Lecture Notes in Networks and Systems

Volume 367

The series "Lecture Notes in Networks and Systems" publishes the latest developments in Networks and Systems—quickly, informally and with high quality. Original research reported in proceedings and post-proceedings represents the core of LNNS.

Volumes published in LNNS embrace all aspects and subfields of, as well as new challenges in, Networks and Systems.

The series contains proceedings and edited volumes in systems and networks, spanning the areas of Cyber-Physical Systems, Autonomous Systems, Sensor Networks, Control Systems, Energy Systems, Automotive Systems, Biological Systems, Vehicular Networking and Connected Vehicles, Aerospace Systems, Automation, Manufacturing, Smart Grids, Nonlinear Systems, Power Systems, Robotics, Social Systems, Economic Systems and other. Of particular value to both the contributors and the readership are the short publication timeframe and the world-wide distribution and exposure which enable both a wide and rapid dissemination of research output.

The series covers the theory, applications, and perspectives on the state of the art and future developments relevant to systems and networks, decision making, control, complex processes and related areas, as embedded in the fields of interdisciplinary and applied sciences, engineering, computer science, physics, economics, social, and life sciences, as well as the paradigms and methodologies behind them.

Indexed by SCOPUS, INSPEC, WTI Frankfurt eG, zbMATH, SCImago.

All books published in the series are submitted for consideration in Web of Science.

For proposals from Asia please contact Aninda Bose (aninda.bose@springer.com).

More information about this series at https://link.springer.com/bookseries/15179

Mykola Nechyporuk · Vladimir Pavlikov ·
Dmitriy Kritskiy
Editors

Integrated Computer Technologies in Mechanical Engineering - 2021

Synergetic Engineering

Springer

Editors
Mykola Nechyporuk
National Aerospace University – Kharki
Kharkov, Ukraine

Vladimir Pavlikov
Kharkov Aviation Institute
National Aerospace University
Kharkov, Ukraine

Dmitriy Kritskiy
Kharkov Aviation Institute
National Aerospace University
Kharkov, Ukraine

ISSN 2367-3370 ISSN 2367-3389 (electronic)
Lecture Notes in Networks and Systems
ISBN 978-3-030-94258-8 ISBN 978-3-030-94259-5 (eBook)
https://doi.org/10.1007/978-3-030-94259-5

This Springer imprint is published by the registered company Springer Nature Switzerland AG
The registered company address is: Gewerbestrasse 11, 6330 Cham, Switzerland

Preface

The International Scientific and Technical Conference "Integrated Computer Technologies in Mechanical Engineering"—Synergetic Engineering (ICTM) was formed to bring together outstanding researchers and practitioners in the field of information technology in the design and manufacture of engines; creation of rocket space systems, aerospace engineering from all over the world to share their experience and expertise. It was established by National Aerospace University "Kharkiv Aviation Institute".

The Conference ICTM-2021 was held in Kharkiv, Ukraine, during October 28–29, 2021. During this conference, technical exchanges between the research community were carried out in the forms of keynote speeches, panel discussions, as well as special session. In addition, participants were treated to a series of receptions, which forge collaborations among fellow researchers. ICTM-2021 received 203 papers submissions from different countries.This book contains papers devoted to relevant topics including

- Information technology in the design and manufacture of engines;
- Information technology in the creation of rocket space systems;
- Aerospace engineering;
- Transport systems and logistics;
- Big data and data science;
- Nano-modeling;
- Artificial intelligence and smart systems;
- Networks and communication;
- Cyber-physical system and IoE;
- Software engineering and IT infrastructure.

All of these offer us plenty of valuable information and would be of great benefit to the experience exchange among scientists in modeling and simulation. The organizers of ICTM-2021 made great efforts to ensure the success of this conference. We hereby would like to thank all the members of ICTM-2021 Advisory Committee for their guidance and advice, the members of program committee and organizing committee, and the referees for their effort in reviewing and soliciting the papers, and all authors for their contribution to the formation of a common intellectual environment for solving relevant scientific problems. Also, we grateful to Springer—Janusz Kacprzyk and Thomas Ditzinger—as the editor responsible for the series "Lecture Notes in Networks and Systems" for their great support in publishing these selected papers.

International Scientific and Technical Conference

Integrated Computer Technologies in Mechanical Engineering—Synergetic Engineering

Kharkiv, Ukraine
October 28–29, 2021

Organized by:

National Aerospace University "Kharkiv Aviation Institute"

Kharkiv Regional State Administration

Patronage:

Ministry of Education and Science of Ukraine

General Schedule (GMT+3)

October 28, 2021 (Thursday)	
08:15	TESTING SESSION: https://meet.google.com/xvk-xhzi-fis
	REGISTRATION OPENS
09:00	OPENINGS CONFERENCE
09:10	PLENARY SESSION
09:50	TECHNICAL BREAK
10:00	Session 1—Mechanical Engineering
	https://meet.google.com/xvk-xhzi-fis
	Topic 1—Integrated Computer Technologies in Aerospace Engineering
12:50	TECHNICAL BREAK
13:10	Topic 2—Nano-Modeling
13:40	Topic 3—Information Technology in Design and Manufacturing of Engines
15:50	Topic 4—Information Technology in The Creation of Rocket Space Systems
16:00	CONCLUSIONS

October 29, 2021 (Friday)		
09:00	Session 2—Software Engineering and IT	Session 3—Project management and business informatics
	https://meet.google.com/cry-xyej-pkm	https://meet.google.com/bmb-fdvs-hoc
	Topic 5—Artificial Intelligence, Smart Systems	Topic 9—Project management
09:40	Topic 6—Software Engineering and IT Infrastructure	
10:30	Topic 7—Control Systems and Engineering	
11:20	TECHNICAL BREAK	
11:40	Topic 8—Information Modeling, Robotics and UAV	
12:20		CONCLUSIONS
14:00	CONCLUSIONS	
14:10	CLOSING OF THE CONFERENCE: LOOKING TO THE FUTURE	

ICTM-2021 Committees

Steering Committee

Honorary Chair

Nechyporuk Mykola	Doctor of Technical Sciences, Professor (National Aerospace University "Kharkiv Aviation Institute", Kharkiv, Ukraine)

Executive Chair

Pavlikov Vladimir	Doctor of Technical Sciences, Senior Research Associate (National Aerospace University "Kharkiv Aviation Institute", Kharkiv, Ukraine)

Program Committee

Bo An	Doctor of Philosophy, Associate Professor (Nanyang Technological University, Nanyang, Singapore)
Bodyanskiy Yevgeniy	Doctor of Technical Sciences, Professor (Kharkiv National University of Radio Electronics, Kharkiv, Ukraine)
Boguslayev Vyacheslav	Doctor of Technical Sciences, Professor (Joint Stock Company Motor Sich, Zaporizhzhia, Ukraine)
Bychkov Sergiy	Doctor of Technical Sciences, Professor (Antonov Company, Kyiv, Ukraine)
Danylov Valeriy	Doctor of Technical Sciences, Professor (National Technical University of Ukraine "Igor Sikorsky Kyiv Polytechnic Institute", Kyiv, Ukraine)

Lukin Vladimir	Doctor of Technical Sciences, Professor (National Aerospace University "Kharkiv Aviation Institute", Kharkiv, Ukraine)
Nikolaev Alexey	Doctor of Physics and Mathematics, Professor (National Aerospace University "Kharkiv Aviation Institute", Kharkiv, Ukraine)
Pashchenko Yuriy	(Scientific and Production Complex Iskra, Zaporizhzhia, Ukraine)
Plankovskyy Sergiy	Professor (O. M. Beketov National University of Urban Economy in Kharkiv, Kharkiv, Ukraine)
Pohudina Olha	Doctor of Philosophy, Associate Professor (National Aerospace University "Kharkiv Aviation Institute", Kharkiv, Ukraine)
Polosukhin Illia	(NEAR.ai, San Francisco, CA, USA)
Ponomarenko Mykola	Doctor of Technical Sciences, Senior Research Associate (Tampere University, Tampere, Finland)
Ponomaryov Volodymyr	Doctor of Technical Sciences, Professor (Instituto Politécnico Nacional, Mexico City, Mexico)
Popov Viktor	Doctor of Philosophy (Joint Stock Company FED, Kharkiv, Ukraine)
Przystalski Karol	Doctor of Philosophy (Jagiellonian University, Kraków, Poland)
Sanin Anatoliy	Doctor of Technical Sciences, Professor (Oles Honchar Dnipro National University, Dnipro, Ukraine)
Shakhovska Nataliya	Doctor of Technical Sciences, Professor (Lviv Polytechnic National University, Lviv, Ukraine)
Shypul Olga	Doctor of Philosophy, Associate Professor (National Aerospace University "Kharkiv Aviation Institute", Kharkiv, Ukraine)
Sokolov Oleksandr	Doctor of Technical Sciences, Professor (Nicolaus Copernicus University, Toruń, Poland)
Stoyan Yuriy	Corresponding Member of the National Academy of Science of Ukraine, Doctor of Technical Sciences, Professor (A. Pidgorny Institute of Mechanical Engineering Problems of the National Academy of Sciences of Ukraine, Kharkiv, Ukraine)
Szalay Tibor	Doctor of Philosophy, Associate Professor (Budapest University of Technology and Economics, Budapest, Hungary)

Turkin Ihor	Doctor of Technical Sciences, Professor (National Aerospace University "Kharkiv Aviation Institute", Kharkiv, Ukraine)
Ugryumov Mykhaylo	Doctor of Technical Sciences, Professor (V. N. Karazin Kharkiv National University, Kharkiv, Ukraine)
Volosyuk Valerii	Doctor of Technical Sciences, Professor (National Aerospace University "Kharkiv Aviation Institute", Kharkiv, Ukraine)
Vozel Benoit	Doctor of Philosophy, Associate Professor (University of Rennes 1, Rennes, France)
Whitehead Charles K.	Doctor of Sciences, Professor (Cornell University, Ithaca, NY, USA)
Yakovlev Sergey	Doctor of Physics and Mathematics, Professor (National Aerospace University "Kharkiv Aviation Institute", Kharkiv, Ukraine)
Yudelson Michael V.	Doctor of Philosophy (Carnegie Mellon University, Pittsburgh, PA, USA)
Zaslavskyi Volodymyr	Doctor of Technical Sciences, Professor (Taras Shevchenko National University of Kyiv, Kyiv, Ukraine)
Zavgorodniy Andrew	Doctor of Philosophy (LinkedIn, Milpitas, CA, USA)

Organizing Team

National Aerospace University "Kharkiv Aviation Institute", Kharkiv, Ukraine

Publication Chairs

Kritskiy Dmitriy
Starovoit Tetiana (Secretary)

Members

Bykov Andrii
Hryhorenko Tetiana
Kalashnikova Vasylysa
Karatanov Oleksandr
Krytska Olha
Morikova Anastasia
Pohudina Olha
Shypul Olga
Tsegelnyk Yevgen

The Official Language of the ICTM-2021 is English

Time for presentations:
Presentation at the plenary session is 20 minutes.
Paper presentation and discussion at the regular session are 10 minutes.

October 28, 2021, Thursday

(GMT + 3)	
8:15	TESTING SESSION: https://meet.google.com/xvk-xhzi-fis
8:30	REGISTRATION OPENS
9:00	OPENINGS CONFERENCE *Vladimir Pavlikov* – Executive Chair of Conference National Aerospace University "KhAI", Ukraine
9:10	PLENARY SESSION **UTM: digitalization of airspace** Dr. Marek Košuda– Keynote Speaker 1 Technical University of Košice, Slovakia
9:30	**Recent advances and future potentials in micro machining** Dr. Márton Takács– Keynote Speaker 2 Budapest University of Technology and Economics, Hungary
09:50	TECHNICAL BREAK
10:00	SESSION 1 – MECHANICAL ENGINEERING https://meet.google.com/xvk-xhzi-fis Dr. Olga Shypul – Chairman
Topic 1—Integrated Computer Technologies in Aerospace Engineering	
10:00	*Sergey Kurennov, Konstantin Barakhov, Olexandr Poliakov and Daria Dvoretska* **Stress Distribution in the Eccentrically Loaded Lapped Adhesive Joint. An Analytical Model**
10:10	*Oleksandr Dveirin, Anton Tsaritsynskyi, Tetyana Nabokina and Andrii Kondratiev* **Assessment of Error of the Experimental Determination of Interlaminar Shear Strength**

(continued)

(continued)

10:20	*Oleksandr Shorinov* **Tensile stress as a factor of deformation and failure of thin films and coatings**
10:30	*Sergei Kurennov and Natalia Smetankina* **Stress-Strain State of a Double Lap Joint of Circular Form. Axisymmetric Model**
10:40	*Vitalii Voronko, Iryna Tkachenko, Yuri Dyachenko, Iryna Voronko and Dmitry Zawistovskiy* **Methodology for Calculating the Productivity of Electrochemical Machining in Stationary Electrolyte**
10:50	*Yurii Shyrokyi and Gennadiy Kostyuk* **Investigation of the influence of crystallization energy on the size of nanostructures during copper ion-plasma treatment**
11:00	*Olena Tachinina, Sergiy Ponomarenko, Viktor Shevchenko, Olexandr Lysenko and Igor Romanchenko* **Synthesis of a Launch Trajectory of Aircraft Optimal in the Minimum Fuel Consumption Based on Sufficient Conditions of Optimal Control**
11:10	*Andrey Volkov* **Analysis of stress development mechanisms in the coating/substrate system**
11:20	*Dmytro Brega, Sergey Eremenko and Maria Surgailo* **The effect of vertical wind turbines position pattern on their aerodynamic characteristics**
11:30	*Ihor Romanchenko, Volodymyr Kotliarov, Andrii Serhienko and Serhii Yenchev* **Simulation of Aircraft Motion in a Corkscrew Taking into Account Nonlinear Effects of Rotation**
11:40	*Wenjie Hu, Kun Tan, Markovych Sergii and Oleksandr Shorinov* **Structural optimization of the special cold spraying nozzle via response surface method characteristics**
11:50	*Kostiantyn Barakhov and Igor Taranenko* **Influence of joint edge shape on stress distribution in adhesive film**
12:00	*Vasilii Makeiev, Yurii Pushkariov, Andrii Raskoshnyi, Iryna Voronko and Svitlana Myronova* **Considering the Meteorological Elements for the Aerial Vehicles Flight by Using "Weight" Functions**
12:10	*Andrii Pohudin, Sergii Gubin, Olha Pohudina and Svitlana Sinchenko* **Development of an algorithm for preparing to start an resistojet propulsion system**
12:20	*Valeriy Sikulsky, Stanislav Sikulskyi, Kateryna Maiorova, Valeriya Suponina and Oleh Komisarov* **The process of forming integral ribbed panels by sequential local phased deformation with information support**
12:30	*Olga Shypul, Oleksiy Pavlenko, Sergiy Zaklinskyy, Vadim Garin and Sergiy Plankovskyy* **Evaluating the mass inside a reservoir during a filling with pressurized component of gas mixture**
12:40	*Iurii Vorobiov, Kateryna Maiorova, Iryna Voronko, Maksym Boiko and Oleh Komisarov* **Creation and Improvement Principles of the Pneumatic Manual Impulse Devices**

(continued)

(continued)

12:50	TECHNICAL BREAK
Topic 2—Nano-Modeling	
13:10	*Andrii Breus, Sergey Abashin, Oleksii Serdiuk and Oleg Baranov* **Linking dynamics of growth of copper oxide nanostructures in air**
13:20	*Andrey Volkov, Oleksandr Shorinov and Sergii Polyviany* **Deposition of titanium-containing coatings by low-pressure cold spraying**
13:30	*Olena Sierikova, Elena Strelnikova, Vasil Gnitko, Anatoliy Tonkonozhenko and Leonid Pisnia* **Nanocomposites Implementation for Oil Storage Systems Electrostatic Protection**
Topic 3—Information Technology in Design and Manufacturing of Engines	
13:40	*Lyudmyla Rozova and Gennadii Martynenko* **The application features of integrated technologies for the design of elements of systems with gas-turbine aircraft engines**
13:50	*Gennadii Martynenko and Volodymyr Martynenko* **Mathematical and Computer Simulation of Rotor Dynamics Phenomena in Electromechanical Systems with Magnetic Bearings**
14:00	*Anatolii Kulik, Konstiantyn Dergachov, Serhii Pasichnik, Vitalii Dzhulgakov and Vasilii Petrenko* **Rational Control of the Operability of the Model Unit of Electric Flywheel Motors under Destabilizing Effects**
14:10	*Volodymyr Martynenko* **Design and Analysis of the Bimetallic Fan Blade**
14:20	*Mykhaylo Ugryumov, Andriy Myenyaylov, Ievgen Meniailov and Irina Trofymova* **Gas turbine engine system multiobjective perfection using the multilevel hierarchy stochastic optimization model and solution searching method**
14:30	*Vyacheslav Merculov, Mykola Kostin, Gennadii Martynenko, Natalia Smetankina and Volodymyr Martynenko* **Peculiarities of the Modelling of the Bird Dynamic Impact on Fan Blades of an Aircraft Turbojet Engine at Operating Modes**
14:40	*Dmytro Konovalov, Halina Kobalava, Mykola Radchenko and Roman Radchenko* **Increasing ecological and energy efficiency of combustion engines by using a thermopressor**
14:50	*Mykhaylo Taranenko and Aleksandr Naryzny* **Modelling the Process of Interaction of a Pulsed Jet with a Workpiece by Electrohydraulic Forming**
15:00	*Victoria Kornienko, Roman Radchenko, Serhiy Forduy and Victor Khaldobin* **Reducing the Harmful Emissions while Water-Fuel Emulsions Combustion**
15:10	*Roman Radchenko, Maxim Pyrysunko, Victoria Kornienko and Mykola Radchenko* **Effect of Utilization Exhaust and Recirculation Gases of Ship Diesel Engine in Absorption Chiller**
15:20	*Andrii Radchenko, Mykola Radchenko, Serhiy Forduy, Veniamin Tkachenko and Felix Tsaran* **Numerical Simulation of an Aerothermopressor with Incomplete Evaporation for Intercooling of the Gas Turbine Engine**

(continued)

(continued)

15:30	*Dmytro Konovalov, Mykola Radchenko, Halina Kobalava, Boris Bileka and Sergey Kantor* **Efficiency of gas turbine circle air by thermopressor**
15:40	*Oleg Savenkov, Andrii Radchenko, Borys Bileka and Ionut Cristian Scurtu* **Improving the Efficiency of Operation of Gas Turbine Plants Based on Turbo-Gear Units**
Topic 4—Information Technology in The Creation of Rocket Space Systems	
15:50	*Igor Taranenko, Anton Tsaritsynskyi, Tetyana Nabokina and Andrii Kondratiev* **Bending-twisting deforming of thin-walled composite sections at thermal-mechanical loading**
16:00	CONCLUSIONS

October 29, 2021, Friday

9:00	SESSION 2 – SOFTWARE ENGINEERING AND IT https://meet.google.com/cry-xyej-pkm Dr. Dmytro Chumachenko – Chairman
Topic 5—Artificial Intelligence, Smart Systems	
9:00	*Galyna Mygal, Valeriy Mygal, Olga Protasenko and Ihor Klymenko* **Cognitive Aspects of Ensuring the Safety, Dependability and Stability of a Dynamic System's Functioning in Extreme Conditions**
9:10	*Mikle Tsuranov, Heorhii Zemlianko, Anatoliy Plakhteev, Kyrylo Leichenko and Vladimir Pevnev* **"Smart City" technology: conception, security issues and cases**
9:20	*Vladimir Lukin, Victoriia Naumenko and Sergey Krivenko* **Analysis of noisy image lossy compression by BPG**
9:30	*Dmitriy Kritskiy, Olha Pohudina, Mykhailo Kovalevskyi, Yevgen Tsegelnyk and Volodymyr Kombarov* **Powder Mixtures Analysis for Laser Cladding using OpenCV Library**
Topic 6—Software Engineering and IT Infrastructure	
9:40	*Olena Zhykhor, Valeriy Ryeznikov, Nataliia Pohribna and Vitalina Delas* **Digital Finance and Fintech Applications in Ukraine: Current Status and Development Prospects**
9:50	*Ekaterina Bataeva and Anastasiia Chumakova-Sierova* **Values in visual practices of Instagram network users**
10:00	*Vladimir Lukin, Sergey Krivenko, Ihor Kaluzhinov, Olha Krylova and Liudmyla Kryvenko* **Lossy compression of remote sensing and dental images corrupted by spatially correlated noise**
10:10	*Dmytro Kolmogorov and Ievgen Meniailov* **Development of Programming Interface for Integrating Assembly Language with Python**
10:20	*Mykola Butkevych, Kseniia Bazilevych, Serhii Krivtsov and Anton Tkachenko* **Classification of Patients with Diabetes Mellitus using Neural Networks**

(continued)

(continued)

Topic 7—Control Systems and Engineering	
10:30	*Nikolay Koshevoy, Oleksandr Zabolotnyi, Oleksii Potylchak and Vitalii Siroklyn* **Development And Modeling Of A Device For Measuring The Moisture Of Oil Products**
10:40	*Oleksandr Zabolotnyi, Vitalii Zabolotnyi and Nikolay Koshevoy* **Classification of Patients with Diabetes Mellitus using Neural Networks**
10:50	*Viktor Makarichev, Irina Vasilyeva, Vladimir Lukin, Nataliia Kussul and Andrii Shelestov* **Classification accuracy of three-channel images compressed by discrete atomic transform**
11:00	*Oleh Pihnastyi, Georgii Kozhevnikov and Olha Ivanovska* **Maxwell-Element Model for Describing Conveyor Belt Stresses**
11:10	*Oleh Pihnastyi, Georgii Kozhevnikov and Pavel Vasyuchenko* **Analysis of Dynamic Stresses During Acceleration and Deceleration of a Conveyor Belt (Maxwell Element Model)**
11:20	TECHNICAL BREAK
Topic 8—Information Modeling, Robotics and UAV	
11:40	*Ivan Ostroumov and Nataliia Kuzmenko* **Performance modeling of aircraft positioning system**
11:50	*Denis Sliusar, Oleksii Isakov, Volodymyr Kolesnyk, Oleg Chugai, Serhii Oliynick* **Computer Simulation Of Abnormal Glow Discharge in Inverse Magnetron Sputtering Systems with Sectioned Cathode Units**
12:00	*Nikolay Koshevoy, Oleksandr Zabolotnyi, Vitalii Siroklyn, Tetiana Rozhnova, Elena Kostenko and Oleh Burlieiev* **Modeling and optimization of photoelectric device positioning for angular displace-ment measurements in aircraft computer control and operation systems**
12:10	*Nikolay Koshevoy, Volodymyr Dergachov, Hanna Pavlyk, Vitalii Siroklyn, Iryna Koshevaya and Olena Hrytsai* **Modified Gray Codes for the Value (Time) Optimization of a Multifactor Experiment Plans**
12:20	*Olexander Betin, Igor Kaluzhynov, Denys Betin and Olena Betina* **Criteria And Similarity Scales In The Task Of Modelling The Operation Of Inflatable Air Shock Absorbers When Landing Parachute Platforms**
12:30	*Alireza Mohammadi and Dmytro Chumachenko* **COVID-19 Epidemic Process Simulation using ARIMA Model**
12:40	*Mariia Pyvovar, Dmitriy Kritskiy, Tetiana Plastun, Vasylysa Kalashnikova, Oleksii Popov* **Takeoff and Landing Model of an Aircraft in the Aviation Simulator**
12:50	*Olha Pohudina, Andrii Bykov, Dmitriy Kritskiy and Mykhailo Kovalevskyi* **The method of flight mission formation for a group autonomous flight of unmanned aerial vehicles**
13:00	*Olena Tachinina, Olexandr Lysenko, Kateryna Nesterenko, Serhii Zybin and Iryna Alekseeva* **Tuning Methodology for Multi-circuit Digital Regulators of Robot Drives with Adjustable Dynamic Characteristics**

(continued)

(continued)

13:10	*Olha Pohudina, Tanya Plastun, Ihor Biletskyi and Igor Vasiliev* **Determination of the Parameters of a Conceptual Quadcopter Model for a Group Outdoor Flights**
13:20	*Zheng Hu and Oleksii Vambol* **Topological design for robotic filament winding of composite beam structures**
13:30	*Mariia Pyvovar, Olha Pohudina, Andrii Pohudin, Olha Kritskaya* **Simulation of Flight Control of Two UAVs Based on the "Master-Slave" Model**
13:40	*Dmitriy Kritskiy, Andrii Bykov, Vladimir Shevel, Olha Krytska and Efosa Mark Guobadia* **Development of a collaborative platform for education in virtual reality**
13:50	*Ihor Honcharenko, Leonid Pisnia, Oleksandr Targonskyi, Hanna Khabarova and Tetiana Klochko* **Information-analytical system of landfill environmental safety assessment**
14:00	CONCLUSIONS
9:00	SESSION 3 – PROJECT MANAGEMENT AND BUSINESS INFORMATICS https://meet.google.com/bmb-fdvs-hoc Karatanov Oleksandr – Chairman
Topic 9—Project management	
9:00	*Andrii Lysenko, Ivan Hrabazii, Iryna Lysenko and Anton Polianskyi* **Violation of Flight Rules and Airspace Use: Current Criminal Situation**
9:10	*Svitlana Gutsu and Andriy Bublikov* **Transformation of Legal Labor Regulation Under Influence of Artificial Intelligence**
9:20	*Olena Syniavska and Olha Pushko* **Some Issues of Corruption Risk Assessment in The Activities of State Bodies, Legal Entities (in Case Study of the Aviation Industry of Ukraine)**
9:30	*Oleksii Lytvynov, Hanna Spitsyna, Yurii Danylchenko and Serafyma Hasparian* **Role and Significance of Situational Analysis in Crime Counteraction at Critical Infrastructure and Aerospace Industry**
9:40	*Anatolii Komziuk, Olena Salmanova and Oleksii Salmanov* **Responsibility for Administrative and Criminal Offenses in the Field of Use of Airspace of Ukraine**
9:50	*Iryna Petrova and Katerina Hovorova* **Offenses Related to Financial and Credit Operations in Aerospace Industry: Causes of Occurrence and Consequences**
10:00	*Olena Savchuk* ***Legal Support of Aerospace Environmental Monitoring***
10:10	*Iryna Petrova and Olha Hetmanets* **Legal Regulation of Public Security and Tasks Related to its Provision in The Aerospace Industry**
10:20	*Ihor Shynkarenko, Vasyl Zakharov, Iryna Shynkarenko, Oleksandra Zakharova and Olha Semenovykh* **Legal and Conceptual Aspects of Securing Space Activity Entities-Facilities in Ukraine**

(continued)

(continued)

10:30	*Mykola Nechyporuk, Oleksandr Kliuiev, Aleksandar Ivanović and Nataliia Filipenko* **Development Strategy of International Cooperation of Forensic Science Institutions of Ukraine with Foreign Experts in Prevention of Terrorist Attacks on Aerospace Industry and Critical Infrastructure**
10:40	*Oleh Uhrovetskyi and Oleg Kurdes* **Certain Aspects of Staffing and Training of Forensic Experts as a Preventive Factor of Safety Ensuring Aerospace Industry Facilities**
10:50	*Nikolay Shelukhin, Ivan Kharaberiush, Oleksandr Shelukhin, Vitalii Tsymbalistyi and Volodymyr Selevko* **Organizational and Legal Support of Economic Security of Aerospace Industry of Ukraine**
11:00	*Ganna Goncharenko, Olena Agapova, Yurii Ryabukha and Oleksandr Shevchuk* **Legal Regulation of Ukrainian Space Industry: Perspective Trends and International Cooperation**
11:10	*Alla Hordeiuk, Andrii Ostropilets and Ihor Bohdaniuk* **Current Issues Of Legal Regulation Of Intellectual Property In Space Field**
11:20	*Ella Simakova-Yefremian and Kateryna Sylenok* **Current Directions of Development of International Cooperation of Forensic Science Institutions of Ukraine in The Fields Historical Forensics, Forensic Archaeology and Aerospace Forensics**
11:30	*Nastasiia Matvieieva, Vasyl Zakharov, Olena Ostapenko, Larisa Yerofyeyenko and Olena Perunova* **Legal regulation of the transport and logistics system**
11:40	*Olena Zhykhor, Olena Iafinovych, Euvgenia Nosova and Nataliia Miedviedkova* **E-waste Handling in the context of Environmental Project Management in Ukraine**
11:50	*Oleh Strelko, Oksana Yurchenko, Halyna Vasilova, Ievgen Gusar and Maryna Rudiuk* **Theoretical Prerequisites for the Development of Mul-timodal Transport-Logistic Technologies in Cooperation With Ten-T Network**
12:00	*Nataliia Fedosenko, Andriy Starodubtsev, Anton Stiebieliev, Valentyn Fesiunin and Svitlana Andrenko* **Contract Structures in The Field of Air Carriage**
12:10	*Natalia Teliura, Nataliia Tsapko, Hanna Khabarova, Olga Lomakina, Olena Pshenichnova and Tetiana Klochko* **Selection Methodology of Ecological Safety Priorities of Sustainable Development Goals of Urban Agglomerations**
12:20	CONCLUSIONS
14:10	CLOSING OF THE CONFERENCE: LOOKING TO THE FUTURE

Contents

Aerospace Engineering

Stress Distribution in the Eccentrically Loaded Lapped Adhesive Joint. An Analytical Model

Sergei S. Kurennov, Konstantin P. Barakhov(✉), D. V. Dvoretskaya, and Olexandr G. Poliakov

National Aerospace University "Kharkiv Aviation Institute", 17 Chkalova Street, Kharkiv 61070, Ukraine
o.poliakov@khai.edu

Abstract. The deflected mode problem for the adhesive joint of two rectangular plates in a simplified two-dimensional formulation is solved. The proposed solution takes into account the bending of the base layers in the joint plane. Base layers are considered as Bernoulli beams. The stress and deformation values are assumed to be linearly distributed along the joint width and uniformly in thickness of the joint elements. The problem is reduced to a system of two linear differential equations of the fourth order with respect to transverse (in the joint plane) shifts of layers. An analytical solution to the problem is obtained. The model problem is solved. Comparison of the results obtained with using the proposed method with the results, obtained using finite element modeling is done. Good correspondence of the results obtained by using two different techniques is shown, and also that the proposed model is adequate and has sufficient precision for engineering tasks.

Keywords: Adhesive joints · Analytical modeling · Interlaminar stresses · Beam theory

1 Introduction

Lapped adhesive joints are widespread in modern technology. There are several basic stress state models for joints, which allow us determine the joint stress state in an analytical form [1–4]. Most of the models are one-dimensional. Base (outer) layers are considered as rods that work in stress-strain, or as beams in the Bernoulli or Timoshenko approximation. The adhesive (connecting) layer is considered as an elastic Winkler base or as a two-parameter elastic base [5–9]. In the latter case, the mathematical model makes it possible to describe with high accuracy the stress state of the adhesive layer at the border of the gluing area. The stress distribution over the base layers thickness is assumed to be linear, and is uniform over the adhesive layer thickness, or stepped, or also linear, depending on the model. In all the listed mathematical models, the stress distribution depends on only one coordinate, i.e. the models are one-dimensional. However, when computing the deflected mode of overlapping adhesive joints, in a number of cases, it is necessary to take into account the nonuniformity of the stress and deformation distribution along the width of the joint. This problem is qualitatively more difficult than

M. Nechyporuk et al. (Eds.): ICTM 2021, LNNS 367, pp. 3–13, 2022.
https://doi.org/10.1007/978-3-030-94259-5_1

constructing a one-dimensional, even though advanced, mathematical model of a three-layer beam. Various numerical and approximate methods for the numerical solution of this problem are proposed [10–17]. An analytical solution to the two-dimensional joint stress state problem is still unknown. However, several simplified mathematical models have been proposed to study the various effects that arise in the joint. To study the effect of base plate transverse deformations, which are caused by Poisson's ratios, on the stress distribution in the adhesive, a model based on hypotheses about the uniformity of the applied load and the shear compliance of the base layers is proposed [16, 18]. Often the structure and the loads applied to it are symmetrical relative to the longitudinal axis. In this case, it is possible to use a simplified model, according to which transverse deformations and shifts in the joint plane are assumed to be zero [19]. In papers [20–22] such approach is developed for the different types of boundary equations. The described mathematical models rank an intermediate position between the models of the elasticity theory and more simple structural mechanics models.

As noted above, in the previously created two-dimensional joint stress state models, which take into account the nonuniform stress distribution along the width of the joint, it is assumed that the applied load is symmetric throughout the longitudinal axis of the structure. I.e. the bending of the joined plates in the plane is absent. In this paper, a joint model is proposed, which takes into account the bending of the base layers in the gluing plane. In this case, the base layers are considered as Bernoulli beams, which bend in the gluing plane. Such loading of the structure occurs when the load is applied to the joint by some eccentricity. This problem was investigated using finite element modeling in [23]. In this paper, an analytical solution to the problem is proposed.

2 Problem Formulation

The force applied to the eccentric joint, at some distance from the point of application, causes the forces in the rods to be linearly distributed over the width. These forces can be considered as a superposition of uniform stress-strain and skew-symmetric forces due to the bending moment, Fig. 1.

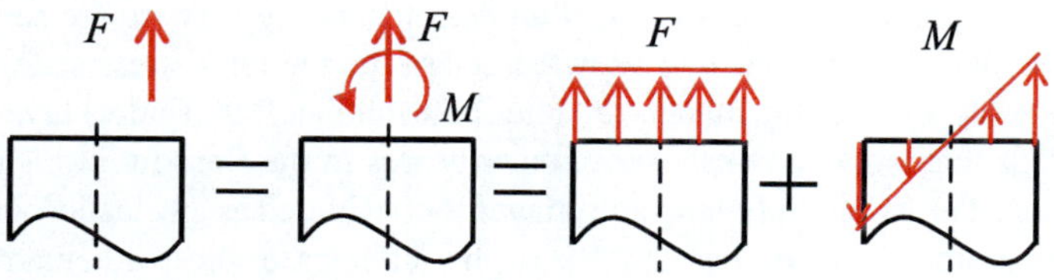

Fig. 1. Load diagram

It is not difficult to find the stress state of the joint, which is due to forces uniformly distributed over the width, since this problem is well known [1, 4]. Whereas the problem of finding the stress state of a joint loaded with a bending moment is new. Therefore, we will focus on this particular task.

Consider an adhesive joint of two rectangular plates (linear dimensions of plates are $L \times 2h$ and thickness δ_1 and δ_2 correspondingly), shown in Fig. 2. The lateral sides of the plates are load free. The plates are deformed only in the joint plane, the adhesive layer acts in shear, the stress distribution is uniform over the layer thickness. Without diminishing the generality of the proposed approach, we assume that the load is applied only to the second layer along the side $x = 0$, and the first layer along the side $x = L$ is rigidly fixed. The load applied to the second layer is assumed to be linearly dependent on the transverse coordinate y.

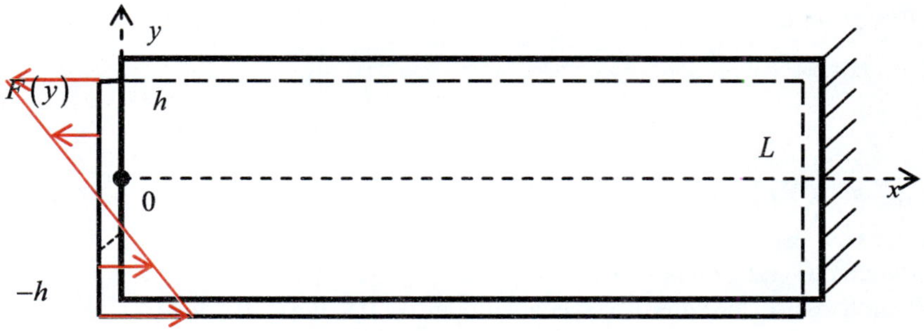

Fig. 2. Diagram of the structure

The applied load is described by the dependence $F(y) = K \cdot y$, where a coefficient $K = \frac{3M}{2h^3\delta_2}$, M is a bending moment applied to the joint.

The forces acting on the differential elements of the joint base layers are shown in Fig. 3.

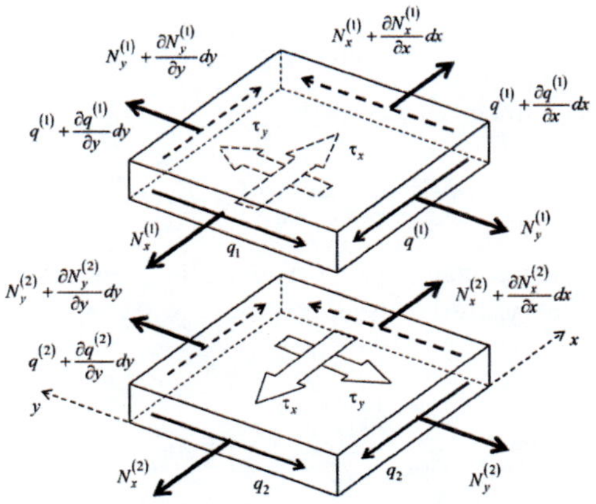

Fig. 3. Equilibrium of the joint differential element

The equilibrium equations for the base layers are

$$\tau_x + \frac{\partial N_x^{(1)}}{\partial x} + \frac{\partial q^{(1)}}{\partial y} = 0, \quad -\tau_x + \frac{\partial N_x^{(2)}}{\partial x} + \frac{\partial q^{(2)}}{\partial y} = 0 \tag{1}$$

$$\tau_y + \frac{\partial N_y^{(1)}}{\partial y} + \frac{\partial q^{(1)}}{\partial x} = 0,\ -\tau_y + \frac{\partial N_y^{(2)}}{\partial y} + \frac{\partial q^{(2)}}{\partial x} = 0 \tag{2}$$

where the superscript is the base layer number and the lower is the direction; $\tau_x,\ \tau_y$ are tangential stresses in the adhesive layer along the corresponding axis; $q^{(1)},\ q^{(2)}$ are tangential forces in the corresponding base layers; $N_x^{(1)}$, $N_y^{(1)}$, $N_x^{(2)}$, $N_y^{(2)}$ are normal forces in the base layers.

Forcess in layers (Poisson deformations are not taken into account)

$$N_x^{(k)} = B_k \frac{\partial u^{(k)}}{\partial x};\quad q^{(k)} = \delta^{(k)} G^{(k)} \left(\frac{\partial u^{(k)}}{\partial y} + \frac{\partial v^{(k)}}{\partial x} \right), \tag{3}$$

where $u^{(k)}$, $v^{(k)}$ are shifts of the $k-th$ layer ($k = 1,\ 2$) in the longitudinal and transverse direction correspondingly; $B_k = \delta^{(k)} E_x^{(k)}$; $E_x^{(k)}$, $G^{(k)}$, $\delta^{(k)}$ are a modulus of elasticity, shear modulus and thickness of the corresponding base layer.

The stresses in the adhesive layer are proportional to the shift difference of the plates [16, 18]

$$\tau_x = \frac{G_0}{\delta_0}\left(u^{(1)} - u^{(2)}\right);\ \tau_y = \frac{G_0}{\delta_0}\left(v^{(1)} - v^{(2)}\right), \tag{4}$$

where G_0 and δ_0 are shear modulus and adhesive layer thickness correspondingly.

Homogeneous boundary conditions:

$$\begin{aligned} &u^{(1)}\big|_{x=L} = 0;\ v^{(1)}\big|_{x=L} = 0;\ N_x^{(2)}\Big|_{x=L} = 0; \\ &N_x^{(1)}\Big|_{x=0} = 0;\ q^{(k)}\Big|_{y=\pm h} = 0;\ N_y^{(k)}\Big|_{y=\pm h} = 0. \end{aligned} \tag{5}$$

Normal force is given on the left lateral side.

$$N_x^{(2)}\Big|_{x-0} = F(y) = K.y \tag{6}$$

We will assume that the bending of the plates in the joint plane is described by the Bernoulli beams theory. In this case, the shifts of the plates are described by the dependences.

$$u^{(k)}(x, y) = -y \cdot \frac{d}{dx} V^{(k)}(x),\ v^{(k)} = V^{(k)}(x),\ k = 1,\ 2. \tag{7}$$

Here $V^{(k)}(x)$ are transversal shifts of $k - th$ layer in the joint plane (i.e. elastic line equation), Fig. 4.

If $V(x)$ is an elastic beam line, then longitudinal point shifts, remote from the beam axis at a distance y are equal $u = y \sin\varphi = -y\frac{dV}{dx}$, since the shifts are assumed to be small (linear theory).

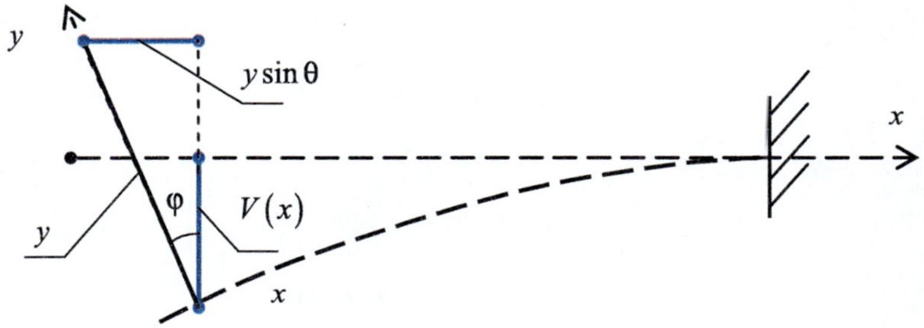

Fig. 4. Shifts of the beam element

3 Constructing the Solution

Substituting Eq. (7) in Eq. (4), we get

$$\tau_x = -\frac{G_0}{\delta_0} y \left(\frac{dV^{(1)}}{dx} - \frac{dV^{(2)}}{dx} \right); \ \tau_y = \frac{G_0}{\delta_0} \left(V^{(1)} - V^{(2)} \right) \tag{8}$$

Substituting longitudinal forces Eq. (3), $N_x^{(k)} = B_k y \frac{d^2 V^{(k)}}{dx^2}$, in Eq. (1), then integrating by y and satisfying zero conditions for tangential stresses at the lateral sides of the joint (5), we get.

$$q^{(k)} = (-1)^k \frac{h^2 - y^2}{2} \left[\frac{G_0}{\delta_0} \left(\frac{dV^{(1)}}{dx} - \frac{dV^{(2)}}{dx} \right) - (-1)^k B_k \frac{d^3 V^{(k)}}{dx^3} \right]. \tag{9}$$

Substituting Eq. (9) in Eq. (2) and integrating by y, we get

$$\begin{aligned} N_y^{(k)} &= (-1)^k \tfrac{G_0}{\delta_0} y \left(V^{(1)} - V^{(2)} \right) + f^{(k)}(x) \\ &+ \tfrac{1}{2} \left(\tfrac{y^3}{3} - h^2 y \right) \left[B_k \tfrac{d^4 V^{(k)}}{dx^4} - (-1)^k \tfrac{G_0}{\delta_0} \left(\tfrac{d^2 V^{(1)}}{dx^2} - \tfrac{d^2 V^{(2)}}{dx^2} \right) \right]. \end{aligned} \tag{10}$$

Functions $f^{(1)}(x)$ and $f^{(2)}(x)$ we find, using conditions (5) $N_y^{(k)}\Big|_{y=h} = 0$:

$$f^{(k)}(x) = -\frac{h^3}{3} B_k \frac{d^4 V^{(k)}}{dx^4} - (-1)^k \frac{G_0}{\delta_0} \left[\frac{h^3}{3} \left(\frac{d^2 V^{(1)}}{dx^2} - \frac{d^2 V^{(2)}}{dx^2} \right) - h \left(V^{(1)} - V^{(2)} \right) \right].$$

Boundary conditions on the second lateral side $N_y^{(k)}\Big|_{y=-h} = 0$ leads us to the differential equations system

$$\boldsymbol{A}_4 \frac{d^4 \boldsymbol{V}}{dx^4} + \boldsymbol{A}_2 \frac{d^2 \boldsymbol{V}}{dx^2} + \boldsymbol{A}_0 \boldsymbol{V} = 0, \tag{11}$$

where

$$\boldsymbol{V} = \begin{pmatrix} V^{(1)} \\ V^{(2)} \end{pmatrix}, \ \boldsymbol{A}_4 = \frac{\delta_0}{G_0} \begin{pmatrix} \delta^{(1)} E_x^{(1)} & 0 \\ 0 & \delta^{(2)} E_x^{(2)} \end{pmatrix}, \ \boldsymbol{A}_2 = \begin{pmatrix} -1 & 1 \\ 1 & -1 \end{pmatrix}, \ \boldsymbol{A}_0 = \frac{3}{h^3} \begin{pmatrix} 1 & -1 \\ -1 & 1 \end{pmatrix}.$$

The general solution to system (11) can be written as

$$\boldsymbol{V} = \sum_{n=1}^{4} S_n x^{(n-1)} \begin{pmatrix} 1 \\ 1 \end{pmatrix} + \sum_{n=1}^{4} S_{n+4}\, e^{\theta_n x} \boldsymbol{H}_n, \tag{12}$$

where θ_n are nonzero roots of a characteristic equation.

$$\det\left(\boldsymbol{A}_4\theta^4 + \boldsymbol{A}_2\theta^2 + \boldsymbol{A}_0\right) = 0.$$

Vectors $\boldsymbol{H}_n$ are solutions of the system

$$\left(\boldsymbol{A}_4\theta_n^4 + \boldsymbol{A}_2\theta_n^2 + \boldsymbol{A}_0\right)\boldsymbol{H}_n = 0$$

and are determined up to an arbitrary factor S_n.

To find eight unknown constants that are included in (12), we formulate eight boundary conditions.

$$V_1(L) = \left.\frac{dV_1(x)}{dx}\right|_{x=L} = \int_{-h}^{h} q^{(1)}(0, y)dy = \left.\frac{d^2V^{(1)}}{dx^2}\right|_{x=0} = \int_{-h}^{h} q^{(2)}(0, y)dy = 0;$$

$$\int_{-h}^{h} q^{(2)}(L, y)dy = 0; \quad \left.\frac{d^2V^{(2)}}{dx^2}\right|_{x=L} = 0; \quad \left.\frac{d^2V^{(2)}}{dx^2}\right|_{x=0} = \frac{K}{E_x^{(2)}};$$

The last condition is similar to the classical Bernoulli beam bending equation $E_x^{(2)} I^{(2)} \left.\frac{d^2V^{(2)}}{dx^2}\right|_{x=0} = M$, where $I^{(2)} = \frac{2h^3\delta_2}{3}$ is a moment of inertia.

The above conditions form a system of linear equations with respect to the unknown constants $S_1, \ldots, S_8$.

4 Model Problem

To analyze the stress state of the joint and verify the proposed analytical model, it was considered the joint of two aluminum ($E^{(1)} = E^{(2)} = 72$ GPa, Poisson coefficient $\mu = 0.28$) plates of length $L = 90$ mm and width $2h = 30$ mm. Thickness of the plates are $\delta_1 = 2,5$ mm and $\delta_2 = 2$ mm. The plates are glued with the adhesive, the elastic parameters of which are $G_0 = 0.34$ GPa, $\mu = 0.32$ and thickness is $\delta_0 = 0.1$ GPa. All materials are isotropic.

To verify the proposed analytical model according to the above parameters, a three-dimensional finite element model was created. The characteristic size of the element in the adhesive layer is equal to the thickness of the adhesive layer.

Stresses in the median plane of the adhesive layer in the longitudinal direction τ_{zx}, calculated using finite element modeling are shown in Fig. 5.

The stresses in the figure are given in dimensionless form, as the ratio of the acting stresses to some hypothetical stresses $\tau^* = \frac{F(h)}{L}$, which would arise in the joint provided

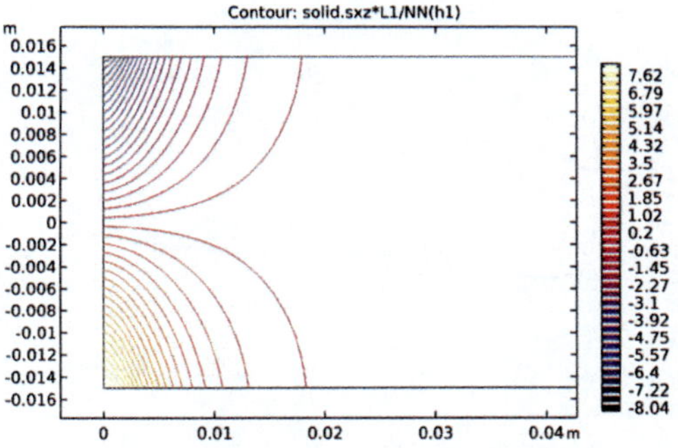

Fig. 5. Stresses in the adhesive (fragment of the structure)

that the applied linear forces are uniformly distributed along the length of the gluing $F(h) = K \cdot h$. These forces represent the maximum values of the forces applied to the joint (5). Therefore, the stress ratio $\frac{\tau_x(x,y)}{\tau^*}$ is a someone similar to the stress concentration coefficient.

The graph shows that on the axis of symmetry the stresses τ_{zx} are equal to zero and increase when approaching the lateral sides. In Fig. 6 it is shown the stresses τ_{zx} along the lateral side of a joint ($y = h$), computed using finite element modeling (FEM) and using proposed analytical model (AM), τ_x, Eq. 4.

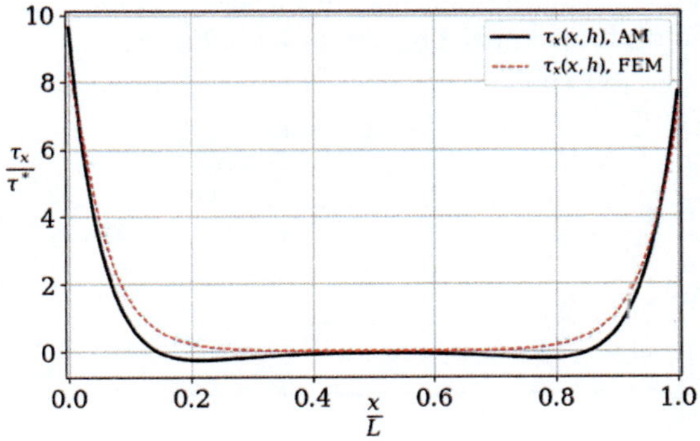

Fig. 6. Stresses τ_x in the adhesive along the lateral side

The stresses in the graph are also given in dimensionless form.

In Fig. 7 the stresses in the transverse direction, τ_{zy}, which appear in the adhesive layer along the lateral side of the joint ($y = h$) are shown. Stresses were calculated using finite element modeling (FEM) and using the proposed analytical model (AM), τ_y, Eq. 4.

As we can see, the stress values calculated using the proposed analytical model and using finite element modeling practically coincide. The most loaded are the ends of the joint, while the analytical model gives slightly overestimated values of stresses in comparison with the finite element model. This phenomenon is well known [5, 8] and is due to a set of simplifying hypotheses underlying the model. However, this does not

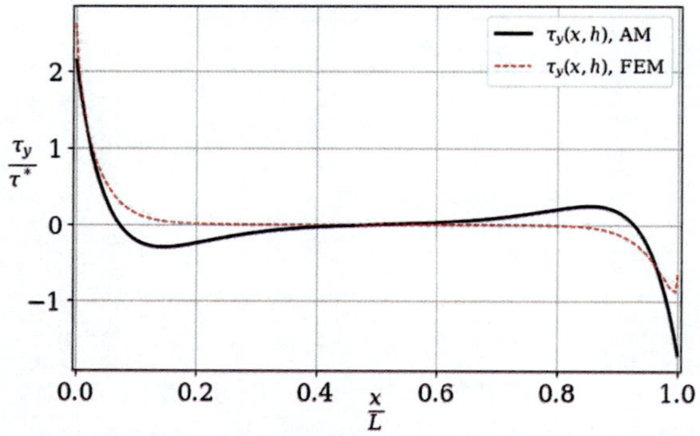

Fig. 7. Stresses τ_y in the adhesive along the lateral side

reduce the possibility of using the proposed model for solving optimization and joint design problems.

An important difference between the finite element and analytical models is that, according to the proposed analytical model, the stresses in the adhesive in the transverse direction are constant throughout the width of the plates, and change only in the longitudinal direction, since they depend only on the coordinate x, Eq. (4). Whereas a three-dimensional finite element model allows us to study the stress distribution in the adhesive layer τ_y in the transverse direction also.

Stresses in the median plane of the adhesive layer in the longitudinal direction τ_{zy}, calculated using finite element modeling are shown in Fig. 8

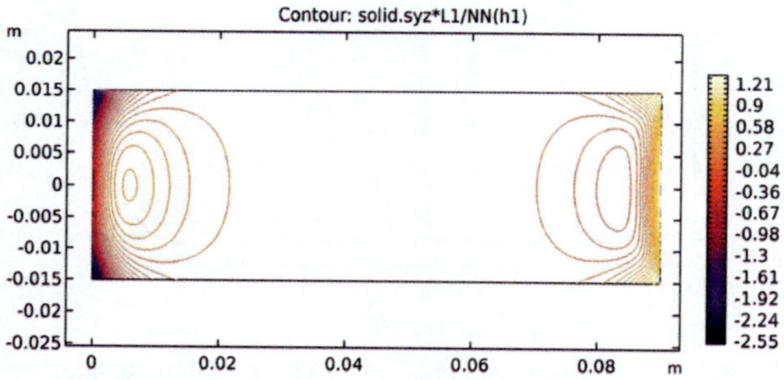

Fig. 8. Stresses τ_{zy} in the adhesive layer

In Fig. 9 graphs that illustrate the distribution of stresses in the adhesive in the transverse direction (τ_y) throughout the joint width are shown. The given graphs are obtained as a result of finite element modeling. The graphs show the stress distribution τ_y throughout the width of the adhesive layer at a distance $\frac{L}{40}$, $\frac{L}{20}$, $\frac{L}{10}$ and $\frac{L}{5}$ from the left edge of the joint.

It should be noted that finite element modeling shows that normal stresses in the base layers are distributed across the width almost linearly. And forces $N_x^{(1)}$ and $N_x^{(2)}$, Eq. (3)

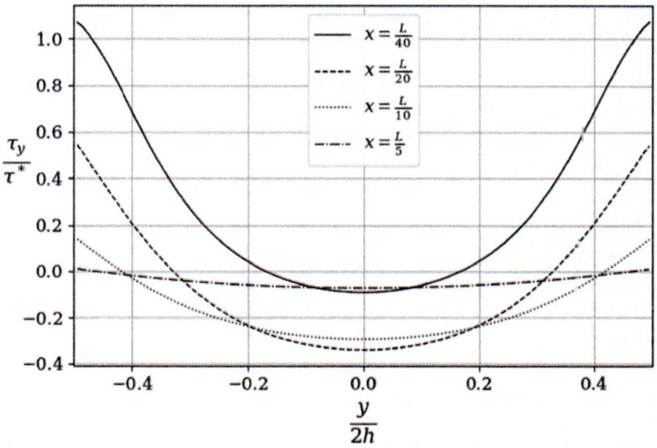

Fig. 9. Stresses in the adhesive in the transverse direction at different distances from the loaded edge

coincide with the results of finite element modeling. Hence, the application of Bernoulli beams theory to describe the shifts and deformations of the base layers is grounded.

5 Conclusions

1) A mathematical model of an overlap joint is proposed, which allows us to describe the three-layer structure stress state, caused by the bending in the gluing plane.
2) The proposed approach is based on the Bernoulli beam mathematical model and is a development of the classical Volkersen joint model [1]. Therefore, the usage of the proposed analytical model is available if the length of the joint is significantly greater than the width.
3) Finite element modeling has shown the high accuracy of the proposed model and the reliability of the hypotheses used.
4) The proposed model expands the class of problems to be solved and, together with the previous results [19], makes it possible to find the stress state of joints with an arbitrary load. To do this, it is necessary to represent the applied load as the sum of a linear load of the form (6) and a load that does not create a bending moment. This will allow us to split the problem into two independent problems, the solutions of which are known.
5) Further development of the proposed model can be aimed at finding the stress state of the coaxial pipes joints, which are loaded with a bending moment [24, 25].

References

1. da Silva, L.F.M., das Neves, P.J.C., Adams, R.D., Spelt, J.K.: Analytical models of adhesively bonded joints. part i: literature survey. Int. J. Adhes. Adhesiv **29**, 319–330 (2009). https://doi.org/10.1016/j.ijadhadh.2008.06.005

2. Budhe, S., Banea, M.D., de Barros, S., da Silva, L.F.M.: An updated review of adhesively bonded joints in composite materials. Int. J. Adhes. Adhes. **72**, 30–42 (2017). https://doi.org/10.1016/j.ijadhadh.2016.10.010
3. Stein, N., Felger, J., Becker, W.: Analytical models for functionally graded adhesive single lap joints: a comparative study. Int. J. Adhes. Adhes. **76**, 70–82 (2017). https://doi.org/10.1016/j.ijadhadh.2017.02.001
4. Wong, E.H., Liu, J.: Interface and interconnection stresses in electronic assemblies – a critical review of analytical solutions. Microelectron. Reliab. **79**, 206—220 (2017). https://doi.org/10.1016/j.microrel.2017.03.010
5. Wang, J., Zhang, C.: Three-parameter elastic foundation model for analysis of adhesively bonded joints. Int. J. Adhes. Adhes. **29**, 495–502 (2009). https://doi.org/10.1016/j.ijadhadh.2008.10.002
6. Karpov, Y.S.: Jointing of high-loaded composite structural components. part 2. modeling of stress-strain state. Strength Mater **38**(5), 481–491 (2006). https://doi.org/10.1007/s11223-006-0067-9
7. Amidi, S., Wang, J.: An analytical model for interfacial stresses in double-lap bonded joints. J. Adhes. 1–25 (2018). https://doi.org/10.1080/00218464.2018.1464917
8. Kurennov, S.S.: Longitudinal-flexural vibrations of a three-layer rod. an improved model. J. Math. Sci. **215**(2), 159–169 (2016). https://doi.org/10.1007/s10958-016-2829-7
9. Kurennov, S.S.: Refined mathematical model of the stress state of adhesive lap joint: experimental determination of the adhesive layer strength criterion. Strength Mater. **52**(5), 779–789 (2020). https://doi.org/10.1007/s11223-020-00231-5
10. Kim, H.S., et al.: Three dimensional stress analysis of a composite patch using stress functions. Int. J. Mech. Sci. **52**, 1646–1659 (2010). https://doi.org/10.1016/j.ijmecsci.2010.08.006
11. Andruet R.H., Dillard, D.A., Holzer, S.M .: Two- and three-dimensional geometrical nonlinear finite elements for analysis of adhesive joints. Int. J. Adhes. Adhes. **21,** 17–34 (2001). https://doi.org/10.1016/S0143-7496(00)00024-5
12. Rapp, P.: Mechanics of adhesive joints as a plane problem of the theory of elasticity. part II: displacement formulation for orthotropic adherends. Arch. Civ. Mech. Eng. **15**(2), 603–619 (2014). https://doi.org/10.1016/j.acme.2014.06.004
13. Kashfuddoja, M., Ramji, M.: An experimental and numerical investigation of progressive damage analysis in bonded patch repaired CFRP laminates. J. Compos. Mater. **49**(4), 439–456 (2014). https://doi.org/10.1177/0021998314521058
14. Kessentini, R., Klinkova, O., Tawfiq, I., Haddar, M.: Transient hygro-thermo-mechanical stresses analysis in multi-layers bonded structure with coupled bidirectional model. Int. J. Mech. Sci. **150**, 188–201 (2019). https://doi.org/10.1016/j.ijmecsci.2018.10.004
15. Kessentini, R., Klinkova, O., Tawfiq, I., Haddar, M.: Theoretical and experimental investigation of bonded patch repairs of a rubber reinforced composite conveyor belt. Polymers **13**(11), 1710 (2021). https://doi.org/10.3390/polym13111710
16. Mathias, J.D., Grédiac, M., Balandraud, X.: On the bidirectional stress distribution in rectangular bonded composite patches. Int. J. Solids Struct. **43**, 6921–6947 (2006). https://doi.org/10.1016/j.ijsolstr.2006.02.016
17. Kim, H.S., Cho, M., Lee, J., Deheeger, A., Grédiac, M., Mathias, J.-D.: Three dimensional stress analysis of a composite patch using stress functions. Int. J. Mech. Sci. **52**(12), 1646–1659 (2010). https://doi.org/10.1016/j.ijmecsci.2010.08.006
18. Kurennov, S.S.: An approximate two-dimensional model of adhesive joints. analytical solution. Mech. Compos. Mater. **50**(1), 105–114 (2014). https://doi.org/10.1007/s11029-014-9397-z
19. Kurennov, S.S.: A simplified two-dimensional model of adhesive joints. nonuniform load. Mech. Compos. Mater. **51**(4), 479–488 (2015). https://doi.org/10.1007/s11029-015-9519-2

20. Kurennov, S.S.: Determining stresses in an adhesive joint with a longitudinal un-adhered region using a simplified two-dimensional theory. J. Appl. Mech. Tech. Phys. **60**, 740–747 (2019). https://doi.org/10.1134/S0021894419040199
21. Kurennov, S.S., Polyakov, OG., Barakhov, K.P.: Two-dimensional stressed state of an adhesive joint. nonclassical problem. J. Math. Sci. **254**(1), 156–163 (2021). https://doi.org/10.1007/s10958-021-05295-5
22. Kurennov, S.S., Barakhov, K.P.: The Stressed state of the double-layer rectangular plate under shift. the simplified two-dimensional model. PNRPU Mech. Bull. **3**, 166–174 (2019). https://doi.org/10.15593/perm.mech/2019.3.16
23. Anyfantis, K.N., Tsouvalis, N.G.: Analysis of an adhesively bonded single lap joint subjected to eccentric loading. Int. J. Adhes. Adhes. **41**, 41–49 (2013). https://doi.org/10.1016/j.ijadhadh.2012.10.007
24. Yang, C., Huang, H., Guan, Z.: Stress model of composite pipe joints under bending. J. Compos. Mater. **36**(11), 1331–1348 (2002). https://doi.org/10.1177/0021998302036011167
25. Kumar, U., Das, R.R.: Adhesion failure analyses of laminated FRP composite made bonded tubular T-joints with axially compressed brace. Compos. Struct. **258**, 113386 (2020). doi:https://doi.org/10.1016/j.compstruct.2020.113386

Assessment of Error of the Experimental Determination of Interlaminar Shear Strength

Oleksandr Dveirin[1], Anton Tsaritsynskyi[2], Tetyana Nabokina[2], and Andrii Kondratiev[3(✉)]

[1] ANTONOV COMPANY, 1 Akademika Tupoleva Street, Kyiv 03062, Ukraine
[2] National Aerospace University "Kharkiv Aviation Institute", 17 Chkalova Street, Kharkiv 61070, Ukraine
[3] O.M. Beketov National University of Urban Economy in Kharkiv, 17 Marshal Bazhanov Street, Kharkiv 61002, Ukraine
andrii.kondratiev@kname.edu.ua

Abstract. Owing to high strength and weight perfection of the polymeric composite materials, they are used in the manufacturing of various equipment. Laminated polymeric materials are the most studied and widespread composite varieties. These materials implement high elastic and strength characteristics in the reinforcement plane; however, they have minimal performance capabilities at interlaminar shear. Assessment of non-uniform distribution of shear stresses over the shear surface is carried out for the specimens intended for determination of the interlaminar strength according to the current standards. Based on the model of adhesive joint, the dependence is obtained which relates the error in assessment of the interlaminar shear strength to the elastic characteristics of the material and geometry of the specimen's working part. The dependence allows controlling the degree of underestimation of interlaminar shear strength of the material when planning the experiment. Modification of specimens recommended by normative documents for interlaminar shear tests is substantiated. The results of experimental study of unidirectional carbon fiber composite interlaminar strength using modified specimens are presented. The typical nature of failure of specimens with 90° and 0° layup patterns after tests is analyzed. According to results of the experiments, values of interlaminar strength are at the level of 53.5 MPa, with the variation coefficient of 3.5% and 5.7% underestimation of the result.

Keywords: Composite specimen · Non-uniform distribution · Strength reduction · Adhesive joint

1 Introduction

At present time, polymeric composite materials (PCM) have widespread application in various industries, successfully replacing metals, which are traditionally used in many branches of technology [1, 2]. Owing to the transition to designing and manufacture of load-bearing elements of structures made of PCM, in particular, in transport, aviation and

M. Nechyporuk et al. (Eds.): ICTM 2021, LNNS 367, pp. 14–25, 2022.
https://doi.org/10.1007/978-3-030-94259-5_2

astronautics, the relevance of the problem of their joints increases [3, 4]. For many reasons, no determination of the interlaminar shear bearing capacity is given in the existing system for calculation of mechanical joints [5, 6]. Firstly, in the bearing-type mechanical joints of thin-walled structures, ratio of diameter of the fastening element to thickness of the package being fastened allows considering the problem as two-dimensional one (in the plane of a part). Secondly, shear characteristics of metals make shear failure in the plane of fastening element practically impossible. Thirdly, mechanical joint design and technology solutions are aimed at reducing the non-uniformity of stresses across the thickness of the package [7, 8]. In case of joining of parts made of PCM, their delamination in the mechanical joint is quite likely, since the interlaminar shear strength of the material is low [9, 10]. Therefore, correct setting of the permissible level of interlaminar shear stresses is essential in order to ensure the bearing capacity and resource of the joints.

2 Literature Review

To determine the interlaminar shear strength, the methods developed for determination of shear characteristics of the package in the laying plane are used [11]. Obtaining of the uniform field of stresses is the main difficulty in the performance of interlaminar shear tests [12]. It gave rise to the large number of test methods, of which there are more than a dozen today [13]. Detailed reviews of these tests are presented in [11, 14]. The features and performance potential of the available shear test methods are compared in the review [14] (Table 1). This paper also gives a conclusion that the Iosipescu method can be recognized the only accurate method, which requires the special equipment, use of strain gauges and specimens of the complex geometry.

Table 1. A comparison of the features and performance potential of available shear test methods [14]

Test method (with ASTM Sid. No., if applicable)	Uniform shear stress state	All three stress states practical	Shear strength obtained	Shear stiffness obtained
Short beam shear (D 2344)	–	–	+	–
Losipescu shear (D 5379)	+	+	+	+
± 45° tensile shear (D3518)	–	–	+	+
Two-rail shear (D 4255)	–	–	+	+
Three-rail shear (D4255)	–	–	+	+

(*continued*)

Table 1. (*continued*)

Test method (with ASTM Sid. No., if applicable)	Uniform shear stress state	All three stress states practical	Shear strength obtained	Shear stiffness obtained
Double-notched shear (D 3846)	+	–	+	+
Torsion of a thin tube (D 5448)	+	–	+	+
Cross-beam sandwich	–	–	+	+
Torsion of a solid rod	–	–	+	+
Four-point shear	–	–	+	–
Picture frame shear	–	–	+	+
Plate twist	–	–	–	+
Off-Axis (Tensile)	–	–	+	+
V-notched rail shear (D 7078)	+	+	+	+

The methods of short beam (three-point bending), two-point plate twist and notched specimen tension are dealt with in [11, 13].

Short beam shear test [13] is the simplest method while giving a fairly good estimate of the ultimate interlaminar shear strength. Peculiar feature of such tests is the need to ensure the failure on shear stresses [15]. More detailed analysis of the features of the short beam failure is given in [16]. In particular, it is indicated that when the ratio of thickness of the sample δ to the distance between supports l is less than 4, there is a probability of collapsing and interlaminar shear (the smaller the ratio, the higher the probability of collapsing). With $4 \leq l/\delta \leq 8$ the failures resulting from the interlaminar shear and bending may occur. Furthermore, as moving closer to the lower limit of the range, probability of the shear increases. Therefore, scatter of results can be explained by the transition from one type of failure to another. The papers [12, 15] present the analysis of influence of the distance between the supports on the value of the ultimate strength. This parameter significantly affects the failure mode: at small distance between the supports the specimen collapses, while the impact of tensile and compressive stresses increases at large distance between supports.

The classical method of plate twist (three-point twist) [13] does not provide the determination of the ultimate interlaminar shear strength, but so-called method of two-point plate twist is described in [11, 16]. In accordance with this method, the specimen should be strictly flat, without initial deflections and curvatures, and its thickness δ should be constant. In plan, the plate should have the shape of a square with side l. The recommended side to thickness ratio is $25 \leq l/\delta \leq 100$, but the results of tests of fiberglass, carbon fiber composite and boron-fiber reinforced plastic mentioned in [11, 16] indicate that already at $l/\delta \geq 15$ there is a good reproducibility of results.

One of the most accurate methods for determination of the interlaminar strength of PCM is the tension of notched specimens (Fig. 1) [13, 16, 17]. The method is quite simple in use, but it has significant drawbacks: in case of testing of specimens with asymmetrical notches (Fig. 1, a), the devices to prevent bending are required; besides, assessment of stress concentration is the laborious procedure, and there is high sensitivity to the quality of making notches. Since an undercut or overcut is unacceptable, it is advisable to generate the notches and hole (Fig. 1, b) in the process of the specimen molding.

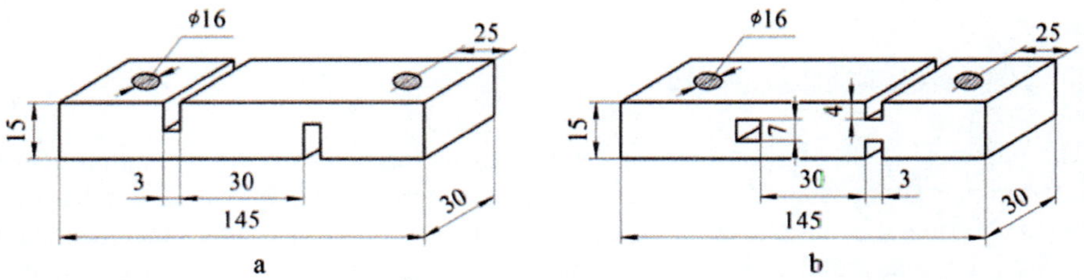

Fig. 1. Notched specimens: a – with asymmetrical notches; b – with symmetrical notches and a hole

There is an open question concerning the determination of distance between notches, where the shear occurs (length of the tested joint). According to the data of [13, 16], the recommended value of the distance is between 10 and 30 mm. The review of test methods [14] shows that at present time, along with tension of notched specimens, the uniform stressed state can be obtained with the use of V-Notched Rail Shear method, but it is also rather difficult to implement in practice.

With regard to mechanical joints, interlaminar shear strength in the system of theoretical calculations has not been sufficiently investigated [18]. In a number of papers [19–21] the laminated composite is considered as a bonded structure. These papers outline the idea (which undoubtedly deserves attention) that interlaminar stresses in the laminated composite are similar to those in the adhesive joint. In this case, the principle of "smearing" of the fastening element is used for its further calculation as the adhesive. This principle has proven itself well in the calculations of compounds with cross-link micro-elements [22] and sandwich structures [23]. The exact solution for analysis of the characteristics of the laminated structures' interlaminar shear is proposed in [24]. The paper [25] describes the function of interlaminar failure, on the basis of which the criteria for delamination failure are established, taking into account the effect of compression across the thickness at the interlaminar shear failure. However, the existing methods are applicable for the laminated structures with simple boundary conditions only. The limitations of the analytical theory and impossibility of its use in the analysis of the interlaminar mechanical properties of PCM are indicated in [26]. This paper proposes to study interlaminar properties of the composite laminates with the use of numerical methods.

Therefore, we can conclude that the problem of assessment of the interlaminar shear strength is multifaceted and controversial one. On the one hand, it is necessary to provide the most accurate description of the behavior of the material in the shear zone of the real joint. On the other hand, during interpretation of the results and their analysis it is necessary to remain within the framework of the engineering methodology for calculation of the joint. The latter requirement restricts the use of numerical methods for the interlaminar shear zones' calculation.

The purpose of the study is to assess non-uniformity of distribution of shear stresses over the shear surface for the specimens meant for determination of the interlaminar strength according to the current standards, and to obtain a dependence relating the error in the assessment of the interlaminar shear strength to elastic characteristics of the material and geometry of the specimen working part.

3 Research Methodology

The papers [11, 13] suggest that for the specimens intended for determination of the interlaminar strength according to the current standards, two approaches can be used when choosing the dimensions of their joints. The first approach assumes that the length of the joint is as small as possible. In this case, the magnitude of stress overshoots is the lowest and stress distribution can be considered uniform along the joint. The second approach consists in choosing the largest possible size of the joint. In this case, the stress peaks will be maximum, but the areas of their manifestation are negligible compared to dimensions of the joint. It allows neglecting the effect of stress non-uniformity in the analysis of results of the experiments. The papers [5, 10, 22] show that in case if the joint is too long, unloaded zone appears in its middle part. It makes the second approach proposed in [11, 13] completely inapplicable.

To resolve the issue of the permissible dimensions of the specimen working zone, we consider it as an adhesive joint of parts of the same stiffness (Fig. 2).

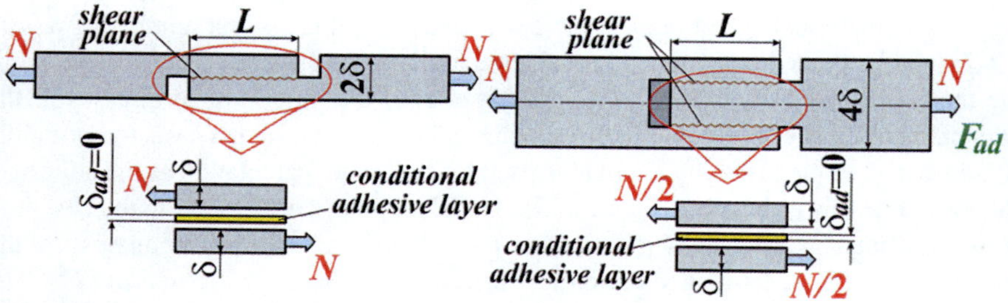

Fig. 2. Modeling of specimens for determination of interlaminar strength as an adhesive joint

The essence of the method is to assess the defect in the bearing capacity of the joint when comparing two one-dimensional models of the joint. The original model is used to calculate the ultimate strength of the package based on the test results.

$$F_{int} = \frac{\overline{N}}{L}, \tag{1}$$

where $\overline{N}$– ultimate linear load (load breaking a specimen and related to its width); L – control length of the joint (distance between notches in the specimen).

From (1) it follows that the first estimated value can be obtained as:

$$\overline{N}^{I} = F_{int}L. \tag{2}$$

The control model is more accurate one, since it takes into account non-uniform distribution of stresses in the connecting layer along the joint. The model, despite its simplicity, describes the adhesive layer stressed state rather accurately [19, 20]. The specimen itself is designed so that under loading the stress distribution curve has two practically identical peaks (see Fig. 3).

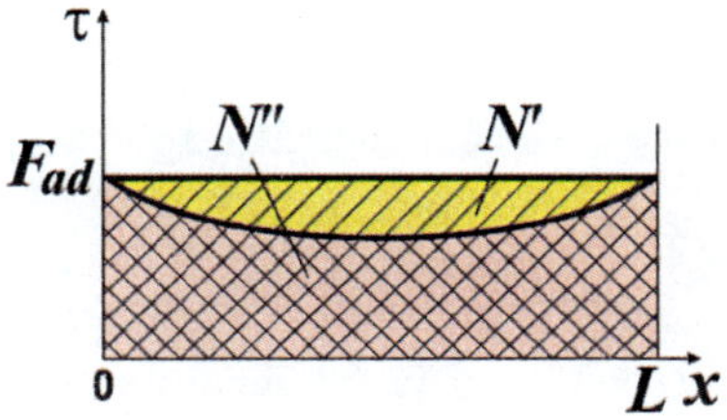

Fig. 3. Stress distribution curve in the connecting layer along the joint

So, it is possible to assess the ultimate load at which the values of stresses in the peaks reach the ultimate interlaminar shear strength and delamination of the specimen occurs

$$\overline{N}^{II} = F_{int}\delta\sqrt{\frac{2E_x}{G_{int}}}th\frac{kL}{2}, \tag{3}$$

where δ – thickness of specimen from its free side to the notch (see Fig. 2); E_x and G_{int} – modulus of elasticity of the material (package) of the specimen along the axis of loading and interlaminar shear module, respectively; k – coefficient of stress concentration along the adhesive joint (according to the Volkersen model [19, 20])

$$k = \frac{1}{\delta}\sqrt{\frac{2G_{int}}{E_x}}. \tag{4}$$

For the further assessment of error of the interlaminar shear, it is sufficient to substitute the load value (3) in (1) taking (4) into account and divide the left and right sides of the equation by the ultimate strength of the package F_{int}. In this case, we obtain the formula to assess the error of the package interlaminar shear strength

$$\Delta = \frac{1}{\overline{L}}\sqrt{2\overline{E}_x}th\left(\frac{\overline{L}}{2}\sqrt{\frac{2}{\overline{E}_x}}\right), \tag{5}$$

where $\overline{L} = L/\delta$; $\overline{E}_x = E_x/G_{int}$.

Now it is sufficient to set the permissible level of error in the assessment of strength, and it is possible to estimate the permissible ratios of the geometry of the specimen for the known ratios of the elastic parameters of the material. For example, when setting the permissible error of 5% we get $\Delta = 0.95$. The relevant solutions of Eq. (5) are given in Table 2.

Table 2. Maximum permissible values of the geometric parameters of the specimen working zone at $\Delta = 0.95$

Elastic properties of the package $\overline{E}_x$	3	50	113	350	546	750	900
Working zone relative length $\overline{L}$	1.0	4.0	6.0	10.56	13.19	15.46	16.94

Based on the data presented, we constructed an approximating dependence

$$\overline{L} \leq 0,5644\sqrt{\overline{E}_x}, \tag{6}$$

Therefore, it is every reason to correct the dimensions of specimens recommended by the normative documents for interlaminar shear tests (Fig. 4).

The papers [11, 13] show the versions of specimens with the relative geometry of the working zone $\overline{L} = 4$ (see Fig. 1, a) and $\overline{L} = 7.5$ (see Fig. 1, b).

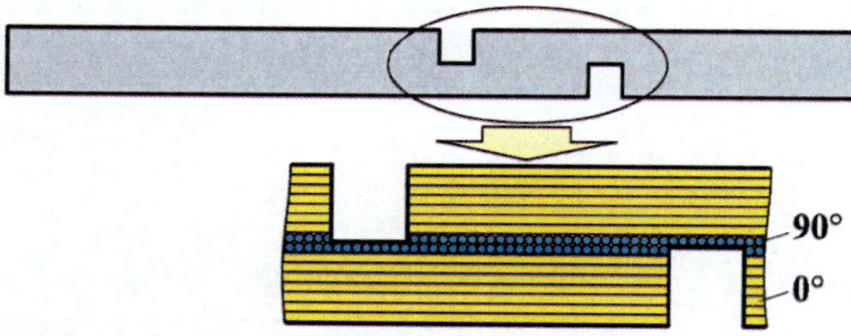

Fig. 4. Modified version of the specimen

Solution of the function inverse to (6) shows that in order to ensure underestimation of the interlaminar strength of up to 5%, the first version of the specimen (Fig. 1, a) is applicable for materials with $\overline{E}_x \geq 50.15$, while the second version (Fig. 1, b) is suitable for the materials with $\overline{E}_x \geq 177$ only. Thus, the first version is applicable for most composites with priority layup pattern of 0°, and the second one can be used only for unidirectional packages made of high-modulus carbon fiber composites and boron-fiber reinforced plastics.

4 Results

According to the recommendations of the standards, dimensions of the specimen should correspond to those indicated in Fig. 5 [11, 13].

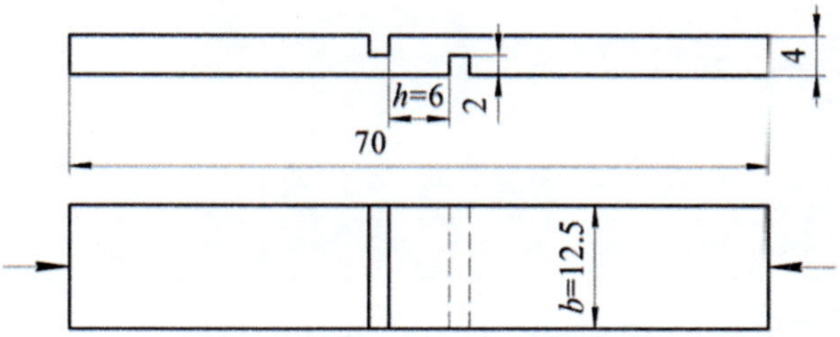

Fig. 5. Interlaminar shear test specimen

For the manufacture of modified specimens, we used unidirectional carbon fiber composite based on ELUR-0.08P tape and 5–211-BN binder with the properties given in Table 3. Two versions of the PCM package were tested with the layup patterns of 0° and 90° to the axis of loading. As shown by Fig. 5, the relative length of the working zone is equal to three. The check according to the proposed method showed that for the carbon fiber composite with 0° layup pattern $\overline{E}_x = 24.64$ (see Table 3). Therefore, according to (5), the ultimate strength estimate will be 0.943 of the real one. It means that underestimation of the strength assessment is 5.7%. For the package with 90° layup pattern – $\overline{E}_x = 1.274$, i.e. $\Delta = 0.507$, and underestimation of the estimate will reach 49.3%.

Tests of specimens with layup pattern of 90° to the axis of loading ended in combined failure in the weakened sections and working zone of Fig. 6, instead of pure shear.

Table 3. Properties of unidirectional carbon fiber composite based on ELUR-008 P/5–211-BN

Parameter	Value	Parameter	Value
Monolayer thickness, mm	0.087	Ultimate tensile strength across the grain, MPa	21
Ultimate tensile strength along the grain, MPa	677	Ultimate compressive strength across the grain, MPa	210
Ultimate compressive strength along the grain, MPa	939	Modulus of elasticity across the grain, GPa	7
Modulus of elasticity along the grain, MPa	136	Ultimate shear strength in the laying plane, MPa	75
Poisson's ratio	0.32	Shear modulus in the laying plane, GPa	5.5

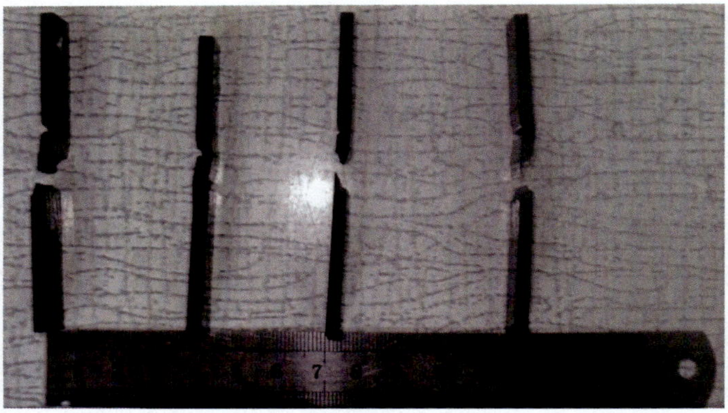

Fig. 6. Specimens with 90° layup pattern after tests

Most likely, the test recommended by the standards [11, 13] in the direction of 90° to the layup can be performed on woven fabrics or unidirectional materials with the dominating layup pattern of 0°, and within the median surface – 90° (Fig. 6).

Tests of specimens with layup pattern of 0° to the axis of loading gave positive results: interlaminar shear failure took place in the area limited by grooves (Fig. 7).

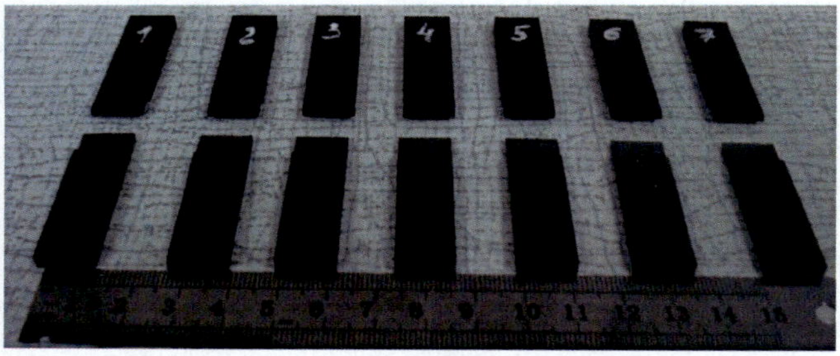

Fig. 7. Specimens with 0° layup pattern after tests

The results obtained are given in Table 4. Statistical processing of the results showed that the arithmetic mean value of the ultimate interlaminar shear strength is 53.5 MPa with the variation coefficient of 3.02%.

According to the obtained values of the ultimate interlaminar shear strength (Table 4), values of the minimum distance to the edge, at which delamination of the composite part is possible, were assessed. The results are shown in Table 5.

Table 4. Interlaminar shear test results for specimens with 0° layup pattern

Specimen number	Dimensions of the specimen working zone (average value)		Area, mm^2	Force, N	Stress, MPa
	Width b, mm	Length h, mm (distance between grooves)			
1	12.52	6.13	76.75	4200	54.72
2	12.61	6.04	76.16	4300	56.46
3	12.64	5.75	72.7	3850	52.96
4	12.55	6.06	76.05	3900	51.28
5	12.51	5.54	69.31	3900	56.27
6	12.58	6.04	75.98	3650	48.02
7	12.51	5.4	67.57	3700	54.76

The relative distance to the edge should be not less than 2.5 diameters in the real structures according to recommendations [5, 7, 9]. It follows from Table 5 that for the carbon fiber composite based on ELUR-008 P/5–211-BN no delamination should occur in almost all cases (except for the use of high-strength fastening elements).

Table 5. Relative distance to the edge at which the composite part may delaminate

Value of ultimate tensile strength of the fastening element, MPa	Relative width of the part per one load-bearing point	Required relative distance to the edge of the part
300	4	1.101
	5	0.881
600	4	2.202
	5	1.762
900	4	3.303
	5	2.642

5 Conclusions

Assessment of non-uniform distribution of shear stresses over the shear surface is carried out for the specimens intended for determination of the interlaminar strength according to the current standards. It is shown that the experimental value of the strength is hardly probable. It is a positive factor; however, underestimation of strength can be unacceptably significant.

Based on the unidirectional model of adhesive joint, the dependence is obtained which relates the error in assessment of the interlaminar shear strength to the elastic characteristics of the material and geometry of the specimen's working part. The dependence allows controlling the degree of underestimation of interlaminar shear strength of the material when planning the experiment.

The analytical dependence is obtained for determination of the limit value of the ratio of specimen dimensions depending on elastic characteristics of the material under condition of 5% error in the assessment of the interlaminar shear strength.

According to the results of the experiments for unidirectional carbon fiber composite based on ELUR-0.08P tape and 5–211-BN binder, the obtained values of interlaminar strength are at the level of 53.5 MPa, with the variation coefficient of 3.5% and 5.7% underestimation of the result.

References

1. Hsissou, R., Seghiri, R., Benzekri, Z., Hilali, M., Rafik, M., Elharfi, A.: Polymer composite materials: a comprehensive review. Compos. Struct. **262**, 15 (2021). https://doi.org/10.1016/j.compstruct.2021.113640
2. Fomin, O., Lovska, A.: Improvements in passenger car body for higher stability of train ferry. Eng. Sci. Technol. Int. J.-Jestech **23**(6), 1455–1465 (2020). https://doi.org/10.1016/j.jestch.2020.08.010
3. Rodichev, Y.M., Smetankina, N.V., Shupikov, O.M., Ugrimov, S.V.: Stress-strain assessment for laminated aircraft cockpit windows at static and dynamic loads. Strength Mater. **50**(6), 868–873 (2019). https://doi.org/10.1007/s11223-019-00033-4
4. Kondratiev, A., Slivinsky, M.: Method for determining the thickness of a binder layer at its nonuniform mass transfer inside the channel of a honeycomb filler made from polymeric paper. East.-Eur. J. Enterp. Technol. **6**(5–96), 42–75 (2018). https://doi.org/10.15587/1729-4061.2018.150387
5. Karpov, Y.S.: Jointing of high-loaded composite structural components. Part 1. Design and engineering solutions and performance assessment. Strength Mater. **38**(3), 234–240 (2006). https://doi.org/10.1007/s11223-006-0036-3
6. Kondratiev, A.: Improving the mass efficiency of a composite launch vehicle head fairing with a sandwich structure. East.-Eur. J. Enterp. Technol. **6**(7–102), 6–18 (2019). https://doi.org/10.15587/1729-4061.2019.184551
7. Li, W., Chen, W., Tang, L.Q., Jiang, Z.Y., Huang, P.Y.: A general strength model for fiber bundle composites under transverse tension or interlaminar shear. Compos. Part Appl. Sci. Manuf. **121**, 45–55 (2019). https://doi.org/10.1016/j.compositesa.2019.03.009
8. Kovalov, A., Otrosh, Y., Rybka, E., Kovalevska, T., Togobytska, V., Rolin, I.: Treatment of determination method for strength characteristics of reinforcing steel by using thread cutting method after temperature influence. Mater. Sci. Forum **1006**, 179–184 (2020). https://doi.org/10.4028/www.scientific.net/msf.1006.179
9. Zhai, Y.N., Li, D.S., Li, X.Q., Wang, L.: An experimental study on the effect of joining interface condition on bearing response of single-lap, countersunk composite-aluminum bolted joints. Compos. Struct. **134**, 190–198 (2015). https://doi.org/10.1016/j.compstruct.2015.08.078
10. Karpov, Y. S.: Jointing of high-loaded composite structural components. part 3. an experimental study of strength of joints with transverse fastening microelements. Strength Mater. **38**(6), 575–585 (2006). https://doi.org/10.1007/s11223-006-0079-5

11. Chaterjee, S., Adams, D., Oplinger, D.: Test Methods for Composites: a Status Report. Materials Science (1993)
12. Boldyreva, A.A., Yarunicheva, Y.A., Dernakova, A.V., Ivashov, I.I.: The strength of the polymer composite (fiberglass) with interlaminar shear. Mag. Civ. Eng. **62**(2), 42–50 (2016). https://doi.org/10.5862/mce.62.5
13. Shah, V.: Handbook of Plastics Testing and Failure Analysis. 3rd ed. Wiley-Interscience. Consultek Brea. California (2007)
14. Adams, D.: A comparison of shear test methods. Delivering the Global Composites Market. (2009) https://www.compositesworld.com/articles/a-comparison-of-shear-test-methods
15. Merzkirch, M., Foecke, T.: Investigation of the interlaminar shear properties of fiber-reinforced polymers via flexural testing using digital image correlation. Mater. Perform. Charact. **9**(5), 666–674 (2020). https://doi.org/10.1520/mpc20190206
16. Vasiliev, V.V., Protasov, V.D., Bolotin, V.V.: Composite materials: Reference (Under total. ed. V.V. Vasilieva, Y.M., Tarnopolsky, Eds.). Mashinostroenie. Moskva (1990)
17. Melin, L.G., Neumeister, J.M., Pettersson, K.B., Johansson, H., Asp, L.E.: Evaluation of four composite shear test methods by digital speckle strain mapping and fractographic analysis. J. Compos. Tech. Res. **22**(3), 161–172 (2000). https://doi.org/10.1520/CTR10636J
18. Mohamed, O.A., Kewalramani, M., Khattab, R.: Fiber reinforced polymer laminates for strengthening of RC slabs against punching shear: a review. Polymers **12**(3), 685 (2020). https://doi.org/10.3390/polym12030685
19. Barakhov, K., Dvoretska, D., Poliakov, O.: One-dimensional axisymmetric model of the stress state of the adhesive joint. In: Nechyporuk, M., Pavlikov, V., Kritskiy, D. (eds.) ICTM 2020. LNNS, vol. 188, pp. 310–319. Springer, Cham (2021). https://doi.org/10.1007/978-3-030-66717-7_26
20. Kurennov, S.S., Polyakov, OG., Barakhov, K.P.: Two-dimensional stressed state of an adhesive joint. nonclassical problem. J. Math. Sci. **254**(1), 156–163 (2021). https://doi.org/10.1007/s10958-021-05295-5
21. Kondratiev, A., Gaidachuk, V., Nabokina, T., Kovalenko, V.: Determination of the influence of deflections in the thickness of a composite material on its physical and mechanical properties with a local damage to its wholeness. East.-Eur. J. Enterpr. Technol. **4**(1–100), 6–13 (2019). https://doi.org/10.15587/1729-4061.2019.174025
22. Beketova, G., Shevtsova, M., Symonov, V.: Static and fatigue characteristics of pinned metal-composite joints. Mech. Compos. Mater. **55**(5), 655–666 (2019). https://doi.org/10.1007/s11029-019-09842-9
23. Slyvynskyi, V.I., Alyamovskyi, A.I., Kondratjev, A.V., Kharchenko, M.E.: Carbon honeycomb plastic as light-weight and durable structural material. In: 63th International Astronautical Congress IAC, vol. 8, pp. 6519–6529. Curran, Red Hook, NY (2012)
24. Chen, W.Q., Cai, J.B., Ye, G.R.: Exact solutions of cross-ply laminates with bonding imperfections. AIAA J. **41**(11), 2244–2250 (2003). https://doi.org/10.2514/2.6817
25. Chen, X.M., Sun, X.S., Chen, P.H., Chai, Y.A.: A delamination failure criterion considering the effects of through-thickness compression on the interlaminar shear failure of composite laminates. Compos. Struct. **241**, 8 (2020). https://doi.org/10.1016/j.compstruct.2020.112121
26. Zhao, L., Gong, Y., Zhang, J.: A survey on delamination growth behavior in fiber reinforced composite laminates. Hangkong Xuebao/Acta Aeronautica et Astronautica Sinica **40**(1), 22200–022200 (2019). https://doi.org/10.7527/S1000-6893.2018.22509

Tensile Stress as a Factor of Deformation and Failure of Thin Films and Coatings

Oleksandr Shorinov(✉)

National Aerospace University "Kharkiv Aviation Institute", Kharkiv 61070, Ukraine
o.shorinov@khai.edu

Abstract. The paper provides analysis of literature data regarding reasons for development of tensile stresses in continuous thin films and coatings, in the process of their formation and under various external influences, as well as study of main mechanisms of their deformation and failure under tensile stresses. Special focus is drawn to the role of film/substrate interface in failure of thin-film structures. Factors controlling initiation and propagation of failure throughout film-substrate system are considered. Main reasons for emergence of the most common models of failure of thin-film structures – edge delamination from the base and film cracking – are demonstrated. In the process of film failure, each of these mechanisms can contribute to the development of the other one, and on the other hand, they both provide relaxation of tensile stresses, thereby competing and suppressing the development of each other. Study of stress relaxation is important for obtaining new structures and certain physical and mechanical properties of coatings. Determining the critical values of stresses that lead to failure is an urgent task, regardless of the method of obtaining coatings and thin films. Development of theoretical basis for predicting and regulating values and sign of stresses that occur during the formation of nanostructures in a plasma environment complement existing models for obtaining these structures in order to further study the plasma methods of growing nanostructures.

Keywords: Tensile stress · Stress-strain state · Thin films and coatings · Nanotechnology · Coating failure

1 Introduction

Carbon and oxide nanostructures are promising materials for use in nanoelectronics and other brunches of industry. Silicon-based nanostructures are considered as potential materials for the production of the next generation of anodes of lithium-ion batteries due to their ultra-high specific capacities [1–3]. Due to their properties, vertically oriented structures of graphene can be used for accumulation and storage of electrochemical energy (in high-performance supercapacitors, fuel cell catalysts, etc.) [4, 5]. In the manufacture of modern semiconductors, field-effect transistors, sensors, oxide nanostructures are widely used, the properties of which largely depend on the methods and conditions of their formation [6, 7]. It should be mentioned that the presence of a boundary between

M. Nechyporuk et al. (Eds.): ICTM 2021, LNNS 367, pp. 26–35, 2022.
https://doi.org/10.1007/978-3-030-94259-5_3

the main material and a layer formed on its surface is a principal feature of all technologies that use the particles (atoms, molecules, or clusters) with energy less than a few keV [8]. A lot of experiments are dedicated to control the parameters of the treating flux to improve the performance of the coatings and thin films [9, 10]. Failure of thin films and coatings is one of the main causes of failure of parts and products on which they are applied caused by combined action of internal mechanical stresses (hereinafter referred to as stresses), stresses arising in the process of their formation (residual), and stresses arising during operation. Failure of films and coatings can be manifested in the form of cracking under tensile stresses [11, 12] or delamination, curvature or swelling at compressive stresses [13–15]. Residual stresses arising in coatings and films affect almost all major performance characteristics of parts. In addition to the possibility of cracking and delamination, residual stresses affect adhesion and cohesion, static and multi-cycle strength, wear resistance, heat resistance [16–18], and can positively affect physical properties of thin layers and nanostructures, such as electrical conductivity [19], dielectric constant [20], piezoelectricity [21], etc.

The reasons for occurrence and development of stresses in coatings and thin films can be as follows:

1) stresses associated with growth of coatings and films. In this case, it is necessary to distinguish between internal stresses due to increasing coverage and stresses caused by geometric characteristics of the surface. The increase in internal stresses is due to chemical reactions, phase transformations, particle bombardment [22–24], etc.;
2) thermal stresses that occur as a result of changes in temperature and various coefficients of thermal expansion of coating materials and substrate [25];
3) external stresses arising from deformation of coating-substrate system [26].

The objective of this article is to analyze the existing models of deformations and failures of thin films and coatings caused by tensile stresses. Understanding the nature of these stresses in thin films and coatings, as well as the possibility to control the stress-strain state of the coating-substrate system is an urgent task, as it directly affects formation, processing and service life of new materials and products.

2 Edge Delamination of Films from Substrate

Under tensile stresses in the film, the effects associated with the stress concentration arising near the free film edge are of particular importance. Figure 1 shows a diagram illustrating the development of shear stresses at film substrate interface as a result of load transfer from the substrate to the film.

If the film is separated from the substrate, then the elastic relaxation of tensile stresses in free state would lead to a decrease in its transverse dimensions (see Fig. 1a). However, under conditions of rigid connection with the substrate, film dimensions need to correspond to substrate dimensions. This situation is equivalent to the application of external forces to unstressed film, which stretch the film to substrate dimensions, leading to its biaxial deformation (see Fig. 1b). If the stretched film is now rigidly fixed on the substrate again, then in order to remove applied external forces in accordance with the principle of superposition, it is necessary to apply forces of the same magnitude – but with the opposite sign – to free edges of the film.

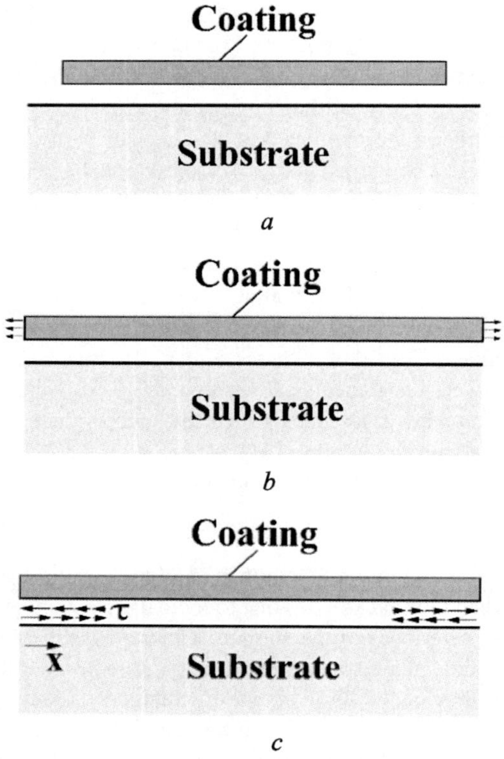

Fig. 1. Schematic diagram of occurrence of shear stresses at film/substrate interface: *a* – unstressed film in a free state; *b* – tensile stresses are applied to the film so that its transverse dimensions correspond to substrate dimensions; *c* – after fixing the film on the substrate, removal of external forces leads to development of shear stresses near the free edge of film.

As a result of the application of these compensating forces at the film edges, normal stresses become equal to zero, and shear stresses appear at film/substrate interface near free edges of the film (see Fig. 1c). These shear stresses provide biaxial deformation of the film, and, thereby, maintain biaxial tensile stresses therein.

In the membrane approximation, when the film is absolutely flexible, and it has no resistance to bending deformation, which makes it possible to ignore the influence of bending moment on the stress-strain state, distribution of shear stresses at the interface $\tau(x)$ is described by the integral equation [27]:

$$\frac{1}{h_f \overline{E}_f} \int_0^x \tau(\xi) d\xi - \frac{2}{\pi \overline{E}_s} \int_0^\infty \frac{\tau(\xi)}{\xi - x} d\xi = (1 + \nu_f)\varepsilon_m, \ 0 < x < \infty \quad (1)$$

where h_f is film thickness, $\overline{E}_f = \frac{E_f}{\left(1-\nu_f^2\right)}$ and $\overline{E}_s = \frac{E_s}{\left(1-\nu_s^2\right)}$ are the moduli of plane deformations, and E_f, E_s, ν_f, and ν_s are Young's moduli and Poisson's ratios of film and substrate, respectively, ε_m is biaxial deformation of film at a distance from the edge ($x \to \infty$).

The exact analytical solution of Eq. 1 is unknown. However, it was shown that, asymptotically at $x \to 0$ shear stresses change as follows:

$$\tau(x) \sim \sigma_m \sqrt{\frac{kh_f}{2\pi x}}, \tag{2}$$

where

$$k = \frac{\overline{E}_s}{E_f} = \frac{E_s}{E_f}\frac{1 - \nu_f}{1 - \nu_s}. \tag{3}$$

Thus, the distribution of shear stresses has a singularity in the limit at $x \to 0$, where the magnitude of these stresses tends to infinity.

Film stiffness, which determines its resistance to bending, leads to the need to consider additional effects arising from occurrence of bending moment. For instance, within the Kirchhoff plate model, the tendency to compress the surface layer of a stretched film causes an internal bending moment M and normal stresses to the interface p near its free edge, which tend to delaminate the film from the substrate (see Fig. 2) [28].

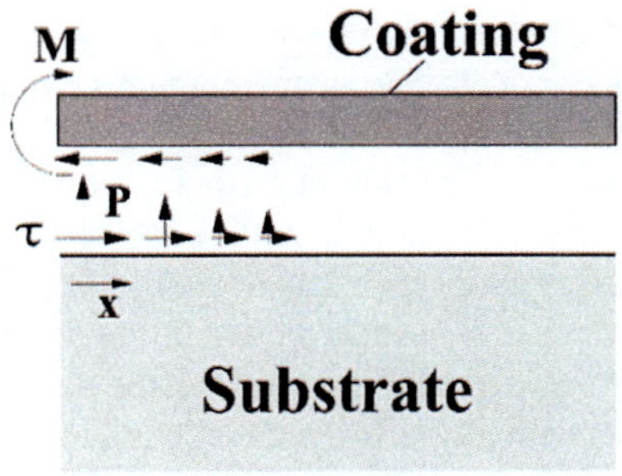

Fig. 2. Diagram of stress distribution near the free edge of film in the framework of Kirchhoff model.

To determine the dependences $\tau(x)$ и $p(x)$ which, as well as the dependence $\tau(x)$ in the model of the membrane film, have a singularity at the edge of the film at $x = 0$, it is necessary to solve the system of integral equations. The results of its numerical solution for the case when the film and the substrate have the same elastic properties ($k = 1$) were obtained in [27].

Occurrence of concentration areas of shear τ and normal stresses p near the free edges of film leads to the fact that one of the most frequent types of failure of thin films and coatings under stress is their edge delamination (see Fig. 3). This mechanism of failure of thin-film structures was observed in many experimental studies [13, 28].

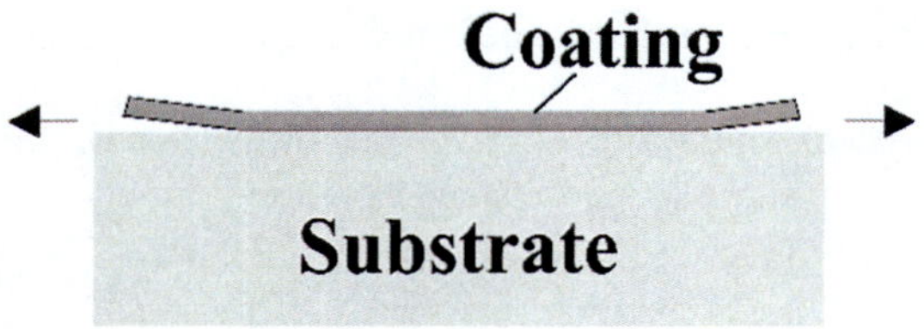

Fig. 3. Film edge delamination from the substrate under tensile stresses.

In theoretical studies, the problem of edge delamination of films and coatings is considered within the framework of two approaches to failure mechanics. The first one assumes that initially the interface between film and substrate is defect-free and delamination starts from the free film edge [29]. Various models are used to study distribution of shear and normal stresses in the immediate vicinity of the film edge and their role in the formation of delamination area. In the general case, the occurrence of edge delamination – failure initiation along the interface – is the result of their joint action, i.e. is a mixed form of failure, which is controlled by a combination of deformation modes I and II. However, limiting cases are also considered when edge delamination develops as a result of only normal [30] or only shear stresses [29] on the interface. In fact, this approach is aimed at developing recommendations for reducing the magnitude of stresses at the interface by selecting the optimal ratio of thicknesses, elastic, thermal, and other characteristics of film and substrate or individual layers in multilayer structures [30]. However, to date, it does not provide an unambiguous criterion to accurately predict the conditions under which edge delamination occurs.

Another approach is based on the assumption that delamination originates in the area of an already existing failure with a sharp tip at the edge of film/substrate interface [31]. This failure can be both the result of separation of the film edge from the substrate under stresses at the interface, which are considered within the first approach, and the initial defect that arose during the film deposition. Stresses at the film/substrate interface tend to cause propagation of the initial failure, while the strength of the material prevents this. This situation can be described using the Griffiths energy failure criterion, according to which the failure propagates if the resulting decrease in potential energy (elastic deformation energy) is greater than or equal to the work required for failure of the material per unit area resulting from this free surface.

The change in the elastic deformation energy of a body ∂U as a result of the crack propagation of length a to an infinitely small distance per unit increment of the crack length is characterized in failure mechanics by the value G, which is called the rate of deformation energy release:

$$G(a) = -\frac{\partial U(a)}{\partial a}. \quad (4)$$

In turn, the energy required for breaking bonds and failure increase, per unit increment of its length, is the specific fracture work Γ. Thus, the Griffiths criterion can be written as follows:

$$G \geq \Gamma. \quad (5)$$

If the film is in a plane deformed state, and the failure front at the interface is a straight line, then the rate of deformation energy release is determined by the elastic deformation energy per unit area of the stressed film. At the stage of stable failure growth, the rate of energy release does not depend on the length of the initial crack:

$$G_s = \frac{\sigma_m^2 h_f}{2\overline{E}_f}, \tag{6}$$

where σ_m is the value of biaxial stresses in the film, h_f is film thickness, and E_f is the modulus of plane deformation. Therefore, from Eq. 5 and Eq. 6 it can be obtained that the film spontaneously delaminates from the substrate under the condition

$$\frac{\sigma_m^2 h_f}{2\overline{E}_f} \geq \Gamma. \tag{7}$$

Taking into consideration that the failure at the interface is constant for a specific film-substrate composition under certain loading conditions, it follows from Eq. 7 that there is a critical value of stresses at which the film edge delaminates from the substrate:

$$\sigma_C = \sqrt{\frac{2\Gamma \overline{E}_f}{h_f}}. \tag{8}$$

As seen from Eq. 8, this critical stress decreases with increasing film thickness, i.e. the thicker the film, the less stress is required to delaminate its edge from the substrate. This also means that a certain stress level corresponds to the critical film thickness at which it delaminates.

It should also be added that the rate of energy release depends significantly on whether the edge of the film coincides with the edge of the substrate or not. The simulation carried out in [31] showed that if film and substrate edges do not coincide, i.e. the film does not cover the entire substrate, the rate of energy release G during failure propagation reaches G_s already at the initial failure length a approximately equal to the film thickness hf. Meanwhile, when film and substrate edges coincide, G approaches G_s only for a failure length of the order of 40 h_f and higher. Moreover, in the latter case, a strong dependence of G on the ratio of the elastic characteristics of film and substrate can be observed.

3 Film Cracking

The second main mechanism that provides partial relaxation of tensile stresses in films and coatings is the formation of transverse cracks therein, i.e. failures normally directed to the film/substrate interface (see Fig. 4).

As in the case of edge delamination of films, the driving force of the cracking process is provided by the elastic energy accumulated in the film. The possibility of forming a transverse failure in a film depends on the ratio of the energy released during its prop-agation per unit length along the film/substrate interface:

$$W_m = \int_0^a G(z)dz, \tag{9}$$

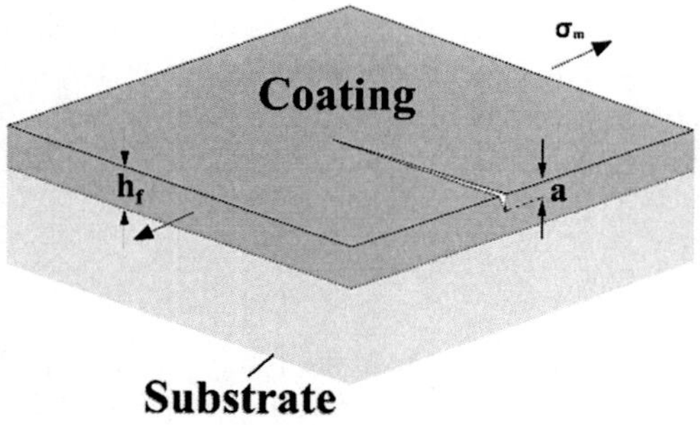

Fig. 4. Generation of a transverse cracks in a thin film under tensile stresses

where.

$$G(z) \approx 1{,}258\pi \frac{\sigma_m^2}{\overline{E}_f} z, \tag{10}$$

is the rate of energy release,

$$W_c = a\Gamma_f. \tag{11}$$

is the energy required for film failure, and a is the depth of crack penetration into the film perpendicular to the film/substrate interface [32].

Freund *et al.* concluded that the failure in the course of its propagation deepens tending to approach the film/substrate interface. If the condition $W_m > W_c$ is still satisfied for $a > h_f$, then the failure penetrates into the substrate and deepens into it until W_m becomes less than W_c. If the substrate has a substantially higher rigidity than the film, then it inhibits the release of elastic energy accumulated in the film, i.e. complicates the failure propagation therein. On the contrary, a pliable substrate easily undergoes elastic deformation and allows a greater release of elastic energy in the film, which results in the easier failure propagation in the latter.

As noted above, the penetration depth of a stable transverse failure into the film/substrate composition, as a rule, reaches the film thickness or overcomes it. The formation of this failure actually leads to the occurrence of new free film edges, where the tensile stresses acting in the film plane are equal to zero. In the areas adjacent to the failure edges, a partial relaxation of these stresses also occurs. However, with distance from the crack, the rigid bond with the substrate prevents the relaxation of tensile stresses in the film, and they increase, approaching the initial level at a distance of several film thicknesses. This contributes to the emergence of new cracks in areas where stresses remain unrelaxed with parallel propagation to the first one. Thus, a system of parallel cracks is formed in the film with each providing the same dissipation of elastic deformation energy in the areas adjacent thereto.

When considering the above mechanisms of failure of thin-film structures, it was assumed that in the case of film delamination, the crack propagates exclusively along the film/substrate interface, while transverse failure occurs only perpendicular to the interface. However, in reality, a crack propagating along the interface can deviate from it, penetrating into the substrate or into the film, and even emerging onto the free edge of the film, leading to shearing of its fragments. In turn, a crack moving normally to the interface can deviate along it, causing the film to delaminate from the substrate. Thus, on the one hand, in the process of film failure, each of these mechanisms can contribute to the development of another, and on the other hand, they both provide relaxation of tensile stresses, and therefore compete with each other and can suppress the development of each other [27, 33, 34].

4 Conclusion

Regardless of the mechanism of stress development, the main factor leading to their occurrence is the rigid connection between the film and the substrate along the interface, which necessitates the compatibility of their deformations. This imposes restrictions on the change in the size of films, and also causes the transfer of stresses from the substrate to the film. Therefore, the stress-strain state of the interface largely determines the nature of deformation and failure of film structures. In this case, one of the key factors influencing the development and distribution of stresses in thin films and coatings, and, consequently, generation and propagation of cracks in the film-substrate system is the interface curvature. Understanding its role in the processes of deformation and failure of thin films and coatings is of great importance in order to increase their reliability and durability.

Acknowledgements. The author acknowledges the support from the project funded by National Research Foundation of Ukraine, under grant agreement No. 2020.02/0119.

References

1. Shi, L., et al.: Vertical graphene growth on SiO Microparticles for stable lithium-ion battery anodes. Nano Lett. **17**(6), 3681–3687 (2017)
2. Mu, Y., Han, M., Li, J., Liang, J., Yu. J.: Growing vertical graphene sheets on natural graphite for fast charging lithium-ion batteries. Carbon **173**, 477–484 (2021)
3. Yang, Q., et al.: Vertically-oriented graphene nanowalls: growth and application in Li-ion batteries. Diam. Relat. Mater. **91**, 54–63 (2019)
4. Zhang, Z., Lee, C.-S., Zhang, W.: Vertically aligned graphene nanosheet arrays: synthesis, properties and applications in electrochemical energy conversion and storage. Adv. Energy Mater. **7**(23), 1700678 (2017)
5. Wang, H., et al.: Facile growth of vertically-aligned graphene nanosheets via thermal CVD: the experimental and theoretical investigations. Carbon **121**, 1–9 (2017)
6. Baranov, O., Romanov, M.: Current distribution on the substrate in a vacuum arc deposition setup. Plasma Process. Polym. **5**, 256 (2008)
7. Tan, K., Markovych, S., Hu, W., Wang, Y., Shorinov, O., Wang, Y.: On the characteristics of cold spray technology and its application in aerospace industries. IOP Conf. Ser. Earth Environ. Sci. **719**, 032023 (2021)
8. Levchenko, I., Romanov, M., Baranov, O., Keidar, M.: Ion deposition in a crossed E×B field system with vacuum arc plasma sources. Vacuum **72**, 335 (2004)

9. Baranov, O., Fang, J., Ostrikov, K., Cvelbar, U.: TiN deposition and morphology control by scalable plasma-assisted surface treatments. Mater. Chem. Phys. **188**, 143–153 (2017)
10. Baranov, O., Romanov, M., Wolter, M., Kumar, S., Zhong, X., Ostrikov, K.: Low-pressure planar magnetron discharge for surface deposition and nanofabrication. Phys. Plasmas **17**(5), 053509 (2010)
11. Skordaris, G., et al.: Effect of PVD film's residual stresses on their mechanical properties, brittleness, adhesion and cutting performance of coated tools. CIRP J. Manuf. Sci. Technol. **18**, 145–151 (2017)
12. Vega-Morón, R.C., et al.: Adhesion and mechanical properties of Ti films deposited by DC magnetron sputtering. Surf. Coat. Technol. **349**, 1137–1147 (2018)
13. Moon, M.-W., Chung, J.-W., Lee, K.-R., Oh, K.H., Wang, R., Evans, A.G.: An experimental study of the influence of imperfections on the buckling of compressed thin films. Acta Mater. **50**(5), 1219–1227 (2002)
14. Boijoux, R., Parry, G., Faou, J.-Y., Coupeau, C.: How soft substrates affect the buckling delamination of thin films through crack front sink-in. Appl. Phys. Lett. **110**(14), 141602 (2017)
15. Faou, J.-Y., Grachev, S., Barthel, E., Parry, G.: From telephone cords to branched buckles: a phase diagram. Acta Mater. **125**, 524–531 (2017)
16. Tang, J.-F., Lin, C.-Y., Yang, F.-C., Chang, C.-L.: Influence of nitrogen content and bias voltage on residual stress and the tribological and mechanical properties of CrAlN films. Coatings **10**(6), 546 (2020)
17. Sarma, B., Sarma, B.K.: Role of residual stress and texture of ZnO nanocrystals on electro-optical properties of ZnO/Ag/ZnO multilayer transparent conductors. J. Alloy. Compd. **734**, 210–219 (2018)
18. Bute, A., et al.: Boron carbide thin films deposited by RF-PECVD and PLD technique: a comparative study based on structure, optical properties, and residual stress. Mater. Chem. Phys. **258**, 123860 (2021)
19. Fluri, A., et al.: Enhanced proton conductivity in Y-Doped $BaZrO_3$ via strain engineering. Adv. Sci. **4**(12), 1700467 (2017)
20. Narayanachari, K.V.L.V., et al.: Growth stress induced tunability of dielectric permittivity in thin films. J. Appl. Phys. **11**(1), 014106 (2016)
21. Fan, Q., Li, D., Li, J., Wang, C.: Structure and piezoelectricity properties of V-doped ZnO thin films fabricated by sol-gel method. J. Alloys Compd. **829**, 154483 (2020)
22. McCluskey, M.D.: Point defects in Ga_2O_3. J. Appl. Phys. **127**(10), 101101, (2020)
23. Cougnon, F.G., Dulmaa, A., Dedoncker, R., Galbadrakh, R., Depla, D.: Impurity dominated thin film growth. Appl. Phys. Lett. **112**(22), 221903 (2018)
24. Cao, S.-G., Li, Y., Wu, H.-H., Wang, J., Huang, B., Zhang, T.-Y.: Stress-induced cubic-to-hexagonal phase transformation in perovskite nanothin films. Nano Lett. **17**(8), 5148–5155 (2017)
25. Abadias, G., et al.: Review article: stress in thin films and coatings: current status, challenges, and prospects. J. Vac. Sci. Technol. **36**(2), 020801 (2018)
26. Zhang, Y., et al.: Strain rate-dependent hardness and deformation behavior in the nanocrystalline/amorphous Ti_2AlNb film. Surf. Coat. Technol. **412**, 127040 (2021)
27. Freund, L.B., Suresh, S.: Thin Film Materials: Stress. Cambridge University Press, Cambridge, England, Defect Formation and Surface Evolution (2003)
28. Xie, K., et al.: Delamination and wrinkling of flexible conductive polymer thin films. Adv. Func. Mater. **31**(21), 2009039 (2021)
29. Alaca, B.E., Saif, M.T.A., Sehitoglu, H.: On the interface debond at the edge of a thin film on a thick substrate. Acta Mater. **50**, 1197–1209 (2002)
30. Klingbeil, N.W., Beuth, J.L.: On the design of debond-resistant biomaterials. part I: free-edge singularity approach. Eng. Fract. Mech. **66**(2), 93–110 (2000)

31. Yu, H.H., He, M.Y., Hutchinson, J.W.: Edge effects in thin film delamination. Acta Mater. **49**, 93–107 (2001)
32. Tada, H., Paris, P., Irwin, G.: The Stress Analysis of Cracks Handbook, 3rd edn. ASME Press, New York, USA (2000)
33. Ayatollahi, M.R., Mirsayar, M.M.: Kinking angles for interface cracks. Procedia Eng. **10**, 325–329 (2011)
34. Mirsayar, M.M.: On fracture of kinked interface cracks – the role of T-stress. Mater. Des. **61**, 117–123 (2014)

Stress-Strain State of a Double Lap Joint of Circular Form. Axisymmetric Model

Sergei Kurennov[1] and Natalia Smetankina[2](✉)

[1] National Aerospace University "Kharkiv Aviation Institute", 17 Chkalova Street, Kharkiv 61070, Ukraine

[2] A. Pidgornyi Institute of Mechanical Engineering Problems of the National Academy of Sciences of Ukraine, 2/10 Pozharsky Street, Kharkiv 61046, Ukraine

nsmetankina@ukr.net

Abstract. The problem of axisymmetrical strain of a structure comprising a plate weakened by a circular cutout and two circular patches lap bonded to both sides of the plate was solved. The patches are bonded to the parent plate with the help of a thin adhesive layer working in shear and peeling. The stresses over the adhesive layer thickness are deemed to be distributed evenly. The Kirchhoff-Love hypotheses were assumed for the patches. Due to structural symmetry, the parent plate is not subjected to bending. The problem considered is the generalization of the classical model of the stress state of an adhesive joint of rods to a domain with radial symmetry. The solution was obtained in analytical form. The model problem was solved. Research shows that tangential stresses achieve a maximum at a distance of about the thickness of the adhesive layer from the joint edge. Normal stresses are of the same order as that of the tangential ones. The results obtained were compared with those of finite-element modelling. The comparison demonstrated high accuracy of the suggested model.

Keywords: Adhesive joints · Interlaminar stresses · Beam theory

1 Introduction

Thin-wall structures, including thin plates, can have defects in the form of holes and cracks occurring during operation, for instance, due to mechanical damage. Local damages to aircraft structures are repaired by using patches. In so doing, to reduce the concentration of stresses around the hole during repair, it is given, as a rule, a circular form. The patch accepts part of the load to relieve the damaged area. The patch can be joined with the parent plate over the entire patch surface (adhesive lap joints [1–4]), along lines (welds [5]) or by a system of points (riveted joints). Adhesive bonding of the patch with the parent plate ensures tightness and aerodynamic effectiveness of the structure. The stress state of such adhesive-bonded structures is analysed by using the finite-element method [6, 7]. Among other reasons, this is because classical analytical models of the stress state of lap joints describe rectangular-form joints and assume a uniform distribution of stresses across the joint width [8, 9]. Existing two-dimensional

M. Nechyporuk et al. (Eds.): ICTM 2021, LNNS 367, pp. 36–46, 2022.
https://doi.org/10.1007/978-3-030-94259-5_4

mathematical models of adhesive-bonded joints, which account for the nonuniformity of stresses across the joint width [10], also assume a rectangular shape of the bonded joint area. In the case considered, the axial symmetry of the structure allows using the polar coordinate system. Mathematical models of adhesive joints are close to the models of the stress state of three-ply (including circular) plates [11–14]. However, the mathematical models of three-ply plates assume identical boundary conditions along the perimeter of both load-bearing layers. The boundary conditions in lap joints are different and most often nonuniform. This is because the joints have to ensure the transmission of forces from one load-bearing layer to the other one. Hence, the mathematical models of bonded joints are more intricate than those of three-ply plates.

The aim of the work is to develop a method for solving the problem of axisymmetrical deformation of a plate weakened by a circular opening, which, in turn, is reinforced on both sides with concentric lap-bonded circular patches. An analytical solution of the problem is presented as the expansion of the sought functions into modified second-order Bessel functions. A similar problem, though without account for the impact of patch bending and peeling stresses in the adhesive, was solved in [15]. Using patches on both sides of the joint, as is well known, is a preferred structural solution because it reduces structure bending and the peeling stresses in the adhesive layer [16, 17].

2 Problem Statement

Let us consider the structure shown in Fig. 1. The radius of the opening in the parent plate is R_1 and the radius of the patches is R_2. Let the parent plate have the thickness $2\delta_2$ and the patches, the thickness δ_1. Between the plate and the patches there is an adhesive layer with the thickness δ_0. The parent plate has the outer radius R_3. If radius R_3 is sufficiently big, the impact of boundary conditions on the plate outer boundary will have a minimal impact on the joint stress state.

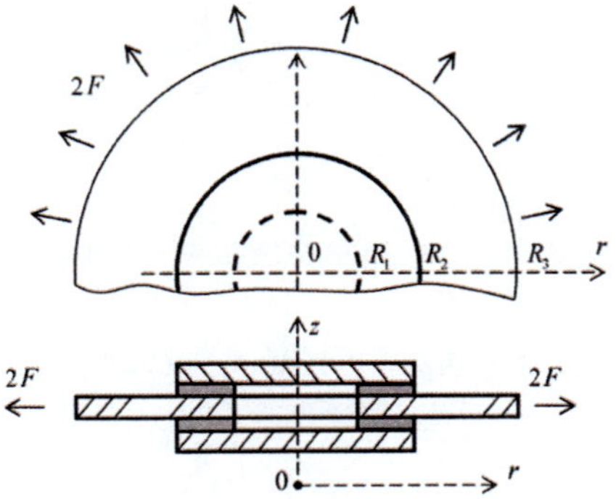

Fig. 1. Structural diagram.

The parent plate is loaded with tensile forces $2F$ applied along radius R_3. Let us consider the bonded joint domain $r \in [R_1; R_2]$. Due to axial symmetry, the forces in the load-bearing layers are independent of the angle coordinate. Due to the symmetry of the structure, the parent plate is not subjected to bending under the action of the load applied to it. However, the patches can bend. Due to symmetry, we can consider one-half of the structure located on one side of the midplane of the parent plate, Fig. 2. Evidently, in this case the thickness of the parent plate is δ_2.

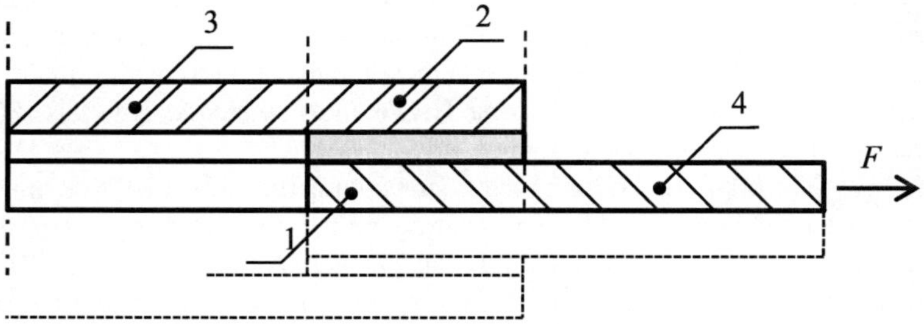

Fig. 2. Structural diagram.

In the bonded joint domain $r \in [R_1; R_2]$, subscript 1 corresponds to the parent plate, and subscript 2, to the circular patch. Beyond the bonded joint $r \in [R_2; R_3]$ the parent plate is designated by subscript 4. The patch on the opening $r \in [0; R_1]$ is designated by subscript 3.

The forces, bending moments and stresses on the adhesive layer side acting on the differential element of the lower patch are shown in Fig. 3.

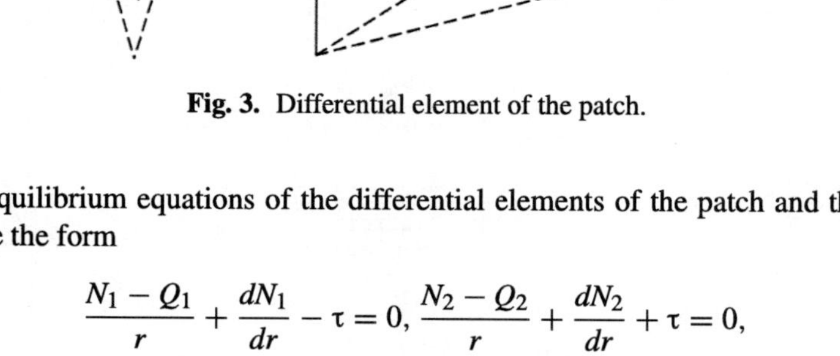

Fig. 3. Differential element of the patch.

The equilibrium equations of the differential elements of the patch and the parent plate have the form

$$\frac{N_1 - Q_1}{r} + \frac{dN_1}{dr} - \tau = 0, \quad \frac{N_2 - Q_2}{r} + \frac{dN_2}{dr} + \tau = 0, \tag{1}$$

$$\frac{M_{2,r} - M_{2,\varphi}}{r} + \frac{dM_{2,r}}{dr} - H_2 + \frac{\delta_2}{2}\tau = 0, \quad \frac{H_2}{r} + \frac{dH_2}{dr} + \sigma = 0, \tag{2}$$

where N_k and Q_k are normal forces in plate k in the radial and circumferential directions, $k = 1,\ 2$; τ and σ are tangential and normal (peeling) stresses in the bonded joint; $M_{k,r}$

is the bending moment in layer k in the radial direction; $M_{k,\varphi}$ is the bending moment in layer k in the circumferential direction; H_k is the shearing force in layer k.

The tangential and normal stresses in the adhesive layer (constant over the adhesive thickness) are proportional to the difference of longitudinal and, respectively, transverse displacements of the sides of the plates facing the adhesive layer [7]

$$\tau = P_1\left(U_1 - U_2 - \frac{\delta_2}{2}\frac{dW_2}{dr}\right), \sigma = P_2 \cdot W_2, \tag{3}$$

where P_1 and P_2 are adhesive layer shear and peeling stiffness, $P_1 = \frac{G_0}{\delta_0}, P_2 = \frac{E_0}{\delta_0(1-\mu_0^2)}$, where respectively E_0, G_0, μ_0 designate the elasticity modulus, shear modulus and the Poisson ratio of the adhesive; W_2 are the transverse displacements of the patch (the transverse displacements of the parent plate are equal to zero); U_k are the longitudinal (radial) displacements of the layers, $k = 1,\ 2$.

Hooke's law is written down as follows

$$N_k = B_k(\varepsilon_{k,r} + \mu_k\varepsilon_{k,\varphi}), Q_k = B_k(\varepsilon_{k,\varphi} + \mu_k\varepsilon_{k,r}), \tag{4}$$

where $B_k = \frac{\delta_k E_k}{1-\mu_k^2}$ is the membrane stiffness of the plates; μ_k is Poisson ratio of the plate k material, $k = 1,\ 2$; E_k is elasticity modulus of the plate k material; $\varepsilon_{k,r}$ and $\varepsilon_{k,\varphi}$ is the layer k strain in the radial and circumferential directions.

The radial displacements are related to strains by relationships

$$\varepsilon_{k,r} = \frac{dU_k}{dr}, \varepsilon_{k,\varphi} = \frac{U_k}{r}. \tag{5}$$

The equations of axisymmetrical bending of the patch (according to the Kirchhoff-Love theory) have the form

$$M_{2,r} = D_2\left(\frac{d^2W_2}{dr^2} + \frac{\mu_2}{r}\frac{dW_2}{dr}\right), M_{2,\varphi} = D_2\left(\mu_2\frac{d^2W_2}{dr^2} + \frac{1}{r}\frac{dW_2}{dr}\right), \tag{6}$$

where $D_2 = \frac{E_2\delta_2^3}{12(1-\mu_2^2)}$ is the patch bending stiffness.

Beyond the bonded joint, in areas 3 and 4, Fig. 2, the stress-strain state of the patch and the parent plate is described by known equations of axisymmetrical strain in circular plates [18].

3 Constructing the Solution

3.1 Adhesive Joint Area

By differentiating Eq. (4) and using Eqs. (1)–(3) and relationships (5)–(7), we obtain a system of equations for stresses in the adhesive layer:

$$\frac{d^3\tau}{dr^3} + \frac{2}{r}\frac{d^2\tau}{dr^2} - \left(a_1 + \frac{1}{r^2}\right)\frac{d\tau}{dr} + \left(\frac{1}{r^3} - \frac{a_1}{r}\right)\tau + Pa_2\sigma = 0; \tag{7}$$

$$\frac{d^4\sigma}{dr^4}+\frac{2}{r}\frac{d^3\sigma}{dr^3}-\frac{1}{r^2}\frac{d^2\sigma}{dr^2}+\frac{1}{r^3}\frac{d\sigma}{dr}+a_3\sigma-Ka_2\left(\frac{\tau}{r}+\frac{d\tau}{dr}\right)=0; \tag{8}$$

where $a_1=P_1\left(\frac{1}{B_1}+\frac{1}{B_2}+\frac{\delta_2^2}{4D_2}\right)$, $a_2=-\frac{\delta_2}{2D_2}$, $a_3=\frac{P_2}{D_2}$.

If we find σ from Eq. (7) and substitute it into Eq. (8), we obtain the differential equation

$$\begin{gathered}\frac{d^7\tau}{dr^7}+\frac{4}{r}\frac{d^6\tau}{dr^6}-\left(a_1+\frac{6}{r^2}\right)\frac{d^5\tau}{dr^5}+\left(\frac{18}{r^3}-\frac{3a_1}{r}\right)\frac{d^4\tau}{dr^4}+\left(\frac{3a_1}{r^2}-\frac{51}{r^4}+a_3\right)\frac{d^3\tau}{dr^3}\\+\left(\frac{126}{r^5}-\frac{6a_1}{r^3}+\frac{2a_3}{r}\right)\frac{d^2\tau}{dr^2}+\left(\frac{9a_1}{r^4}-\frac{225}{r^6}-\frac{a_3}{r^2}+a_2^2P_1P_2-a_1a_3\right)\frac{d\tau}{dr}\\+\left(\frac{225}{r^7}-\frac{9a_1}{r^5}+\frac{a_3}{r^3}+\frac{a_2^2P_1P_2-a_1a_3}{r}\right)\tau=0.\end{gathered} \tag{9}$$

The general solution of (9) is found in the form

$$\tau=\sum_{n=1}^{6}C_nK_1(\lambda_nr)+\frac{C_7}{r}; \tag{10}$$

where $K_1(\lambda_nr)$ are modified second-order Bessel functions (Macdonald's functions); C_n are arbitrary constants.

Substituting (9) into (8) yields the characteristic equation

$$\lambda^6-a_1\lambda^4+a_3\lambda^2+a_2^2P_1P_2-a_1a_3=0,$$

whose roots are coefficients λ_n $n=1,\ldots,6$ included in Eq. (10).

From Eq. (7) we easily find the normal (peeling) stresses in the adhesive layer

$$\sigma=\sum_{n=1}^{6}C_n\alpha_nK_0(\lambda_nr), \tag{11}$$

where $\alpha_n=\frac{\lambda_n^3-a_1\lambda_n}{P_1a_2}$.

Note that the stress state of rods and beams, which are lap joined with the help of an elastic bonding layer [7, 8], is described using linear combinations of exponential functions. In this problem, due to circular symmetry, the analogues of exponential functions are unbounded and nonperiodic modified Bessel functions.

From Eq. (1), using Eqs. (4) and (5), we obtain

$$\frac{d^2U_1}{dr^2}+\frac{1}{r}\frac{dU_1}{dr}-\frac{U_1}{r^2}-\frac{\tau}{B_1}=0,\ \frac{d^2U_2}{dr^2}+\frac{1}{r}\frac{dU_2}{dr}-\frac{U_2}{r^2}+\frac{\tau}{B_2}=0$$

The solution of these equations has the form

$$U_1=\frac{1}{B_1}\sum_{n=1}^{6}\frac{C_n}{\lambda_n^2}K_1(\lambda_nr)+\frac{C_7}{4B_1}r(2\ln r-1)+C_8r+\frac{C_9}{r}$$

$$U_2 = \frac{1}{B_2}\sum_{n=1}^{6}\frac{C_n}{\lambda_n^2}K_1(\lambda_n r) - \frac{C_7}{4B_2}r(2\ln r - 1) + C_{10}r + \frac{C_{11}}{r}$$

From Eq. (3) it is easy to find the transverse displacements W_2 of the patch. Using the relationships for stresses in the adhesive layer (10) and (11), and the longitudinal and transverse displacements of the patch and parent plate, it is easy to find from Eqs. (4)–(6) the radial and circumferential forces in the plates and the bending moments in the patch. The shearing forces H_2 in the patch are found from Eq. (2).

3.2 Areas Beyond the Adhesive Joint

Beyond the adhesive joint there are two areas – the parent plate ($R_2 \le r \le R_3$) and the patch on the cutout ($0 \le r \le R_1$). Subscript 4 designates the displacements and force factors in the parent plate beyond the adhesive joint. Subscript 3 designates the displacements and force factors in the patch on the cutout. The displacements in these areas are described by known equations of strain in circular plates with no shearing forces [18]

$$\frac{d^2U_m}{dr^2} + \frac{1}{r}\frac{dU_m}{dr} - \frac{U_m}{r^2} = 0,\ \frac{d^3W_3}{dr^3} + \frac{1}{r}\frac{d^2W_3}{dr^2} - \frac{1}{r^2}\frac{dW_3}{dr} = 0,$$

where $m = 3,\ 4$.

The solutions of these equations have the form

$$U_3 = c_1 r + \frac{c_2}{r},\ U_4 = c_3 r + \frac{c_4}{r},\ W_3 = s_1 + s_2\ln r + s_3 r^2.$$

Using Eqs. (4)–(6), we find the bending moments, and the radial and circumferential forces.

3.3 Boundary Conditions

Unknown coefficients are found from the boundary conditions and the conjugation conditions on the boundaries of the domains.

In point $r = 0$, the transverse displacements should be bounded by magnitude, and the radial displacements should be equal to zero. It follows that

$$c_2 = 0, s_2 = 0.$$

The tension radial forces are specified on the outer boundary of the parent plate

$$N_4(R_3) = F.$$

On the adhesive joint domain boundaries, we have the conditions

$$N_4(R_2) = N_1(R_2), U_4(R_2) = U_1(R_2), N_2(R_2) = M_{2,r}(R_2) = N_2(R_2) = 0,$$

$$N_1(R_1) = 0, U_2(R_1) = U_3(R_1), W_2(R_1) = W_3(R_1), \left.\frac{dW_2}{dr}\right|_{r=R_1} = \left.\frac{dW_3}{dr}\right|_{r=R_1},$$

$$M_{2,r}(R_1) = M_{3,r}(R_1), N_2(R_1) = N_3(R_1) H_2(R_1) = H_3(R_1),$$

Besides this, radial displacements $U_1(r)$, $U_2(r)$ and transverse displacements $W_2(r)$ of the patch should reduce Eq. (3) to an identity. It follows that

$$C_7 = 0, -C_9 + C_{11} + \frac{C_7}{P_1} = 0, C_{10} - C_8 = 0.$$

Hence, we obtain a closed system of linear equations for unknown coefficients.

4 Model Problem

To analyse the stress state of the joint and verify the suggested analytical model, we consider the joining of two aluminium plates ($E^{(1)} = E^{(2)} = 70$ GPa, Poisson ratio is $\mu = 0.28$). The thicknesses of the plates $\delta_1 = \delta_2 = 3$ mm. The plates are bonded with an adhesive whose elastic properties are $G_0 = 0.3125$ GPa, $E_0 = 0.8$ GPa, and the adhesive layer thickness $\delta_0 = 0.1$ mm. The radius of the hole in the plate $R_1 = 20$ mm, and the radius of the patches $R_2 = 50$ mm. During analysis, we will assume that the parent plate has an infinitely big radius $R_3 = \infty$ mm. This assumption can be explained by that the effect of the hole and the patches on the stress-strain state of the parent plate has a local character. In the case of a relatively big plate of any form with a significant distance of the cutout from its edge, we can assume that the parent plate is infinitely big. Radial tension forces F are specified on the outer boundary of the parent plate. Obviously, in this case, the radial forces in the parent plate at infinity are also equal to F.

An axisymmetrical finite-element model was built to verify the suggested analytical model of the stress-strain state of the joint. The geometrical form of the structure is shown in Fig. 4. Outer radius $R_3 = 3R_2$.

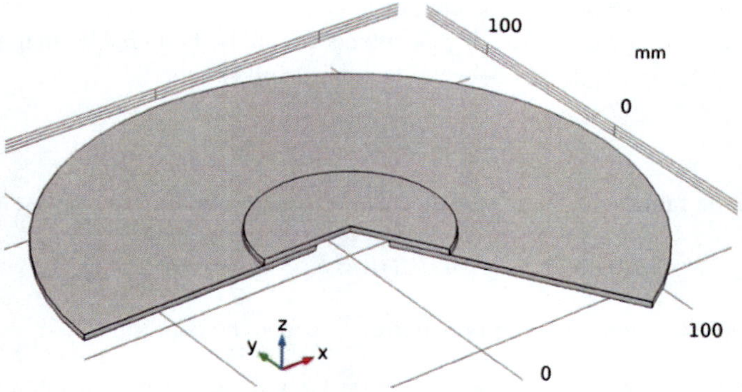

Fig. 4. Design of the patch and plate joint (fragment).

Edge chamfers and squeezed out adhesive were included in the finite-element model as shown in Fig. 5. It shows the model fragment in the vicinity of the adhesive joint edge. Since stress gradients are extremely big in the adhesive layer, the biggest size of the adhesive layer element was chosen fairly small, equal to one-half of the adhesive layer thickness.

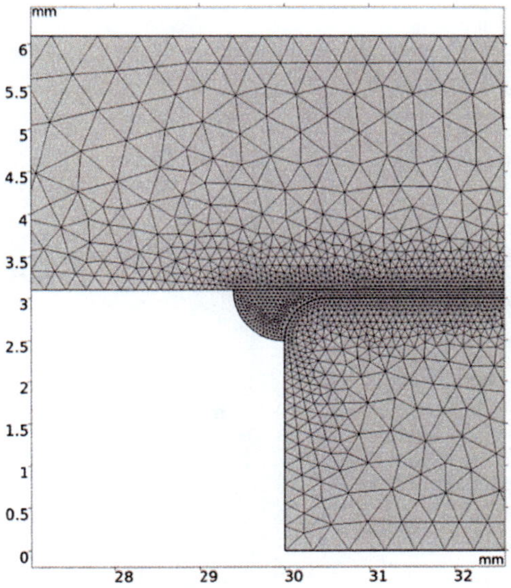

Fig. 5. Fragment of the finite-element model.

The tangential stresses τ in the adhesive layer are shown in Fig. 6. The solid line designates the stresses calculated by the analytical model (AM) suggested in the paper. The dashed line shows the tangential stresses in the adhesive joint midplane calculated using the finite-element model (FEM). The stresses are shown in dimensionless form as ratios of acting stresses and some stresses $\tau^* = \frac{F}{R_2 - R_1}$. Here F are the line forces applied to the butt end of the parent plate, $R_2 - R_1$ is the width of the adhesive joint.

Normal stresses σ in the adhesive layer are shown in Fig. 7.

As in the previous graph, the solid line designates the stresses calculated using the analytical model (AM) suggested in the paper. The dashed line shows the tangential stresses in the adhesive joint midplane calculated using the finite-element model (FEM).

Note that in the graphs the tangential stresses achieve maximum values on the edges of the adhesive joint domain. However, on the other hand, if the outer edge of the adhesive joint is not loaded, then, by virtue of the duality law of tangential stresses, the tangential stresses on the boundary of the adhesive layer should be equal to zero. More accurate mathematical models of the stress state of the adhesive layer [19–23], as well as the division of the adhesive layer near the joint boundary into finite elements with a size much less than the adhesive layer thickness, demonstrate this effect. Research shows that tangential stresses achieve a maximum at a distance of about the thickness of the

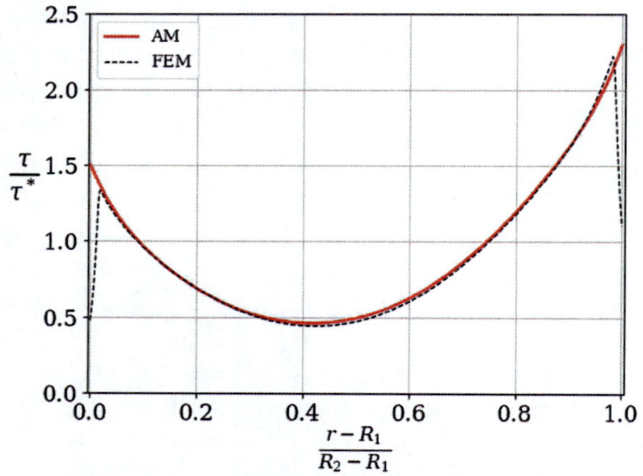

Fig. 6. Tangential stresses in the adhesive layer.

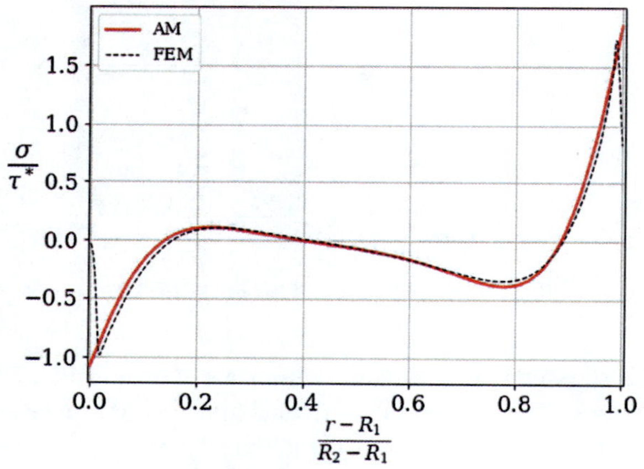

Fig. 7. Normal stresses in the adhesive layer.

adhesive layer from the joint edge. However, here the maximum values of stresses are little different from stresses calculated using simpler models.

Besides, the adhesive layer has normal stresses not accounted for in [15]. Analysis has shown that in this case these stresses are of the same order as that of the tangential ones.

5 Conclusions

An axisymmetrical mathematical model is proposed for a double lap joint of a plate, which has a circular cutout, with two circular patches arranged on both sides of the plate. The approach suggested is based on the Kirchhoff-Love classical theory of plates. The adhesive layer is considered as a Winkler's elastic foundation. The problem solution was obtained in the analytical form.

In spite of the presence of patches on both sides of the plate, due to which the parent plate does not bend under loading, normal stresses occur in the adhesive layer. These stresses are caused by bending of the patches. To reduce the bending of the patches, it makes sense to fill the cutout space between the patches with a stuffing.

Finite-element modelling has demonstrated the high accuracy of the suggested model. In the future, a problem can solved in optimal design of minimal weight patches [24, 25] with specified strength parameters of structure materials.

References

1. Okafor, A.C., Singh, N., Enemuoh, U.E., Rao, S.V.: Design, analysis and performance of adhesively bonded composite patch repair of cracked aluminum aircraft panels. Compos. Struct. **71**(2), 258–270 (2005). https://doi.org/10.1016/j.compstruct.2005.02.023
2. Tomblin, J.S., Salah, L., Welch, J.M., Borgman, M.D.: Bonded repair of aircraft composite sandwich structures. Final Report DOT/FAA/AR-03/74. Federal Aviation Administration, Office of Aviation Research, Washington (2004)
3. Gui-Fang, W.: Stress analysis of plates with a circular hole reinforced by flange reinforcing member. Appl. Math. Mech. **8**(6), 569–588 (1987). https://doi.org/10.1007/bf02017406
4. Bakuckas, J.G., et al.: Bonded repairs of composite panels representative of wing structure. In: Niepokolczycki, A., Komorowski, J. (eds.) ICAF 2019. LNME, pp. 565–580. Springer, Cham (2020). https://doi.org/10.1007/978-3-030-21503-3_45
5. Zemlyanova, A.Y.: Reinforcement of a plate weakened by multiple holes with several patches for different types of plate-patch attachment. Math. Mech. Solids **21**(3), 281–294 (2016). https://doi.org/10.1177/1081286513519812
6. Khan, M.A., Aglietti, G.S., Crocombe, A.D., Viquerat, A.D., Hamar, C.O.: Development of design allowables for the design of composite bonded double-lap joints in aerospace applications. Int. J. Adhes. Adhes. **82**, 221–232 (2018). https://doi.org/10.1016/j.ijadhadh.2018.01.011
7. Ramalho, L.D.C., Campilho, R.D.S.G., Belinha, J., da Silva, L.F.M.: Static strength prediction of adhesive joints: a review. Int. J. Adhes. Adhes. **96**, 102451 (2020). https://doi.org/10.1016/j.ijadhadh.2019.102451
8. Da Silva, L.F.M., das Neves, P.J.C., Adams, R.D., Spelt, J.K.: Analytical models of adhesively bonded joints – part I: literature survey. Int. J. Adhes. Adhes. **29**(3), 319–330 (2009). https://doi.org/10.1016/j.ijadhadh.2008.06.005
9. Wong, E.H., Liu, J.: Interface and interconnection stresses in electronic assemblies – a critical review of analytical solutions. Microelectron. Reliab. **79**, 206–220 (2017). https://doi.org/10.1016/j.microrel.2017.03.010
10. Kurennov, S.S.: Determining stresses in an adhesive joint with a longitudinal unadhered region using a simplified two-dimensional theory. J. Appl. Mech. Tech. Phys. **60**(4), 740–747 (2019). https://doi.org/10.1134/s0021894419040199
11. Starovoitov, E.I., Leonenko, D.V., Tarlakovskii, D.V.: Thermoelastic deformation of a circular sandwich plate by local loads. Mech. Compos. Mater. **54**(3), 299–312 (2018). https://doi.org/10.1007/s11029-018-9740-x
12. Kudin, A.V., Choporov, S.V., Gomenyuk, S.I.: Axisymmetric bending of circular and annular sandwich plates with nonlinear elastic core material. Math. Models Comput. Simul. **9**(5), 601–612 (2017). https://doi.org/10.1134/S2070048217050076

13. Shupikov, A.N., Smetankina, N.V., Svet, Ye.V.: Nonstationary heat conduction in complex-shape laminated plates. Trans. ASME. J. Heat Transfer **129**(3), 335–341 (2007). https://doi.org/10.1115/1.2427073
14. Smetankina, N., Kravchenko, I., Merculov, V., Ivchenko, D., Malykhina, A.: Modelling of bird strike on an aircraft glazing. In: Nechyporuk, M., Pavlikov, V., Kritskiy, D. (eds.) Integrated Computer Technologies in Mechanical Engineering. AISC, vol. 1113, pp. 289–297. Springer, Cham (2020). https://doi.org/10.1007/978-3-030-37618-5_25
15. Barakhov, K., Dvoretska, D., Poliakov, O.: One-dimensional axisymmetric model of the stress state of the adhesive joint. In: Nechyporuk, M., Pavlikov, V., Kritskiy, D. (eds.) ICTM 2020. LNNS, vol. 188, pp. 310–319. Springer, Cham (2021). https://doi.org/10.1007/978-3-030-66717-7_26
16. Campilho, R.D.S.G., Banea, M.D., Pinto, A.M.G., da Silva, L.F.M., de Jesus, A.M.P.: Strength prediction of single- and double-lap joints by standard and extended finite element modelling. Int. J. Adhes. Adhes. **31**(5), 363–372 (2011). https://doi.org/10.1016/j.ijadhadh.2010.09.008
17. Osnes, H., McGeorge, D.: Analysis of overlaminated double-lap joints. Compos. Part B: Eng. **36**(6–7), 544–558 (2005). https://doi.org/10.1016/j.compositesb.2005.01.002
18. Timoshenko, S.P., Woinowsky-Krieger, S.: Theory of Plates and Shells, 2nd edn. McGraw-Hill, New York (1959)
19. Kurennov, S.S.: Longitudinal-flexural vibrations of a three-layer rod. An Improved Model. J. Math. Sci. **215**(2), 159–169 (2016). https://doi.org/10.1007/s10958-016-2829-7
20. Kurennov, S.S.: Refined mathematical model of the stress state of adhesive lap joint: experimental determination of the adhesive layer strength criterion. Strength Mater. **52**(5), 779–789 (2020). https://doi.org/10.1007/s11223-020-00231-5
21. Wang, J., Zhang, C.: Three-parameter elastic foundation model for analysis of adhesively bonded joints. Int. J. Adhes. Adhes. **29**, 495–502 (2009). https://doi.org/10.1016/j.ijadhadh.2008.10.002
22. Frostig, Y., Thomsen, O.T., Mortensen, F.: Analysis of adhesive-bonded joints, square-end, and spew-fillet - high-order theory approach. J. Eng. Mech. **125**, 1298–1307 (1999). https://doi.org/10.1061/(ASCE)0733-9399(1999)125:11(1298)
23. Amidi, S., Wang, J.: An analytical model for interfacial stresses in double-lap bonded joints. J. Adhes. **95**(11), 1031–1055 (2019). https://doi.org/10.1080/00218464.2018.1464917
24. Kondratiev, A.: Improving the mass efficiency of a composite launch vehicle head fairing with a sandwich structure. Eastern-Eur. J. Enterp. Technol. **6**(7), 6–18 (2019). https://doi.org/10.15587/1729-4061.2019.184551
25. Kondratiev, A., Potapov, O., Tsaritsynskyi, A., Nabokina, T.: Optimal design of composite shelled sandwich structures with a honeycomb filler. In: Ivanov, V., Trojanowska, J., Pavlenko, I., Zajac, J., Peraković, D. (eds.) DSMIE 2021. LNME, pp. 546–555. Springer, Cham (2021). https://doi.org/10.1007/978-3-030-77719-7_54

Methodology for Calculating the Productivity of Electrochemical Machining in Stationary Electrolyte

Vitalii Voronko[1], Iryna Tkachenko[1], Yuri Dyachenko[2], Iryna Voronko[2](✉), and Dmitry Zawistovskiy[2]

[1] O. M. Beketov National University of Urban Economy in Kharkiv, 17 Marshala Bazhanova Street, Kharkiv 61002, Ukraine

[2] National Aerospace University "Kharkiv Aviation Institute", 17 Chkalova Street, Kharkiv 61070, Ukraine

i.voronko@khai.edu

Abstract. The technologies that make it possible to manufacture parts with high quality surfaces and with high accuracy of dimensions and configuration of the part are especially actual now. The technological capabilities of electrochemical machining (ECM) (no tool wear, high quality of the machined surface, etc.) have ensured widespread use of this method in the manufacture of parts from difficult-to-machine materials of complex-profiled surfaces, for example, in the production of dies, press forms, and turbine blades. Some manufacturing requirements cannot be met without the use of electrochemical machining equipment. The method of machining metals and alloys is considered based on the principle of anodic dissolution of the machined workpiece in the electrolyte. It is very important to know the processes occurring during the machining on the electrodes in order to understand the basic regularity and fundamental capabilities of the ECM. The machining of a specific metal must be carried out in the correct electrolyte as this has a significant effect on productivity, surface roughness, machining accuracy, current efficiency and energy consumption of the ECM.

Keywords: Electrochemical machining · Electrolyte · ECM in stationary electrolyte · Productivity · Current density

1 Introduction

Electrochemical machining has demonstrated several undeniable advantages when working with difficult-to-machine materials [1], which are often used in the manufacture of aircraft parts.

During operation, various unforeseen situations may arise, which can cause defects on the surface of the part [2]. The chemical nature of the ECM makes it difficult to predict results. The researchers think that mathematical and computer simulations are characterized by limited accuracy due to random behavior during the process. That is,

M. Nechyporuk et al. (Eds.): ICTM 2021, LNNS 367, pp. 47–56, 2022.
https://doi.org/10.1007/978-3-030-94259-5_5

in the manufacture of a finished part, it is possible to obtain parameters that differ from the required ones.

Numerous physical and chemical phenomena occurring in the processing area can cause difficulties in the implementation of the process and give insufficiently accurate simulation results [3].

However, electrochemical processing is characterized by a variety of schemes. For example, the authors of work [4] made a comparative analysis of two types of the ECM: impulse and jet, successfully selected the conditions for the processing of samples. This made it possible to obtain good surface quality and high dimensional accuracy. The changes that took place in the samples were calculated by simulations and confirmed by various experiments. The deviation of the obtained geometric shapes from the required ones was several micrometers.

ECM processing method shows high accuracy of product parameters. That is why it is widely used in aircraft production. The scientific team describes a new strategy for the manufacture of a critical part [5], which is an important component of an aircraft engine. The study of electrochemical treatment of a turbine blade was carried out in the form of a full-scale experiment and using computer simulation. The high accuracy of the obtained product proves the correctness of the strategy, and the insignificant discrepancy between the research results is the verification of the simulation.

Scientists of this scientific team [6] also analyzed the relationship between the initial parameters of the ECM process and the roughness of the finished part. The authors believe that the most influential parameter is the frequency of the switching mode power supply.

By influencing various factors during the process, it is possible not only to reduce the roughness, but also to change the quality of the surface layer as a whole. For example, the feed of compressed air to the treatment area can reduce the phenomenon of accidental corrosion [7]. Or regulation of the electrolyte flow rate and vortex distribution can optimize the overall parameters of the ECM [8].

In general, the ECM processing method has many advantages, which have become the reasons for its widespread use in the space, aviation, automobile and electromechanical industries, as well as the activity of researchers in the development of new methods and strategies for this process [3].

Electrochemical machining (ECM) – a method of processing electrically conductive materials, consists in changing the shape, dimension and (or) roughness of the surface of the workpiece as a result of the anodic dissolution of its material in the electrolyte under the action of an electric current. The peculiarities of electrolysis are the spatial dissolution of the anode and the deposition of metal on the cathode surface [3, 9, 10].

Methods of electrochemical machining of metals in production are becoming more widespread due to the relatively low energy consumption, low noise, vibration, high accuracy and repeatability of the result with strict adherence to processing technology. Permissible processing of any conductive materials resistant to the chemical constituents of the solution [5].

Electrochemical machining of parts is sometimes the only possible method of shaping a metal product [6]. In some cases, such types of processing as thermal or mechanical can cause changes in the structure of the surface layer [11, 12]. The properties of the metal will change: fragility will increase, elasticity and resilience will be lost, and corrosion resistance will decrease. This will make it impossible for the further use of the part in the mechanism assembly, especially if it is a critical design [13].

There are several ECM schemes [1, 3, 14]:

- Machining with a stationary electrode-tool (ET). Using this scheme, holes are made in sheet metals, information is engraved, burrs are removed, sharp edges are rounded, and the tool is sharpened.
- Drilling of recesses, cavities, holes. According to this scheme, processing cavities of forging dies are made, holes, grooves, and turbine blade feathers are drilled.
- Broaching outside and inside surfaces in workpieces that have pre-machined surfaces on which the ET can be based. According to this scheme, finishing machining of cylindrical holes, splines, screw grooves is performed.
- Cutting workpieces. The ET is a rotating disc. According to this scheme, grooves, slots are made, springs are cut.
- Grinding or polishing. A cylindrical ET is used that rotates and moves progressively along the workpiece. This scheme is used as the final operation in the manufacture of thin plates, as well as parts from tough and strong alloys.

The main technological properties of the ECM processes are productivity, dimensional and shape accuracy and roughness of the treated surfaces.

Various metals are actively dissolved only in electrolytes of a certain composition and concentration. However, the technological properties of the processes are influenced not only by the composition of the electrolyte and concentration, but also by the temperature, hydrogen pH, which characterizes the concentration of hydrogen ions in the ECM or acidity, and the rate of its pumping in the inter-electrode gap (IEG) [15].

The concentration of components in the electrolyte is an indicator of their quantitative content in the solvent (water). It is expressed in relative, weight, or volumetric values. In the technological documentation, to control the concentration of the electrolyte, the required density is additionally indicated, which makes it possible to perform a more accurate dosing of the components. The composition of the electrolyte is determined based on the required productivity and the purpose of this ECM process, the accuracy and quality of processing. For example, in the electrochemical polishing of steel parts, aqueous solutions of phosphoric or sulfuric acids are used with the addition of chromic anhydride CrO_3, since this will provide a low roughness of the polished surface with a relatively minimum productivity of the process. And in the electrochemical shaping of products from the same steel, when it is necessary to maintain the dimension and shape with a given accuracy, aqueous solutions of sodium nitrate $NaNO_3$ are used. Because it is necessary to achieve high productivity with a relatively low roughness of the processed surface.

The ECM varieties are combined into two groups, depending on the physicochemical characteristics of the removal of the workpiece material. The first group includes all types of the ECM, during which the allowance from the workpiece is removed due to electrochemical dissolution, and the second group includes processing methods in which, together with the electrochemical dissolution of the removed allowance, mechanical or electrothermal action is carried out.

The electrochemical polishing process can be performed in a flowing or stationary electrolyte. The best results are obtained when electropolishing homogeneous metals and alloys [16].

Most often, surface cleaning and degreasing, marking, stock removal, edge rounding, blade sharpening, making engravings, changing the physical and technical properties of the surface (removing surface stresses after machining, increasing corrosion resistance), polishing gear teeth are performed in a stationary electrolyte.

The ECM at a low current density (from 0.02 to 0.03 A/mm^2) is carried out in a stationary electrolyte. For example, a diagram of the most typical operation is shown – electrolytic grinding or polishing (Fig. 1).

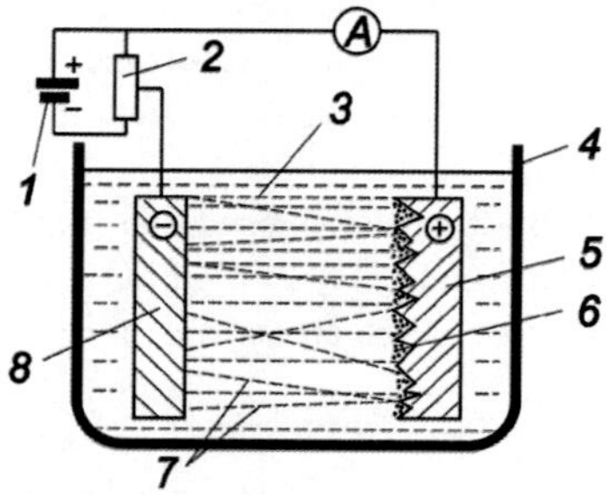

Fig. 1. Schematic diagram of the ECM in a stationary electrolyte: power supply (1), resistor (2), electrolyte (3), bath (4), electrode-workpiece/anode (5), film of dissolved metal products (6), power lines (7), electrode-tool / cathode (8)

When processing the ECM in a stationary electrolyte with fixed electrodes, the average rate of metal dissolution is determined by the formula (1):

$$V_{avg} = {}^{Z}/_{\tau} \quad (1)$$

where **Z** is machining allowance; τ is process time.

The productivity of the ECM in a stationary electrolyte can be determined by the process time, which ranges from 5 to 10 min for polishing carbon steels and from 2 to 3 min for aluminum. The productivity of dimensional electrochemical shaping is characterized by the rate of anodic dissolution of the metal, which is expressed in linear (mm/min) or volume (mm^3/min) units.

The study of the influence of the current density on the productivity of the ECM process in a stationary electrolyte has been carried out. The influence of the current density on the roughness of the processed samples was not included in the objectives of this study.

2 Methodology for Calculating the Productivity of the ECM

Samples with dimensions of 25 × 40 × 1 mm made of 1X18H9T stainless steel sheet were used for the experiment. (Fig. 2). They were immersed in the solution to a depth of 25 mm. The immersion time was 5 min.

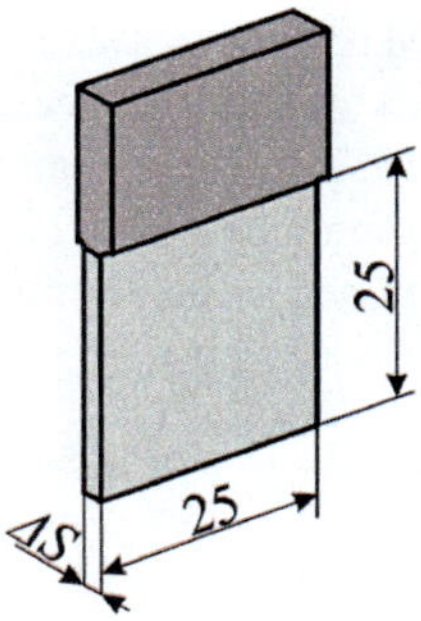

Fig. 2. The workpiece-sample (anode) for the ECM in a stationary electrolyte.

The prepared electrolyte has the following composition: phosphoric acid – 73–74%; chromic anhydride – 9–10%; water – 15–16%.

The study was carried out on a serial machine ЭЗИ–2M for electrolytic sharpening of a tool, the main structural elements of which are shown in Fig. 3.

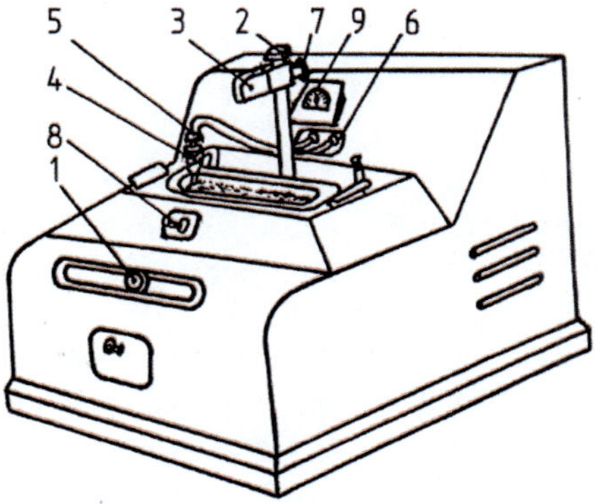

Fig. 3. Machine for the ECM in stationary electrolyte: movable contact of the rheostat (1), rack (2), anode clamp (3), bracket (4), thermal contactor (5), panel (6), screw (7), toggle switch (8), ammeter (9)

Contact occurs on both sides; the end face machining has been neglected. Then the contact area of the metal with the electrolyte can be calculated in advance (2):

$$F = 2a^2 = 2 \cdot 25 \cdot 25 = 1250\,\text{mm}^2 = 0.125\,\text{dm}^2; \qquad (2)$$

where ***F*** is the contact surface area; ***a*** is the workpiece-sample side.

The known relationship between density and current (3):

$$i = {}^{I}/_{F}; \qquad (3)$$

where ***F*** is the area of the processed surface; ***I*** is the current.

The current density in the ECM process in a stationary electrolyte is set according to the technological recommendations for stainless steel: $\boldsymbol{i} = 15, 20, 25, 30, 35$ A/dm^2. The dimension of the current density, which differs from the designations in the SI system, is generally accepted in mechanical engineering technology for the ECM processes.

The current $\boldsymbol{I}$ [A] can be determined by the formula (4), knowing the contact surface area $\boldsymbol{F}$ and the density formula $\boldsymbol{i}$:

$$I = i \cdot F = 0.125 \cdot i. \tag{4}$$

It is possible to determine the volume $\boldsymbol{V}$ [mm^3] using the formula (5):

$$V = F \cdot \Delta S = 25 \cdot 25 \cdot \Delta S = 625\Delta S; \tag{5}$$

where $\boldsymbol{F}$ is the area of the processed surface; $\boldsymbol{\Delta S}$ is the difference in sample thickness.

The productivity of the ECM process in a stationary electrolyte $\boldsymbol{Q}$ [mm^3/min] can be defined as the volume of dissolved metal lost over a certain time (6):

$$Q = {}^{V}\!/_{\tau}; \tag{6}$$

where $\boldsymbol{V}$ is the volume of dissolved metal; $\boldsymbol{\tau}$ is time of the process.

2.1 Procedure for the Experiment

1. Preparing the machine for work:

- the movable contact 1 of the rheostat was set in the middle position, otherwise, when the current is turned on, the ammeter may break down;
- the anode clamp 3 was fixed on the rack 2;
- the thermal contactor 5 was inserted into the bracket 4 and connected to the machine circuit using two pins inserted into the sockets of the panel 6;
- the thermal contactor 5 was lowered into the electrolyte to a depth of 15 mm;
- the machine was connected to an alternating voltage of 220 V.

2. The order of the experiment:

- each of the samples was installed in the anode clamp 3, lowered into the bath to the line of the immersion mark, then the anode clamp was fixed with the screw 7;
- the current was switched on by toggle switch 8;
- the required current was created using the movable contact of the rheostat 1 by the ammeter 9 for a given current density;
- each sample was processed within 5 min;
- the current was switched off by the toggle switch 8;
- the movable contact of the rheostat 1 was set to the middle position;
- the sample was taken out of the bath, was removed from the anode clamp 3 using tweezers;
- the samples were washed with running water.

The processing of samples on the ЭЗИ-2M machine was carried out in the following sequence:

After washing the sample and wiping it with dry napkins, the thickness of the plates was measured with a device (lever bracket).

Each sample was processed under specific conditions (Fig. 4).

(a) (b)

Fig. 4. Samples for studying the ECM process in a stationary electrolyte: (A) untreated part of the sample, (B) sample after ECM processing

Measurements of the difference in the thickness of the material of the initial and processed samples ***ΔS*** were performed using the lever bracket CP 0–25 (0002) ГОСТ 11098–64 with an absolute measurement error of 2 μm (Fig. 5).

Fig. 5. Measuring the thickness of the sample-plate

The results of measurements of samples and calculations are presented in tabular form (Table 1):

Table 1. Parameters of the ECM process in a stationary electrolyte.

No. of sample plate	Current density *i*, A/dm^2	Current *I*, A	Difference in sample thickness **Δ*S***, mm	Volume of dissolved metal *V*, mm^3	Productivity *Q*, mm^3/min
1	14	1.75	0.033	20.625	4.125
2	15	1.87	0.040	25.00	5.0
3	16	2	0.045	28.125	5.625
4	18	2.25	0.053	33.125	6.625
5	20	2,5	0.060	37,50	7,5
6	22	2.75	0.061	38.125	7.625
7	24	3	0.065	40.625	8.125
8	26	3.25	0.066	41.25	8.25
9	28	3.5	0.069	43.125	8.625
10	30	3.75	0,070	43.75	8.75
11	32	4	0.069	43.125	8.625
12	34	4.25	0.070	43.75	8.75
13	35	4.37	0.070	43.75	8,75
14	36	4.5	0.069	43.125	8.625

The figure graphically shows the dependence of productivity on current density (Fig. 6).

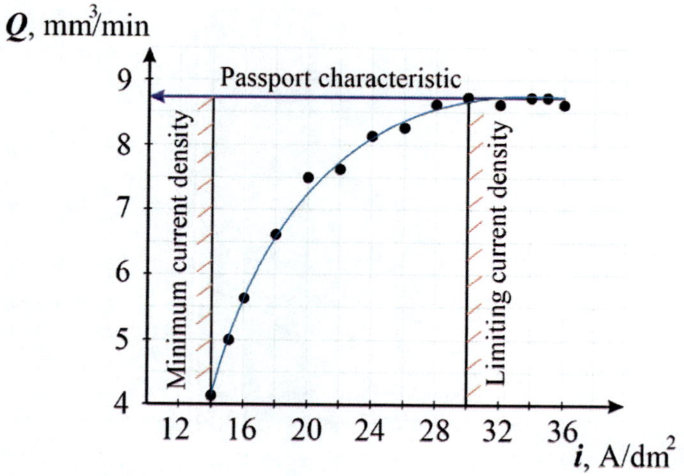

Fig. 6. Productivity of the ECM process in a stationary electrolyte

3 Conclusions

The proposed methodology for calculating the productivity of the ECM in a stationary electrolyte is illustrative and can be used for production and educational purposes.

The productivity of the ECM in a stationary electrolyte, depending on the current density, has a non-linear character.

There is a limiting current density, which is the passport characteristic of a specific machine model for the ECM in a stationary electrolyte.

Applying a current density in excess of the limit shown in the graph does not improve the surface quality of the sample. Only overheating of the electrolyte occurs.

References

1. Zhengyang, X., Wang Y.: Electrochemical machining of complex components of aero-engines: developments, trends, and technological advances. Chin. J. Aeronaut. **34**(2), 28–53 (2019). https://doi.org/10.1016/j.cja.2019.09.016
2. Ruszaj, A., Cygnar, M., Grabowski, M.: The state of the art in electrochemical machining process modeling and applications (2018). https://doi.org/10.1063/1.5056292
3. Ruszaj, A.: Electrochemical machining – state of the art and direction of development. Mechanik **90**(12), 1102–1109 (2017). https://doi.org/10.17814/mechanik.2017.12.188
4. Schuberta, A., et al.: Generation of complex surfaces by superimposed multi-dimensional motion in electrochemical machining. Procedia CIRP, 42, 384 – 389 (2016). https://doi.org/10.1016/j.procir.2016.02.216
5. Dong, Z., Di, Z., Zhengyang, X., Laishui, Z.: Trajectory control strategy of cathodes in blisk electrochemical machining. Chin. J. Aeronaut. (2013). https://doi.org/10.1016/j.cja.2013.06.012
6. Xuezhen, C., Zhengyang, X., Dong, Z., Zhongdong, F., Di, Z.: Experimental research on electrochemical machining of titanium alloy Ti60 for a blisk. Chin. J. Aeronaut. **29**(1), 274–282 (2016). https://doi.org/10.1016/j.cja.2015.09.010
7. Xingyan, H., Dong, Z., Jiabao, L., Zhouzhi, G.: Flow field research on electrochemical machining with gas film insulation. J. Mater. Process. Technol. **267**, 247–256 (2018). https://doi.org/10.1016/j.jmatprotec.2018.12.019
8. Mingxia, C., Zhiyong, L., Hongjuan, Y., Xiaoyu, S.: Experimental investigations on aircraft blade cooling holes and CFD fluid analysis in electrochemical machining. Adv. Mater. Sci. Eng. (2019). https://doi.org/10.1155/2019/4219323
9. Воронько, И.А.: *Кондуктивное шлифование как разновидность формообразующего процесса электрохимического метода обработки металлов и сплавов*. Coleção de trabalhos científicos «ΛΌΓΟΣ» com materiais da conferência científicoprática internacional. Lisboa: Plataforma Científica Europeia, vol. 2, 29–31 (2020). https://doi.org/10.36074/09.10.2020.v2.07
10. Taran, A., Plankovskyy, S., Voronovich, D., et al.: Emission properties of Re-W dispenser cathodes. In: 2009 IEEE International Vacuum Electronics Conference, pp. 407–408, (2009). https://doi.org/10.1109/IVELEC.2009.5193581
11. Plankovskyy, S., Myntiuk, V., Tsegelnyk, Y., Zadorozhniy, S., Kombarov, V.: Analytical Methods for Determining the Static and Dynamic Behavior of Thin-Walled Structures During Machining. In: Shkarlet, S., Morozov, A., Palagin, A. (eds.) MODS 2020. AISC, vol. 1265, pp. 82–91. Springer, Cham (2021). https://doi.org/10.1007/978-3-030-58124-4_8

12. Plankovskyy, S., Breus, V., Voronko, V., Karatanov, O., Chubukina, O.: Review of Methods for Obtaining Hardening Coatings. In: Nechyporuk, M., Pavlikov, V., Kritskiy, D. (eds.) ICTM 2020. LNNS, vol. 188, pp. 332–343. Springer, Cham (2021). https://doi.org/10.1007/978-3-030-66717-7_28
13. Krivtsov, V.S., Voronko, V.V., Zaytsev, V.Y.: Advanced prospect for the development of aircraft assembly technology. Sci. Innov. **11**(3), 11–18 (2015). https://doi.org/10.15407/scine11.03.011
14. Сикульский, В.Т., Дьяченко, Ю.В., Божко, В.П., Воронько, В.В., Борисевич, В.В., Проскурин, С. Д., Воронько, И.А.: *Технология производства деталей летательных аппаратов размерной обработкой / Technology of Aircraft Parts Manufacturing by Dimensional Machining.* Харьков: НАУ «ХАИ» (2017). http://library.khai.edu/library/fultexts/metod/Sikulskij_TECHNOLOGY_OF_AIRCRAFT.pdf
15. Невский, О.И., Бурков, В.М., Гришина, Е.П., Гаврилова, Е.Л., Балмасов А.В., Носков, А.В., Донцов, М.Г.: *Электрохимическая размерная обработка металлов и сплавов. Проблемы теории и практики.* Иваново: ИГХТУ (2006). http://main.isuct.ru/files/publ/PUBL_ALL/159.pdf
16. Рахимянов, Х.М., Василевская, С.И.: *Технологические возможности электрохимической обработки отверстий неподвижным катодом-инструментом.* Обработка металлов (технология, оборудование, инструменты). № 2 (71). Новосибирск: НГТУ, 12–20 (2016). https://doi.org/10.17212/1994-6309-2016-2-12-20

Investigation of the Influence of Crystallization Energy on the Size of Nanostructures During Copper Ion-Plasma Treatment

Yurii Shyrokyi(✉) and Gennadiy Kostyuk

National Aerospace University "Kharkiv Aviation Institute", Chkalov Street, 17, Kharkiv 61070, Ukraine
i.shyrokyi@khai.edu

Abstract. The paper presents the results of the calculation of the crystallization energy and investigates its effect on the size of the nanostructured grain during ion-plasma treatment of copper with oxygen and nitrogen ions. It is shown that the crystallization energy increases the energy of ions required to obtain nanostructures. At energies near 300 eV, it ranges from 0.1 to 7 eV, ie can be ignored, but at energies close to 15,000 eV, the crystallization energy is from 200 to 11,000 eV, and the energy for oxygen ions is greater. Calculations also showed that the ion charge significantly affects the crystallization energy for large ion charges in the direction of increase. All these results confirm that it is necessary to take into account the crystallization energy only at energies of $2 \cdot 10^3$–$2 \cdot 10^4$ eV, and this allows to clarify the technological parameters of ion-plasma treatment of copper to increase the probability of obtaining nanostructures.

Keywords: Crystallization energy · Surface hardening · Micro-coarseness · Wear resistance · Diffusion layer

1 Introduction

In modern research, many scientists are studying the issue of obtaining nanostructures in various structural materials. This gives not only new opportunities for their use, but also quite interesting new properties for known materials. The study of copper is especially interesting because it has a very wide range of possibilities for using both as mono material and in combination with others. For example, by deposition of silver nanostructures on copper substrates [1] or its oxidation followed by heat treatment [2], superhydrophobic anti-corrosion coatings are obtained. Developed in the [3] hydrothermal method, a superhydrophobic copper surface is obtained, which has not only resistance to chemical exposure and UV radiation, but also a self-cleaning effect. With the development of technology, researchers are particularly interested in nanostructures in copper and its combinations. Thus, in [4], copper nanostructures are used for chemical analysis using Raman spectroscopy. And [5] developed a resistive memory device based on copper oxide synthesized by ion implantation.

M. Nechyporuk et al. (Eds.): ICTM 2021, LNNS 367, pp. 57–66, 2022.
https://doi.org/10.1007/978-3-030-94259-5_6

It is quite natural that with the spread of interest in the use of nanostructured copper, the interest in the methods and technologies of its production also spreads. A lot of research is being done to expand them. For example, [6] investigated the effect of nitrogen ion implantation on the properties of thin films of copper oxide obtained by direct current magnetron sputtering. And in [7] the influence of radiation on the submicrostructures of copper and copper alloys is considered. The effect of implantation of nitrogen ions of different charges into copper was also studied in [8]. In [9], the structural characteristics of nanocrystalline copper after implantation of carbon ions were studied. Particular physical properties were also studied. Thus, [10] studied the control of residual stress gradients in copper films by implantation of inert ions.

But there are far fewer studies of ion-plasma methods for obtaining nanostructures in copper. Thus, the effect of ion-plasma treatment of the copper surface on its resistance to high-vacuum breakdowns was studied in [11]. In [12, 13] the evolution of the charge positions of ions in the plasma of the cathode vacuum arc was studied. And [14, 15] are practical works using vacuum-arc plasma to obtain nanostructured films. Practical work on measuring the dependence of the erosion rate on the parameters of the arc on nanostructured copper cathodes was also carried out in [16, 17].

As we can see, a lot of practical works [12–15] are focused on obtaining nanostructured materials, but much less work is devoted to theoretical research [18, 19]. Although it should be noted that in [20, 21] the study of energy required to create nanostructures in some structural materials was given and the evaluation of the action of ions of a certain energy, charge and grade was carried out [22, 23], which allows to assess the possibility of formation of nanostructures. However, the crystallization processes were not considered, which means that the energy expended on crystallization when obtaining nanostructures by the ion-plasma method was not estimated. This means that the accuracy of calculations of the energy required to obtain nanostructures by the above method decreased. All this indicates the need to assess the impact of crystallization energy on the energy of nanostructures, which will determine the technological parameters of ion-plasma treatment where its accounting is mandatory and were not. In general, this will make it possible to more accurately determine the technological modes of processing at which there is a high probability of obtaining nanostructures.

2 The Main Part

To solve this problem, we take in [22] a model where the solution of the joint problem of thermal conductivity and thermoelasticity in the area of action of an individual ion of the corresponding energy, charge and grade, which allows to estimate the possible number of particles in the nanostructure, then calculates the energy and chemical compounds. Taking into account these energies and the number of atoms and chemical compounds in the grain, we find the total energy of atomization of the grain E_{ac}.

Further, taking the energy of grain formation as $E_c = 1.1E_{ac}$, we can find the energy required for grain formation. Given this value, we determine all the characteristics in the ion's range: temperature, rate of temperature rise, temperature stresses and strain rate, grain size, volume of nanostructure and depth of its occurrence and the actual number of particles in the nanostructure.

2.1 Research Methodology

Using the above, we obtain new values of the considered quantities, which will make it possible to determine their influence on the occurrence of nanostructures.

Having determined the number of particles in the nanostructure, we need to determine the atomization energy of particles in the structure, which can be determined taking into account two energies: the Coulomb repulsion energy E_k and the ionic bond energy E_{ib} by the expression:

$$\begin{aligned} \mathrm{E}_c &= \mathrm{E}_k + \mathrm{E}_{ib} = \varepsilon^2 U + \left(1-\varepsilon^2\right)^{\frac{1}{2}} D_0 N \\ &= -\frac{\varepsilon^2 k \mathrm{M} \mathrm{e}^2 \mathrm{Z}^2 \mathrm{N}}{R_0}\left(1-\frac{\rho}{R_0}\right) + \left(1-\varepsilon^2\right)^{\frac{1}{2}} D_0 N \end{aligned} \tag{1}$$

where ε is the degree of ionicity; U is the energy of interaction of ions; k is electrostatic constant,

$$k = \frac{1}{4\pi\varepsilon_0} \approx 8,987742438 \cdot 10^9 \left[\frac{H * \mathrm{M}^2}{\text{к}\pi^2}\right];$$

M *is* Madelung constant; e is electron charge;

z is the ion charge; N is the number of particles;

R_0 is the shorter interatomic distance; ρ is the degree of repulsion; D_0 is the dissociation energy of one particle of matter.

The degree of ionicity ε for substances AB is determined by the expression:

$$\varepsilon = \frac{X_A - X_B}{X_A + X_B}, \tag{2}$$

where X_A, X_B are oxidation states for elements A and B on the Pauling scale.

The Madelung constant M is defined as follows:

$$M = \sum\nolimits_j^n \frac{z_i z_j}{P_{ij}}; \tag{3}$$

where z_i, z_j are charges of interacting ions.

The energy of the electrostatic Coulomb interaction is defined as follows:

$$P_{ij} = \frac{e^2 z_i z_j}{R_{ij}}, \tag{4}$$

where R_{ij} is the distance between charges i, j, which can be expressed as $R_{ij} = P_{ij} * R_0$.

Then the energy of interaction of the *i*-th ion with all other ions (n) is represented as:

$$U = 1/2\left(n\sum_{i}^{n}\frac{e^2 z_i z_j}{R_0} * \frac{1}{P_i j} + n\sum_{j}^{n}\frac{e^2 z_i z_j}{R_0} * \frac{1}{P_i j}\right) = \frac{ne^2}{R_0}\sum_{j}^{N}\frac{z_i z_j}{P_i j}; \quad (5)$$

The use of the factor $\frac{1}{2}$ in (1) is due to the fact that we will use the interacting pair of ions only once.

The degree of repulsion ρ depends on the magnitude and characteristics of grain deformation. As the compression of the crystal increases, its resistance increases rapidly. The ions repel each other, and the closer the ions get closer to each other, the greater the repulsive forces. Since the electronic shells of ions that penetrate each other are repelled, the correct description of these forces is possible on the basis of quantum mechanics or by the Bourne method [20] by the expression:

$$\frac{1}{\rho} = \frac{2}{R_0} + \frac{9V}{M\beta}; \quad (6)$$

where V is the mole volume of the crystal; M is the Madelung constant; β is compressibility of crystals.

There is a relationship between repulsion indices and orbital exponents that connect atomic orbitals [20]. For oxides and chalcogenides this relationship can be represented as follows:

$$\frac{1}{\rho} \approx (1 + \varepsilon)(\xi_M + \xi_X); \quad (7)$$

Where ξ_M Ta ξ_X are the orbital exponents of the atomic orbitals of metal (m) and halogen (x).

Estimation of the value of the orbital exponent can be obtained using the first ionization potential (I), determined in Ridbergs *(1Ry = 13.6 eV = 313 kcal)*:

$$\xi = \sqrt{I}; \quad (8)$$

The interatomic distanceR_0 for substances A and B is defined as the sum of covalent radii:

$$R_0 = R_\kappa(A) + R_\kappa(B); \quad (9)$$

The dissipation energy D_0 has been experimentally determined for many substances, but to avoid limitations in estimating the energies of covalent bonds for D_0 crystals, we use Pauling's general postulate:

$$D_0 = \sum_{i=1}^{m}\Delta H_{aT}(i); \quad (10)$$

To test the developed model, consider the effect of nitrogen and oxygen ions in the plasma medium on copper. Theoretical calculations of the crystallization energy depending on the ion energy, which takes into account the number of particles in the cluster, obtained without taking into account the crystallization energy, are given in [22]. In the future, on the basis of a refined theoretical model, we will compare the size of the theoretically calculated nanograin, taking into account and without taking into account the crystallization energy. If we assume that the crystallization energy is equal to or close to the atomization energy, then as a result of calculations we obtain the atomization energy for chemical compounds at different charge numbers, degrees of repulsion and dissociation energies.

3 The Results of Calculations and Their Discussion

Using the calculated values of the atomization energy and knowing the number of particles in the nanocluster, which are taken from [22], we obtain the energy required to obtain grain. To ensure a greater probability of its creation, increase this value by 10% and use in further calculations.

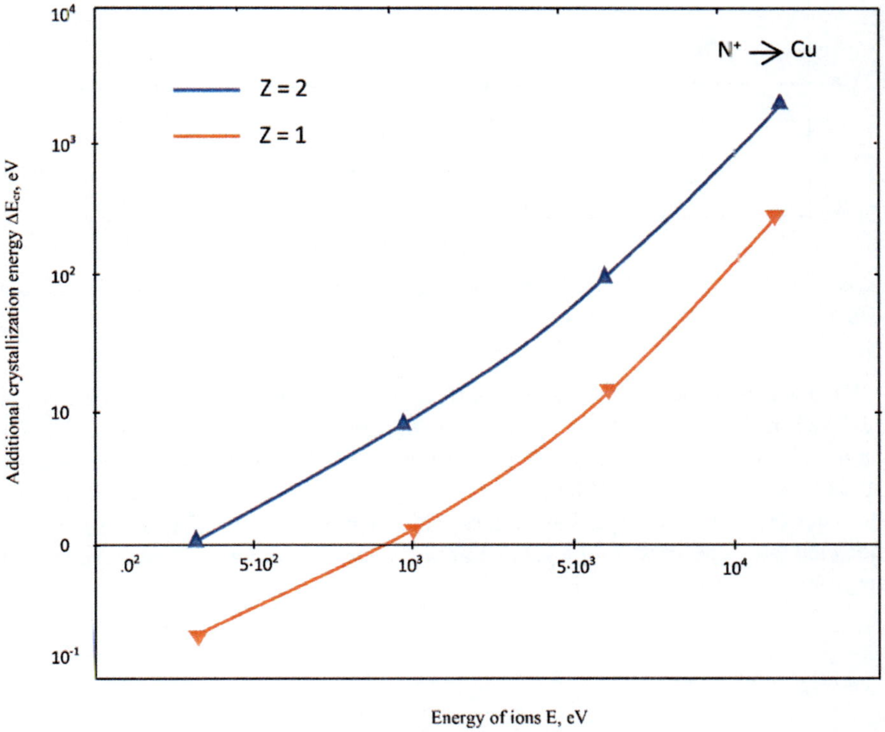

Fig. 1. Dependence of additional crystallization energy on the energy of ions at different ion charges N^+ when it acts on copper.

The results of the study of the dependence of the crystallization energy on the energy of ions acting on copper are presented in Fig. 1, 2 for the case of action of nitrogen ions (see Fig. 1) and oxygen (see Fig. 2).

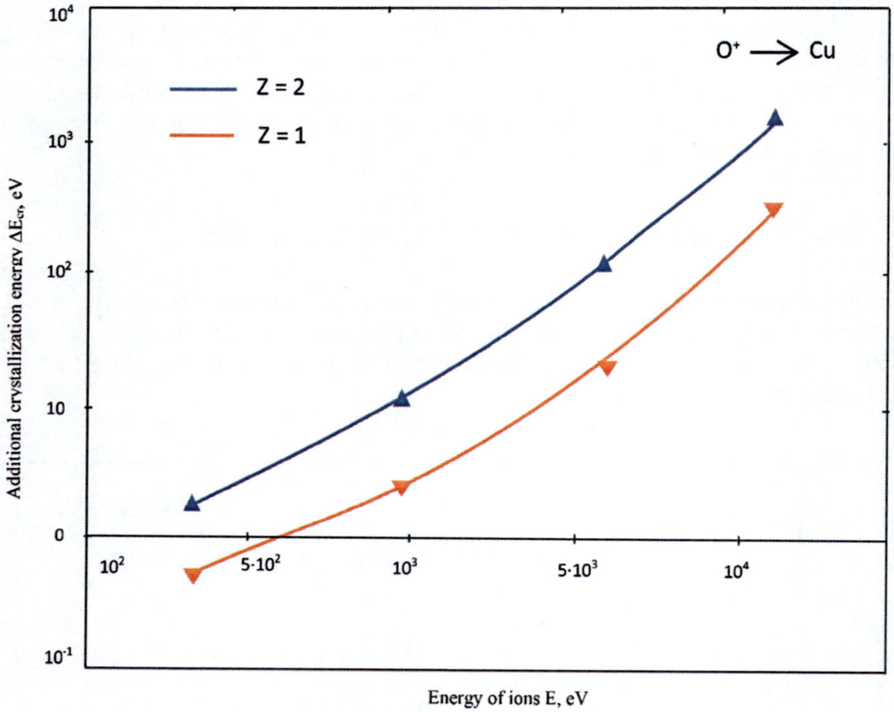

Fig. 2. Dependence of additional crystallization energy on the energy of ions at different ion charges O^+ when it acts on copper.

We can see that at the energy of ions ($E = 3{\cdot}10^3$ eV), the crystallization energy is (0.1–3) eV for N^+ and (0.25–9) eV for O^+, with large values corresponding to a larger charge of ions (z = 2). At the energy of ions E = 1, $5{\cdot}10^4$ eV, the crystallization energy is already ($2.7{\cdot}10^3 - 9{\cdot}10^4$) eV for N^+ ions and ($10^3 - 1.1{\cdot}10^3$) eV for O^+ ions, with large values corresponding to larger ion charges. This change in the crystallization energy is associated with a decrease in the volume of the nanocluster, with an increase in the ion energy and, accordingly, with the crystallization energy.

Obviously, the crystallization energy will affect the technological parameters necessary for the formation of nanostructures. For this purpose, the dependences of the grain size on the ion energy, taking into account and without taking into account the crystallization energy, were calculated according to the model [22]. Such dependences are presented in Fig. 3 and Fig. 4 for the case of action of nitrogen and oxygen ions.

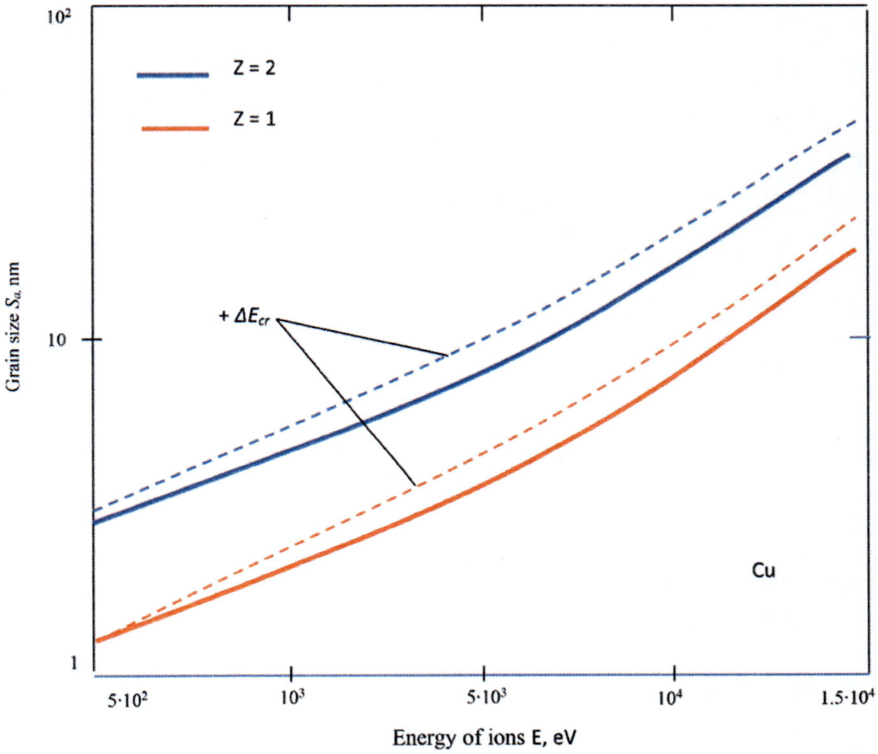

Fig. 3. Dependence of grain size on the energy of ions N^+ (with charge: z = 1; z = 2) when they act on copper, taking into account (ΔE_{cr}) and without taking into account (–) the energy expended on crystallization.

Thus, in the case of the activity of nitrogen ions with an energy of $3{\cdot}10^2$ eV, the grain size can be 1.7–6 nm, whereas at an ion energy of 1, $5{\cdot}10^4$ eV, they can grow up to 15–30 nm. For oxygen ions with an energy of $3{\cdot}10^2$ eV, they are in the range of 1.5–5.5 nm, while the grain size increases to 10–50 nm under the action of oxygen ions with an energy of 1, $5{\cdot}10^4$eV.

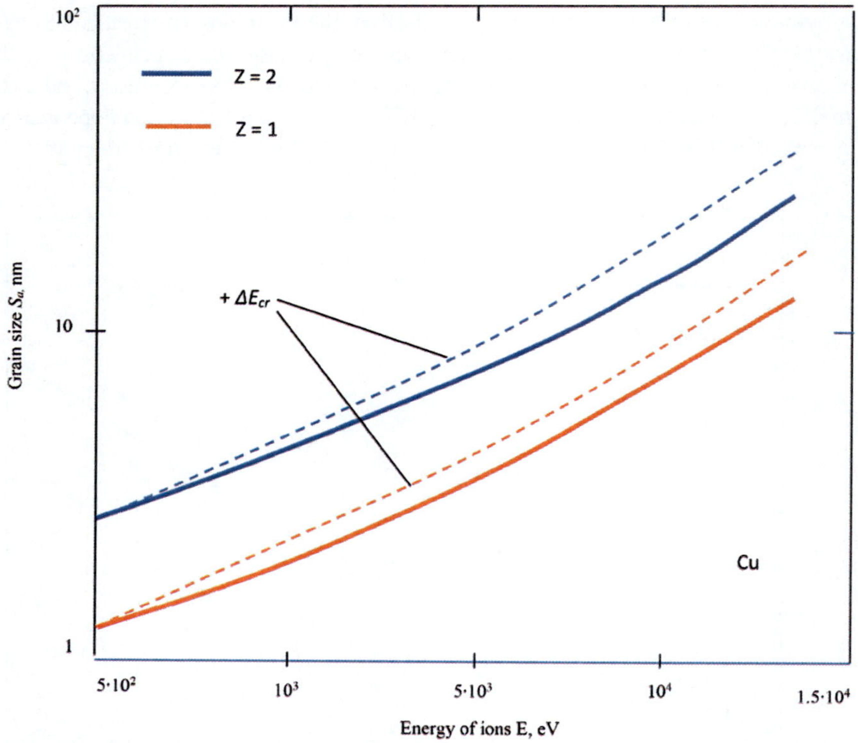

Fig. 4. Dependence of grain size on the energy of ions O^+, (with charge: z = 1; z = 2) when they act on copper, taking into account (ΔE_{cr}) and without taking into account (–) the energy expended on crystallization.

4 Conclusions

Theoretical studies have shown that the crystallization energy increases the energy of ions required to obtain nanostructures. The crystallization energy, at values ranging from 0.1 to 7 eV, which corresponds to ion energy of about 300 eV, can be neglected. But at ion energies close to $1.2 \cdot 10^4$ eV, the crystallization energy is from $2 \cdot 10^2$ to $1.1 \cdot 10^3$ eV. Also, calculations have shown that the ion charge significantly affects the crystallization energy for large ion charges in the direction of increase.

All this suggests that it is necessary to take into account the crystallization energy only at ion energies $2 \cdot 10^3$–$2 \cdot 10^4$ eV. Its consideration will allow to clarify the technological parameters in the ion-plasma treatment of copper to increase the probability of obtaining nanostructures. It may be interesting for professionals engaged in developing devices for nanostructures.

Acknowledgment. The authors would like to acknowledge financing of National Research Foundation of Ukraine under grant agreement No. 2020.02/0119.

References

1. Zhu, J., Dai, X., Zhou, B.: One-step fabrication of a superhydrophobic copper surface by nano-silver deposition. AIP Adv. **10**, 95–111 (2020)
2. Feng, L., Zhao, L., Qiang, X., Liu, Y., Sun, Z., Wang, B.: Fabrication of superhydrophobic copper surface with excellent corrosion resistance. Appl. Phys. **119**(1), 75–83 (2014). https://doi.org/10.1007/s00339-014-8959-1
3. Feng, L., Wang, J., Shi, X., Chai, C.: Superhydrophobic copper surface with mechanical, chemical, and UV durability along with corrosion resistance and self-cleaning effect. Appl. Phys. **125**(4), 1–12 (2019). https://doi.org/10.1007/s00339-019-2562-4
4. Markin, A.V., Markina, N.E., Popp, J., Cialla-May, D.: Copper nanostructures for chemical analysis using surface-enhanced Raman spectroscopy, TrAC Trends Anal. Chemi. **108**, 247–259 (2018)
5. Bishop, S.M., Bakhru, H., Novak, S.W., Briggs, B.D., Matyi, R.J., Cady, N.C.: Ion implantation synthesized copper oxide-based resistive memory devices. Appl. Phys. **99**, 202–215 (2011)
6. Jafari, A., et al.: Ion implantation of copper oxide thin films; statistical and experimental results Hamid Garmestani, Ştefan Ţăl. Surf. Interfaces **18**, 445–463 (2020)
7. Li, M., Zinkle, S.J., Physical and mechanical properties of copper and copper alloys. Compr. Nucl. Mater. **4**, 667–690 (2012)
8. Sari, A.H., Salem, M.K., Shoorche, A.: Effect of nitrogen ion implantation in copper. J. Fus. Energy **30**(4), 323-327 (2011)
9. Lin, W.-M., et al.: Structural characteristics of nanocrystalline copper after carbon ion implantation. Micron **42**(7), 691–694 (2011)
10. Jeon, J.G., Kim, J.H., Kang, T.J.: Control of the residual stress gradients in copper films by inert ion implantation. Mater. Lett. IF 3.204, 222, 54–57 (2018)
11. Karpenko, A., Storizhko, V.E., Shutko, V.A., Baturin, V.A.: Investigation of copper samples with ion-plasma treatment on the high voltage breakdowns. Atom. Sci. Technol. Issues **4**, 297–301 (2018)
12. Anders, A.: The evolution of ion charge states in cathodic vacuum arc plasmas: a review. Plasma Sour. Sci. Technol. **21**(3), 151–162 (2012)
13. Baranov, O., Romanov, M., Fang, J., Cvelbar, U., Ostrikov, K.: Control of ion density distribution by magnetic traps for plasma electrons. J. Appl. Phys. **112** (7), 073302 (2012)
14. Medhisuwakul, N., Pasaja, S., Sansongsiri, J., Kuhakan, S., Intarasiri, L.D.: Yu, Development and application of cathodic vacuum arc plasma for nanostructured and nanocomposite film deposition. Surf. Coat. Technol. **229**, 36–41 (2013)
15. Glukhov, O.V., et al.: Nanostructured coatings based on amorphous carbon and gold nanoparticles obtained by the pulsed vacuum-arc method. J. nano- Electron. Phys. **11**(4), 04019 (2019)
16. Rao, L., Munz, R.J., Meunier, J-L.: Vacuum arc velocity and erosion rate measurements on nanostructured plasma and HVOF spray coatings. J. Phys. Appl. Phys. **14**(40), 4192 (2007)
17. Kajita, S., Ohno, N., Takamura, S., Tsuji, Y.: Direct observation of cathode spot grouping using nanostructured electrode. Phys. Lett. **373**(46), 4273–4277 (2009)
18. Baranov, O., Romanov, M.: Current distribution on the substrate in a vacuum arc deposition setup. Plasma Process. Polym. **5**, 256 (2008)
19. Popov, V., Kostyuk, G., Nechyporuk, M., Kostyk, K.: Study of ions energy, their varieties and charge on temperature, rate of temperature rise, thermal stresses for nanostructures on construction materials. In: Grabchenko's International Conference on Advanced Manufacturing Processes. Advanced Manufacturing Processes. Inte rPartner 2019. Lecture Notes in Mechanical Engineering. Springer, Cham, pp. 107–121 (2019)

20. Kostyuk, G.I.: Nanotechnology: theory, experiment, technology, perspective. Ed. center Int. academies of sciences and innov. technologies, 648 (2012)
21. Baranov, O.O., Fang, J., Rider, A.E., Kumar, S., Ostrikov, K.: Effect of ion current density on the properties of vacuum arc-deposited TiN coatings. IEEE Trans. Plasma Sci. **41**(12), 3640–3644 (2013). https://doi.org/10.1109/TPS.2013.2286405
22. Kostyuk, G., Melkoziorova, O., Kostyuk, E., Shirokiy, I.: Prospects for producing nanostructures in the volume of parts under the action of plasma flows. Dev. Tools Technol. Syst. KhNTU "KhPI" **92**, 107–121 (2020)
23. Kostyuk, G., Popov, V., Shyrokyi, Y., Yevsieienkova, H.: Efficiency and performance of milling using cutting tools with plates of a new class. In: Tonkonogyi, V., (eds.) et al., InterPartner 2020. LNME, pp. 598–608. Springer, Cham (2021). https://doi.org/10.1007/978-3-030-68014-5_58

Synthesis of a Launch Trajectory of Aircraft Optimal in the Minimum Fuel Consumption Based on Sufficient Conditions of Optimal Control

Olena Tachinina[1(✉)], Sergiy Ponomarenko[2], Victor Shevchenko[3], Olexandr Lysenko[2], and Igor Romanchenko[4]

[1] National Aviation University, Kyiv, Ukraine
[2] National Technical University of Ukraine «Igor Sikorsky Kyiv Polytechnic Institute», Kyiv, Ukraine
sol_@ukr.net
[3] Taras Shevchenko National University of Kyiv, Kyiv, Ukraine
gii2014@ukr.net
[4] Central Research Institute of the Armed Forces of Ukraine, Kyiv, Ukraine

Abstract. In the theory of automatic control of complex dynamic objects, the necessary conditions for optimality of control are widely used on the basis of the method of L.S. Pontryagin, necessary and sufficient conditions for optimality of control based on the method of R. Bellman and sufficient conditions for optimality of control based on the method of V.F. Krotova. To date, control objects are becoming more complex and the development of optimal control methods themselves continues. Of particular interest for complex objects is the development of control synthesis methods based on sufficient optimality conditions. A modification of V.F. Krotova-M.M. Khrustaleva method, which consists in a conditional change in the characteristics of the object under study. This technique makes it possible, with minor changes, to apply the already known technique for synthesizing the suboptimal launch trajectory, obtaining better estimates of the final mass in comparison with the known method. In this article, this approach is used to construct an optimal launch-acceleration trajectory of an aircraft under the influence of gravity, aerodynamic forces, and engine thrust that is optimal in terms of fuel consumption. Using the method of mathematical modeling, numerical results have been obtained that confirm the efficiency of the proposed modification of the V.F. Krotova-M.M. Khrustalev method.

Keywords: Optimal control · Sufficient conditions of optimal · Launch trajectory · Minimum fuel consumption · Angle of attack

1 Introduction

In the modern theory of automatic control, fundamental results are actively used, such as the necessary conditions for optimality of control based on the method of L.S. Pontryagin

M. Nechyporuk et al. (Eds.): ICTM 2021, LNNS 367, pp. 67–77, 2022.
https://doi.org/10.1007/978-3-030-94259-5_7

[1], necessary and sufficient conditions for optimality of control based on the method of R. Bellman [2], as well as sufficient conditions for optimality of control based on the method of V.F. Krotov [3, 4, 5]. The development of these methods is reflected in the works of many domestic and foreign authors [6–15]. All existing methods for solving optimal problems are focused on certain classes of problems in a formulation that presupposes certain assumptions.

One of the successful options for the development of control theory with the help of sufficient optimality conditions of V.F. Krotov [1, 2, 3] is the method of V.F. Krotov-M.M. Khrustalev [4, 5]. It allows you to find the optimal ascent-acceleration trajectory $V(h)$ in terms of minimum fuel consumption, as well as an estimate of the upper limit of the final mass of the aircraft. By virtue of the latter property, the found trajectory is called estimated. In [4], it was proved that when flying along an optimal trajectory, the final mass turns out to be no more than the estimated one. For some cases of the problem ([4] problem A'), the found estimate corresponds to the exact solution of the boundary value problem.

Method V.F. Krotov-M.M. Khrustaleva has its own limitations. In particular, when substantiating the method [4], the differential connection by the angle of inclination of the trajectory is excluded θ and further θ acts as a control. The dependencies with which the estimated trajectory can be found does not take into account the feasibility of the processes, which, for the known ones h, V, m, β, is determined by the relationship between α and θ. Although, in the same work of V.F. Krotov and M.M. Khrustalev, a possible method for determining α and θ is proposed, which allows one to realize the found optimal trajectory. But at the same time, the search for the very trajectory $V(h)$ to be implemented is based on the dependencies that assume $\alpha = \theta$. That is, first a trajectory close enough to the optimal one is synthesized, and then it is refined.

The undoubted advantage of the method lies in the possibility of obtaining solutions that meet the requirements of design calculations in terms of accuracy with a relatively low cost of computer time, which is especially important when forming the appearance of a future object at the preliminary design stage. It is this quality that prompts researchers [14, 15] to turn to the method of V.F. Krotov-M.M. Khrustaleva. The breadth of application of the method largely depends on the degree of closeness of the results of the synthesis of the estimated trajectory and the exact solution of the boundary value problem. Therefore, it is understandable to strive to refine the results obtained with low additional costs of computer time.

In the proposed work:

- an attempt was made to refine the method of V.F. Krotov-M.M. Khrustalev by introducing into the calculation formulas the reference values of the angle of attack.
  $\alpha_{оп} = \alpha_{min}$ at the stage of trajectory synthesis; at the same time, additional costs of processor time are no more than 0.1%;
- an iterative algorithm for the synthesis of a realizable injection trajectory based on the method of V.F. Krotov-M.M. Khrustalev, which allows one to approach the solution of longitudinal and spatial problems of injection from a single point of view, is proposed.

2 Application of the V.F. Krotov-M.M. Khrustalev Method for Finding the Estimated Launch Trajectory

Let us clarify the problem statement [5]. The longitudinal motion of the center of mass of an aircraft in a vertical plane passing through the center of the Earth is considered.

In accordance with the differential equations of motion (1–3), the object is acted upon by the force of gravity, aerodynamic forces and engine thrust. It is necessary to find an ascent-acceleration trajectory that is optimal in terms of the minimum fuel consumption. Initial conditions:

$$t = 0;\quad V_0 = V(0);\quad \theta_0 = \theta(0);\quad h_0 = h(0);\quad L_0 = L(0);\quad m_0 = m(0).$$

End conditions:

time t_1 is not fixed; $V_1 = V(t_1)$; $\theta_1 = \theta(t_1)$; $h_1 = h(t_1)$.

Constraints on control α, β and phase coordinates $h,\ V,\ \theta,\ m,\ L$:

$$0 < V_{min}(h) \le V \le V_{max}(h),\ \ where\ \ h \ge 0;$$

$$0 < \beta_{min}(h, V) \le \beta \le \beta_{max}(h, V),\ \ where\ 0 < \beta_{max}(V, h) < \infty; \tag{1}$$

$$\alpha_{min}(h, V) \le \alpha \le \alpha_{max}(h, V),\quad \text{where } \alpha_{min} \le 0;\ \ \alpha_{max} \ge 0.$$

The functions $a_{min},\ a_{max},\ \beta_{min},\ \beta_{max}$ are continuously differentiable. Also assume that

$$0 \le \theta \le \pi. \tag{2}$$

The need for this requirement near the Earth's surface is obvious [5], and for many types of aircraft this condition must be fulfilled throughout the entire lift-acceleration section. The research carried out by the author of the thesis on the synthesis of optimal launch trajectories shows that the need to violate the requirement (2) is most likely at transonic speeds in the process of breaking the sound barrier, which is caused by a sharp increase in the profile resistance coefficient of the value $P\cdot\cos\alpha - X(\alpha)$. In the remaining sections of the trajectory, the need to violate the requirement (2) is less likely, but may be due to a sharp change in the traction characteristics or the operating mode of the aircraft power plant. In the overwhelming part of the extraction stage, the optimal trajectory satisfies condition (2). Consequently, this assumption does not lead to significant errors in the synthesis of the optimal launch trajectory.

The functions $g = g(h)$; $P = P(V, h, \beta)$; $X = X(V, h,\ \alpha)$; $Y = Y(V, h,\ \alpha)$ are defined and continuously differentiable on the set of admissible values $\Omega_{h,V,\alpha,\beta}$ for all time values. The physical meaning of the problem implies the properties of the listed functions

$$\begin{gathered} P(V, h,\ 0) = 0;\ \frac{\partial P}{\partial \beta} \ge 0;\ \frac{\partial Y}{\partial \alpha} > 0;\ \frac{\partial X}{\partial V} > 0; \\ X(V, \infty, \alpha) = 0;\ \ X_0(V, h) = X(V, h,\ 0) \le X(V, h, \alpha). \end{gathered} \tag{3}$$

For existing aircraft, the following conditions are characteristic

$$\frac{\partial}{\partial h}G(V,h) > 0;\quad 0 > \frac{\partial}{\partial V}G(V,h) > -C. \tag{4}$$

where $C = \text{const}$ for all $(h, V) \in \Omega_{V,h}$ satisfying

$$G(V,h) = P(V,h,\ \beta_{max}(V,h)) - X_0(V,h) \geq 0 \tag{5}$$

The boundary conditions satisfy the inequalities

$$G(V,h) > 0; G(V_1,h_1) > 0. \tag{6}$$

Inequalities (6) are certainly valid when the engine thrust does not depend on V and h.

Trajectories satisfying the above conditions will be called admissible, and the set of elements $z = (t_1,\ x(t),\ u(t))$, corresponding to these trajectories will be denoted D. The vector function of the phase coordinates $x(t) = (V(t),\ \theta(t),\ h(t),\ L(t),\ m(t))$ is assumed to be continuous and piecewise differentiable, and the control $u(t) = (\alpha(t),\ \beta(t))$ is piecewise continuous. Both functions are defined on the segment $[0,\ t_1]$. It is required from the number of admissible injection trajectories to choose the one on which the mass of the consumed fuel is minimal, i.e., the functional

$$J(z) = -m(t_1)$$

takes the smallest value on the set of valid elements D. To solve the problem of finding an estimate for the maximum finite mass, we define on the set D the set of triples of arguments D_0 satisfying the inequality

$$P(V,h,\ \beta) - X_0(V,h) > 0. \tag{7}$$

Let us make some remarks about the physical meaning of inequality (7). For this, we write down the expression for the derivative of the energy height with respect to time

$$\frac{dh_\varepsilon}{dt} = Vn_x = \frac{V\Big(P(V,h,\beta)\cos\alpha - \Big(X_0(V,h) + AC_y^{\alpha^2} qS\alpha^2\Big)\Big)}{mg} \tag{8}$$

The energy altitude is the reduced total energy of the aircraft

$$h_\varepsilon = \frac{E}{m} = h + \frac{V^2}{2g},$$

that is expression (8) characterizes the rate of energy growth. It is easy to see that the extremum (8) with respect to the angle of attack is attained at $\alpha = 0$. Moreover, the variables $V,\ m,\ g$ are nonnegative. Consequently, condition (7) indicates the requirement that the sign of the derivative (8) be positive, that is, the need for a constant increase in energy in the case of a zero angle of attack. Simulation shows. That condition (7) is satisfied for the optimal trajectory of extraction, as well as for the set of trajectories lying in some of its neighborhood.

Let us define on the set additional functions of the form [5]

$$Q(V,h,\beta)=\frac{\beta}{P(V,h,\beta)-X_0(V,h)} \tag{9}$$

$$Q(V,h)=\min_{\beta} Q(V,h,\beta) \tag{10}$$

$$S(V,h)=\int\limits_{V^*}^{V}\frac{\partial}{\partial h}\cdot Q_{\min}(V,h)\cdot dV-\frac{g}{V}\cdot Q_{\min}(V,h) \tag{11}$$

$$\gamma(h)=Q_{\min}\big(V^*(h),\ h\big)\frac{dV^*(h)}{dh}\ . \tag{12}$$

Here $V^*(h)$ is a function satisfying (7), continuously differentiable. In addition, we require that for all $(h,\ V)$ satisfying the set D_0, the function $Q_{min}(V,h)$ is continuously differentiable. This requirement is being met. $P(V,h,\beta)=C(V,h)\beta$ where C is the outflow rate. In other cases, as a rule, a slight change in the dependency $P(V,h,\beta)$ allows you to achieve its implementation.

An expression similar to (9) for the dynamic relations of the mechanical energy of the aircraft has the following form:

$$Q_\varepsilon(V,h,\alpha,\beta)\ =-\frac{V(P\cos\alpha-X(\alpha))}{\beta}\ =\frac{dE}{dm}\ -gh_\varepsilon$$

Taking into account (9), we can write

$$Q_\varepsilon(V,h,\beta)\ =\frac{-V}{Q_\varepsilon(V,h,\alpha=0,\beta)}\ .$$

The extremum of the function Q_ε with respect to α is achieved at $\alpha=0$, therefore, expression (9) can be interpreted as an indirect indicator of the efficiency of motion in the case when one of the controls α is equal to its optimal value $\alpha=0$.

According to Theorem 2 from [5], for any admissible trajectory, the optimal value of the final mass $m(t_1)$ does not exceed the estimated value

$$m^*=m_0\exp\left[\int\limits_{h_0}^{h_1}\max_V S(V,h)\cdot dh-\int\limits_{V_0}^{V_1}Q_{\min}(V,h)\cdot dV\right]. \tag{13}$$

There, it is also proved that the optimal values satisfy $G(V,h,t)$ for all $t\in[0,\ t_1]$, if the condition is satisfied $G(V_0,\ h_0)>0$, and it is also shown that it is necessary to consider only the trajectories lying in the domain $G(V,\ h)>0$.

In [5], an estimate of the optimal value of the final mass was obtained in a refined form

$$m^*=m_0\exp\int\limits_{h_0}^{h_1}\left[\max_V S(V,h)-\gamma(h)\right]dh\ . \tag{14}$$

If we assume that $V^*(h_0) = V_0,\ V^*(h_1) = V_1$, by substituting expression (12) for $\gamma(h)$, dependence (14) can be easily transformed to form (13). Consequently, in the analytical sense, dependencies (13) and (14) are similar, however, the numerical implementation of (14) is much simpler, since it requires fewer calls to the subroutine for finding the mass estimate. In the absence of aerodynamic and gravity forces (14) takes the form of K.E. Tsiolkovsky's formula

$$m^* = m_0 \exp\left[-\frac{1}{C}(V_1 - V_0)\right]. \tag{15}$$

A special case of the formula is also possible, which does not take into account aerodynamic forces, but makes it possible to take into account the energy costs for overcoming gravitational forces

$$m^* = m_0 \exp\left[-\frac{1}{C}\left(\frac{g(h_1 - h_0)}{V_1} + (V_1 - V_0)\right)\right]. \tag{16}$$

Dependences (14–16) were used to synthesize the lift-acceleration trajectory and estimate the final mass of specific aircraft. Note that the above-described method of V.F.Krotov and M.M. Khrustalev [3], as well as a refined version of the method (14) can be used to search not only the final mass, but also the mass of the aircraft at all current points of the trajectory. To do this, it is enough to take the current one $(V,\ h)$ as the final values in the above dependencies (V_1, h_1).

3 Modeling the Method of V.F. Krotov-M.M. Khrustalev for the Synthesis of the Estimated Launch-Acceleration Trajectory

In the works of V.F. Krotov-M.M. Khrustalev [4– 7], two problems of launching an aircraft in a longitudinal channel are considered: A and A'. The problem A' differs from the A fact that it accepts the assumptions $\cos\alpha = 1$ and $C_{xi} = 0$ (equality to zero inductive resistance). To perform a modification of the method of V.F. Krotov-M.M. Khrustalev in order to improve its accuracy, we defined the problem A''. To do this, on the considered flight segment, from the set of admissible angles of attack Ω_α, we select the set of angles that make it possible to realize trajectories belonging to a certain neighborhood of the optimal trajectory. Let's call such values $\alpha \in \Omega_\alpha$ realizable.

Analysis of the behavior of α for a trajectory lying in some neighborhood of the optimal trajectory $V^*(h)$ shows that on most sections of the launch trajectory there is some $\alpha_{\min}$ such that for all $h \in \Omega_h$

$$0 < \alpha_{\min} < \alpha_{\text{оПТ}} < \alpha_{\text{оПТ}}\ . \tag{17}$$

If in our formulation of the problem it was necessary to accept constraint (2), then for steady flight modes it is possible to limit the possible values of the angles of attack from below

$$\alpha_P < \alpha_{\Pi\Gamma} < \alpha_{\min}\ . \tag{18}$$

where $\alpha_{\Gamma\Pi}$ is the angle of attack required for the implementation of steady horizontal flight

$$\alpha_{\Pi\Gamma} = \frac{mg}{P + C_y^{\alpha} qS}\left[1 - \frac{V^2}{rg}\right]. \tag{19}$$

In the case of a choice of altitude or a positive change in the angle of inclination of the trajectory, the angle of attack α_P will be greater than $\alpha_{\Gamma\Pi}$. Violation of conditions (18) is possible only in cases of a sharp decrease in the angle of inclination of the trajectory or an arbitrary decrease θ less than a certain one $\theta_{\min} < 0$, which is unacceptable in the presence of constraint (2). Note that in the injection site, the modes associated with a sharp change θ are extremely rare and short-lived. Thus, with an insignificant distortion of the physical meaning and insignificant errors in the estimate of the mass, it is possible to determine some according $\alpha_{\min}$ to the formula (19) (or another similar dependence) that satisfies the conditions (17, 18). If the specified $\alpha_{\min}$ exists, then it is possible to define the task A'', which differs from the task A by the presence of an assumption $\alpha = \alpha_{\min} = \text{const}$. Let us systematize the signs of tasks $A,\ A',\ A''$ in Table 1.

Table 1. The systematize the signs of tasks $A,\ A',\ A''$

Task	Attack angle	Longitudinal overload
A	$\alpha = \alpha_P = v\alpha r$	$(P\cos\alpha - X(\alpha))/mg$
A'	$\alpha = 0$	$(P - X_0)/mg$
A''	$\alpha = \alpha_{\min} = \text{const}$	$(P' - X_0')/mg$

The problem A'' easily turns into A' if, instead of the original object, we consider an object with new traction and aerodynamic characteristics found from the relations

$$P' = P\cos\alpha_{\min}, \quad X_0' = X_0 + C_{=1}(\alpha_{\min})qS \tag{20}$$

where $\alpha_{\min} = \text{const}$.

In this case, all conclusions that are correct for the problem A' will also be true for A''.

In [5], an exact solution A' to the problem was obtained. In this case, the solution to the problem A' can be considered as an approximate solution to the problem A [5]. Due to the fact that $\alpha_{\min}$ it is closer to $\alpha_{o\Gamma\Pi} = \alpha_P$ than $\alpha = 0$, the solution to the problem A'' will be closer to the solution to the problem A than the solution A'. Of all $\alpha \in \Omega_\alpha$, the minimum fuel consumption of the task would be achieved at $\alpha = 0$, if such a regime were possible. However, in real flight during the launching phase $\alpha > 0$.

The proposed modification makes it possible to reduce the difference between the problem A' and the solution of the problem A by passing to the problem A''. The $\alpha_{\min}$ can be determined directly at each step of the trajectory synthesis $V^*(h)$. In general, the methodology for finding the optimal estimated trajectory and searching for the mass estimate remains the same. Constraints and boundary conditions are also unchanged. Let us note the existing differences in the form of the used dependences (5), (7)

$$G(V, h) = P(V, h, \beta_{\max}(V, h)) \cos \alpha_{\min} - X(V, h, \alpha_{\min}) \geq 0 \,, \tag{21}$$

where $X(V, h, \alpha_{\min}) = X_0(V, h) + AC_y^{\alpha^2} qS\alpha_{\min}^2$.

The set D_0 is defined as the set of arguments satisfying

$$P(V, h, \beta) \cos \alpha_{\min} - X(V, h, \alpha_{\min}) > 0 \,. \tag{22}$$

Function (9) is defined as

$$Q(V, h, \beta) = \frac{\beta}{P(V, h, \beta) \cos \alpha_{\min} - X(V, h, \alpha_{\min})} \,. \tag{23}$$

The modified method retains the ability to find the mass estimate at the final and intermediate points. In addition, the described modification can be used to organize an iterative search for a realized suboptimal trajectory and the corresponding mass estimate.

The presented procedure is repeated for each current height value. For the task A'' it is supposed to $\alpha_{\min} = \alpha_{o\Pi}$. For each controversial value $\alpha_{o\Pi} \in \Omega_\alpha$, the optimal values of speed $V_{o\Pi T}$ and fuel consumption per second are found $\beta_{o\Pi T}$. From the differential equations written in finite differences, θ and α_P, are found, which make it possible to realize the transition of the aircraft to a new point from the previous one.

We are interested in such values $V_{o\Pi T}$, α_P for which the reference value α is simultaneously realizable $\alpha_p = \alpha_{o\Pi}$, that is, when the dependence graph $\alpha_p(\alpha_{o\Pi})$ intersects with the bisector of the angle formed by the coordinate axes.

If there are several intersection points, then we are interested in the one in which the value of α is minimal. It is possible that there is no intersection of the curve $\alpha_p(\alpha_{o\Pi})$ with the bisector of the angle. In this case, it is necessary to go back one step in height, exclude the value found at the previous step $V_{oT\Pi}$ from the number of possible solutions, and repeat the search procedure.

The trajectory synthesis problem can be divided into the following stages:

1. analysis of the feasibility of a particular launching task;
2. synthesis of the implementation trajectory of launch;
3. synthesis of control for the flight of wills of the found reference trajectory.

If at the first stage it turns out that among the admissible trajectories there is no one that will allow the inference task to be completed, then it makes no sense to proceed to stages 2, 3. At the first stage, as a rule, the problem is considered in a somewhat simplified form, but the more accurately the problem is solved on at this stage, the easier it is to perform items 2, 3. The feasibility of the lifting-acceleration task is defined as the ability of an object to bring the payload to a given point in space no less than a given one.

In our case, at the first stage, the method of V.F. Krotova-M.M. Khrustalev or one of its modifications that give an estimate by mass (which makes it possible to judge the feasibility of the injection task) and an optimal trajectory $V(h)$ synthesized without taking into account its feasibility. Refinement of the initial trajectory $V(h)$ in the course of bringing it to a realizable form simultaneously with refining the estimate is carried out at the second stage. Modifications of the V.F. Krotova-M.M. Khrustaleva allows:

1. to clarify the solution sought at the first stage (task A'');

2. choose a compromise between the real and joint solution of the tasks of stages 1 and 2 by establishing the degree of proximity α_P and $\alpha_{o\Pi}$, and, consequently, the number of necessary iterations, depending on the formulation of the research problem.

The validity of the theoretical studies described is confirmed by the results of numerical modeling. The synthesis of the estimated launch trajectory was performed with different reference values $\alpha_{\min} = \text{const}$, $\alpha_{\min} \in [0°,\ 10°]$ for an aircraft with a turbojet engine in the altitude range from 0 to 19 km.

Let us dwell on individual results and features of the algorithm.

Beginning with $\alpha_{\min} = 7°$, a range of heights appears (Table 2) within which the set turns out to be empty, ie, there are no such V for which condition (22) would remain valid.

Table 2. The range of heights appears with different reference values $\alpha_{\min}$

$\alpha_{\min}$	The h range in which $D_o = 0$
7°	4–8 КМ
10°	1–13 КМ

This indicates that the use of the same $\alpha_{\min}$ at different heights is not always acceptable and gives obviously worse results than the flexible change algorithm $\alpha_{\min}$ as the altitude changes. In the latter case, there is a real opportunity to obtain an estimate of the mass that completely coincides or is close enough to the exact solution of the boundary value problem. As a result of calculations, a functional dependence of the relative mass on the height and the reference value of the angle of attack $\mu(h,\ \alpha_{o\Pi})$ was found.

The magnitude of the improvement in the weight estimate compared to the unmodified method was also determined

$$\Delta\mu(h,\ \alpha_{o\Pi}) = \mu(h,\ 0) - \mu(h,\ \alpha_{o\Pi}) \tag{24}$$

The studies indicate that in the considered range of heights $h \in [0, 19]$ KM , the realizable value of the angle of attack in the vicinity of the optimal trajectory is not less 3°–4°. If we focus on such a minimum possible value $\alpha_{\min}$, then the degree of improvement of the estimate in comparison with the base version of the V.F. Krotova-M.M. Khrustalev method fluctuates in a range $\Delta\mu = 0{,}1 - 0{,}2\%$ that corresponds to the absolute mass $\Delta m = 300 - 1200$ КГ for objects with a starting mass of 300 to 600 tons. When the reference value of $\alpha_{O\Pi}$ is varied within the limits $3° - 5°$, an improvement in the estimate for the final mass at an altitude of 19 km can reach, respectively, the values

$$\Delta\mu = 0{,}1{-}0{,}4\%, \quad \Delta m = 300{-}2400 \text{ kg}$$

When driving to high altitudes, the difference $\Delta\mu$ will be significant and can reach a few percent of the relative mass.

It is easy to see that $\Delta\mu(h,\ \alpha_{o\Pi})$ it changes non-monotonically for all values $\alpha_{on} = const$. In addition, for different $\alpha_{o\Gamma}$ types of dependence $\Delta\mu(h,\ \alpha_{o\Pi})$, the quality is also different. In a local sense, the reasons for this phenomenon lie in the imperfection of the interpolation of the original data arrays during printing. In a global sense, the general view of the dependency $\Delta\mu(h,\ \alpha_{o\Pi})$ is inextricably linked with the type of dependency $V(h)$ corresponding to the value used $\alpha_{o\Pi}$.

$$\frac{d\mu}{dh} = \frac{d\mu}{dt}\frac{dh_\theta}{dh}\left[\frac{dh_\theta}{dt}\right]^{-1} = \frac{-\beta\left[1+\frac{V}{g}\frac{dV}{dh}\right]}{m_0\ V\ n_x} = \frac{-\beta\ \mu\ g\left[1+\frac{V}{g}\frac{dV}{dh}\right]}{V[P\ \cos\ \alpha_{o\Pi} - X(\alpha_{o\Pi})]} \tag{25}$$

The change causes a change in the type of dependence, which, in accordance with (24), (25), leads to the non-monotonicity of the dependence $\Delta\mu(h,\ \alpha_{o\Pi})$.

4 Conclusion

1. The proposed modification of the methods of V.F. Krotova-M.M. Khrustalev consists in a conditional change in the characteristics of the object under study, which makes it possible, with minor changes, to apply the already known [5] method for synthesizing the suboptimal launch trajectory, while obtaining better estimates of the final mass in comparison with the known method. To apply the modification, information is needed on the possible realizable angles of attack when withdrawing along a trajectory belonging to a certain neighborhood of the optimal one. Based on conditions (17), (18), the dependences for the smallest realizable values of the angle of attack were found by numerical methods on the basis of dependence (19).
2. Based on the modification of the method of V.F. Krotova-M.M. Khrustalev developed an iterative algorithm for synthesizing a realizable injection trajectory, which allows one to approach the solution of the longitudinal and spatial injection problem from one point of view. In addition, the algorithm provides the ability to choose a compromise solution between the accuracy and speed of the algorithm, depending on the formulation of the research problem.
3. At the smallest realizable angles of attack of the aircraft $\alpha = 3°–4°$ (which corresponds to reality), the refinement of the estimate for the relative final mass at an altitude of 19 km varies within the limits $\Delta\mu = 0{,}1–0{,}2\%$, which corresponds to the absolute mass $\Delta m = 300–1200$ kg for objects with a launch mass of 300 to 600 tons. By varying the smallest realizable angles of attack within the limits $\alpha = 3°–5°$, an improvement in the estimate for the final mass at an altitude of 19 km, respectively, can reach $\Delta\mu = 0{,}1–0{,}4\%$, $\Delta m = 300–2400$ kg.
4. When fixed $\alpha_{o\Pi} =$ const, the dependence $\Delta\mu(h)$ has a non-monotonic character, which is explained by a change in the slope of the dependence $V(h)$, the type of which depends on the method of redistribution of the energy entering the system between the kinetic potential components, as well as on the indicators of the efficiency of the excretion process.
5. The change in the qualitative and quantitative characteristics of the dependence $\Delta\mu(h)$ when changing $\alpha_{o\Pi} =$ const is associated with a change in the type of trajectory $V(h)$ when changing $\alpha_{o\Pi}$. The impact $V(h)$ on efficiency indicators is described above.

6. For the object under study, there is some $\alpha_{\min}$, starting from which the set $D_0 = 0$ turns out to be empty (Table 1). Consequently, for a more complete study with a more accurate estimate of the mass, it is necessary to connect the apparatus of flexible change $\alpha_{o\Pi}$ when changing h.

References

1. Pontryagin, L., Boltyansky, V., Gamkrelidze, R., Mishchenko, E.: Matematicheskaya teoriya optimal'nykh protsessov [mathematical theory of optimal processes]. Science, Moscow (1969)
2. Bellman, R.: Dinamicheskoye programmirovaniye [Dynamic programming]. Foreign literature, Moscow (1960)
3. Krotov, V.: Ob optimal'nom upravlenii trayektoriyami poleta. Absolyutnyy optimum, analiticheskiye resheniya, algoritmy [Optimal control of flight trajectories. Absolute optimum, analytical solutions, algorithms]. Automation and telemechanics (1996)
4. Krotov, V., Gurman, V.: Metody i zadachi optimal'nogo upravleniya [optimal control methods and problems]. Science, Moscow (1973)
5. Khrustalev, M.: Neobchodimue i dostatochnue uslovija dlja zadachi optimalnogo upravlenija [Necessary and sufficient conditions for the optimal control problem]. Daclads of the Academy of Sciences, Moscow (1973)
6. Khrustalev, M.: Exact description of reachable sets and global optimality conditions for dynamic systems. P. I: Estimates and description of reachability and controllability sets optimality conditions. Avtomatika i Telemekhanika (1988)
7. Khrustalev, M.: Exact description of reachable sets and global optimality conditions. P. I: Global optimality conditions. Avtomatika i Telemekhanika (1988)
8. Krasovsky, A.: Sistemy avtomaticheskogo upravleniya poletom i ikh analiticheskoye konstruirovaniye Automatic flight control systems and their analytical design. Science, Moscow (1973)
9. Krasovsky, A.: Spravochnik po teorii avtomaticheskogo upravleniya [Automatic Control Theory Handbook] Science, Moscow (1987)
10. Letov, A.: Dinamika poleta i upravleniya [Flight dynamics and control]. Science, Moscow (1969)
11. Bryson E., Y.-C., Ho.: Applied Optimal Control: Optimization, Estimation, and Control. Mir, Moscow (1972)
12. Appazov, R., Sytin, O.: Metody proyektirovaniya trayektoriy nositeley i sputnikov Zemli [Methods for designing trajectories of carriers and satellites of the Earth]. Nauka, Moscow (1987)
13. Kazakov, I., Gladkov, D., Kriksunov, L., et al.: Sistemy upravleniya i dinamika navedeniya raket [Control systems and dynamics of missile guidance]. Zhukovsky Air Force Engineering Academy, Moscow (1973)
14. Lysenko, O., Ponomarenko, S., Tachinina, O., et al.: Feasibility reasoning of creating ultra-low orbit communication systems based on small satellites and method of their orbits designing. Inf. Telecommun. Sci. (2020)
15. Lysenko, O., Tachinina, O.: Method of path constructing of information robot on the basis of unmanned aerial vehicle. Proc. Natl. Aviat. Univ. (2017)

Analysis of Stress Development Mechanisms in the Coating/Substrate System

Andrey Volkov(✉)

National Aerospce University "Kharkiv Aviation Institute",
17 Chkalova Street, Kharkiv 61070, Ukraine
a.volkov@khai.edu

Abstract. The development of nanotechnologies and the general tendency to increase the areas of use of thin coatings with the simultaneous high cost of high-tech materials makes us pay attention to the possibility of predicting the properties of the obtained coatings. This paper presents an analysis of the mechanisms of stress development of thin coatings for a deeper understanding of the possibilities of modeling the stress state of the obtained coatings with subsequent prediction of the required properties. The magnitude and sign of the internal stress in the coatings depend on many factors related to the conditions and method of coating the substrate, as well as the nature of their growth. Thermal stresses make an important contribution to the overall level of stresses in thin coatings, as well as the development of stresses in current-carrying coatings is significantly influenced by electromigration processes. No less significant contribution to the development of stresses in thin coatings can have the effects associated with the influence of the electromagnetic field: the piezoelectric effect, electro- and magnetostriction.

Keywords: Nanostructures · Thin coatings · Stress · Substrate · Coating properties

1 Introduction

The need for new functional properties in coatings and tendency to use of nanomaterials in our time leads to the emergence of a large number of new compositions of materials. As a consequence of the introduction of increasingly complex technologies for obtaining coatings requires further development of both models of their development and forecasting the properties of the obtained coatings.

The stresses that occur in the coatings can be divided into two large groups [1]. The first is the internal stresses associated with changes in the composition and structure that occur inside the coatings during their application. The second - stresses that are the result of exposure to the coating/substrate system of any external forces and occur after coating on the substrate. In addition, the relaxation processes that occur in the coatings can lead to a change in their stress state, even in the absence of any external influences. However, such a classification allows a deeper understanding of the causes and mechanisms of stress in coatings, as well as to develop methods for managing them.

M. Nechyporuk et al. (Eds.): ICTM 2021, LNNS 367, pp. 78–88, 2022.
https://doi.org/10.1007/978-3-030-94259-5_8

2 Stress of Nonequilibrium in Epitaxial Coatings

The size and sign of internal stress in the coatings depend on many factors associated with the conditions and method of applying coatings on the substrate, as well as the nature of their growth. Thus, with heteroepitaxial growth, when the coating materials and substrates are different, the appearance of internal stresses is mainly caused by the nonequilibrium of their constant crystal lattices. At the initial stage of growth, a continuous heteroepitaxial coating is a biaxially stressed layer coherently coherently linked to the substrate interface. If the parameters of the coating and substrate lattices differ slightly, the pseudomorphic growth of the coating can take a long time [2, 3]. However, since the energy of elastic deformation of the coating W is proportional to its volume V:

$$W = \frac{2 \cdot \mu \cdot (1 + \nu)}{1 - \nu} \cdot \varepsilon^2 \cdot V \quad (1)$$

where μ – shear module, ν – Poisson's ratio of coating, ε – flat deformation that occurs under the action stresses of nonequilibrium, the increase in the thickness of the applied layer leads to a rapid increase in W, resulting in the achievement of a certain critical thickness begins stress relaxation.

Depending on the application conditions and the characteristics of the coating and the substrate, this process may be accompanied by both plastic and elastic deformation of the coating [4].

The mechanism of surface coarsening of coatings depends on the kinetic behavior of atoms on the surface of the growing layer, which is regulated by deformation. At the boundaries of atomic clusters there is an almost complete relaxation of stresses, while in their central part the deformation energy increases significantly. Since in monocrystalline epitaxial coatings there are no internal interface, the main mechanism of mass transfer in them is diffusion on the surface. Atoms move from one place on the surface to another, leading to a change in free energy, which is determined by the gradient of chemical potential along the surface. Thus, there is a movement of growth centers along the surface in places with less free energy, ie to the tops of the protrusions, which causes the transition to the island mechanism of growth [4].

3 Internal Stresses in Polycrystalline Coatings

One of the main reasons for the development of internal stresses in polycrystalline thin coatings is the compaction of their structure. The nonequilibrium of polycrystalline coatings is most significant at low temperatures and/or at high rates of application, when low mobility of adatoms on the surface of the substrate leads to the formation of a large number of nuclei and does not allow accelerated growth of the crystallites with energetically favorable orientation.

At the initial stage of formation, when the islands do not touch each other, in the coatings, compressive stresses develop, the magnitude of which can be quite high.

It is currently assumed that the main cause of compressive stresses in the island thin coatings is associated with the action of surface tension forces [7].

The increase in the size of the island during the growth of the coating should lead to a decrease in capillary pressure and, accordingly, to the gradual relaxation of the permanent lattice to the volumetric values. However, after the island reaches a certain critical size, it is rigidly fixed on the substrate, which prevents further changes within the interatomic distances in the plane parallel to the interface of the coating / substrate [8, 9]. Therefore, the long-term growth of the islands contributes to their deformation along a given plane, which according to Hooke's law is equal to

$$\varepsilon = -\frac{2 \cdot s \cdot (1 - \nu)}{E \cdot R}, \tag{2}$$

where s – is the force of surface tension, R – the radius of the island (the average curvature of its surface), E and ν – modulus of elasticity and Poisson's ratio. Since for most low-index surfaces of metals and dielectrics $s > 0$, then $\varepsilon < 0$, that is, compressive stresses develop in the islands.

Also a significant contribution to the development of compressive stresses can be made by the interaction of adatoms with the surface of the substrate and the islands that are formed [10]. This assumption is based on the fact that the interaction of the adatom with the surface of the substrate leads to the formation in its surface layer of a field of elastic deformation, which can be considered as a "power dipole". This field causes the development of compressive stresses in the islands, the growth of which is accompanied by the continuous deposition of a significant number of new adatoms.

As the size of the islands increases, they cover most of the surface of the substrate, and neighboring islands begin to interact with each other. The result of this interaction is the fusion of the islands, which is usually accompanied by the development of tensile stresses in the coatings [11]. The maximum value of these stresses $\sigma_{\max}$, significantly depends on the grain size during coalescence d_0 ($\sigma_{\max} \sim 1/d_0$). In particular, at $d_0 \leq$ 10 nm stresses can reach several GPa [12].

The driving force of coalescence is the reduction of the total energy of the coating during the transformation of the two free surfaces of the islands into one grain boundary. The change in energy in this case can be written as

$$\Delta\gamma = 2\gamma_s - \gamma_{gb}, \tag{3}$$

where γ_s – surface energy, γ_{gb} – grain boundary energy. Because usually $2\gamma_s > \gamma_{gb}$, then coalescence of islands is energetically favorable.

At the initial stage of coalescence, the growth of the islands and the filling of the free space between them is carried out by diffusion of adatoms on the surface of the islands and the surface of the substrate. However, when the distance between the islands becomes less than some critical value, the diffuse mechanism of filling the cavities is difficult. It has been suggested that the final stage of coalescence (corresponding to the thickness of the applied layer 10–50 nm), during which the islands become grains of solid coating and boundaries are formed between them, is accompanied by contraction of space between the side faces of neighboring islands by their spontaneous elastic deformation [13]. This mechanism allows to explain the development in the coating of tensile stresses.

According to the model first proposed in [13], the maximum tensile stress in the coating can reach

$$\sigma_{\max} = \frac{E}{(1-\nu)} \cdot \varepsilon_{\max} = \frac{E}{(1-\nu)} \cdot \frac{\Delta_{\max}}{2a} = \left[\frac{E}{a \cdot (1-\nu)} \cdot \left(2 \cdot \gamma_s - \gamma_{gb}\right)\right]^{\frac{1}{2}} \tag{4}$$

The above model is very simplified and contains a number of assumptions that do not correspond to reality. For example, it assumes that the walls of the islands are located vertically and can interact with each other through the gap.

A more complex and physically realistic model was proposed in [4]. In it, the filling of the gap between adjacent islands and the formation of the grain boundary were considered as a process of closing the crack from the standpoint of the Griffiths destruction criterion. This model assumes that the islands have a rounded shape and their coalescence begins with the occurrence of contact at one point. Next, the area between adjacent islands is considered as a crack, the closure of which is energy efficient, because it reduces the total free energy of the system. In contrast to the above model [13], in [4] it is assumed that the process of crack closure occurs gradually, accompanied by elementary acts of elastic deformation of the islands, as long as the decrease in surface energy due to grain boundary exceeds the corresponding increase in elastic deformation. Despite the significant differences of this mechanism of coalescence of the islands from the model proposed in [13], obtained in [4] the expression for the average tensile stresses developing in the coating

$$\sigma_{aver} = \left[\frac{(1+\nu)}{a} \cdot \frac{E}{(1-\nu)} \cdot \left(2 \cdot \gamma_s - \gamma_{gb}\right)\right]^{\frac{1}{2}}, \tag{5}$$

differs from the ratio (4) only by the factor $1 + \nu$. Thus, the average stresses, which are determined by relation (5), exceed the maximum stresses in (4), ie are even more inflated. Among the possible reasons leading to such overestimation, the authors [13] note the lack of consideration in the model of shear stresses occurring at the interface/substrate during the deformation of the islands, as well as the assumption that the coalescence of all islands begins simultaneously.

Another approach to the description of the mechanism of development of tensile stresses in thin coatings at the stage of coalescence of islands was proposed in [14]. The authors of this work considered the coalescence of the islands as the contact of elastic bodies between which the force of adhesion acts. This model is based on the classical Hertz theory of elastic contact of solids, taking into account the cohesive interaction between the contact surfaces [15]. For three-dimensional islands of hemispherical shape in [14] the following expression is obtained for tensile stresses averaged over the volume of the island

$$\sigma_{aver} = \frac{2 \cdot \left(2 \cdot \gamma_s - \gamma_{gb}\right)}{a} \tag{6}$$

The results of this model are in better agreement with the experimentally obtained stress values. However, this model does not allow to explain the large scatter of experimental data for different coatings, as the value determined by relation (6), in fact, can change only by varying the size of the islands.

After the coalescence of the islands and the formation of a continuous thin coating, one of the main factors determining the magnitude of internal stresses in it is the diffusion mobility of adatoms, which depends on the deposition rate and homologous temperature (the ratio of the deposition temperature of the coating to its melting point *Td/Tm*). In the general case, thin-film materials can be divided into two types [11, 16]. Coatings of the first type, applied at low homologous temperature (*Td/Tm* < 0,2) and/or high speed. In these coatings, the tensile stresses that occur during the coalescence stage of the islands are maintained after its completion and increase with increasing thickness of the applied layer, eventually reaching saturation. Coatings of the second type are formed at higher homologous temperatures and/or low deposition rates. In these coatings with increasing thickness there is a transition of tensile internal stresses in compressive.

Another mechanism for the development of tensile stresses is associated with recrystallization and growth of grains flowing in a continuous coating, both during its deposition and after its completion [11, 16, 17]. This mechanism has a low effect on coatings of the first type, because in this case the low diffusion mobility of atoms does not contribute to significant grain growth. However, it contributes to the relaxation of compressive stresses and the re-emergence of tensile stresses in coatings of the second type, the mobility of atoms in which is high.

Since the substrate does not have a significant orienting effect on the growth of polycrystalline coatings, the location and crystallographic orientation of neighboring islands are random, which does not contribute to the formation of equilibrium boundaries when combined. In the process of grain growth there is a decrease in the number of grain boundaries and, consequently, the redistribution of free volume. As a result, the coating is compacted, which in the free state would reduce its size. The transverse dimensions of the coating remain unchanged, and it undergoes biaxial deformation, which with increasing average grain size from d_0 to d is [12]:

$$\varepsilon_d = \Delta V \cdot \left(\frac{1}{d} - \frac{1}{d_0}\right) \tag{7}$$

If this deformation accommodates elastically, then the compaction of the coating during grain growth causes the development of biaxial tensile stresses in it.

$$\sigma = \frac{E}{(1-\nu)} \cdot \varepsilon_d, \tag{8}$$

and leads to an increase in its average energy density (per unit volume of coating) by

$$\Delta F_d = \frac{E}{(1-\nu)} \cdot \varepsilon_d^2 \tag{9}$$

From the above, it follows that the tensile stresses in the coatings can inhibit the growth of grains in them, because it is energetically unfavorable due to the additional increase in stresses and the growth of elastic energy. At rather small d_0 ($d_0 \leq 4$ HM) nm) decrease in energy density associated with a decrease in the number of grain boundaries per unit area of coating [12]:

$$\Delta F_{gb} = \gamma_{gb} \cdot \left(\frac{2}{d_0} - \frac{2}{d}\right), \tag{10}$$

may be less than the value ΔF_d. In this case, each stress value corresponds to a certain grain size limit, at which time the growth of grains in the coating stops. If $\Delta F_{gb} \geq \Delta F_d$ then the growth of grains will continue, leading to the development of strong deformations. In the general case, the presence of tensile stresses promotes the formation of fine-grained coatings, and compressive stresses, on the contrary, promote the growth of large grains.

In contrast to tensile stresses, the main mechanism of which in polycrystalline coatings is currently considered to be the above-described process of grain boundary formation, today there is no complete understanding of the mechanisms of development of compressive stresses. Many studies suggest that compressive stresses in solid coatings are inherited from island coatings, in which they are the result of capillary forces [7, 16, 18]. In other work [10], it is believed that, as in island coatings, in the process of growth of solid coatings there is a mechanism of " power dipole", which leads to compression of the coating due to the interaction of adatoms with its surface. It is also assumed that the compressive stresses in thin coatings may be the result of the introduction of excess atoms between adjacent layers on the growing surface [19]. The recently proposed model assumes that stresses in solid polycrystalline coatings arise as a result of the interaction of adjacent grains, which causes their deformation, including grain boundary slippage [20].

It should be noted that the model that linked the mechanism of development of compressive stresses with the diffusion flux of atoms from the free surface of the growing coating at the grain boundary [7, 21–23].

The advantage of this model is the fact that it directly links the development of compressive stresses in polycrystalline coatings with the process of their growth, as it considers as the main reason for this the nonequilibrium nature of the surface of the growing coating.

Another advantage of this approach is that it also explains the relaxation of compressive stresses in the coatings after the termination of the application process. As soon as the flux of atoms deposited on the free surface of the coating becomes zero, the chemical potential there decreases to equilibrium. At the same time, the chemical potential at the grain boundaries remains high due to the accumulation of excess atoms. As a result, there is an oppositely directed gradient of chemical potential, which causes the reverse flow of atoms from the grain boundaries to the free surface, which leads to the relaxation of compressive stresses. Restoration of coating deposition again causes an increase in the chemical potential on the free surface of the coating and the flow of atoms at the grain boundary.

4 Factors Influencing the Occurrence of Internal Stresses in Thin Coatings

Along with the above processes underlying the formation of thin coatings, the development of internal stresses in them can also be caused by the formation and annihilation of point defects, the introduction of impurities, phase transformations, the formation of precipitates, absorption and desorption of moisture and volatile substances, and structural damage due to the effects of high-energy processes (for example, ion bombardment).

Point defects, which are formed in the process of applying thin coatings, contribute to the development of internal stresses. One of the important mechanisms for the development of stresses in coatings obtained by spraying is the so-called "atomic peening" [24]. The introduction of atoms sprayed into the internodes of the crystal lattice leads to distortion of the crystal lattice and excessive compaction of the coating. As a result, it has compressive stresses, the magnitude of which is directly proportional to the molar volume and biaxial modulus of elasticity of the coating [25]. In addition to the atoms of the atomized material in the interstitial of the crystal lattice can be introduced and impurity atoms and ions, such as inert gas ions used to bomb the target. Another important parameter that affects the concentration of internodal atoms is the pressure in the working chamber [26]. Under conditions of low pressure, most of the deposited atoms do not have collisions with each other and when in contact with the surface of the coating have a high energy, which facilitates their introduction into the internodes. Also, a significant influence on the development of compressive stresses in the coatings due to "atomic peening" has a shear stress applied to the substrate [27]. At high shear stresses, a strong electromagnetic field is created near the surface of the substrate, which accelerates the ions bombarding it. This leads to an increase in their kinetic energy and, as a consequence, increases the concentration of point defects in the coating.

High-energy atoms and ions can lead to the appearance in coatings not only of internodal atoms, but also of substitution atoms [27]. In this case, if the substitution occurs by an atom having a larger radius than the lattice atom, then compressive stresses develop in the coating. Otherwise, tensile stresses occur.

Another type of point defects - vacancies - can also be a source of internal stresses in thin coatings. The development of internal stresses, sign and magnitude depend on the partial molar volume of vacancies, as well as the location of the annihilation [17].

In practice, different types of point defects are present in the coatings at the same time. Therefore, their influence on the development of internal stresses is complex, as different types of defects not only cause different mechanisms of stress, but also interact with each other, which can also change the stress-strain state of the material.

Internal stresses in coatings can also occur due to phase transformations that occur inside them in the process of deposition, the formation of new phases and precipitates, as well as the introduction of impurities.

Another type of phase transformations that occur during the deposition of coatings is associated with the course of chemical reactions that lead to the formation of new phases. These can be reactions caused by mutual diffusion of the coating atoms and the substrate, or diffusion between different layers. In addition, chemical reactions of coating atoms with residual gas molecules or with embedded impurity atoms are possible.

The result of the interaction of the coating with the residual gases in the working chamber in a complex way depends on their partial pressure.

5 Stresses that Develop in Thin Coatings Under External Influences

Thermal stresses make an important contribution to the overall stress level in thin coatings. Thermal stresses are external because they are caused by changes in temperature after the deposition of coatings. But they are also directly related to the deposition process of coatings, as many technologies for their application require high temperatures.

In addition, thermal stresses can develop during heat treatment of coatings after their application, as well as due to changes in temperature during operation of coatings.

The reason for the development of stresses in thin-film structures when the temperature changes is the difference between the coefficients of thermal expansion of the coating and the substrate.

If the composition coating/substrate is cooled or heated from temperature T_1 to temperature T_2, the biaxial deformation of the coating that develops is determined as follows:

$$\varepsilon = -(\alpha_d - \alpha_s) \cdot (T_2 - T_1) = -\Delta\alpha \cdot \Delta T, \tag{11}$$

where α_d and α_s – coefficients of thermal expansion of the coating and the substrate, respectively. If the temperature changes during the deposition of the coating, then different layers of the coating will experience different deformation. In this case, its average biaxial deformation can be calculated using the ratio [17]:

$$\varepsilon \approx -\Delta\alpha \cdot \left(T_2 - \frac{1}{h_f} \cdot \int\limits_0^h T(z)dz \right), \tag{12}$$

where h – coating thickness, dz – thickness increase. Since biaxial stresses are associated with deformation as

$$\sigma = \frac{E}{(1-\nu)} \cdot \varepsilon, \tag{13}$$

then from (11) and (12) we obtain that the thermal stresses are directly proportional to the temperature changing:

$$\sigma = -\frac{E}{(1-\nu)} \cdot \Delta\alpha \cdot \Delta T \tag{14}$$

Since, in thin coatings both thermal and internal stresses develop at the same time as a result they can both strengthen, and weaken each other.

A significant influence on the development of stress in current-carrying coatings, have electromigration processes.

Thus, according to [28], under the action of current, the interatomic distances in Al conductors increased near the anode and decreased near the cathode.

It should be noted that thermal stresses are always present in conductive coatings, because high-density electric current causes intense release of Joule heat, which leads to heating of the coatings. This complicates the precise determination of the magnitude of the stresses caused directly by electromigration in the coatings applied to the substrate.

Significant contribution to the development of stresses in thin coatings can have effects associated with exposure to electromagnetic fields. First of all, they include the piezoelectric effect, electro- and magnetostriction.

The piezoelectric effect under the influence of an external electric field causes the deformation of the coating crystals. And since the coatings can not be freely deformed due to the restrictions imposed by the substrate, they have mechanical stresses.

In contrast to the piezoelectric effect, electrostriction is observed in absolutely all dielectrics and is also due to their polarization in an external electric field. At the same field strength, the deformation of bulk materials caused by electrostriction is several orders of magnitude smaller than in the piezoelectric effect. But in thin-film materials, electrostriction can be an important source of stress. In particular, strong stresses caused by electrostriction develop in coatings of anode oxides of metals, the thickness of which can be several nanometers [29, 30].

It should be noted that the development of stresses in dielectric coatings under the action of an external electric field is due to two different factors. First, as a result of polarization on the surface of the coating and at its interface with the substrate, uncompensated charges appear. Which leads to the exploration of the so-called Maxwell stresses [29]:

$$\sigma_M = -\frac{\nu}{1-\nu} \cdot \frac{\varepsilon_0 \cdot \varepsilon_{di}}{2} \cdot E_e^3, \tag{15}$$

where ε_{di} – dielectric permittivity, ε_0 – dielectric constant, E_e – electric field strength. As can be seen from expression (15), Maxwell's stresses have a minus sign, ie they are always compressive, which follows from the geometry of the Coulomb interaction.

Secondly, under the action of an external electric field, molecular dipoles tend to line up along its lines of force, which also causes the deformation of the coating and the occurrence of stresses in it. This effect is sometimes called dielectrostriction. According to [29, 31], the magnitude of these stresses can be recorded

$$\sigma_{ES} = -\frac{\nu}{1-\nu} \cdot \frac{\varepsilon_0}{2} \cdot [\varepsilon_{di} - (\beta_1 + \beta_2)] \cdot E_e^3, \tag{16}$$

where β_1 and β_2 – electrostriction parameters that depend on the dielectric permittivity of the coating. As can be seen from expression (16), the sign of the stresses caused by electrostriction is determined by the ratio ε and amounts $\beta_1 + \beta_2$, that is, depending on the dielectric properties of the coating, these stresses can be both compressive and tensile. The final value of the stresses caused by electrostriction depends on the ratio σ_M and σ_{ES}.

This effect of deformation of the material when changing the state of its magnetization under the action of an external magnetic field is called magnetostriction. The mechanism of stress development in thin-film structures due to magnetostriction is similar to the cases of piezoelectric effect and electrostriction described above. [32, 33].

Finally, thin coatings in the process of their operation can be directly exposed to various mechanical influences, which lead to the development of stresses in them. Thus, in recent years, researchers are of great interest to thin-film structures on flexible substrates, which have broad prospects for use in the production of flexible LED screens, solar panels, various microelectromechanical systems, etc. [34, 35]. During operation, such flexible thin-film structures experience multiple bending deformation, which leads to the development of inhomogeneously distributed mechanical stresses, which causes their high gradients.

6 Conclusions

The analysis showed that the study of the mechanism of stress development in the coating/substrate system and their subsequent relaxation is of great interest. The high level of mechanical stresses in thin coatings is primarily due to the rigid connection with the substrate, which prevents changes in the size of the coating in the plane parallel to the interface between them and thus prevents elastic stress relaxation. In turn, the small thickness of the coatings and the presence of the interface with the substrate impose restrictions on the sliding of dislocations, complicating the development of plastic deformation. Therefore, either there is a coherent elastic deformation of the coating/substrate system due to its bending, which provides partial relaxation of stresses, or there is a destruction of the system.

To date, there is no consensus on the adequacy of the proposed models and their agreement with the available experimental results.

Obviously, to understand and take into account all the mechanisms of occurrence and evolution of stresses in thin coatings requires further experimental and theoretical studies.

Further research is needed to fully understand the wide range of factors that control the evolution of internal and external stresses in thin coatings. The contribution of each of the mechanisms to the total amount of stresses significantly depends on the growth conditions of the coatings. Therefore, to obtain coverage with the required level of stress, it is necessary to conduct research in a wide range of parameters of their deposition.

References

1. Freund, L.B., Suresh, S.: Thin Film Materials: Stress. Cambridge University Press, Cambridge, Defect Formation and Surface Evolution (2003)
2. Frank, F.C., van der Merwe, J.H.: One-dimensional dislocations. I. Static theory, Proc. R. Soc. London **198** (1053), 205–216 (1949)
3. Frank, F.C., van der Merwe, J.H.: One-dimensional dislocations. II. misfitting monolayers and oriented overgrowth. Proc. R. Soc. London, **198** (1053), 216–225 (1949)
4. Shugurov, A.R., Panin, A.V.: Mechanisms of stress generation in thin films and coatings. Tech. Phys. **65**(12), 1881–1904 (2020). https://doi.org/10.1134/S1063784220120257
5. Pang, Y., Huang, R.: Nonlinear effect of stress and wetting on surface evolution of epitaxial thin films. Phys. Rev. B., **74**, 075413–1–11 (2006)
6. Tekalign, W., Atena, A.: Thin film evolution equation for a strained anisotropic solid film on a deformable isotropic substrate. J. Appl. Math. Phys. **6**(4), 864–879 (2018)
7. Chason, E., Guduru, P.R.: Tutorial: Understanding residual stress in polycrystalline thin films through real-time measurements and physical models. J. Appl. Phys. **119** (19), 191101 (2016)
8. Laugier, M.: Intrinsic stress in thin films of vacuum evaporated LiF and ZnS using an improved cantilevered plate technique. Vacuum **31**(3), 155–157 (1981)
9. Cammarata, R.C.: Surface and interface stress effects in thin films. Prog. Surf. Sci. **46**(1), 1–38 (1994)
10. Friesen, C., Thompson, C.V.: Reversible stress relaxation during precoalescence interruptions of volmer-weber thin film growth, Phys. Rev. Lett. **89**(12), 126103–1–4 (2002)
11. Floro, J.A., Hearne, S.J., Hunter, J.A., et al.: The dynamic competition between stress generation and relaxation mechanisms during coalescence of Volmer-Weber thin films. J. Appl. Phys. **89**(9), 4886–4897 (2001)

12. Thompson, C.V., Carel, R.: Stress and grain growth in thin films. J. Mech. Phys. Solids. **44**(5), 657–673 (1996)
13. Abadias, G., Chason, E., Keckes, J., et al.: Review Article: Stress in thin films and coatings: Current status, challenges, and prospects. J. Vac. Sci. Technol. **36**, 020801 (2018)
14. Freund, L.B., Chason, E.: Model for stress generated upon contact of neighboring islands on the surface of a substrate. J. Appl. Phys. **89**(9), 4866–4873 (2001)
15. Johnson, K.L., Kendall, K., Roberts, A.D.: Surface energy and the contact of elastic solids. Proc. R. Soc. Lond. **324**(1558), 301–313 (1971)
16. Koch, R.: The intrinsic stress of polycrystalline and epitaxial thin metal films. J. Phys.: Condens. Matter. **6**, 9519–9550 (1994)
17. Doerner, M.F., Nix, W.D.: Stresses and deformation processes in thin films on substrates. Crit. Rev. Solid State Mater. Sci. **14**(3), 225–268 (1988)
18. Abermann, R., Koch, R., Kramer, R.: Electron microscope structure and internal stress in thin silver and gold films deposited onto MgF2 and SiO substrates. Thin Solid Films **58**(2), 365–370 (1979)
19. Spaepen, F.: Interfaces and stresses in thin films. Acta Mater. **48**, 31–42 (2000)
20. Gonzalez-Gonzalez, A., Polop, C., Vasco, E.: Postcoalescence evolution of growth stress in polycrystalline films, Phys. Rev. Lett., **110** (5), 056101–1–5 (2013)
21. Floro, J.A., Chason, E., Cammarata, R.C., Srolovitz, D.J.: Physical origins of intrinsic stresses in Volmer-Weber thin films. MRS Bull. **27**, 19–25 (2002)
22. Chason, E.: A kinetic analysis of residual stress evolution in polycrystalline thin films. Thin Solid Films **526**, 1–14 (2012)
23. Chason, E., Shin, J.W., Hearne, S.J., Freund, L.B.: Kinetic model for dependence of thin film stress on growth rate, temperature, and microstructure, J. Appl. Phys., **111** (8), 083520 (2012)
24. Mwema, F.M., Akinlabi, E.T., Oladijo, O.P.: Micromorphology of sputtered aluminum thin films: a fractal analysis. Mater. Today Proc. **18** (7), 2430–2439 (2019)
25. Mbam, S.O., Nwonu, S.E., Orelaja, O.A., et al.: Thin-film coating; historical evolution, conventional deposition technologies, stress-state micro/nano-level measurement/models and prospects projection: a critical review. Mater. Res. Expr. **6** (12), 122001 (2019)
26. Koch, R.: Stress in evaporated and sputtered thin films – a comparison. Surf. Coat. Technol. **204**, 1973–1982 (2010)
27. Al-masha'al, A., Bunting, A., Cheung, R.: Evaluation of residual stress in sputtered tantalum thin-film. Appl. Surf. Sci. **371**, 571–575 (2016)
28. Wang, P.-C., Cargill, G.S., III., Noyan, I.C., Hu, C.-K.: Electromigration-induced stress in aluminum conductor lines measured by x-ray microdiffraction. Appl. Phys. Lett. **72**(11), 1296–1298 (1998)
29. Vanhumbeeck, J.-F., Proost, J.: On the contribution of electrostriction to charge- induced stresses in anodic oxide films. Electrochim. Acta. **53**, 6165–6172 (2008)
30. Heuer, A.H., Kahn, H., Natishan, P.M., et al.: Electrostrictive stresses and breakdown of thin passive films on stainless steel. Electrochim. Acta. **58**, 157–160 (2011)
31. McMeeking, R.M., Landis, C.M.: Electrostatic forces and stored energy for deformable dielectric materials. J. Appl. Mech. **72**(4), 581–590 (2005)
32. Jay, J.-P., Le Berre, F., Pogossian, S.P., Indenbom, M.V.: Direct and inverse measurement of thin film magnetostriction. J. Magn. Mag. Mater. **322**, 2203–2214 (2010)
33. Varghese, R., Viswan, R., Joshi, K., et al.: Magnetostriction measurement in thin films using laser doppler vibrometry. J. Magn. Mag. Mater. **383**, 179–187 (2014)
34. Fan, Z., Ho, J.C., Takahashi, T., et al.: Toward the development of printable nanowire electronics and sensors. Adv. Mater. **21**(37), 3730–3743 (2009)
35. Rogers, J.A., Someya, T., Huang, Y.: Materials and mechanics for stretchable electronics. Science **327**, 1603–1607 (2010)

The Effect of Vertical Wind Turbines Position Pattern on Their Aerodynamic Characteristics

Dmytro Brega(✉), Maria Surgailo, and Sergey Eremenko

National Aerospace University "Kharkiv Aviation Institute", 17 Chkalova Street, Kharkiv 61070, Ukraine

Abstract. The article is devoted to the actual problem of increasing efficiency in the field of wind energy. Today, in order to overcome the energy crisis in Ukraine, more and more attention is paid to renewable energy sources. Traditionally, in our country, horizontal wind generators are used. Total power which is produced by wind turbines in Ukraine is nearly 1170 MW this is approximately 18% of whole green power produce in our country. The number of «wind farm» gives a lot of options for optimization. Horizontal wind generators characteristics are well studied, and the existing engineering methodologies and design methods make it possible to create efficient wind power plants. However, the efficiency of vertical wind turbines remains underestimated. Especially in matters of their mutual influence. In this regard, the article devoted to the study of the wind generator pattern position influence on their aerodynamic characteristics is relevant. Both analytical and numerical approach were used to carry calculation with different rotors position. It is shown that with some variation of rotors position their total efficiency can be increased.

Keywords: Vertical wind turbines · Rotor · Position pattern · Synchronous operation · Numerical approach · Management system

1 Introduction

Today, special attention is paid to the wind turbine rotors position, which is vital in the design of a wind turbine management system. Moreover, in most of the works devoted to both experimental and theoretical studies, the joint operation of wind turbine rotors with a horizontal axis of rotation (Fig. 1, a) is considered [1, 2]. Therefore, the study of the effect of the distance between the rotors of wind turbines with a vertical axis (Fig. 1, b) and their relative position is of current interest. This can help to find an optimal position and number of wind turbines installed together in order to increase total productivity. Special attention should be paid for wind turbines which are installed along the highways. Their position affects directly on total efficiency as the use not only wind energy but also the turbulence from the vehicle with certain velocity.

Several factors should be considered – rotors mutual location, delay time for each rotor, number of blades for each turbine. In current work 2d approach was used for both analytical method and numerical simulation. CFD method was used for verification purpose.

M. Nechyporuk et al. (Eds.): ICTM 2021, LNNS 367, pp. 89–97, 2022.
https://doi.org/10.1007/978-3-030-94259-5_9

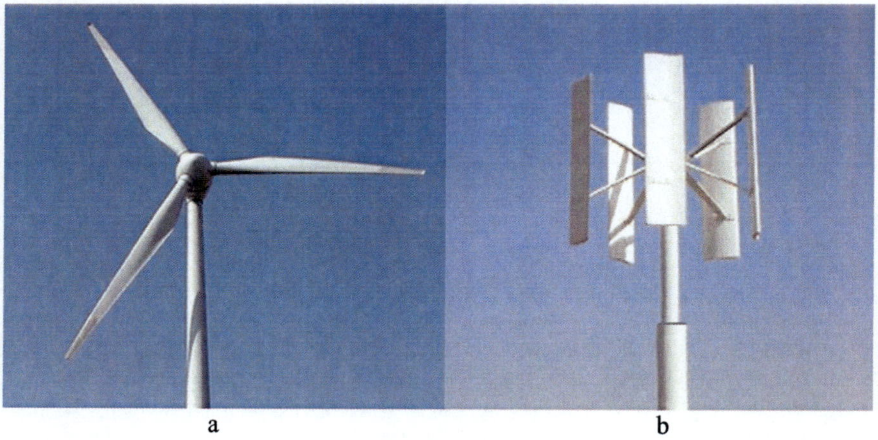

Fig. 1. Horizontal (a) and vertical (b) wind turbine

2 Analytical Model

Current study considers the synchronous operation of two wind turbine rotors with a vertical axis of rotation, located one after the other along the axis ***0x*** (Fig. 2, a) and along the axis ***0y*** (Fig. 2, b).

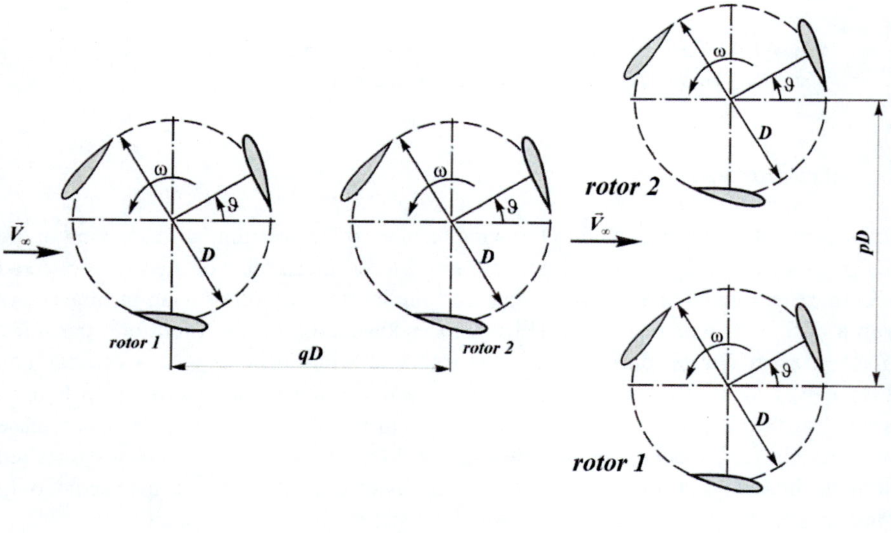

Fig. 2. Variants of mutual arrangement of wind turbine rotors

The distance between the centers of the rotors was varied and its influence on the dependence of the torque coefficient $\mathbf{c_m(z)}$ on the values of the maximum torque $\mathbf{c_{m_{max}}}$ and maximum speed $\mathbf{z}_{max}$ (with a wind energy utilization factor equal to zero) was investigated as well as the dependence of the angles of attack $\boldsymbol{\alpha}(\boldsymbol{\vartheta})$ and the relative resultant speed $\overline{\mathbf{W}}(\boldsymbol{\vartheta})$ on the swept circle with the azimuthal angle ϑ.

The study of the joint operation of wind turbine rotors with a vertical axis of rotation was carried out using a model of potential flow around a rotating permeable cylinder [3].

Expressions for the relative normal and tangential velocity components on the swept circles of the rotors are given with equations:

$$\overline{V}_{r1R_i} = \frac{V_{r1R_i}}{V_\infty} = \left(1 - R^2\frac{1-K}{1+K}\sum_{j=1}^{k}\frac{1}{\left(x_{R_i}-x_{c_j}\right)^2+\left(y_{R_i}-y_{c_j}\right)^2}\right)\cos\vartheta,$$

$$\overline{V}_{\vartheta 1R_i} = \frac{V_{\vartheta 1R_i}}{V_\infty} = -\left(1 + R^2\frac{1-K}{1+K}\sum_{j=1}^{k}\frac{1}{\left(x_{R_i}-x_{c_j}\right)^2+\left(y_{R_i}-y_{c_j}\right)^2}\right)\sin\vartheta$$

$$+ zR\left(1+\frac{1-K}{1+K}\right)\sum_{j=1}^{k}\frac{1}{\sqrt{\left(x_{R_i}-x_{c_j}\right)^2+\left(y_{R_i}-y_{c_j}\right)^2}},$$

$$\overline{V}_{r2R_i} = \frac{V_{r2R_i}}{V_\infty} = \overline{V}_{r1R_i},$$

$$\overline{V}_{\vartheta_2 R_i} = \frac{V_{\vartheta_2 R_i}}{V_\infty} = -\left(1 - R^2\frac{1-K}{1+K}\sum_{j=1}^{k}\frac{1}{\left(x_{R_i}-x_{c_j}\right)^2+\left(y_{R_i}-y_{c_j}\right)^2}\right)\sin\vartheta$$

$$+ zR\left(1+\frac{1-K}{1+K}\right)\sum_{j=1}^{k}\frac{1}{\sqrt{\left(x_{R_i}-x_{c_j}\right)^2+\left(y_{R_i}-y_{c_j}\right)^2}},$$

where $x_{R_i} = x_{c_i} + qD$ and $y_{R_i} = y_{c_i} + pD$ – the coordinates of the swept circle of the $\boldsymbol{i}$-th rotor;

$\mathbf{x}_{c_j}$ and y_{c_j} – coordinates of the center of the **j**-th rotor;
k – number of rotors;
q and **p** – coefficients that determine the distance between the rotors, proportional to their diameter;
D and **R** – diameter and radius of the rotors, m;
$K = \left(1-\sigma_{o\kappa p}\right)e^{-z(0,823+0,628\lg\sigma_{o\kappa p})}$ – permeability coefficient [3];
$\sigma_{o\kappa p} = \frac{i b_\pi}{\pi D}$ – "Circumferential" filling factor of the rotor, which is the ratio of the sum of the chords of the blades to the length of the swept circle $l_{o\kappa p} = \pi D$ [3];
z – speed factor;

For each rotor on the windward side, i.e. at $\frac{\pi}{2} \le \vartheta \le \frac{3\pi}{2}$, we work with the velocity components with the index "1" ($\overline{V}_{r1R_i}$ and $\overline{V}_{\vartheta 1R_i}$), and on the leeward side, i.e. for $-\frac{\pi}{2} < \vartheta < \frac{\pi}{2}$ - with the index "2" ($\overline{V}_{r2R_i}$ and $\overline{V}_{\vartheta 2R_i}$) (Fig. 3).

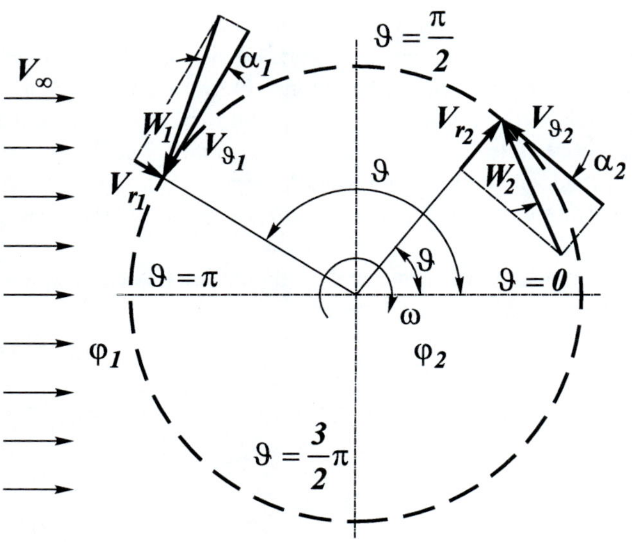

Fig. 3. Determination of velocity and angles of attack on the windward and leeward parts of the rotor.

On the swept circles of the rotors, knowing the velocity at each point of the surface, it is possible to determine the angles of attack and the resulting velocity approaching the blade from the expressions:

$$\alpha(\vartheta) = arctg\frac{V_r(\vartheta)}{V_\vartheta(\vartheta)},$$

$$W(\vartheta) = \sqrt{(V_r(\vartheta))^2 + (V_\vartheta(\vartheta))^2}.$$

In Figs. 4–7 show the dependences of the angles of attack $\alpha(\vartheta)$ and the relative resultant speed $\overline{W}(\vartheta)$ for an isolated rotor and two rotors of a wind turbine with a vertical axis, operating synchronously at a speed coefficient $z = 2, 5$.

The angles of attack were obtained for three-blade rotors with a vertical axis, having a diameter $D = 1\ M$, a chord of the blade $b = 0, 1\ M$, a height of the rotor $H = 0, 7\ M$, and an angle of installation of the blade $\varphi_\pi = 0°$.

1 and 2 - rotors 1 and 2 with a distance between them equal ***2D***; 3 - rotors 1 and 2 with a distance between them equal ***5D***; 4 - rotors 1 and 2 with a distance between them equal ***10D***; 5 - rotors 1 and 2 with a distance between them equal ***60D*** and an insulated rotor.

1 and 2 - rotors 1 and 2 with a distance between them equal ***2D***; 3 - rotors 1 and 2 with a distance between them equal ***5D***; 4 - rotors 1 and 2 with a distance between them equal ***10D***; 5 - rotors 1 and 2 with a distance between them equal ***60D*** and an insulated rotor.

From Fig. 4 and 5, it can be seen that with the synchronous operation of two rotors of a wind turbine with a vertical axis, both when arranged sequentially along the axis ***Ox*** and along the axis ***Oy***, the range of variation of the angles of attack on the swept surfaces is narrowed in comparison with an isolated rotor.

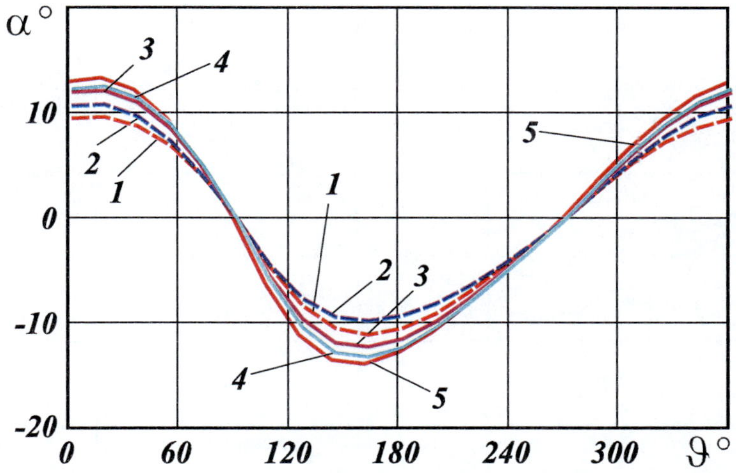

Fig. 4. Comparison of dependencies $\alpha(\vartheta)$ for isolated and synchronously operating rotors when they are located along the axis ***0x***.

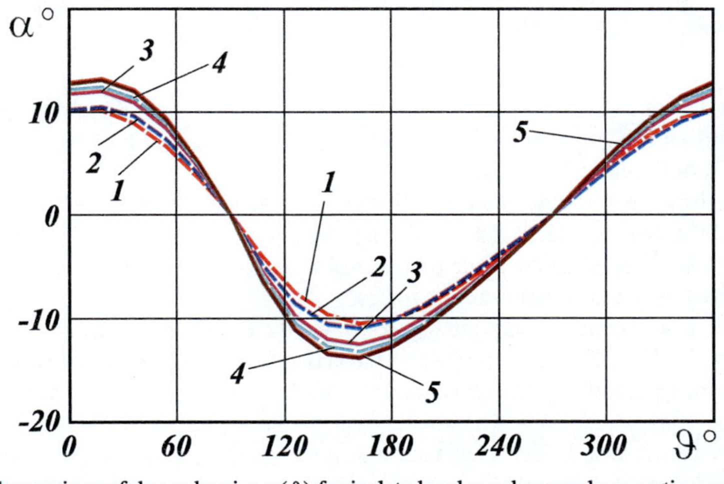

Fig. 5. Comparison of dependencies $\alpha(\vartheta)$ for isolated and synchronously operating rotors when they are located along the axis ***0y***.

When the distance between the centers of the rotors is from ***2D*** to ***5D***, the dependences $\alpha(\vartheta)$ for the first and second rotors differ significantly. Starting from a distance of equal ***5D*** dependences $\alpha(\vartheta)$ practically merge, but the range of variation of the angles of attack on the swept circles remains smaller than for an isolated rotor up to a distance ***(50...60) D*** where the mutual influence of the rotors on each other decreases. When the distance between the centers of the rotors is equal ***(50...60) D***, each of the rotors works as an isolated one.

1 and 2 - rotors 1 and 2 with a distance between them equal ***2D***;

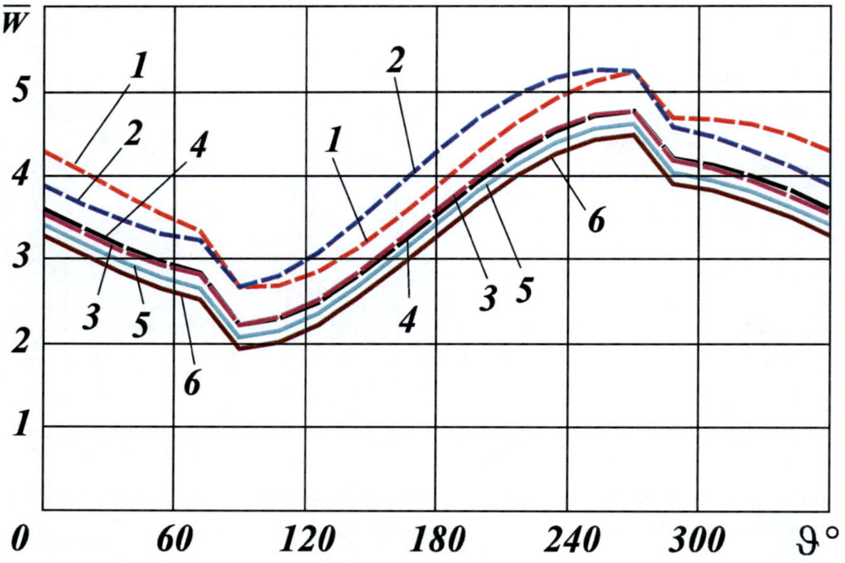

Fig. 6. Comparison of dependencies $\overline{W}(\vartheta)$ for isolated and synchronously operating rotors when they are located along the axis ***0x***.

3 and 4 - rotors 1 and 2 with a distance between them equal ***5D***; 5 - rotors 1 and 2 with a distance between them equal ***10D***; 6 - rotors 1 and 2 with a distance between them equal ***60D*** and an insulated rotor.

The narrowing of the range of variation of the angles of attack leads to the fact that in most of the swept surface of the rotor, the blades will operate in the range of angles of attack that are less than the angle of attack of the start of stall, which should improve the aerodynamic characteristics of the rotors.

With the synchronous operation of two rotors, the centers of which are located at a distance from ***2D*** to ***5D***, the relative net speed on the swept surfaces increases. Moreover, when the rotors are placed along the axis ***0x***, on the swept circle of the first rotor, the relative resulting velocities will be greater than for the second, on the leeward side, and smaller on the windward side. When the rotors are located along the axis ***0y*** on the swept circle of the first rotor, the relative resulting velocities are greater than on the second, in the range of azimuthal angles $0 < \vartheta < \pi$, and smaller for $\pi < \vartheta < 2\pi$.

The same regularities are preserved for the distances between the centers of rotors larger then 5*D*, but the mutual influence of the rotors on their kinematic decreases. At a distance from ***(50...60)D*** each of the rotors can be considered as working in isolation, and their kinematic parameters will fully coincide with the parameters of an isolated rotor.

1 and 2 - rotors 1 and 2 with a distance between them equal ***2D***;

3 and 4 - rotors 1 and 2 with a distance between them equal ***5D***; 5 - rotors 1 and 2 with a distance between them equal ***10D***; 6 - rotors 1 and 2 with a distance between them equal ***60D*** and an insulated rotor.

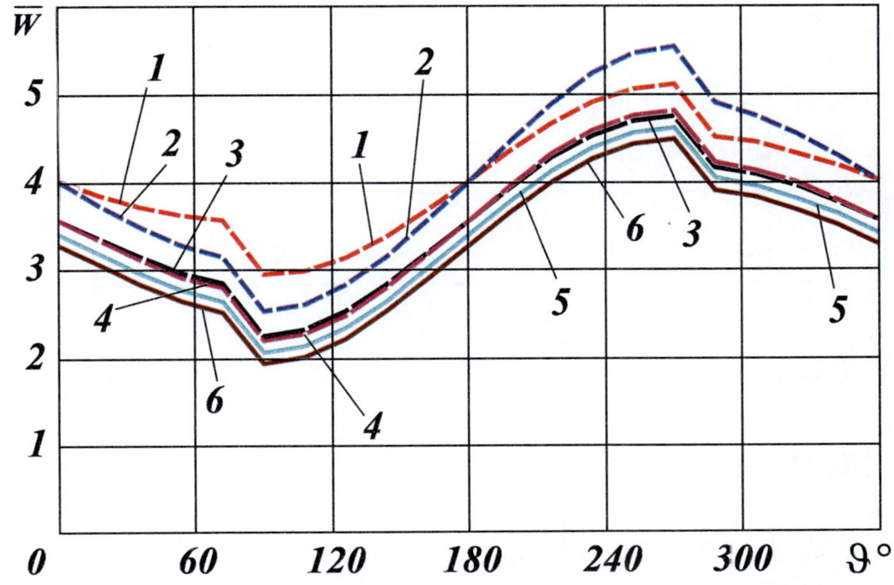

Fig. 7. Comparison of dependencies $\overline{W}(\vartheta)$ for isolated and synchronously operating rotors when they are located along the axis ***Oy***.

Changes in the dependencies $\overline{W}(\vartheta)$ will affect the aerodynamic characteristics of each of the synchronously operating rotors.

The rotor torque coefficient is determined from the expression [3]:

$$c_m = \frac{\sigma_{o\kappa p}}{2} \int_0^{2\pi} c_x(\vartheta)\overline{W}(\vartheta)^2 d\vartheta,$$

where $c_x(\vartheta)$ – longitudinal force coefficient of the blade, determined depending on the angle of attack $\alpha(\vartheta)$.

In Fig. 8 shows the dependence of the longitudinal force coefficient on the angle of attack $c_x(\alpha)$ for the NACA-0018 profile. Experimental studies of an isolated blade were carried out at the Reynolds number $Re = 0,3 \cdot 10^6$, blade elongation $\lambda = 6{,}36$ [4].

The dependences show that the placement of two synchronously rotating rotors with a vertical axis gives an increase in the maximum value of the torque coefficient in comparison with the isolated one. There are some related studies [5–7] which confirm the positive effect of using several wind turbines, however, they are more aimed at studying the mutual arrangement on the self-starting torque.

Let us consider the dependences of the maximum value of the torque coefficient for two synchronously operating wind turbine rotors with a vertical axis, located along the axis ***Ox*** and ***Oy***, on the distance between their centers.

In order to check the analytical model results a numerical simulation for the case 1 (Fig. 2 a) was carried. Numerical model is based on Navier-Stokes equations describing the conservation of mass, momentum and energy.

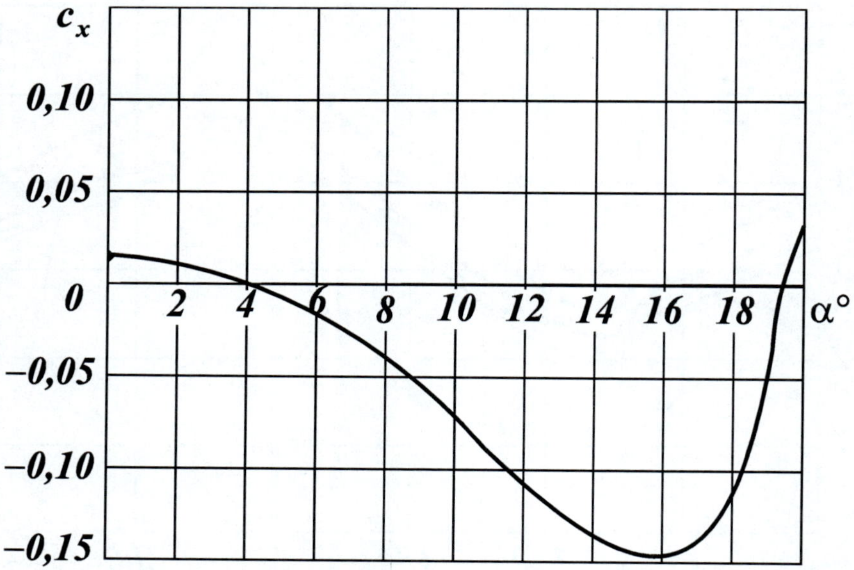

Fig. 8. Dependency $c_x(\alpha)$ for profile $NACA - 0018$

Figure 9 shows the comparative analysis of the results. As can be seen both approaches gives a very close results.

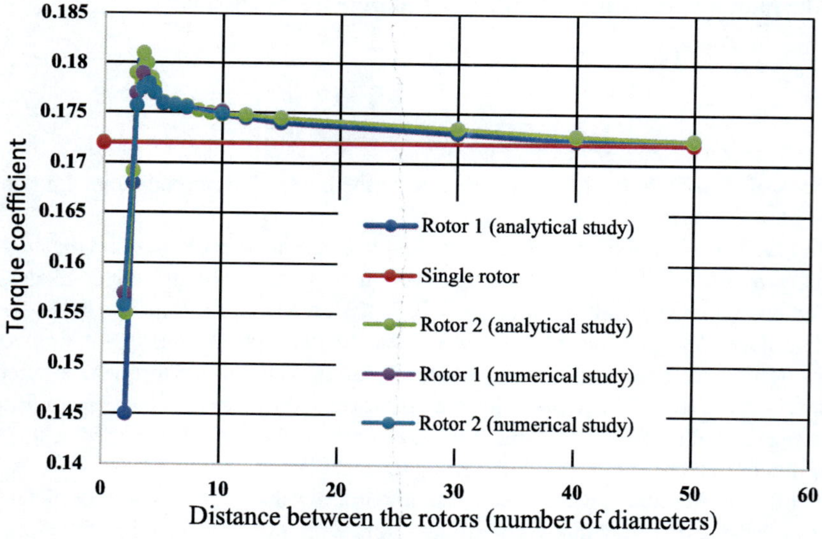

Fig. 9. Comparative analysis for analytical and numerical model for a set of rotors

3 Conclusions and Future Work

The specified research direction is rather promising. Obtained results shows a positive impact from optimum mutual position of wind turbines. It was found that the distance between rotors pattern equal to 5D gives an increase in power. Also mutual arrangement can reduce a torque required for self-start of a rotor. Initial comparative analysis of the analytical model and numerical shows good agreement between results. However further research is vital.

Several direction are left for the further research – parametric study on flow velocity vector angle effect on wind farm efficiency. Another important thing is desynchronization of the rotors which is also not studied, thus can give a positive boost in performance.

Promising results can be obtained with a combination of different geometrical parameters for individual turbine and their amount.

References

1. Okulov, V.L., Naumov, I.V., Tsoi, M.A., Mikkelsen, R.F.: Efficiency losses with coaxial arrangement of a pair of wind generators. Thermophys. Aeromech. **24**(4), 561–567 (2017)
2. Lundquist, J.K., DuVivier, K.K., Kaffine, D., et al.: Costs and consequences of wind turbine wake effects arising from uncoordinated wind energy development. Nat. Energy 4, 26–34 (2019)
3. Potential flow of a permeable cylinder in relation to the modeling of the rotor of a vertical-axial wind turbine [Text]: dissertation abstract. cand. tech. Sciences: 05.14.08 / Maria Leonidovna Surgailo; Inst. Of Renewable Energy. 20 p . fig. Bibliogr. pp. 15–17 (2014)
4. Chmovzh, V.V., Eremeev, I.D., Solyanik, P.N., Surgailo, M.L., Chmovzh, V.V., Usik, Y.F.: Aerodynamics of a wind turbine with a vertical axis of rotation: research report/Nat. aerospace. un-t "KhAI"; hands, p. 109, No. GR 0107U001786. Inv. No. 0207U006273 (2007)
5. Hosseini, A., Goudarzi, N.: Design and CFD study of a hybrid vertical-axis wind turbine by employing a combined Bach-type and H-Darrieus rotor systems. Energy Convers. Manag. **189**, 49–59 (2019)
6. Kyozuka, Y.: An experimental study on the darrieus-savonius turbine for the tidal current power generation. J. Fluid Sci. Technol. **3**, 439–449 (2008)
7. Kumar, P.M., et al.: On the self starting of darrieus turbine: an experimental investigation with secondary rotor. In: 2017 Asian Conference on Energy, Power and Transportation Electrification (ACEPT). IEEE, pp. 1–7 (2017)

Simulation of Aircraft Motion in a Corkscrew Taking into Account Nonlinear Effects of Rotation

Ihor Romanchenko[1], Volodymyr Kotliarov[1], Andrii Serhiinko[1], and Serhii Yenchev[2(✉)]

[1] Central Research Institute of the Armed Forces of Ukraine, Kyiv, Ukraine
{info-cvni,sergiienko_sao}@ukr.net
[2] National Aviation University, Kyiv, Ukraine
esw@ukr.net

Abstract. The issues of computer modeling of an aircraft corkscrew as a complex rotational-oscillating (self-oscillating) motion on the basis of a refined nonlinear model of aerodynamic coefficients are considered. The main idea of the refinement is to obtain a general approximate analytical solution (the small parameter method is used) for a nonlinear vibrational dynamical system, where the parameters of the self-oscillating process (amplitudes and frequencies) are matched with their groups of aerodynamic coefficients, which are purposefully varied to match. Based on the physics of the process (the presence of a "curve" of self-rotation), the emphasis is on a substantially nonlinear functional dependence, which is explicitly included not in the equations of vibrational forms of motion, but in the equation of rotational coordinates. An approach to determining the quantitative values of polynomial components of the selected nonlinearity, the calculation of which by other methods is almost impossible. At the end of the article, the modes of corkscrew motion are modeled. The comparative analysis of this movement with real records of registrars of flight parameters is resulted.

Keywords: Aircraft corkscrew · Computer simulation · Self-oscillation process · Nonlinear aerodynamic model · Dynamic system

1 Introduction

The relevance of the article is determined by the insufficient study of the complicated dimensional motion of the aircraft (A), which occurs in the extraterrestrial (angle of attack) area of change of flight parameters, the unintentional occurrence of which often leads to catastrophic consequences.

Such regimes have caused catastrophes, such as Tu-154 aircraft near Irkutsk in 2001 and in the Donetsk region in 2006, as well as possible causes of the crash of Airbus 330 in 2009 over the Atlantic Ocean [1]. It is no exception to the emergence of such modes of maneuverable aircraft. In particular, according to [2], the crashes of L-39M1 aircraft in 2017 and Su-27C in 2018 were associated with their falling into difficult

M. Nechyporuk et al. (Eds.): ICTM 2021, LNNS 367, pp. 98–109, 2022.
https://doi.org/10.1007/978-3-030-94259-5_10

spatial positions. Thus, in general, the study of aircraft motion at large angles of attack is not complete.

In this multifaceted subject area, along with the practical part, we can distinguish the theoretical, which contains the whole complex of research - from compiling a mathematical image as a whole, covering all aspects of the phenomenon, to the implementation of this image in the form of analytical estimates of process parameters in nonlinear dynamic system order, which describes the movement of the aircraft at large supercritical angles of attack.

It is important, first of all, from the point of view of further modeling of these modes in aviation simulators [3]. Of undoubted interest is also the process of possible improvement of the existing algorithms of the automatic control system of the aircraft in order to prevent dangerous modes of the corkscrew, and, if necessary, - to get out of it.

2 Problem Statement

It should be emphasized that the problem of corkscrew research is closely related to unresolved issues in general theory related to both structural identification of systems (identification in the broadest sense of the word, when determining the model of process based on registered records of object operation) and obtaining a general analytical solution in a nonlinear high-order dynamic system for further parametric identification of aerodynamic coefficients, when the analytical solution found in their variation is compared with the results of flight recorders. The coincidence of indicators determine the desired values of the coefficients.

In the special literature concerning the research of this subject area, is concentrated mainly in two areas. For the first of them [4, 5] and others the analysis of nonlinear dynamics of the plane by qualitative methods on the basis of the created mathematical model of aerodynamic characteristics with definition of the established decisions and their stability is carried out.

In the second direction, for example, [5], mathematical modeling of corkscrew modes is carried out with the subsequent comparison of its results both with the data of tests in wind tunnels, and with records of onboard flight recorders on these modes. The purpose of comparative analysis is to refine the mathematical model of the aircraft. Close to this approach are the directions of identification of the aerodynamic model based on the results of flight tests, as well as others.

It should be noted that the narrowest point in all studies of the corkscrew is the aerodynamic model of this type of movement. And, if the general form of the model, in principle, is not in doubt, then its content in the form of specific numerical values of some really nonlinear characteristics (which, incidentally, can not always be detected by conventional experimental approaches) needs to be clarified. The physics of the process, of course, implies their presence, but only gives an idea of the quality of such dependencies.

The article will focus on the possibility of determining the quantitative values of these nonlinear dependences by analogy based on the results of the theory of nonlinear oscillations, where the most characteristic types of nonlinearity are already used and are important factors in the self-oscillating process.

A research algorithm is proposed, where, firstly, the initial system of differential equations is transformed into a form convenient for the analysis of rotational-oscillating motion, then their analytical solution is carried out taking into account the purposeful variation of nonlinear aerodynamic factors. In order to verify the correctness of the transformation, the obtained results are compared with the results of numerical integration of complete equations. At the end of the article, the graphs of real corkscrew motion of the Tu-154M aircraft and the simulated motion of a hypothetical aircraft are compared with the basic model of aerodynamic force coefficients and moments, which was refined on the basis of the above approach.

3 Problem Solving

To describe the motion of the aircraft, a system of differential equations has become widespread, which in the conventional notation [4, 6] has the form of:

$$\begin{aligned}
\dot{\alpha} &= \omega_z - sec\beta\big((F_x - \omega_y sin\beta)sin\alpha + (F_y + \omega_x sin\beta)cos\alpha\big);\\
\dot{\beta} &= F_z cos\beta - (F_x sin\beta - \omega_y)\, cos\alpha + (F_y\, sin\beta + \omega_x)\, sin\alpha;\\
\dot{\omega}_z &= (I_x - I_y)\, I_z^{-1}\omega_x\omega_y + 0,5\rho V^2 b_A S\, I_z^{-1} m_z;\\
\dot{\omega}_y &= (I_z - I_x)\, I_y^{-1}\omega_x\omega_y + 0,5\rho V^2 Sl\, I_y^{-1} m_y;\\
\dot{\omega}_x &= (I_y - I_z) I_z^{-1}\omega_y\omega_z + 0,5\rho V^2 Sl\, I_x^{-1} m_x;\\
\dot{V} &= V\big(F_x\, cos\beta\, cos\alpha - F_y\, cos\beta\, sin\alpha + F_z\, sin\beta\big);\\
\dot{\vartheta} &= \omega_y\, sin\alpha + \omega_z\, cos\gamma;\\
\dot{\gamma} &= \omega_x - tg\vartheta\,(\omega_y cos\gamma - \omega_z sin\gamma);\\
F_x &= -\tfrac{\rho VS}{2M}(C_x - C_\rho) - \tfrac{g}{V}\, sin\vartheta;\\
F_y &= \tfrac{\rho VS}{2M} C_y - \tfrac{g}{V}\, cos\vartheta\, cos\gamma;\\
F_z &= \tfrac{\rho VS}{2M} C_z + \tfrac{g}{V}\, cos\vartheta\, sin\gamma.
\end{aligned} \tag{1}$$

For further analysis, the model of aerodynamic force coefficients $C_{x,y,z}$ and moments $m_{x,y,z}$ [5, 7, 8] is chosen as the base, with its further refinement.

The study of system (1) in general is a rather difficult task. At the same time, the analysis of materials in which the corkscrew motion of aircraft is studied shows that its modes reveal characteristic features related to the dynamics of motion realized in a multidimensional nonlinear system, which usually has a rotational-oscillating (self-oscillating) character. The main of these features are the periodicity (quasi-periodicity) of graphs of parameters in the process of movement, as well as in some cases the same (multiple) period of their change [9]. Moreover, the transition from one type of process to another can be carried out under the condition of constant control parameters.

It is well known that in the study of nonlinear systems of differential equations an important place is occupied by the hypothesis of the type of solution sought. Thus, having the existing appearance of the transient process in the corkscrew (as a rotational-oscillating motion), it is possible to correctly approach, according to this process, the choice of the general structure of the model in the form of two oscillating (self-oscillating) contours for generalized longitudinal X and lateral Y coordinates related by rotation of a separate nonlinear differential equation for the parameter Z, which is very important from the point of view of conducting the following analytical analysis of the behavior of the aircraft at large angles of attack.

We write these equations as follows:

$$\begin{gathered}\Delta\dot{Z} = A_z\Delta Z + f_z\left(\Delta X,\, \Delta\dot{X},\, \Delta X^2,\, \Delta\dot{X}^2, \ldots,\, \Delta X^4,\, \Delta\dot{X}^4, \ldots\right);\\ \Delta\ddot{X} + \omega_1^2\Delta X + A_{xy}\Delta Y = \mu\big(f_x\left(\Delta Z,\, \Delta\dot{Z}\right) + A_z\Delta\dot{X}+, \ldots\big);\\ \Delta\ddot{Y} + \omega_2^2\Delta Y + A_{yx}\Delta X = \mu\big(f_y\left(\Delta Z,\, \Delta\dot{Z}\right) + A_y\Delta\dot{Y}+, \ldots\big).\end{gathered} \tag{2}$$

Such equations are obtained from system (1) by a special analytical tansformation (see, for example, [10]) in predicting the almost vertical trajectory of the plane in the corkscrew, when the resulting forces (aerodynamic and traction) are compensated by gravity $\left(F_{x,y,z} \approx 0\right)$. Then V≈const, which corresponds to the motion in the corkscrew that has developed.

The system (2) is written in increments relative to the balancing (stationary) values of the parameters, which are generally determined from the system (1) when equating the derivatives to zero. A feature of system (2) is also its study and possibility to introduce a small parameter $\mu(\mu > 0)$. The basis for its introduction are the results of full-scale flight experiments, as well as the values of individual coefficients of aerodynamic moments. The dependences $f_{x,y,z}$ in the general case are nonlinear functions of their arguments.

Thus, a fifth-order dynamic system is chosen for the study, which is also widely used to analyze the demensional motion of the aircraft in [4]. According to this work, one of the possible directions of research of such systems is consideration of separate partial cases with the subsequent generalization of the received results on a general case of movement.

Later in the article we will consider one of the following partial cases (model case), when the degree of stability of the aircraft at the sliding angle is large enough and we can assume in the first approximation that for the entire time of the aircraft the sliding ratio $\dot{\beta} \approx \beta \approx 0$ [4]. Taking into account this remark and following the general direction of the article on the study of rotational-oscillating motion, we write the equation for this motion of the aircraft in the following form:

$$\begin{aligned}&\Delta\dot{Z} = A_z\Delta Z + f_z\left(\Delta X,\, \Delta\dot{X},\, \Delta X^2,\, \Delta\dot{X}^2, \ldots,\, \Delta X^4,\, \Delta\dot{X}^4, \ldots\right);\\ &\Delta\ddot{X} + \omega_1^2\Delta X = \mu\left(A_{xz}\Delta Z + A_{x\dot{z}}\Delta\dot{Z} + A_x\Delta\dot{X}\right).\end{aligned} \tag{3}$$

The main feature of expressions (3) is the possibility of using the method of small parameter for their study. Then we introduce the following dependencies:

$$\Delta X = b\cos(\omega t + \varphi); \quad \Delta\dot{X} = -b\omega\sin(\omega t + \varphi), \tag{4}$$

where the variables amplitude b and phase φ are slowly changing functions of time.

The algorithm of further analytical research is as follows: Eqs. (4) are substituted into the first Eq. (3), after which it is integrated (here it is assumed that $b \approx const$). In the future (taking into account the obtained solution), following the method of small parameter, abbreviated equations for amplitude and phase are compiled and the procedure of their averaging is carried out.

Finally, these equations will take the form of:

$$\begin{aligned} \dot{b} &= -f_b\left(b, b^2, b^3, b^4, \omega_1, \omega, \ldots..\right); \\ \dot{\varphi} &= \omega_1 - \omega - f_\varphi\left(b, b^2, b^3, b^4, \omega_1, \omega, \ldots..\right). \end{aligned} \quad (5)$$

The functions f_b and f_φ are polynomial dependences of the argument b with corresponding coefficients at its different degrees, which, in turn, depend, inter alia, on the parameters of the nonlinear function f_z.

From Eqs. (5) for $\dot{b} = 0$ and $\dot{\varphi} = 0$ we find the stationary value of the amplitude b_c of the oscillatory coordinate ΔX, as well as the frequency of the oscillating motion ω. After that, it is easy to find the amplitude of oscillation a_z of the parameter ΔZ.

Leaving outside the article of the study of the stability of the general system (3), however, we show that the stable self-oscillating processes in it for $A_z < 0$ possible (coefficient A_z to a greater extent determines the aperiodic stability of motion). So, as an example, in Fig. 1 shows a graph of such a process for the coordinate ΔZ for different values of the nonlinear function f_z and for constant values of all other coefficients.

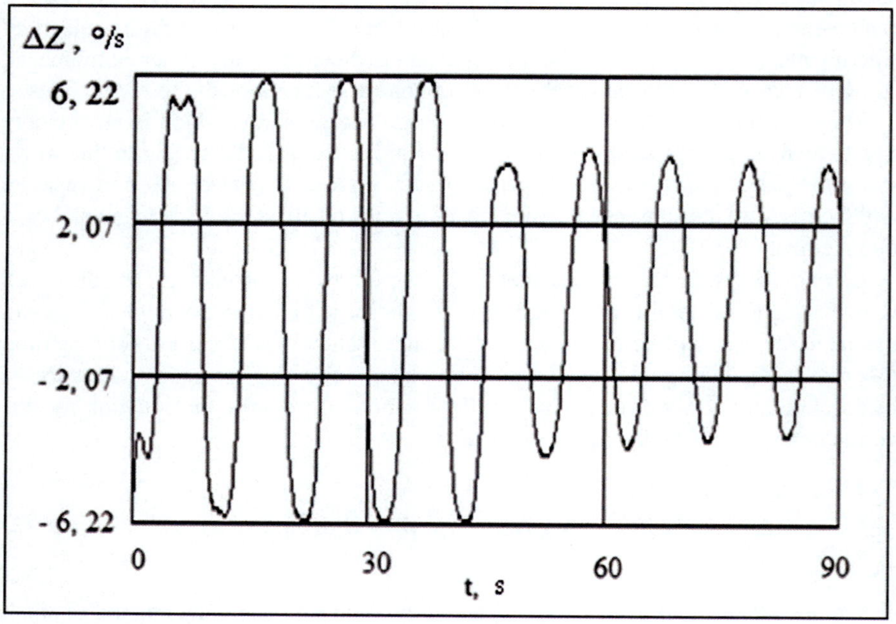

Fig. 1. Graph of increment of rotation coordinates ΔZ.

The Fig. 1clearly shows the amplitude of oscillations significantly depends on the form of the nonlinear dependence f_z, the change of which was carried out at t = 45s. For the considered cases amplitudes are stabilized.

Thus, other things being equal, if we set as independent variable components of the function f_z, it is possible to quantify the parameters of its nonlinearity. The criterion of truth here will be the coincidence of the calculated value of the amplitude of the parameter ΔZ (or ΔX) with the real ones, which was obtained, for example, by means of a field experiment. It should be emphasized that the physics of corkscrew motion presupposes that the nature of the function f_z (judging by the self-rotation curve) does have a pronounced nonlinear dependence, but it is almost impossible to calculate its quantitative parameters using other approaches. It should also be noted that most often during the analysis of the basic model of aerodynamic moment coefficients in direct formulation the attention to the peculiarities of changes in nonlinear functions of aerodynamic derivatives of rotation, from the point of view of their main role in self-oscillating process, is practically not emphasized.

To test the proposed approach while preserving the previous constraints $\left(F_{x,y,z} \approx 0; \dot{\beta} \approx \beta \approx 0\right)$, a numerical modeling of the system of differential equations was performed not in increments ΔX and ΔZ, but in the form of complete coordinates $\dot{\alpha}, \dot{\omega}_z, \dot{Z} = f\left(\dot{\omega}_x, \dot{\omega}_y\right)$ from expressions (1). Graphs for parameters α are presented in Fig. 2, and for Z - in Fig. 3. These curves are shown by solid lines (without points).

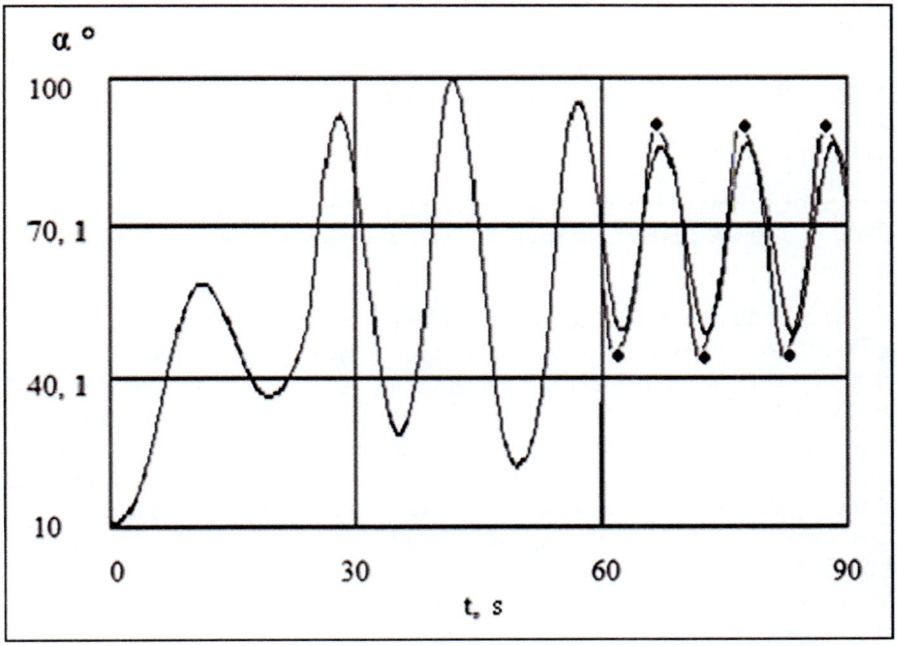

Fig. 2. Changing the angle of attack in model experiments.

Here, as in the previous case, the type of nonlinear functional dependence f_z(при $t=60$ c)was changed. As can be seen in Fig. 2 and Fig. 3, for full coordinates, the change in the type of nonlinearity also significantly affects the transient process, the amplitude of which decreases and stabilizes over time (t > 60 s) (there is an self-oscillating mode). Note that the parameters of the self-oscillating mode are determined by the type of nonlinear function that is included in the equation of rotation.

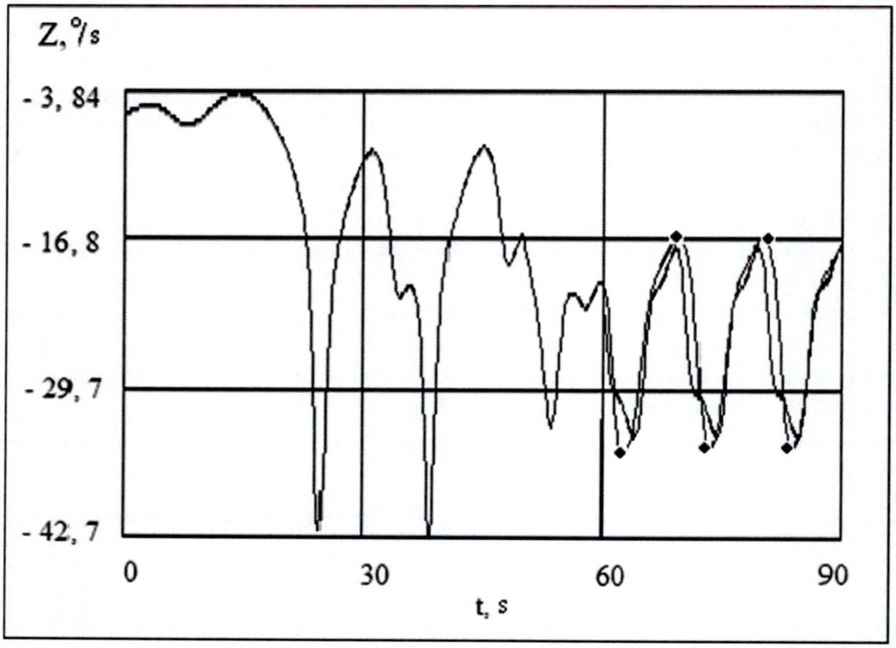

Fig. 3. Changing the coordinates of rotation in model experiments.

Now we calculate this model type of corkscrew motion $t > 60\ s$ first analytically by Eqs. (5), and then by numerical integration of system (3). Known there are the values of the frequency $\omega_I 1$ and the parameters A_x, A_z, A_{xz}, which are functions of the balancing values α_δ, Z_δ (for the calculation case from Fig. 2 $\alpha_\delta = 67, 7°$, and from Fig. 3 $Z_\delta = -25, 7°/\text{s}$) and aerodynamic coefficients of the base model, as well as nonlinear values f_z used in system (3) at $t > 60\ s$. As a result of analytical calculations, the calculated value of b_c was the value of $b_c = 0, 205°/\text{s}$, and the correction to the frequency ω_I was the value of $\Delta\omega = 0, 05°/\text{s}$. Taking into account b_c and the frequency ω, the amplitude of the first harmonic of the analytical solution of the first Eq. (3) was $a_z = 9, 17°/\text{s}$. The graph in Fig. 3 shows that this value relative to the value of $Z_\delta = -25, 7°/\text{s}$ was approximately the same value $a_z = \Delta Z = 8, 9°/\text{s}$.

Graphs of numerical integration of system (3) on the segment $t > 60\ s$ in Fig. 2 and Fig. 3 are marked by curves with points. Here, the calculation of the coordinate $\Delta\alpha$ (relative to the balancing value α_δ) was carried out on the value of ΔX. As can be seen in Fig. 2 and Fig. 3, the obtained transients coincide in frequency and amplitude with the corresponding graphs for the full coordinates $\alpha(\alpha = \alpha_\delta + \Delta\alpha)$ and $Z(Z = Z_\delta + \Delta Z)$.

Note that if the inverse problem of finding the function f_z was solved, its components (independent variables) would be selected so that the calculated (with their participation) value, such as the amplitude a_z, coincided with the real one. The value of a_z would play the role of a dependent parameter here. As part of solving the same inverse problem, we could consider refining the quantitative values of other aerodynamic coefficients that are included in the expression for the frequency ω_I. Here they would already act as independent variables, and the frequency ω1 itself would play the role of a dependent parameter in this case. The coincidence of frequencies (calculated and real) and would determine the true value of these coefficients.

In the general case, all the above approach can be extended to a more complex case of spatial motion of the aircraft at large angles of attack, when the simulation uses two oscillatory circuits $\left(\dot{\beta} \neq \beta \neq 0\right)$, which are connected by rotation by a nonlinear differential equation. Such graphs for a model example are presented in Fig. 4, where at t > 60 c the second oscillating circuit was included.

In this case, the process has a pronounced biharmonic character, and there is a change even in the sign of rotation. In practice, this type of corkscrew is called "falling leaf".

The authors of the article did not aim to obtain a complete picture of the movement of the aircraft at large supercritical angles of attack. However, extending this approach to the whole area of complex rotational-oscillating motion, where each parameter of this process corresponds to its own group of coefficients of aerodynamic forces and moments, it is still possible to obtain acceptable results for practice.

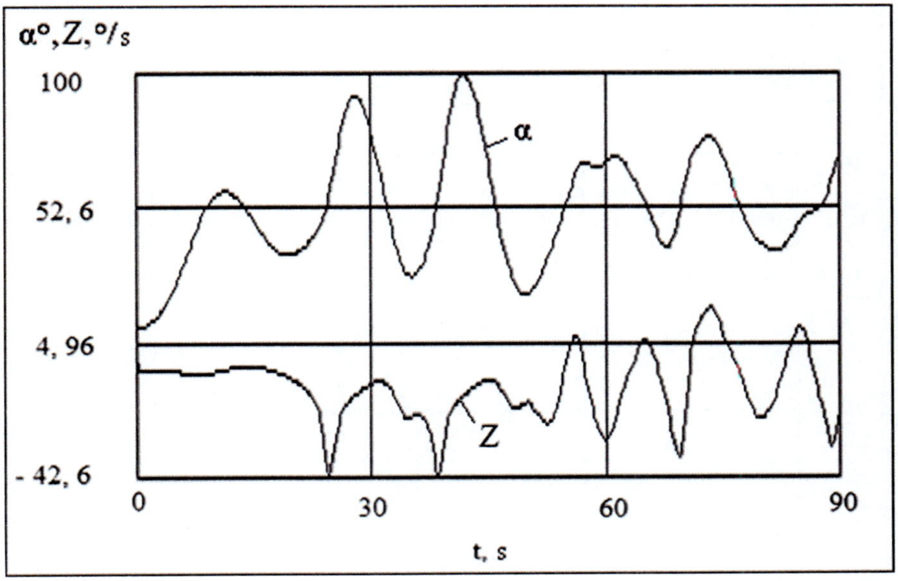

Fig. 4. A model example of a biharmonic process at t > 60 s.

To confirm these words, the graphs of the transition process in the corkscrew for the complete system of Eqs. (1) of the hypothetical aircraft for angles of attack α (Fig. 5), ϑ (Fig. 6), γ (Fig. 7) and ψ are given below. 8). System (1) additionally takes into account the gyroscopic forces from the rotation of the rotors of aircraft engines (especially manifested in the initial period of dumping into the corkscrew), as well as changes in air density ρ depending on the altitude.

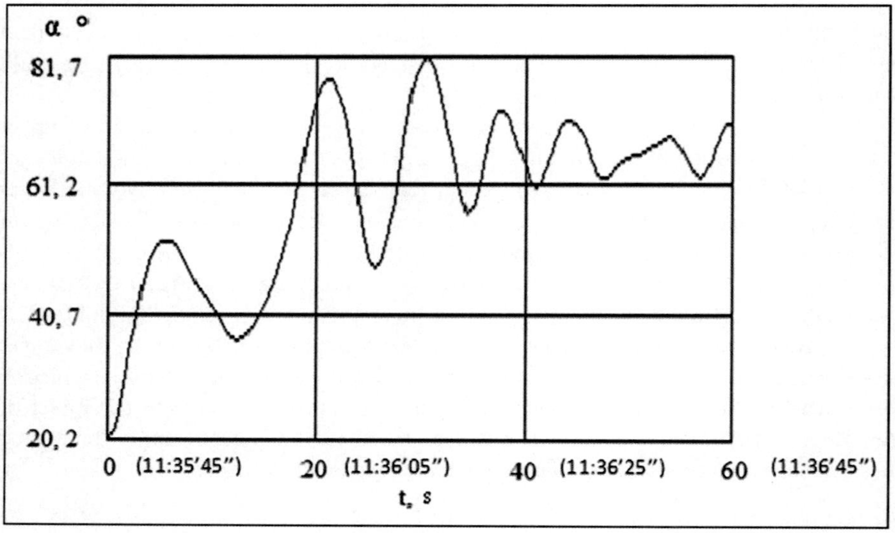

Fig. 5. Transition process by angle of attack (simulation).

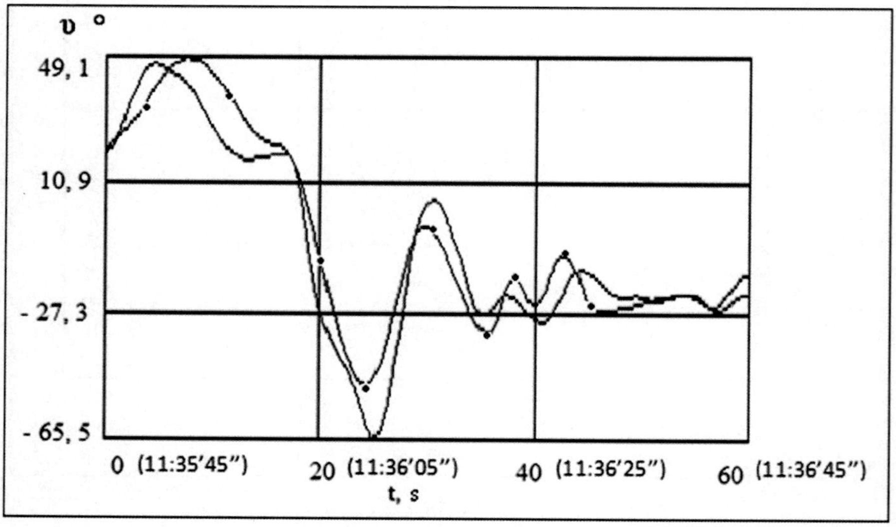

Fig. 6. Comparison of simulation results and flight recorder (pitch angle).

In Figs. 6, 7 and 8 the separate graphs show the curves of the parameters ϑ, γ and ψ (marked by dots) of the real process of the plane Tu-154M crash in a corkscrew near Donetsk [12]. In parentheses along the abscissa axis in Figs. 5, 6, 7 and 8 indicates the actual flight time (hours, minutes, seconds).

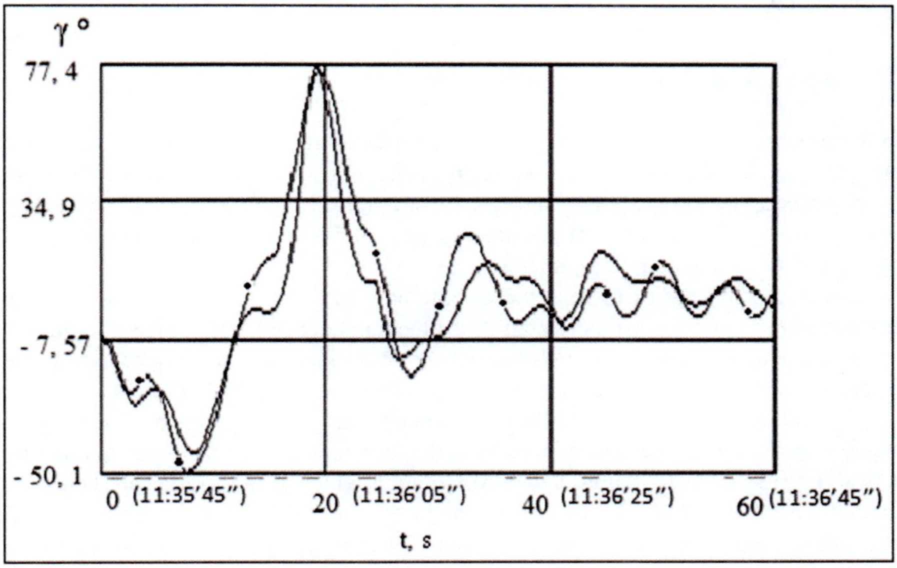

Fig. 7. Comparison of simulation results and flight recorder recording (roll angle).

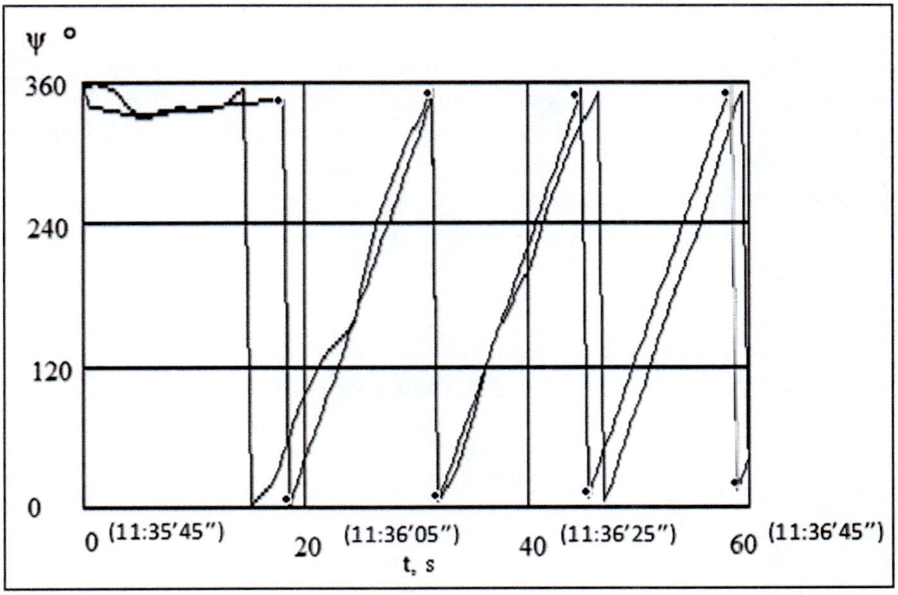

Fig. 8. Comparison of simulation results and flight recorder recording (course angle).

The sensor of an angle of attack through the design features at big values of this parameter does not give real idea of size α (the weather vane is put "to the stop"). Therefore, the signal of this parameter in Fig. 5 not listed. The identity of the corresponding graphs for other parameters, such as pitch, roll and course, suggests that the proposed approach can be used as a basis for the analysis of complex rotational-oscillating motion of the aircraft at large angles of attack.

4 Conclusions

In the process of computer modeling of rotary-oscillating corkscrew motion, a refined analytical model of aerodynamic coefficients is key. The main idea of their refinement is to obtain a general analytical solution for a nonlinear oscillatory (self-oscillating) system, where the parameters of this process (amplitudes and frequencies) are matched by a group of aerodynamic coefficients.

Based on the physics of the process (the emergence of the self-rotation mode), emphasis is placed on essentially nonlinear functional dependence, which is explicitly included not in the equations of oscillatory forms of motion, as is most often used in the theory of nonlinear oscillations, but in the equation of rotational coordinates.

The algorithm for determining the desired nonlinearity involves first representing the expressions for the oscillating coordinates in the form of periodic functions, followed by substituting them into a dynamic equation of rotation, which in this case can be integrated. In the future (taking into account the obtained solution), following the method of small parameter, abbreviated equations (for amplitudes and phases) are made, to which the averaging procedure is applied. From these equations are the stationary values of the parameters of self-oscillations.

When solving the inverse problem, iteratively selects such quantitative values of the polynomial nonlinear function, according to which the calculated values of amplitudes and frequencies coincide with the real ones. Obtaining quantitative values of such functions with other approaches is almost impossible.

The article presents model variants of corkscrew movement, which coincide with real processes, which testifies to the efficiency of the considered approach and the possibility of its practical use.

References

1. Mikoyan, S.: Bol'shiye oshibki v polete na bol'shikh uglakh ataki [Big errors in flight at high angles of attack]. «Aviapanorama». № 1 (2012)
2. Kulba, P.P., Lampik, M.M., Bursala, O.O.: Obgruntuvannya dotsil′nosti ta bezpeky pidhotovky l′otnoho skladu vykonannyu pol′otiv na doslidzhennya shtoporu ta yikh vprovadzhennya u boyovu pidhotovku [Substantiation of expediency and safety of flight crew training for corkscrew research and their introduction into combat training]. ollection of scientific works of the State Research Institute of Testing and Certification of Arms and Military Equipment. Kyiv, vol. 2 (2019)
3. Kotlyarov, V.P., Seryogin, G.N., Shihaleyev, V.N.: Simulation of aircraft autooscillatory motion in high-angel-of attack. In: Book of Abstracts in Mosaeroshow. Central Aaerohydrodynamic Iinsntitute Moskow, pp. 59–60(1992)

4. Biushhens, H.S., Studnev, R.V.: Dynamyka samoleta. Prostranstvennoe dvyzhenye [Aircraft dynamics. Spatial movement]. Mechanical Engineering Moskow (1983)
5. Voronyn, A.Y., Dubov, Y.B., Zhelonkyn, M.V., Zhyvov, Y.H., Poedynok, A.M.: Metodyka yssledovanyia shtopora samoleta s yspolzovanyem pylotazhnoho stenda [Technique for the study of the aircraft spin using the aerobatic stand]. In: Proceedings of the MAI, vol. 58, p. 20 (2012)
6. Homan, M.H., Khramtsovskyi, A.V.: Byfurkatsyy ustanovyvshykhsia rezhymov samoleta. Yssledovanyia po dynamyke poleta letatelnykh apparatov [Bifurcations of steady-state airplane spin modes. Research on the dynamics of the flight of aircraft]. Moscow Institute of Physics and Technology, pp. 17–25 (1986)
7. Studnev, R.V., Shybaev, V.M., Kostrykyn, V.N.: Raschet kharakterystyk ustanovyvshehosia shtopora samoleta [Calculation of the characteristics of the aircraft's steady spin]. In: Proceedings of prof. N.E. Zhukovsky. Department of TsAGI Moscow, 1985, vol. 2270, p. 3 (1985)
8. Studnev, R.V., Shybaev, V.M., Zhelonkyn, V.Y.: Otsenka lokalnykh kharakterystyk ustoichyvosty y upravliaemosty samoleta na rezhymakh ustanovyvshehosia shtopora. [Evaluation of the local characteristics of stability and controllability of the aircraft in the steady-state spin modes]. In: Proceedings of prof. N.E. Zhukovsky. Department of TsAGI Moscow, 1985, vol. 2270, p. 23 (1985)
9. Hryshyn, V.V., Mandelbaum, Y.V., Chetverhov, V.N.: Yssledovanyia po dynamyke poleta letatelnykh apparatov [Research on the dynamics of the flight of aircraft]. Moscow Institute of Physics and Technology, pp. 4–10 (1986)
10. Kotliarov, V.P.: Pobudova struktury dynamichnoi systemy, shcho opysuie avtokolyvalnyi rukh obiekta, za materialamy naturnykh vyprobuvan [Construction of the structure of a dynamic system that describes the self-oscillating motion of an object, based on field tests]. Coll. Science. etc. Central Research Institute of the Armed Forces of Ukraine. Kyiv, no 3(8), pp. 165–169 (1999)
11. Demchenko, V.: Za try s polovynoi mynuty do padenyia samoleta eshche mozhno bylo spasty [Three and a half minutes before the plane crashed, it was still possible to save] Izvestia in Ukraine, no 49/208, p. 4 (2006)

Check for updates

Structural Optimization of the Special Cold Spraying Nozzle via Response Surface Method

Wenjie Hu[1,2](✉), Kun Tan[1], Sergii Markovych[1], and Tingting Cao[2,3]

[1] National Aerospace University "Kharkiv Aviation Institute", Kharkiv, Ukraine
[2] School of Aeronautics and Astronautics, Nanchang Institute of Technology, Nanchang, China
[3] Nanchang Hangkong University, Nanchang, China

Abstract. This paper studies a new 90° cold spraying nozzle by response surface method (RSM), which can spray in a limited space. A multi-factor and multi-level response surface regression model with divergent section length, spraying distance, and fillet radius (throat size) as independent variables and powder collision speed as dependent variables are established. The divergent section of the optimized new nozzle is only 12 mm. The results show that the length of the divergent section and spraying distance has a significant effect on the powder impact velocity, but the interaction between the three factors is not obvious. Compared with the numerical simulation data, the error of the optimized parameters is 0.3%; In the range of 400 k–1200 k of propulsion gas, the new 90° nozzle can meet a variety of powder sprayings, such as aluminum (Al), titanium (Ti), nickel (Ni), copper (Cu), magnesium (Mg), and zinc (Zn), which has a good application prospect.

Keywords: Cold spraying nozzle · Multi-factor · Multi-level · Regression model · RSM

1 Introduction

Cold spraying technology (CS) can obtain coatings in the solid-state, which deposited a process of high-speed collision with substrate and formation of coating [1, 2], and it is mainly deposited for restorative [3] and protective coating [4, 5], or additive manufacturing applications [6–9], etc.

In the current study, most researchers focus on straight line nozzles [10–13], resulting in the limited or inconvenient spraying operation in limited Spaces. Hence, to facilitate spraying in a limited space, or areas where linear nozzles cannot be sprayed, this study used the RSM to optimize the design of the 90° nozzle with a rectangular cross-section, the outlet section is rectangular and can be applied to thinner rotating components. After that, three key structural parameters (divergent section length, spraying distance and fillet radius) that affect the powder collision speed are used as independent variables, and the powder collision speed is the dependent variable, a response surface quadratic regression equation is established to analyze the interaction between various factors, and the optimal structure parameters are obtained.

M. Nechyporuk et al. (Eds.): ICTM 2021, LNNS 367, pp. 110–122, 2022.
https://doi.org/10.1007/978-3-030-94259-5_11

2 Theoretical Details of the Nozzle

As shown in Fig. 1a, although the flow field distribution of the circular cross-section is uniform and the turbulence can be minimized, it is not suitable for spraying fine rotating specimens and workpieces.

Therefore, the rectangular nozzle design is considered in this study [14]. Considering the convenience of installation, the upper and lower structures are adopted, and the initial model is shown in Fig. 2.

It is found [15] that when the throat diameter $D_{throat} \geq 6$ mm of the circular section nozzle, a large fluid outlet velocity can be obtained. The rectangular section width is preliminarily 3 mm, so the throat section length $L \geq 9.42$ mm. This study takes the throat with the same cross-sectional area as the benchmark. To achieve the compression effect, the rectangular nozzle in this paper uses the fillet method to compress the cross-sectional area at the 90° corner, and the larger the fillet, the better the compression effect. Through numerical simulation, it is found that when the inlet width of propulsion gas is 16 mm and the fillet exceeds 30 mm, the throat compression radius is close to 10 mm. The primary fillet radius $R = 18$ mm and the cross-section width of the throat inlet is 16 mm, At this time, the throat has a certain compression effect.

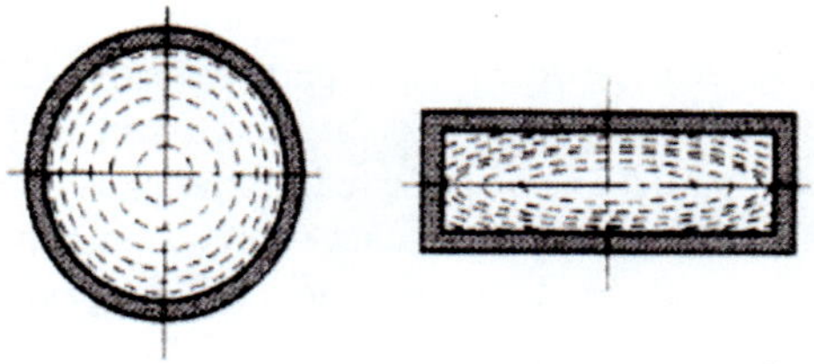

a b

Fig. 1. Three different nozzle cross-sections: a - circular; b - rectangular [14]

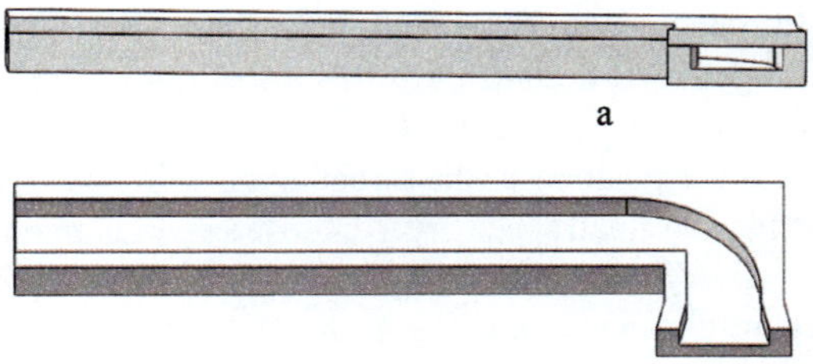

a

b

Fig. 2. The actual picture of the initial model of the 3D model: a - the combination of the upper and lower parts; b - only the lower part

3 Theoretical Details of the Single Factor

Since the particle acceleration process is mainly affected by the main drag force of the air flow, this paper assumes that the powder are spherical particles with smooth surface (Fig. 3), and the aerodynamic drag force formula is shown in Eq. 1:

$$F = \frac{1}{2} C_D \rho_g (v_g - v_p)^2 S \tag{1}$$

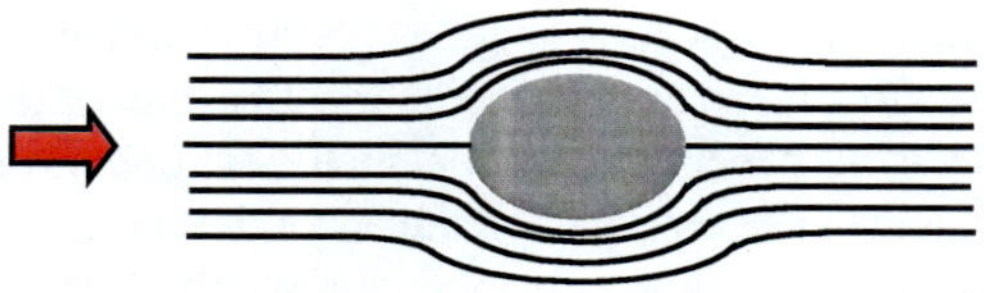

Fig. 3. Particles are subjected to fluid action

where, F is the aerodynamic drag force on the particles; CD is the drag coefficient; ρ_g is the density of the air flow; Vg and VP are gas flow velocity and particle velocity respectively; S is the upwind area of particles.

When the injection speed of powder is 0, the particles can obtain a large acceleration. If it is necessary to obtain a large acceleration, it is necessary to increase the difference between air velocity and particle velocity or reduce the powder particle size. In order to reduce the energy loss, try to avoid contact or collision between the powder and the inner wall of the nozzle. In addition, the factors affecting powder collision include powder characteristics, propulsion gas characteristics, divergent section length, spraying distance, and nozzle throat size, etc. [16]. This study mainly considers the structure. Therefore, the spraying distance, throat size, and divergent section length are selected for optimization.

Solidworks/Flow simulation module for fluid analysis is used, the influence of turbulence is considered (turbulence intensity 2%, and the inner wall conditions are adiabatic and smooth). The nitrogen (N2) is selected as the propelling gas, the internal cavity and excludes the internal non-flowing area are selected. The powder injection is selected in the low-pressure area at the junction of the throat and the divergent section.

To facilitate spraying in a small space, try to choose a shorter divergent section length. Although the shorter divergent section length is unfavorable to powder acceleration, optimizing the spraying distance can make up for the defect of insufficient divergent section length. Therefore, 4 mm, 8 mm, 12 mm, and 16 mm are selected in this paper. The spraying distance is 50 mm, nitrogen is selected as propulsion gas, and other specific fluid parameters are shown in Table 1.

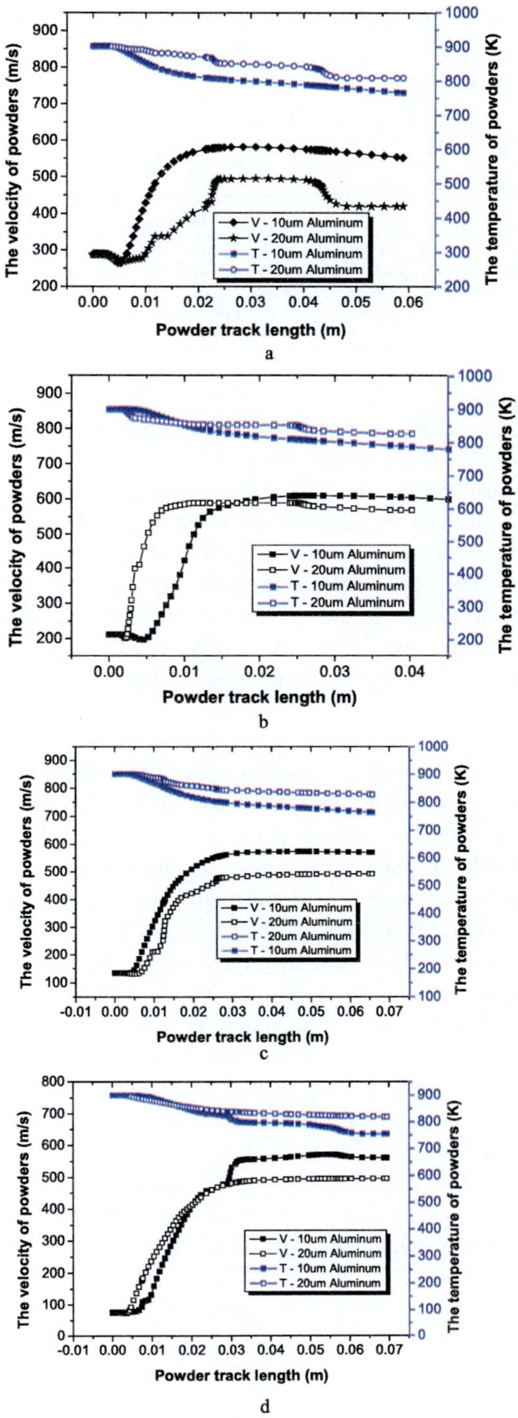

Fig. 4. The impact velocity and temperature changes of powder with different lengths of divergent section: a-4 mm; b-8 mm; c-12 mm, d-16 mm

Table 1. Numerical simulation parameters for different lengths of divergent segments.

Length mm	Parameters			
	Propulsion gas pressure, MPa	Propulsion gas temperature, K	Powder injection pressure, MPa	Maximum collision velocity of powder, m/s
4	5	900	0.8	1039
8	5	900	0.8	1065
12	5	900	1.2	955
16	5	900	1	1015

If the spraying distance is too large, the powder speed and powder temperature will be reduced. Figure 4 shows that the spraying distance is between 10 mm and 20 mm, and the powder has reached the maximum. The spraying distance is 10 mm (injection pressure 0.8 MPa), 15 mm (injection pressure 0.9 MPa) and 20 mm (injection pressure 0.95 MPa), the propulsion gas temperature is 900 K and the powder is 10 μm aluminum. In addition, the throat size is the key factor affecting the fluid, thus affecting the particle acceleration. According to Fig. 5, the spraying distance 15 mm is selected. Since the width of the propulsion gas inlet is 16 mm, in order to achieve the throat compression effect, the fillet is 18 mm, 22 mm and 26 mm (Fig. 6), and the powder injection pressure is 0.9 MPa.

Hence, the powder collision velocity increases with the increase of the length of the divergent section, while the powder velocity reaches the maximum when the spraying distance is basically within 20 mm, while the collision temperature decreases with the increase of particle trajectory, and the smaller the particle size, the more it decreases. However, 10 μm aluminum powder can obtain greater collision velocity than 20 μm aluminum powder under the same conditions, mainly because the obtained acceleration is greater. According to Formula (1), it is further deduced that the powder acceleration a = F/m, and the smaller the particle size, the greater the acceleration, which also explains the greater collision velocity of 10 μm powder. Figure 5 shows that the powder speed reaches the maximum when the spraying distance is 15 mm, indicating that the spraying distance should not be too short or too long. Too short will lead to the powder can not accelerate effectively, and too long will lead to the loss of powder speed. Figure 6 shows that the throat is also a key factor affecting the powder collision speed. When the fillet 18 mm increases to 22 mm, the maximum powder speed increases by about 23 m/s, while when it increases from 22 mm to 26 mm, the growth rate is less, and the maximum powder speed only increases by about 8 m/s. therefore, with the increasing fillet radius (i.e. the decreasing throat size), the growth rate of the maximum powder speed will be less and less.

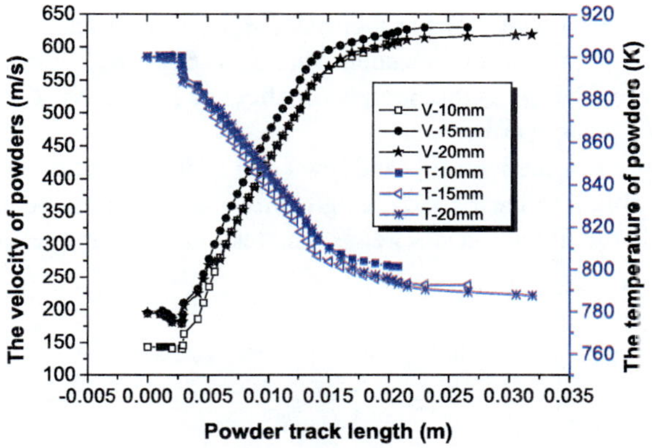

Fig. 5. Impact velocity and temperature variation of aluminum powder at different spraying distances

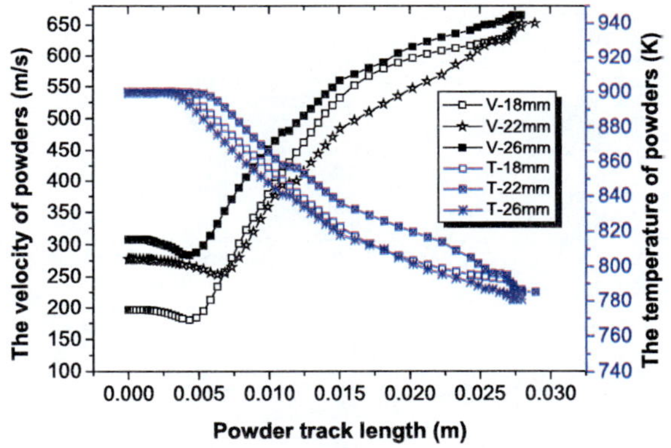

Fig. 6. Impact velocity and temperature variation of aluminum powder with different throat sizes

4 Multivariate Results and Discussion

The single factor method is commonly used by most researchers to seek rules [17], but the actual operation process is often multi-factor interaction. Therefore, multi-factor interaction analysis is more scientific and realistic.

RSM is a product of a combination of mathematics and statistical methods. It is often used to find the optimal process parameters in a multi-parameter system [18]. This study uses the Box Behnken principle to design, and the comprehensive numerical results of the single factors, three independent variables (Diffusion L, Spraying D and Fillet radius of throat R) are set through the Design-Expert software, and the powder speed is the dependent variable. A three-factor, three-level response surface quadratic regression equation is constructed. The model is Eq. 2:

$$y = \beta_0 + \sum_{i=1}^{m} \beta_i x_i + \sum_{i=1}^{m} \beta_{ij} x_j + \sum_{i=1}^{m} \beta_{ii} x_i^2 + \varepsilon \tag{2}$$

where y is the response value of the regression equation, Xi and Xj are independent variables; m is the number of independent variables, β_0 is the regression intercept; β_i is the linear effect of Xi; β_{ij} is the interaction effect of Xi and Xj; β_{ii} is the secondary effect of Xi; ε is a random error.

Input the high (+1), medium (0), and low (−1) codes and actual parameters of the independent variable into the Expert Design software, which can reduce unnecessary testing and obtain the optimal test scheme. The test methods and results are shown in Table 2.

Table 2. Test scheme and results.

Run	High and low level code			Actual value			Impact V (m/s)
	Diffusion L (mm)	Spraying D (mm)	Fillet radius of throat (mm)	Diffusion L (mm)	Spraying D (mm)	Fillet radius of throat (mm)	
1	0	0	0	8	15	22	666
2	−1	+1	0	4	20	22	578
3	0	0	0	8	15	22	666
4	+1	+1	−1	12	15	18	677
5	+1	+1	0	12	20	22	664
6	0	0	−1	8	20	18	599
7	0	0	0	8	15	22	666
8	+1	+1	+1	12	15	26	684
9	+1	+1	0	12	10	22	700
10	0	0	+1	8	20	26	630
11	0	0	0	8	15	22	666
12	−1	−1	+1	4	15	26	645
13	0	0	+1	8	10	26	675
14	−1	−1	−1	4	15	18	637
15	0	0	−1	8	10	18	643
16	−1	−1	0	4	10	22	604
17	0	0	0	8	15	22	666

As shown in Table 3, the model F-value of 7.04 and P-value of 0.0087 (less than 5%) imply the model is significant. The P-values of A and B are both less than 0.01 (less than 0.05), indicating that a single factor has a significant influence on the powder velocity, however, the P-values of C is greater than 0.05, indicating that a single factor has not a significant influence on the powder velocity. The P-value of AB, AC, and BC are both greater than 0.05, hence, the interaction between diffusion length and spraying distance, diffusion length and Fillet radius of throat, and spraying distance and fillet radius of

Table 3. Analysis of variance.

Source	Sum of squares	df	Mean square	F value	P-value
Model	15505.76	9	1722.86	7.04	0.0087
A-Diffusion L	8580.5	1	8580.5	35.08	0.0006
B-Spraying D	2964.5	1	2964.5	12.12	0.0102
C-Fillet radius of throat R	760.5	1	760.5	3.11	0.1212
AB	12.25	1	12.25	0.05	0.8293
AC	0.25	1	0.25	1.022e−3	0.9754
BC	0.25	1	0.25	1.022e−3	0.9754
A2	34.8	1	34.8	0.14	0.7172
B2	3041.12	1	3041.12	12.43	0.0096
C2	23.75	1	23.75	0.097	0.7644
Residual	1712	7	244.57		
Lack of fit	1712	3	570.67		
Error	0	4	0		
Total	17217.76	16			

throat are not obvious. The regression equation at this time is: Y = 666 + 32.75A − 19.25B + 9.75C − 1.75AB − 0.25AC − 0.25BC − 2.88A2 − 26.88B2 − 2.37C2. The lack of fit error is not significant. The determination coefficient of the regression equation is R2 = 0.9006 and the correction coefficient R2 = 0.754. These results indicate that the regression model can explain 90.06% of the change in powder velocity response.

It can be seen from Fig. 7 that the order of the influence of the three factors is that diffusion length is greater than spraying distance, and spraying distance is greater than fillet radius of the throat, and the maximum powder velocity is taken as the target, the optimal velocity is predicted to be 707.2 m/s (Fig. 8).

To verify the optimized parameters, import the optimized parameters into the Solidworks flow simulation module. The simulation results (Fig. 9) show that the powder velocity is 705 m/s, with an error of 0.3%. Therefore, the response surface analysis is highly accurate.

The application range of the optimized new nozzle structure is further discussed. Taking the critical velocity of powder as the standard, the critical velocity can be calculated theoretically by Eq. 3. The 10-micron size powder particles are selected, the powder is injected at room temperature, and the temperature of propulsion gas ranges from 400 k to 1200 k. Through the summary of theoretical critical velocity and numerical simulation velocity data (Table 4), the new cold spraying nozzle of this study can spray Ni, Ti, Cu, Al, Zn and Mg, which has a wide application prospect.

$$V_{\text{crit}} = \sqrt{C_P(0.7T_{\text{m}} - T_{\text{i}})} \tag{3}$$

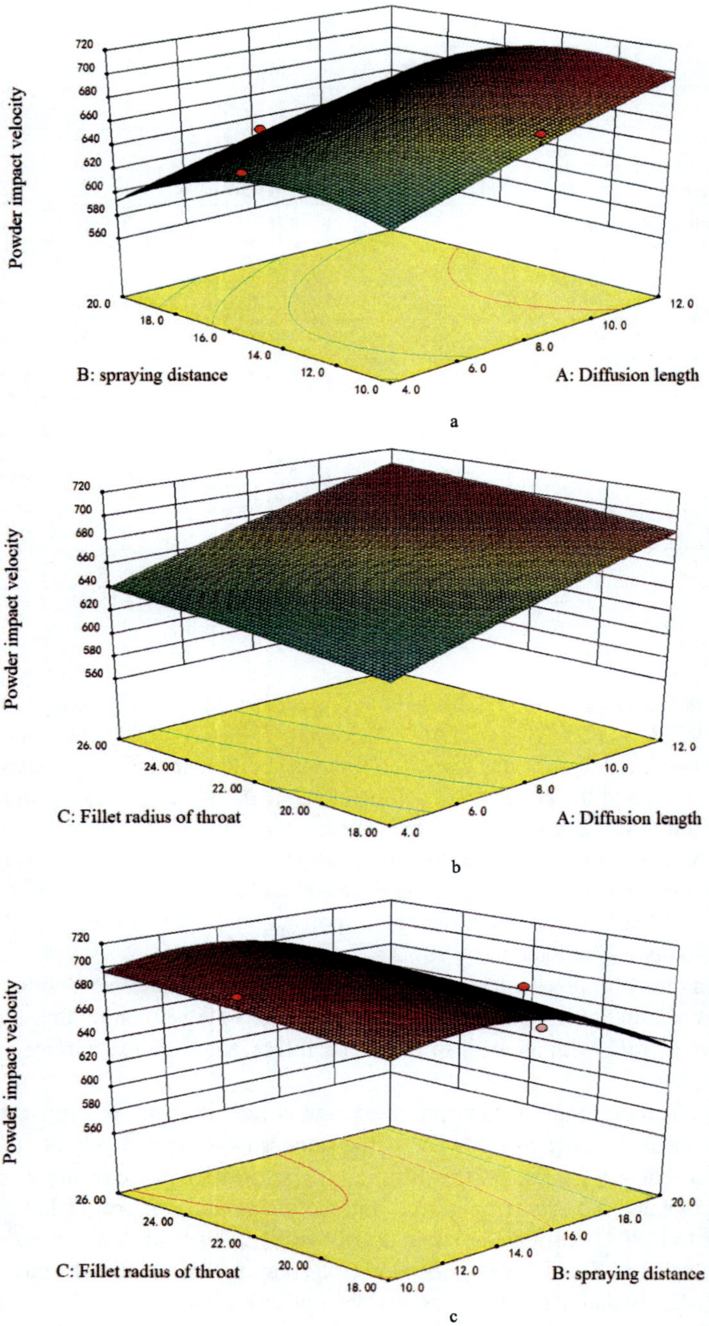

Fig. 7. Interaction effect of three factors on response under N2 conditions: a - A and B factors interaction influence; b - A and C factors interaction influence; c - B and C factors interaction influence

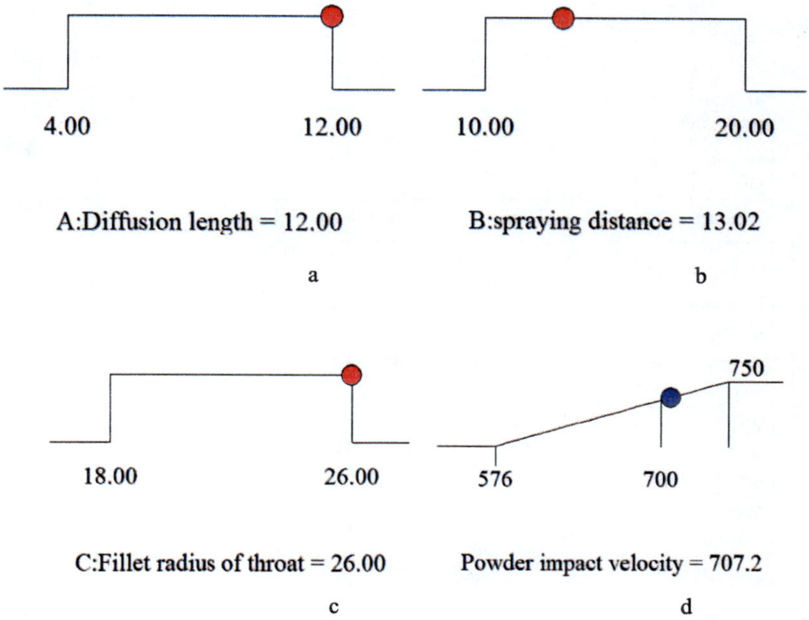

Fig. 8. Numerical simulation collision velocity under optimal parameter conditions under N2: a-optimum diffusion length; b-optimum spraying distance; c-optimum fillet radius of throat; d-response speed

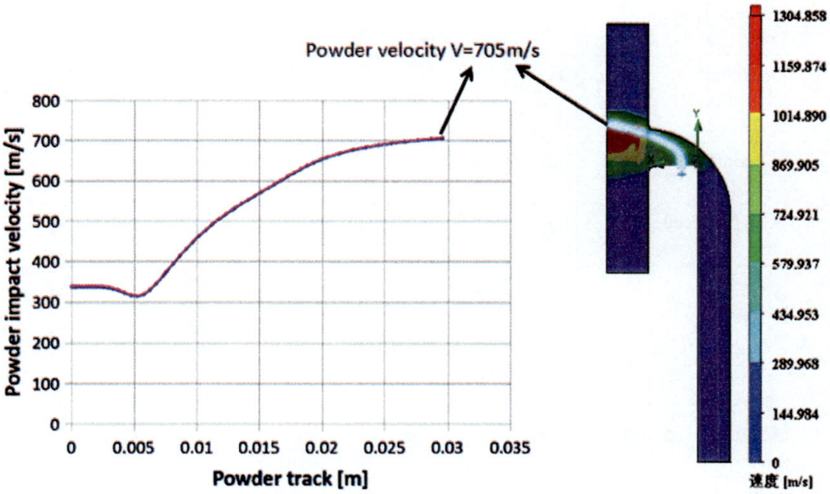

Fig. 9. Powder velocity trajectory and velocity nephogram after structure optimization

Table 4. The optimized spraying is applicable to different powder materials.

Powder parameters		Gas temperature, K								
		400	500	600	700	800	900	1000	1100	1200
Ni	Maximum speed, m/s	–	–	–	–	–	533	550	570	585
	Critical speed, m/s	–	–	–	–	–	587	574	564	552
	Whether deposition? (yes or no)	–	–	–	–	–	No	No	Yes	Yes
Ti	Maximum speed, m/s	–	–	–	587	615	640	658	680	698
	Critical speed, m/s	–	–	–	632	619	605	590	576	560
	Whether deposition? (yes or no)	–	–	–	No	No	Yes	Yes	Yes	Yes
Cu	Maximum speed, m/s	–	–	464	488	512	532	–	–	–
	Critical speed, m/s	–	–	480	467	452	437	–	–	–
	Whether deposition? (yes or no)	–	–	No	Yes	Yes	Yes	–	–	–
Al	Maximum speed, m/s	525	568	608	–	–	–	–	–	–
	Critical speed, m/s	570	544	517	–	–	–	–	–	–
	Whether deposition? (yes or no)	No	Yes	Yes	–	–	–	–	–	–
Mg	Maximum speed, m/s	562	608	652	–	–	–	–	–	–
	Critical speed, m/s	630	600	570	–	–	–	–	–	–
	Whether deposition? (yes or no)	No	Yes	Yes	–	–	–	–	–	–
Zn	Maximum speed, m/s	432	463	–	–	–	–	–	–	–
	Critical speed, m/s	299	274	–	–	–	–	–	–	–
	Whether deposition? (yes or no)	Yes	Yes	–	–	–	–	–	–	–

5 Conclusion

The RSM has certain guiding significance for multi-factor parameter optimization. The response surface model obtained by this study is reliable with an error of 0.3% and high accuracy. By analyzing the structure of cold spraying 90° rectangular nozzle, some meaningful conclusions are draws:

1. When the length of the inlet cross-sectional area of the propulsion gas is 16 mm and the width is 3 mm, that is, the cross-sectional area of the inlet is 48 mm^2, the optimal structural parameters are as follows: the length of the divergent section is 12 mm, the spraying distance is 13 mm, and the fillet radius is 26 mm. At this time, the maximum speed of 10 μm aluminum powder can reach 705 m/s.
2. The optimized 90° rectangular nozzle can meet the spraying of a variety of common metal powders with a size of 10 μm between 400 k–1200 k, which has a certain process reference value.

Acknowledgment. The authors would like to thank the China Scholarship Council for its support (NO. 202008100011).

References

1. Li, W.Y., Cao, C.-C., Yin, S.: Solid-state cold spraying of Ti and its alloys: a literature review. Prog. Mater. Sci. **110**, 1–53 (2019). https://doi.org/10.1016/j.pmatsci.2019.100633
2. Assadi, H., Gartner, F., Stoltenhoff, T., Kreye, H.: Bonding mechanism in cold gas sparing. Acta Mater. **51**, 4379–4394 (2003)
3. Hu, W.J., Markovych, S., Tan, K., Shorinov, O., Cao, T.T.: Surface repair of aircraft titanium alloy parts by cold spraying technology. Aerosp. Tech. Technol. **163**, 30–42 (2020)
4. Sun, W., Tan, A.-W.-Y., Marinescu, L., Toh, W.Q., Liu, E.: Adhesion, tribological and corrosion properties of cold-sprayed CoCrMo and Ti6Al4V coatings on 6061-T651 Al alloy. Surf. Coat. Technol. **326, part A**, 291–298 (2017). https://doi.org/10.1016/j.surfcoat.2017.07.062
5. Han, X.J.: Research on preparation of Zn-Al alloy coating on the surface of TC4 Alloy and its resistance to titanium/aluminum contact corrosion. Nanjing University of Aeronautics and Astronautics (2016)
6. Vargas-Uscategui, A., King, P.C., Styles, M.J., Saleh, M., Luzin, V., Thorogood, K.: Residual stresses in cold spray additively manufactured hollow titanium cylinders. J. Therm. Spray Technol. **29**(6), 1508–1524 (2020). https://doi.org/10.1007/s11666-020-01028-3
7. Garmeh, S., Jadidi, M., Dolatabadi, A.: Three-dimensional modeling of cold spray for additive manufacturing. J. Therm. Spray Technol. **29**, 38–50 (2020)
8. Li, W.Y., Yang, K., Yin, S., Yang, X.W., Xu, Y.X., Lupoi, R.: Solid-state additive manufacturing and repairing by cold spraying: a review. J. Mater. Sci. Technol. **34**(3), 440–457 (2017). https://doi.org/10.1016/j.jmst.2017.09.015
9. MacDonald, D., Fernandez, R., Delloro, F., Jodoin, B.: Cold spraying of Armstrong process titanium powder for additive manufacturing. Therm. Spray Technol. **26**, 598–609 (2017)
10. Wu, Z.L.: Numerical simulation research of the internal flow field cold of the spray gun nozzle and structural optimization. Henan Polytechnic University (2011)

11. Li, W.Y., Li, C.J.: Optimal design of a novel cold spray gun nozzle at a limit space. J. Therm. Spray Technol. **14**, 391–396 (2005)
12. Canales, H., Litvinov, A., Markovych, S., Dolmatov, A.: Calculation of the critical velocity of low pressure cold sprayed materials. Aircr. Des. Manuf. Issues **3**, 86–91 (2014). http://nbuv.gov.ua/UJRN/Pptvk_2014_3_11
13. Hu, W.J., Tan, K., Markovych, S., Cao, T.T., Liu, X.L.: Optimization of cold spraying 90° rectangular nozzle technological parameter via response surface analysis. Metallofizika i Noveishie Tekhnologii (2021)
14. Zho, X.L., Zhang, J.S., Wu, X.K.: Advanced Cold Spray Technology and Application. China Machinery Industry Press, Beijing (2011)
15. Hu, W.J., Tan, K., Markovych, S., Liu, X.L.: Study of cold spray nozzle throat on acceleration characteristics via CFD. J. Eng. Sci. **8**(1), 8–12 (2021)
16. Alhulaifi, A.S., Buck, G.A.: A simplified approach for the determination of critical velocity for cold spray processes. J. Therm. Spray Technol. **23**(8), 1259–1269 (2014). https://doi.org/10.1007/s11666-014-0128-8
17. Yin, S., Meyer, M., Li, W., Liao, H., Lupoi, R.: Gas flow, particle acceleration, and heat transfer in cold spray: a review. J. Therm. Spray Technol. **25**(5), 874–896 (2016). https://doi.org/10.1007/s11666-016-0406-8
18. Zhou, X.H., Wang, Y., Song, D.P., Bai, G., Li, A., Dong, Q.: Analysis and prediction on thermal conductivity of coal based on Box Behnken design. J. Saf. Sci. Technol. **13**(9), 109–115 (2017)

Influence of Joint Edge Shape on Stress Distribution in Adhesive Film

Konstantin P. Barakhov(✉) and Igor M. Taranenko

National Aerospace University "Kharkiv Aviation Institute", 17 Chkalova Street, Kharkiv 61070, Ukraine

Abstract. Adhesive overlapped joints are widely used in up-to-date aviation and aerospace engineering. Models for joint stress state are based on definite hypothesis about stress distribution and translations in adhesive film and load-carrying layers. Known papers generally use hypothesis about uniform shear and normal stress distribution in adhesive film. Mathematica models based on mentioned hypothesis describe joint stress state in regular zone far from adhesive film edges. But near edge of joint adhesive film withstandss complicated stress state which depends on many different factors. Analysis of stress near zone of joint is conducted and influence of such technological factors as fillets and adhesive spew-fillet on adhesive stress is shown. Comparison of analytical calculations with finite-element methods modelling is done. It is shown that presence of fillet and spew-fillet reduces maximum adhesive stress and as a result increases structural strength.

Keywords: Adhesive joints · Analytical model · Finite elements method

1 Introduction

Wide expansion of composites in aviation and aerospace engineering demands developing novel types of approaches for joints design and analysis, which can realize all advantages of composites. Overlapped adhesive joints belong to such type of joints. This is stipulated by fact that overlapped adhesive joints doesn't violate composite structure integrity, ensuring loads transferring from structure surface (but not from points like in bolt and rivet joints), and guarantee air tightness, low weight, and other technological advantages. There are many structural-manufacturing solutions of overlapped joints for plates which loaded with longitudinal load [1–4]. Adhesive film in overlapped joint withstands shear and peeling off. And stress distribution over joint length is not uniform. Stress concentration near joint edges is observed [2–4] (Fig. 1).

In symmetrical double-sided joints and in scarf joints influence of peeling off stress stipulated by structural bending is negligible. In these cases following model can be applied: adhesive film withstands shear only and shear stress distribution over film thickness is uniform and proportional to difference of load-carrying layers translations (Vőlkersen model and its generalization) [2, 5–10].

M. Nechyporuk et al. (Eds.): ICTM 2021, LNNS 367, pp. 123–132, 2022.
https://doi.org/10.1007/978-3-030-94259-5_12

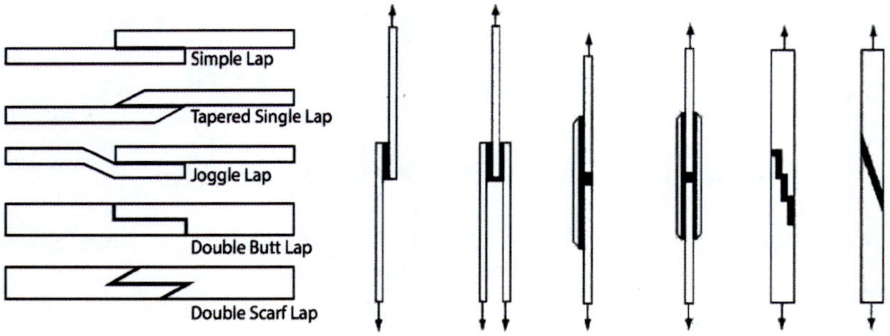

Fig. 1. Typical arrangement of adhesive overlapped joints.

More general is Holland-Reissner model [2] in which load-carrying layers are considered as beams in Bernoulli approximation and adhesive film is considered as Winkler elastic basement. I.e. normal stress in adhesive is proportional to the difference of lateral translations of load-carrying layers and shear stress – proportional to difference of translations of load-carrying layers oriented to adhesive film. It is assumed that distribution of shear and normal stress in adhesive film is uniform through thickness. In both Vőlkersen model and Holland-Reissner model shear stress reach maximum at edges of adhesive film. However, if consider adhesive film as elastic medium having free of load boundary at edges then shear stress must be equal to zero at the edges of adhesive film according to the law of shear stress parity. To remove this contradiction several more precise mathematical models for stress state of adhesive film are created. These models allow to consider boundary conditions at the edge of adhesive film [11–17]. Shear stress is assumed to be uniformly distributed through adhesive thickness and normal one varies through adhesive thickness. More precise models for stress state of adhesive film allows to establish that normal stress change significantly through adhesive thickness near edges of joint and at the boundaries with load-carrying layers can have even different signs. In their turn shear stress reach maximum at spacing near thickness of adhesive film from the joint edge. High values of stress near edge of joint can lead to failure of both adhesive film and entire structure. Therefore, to reduce stress near adhesive film edge following structural measures as creation of a fillet, spew-fillet of adhesive etc. [18]. Paper [19] deals with influence of other factors governing strength of composite structures. Analysis of stress state of overlapped adhesive joints is also conducted by numerical methods [18–24]. Different researchers have found that the most dangerous area is joint edge from the strength point of view (Fig. 2).

However, stress state of adhesive at this area depends significantly on the shape of joining layers and outer edge of adhesive film.

The goal of this paper is analysis of influence of structural distinctions of joint edge on stress state of adhesive joint. To achieve the goal let's consider definite joint, find stress state of a structure by means of analytical model, create finite-element model of a structure to which outer fillet, inner fillet and spew-fillet of adhesive will be embedded. Then influence of all mentioned structural solutions on stress state of adhesive film will be conducted.

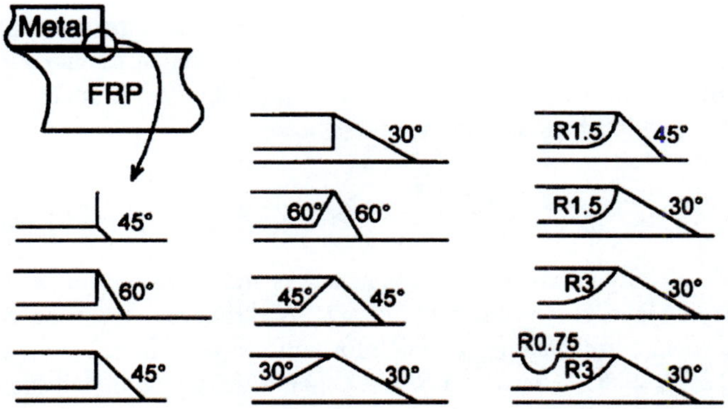

Fig. 2. Arrangement of joint edge geometry.

2 Analytical Model

To analyze stress state of adhesive film we can use one of the most precise analytical models allowing to consider boundary conditions at edges of adhesive film [13].

Differential element and force factors in layers and their edges are shown on Fig. 3.

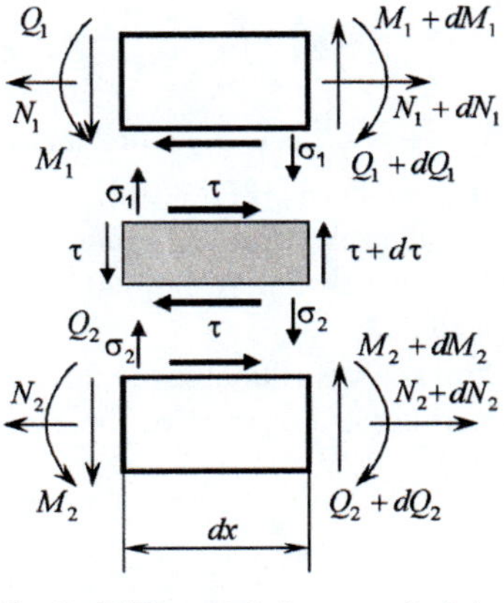

Fig. 3. Differential element of a joint.

Equilibrium equations can be written as.

$$\frac{dN_1}{dx} = -\tau,\ \frac{dQ_1}{dx} = \sigma_1,\ \frac{dM_1}{dx} = Q_1 - \frac{\delta_1}{2}\tau,$$

$$\frac{dN_2}{dx} = \tau,\ \frac{dQ_2}{dx} = -\sigma_2,\ \frac{dM_2}{dx} = Q_2 - \frac{\delta_2}{2}\tau,$$

$$\frac{d\tau}{dx} = \frac{\sigma_2 - \sigma_1}{\delta_0},$$

where $N_i,\ Q_i,\ M_i$ – longitudinal, lateral forces and bending moment per unit length in i-ом load-carrying layer, $i = 1,\ 2$; $\tau,\ \sigma_i$ – shear and normal stress in adhesive; $\delta_1,\ \delta_2$ – thickness of 1-st and 2-nd load-carrying layer; δ_0 – adhesive film thickness.

Mechanics relationships for Timoshenko's beams.

$$\frac{du_i}{dx} = \frac{N_i}{B_i};\ D_i\frac{d\varphi_i}{dx} = M_i;\ \frac{dw_i}{dx} + \varphi_i = \frac{Q_i}{H_i},$$

where $u_i,\ w_i,\ \varphi_i$ – longitudinal, lateral translations and cross-section twisting angle of i-th load-carrying layer correspondingly; $B_i,\ D_i,\ H_i$ – rigidities at tension-compression, bending and shear correspondingly, which for uniform layer have the view $B_i = E_i\delta_i$, $D_i = \delta_i^3 E_i/12$, $H_i = 5G_i\delta_i/6$, where E_i and G_i – elasticity moduli and shear moduli of load-carrying layers.

Adhesive film is modelled by multi-parametric elastic basement in membrane interpretation by M. N. Filonenko-Borodich. According to the model adhesive film is considered as membrane located in mid-plane of a film. Elastic elements are located between membrane and load-carrying layers. Normal stress according to the model can be found as.

$$\sigma_1 = -K_1(w_0 - w_1) + K_2\left(\frac{d^2w_0}{dx^2} - \frac{d^2w_1}{dx^2}\right),$$

$$\sigma_2 = K_1(w_0 - w_2) - K_2\left(\frac{d^2w_0}{dx^2} - \frac{d^2w_2}{dx^2}\right).$$

where w_0 - membrane lateral translation; $K_1,\ K_2$ – first and second bedding coefficients, $K_1 = 2E_0\left[\delta_0\left(1 - \mu_0^2\right)\right]^{-1}$; $K_2 = E_0\delta_0\left[12(1 + \mu_0)\right]^{-1}$. Here E_0 and μ_0 - elasticity modulus and Poisson's ratio of adhesive film. Difference of the joint model stress state [13] from [11] is in formulas for calculation normal stress in adhesive film second bedding coefficient K_2 is considered.

Shear stress in adhesive film is proportional to the difference of longitudinal translations of inner sides of load-carrying layers, relative shear of adhesive film and constant through adhesive film thickness.

$$\tau = P_0\left(u_1 - u_2 - \frac{\delta_1}{2}\varphi_1 - \frac{\delta_2}{2}\varphi_2\right) + G_0\frac{dw_0}{dx},$$

where P_0 – shear rigidity of adhesive film; G_0 – shear modulus of adhesive film. If one assumes that shear is concentrated in adhesive film only, then $P_0 = \frac{G_0}{\delta_0}$. However, such model gives slightly overrated values of shear stress. More precise approach assumes uniform shear stress distribution through adhesive thickness and linear distribution in load-carrying layers. In this case $P_0 = \left(\frac{\delta_0}{G_0} + \frac{1}{2}\frac{\delta_1}{G_1} + \frac{1}{2}\frac{\delta_2}{G_2}\right)^{-1}$.

Above-mentioned equations can be reduced to the system of linear ordinary differential equations with respect to longitudinal and lateral translations of load-carrying layers and also twisting angles of cross-sections. The system has analytical solution. Unknown coefficients can be found from boundary conditions. Boundary conditions include translations and forces in load-carrying layers at boundaries of joint and also two boundary conditions for shear force at the edge of joint. As boundary condition one can consider equality to zero of shear stress in adhesive at the edge of joint in the case of absence of spew-fillet. If adhesive film has spew-fillet one can consider as boundary condition reaching of shear stress its maximum value at the edge of joint, i.e. equality to zero the derivative from shear stress in adhesive film [13, 16].

3 Numerical Analysis and Finite-Element Modelling

Let's consider joint of two the same aluminum plates shown on на Fig. 4 to study distinctions of stress state for adhesive film at joint edges.

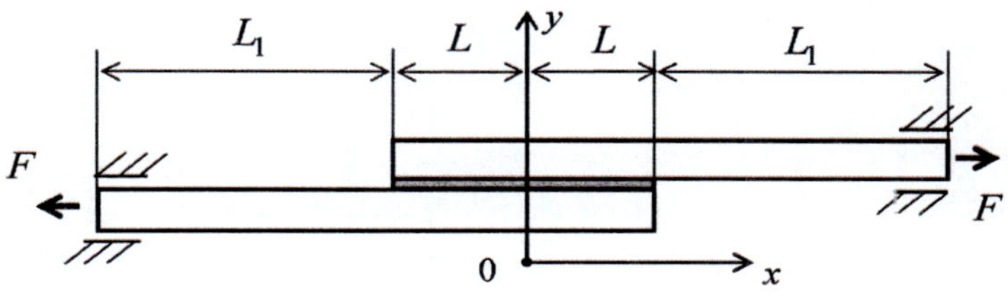

Fig. 4. Design of adhesive joint.

Parameters of load-carrying layers: $E_1 = E_2 = 70$ GPa, $\delta_1 = \delta_2 = 4$ mm, $\mu_1 = \mu_2 = 0.28$, $L = 10$ mm, $L_1 = 100$ mm. Adhesive film parameters: $G_0 = 0.25$ GPa, $E_0 = 0.65$ GPa, $\delta_0 = 0.1$ mm.

Let's consider two cases described in papers [13, 16]. In the first case (a) spew-fillet at the edge of joint is absent therefore adhesive film has shape of thin rectangle with rectangular outer boundary (square-end, «S-E»). In this case one can add condition of free boundary at the edge of joint $\tau(\pm L) = 0$ to boundary conditions and conditions of conjugacy at joint boundary, to conditions of sliding clamping at points of loading with longitudinal tension force F. In the second case (b) spew-fillet of adhesive is presented on adhesive film edge (spew-fillet, «S-F»), therefore shear stress at the joint boundary reaches maximum. This corresponds to condition at the boundary $\left.\frac{d\tau}{dx}\right|_{x=\pm L} = 0$.

Figure 5 shows diagrams of shear stress τ distribution in adhesive film and peeling off stress σ_1, σ_2 at the boundary «adhesive – load-carrying layer» for both cases. Stress is given in dimensionless form as relationship to some hypothetical shear stress τ^*, which could appear at uniform distribution of longitudinal force F transferred by joint along the length of joining area, i.e. $\tau^* = \frac{F}{2L}$.

It can be seen that presence of adhesive spew-fillet no influence on values of maximum shear stress (changing position of maximum point negligibly) but changes normal stress distribution in surrounding of joint edge significantly. Moreover, peeling off stress in the first case at the boundary of one layer practically two times overexceeds analogous stress in the second case. This reduces structural strength because can lead to both adhesive failure and adhesive tearing out from joining surface. If joining structure is laminated composite, then high values of peeling off stress can lead to composite delamination [20].

Up-to-date methods of finite-element analysis allow to analyze stress state of a joint with high precision. Let's consider several finite-element models which differ by geometry of joint edge.

First of all, consider model which has large spew-fillet of adhesive excess at joint edge and load-carrying layers have fillet from the adhesive layer side. Let's consider two cases in which fillet size is equal to $\frac{1}{3}$ (a) and $\frac{2}{3}$ (b) of outer layer thickness. All other parameters are analogous to above-mentioned. Figure 6 shows fragments of both models where relative sizes of fillets are designated.

Figure 7 shows diagrams of tangent and normal stress distribution in mid-plane of adhesive layer calculated by means of finite element method (FEM).

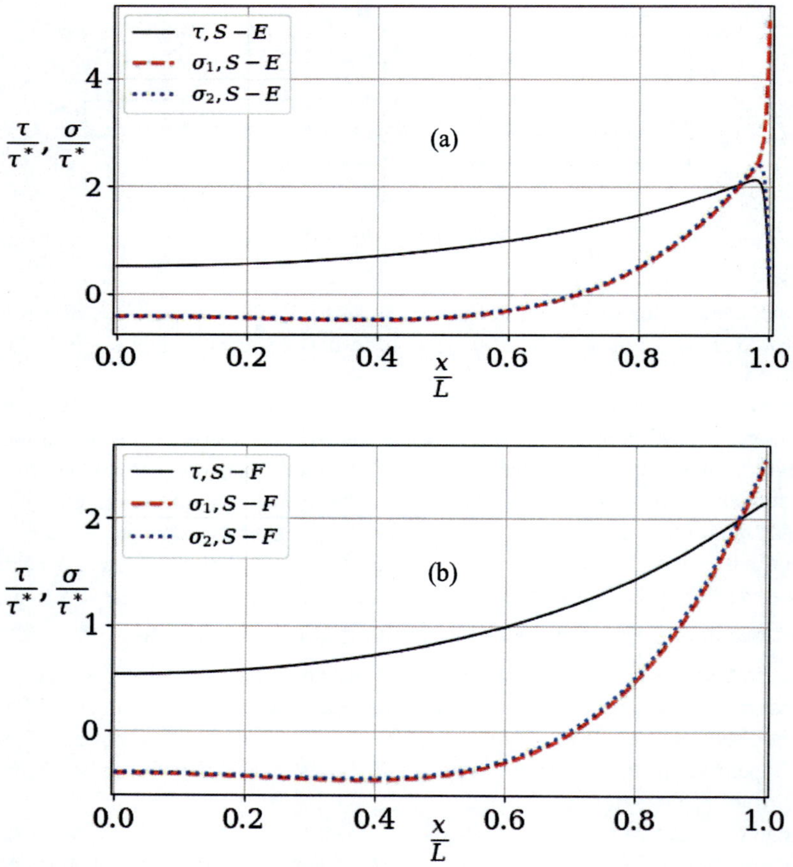

Fig. 5. Stress distribution in adhesive layer in case of absence (a) and presence of adhesive excess on boundary (b).

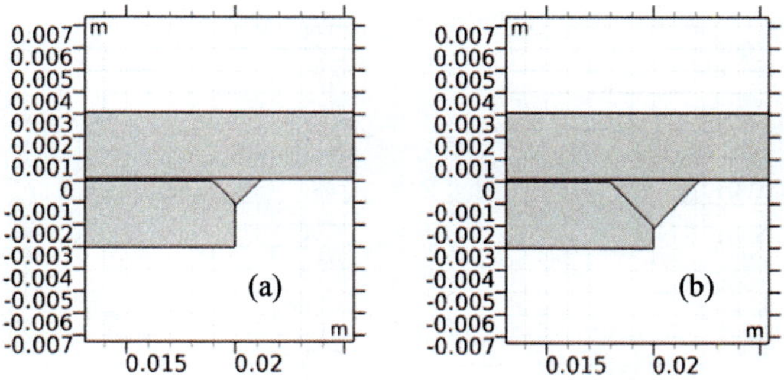

Fig. 6. Geometry of joint edge in finite-element models.

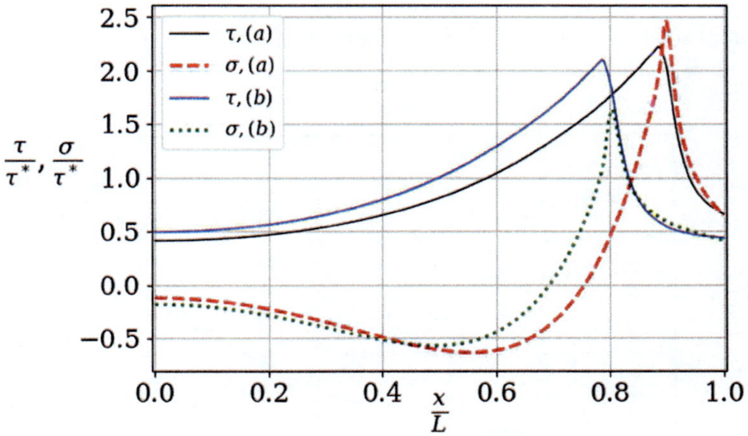

Fig. 7. Stress distribution in adhesive joint.

Pikes on the diagram correspond to beginning of a fillet. More thickness of adhesive film behind a fillet less stress in adhesive. The diagram shows that increasing of a fillet size practically doesn't reduce maximum values of shear stress in adhesive but normal (peeling off) stress in adhesive are reduced significantly. Moreover, comparison of results shown with ones obtain by analytical model and shown on the Fig. 5 (S-F) reveals that analytical model guarantees quite good description of stress state of a joint having relatively small fillet and spew-fillet of adhesive excess. Maximum values of stress in adhesive film calculated by analytical model and shown on Figs. 5(a) and 7(a) coincide practically.

Let's consider joints which have scarf of load-carrying layer at the edge of a joint. Let's analyze two cases in which load-carrying layer are cut at the edge up to $\frac{1}{3}$ (a) and $\frac{1}{10}$ (b) of layer thickness. Scarf angle is done 30° with respect to joining plane. Figure 8 shows fragments of both models.

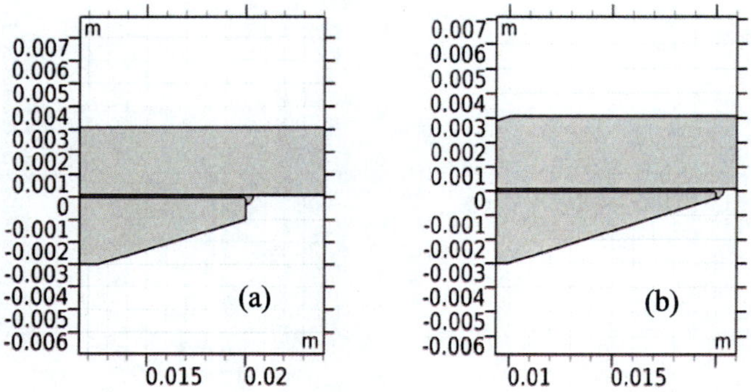

Fig. 8. Geometry of joint edge in finite-element models.

Figure 9 shows diagrams of shear and normal stress in mid-plane of adhesive film for both cases considered.

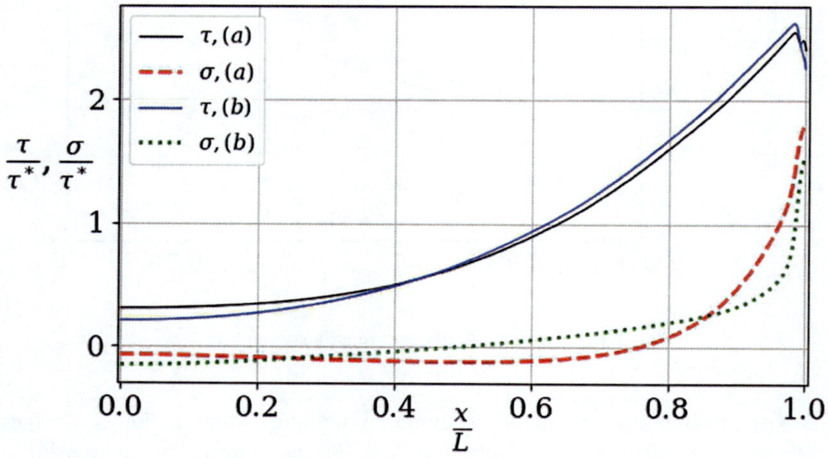

Fig. 9. Stress distribution in adhesive joint.

Diagrams shown allow to make conclusion that scarf of load-carrying layer has negligible influence on shear stress distribution in adhesive film but reduce normal (peeling off) stress in adhesive. But in small zone at the edge of joint there is the surrounding in which normal stress in adhesive stay quite high.

One can see that the first methods of reduction stress at the edge of a joint is preferable that second one. It can be explained by following: creation of a fillet from the adhesive side (and filling free space with adhesive) not only reduces rigidity of load-carrying layers but also increases compatibility of adhesive film that in its turn reduces stress in adhesive.

4 Conclusions

Numerical modelling of stress state of overlapped adhesive joint has shown that:

1) Analytical model suggested in papers [12, 13] possesses high precision.
2) Such structural solutions as creation of a fillet at the edge of load-carrying layer and beard of adhesive excess allow to reduce stress at the end of joint.
3) Fillet created in load-carrying layer from the side attracted to adhesive is preferable because allows both to reduce rigidity of load-carrying layer and increase compatibility of adhesive film. However, in this case all space between outer layers has to be filled with adhesive.

References

1. Mittal, K.L., Panigrahi, S.K. (eds.): Structural Adhesive Joints: Design, Analysis and Testing. Scrivener Publishing LLC, Beverly (2020)
2. da Silva, L.F.M., das Neves, P.J.C., Adams, R.D., Spelt, J.K.: Analytical models of adhesively bonded joints. Part I: literature survey. Int. J. Adhes. Adhes. **29**, 319–330 (2009). https://doi.org/10.1016/j.ijadhadh.2008.06.005
3. Wong, E.H., Liu, J.: Interface and interconnection stresses in electronic assemblies – a critical review of analytical solutions. Microelectron. Reliab. **79**, 206–220 (2017). https://doi.org/10.1016/j.microrel.2017.03.010
4. Budhe, S., Banea, M.D., de Barros, S., da Silva, L.F.M.: An updated review of adhesively bonded joints in composite materials. Int. J. Adhes. Adhes. **72**, 30–42 (2017). https://doi.org/10.1016/j.ijadhadh.2016.10.010
5. Karpov, Ya.S.: Jointing of high-loaded composite structural components. Part 2. Modeling of stress-strain state. Strength Mater. **38**(5), 481–491 (2006). https://doi.org/10.1007/s11223-006-0067-9
6. Kurennov, S.S., Barakhov, K.P.: The stressed state of the double-layer rectangular plate under shift. The Simplified two-dimensional model. PNRPU Mech. Bull. **3**, 166–174 (2019). https://doi.org/10.15593/perm.mech/2019.3.16
7. Kurennov, S.S.: An approximate two-dimensional model of adhesive joints. Analytical solution. Mech. Compos. Mater. **50**(1), 105–114 (2014). https://doi.org/10.1007/s11029-014-9397-z
8. Kurennov, S.S.: A simplified two-dimensional model of adhesive joints. Nonuniform load. Mech. Compos. Mater. **51**(4), 479–488 (2015). https://doi.org/10.1007/s11029-015-9519-2
9. Kurennov, S., Barakhov, K., Dvoretska, D., Poliakov, O.: Stress state of two glued coaxial tubes under nonuniform axial load. In: Nechyporuk, M., Pavlikov, V., Kritskiy, D. (eds.) ICTM 2020. LNNS, vol. 188, pp. 389–400. Springer, Cham (2021). https://doi.org/10.1007/978-3-030-66717-7_33
10. Kurennov, S.S.: Determining stresses in an adhesive joint with a longitudinal un-adhered region using a simplified two-dimensional theory. J. Appl. Mech. Tech. Phy. **60**, 740–747 (2019). https://doi.org/10.1134/S0021894419040199
11. Amidi, S., Wang, J.: An analytical model for interfacial stresses in double-lap bonded joints. J. Adhes. 1–25 (2018). https://doi.org/10.1080/00218464.2018.1464917
12. Kurennov, S.S.: Longitudinal-flexural vibrations of a three-layer rod. An improved model. J. Math. Sci. **215**(2), 159–169 (2016). https://doi.org/10.1007/s10958-016-2829-7

13. Kurennov, S.S.: Refined mathematical model of the stress state of adhesive lap joint: experimental determination of the adhesive layer strength criterion. Strength Mater. **52**(5), 779–789 (2020). https://doi.org/10.1007/s11223-020-00231-5
14. Kim, H.S., Cho, M., Lee, J., Deheeger, A., Grédiac, M., Mathias, J.D.: Three dimensional stress analysis of a composite patch using stress functions. Int. J. Mech. Sci. **52**, 1646–1659 (2010). https://doi.org/10.1016/j.ijmecsci.2010.08.006
15. Dua, Y., Liu, Y., Zhou, F.: An improved four-parameter model on stress analysis of adhesive layer in plated beam. Int. J. Adhes. Adhes. **91**, 1–11 (2019). https://doi.org/10.1016/j.ijadhadh.2019.02.005
16. Frostig, Y., Thomsen, O.T., Mortensen, F.: Analysis of adhesive-bonded joints, square-end, and spew-fillet—high-order theory approach. J. Eng. Mech. **125**(11), 1298–1307 (1999). https://doi.org/10.1061/(asce)0733-9399(1999)125:11(1298)
17. Radice, J.: On the decoupled biharmonic airy stress function for the square-end adhesive layer and sandwich structure core. J. Sandw. Struct. Mater. (2018). https://doi.org/10.1177/1099636218818624
18. Harris, J.A., Adams, R.A.: Strength prediction of bonded single lap joints by non-linear finite element methods. Int. J. Adhes. Adhes. **4**(2), 65–78 (1984). https://doi.org/10.1016/0143-7496(84)90103-9
19. Kondratiev, A., Gaidachuk, V., Nabokina, T., Kovalenko, V.: Determination of the influence of deflections in the thickness of a composite material on its physical and mechanical properties with a local damage to its wholeness. East.-Eur. J. Enterp. Technol. **4**(1) (100), 6–13 (2019). https://doi.org/10.15587/1729-4061.2019.174025
20. Cheuk, P.T., Tong, L.: Failure of adhesive bonded composite lap shear joints with embedded precrack. Compos. Sci. Technol. **62**(7–8), 1079–1095 (2002). https://doi.org/10.1016/s0266-3538(02)00054-4
21. Wang, C.H., Rose, L.R.F.: Compact solutions for the corner singularity in bonded lap joints. Int. J. Adhes. Adhes. **20**(2), 145–154 (2000). https://doi.org/10.1016/s0143-7496(99)00032-9
22. Kairouz, K.C., Matthews, F.L.: Strength and failure modes of bonded single lap joints between cross-ply adherends. Composites **24**(6), 475–484 (1993). https://doi.org/10.1016/0010-4361(93)90017-3
23. Goglio, L., Rossetto, M.: Precision of the one-dimensional solutions for bonded double lap joints. Int. J. Adhes. Adhes. **31**(5), 301–314 (2011). https://doi.org/10.1016/j.ijadhadh.2010.10.004
24. Kurennov, S.S., Barakhov, K.P., Poliakov, A.G.: Stressed state of the axisymmetric adhesive joint of two cylindrical shells under axial tension. Mater. Sci. Forum **968**, 519–527 (2019). https://doi.org/10.4028/www.scientific.net/msf.968.519

Considering the Meteorological Elements for the Aerial Vehicles Flight by Using "Weight" Functions

Vasilii Makeiev[1], Yurii Pushkariov[1], Andrii Raskoshnyi[1], Iryna Voronko[2](✉), and Svitlana Myronova[2]

[1] Sumy State University, 2 Rimskoho-Korsakova Str., Sumy 40007, Ukraine
[2] National Aerospace University "Kharkiv Aviation Institute", 17 Chkalova Str., Kharkiv 61070, Ukraine
i.voronko@khai.edu

Abstract. Meteorological preparation of aerial vehicles with empennage launching is one of the most important elements of the whole launch preparation. It's task to determine the deviation of launch meteorological conditions from normal (tabular) values, as they are necessary for the flights calculating. Nevertheless, the distribution of meteorological elements deviation through the atmosphere heights doesn't go in parallel with tabular ones. This is resulted in need to fulfill not only the ground measurements but also the altitudinal sounding of atmospheric temperature and wind. According to the results of these admeasurements some conditional deviations of meteorological elements, which influent on the aerial vehicle flight, can be calculated in correspondence with trajectory height, in the same way as real ones, which are not equal on the different heights. The article offers to consider the influence of meteorological elements on the flight of the aerial vehicles with empennage by using the "weight" functions of these elements. "Weight" function can be represented analytically or graphically. The way of determining the average ballistic meteorological elements for aerial vehicles within powered trajectory, passive trajectory and within flight trajectory of the falling elements by using the weight functions of meteorological elements distribution is proposed.

Keywords: Meteorological elements · Weight functions · Powered and passive trajectory · Flight trajectory of falling elements · Ballistic temperature of air · Ballistic wind · Correction coefficients · Jet acceleration · Acceleration of drag · Longitudinal (lateral) wind

1 Introduction

The rate of meteorological preparation errors in the total preparation error is 60–70% [1–6].

M. Nechyporuk et al. (Eds.): ICTM 2021, LNNS 367, pp. 133–144, 2022.
https://doi.org/10.1007/978-3-030-94259-5_13

In scientific articles about meteorological preparation to the launching of aerial vehicles with empennage are usually considered the theoretical methods for determining of meteorological elements ballistic deviations [1, 2, 7, 8].

The influence of meteorological elements on flight of aerial vehicles with empennage (EAV) is more complicated and it's necessary to consider separately the correction coefficients for flight within powered trajectory, passive trajectory and within flight trajectory of the falling elements. Thus, there is a need to consider the influence of meteorological elements on the EAV flight [1, 9–12].

2 The Main Part

2.1 Meteorological Conditions of Launch of Aerial Vehicles with Empennage

When determining the launching directions is necessary to calculate the next corrections of meteorological elements deviation from tabular ones:

a) Range corrections:

$$\Delta X_m = \Delta X_{Wax} Wax + \Delta X_{Wnx} Wnx + \Delta X_{Wex} Wex + \Delta X_H \Delta H_0 + \Delta X_{HH} (\Delta H_0)^2 \\ + \Delta X_t \Delta T + \Delta X_{tt} (\Delta T)^2 + \Delta X_{tH} \Delta T \Delta H_0, \quad (1)$$

where ΔX_{Wax}, ΔX_{Wnx}, ΔX_{Wex} are correction coefficients for longitudinal component of ballistic wind within powered and passive trajectory and flight trajectory of the falling elements accordingly, 1000/mps; ΔX_H, ΔX_{HH} are correction coefficients of ground atmosphere pressure deviation from tabular ones, 1000/mmHg; 1000/mmHg^2; ΔX_t, ΔX_{tt} are correction coefficients of air temperature ballistic deviation from tabular ones, 1000/(°C); 1000/(°C)^2; ΔX_{tH} is correction coefficient of common influence of the deviations of air temperature and ground atmosphere pressure 1000/(°C), mmHg. These correction coefficients can be calculated with the help of differential equations system (DES) (9) [13, 14].

Wax, *Wnx*, *Wex* are longitudinal components of ballistic wind within powered and passive trajectory and flight trajectory of the falling elements accordingly, mps; $\Delta H0$ is deviation of ground atmosphere pressure from tabular one on the height of launching position, mmHg; ΔT is ballistic deviation of air temperature within the whole trajectory, °C.

b) correction coefficient of corrective turn with respect to the main flight direction:

$$\Delta \partial_m = \Delta Z_{Waz} Waz + \Delta Z_{Wnz} Wnz + \Delta Z_{Wez} Wez; \quad (2)$$

where *Waz*, *Wnz*, *Wez* are side components of ballistic wind within powered and passive trajectory and flight trajectory of the falling elements accordingly; ΔZ_{Waz}, ΔZ_{Wnz}, ΔZ_{Wez} are correction coefficients of ballistic wind lateral component within powered and passive trajectory sector and flight trajectory of the falling elements accordingly, calculated with the help of differential equations system (DES) (9).

The need for meteorological preparation of EAV launches is confirmed by the quantitative characteristics of the meteorological factors influence on their flight.

Table 1 shows the deviations of EAV due to disregard for the influence of meteorological factors.

From Table 1 it follows that without taking into account the deviations of real meteorological values from the tabular ones, the deviations of the EAV impact point from the calculated tabular impact point may reach several hundred meters. It is obvious that with such deviations the aerial vehicles' task may be failed. The data in Table 1 are calculated using the DES (9) of EAV movement.

Table 1. Possible deviations of EAV in range and direction due to meteorological elements significant deviations

D, km	Range deviations, m					Direction deviations, m		
	$Wax =$ 15 mps	$Wnx =$ 20 mps	$Wex =$ 15 mps	$\Delta Ho =$ 20 mmHg	$\Delta T\text{в} =$ 20 °C	$Waz =$ 15 mps	$Wnz =$ 20 mps	$Wez =$ 15 mps
30	1140	440	1050	980	1200	225	220	840
40	690	320	615	900	680	330	520	630
50	735	960	525	1240	280	450	720	555
60	885	1120	630	1560	320	585	840	500
70	1020	1100	780	1800	1060	705	860	675

Thus, the data in Table 1 confirms the need of carrying out the meteorological preparation of EAV launches [9, 14]. Initial data for calculating the results, which shown in Table 1, are taken in [1, 2].

3 Fundamentals of Meteorological Conditions Determination Methods

3.1 Determining of a Ballistic Deviation of Air Temperature Within Full Trajectory and Powered Trajectory

The method of determination of air temperature ballistic deviation includes the replacement of true weight function r_T^τ with approximate function r_p^τ (Fig. 1). Approximate weight functions consist of two-three linear segments (0-1, 1-2, 2-3). The replacement is done under condition that areas confined by true and approximate weight functions are equal by values and have opposite signs. In this case the error, which appears due to replacement of true weight function by approximate one, is insignificant. The true weight function r_T^τ is calculated with the help of the equation system (9).

Then the formula for calculation of ballistic air temperature deviation ΔT is:

$$\Delta T = \sum_{i=1}^{n} K_i^{\tau} \Delta \tau_{Yi} \tag{3}$$

where K_i^{τ} is the weight coefficient for i air layer, τ_{Yi} is average air temperature deviation in the air layer from ground to the altitude Y_i (these values should be taken from the "Метеосредний" bulletin).

For calculation of the weight coefficients will be used the next formula:

$$K_i = (tg\beta_i - tg\beta_{i+1})Y_i \tag{4}$$

where β_i is an angle of approximate weight function i-segment declination to abscissa axis (Fig. 1).

3.2 The Replacement of True Weight Function by Approximate Weight Function

Let's consider the procedure of the weight coefficients calculation during the replacement of true weight function by approximate one, which consists of three segments (see Fig. 1).

Using the expression (4) the weight coefficients will be represented as:

$$\begin{gathered} K_1^{\tau} = (tg\beta_1 - tg\beta_2)Y_1 = (q_1/\Delta Y_1 - q_2/\Delta Y_2)Y_1 = ((r_1 - r_0)/(Y_1 - Y_0) - (r_2 - r_1)/(Y_2 - Y_1))Y_1 \\ K_2^{\tau} = tg\beta_2 Y_2 = (q_2/\Delta Y_2)Y_2 = ((r_2 - r_1)/(Y_2 - Y_1))Y_2 \\ K_3^{\tau} = tg\beta_3 Y_3 = 0 \end{gathered} \tag{5}$$

where q_1, q_2 are the "weight" of the first and the second layer; r_0, r_1, r_2, r_3 are the coordinates of the points of replacement of true weight function by the approximate one.

The mathematical expression for calculation of ballistic air temperature deviation within the whole trajectory with replacement of true weight function by approximate one as shown in the Fig. 1 is:

$$K_1^{\tau} \Delta T = \sum_{i=1}^{n} K_i^{\tau} \Delta \tau_{Y_i} = K_1^{\tau} \Delta \tau_{Y_1} + K_2^{\tau} \Delta \tau_{Y_2} = K_{bul}^{7} \Delta \tau_{Y_7} + K_{bul}^{8} \Delta \tau_{Y_8}. \tag{6}$$

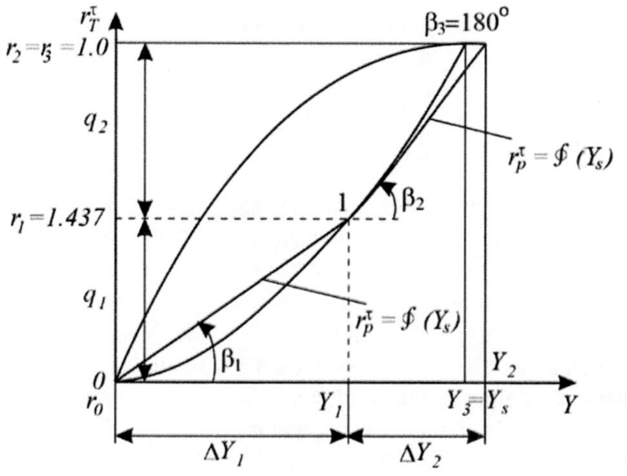

Fig. 1. The weight functions, which characterize the influence of air temperature within the whole trajectory of aerial vehicle with empennage (EAV)

Besides, the range of the flight includes the correction factor for ballistic temperature deviation ΔT_a within powered trajectory. The true function $r_T^{\tau\alpha}$, which characterizes the influence of air temperature within powered trajectory, is shown in the Fig. 2.

The true weight function is replaced by the approximate one $r_p^{\tau\alpha}$, which consists of two segments: 0-1 and 1-2. In this replacement n = 2; $\beta_1 > 0$; $Y_1 < Y_\alpha$; $Y_2 = Y_\alpha$ (see Fig. 2).

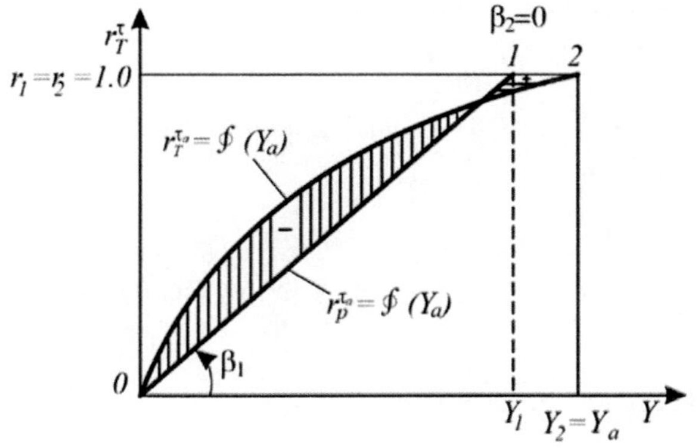

Fig. 2. The replacement of the true weight function of air temperature influence within the powered trajectory with the approximated one

The weight coefficients will be found from the following expressions:

$$
\begin{aligned}
K_1^{\tau\alpha} &= (tg\beta_1 - tg\beta_2)Y_1 = (1.0/Y_1 - 0)Y_1 = 1.0 \\
K_2^{\tau\alpha} &= tg\beta_2 Y_2 = 0
\end{aligned}
\tag{7}
$$

Then the deviations of air temperature will be found from the expressions:

$$\begin{aligned} \Delta T_a &= \sum_{i=1}^{n=2} K_i^{\tau\alpha} \Delta\tau_{Y_i} = K_1^{\tau\alpha}\,\Delta\tau_{Y_1} + K_2^{\tau\alpha}\,\Delta\tau_{Y_2} = \Delta\tau_{Y_1}^a, \\ \Delta T_a &= \Delta\tau_{Y_1}^a = \Delta\tau_{Y_9}^a \end{aligned} \quad (8)$$

The weight coefficients $K_1^{\tau\alpha}$, $K_2^{\tau\alpha}$ can be calculated, depending on altitudes Y_1, Y_2, by the expressions (7) and saved as the appendices to the flight tables of considered EAV. In these table $K_1^{\tau\alpha}$ are denoted K_{bul}^7; $K_2^{\tau a} - K_{bul}^8$; $Y_1 - Y_7$; $Y_2 - Y_8$; $Y_1^\alpha - Y_9$.

The temperature deviations in the layers $\Delta\tau_{Y7}, \Delta\tau_{Y8}, \Delta\tau_{Y9}$ are determined according to the "Метеосредний" bulletin for altitudes Y_7, Y_8, Y_9 accordingly. The true weight function rT^τ is calculated by using the system of differential Eq. (9).

Thus, ballistic air temperature deviation within powered trajectory ΔT_a, calculated by (8), equal to average deviation of air temperature $\Delta\tau_{Y1}^a$ in the layer from ground to the altitude Y_1. The altitude $Y_1 = f(Ya) = f(Dg)$.

In order to solve the problems of building the exact "weight" functions and calculating the weight coefficients for meteorological elements, the authors of this article proposed the following DES for the EAV flight [13].

System of differential equations of EAV spatial motion (ballistic model) is:

1. $\dot{X} = (V \cdot \cos\theta \cdot \cos\Psi)/(1 - 2Y/Rz)$.
2. $\dot{Y} = V \cdot \sin\theta$.
3. $\dot{Z} = V \cdot \cos\theta \cdot \sin\Psi$.
4. $\dot{V} = a(t) - a_x - g\sin\theta \cdot R_z^2/(1 - 2Y/R_z)$.
5.

$$\begin{aligned} \dot{\theta} =& a \cdot \delta + 2\lambda \cdot \delta + a\delta - (g \cdot \cos\theta/V) - \Omega_z(\cos B \cdot \sin\alpha_g) \\ &+ (V \cdot \cos\theta/R_z + Y) \end{aligned} \quad (9)$$

6. $\dot{\Psi} = (1/\cos\theta)[a \cdot \delta + 2\lambda \cdot \delta - a \cdot \delta] + \Omega_z(\sin B + \cos B \cdot tg\theta \cdot \cos\alpha_g)$.
7. $\dot{\delta}_1 = Z_1$.
8. $\dot{\delta}_2 = Z_2$.
9. $\dot{Z}_1 = -a \cdot Z_2 - \tilde{b}^* \cdot Z_1 + c^* \cdot \delta_1 + e^* \cdot \delta_2 + (b_0 \cdot g_0\cos\theta/V)$.
10. $\dot{Z}_2 = a \cdot Z_1 - \tilde{b}^* \cdot Z_2 + c^* \cdot \delta_2 - e^* \cdot \delta_1 - (\alpha \cdot g_0 \cdot \cos\theta/V)$.
11. $\dot{\pi}(y) = -(\pi(y) \cdot \dot{y})/(R \cdot \tau_y)$.

The righthand sides of the DES are:

12. $a(t) = \omega_0(I_{1N} + K_1 \cdot \Delta Tr)/m_0[\tau_{aN} - K_2 \cdot \Delta Tr] \cdot (1 - \mu_y)$.
13. $a_x = -0.474i \cdot d^2/(q_0 + \Delta q) \cdot \pi(y) \cdot V_{r\tau}^2 \cdot C_X(M_{\tau,\delta})$.
14. $a_N = 0.474 \cdot d^2/(q_0 + \Delta q) \cdot \pi(y) \cdot V_{r\tau} \cdot C_y(M_{\tau,\delta})$.
15. $\lambda = a(t)/V_{r\tau}$.
16. $\alpha = 0.474 \cdot d^3(q_0 + \Delta q) \cdot \pi(y) \cdot p \cdot C_z^{p\delta}(M_\tau)$.
17. $a = \alpha + a_L$.
18. $\alpha = (A/2B)p$.

19. $p0 = 2\pi V0/\eta d$.
20. $b* = 0.474 \cdot (d^2 \cdot L^2/2Bg_0) \cdot V_{r\tau}\pi(y) \cdot [m_z^{\omega_x}(M_\tau) + B/m_0 L^2 \cdot C_y^\delta(M_\tau)]$.
21. $c* = \pm b_R + \alpha \cdot a_L - b_d(a_n + \lambda)$.
22. $b_R = 0.474(d^2 \cdot L)/(B \cdot g) \cdot \pi(y) \cdot V_{r\tau}^2 \cdot m_z^\delta(M_\tau)$.
23. $b_d = 0.474(d^2 \cdot L^2)/(B \cdot g) \cdot \pi(y) \cdot V_{r\tau} \cdot m_x^{\varpi z}(M_\tau)$.
24. $b_r = 0.474(d^3 \cdot L)/(A \cdot g) \cdot \pi(y) \cdot V_{r\tau} \cdot m_z^{\varpi x}(M_\tau)$.
25. $$\begin{matrix} e* = 0.474(d^3 \cdot L)/(B \cdot g) \cdot \pi(y) \cdot p \cdot V_{r\tau}[\overline{m_z^{\rho\delta}}(M_\tau) - \\ -(d \cdot Ag \cdot C_y^\delta)/(L \cdot q_0 \cdot d^2)] - (A \cdot p \cdot a_p)/(B \cdot V) \end{matrix}.$$

"Virtual" temperature distribution by altitude (τy), °C:

26. $$\tau(y) = \tau + \begin{cases} \tau = 288.9 - 6.328 \cdot 10^{-3}y;\ y \le 9300 \\ \tau = 230 - 6.328 \cdot 10^{-3}y(y - 9300) + 1.172 \cdot 10^{-6}(y - 9300)^2; \\ 9300 < y \le 12000 \\ \tau = 221°C;\ y > 12000 \end{cases}.$$

Considering the wind influence may be done as following.
Initial conditions are:

$$x = 0;\ y = 0;\ z = 0;\ V \ne 0 = V_{or};\ \theta \ne 0 = \theta_{or};$$
$$\Psi = 0;\ \delta_1 = 0;\ \delta_2 = 0;\ \dot{\delta}_1 = 0;\ \dot{\delta}_2 = 0;\ p = p_0$$

Stage 1: before the calculation wind correction factors are included into the initial conditions (i.e., the transition to the relative coordinate system is carried out).

1. $V_{or} = \sqrt{(V_{ox} - W_x)^2 + (V_{oy} - W_y)^2 + (V_{oz} - W_z)^2}$.
2. $V_{ox} = \dot{x} = V_0 \cdot \cos\theta_0 \cdot \cos\psi_0$.
3. $V_{oy} = \dot{y} = V_0 \cdot \sin\theta_0$.
4. $V_{oz} = \dot{z} = V_0 \cdot \cos\theta \cdot \sin\psi_0$.
5. $\theta_{or} = \theta_0(1 + W_x/V_{ox}) - W_y/V_{ox}$.
6. $\psi_{or} = \psi_0(1 + W_y/V_{ox}) - W_z/V_{ox}$.
7. $\delta_{1or} = \delta_{10} - \theta_0(W_x/V_{ox}) + (W_y/V_{ox})$.
8. $V_{2or} = \delta_{20} - \psi_0(W_x/V_{ox}) + (W_z/V_{ox})$.
9. $\dot{\delta}_{1or} = \dot{\delta}_{10} - \dot{\theta}_0(W_x/V_{ox})$.

Stage 2: after integrating the DES under the initial conditions, the transition to the absolute kinematic formulas is carried out.

10. $\dot{\delta}_{2or} = \dot{\delta}_{20} - \dot{\psi}_0(W_x/V_{ox})$.
11. $$\begin{cases} V_{rx} = V_{tec} \cdot \cos\theta_{tec} \cdot \cos\psi_{tec} \\ V_{ry} = V_{tec} \cdot \sin\theta_{tec} \\ V_{rz} = V_{tec} \cdot \cos\theta_{tec} \cdot \sin\psi_{tec} \end{cases}.$$
12. $$\begin{cases} V_x = V_{rx} + W_x \\ V_y = V_{ry} + W_y \\ V_z = V_{rz} + W_z \end{cases}.$$

13. $V = \sqrt{V_x^2 + V_y^2 + V_z^2}$.

$x = x_r + W_x \cdot t$

14. $y = y_r + W_y \cdot t$.

$z = z_r + W_z \cdot t$

15. $\theta = \theta_r \cdot (1 - W_x/V_x) + W_z/V_x$.
16. $\psi = \psi_r(1 - W_x/V_x) + W_z/V_x$.
17. $\delta_1 = \delta_{1r} + (\theta_r \cdot W_x/V_x) - (W_y/V_y)$.
18. $\delta_2 = \delta_{2r} + (\psi_r \cdot W_x/V_x) - (W_z/V_x)$.
19. $\dot{\delta}_1 = \dot{\delta}_{1r} + \dot{\theta}_r \cdot (W_x/V_x)$.
20. $\dot{\delta}_2 = \dot{\delta}_{2r} + \dot{\psi}_r \cdot (W_x/V_x)$.

All the designation are generally accepted in the external ballistics [15, 16].

3.3 Determining the Ballistic Wind Within Passive Trajectory

The influence of the longitudinal Wx and lateral Wz wind on the EAV flight within passive section of trajectory have insignificant differences. That's why the "weight" functions of the longitudinal r^{Wx} and lateral wind r^{Wz} are differ from each other only a little, and they can be averaged while determining the components of ballistic wind within passive trajectory.

True "weight" functions of longitudinal (lateral) wind $r^{Wx(z)} = \oint (Ys)$ which are received from system of Eq. (9) are replaced by approximate ones.

If the true "weight" function $r_T^{Wx(z)}$ is replaced by the approximate one $r_p^{Wx(z)}$, as it's shown in Fig. 3, then, $n = 3$; $\beta_1 > 0$; $\beta_2 > 0$; $\beta_3 = 180°$; $Y_1 \ll Ys$; $Y_2 > Ys$; $Y_3 = Ys$.

The expressions (9) are used for such approximation and finally formulas for the weight coefficients calculations are:

$$\begin{aligned} K_1^{Wx(z)} &= (tg\beta_1 - tg\beta_2)Y_1 \\ K_2^{Wx(z)} &= (tg\beta_2 - tg\beta_3)Y_2 = (tg\beta_2 - 0)Y_2 = tg\beta_2 Y_2 \\ K_2^{Wx(z)} &= (tg\beta_2 - tg\beta_3)Y_2 = (tg\beta_2 - 0)Y_2 = tg\beta_2 Y_2 \\ K_3^{Wx(z)} &= tg\beta_3 Y_3 = 0. \end{aligned} \tag{10}$$

Then, the formulas for determining the longitudinal and lateral components of ballistic wind within passive trajectory are:

$$\begin{aligned} W_{nx} &= K^{Wx} \cdot Wx_{Y_1} + K_2^{Wx} \cdot Wx_{Y_2} \\ W_{nz} &= K^{Wz} \cdot Wx_{Y_1} + K^{Wz} \cdot Wx_{Y_2}, \end{aligned} \tag{11}$$

where $Wx(z)_{Y_1}$ are the longitudinal (lateral) components of average wind within the layer $0 \ldots Y_1$, which is determined according to the bulletin "Метеосредний"; $Wx(z)_{Y_2}$ are longitudinal (lateral) components of average wind within the layer $0 \ldots Y_2$, which is determined according to the bulletin "Метеосредний".

The "weight" coefficients $K_1^{Wx(z)}$, $K_2^{Wx(z)}$, $K_3^{Wx(z)}$ depend on the altitudes Y_1, Y_2, Y_3, and after they are calculated according to dependences (10), they may be saved as the appendices to the flight tables of considered EAV.

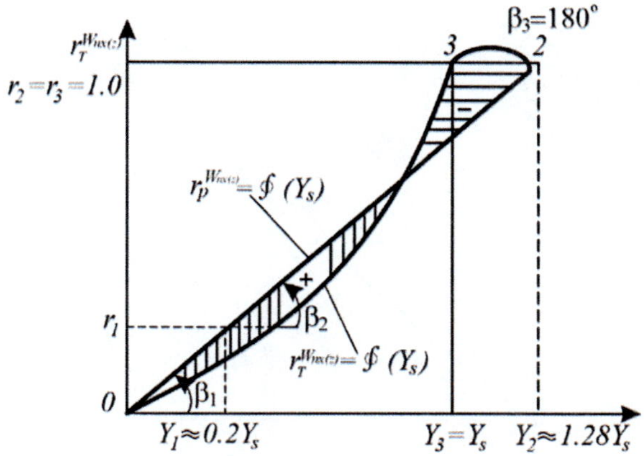

Fig. 3. The variant of replacement of true "weight" function of longitudinal (lateral) wind influence within passive trajectory by the approximate function

In the appendices to the EAV flight tables the weight coefficients $K_1^{Wx(z)}$ are designated as K_{bul}^5, $K_2^{Wx(z)} = K_{bul}^6$, and the altitudes in the bulletin Y_1 and Y_2, are under numbers Y_5 and Y_6.

Then, the values of the ballistic wind components within passive trajectory can be determined by the next dependences:

$$\begin{aligned} W_{nx} &= K_{bul}^5 \cdot Wx_{Y_5} + K_{bul}^6 \cdot Wx_{Y_6} \\ W_{nz} &= K_{bul}^5 \cdot Wz_{Y_5} + K_{bul}^6 \cdot Wz_{Y_6}. \end{aligned} \tag{12}$$

3.4 Determining the Ballistic Wind Within the Flight Trajectory of an EAV Falling Elements

The peculiarity of the meteorological preparation of the EAV of cassette type is the need to determine the ballistic wind within flight trajectory of the falling elements.

The speed of the falling elements flight varies insignificantly; in the top layers of atmosphere, it's less than in the bottom ones. In accordance with this, the flight time of falling elements in the top layers is longer, than in bottom ones. This is proved by the true weight function of longitudinal (lateral) wind influence on the flight of falling elements $r_T^{Wex(z)}$ (Fig. 4).

During the calculation of ballistic wind within the flight trajectory of the falling elements the true weight function $r_T^{Wex(z)}$, which was calculated by the dependence (9), is replaced by the approximate one $r_p^{Wex(z)}$, which consists of two segments: 0-1 and 1-2.

After such approximation the weight coefficients are:

$$\begin{aligned} K_1 = (tg\beta_1 - tg\beta_2)Y_1 = (r_1/Y_1 - 0)Y_1 = r_1 = 1.0; \\ K_2 = tg\beta_2 Y_2 = 0 \end{aligned} \tag{13}$$

The coefficients K_1 and K_2, which calculated by the dependence (13), are saved in appendices to the EAV flight tables, where $Y_1 = Y_{10}$.

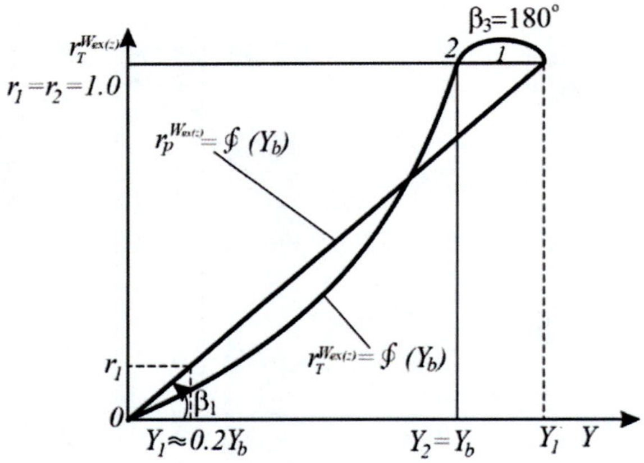

Fig. 4. Replacement of the true "weight" function of a longitudinal (lateral) wind influence within flight trajectory of EAV falling elements by the approximate one

The mathematical expression for ballistic wind calculation, when n = 2, is:

$$\overline{W}_{eYp} = \sum_{i=1}^{n=2} K_i \overline{W}_{Yi} = K_1 \overline{W}_{Y1} + K_2 \overline{W}_{Y2} = Y_p \overline{W}_{Y1} \quad (14)$$

Thus, a ballistic wind within the falling elements flight sector W_{eYp} is equal to an average wind W_{Y1} in the layer from ground to the altitude $Y = f(Y_p)$.

When determining the ballistic wind in the falling elements flight sector, it's considered a scalar value, not a vector one:

$$\begin{aligned} W_{exYp} &= W_{xY10} \\ W_{ezYp} &= W_{zY10} \end{aligned} \quad (15)$$

where W_{xY10}, W_{zY10}, Y_{10} are calculated by the dependences (15) and saved in appendices to the considered EAV flight tables.

3.5 The Calculation of a Ballistic Wind Within Powered Trajectory

The influence of a wind on the EAV flight in the near-the-ground atmospheric layers, i.e., within the powered trajectory, is much more significant, than within passive trajectory [1, 9]. So, according to the bulletin "Метеосредний", the influence of a wind within the powered trajectory should be taken into account as well as within passive trajectory.

The "weight" functions which characterize the longitudinal and lateral wind influence on the EAV flight within the powered and passive trajectories for the same flight range differ significantly. As a result, the "weight" coefficients and altitudes can't be averaged during the determining of longitudinal and lateral components of a ballistic wind, as it was done whilst calculating of passive trajectory. That is why, the calculation of a ballistic wind within the powered trajectory is much more complicated procedure than for the passive trajectory.

As a result of rational replacement of the true weight functions $r_T^{W^a_{x(z)}}$ by the approximate ones $r_p^{W^a_{x(z)}}$ there was obtained the mathematical expressions for calculation of longitudinal and lateral ballistic wind components within powered trajectory [1]:

$$\begin{aligned} W_{ax} &= K^1_{bul} \cdot W_{xY1} + K^2_{bul} \cdot W_{xY2} \\ W_{az} &= K^3_{bul} \cdot W_{zY3} + K^4_{bul} \cdot W_{zY4} \end{aligned} \tag{16}$$

where $K^{1,2,3,4}{}_{bul}$ are the "weight" coefficients calculated by the dependence (5) which are placed as appendices to the EAV flight tables.

W_{xY1}, W_{xY2} are longitudinal components of an average wind within altitudes Y_1 and Y_2, which are determined according to the appendices to the EAV flight tables; W_{zY3}, W_{zY4} are the lateral components of an average wind within altitudes Y_3 and Y_4 are determined according to the appendices to the EAV flight tables; Y_1, Y_2, Y_3, Y_4 are the altitudes in the bulletin "Метеосредний", which are shown in the appendices to the EAV flight tables.

The true "weight" functions have been calculated by DES (9).

Thus, having the DES of the EAV flight, it is possible to build the exact "weight" functions of the meteorological elements distribution within the whole flight trajectory, and using the proposed methods of approximation – to calculate "weight" coefficients for all meteorological factors. With the help of "weight" coefficients, it is possible to calculate corrections of flight range and direction while preparing the EAV launches. The accuracy of EAV flight prepared in this way is considered as 0,8–0,9% Dg [1, 2, 16] vs the currently existing method which gives 1,3–1,5 Dg [9, 14].

4 Conclusions

In this article the calculation methods of "weight" coefficients for flight of the aerial vehicles with empennage (EAV) have been explained.

It's proposed to save the calculated weight coefficients for considering the influence of the air temperature and ballistic wind within passive trajectory, powered trajectory and within flight sector of the falling elements as the appendices to the EAV flight tables.

Using the weight coefficients tables, it's possible to calculate the corrections of flight conditions deviations from the tabular ones during the launch preparation, taking into the account the ballistics of particular type of the aerial vehicle.

True "weight" functions are calculated using DES of the EAV flight what is resulted in significant improving their accuracy.

The mathematical model of flight of the aerial vehicles with empennage, which was proposed by the authors of the article, allows to take into the consideration the influence of geophysical conditions on their flight.

References

1. Kovalenko, V.V., Shevkunov, V.I.: Meteorological preparation of artillery launch. Leningrad (1975)

2. Volobuyeva, V.I.: The preparation of firing and artillery fire control. Military Publishing, Moscow (1987)
3. Krivtsov, V.S., Voronko, V.V., Zaytsev, V.Y.: Advanced prospect for the development of aircraft assembly technology. Sci. Innov. **11**(3), 11–18 (2015). https://doi.org/10.15407/scine11.03.011
4. Thibbotuwawa, A., Bocewicz, G., Zbigniew, B., Nielsen, P.: A solution approach for UAV fleet mission planning in changing weather conditions. Appl. Sci. **9**(19), 25 (2019). https://doi.org/10.3390/app9193972
5. Mattei, A.L., Orbital, E.S., Toledo, C.F., da Silva Arantes, J., Trindade, O., Jr.: Unmanned aerial vehicles flight safety improvement using in-flight awareness. Intell. Inf. Manag. **13**, 97–123 (2021). https://doi.org/10.4236/iim.2021.132005
6. Radzki, G., Nielsen, I., Golinska, P., Bocewicz, G., Banaszak, Z.: Reactive UAV fleet's mission planning in highly dynamic and unpredictable environments. Sustainability **13**(9), 5228 (2021). https://doi.org/10.3390/su13095228
7. Romanova, N., Crosby, N., Pilipenko, V.: Relationship of worldwide rocket launch crashes with geophysical parameters. Int. J. Geophys. **2013**, 1–15 (2013). https://doi.org/10.1155/2013/297310
8. Houston, J.: Liftoff and time equivalent duration data evaluation of exploration flight test 1 Orion multi-purpose crew vehicle. J. Acoust. Soc. Am. **140**(4), 3096–3096 (2016). https://doi.org/10.1121/1.4969653
9. Trophimenko, P.E., Makeev, V.I., Raskochniy, A.F.: Accuracy assessment account meteorological factors firing for long range. Syst. Arms Mil. Equipment **3**(23), 85–89 (2010)
10. Caruzzo, A., Belderrain, M.C.N., Fisch, G., Manso, D.F.: The mapping of aerospace meteorology in the brazilian space program: challenges and opportunities for rocket launch. J. Aerosp. Technol. Manag. **7**(1), 7–18 (2015). https://doi.org/10.5028/jatm.v7i1.461
11. Caruzzo, A., Belderrain, M.C.N., Fisch, G., Young, G.S., Hanlon, C.J., Verlinde, J.: Modelling weather risk preferences with multi-criteria decision analysis for an aerospace vehicle launch: modelling weather risk preferences with MCDA. Meteorol. Appl. **25**(3), 456–465 (2018). https://doi.org/10.1002/met.1713
12. Ceotto, G.H., Schmitt, R.N., Alves, G.F., Pezente, L.A., Carmo, B.S.: RocketPy: six degree-of-freedom rocket trajectory simulator. J. Aerosp. Eng. **34**(6), 04021093 (2021). https://doi.org/10.1061/(ASCE)AS.1943-5525.0001331
13. Makeev, V.I., Lyapa, M.M., Nazarenko, L.D.: The mathematical model of simple motion of solid fuel aircrafts in the atmosphere. Bull. Sumy State Univ. **2**, 5–15 (2008)
14. Makeiev, V.I., Petrenko, V.M., Zhytnyk, V.Ye., Hridin, V.I.: Determination of launching installations based on the solution of the system of differential equations of motion of aircraft. Syst. Arms Mil. Equipment **34**, 34–38 (2013)
15. Dmitrievsky, A.A., Lysenko, L.N.: External Ballistics, 4th edn. Mechanical Engineering, Moscow (2005)
16. Lysenko, L.N., Lipanov, A.M. (eds.): Barrel Systems Ballistics. Mechanical Engineering, Moscow (2006)

Development of an Algorithm for Preparing to Start an Resistojet Propulsion System

Andrii Pohudin(✉), Sergii Gubin, Olha Pohudina, and Svitlana Sinchenko

National Aerospace University Kharkiv Aviation Institute, Chkalova Street, 17, Kharkiv 61070, Ukraine

Abstract. The object for the creation of algorithms is the process of preparation for the start of the resistojet propulsion system (RJPS) in terms of duration and low energy consumption. The aim is to test the algorithm for reducing energy consumption and preparation time for the start of RJPS. The methods used are mathematical and simulation modeling of power supply processes of an resistojet propulsion system. The obtained results: simulation models of the primary converter of solar energy into electricity taking into account the change of light and temperature, chemical energy storage taking into account the operating modes and the dynamic model of changing the load of RJPS. The scientific novelty of the obtained results is as follows: the obtained limit values of current-voltage characteristics of the solar cell unit in terms of lighting and temperature, the charging and discharge characteristics of the chemical battery are obtained, the algorithm of preparation for RJPS start is obtained.

Keywords: Spacecraft · Electric heating propulsion system · Photovoltaic battery · Chemical battery · Start-up algorithm

1 Introduction

At present, small spacecraft (SSC) groups have become widespread [1]. You can use a space tug (ST) to create an SSC group. Electric rocket engines are installed on the ST to perform orbital maneuvers. The simplest in design, and the cheapest in the price of traction is an electric heating propulsion system (RJPS). But it has some peculiarities of use. Thus, for the operation of RJPS it is necessary to make a preliminary heating of the working fluid and units of RJPS. This time is very long, so the urgent scientific task is to reduce the time of preparation of RJPS for launch.

The power supply system of the ST is limited [2], so a pre-limited amount of energy is supplied which should be enough for the operation of the RJPS.

In general, the operating conditions of the RJPS are to limit the initial temperature of the structure of the RJPS and the propellant, the heating temperature range of the RJPS in the pauses between switch-on, the voltage range of the power supply. The system under consideration uses a propellant – gaseous ammonia [3] (thrust is created by the products of its decomposition – nitrogen and hydrogen). The immediate problem is that the preparation time for such an RJPS before the first start of the thruster is several hours.

M. Nechyporuk et al. (Eds.): ICTM 2021, LNNS 367, pp. 145–156, 2022.
https://doi.org/10.1007/978-3-030-94259-5_14

If we consider the start of one thruster with a duration of one start – up to 20 min and limiting the DC supply voltage from 24 to 34 V and the temperature of normal operation of the RJPS from 273 K to 313 K, it is possible to build and model an algorithm that will limit the energy required to prepare for the launch, and the duration itself.

2 Computational Simulation of Experiment

Since RJPS is installed on the ST, the initial start and stop points of the propulsion system are not initially determined (this is due to the location of the ST in the initial orbit, where it has not yet been detected when the SSC withdrawal maneuver begins). This imposes a number of power and current voltage restrictions on the central bus of the RJPS power supply system, (PSS). In this case, the voltage and the amount of electricity supplied by the source (solar array, SA) and storage (chemical battery, CU) can be expressed within the limits of discharge/charge characteristics of CU. Which in turn are determined by the magnitude of the discharge/charge current, the temperature of the electrochemical system and the current capacity of the CU. The output voltage of the SA and, accordingly, its output current is limited by the level of illumination, the value of temperature and the level of degradation of the elements. It can be concluded that it is difficult for ST to determine the required dynamic characteristics of CU and SA, so use ranges of random values in the range of limit values of CU and SA voltage. $24B < V_{PSS} < 34B$, and $10,5A < I_{PSS} < 16,3A$ current strength, respectively.

For the power supply of RJPS we give the energy characteristics of the SA taking into account the solar energy, chemical energy storage and program the cyclogram of energy consumption of RJPS for the test problem of orbital maneuvering. The main elements will be described by substitution schemes. SA is conveniently described in the form of the Shockley model for a photovoltaic converter [4], which is shown in Fig. 1.

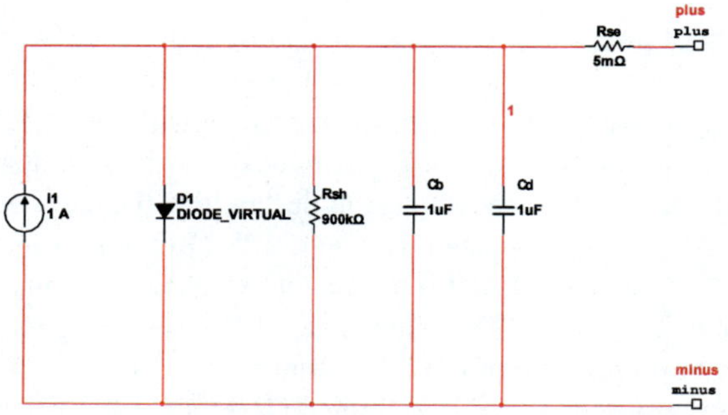

Fig. 1. SA substitution scheme in the form of the Shockley model, where I1 – DC generator, D1 – diode operating in direct bias mode, characterizes the diode properties of the p-n junction, Rsh – shunt resistance, Rse – series resistance, Cb, Cd – barrier and diffusion capacities, respectively, they are required when high switching speed, do not work in static mode, plus, minus – power outputs from the SA.

The shape and size of the current-voltage characteristic depend on the parameter that characterizes the quality of the transition, the values of the shunt and series resistances and the total radiation energy converted into electrical energy regardless of the spectral distribution of radiation intensity.

When modeling the SA substitution scheme in the form of the Shockley model, the limiting conditions of the current-voltage characteristics of the SA operation are obtained. Figure 2 shows the characteristics of lighting, and Fig. 3 presents a characteristic of temperature.

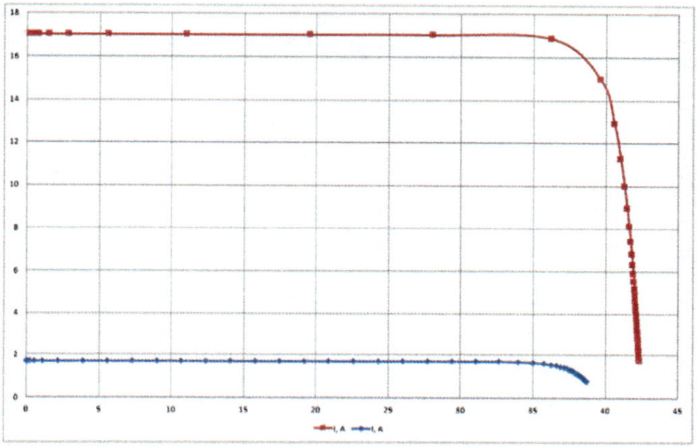

Fig. 2. Limit values of C-V characteristics of SA on illumination

The upper edge of the range corresponds to the power supplied at an average value of out-of-atmosphere lighting of 1360 W/m^2. The lower edge of the range corresponds to the minimum illumination, which for the task is 10% of the maximum value, namely 136 W\m^2.

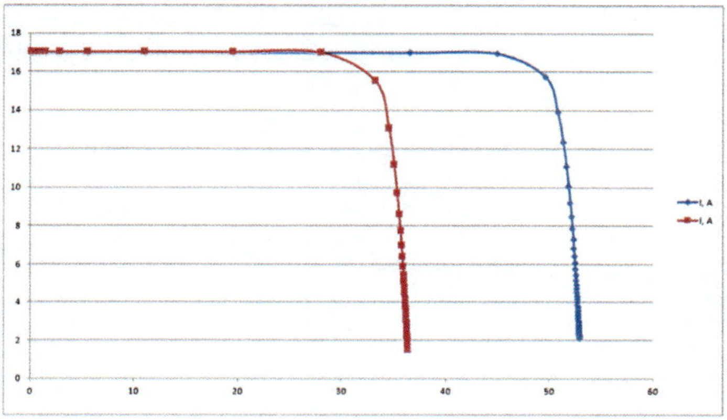

Fig. 3. Limit values of C-V characteristics of SA by temperature

At the maximum heating of SA we have red area of functioning, accordingly at the minimum heating – blue area of functioning.

To provide 34 V, it is necessary to sequentially switch 17 photoelectrical converters (FEC) [5], because each converter produces 2.5 V. To provide 16.5 A current, it is necessary to connect 28 strings in parallel (line FEC, mounted in length). That is, the SA is a parallelogram with sides 17 by 28 FEC. Each FEC has a size of 8 × 4 cm, so we have a FEC size 135 × 112 cm.

The CU substitution scheme (Fig. 4) is based on resistive capacitive cells that emulate activation and concentration polarization. In addition, the circuit includes a varistor (resistor, which resistance depends on the applied voltage) [6] to describe the final concentration polarization, internal resistance and the basic source of electromotive force.

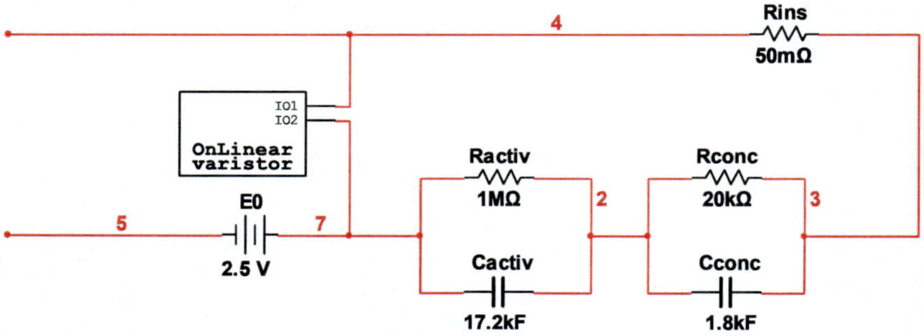

Fig. 4. CU substitution scheme, where OnLinervaristor – varistor, Ractiv – resistor cell activation polarization, Rconc – resistor cell concentration polarization, Rins – internal resistance of CU, Cactiv – capacitor cell activation polarization, Cconc – capacitor of concentration polarization cell, E0 – element of constant voltage 5 V.

When modeling the CU substitution scheme, charging and discharge characteristics are obtained. Figure 5 shows the characteristics of the CU charge depending on the current, and in Fig. 6 presents the discharge characteristic of CU depending on the discharge current.

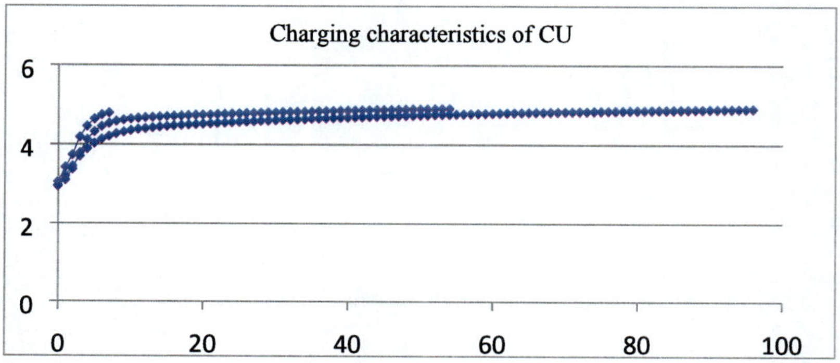

Fig. 5. Charging characteristics of CU

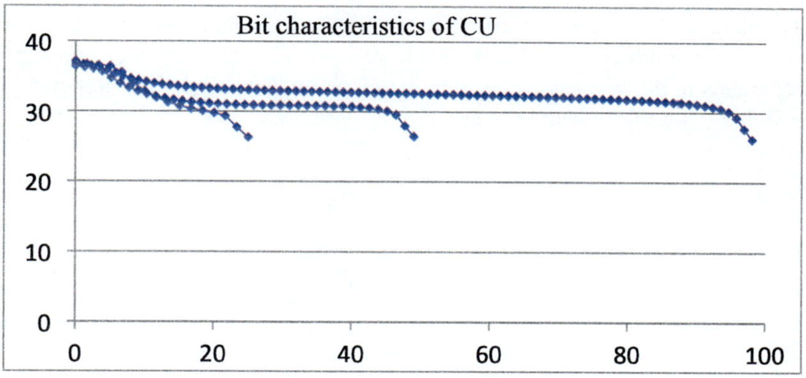

Fig. 6. Bit characteristics of CU

According to the flight dynamics, a CU capacity of 7.5 A-h is sufficient for RJPS power supply in the shadow area. To simplify this, choose a standard CU at 8 A × h. In this case, the SA is formed in such a way that at the point of maximum power provides a voltage greater than the limit value.

According to the developed mathematical model of the process of heating the elements of RJPS, and taking into account the capabilities of the power supply system above, it is proposed to create an algorithm for preparation for the launch of RJPS. Let's start with the list of RJPS commands, which is presented in Table 1:

Table 1. List of RJPS commands

Marking	Functional purpose of the command
RJPS CU1on	Power on CU RJPS
RJPS CUoff	Power off CU RJPS
Interface commands	
InstArrPrev	Use an array of previous values
OffRP	Disable all running procedures
OnPPAT	Activation of the preparation procedure and operation of the acceleration thruster
OnPPDT	Switch-on of the procedure of preparation and operation of the deceleration engine

Next, procedures should be developed to prepare the elements of the RJPS for start. The algorithm of preparation for the first start of RJPS will have the form presented in Fig. 7. This algorithm is represented by UML (Unified Modeling Language) [7].

The end time of the algorithm is due to the completion of the heating of the balloon, steam generator (SG) and thruster.

According to the created algorithm we will build the cyclogram of work of power consumption of RJPS realized by algorithm of start. The cyclogram of preheating of RJPS elements in the lossless mode from the moment of receipt of the command from CU on switching-on is presented in Fig. 8.

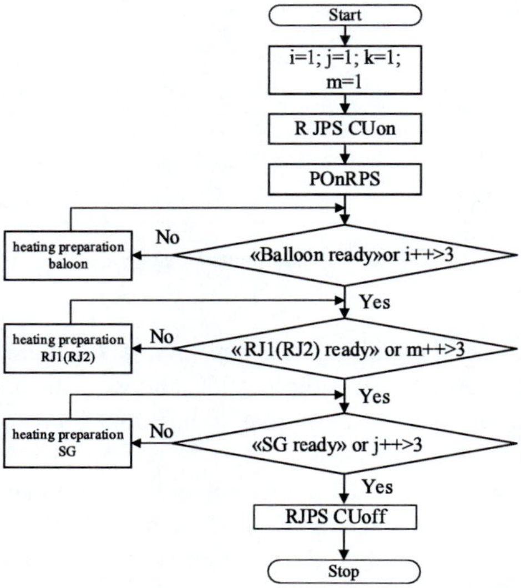

Fig. 7. Algorithm of preparation for the first start of RJPS

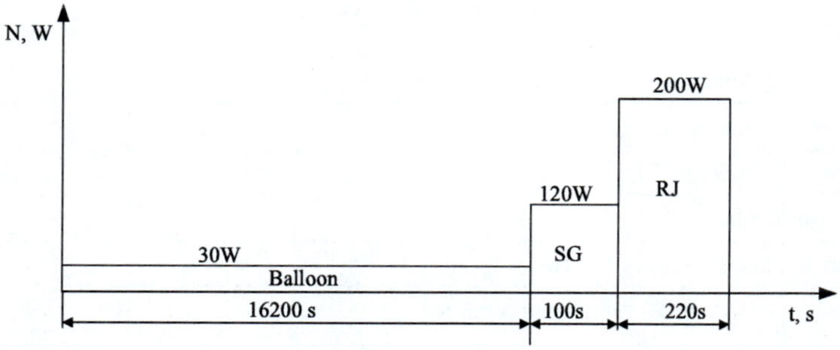

Fig. 8. Cyclogram of heating of RJPS elements and pressure rise in the receiver.

Thus, the time to prepare RJPS for start is 16520 s.

In flow mode, the balloon heater consumes 30 W, while the thruster heater constantly consumes 200 W, and the SG heater 120 W, 15 W are used for CU and solenoid valves. The peak load on the space tug EPS from the RJPS does not exceed 350 W.

To reduce the start-up time of the RJPS, someone can simultaneously perform processes occurring in the units of the RJPS. This makes it clear that it does not make sense to open the solenoid valves at the same time. Therefore, to reduce the start-up time, it is necessary to heat simultaneously the balloon, thruster and SG, which will make the return of EPS ST increased requirements [8]. Thus, we have the following options:

- the balloon and the SG are heated simultaneously, then the thruster is heated;
- the balloon and the thruster are heated simultaneously, then the SG is heated;
- balloon, SG, thruster are heated simultaneously.

The most efficient option is option 3, but it in turn will be the most energy consuming.

Next, it should be noted that the heating of the SG and the thruster in the heating mode must begin with the heating of the balloon, so that by the end of the balloon heating the SG and the thruster were also in the operating temperature range.

Based on the above proposed cyclogram will take the form shown in Fig. 9.

Another way to reduce the preparation time of the RJPS for the first start is to increase the capacity of the balloon heater. Thus, increasing the power supplied to the balloon heater. Without exceeding the consumption of the provided 350 W PSS, it is necessary to supply 130 W to the balloon heater, while heating the SG and then the thruster. Thus, the cyclogram will take the form shown in Fig. 10.

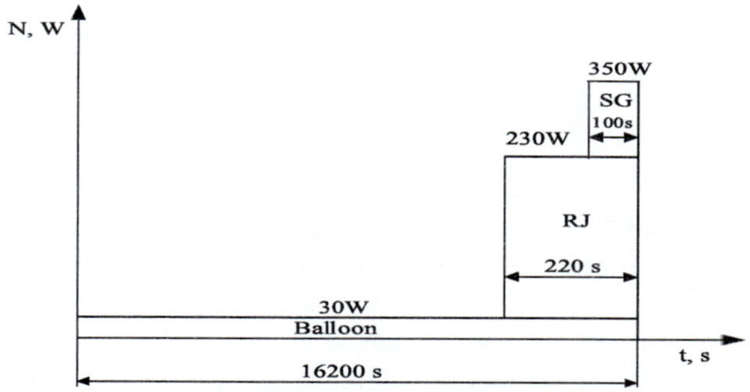

Fig. 9. Cyclogram of RJPS start-up with simultaneous heating of elements.

But there is another way to ensure a minimum time to start the RJPS. To do this, all the power of 350 W emitted by PSS can be applied to each of the elements of the RJPS, which are pre-heated separately. The most difficult element of heating is the balloon, so first it is heated. After that, heat the next mass element, namely the thruster. Finally, heat the SG. We will supply 350 W for each element in series. We obtain graphs of heating of the balloon (Fig. 11), SG (Fig. 12), thruster (Fig. 13).

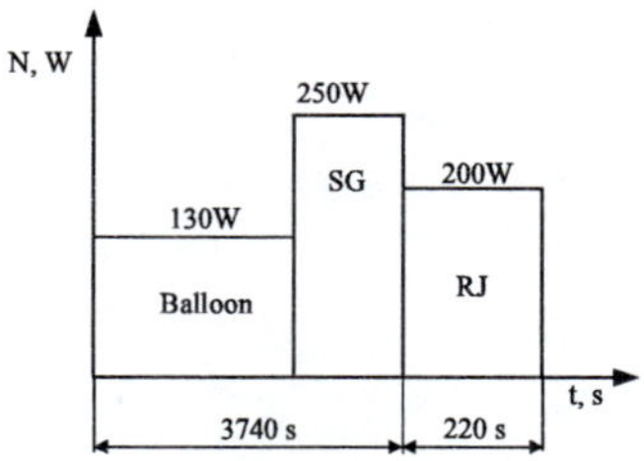

Fig. 10. Cyclogram of RJPS startup with maximum power balloon heater

The balloon will be heated to a set temperature of 308 K. Figure 11 shows the heating of the balloon with a heater at a power of 30 W – black line, 130 W – red line, and the blue line – 350 W.

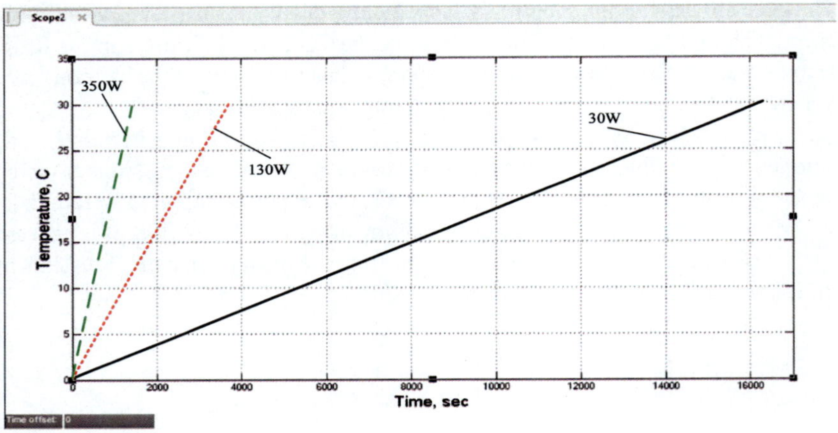

Fig. 11. Balloon heating

When the power supplied to the balloon heater increases, the time of the set temperature 308 K decreases. When all the energy provided by the PSS supplied to the balloon heater, the heating time is 1388 s.

The SG will be heated to the set temperature of 333 K. Figure 12 shows the heating of the SG by a heater at a power of 120 W – black line and heating 350 W – red line.

When the power supplied to the SG increases, the time of the set temperature 333 K decreases. When all the energy provided by the PSS is supplied to the SG, the heating time is 34 s.

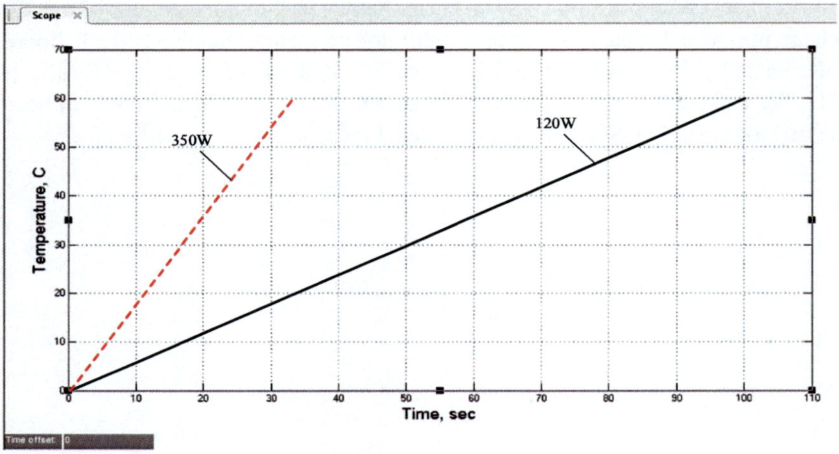

Fig. 12. SG heating.

The thruster will be heated to a set temperature of 1373 K. Figure 13 shows the heating of the thruster by the heater at a power of 200 W – black line, 150 W – green line and heating 350 W – red line.

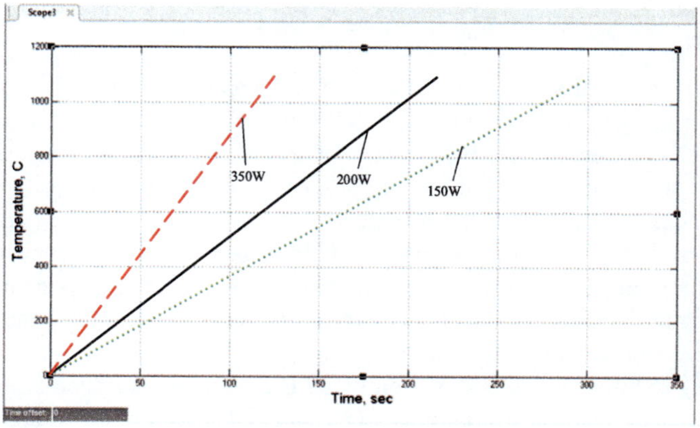

Fig. 13. Thruster heating.

When the power supplied to the thruster increases, the time of the set temperature 1373 K decreases. When all the energy provided by the PSS is supplied to the thruster, the warm-up time is 125 s.

According to the received schedules we will build the cyclogram with the minimum time of preparation of RJPS (Fig. 14). After all elements of RJPS are ready for start we pass to an mass flow rate operation mode.

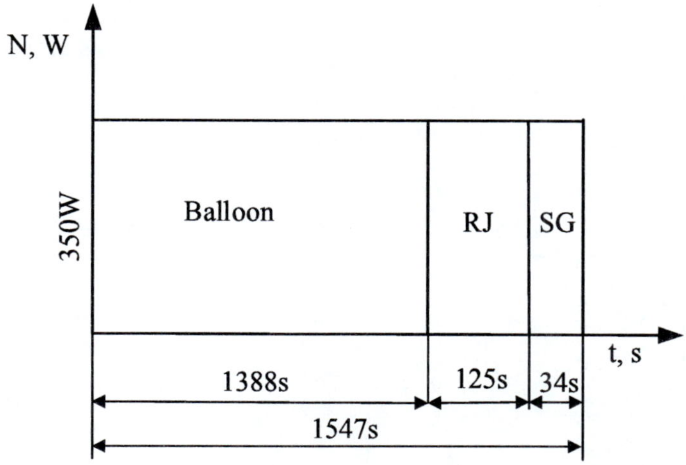

Fig. 14. Cyclogram of heating of each element of RJPS separately

After simulation, it was found that the time to prepare the RJPS for start-up was reduced from 16520 s to 1547 s, due to the fact that the total power provided by the PSS is supplied in the selected sequence by weight from heaviest to lightest.

In the flow mode, the power of the PSS is distributed as follows:

$$W_{PSS} = W_B + W_{SG} + W_{RJ} = 30 + 120 + 200 = 300W$$

The disadvantage of such heating may be that the power of the PSS is not enough to perform the heating of the elements of the RJPS with maximum power. This requires a preliminary calculation of the orbit, which will provide the PSS charging with SA.

Thus, the algorithm of the first start-up of RJPS with redistribution of power supply among RJPS elements is received. The algorithm takes into account the partial or complete failure of one or all heating elements. If all heaters fail, the RJPS switches to the gas jet thruster mode, i.e. the propellant is no longer divided into components, but flows down through the thruster nozzle. This increases the cost of the propellant, but the ST can continue to perform tasks in space.

According to the results of simulation in the ElectronicWorkBench environment of RJPS according to the task, cyclograms were obtained showing the compliance of the algorithm according to which the preparation for start can be realized in one orbit of the spacecraft from the available power of the power system. The simulation was performed according to the scheme shown in Fig. 15. In the process of modeling the algorithms of Figs. 9 and 11 were considered. The initial condition was that the chemical battery on the first orbit has zero capacity. Under such conditions, according to the first cycle, normal operation is possible only at the fourth orbit, and according to the second cycle, normal operation is possible from the second one, and at the first orbit the thruster consumption does not go beyond the normal range.

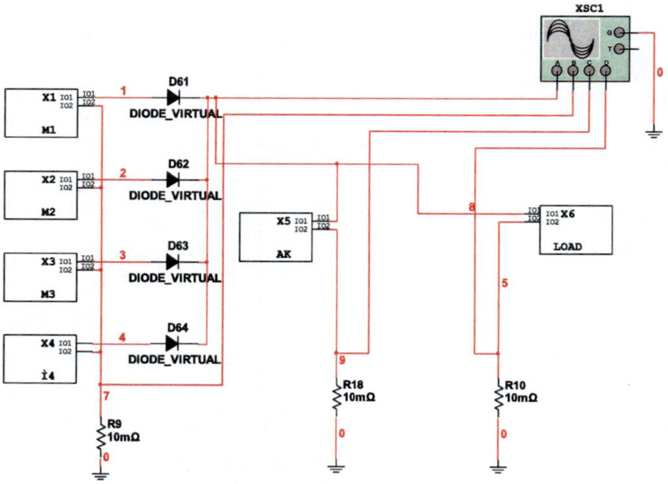

Fig. 15. Scheme for modeling RJPS, where X1–X4 units of the photovoltaic battery; X5 – chemical battery; X6 – EJPS; XSC1 four-beam oscilloscope.

According to the cyclogram of Fig. 9 the simulation results obtained shown in Fig. 16.

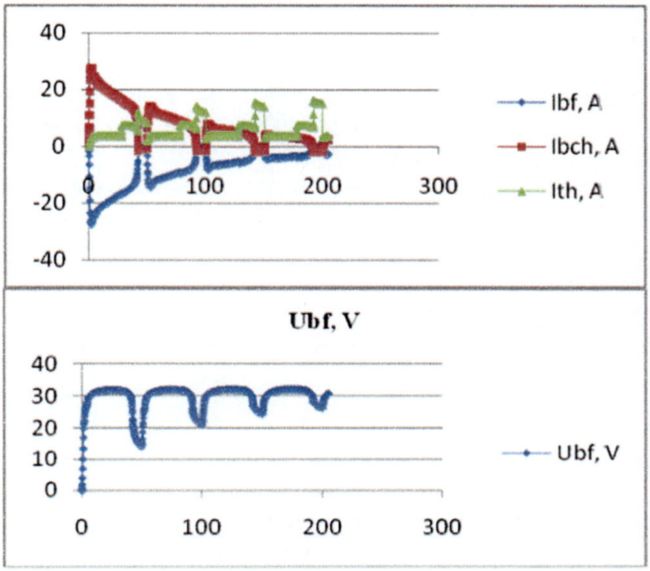

Fig. 16. The results of modeling the work of RJPS with step energy consumption

According to the cyclogram of Fig. 11 obtained the simulation results shown in Fig. 17.

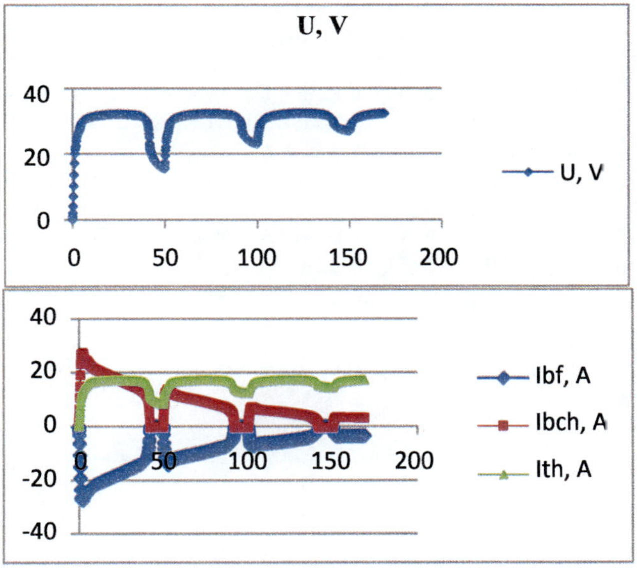

Fig. 17. The results of modeling the work of RJPS with constant energy consumption

3 Conclusion

Having conducted a study of the dynamic characteristics of RJPS, we can give the following recommendations for its use:

- in order to reduce the time of preparation of the RJPS for the first start, it is necessary to increase the requirements for EPS in the ST, namely it is necessary to increase the power provided to the RJPS over 350 W;
- in order to reduce the duration of preparation of the RJPS for the first start, it is further necessary to increase the capacity of the balloon, SG and thruster heaters;
- in order to increase the time of operation of the ST in orbit, it is necessary to increase the store of propellant;
- for the mode of complete or partial failure of the elements of the RJPS to develop and test separate algorithms of the RJPS operation in order to continue the task.

References

1. Ganswind, I.N.: Small spacecraft in earth remote sensing. Explor. Earth Space **5**, 82–88 (2019)
2. Blinov, V.N., et al.: Experimental studies of arcjet thruster in pulsed mode of operation for corrective propulsion systems of small spacecraft. J. Phys.: Conf. Ser. **1441**(1) (2020)
3. Zółtowski, A., Gis, W.: Ammonia emissions in SI engines fueled with LPG. Energies **14**, 691 (2021). https://doi.org/10.3390/en14030691
4. Komilov, M.: Study of the effect of optical layer thickness on the photoelectric parameters of silicon-based solar cells using Sentaurus TCAD. Sci. Eur. **72–1**, 52–54 (2021)
5. Dhinesh, V., Vijayakumar, G., Saravanan, S.: A photovoltaic modeling module with different converters for grid operations. Int. J. Innov. Res. Technol. **6**(8), 89–95 (2020)
6. Martzloff, F.D., Levinson, L.M.: Surge-protective devices. In: Electronic Ceramics, pp. 275–305. CRC Press (2020)
7. Oberhauser, R.: VR-UML: the unified modeling language in virtual reality – an immersive modeling experience. In: Shishkov, B. (ed.) BMSD 2021. LNBIP, vol. 422, pp. 40–58. Springer, Cham (2021). https://doi.org/10.1007/978-3-030-79976-2_3
8. Pohudin, A.V., Gubin, S.V., Sinchenko, S.V., Shulga, I.M., Pohudina, O.K.: A mathematical model of electric Ohmic propulsion system dynamics. In: 2020 IEEE 15th International Conference on Computer Sciences and Information Technologies (CSIT), vol. 1, pp. 1–4 (2020). https://doi.org/10.1109/CSIT49958.2020.9322028

The Process of Forming Integral Ribbed Panels by Sequential Local Phased Deformation with Information Support

Valeriy Sikulskyi[1], Stanislav Sikulskyi[2], Kateryna Maiorova[1](✉), Valeriya Suponina[1], and Oleh Komisarov[3]

[1] National Aerospace University "KhAI", Chkalova Street, 17, Kharkiv, Ukraine
kate.majorova@ukr.net
[2] Embry-Riddle Aeronautical University, Daytona Beach, FL, USA
[3] Motor Sich JSC, Motorostroiteley Avenue, Zaporozhye, Ukraine

Abstract. The study considers a shaping method for the stiffened monolithic panels by local deformation of the panel ribs. The local deformation is caused by bending moments along a rib towards each other. The purpose of the work is to investigate form-shape and finish ribbed panels by multi-point local deformation using devices for controlling the shape and processing of data in the information technology environment for increase accuracy. The data from the 3D-scanner enters the computer, where it is compared with the data describing the desired shape of the panel, and then in accordance with the technological recommendations are assigned additional corrective actions aimed to obtaining high accuracy of the panel shape. Simulation of deformation, stress, and strain distributions along the panel at loaded and unloaded states was obtained. The modeling allows to foresee the panel shape deviations after the shaping process and to correct the impact location, which is hard to accomplish in case of a conventional stamp forming. During the implementation of the technological process by specified way the following results were achieved: the time of formation of double curvature panels 5.5 m long and 1.1 m wide with shape control using a 3D-scanner – 5 – 6 h; time to prove the shape of the panel – 1 – 2 h. During the formation, deviations from the specified surface of the form are achieved – 0.7…0.9 mm, and after proving the form – 0.3… 0.5 mm.

Keywords: Monolithic panels · Finishing · Bending moment · Modeling · Deflection · Residual deflection · Finite element method

1 Introduction

The group of processes in which the assignment of initial technological modes cannot be made unambiguously to obtain a given result is distinguished among different technological processes in mechanical engineering. First of all, these processes include forming of parts from a sheet by multi-point deformation, for instance, punching, bending of parts in advance increment, bending of profiles with an asymmetric section profile, forming

M. Nechyporuk et al. (Eds.): ICTM 2021, LNNS 367, pp. 157–165, 2022.
https://doi.org/10.1007/978-3-030-94259-5_15

of shells with anisotropy, for example, ribbed shells. In order to achieve an acceptable result, these processes shall be performed in accordance with the sequence algorithm for conducting individual technological operations. The accuracy of products obtained is of great importance, which is difficult to predict for these processes. The principal method to achieve the accuracy of the product form is sequential forming with periodic control of surface form, which arises the task to compare a given and obtained form and develop solutions for carrying out subsequent technological operations. An example of organization of this process is forming integral panels, especially in regards to the tendency to increase the overall dimensions of panels, as well as the need to ensure high accuracy of forming large-sized integral panels. Hence, the requirements for both the accuracy of the product and direct control of panel form at all stages of its manufacture increase.

2 Analysis of Basic Research and Publications

The methods of forming and control used in production have low productivity and require the usage of sophisticated equipment. The surface form control after forming is also imperfect [1]. Currently control and measuring machines are used in the forming technology for form measurements, however, in this case obtaining a quick measurement result during the forming process is impossible. In the applied technological methods, the challenging issue is the modeling of joint deformation of the panel surface and edges at the stages of the technological process that has several transitions [2]. In the manufacture of panels and cladding, technological processes of multipoint local deformation [3, 4], widely used at the stages of preliminary (rough) forming and final panel refinement, are widely used, which can significantly improve the quality of obtained profile-shaping surface of the integral panel and increase the manufacture accuracy thereof. These processes include local deformation shown in Fig. 1.

(a) (b) (c)

Fig. 1. Appearance of devices for sequential deformation of ribbed panels by stretching with panel section bending (a), panel rib rolling (b) and local rib bending (c).

These processes are implemented according to sequential deformation scheme for ribbed panels when individual sections of the panel are deformed by local deformation and the final panel form is achieved due to a defined distribution of local effects. Because of elasticity property of material, the spring back affects to deformed state and it is very important to control the form at the deformation stages and adjust the distribution of subsequent effects. The authors used the following methods of local action: stretching with panel section bending, panel rib rolling, local rib bending.

Local point-forming enabled significant reduction of the deformation area and the bending angle by using a scheme with two torques applied to two sections of the rib and directed towards each other. This allows significant reduction of the deformation area, and therefore, of bending angle under load and spring angle [5].

Productivity increase can be achieved by using devices that allow panel form control directly during forming after the end of the successive intermediate stage [6].

At the same time, the use of systems to control the form during forming and finishing integral panels in blanking and stamping shops remains quite limited due to the lack of methods and devices for implementing the process.

3 Problem Statement

Currently, industrial enterprises need technological means and equipment for form-shaping, finishing, and form control of integral panels at all stages of their production. Technological modes of forming integral panels are also not fully understood and therefore new methods are used in industry with a great caution.

One of the problematic issues is the implementation of the panel of double curvature from a flat blank with ribs. As a rule, aircraft manufacturing enterprises make this blank from a slab or profile on 3-coordinate CNC machines. To obtain this panel the local deformation is carried out sequentially, and local point effects are superimposed in the form of successive stages, covering the entire surface of the panel [5, 7] (Fig. 2).

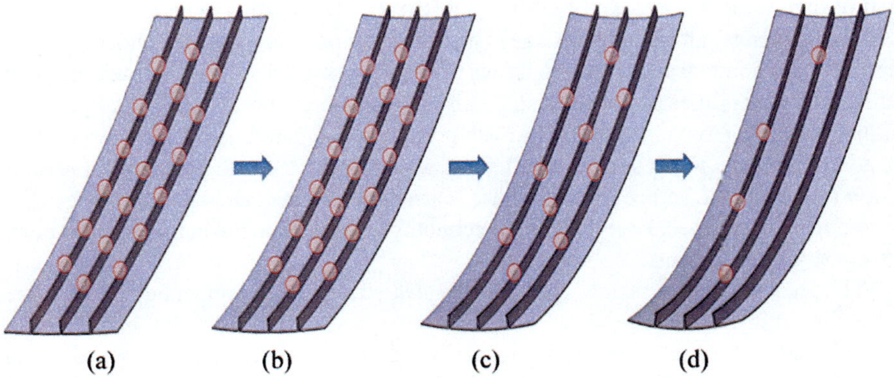

Fig. 2. Scheme of a sequential stepwise method for obtaining a panel of double curvature by multipoint local deformation.

In this case local point effects were applied sequentially at pre-calculated points using a special original device shown in Fig. 1,a. This device is installed in a cantilever press and provides local stretching with panel section bending. However, due to a large number of factors, it is impossible to obtain an accurate form due to preliminary calculation of parameters.

To obtain the exact form of the product in this example, it is necessary to finish the form in several technological transitions using intermediate measurements of the panel form.

The capabilities of computer technology lead to creation of modern industrial measurement systems that allow the user to receive, process and analyze the information necessary to control the quality of measured parts of complex form for compliance with obtained surfaces of their aerodynamic standards.

4 Purpose and Objectives of Research

The purpose of the paper is to increase product form accuracy by form-shaping and finishing ribbed panels by multi-point local deformation using devices for form control and data processing in the information technology environment.

To achieve this, the following tasks were solved:

- form-shaping by local forces during the formation of double curvature of the panel;
- construction of the portraits of panel geometry in real time at individual technological transitions;
- implementation of procedures for comparing the analytical standard and the form of ribbed panels.

5 Experimental Procedures

This paper provides the sequence of digital panel standards after each stage of deformation of digital mock-up (DMU): DMUa, DMUb, DMUc, DMUd, where DMUd is the DMU of the panel after the final stage. The formation of each of these standards is carried out taking into account the properties of the panel material, its design features and equipment capabilities. After each stage of deformation is completed, digital portraits of the panels Pa, Pb, Pc, Pd are built. Each portrait is compared with the corresponding standard. If the deviation at all controlled points does not exceed the tolerance, the transition to the next stage occurs. Otherwise, corrective actions calculated by the control program are performed. Laser scanning technology was used in the considered problem for constructing portraits.

The process diagram is presented in Fig. 3, which depicts an application of real-time form control [5].

Volumetric scanners were used to control the form, which made it possible to create their three-dimensional analogues with real volumetric models in the format of stereo lithography (point diffusion in STL format).

The experiments included the use of scanners that allow digitizing sufficiently large geometric objects with an area of 50–100 m^2 in a fairly short time (6–10 h) and with high accuracy (5… 10 μm) [8]. The portraits were obtained using Artec EVA 3D scanner and processed in the Artec-Studio software environment.

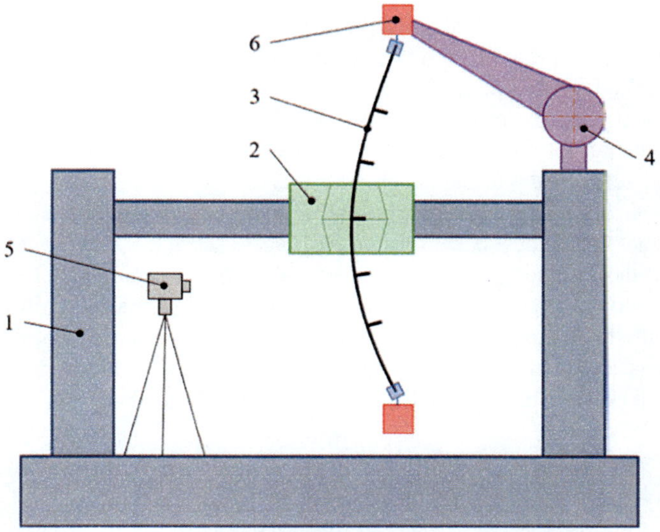

Fig. 3. Schematic diagram of the automated forming of integral panels with form control using a 3D scanner: cantilever press (1), breeding device with a bend (2), deforming panel (3), robotic arm with five degrees of freedom (4), 3D scanner (5), supporting device (6).

The rather quick calculation of deviations from a given form for all surface points was carried out using special software. Surface point deviation was visualized using deviation charts or graphical displays. The scheme for form-shaping and finishing the panel using a 3D scanner in the CAD/CAM/CAE/ system can be displayed in the form of block diagram shown in Fig. 4.

The experiments carried out provided for the measurement and quality control of large-sized products of various configurations in real time, including parts of blank-stamping production.

The 3D scanning results were processed in the following stages:

- creation of data on the object surface for a computer model using CAD;
- comparison of obtained data with CAD, drawings or standard reference;
- quick decision on changing a part form;
- additional technological impact performance.

Measurement data can be output in several standard formats, such as IGES, and VDA, ASCII, STL point set. Standard software also supports the creation of cross-sectional data for other types of laser systems compared to mechanical systems that are based on the principle of non-contact measurements.

Data from a 3D scanner in portraits form is sent to a computer, where it is compared with data describing the required panel form, and then, in accordance with technological recommendations, additional corrective actions are assigned to obtain high accuracy of the panel form [9, 10].

The principal element of this process is a device for editing and finishing panels, which is installed in the arm of a robot manipulator.

It was applied a position the device at the desired point on the panel normal to the tangent plane at the point of adding a specific action on the front surface of the panel. In this case, the purpose of the supporting device is to exclude the action of mass forces on the elastic deformation of the integrated panel.

Figure 5 shows a graphical comparison of the portrait of the outer side (Pa, Pb, Pc, Pd) of the panel with a ideal panel form in the process (DMUa, DMUb, DMUc, DMUd) of successive deformation at one of the stages of double curvature panel forming.

Special markers were applied to the panel due to its geometric uniformity and in order to assure the scanning accuracy. The result of scanning and processing is an STL file, which was used to measure the control sections and compare them with the reference standard (mathematical model DMUa, DMUb, DMUc, DMUd).

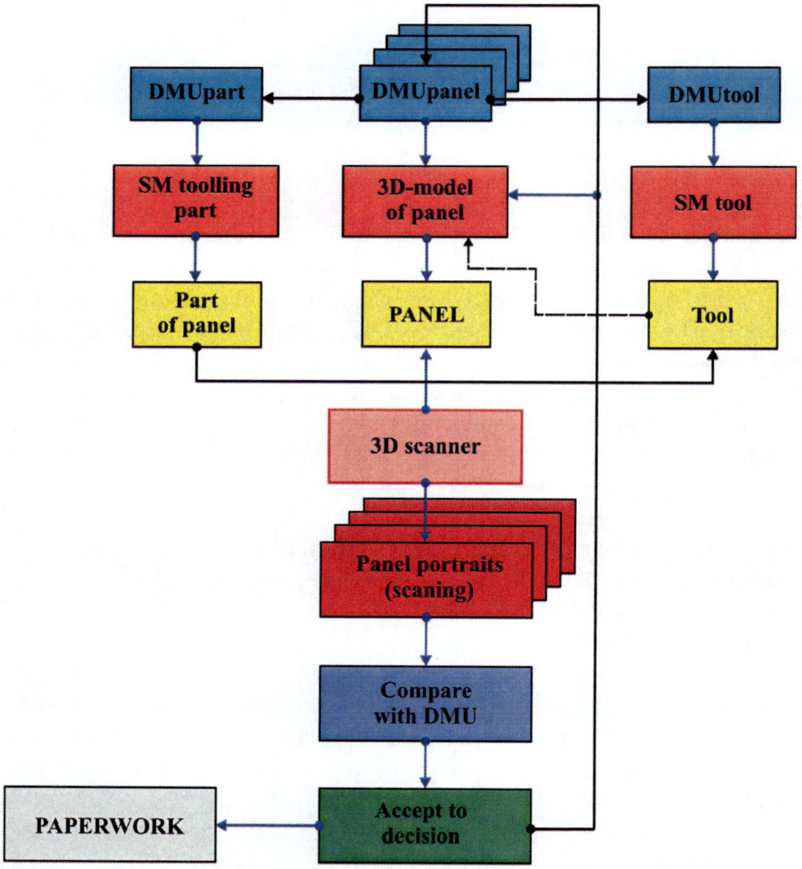

Fig. 4. Scheme for form-shaping and finishing the panel using a 3D scanner in a CAD/CAM/CAE/ system.

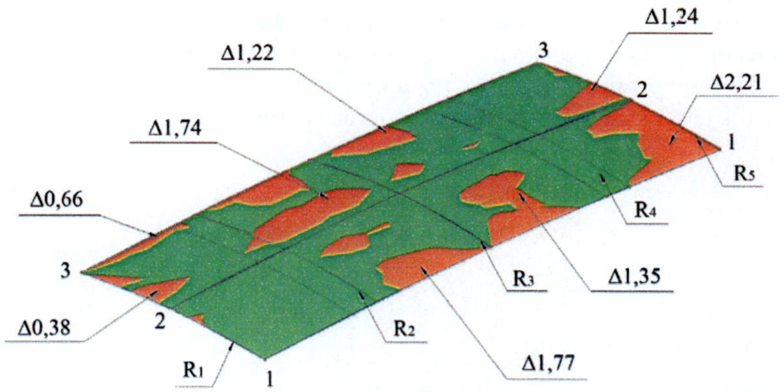

Fig. 5. Graphical comparison of a panel portrait with a given form in the process of successive deformation.

Areas that differ from the reference panel (mathematical model DMUa, DMUb, DMUc, DMUd) within the tolerance are highlighted in green, whereas areas of the panel that have large deviations and require revision are highlighted in orange. It can be seen that the greatest deviations are observed at the panel edges and the surface torsion is observed in the area with the greatest deviation (Δ2.21).

The calculated values of deviations were obtained by comparing the given radii of panel curvature and those obtained by scanning (Table 1). For the calculation method, deviations along the radii of curvature of transverse ($R_1 \ldots R_5$) and longitudinal ($R_{1\text{-}1} \ldots R_{3\text{-}3}$) panel sections were measured.

Table 1. Calculated deviations along the radii of curvature.

Radius of curvative	Deviation Δ, mm
R_1	0.38
R_2	1.77
R_3	1.35
R_4	2.21
R_5	1.24
$R_{1\text{-}1}$	1.74
$R_{2\text{-}2}$	1.22
$R_{3\text{-}3}$	0.66

In Table 1, the number in the index indicates the number of longitudinal or transverse section.

These calculated deviations are approximated in percentage values of deviations along the radii of the control sections of the panel under study. In contrast to the calculated graphical method, it provides a visual representation of deviation distribution over the surface. The deviation distribution is caused by the appearance of the total measurement error, random component whereof is the noise that occurs when obtaining portraits. In regards of scanning, this panel has a complex geometry due to its homogeneity, and the greatest difficulties arise when comparing portraits during scanning.

Fig. 6. A sample of integral panel of double curvature, which is formed with shape control using portraits and analytical standards.

Experimental verification of the stated provisions was carried out on samples of full-size cross-section panels and large panels (Fig. 6). With the introduction of the technological process, the following results were achieved: the forming time of double curvature panels with form control using a 3D scanner amounts to 5–6 h; panel finishing time amounts to 1–2 h. During forming, deviations of 0.7…0.9 mm from a given surface form, and 0.3…0.5 mm after finishing the form are achieved.

6 Conclusion

Increasing the accuracy of the outer surface of shallow large-sized integral panels, regardless of deviations in the cross-sectional dimensions of the ribs and surface after their machining, is possible by implementing information support for the process of measuring the product form in real time. For shallow parts of large dimensions, it is preferable to use 3D scanners, when the measuring part is mounted on a manipulator and moves along the part. In this case, sufficiently large parts can be measured with the appropriate division into sections using special markers.

The use of integral panels as initial data for the sequence of the analytical standard at the stages DMUa, DMUb, DMUc, DMUd enabled the development of a software module for automated forming of integrated panels with control of their forms using portraits of Pa, Pb, Pc, Pd. The method can be used to control geometries in all studies of panel life cycle.

References

1. Belykh, S., et al.: Numerical and experimental study of multi-point forming of thick double-curvature plates from aluminum alloy 7075. In: IV Sino-Russian ASRTU Symposium on Advanced Materials and Materials and Processing Technology, KnE Materials Science, pp. 17–23 (2016)
2. Ramati, S., Levasseur, G., Kennerknecht, S.: Single piece wing skin utilization via advanced peen forming technology. http://www.aerosphere.ca/docs/ICSP-7.pdf. Accessed 13 Feb 2019
3. Tan, J., Zhan, M., Liu, S.: Guideline for forming stiffened panels by using the electromagnetic forces. Metals **6**(11), 2–24 (2016)
4. Liu, C.G., Li, J., Dong, Y.N., Zhang, X.G., Yue, T.: Fracture prediction in the forming of aircraft Al stiffeners using multi-point dies. Int. J. Adv. Manuf. Technol. **90**, 3109–3118 (2017)
5. Sikulskiy, V., Kashcheyeva, V., Romanenkov, Yu., Shapoval, A.: Study of the process of shape-formation of ribbed double-curvature panels by local deforming. East.-Eur. J. Enterp. Technol. **4/1**(88), 43–49 (2017). https://doi.org/10.36074/24.04.2020.v2.27
6. Sikulskiy, V., Sikulskyi, S., Garin, V.: Investigation into the forming process of wing panel oblique bending by means of rib rolling. In: Nechyporuk, M., Pavlikov, V., Kritskiy, D. (eds.) ICTM 2020. LNNS, vol. 188, pp. 598–608. Springer, Cham (2021). https://doi.org/10.1007/978-3-030-66717-7_51
7. Yoon, J.S., Kim, J., Kim, H.H., Kang, B.S.: Feasibility study of flexibly reconfigurable roll forming process for sheet metal and its implementation. Adv. Form. Technol. Lightweight Compon.: Theory Appl. **6**, 45–63 (2014)
8. Luo, H., Li, W., Li, C., Wan, M.: Investigation of creep-age forming of aluminum lithium alloy stiffened panel with complex structures and variable curvature. Int. J. Adv. Manuf. Technol. **91**, 3265–3271 (2017)
9. Zhang, M., Tian, X., Li, W., Shi, X.: An equivalent calculation method for press-braking bending analysis of integral panels. Metals **8**, 2–18 (2018). https://doi.org/10.3390/met8050364
10. Bychkov, I., Maiorova, K., Suponina, V., Riabikov, S.: Reengineering based on 3D-scanning in the process of propeller digital mock-up constructing for an ultra-light twin-seat aircraft. In: Integration of Fundamental and Applied Sciences in the Paradigm of Post-Industrial Society: Collection of Scientific Documents "ΛΌΓΟΣ" with Proceedings of the International Scientific and Practical Conference, vol. 2, pp. 32–38. The European Science Platform, Spain (2020). https://doi.org/10.36074/24.04.2020.v2.09

Determination of the Mass of Gas in a Reservoir at Filling with a Mixture Component Under the Pressure

Olga Shypul[1], Sergiy Plankovskyy[2], Sergiy Zaklinskyy[1], Oleksiy Pavlenko[1](✉), and Vadim Garin[1]

[1] National Aerospace University "Kharkiv Aviation Institute", 17 Chkalova Street, Kharkiv 61070, Ukraine
alexey.pavlenko@khai.edu

[2] O. M. Beketov National University of Urban Economy in Kharkiv, 17 Marshala Bazhanova Street, Kharkiv 61002, Ukraine

Abstract. The subject of the study is a method for ensuring the mass fraction of a gas mixture when filling a reservoir according to the parameters of a gas component using a mixture generator and a numerical control system for dosing. The aim of the work is the experimental verification of the mathematical model of the unsteady gas-dynamic process of filling the reservoir for use in the method for determining the composition of the mixture. Thus, the Computational Fluid Dynamics (CFD) code CFX-based simulation of filling a reservoir with nitrogen is compared to full-scale experiments. The following results were obtained. The full-scale experiments to determine the pressure and temperature of the gas being filled were carried out with high-frequency monitoring. Models of a gas-dynamic unsteady process for various inlet boundary conditions had been built and compared with each other and experimental data. The difference in pressure between filling models with mass flow control and pressure boundary conditions is up to 10%. The discrepancy between the simulation results and full-scale experiments does not exceed 8% in pressure and 1% in temperature at the monitor point. The use of the SAS SST turbulence model and constant temperature on the wall as a heat transfer condition has been substantiated. It was found that the model with pressure boundary condition on inlet closely matches the reservoir pressure trend over time observed in full-scale experiments, with a typical error of less than 3%, which can provide accurate predictions of gas dosing.

Keywords: The mathematical model of a gas-dynamic process · Reservoir filling parameters · Numerical modeling · Dosing accuracy · The components of the gas mixture

1 Introduction

The requirements of the fourth industrial revolution force modern high-tech enterprises to use fully automated production facilities that are part of cyber-physical systems [1, 2]. Therefore, the most pressing issue in the technology of processing parts, along with

M. Nechyporuk et al. (Eds.): ICTM 2021, LNNS 367, pp. 166–177, 2022.
https://doi.org/10.1007/978-3-030-94259-5_16

flexibility, productivity, automation, is the digitalization of processes. It is the construction of digital twin processes that guarantee quick readjustment of the system for specific tasks with automatic determination of the optimal mode to achieve a given processing accuracy [3].

As for the technologies of heat-pulse finishing, the construction of these digital twins requires a comprehensive model of all equipment workflows, including the gas-dynamic processes of filling the working chamber with a gas mixture, which is the purpose of this study. Requirements for dosing accuracy of fuel mixture components for precision heat pulse treatment are given in [4]. It is proved that taking into account the qualimetric parameters of the edges of precision parts, the permissible error range of the component composition of the fuel mixture should not exceed 0.1...0.5%. In addition, in [5] it was shown that to ensure a stable value of the specific heat flux during cyclic heat-pulse treatment, it is necessary to simultaneously control both the charge mass and the pressure in the chamber during the generation of the mixture during filling with subsequent correction of the processing time at a certain initial temperature of the fuel mixture. The dynamic method for the generation of gas mixtures proposed in [6] is based on the effect of supercritical gas leakage from tanks through calibrated holes, which provides high accuracy of given mixture composition and can be used to quickly measure the volume of complex vessels [7].

Existing studies of gas-dynamic filling of vessels are mostly conducted on the example of hydrogen and determine the critical values of the following parameters: gas temperature, thermal stresses in the vessel walls, etc. [8–12]. In addition, in the aforementioned works and works [13–15], the time of filling the vessels with gas was several minutes, in contrast to the process under study. The difference lies, firstly, in its transience (about 2 s), and secondly, in the need for high-frequency monitoring of pressure and temperature in dynamics as control parameters of gas filling.

To test the automated system for generating a fuel mixture in dynamics and further creation of its digital twin, the task of this study is to conduct a series of experimental tests on filling the tank with gas and create adequate mathematical models of the process.

The construction of this mathematical model should be a further development of a simplified model of the gas mixture generator, investigated by the authors of this work [16].

2 Experimental Study of Filling a Reservoir with a Gas Mixture Component Under Pressure

Experimental studies were carried out on the developed gas mixture generator of the automated thermal pulse equipment [17], in particular, the node of filling the intermediate tank with one component of the gas mixture, namely nitrogen, was used. A diagram of the mixture generator experimental setup, which includes: the supply line of the component to the generator, the gas supply line to the tank with an actuator valve is shown in Fig. 1.

A temperature sensor (thermocouple) and a pressure sensor (type TSZ-6002) are mounted inside the tank. The registration of the readings of the temperature and pressure sensors 8 with a frequency of 1 kHz, as well as the control of valve 6, were carried out using the CNC system.

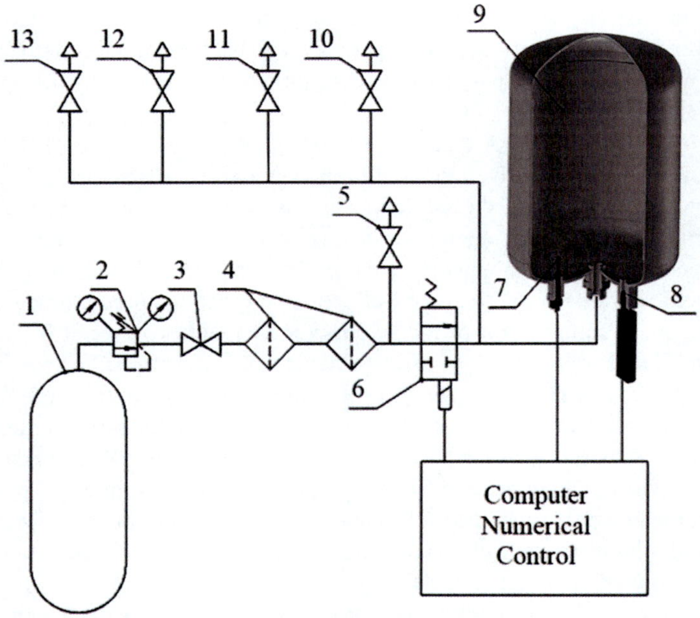

Fig. 1. A diagram of the experimental setup: 1 - high pressure storage tank; 2 - reducer; 3, 5, 10, 11, 12, 13 - valves; 4 - filters; 6 - solenoid valve type Burkert; 7 - temperature sensor (thermocouple); 8 - pressure sensor type TSZ-6002; 9 - reservoir.

All experiments began with ensuring thermal equilibrium between the internal gas temperature and the ambient temperature. To remove residues of other gases from the reservoir cavity and ensure the presence of only the investigated gaseous nitrogen in it, a preliminary five-fold purging of the system was carried out [18].

Field experiments were performed with the initial conditions listed in Table 1, and according to the following algorithm:

1. Starting registration of readings and control signals on the CNC system.
2. Setting the initial filling pressure through the pressure regulator with reducing valve 2.
3. Opening the tap 3.
4. Setting the filling time of the tank - opening the valve 6 through the interface of the CNC system.
5. Filling the tank with gas.
6. Completion of filling the tank with gas at the specified time - closing the valve 6 through the CNC system.
7. Closing the tap 3.
8. Continuation of data recording for 40 s to equalize the temperature of the gas in the tank.

The experimental data were recorded for 40s after completion of the filling to reduce the error in determining the final mass and the average flow rate due to a significant deviation of the current temperature from the average in the reservoir.

Table 1. Initial conditions to the experiments.

Parameter	Experiment's mode 1	Experiment's mode 2
The initial pressure in the reservoir, Pa	476569,32	455444,58
Pressure through a reducer, Pa	111457,5	111457,5
The temperature in the reservoir, K	289,21	288,78
The reservoir volume, m^3	0,009605	0,009605
The initial mass of the component inside the reservoir (nitrogen), g	53,333	51,045
Filling time, sec	2	2,8
Molar mass of nitrogen, g/mol	28	28
Gas constant of nitrogen, J/(mol × K)	8,314	8,314
Prandtl number (for T0 = 290 K)	0,716	0,716

The current mass of gas was determined by the equation:

$$m(t) = \frac{P(t)VM}{RT(t)}, \tag{1}$$

where *t* – time, sec; *P(t)* – current pressure, Pa; *V* - a volume of the reservoir, m^3; *M* – a molar mass of gas, g / mol; *R* – gas constant, J / (mol × K); *T(t)* – current temperature, K.

3 Numerical Determination of the Mass of Gas in the Reservoir During Filling

Model Description. To determine gas mixture parameters during filling a reservoir to assure its mass fraction the numerical model of the process was built. This study does not consider the solid bodies of the objects of the mixture generator, the only fluid of their internal volumes shape the geometrical model. The developed model is an advanced model built earlier [16] with the following features.

Firstly, the geometry of the present model exactly matches the experimental setup, which allows us to get away from the assumptions about intermediate volumes and take into account well the peculiarities of gas outflow in places with variable cross-sections of the pipeline. Of course, this improvement entailed a several-fold increase in the number of finite elements, and, accordingly, in computational resources.

Secondly, the initial conditions for the gas flow are set by the condition of constant pressure at the inlet, for the two options both the prohibition of overflow from the inlet and its possibility. In contrast to the previously presented model, where a constant flow rate through the inlet was set, the stability of the solution for this model is worse, however, the constant pressure condition exactly corresponds to the experiments, namely, the gas flow through the reducer. The difference in the resulting pressure between the mass flow filling models [16] and the developed model with pressure boundary conditions is up to 10%.

And, finally, since the calculation of the problem with the indicated initial conditions is poorly stable, it was required to significantly reduce the time step, which entailed a significant increase in computing resources.

Thus, the calculation model (Fig. 2) consists of three domains: the reducer section, the valve section, the reservoir section. The connections of these sections are specified by the corresponding interface conditions. Appropriate monitoring points are located in the places where pressure and temperature sensors are installed (Fig. 2).

Simulating of filling with one-component nitrogen gas was prepared using the ideal gas model. The calculated model is divided into 3 345 434 finite elements with a maximum linear size of 3 mm, which provides the value of the Courant number C < 5.

To solve the defining equations of the model, the finite volume method was used [19], according to which the values of physical variables were calculated (and stored) only in the centers of finite elements, and mass, momentum, and energy flows were calculated on the faces of these elements. At the same time, the spatial derivatives were approximated using implicit difference operators of the second order of accuracy. Five layers of finite elements have been created in the near-wall region for the correct display of the turbulent spectrum.

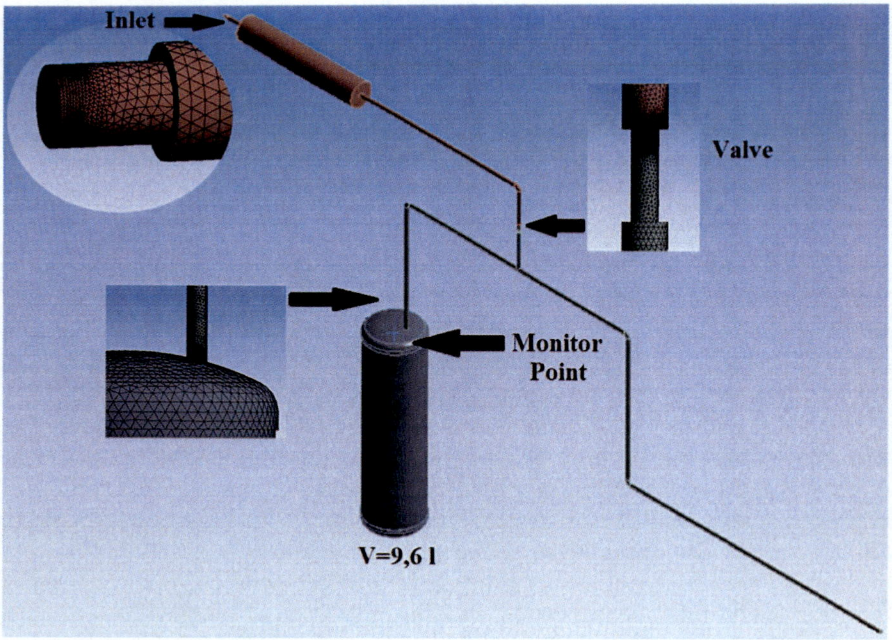

Fig. 2. A computational model of a real experimental setup.

Boundary and Initial Conditions. Corresponding to the full-scale tests it was set the pressure boundary conditions – the flow direction is normal to the inlet, the flow regime is subsonic and zero gradient. Both pressure boundary conditions as allowed as prohibited backflow through the inlet were simulated. The initial temperature of 289 K and the initial pressure of 1.11 MPa were set. Other initial values of gas correspond to tests shown in Table 1. The SAS SST turbulence model that can resolve turbulent structures in unstable regions of complex engineering flows was applied.

Modeling the heat transfer between the gas and the rigid reservoir wall is a very important parameter of the filling model, but the calculation of this process critically increases the resource requirements of the solver, so it is not included in the scope of this study. So based on previously studied models [16, 20] it was chosen a static temperature on the wall is a type of boundary condition of heat exchange for the developed model.

Defining Equations. The problem of filling a reservoir is described by a system of equations that includes the laws of conservation of mass (the continuity equation), momentum (Navier-Stokes equation), and energy of nonstationary spatial flow according to Euler's approach in the Cartesian system of coordinates (x_i, $i = 1, 2, 3$) [21]:

$$\frac{\partial \rho}{\partial t} + \frac{\partial}{\partial x_i}(\rho u_i) = 0, \tag{2}$$

$$\frac{\partial(\rho u_i)}{\partial t} + \frac{\partial}{\partial x_j}(\rho u_i u_j) + \frac{\partial P}{\partial x_i} = \frac{\partial}{\partial x_j}\left(\tau_{ij} + \tau_{ij}^R\right), \tag{3}$$

$$\frac{\partial \rho H}{\partial t} + \frac{\partial \rho u_i H}{\partial x_i} = \frac{\partial}{\partial x_i}\left(u_i\left(\tau_{ij} + \tau_{ij}^R\right) + q_i\right) + \frac{\partial P}{\partial t} - \tau_{ij}^R\frac{\partial u_i}{\partial x_j} + \rho\varepsilon, \tag{4}$$

where u_i – velocity vector components, [m/sec]; $\rho = \frac{P}{RT}$ – gas density (for ideal gas), [kg/m^3]; $H = h + \frac{\rho u^2}{2}$ – enthalpy, [J]; τ_{ik} – tensor of viscous shear stresses, [Pa]; q_i – specific heat flux, [W/m^2].

The tensor of viscous shear stresses is determined by the following equation:

$$\tau_{ij} = \mu\left(\frac{\partial u_i}{\partial x_j} + \frac{\partial u_j}{\partial x_i} - \frac{2}{3}\frac{\partial_k}{\partial x_k}\delta_{ij}\right). \tag{5}$$

where $\mu = \mu_1 + \mu_t$ – viscosity coefficients; μ_t – coefficient of turbulent viscosity; μ_1 – molecular viscosity coefficient; δ_{ij} – Kronecker delta function (if $i = j$, then $\delta_{ij} = 1$; if $i \neq j$, then $\delta_{ij} = 0$); k – kinetic energy of turbulence, [J].

Reynolds stress tensor in the Boussinesq approximation:

$$\tau_{ij}^R = m_t\left(\frac{\partial u_i}{\partial x_j} + \frac{\partial u_j}{\partial x_i} - \frac{2}{3}\frac{\partial u_k}{\partial x_k}\right) - \frac{2}{3}rkd_{ij} \tag{6}$$

To determine the coefficient of turbulent viscosity μ_t in this paper we use the equations of the SAS SST-model of turbulence, which shows high accuracy in modeling near-wall flows [22]:

$$m_t = \frac{\rho a_1 k}{\max(a_1\omega; \Omega F_2)}, \tag{7}$$

where $F_2 = \tan(\arg_2^2)$; $\arg_2 = \max\left(2\frac{\sqrt{k}}{0{,}09\omega y}; \frac{500v}{y^2\omega}\right)$ – a function equal to one for the wall layer and zero for the free layers; $\Omega = (\partial u/\partial n)$ – a derivative of the flow velocity in the direction normal to the wall.

To determine the kinetic energy and its dissipation used the equations:

$$\frac{\partial \rho k}{\partial t}\frac{\partial}{\partial x_i}(\rho u_i k) = \tau_{ij}\frac{\partial u_i}{\partial x_j} - \beta^* \rho \omega k + \frac{\partial}{\partial x_i}\left((\mu_1 + \sigma_k \mu_t)\frac{\partial k}{\partial x_i}\right) \tag{8}$$

$$\frac{\partial \rho \omega}{\partial t} + \frac{\partial}{\partial x_i}(\rho u_i \omega) = \frac{\gamma \rho}{\mu_T}\tau_{ij}\frac{\partial v_i}{\partial x_j} - \beta \rho \omega^2 +$$

$$\frac{\partial}{\partial x_i}\left((\mu_1 + \sigma_\omega \mu_t)\frac{\partial \omega}{\partial x_i}\right) + 2\rho(1 - F_1)\sigma_{\omega 2}\frac{1}{\omega}\frac{\partial k}{\partial x_j}\frac{\partial \omega}{\partial x_j}, \tag{9}$$

where β, β^*, σ_k, σ_ω – empirical constants; F_1 – a function that acts as a switch between models, so that near the wall it is equal to one, and at a distance from the wall - zero.

The specific heat flux is determined by the equation:

$$q_k = -\left(\frac{\mu_1}{Pr} + \frac{\mu_t}{\sigma_c}\right)c_p\frac{\partial T}{\partial x_k}, \tag{10}$$

where $\sigma_c = 0.9$ – empirical constant; Pr – Prandtl number; c_p – specific heat at constant pressure, [J / (kg × K].

Solver Settings. The step of solving equations overtime is set adaptive to the Courant number ($C < 5$). Timeline - Second Order Backward Euler. Others - High Resolution. The CEL language was used to determine the dependences of the calculated process parameters over time. For instance, the mass of the gas component at the estimated time in the volume of the reservoir was determined by integrating the value of the current gas density:

$$m(t) = \text{volumeInt}(\rho(t))@\text{Volume}. \tag{11}$$

4 Analysis of the Results Obtained

Following the objectives of the study, the results of the tests were compared with the numerical solutions. It should be emphasized that the temperature distribution field in the tank cavity is extremely uneven, which is expected (Fig. 3). At the same time, the calculation of the mass of the filled component in the volume strictly depends on the temperature. The difference in the values between the volume-averaged temperature and the temperature determined at a point (thermocouple) systematically leads to an error in determining the gas mass.

Comparison of Numerical Models with Each Other and with the Experiments. Figure 4 shows the graphical presentation of temperature change in the reservoir according to the full-scale experiments № 1 and № 2 (Table 1) as well as the results of calculations for models with two types of boundary pressure conditions through the inlet:

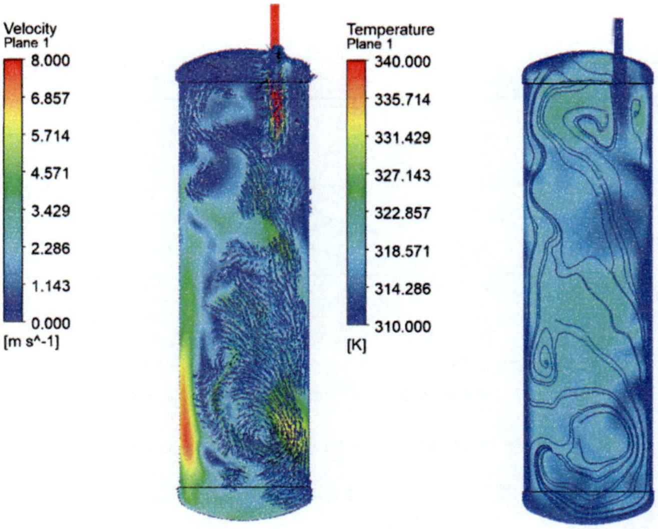

Fig. 3. The field of velocity and temperature in the intermediate stages of filling.

allowed backflow– "opening" and prohibited backflow – "inlet". Analysis of the results reveals the following.

First, the obtained values of temperatures at the control point for different cases of the boundary conditions of the model differ from each other to 1.5%. This difference in the calculated results makes it possible to establish a preference for using the "opening" boundary over the "inlet" because of a more stable solution that requires a lower degree of discretization with an equal solution resolution.

In addition, the discrepancy in temperature values at the control point of the calculation model and experimental data is 1% up to 2.5%, which verifies the calculation models well. Comparison of the volume-averaged temperature values in the reservoir shows an error of up to 1% for "inlet" and "opening" models between each other, and up to 2% between them and experiments. Despite the qualitative difference between temperature curves of the control point from the volume-averaged, they can be used to estimate the filled gas mass with the condition of determining the systematic deviation.

Next, the volume-averaged pressure in the reservoir over the filling time was calculated. Comparison of pressure change for models with different boundary conditions on the wall establishes a difference between their values of 5%. The discrepancy between the simulation results and the full-scale experiments does not exceed 8% (see Fig. 5).

Similarly, the graphs of gas mass change in the reservoir for mentioned cases (Fig. 6) were obtained. Despite the above discrepancy between the values of pressure and temperature, the error in calculating the mass does not exceed 3%.

In a real case, when it is not possible to measure gas parameters in a real reservoir, for instance, in combustion chambers, the use of verified numerical models of the process (and even better, digital twins built on their basis) makes it possible to control the filling process with predictable accuracy, based only on the gas state in the original tanks. This is a direction for further research.

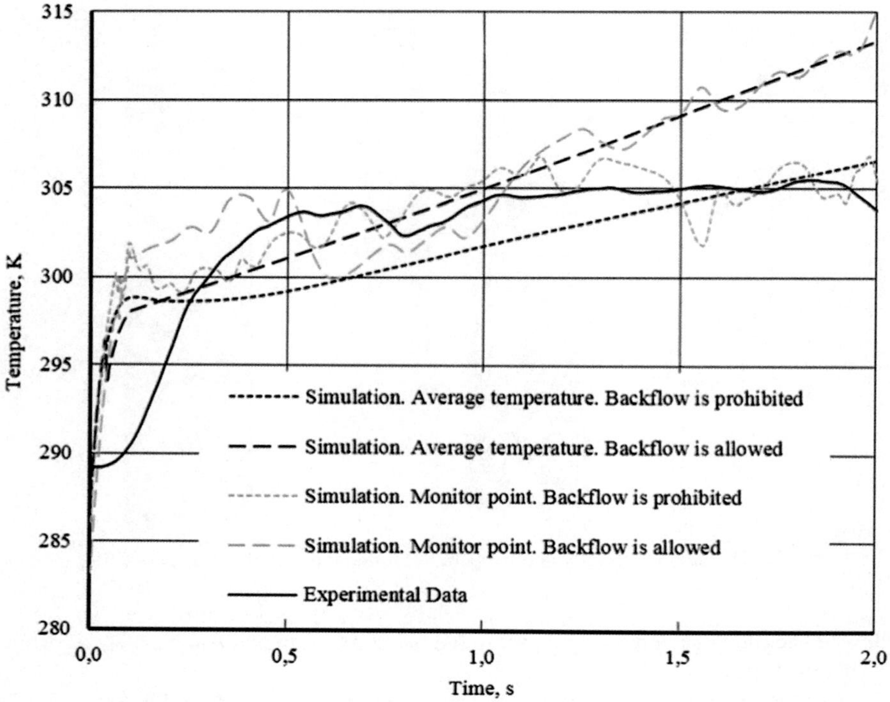

Fig. 4. Changing the temperature of the gas in the reservoir during filling time.

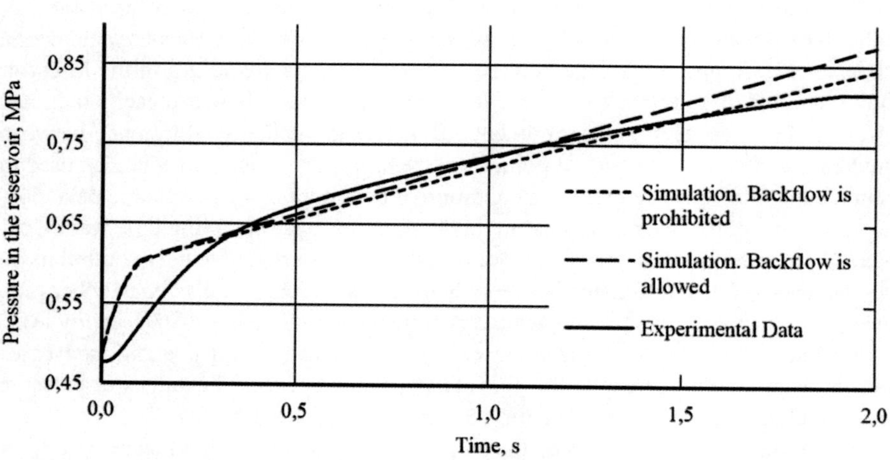

Fig. 5. The calculated volume-averaged and experimentally measured pressure in the reservoir during the filling time.

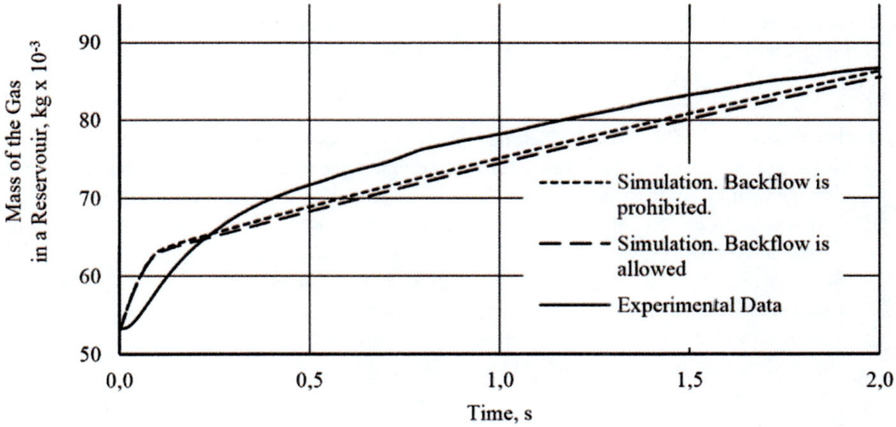

Fig. 6. The mass of gas in the reservoir, determined during the filling time.

5 Conclusion

As to fill a reservoir with a guaranteed composition of the gas mixture mass of gas component (here is nitrogen) by full-scale tests and corresponds CFD simulations were evaluated. During the full-scale tests, the mass of the filled gas was estimated from the gas parameters by CNC monitoring of high-frequency sensors. Models of the gas-dynamic unsteady process for various inlet boundary conditions had been built and compared with each other and experimental data.

It was substantiated the use of SAS SST turbulence model and the constant temperature on the wall as heat transfer conditions. The obtained difference in pressure values ranges up to 10% for the model of filling with a mass flow rate and the developed model with pressure boundary conditions.

It was established the following deviation of the calculated data when using the model with pressure boundary conditions both allowed and prohibited backflow through the inlet: for pressure - not more than 5%, for the average temperature up to 1%, for the temperature at the control point – 1.5%, for the mass - 2.5%. Besides that, the discrepancy between the simulation results and the full-scale experiments does not exceed 8% in pressure and 1% in temperature at the monitor point.

The model with pressure boundary condition on the inlet closely matches the trend in the reservoir pressure over time seen in laboratory experiments, with a typical error of less than 3% in the mass of the component that can provide an accurate prediction of gas dosing.

References

1. Xu, L.D., Xu, E.L., Li, L.: Industry 4.0: state of the art and future trends. Int. J. Prod. Res. **56**(8), 2941–2962 (2018). https://doi.org/10.1080/00207543.2018.1444806
2. Mubarak, M.F., Petraite, M.: Industry 4.0 technologies, digital trust and technological orientation: what matters in open innovation? Technol. Forecast. Soc. Change **161**, 120332 (2020). https://doi.org/10.1016/j.techfore.2020.120332

3. Plankovskyy, S., Popov, V., Shypul, O., et al.: Advanced thermal energy method for finishing precision parts. In: Gupta, K., Pramanik, A. (eds.) Advanced Machining and Finishing, pp. 527–575. Elsevier, Amsterdam (2021). https://doi.org/10.1016/B978-0-12-817452-4.00014-2
4. Plankovskyy, S.I., Shypul, O.V., Zaklinskiy, S.A.: Application perspectives of modern methods for gas mixtures generating to precision thermal pulse treatment. Aerosp. Tech. Technol. **3**, 85–93 (2017). [in Russian]
5. Plankovskyy, S.I., Shypul, O.V., Tryfonov, O.V., Zaklinskiy, S.O.: Algorithm of mixture generation control system for precision thermal pulse treatment. Aerosp. Tech. Technol. **5**, 58–66 (2018). https://doi.org/10.32620/aktt.2018.5.09. [in Russian]
6. Plankovskyy, S., Shypul, O., Zaklinskyy, S., Tryfonov, O.: Dynamic method of gas mixtures creation for plasma technologies. Probl. At. Sci. Technol. **6**(118), 189–193 (2018)
7. Plankovskyy, S., Shypul, O., Zaklinskyy, S., Tsegelnyk, Y., Kombarov, V.: A method of rapid measurement of vessels volume with complex shape by critical nozzles. In: Nechyporuk, M., Pavlikov, V., Kritskiy, D. (eds.) ICTM 2020. LNNS, vol. 188, pp. 247–255. Springer, Cham (2021). https://doi.org/10.1007/978-3-030-66717-7_20
8. Bourgeois, T., Ammouri, F., Weber, M., Knapik, C.: Evaluating the temperature inside a tank during a filling with highly-pressurized gas. Int. J. Hydrogen Energy **40**(35), 11748–11755 (2015). https://doi.org/10.1016/j.ijhydene.2015.01.096
9. Elgin, R.C., III., Hagen, C.L.: A semiempirical compressed-natural-gas tank-filling model created for a novel self-refueling vehicle system. Proc. Inst. Mech. Eng. Part D: J. Automob. Eng. **230**(12), 1719–1726 (2016). https://doi.org/10.1177/0954407015623409
10. Striednig, M., Brandstätter, S., Sartory, M., Klell, M.: Thermodynamic real gas analysis of a tank filling process. Int. J. Hydrogen Energy **39**(16), 8495–8509 (2014). https://doi.org/10.1016/j.ijhydene.2014.03.028
11. Li, J.Q., Myoung, N.S., Kwon, J.T., et al.: A study on the prediction of the temperature and mass of hydrogen gas inside a tank during fast filling process. Energies **13**(23), 6428 (2020). https://doi.org/10.3390/en13236428
12. Bourgeois, T., Ammouri, F., Baraldi, D., Moretto, P.: The temperature evolution in compressed gas filling processes: a review. Int. J. Hydrogen Energy **43**(4), 2268–2292 (2018). https://doi.org/10.1016/j.ijhydene.2017.11.068
13. Heitsch, M., Baraldi, D., Moretto, P.: Numerical investigations on the fast filling of hydrogen tanks. Int. J. Hydrogen Energy **36**(3), 2606–2612 (2011). https://doi.org/10.1016/j.ijhydene.2010.04.134
14. Suryan, A., Kim, H.D., Setoguchi, T.: Three dimensional numerical computations on the fast filling of a hydrogen tank under different conditions. Int. J. Hydrogen Energy **37**(9), 7600–7611 (2012). https://doi.org/10.1016/j.ijhydene.2012.02.019
15. Liu, J., Zheng, S., Zhang, Z., et al.: Numerical study on the fast filling of on-bus gaseous hydrogen storage cylinder. Int. J. Hydrogen Energy **45**(15), 9241–9251 (2020). https://doi.org/10.1016/j.ijhydene.2020.01.033
16. Shypul, O., Zaklinskyy, S., Kombarov, V., et al.: Numerical and experimental study of filling a vessel with a component of gas mixture. Aerosp. Tech. Technol. **4**, 63–72 (2021). https://doi.org/10.32620/aktt.2021.4.09. [in Ukrainian]
17. Plankovskyy, S.I., Shypul, O.V., Tsegelnyk, Y.V., et al.: Development of an automated complex for precision thermal pulse treatment with detonating gas mixtures: scientific materials. National Aerospace University "Kharkiv Aviation Institute", Kharkiv (2020). http://dspace.library.khai.edu/xmlui/handle/123456789/368. [in Ukrainian]
18. Sysoiev, Y.A.: The creation of multicomponent gas mixtures for ion-plasma technologies. Probl. At. Sci. Technol. **90**(2), 137–142 (2014)
19. ANSYS CFX Reference. Ansys, Inc. (2020)

20. Plankovskyy, S., Shypul, O., Tsegelnyk, Y., Pankratov, A., Romanova, T.: Amplification of heat transfer by shock waves for thermal energy method. In: Nechyporuk, M., Pavlikov, V., Kritskiy, D. (eds.) ICTM 2020. LNNS, vol. 188, pp. 577–587. Springer, Cham (2021). https://doi.org/10.1007/978-3-030-66717-7_49
21. Menter, F.R.: Two-equation eddy-viscosity turbulence models for engineering applications. AIAA J. **32**(8), 1598–1605 (1994). https://doi.org/10.2514/3.12149
22. Menter, F., Kuntz, M., Bender, R.: A scale-adaptive simulation model for turbulent flow predictions. In: 41st Aerospace Sciences Meeting and Exhibit, AIAA 2003-0767. AIAA, Reno, Nevada (2003). https://doi.org/10.2514/6.2003-767

Creation and Improvement Principles of the Pneumatic Manual Impulse Devices

Iurii Vorobiov[1], Kateryna Maiorova[1], Iryna Voronko[1](✉), Maksym Boiko[1], and Oleh Komisarov[2]

[1] National Aerospace University "Kharkiv Aviation Institute", 17 Chkalova Street, Kharkiv 61070, Ukraine
kate.majorova@ukr.net, i.voronko@khai.edu

[2] Motor Sich JSC, 15, Motorostroiteley Avenue, Zaporozhye 69068, Ukraine

Abstract. The aircraft structures complexity and the inaccessibility of rivets setting areas create the need for manual rivet hammers. Many developments in this area provide a wide range of tools for the assembly work. During the assembly work, the multiple-hitting pneumatic impulse hammers are used. Such kind of tools create noise and vibrations that affect a fitter. Moreover, such equipment can be used only by a highly qualified specialist. The purpose of the study is to determine the pneumatic manual impulse devices (PMID) blow energy required to auto-mate the devices design process, taking into account their design features. The engineering process of the PMID operation is simulated using the ANSYS CFX Software. For the numerical study of the problem, a system of Navier-Stokes equations is used. To determine the optimal configuration of the devices gas tracts, energy losses and stress-stain state of structural elements in their interaction, the finite element method is used. As a result of the study, the simulation with various geometry of computational domain has been carried out. The results of different approaches to reduce the calculation time when using 2D geometry are compared. Using the results of gas-dynamic calculations of the devices energy parameters, the task of designing the PMID with the required blow energy has been solved. The comparison of the simulation results made in the ANSYS CFX software with the theoretical data revealed the insignificant error, the maximum value of which is 12%. It may be caused by using many simplifications during the theoretical calculations.

Keywords: Pneumatic manual impulse device · ANSYS CFX system · Blow energy · Striker · Finite element method

1 Introduction

Basic technical requirements for the PMID are formulated according to the design analysis and experience of using assembly devices in the aviation industry [1–4]. As a result of the analysis of re-cent research and publications, it has been stated that pneumo-impulse rivet hammers PIH-90H and pneumo-impulse devices (PID) for holes mandrelling PID-90, developed at the Aircraft Manufacturing Technology department of Kharkiv Aviation

M. Nechyporuk et al. (Eds.): ICTM 2021, LNNS 367, pp. 178–191, 2022.
https://doi.org/10.1007/978-3-030-94259-5_17

Institute in the mid-1980s have proved to be the most successful in terms of design and reliable in terms of manufacturing requirements. These devices differ favorably from their analogs in improved working conditions of the operators due to minimum pressure load, minimization of the force of pressing the trigger and damping the recoil energy of the device during impact. They have higher impact energy stability due to the pressure relief valve, shut-off valve and receiver that provides the accumulation of compressed air in the required amount.

However, the existing experience of operating these models revealed a number of their short-comings. The PIH-90H crimp that has a step form to ensure self-rotation at the end of working cycle has an extremely low stability. Its best specimens made of titanium alloy Ti-5Al-5Mo-5V-3Cr (Ti5553) have maximum stability of 5–7 thousand blows. The stability of the crimps made of carbon steel W1, W108, W110, W112 and heat treated to HRC 44–46 is only 300–500 blows. Failure occurs at the point of transition from maximum diameter to minimum.

Violations of the smooth functioning of the quick release valve not only led to instability of the energy parameters of the devices, but sometimes to failures. Such a disadvantage is the most characteristic one during the long breaks in the work of PMID.

In the device design, the blow energy is regulated only by the external gearbox built into the compressed air preparation unit. Although adjusting the blow energy in this way is effective, it doesn't meet the needs for efficiency and convenience in operation. The blow energy significantly relies on the quality of the lock (the fit of the striker shank in the fluoroplastic ring), which deter-mines the pressure drop magnitude in the cavities in front of and behind the striker. Moreover, the presence of the back pressure in the cavity in front of the striker during its acceleration significantly reduces the blow energy.

In some specific cases, during the assembly in a jig, the technological operations performance by the devices with a series arrangement of the chassis and the handle (forward type) in places with constrained approaches became impossible, and therefore the pistol arrangement type of the PMID was needed.

The blow energy significantly depends on the quality of the lock (the fit of the striker shank in a fluoroplastic ring), which determines the magnitude of the pressure drop in the cavities in front of and behind the striker. The presence of back pressure in the cavity in front of the striker during its acceleration, in addition, significantly reduces the blow energy.

In some specific cases, during the assembly in a jig the performance of the technological operations by devices with a series arrangement of chassis and handle (forward type) in places with con-strained approaches became impossible, and therefore the pistol arrangement type of the PMID was asked for [5].

The blow energy damping only by a rubber damper on the tappet is also cannot be considered effective enough. The PMID reserves regarding the noise reduction parameters, ergonomic parameters, etc. are also not exhausted.

The disadvantages of PID-90 include the necessity for precise strokes coordination of the striker and mandrel with the holder and shank in the form of a rod with a flange at the end. In addition, during the forward stroke the rod will be subject to vibration and quick fatigue failure in the point of attachment to the holder due to the impulse load on the holder through the long cantilever.

In this regard, the following areas of the PMID design improvements were formulated:

- increasing the reliability of the device;
- increasing the energy stability of a single blow;
- providing the quick and convenient adjustment of the blow energy;
- extending technological capabilities of the device through the reduction of dimensional and mass parameters and new arrangement solutions;
- increasing the energy conversion efficiency of the device;
- improving hygienic and ergonomic criteria.

When implementing these improvements, the simulation of gas-dynamic processes in the PMID with finite element method (FEM) was carried out to determine the optimal configuration of the gas tracts of the devices, energy losses and stress-strain state of structural elements in their interaction.

2 Description of the Mathematical Models Applied During the Gas-Dynamic Processes Simulation in the PMID

The energy parameters of the PMID are largely determined by the gas-dynamic processes in the channels and tracts of the device, as well as conditions of energy transfer during the interaction of structural elements. Therefore, the creation of the mathematical model (MM), which allows a comprehensive description of these processes, will be especially important because it correctly predicts the blow energy, taking into account all the factors that determine the behavior of the striker and the crimp of the pneumatic tool. The ANSYS CFX system is chosen as a computer simulation system for the numerical study of the gas-dynamic processes, as its solver contains a large number of models that allow calculating the turbulent gas flow taking into account the presence of the moving solids and their interaction.

Description of the Mathematical Models Applied During the Simulation of the Gas-Dynamic Processes in the PMID. For the numerical study of this problem, a system of Navier–Stokes equations (direct numerical simulation – DNS) [6] is used, which involves the laws of conservation of mass, momentum and energy of unsteady spatial flow in the Cartesian coordinate system (x_i, $i = 1, 2, 3$):

$$\begin{aligned} &\frac{\partial \rho}{\partial t} + \frac{\partial}{\partial x_k}(\rho u_k) = 0; \\ &\frac{\partial (\rho u_i)}{\partial t} + \frac{\partial}{\partial x_k}(\rho u_i u_k - \tau_{ik}) + \frac{\partial P}{\partial x_i} = S_i; \\ &\frac{\partial (\rho E)}{\partial t} + \frac{\partial}{\partial x_k}((\rho E + P)u_k + q_k - \tau_{ik} u_i) = S_k u_k + Q_H, \end{aligned} \tag{1}$$

where u_i are the gas velocity vector components; ρ, P are the gas density and pressure; S_i are the external volumetric forces; E is the total energy per unit mass of gas; Q_H is the heat released in a unit volume of gas; τ_{ik} is the tensor of viscous shear stresses; q_i is the heat flow.

Tensor of viscous shear stresses is determined as follows:

$$\tau_{ik} = \mu \cdot (\frac{\partial u_i}{\partial x_j} + \frac{\partial u_j}{\partial x_i} - \frac{2}{3}\frac{\partial u_l}{\partial x_l}\delta_{ij}) - \frac{2}{3}\rho k \delta_{ij}, \quad (2)$$

where $\mu = \mu_l + \mu_t$ is the viscosity coefficient; μ_l is the molecular (dynamic) viscosity coefficient; μ_t is the turbulent viscosity coefficient; δ_{ij} is the Kronecker delta; k is the kinetic turbulence energy.

Certainly, the gas flow is turbulent. Obviously, the direct solution to three-dimensional unsteady DNS using spatial grids can be used to describe all significant flows, including spatiotemporal inhomogeneities. However, it is not less obvious that for its numerical implementation, it is necessary to use very small grids, where the nodes number should significantly increase with increasing Reynolds number.

As DNS requires three-dimensional unsteady calculation, the cost of the calculation is proportional to the total number of grid nodes, the number of time steps, and proportional to Re^3. This means that when the Reynolds number doubles, the costs increase by about an order of magnitude. Therefore, these methods are not used to calculate flows with complex geometry, and in practice, DNS is used only to calculate simple turbulent flows at low Reynolds numbers (about 103 and below).

Thus, as an alternative approach, the Reynolds-averaged method has now become the most widespread.

It is believed [7] that the average values of the pulsation components are zero, but the averaging from the multiplication of the pulsation components by each other is not zero. In the averaged equations of motion, additional terms (Reynolds stress) equal to $-\rho\overline{u_i'u_l'}$ appear, which describe the change in momentum caused by the interaction of turbulent velocity fluctuations.

To determine the Reynolds stresses, the Boussinesq hypothesis [7], which joins these stresses and the mean velocity gradients, is usually used:

$$-\rho\overline{u_i'u_l'} = \mu_T\left(\frac{\partial u_i}{\partial x_j} + \frac{\partial u_j}{\partial x_i}\right) - \frac{2}{3}\left(\rho k + \mu_T\frac{\partial u_i}{\partial x_i}\right)\delta_{ij} \quad (3)$$

The obtained equations describe the behavior of the averaged characteristics of turbulent flow, if the coefficients of viscosity and thermal conductivity are taken to be effective values equal to the sum of molecular (laminar dynamic) and turbulent viscosity and thermal conductivity, respectively:

$$\mu = \mu_l + \mu_T, \; \lambda = \lambda_l + \lambda_T \quad (4)$$

To determine the μ_T and λ_T, there are numerous semi-empirical theories of turbulence, which can be divided into algebraic and differential models. In this paper, the equation of the shear-stress transport (SST) model of turbulence is used [8], which shows high accuracy in the near-wall flows simulation.

To specify the turbulent viscosity value in this model, the following equation is used:

$$\mu_T = \frac{\rho a_1 k}{\max(a_1\omega; \Omega F_2)}, \tag{5}$$

where $F_2 = \tanh(\arg_2^2)$; $\arg_2 + \max(2\frac{\sqrt{k}}{0.09\omega y}; \frac{500\nu}{y^2\omega})$ is the function equal to one for the boundary layer and zero for the free layers; $\Omega = (\partial u/\partial n)$ is the derivative of the flow velocity in the direction normal to the wall. To determine the kinetic energy and its dissipation, the following equation is used:

$$\frac{\partial \rho k}{\partial t} + \frac{\partial}{\partial x_i}(\rho u_i k) = \tau_{ij}\frac{\partial u_i}{\partial x_j} - \beta^* \rho\omega k + \frac{\partial}{\partial x_i}((\mu_l + \sigma_k \mu_t)\frac{\partial k}{\partial x_i});$$

$$\frac{\partial \rho\omega}{\partial t} + \frac{\partial}{\partial x_i}(\rho u_i \omega) = \frac{\gamma\rho}{\mu_T}\tau_{ij}\frac{\partial v_i}{\partial x_j} - \beta\rho\omega^2 + \frac{\partial}{\partial x_i}((\mu_l + \sigma_\omega \mu_t)\frac{\partial \omega}{\partial x_i}) +$$

$$+ 2\rho(1 - F_1)\sigma_{\omega 2}\frac{1}{\omega}\frac{\partial k}{\partial x_j}\frac{\partial \omega}{\partial x_j}, \tag{6}$$

where β, β^*, σ_k, σ_ω are the empirical constants, which are calculated by the formula $\varphi = F_1\varphi_1 + (1 - F_1)\varphi_2 k - \varepsilon k - \omega$,

where φ_1, φ_2 are the corresponding empirical coefficients $k - \varepsilon$ and $k - \omega$ of turbulence models; F_1 is the function that acts as a switch between models so that near the wall $F_1 = 1$, and away from the surface $F_1 = 0$.

German scientist F. R. Menter was the first who proposed the formulation of the SST model from the condition of calculations stability [8], justifying the choice of the F_1 function in the form

$$F_1 = \tanh(\arg_1^4) \tag{7}$$

where $\arg_1 = \min[\max(\frac{\sqrt{k}}{0.09\omega y}; \frac{500\nu}{y^2\omega}); \frac{4\rho\sigma\omega_2 k}{CD_{k\omega}y^2}]$; y is the distance to the nearest wall.

$$CD_{k\omega} = \max(2\rho\sigma_{\omega 2}\frac{1}{\omega}\frac{\partial k}{\partial x_j}\frac{\partial \omega}{\partial x_j}, 10^{-20}). \tag{8}$$

3 Description of the Mathematical Models Used to Simulate the PMID Striker Movement

3.1 Determination of the Forces Acting on the Striker

Acceleration of the striker is a complex multifactorial process, which is significantly influenced by both the parameters of the gas environment and the conditions of gas leakage associated with the design features of the tool. As mentioned above, this process is described numerically using the system of Navier–Stokes equations, which is solved using FEM. Solving such a problem with complex spatial geometry requires large computational resources, which is enhanced by the need to rearrange the finite elements (FE) grid in real time during the calculation when its quality deteriorates. In this regard, two approaches are proposed to solve the problem of the PMID gas dynamics taking into account the movement of the striker:

- multiconfiguration approach in a flat axisymmetric setting;
- axisymmetric approach with the elementary sector selection and the use of the mechanism of the FE grid automatic rearrangement.

Acceleration of the striker is determined by the net force – the resultant of forces acting on the striker [9]:

$$R = P_3 - P_n - F_T, \tag{9}$$

where R is the resultant force; P_3 is the force developed by the gas expanding behind the striker; P_n is the force of air resistance in front of the striker; F_T is the force of friction of the striker against the barrel walls.

Sealing with O-rings has become widespread in hydraulic and pneumatic systems. The tightness in the seals of this type is achieved when there is no pressure, due to the previous (mounting) compression of rubber ring in the groove. Whenever the pressure in the system appears, the ring is further deformed, creating a dense contact with a sealed surface.

When performing design, calculation, research works and the construction of MM of various hydraulic and pneumatic devices, there are some difficulties in taking into account friction forces because of the lack of analytical dependencies for them. Known methods for determining friction are mainly based on tabular experimental data. In the work [9], it is recommended to calculate the friction forces in seals with rubber O-rings using the formula

$$F_T = q \cdot \pi \cdot D \tag{10}$$

where q is the specific friction force (force per unit length of contact of the O-ring with the moving surface); D is the diameter of the sealing surface.

The value of the specific friction force depends on the cross-sectional diameter of the ring d, the relative compression of the ring W, and the hardness of the O-ring material.

For the inertial forces, the striker was considered as a rigid body with a mass corresponding to the real mass of the striker, the boundary of which is assigned to the boundary conditions of the wall. During the development of the MM, a special 6 Degree of Freedom (6DOF) solver was connected, which calculates the position and orientation of the striker as a rigid body using the equations of motion. These equations describe six degrees of freedom: three translational and three rotational. In this case, the position of the body is determined by the three Euler angles and three translational coordinates [10].

When solving problems with deformed areas, the distortion of the FE grid leads to the appearance of the degenerate FE or significant deterioration of their quality. Periodic rearrangement of the FE grid for the computational domain is a very important part of the analysis, which contains significant displacements of the computational domain boundaries and deformation of the FE grid.

Therefore, in order to obtain an adequate result of the problem to be solved, the integration of the FE grid rearrangement cycle into the general problem solving cycle is applied.

As mentioned above, two approaches have been proposed to solve the PMID gas dynamics problem taking into account the striker motion.

3.2 Approach 1. Multiconfiguration Approach to Solving Problems with Moving Boundaries

The computational domain is a flat area that corresponds to the cross-section of the pneumo-impulse hammers with a plane passing through the axis of the symmetry (Fig. 1). The displacement of the striker under the pressure drop in cavities in front of and behind the striker was studied.

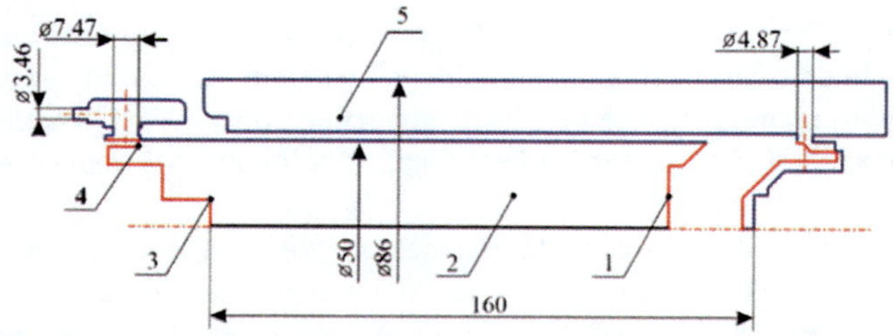

Fig. 1. Geometry of the computational domain with the striker (1), the barrel (2), the crimp (3), the quick release valve (4) and the receiver (5)

It is assumed that at the time before the opening of the relief valve, the air in the barrel and receiver was at the temperature of 300 K and the initial pressure of 0.5 MPa. The pressure of the external environment is taken to be equal to 0.1 MPa, the temperature is 300 K. It is assumed that the relief valve is opened in 0.01 s. The weight of the striker is 0.139 kg. The condition for stopping the calculation is the moment of contact of the striker and the crimp.

The simulation was performed for cases with different ratio of the striker V_x and the volume of the receiver V_p. The influence of the gas leakage conditions on the velocity and energy of the striker has also been studied. The multiconfigurational approach assumes the presence of several computational domain configurations depending on the striker position during the task performing. In this case, seven configurations were used, which corresponded to the stroke of the striker at 0.001, 0.008, 0.028, 0.058 and 0.088 m. Within a specific configuration, the FE calculation grid is deformed at a speed that is characteristic to the instantaneous velocity of the striker. Thus, switching to a new grid occurs automatically when the striker moves on the set distance. The maximum cells number of the FE grid was 15643.

The gas velocities distributions obtained from the simulation results, as well as the calculated striker velocities for different cases are shown in Figs. 2 and 3.

In Fig. 3, the computational domain was supplemented by a reflection relative to the axis of the symmetry for better clarity. Velocity and energy of the striker for cases with different V_x/V_p ratio is shown in Fig. 3 (a), with different conditions of gas leakage is shown in Fig. 3 (b).

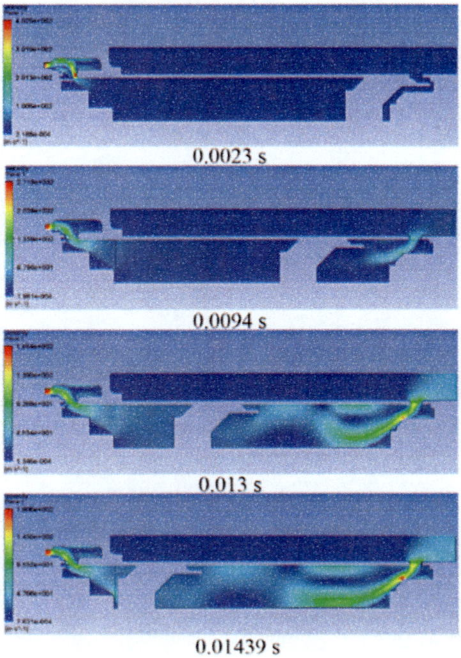

Fig. 2. Fields of gas flow velocities in different periods of time

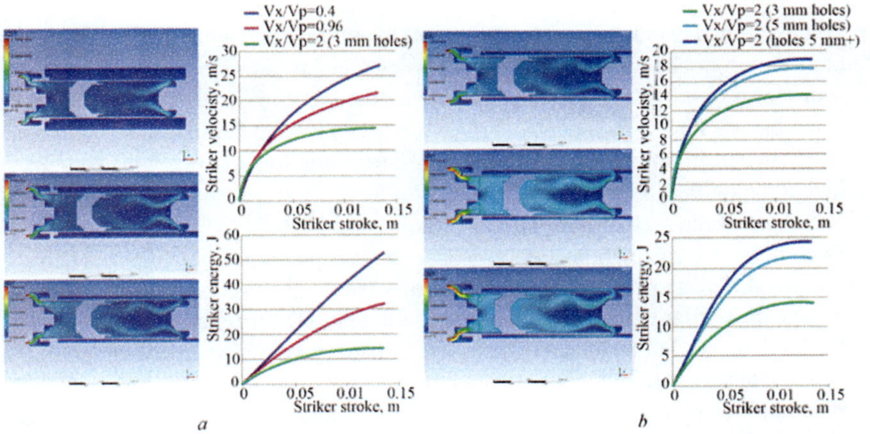

Fig. 3. Velocity and energy of the striker for cases: (a) with different V_x/V_p ratio; (b) with different conditions of gas leakage

The simulation results show that with the increasing V_x/V_p ratio, the velocity and energy of the striker decrease, and at the $V_x/V_p \geq 2$ the braking of the striker is happening. Thus, when the striker passes 0.13 m, which is equivalent to 2.7 d_s, its velocity reaches a peak value of 14 m/s, and then begins to decline. With the increasing diameter of the outlets on the PIH-90 hammer front cover, the striker begins to brake when passing the distance of 0.135 m, but the maximum velocity reaches 19 m/s. A summary graph of the simulation results for different cases is shown in Fig. 4. The calculation time for each individual case was not more than 1 h depending on the size of the computational domain.

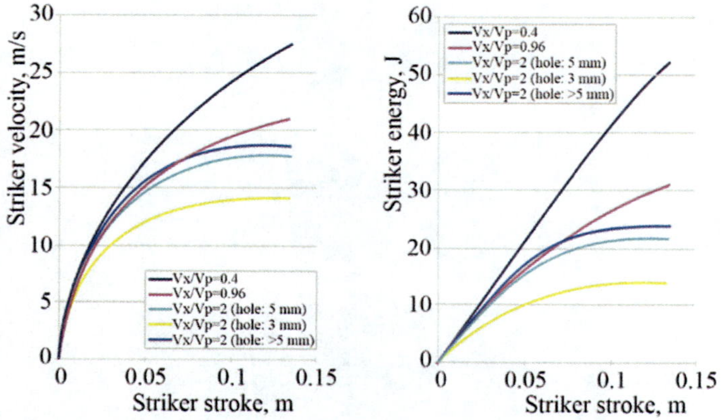

Fig. 4. Comparison of velocity and energy of the striker for different conditions

3.3 Approach 2. Axisymmetric Approach with the Elementary Sector Selection

The computational domain is a sector. To be able to build a sector with a regular placement of holes, the original geometry of the tool was changed to provide the same number of holes in the gas outlet areas on the front cover, the gas flow when opening the relief valve and the gas flow from the receiver into the cavity behind the striker. The holes size was recalculated in compliance with the holes cross-sectional area equivalence in the original and modified geometry of the tool.

The angle of the sector with one row of holes was 17.14°.

To compare the calculations performance, two computational domains were considered, which differed in the number of holes rows (see Fig. 5). The initial and boundary conditions are similar to the solution of the problem in 2D formulation.

To implement the formulated task, the mechanism of the FE grid automatic rearrangement was used, which is shown in Fig. 6.

When implementing this approach, the minimum angle of the FE grid element over the entire computational domain was chosen as a control parameter. The connection of the grid rearrangement cycle occurred when an element with an angle at the vertex less than 5° appeared in the computational domain. Then the grid was rearranged, and the calculation continued on the new grid, where, after interpolation of the results from the previous iteration, the initial conditions for further calculations were added.

The simulation results showed that the sector size does not affect the flow nature in the PMID channels.

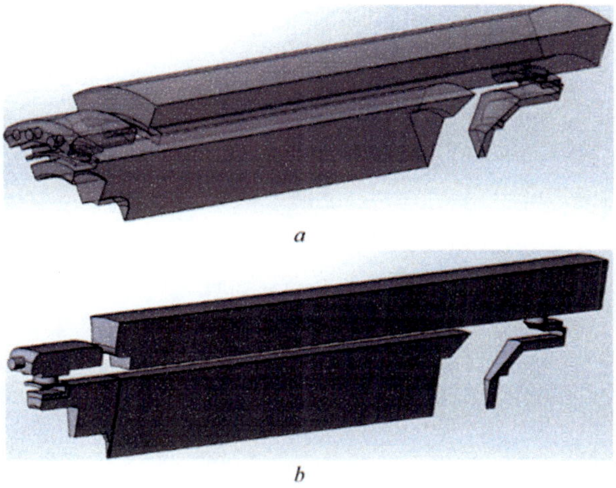

Fig. 5. Computational domain geometry in the form of an elementary sector, where three rows (a) and one row (b) are shown

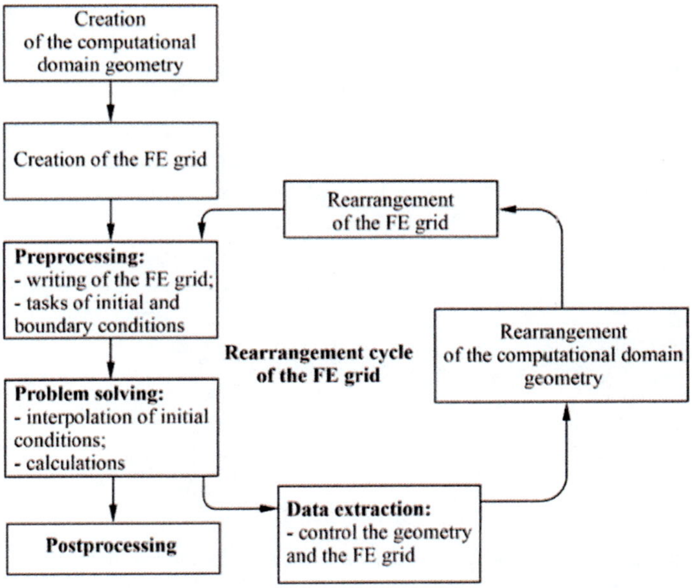

Fig. 6. Integration of the FE grid rearrangement cycle into the general problem solving cycle

The graph of the striker velocity change with its stroke increase for the cases with different sectors is shown in Fig. 7. The maximum relative error was 4%. At the same time, the calculation time was reduced from 13 to 6 h, which was due to a significant reduction in the FE grid cells number. Thus, for the sector with three rows of holes, the calculation grid consisted of 263,326 cells, and the sector with one row of holes consisted of 102,359 cells. This confirmed the expediency of using a sector with a smaller angle to speed up the calculations.

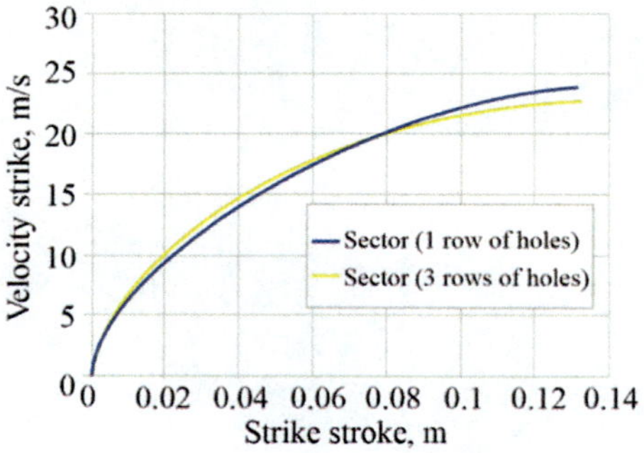

Fig. 7. Striker velocity comparison for different computational domain geometries

In order to verify the possibility of implementing the coherent problem of the PMID gas dynamics and the striker and crimp mechanics, the possibility to use the laminar flow in the gas dynamics calculations was verified (which was due to the solver limitations in the coherent problems solving). It is necessary to compare the crimp velocity obtained during the simulation with experimental data in this calculation in order to confirm the possibility of using the developed MM for designing a new PMID.

The comparison results are shown in Fig. 8 (a), according to which maximum relative error was 2% (for convenience, the figure shows the laminar flow part of the graph).

Figure 8 (b) shows the comparison of simulation results with theoretical data.

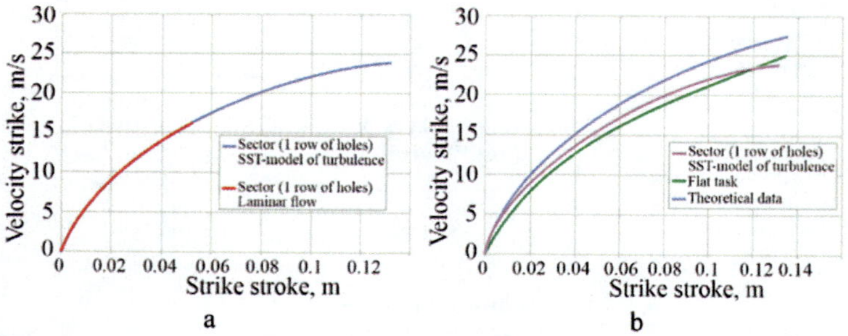

Fig. 8. Comparison of the turbulent and laminar flows (a) and comparison of the simulation results with known theoretical data (b)

The maximum error was 8% for a plain problem and 12% for a sector. The error is explained by a large number of assumptions taken in theoretical calculations. Thus, in work [11], the following assumptions were adopted:

- the process of gas expansion occurs not continuously, but discreetly, so the entire gas tract is divided into several sections and the pressure inside the section is considered a constant value for the entire section;

- the process is considered one-dimensional, and the inhomogeneity of the velocity field and losses at the boundaries and within the sections are taken into account using the coefficients of local and path losses;
- the interaction between the sections occurs according to the decay scheme of the initial gap;
- local sound speed is considered a constant value for all sections.

This leads to the inflated calculated values of the PMID energy parameters.

Considering the influence of the boundary layer on the energy parameters of the tool, it should be noted that the regular FE grid does not allow to accurately simulate the boundary layer near the walls. Simulation with the Free Slip and No Slip boundary conditions showed identical results. The striker velocity at the time of collision with the crimp was 33 m/s. The simulation was performed taking into account the supply of the receiver with compressed air with a pressure of 0.5 MPa.

Assuming the losses caused by gas inhibition near the walls, it was decided to improve the FE grid by creating prismatic layers (10 layers at a distance of 1 mm from the wall).

Solving the problem using the gas laminar flow model showed the limitations of this approach. At a distance of 0.125 m from the initial position of the striker (see Fig. 9), the calculation was stopped with the error of the Reynolds number value going outside the laminar flow model.

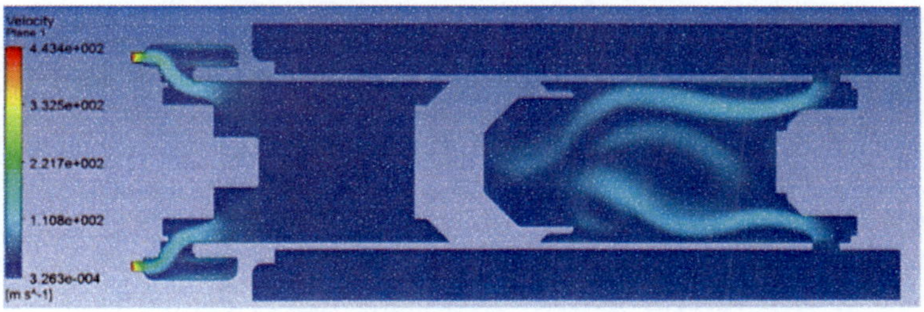

Fig. 9. Gas flows in the pneumo-impulse hammer channels

Thus, for a more accurate gas-dynamic calculation, it is necessary to apply the turbulence model, which would take into account the presence of a boundary layer under the walls. For this purpose, the SST model of turbulence was chosen.

According to the calculation results, the striker velocity at the time of contact with the crimp was 24 m/s.

According to the results of the numerous simulations using an improved grid, additional energy losses were revealed. Consequently, the striker velocity decreased by 9 m/s. The relative deviation was 27%.

Summary data for all calculation cases is shown in Fig. 10.

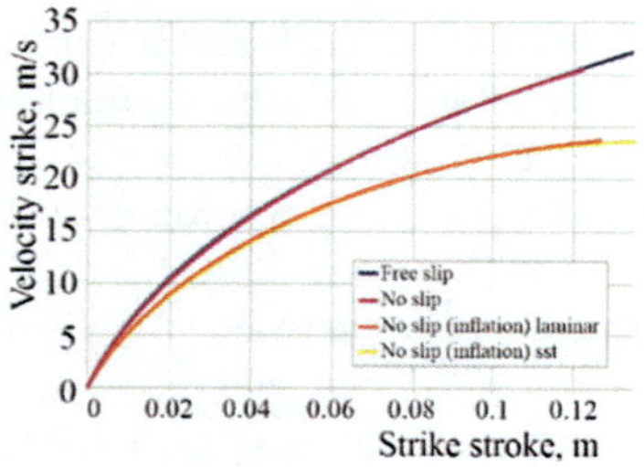

Fig. 10. Simulation results

A significant limitation of the ANSYS CFX system is inability to take into account the contact of solids. This significantly narrows the range of the system application and makes it impossible to solve the problem of structural-fluid interaction in a single computing environment. Therefore, it is necessary to use additional systems of structural analysis, which will lead to a significant increase of calculation time and to an additional error in data transmission between different packages.

4 Conclusions

The principles of creation and improvement of PMID were developed.

The direct problem of PMID design was solved using the results of gas-dynamic calculations of the devices energy parameters in the ANSYS CFX software package, which allow to determine the velocity and energy of the striker taking into account its mass characteristics, friction losses and design features of the device elements.

It is shown that increasing the V_x/V_p ratio from 0.4 to 2 leads to a decrease in the striker velocity from 28 to 14 m/s and in the blow energy from 52 to 14 J. At $V_x/V_p = 2$, the braking area appears after striker passes the distance equivalent to 2.7 d_s. Increasing the diameter of the holes at the outlet of the hammer front cover from 3 to 5 mm allows to increase the striker velocity from 14 to 19 m/s, and the blow energy from 14 to 24 J.

The simulation with different geometry of the computational domain was performed: two-dimensional formulation, a sector with three rows of holes and a sector with one row of holes. The calculation time for a sector with three rows of holes was 13 h; for a sector with one row of holes – 6 h; for a plane problem –1 h. The maximum relative error of the results in the case of using 2D geometry and the sector does not exceed 5%. Thus, for design calculations, it is expedient to use a 2D formulation, as it significantly reduces the calculation time. After completing the design, it is expedient to perform a test calculation, using the geometry in the form of a sector.

References

1. Krivtsov, V.S., Nechyporuk, N.V., Vorobiov, I.A., Voronko, V.V., Vorobiov, A.I.: Development of a technological process and a tool for impulse riveting of aviation parts made of carbon fiber. KhAI, Kharkov (2012)

2. Voronko, V.V., Vorobiov, I.A., Voronko, I.A., Kruglov, V.V.: Development of control systems for the pneumo-pulse energy unit for holes mandrelling of aviation parts as a robotic complex. Open Inf. Comput. Integr. Technol. (74), 88–98 (2016)
3. Krivtsov, V.S., Vorob'ev, Yu.A., Voron'ko, V.V.: Advanced devices for mandreling bores. Kuznechno-Shtampovochnoe Proizvodstvo (Obrabotka Metallov Davleniem) (12), 18–30 (2004)
4. Vorobiov, I.A.: The scientific basis for the creation of a complex of impulse technologies and equipment for the aggregate assembly of airframes (Doctoral dissertation) (Engineering). Kharkiv (2020)
5. Krivtsov, V.S., Voronko, V.V., Zaytsev, V.Ye.: Advanced prospects for development of aircraft assembly technology. Sci. Innov. **11**(3), 12–20 (2015). https://doi.org/10.15407/scin11.03.012
6. Garbaruk, A.V., Strelets, M., Shur, M.L.: Modeling of Turbulence in Calculations of Complex Flows. Publishing by Politechnic University, St. Petersburg (2012)
7. Plankovskyy, S., Shypul, O., Tsegelnyk, Ye., Tryfonov, O., Golovin, I.: Simulation of surface heating for arbitrary shape's moving bodies/sources by using R-functions. Acta Polytechnica **56**(6), 472–477 (2016). https://doi.org/10.14311/AP.2016.56.0472
8. Menter, F.R., Kuntz, M., Bender, R.: A scale-adaptive simulation model for turbulent flow predictions. In: 41st Aerospace Science Meeting & Exhibit, Nevada, Reno, 11 p. (2003)
9. Mitar, J., Šević, D., Karanovic, V., Beker, I., Dudić, S.: Increased efficiency of hydraulic systems through reliability theory and monitoring of system operating parameters. Strojniski Vestnik **584**(4), 281–288 (2011). https://doi.org/10.5545/sv-jme.2011.084
10. Plankovskyy, S., Tedorczyk, A., Shypul, O., Tryfonov, O., Brega, D.: Determination of detonable gas mixture heat fluxes at thermal deburring. Acta Polytechnica **59**(2), 162–169 (2019). https://doi.org/10.14311/AP.2019.59.0162
11. Encyclopedia of Mechanical Engineering XXL. http://mash-xxl.info/info/687955/. Accessed 09 Sept 2012

Artificial Intelligence and Smart Systems

Cognitive Aspects of Ensuring the Safety, Dependability and Stability of a Dynamic System's Functioning in Extreme Conditions

Galyna Mygal[1](✉), Valeriy Mygal[1], Olga Protasenko[2], and Igor Klymenko[1]

[1] The National Aerospace University "Kharkiv Aviation Institute", 17 Chkalov Street, Kharkiv 61070, Ukraine
{g.mygal,valeriy.mygal,igor.klymenko}@khai.edu

[2] Simon Kuznets Kharkov National University of Economics, 9a Science pr., Kharkiv 61166, Ukraine
olha.protasenko@hneu.net

Abstract. Digitalization leads to the necessity of taking into account the cognitive aspects of human-computer interaction. Research of the complex dynamic system's functioning in extremal conditions are limited by the hidden interconnections of elements. The influence of the increasing complexity of interaction on ensuring the safety, dependability and stability of a complex dynamic system in extreme conditions is considered. The analysis of dynamic complexity from the perspective of interdisciplinary principles and laws made it possible to establish that taking into account cognitive aspects, as well as the person's psychophysiological state are critically important. Attention is drawn to the functional asymmetry of the cerebral hemispheres, the features of which lead to latent cognitive problems. Features of asymmetry are inherited or acquired during life. Their consequence is the violation of the basic communication interrelations in a complex dynamic system. This is reflected in the system's cognitive model, in which the influence of the human factor prevails. The key element of the model is the spatio-temporal inconsistency of relevant information flows, which complicates the analysis and selection of risk mitigation strategies. It is proposed to eliminate the actions of systematic security threats by applying an interdisciplinary approach. It is developed a structural-functional approach to the study and modelling of the elements functioning of the complex dynamic system. It is shown that such an approach to the modelling allows taking into account the person's psychophysiological state in real-time. The complex dynamic system's viability in extreme conditions depends on the spatio-temporal coordination of the elements functioning.

Keywords: Complex dynamic systems · Dependability · Viability · Human factors · Cognitive aspects · Spatio-temporal patterns · Parametric geometrization · Visual analytics

1 Introduction

The relevance of complex dynamic systems (CDS) dependability forms a significant increase in the complexity of systems (information, technical, physical, biological),

M. Nechyporuk et al. (Eds.): ICTM 2021, LNNS 367, pp. 195–206, 2022.
https://doi.org/10.1007/978-3-030-94259-5_18

which leads to an increase in the number, significance and consequences of the realization of the risks. The interdisciplinarity of the problem is caused by the complex influence on the CDS dependability and safety of a person, society and the environment. It determinates the search for new ways, methods and means of risk mitigation to improve the dependability of the systems functioning taking into account the cognitive aspects of digitalization [1]. The interest in man-machine interaction increased with the rising integration of cyber-physical systems and Industry 4.0, which develops according to exponential laws. Industry 4.0 was marked by the rapid adoption of digital technologies that removed the barriers between the physical and biological areas. The next technological revolution (Industry 5.0) links system dependability to man-machine communication. A key trend of Industry 4.0 is technology and dehumanization, will be changed. Industry 5.0 aims to increase the role of a person, his cognitive and intellectual abilities. The unification of human and machine, the emergence of cobots and the complication of human perception of new robotic systems led to the appearance of interest in humans' cognitive and psychophysiological abilities [1]. Statistical data on man-made incidents evidence the influence of environment and activity stress factors on them. Their analysis indicates that the main reasons for the occurrence of unpredictable events are the features of man-machine interaction, which are associated with the diversity, nonlinearity and blurring of information flows [2]. Accordingly, the account of human cognitive capabilities in risk management getting actual [3].

In this regard, the research aims to the problem of the complex dynamic systems functioning from the standpoint of individuality influence of human factor on their dependability, safety and viability.

The goal is to explore the influence of the human factor on the complex dynamic systems dependability under extreme conditions, to develop means for declining and estimating the human factor.

The objectives are: a) to show possible ways to take into account the influence of the human factor on the complex dynamic systems' dependability at the design stage; b) to develop unified means for the study of cognitive aspects of complex dynamical systems' dependability under extreme conditions.

2 Research Methods

Despite the significant resources that are invested in ensuring safety, the negative consequences of man-made disasters are growing. One of the dominant reasons is the local features of the CDS elements' functional characteristics manifest in extreme conditions caused by external and internal factors. At the same time, the problem of ensuring complex dynamic systems' stability in different ways and approaches are tried to solve [4], which combine intellectual analysis and processing of scientific data based on system analysis and methodologies of foresight and cognitive modelling [5–7].

2.1 Viability as CDS Emergent Property

CDS include a large number of elements, which determines the variety of emergent properties of the system. These properties are advisable to consider from the standpoint of their influence on the system's dependability and safety. The CDS' emergent properties and the hidden relationships in it determine its dynamic stability, which A. Bogdanov called vitality [8]. He showed that unstable or unreliable elements can form a viable, stable and dependable whole, due to their emergent properties. Further, this idea was developed in the works of W.R. Ashby, S. Beer, R. Espejo, R. Harden, Razumovsky, Khazov [9–15]. Thanks to them, the theory of viability was developed for most types of CDS. Viability as a system's emergent property determines its functioning in extreme conditions and explains the existence of the "human factor" phenomenon and a large number of failures in complex systems controlled by a human. Thus, improving the CDS dependability depends on understanding the role of a person, his cognitive, behavioural and psychophysiological characteristics. Viability determines the system functioning in extreme conditions and explains the existence of the "human factor" phenomenon and a large number of failures in complex systems controlled by a human. Thus, improving the CDS dependability depends on understanding the role of a person, his cognitive, behavioural and psychophysiological characteristics. Note that taking into account the cognitive aspects of the CDS' viability in extreme conditions is important for the development of systems for intelligent support for systemic decision making and predictive analytics [16–20]. Thus, **the set of methods that we used to study CDS are the extreme principles of natural science** that underlie the theory of CDS viability.

2.2 Strategies for Managing CDS' Dependability

At the moment, CDS dependability management strategies are based on standardization. Updates and the appearance of new international standards demonstrate the importance of certain procedures to ensure dependability. So, additions appeared in the ISO 9000 standards series related to the control and management of the human factor. Along with the changes in the employe's working place, ergoecology develops, it is introduced the standards indicating the need to control and take into account the human factor – ISO 45001 (Health & Safety), ISO 31000 (Risk management). Finally, the IEC 60300 (Dependability management) standard combines management systems and system dependability analysis and is designed to reduce the risks of design (systematic) errors that are directly related to human activities. It is devoted to the procedures, the implementation of which should increase the dependability indicators of technical systems, most of which are directly related to human and the manifestation of the human factor. Thus, the lack of attention to the problems of man-machine interaction leads to an increase in the negative consequences of the human factor manifestation, which led to the need to introduce changes in safety standards. It should be noted that the dependability management strategies are based on the Japanese concepts of quality assurance: Total Quality Management, lean production, Six Sigma, Kaizen and KAIRYO, Gemba kaizen, Method 5S (Seiri, Seiton, Seiso, Seiketsu, Shitsuke), structured function QFD - (Quality Function Deployment), Re-engineering and EFQM Excellence Model. All of them directly use the principle of total control over human actions to avoid mistakes. And this is the problem because a human in the system should not make decisions that are not provided by the regulations. And it does not work in difficult and extreme conditions. Human activity according to clear instructions and the lack of opportunity to make independent decisions are rigidly ordered system, which over time tends to degrade. That is why the above models of man-machine systems do not work today, do not give the same result as it was several decades ago. **We used a number of standards ISO 9000, ISO 45001, ISO 31000, IEC 60300 to study the CDS.** It allowed us to analyse the operation of the CDS both from the standpoint of safety theoretical provisions and in real work conditions.

2.3 Account of the Influence of the Cognitive Distortions

The CDS dependability today directly depends on the human: his knowledge, perception of information, the presence of cognitive biases or distortions [21, 22]. There is a causal relationship between the CDS dependability and the human factor. According to the theory of practical bias by Scott A. Snook, the expected properties of a system always differ significantly from the obtained result [23]. The calculated dependability of the system is based on three basic assumptions: the absence of problems with software and hardware, the necessary training of a human and the observance of the rules by him. However, it is unreal to ensure three conditions at the same time. In this case, an inevitable shift towards real indicators occurs, the main reason for which is the influence of the human factor. At the same time, the application of human factor engineering to the development of CDS can reduce this "practical bias" and bring the result closer to the predicted one. Thus, the third method that **we used to study CDS is Scott A. Snook's "practical drift" theory**.

2.4 Systemic Problems that are Induced by the Human Factors

The human factor as a source of problems is present in all aspects and stages of the life cycle of complex systems. Human is a source, receiver and analyser of information, and he ensures the reliability, security and stability of the system. The essential challenges of digitization that directly affect complex systems and technologies safety are: features of man-machine interaction in extreme conditions; increasing information complexity of systems (technical, physical, biological); human cognitive abilities (individuality of perception, imagination and thinking), as well as cognitive distortions under stress. The human factor takes place on the stages: creation of complex systems and technologies; selection of relevant information; information processing, analysis and generalization based on previously set or formed evaluation criteria; decision-making based on the analysis of information and image-conceptual model; execution of the decision using a certain sequence of actions. To study the safety and reliability of CDS functioning in extreme conditions, **we adapted methods of systems analysis of the structure of the relationships based on Ishikawa diagrams**. This is the fourth method we used to study CDS.

3 Results and Discussion

3.1 Identification of Dominant Factors Influencing the CDS Reliability and Safety in Extreme Conditions

Thus, the human factor is a cause and a result of system problem in CDS operation at the same time. It is the reason because of systems and technologies are developed without taking into account certain human characteristics. It is the result since the limitation of human cognitive abilities leads to dangerous situations. Analysis of standards ISO 9000, ISO 45001, ISO 31000, IEC 60300 and statistics of the results of their practical application showed that the safety and reliability of CDS depend on standards, rules, algorithms and programs **only theoretically**. In reality, the neuroergonomic, psychophysiological and psychological aspects of human perception of information play a significant role. An adapted version of the Ishikawa diagram demonstrates this statement. (see Fig. 1). **The adapted diagram is the first result of the research.**

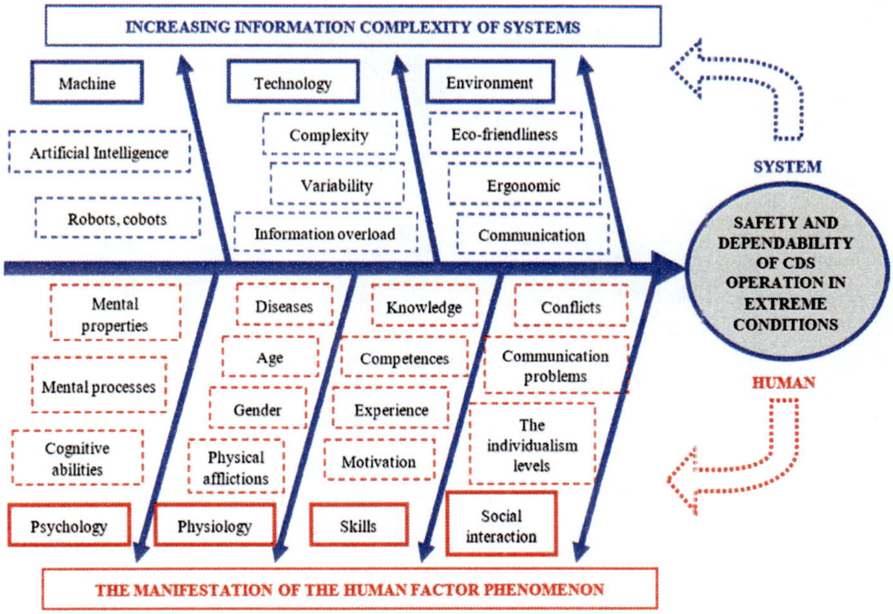

Fig. 1. Factors affecting CDS dependability and safety in extreme conditions (an adapted version of the Ishikawa diagram).

3.2 CDS Cognitive Model

Functional asymmetry of the cerebral hemispheres leads to latent cognitive problems that are genetically inherited or acquired during learning and activity. They manifest in difficult conditions of functioning and cause the manifestation of the "human factor" phenomenon. As a consequence, there is a violation of the basic communication relationships "human – human" and "human – hardware, software and environment" in complex systems controlled by a human. A cognitive model of a complex system in which the influence of the human factor prevails reflects this (transformed SHEL model) (see Fig. 2).

The presented model is built on the basis of practical results of the application of the CDS viability theory principles and provisions of Scott A. Snook's "practical drift" theory for the research of CDS.

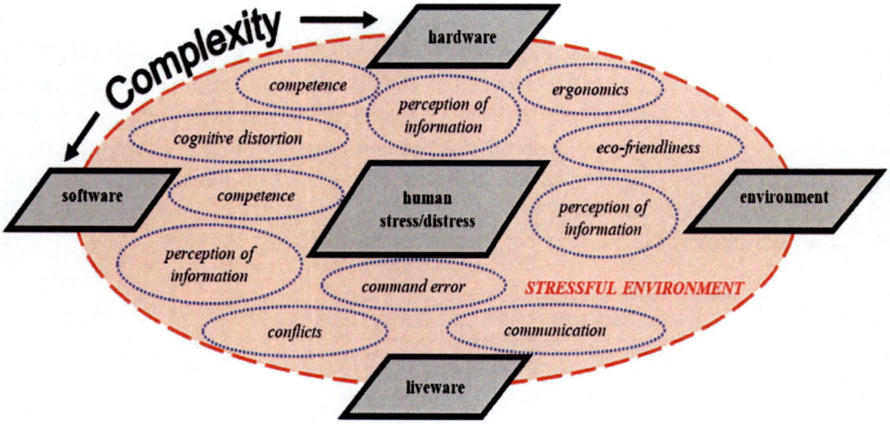

Fig. 2. CDS cognitive model.

The basis of the problem is CDS, which often finds itself in extreme conditions. A key element of the problem is the spatio-temporal inconsistency of all information flows, which significantly complicates human decision-making. Thus, the main reasons for the decrease in the complex dynamic systems viability in extreme conditions are the problems of perception of nonlinearity, ambiguity and increasing complexity of information interaction. **The cognitive model of hidden relationships, in which the human factor predominates, is the second result of the research.**

3.3 The Strategies of Risk Management and Improving CDS Reliability

1. Distribution of random and systematic threats and risks. Eliminating systematic threats by using standards and creating viable systems based on human factors.
2. Training of future complex dynamical systems developers. A transition to convergent learning with an emphasis on understanding the features of man-machine interaction is needed. The transition model can be represented as follows (see Fig. 3).
3. Interdisciplinary view on the "human factor" problem. An important moment in the "human factor" management is the collection and analysis of data on the CDS functioning. At the same time, when studying the CDS behaviour in extreme conditions, an additional question arises: the unification of the means of obtaining, processing and displaying information flows from each of its elements in one space. For today, using of a significant number of unrelated parameters, indicators and criteria from different areas in the study of the CDS functioning proved ineffectiveness. Thus, because the CDS dependability and safety directly depend on the human factor at all stages of the system's life cycle, then only an interdisciplinary view on the human capabilities in CDS creates the necessary basis for improving CDS' dependability and safety. In this case, interdisciplinarity means the intersection of such areas in the CDS study, as a person's psychological and physiological capabilities, his social capital, the technical perfection level of equipment and so on. This approach allows reducing the human factor impact through: study of the human perception cognitive features of information, according to the results of which it is possible to carry

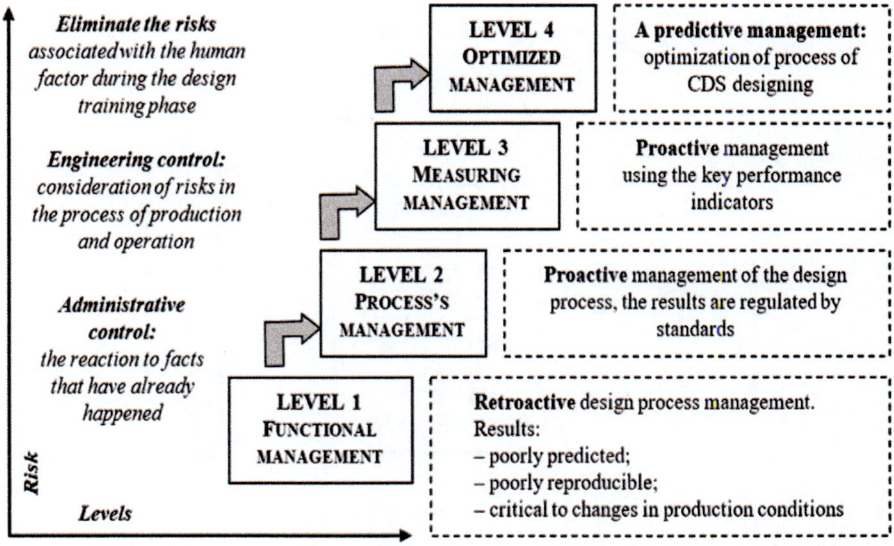

Fig. 3. An engineering training levels of a complex systems designer.

out a rational redistribution of the CDS functions; Determining the characteristics of the CDS behaviour in extreme functioning conditions; Taking into account a person's potential opportunities and existent limitations; Development of a person's professional skills based on an individual approach to learning. The interdisciplinary approach to the study of the "human factor" creates the conditions of managing by him [23]. From the above, the third result of the research follows: the use of methods of the human factor engineering allows to influence the most unpredictable factor in the functioning of the CDS. This means the possibility to manage risks, which will increase the reliability and security of the system as a whole.

4. An application of strategic proactive management principles by the "human factor". Today, for human factor management individual and system approaches are used. According to the individual approach, dangerous actions occur primarily due to aberrant mental processes and countermeasures aimed primarily at reducing unwanted changes in human behaviour. According to the system approach, errors are a consequence of system problems in the system's organisational structure. Obviously, the most effective mechanism for the "human factor" managing can be created using both approaches. In this case, the role of the individual approach is to study the human activity cognitive aspects, and the system one is in the study of CDS behaviour in extreme conditions [23]. In Fig. 4 shows the interrelations in the case of the implementation of the human factor proactive management and ensuring the CDS viability.

The interconnectedness and cyclicity of the stages research and analysis of CDS dependability is shown in Fig. 5.

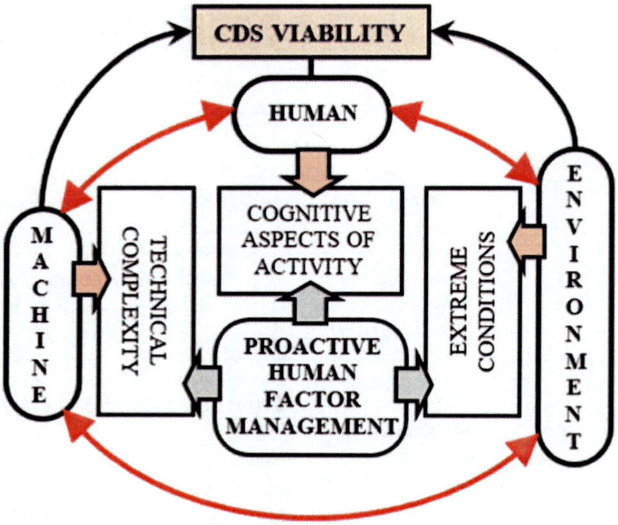

Fig. 4. Interrelations in the CDS in the case of human factor's proactive management.

The starting point is the study and analysis of the human activity cognitive aspects. The research results determine the choice of methods for the "human factor" proactive management. The practical implementation of the selected management methods determines the degree of CDS dependability, which, in turn, leads to risks mitigation arising in the extreme conditions of CDS functioning. Improving the CDS dependability and risks mitigating arising during its functioning determine the level of system safety. The

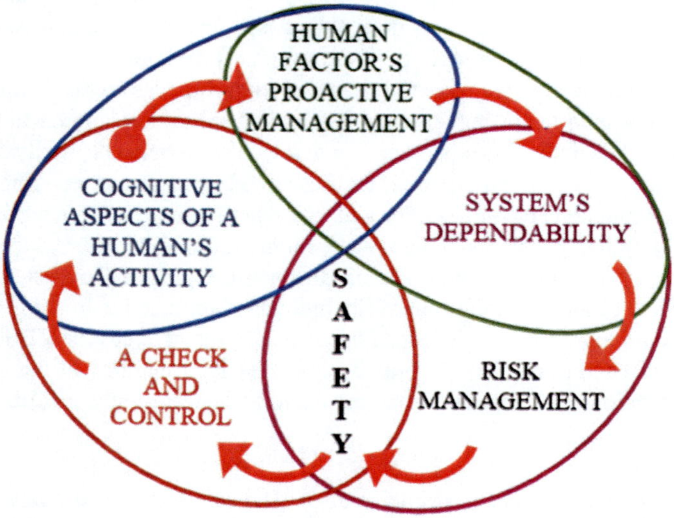

Fig. 5. Cyclicity of the stages of research and analysis of CDS dependability.

final stage is a check and control of the obtained results. Then the cycle is repeated until the required level of CDS dependability and safety is achieved. **The application of the principles of strategic proactive management of the "human factor" is the fourth result of the research.**

Structural-functional approach to the complex dynamic systems studying and modelling. Currently, there is no universal approach to the study and modelling of CDS functioning in extreme conditions. A specific feature of any approach is that it must rely on the notion of system, as well as use the general patterns of CDS' construction, functioning and development. However, common for all approaches to systems research is the determination of the regularities of the system's functioning and the system structure's models synthesis through decomposition. Analysis and synthesis of the model the researched system should solve problems in practice. For this, it is necessary to analyse the functioning of all CDS elements including the operator in the polydisciplinary cognitive space that is proposed in the study [24]. The visualisation in the dynamic events space presented in the study made it possible to connect the following ideas: N. Wiener on the spatio-temporal ordering of the signal structure [25]; Packard and Takens on the reconstruction of the functioning model from a one-dimensional measured signal [26, 27]; Forester on feedback loops and response delays in system dynamics [28].

The synergy of these ideas is most evident in the uniqueness of the dynamic events space [29]. It is due to: interconnected dynamic parameters (state-speed-acceleration); geometric interpretation of the least action principal formulations; energetic interpretation of the least action principle.

Their complementarity simplifies the synthesis of models, the analysis of which can use fundamental laws and principles, as well as thermodynamic criteria. Information about the CDS viability is contained exactly in the individual features of its functioning [29]. The structural-functional approach is based on the connectiveness of the dynamic events space [29, 30]. It is due to the following facts: morphologically different dynamic systems are functionally subordinate to the same principles of physics and biomimicry; different dynamic systems operate according to the same laws; cognitive simplicity is inherent in models in the space of dynamic events; it is possible to visualise dynamic processes that obey the principle of detailed balance; reconstruction of the transformation of the 3D model is accompanied by natural decomposition.

Therefore, the approach is based on the interrelation between local distortions of the fractal signal of the object functioning of study and the features of its dynamics. That is why cognitive visualisation of different nature and scale dynamic processes in the same space underlies the cognitive approach to predictive analytics for safety purposes [30]. In the space of dynamic events, the latent spatio-temporal features of fractal signals become available for cognitive analysis and synthesis. This allows you to identify relevant sources of information and predict their work in difficult conditions.

Cognitive graphics of the CDS elements functioning. It is proposed to visualise the dynamic structure of information flows of various nature in the parametric space of dynamic events to ensure physical, functional and information safety. In this space, a scalar time series (signal) of any nature is transformed into a topological 3D model. Its orthogonal projections are space-time signatures. Our researches [29, 30] evident that **these signatures can be transformed into patterns. This is the fifth result of the**

current research. The use of extreme principles of dynamics and biomimicry for the synthesis and analysis of 3D models of the CDS functioning of different nature has a high innovative potential.

4 Conclusion

The article analyses the interdisciplinarity and cognitive aspects problems of safety, dependability and stability CDS functioning in extreme conditions that are caused by digitalization and increasing dynamic complexity. A variety of concepts that reflect different aspects of dependability are analysed. It is shown that the system viability is the emergence property, which is important for modelling the CDS elements functioning of a different nature, including a human under external influence.

The leading role of the human factor in CDS safety, as well as its dependence on the increasing dynamic complexity of modern systems, lead to insufficient predictability of man-machine interaction. Predictive analytics directly depends on the diversity, nonlinearity and blurring of information flow, the perception of which depends on the person's psycho-functional state. It is shown that increasing the CDS viability in extreme conditions is in the plane of using the methods of human factors engineering in the design of the dynamic system, as well as in the synthesis and analysis of their models. The viability and dependability of dynamic systems are most dependent on various aspects of the human factor that manifest themselves at all stages of the CDS life cycle. At the same time, it is important to eliminate the action of systematic safety threats by applying an interdisciplinary resource approach based on proactive management.

It is shown that for the implementation of monitoring and forecasting the CDS functioning there is a necessity for the development of interdisciplinary tools for predictive analytics. These include a structural-functional approach to modelling the complex system behaviour and state. It was proposed a smart technology in the basis of which 3D visualisation of information flows in the parametric space of dynamic events. This technology makes possible to study and model the CDS functioning, including the assessment of the person's psychophysiological abilities in real-time. It is shown that the methodology of human factors engineering and the development of structural-functional approach to modelling make possible to foresee the problems of physical, functional and information safety, which reduces the risks of cognitive distortions.

References

1. Nahavandi, S.: Industry 5.0—a human-centric solution. Sustainability **1,** 4371 (2019). https://doi.org/10.3390/su11164371. https://www.researchgate.net/publication/335148344_Industry_50-A_Human-Centric_Solution
2. Gevins, A., Smith, M.: Neurophysiological measures of cognitive workload during human-computer interaction. Theoret. Issues Ergon. Sci. **4**(1, 2), 113–131 (2003)
3. Mygal, V., Mygal, G.: Cognitive and ergonomic aspects of human-computer interaction. Radioelectron. Comput. Syst. **1**(93), 90–102 (2020)
4. Zgurovskiy, M., Boldak, A., Efremov, K.: Intelligent analysis and systemic adjustment of scientific data in interdisciplinary research. Cybern. Syst. Anal. **49**(4), 62–75 (2013)

5. Mudla, B.G., Yefimova, T.I., Rudko, R.M.: Dependability as a fundamental generalizing and integrating approach. Math. Mach. Syst. **1**(2), 148–165 (2010)
6. Smith, P.: Network resilience: a systems approach. IEEE Commun. Mag. **49**(7), 88–97 (2011)
7. Mitchell, S.M., Mannan, M.S.: Designing resilient engineered systems. Chem. Eng. Prog. **102**(4), 39–45 (2006)
8. Bogdanov, A.: Tectology: a universal organizational science. In: Economics, Moscow, vol. 1, pp. 213–215 (1989)
9. Ashby, W.R.: Introduction to Cybernetics, Moscow (1959)
10. Beer, S.: Cybernetics and Management, London (1959)
11. Espejo, R., Harnden, R.: The Viable System Model, Chichester (1989)
12. Khazov, M.: Learning to Be a God: Creating Sustainable Systems. Science, Novosibirsk (1992)
13. Razumovskij, O., Khazov, M.: The problem of the viability of systems. Humanit. Siberia **1**, 3–7 (1998)
14. Baksansky, O.: Convergence of knowledge, technology and society: beyond convergent technologies. Philos. Cult. **7**(79), 1061–1068 (2013)
15. Kovalchuk, M.: Convergence of sciences and technologies – a breakthrough in the future. Russ. Nanotechnol. **6**(1, 2), 13–23 (2011)
16. Mygal, V., Mygal, G.: The viability of dynamic systems in difficult conditions: cognitive aspects. In: 11th International IEEE Conference on Dependable Systems, Services and Technologies, pp. 224–229 (2020)
17. Fedota, J.R., Parasuraman, R.: Neuroergonomics and human error. Theor. Issues Ergon. Sci. **11**(5), 402–421 (2010)
18. Lee, J.D., Wickens, C.D., Liu, Y., Boyle, L.N.: Designing for People: An Introduction to human Factors Engineering. SC: CreateSpace, Charleston (2017)
19. Wickens, C.D., Lee, J.D., Liu, Y., Becker, G.: An Introduction to Human Factors Engineering. Prentice Hall, Hoboken (2004)
20. Dul, J., Bruder, R.: A strategy for human factors/ergonomics: developing the discipline and profession. Ergonomics **55**(4), 377–395 (2012)
21. Hettinger, L.J., Branco, P., Encarnacao, L.M., Bonato, P.: Neuroadaptive technologies: applying neuroergonomics to the design of advanced interfaces. Theoret. Issues Ergon. Sci. **4**(1, 2), 220–237 (2003)
22. Parasuraman, R., Mehta, R.: Neuroergonomics: a review of applications to physical and cognitive work. Front. Hum. Neurosci. **7**(889), 1–10 (2013)
23. Mygal, G., Protasenko, O.: The role of human factor in manufacturing safety management. NTU "KhPI" Bull.: Ser. "New Solut. Mod. Technol." **1**(3), 60–65 (2020)
24. Mygal, V., Mygal, G.: Convergent approach to identification of transient states of a dynamic system. J. Nano- Electron. Phys. **12**(6), 06018–06024 (2020)
25. Wiener, N.: Cybernetics: Or Control And Communication in the Animal and the Machine. MIT Press, Moscow (1961)
26. Packard, N.H., Shaw, R.S.: Geometry from a time series. Phys. Rev. Lett. **45**, 712–716 (1980)
27. Takens, F.: Detecting strange attractors in turbulence. In: Rand, D., Young, L.-S. (eds.) Dynamical Systems and Turbulence, Warwick 1980. LNM, vol. 898, pp. 366–381. Springer, Heidelberg (1981). https://doi.org/10.1007/BFb0091924
28. Forrester, J.W.: System dynamics – a personal view of the first fifty years. Syst. Dyn. Rev. **23**(2), 345–358 (2007)

29. Mygal, V., But, A., Mygal, G., Klimenko, I.: An interdisciplinary approach to study individuality in biological and physical systems functioning. In: Scientific Reports, vol. 6, pp. 387–391. Nature Publishing Group (2016)
30. Mygal, V.P., Mygal, G.V., Illiashenko, O.: Intelligent decision support—Cognitive aspects. In: Tagarev, T., Atanassov, K.T., Kharchenko, V., Kacprzyk, J. (eds.) Digital Transformation, Cyber Security and Resilience of Modern Societies. Studies in Big Data, vol. 84, pp. 395–411. Springer, Cham (2021). https://doi.org/10.1007/978-3-030-65722-2_25

"Smart City" Technology: Conception, Security Issues and Cases

Vladimir Pevnev(✉), Anatoliy Plakhteev(✉), Mikhail Tsuranov(✉), Heorhii Zemlianko(✉), and Kyrylo Leichenko(✉)

National Aerospace University "Kharkiv Aviation Institute", Kharkiv, Ukraine
{v.pevnev,a.plakhteev,m.tsuranov,g.zemlynko}@csn.khai.edu

Abstract. Before all "smart" technologies were considered as tools to make the city and its objects more effective for work of inhabitants, the plants, factories and other infrastructure and housing estates. Phones, laptops, tablets and other techniques have become key to the city, putting in millions of hands instant information about transit, traffic congestion, health services, community safety and news alerts, catch and search for bouts, currency analyses and quality of life improvements. However, the rough side of intensive industrialization and population growth in urban areas have become a big problem for administrators, architects and urban planners. This article will discuss the concepts of "smart city" and will show for what the introduction of technologies is and what consequences are expected and cases of realizes of "Smart City".

Keywords: Cybersecurity · Smart City · Urbanization · User experience · Smart grid · Smart network · Wi-Fi · Safety

1 Introduction

Today, more than ever, cities need to develop strategic planning processes, because only then will they be able to chart ways to innovate and prioritize the aspects that are most important to their future.

The megacities of the world are overcrowded, and by 2050, according to some estimates, almost 70% of the total population of the Earth will concentrate. This development makes us doubly responsible for how resources are distributed and processed within such individual territories [1].

The strategic planning process must be participatory and flexible, and a central goal must be set to define a sustainable plan of action that would make the metropolitan area unique and known. Just as the two companies do not have one recipe for success, each city must seek its own model based on a series of shared reflections and considerations [1].

Experience shows that cities must avoid short termism and expand their field of view. They should turn more often to innovation to improve the efficiency and sustainability of their services. They should also promote communication and ensure public and business participation in their projects.

M. Nechyporuk et al. (Eds.): ICTM 2021, LNNS 367, pp. 207–218, 2022.
https://doi.org/10.1007/978-3-030-94259-5_19

It is time to practice sound governance that takes into account all the factors and social actors of the global vision. In fact, over the past few decades, various national and international organizations have produced studies that focus on the definition, creation and use of indicators with a variety of objectives, although they mainly contribute to the diagnosis of cities. The definition of indicators and the process of their creation are the result of the peculiarities of each study, statistical and econometric methods best suited to the theoretical model and available data, as well as the preferences of analysts.

Today we have many "urban" indicators, although many of them are not standardized and consistent and cannot be used to compare cities. In fact, despite numerous attempts to develop urban indicators at the national, regional and international levels, few have been sustainable in the medium term because they have been designed for research designed to meet the specific information needs of some bodies whose lifespans depend on the duration of funding. In other cases, the system of indicators depended on political desire in specific circumstances, so they were abandoned when political priorities or the authorities themselves changed. With regard to indicators developed by international organizations, they did seek to ensure the consistency and sustainability necessary to compare cities; however, in most cases they tend to be biased or focus on a specific area (particularly technology, economics and environment) [2].

The absence of standardized indicators of "Smart Cities" leads to the impossibility of unified development of platforms for implementation of smart functions in cities, which in turn can bring to the developing technology big problems with cybersecurity of data of citizens and key systems of the city. And the impossibility of clearly planning the functional security of critical urban infrastructures.

We must not forget that the human factor is fundamental to urban development. Without a participatory and active society, any strategy, if reasonable and comprehensive, would be doomed to failure. In addition to technological and economic development. The public owns the key to the transition of cities from "smart" to wise. It is goal that every city must strive: that the people who live there and rulers deploy all their talent in favor of progress.

The concept of "smart city" is used not only for the rational use of natural resources, but also for improving the safety of the city. Cameras monitor public order on the streets of the city, and special sensors monitor the state of critical infrastructure. Errors in the design of these systems can lead to their unlawful use and, instead of increasing the level of security of citizens, lead to the disclosure of sensitive private information of citizens.

The purpose of the article is to assess ways of developing smart city technologies with the possibility of maintaining a high level of privacy and security of citizens.

2 Analize of Conception of "Smart City"

A smart city is an interconnected system of communication and information technologies with the Internet of Things (IoT), which simplifies the management of the city's internal processes and improves the standard of living of the population.

Building Smart City is a set of measures aimed at modernizing all spheres of city management. Every city usually has its own tasks. Some solve security problems, others improve public transport, and others reduce energy in homes. In its essence, it resembles an advanced city telecom operator [3]. As we can see most important part of telecom "Smart city" infrastructure is cybersecurity.

The advantages of smart city are to improve the standard of living of citizens and to reduce the costs of business processes by automating activities that do not require the use of analytical skills.

The term "smart city" has been introduced relatively recently, and there is still no explicit interpretation of this concept. However, experts agreed that the main source of smart city management is population data [3].

The British Standard Institution (BSI) describes a smart city as "effectively integrating physical, digital and human systems in an artificially created environment with the aim of ensuring a sustainable, prosperous and comprehensive future for citizens" [4].

Information and communication technologies (ICT) allow city authorities to directly interact with communities and urban infrastructure, and monitor what is happening in the city, how the city is developing, and what ways to improve the quality of life. By using real-time integrated sensors, accumulated data from urban residents and devices is processed and analyzed. The collected information is the key to solving the problems of inefficiency [5].

Digital cities are constantly improving their functions by continuously processing and updating information. Integrated sensors collect information received from residents of the city and using electronic devices. After analyzing the collected data, optimization occurs, solving the problems of inefficiency.

Smart Cities Components: video surveillance and photo fixing, intelligent transport systems (ITS), unified emergency call, unified dispatching service and situation centers, internet of prophetic (IoT), fifth generation mobile (5G).

These information collection and analysis tools are used to improve the functioning of the transport interchange, medicine, industry and other spheres that form the digital city model. And yet their language converges in one thing: a smart city is managed by data, and data management allows municipal services to improve the quality of life of the population. The data cover such areas of life of citizens as security, transport, medical services, public utilities, improvement, etc. Video cameras, various sensors, sensors, information systems and other are data sources [6].

The world experience of creating information systems in various spheres of human activity, recorded in the methods of world grants of management consulting (Big Four, IBM, McKinsey, Microsoft, etc.). The standards define a list of targets whose measurement and monitoring allows cities to assess their development. ISO 37120:2014 "Sustainable community development. Indicators of urban services and quality of life" regulates 46 mandatory and 56 auxiliary indicators in 17 areas [7].

An international team of scientists from Canada, the United States and Mexico identified eight clusters of critical factors that form the content of smart city initiatives [8]: management and organization, ICT technology, management, political context, people and communities, economy, built infrastructure and environment.

These factors form the basis of an integrative smart city model that can be used by city authorities to shape and implement these initiatives, and which is essentially the framework of the initiative. In this case, the framework refers to a large block overview, outline, or block relationship diagram that supports a specific approach to achieving a particular goal, and serves as a guide that can be modified as needed by adding or removing blocks. Moreover, integrativeness refers to the use in one model of ideas about a smart city from various disciplines, for example, science of environmental protection, science of urban planning and architectural design of cities, etc. [8].

The main goal of any city should be to improve its human capital. A city with smart governance should be able to attract and retain talent, create plans to improve education, promote creativity and research.

While human capital includes factors that make it more extensive than what can be measured with these indicators, there is international consensus that educational attainment and access to culture are indispensable components of measuring human capital. A core element of human development is human capital, and given that the Human Development Index, published annually by the United Nations Development Programmed (UNDP), includes education and culture as aspects, these indicators should be considered as factors explaining differences in human capital in the city. Because of this, the speed of smart technology development in different cities is different, so not all stages or elements of the smart city system may be identified [9].

Smart management represents the well-functioning of the communication system between residents of the city and representatives of the executive power, information openness of the city administration, activity of citizens in the management of the city, relevance of documentation of strategic planning, high attendance of official sites of the city administration. In addition, this aspect covers all activities aimed at improving the efficiency of the administration, including the development of new organizational and management models. In this area, there are great opportunities for private initiative, which can bring greater efficiency [10].

Social cohesion is a sociological aspect of cities that can be defined as the degree of consensus among members of a social group or as the perception of belonging to a general situation or project. It is a measure of the intensity of social interaction within a group. Taking care of the social conditions of the city requires an analysis of factors such as immigration, community development and care for the elderly, the effectiveness of the health system, social integration and the safety of the population.

Cities of the future must address two major mobility and transport challenges: facilitating movement through cities (often large ones) and access to public services.

Mobility and transport - both for road and route infrastructure, vehicle fleet and public transport, and for air transport - affect the quality of life of city residents and can be vital to the sustainability of cities over time. However, perhaps the most important aspect is the external factors that arise in [9].

Smart infrastructure is the well-functioning of Internet services for calling and paying for taxis, the ability to monitor road traffic online, the availability of a network of filling stations for electric vehicles, a service to provide carsharing services.

Sustainable city development can be defined as "development that meets the needs of the present without compromising the ability of future generations to meet their own needs." In this regard, improved environmental sustainability through pollution abatement plans, support for green buildings and alternative energy, effective water management and policies that help to address the impacts of climate change are important to ensure the long-term sustainability of cities [10].

A smart environment is a developed system of environmental security monitoring, participation of citizens and administration in eliminating the consequences of unauthorized release of garbage. Cities that want progress must provide a privileged place in the world.

While ICT are not the only important aspect for cities, they are part of the foundation of any society seeking smart status. Technology, an inherent dimension, is an aspect of society that improves the current quality of life, and its level of development or diffusion is an indicator of the quality of life achieved or the potential quality of life. In addition, technological development is a dimension that allows cities to be resilient over time and maintain or expand the competitive advantages of their production system and employment quality.

The covid-19 pandemic showed that the infrastructure of a smart city is extremely necessary in critical situations, but for the effective operation of all subsystems of a smart city, a single standard system is needed that can ensure reliable communication of all devices and sensors, at the moment the most promising is the use of fifth-generation communication technologies, since these technologies allow for scalability and security of the system in the future.

Based on the considered concepts of the smart city, the authors offer their vision of the development of this technology (Fig. 1). The concept of a smart city is built around the communication infrastructure, currently the most secure networks: 5G, Wi-Fi 6 for the wireless segment and fiber optics for the wired. Since the number of connected Smart City subsystems will only increase, it is proposed to make the backbone communication channels fiber-optic in the current concepts. At the moment, 4 main subsystems of the smart city can be distinguished:

- "Mobile City" - a monitoring system for road and public transport, which necessarily includes geo-beacons based on GPS and other positioning systems and smart traffic lights, for example, in Kharkov, a traffic light is used where special cameras monitor cars in a certain area for switching the signal and if pedestrians or vehicles appear there, the main motorway is blocked to allow pedestrians or vehicles from a secondary road to manoeuvre;
- "Security City" - the subsystem consists of a large number of surveillance cameras, as well as sensors for warning emergency services, directly interacts with the Mobile City subsystem to ensure the passage of emergency vehicles.
- "Illuminated City" is a subsystem for lighting city streets depending on external factors, which works on renewable energy sources, for example, in Kharkov, all the main areas and parks for lighting use a similar system based on solar panels and lighting sensors.
- "Comfortable City" is a system for monitoring public services, closely integrated with the networks of suppliers of electricity, gas, heat, water, waste removal and cleaning of house areas, which allows you to monitor not only the quality of services, but also the state of the networks, which will significantly increase the functional safety of urban infrastructure.

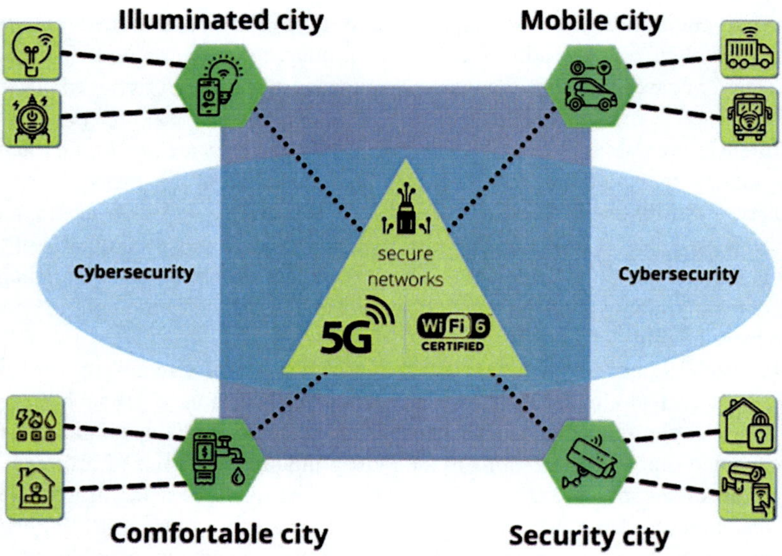

Fig. 1. Smart city functioning concept

3 Cases of "Smart City"

3.1 "Smart City" or "IoT" City of Kharkiv

Since 2016, Kharkiv has continued to work on the creation of a concept and plan for the introduction of smart city technologies. The International Forum "Kharkiv: Development and Investment" discussed cooperation with major world companies on investment and development of the city to international standards on the smart-city system [12].

According to the city head adviser on information technologies, work is under way in the areas of video analytics for a safe city, automation of rapid response services, monitoring of water supply systems, garbage removal.

The city authorities also cooperate in many areas with Ukrainian telecommunication companies. Among such directions - identification of the citizen when receiving electronic services, creation of maps for analysis of traffic flows to optimize traffic in the city [11].

To Kharkiv management of investment development and image projects is created today. For 2 years of activity of this management it is possible to observe certain achievements [12]: the electronic system of data created on the basis of the principle of a cloud computing (it allows to provide services to the population in electronic form); draft of the QR codes "7 miracles of Kharkiv" (the code is on the plastic plate, in these three technologies are put); the program of video surveillance behind the major infrastructure facilities "The safe city" (cameras with high resolution the 28th ability will fix any violations, penetrations on objects, the facts of damage or theft of municipal property).

3.2 Green Technology as Part of a "Smart City"

One of the components in this work is the "green" direction. This requires an interdisciplinary approach and horizontal cooperation in urban development design.

In any case, it is an increase in energy efficiency, energy and resource conservation, improvement of monitoring (air quality sensors, water, climate monitoring, automated environmental control at enterprises, output of data to the control room and distinction of access rights, total accounting and possibility to control consumption regimes in buildings). Accounting (including the introduction of remote readings from metering devices in utilities and the creation of cloud accounting ecosystems). Emission and discharge treatment technologies to reduce environmental impacts.

Green technologies are closely linked to digitalization, as technological advances can produce a new round of efficiency in all areas. A simple example - repairs not according to schedule, but according to the actual condition of equipment, replacement of garbage containers according to actual filling and strength, which helps to save labor, departure of teams and equipment. All planned industry and cross-industry information technology guides for best available technologies (BAT) have already been developed, and a regulatory framework will be established to confirm compliance with BAT and to obtain an integrated environmental permit from enterprises [12]. For energy, it is the emergence of a mix of renewable sources and fossil-based fuels, the transition to cleaner fuels, new models of consumer behavior, the emergence of active users, who can also be suppliers of energy to the network through microgeneration, and thus new mechanisms of balancing and optimization in networks (smart grid).

For housing and utilities, it is already called total digital accounting of consumption of utilities. New models of waste management, the emergence of green, environmentally neutral buildings and passive houses in terms of annual energy balance, smart house technology, and therefore construction, insulation materials, engineering networks, controllers, etc.

The capabilities of "green" solutions supported by digital technologies are available for transport. The theory of green computing for critical infrastructure is presented in [12].

3.3 System of Smart Watering for "Smart City"

A smart watering system can be used in smart city technology to ensure plant vitality in parks and green plantation areas. The authors developed a smart watering system for smart cities based on ARM microcontrollers.

Thus, it was decided to develop an automated watering system based on STM32L1, in connection with ESP01S and WeMos D1. The system consists of at least two subsystems: control and actuating. There may be several executive subsystems, which provide system scalability. Communication of control and actuating subsystems is provided through wireless communication channel WiFi. STM32L1acts as the controller of the control part of the system, providing basic functionality, namely monitoring of climate indicators; Setting of watering modes, time and duration; Providing access to control panel and output of information on current weather conditions and other information. The algorithm system of smart watering is presented Fig. 2. The algorithm demonstrates the operation of the initialization and watering module according to the schedule, the condition for leaving the watering cycle is the user's command, which the device checks after each watering cycle. The command can be set in advance and limited to the number of cycles, and can also focus on the weather forecast when there is a network module or depending on the soil moisture when there is a humidity sensor.

Weather forecast data make it possible to plan water consumption, which in turn allows to make observations of the site, evaporation of moisture. This approach allows eliminating humidity sensors, as this approach is cumbersome and expensive. ESP01S works in conjunction with STM32 to communicate with the executive sub-systems. Monitoring of weather indicators is provided by digital temperature and humidity sensor, illumination sensor and rain sensor. Temporary measurements are made using the real-time clock module. The system is controlled by a joystick and display. The actuating subsystem is a WeMos D1 to which relays having an output to the pumps are connected. Up to 9 relays can be installed per actuator subsystem.

Thus, a new approach is proposed in the architectural feature of a smart watering system, namely, the division of the system into subsystems, and their communication through a wireless communication channel, which provides scalability of the system. The experiment showed that the prototype is capable of performing the tasks and solves the problems present in the known solutions, in particular the problems of scaling and tuning for different tasks. The system architecture provides portability to different scales. Experimentally confirmed that this development can be easily integrated into existing smart city management systems, both at the software and hard-ware levels. The introduction of the considered technology in drylands, such as Dubai, had significantly reduced water consumption, while improving the survival of green spaces, resulting in significant savings in both monetary and water resources. By using the microcontroller, the STM32L1was able to significantly improve the system's resistance to external influences, as well as protect it from cyber-attacks, also have low power consumption and scale architecture, the STM32L1 family contains a fairly large number of different microcontrollers, the specific choice of which depends on the size of the designed system.

4 Security and Safety of "Smart City"

Ensuring the safety of the residents of a "reasonable city" is one of its fundamental functions. Tens and hundreds of thousands of sensors and video cameras collect information on the situation on roads, in public places and buildings, at sports events and territories with limited access in order to notice the threat in time, analyze it and transmit the information to the relevant services. There are a number of major threats (crimes, terrorism and natural disasters) that need to be taken into account in the development of urban security systems. By investigating threats, funding and the possibility of technology adoption, we can classify cities by their degree of readiness to introduce modern urban security systems. Various factors should be taken into account, including the cultural, social and political characteristics of the city [10].

Depending on the quality and number of threats faced by the city, the authorities are ready to allocate budgetary funds to maintain the security of citizens. In cities with high crime rates, surveillance cameras help to contain the number of offences in the most

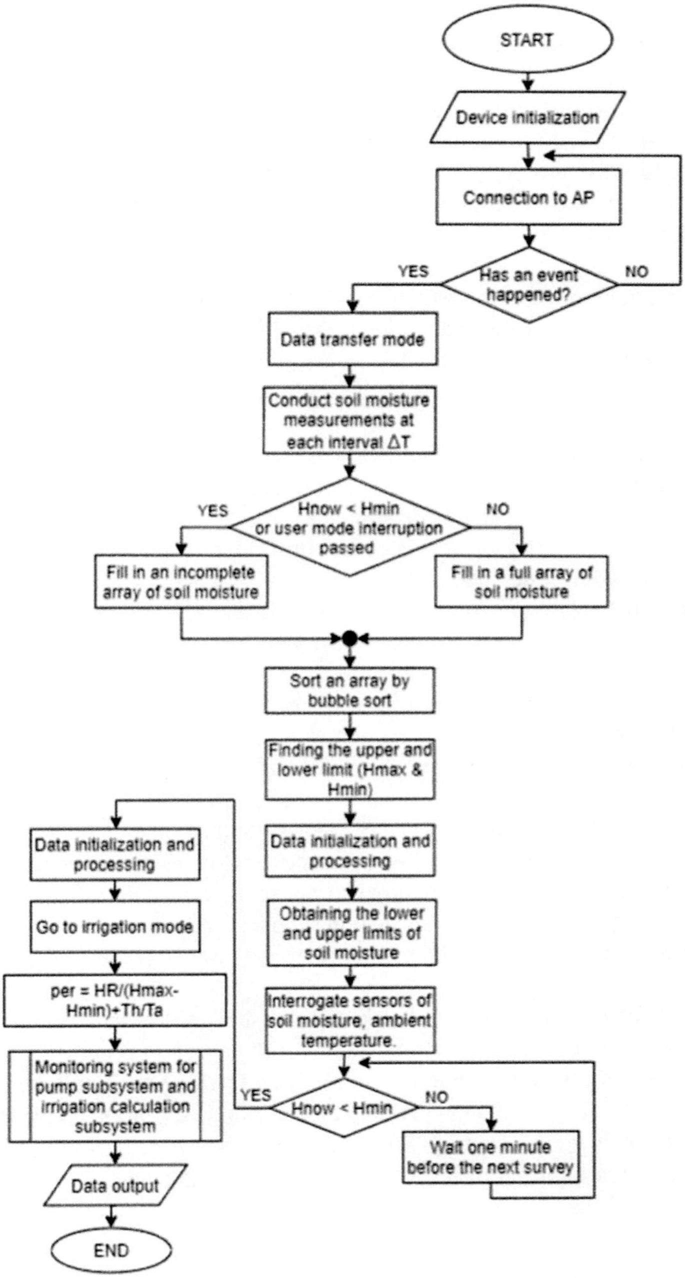

Fig. 2. The algorithm system of smart watering

dangerous neighborhoods, as well as quickly monitor what is happening in the rest of the districts. If security personnel receive information about the type and location of a possible threat in time, they can prevent it. The more dangerous the potential threat, whether it is terrorist or natural, the more reliable the protection must be. In order for the city to be prepared to confront any situation, it is necessary to carefully plan the security system, establish uninterrupted communication channels and introduce technology that will allow the rapid response services to take the necessary action immediately. Functionality of integrated safety systems [10]:

- Continuous monitoring of infrastructure facilities, residential and public buildings and structures, places of mass accumulation of people and carrying out cultural and public events using video cameras and intelligent sensors and video analysis technologies, intelligent video processing in real time with detection of offences and emergency situations.
- Monitoring and quality control of public services: collection and processing of information from sensors and survey video cameras, accounting of up-to-date data on the state of municipal (communal) infrastructure.
- Ensuring industrial safety: operational monitoring of the state of hazardous production facilities, hydraulic structures.
- Situational analysis: automatic generation of incident cards and calculation of optimal response scenarios, maintenance of incident database and their solutions.
- Modeling of possible scenarios: construction of basic models based on accumulated data and their analysis, formation of proposals for preventive organizational and technical measures.
- Communication with the population: operational communication through terminals "Citizen - Police" and "Citizen - Situation Center," use of media devices (info kiosks, graphical signs) for warning about emergency situations.
- Organization of a single operational response center to collect and process information on the current situation, create optimal response scenarios, coordinate the work of services and agencies, provide information support for their interaction.
- Transparency of operation of "Smart City" subsystems with personal data of citizens.

Particular, smart cities may be at risk. Among the key factors contributing to the growth of cyber threats in smart city ecosystems will be lack of cryptographic security, improper management of encryption keys, machine learning for a malicious purpose, lack of knowledge about social engineering methods and lack of DDoS protection.

Based on the above concept of "Smart City" and subsystems of which it consists, taking into account the need to involve citizens in the implementation and control of the work of these subsystems, when designing the security subsystems of "Smart City," it is necessary to pay attention to the following aspects: openness of source code, to involve citizens in the process of testing security systems; audit of security systems by various public audit companies; the presence of a large amount of illustrative material, in a language understandable to both citizens and experts, about the principles of working security systems "Smart City".

5 Conclusion

Urbanization is one of the most important challenges of the twenty-first century. As the world's population moves towards cities, problems are growing and new ones are emerging, which are also heavily affected by globalization. This tendency means a relationship between global dynamics and cities, generating local impacts: economic impacts, demographics, social divisions, or environmental impacts. Despite these challenges, cities and their leaders must understand the positive aspect that cities have. In our view, the city offers a much more distinct sphere of action that allows us to work more directly for the benefit of people. But city leaders must step back and analyze their problems, try to discover what other cities are doing, and learn what best practices are being carried out elsewhere in the world. The day-to-day management of the city makes it difficult for cities to identify how to contribute to the positive impact of urbanization and how to reduce the negative impacts. According to analysts, AI will be one of the cornerstone technologies used in building smart cities, along with personalized health care, robotics, distributed energy and advanced driver assistance systems. In the process of building smart cities, most of its users, including private ones, will be centrally stored in the smart city system, which will lead to the emergence of new and updated threats in the cybersecurity system of this technology.

Problems with the privacy and accessibility of these citizens need to be solved now, having prepared the necessary normative documents and standards for this purpose, as well as it is necessary to develop a set of strict technical measures to ensure functional and information security of smart cities. The safety level of which should not be lower than nuclear power plants, which in the process of implementation can lead to significant contradictions between various subsystems of Smart City technology.

First, a man with his needs and safety is at the head of the corner. Smart technologies are used to serve the well-being and comfort of people in the city and are not an end in themselves. Second, it is more correct to implement solutions on a single platform, rather than to break down individual pieces. Third, the principle of data openness is important. Fourthly, it is important to provide consensus and acceptance of the events by all interested parties, stakeholders. Synchronization of "green" technologies and "smart" means new economic, motivational models, innovations in management and communication system, contractual capacity of all involved participants. In fact, today there is a design of the future of cities for several decades to come, and no one will give it to the city outside, it is a product of their own intellectual and organizational work, search for compromise and consent of all parties, formation of a technical assignment and own solutions.

References

1. Kulkarni, S., Kulkarni, S.: Communication models in IoT: a survey. IJSTE – Int. J. Sci. Technol. Eng. **3**(11), 2017 (2017)
2. Kienzler, R.: Digital twins and the Internet of Things (2019). https://developer.ibm.com/articles/digital-twins-and-the-internet-of-things/. Accessed 25 June 2019
3. Kroes, N.: European Commissioner for Digital agenda, The Critical Role of Cities in Making the Digital Agenda a Reality. Closing Speech to Global Cities Dialogue Spring Summit of Mayors Brussels, 28 May 2010 (2010)

4. Smart Cities|BSI: Smart Cities Current Standarts Portfolio (2020). https://www.bsigroup.com/en-GB/smart-cities/. Accessed 12 Feb 2020
5. European Commission, DG INFSO: Future Internet Research and Experimentation, September 2010
6. Mora, L., Bolici, R.: How to become a smart city: learning from Amsterdam. In: Bisello, A., Vettorato, D., Stephens, R., Elisei, P. (eds.) SSPCR 2015. GET, pp. 251–266. Springer, Cham (2017). https://doi.org/10.1007/978-3-319-44899-2_15
7. Florida, R.: The Rise of the Creative Class: And How It's Transforming Work, Leisure, Community and everyday Life. Basic Books, New York (2007). 432 p.
8. Bakici, T., Almirall, E., Wareham, J.: A smart city initiative: the case of Barcelona. J. Knowl. Econ. **4**(2), 135–148 (2013)
9. Estevez, E., Lopes, N.V., Janowski, T.: Smart sustainable cities – Reconnaissance study, Canada (2017). https://joinup.ec.europa.eu/sites/default/files/document/2016-04/smart_cities_report.pdf. Accessed 18 Feb 2020
10. Kaufman, T., Bothe, D., Gawlik, W., Ponweiser, K.: Optimization of load flows in urban hybrid networks. In: Bisello, A., Vettorato, D., Stephens, R., Elisei, P. (eds.) SSPCR 2015. GREEN, pp. 3–13. Springer, Cham (2017). https://doi.org/10.1007/978-3-319-44899-2_1
11. The city of Kharkov: Image projects of Kharkov - Smart City (2019). https://invest.kh.ua/ru/kharkiv-strategy-success/image-projects/641-kharkiv-smart-city. Accessed 20 Feb 2020
12. Smart Cities: Securing the Smart City (2019). https://www.govtech.com/security/Securing-the-Smart-City.html. Accessed 15 Feb 2020

Control Systems and Engineering

Development and Modeling of a Device for Measuring the Moisture of Oil Products

Nikolay Koshevoy, Oleksandr Zabolotnyi, Oleksii Potylchak(✉), and Vitalii Siroklyn

National Aerospace University "Kharkiv Aviation Institute", 17 Chkalova Str., Kharkiv 61070, Ukraine
kafedraapi@ukr.net, o.potylchak@khai.edu

Abstract. A portable device for measuring the moisture of oil products in the range from 0 to 30% by volume has been developed. Using the cost-optimal planning of the experiment, a research was carried out to obtain an analytical dependence that would allow calibrating the device with a sufficient level of accuracy without making test samples with different moisture content. To implement the calibration process, mathematical models were obtained in coded and natural values of factors.

Keywords: Calibration · Moisture meter · Oil products · Device · Experiment planning · Cost · Mathematical model · Humidity · Accuracy

1 Introduction

The process of measuring the moisture content of materials is widely used in various branches of science and technology [1–13]. To automate this process, it is necessary to develop new methods and devices that make it possible to measure this parameter with sufficient accuracy. This raises the problem of calibrating the developed devices for measuring the moisture content of various materials.

Calibration problems for dielcometric moisture meters, which are used to measure the moisture content of various materials, are considered in many literary sources.

Some of them are devoted to the calibration of instruments of foreign companies for measuring the moisture content of domestic materials. For example, in [14], the calibration of the WILE200 device of a Finnish company on domestic crops and varieties in the ranges of measuring their main properties characteristic of Ukrainian conditions is considered.

In the dissertation work [15] methods of moisture meter calibration are developed to control the moisture content of grain products.

In the collection of scientific papers [16], the issues of the refined calibration of capacitive moisture meters for a specific soil are considered.

For moisture meters of wood "Kedr-1" and "Kedr-2", which were developed at the Engineering Bureau of the Aviation Institute (Kharkiv, Ukraine), calibration for various materials were provided [17].

M. Nechyporuk et al. (Eds.): ICTM 2021, LNNS 367, pp. 221–228, 2022.
https://doi.org/10.1007/978-3-030-94259-5_20

In [18], a pulse digital dielcometric converter of the moisture content of industrial oil with automatic error correction is considered, but the results of its calibration are not given.

When calibrating moisture meters for measuring the moisture content of emulsions, one can use the analytical dependences of Piekar, Kubo-Nakamura, Wiener, but they have a complex form of the function $\varepsilon = f(\varepsilon_{H}, W)$ [19].

In this regard, the problem arises of constructing a simpler mathematical model $\varepsilon = f(\varepsilon_{H}, W)$, which allows the calibration of moisture meters for viscous oil products.

2 Main Results of Research

Employees of the Department of Intelligent Measuring Systems and Quality Engineering of the National Aerospace University "Kharkov Aviation Institute" has developed a portable device for measuring the moisture content of oil products in the range from 0 to 30% by volume.

Figure 1 shows the functional diagram of this device. The principle of the device operation is based on the formation of a width-modulated signal at the output of the operational amplifier 5, whose characteristics depend on the measured humidity.

Before measuring the moisture content of materials, it is necessary to check the correct functioning of the device. To do this, the operator closes the switch 8 in an electrical circuit containing a series-connected stabilized power supply 9 and a loop of wire 2, through which the reference current I_{et} passes. This current induces a magnetic flux in the core of the magnetically sensitive transducer

$$F_{et} = \frac{I_{et}}{R_m},$$

where R_m – the magnetic resistance of the core, characterized by its geometric dimensions and magnetic permeability.

Due to the influence of this flow, a width-modulated signal is formed at the output of the operational amplifier. When the device functions correctly, the indicator 20 shows the constant value.

To measure moisture, the operator breaks with a switch 8 the electrical circuit, consisting of a series-connected stabilized power supply 9 and a turn of wire 2.

During the measurement of the moisture content of a particular material, the capacitance of the measuring transducer 11, filled with this material, is compensated by a corresponding capacitor from the capacitance box 12. In the case when there is no moisture in the material under research, the device fixes a zero moisture value.

When there is a certain amount of water in the material, the capacitance of the measuring transducer with the material 11 increases relative to the capacitance of the corresponding capacitor from the reference capacitance box 12. The duty cycle of rectangular pulses at the corresponding output of the multivibrator 10 increases in proportion to the moisture contained in the material. This leads to an increase in the constant component of the signal at the output of the corresponding signal rectifier, and a difference current I appears in the magnetosensitive converter 1, obtained by subtracting the signals from the windings 3 and 4. Due to the appearance of such a current, the duration of the pulses of positive polarity at the output of the operational amplifier 5 will increase in proportion to the moisture contained in the material, with respect to the duration of the pulses of negative polarity (the value of the signal period at the output of the operational amplifier remains unchanged).

Thus, a certain constant signal appears at the output of the low-pass filter 16, the value of which is proportional to the moisture contained in the material. After amplification, this signal is fed to an analog-to-digital converter 19 and is fixed by indicator 20. A more detailed description of the operation of the device is given in [20].

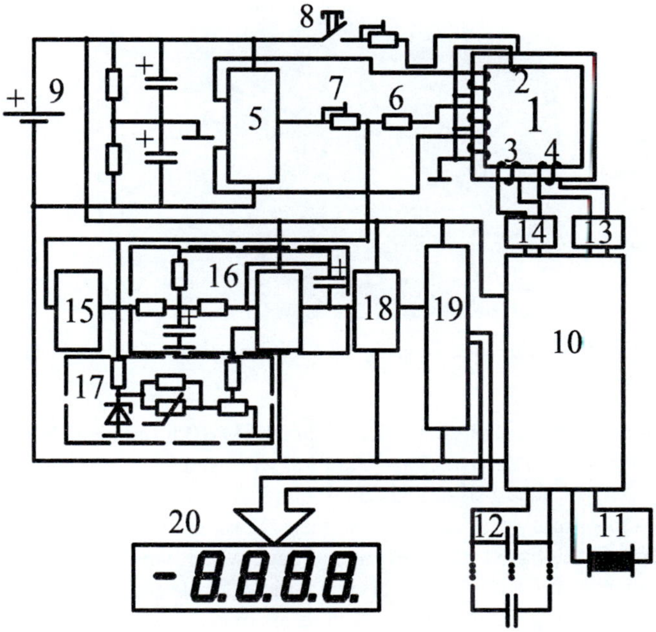

Fig. 1. Functional diagram of the device for moisture measuring.

Notation in Fig. 1: 1 – magnetosensitive converter; 2 – wire loop; 3.4 – additional windings; 5 – operational amplifier; 6.7 – resistors; 8 – switch; 9 – power supply; 10 – multivibrator; 11 – capacitive measuring transducer; 12 – store of reference containers; 13,14 – rectifiers; 15 – signal generator; 16 – low-pass filter; 17 – temperature compensation circuit; 18 – information signal amplifier; 19 – analog-to-digital converter; 20 – digital indicator.

This device [20] consists of an electronic unit and a measuring transducer located in a container. The electronic unit is equipped with an analog or digital indicator (Fig. 2).

To perform measurements the container is filled with the research oil product (fuel oil, heating oil, etc.), then the moisture value for this product appears on the indicator of the electronic unit. It is also possible to install a measuring transducer in a pipe elbow (Fig. 3).

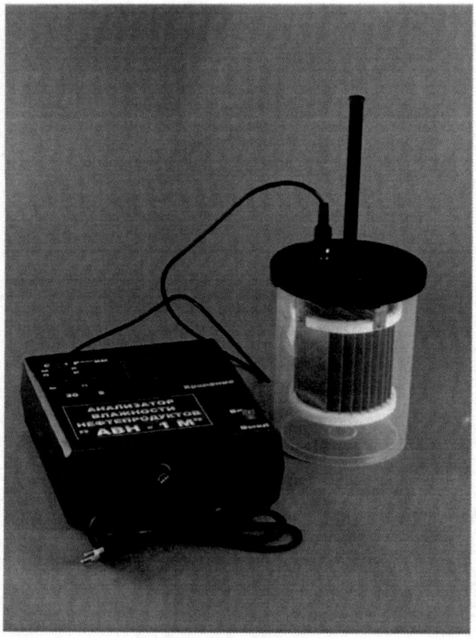

Fig. 2. Device for measuring the moisture content of oil products.

To calibrate the device, a reference fuel oil of the "M-100" brand of Closed Joint-stock company "Ukrtatnafta" (Kremenchuk, Ukraine) was used.

In order not to produce a significant number of samples with different water content, it was decided to simplify the calibration process. For this, it is proposed to construct an analytical dependence based on experimental data, which would allow to calibrate the developed device with a sufficient level of accuracy. To synthesize such a dependence, the experiment planning that was optimal in terms of costs was used [21].

For experimental studies, the following factors were chosen as the dominant factors: X_1 – relative dielectric constant $\varepsilon_{н}$ of dehydrated oil product; X_2 – water content W in percentage. The output index is the relative permittivity ε of the internal phase.

The initial plan of a full factorial experiment to obtain a mathematical model $\varepsilon = f(\varepsilon_{н}, W)$ is given in Table 1.

The cost of changes in the values of the factors levels are given in Table 2.

Let us optimize the initial experiment according to the criterion of the total cost of its implementation using the method based on the Gray code [22]. The matrix of the optimal experimental design is shown in Table 1.

The cost of implementing the experiment according to the optimal plan is 70 conventional units, while the cost of the initial plan is equal to 80 conventional units.

Thus, we have a gain in the cost of implementing the experiment by 1.14 times in comparison with the initial plan. The values of the selected factors levels and their variation intervals are presented in Table 3.

Fig. 3. Positioning the transmitter in pipe elbow.

Table 1. Initial and optimal plans of experiment.

Initial plan			Optimal plan		
Number of experiment	X_1	X_2	Number of experiment	X_1	X_2
1	– 1	– 1	1	– 1	– 1
2	+1	– 1	2	+1	– 1
3	– 1	+1	3	+1	+1
4	+1	+1	4	– 1	+1

Table 2. The cost of changes in the values of the factors levels.

Factor	Cost of changes in the values of the factors levels, conv. units			
	from "0" to "+1"	from "0" to " –1"	from " –1" to " +1"	from " +1" to " –1"
X_1 (ε_H)	10	10	10	10
X_2 (W)	60	20	20	60

Table 3. The values of the selected factors levels and their variation intervals.

Factor	Level of the factor			Variation interval	Dimension
	– 1	0	+1		
X_1 (ε_H)	1,6	2,8	4,0	1,2	-
X_2 (W)	0	15	30	15	%

The lower and upper levels of the factor $X_2(W)$ were selected according to the measurement range of the device (from 0 to 30%). The values of the lower and upper levels of the factor $X_1(\varepsilon_H)$ were established by determining the dielectric constants for fuel oil of the "M-100" brand and a previously dehydrated heavier oil product with a significant content of sulfur, paraffin and asphalt resins. The dielectric constant ε_H in both cases was determined from the ratio of the capacity of the measuring transducer filled with oil to the capacity of the empty measuring transducer (filled with air). The measurement results are shown in Table 4.

Table 4. Planning matrix and measurement results.

Number of experiment	X_1	X_2	Y_1	Y_2	Y_3	Y_4	Y_5	Y_6	Y_7
1	– 1	– 1	1,600	1,585	1,586	1,556	1,579	1,594	1,602
2	+1	– 1	4,012	3,977	4,023	3,990	3,969	4,012	4,001
3	+1	+1	7,833	7,844	7,820	7,787	7,800	7,864	7,800
4	– 1	+1	2,917	2,932	2,895	2,917	2,910	2,895	2,917

Number of experiment	Y_8	Y_9	Y_{10}	Υ
1	1,571	1,579	1,586	1,584
2	4,023	3,980	4,012	3,999
3	7,825	7,843	7,832	7,823
4	2,902	2,902	2,887	2,907

As a result of processing the experimental data, mathematical models were obtained in coded and natural values of factors that characterize the dependence of the dielectric constant of a two-phase emulsion on the dielectric constant of a dehydrated oil product and the water content in this emulsion:

- in coded values of factors

$$Y = 4.079 + 1.833X_1 + 1.287X_2 + 0.625X_1X_2; \quad (1)$$

- in natural values of factors

$$\varepsilon = -0.015 + 0.988\varepsilon_H - 0.012W + 0.035\varepsilon_H W. \quad (2)$$

The mathematical model (2), synthesized using the theory of cost-optimal experiment planning [21, 22], will allow to calibrate viscous oil products moisture meters for the specified absolute error and measurement range only by determining the dielectric constant $\varepsilon_н$ of the oil product under research.

3 Conclusion

For the developed portable dielectric moisture meter using the theory of cost-optimal planning of the experiment, a mathematical model was obtained that characterizes the dependence of the two-phase emulsion ε on the dielectric constant $\varepsilon_{н}$ of the oil product under research and the water content in this emulsion. The obtained mathematical model will make it possible to calibrate moisture meters, only by determining the dielectric constant $\varepsilon_{н}$ of the product under research.

References

1. Sharma, P., Yeung, H.: Recent advances in water cut sensing technology. In: Yurish, S.E. (ed.) Advances in measurements and Instrumentation: Reviews, vol. 1, pp. 147–175. IFSA Publishing, Barcelona (2018)
2. Kadota, T., Yamasaki, H.: Recent advances in the combustion of water fuel emulsion. Prog. Energy Combust. Sci. **28**(5), 385–404 (2002). https://doi.org/10.1016/S0360-1285(02)00005-9
3. Shinjo, J., Xia, I., Ganippa, L.C., Megaritis, A.: Physics of puffing and micro-explosion of emulsion fuel droplets. Phys. Fluids **26**(10), 103302 (2014). https://doi.org/10.1063/1.4897918
4. Garvey, R., Fogel, G.: Estimating water content in oils: moisture in solution, emulsified water, and free water. In: Technology Showcase: Integrated Monitoring Diagnostics and Failure Prevention. Proceedings of a Joint Conference, Mobile, Alabama, pp. 1–14, 22–26 (1996)
5. Zabolotny, A.V.: Improving efficiency of the quality control of substances with dielectric properties. In: Zabolotny, A.V., Koshevoi, N.D., Gordienko, V.A. (eds.) Telecommunications and Radio Engineering, vol. 57, № 2–3, pp. 184–196 (2002). https://doi.org/10.1615/TelecomRadEng.v57.i2-3.200
6. Zabolotnyj, A.V., Koshevoj, N.D.: Development, examination and optimization of the device for quality control of dielectric materials, №. 1, pp. 39–43 (2004)
7. Zabolotnyi, O.V.: Conditionality examination of the new testing algorithms for coal-water slurries moisture measurement. In: Zabolotnyi, O.V., Zabolotnyi, V.A., Koshevoi, N.D. (eds.) Scientific Bulletin of National Mining University. Scientific and Technical Journal, № 1(163), pp. 51–59. https://doi.org/10.29202/nvngu/2018-1/21
8. Koshevoy, N., Zabolotny, A., Koshevaya, I., Kostenko, E., Rozhnova, T.: Research of moisture-meter device for bulk and liquid materials. In: 29th International Scientific Symposium “Metrology and Metrology Assurance”. Proceeding. Institute of Electrical and Electronics Engineers Inc. (2019). https://doi.org/10.1109/MMA.2019.8935983
9. Zabolotnyi, O., Koshevoi, M.: An effective method of bulk materials moisture measurement using capacitive sensors. J. Stored Prod. Res. **89** (2020). https://doi.org/10.1016/j.jspr.2020.101733
10. Zabolotnyi, O., Zabolotnyi, V., Koshevoy, N.: Oil products moisture measurement using adaptive capacitive instrument measuring transducers. In: Nechyporuk, M., Pavlikov, V., Kritskiy, D. (eds.) ICTM 2020. LNNS, vol. 188, pp. 81–91. Springer, Cham (2021). https://doi.org/10.1007/978-3-030-66717-7_7
11. Koshovyi, O., Raal, A., Kovaleva, A., Myha, M., Ilina, T., Borodina, N., Komissarenko, A.: The phytochemical and chemotaxonomicstudy of Salvia spp. growing in Ukraine. J. Appl. Biol. Biotech. **8**(03), 29–36 (2020). https://doi.org/10.7324/JABB.2020.80306
12. Starchenko, G., Hrytsyk, A., Raal, A., Koshovyi, O.: Phytochemical profile and pharmacological activities of water and hydroethanolic dry extracts of Calluna vulgaris (L.) Hull. Herb. Plants **9**, 751 (2020). https://doi.org/10.3390/plants9060751
13. Shinkovenko, I.L., Kashpur, N.V., Ilyina, T.V., Kovalyova, A.M., Goryacha, O.V., Koshovyi, O.M., Toryanyk, E.L., Kryvoruchko, O.V.: The immunomodulatory activity of the extracts and complexes of biologically active compounds of Galiumverum L. herb. Ceska a Slovenska Farmacie **67**(1), 25–29 (2018)

14. Sekanov, Yu.P., Stepanov, M.A., Pavlov, E.L.: Moisture tester WILE200: research results. Veg. Russ. **4**, 94–97 (2018). https://doi.org/10.18619/2072-9146-2918-4-94-97
15. Afonin, V.S.: Development of the device for control of humidity of grain products on the basis of multi-electrode capacitive converters: dissertation .... Candidate of Technical Sciences, 05 November 2013, Barnaul (2007). 134 p.
16. Collection of Scientific Papers of the International Scientific Conference: Fundamental Concepts of Soil Physics: Development, Modern Applications and Prospects, Moscow, 27–30 May 2019: Collection of Articles. "KDU", "Dobrosvet", Moscow (2019). 795 p.
17. Koshevoy, N.D.: Direction and prospects of scientific research of the department of aviation instruments and measurements. Radioelectron. Comput. Syst. **5**(24), 22–28 (2007)
18. Brago, Velikanov, Martynov: Comprehensive quality control of industrial oil preparation at the stages of dehydration and desalination. Autom. Telemech. Commun. Oil Ind. (6), 5–8 (2016). https://rucont.ru/efd/383740
19. Krichevsky, E.S., Bondar, V.K., Benediktov, M.V., et al.: Theory and practice of express control of moisture content of solid and liquid materials. Energy, Moscow (1980). 240 p.
20. Zabolotny, O., Koshovy, M.D., Zabolotny, V.A.: Pat. 53186 Ukraine, МПК G01N 27/22. Capacitive moisture meter of materials. No. 2002042547; app. 04/01/02; publ. 15.01.03, Bul. №1, 4 p.
21. Koshevoy, N.D., Kostenko, E.M., Gordienko, V.A., Syroklyn, V.P.: Optimum planning of experiment in manufacturing the electronic equipment. Telecommun. Radio Eng. **70**(8), 731–734 (2011). https://doi.org/10.1615/TelecomRadEng.v70.i8.60
22. Koshevoy, N., Zabolotnyi, A., Koshevaya, I., Rozhnova, T., Tsekhovskoi, M., Bekirov, A.: Methods and devices for angular measurement movement. In: 30th International Scientific Symposium Metrology and Metrology Assurance, MMA, 9254257 (2020)

Method of Grain Moisture Measurement with Application of Testing Influences on a Substance Under Research

Oleksandr V. Zabolotnyi(✉), Vitalii A. Zabolotnyi, and Nikolay D. Koshevoy

National Aerospace University "Kharkiv Aviation Institute", Kharkiv 61070, Ukraine
pretorian14@ukr.net

Abstract. When we speak about moisture measurement in grain using capacitive sensors we face with different values of dielectric permittivity for different grain types in dehydrated state, what causes a method error that can be named 'type uncertainty'. That problem is even more significant, because different varieties of the same grain have different values of dielectric permittivity, which depend from geographical origin, processing conditions etc. These situations can be hardly predicted automatically and type uncertainty can be compensated only in separate situations with the help of preliminary calibration. The main task of the research is to decrease the type uncertainty value by developing new method of moisture measurement with testing influences on the substance under research. To estimate the relevant level of testing influences a group of different materials with 2, 2,5, 3 and 3,5 dielectric permittivity values was taken for 0, 10, 20 and 30% of moisture content. It was suggested to apply two additive and two multiplicative testing influences on a substance under research. This was enough to build a system of equations, which solvation would give us a possibility to get almost linear static function of grain moisture meter and calculate initial moisture content of a substance. During the process of experimental researches nominal moisture content values 0%, 10%, 20% and 30% were reproduced by certain types of grain with different dielectric permittivity values: pearl barley ($\varepsilon = 3{,}68$); poppy ($\varepsilon = 3{,}56$); millet ($\varepsilon = 3{,}17$); pea ($\varepsilon = 2{,}97$); wheat cereals ($\varepsilon = 2{,}55$). Experimental setup for measuring moisture content was assembled using the substitution method to provide good accuracy for the conditions of capacitance measurement in substances with significant dielectric losses. Obtained experimental results helped to estimate the value of related full scale error, equal to $\gamma = 1.7\%$, what corresponds the restrictions of OIML R59.

Keywords: Moisture measurement · Grain · Capacitive sensor · Type uncertainty method error · Testing influence · Related full scale error

1 Introduction

Grain farming of Ukraine is a strategic and one of the most effective branches of economy. Grain itself and grain products have always been liquid as they form the basis of the state food safety. First of all it is caused by special significance of bread and other bakery

M. Nechyporuk et al. (Eds.): ICTM 2021, LNNS 367, pp. 229–245, 2022.
https://doi.org/10.1007/978-3-030-94259-5_21

products for native system of population feeding, their comparatively moderateness of prices and availability. Ukrainian grain belongs to rather short list of products that bring good benefits on the internal market and can be competitive abroad.

It is necessary to emphasize that Ukraine is among the top 10 world grain producers in accordance with FAOSTAT. Local grain farming produces approximately 25 million tons of grain and demonstrates stable growth of grain production index. Structure of local grain production is relatively stable: biggest part traditionally belongs to the hard, red variant of winter wheat which is used in bread making (up to 90%), part of rye is not more than 10% and part of other grain types is up to 1%.

In spite of that approximately 50% of produced grain requires quality improvement. It is reasonable that grain producers always try to use its' food potential most fully, especially peripheral components and fetus as a source of useful nutrients. Grain processing industry implements a list of technologies for that purpose. Among them we have long term hydrothermal grain treatment with further sprouting and its usage as basic foodstuffs. It is possible to use one or two times humidification with further binning, steaming with humid saturated steam for 20–30 s and 44–46 °C temperature control during 20 min for a grain with weak gluten to improve its bakery properties.

Long term hydrothermal treatment (cold conditioning) during the process of traditional wheat preparation to refining, in particular one after another cycles of grain humidification and binning during 28 h and its germination during 48 h, increases the vitamins concentration: thiamine – on 66%; chlorine – on 67%; inositol – on 22%; niacin – on 18%; riboflavin, pyridoxine, ascorbic acid – in more than two times, and concentration of tocopherols increases in 55 times [1–3]. During the process of grain germination change in amino acid composition is observed, concentration of indispensable amino acids increases: lysine – from 48% to 65%; isoleucine – from 74 to 86%. It is proven that during the process of grain hydrothermal treatment and further germination at 22 °C temperature intensity of E, C and B vitamins formation is maximal in a period of sprout development from the beginning of germination up to 2–3 mm length.

From other hand, modern performance indexes of grain are regulated with a group of national and international standards, where it is mentioned that grain moisture content shouldn't be more than 14,5…15,5% (in general) to provide its long – term storage [4].

All it makes possible to deduce that moisture measurement with high accuracy is really necessary for all these examples.

Big number of factors influence on moisture measurement accuracy. Among them we traditionally have physicochemical composition, granulometric composition and temperature variation. Not only material's type or sort has an effect on its physicochemical properties, but conditions of extraction, processing, etc., and such group of factors is complicated for analytic forecasting. Granulometric composition of different bulk materials is usually far from ideal what decreases the repeatability of measurements. Temperature influence, in its turn, can be taken into account by the introduction of correction coefficients into the result of moisture measurement.

A lot of analytical methods were developed to define the moisture content of foods, which are based on the fact that water has appreciably different bulk physical characteristics than the food matrix, e.g., density, electrical conductivity or refractive index. These methods are usually suitable only for analysis of foods in which the composition of food matrix does not change significantly, but the ratio of water-to-food matrix changes. For example, the water content of oil-in-water emulsion can be defined by measuring its' density or electrical conductivity, because density and electrical conductivity of water are significantly higher than those of oil. If composition of the food matrix changes as well as the water content, then it may not be possible to determine the moisture content of food accurately because more than one food composition may give the same value for the physical property being measured [5].

Unfortunately, it is impossible to get analytic forecast of chemical composition and different features of all possible food substances, being under research. That's why most of indirect methods of moisture measurement have local effectiveness and are not versatile, and in-depth analysis of grain moisture control modern approaches with a purpose of further development and improvement is still a relevant task.

Problem Statement. Now we have approximately 33% of moisture meters, represented by dielcometer measuring instruments with capacitive sensors [6, 7]. Moisture meters of this type have method error (type uncertainty), connected with different values of dielectric permittivity for the substances under research in dehydrated state. Influence of this error is usually significant and traditional ways of compensation can be effective when we know the composition of a substance. In other case effectiveness of traditional compensation methods essentially decreases. Mainly because of that further improvement of the existing methods of moisture measurement with a task to solve the problem of type uncertainty compensation is a relevant and perspective mission.

2 Materials and Methods

As we pretend grain moisture meter to be versatile for the wide range of grain types, it is necessary to reduce the influence of "type uncertainty" as much as possible, and one of the perspective solutions had been detected for that purpose. The idea was to use special testing influences, which help to increase the accuracy of measurements [8, 9]. Essentiality of this method consists in determining the parameters of a static function for the moisture meter with a help of special testing influences on the object under control. For the purposes of moisture measurement testing influences should be performed as a number of water injections into the substance under consideration. Using the values of dielectric permittivity (or electric capacitance) after each testing influence it is possible to calculate the initial moisture content of dielectric substance.

In [8, 9] three measurements of dielectric permittivity are used to determine moisture content: of initial sample with unknown moisture level; of initial sample after injection of certain amount of water (first testing influence); of initial sample after first testing influence, where fixed volume of preliminary prepared initial sample with known moisture level should be added (sample with known moisture level), or of initial sample after first testing influence, where certain volume of dehydrated initial sample should be introduced. Authors suggest three ways to receive third measurement (second testing influence on a substance under research).

In the first variant we have to add sample with known moisture level (similar to initial sample in dehydrated state) to a substance, received after first testing influence. In this case we can build such a system of three equations:

$$\begin{cases} \varepsilon_1 = a + b \cdot W_1, \\ \varepsilon_2 = a + b \cdot k_1 \cdot (W_1 + \Delta W_1), \\ \varepsilon_3 = a + b \cdot k_2 \cdot (W_1 + \Delta W_1 + \Delta W_2 \cdot W_{add}), \end{cases} \tag{1}$$

where $\Delta W_1 = \Delta V_1/V_0$ – relation of added volume of water to the full volume of initial sample, $k_1 = 1/(1 + \Delta W_1)$ – coefficient, applied for simplification, $\Delta W_2 = \Delta V_2/V_0$ – relation of the samples' with known moisture level volume to the full volume of initial sample, $k_2 = 1/(1 + \Delta W_1 + \Delta W_2)$ – coefficient, applied for simplification, W_{add} – moisture content in the sample with known moisture level, added to a substance, received after first testing influence.

Solvation of system (1) relatively to initial moisture content W_1 would help to receive appropriate equation:

$$W_1 = \frac{2(\varepsilon_1 - \varepsilon_2) \cdot k_1 \cdot \Delta W_1 - (\varepsilon_1 - \varepsilon_2) \cdot \Delta W_1 \cdot \Delta W_2 \cdot W_{add} \cdot k_2}{(\varepsilon_1 - \varepsilon_2) \cdot (k_2 - k_1) - (\varepsilon_1 - \varepsilon_3) \cdot (k_1 - 1)}. \tag{2}$$

Second way to receive third measurement (second testing influence on a substance under research) can be performed by introducing sample with known moisture level directly into initial sample with unknown moisture level. Then we can have another system of three equations:

$$\begin{cases} \varepsilon_1 = a + b \cdot W_1, \\ \varepsilon_2 = a + b \cdot k_1 \cdot \left(W_1 + \Delta W_1\right), \\ \varepsilon_3 = a + b \cdot k_2^{I} \cdot \left(W_1 + \Delta W_2 \cdot W_{add}\right), \end{cases} \tag{3}$$

where $k_2^{I} = 1/\left(1 + \Delta W_2\right)$– coefficient, applied for simplification. After solving system (3) relatively to W_1 we'd get another equation:

$$W_1 = \frac{\left(\varepsilon_1 - \varepsilon_2\right) \cdot \Delta W_2 \cdot W_{add} \cdot k_2^{I} - \left(\varepsilon_1 - \varepsilon_3\right) \cdot \Delta W_1 \cdot k_1}{\left(\varepsilon_1 - \varepsilon_2\right) \cdot \left(1 - k_2^{I}\right) - \left(\varepsilon_1 - \varepsilon_3\right) \cdot \left(1 - k_1\right)}. \tag{4}$$

In a third variant of third measurement it is suggested to use a certain volume of dehydrated initial sample. Appropriate system of equations would be:

$$\begin{cases} \varepsilon_1 = a + b \cdot W_1, \\ \varepsilon_2 = a + b \cdot k_1 \cdot \left(W_1 + \Delta W_1\right), \\ \varepsilon_3 = a + b \cdot k_2^{II} \cdot W_1, \end{cases} \tag{5}$$

where $k_2^{II} = 1/\left(1 + \Delta W_2^{II}\right)$, $\Delta W_2^{II} = \Delta V_d / V_0$, ΔV_d – volume of dehydrated initial sample.

Solvation of system (5) would give us appropriate equation to calculate moisture content:

$$W_1 = \frac{(\varepsilon_1 - \varepsilon_2)\cdot \Delta W_1 \cdot k_1}{(\varepsilon_1 - \varepsilon_3)\cdot(1 - k_1) - (\varepsilon_1 - \varepsilon_2)\cdot\left(1 - k_2^{II}\right)}. \tag{6}$$

3 Theory/Calculation

Checking the "type uncertainty" Influence on the Result of Moisture Content Calculation for Eqs. (2), (4) and (6). To estimate three values of dielectric permittivity ε_1, ε_2 and ε_3, we took initial data from patent [8], given in Table 1, four different dielectric bulk substances with values of dielectric permittivity in dehydrated state, equal to $\varepsilon_d = 2.0$, $\varepsilon_d = 2.5$, $\varepsilon_d = 3.0$, and $\varepsilon_d = 3.5$, besides we took four control points of nominal moisture content W = 0%, W = 10%, W = 20% and W = 30%.

Table 1. Initial data for moisture content calculation

V_0, l	ΔV_1, l	ΔV_2, l	W_{add}, %	Dielectric permittivity of water, ε_w
0.5	0.05	0.25	0.1	80

To calculate values of dielectric permittivity for different substances with different moisture content, we used universal Wiener equation [10–15]:

$$\varepsilon_1 = \varepsilon_d\left(1 + \frac{3W_1}{\frac{\varepsilon_w + 2\varepsilon_d}{\varepsilon_w - \varepsilon_d} - W_1}\right),\ \varepsilon_2 = \varepsilon_d\left(1 + \frac{3W_2}{\frac{\varepsilon_w + 2\varepsilon_d}{\varepsilon_w - \varepsilon_d} - W_2}\right),$$

$$\varepsilon_3 = \varepsilon_d\left(1 + \frac{3W_3}{\frac{\varepsilon_w + 2\varepsilon_d}{\varepsilon_w - \varepsilon_d} - W_3}\right),$$

where $W_2 = W_1 + \Delta W_1$, $W_3 = (W_2 + W_{add})/2$ for Eqs. (2) and (4), $W_3 = W_1/2$ for Eq. (6). Values of dielectric permittivity, calculated for mentioned above conditions with a help of universal Wiener equation, can be found in Table 2.

Calculated and normalized values of moisture content W_1, defined with a help of Eqs. (2), (4) and (6), are given in Table 3.

Graphs, where we can compare the difference between nominal and calculated static functions (calculated with a help of Eqs. (2), (4) and (6)), are given on Fig. 1. If we analyze the results from Table 3 and graphs from Fig. 1, it would be possible to conclude next: all three equations, suggested in patent [8], perform almost complete reduction

Table 2. Calculated values of dielectric permittivity

W, %	ε_d											
	2.0			2.5			3.0			3.5		
	Values of dielectric permittivity, calculated for Eq. (2)											
	ε_1	ε_2	ε_3	ε_1	ε_2	ε_3	ε_1	ε_2	ε_3	ε_1	ε_2	ε_3
0	2.0	2.614	2.614	2.5	3.252	3.252	3.0	3.885	3.885	3.5	4.512	4.512
10	2.614	3.368	2.971	3.252	4.173	3.688	3.885	4.963	4.396	4.512	5.741	5.095
20	3.368	4.317	3.368	4.173	5.324	4.173	4.963	6.305	4.963	5.741	7.262	5.741
30	4.317	5.545	3.814	5.324	6.806	4.714	6.305	8.022	5.596	7.262	9.197	6.459
	Values of dielectric permittivity, calculated for Eq. (4)											
	ε_1	ε_2	ε_3	ε_1	ε_2	ε_3	ε_1	ε_2	ε_3	ε_1	ε_2	ε_3
0	2.0	2.614	2.292	2.5	3.252	2.858	3.0	3.885	3.422	3.5	4.512	3.983
10	2.614	3.368	2.614	3.252	4.173	3.252	3.885	4.963	3.885	4.512	5.741	4.512
20	3.368	4.317	2.971	4.173	5.324	3.688	4.963	6.305	4.396	5.741	7.262	5.095
30	4.317	5.545	3.368	5.324	6.806	4.173	6.305	8.022	4.963	7.262	9.197	5.741
	Values of dielectric permittivity, calculated for Eq. (6)											
	ε_1	ε_2	ε_3	ε_1	ε_2	ε_3	ε_1	ε_2	ε_3	ε_1	ε_2	ε_3
0	2.0	2.614	2.0	2,5	3.252	2.5	3.0	3.885	3.0	3.5	4.512	3.5
10	2.614	3.368	2.292	3.252	4.173	2.858	3.885	4.963	3.422	4.512	5.741	3.983
20	3.368	4.317	2.614	4.173	5.324	3.252	4.963	6.305	3.885	5.741	7.262	4.512
30	4.317	5.545	2.971	5.324	6.806	3.688	6.305	8.022	4.396	7.262	9.197	5.095

of "type uncertainty" for different types of grain; from the other hand, equations are strongly nonlinear, what would be a reason for significant method error if being applied for the process of moisture measurement; another disadvantage is that second testing influence on a substance under research is rather difficult for practical implementation because it requires preliminary dehydrated sample of initial substance or preliminary dehydrated sample with further addition of certain water volume to be prepared, and, if we apply dehydration it would be evident to use thermogravimetric method of moisture measurement, which is far more accurate.

Method of Moisture Measurement with Application of One Additive and One Multiplicative Testing Influences on a Substance Under Research. Like it was told before, we need to have at our disposal value of electric capacitance C_1 of a sensor, filled with a substance, being under research. Besides, we need the value of electric capacitance C_2, taken from a sensor, filled with a substance, being under research, where the certain volume of water was introduced as an additive testing influence (first testing influence on a substance under research). Value of electric capacitance C_3 would be received after implementation of multiplicative testing influence (capacitance C_1 should be multiplied into factor k, $k = 2$ for example). This would be enough to build a system of three

Table 3. Calculated and normalized values of moisture content

W, %	ε_d							
	2.0		2.5		3.0		3.5	
	Values of moisture content, calculated with a help of Eq. (2)							
	W_{1calc}	W_{1norm}	W_{1calc}	W_{1norm}	W_{1calc}	W_{1norm}	W_{1calc}	W_{1norm}
0	−0.925	0	−0.925	0	−0.925	0	−0.925	0
10	−0.741	14.9	−0.741	14.9	−0.742	14.8	−0.742	14.8
20	−0.629	24.0	−0.629	24.0	−0.629	24.0	−0.629	24.0
30	−0.556	29.9	−0.556	29.9	−0.555	30.0	−0.555	30.0
	Values of moisture content, calculated with a help of Eq. (4)							
	W_{1calc}	W_{1norm}	W_{1calc}	W_{1norm}	W_{1calc}	W_{1norm}	W_{1calc}	W_{1norm}
0	−0.034	0.1	−0.034	0.1	−0,034	0.1	−0.035	0.0
10	0.100	13.8	0.100	13.8	0.100	13.8	0.100	13.8
20	0.192	23.2	0.193	23.3	0.193	23.3	0.193	23.3
30	0.257	29.8	0.257	29.8	0.258	29.9	0.259	30.0
	Values of moisture content, calculated with a help of Eq. (6)							
	W_{1calc}	W_{1norm}	W_{1calc}	W_{1norm}	W_{1calc}	W_{1norm}	W_{1calc}	W_{1norm}
0	−0,273	0.0	−0.273	0.0	−0.273	0.0	−0.273	0.0
10	−0,244	13.6	−0.244	13.6	−0.244	13.6	−0.244	13.6
20	−0,224	23.0	−0.224	23.0	−0.224	23.0	−0.224	23.0
30	−0,21	29.5	−0.210	29.5	−0.209	30.0	−0.209	30.0

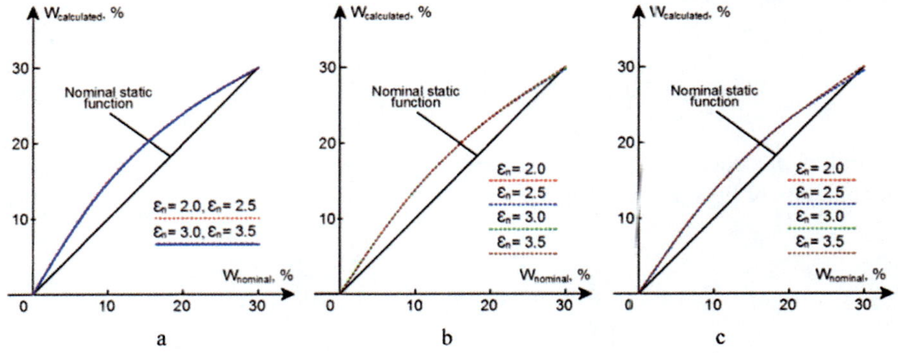

Fig. 1. Calculated and nominal values of moisture content: a – for Eq. (2); b – for Eq. (4); c – for Eq. (6)

equations, which solvation would give us an equation, similar to (2), (4) or (6). We'll use simple (linear) dependence between electric capacitance and moisture content in this system (7).

$$\begin{cases} C_1 = \varepsilon_d(1 + 3W) \cdot g, \\ C_2 = \varepsilon_d(1 + 3(W + \Delta W)) \cdot g, \\ C_3 = k \cdot \varepsilon_d(1 + 3W) \cdot g, \end{cases} \tag{7}$$

where C_1, C_2 and C_3 – electric capacitances of capacitive sensor;

g – spatial characteristic of electric field between capacitive sensors' electrodes, g = 10 m;

ΔW – water volume, added into a substance as an additive test ($\Delta W = 10\%$).

Solving system (7) we can get formula for initial moisture content of a substance under research to be calculated:

$$W = \Delta W(C_3 - C_1)/[(k - 1)(C_2 - C_1)] - 1/3. \tag{8}$$

We can see that parameter ε_d is absent in this formula, but it influences electric capacitances C_1, C_2 and C_3 indirectly. To eliminate this influence it's necessary to build another system of three equations with bigger water supplement ΔW^l (let it be $\Delta W^l = 20\%$) and bigger factor k^l ($k^l = 4$).

$$\begin{cases} C_1 = \varepsilon_d\left(1 + 3W_1\right) \cdot g, \\ C_2^l = \varepsilon_d\left(1 + 3\left(W_1 + \Delta W^l\right)\right) \cdot g, \\ C_3^l = k^l \cdot \varepsilon_d\left(1 + 3W_1\right) \cdot g, \end{cases} \tag{9}$$

where C^l_2 and C^l_3 – new values of electric capacitances after second pair of testing influences implementation.

Solution for system (9) is given below:

$$W_1 = \Delta W^l\left(C_3^l - C_1\right) / \left[\left(k^l - 1\right)\left(C_2^l - C_1\right)\right] - 1/3. \tag{10}$$

Initial moisture content of a substance under research can be expressed as a combination of Eqs. (8) and (10):

$$W_{initial} = \frac{W_1 + W}{W_1 - W} = \frac{\dfrac{\Delta W^l\left(C_3^l - C_1\right)}{\left(k^l - 1\right)\left(C_2^l - C_1\right)} + \dfrac{\Delta W\left(C_3 - C_1\right)}{\left(k - 1\right)\left(C_2 - C_1\right)}}{\dfrac{\Delta W^l\left(C_3^l - C_1\right)}{\left(k^l - 1\right)\left(C_2^l - C_1\right)} - \dfrac{\Delta W\left(C_3 - C_1\right)}{\left(k - 1\right)\left(C_2 - C_1\right)}} \tag{11}$$

Values of moisture content $W_{initial}$, calculated in accordance with formula (11), are given in Table 4.

Table 4. Calculated values of moisture content $W_{initial}$

$W_{nominal}$, %	$\varepsilon_d = 2$	$\varepsilon_d = 2.5$	$\varepsilon_d = 3$	$\varepsilon_d = 3.5$
	$W_{initial}$, not normalized			
0	−18.543	−18.799	−19.342	−19.654
10	−16.467	−17.017	−17.338	−17.836
20	−14.606	−14.909	−15.315	−15.696
30	−12.504	−12.952	−13.353	−13.772

As we can see from Table 4, influence of dielectric permittivity ε_d on the calculated values of moisture content remains significant. To fix this situation, appropriate researches were carried out separately for numerator and denominator of Eq. (11). It helped to see that denominator of Eq. (11) should be adjusted by adding a multiplying factor $(1 + a \cdot C_1)$, where 'a' is a constant, equal to $a = 0.0029$. Constant 'a' was defined by selection for the conditions, when the difference of moisture content values for $\varepsilon_d = 2$ and $\varepsilon_d = 2.5$ is minimal.

$$W_{initial} = \frac{W_1 + W}{W_1 - W} = \frac{\dfrac{\Delta W^I (C_3^I - C_1)}{(k^I - 1)(C_2^I - C_1)} + \dfrac{\Delta W (C_3 - C_1)}{(k - 1)(C_2 - C_1)}}{\left(\dfrac{\Delta W^I (C_3^I - C_1)}{(k^I - 1)(C_2^I - C_1)} - \dfrac{\Delta W (C_3 - C_1)}{(k - 1)(C_2 - C_1)}\right) \cdot (1 + 0.0029 \cdot C_1)} \tag{12}$$

New values of moisture content, recalculated with application of formula (12), are given in Table 5.

Received results are satisfactory, so, formula (12) can be normalized in a way: $W_{norm} = (W_{initial} + 17.843)/22.08$. Normalized values of moisture content, calculated with application of formula (12), can be found in Table 6.

Table 5. Calculated values of moisture content $W_{initial}$

$W_{nominal}$, %	$\varepsilon_d = 2$	$\varepsilon_d = 2.5$	$\varepsilon_d = 3$	$\varepsilon_d = 3.5$
	$W_{initial}$, not normalized			
0	−17.526	−17.528	−17.794	−17.843
10	−15.306	−15.551	−15.582	−15.772
20	−13.306	−13.300	−13.388	−13.455
30	−11.112	−11.219	−11.288	−11.376

Table 6. Normalized values of moisture content $W_{initial}$

$W_{nominal}$, %	$\varepsilon_d = 2$	$\varepsilon_d = 2.5$	$\varepsilon_d = 3$	$\varepsilon_d = 3.5$
	$W_{initial}$, normalized			
0	1.4	1.4	0.2	0.0
10	11.5	10.4	11.1	9.4
20	20.5	20.6	20.2	19.9
30	30.5	30.0	29.7	29.3

As we can see, addition of multiplying factor into formula (11) fulfills the expectations and provides satisfactory results (Fig. 2).

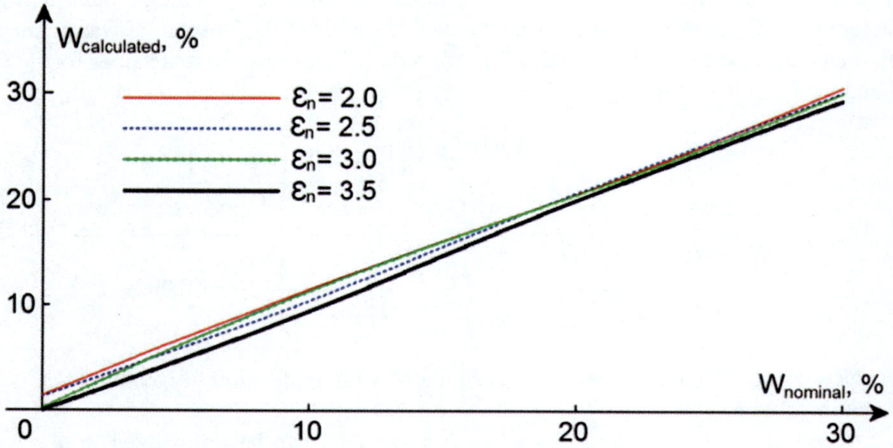

Fig. 2. Calculated values of moisture content after normalization

It was necessary to check, which equation would be more relevant as a static function of grain moisture meter. Dispersion analysis had been used for that purpose, where dispersions of repeatability D_{rep_j} should be compared with dispersions of adequacy D_{ad_j} for Eqs. (2), (4), (6) and (12) with a help of F-test.

$$D_{rep_j} = \sum_{j=1}^{m}\sum_{i=1}^{n}\left(W_{i,j} - \overline{W_j}\right)^2 \Big/ (n \cdot m - 1), \tag{13}$$

$$D_{ad_j} = \sum_{j=1}^{m}\left(\overline{W}_j - \overline{W}_{jnominal}\right)^2 \Big/ (N - (m + 1)), \tag{14}$$

where n – number of substances being under consideration;

m – number of nominal points of moisture content $W_{nominal}$;

Table 7. Mean values of moisture content

$W_{nominal}$, %	Mean values $\overline{W}_j$, %			
	For Eq. 2	For Eq. 4	For Eq. 6	For Eq. 12
0	0.000	0.075	0.000	1.000
10	14.850	13.800	13.600	10.600
20	24.000	23.275	23.000	20.300
30	29.950	29.875	29.875	29.875

$N = n \cdot m.$

Both dispersions require mean values to be calculated (Table 7).

In accordance with formula (13), dispersion of repeatability D_{rep_1} for values of moisture content, calculated with a help of Eq. (2) (placed in Table 3), can be received like:

$$D_{rep_1} = \frac{(0-0)^2 + (14.9 - 14.85)^2 + (24.0 - 24.0)^2 + (29.9 - 29.95)^2}{5 \cdot 4 - 1} +$$

$$+ \frac{(0-0)^2 + (14.9 - 14.85)^2 + (24.0 - 24.0)^2 + (29.9 - 29.95)^2}{5 \cdot 4 - 1} +$$

$$+ \frac{(0-0)^2 + (14.8 - 14.85)^2 + (24.0 - 24.0)^2 + (30.0 - 29.95)^2}{5 \cdot 4 - 1} +$$

$$+ \frac{(0-0)^2 + (14.8 - 14.85)^2 + (24.0 - 24.0)^2 + (30.0 - 29.95)^2}{5 \cdot 4 - 1} = 0.00105.$$

As for the dispersion of adequacy D_{ad_1} for Eq. (2) we would have:

$$D_{ad_1} = \frac{(0-0)^2 + (14.85 - 10)^2 + (24.00 - 20)^2 + (29.95 - 30)^2}{20 - (4+1)} = 2.635.$$

Calculations for the rest of dispersions and F-test values are given in Table 8.

Table 8. Repeatability and adequacy dispersions and values of F-test

Statistical parameter	For Eq. 2	For Eq. 4	For Eq. 6	For Eq. 12
D_{rep_j}	0.00105	0.00224	0.01645	0.29293
D_{ad_j}	2.635	1.688	1.465	0.098
$F_j = D_{ad_j} / D_{rep_j}$	2252.38	753.57	89.06	0.33

It can be seen that Eq. (12) has biggest dispersion of repeatability, but smallest dispersion of adequacy and F-test value. It means that influence of 'type uncertainty' on

Eqs. (2), (4) and (6) is much smaller than on Eq. (12). But, from the other hand, Eq. (12) corresponds nominal static function in the best way, because its' shape is close to linear and it has smallest dispersion of adequacy. And if we take the values of F-test, only Eq. (12) can be called adequate. That's why Eq. (12) would be the best we can choose as a static function for grain moisture meter.

4 Results/Discussion

To fulfill experimental researches it was necessary to create moist samples with 0%, 10%, 20% and 30% of water content. To prepare such samples it was necessary to fulfill preliminary drying of a sample under research for maximal extraction of water with a help of air-oven reference method [16]. Seeds of a pearl-barley, poppy, millet, pea and wheat had been chosen for that purpose. As a first moisture sample (with 0% of moisture content) we can use any of just mentioned substances, but preliminary dehydrated (we took dehydrated samples of 600 g mass). Second moist sample had been received by adding 60 g of water (10% from total volume) into the first sample. To get third moisture sample we took 120 g of water (20%), fourth – 200 g of water (30%). After that each moisture sample should be carefully mixed to provide moisture distribution uniformity. When working with seeds it's necessary to take into account that presence of water activates physiological, physical and chemical processes: absorption of water, swelling, sprouting etc. That's why free water in seed is usually absent. To simulate real conditions it was necessary to place just prepared moist samples in desiccators, cover with lids and wait for an hour till seed soaks free water.

Process of moisture measurement was performed in accordance with substitution method, when the value of the parameter being measured is not found directly from a reading of the measuring instrument, but from the magnitude of the standard, being regulated in such a way that the reading of the measuring instrument remains the same when the parameter being measured is replaced by the standard. Application of this method eliminates systematic errors and provides high accuracy, and substitution method is extensively used in measuring electrical quantities, such as resistance, capacitance and inductance. Experimental setup, assembled for the process of moisture measurement, is illustrated on Fig. 3.

Secondary measuring transducer converts capacitance values, taken from sensors' output, into corresponding dc voltages and consists of two 555 or 777 interval timers and low-pass RC filter. First timer functions as a multivibrator and generates rectangular pulses. Second timer, to which capacitive sensor is connected, operates like monostable multivibrator, and duration of its' output pulses is in straight proportion with the capacitance of a sensor. Low-pass RC filter, connected to the output of a monostable multivibrator, operates like pulse duration into dc voltage converter in a range from 0,000 to 2,000 V.

Process of measurement included several steps. At first capacitive sensor should be merged into the substance under consideration into 10 cm depth to receive capacitance value C_1 (on Fig. 3, b, c, we can see dehydrated poppy seed being under research). Appropriate dc voltage value, proportional to C_1 and taken from the indicator of a digital multimeter, should be remembered by operator. Another capacitance value, necessary to

a b c

Fig. 3. Experimental setup for moisture measurement: a – capacitive sensor; b – secondary measuring transducer with capacitive sensor and reference capacitor; c – reference capacitor, connected to the accurate RLC meter

be measured, is C_2 (first additive test). It can be received after direct addition of the first portion of water (10%) into the substance under research. Second value of dc voltage is to be remembered by operator like it was with C_1. Third value, necessary to calculate moisture content with a help of formula (12), is C'_2 (second additive test), received after adding another (20%) portion of water. Corresponding value of dc voltage should be remembered by operator too. Rest of the capacitances can be received with a help of simple calculations (see systems (7) and (9)) without any measurements. After receiving three values of dc voltage capacitive sensor was substituted with variable standard capacitor. Its capacitance had been slowly changed in correspondence with method of substitution up to the moment till dc voltage value on the multimeters' indicator became equal to the value, detected before (dc voltage value, proportional to C_1). After that standard capacitor vas disconnected from the secondary transducer and its capacitance had been measured with high accuracy by UNI-T UTM1612 RLC-meter on 10 kHz frequency (Fig. 3, c). Arithmetic mean value for ten capacitance measurements was taken for future calculation of moisture content. Mean values for C_2 and C'_2 had been defined in the same way. All mean values are given in Table 9.

Table 9. Mean values of C_1, C_2 and C^{l}_2 capacitances

$\overline{C}_i$, pF	$W_{nominal}$, %			
	0	10	20	30
Pearl barley				
$\overline{C}_1$	28.56	36.88	46.97	58.75
$\overline{C}_2$	36.70	43.48	55.19	69.31
$\overline{C}^{l}_2$	46.75	51.71	65.73	83.35
Poppy				
$\overline{C1}$	27.05	34.77	44.57	56.55
$\overline{C}_2$	34.64	41.13	52.21	65.81
$\overline{C}^{l}_2$	43.90	48.81	61.89	78.15
Millet				
$\overline{C}_1$	24.08	31.44	39.93	51.16
$\overline{C}_2$	31.07	36.88	46.85	59.41
$\overline{C}^{l}_2$	39.61	43.67	55.76	70.43
$\overline{C}_i$, pF	$W_{nominal}$, %			
	0	10	20	30
Pea				
$\overline{C}_1$	22.56	29.68	37.79	48,29
$\overline{C}_2$	29.35	34.73	44.37	56.14
$\overline{C}^{l}_2$	37.65	41.05	52.84	66.67
Wheat cereals				
$\overline{C}_1$	19.11	25.26	32.88	42.26
$\overline{C}_2$	25.94	29.97	38.25	48.78
$\overline{C}^{l}_2$	34.30	35.50	45.19	57.83

Experimental values of moisture content, calculated for mean values of corresponding capacitances using formula (12) are given in Table 10.

Table 10. Experimental values of moisture content

$\overline{C}_i$, pF	$W_{nominal}$, %			
	0	10	20	30
Pearl barley				
$W_{experim}$, %	0.120	10.454	20.333	29.971
Poppy				
$W_{experim}$, %	0.265	10.041	20.292	30.128
Millet				
$W_{experim}$, %	0.213	9.759	20.293	29.882
Pea				
$W_{experim}$, %	0.100	10.314	19.888	30.293
Wheat cereals				
$W_{experim}$, %	0.105	10.185	20.107	30.514

Graphs of experimental moisture content values are given on Fig. 4, and the most probable related full scale error can be calculated with a help of formula (15):

$$\gamma = \frac{\Delta W_{max}}{W_{normalized}} \cdot 100\ \% = \frac{30.514 - 30}{30} \cdot 100\ \% = 1.71\ \%. \tag{15}$$

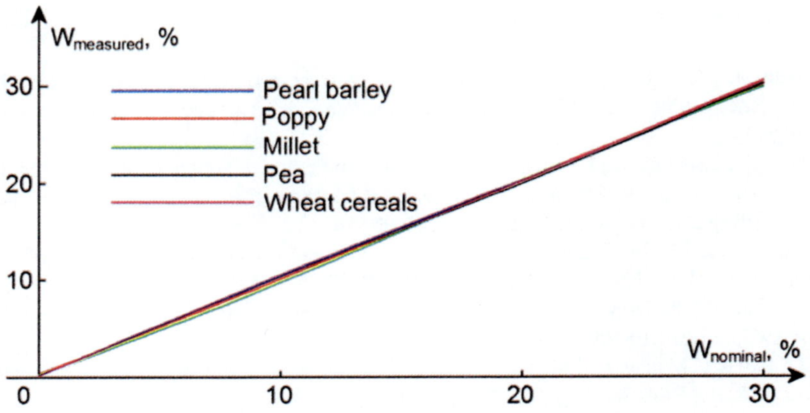

Fig. 4. Measured values of moisture content for different types of grain

5 Conclusions

Big number of analytical methods for moisture measurement in bulks and solids are based on the fact that water has appreciably different bulk physical characteristics than the substance matrix, e.g., density, electrical conductivity or refractive index. These methods can give us good results when the composition of substance matrix does not change significantly, but the ratio of water-to-substance matrix changes. If composition of the substance matrix changes as well as the water content, then it may be a problem to measure the moisture content accurately. It is impossible to get analytic forecast of chemical composition and different features of all possible food substances, being under research. That's why most of indirect methods of moisture measurement have local effectiveness.

In the article we used special testing influences on moist bulk substances, being under research with a purpose to define the parameters of a static function for the moisture meter. For the purposes of moisture measurement testing influences should be performed as a number of water injections into the substance under consideration (four different dielectric bulk substances with values of dielectric permittivity in dehydrated state, equal to $\varepsilon_d = 2.0$, $\varepsilon_d = 2.5$, $\varepsilon_d = 3.0$ and $\varepsilon_d = 3.5$, were taken for theoretical researches, besides we took four control points of nominal moisture content W = 0%, W = 10%, W = 20% and W = 30%). To build an adequate static function for the capacitive grain moisture meter with significantly reduced 'type uncertainty' value we used two additive and two multiplicative testing influences on a substance under consideration. During the process of experimental researches we used seeds of a pearl-barley, poppy, millet, pea and wheat with 0%, 10%, 20% and 30% of moisture content respectively. Obtained experimental results helped to estimate the value of related full scale error, equal to $\gamma = 1.7\%$. OIML organization in its' normative OIML R59 [17] restricts maximal permissible value of moisture meters uncertainty to not more than 3% of relative full scale error, so we can say that the suggested method of moisture measurement is workable.

References

1. Cozma, D., Tenu, I., Zapodeanu, C.: Influence of hydrothermal treatment on wheat for milling. Appl. Mech. Mater. **659**, 511–514 (2014). https://doi.org/10.4028/www.scientific.net/amm.659.511
2. Pandey, S., Senthil, A., Fatema, K.: Effect of hydrothermal treatment on the nutritional and functional properties of Husked and Dehusked Buckwheat. J. Food Process. Technol. **6**, 461 (2015). https://doi.org/10.4172/2157-7110.1000461
3. Panasiewicz, M.: Influence of hydrothermal processes on final moisture content of barley grain. Pol. J. Food Nutr. Sci. **57**(2A), 129–135 (2007)
4. Pomeranz, Y., Meloan, C.E.: Food Analysis. Theory and Practice. Springer, Boston (1994). https://doi.org/10.1007/978-1-4615-6998-5
5. El-Sayd, N.I., Makawy, M.M.: Comparison of methods for determination of moisture in food. Res. J. Agric. Biol. Sci. **6**(6), 906–911 (2010)
6. Zabolotnyi, O., Koshevoi, M.: An effective method of bulk materials moisture measurement using capacitive sensors. J. Stored Prod. Res. **89**, 101733 (2020). https://doi.org/10.1016/j.jspr.2020.101733

7. Zabolotny, A.V., Koshevoi, M.D.: Improving efficiency of the quality control of substances with dielectric properties. Telecommun. Radio Eng. **57**(2–3), 177–190 (2002)
8. Zabolotnyi, O., Zabolotnyi, V., Koshevoy, N.: Oil products moisture measurement using adaptive capacitive instrument measuring transducers. In: Nechyporuk, Mykola, Pavlikov, Vladimir, Kritskiy, Dmitriy (eds.) ICTM 2020. LNNS, vol. 188, pp. 81–91. Springer, Cham (2021). https://doi.org/10.1007/978-3-030-66717-7_7
9. Wobschall, D.: A theory of the complex dielectric permittivity of soil containing water: the semidisperse model. IEEE Trans. Geosci. Electron. Ge-**15**(1), 49–58 (1977). https://doi.org/10.1109/TGE.1977.294513
10. Hanai, T., Koizumi, N., Goto, R.: Dielectric constants of emulsions. Bull. Inst. Chem. Res. Kyoto Univ. **40**(4), 240–271 (1962)
11. Nandi, N., Bhattacharyya, K., Bagchi, B.: Dielectric relaxation and solvation dynamics of water in complex chemical and biological systems. Chem. Rev. **100**(6), 2013–2046 (2000). https://doi.org/10.1021/cr980127v
12. Josh, M., Clennell, B.: Broadband electrical properties of clays and shales: comparative investigations of remolded and preserved samples. Geophysics **80**(2), 129–143 (2015). https://doi.org/10.1190/2015-0225-TIOGEO.1
13. Kausik, B., Kothari, V.K.: Permittivity of woven fabrics: a comparison of dielectric formulas for air-fiber mixture. IEEE Trans. Dielectr. Electr. Insul. **17**(3), 881–889 (2010). https://doi.org/10.1109/TDEI.2010.5492262
14. Enis, T.: Dielectric mixtures-importance and theoretical approaches. IEEE Electr. Insul. Mag. **29**(6), 1–22 (2009). https://doi.org/10.1109/MEI.2013.6648753
15. ISO: Cereals and Cereal products – Determination of moisture content – Reference method. ISO 712:2009. Switzerland, 20 p. (2009)
16. OIML: Moisture Meters for Cereal Grain and Oilseeds. OIML R59, 2006. Paris – France, 40 p. (2006)

Classification Accuracy of Three-Channel Images Compressed by Discrete Atomic Transform

Viktor Makarichev[1], Irina Vasilyeva[1], Vladimir Lukin[1(✉)], Nataliia Kussul[2,3], and Andrii Shelestov[2,3]

[1] National Aerospace University "Kharkiv Aviation Institute", 17 Chkalov Street, Kharkiv 61070, Ukraine
i.vasilieva@khai.edu, lukin@ai.kharkov.com

[2] National Technical University of Ukraine "Igor Sikorsky Kyiv Polytechnic Institute", 37 Prosp. Peremohy, Kyiv 03056, Ukraine

[3] Space Research Institute of National Academy of Sciences of Ukraine and State Space Agency of Ukraine, 40 Glushkov Aveune, 4/1, Kyiv 03680, Ukraine

Abstract. Acquired images often have a large size while there can be limitations on communication line capacity and/or storage memory. Then, there is a need to compressed them. If lossy compression is applied, compressed images should have quality enough high for solving the tasks of their further processing as segmentation, classification, object detection. Here, we consider the influence of lossy compression on classification accuracy of three-channel remote sensing images. A specific feature of our analysis is that discrete atomic transform is studied as the basis of lossy compression and maximum likelihood method is applied at classification stage. It is shown that classification accuracy depends on both compression degree that can be characterized in different ways and image complexity. Under certain conditions, classification accuracy remains practically the same as in case of classifying an original (uncompressed) image. Then, it starts to worsen. We show how to provide quite large compression ratio with avoiding sufficient degradation of classification accuracy.

Keywords: Image lossy compression · Discrete atomic transform · Classification accuracy · Parameter adaptation

1 Introduction

Images of different origin are used in numerous applications nowadays including medicine, non-destructive control, remote sensing (RS) and so on [1, 2]. In all these areas, there is an obvious tendency to increasing the total number of acquired images and average size of images. This is due to many factors: improved resolution of imaging systems, using multichannel imaging mode (multispectral and hyperspectral data are good examples of such data), acquiring images more often (with a higher periodicity) to control sensed terrain state and its changes, in particular, fires, floods, etc. State-of-the-art

M. Nechyporuk et al. (Eds.): ICTM 2021, LNNS 367, pp. 246–258, 2022.
https://doi.org/10.1007/978-3-030-94259-5_22

agriculture monitoring systems require a huge amount of images in the geospatial time-series formats [3]. The same situation takes place in other applied geography science related spheres – multispectral time-series of remote sensing data are already deeply integrated in the urbanization, economy and environmental assessment tasks [4]. The total amount of satellite-based image volume published by the Sentinel mission of Copernicus programme until 2019 is 17.23 PiB, while 7.54 PiB of images were published only before 2019. The global trend of image generation indicates a significant growth for last few years. These images have to be stored, transferred from a point of their acquiring to a point of their processing or storage, disseminated between customers who might have interest to them. Certainly, characteristics of communicartion lines and memory means improve, but, anyway, necessity to carry out image compression remains for many applications [5–7].

Image compression techniques can be divided into lossless and lossy ones. Lossless techniques do not introduce distortions into compressed data but compression ratio (CR) provided by them is usually not large enough and it often occurs inacceptable for practice [5, 6]. Then, lossy compression methods have to be used. A lot of lossy compression methods have been proposed so far [5, 6]. Probably, the most known among them are the standards JPEG and JPEG2000 based on discrete cosine and wavelet transforms, respectively.

Meanwhile, there are also other methods employing similar principles. One of them is the method based on discrete atomic transform (DAT) [8, 9]. Discrete atomic compression (DAC) has several advantages. It allows compression single and multichannel images with larger compression ratios compared to JPEG for the same quality. DAC is fast enough and it is able to provide data privacy. An interesting property of DAC is that it can easily (in non-iterative way [10]) provide a desired quality of compressed images characterized by maximal absolute deviation (MAD). This property is important since in near-lossless and lossy compression allows characterizing degree of spectral portrait degradation [11] connected with RS image classification accuracy. However, performance of DAC has not been analyzed from the viewpoint of classification of compressed images.

The paper structure is the following. Section 2 given a brief description of DAC and presents some properties of DAC with application to two real-life three-channel images of different complexity acquired bySentinel-2. Section 3 briefly presents background of maximum likelihood classification and gives the results of image classification for compressed data. Finally, the conclusions follow.

2 Discrete Atomic Compression and Its Main Characteristics

It is known that most modern methods of lossy image compression are based on orthogonal transforms, in the first order, discrete cosine and wavelet ones. The DAT procedure employs the system of atomic wavelets.

$$\left\{ w_k\left(x - j \cdot \frac{2^{k+1}}{N} \right), v_m\left(x - j \cdot \frac{2^{m+1}}{N} \right) \right\}_{j \in Z, k=1,2,\ldots,m},$$

where N is a fixed non-zero number; m denotes a positive integer and the functions $w_k(x)$, $v_m(x)$ are constructed using $up_s(x)$. In our research, we use s = 32 and N = 64. Let us consider a function d(x) that represents some array $D = [d_1, d_2, ..., d_n]$. Then, the atomic wavelet expansion of this function is defined as:

$$d(x) = \sum_{k=1}^{m} \sum_{j} \omega_{kj} w_k \left(x - j \cdot \frac{2^{k+1}}{N} \right) + \sum_{j} \upsilon_j v_m \left(x - j \cdot \frac{2^{m+1}}{N} \right).$$

The set of atomic wavelet coefficients $W = \{\omega_{kj}, \upsilon_j\}$ describes the orthogonal components of the data D. A procedure of computation of atomic wavelet coefficients is called DAT. Details concerning this procedure can be found in [12]. The transform has the parameter m that defines a number of orthogonal layers, i.e. the depth of the transform. As a starting point of our research, we take m = 1.

Now let M be some matrix. There are several possible ways to construct DAT for the matrix M. Here we employ the approach considered classical in wavelet processing of digital images. If m = 1, the transform is applied to all rows of the matrix M. Then, we apply it to all columns of the matrix calculated at the previous step (Fig. 1). Since each time the array transform of the depth 1 is used, we call the constructed matrix DAT of the depth 1. Thus, DAT of the depth 1 applied to the matrix M produces the matrix W of atomic wavelet coefficients. The matrix W consists of four blocks, namely B_{11}, B_{12}, B_{21}, B_{22}. Each of them contains wavelet coefficients that are determined by the corresponding orthogonal component of M. As in other wavelets, the block B_{11} presents a small averaged copy of the matrix M.

Compression of grayscale digital images based on DAT is performed as follows. First, DAT is applied to the matrix M (a processed image) with obtaining a set of wavelet coefficients W. Second, each element of W is quantized (losses are introduced just at this step). By varying the quantization coefficients, different losses can be introduced. Third, quantized coefficients are encoded using lossless compression algorithm. In our study, we apply Huffman coding combined with run-length encoding. Decompression is carried out in inverse order.

If DAC is applied to color or three-channel image, RGB-to-YCrCb conversion is carried out for the matrix M of RGB-components (three compoents of RS image) at the first step. After such a conversion, all three matrices Y, Cr, and Cb are independently compressed using the procedure described above.

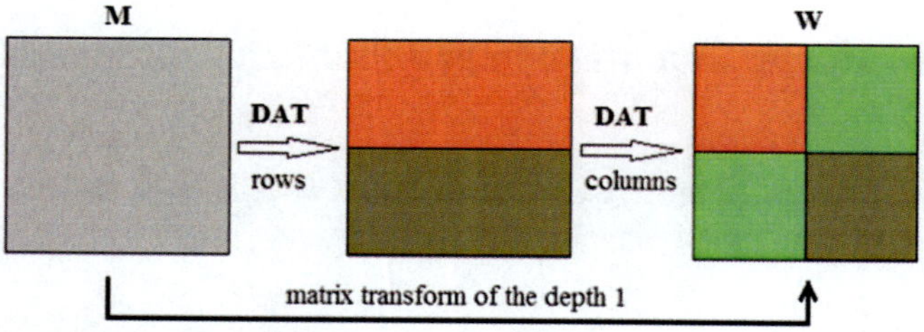

Fig. 1. Discrete atomic transform of the matrix M

As it is known, performance of a method of lossy compression can be described from different viewpoints. It is quite common to analyze several test images and the so-called rate-distortion curves that are dependences of some metric that characterizes distortions or image quality on some parameter connected with CR or CR itself. As metrics, one can use standard metrics as peak signal-to-noise ratio (PSNR) or mean square error (or root mean square error (RMSE)). Following this approach, we analyze PSNR and RMSE. However, in addition to them, we also analyze maximal absolute deviation (MAD). The reason for this has been mentioned earlier. Besides, there is an easy algorithm that allows providing a desired MAD at image compression stage. Then, just MAD serves as parameter that controls compression.

Lossy compression has to be tested for images of different complexity to see the range of possible rate-distortion curves and their peculiarities. Because of this, we have used two real-life Sentinel-2 three-channel images in our experiments. Both images are for Kharkov region in Ukraine obtained in August of 2019. They are shown in Fig. 2 where the image in Fig. 2,a is for country-side region and it is quite simple (cobtains large homogeneous regions). The second image (Fig. 2,b) is for city region and it is quite complex (contains a lot of small-sized objects and textures).

a

b

Fig. 2. Real life images used in experiments

The data that characterize lossy compression for these images are collected in Table 1. We have collected data for the following four typical situations.

First, near-lossless compression is carried out. In this case, MAD should be small and this corresponds to RMSE smaller than unity and PSNR > 50 dB (about 51 dB in the considered case). Then, CR is about 2, i.e. rather small. The second situation relates to compression images with providing PSNR about 42 dB when it is guaranteed that distortions are invisible. This corresponds to MAD about 13 and RMSE about 2. CR has increased but it is still quite small. The third situation corresponds to lossy compression "at the edge" of distortion visibility. The provided PSNR is about 35.5 dB, MAD is about

28, CR is considerably larger than for lossless and near-lossless compression. Finally, for PSNR < 35 dB, distortions are visible and this happens for MAD > 30 and RMSE > 5.

Table 1. Data for lossy compression of the images in Fig. 2.

Compressed image	MAD	RMSE	PSNR, dB	CR
Figure 2a	4	0.730	50.86	2.072
Figure 2a	12	1.950	42.33	3.705
Figure 2a	26	4.047	35.99	7.750
Figure 2a	35	4.842	34.43	9.932
Figure 2b	4	0.730	50.83	1.757
Figure 2b	14	2.026	42.00	2.715
Figure 2b	30	4.642	34.79	4.636
Figure 2b	35	5.724	32.97	5.614

So, now we know what values of MAD have to be used in simulations. The results given in Table 1 also show the following. The image with the more complex structure is compressed with a smaller CR values compared to the simple structure image for the same MAD. One can also assume that MAD and RMSE are highly correlated (this assumption will be checked in future studies).

3 Maximum Likelihood Classification of Three-Channel Images

Maximum likelihood (ML) classification is known to be a rather efficient classification approach. It is based on an assumption that there is some feature space. In our case of three-channel images, these features are just pixel values in the spectral bands (we do not consider more complex cases of exploiting, e.g., texture features estimated for a group of neighboring pixels). In other words, we consider the simplest variant of ML classification [13] which is pixel-wise. ML classification relies on preliminary training usually performed in a supervised manner. One has to choose (prepare) sample data for chosen classes and obtain distributions for them. These distributions are estimated or modeled where we prefer to employ Johnson's SB distribution for modeling [14]. The reason is that it is able to approximate rather complex distributions using a limited (four) number of adjustable parameters. Then, the classifier can be applied to RS images under interest. Also, it is possible to use other common pixel-wise classification approaches, such as Classification by Regression Trees, random forest, Support Vector Machine or multi-layer perceptron [15, 16]. These models could have higher accuracy, due to fitting of more complex distribution model, especially if data have more complex feature space in multispectral and temporal context. However, in the experiment with one image, the overall result will be the same – if ML accuracy is degraded by the image compression, other classification accuracies will be decreased too and vice versa.

ML training has been carried out for original (uncompressed) images under assumption that such images are available. Analysis of feature distributions has clearly shown that they are non Gaussian for all classes [13]. It has been also supposed that images contain four classes: Water, Vegetation, Bare Soil, and Urban. Image fragments are of the size 512 × 512 pixels. Training samples contained thousands of pixels. Verification samples were of several times larger sizes. Fragments used for training and verification are given in Fig. 3. Here, Water is shown by blue, Vegetation – by green, Bare Soil – by black, and Urban – by yellow.

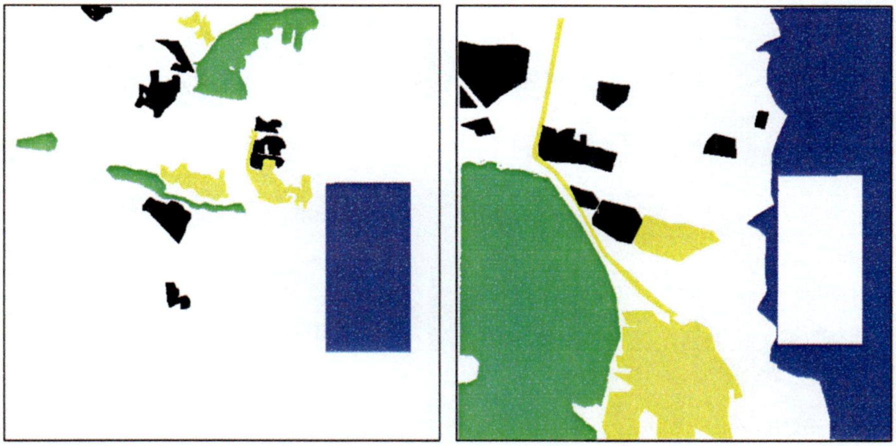

a b

Fig. 3. Training (a) and verification (b) maps for the image in Fig. 2,a

Let us start our analysis from the results obtained for the image in Fig. 2a. Examples of three confusion matrices for original image and two compressed images with different values of MAD are presented in Tables 2, 3, and 4.

As one can see (Table 2), three classes, namely, Urban, Water, and Vegetation are recognized well enough. Meanwhile, Bare Soil is often misclassified with the class Urban. As one can see, probabilities change a little if MAD changes but not a lot. For some classes, probabilities of correct classification in compressed images can be even slightly larger than in original image (see the results for the class Vegetation). Such effects have been earlier reported for other methods of lossy compression [13, 17, 18].

Finally, Table 5 presents data for probabilities of correct classification for particular classes in aggregate (P_{total}) depending upon allowed MAD of introduced distortions.

As one can see (Table 5), aggregate classification accuracy has reduced a little but such a reduction can be considered acceptable.

The classification results are presented in Fig. 4 for the original image and three compressed images that are characterized by different MAD values.

It is seen that point-wise misclassifications occur in the classification map for the original images (Fig. 4a). Similar misclassifications take place for compressed images if MAD = 4 and MAD = 12. For the case of MAD = 26, some misclassifications become

Table 2. Confusion matrix for original image, $P_{total} = 0.876$.

Class	Probability of decision			
	Urban	Water	Vegetation	Bare soil
Urban	0.912	1.71E−04	5.05E−03	0.083
Water	4.39E−03	0.995	7.67E−04	0
Vegetation	0.02	0.056	0.889	0.035
Bare soil	0.294	0	1.57E−04	0.706

Table 3. Confusion matrix for compressed image with MAD = 4, $P_{total} = 0.870$.

Class	Probability of decision			
	Urban	Water	Vegetation	Bare soil
Urban	0.916	1.71E−04	5.39E−03	0.079
Water	0.018	0.972	0.011	0
Vegetation	0.02	0.051	0.902	0.026
Bare soil	0.31	0	7.85E−05	0.69

Table 4. Confusion matrix for compressed image with MAD = 26, $P_{total} = 0.860$.

Class	Probability of decision			
	Urban	Water	Vegetation	Bare soil
Urban	0.915	1.71E−04	6.46E−03	0.078
Water	0.023	0.964	0.013	1.47E−04
Vegetation	0.028	0.046	0.895	0.031
Bare soil	0.333	0	0	0.667

Table 5. Probabilities of correct classification of particular classes and in aggregate for compressing with different MAD.

Classes	MAD = 4	MAD = 12	MAD = 26	MAD = 35
Urban	0.916	0.921	0.915	0.916
Water	0.972	0.954	0.964	0.975
Vegetation	0.902	0.908	0.895	0.896
Bare soil	0.690	0.659	0.667	0.665
P_{total}	0.870	0.861	0.860	0.863

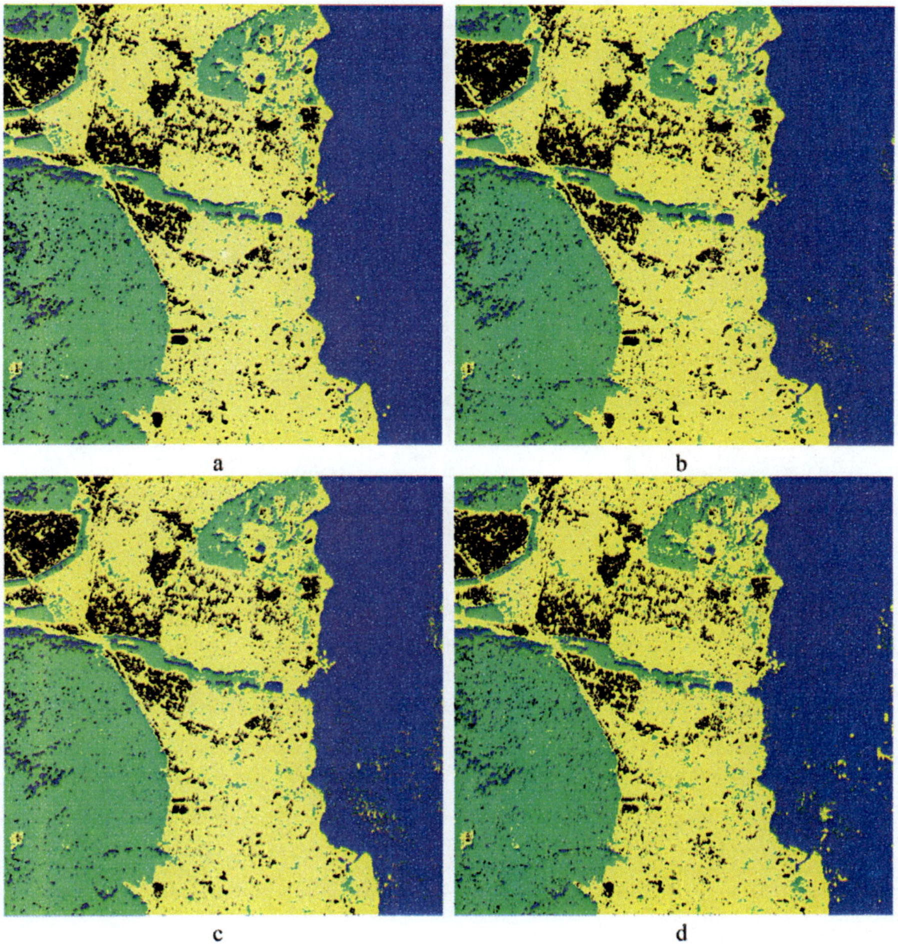

Fig. 4. Classificaion maps for original image (a) and images compressed with MAD equal to 4 (b), 12 (c), and 26 (d).

be grouped (yellow "urban" and 'vegetation" objects on "water" surface). This is due to distortions introduced by the DAC.

Consider now the results for the second analyzed three-channel image (Fig. 2b). They are presented in Tables 6, 7 and 8. The main observations are the following.

First, P_{total} for this image is smaller than for the image in Fig. 2a. This can be explained by the fact that the image in Fig. 2b has a more complex structure; in particular, this image contains a large number of linear and small-sized objects, as well as sharp changes in intensity. Second, the classes "Urban" and "Bare Soil" are not classified well and there are quite many false classifications between them. This is quite natural for the city territory. The class "Water" is recognized less accurately than in the previous case. It should be noted that classes similar in spectral characteristics (highly moistened soils, vegetation, shadows, and some anthropogenic objects) could be mistakenly assigned to the class "Water". Only the class "Vegetation" is recognized well enough. In addition, the "salt and pepper"-like noise on the classification map (misclassification of random pixels) is common for such products, due to the high variativity of real-life spectral values of land pixels. The only way to solve this problem is expansion of feature space to increase the distances between features vectors. It can be done by adding of new images and forming of time-series [19].

Table 6. Confusion matrix for original image, $P_{total} = 0.811$.

Class	Probability of decision			
	Urban	Water	Vegetation	Bare soil
Urban	0.645	2.90E−02	9.60E−02	0.23
Water	4.30E−02	0.860	8.70E−02	9.70E−03
Vegetation	4.87E−03	0.052	0.943	1.80E−04
Bare soil	0.16	7.54E−03	3.80E−02	0.795

Compression leads to additional degradation of classification accuracy. Sufficient degradation is observed for the classes "Water" and "Vegetation".

Meanwhile, probability of correct classification for the class "Urban" slightly increases although remains quite low due to sufficient intersection of its feature distribyution with other classes.

Table 7. Confusion matrix for compressed image with MAD = 4, $P_{total} = 0.798$.

Class	Probability of decision			
	Urban	Water	Vegetation	Bare soil
Urban	0.647	2.68E−02	9.81E−02	0.2282
Water	0.045	0.842	0.1034	9.15E−03
Vegetation	5.33E−03	0.0887	0.906	1.80E−04
Bare soil	0.1579	6.60E−03	0.0391	0.796

Table 8. Confusion matrix for compressed image with MAD = 30, $P_{total} = 0.780$.

Class	Probability of decision			
	Urban	Water	Vegetation	Bare soil
Urban	0.654	0.027	0.096	0.223
Water	0.072	0.814	0.105	9.42E−03
Vegetation	0.021	0.09	0.888	3.61E−04
Bare soil	0.186	6.60E−03	0.042	0.765

Table 9 gives data for probabilities of correct classification of particular classes and aggregate probability P_{total} for original image and its compressed versions with different MAD.

Table 9. Probabilities of correct classification of particular classes and in aggregate for compressing with different MAD.

Classes	MAD = 4	MAD = 14	MAD = 30	MAD = 35
Urban	0.647	0.647	0.654	0.667
Water	0.842	0.821	0.814	0.808
Vegetation	0.906	0.900	0.888	0.895
Bare soil	0.796	0.772	0.765	0.760
P_{total}	0.798	0.785	0.780	0.782

With a general tendency of decreasing for some particular and aggregate probabilities, classification of compressed images remains almost the same as for the original image even for MAD = 35.

Consider now the classification maps. They are given in Fig. 5.

It is quite difficult to carry out comparison of the obtained classification maps because there is a large number of small sized objects.

Note that there are quite many pixels classified as "Water" in forest regions that have to be classified as "Vegetation" (left upper part of the image). This is, probably, due to some shadowed zones that can be easily misclassified with water surface.

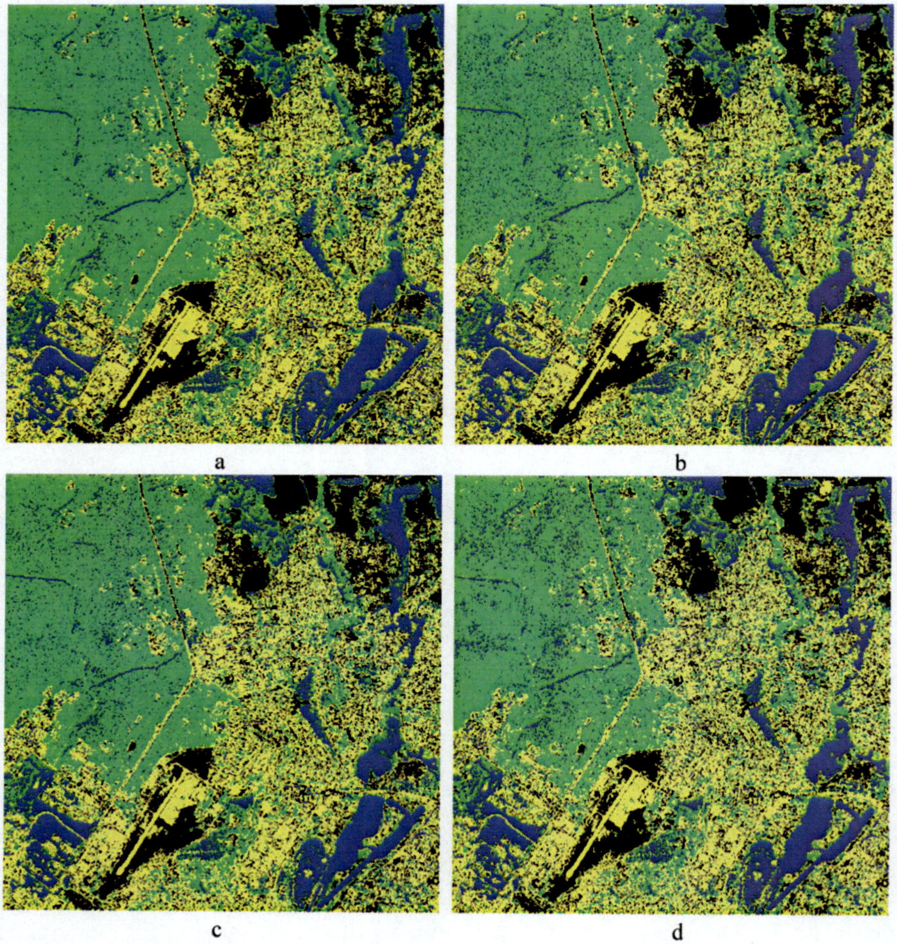

Fig. 5. Classificaion maps for original image (a) and images compressed with MAD equal to 4 (b), 14 (c), and 30 (d).

Compression leads to some blurring of the image, the appearance of noise halos around the sharp edges of objects, colour noise, therefore, the original distributions of class features change their shape. Since the statistical classifier was trained on original (uncompressed) images and then applied to compressed images, this affected the classification results.

4 Conclusions

We have considered DAC of three-channel RS images for ML classifier trained for original (uncompressed) images. It has been established that lossy compression leads to some degradation of images of both simple and complex structure. This degradation expressed in reduction of aggregate and particulat probabilities of correct classification is not too large (a few percent). To our opinion, reduction of P_{total} by 2% (0.02) is acceptable and it seems that it occurs if MAD is less than approximately 20 (PSNR is larger than 39 dB that corresponds to invisibility of introduced distortions). Meanwhile, to be confident in such conclusions and recommendations, more images have to be analyzed.

The obtained results show that it is worth continuing research of DAC. In particular, DAC with other depths have to be studied since they are able to produce slightly better CR and, possibly, a better adaptation to image content. It is also worth considering the ML classifier training for the compressed images since this approach occurred useful for other coders [11]. Other classifiers can be tested as well. Finally, post-classification can be applied.

Acknowledgments. The authors acknowledge the funding received from the National Research Foundation of Ukraine within the grant support 2020/01.0273 "Intelligent models and methods for determining land degradation indicators based on satellite data" and 2020.01/0268 "Information technology for fire danger assessment and fire monitoring in natural ecosystems based on satellite data" (NRFU competition "Science for the safety of human and society").

References

1. Pillai, D.K.: New computational models for image remote sensing and big data. In: Big Data Analytics for Satellite Image Processing and Remote Sensing, pp. 1–21. IGI Global (2018)
2. Taubman, D., Marcellin, M.: Standards and Practice JPEG 2000: Image Compression Fundamentals. Kluwer, Boston (2002)
3. Kussul, N., Mykola, L., Shelestov, A., Skakun, S.: Crop inventory at regional scale in Ukraine: developing in season and end of season crop maps with multi-temporal optical and SAR satellite imagery. Eur. J. Remote Sens. **51**(1), 627–636 (2018)
4. Kussul, N., Lavreniuk, M., Kolotii, A., Skakun, S., Rakoid, O., Shumilo, L.: A workflow for Sustainable Development Goals indicators assessment based on high-resolution satellite data. Int. J. Dig. Earth **13**(2), 309–321 (2020)
5. Salomon, D., Motta, G., Bryant, D.: Handbook of Data Compression. Springer-Verlag, London (2010)
6. Sayood, K.: Introduction to data compression. Morgan Kaufman, Burlington (2017)
7. Blanes, I., Magli, E., Serra-Sagrista, J.: A tutorial on image compression for optical space imaging systems. IEEE Geosci. Remote Sens. Mag. **2**(3), 8–26 (2014)
8. Lukin, V., Brysina, I., Makarichev, V.: Discrete atomic compression of digital images: a way to reduce memory expenses. In: Nechyporuk, M., Pavlikov, V., Kritskiy, D. (eds.) Integrated Computer Technologies in Mechanical Engineering. AISC, vol. 1113, pp. 492–502. Springer, Cham (2020). https://doi.org/10.1007/978-3-030-37618-5_42
9. Makarichev, V., Lukin, V., Brysina, I., Vozel, B., Chehdi, K.: Atomic wavelets in lossy and near-lossless image compression. In: Proceedings of SPIE 11533, Image and Signal Processing for Remote Sensing XXVI, vol. 11533, pp. 1–12 (2020)
10. Zemliachenko, A., Ponomarenko, N., Lukin, V., Egiazarian, K., Astola, J.: Still image/video frame lossy compression providing a desired visual quality. Multidimension. Syst. Signal Process. **27**, 697–718 (2015)
11. Aiazzi, B., Alparone, L., Baronti, S.: Near-lossless compression of 3-D optical data. IEEE Trans. Geosci. Remote Sens. **39**(11), 2547–2557 (2001)

12. Brysina, I.V., Makarichev, V.A.: Atomic functions and their generalizations in data processing: function theory approach. Radioelectron. Comput. Syst. **87**(3), 4–10 (2018)
13. Lukin, V., et al.: Lossy compression of multichannel remote sensing images with quality control. Remote Sens. **12**(22), 3840 (2020)
14. Parresol, B.R.: Recovering parameters of Johnson's SB distribution. US Department of Agriculture, Forest Service, Southern Research Station (2003)
15. Shelestov, A., et al.: Cloud approach to automated crop classification using Sentinel-1 imagery. IEEE Trans. Big Data **6**(3), 572–582 (2020)
16. Kussul, N., Lavreniuk, M., Skakun, S., Shelestov, A.: Deep learning classification of land cover and crop types using remote sensing data. IEEE Geosci. Remote Sens. Lett. **14**(5), 778–782 (2017)
17. Zabala, A., Pons, X.: Impact of lossy compression on mapping crop areas from remote sensing. Int. J. Remote Sens. **34**(8), 2796–2813 (2013)
18. Garcia-Sobrino, J., Laparra, V., Serra-Sagristà, J., Calbet, X., Camps-Valls, G.: Improved statistically based retrievals via spatial-spectral data compression for IASI data. IEEE Trans. Geosci. Remote Sens. **57**(8), 5651–5668 (2019)
19. Lavreniuk, M., Shelestov, A., Kussul, N., Rubel, O., Lukin, V., Egiazarian, K.: Use of modified BM3D filter and CNN classifier for SAR data to improve crop classification accuracy. In 2019 IEEE 2nd Ukraine Conference on Electrical and Computer Engineering (UKRCON), pp. 1–6 (2019)

Maxwell-Element Model for Describing Conveyor Belt Stresses

Oleh Pihnastyi[1], Georgii Kozhevnikov[1(✉)], and Olha Ivanovska[2]

[1] National Technical University "Kharkiv Polytechnic Institute", Kharkiv, Ukraine
[2] National Aerospace University "Kharkiv Aviation Institute", Kharkiv, Ukraine
o.ivanovska@khai.edu

Abstract. The wave equation which allows researching the occurrence and the dynamic stress propagation in the conveyor belt is obtained for the conveyor belt, the material of which corresponds to the Maxwell element model. The boundary and initial conditions were written for power switching modes to consider the mechanical characteristics of the asynchronous engine with the phase rotor, which determine the dependencies between the traction torque and the rotational speed of an asynchronous electric engine with a phase rotor. The estimate is given to separate wave equation terms. The expression is obtained for the calculation propagation belt speed of the dynamic stress along the conveyor belt. The conditions are shown by which the wave equation will correspond to the model of Hooke's element. By designing the dependencies between the traction torque and the rotational speed for a specified interval of the mechanical characteristic, the linear approximation is used. It is shown that the change of the material flow value coming into a section input doesn't render a special influence on the dynamic stresses propagation process along the conveyor belt. The expressions are obtained for the dynamic stress propagation speed calculation. By deriving the wave equation it is assumed the uneven material distribution along the conveyor section.

Keywords: Dynamic stress · Conveyor belt · Asynchronous electric engine

1 Introduction

The transport conveyor is used as the main method of transporting materials in the mining industry [1, 2]. As rule, this transportation method allows designing the transport route the shortest length [3], including through difficult terrain. Transport costs are directly related to the material load factor of the transport system and reach 20% of the material cost at a material load factor of 0.5–0.7 [5]. The unevenness of the incoming flow material at the section input and the transportation section length significantly affect the material load factor of the transport system [6]. To increase the load factor, the transport route is separated into sections, each of them uses the control system of the belt speed [7, 8], of the value incoming flow from the bunker [9, 10], as well as the energy management methodology [11]. The material load factor value change as a result of the speed regulation or the value incoming flow from the bunker led to a change

M. Nechyporuk et al. (Eds.): ICTM 2021, LNNS 367, pp. 259–269, 2022.
https://doi.org/10.1007/978-3-030-94259-5_23

in the consume power value of the transport system [12–14], which in turn causes the acceleration (deceleration) conveyor belt and, as consequently is the cause of the dynamic stresses occurrence in the belt. Depending on the algorithm used, dynamic stresses can reach critical values, which leads to the failure of the conveyor belt.

2 Problem Statement

The qualitative mechanical characteristics determining the dependency between the traction torque M_{eng} and the rotational speed n_{eng} of an asynchronous electric engine with a phase rotor are shown in the Fig. 1 [15], where M_{eng0}, n_{eng0} are nominal values of the traction torque and the rotational speed. Each working mode of the electric engine is characterized the mechanical characteristic. With stepwise engine power regulation, a sharp change in the traction moment occurs, which leads to the appearance of acceleration (deceleration) of the belt in the belt and, accordingly, the appearance of dynamic stresses in the belt. The short excess of the maximum allowable stress value can lead to the conveyor belt demang. Thus, the restrictions are imposed on the acceleration (deceleration) process of the belt, which must be taken into account when designing algorithms for controlling the flow parameters of a conveyor section. In addition, the properties of the material from which the belt is made affect the process of dynamic stress propagation in the belt. Depending on the properties, the composite material can be represented by a certain model of an elastic element. In this article, we will focus on the process of dynamic stress propagation in a belt, the material of which corresponds to the Maxwell element model.

To calculate the traction torque moment required to move the belt with material, it is necessary to know the mass of the material on the belt. The mass of the material can be calculated through the linear density of the material $[\chi]_0(t, S)$ in the moment time t in the point of the conveyor section with the coordinate $S \in [0, S_d]$ [6]:

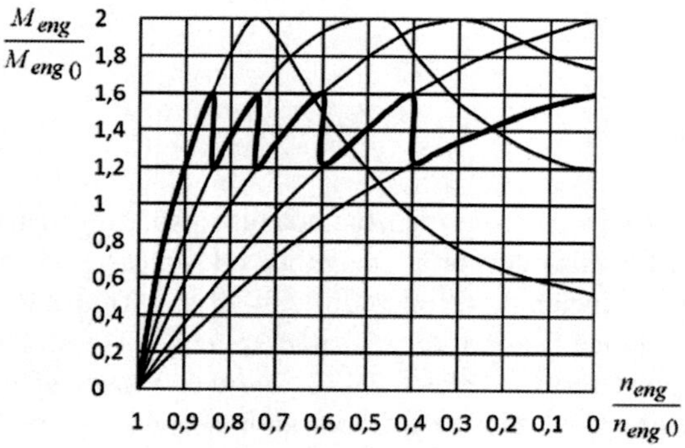

Fig. 1. Qualitative mechanical characteristics of a phase rotor induction motor

$$\theta_0(\tau, \xi) = (H(\xi) - H(\xi - G(\tau)))\frac{\gamma_1(\tau - \Delta\tau_\xi)}{g(\tau - \Delta\tau_\xi)} + H(\xi - G(\tau))\psi(\xi - G(\tau)), \quad (1)$$

$$\tau = \frac{t}{T_d}, \xi = \frac{S}{S_d}, H(S_d\xi) = H(S), \frac{\delta(\xi)}{S_d} = \delta(S), \psi(\xi) = \frac{\Psi(S)}{[\chi]_{0\max}}, \tag{2}$$

$$\gamma_1(\tau) = \lambda_1(t)\frac{T_d}{S_d[\chi]_{0\max}}, g(t) = \mu_\psi(t)\frac{T_d}{S_d}, \theta_0(\tau, \xi) = \frac{[\chi]_0(t, S)}{[\chi]_{0\max}}. \tag{3}$$

The expression (1) determines the dependence of the linear density $[\chi]_0(t, S)$ from the belt speed $\mu_\psi(t)$ and the material flow from the bunker $\lambda_1(t)$ with initial material distribution with density $\Psi(S)$. The average time T_d of the material passing along the transport route by the length S_d of the conveyor section is taken as the characteristic time of the process. The value $[\chi]_{0\max}$ corresponds to the limit allowable linear density material value for the conveyer belt. The choice of characteristic values T_d, S_d, $[\chi]_{0\max}$ is conditional. The rigidity of conveyor belts generally depends on the running time. Due to this the dynamic stress propagation mechanism let's consider for the belt, the material of which corresponds to the Maxwell-element model [16] (Fig. 2)

$$\frac{1}{E}\frac{d\sigma(t, S)}{dt} + \frac{\sigma(t, S)}{\eta} = \frac{d\varepsilon(t, S)}{dt}, \tag{4}$$

where E is the elastic modulus of the element; η is the viscosity of the element.

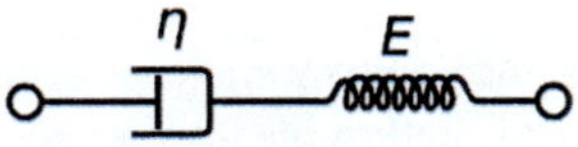

Fig. 2. Maxwell element

These materials include rubber, nylon, and polyester [17]. The model consists of two elements. With a sudden change in the traction load as a result of power switching, the Hooke-element suddenly deforms by an amount σ_0/E, and the viscous element deforms at a constant speed σ_0/η. The general deformation $\varepsilon(t, S)$ is determined by integrating the Eq. (4)

$$\frac{\sigma_0}{E} + \frac{\sigma_0}{\eta}t = \varepsilon(t, S), \varepsilon(0, S) = \sigma_0/E. \tag{5}$$

With a long time during which the conveyor belt is in a stressed state, the deformation $\varepsilon(t, S)$ becomes quite significant. The belt is stretched. The resulting slack in the belt is eliminated using a tensioner. For the constant deformation speed $d\varepsilon(t, S)/dt = v_\varepsilon \approx$ *const*, the solution of the Eq. (5) has the form

$$\sigma(t, S) = \eta v_\varepsilon(1 - \exp(-t/t_0)), t_0 = \eta/E. \tag{6}$$

If the deformation process is long, then the stress in the belt tends to a constant value

$$\lim_{t\to\infty} \sigma(t, S) = \eta v_\varepsilon. \tag{7}$$

The model of the transport system containing Eqs. (1), (5), taking into account the relationship between the traction torque and the belt speed (Fig. 1), is used to analyze the mechanism of the propagation of dynamic stresses in the conveyor belt.

3 Literature Review

A comparative analysis of the models of elastic elements (Hooke-element, Newtonian-element, Maxwell element, Kelvin-element, CDI geometric beam-element, CDI five-element) used for the propagation of dynamic disturbances along the conveyor belt is presented in the paper [17]. In article [18] the models of elastic elements, Voigt-element and Maxwell-element are considered. The Maxwell-element and Winkler foundation models are given in the paper [11]. The elastic Kelvin-Voigt-element was used to calculate the belt stretch curve, speed curves in the papers [20, 21]. Combinations of the Kelvin-Voigt-element, the Hooke-element, and the Kelvin-Voigt-element for analyzing the propagation of dynamic disturbances are considered in paper [22]. The generalized Maxwell model is studied in the paper [23], the Kelvin-Voigt element in the paper [24]. The use of the Hooke element model is discussed in detail in the paper [25]. The results of experimental studies of the mechanical properties of composite materials for the manufacture of conveyor belts in [26]. To calculate the parameters of the system, the finite element method (FEM) was used in [17, 24], and the Lagrange equations were used in [18, 20]. The solution of the wave equation is given in [21, 25].

4 Dynamic Stress Propagation Equation

The movement of an element dS at a point S on the conveyor belt with an effective mass dm and acceleration $d\mu(t,S)/dt$ is determined by the equation

$$\frac{d\mu(t,S)}{dt}dm = \sigma(t, S + dS)Bh - \sigma(t,S)Bh - dF_W. \tag{8}$$

The force of resistance to the belt movement F_W is the sum of the resistances [24, 27]

$$dF_W = dS\ C\, f_C\ g_m\big([\chi]_{0R} + [\chi]_0(t,S) + [\chi]_{0C}\big),\ g_m = 9.81\ (\text{m/s}^2),\ C \approx 1,05, \tag{9}$$

where f_C is the coefficient of resistance to movement; $[\chi]_{0C}$ is linear density of the belt; $[\chi]_{0R}$ is linear load of rotating parts, Let's represent Eq. (8) in the form

$$\frac{d\mu(t,S)}{dt} = \frac{Bh}{[\chi]_{0ef}(t,S)}\frac{\partial\sigma(t,S)}{\partial S} - C\, f_C\ g_m,\ [\chi]_{0ef}(t,S) = [\chi]_{0R} + [\chi]_0(t,S) + [\chi]_{0C}. \tag{10}$$

The ratio of the elongation at a point $dW(t,S)$ in time t for the coordinate to the length of the segment dS is the relative deformation of the element

$$\varepsilon(t,S) = \frac{\partial W(t,S)}{\partial S},\ \varepsilon(t,S) \approx 10^{-2},\ dS >> dW(t,S), \tag{11}$$

$$\mu(t,S) = \mu_\psi(t) + \frac{dW(t,S)}{dt},\ \frac{dW(t,S)}{dt} = \frac{\partial W(t,S)}{\partial t} + \mu(t,S)\frac{\partial W(t,S)}{\partial S}. \tag{12}$$

where $\mu_{\psi}(t)$ is the speed of the belt in the equilibrium state; $dW(t,S)/dt$ is the belt speed in the vicinity of the equilibrium state. Let us write the last equation taking into account (11) under the assumption that the relative elongation value $\varepsilon(t,S)$ is small

$$\frac{dW(t,S)}{dt}=\frac{\partial W(t,S)}{\partial t}+\mu(t,S)\varepsilon(t,S), \tag{13}$$

$$\mu(t,S)=\mu_{\psi}(t)+\frac{\partial W(t,S)}{\partial t}+\mu(t,S)\varepsilon(t,S)\approx\mu_{\psi}(t)+\frac{\partial W(t,S)}{\partial t}, \tag{14}$$

$$|\mu(t,S)\varepsilon(t,S)|=\left|\mu(t,S)\frac{\partial W(t,S)}{\partial S}\right|<<\left|\frac{\partial W(t,S)}{\partial t}\right|. \tag{15}$$

Let the length of the segment dS change and become equal $(dS+dW(t,S))$. In this case, the effective density $[\chi]_{0ef}(t,S)$ will change and become equal $\left([\chi]_{0ef}(t,S)+\Delta[\chi]_{0ef}\right)$, whence it follows

$$dS[\chi]_{0ef}(t,S)=(dS+dW(t,S))\left([\chi]_{0ef}(t,S)+\Delta[\chi]_{0ef}\right). \tag{16}$$

Neglecting a quantity of the order of smallness $dW(t,S)\Delta[\chi]_{0ef}(t,S)$, let obtain

$$\left|\frac{\partial W(t,S)}{\partial S}\right|\approx\left|\frac{\partial\Delta[\chi]_{0ef}}{[\chi]_{0ef}}\right|\approx\varepsilon \tag{17}$$

$$\begin{aligned}[\chi]_{0ef}(t,S)&=[\chi]_{0\psi\,ef}(t,S)+\Delta[\chi]_{0ef}(t,S)\approx[\chi]_{0\psi\,ef}(t,S)+\varepsilon[\chi]_{0ef}(t,S)\\&\approx[\chi]_{0\psi\,ef}(t,S)\end{aligned} \tag{18}$$

where $[\chi]_{0\psi\,ef}$ is the linear effective density in the absence of dynamic stresses. Taking into account (14), (15), (18), the Eq. (10) has the form

$$\frac{\partial^2W(t,S)}{\partial t^2}=\frac{Bh}{[\chi]_{0\psi\,ef}(t,S)}\frac{\partial\sigma(t,S)}{\partial S}-C\,f_C\,g_m-\frac{d\mu_{\psi}(t)}{dt}. \tag{19}$$

Equation (4) can be represented as

$$\frac{\partial\sigma(t,S)}{\partial t}+\frac{\partial\sigma(t,S)}{\partial S}\mu+\frac{E}{\eta}\sigma(t,S)=E\frac{\partial\varepsilon(t,S)}{\partial t}+E\frac{\partial\varepsilon(t,S)}{\partial S}\mu. \tag{20}$$

In this paper, we will consider perturbations whose propagation satisfies the condition

$$\frac{\partial\sigma(t,S)}{\partial S}\mu<<\frac{\partial\sigma(t,S)}{\partial t},\ \frac{\partial\varepsilon(t,S)}{\partial S}\mu<<\frac{\partial\varepsilon(t,S)}{\partial t}. \tag{21}$$

For the case when the functions $\sigma(t,S)$,$\varepsilon(t,S)$ have a large gradient, the destruction of the conveyor belt occurs. Let represent the functions $\sigma(t,S)$, $\varepsilon(t,S)$ in the form of a plane wave $\sigma(t,S)\sim e^{i(\omega t-kS)}$, $\varepsilon(t,S)\sim e^{i(\omega t-kS)}$, $\varepsilon(t,S)\sim e^{i(\omega t-kS)}$, and write condition (21) in the form

$$|\mu|<<\left|\frac{\omega}{k}\right|=\frac{1}{2\pi}|\lambda_k\omega|=\mu_{ph} \tag{22}$$

where ω is the oscillation frequency; k is the wave vector; λ_k is the wavelength of the disturbance; μ_{ph} is the phase velocity of the disturbance wave propagation. Considering (21), (22), let's represent Eq. (4) in the form followed by integration over t

$$\frac{\partial\sigma(t,S)}{\partial t}+\frac{E}{\eta}\sigma(t,S)=E\frac{\partial\varepsilon(t,S)}{\partial t},\ \ \sigma(t,S)+\frac{E}{\eta}\int\sigma(t,S)dt=E\varepsilon(t,S)+\sigma_0(S). \tag{23}$$

The limiting case $E/\eta\to 0$ corresponds to Hooke's law

$$\sigma(t,S)=E\varepsilon(t,S), \tag{24}$$

whence it follows $\sigma_0(S)=0$. Taking this into account, we substitute (23) into the Eq. (19)

$$\frac{\partial^2 W(t,S)}{\partial t^2}+\frac{C_\psi^2(t,S)}{\eta}\int\frac{\partial\sigma(t,S)}{\partial S}dt+C\,fc\;g_m+\frac{d\mu_\psi(t)}{dt}=C_\psi^2(t,S)\frac{\partial\varepsilon(t,S)}{\partial S}, \tag{25}$$

$$C_\psi^2(t,S)=\frac{BhE}{[\chi]_{0\psi\ ef}(t,S)} \tag{26}$$

where $C_\psi(t,S)$ is the speed of propagation of disturbances along the conveyor belt [26]. Let us calculate the term in (23) represented by the integral

$$\frac{E}{\eta}\frac{Bh}{[\chi]_{0\psi\ ef}(t,S)}\int\frac{\partial\sigma(t,S)}{\partial S}dt=\frac{E}{\eta}\int\left(\frac{\partial^2 W(t,S)}{\partial t^2}+C\,fc\;g_m+\frac{d\mu_\psi(t)}{dt}\right)dt \tag{27}$$

$$=\frac{E}{\eta}\left(\frac{\partial W(t,S)}{\partial t}-\frac{\partial W(0,S)}{\partial t}\right)+\frac{E}{\eta}C\,fc\;g_m t+\frac{E}{\eta}\left(\mu_\psi(t)-\mu_\psi(0)\right),$$

taking into account (10), (23), we obtain the wave equation for a conveyor belt with material distributed over it

$$\frac{\partial^2 W(t,S)}{\partial t^2}+\frac{1}{t_0}\left(\frac{\partial W(t,S)}{\partial t}-\left.\frac{\partial W(t,S)}{\partial t}\right|_{t=0}\right) \tag{28}$$

$$=-C\,fc\;g_m\left(\frac{t}{t_0}+1\right)-\frac{\mu_\psi(t)-\mu_\psi(0)}{t_0}-\frac{d\mu_\psi(t)}{dt}+C_\psi^2(t,S)\frac{\partial^2 W(t,S)}{\partial S^2}.$$

5 Analysis of Results

Let us supplement Eq. (28) with the boundary conditions

$$\left.\frac{\partial W(t,S)}{\partial S}\right|_{S=0}=\frac{k_s}{k_s+1}\frac{M_w}{EBh}\left(g_m+\left.\frac{\partial^2 W(t,S)}{\partial t^2}\right|_{S=0}\right). \tag{29}$$

If the acceleration of the suspended load is small and assuming $k_s/(k_s+1) \approx 0.5$, the boundary condition (29) can be represented in the form

$$\left.\frac{\partial W(t,S)}{\partial S}\right|_{S=0} = \frac{M_w g_m}{2EBh}, \; g_m >> \left.\frac{\partial^2 W(t,S)}{\partial t^2}\right|_{S=0}. \tag{30}$$

To construct the second boundary condition, we will use the relationship between the tractive torque and the angular speed of rotation. Let us consider the initial acceleration of the conveyor belt for the range of angular speed $n_{eng}/n_{eng0} \in [0.0; 0.4]$, for which we write down the relationship between the tractive torque and the rotor speed with the mechanical characteristic shown in Fig. 1:

$$\frac{M_{eng}}{M_{eng0}} = a - b\frac{n_{eng}}{n_{eng0}}, \; a = 1.6, \; b = 1.0, \; \mu_{eng} = \mu_{\psi}(t). \tag{31}$$

where the standard speed of the belt μ_0 corresponds to the standard speed of rotation of the rotor n_{eng0} with a drum of radius r, $\mu_0/2\pi r = n_{eng0}$. Taking into account Eq. (31), we write down the second boundary condition

$$\begin{aligned}\left.\frac{\partial W(t,S)}{\partial S}\right|_{S=S_d} &= \frac{1}{EBh}\frac{M_{eng}}{r}\frac{\exp(k_b\alpha)}{\exp(k_b\alpha)-1} \\ &= \frac{M_{eng0}}{r}\frac{\exp(k_b\alpha)}{\exp(k_b\alpha)-1}\frac{a}{EBh}\left(1-\frac{\mu_{\psi}(t)}{a}\frac{b}{\mu_0}\right)\end{aligned} \tag{32}$$

Boundary condition (32) depends on the linear speed of rotation of the drum $\mu_{\psi}(t)$. Taking into account expression (31), let obtain an equation for determining the speed:

$$\frac{d\mu_{\psi}(t)}{dt} = a_0 - b_0\mu_{\psi}(t), \tag{33}$$

$$d_{w1} = \frac{k_s+1}{EBh}\frac{\exp(k_b\alpha)}{\exp(k_b\alpha)-k_s}\left([\chi]_{0R}+[\chi]_{0C}\right)S_d, \; d_{w2} = \frac{1}{EBh}\frac{\exp(k_b\alpha)}{\exp(k_b\alpha)-k_s}M, \tag{34}$$

$$\begin{aligned}c_{w1} &= \frac{k_s+1}{EBh}\frac{\exp(k_b\alpha)}{\exp(k_b\alpha)-k_s}C\,f_C\,g_m\left([\chi]_{0R}+[\chi]_{0C}\right)S_d, c_{w2} \\ &= \frac{1}{EBh}\frac{\exp(k_b\alpha)}{\exp(k_b\alpha)-k_s}C\,f_C\,g_m\,M,\end{aligned} \tag{35}$$

$$a_w = \frac{M_{eng0}}{r}\frac{\exp(k_b\alpha)}{\exp(k_b\alpha)-1}\frac{a}{EBh}, \; a_0 = \frac{a_w - c_{w1} - c_{w2}}{d_{w1}+d_{w2}}, \; b_0 = \frac{b_w}{d_{w1}-d_{w2}}. \tag{36}$$

Consider the case of a disturbance propagation with the initial condition $\mu_{\psi}(0) = 0$ for the Eq. (33). During the characteristic time ΔT_f of the belt acceleration, the material passes a path $S_{\Delta T_f} << S_d$, which makes it possible to consider the material mass M as slowly changing over time:

$$\frac{M}{t_{\mu}} >> \frac{dM}{dt}, \; S_{\Delta T_f} = \int_0^{\Delta T_f}\mu_{\psi}(t)dt << S_d, \; \int_0^{S_d}[\chi]_0(t,S)dS = M. \tag{37}$$

For longer conveyor systems, due to inequality (37) for the characteristic acceleration (deceleration) time t_μ the mass M is a quasi-stationary quantity, and therefore the coefficients a_0, b_0 constant during the acceleration process, which allows us to represent the solution of Eq. (33) in the form

$$\mu_\psi(t) = \frac{a_0}{b_0}\left(1 - e^{-b_0 t}\right),\ \mu_\psi(0) = 0. \tag{38}$$

Let estimate the characteristic acceleration (deceleration) time $t_\mu = b_0^{-1}$ by $k_s \to 1$

$$b_0 = \frac{b_w}{d_{w1} + d_{w2}} = \frac{\frac{M_{eng0}}{r}\frac{\exp(k_b\alpha)}{\exp(k_b\alpha)-1}\frac{b}{2\pi\ r\ n_{eng0}}\frac{1}{EBh}}{\frac{1}{EBh}\frac{\exp(k_b\alpha)}{\exp(k_b\alpha)-k_s}2\left([\chi]_{0R} + [\chi]_{0C}\right)S_d + M} \tag{39}$$

$$= \frac{M_{eng0}}{r}\frac{\exp(k_b\alpha) - k_s}{\exp(k_b\alpha) - 1}\frac{b}{2\pi\ r\ n_{eng0}}\frac{1}{M_\mu},\ M_\mu = 2\left([\chi]_{0R} + [\chi]_{0C}\right)S_d + M.$$

$$b_0|_{k_s\to 1} = \frac{M_{eng0}}{r}\frac{b}{2\pi\ r\, n_{eng0}}\frac{1}{M_\mu} \tag{40}$$

The characteristic time t_μ is the time, for which the belt with the material achieves the value of the normative speed μ_0 for the normative traction torque M_{eng0}. The speed value $\mu_\psi(t)$ for the steady-state (at $t \to \infty$) is determined by the expression

$$\mu_\psi(\infty) = \frac{a_0}{b_0}\left(1 - e^{-b_0 t}\right)\Big|_{t\to\infty} = \frac{a_0}{b_0}, \tag{41}$$

$$\frac{a_0}{b_0} = \frac{a_w - c_{w1} - c_{w2}}{b_w} = \frac{a}{b}\mu_0 - \frac{C\, f_C\, g_m}{b_0} = \mu_{st} \tag{42}$$

$$\mu_{st} = \frac{a}{b}\mu_0 - \frac{C\, f_C\, g_m}{b_0},\ \frac{a_0}{b_0}\Big|_{k_s\to 1} = \frac{a}{b}\mu_0 - \frac{C\, f_C\, g_m}{b_0}\Big|_{k_s\to 1} = \frac{\mu_0}{b}\left(a - \frac{C\, f_C\, g_m}{f_{\mu 0}}\right). \tag{43}$$

The solution $\mu_\psi(t)$ is represented in Fig. 3. For a sufficiently long period of time $t >> t_\mu$ the speed $\mu_\psi(t)$ tends to a value $\mu_\psi(t)\big|_{t>>t_\mu} \to \mu_{st}$, corresponds to the tractive torque

$$M_{eng}/r = C\, f_C\, g_m\, M_\mu. \tag{44}$$

The conveyor belt reaches speed μ_{st} over time $t \approx 5t_\mu$. The acceleration of the belt has a maximum value in the interval $t \in [0, t_\mu]$. After a period of time $t \approx 5t_\mu$ the movement of the belt can be considered uniform. If the speed μ_{st} is less or more than the required value of the standard speed of the conveyor line, then the change in the value μ_{st} can be achieved through a decrease or increase in engine power as a result of switching the power mode (Fig. 1).

Let us supplement Eq. (28) with the initial conditions

$$\frac{d\mu_\psi}{dt}\Big|_{t=0} = 0, \tag{45}$$

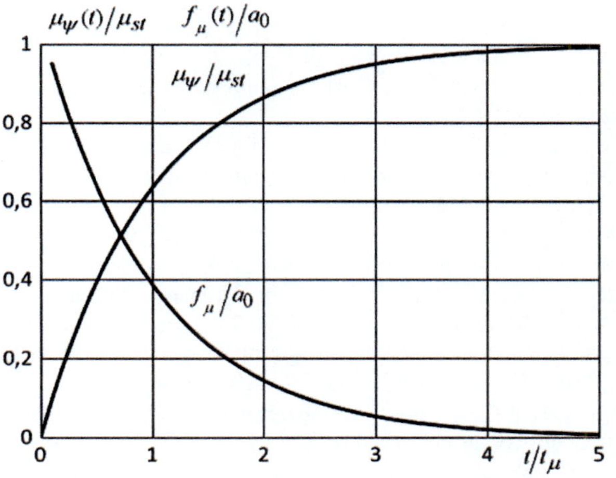

Fig. 3. Conveyor belt speed and acceleration

$$\left.\frac{\partial W(t,S)}{\partial S}\right|_{t=0} = \frac{k_s}{k_s+1}\frac{M_W g_m}{EBh} + \left(a_W - b_W\mu_\psi(t) - \frac{k_s}{k_s+1}\frac{M_W g_m}{EBh}\right)\left.\frac{S}{S_d}\right|_{t=0},$$
$$\left.\frac{\partial W(t,S)}{\partial t}\right|_{t=0} = 0. \quad (46)$$

With a uniform distribution of material along the transport route, the speed of propagation of dynamic disturbances $C_\psi(t,S)$ (26) s constant. The solution of the wave Eq. (28) with boundary (29), (32), and initial conditions (45), (46) makes it possible to calculate the value of dynamic stresses in the conveyor belt.

6 Conclusion

In this paper, the mechanism of the occurrence of dynamic stresses in a conveyor belt, the material of which corresponds to the model of the elastic Maxwell-element, is considered. When deriving the wave equation, the restriction between the belt speed and the traction torque was used, which is determined by the mechanical characteristics of the induction motor. The time interval is estimated during which, after switching power modes, significant dynamic disturbances are observed in the conveyor belt. The dynamic stresses magnitude significantly depends on the magnitude of the acceleration of the belt. Dynamic stresses, which have arisen as a result of switching power modes and have not reached a critical value, fade out for a short time. The speed of the conveyor belt changes in accordance with the mechanical characteristics of the asynchronous electric motor, reaching the standard value at which there is no belt acceleration. Prospects for further research are 1) analysis of the solution of the wave equation for the case when the tape material corresponds to the Maxwell-element model; 2) assessment of the influence of boundary conditions on the process of propagation of dynamic stresses in the belt.

References

1. SIMINE for Conveyors. Siemens. Innovative solutions for the mining industry. www.siemens.com/mining. Accessed 21 Jan 2021
2. Alspaugh, M.: Longer Overland Conveyors with Distributed Power, Rockwell Automation Fair (2005). http://www.overlandconveyor.cn/uploadfile/pdf/13-Longer_Overland_Conveyors_with_Distributed_Power[1].pdf
3. Conveyor Dynamics, Inc. Curragh Project – World Record (2020). http://conveyordynamics.com/index.php/project/curragh/. Accessed 21 Jan 2021
4. Razumnyj, Ju., Ruhlov, A., Kozar, A.: Improving the energy efficiency of conveyor transport of coal mines. Min. Electromech. Autom. **76**, 24–28 (2006). https://docplayer.ru/64655888-Povyshenie-energoeffektivnosti-konveyernogo-transporta-ugolnyh-shaht.html
5. Mathaba, T., Xia, X.: A parametric energy model for energy management of long belt conveyors. Energies **8**(12), 13590–13608 (2015). https://doi.org/10.3390/en81212375
6. Pihnastyi, O.: Control of the belt speed at unbalanced loading of the conveyor. Sci. Bull. Natl. Min. Univ. **6**, 122–129 (2019). https://doi.org/10.29202/nvngu/2019-6/18
7. Antoniak, J.: Energy-saving belt conveyors installed in polish collieries. Transp. Prob. **5**(4), 5–14 (2010). http://transportproblems.polsl.pl/pl/Archiwum/2010/zeszyt4/2010t5z4_01.pdf
8. Halepoto, I., Shaikh, M., Chowdhry, B.: Design and implementation of intelligent energy efficient conveyor system model based on variable speed drive control and physical modeling. Control and physical modeling international. J. Contr. Autom. **9**(6), 379–388 (2016)
9. Pihnastyi, O.M.: The optimal control problem for output material flow on conveyor belt with input accumulating bunker. In: Pihnastyi, O.M., Khodusov, V.D. (eds.) Bulletin of the South Ural State University. Series Mathematical Modelling, Programming & Computer Software (Bulletin SUSUMMCS), vol. 12, no. 2, pp. 67–81 (2019). https://doi.org/10.14529/mmp190206
10. Bardzinski, P., Walker, P., Kawalec, W.: Simulation of random tagged ore flow through the bunker in a belt conveying system. Int. J. Simulat. Model. **4**, 597–608 (2018). https://doi.org/10.2507/IJSIMM17(4)445
11. Marais, J., Mathews, E., Pelzer, R.: Analysing DSM opportunities on mine conveyor systems. In: Industrial and Commercial Use of Energy Conference, Cape Town, South Africa, 28–30 May 2008. http://citeseerx.ist.psu.edu/viewdoc/download?doi=10.1.1.469.574&rep=rep1&type=pdf
12. Kiriia, R., Shyrin, L.: Reducing the energy consumption of the conveyor transport system of mining enterprises. In: International Conference Essays of Mining Science and Practice, vol. 109 (2019). https://doi.org/10.1051/e3sconf/201910900036
13. Marasova, D., Andrejiova, M., Grincova, A.: Creation of the project of a logistic system for transportation of minerals - case study. TEM J. **6**(2), 205–213 (2017). https://doi.org/10.18421/TEM62-03
14. Pihnastyi, O.M., Khodusov, V.D.: Calculation of the parameters of the composite conveyor line with a constant speed of movement of subjects of labour. Sci. Bull. Natl. Min. Univ. **4**(166), 138–146 (2018). https://doi.org/10.29202/nvngu/2018-4/18
15. Spivakovsky, A.O., Dyachkov, V.A.: Transporting machines. In: Mechanical Engineering, Moscow, p. 487 (1983)
16. Roylance, D.: Engineering Viscoelasticity, p. 37. Massachusetts Institute of Technology, Cambridge (2001). http://web.mit.edu/course/3/3.11/www/modules/visco.pdf
17. Nordell, L.K., Ciozda, Z.P.: Transient belt stresses during starting and stopping: elastic response simulated by finite element methods. Bulk Solids Hand. **4**(1), 99–104 (1984). http://www.ckit.co.za/secure/conveyor/papers/troughed/transient/transient-belt-stresses.htm

18. Yang, G.: Dynamics analysis and modeling of rubber belt in large mine belt conveyors. Sens. Transduc. **81**(10), 210–218 (2014). https://www.sensorsportal.com/HTML/DIGEST/P_2492.htm
19. Yan, L., Lin, F.-Y., Wang, Y.-C.: Investigation on influence of speed on rolling resistance of belt conveyor based on viscoelastic properties. J. Theor. Appl. Mech. **45**(3), 53–68 (2015). https://doi.org/10.1515/jtam-2015-0017
20. Sakharwade, S.G., Nagpal, S.: Analysis of transient belt stretch for horizontal and inclined belt conveyor system. Int. J. Math. Eng. Manag. Sci. **4**(5), 1169–1179 (2019). https://doi.org/10.33889/IJMEMS.2019.4.5-092
21. Pihnastyi, O., Khodusov, V.: Hydrodynamic Kelvin-Voigt model transportation system. East Eur. J. Phys. (4), 95–109 (2020). https://doi.org/10.26565/2312-4334-2020-4-13
22. Kulinowski, P.: Simulation method of designing and selecting tensioning systems for mining belt conveyors. Arch. Min. Sci. **59**(1), 123–138 (2014). https://doi.org/10.2478/amsc-2014-0009
23. Thomas, R., Allen, R.: Viscoelastic indentation and resistance to motion of conveyor belts using a generalized Maxwell model of the backing material. Rubber Chem. Technol. **79**(2), 307–319 (2006). https://doi.org/10.5254/1.3547939
24. Kurt, S., Gerdemeli, I., Cengiz, C.: Analysis of belt conveyor using finite element method. In: Scientific Proceedings IX International Congress "Machines, Technologies, Materials", pp. 111–113 (2012) http://mtmcongress.com/proceedngs/2012/2/31.ANALYSIS%20OF%20BELT%20CONVEYOR%20USING%20FINITE%20ELEMENT%20METHOD.pdf
25. Pihnastyi, O.M., Khodusov, V.D.: Hydrodynamic model of transport system. East Eur. J. Phys. **1**, 121–136 (2020). https://doi.org/10.26565/2312-4334-2020-1-11
26. Manjgo, M., Piric, E., Vuherer, T., Burzic, M.: Determination of mechanical properties of composite materials-the rubber conveyor belt with cartridges made of polyester and polyamide. Ann. Facult. Eng. Hunedoara **16**(1), 141–144 (2018). http://annals.fih.upt.ro/pdf-full/2018/ANNALS-2018-1-22.pdf
27. DIN 22101:2002-08. Continous conveyors. Belt conveyors for loose bulk materials. Basics for calculation and dimensioning. [Normenausschuss Bergbau (FABERG), DIN Deutsches Institut für Normung e.v. Normenausschuss Maschinenbau (NAM)], p. 51 (2002)

Analysis of Dynamic Stresses During Acceleration and Deceleration of a Conveyor Belt (Maxwell Element Model)

Oleh Pihnastyi[1], Georgii Kozhevnikov[1](✉), and Pavel Vasyuchenko[2]

[1] National Technical University “Kharkiv Polytechnic Institute”, Kharkiv, Ukraine
[2] Ukrainian Engineering Pedagogical Academy, Kharkiv, Ukraine

Abstract. This paper presents a solution to the equation of longitudinal vibrations in a conveyor belt, the material of which corresponds to the Maxwell elastic element model. The method of consecutive approximations was used to solve the wave equation. The boundary and initial conditions are recorded taking into account the mechanical characteristics of the phased rotor induction motor, which determine the relationship between the tractive torque and the rotational speed of the phased rotor induction motor. The propagation of dynamic stresses along the conveyor belt is analyzed. An estimate of the magnitude of the stresses in the belt is presented as a function of the characteristic time of the acceleration process. The urgency of the problem is associated with the high cost of the conveyor belt in comparison with the cost of the conveyor, the damage of which can be caused by dynamic stresses, the values of which exceed the maximum permissible value.

Keywords: Dynamic stress · Conveyor belt · Asynchronous electric engine

1 Introduction

The conveyor plays an important role in reducing the cost of extracting material in the mining industry [1, 2], acting as a means of direct conversion of electrical energy into mechanical energy [3]. Uniform distribution of material along the route makes it possible to increase the load factor of the transport system with material [4], which directly leads to a decrease in transportation costs [5, 6]. For these purposes, the belt speed control [7, 8], material flow control from the bunker [9, 10], the methodology of the energy management [11] are effectively used. The industry-standard DIN 22101 gives recommendations for reducing energy consumption, which is based on the achievement of the maximum value of the upload factor of the conveyor [12] by regulation flow parameters value [13, 14]. The presence of transient modes leads to acceleration of the belt, which is the reason for the occurrence of longitudinal vibrations in the belt [15].

2 Problem Statement

Dynamic stresses in the belt arise when switching the power modes of the conveyor, which occurs when regulating the flow parameters, and is also typical for the start and

M. Nechyporuk et al. (Eds.): ICTM 2021, LNNS 367, pp. 270–280, 2022.
https://doi.org/10.1007/978-3-030-94259-5_24

stop modes. In this paper, the mechanism of occurrence and propagation of dynamic disturbances will be considered for transport systems equipped with an asynchronous electric engine with a phase rotor. For a given power mode, the qualitative mechanical characteristics determining the dependency between the traction torque M_{eng} and the rotational speed n_{eng} can be represented by a linear relationship [16]

$$\frac{M_{eng}}{M_{eng0}} = a - b\frac{n_{eng}}{n_{eng0}}, a = 1.6, b = 1.0, n_{eng}/n_{eng0} \in [0.0; 0.4], \tag{1}$$

where M_{eng0}, n_{eng0} are nominal values of the traction torque and the rotational speed. In the general case, the acceleration/deceleration of the belt is characterized by sequential switching of power modes, which corresponds to a sawtooth change in the magnitude of the traction torque M_{eng} depending on the change in the rotational speed of the engine [16]. The propagation analysis of dynamic stresses in the conveyor belt is carried out for the Maxwell model of the elastic element [17]

$$\frac{1}{E}\frac{d\sigma(t,S)}{dt} + \frac{\sigma(t,S)}{\eta} = \frac{d\varepsilon(t,S)}{dt}, \tag{2}$$

where E,η is the elastic modulus and the viscosity of the element; $\sigma(t, S)$ stress in the belt at the moment in time t at the point of the route, characterizing the coordinate S, $S \in [0, S_d]$; S_d is length of the section. The Maxwell model of an elastic element describes the propagation of dynamic stresses for conveyor belts made of a material (rubber, nylon), the rigidity of which depends on time [18]. Equation (2) mutually with the equation of motion of a conveyor belt element dS with width B and thickness h

$$\frac{d\mu(t,S)}{dt}dm = \sigma(t, S+dS)Bh - \sigma(t,S)Bh - dF_W \tag{3}$$

$$\frac{dm}{dS} = [\chi]_{ef}(t,S) = [\chi]_{0R} + [\chi]_0(t,S) + [\chi]_{0C}, \quad \frac{dF_W}{dS} = C\,f_C\,g_m[\chi]_{ef}(t,S), \tag{4}$$

$$\mu(t,S) = \mu_\psi(t) + \frac{dW(t,S)}{dt}, \frac{dW(t,S)}{dt} = \frac{\partial W(t,S)}{\partial t} + \mu(t,S)\varepsilon(t,S), \varepsilon(t,S) = \frac{\partial W(t,S)}{\partial S}, \tag{5}$$

allows you to construct a wave equation describing the absolute elongation of the conveyor belt $W(t, S)$ at the time moment t for the coordinate S

$$\frac{\partial^2 W(t,S)}{\partial t^2} + \frac{1}{t_0}\left(\frac{\partial W(t,S)}{\partial t} - \left.\frac{\partial W(t,S)}{\partial t}\right|_{t=0}\right) = C\,f_C\,g_m\left(\frac{t}{t_0}+1\right) - \frac{\mu_\psi(t)-\mu_\psi(0)}{t_0} - \frac{d\mu_\psi(t)}{dt} + C_\psi^2(t,S)\frac{\partial^2 W(t,S)}{\partial S^2} \tag{6}$$

with boundary, initial conditions

$$\left.\frac{\partial W(t,S)}{\partial S}\right|_{S=0} = \frac{k_s}{k_s+1}\frac{M_w}{EBh}\left(g_m + \left.\frac{\partial^2 W(t,S)}{\partial t^2}\right|_{S=0}\right),$$

$$\left.\frac{\partial W(t,S)}{\partial S}\right|_{S=S_d} = a_w - b_w \mu_\psi(t), \tag{7}$$

$$a_w = \frac{M_{eng0}}{r}\frac{\exp(k_b\alpha)}{\exp(k_b\alpha)-1}\frac{a}{EBh}, \quad b_w = \frac{a_w}{a}\frac{b}{\mu_0}, \quad \frac{\mu_0}{2\pi r} = n_{eng0}, \tag{8}$$

$$\left.\frac{\partial W(t,S)}{\partial S}\right|_{t=0} = \frac{k_s}{k_s+1}\frac{M_w g_m}{EBh} + \left(a_w - \frac{k_s}{k_s+1}\frac{M_w g_m}{EBh}\right)\frac{S}{S_d}, \quad \left.\frac{\partial W(t,S)}{\partial t}\right|_{t=0} = 0, \tag{9}$$

and the law of changing the speed of the belt

$$\mu_\psi(t) = \mu_{st}(1-e^{-b_0 t}), \ \mu_\psi(0) = 0, \ \mu_{st} = \frac{a}{b}\mu_0 - \frac{C f_C \, g_m}{b_0}, \ C \approx 1.05, \ g_m = 9.81, \tag{10}$$

$$b_0 = \frac{\exp(k_b\alpha)-k_s}{\exp(k_b\alpha)-1}\frac{b}{2\pi \ n_{eng0}}\frac{M_{eng0}}{2([\chi]_{0R}+[\chi]_{0C})S_d+M}, \quad \int_0^{S_d}[\chi]_0(t,S)dS = M. \tag{11}$$

The resistance movement force of the belt F_W is expressed through the linear density of the transported material $[\chi]_0(t,S)$, linear density of the belt $[\chi]_{0C}$ the load from the rotating parts $[\chi]_{0R}$, takes into account the coefficient of resistance to movement f_C [12]. The belt speed for the steady-state mode $\mu_\psi(t)$ for the start of the conveyor tends to a steady-state value μ_{st} with the characteristic time of the acceleration process $\sim 1/b_0$. A mass M_w is used to pre-tension the belt. For transport conveyors, the drum loss factor $k_s \approx 1,03$ [16]. The value $k_b\alpha$ specifies the quality of the adhesion of the belt to the leading drum of the radius r(α is angle of adhesion). Dynamic stresses in the belt propagate at a speed C_ψ. When constructing algorithms for controlling, a common case is the assumption of instantaneous switching of the modes of regulation of the flow parameters of a section or the synthesis of optimal controls without taking into account restrictions on the amount of belt acceleration [7–10].

3 Literature Review

The elastic element Maxwell model was considered in the paper [19] to analyze the rolling friction coefficient. In discrete modeling of a conveyor system using the Lagrange equations and the Voigt-element model, a comparative analysis of the Voigt-element and Maxwell-element models is presented [20]. The influence of the belt speed, made from a rubber compound, on the rolling resistance is investigated using the three-parametric viscoelastic Maxwell model [21]. Based on theoretical analysis and experiment, the relationship between belt speed and rolling resistance is shown. Belt elongation is an obvious dynamic stress value that can be demonstrated for horizontal and inclined belt conveyors using the Lagrange equation [22]. The belt conveyor system is considered as a series of oscillating masses, and its unit is a viscoelastic segment with material properties corresponding to the Kelvin-Voigt model [23, 24]. The propagation speed of longitudinal waves is studied in a conveyor belt made of Kelvin-Voigt Model material in [25]. The rheological properties of conveyor belts must also be taken into account when designing tensioning devices, taking into account their effect on the mechanism of dynamic stress propagation. Using the example of the Kelvin-Voigt Model [24], the paper demonstrates the possibility of starting the conveyor with a significantly reduced initial tension, which will increase the service life of the belt and improve the reliability of the transport system. Hooke's elastic element model is a limiting case of the Maxwell-element model [15]. A comparative analysis of the models of elastic elements and the results of experimental studies of the properties of composite materials are presented in papers [18, 26]. The study's relevance considered is due to the further improvement of control systems for the parameters of conveyor sections, on the one hand, and the fact that the CEMA and DIN 22101 standards use static analysis for recommendations for the design of conveyor systems [25].

4 Equation of Propagation of Dynamic Stresses

For a constant strain speed $d\varepsilon(t, S)/dt \approx const$ he solution to Eq. (2) has a form $\sigma(t, S) \sim (1 - \exp(-t/t_0))$, that defines t_0 as the characteristic deformation time reduction in a belt, the material of which corresponds to the Maxwell element model. When constructing the solution, let assume that the characteristic time of the development of dynamic stresses t_w is much less than the characteristic time t_0, during which there is a significant decrease in stress for the Maxwell element

$$\varepsilon_w = t_w/t_0 << 1, \;\; t_0 = \eta/E, \tag{12}$$

the material is evenly distributed along the transport route and there is no acceleration of the belt at the initial moment of time. Taking into account the accepted assumptions, the dynamic stress propagation Eq. (6) can be represented in the form

$$\begin{aligned} &\frac{\partial^2 W(t,S)}{\partial t^2} - C_\psi^2 \frac{\partial^2 W(t,S)}{\partial S^2} + \frac{d\mu_\psi(t)}{dt} + C\, fc\; g_m \\ &= -\varepsilon_w \left(\frac{1}{t_w} \frac{\partial W(t,S)}{\partial t} + C\, fc\; g_m \frac{t}{t_w} + \frac{\mu_\psi(t)}{t_w} \right). \end{aligned} \tag{13}$$

Equation (13) does not contain terms $\partial W(t, S)/\partial t\big|_{t=0}$ due to their equality to zero. We represent the solution of Eq. (13) in the form

$$W(t,S) = W_0(t,S) + \varepsilon_w W_1(t,S) + \ldots + \varepsilon_w^n W_n(t,S) + \ldots \tag{14}$$

Substituting series (14) into Eq. (13), we obtain the system of equations:

$$\frac{\partial^2 W_0(t,S)}{\partial t^2} - C_{\psi 0}^2 \frac{\partial^2 W_0(t,S)}{\partial S^2} = -C\, fc\; g_m - \frac{d\mu_\psi(t)}{dt}, \tag{15}$$

$$\frac{\partial^2 W_1(t,S)}{\partial t^2} - C_{\psi 0}^2 \frac{\partial^2 W_1(t,S)}{\partial S^2} = -\frac{1}{t_w} \frac{\partial W_0(t,S)}{\partial t} - C\, fc\; g_m \frac{t}{t_w} - \frac{\mu_\psi(t)}{t_w}, \tag{16}$$

$$\frac{\partial^2 W_\kappa(t,S)}{\partial t^2} - C_{\psi 0}^2 \frac{\partial^2 W_\kappa(t,S)}{\partial S^2} = -\frac{\partial W_{\kappa-1}(t,S)}{t_w \partial t}, k = 2..n. \tag{17}$$

Let us stop our attention on the solution of the first equation, which is determined by the zero approximation in parameter ε_w. We will look for a solution in the form

$$W_0(t,S) = V_S(t,S) + V(t,S). \tag{18}$$

$$V_S(t,S) = \frac{k_s}{k_s+1}\frac{M_w g_m}{EBh} S + \left(a_w - b_w \mu_\psi(t) - \frac{k_s}{k_s+1}\frac{M_w g_m}{EBh}\right)\frac{S^2}{2S_d} + V_{S0}. \tag{19}$$

The function $V_S(t,S)$ determines the static deflection of the belt in the absence of dynamic stresses. The constant V_{S0} is selected from the requirements to ensure the minimum deflection of the belt. The function $V(t,S)$ is a deviation from the undisturbed state as a result of the occurrence of dynamic stresses. Let us introduce the notation

$$A_0 = \frac{k_s M_w g_m}{k_s+1} \Big/ \frac{M_{eng0}}{r}\frac{\exp(k_b\alpha)}{(\exp(k_b\alpha)-1)}, \tag{20}$$

and let's write down the expression for stresses in the unperturbed state

$$\frac{a}{a_w}\frac{\partial V_S(t,S)}{\partial S} = A_0\left(1-\frac{S}{S_d}\right) + \left(a - b\frac{\mu_{st}}{\mu_0}\left(1-e^{-b_0 t}\right)\right)\frac{S}{S_d},$$
$$\frac{\mu_{st}}{\mu_0} = \frac{a}{b} - \frac{C\,f_C\,g_m}{\mu_0 b_0}, t_\mu = \frac{1}{b_0}. \tag{21}$$

5 Analysis of Results

The graph of the function $Q(t/t_\mu)$ characterizing the stresses at the point $S = S_d$ for the values μ_{st}/μ_0 ={0.1; 0.2; 0.3; 0.4} with $a = 1.6$, $b = 1.0$ is shown in Fig. 1

$$Q(t/t_\mu) = \left(\frac{a}{a_w}\right)\frac{\partial V_S(t,S)}{\partial S}\bigg|_{S=S_d} = a - b\frac{\mu_{st}}{\mu_0}\left(1-e^{-t/t_\mu}\right). \tag{22}$$

The relative stresses for the steady motion correspond to the range of variation of the belt speed $\mu_{eng}/\mu_{eng0} \in [0.0; 0.4]$ and tractive torque $M_{eng}/M_{eng0} \in [1.6; 1.2]$. We substitute solution (18) into the zero approximation Eq. (15) and considering that

$$C_{\psi 0}^2 \frac{\partial^2 V_S(t,S)}{\partial S^2} = C\,f_C \cdot g_m + \frac{d\mu_\psi(t)}{dt}, \tag{23}$$

$$\frac{\partial V_S(t,S)}{\partial S} = \frac{k_s}{k_s+1}\frac{M_w g_m}{EBh} + \left(a_w - b_w\mu_\psi(t) - \frac{k_s}{k_s+1}\frac{M_w g_m}{EBh}\right)\frac{S}{S_d}, \tag{24}$$

$$\frac{\partial^2 V_S(t,S)}{\partial t^2} = \frac{\partial^2}{\partial t^2}\left(a_w - b_w\mu_\psi(t) - \frac{k_s}{k_s+1}\frac{M_w g_m}{EBh}\right)\frac{S^2}{2S_d} \approx \left(\frac{S_d/t_\mu}{C_{\psi 0}}\right)^2\frac{d\mu_\psi}{dt} \tag{25}$$

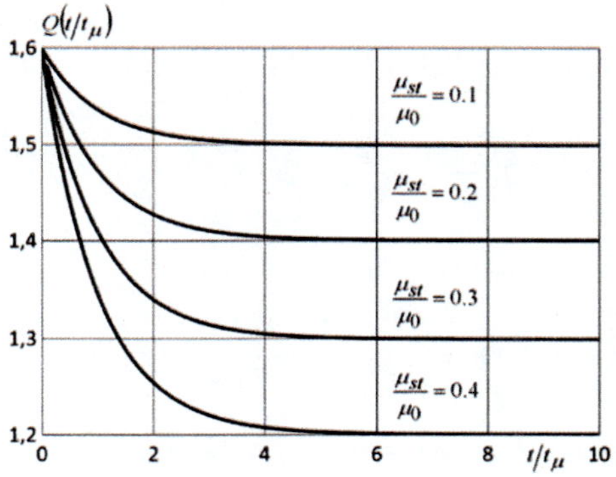

Fig. 1. Relative stress of the conveyor belt

let obtain the system of equations for the function $V(t, S)$

$$\frac{\partial^2 V(t,S)}{\partial t^2} = C^2_{\psi 0} \frac{\partial^2 V(t,S)}{\partial S^2}, \tag{26}$$

$$\left.\frac{\partial V(t,S)}{\partial S}\right|_{S=0} = \frac{k_s}{k_s+1} \frac{M_w}{EBh} \left.\frac{\partial^2 V(t,S)}{\partial t^2}\right|_{S=0}, \quad \left.\frac{\partial V(t,S)}{\partial S}\right|_{S=S_d} = 0, \tag{27}$$

$$\left.\frac{\partial V(t,S)}{\partial S}\right|_{t=0} = 0 \left.\frac{\partial V(t,S)}{\partial t}\right|_{t=0} = b_w\big(a_0 - b_0\mu_\psi(t)\big)\left.\frac{S^2}{2S_d}\right|_{t=0} = b_w a_0 \frac{S^2}{2S_d}. \tag{28}$$

When deriving Eq. (26), it was taken into account that

$$C^2_{\psi 0}\frac{\partial^2 V_S(t,S)}{\partial S^2} >> \frac{\partial^2 V_S(t,S)}{\partial t^2} \approx \left(\frac{S_d/t_\mu}{C_{\psi 0}}\right)^2 \frac{d\mu_\psi}{dt}, \ \left(\frac{S_d/t_\mu}{C_{\psi 0}}\right)^2 \approx \left(\frac{60}{2000}\right)^2 \approx 10^{-3}. \tag{29}$$

The ratio S/t_μ of the operating conveyors is presented in Table 1. The propagation of disturbances speed, depending on the material of the conveyor belt $C_{\psi 0} \approx 2000$ m/s.

Table 1. Estimated conveyor belt acceleration mode time

Name	S_d, m	t_μ, s	S_d/t_μ
Conveyor at the Zimbabwe Iron & Steel Co [27]	15600	500	31.2
Conveyor C3, Indo Kodeco System Layout [27]	8600	250	34.4
North Shaft decline conveyor [28]	1350	30	45.0
Experimental conveyor [29]	3620	60	61.0
Conveyor CV002B, El Brocal [30]	2781	80	34.8

Let us investigate the natural frequencies of stress oscillations arising in the conveyor belt. The solution to Eq. (26) will be sought in the form

$$V(t,S) = T(t)X(S), \tag{30}$$

$$\frac{d^2X_n(S)}{dS^2}+\lambda_n^2X_n(S)=0,\ X_n(S)=\sin(\lambda_nS_d)(\sin(\lambda_nS)+B_n\cos(\lambda_nS)), \tag{31}$$

$$\left.\frac{\partial X_n(S)}{\partial S}\right|_{S=0}=-C_{\psi 0}^2\frac{k_s}{k_s+1}\frac{M_w}{EBh}\lambda_n^2X(t,S)\Big|_{S=0},\ \left.\frac{\partial X_n(S)}{\partial S}\right|_{S=S_d}=0. \tag{32}$$

From the boundary conditions we find

$$1=-c_0\lambda_nB_n,\ c_0=C_{\psi 0}^2\frac{k_s}{k_s+1}\frac{M_w}{EBh},\ tg(\lambda_nS_d)=-c_0\lambda_n, \tag{33}$$

$$X_n(S)=\cos(\lambda_nS-\lambda_nS_d). \tag{34}$$

The orthogonality conditions for functions $X_n(S)$ have the form

$$\int_0^{S_d}X_n(S)X_m(S)dS+c_0X_n(0)X_m(0)=0,\ \left\|X_n^2\right\|=\int_0^{S_d}X_n^2(S)dS+c_0X_n^2(0). \tag{35}$$

We will seek a solution to problem (26) in the form

$$V(t,S)=\sum_{n=1}^{\infty}T_n(t)\cos(\lambda_nS-\lambda_nS_d), \tag{36}$$

considering time t as a parameter. The value λ_n is found from the solution of the transcendental Eq. (33). Using the initial conditions

$$\sum_{n=1}^{\infty}T_n(0)\lambda_n\sin(\lambda_nS-\lambda_nS_d)=0,\ b_wa_0\frac{S^2}{2S_d}=\sum_{n=1}^{\infty}\frac{dT_n(0)}{dt}\cos(\lambda_nS-\lambda_nS_d), \tag{37}$$

let us define

$$T_n(0)=0,\ \frac{dT_n(0)}{dt}=\frac{1}{\|X_n^2\|}\int_{-S_d}^{0}\cos(\lambda_ny)(y+S_d)^2dy,\ y=S-S_d. \tag{38}$$

For the case where the weight of the gravity weight is very large in comparison with the mass $M_w/[\chi]_{0\psi\, ef0}S_d >> 1$ to be moved, it follows

$$c_0\lambda_n=C_{\psi 0}^2\frac{k_s}{k_s+1}\frac{M_w}{EBh}S_d\approx\frac{BhE}{[\chi]_{0\psi\, ef0}}\frac{M_w}{EBh}\frac{1}{S_d}\approx\frac{M_w}{[\chi]_{0\psi\, ef0}S_d}>>1, \tag{39}$$

$$ctg(\lambda_nS_d)\to 0,\ \lambda_n=\frac{\pi}{S_d}\left(n-\frac{1}{2}\right),\ B_n=0,\ X_n(0)=0. \tag{40}$$

The resulting boundary condition corresponds to the case of rigid fixation at a point $S=0$. The solution $X_n(S)$, $V(t,S)$ for the considered case will be sought in the form

$$X_n(S)=\sin(\lambda_nS),\ V(t,S)=\sum_{n=1}^{\infty}T_n(t)\sin(\lambda_nS). \tag{41}$$

$$T_n(0) = 0, \quad \frac{dT_n(0)}{dt} = b_w a_0 S_d \left(\frac{(-1)^{n-1}}{\pi n - \pi/2} - \frac{1}{(\pi n - \pi/2)^2} \right). \tag{42}$$

The solution $T_n(t)$ to the given initial conditions has the form

$$T_n(t) = c_n \sin(\omega_n t), \ \omega_n = C_{\psi 0}\lambda_n = \sqrt{\frac{BhE}{[\chi]_{0\psi\, ef0}}} \frac{\pi}{S_d}\left(n - \frac{1}{2}\right), \ c_n = \frac{1}{\omega_n}\frac{dT_n(0)}{dt}. \tag{43}$$

The solution that determines the dynamic tension of the belt can be represented as

$$V(t, S) = \sum_{n=1}^{\infty} c_n \sin(\omega_n t) \sin(\lambda_n S). \tag{44}$$

The total tension in the conveyor belt is determined by the expression

$$\frac{\partial W_0(t, S)}{\partial S} = \frac{\partial V_S(t, S)}{\partial S} + \frac{\partial V(t, S)}{\partial S}, \tag{45}$$

$$\frac{a}{a_w}\frac{\partial W_0(t, S)}{\partial S} = A_0\left(1 - \frac{S}{S_d}\right) + \left(a - b\frac{\mu_{st}}{\mu_0}\left(1 - \exp(-t/t_u)\right)\right)\frac{S}{S_d} \tag{46}$$

$$+ \frac{b_0}{C_{\psi 0}} S_d b \frac{\mu_{st}}{\mu_0} \sum_{n=1}^{\infty} \left(\frac{(-1)^{n-1}}{\pi n - \pi/2} - \frac{1}{(\pi n - \pi/2)^2} \right) \sin(\omega_n t) \cos(\lambda_n S).$$

Consider the propagation of dynamic stresses for n = 1. Estimated calculation shows that for the acceleration modes presented in Table 1, $A_0 >> B_0$

$$B_0 = 0.64 \frac{S_d b}{C_{\psi 0} t_\mu} \frac{\mu_{st}}{\mu_0}, \ \omega_w = \omega_1 t_\mu \approx 1.57 C_{\psi 0} \frac{t_\mu}{S_d}, \ \lambda_w = \lambda_1 S_d \approx 1.57, \tag{47}$$

and therefore the dynamic stresses in the conveyor belt will be significantly less than the static stress. Estimation of the characteristic acceleration time at which $A_0 \approx B_0$, is determined by the expression

$$t_\mu \approx \frac{1}{A_0}\frac{S_d}{C_{\psi 0}}\frac{\mu_{st}}{\mu_0}, \ \omega_w = \omega_1 t_\mu \approx \left(1.57 C_{\psi 0}\frac{1}{S_d}\right)\left(\frac{1}{A_0}\frac{S_d}{C_{\psi 0}}\frac{\mu_{st}}{\mu_0}\right) \approx \frac{1.57}{A_0}\frac{\mu_{st}}{\mu_0}. \tag{48}$$

For values $A_0 \approx 10$,$b \approx A_0$, $a \approx 2.5A_0$, $S_d \approx 10^4$,$C_{\psi 0} \approx 2000$, $\mu_{st}/u_0 \approx 0.4$ we get $t_\mu \approx 0.1$, $\omega_w = \omega_1 t_\mu \approx \pi$, $\lambda_w = \pi/2$ and correspondingly ($\xi = S/S_d$, $\tau = t/t_\mu$)

$$Q_1(\tau, \xi) = \frac{a}{A_0 a_w}\frac{\partial W_0(t, S)}{\partial S} \approx 1 + \xi - 0.4(1 - \exp(-\tau))\xi + \sin(\pi\tau)\cos\left(\frac{\pi}{2}\xi\right), \tag{49}$$

When $S = S_d$, the function $Q_1(\tau, \xi)$ goes to the function $Q(t/t_\mu)/A_0$ (Fig. 1)

$$Q_1(\tau, 1) \approx \left(2 - 0.4\left(1 - \exp(t/t_\mu)\right)\right)\frac{S}{S_d}. \tag{50}$$

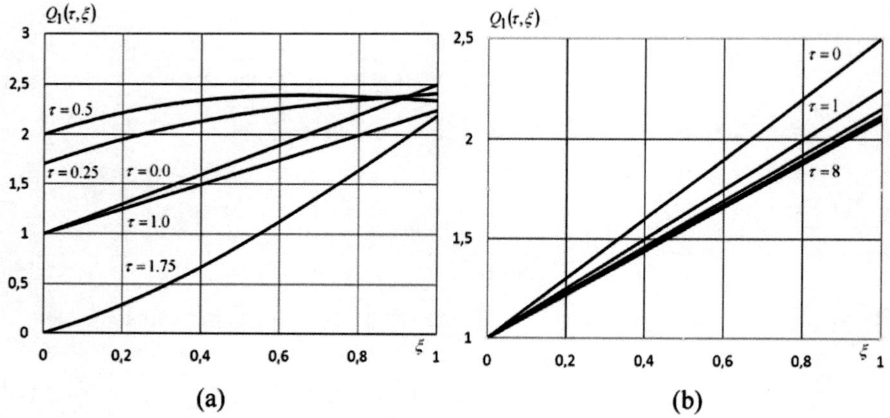

Fig. 2. Relative tension of the conveyor belt: a)$B_0 = A_0$; b) $B_0 = 1.25A_0$

The dynamics of the stresses in the conveyor belt for the case determined by the function $Q_1(\tau, \xi)$ are shown in Fig. 2a ($B_0 = A_0$). A family of curves characterizes the magnitude of the stresses in the belt at different points in time. At $B_0/A_0 = 1$ (Fig. 2a) the stress value does not exceed the maximum stress value $Q_1(0, 1)$ at the moment of the start of the belt. The maximum amplitude of fluctuations in the magnitude of stresses is achieved at the beginning of the belt $\xi = 0$, which can lead to the effects of sagging of the conveyor belt. An increase in the ratio B_0/A_0 leads to the occurrence of stresses with a maximum value in the middle of the transport section. The analysis of the function $Q_1(\tau, \xi)$ allows us to conclude that when accelerating the belt, the most dangerous is the initial moment of acceleration, determined by the gap $\tau \in [0, 1[$. In subsequent time intervals, a decrease in the stresses value occurs. An estimate of the value and rate of stress reduction during the transition to the standard value of the belt speed μ_{st} can be obtained from the analysis of the family of dependencies shown in Fig. 2b. The stresses in the conveyor belt are set by the ratio

$$Q_1(\tau, \xi) \approx 1 + \xi - 0.4(1 - \exp(-\tau))\xi, \tag{51}$$

which is written taking into account the fact that for the moments of time $\tau = 0, 1, 2, ...$ there are no dynamic stresses in the conveyor belt due to the equality $\sin(\pi\tau) = 0$.

6 Conclusion

The article discusses the problems associated with the functioning of a long transport conveyor in transient modes, driven by an asynchronous motor. The distribution of dynamic stresses in a belt is analyzed taking into account the mechanical characteristics of an induction motor. When constructing a solution to the wave equation of longitudinal vibrations in a conveyor belt, it was assumed that the characteristic time of the development of dynamic stresses is much less than the characteristic time during which there is a significant decrease in the stress of the belt material. This approach made it possible, using the method of successive approximation to solve the wave equation, to determine the magnitude of the stress in the belt, taking into account the acceleration of the conveyor belt with the material. The time interval is estimated during which, after switching power modes, significant dynamic disturbances are observed in the conveyor belt. The relationship between the mode of acceleration of the conveyor belt and the mechanism of the propagation of dynamic stresses in the conveyor belt is analyzed. Analysis of the solution shows that the initial period of time after switching the power mode is the most probable in terms of exceeding the critical value for dynamic stresses. At the subsequent stages of acceleration, a gradual attenuation of dynamic stresses occurs, which is explained by the dependence of the traction torque of the engine on the angular speed of the rotor on the mechanical characteristics of the induction motor. Prospects for further research: 1) analysis of the mechanism of propagation of dynamic stresses in the belt associated with the presence of shock loads caused by switching the power mode; 2) assessment of the influence of boundary conditions on the process of propagation of dynamic stresses in the belt.

References

1. SIMINE for Conveyors. Siemens. Innovative solutions for the mining industry. www.siemens.com/mining
2. Alspaugh, M.: Longer Overland Conveyors with Distributed Power, Rockwell Automation Fair (2005). http://www.overlandconveyor.cn/uploadfile/pdf/13-Longer_Overland_Conveyors_with_Distributed_Power[1].pdf
3. Conveyor Dynamics, Inc. Curragh Project – World Record (2020). https://www.cvdyn.com/experience/curragh-project/
4. Pihnastyi, O.: Control of the belt speed at unbalanced loading of the conveyor. Sci. Bull. Natl. Min. Univ. **6**, 122–129 (2019). https://doi.org/10.29202/nvngu/2019-6/18
5. Razumnyj, Ju., Ruhlov, A., Kozar, A.: Improving the energy efficiency of conveyor transport of coal mines. Min. Electromech. Autom. **76**, 24–28 (2006). https://docplayer.ru/64655888-Povyshenie-energoeffektivnosti-konveyernogo-transporta-ugolnyh-shaht.html
6. Mathaba, T., Xia, X.: A parametric energy model for energy management of long belt conveyors. Energies **8**(12), 13590–13608 (2015). https://doi.org/10.3390/en81212375
7. Antoniak, J.: Energy-saving belt conveyors installed in polish collieries. Transp. Prob. **5**(4), 5–14 (2010). http://transportproblems.polsl.pl/pl/Archiwum/2010/zeszyt4/2010t5z4_01.pdf
8. Halepoto, I., Shaikh, M., Chowdhry, B.: Design and implementation of intelligent energy efficient conveyor system model based on variable speed drive control and physical modeling. Cont. Phys. Model. Int. J. Cont. Autom. **9**(6), 379–388 (2016)
9. Pihnastyi, O.M., Khodusov, V.D.: The optimal control problem for output material flow on conveyor belt with input accumulating bunker. Bull. South Ural State Univ. Ser. Math. Model. Prog. Comput. Softw. **12**(2), 67–81 (2019). https://doi.org/10.14529/mmp190206
10. Bardzinski, P., Walker, P., Kawalec, W.: Simulation of random tagged ore flow through the bunker in a belt conveying system. Int. J. Simulat. Model. **4**, 597–608 (2018). https://doi.org/10.2507/IJSIMM17(4)445
11. Marais, J., Mathews, E., Pelzer, R.: Analysing DSM opportunities on mine conveyor systems. In: Industrial and Commercial use of Energy Conference, Cape Town, South Africa, 28–30 May 2008. http://citeseerx.ist.psu.edu/viewdoc/download?doi=10.1.1.469.574&rep=rep1&type=pdf

12. DIN 22101:2002-08. Continous conveyors. Belt conveyors for loose bulk materials. Basics for calculation and dimensioning. [DIN Deutsches Institut für Normung e.v. Normenausschuss Maschinenbau (NAM)], p. 51 (2002)
13. Kiriia, R., Shyrin, L.: Reducing the energy consumption of the conveyor transport system of mining enterprises. Int. Conf. Essays Min. Sci. Pract. **109**, 00036 (2019). https://doi.org/10.1051/e3sconf/201910900036
14. Marasova, D., Andrejiova, M., Grincova, A.: Creation of the project of a logistic system for transportation of minerals - case study. TEM J. 6(2), 205–213, https://doi.org/10.18421/TEM62-03. ISSN 2217–8309
15. Pihnastyi, O.M., Khodusov, V.D.: Hydrodynamic model of transport system. East Eur. J. Phys. **1**, 121–136 (2020). https://doi.org/10.26565/2312-4334-2020-1-11
16. Spivakovsky, A.O., Dyachkov, V.A.: Transporting machines. Mechanical Engineering, Moscow p. 487 (1983)
17. Roylance, D.: Engineering viscoelasticity. Massachusetts Institute of Technology, p. 37. (2001). http://web.mit.edu/course/3/3.11/www/modules/visco.pdf
18. Nordell, L.K., Ciozda, Z.P.: Transient belt stresses during starting and stopping: elastic response simulated by finite element methods. Bulk Solids Hand. **4**(1), 99–104 (1984). http://www.ckit.co.za/secure/conveyor/papers/troughed/transient/transient-belt-stresses.htm
19. Thomas, R., Allen, R.: Viscoelastic indentation and resistance to motion of conveyor belts using a generalized Maxwell model of the backing material. Rubber Chem. Technol. **79**(2), 307–319 (2006). https://doi.org/10.5254/1.3547939
20. Yang, G.: Dynamics analysis and modeling of rubber belt in large mine belt conveyors. Sens. Transd. **81**(10), 210–218 (2014). https://www.sensorsportal.com/HTML/DIGEST/P_2492.htm
21. Yan, L., Lin, F.-Y., Wang, Y.-C.: Investigation on influence of speed on rolling resistance of belt conveyor based on viscoelastic properties. J. Theor. Appl. Mech. **45**(3), 53–68 (2015). https://doi.org/10.1515/jtam-2015-0017
22. Sakharwade, S.G., Nagpal, S.: Analysis of transient belt stretch for horizontal and inclined belt conveyor system. Int. J. Math. Eng. Manag. Sci. **4**(5), 1169–1179 (2019). https://doi.org/10.33889/IJMEMS.2019.4.5-092
23. Pihnastyi, O., Khodusov, V.: Hydrodynamic Kelvin-Voigt model transportation system. East Eur. J. Phys. **4**, 95–109 (2020). https://doi.org/10.26565/2312-4334-2020-4-13
24. Kulinowski, P.: Simulation method of designing and selecting tensioning systems for mining belt conveyors. Arch. Min. Sci. **59**(1), 123–138 (2014). https://doi.org/10.2478/amsc-2014-0009
25. Kurt, S., Gerdemeli, I., Cengiz, C.: Analysis of belt conveyor using finite element method. Sci. Proc. Int. Congr. Mach. Technol. Mater. **2**, 111–113 (2012)
26. Manjgo, M., Piric, E., Vuherer, T., Burzic, M.: Determination of mechanical properties of composite materials-the rubber conveyor belt with cartridges made of polyester and polyamide. Ann. Facult. Eng. Hunedoara **16**(1), 141–144 (2018)
27. Cornet, J.: Head and tail controls in long overland conveyors. Bulk Mater. Handl. Conveyor Belt **4**, 55–67 (2002)
28. Lodewijks, G.: Two decades dynamics of belt conveyor systems. Bulk Solids Handl. **22**(2), 124–132 (2002)
29. Karolewski, B., Ligocki, P.: Modelling of long belt conveyors. Mainten. Reliab. **16**(2), 179–187 (2014). http://yadda.icm.edu.pl/yadda/element/bwmeta1.element.baztech-ce355084-3e77-4e6b-b4b5-ff6131e77b30
30. Lawson, B.R.: Overland Conveyor: Control System Re-design and Implementation (2017). https://www.semanticscholar.org/paper/Overland-Conveyor-%3A-Control-System-Re-design-and-Lawson/a4d3dda013402e1e5b98aa5c4547d886b84a4983

Hyper Reality

Development of a Collaborative Platform for Education in Virtual Reality

Dmitriy Kritskiy(✉), Andrii Bykov, Vladimir Shevel, Krytska Olha, and Guobadia Efosa Mark

National Aerospace University "Kharkiv Aviation Institute", 17 Chkalova Str., Kharkiv 61070, Ukraine
d.krickiy@khai.edu

Abstract. The aim of the work is to study the possibility of creating and using virtual reality to develop a collaborative platform for virtual meetings in a pandemic. The aim of the study is to create an affordable solution that would provide access to the platform to as many people as possible, without complicating the development process. As a solution, an application was developed using the Unity Engine game engine for the family of Windows Mixed Reality and Google Cardboard platforms. The resulting prototype has the necessary set of functions for creating group meetings and training sessions. Users using Windows Mixed Reality devices have the ability to navigate with six degrees of freedom and interact with physical objects on the playing field, while users of Cardboard-compatible devices have only three degrees of freedom and no controllers to interact directly with the virtual environment. To solve this problem, the playing field has special interaction markers using the screen focus on the object or trigger.

Keywords: Virtual reality · Collaborative platform · Unity Engine · Windows Mixed Reality · Learning platform

1 Introduction

1.1 Formulation of the Problem

In 2019 the world faced a new virus called COVID-19, which swiftly spread across the world, leaving people around the world in social isolation.

Amongst many others, students felt detrimental damage due to inability to study in person, taking away an important part of education – personal contact. Many scholars were transferred home from abroad, others – were stuck at home. With new restrictions schools, colleges and universities were unable to host public classes and started to take a big step towards new ways of online education [1].

While before that the most popular online educational platform type were platforms like Moodle. Moodle is a web-based Learning Content Management System (LCMS), a Course Management System (CMS) and a Virtual Learning Environment designed to project pedagogical principles, namely a social constructivist philosophy using the

M. Nechyporuk et al. (Eds.): ICTM 2021, LNNS 367, pp. 283–294, 2022.
https://doi.org/10.1007/978-3-030-94259-5_25

possibilities of the social collaboration on the Internet. It gives an opportunity for teachers to provide and share files, provides functionality for graded assignments, quizzes, etc. with students in an easy-to-learn way, and to create high quality on-line courses [2].

Moodle is a free OSS and users are free to download, use, modify and distribute it under the terms of GNU. Effectiveness of Moodle comes from its modularity, flexibility and ability to be freely integrated with other web-applications.

Using Moodle-like platforms teachers were able to create, host and share classes during pandemic, and while sharing information using pre-recorded videos were an effective way to give an access to information no matter where attendee is located and what time it is, the inability to communicate with teachers was quickly brought up [3].

In order to mitigate this issue, universities combined this approach with the usage of video-conference software. By utilizing such programs as Zoom, WeChat, QQ, Skype, etc. teachers were able to combine both methods of synchronized and unsynchronized learning, hosting online meetings with video presentations.

And while this approach works for the purpose of knowledge transfer, the socialization factor takes a strong hit. Being unable to socialize with people in a familiar environment and to communicate using voice, face expressions and gestures left many scholars with feelings of unfairness.

Combined with travel and lodging restrictions, imposed both by government and universities, many students expressed doubts where is reasonable to continue an academical study or is to drop out and self-educate is more preferable [4].

With both educational institutes and scholar looking for the solutions to all the occurred problems, many technological instruments found their use in educational process. Amongst others arose the idea to use Virtual Reality to conduct online classes.

The idea of creating such systems has existed for a long time, but it has become especially relevant against the background of the restrictions imposed by the advent of COVID-19 in 2020. With the need for social distancing, the issues of knowledge transfer, information exchange and communication in general have become urgent for a large number of people.

1.2 Analysis of Recent Research and Publications

A study, conducted by Song, regarding online education draws attention to difficulty in understanding the objectives, instructional goals, and technical problems of online courses, which indicates that online education has significant drawbacks compared to regular education process [5]. However, the concept of online education, reviewed by Song can be improved by adding an interactive layer between the scholar and the knowledge itself.

While many educational institutions have switched to the use of pre-recorded video lectures, online conferences and other types of distance learning, the idea arose to use advanced methods and computer technology to create conditions for collaborative educational work.

In medical field, for example, the ability to eliminate the cost of travel and lodging and the loss of patient care opportunities for the visiting professor along with the ability for residents to attend across a wide geographic area all are strengths of the online meeting space platform [6].

The main limitations of all existing platforms are the lack of presence and limited opportunities for interaction between interlocutors.

The task is to create a platform that immerses the user in what is happening and uses the effect of presence to improve the assimilation and transmission of information. Virtual and augmented reality systems are used to solve such problems.

Virtual reality (VR), sometimes called immersive computing (ICT) technology, is a system that provides a unique way to interact with the ever-growing digital landscape. VR is often described as a set of technologies that allow people to immerse themselves in a world beyond reality [7].

Virtual reality devices and systems have received a new round of development with their popularization as a new type of controllers for video games and entertainment content. According to CNBC, in 2020, 26 million virtual reality devices were in the hands of private buyers, and 5.5 million are planned to be delivered to consumers during 2020.

Virtual reality devices often consist of a head display (HMD - Head Mounted Display) and input devices (controllers or others) and tracking stations or tracking marks [8]. The main display provides the transfer of the image from the information processing device (PC or other) to the user in a form that creates the effect of a stereoscopic image [9].

The picture quality largely depends on the resolution of the HMD screen, which should also provide the ability to adjust the distance between the lenses and the distance from the eyes to the monitor.

Virtual reality systems are divided into autonomous and semi-autonomous. By combining different tools and techniques of virtual reality it is possible to achieve high levels of immersion.

The purpose of this research project is to study the methods of virtual reality and develop with their help a virtual collaborative environment for learning.

The purpose of this research project is to study the methods of virtual reality and develop with their help a virtual collaborative environment for learning.

2 Building a Virtual Environment

Building a virtual environment for conducting training sessions requires a preliminary analysis of not only hardware solutions, but also the software products that provide support.

Considering modern game engines, Unity, Unreal Engine and Godot are the most advanced and well supported. These three software packages provide broad support for augmented and virtual reality, both native and through extensions.

While Godot is the fresh new engine with low community support, which is detrimental for continuous development, Unreal and Unity both provide good capabilities for the purpose of study 3D-game development. In conclusion, it was pointed out, that Unity provides easier experience, requires less preliminary knowledge, is more stable and provides vaster range of programming languages available [10]. Since stability and scalability are important points of this project, it was decided to use Unity Engine.

Comparative study, conducted by Dickson, Block and others, puts those two engines against each other in having decided on the development environment, it is necessary to select the target platforms. The native Unity XR library [11] supports augmented and virtual reality systems from various manufacturers: Microsoft HoloLens, Windows Mixed Reality, Magic Leap, Oculus, PlayStation VR.

In addition to built-in engine support, Unity allows you to expand the list of supported platforms using assets. Despite the fact that Unity has stopped official support for VR-solution for Android - Google Cardboard, the developer company - Google is engaged in support of this product.

Google Cardboard is a low-cost solution for mobile virtual reality in a form of a box with embedded in it lenses. Smartphone has to be located in the special chamber. The gyroscopic systems that are built-in into the smartphone are used to track user's positional data, such as head movements.

Google Cardboard support is built into all Android smartphones by default, which greatly simplifies the development process by unifying the program code. At the same time, Unity supports Android development out of the box, thus creating a convenient ecosystem for developing VR applications [12].

Based on the above comments, it was decided to develop application for the Windows Mixed Reality device family and the Google Cardboard platform using the Unity Engine in the C# programming language.

The development process begins with creating a scene: a virtual space on which planes and objects will be placed. When creating levels in virtual reality, you must take into account that the sizes of objects may not coincide with those expected in reality. To solve this problem, a certain object is created, the side of which will be equal to 1 m in virtual space [13].

Using the built-in Unity plugins, a XR Rig is placed on the stage - a prefab consisting of the necessary components to initialize the camera and controllers in virtual reality. The hierarchy of VR Rig objects is shown in Fig. 1. By using and modifying these components it is possible to create an immersive experience for the player by transferring the model, movements and gestures to the game models.

Fig. 1. The hierarchy of VR Rig objects

For hand gestures model's compatibility this project utilizes Oculus Hands. The Oculus Hand Models package includes left and right hands models and animations for use with Autodesk Maya. These models can assist in creating custom hand poses for use with the hands included with the Oculus Avatar SDK, or they can serve as a starting point for creating custom hands (Fig. 2).

The version of assets without modifications is included in the Unity Sample Framework or with Unreal Engine 4 source distribution. These contain ready-to-use models and animations for use with Oculus Touch controllers.

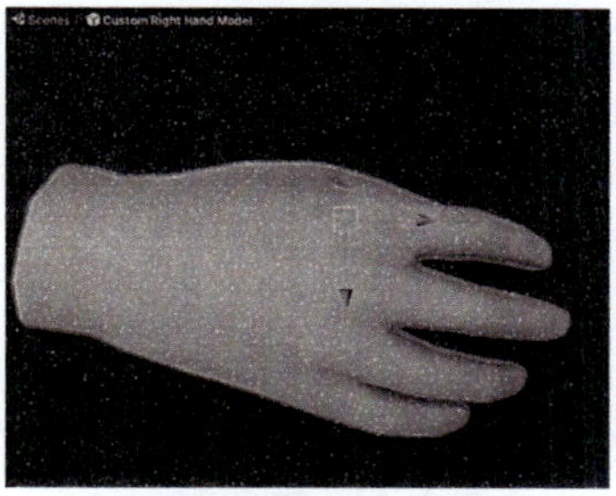

Fig. 2. Oculus hands

Despite focus on the Oculus hardware, this package provides great capabilities for customizing hand models compatible with most of the types of HMD-compatible proprietary controllers.

With the idea of multiplayer comes another obstacle: in order to offer a user immersive experience together with other players we need to synchronize both positions and rotation for all the objects, including interactable whiteboard and parts of the model's body.

Unfortunately, Unity's own framework for networking is being deprecated in the current version. Plus, it requires one user to be a host of the session, which in terms means that the person starting the meeting should have persistent internet connection, which cannot be guaranteed.

In order to mitigate this issue, we have explored the possibility of using a third-party networking framework, and chose Photon for multiplayer. It provides flexible API for developer to use their network service with stable and affordable cloud-based networking. It also provides Voice Over IP capabilities (VOIP) to translate voice using networking cloud [14].

The Photon Unity Networking or PUN is a library that used to provide access to cloud-based networking for multiplayer in Unity.

Photon allows us to use dedicated servers to transfer data between users in the same session. As a downside, the amount of data that can be reliably transferred using this system, but it has enough capability to be used as a protocol connection.

Using Photon, we can synchronize the absolute position of objects in the scene, changes in texture of the whiteboard, timestamps of shown videos on the screen and slide pages, all in real time. Photon also provides detailed statistics on resource management, example of which is shown on Fig. 3.

In order to stream in real-time actions of other players in given environment we have to determine what data and in which format is going to be transmitted to other players. The basis for an immersive VR experience is described in two concepts: place illusion and plausibility illusion [15].

While place illusion is achieved through using VR-googles that cover entire field of view, virtually replacing the environment, the plausibility illusion is harder to achieve.

All physical objects inside the scene have to either react to player's action or have set of strict and understandable rules of behavior. For example, whiteboard and marker are both real-life physical objects, rules of behavior are known to user [16]. To provide user

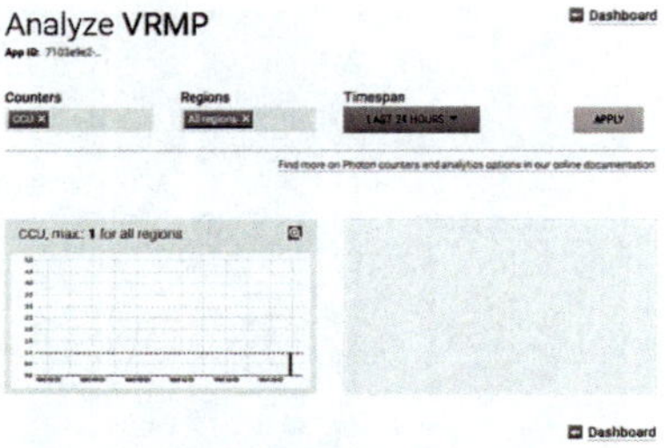

Fig. 3. Photon dashboard

with plausible versions of those objects, we allowed marker to be moved by player and to interact with other objects using RigidBody physics, while whiteboard is a stationary object, that can interact with other objects, but cannot be moved by player.

The information that is given to Photon, shown on Fig. 4, includes:

- Player Position
- Equipment State
- Player Model (including gestures, posing, etc.)
- World State (light, video and texture synchronization)

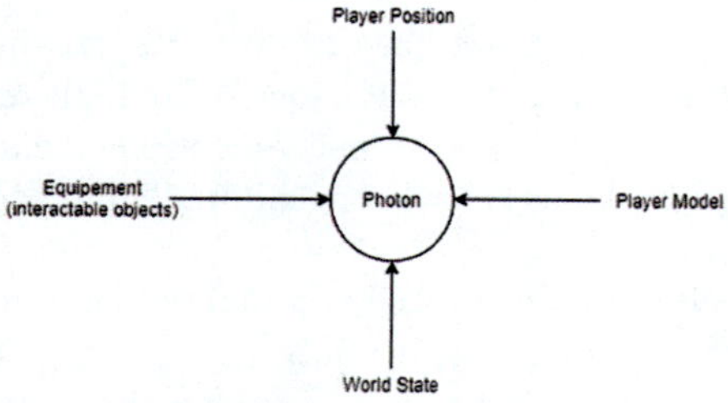

Fig. 4. Photon interaction diagram

The general process-flowchart of network interaction is shown in Fig. 5 Note, that some processes, for example world state synchronization, consists from multiple subroutines, some of which will be described in more detail further.

Using Unity's app analytics profiler, we have been able to get detailed graph of CPU, GPU and memory usage information as well as information about scripts execution runtime, example of which is shown in Fig. 6.

On the example we can see that most of the tasks work under 200 FPS, with some spikes up to 100 FPS, which is still a good result with industry standard being 60 FPS.

In order to release the need for session host to upload and share all media files used in the environment, which in turn would require stable connection with good bandwidth (those requirements would get bigger the more users are requesting those media files at a time), all media is shared through remote host servers.

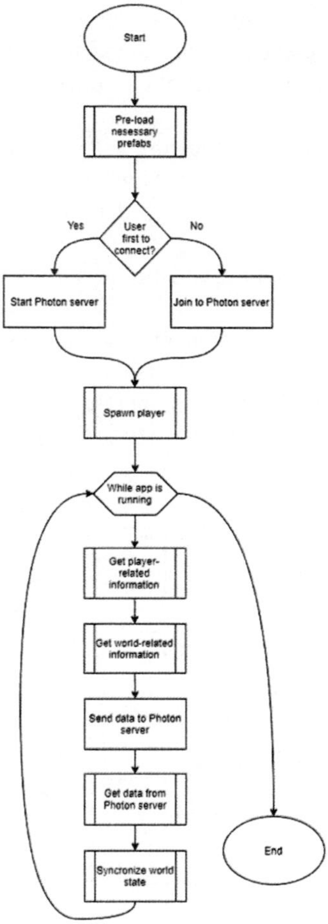

Fig. 5. General process-flowchart

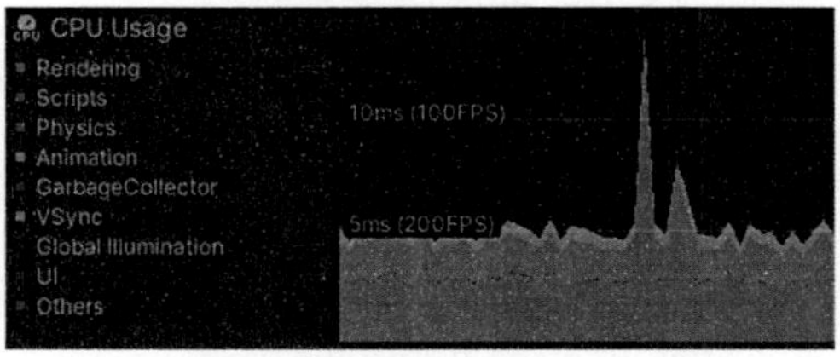

Fig. 6. CPU profiler information

The host will enter the link in the required field if needed, and tile itself is going to be downloaded via the developed application at the log-in phase by each individual user. This solution makes it easier both to host and to connect by avoiding creation of bottlenecks in the design.

One of the essential tools for any online class is a whiteboard: the easiest way to display information to the audience. Thanks to the ability to create objects of any shape and size, the whiteboard created in the developed environment can be as wide and as tall as needed.

Using the advantages of digital modeling and modern programming tools we can give the whiteboard other abilities to widen its area of use. For example, we implemented the ability to use the whiteboard as a multimedia screen to display photos, videos and PowerPoint presentations.

On Fig. 7 shown the class diagram, which shows from which basic components such a whiteboard consists of.

Using Unity as development environment gives an opportunity to operate with objects as classes. To create object/class, that will behave like real-life whiteboard we need 3 child classes-entities: the one to draw, one to erase and one display those changes. All of those entities are presented as object prefabs: prebaked objects, that can be serialized on demand.

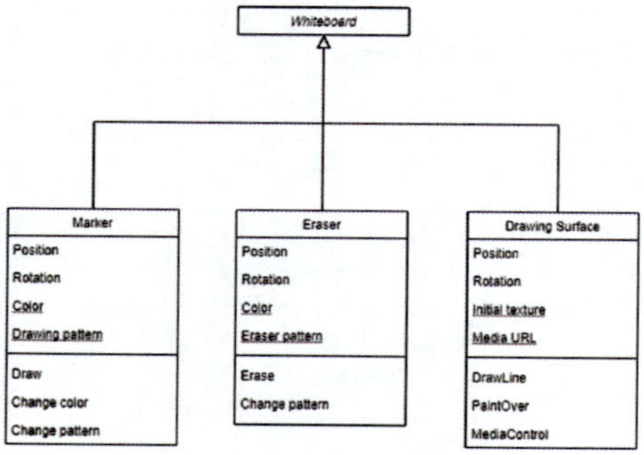

Fig. 7. Whiteboard class diagram

In order to process an input from the marker we need to define the surface that will work with it. The script works as follows:

- Loads an initial texture;
- Waits for the signal from the marker to process changes;
- Applies the difference between the initial texture's color and new color placed.

The full process flow-chart is shown on Fig. 8.

After the initialization, the surface waits for the marker/eraser to send the command to edit the texture. There are two types of commands: to create a splash, which is a single change of receiver's texture or to create a line, which, for the sake of the optimization, requires some additional processing.

The general idea of such approach is to replace only the relevant part of the texture and not to reload the whole texture with every edit required. To implement this idea sender (marker or eraser) will send the data about its position to the receiver and receiver will calculate the difference only to the relevant pixels of the texture.

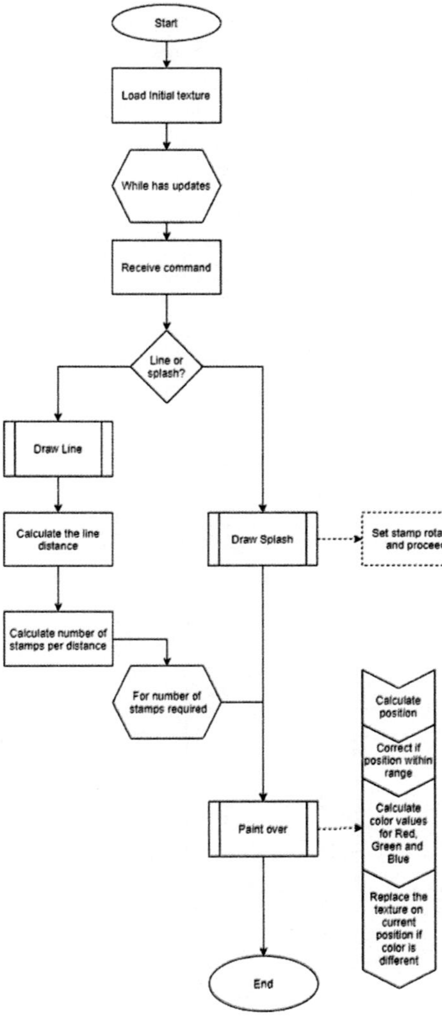

Fig. 8. Whiteboard drawing surface processing flow-chart

In order to get the position of a marker and the direction the marker is pointing towards, as well as the distance to the receiver we are using raycasting – a set of interaction techniques for immersive environments, such as VR. This approach resembles a laser pointer where user interacts with the environment by pointing an object with a light source towards an object of interest, that way confirming the selection.

Raycast – is a built-in Unity class, instance of which casts a ray, from an origin point in a specified direction of a specified length against all colliders in the Scene. Using a short-lengths collider we can achieve the illusion of a marker's painting head leaving the trace on the surface.

After the raycast registers a hit against a surface, the script sends a Draw Line or a Create Splash command depending on whether the raycast hit multiple different coordinates in a span of one frame or not.

Unity Engine then processes the command sent: if the object-receiver is the receiver and has the command-processing script required – it will process it accordingly to it; otherwise the command will not be generated.

The general flow-chart presented on Fig. 9.

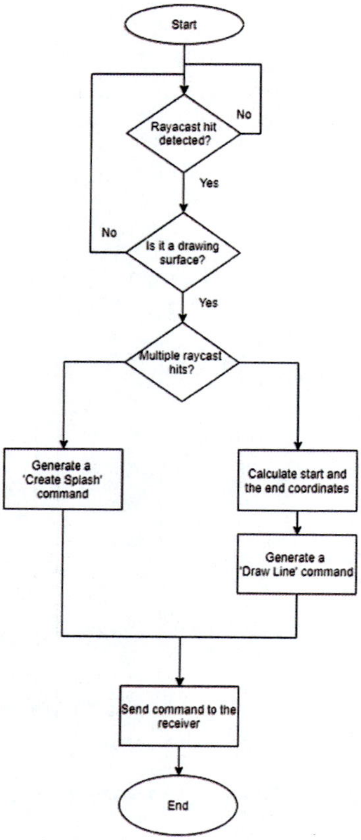

Fig. 9. Marker drawing surface processing flow-chart

Because the host-entity for Drawing Surface script can be any 3D-surface, we can use a plane for that purpose. We used the plane as a base object for media projection which gives us opportunity to include media controls into the same script during the design.

As a result, the created environment is capable of hosting classes in different subjects. Created environment can be run on PC, using Windows Mixed Reality family of headsets, and on Android devices via Google Cardboard or similar systems (Fig. 10).

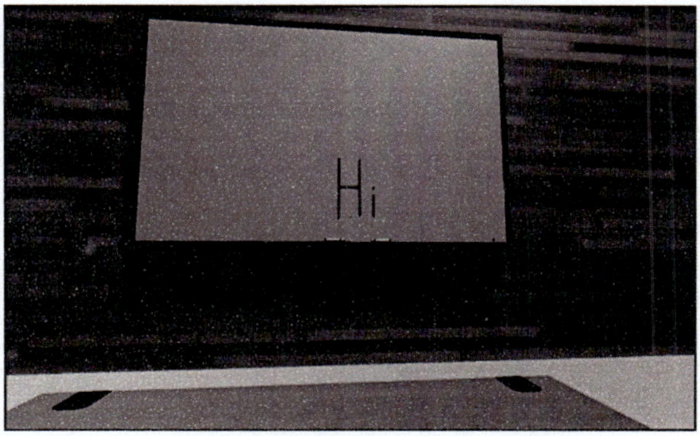

Fig. 10. First-person view

3 Conclusion

As a result, the created environment is capable of hosting classes in different subjects. Created environment can be run on PC, using Windows Mixed Reality family of headsets, and on Android devices via Google Cardboard or similar systems.

Using described in this paper system users can host online lectures and classes on dedicated servers, preventing lags, on which to show presentations, share videos or use interactive elements of the environment.

The proposed system is going to be used for the Master's thesis proposal presentation in Kharkiv National Aerospace University in 2021 due to COVID-19 restrictions.

Compared to existing solutions in the VR market, the scalability, the accessibility and the price of compatible hardware are the biggest advantages of our system achieved by creating the application that is compatible with wide-ranged families of VR hardware.

References

1. Karatanov, O., Bykov, A., Serginko, M., Miroshnichenko, D.: Implementation of augmented reality technologies in the training process with the design of aircraft equipment. Radioelectronic Comput. Syst. **1**, 110–118 (2021). https://doi.org/10.32620/REKS.2021.1.10
2. Mallampalli, S.S., Goyal, S.: Virtual learning of engineering students from the low-resource environment during the pandemic. In: Auer, M.E., Bhimavaram, K.R., Yue, X.-G. (eds.) REV 2021. LNNS, vol. 298, pp. 557–564. Springer, Cham (2022). https://doi.org/10.1007/978-3-030-82529-4_52
3. Dascalu, M., et al.: Before and during COVID-19: a cohesion network analysis of students' online participation in moodle courses. Comput. Hum. Behav. **121** (2021). https://doi.org/10.1016/j.chb.2021.106780
4. Ozturk, P., Avci, C., Kaya, C.: The effect of remote collaborative work on design processes during the pandemic. Strateg. Des. Re. J. **14**(1), 114–123 (2021). https://doi.org/10.4013/sdrj.2021.141.10
5. Simonova, I., Faltynkova, L., Kostolanyova, K.: Students' reflection on online distance learning: advantages, disadvantages, recommendations (2021). https://doi.org/10.1007/978-3-030-80504-3_23

6. Egarter, S., Mutschler, A., Brass, K.: Impact of COVID-19 on digital medical education: compatibility of digital teaching and examinations with integrity and ethical principles. Int. J. Educ. Integr. **17**(1), 1–19 (2021). https://doi.org/10.1007/s40979-021-00084-8
7. Geng, J., Wu, X.: Application of virtual reality technology in university education. J. Phys. Conf. Ser. **1972**(1) (2021). https://doi.org/10.1088/1742-6596/1972/1/012023
8. Pohudina, O., Kritskiy, D., Karatanov, A.V., Bykov, A.N.: Possibilities of position determination. In: Nechyporuk, M., Pavlikov, V., Kritskiy, D. (eds.) Integrated Computer Technologies in Mechanical Engineering. AISC, vol. 1113, pp. 523–537. Springer, Cham (2020). https://doi.org/10.1007/978-3-030-37618-5_45
9. Huang, W., Roscoe, R.D.: Head-mounted display-based virtual reality systems in engineering education: a review of recent research. Comput. Appl. Eng. Educ. **29**(5), 1420–1435 (2021). https://doi.org/10.1002/cae.22393
10. Bikmullina, I., Garaeva, E.: The development of 3D object modeling techniques for use in the unity environment. Paper presented at the 2020 International Multi-Conference on Industrial Engineering and Modern Technologies, FarEastCon 2020 (2020). https://doi.org/10.1109/FarEastCon50210.2020.9271568
11. Cook, J., Brown, M., Sellwood, M., Campbell, C., Kouppas, P., Poronnik, P.: XR game development as a tool for authentic, experiential, and collaborative learning. Biochem. Mol. Biol. Educ. (2021). https://doi.org/10.1002/bmb.21572
12. Chen, M.: The cardboard VR game development tool. Paper presented at the 2019 IEEE Eurasia Conference on IOT, Communication and Engineering, ECICE 2019, pp. 438–441 (2019). https://doi.org/10.1109/ECICE47484.2019.8942667
13. Pohudina, O., Kritskiy, D., Bykov, A.N., Szalay, T.: Method for identifying and counting objects. In: Nechyporuk, M., Pavlikov, V., Kritskiy, D. (eds.) Integrated Computer Technologies in Mechanical Engineering. AISC, vol. 1113, pp. 161–172. Springer, Cham (2020). https://doi.org/10.1007/978-3-030-37618-5_15
14. Photon Unity Networking (PUN). https://www.photonengine.com/pun. Accessed 8 Oct 2021
15. Kritskiy, D., Alexander, K., Juliia, P., Koba, S.: Modeling the characteristics of complex projects using parallel computing. Paper presented at the International Scientific and Technical Conference on Computer Sciences and Information Technologies, vol. 1, pp. 160–163 (2018). https://doi.org/10.1109/STC-CSIT.2018.8526667
16. Zheng, J., Liu, Q.: Design of 3D scene visual communication modeling based on virtual reality graphics rendering framework. J. Phys. Conf. Ser. **1982**(1) (2021). https://doi.org/10.1088/1742-6596/1982/1/012183

Information Modeling

Performance Modeling of Aircraft Positioning System

Ivan Ostroumov(✉) and Nataliia Kuzmenko

National Aviation University, Kyiv, Ukraine
{ostroumovv,nataliiakuzmenko}@ukr.net

Abstract. Navigational aids are widely used in the case of primary on-board positing system lock. An accuracy of airplane location detection depends on a variety of factors including performance and configuration of ground navigational aids network. A cell-based algorithm for accuracy estimation of positioning by navigational aids within defined airspace volume is proposed in the article. The algorithm is based on navigation system error evaluation for each point of airspace taking into account performance and availability of ground navigational aids network. Service volume of each navigational aid is estimated based on maximum range of a radio communication link taking into account technical characteristics of the ground and on-board equipment. Accuracy distribution within Ukrainian airspace was estimated with the help of proposed algorithm. Availability of a certain level of error was estimated for investigated airspace volume. Also, changes in availability levels were studied due to the national navigational aids network development process.

Keywords: Transportation · Performance · Distance Measuring Equipment · Positioning · Aviation · Ukrainian airspace · Service volume

1 Introduction

Modern air traffic system is based on precise airspace usage. Each airspace user measures its location and has to follow the predefined trajectory precisely. Performance-Based Navigation (PBN) considers a specific requirement for measured airplane coordinates [1]. In a common case, Global Navigation Satellite Systems (GNSS) is considered as a primary positioning sensor on board of airplane. Inertial Reference System (IRS) and positioning algorithms by pairs of navigational aids are used only in case of GNSS lock or poor accuracy [2]. Time of IRS operation is limited due to the additive noise of accelerometers and gyroscopes [3]. Algorithms of positioning by data from navigational aids are initiated inside of Flight Management System (FMS) of civil airplane [4, 5]. These algorithms use only a pair of simultaneously measured ranges from Distance Measuring Equipment (DME), angles measured by VHF Omni-directional Range (VOR), or their combinations [6]. Also, some FMS may use sequential operations [7] or data extrapolation by spline [8] to extend input data and getting some benefits in the accuracy

M. Nechyporuk et al. (Eds.): ICTM 2021, LNNS 367, pp. 297–310, 2022.
https://doi.org/10.1007/978-3-030-94259-5_26

of positioning. However, application of such approaches requires valuable computation power [9].

A quality of navigational determinations is estimated by characteristics of availability, accuracy, and integrity of aeronautical information [10]. According to PBN, navigational characteristics are limited to certain values defined in navigation specifications for particular airspace [1]. Availability of radio navigational aids is limited by service volume due to the radio propagation model in the space. Compliance with the regulative documents of area navigation is the necessary requirement for any flight operation.

The problem of alternative positioning to GNSS is quite important today due to degradation influence of interference and unintentional jamming of radio waves in space [11, 12]. Therefore, alternative positioning methods are extremely important for safety of aviation.

On-board DME interrogator provides range measuring from aircraft to ground transponder. Transponders usually are located at waypoints. Measured ranges to waypoints are used to navigate an aircraft in space. The Time Of Arrival (TOA) method is used to determine airplane coordinates based on a known database of navigational aids [14].

VOR provides a pilot with bearing information according to magnetic North of VOR ground station location. This angle is called Magnetic bearing from a station or radial (QDR). Magnetic heading to a station (QDM), also can be provided to a pilot. A pair of VOR receivers can be used to determine the coordinates of airplane location by Angle of Arrival (AOA) method [8]. Airplane is located in the point of intersection of two radials carried out from the point of ground VORs location.

In addition, VOR/DME pair located at one point can be used for aircraft positioning in space. In this case, the coordinates of location are determined from simple trigonometric relations in right triangle.

Each of positioning methods supports airplane localization in the lateral plane only, due to a pair of input data usage. Airplane position is calculated in the horizontal plane of the local Cartesian reference frame with subsequent transformation to the geodetic reference frame without height consideration.

Network of navigational aids includes ground stations located at a known position. Technical characteristics of navigational aids' ground equipment are significantly different, because the ground network is represented by various equipment, which is the result of a long process of its development [13].

The performance study of positioning by navigational aids is an important task for the safety of civil aviation. Results of network performance evaluation together with a traffic data are used for capacity study [15]. In particular, the Ukrainian navigational aids ground network is constantly evolving and developing. Previous studies [16, 17] showed that in the major part of the airspace, DME/DME, VOR/VOR, and VOR/DME positioning services are available for aircraft navigation. In particular, DME/DME positioning meets RNAV 1 requirements of area navigation and may be used as a stand-by source of coordinate data during airplane operation. However, the territorial conflict in the East and South of Ukraine in 2014 led to a restriction on the use of a certain part of navigational aids ground equipment, that negatively affected positioning performance and safety of air transportation as well. Therefore, the main objective of the article is to develop a cell-based algorithm for estimating the accuracy distribution of positioning by navigational aids methods within investigated airspace.

2 Service Volume

Service of navigational aid can be provided only within a certain three-dimensional volume which is defined by the radio propagation model. Service volume can be estimated by the maximum range of radio waves propagation limited by line-of-radio-sight. Maximum operational range is estimated taking into account technical characteristics of the ground and on-board equipment within the area of line-of-sight visibility for the frequency range of VHF and UHF [18, 19]. The maximum range (R_{max}) of a radio communication link is limited by the threshold of receiver sensitivity (P_r):

$$R_{max} = \frac{\lambda}{4\pi}\sqrt{\frac{P_t G_t G_r}{P_r}}, \quad (1)$$

where λ is a wavelength; P_t is a power of transmitter; G_t and G_r are antenna gain function of transmitter and receiver.

A gain depends on technical characteristics of antenna system [20, 21]. Service volume of common DME transponder estimated by (1) taking into account gain function for "dbs5100" antenna in free space is represented in Fig. 1. Also, propagation loss during traveling in troposphere should be taking into account in model (1) [22].

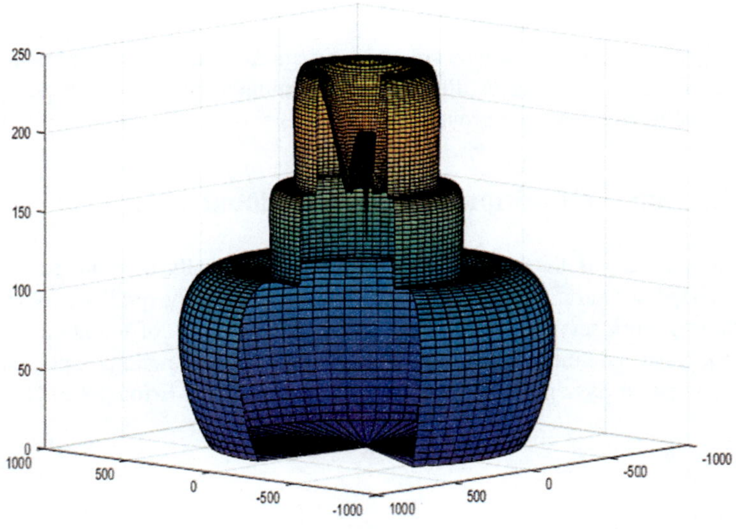

Fig. 1. Service volume model of "dbs5100" antenna in free space.

Model (1) can not be used in FMS due to required high level of computation power. In this case, a simple cylindrical model of Standard Service Volume (SSV) is used. SSV represents a volume within which services of navigational aid can be guaranteed with a certain probability level [23]. The SSV is fixed at the location of navigational aid and is characterized by a certain radius R and height H limited by terrain from the bottom and Elevation Angle (α_{el}) in the vertical plane (40° for DME and 60° for VOR [23]). Boundaries of elevation angle create a cone of unavailability above navigational aid within which its services can not be guaranteed (see Fig. 2). According to international regulative documents, there are three main types of cylindrical SSV [23]: terminal (maximum horizontal range is 25 NM for the heights from 1000 ft up to 12000 ft), low altitude (maximum horizontal range is 40 NM for the heights from 1000 ft up to 18000 ft), and high altitude (maximum horizontal range is 130 NM, maximal height up to 45000 ft).

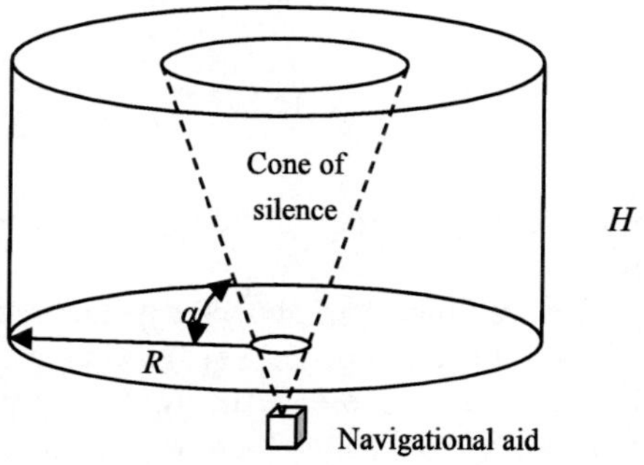

Fig. 2. Cylindrical standard service volume model.

The cylindrical model of service volume estimation is a rough approximation of the maximum range, but its use is justified when performing a comprehensive performance analysis of the navigational aid ground network.

3 Performance of Positioning by Navigational Aids

Positioning accuracy is the main characteristic that is usually determined within the confidence band of a certain probability level when analyzing the quality of navigational aids ground network service. In the general case, the accuracy of maintaining a given aircraft trajectory is defined as Total System Error (TSE). The main components of the TSE are Navigation System Error (NSE) and Flight Technical Error (FTE) [1]:

$$TSE^2 = NSE^2 + FTE^2, \tag{2}$$

FTE values depend on airplane model [24] and type of piloting, in particular: in case of manual guidance – 0.5 NM; flight director – 0.25 NM; autopilot – 0.125 NM [25]. Also, FTE includes influence of human factor on pilot and air traffic control sides [26]. NSE value is chosen from the assumption that aircraft positioning system error has a normal distribution, equal to two root-mean-square deviations of the radial error to ensure a 95% confidence band. An accuracy of aircraft coordinates determination by DME/DME, VOR/VOR, or VOR/DME pairs varies according to the positioning method.

NSE can be estimated, based on the weighted mean least squared method as follows [8]:

$$NSE^2 = 4trace\left(\sigma_0^2\left(H^T WH\right)\right), \tag{3}$$

where σ^2_0 is a mean-squared deviation of error introduced by ground navigational aid; W is a diagonal matrix of weight coefficients; H is a matrix of partial derivatives by coordinates.

Matrix of partial derivatives and weight matrix depends on positioning method:

- in case of positioning by pair of DME/DME [27]:

$$H_{DME} = \begin{bmatrix} \frac{x - x_{GS\,A}}{d_A} & \frac{y - y_{GS\,A}}{d_A} \\ \frac{x - x_{GS\,B}}{d_B} & \frac{y - y_{GS\,B}}{d_B} \end{bmatrix}, \; W_{DME} = diag\left(\left[\sigma_{DME\,A}^{-2}, \sigma_{DME\,B}^{-2}\right]\right) \tag{4}$$

- in case of positioning by pair of VOR/VOR [28]:

$$H_{VOR} = \begin{bmatrix} \frac{y_{GS\,A} - y}{d_a^2} & \frac{x - x_{GS\,A}}{d_a^2} \\ \frac{y_{GS\,B} - y}{d_B} & \frac{x - x_{GS\,B}}{d_B} \end{bmatrix}, \; W_{WOR} = diag\left(\left[\sigma_{VOR\,A}^{-2}, \sigma_{VOR\,B}^{-2}\right]\right) \tag{5}$$

where x and y airplane position in the local Cartesian coordinate system; x_{GS} and y_{GS} are coordinates of the ground station of navigational aid location; d_A and d_B are ranges to navigational aids in pair of AB.

4 An Accuracy Estimation

The geometry of ground navigational aids location creates a unique accuracy distribution within the area of airspace. The study of accuracy distribution is an important part of performance support and validation of airspace compliance with area navigation requirements.

An analysis of airspace is carried out using an iterative approach with space division into elementary cells with a further estimation of navigation performance. Navigation characteristics are considered to be constant within the boundaries of each elementary cell. If necessary, scaling or change of the size of an elementary cell can be used in order to increase the accuracy of the study. Accuracy of the approach is defined by geometric size of the elementary cell and in two-dimensional space can be represented as an area of a cell. Performance estimation is carried out for a specific altitude level in which, the whole investigated area is divided into elementary cells with a certain constant step in the horizontal plane of local Cartesian system. According to that, it is necessary to calculate a matrix of cells centers location (φ_{cell}, λ_{cell}). Coordinates of elementary cells are used to estimate navigation performance at the given point of space by the ground infrastructure.

Proposed algorithm performs estimation of positioning characteristics for navigational aids ground network within defined airspace at a given altitude level. As input data, the algorithm uses a global database of navigational aids in ARINC 404 format. Since global database includes many entries, then at the beginning of algorithm it is important to roughly detect potentially available DMEs (D_{DME}) and VORs (D_{VOR}) in investigated space for saving computation time. An algorithm implements the parsing of each elementary cell in space. For each of n points in space, a logical operation for estimation of navigational aids availability, a model of maximum range (1), or a simplified cylindrical model may be used, depending on the desired accuracy of the assessment. Each of the models should be applied taking into account the break of line-of-sight by high-altitude elements of the terrain. Therefore, a digital map of terrain should be used in the analysis of availability of navigational aids services at the point of investigation.

In particular, digital elevation model obtained from the Shuttle Radar Topography Mission of the space shuttle Endeavor that has global coverage with 1 arc-second resolution [29] was used. Based on the received binary availability matrices A_{DME} and A_{VOR}, which contain logical operators for each of the elementary cell for all the studied navigational aids, the total amount of available navigational aids (N_{DME}, N_{VOR}) and the amount of available navigational aids pairs for positioning ($N_{DMEpairs}$, $N_{VORpairs}$) have been estimated. If more than one pair is available, the choice of the most optimal one is performed by comparing the estimated values of accuracy and the internal angle between them [1]. For the optimal pair (P_{DME}, P_{VOR}, $P_{VOR/DME}$) [30], the values of NSE and TSE are estimated separately for each of the positioning methods by (2), (3), (4), (5).

Algorithm of performance estimation

```
Input: limits of analyzed airspace (φ_min, φ_max, λ_min, λ_max),
Flight Level (H_FL), Navigational aids database (φ_GS, λ_GS,
h_GS, Type)
Output: Elementary cell data (φ_cell, λ_cell), amount of navi-
gational aids (N), Amount of pairs (N_pairs), NSE, TSE

(φ_cell, λ_cell) ←cellcalculation(φ_min, φ_max, λ_min, λ_max)
D_DME←DMEdataselection(φ_GS, λ_GS, h_GS, Type, φ_cell, λ_cell)
D_VOR←VORdataselection(φ_GS, λ_GS, h_GS, Type, φ_cell, λ_cell)
h←FLtometers(H_FL)
foreach element step i ∈ 1, …, n
    A_DME[i]←DMEAvailability(D_DME, h, φ_cell[i], λ_cell[i])
    N_DME [i]←numberoftrue(A_DME[i])
    N_DMEpairs [i]←numberofpairs(N_DME[i])
    if N_DMEpairs [i]≥1
        P_DME[i]←optimalpairdetection(N_DMEpairs[i])
        NSE_DME[i]←nsedme(P_DME[i])
        TSE_DME[i]←tsedme(P_DME[i])
    end
    A_VOR[i]←VORAvailability(D_VOR, h, φ_cell[i], λ_cell[i])
    N_VOR [i]←numberoftrue(A_VOR[i])
    N_VORpairs [i]←numberofpairs(N_VOR[i])
   if N_VORpairs [i]≥1
        P_VOR [i]←optimalpairdetection(N_VORpairs[i])
        NSE_VOR[i]←nsevor(P_VOR[i])
        TSE_VOR[i]←tsevor(P_VOR[i])
    end
   if N_VOR [i]≥1 and N_DME [i]≥1
        N_VOR/DMEpairs [i]←numberofpairs(N_VOR[i], N_DME[i]))
        P_VOR/DME [i]←optimalpairdetection(N_VOR/DMEpairs[i])
        NSE_VOR/DME[i]←nsevor(P_VOR/DME [i])
        TSE_VOR/DME[i]←tsevor(P_VOR/DME [i])
    end
end
return(φ_cell, λ_cel , N, N_pairs, NSE, TSE)
```

Obtained results of performance estimation can be visualized in the form of pixel mapping or represented as contour graphs for a case study of space with a poor performance level. In addition, results can be represented as a dependence of certain characteristics on the percentage of their coverage area. Such a representation of the results of study allows us to estimate performance within the area of the investigated airspace in a general form.

5 An Accuracy Estimation of Ukrainian Navigational Aids Ground Network

In the numerical demonstration, we use navigational aids ground network of Ukrainian air navigation service provider. Ukrainian navigational aids ground network, on May 2020, includes 9 DMEs (BAH, KSN, KVR, RVN, STB, TER, UZH, VIN, YHT) and 8 VOR/DMEs (BRP, DNP, IVF, KHR, KVH, LIV, ODS, SLV) [31]. Result of NSE estimation by (3) for the case of positioning by pairs of DME/DME within Ukrainian airspace is represented in Fig. 3.

Ukrainian navigational aids network has been changing rapidly over the last decade. In 2013 the network included 8 DMEs and 10 VORs. Territorial conflict in the East and South of the country (2014) led to the unavailability of two VOR/DMEs ("DON" and "SMF") and two DMEs ("MRP" and "KRH"). In January 2018 ground network contained 6 DMEs (in particular: BAH, KSN, KVR, STB, VIN, YHT) and 8 VOR/DMEs (BRP, DNP, IVF, KHR, KVH, LIV, ODS, SLV). The loss of four navigational aids negatively affected positioning accuracy by pairs of navigational aids within airspace of the country.

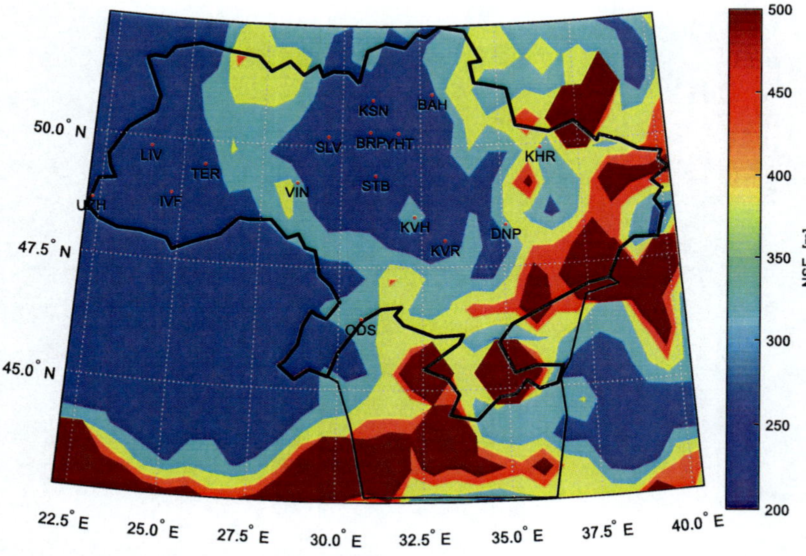

Fig. 3. NSE for DME/DME positioning within Ukrainian airspace

We perform a comparative analysis of the state of the navigational aids ground network in January 2018 and at the end of 2013 for the FL 195, based on the developed algorithm for estimation of navigational aids performance. Investigated part of the airspace was located between 42.75°–52.44° latitude and 22.12°–40.27° longitude with an elementary cell area equal to 3 km^2. The total amount of cells within the investigated range was $n = 5.34 \times 105$, within the airspace of Ukraine $n_{ua} = 2.84 \times 105$. Results of changes estimation in DMEs availability are presented in Fig. 4.

From Fig. 4 it can be seen that the operational areas of investigated navigational aids intersect over the most of the Eastern region. An area above the Sea of Azov, where all the lost navigational aids were available in 2013, suffered the greatest degradation in availability.

In particular, the area of NSE less than 800 m has decreased for 142,000 km^2, which corresponds to 16.65% of the total airspace. The territory with an error less than 2 km decreased for 98,700 km^2 (11.57%). Results of changes estimation in the areas of compliance with RNAV 1 navigation characteristics are shown in Fig. 5.

In the general case, RNAV 1 compliance area decreased from 95.49% (814,900 km^2) in 2013 to 83.28% (710,700 km^2) in 2018. The total reduction of space corresponds to 12.21% (104,200 km^2). The loss of airspace for RNAV 2 performance is 11.53% (98,400 km^2).

At the next step of the study, we will analyze changes in the size of areas in percentage of the total airspace for different DME/DME, VOR/DME, VOR/VOR positioning methods for FL 195. Results of estimation of available amount of navigational aids in the airspace metrics are represented in Fig. 6. It should be noted that the availability of VORs and VOR/DMEs is the same because Ukrainian navigational aids ground network does not contain single VORs. The degradative changes in availability are significantly expressed for a range of 2 to 10 navigational aids. Results in Fig. 6 indicate that the greatest loss of space for positioning by DME at 19.81% was reached at the level of 5 DMEs. On the other hand, the largest decrease in the area of 19.26% of the angular navigational aids is observed for availability of more than three VORs.

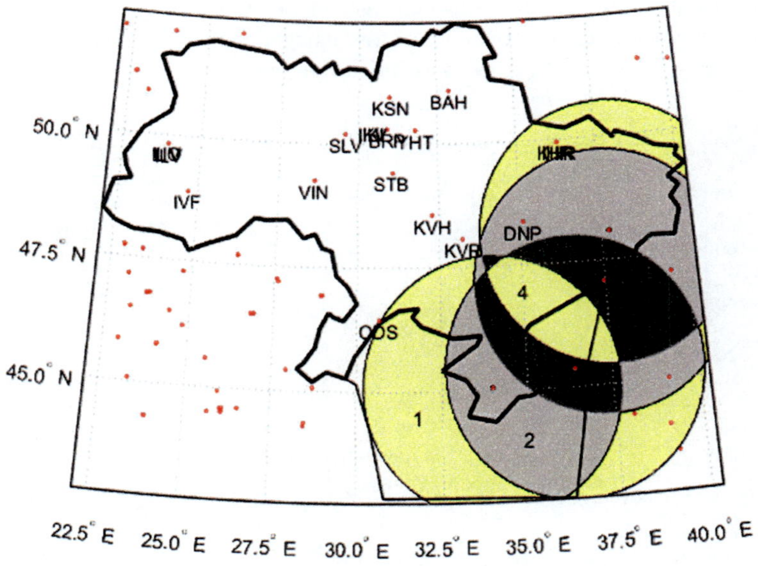

Fig. 4. Changes in DMEs availability during 2013/2018 for FL 195

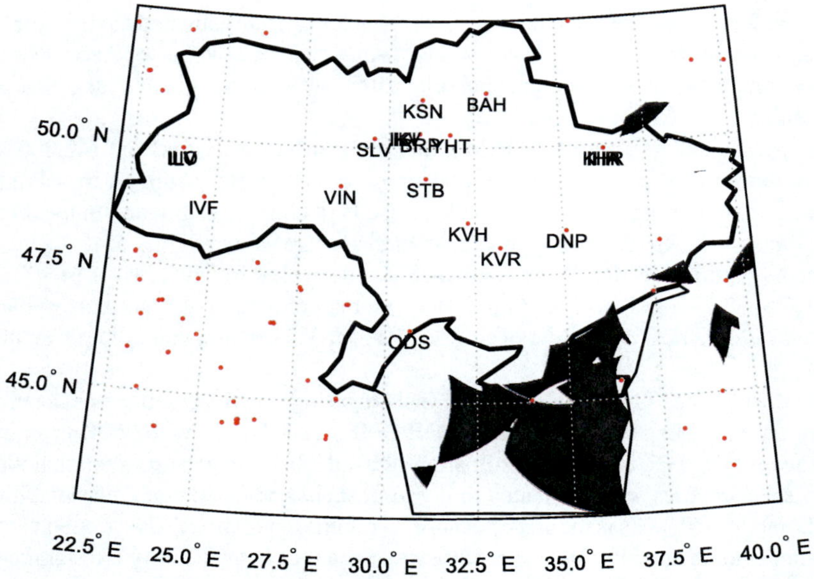

Fig. 5. Area of RNAV 1 degradation at DME / DME positioning during 2013/2018 for FL 195

Results of estimation of degraded changes in the availability of navigational aids pairs, together with a corresponding decrease in the area of space, are shown in Fig. 7. Obtained results in Fig. 7 indicate that changes in the number of pairs are much larger than six combinations created by loose of four navigational aids and are the result of a change in combinations with other available navigational aids.

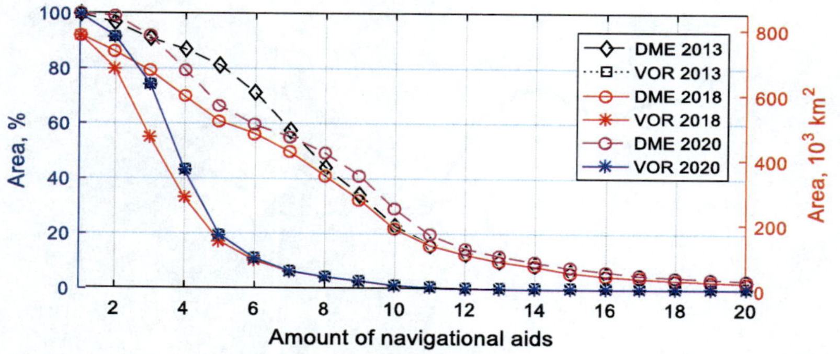

Fig. 6. The coverage area of navigational aids during 2013/2018/2020 for FL 195.

An important task is to estimate changes in positioning accuracy, results of which are shown in Fig. 8.

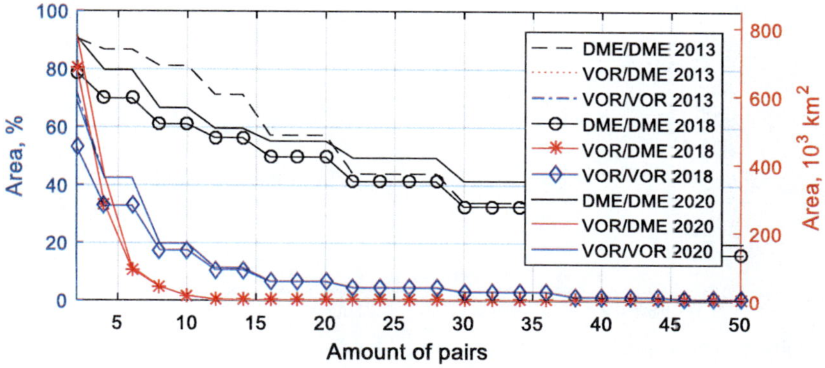

Fig. 7. Availability area of navigational aids pairs during 2013/2018/2020 for FL 195.

Results in Fig. 8 show that the coverage area has decreased significantly for DME/DME navigation. Significant changes occurred in areas with an accuracy of 1 km (the largest changes for NSE = 800m is 16.65% (142,000 km^2)). Changes in the area for VOR/DME and VOR/VOR navigation have a growing trend with an increase in NSE.

Results of estimating changes in the airspace under RNAV requirements are shown in Fig. 9. However, the loss of four navigational aids affected only the reduced RNAV areas for DME/DME positioning method. At that time, the loss of two VOR/DMEs in the general approach did not affect the size of RNAV availability areas for VOR/DME and VOR/VOR positioning methods.

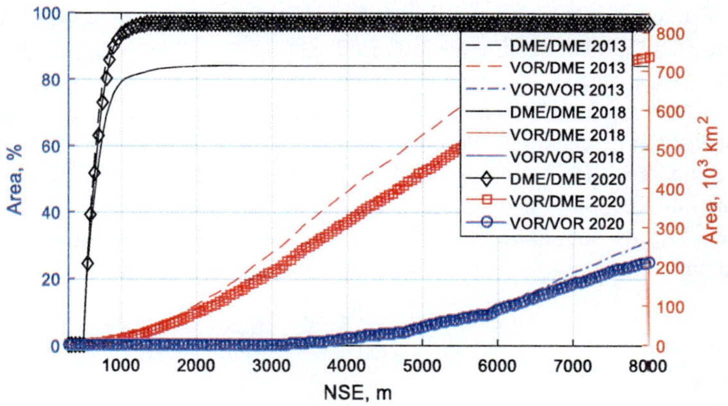

Fig. 8. Changes in positioning accuracy by navigational aids during 2013/2018 for FL 195

Obtained result in Fig. 9 shows that the area of RNAV 1 compliance for DME/DME positioning algorithm decreased from 95% in 2013 to 83% in 2018. Introduced three new DME transponders increase coverage area to 96% in 2020.

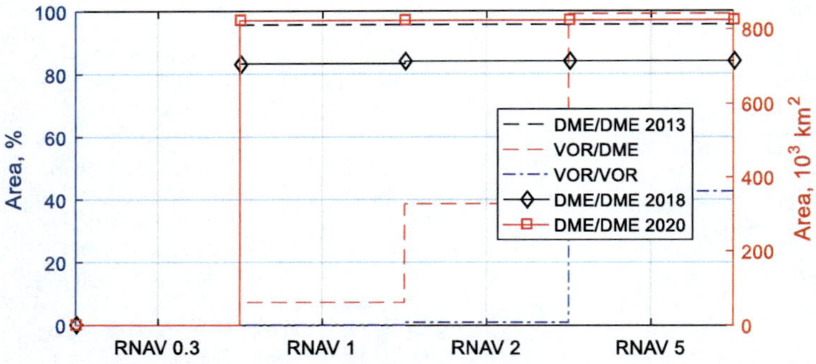

Fig. 9. Changes in airspace by RNAV requirements during 2013/2018 for FL 195

6 Conclusions

Performance assessment of navigational aids ground network is an important part of air navigation service support for safe airspace usage. Proposed cell-based algorithm helps to evaluate navigational aids network service performance, in particular: number of available navigational aids, number of pairs for positioning, the accuracy of positioning, and PBN compliance for each navigation method by DME/DME, VOR/DME, VOR/VOR pairs. Proposed algorithm is used for performance estimation of Ukrainian network of navigational aids, which has been developing rapidly during the last decade. Result of performance analysis helps to provide a comparative study of network performance at the end of 2013 with January 2018 and April 2020. Reducing of navigational aids network in 2014 as a result of the terrestrial conflict in the Eastern and Southern regions of Ukraine decreased availability, amount of pairs, and accuracy in comparison with 2013. In particular, DME/DME positioning reduced area for NSE > 500m degraded more than 12% of the total airspace. As a result, in the North-Eastern direction, above the Sea of Azov and in the southern direction over the Black Sea, areas with a significant degradation in positioning accuracy that did not meet RNAV 1 requirements have appeared (Fig. 5). An area of RNAV 1 airspace for FL 195 decreased by 12% (104,200 km^2), despite the introduction of new navigational aids in comparison with 2013.

Obtained results should be taken into account during planning the air traffic within discussed parts of the airspace and should be considered to improve the safety of civil aviation.

References

1. Performance-based Navigation (PBN) Manual. Doc 9613. ICAO (2008)
2. Solomentsev, O., Zaliskyi, M.: Correlated failures analysis in navigation system. In: Intertational Conference on Methods and Systems of Navigation and Motion Control, MSNMC, pp. 123–126, Kyiv, Ukraine (2018)
3. Sushchenko, O.A., Bezkorovainyi, Y.M., Golitsyn, V.O.: Fault-tolerant inertial measuring instrument with neural network. In: 40th International Conference on Electronics and Nanotechnology, ELNANO, pp. 797–801. IEEE. Kyiv (2020)

4. Han, S., Gong, Z., Meng, W., Li, C., Gu, X.: Future alternative positioning, navigation, and timing techniques: a survey. IEEE Wireless Commun. **23**(6), 154–160 (2016)
5. Eldredge, L., et al.: Alternative Positioning, Navigation & Timing (PNT) Study. ICAO Navigation Systems Panel (NSP). Working Group Meetings, Montreal (2010)
6. Ostroumov, I.V., Kuzmenko, N.S.: An area navigation (RNAV) system performance monitoring and alerting. In: First International Conference on System Analysis & Intelligent Computing, SAIC, pp. 1–4. IEEE, Kyiv (2018)
7. Jalloul, T., Ajib, W., Yeste-Ojeda, O.A., Landry, R., Thibeault, C.: DME/DME navigation using a single low-cost SDR and sequential operation. In: First International Conference on Digital Avionics Systems Conference, DASC, pp. 3C2–1 (2014)
8. Ostroumov, I.V., Kuzmenko, N.S.: Accuracy improvement of VOR/VOR navigation with angle extrapolation by linear regression. Telecommun. Radio Eng. **78**(15), 1399–1412 (2019)
9. Kulik, A., Dergachev, K.: Intelligent transport systems in aerospace engineering. Study Syst. Decis. Control **32**, 243–303 (2016)
10. Global Navigation Satellite System (GNSS) Manual, Doc 9849, ICAO (2012)
11. Narins, M., Eldredge, L., Enge, P., Harrison, M., Kenagy, R., Lo, S.: Alternative position, navigation, and timing – the need for robust radio navigation global navigation satellite systems. In: Report of a Joint Workshop of the National Academy of Engineering and the Chinese Academy of Engineering, pp. 119–136. The National Academies Press (2012)
12. Elghamrawy, H., Karaim, M., Tamazin, M., Noureldin, A.: Experimental evaluation of the impact of different types of jamming signals on commercial GNSS receivers. Appl. Sci. **10**(12), 4240 (2020)
13. Lo, S., Chen, Y.H., Enge, P., Peterson, B., Erikson, R., Lilley, R.: Distance measuring equipment accuracy performance today and for future alternative position navigation and timing (APNT). In: Proceedings of the Technical Meeting of the Satellite Division of the Institute of Navigation, pp. 711–721. ION GNSS (2013)
14. Radio Navigation Aids: International Standards and Recommended Practice. Aeronautical Telecommunications, Volume I. Annex 10 to the Convention on International Civil Aviation. ICAO (2006)
15. Ostroumov, I., Kuzmenko, N.: Interrogation rate measurements of distance measuring equipment in air navigation system. In: 2nd International Conference on System Analysis & Intelligent Computing, SAIC, pp. 1–5. IEEE, Kyiv (2020)
16. Ostroumov, I.V., Kuzmenko, N.S.: Risk analysis of positioning by navigational aids. In: Signal Processing Symposium, SPSympo, pp. 92–95. IEEE, Krakow (2019)
17. Ostroumov, I., Kharchenko, V., Kuzmenko, N.: An airspace analysis according to area navigation requirements. Aviation **23**(2), 36–42 (2019)
18. Pavlikov, V.V., Zhyla, S.S., Odokienko, A.V., Antonov, M.O.: Radiometer with signal energies ratio. In: Proceedings of the Radar Methods and Systems Workshop (RMSW), pp. 99–102. IEEE, Kyiv (2016)
19. Merzlikin, A., Pavlikov, V., Ruzhentsev, N., Sobkolov, A., Tsopa, O., Zhyla, S.: Peculiarities of microwaves attenuation statistics for Ukrainian region. In: Proceedings of the Ukrainian Microwave Week (UkrMW), pp. 761–764. IEEE, Kharkiv (2020)
20. Shcherbyna, O., Kozhokhina, O.: Construction principles of quadrifilar helical antennas. Telecommun. Radio Eng. **79**(16), 1441–1453 (2020)
21. dB Systems Inc. Homepage. https://www.dbsant.com
22. Barrios, A.E.: Considerations in the development of the advanced propagation model (APM) for US Navy applications. In: International Conference on Radar, pp. 77–82 (2003)
23. AC 00-31A U.S. National Aviation Standard for the Very High Frequency Omnidirectional Radio Range (VOR)/Distance Measuring Equipment (DME)/Tactical Air Navigation (TACAN) Systems, FAA (1982)

24. Lo, S., Enge, P., Niles, F., Loh, R., Eldredge, L., Narins, M.: Preliminary assessment of alternative navigation means for civil aviation. In: Proceedings of the Technical Meeting at the Institute of Navigation, pp. 484–492. ITM (2010)
25. Minimum Operational Performance Standards for Airborne Supplemental Navigation Equipment Using Gloioal Positioning System. RTCA/DO-208. RTCA Inc. (1993)
26. Rizun, N., Shmelova, T.: Decision-Making models of the human-operator as an element of the socio-technical systems. In: International Workshop on Strategic Imperatives and Core Competencies in the Era of Robotics and Artificial Intelligence, pp. 167–204 (2016)
27. Ostroumov, I.V., Kuzmenko, N.S.: Accuracy estimation of alternative positioning in navigation. In: 4th International Conference on Methods and Systems of Navigation and Motion Control, MSNMC, pp. 291–294. IEEE, Kiev (2016)
28. Lv, X., Liu, K., Hu, P.: Geometry influence on GDOP in TOA and AOA positioning systems. In: International Conference on Networks Security Wireless Communications and Trusted Computing, NSWCTC, vol. 2, pp. 58–61 (2010)
29. Shuttle Radar Topography Mission. U.S. Geological Survey's, Long Term Archive at the National Center for Earth Resource Observations and Science. http://wwwlta.cr.usgs.gov
30. Ostroumov, I.V., Kuzmenko, N.S., Marais, K.: Optimal pair of navigational aids selection. In: 5th International Conference on Methods and Systems of Navigation and Motion Control, pp. 32–35. IEEE, Kiev (2018)
31. Aeronautical Information Publication (AIP) of Ukraine Ukrainian State Air Traffic Services Enterprise (2020)

Computer Simulation of Abnormal Glow Discharge in Inverse Magnetron Sputtering Systems with Sectioned Cathode Units

Denis Slusar(✉), Oleksii Isakov, Volodymyr Kolesnyk, Oleg Chugai, and Serhii Oliynick

National Aerospace University "Kharkiv Aviation Institute", Chkalova Street, 17, Kharkiv 61070, Ukraine
d.sliusar@khai.edu

Abstract. A brief review of methods for the formation of coatings from high-entropy alloys is presented. It is concluded that the multi-target co-deposition method is promising for the formation of coatings from high-entropy alloys. The advantage of inverted magnetron sputtering systems with sectioned cathode units developed at the National Aerospace University "Kharkiv Aviation Institute" over the currently used technological ion-plasma generators of the coating material flow for the formation of coatings from high-entropy alloys is shown. It is concluded that it is necessary to develop a mathematical model that allows one to calculate the operating parameters of the sputtering system. These parameters are to be known to obtain coatings of a given composition and thickness. It is mentioned that to solve this problem, first of all, a method was developed for calculating the distribution of local plasma parameters in the discharge gap of an inverse magnetron sputtering system with sectioned cathode nodes and axial plasma flows. When constructing a model of the investigated sputtering system, a fluid plasma model was used. Further, using the proposed model, the current-voltage characteristics of the investigated sputtering systems were obtained. They were compared with experimental data. The character of the calculated and experimental volt-ampere characteristics is the same. Based on comparison of these calculated parameters with experimental data, the conclusion was made about the correctness of the developed mathematical model.

Keywords: Plasma fluid model · Magnetron sputtering system · Volt-ampere characteristic

1 Introduction

Recently a new class of materials has become widespread as functional coatings on parts for various purposes - high-entropy alloys (HEAs) [1], the main creators of which are Professors J. W. Yeh, B. Cantor and colleagues. There are several widespread methods of forming coatings based on HEA.

M. Nechyporuk et al. (Eds.): ICTM 2021, LNNS 367, pp. 311–320, 2022.
https://doi.org/10.1007/978-3-030-94259-5_27

K. Yu. Yakovchuk and colleagues [2] developed composite ingots to form gradient coatings of both metallic and non-metallic components by the methods of electron-beam evaporation. In the articles by L. G Shaginyan and colleagues [3], to create HEA coatings the method of sputtering a target cathode of preformed alloys was used. In turn M. Beresnev and colleagues in the works [4] also used cathodes of preformed alloys to create HEAs coatings by a vacuum-arc method of generating a coating material flow.

Many other scientists have followed a similar path [5, 6]. Cathodes were manufactured by rather laborious and expensive method of multiple (at least 5–7 times) remelting of a mixture from individual components of an alloy. Due to significant complexity, and sometimes impossibility of creating an intended component composition, this way of solving the problem of creating multicomponent HEAs coatings significantly limits the possibilities during developing a new coating composition.

The possibility of applying coatings by generating streams of particles of pure metals or partial alloys, which forms particular coating, has been shown quite a long time ago. The method of multi-target co-deposition is discussed in detailed reviews by Yan X. H. and colleagues [5], Wei Li and colleagues [6], Zhang W. and colleagues [7].

Among many methods for forming coatings, the most common methods are vacuum ion-plasma technologies. Specific of currently used equipment for ion-plasma coating limits the number of plasma sources used simultaneously in the working area of a technological unit. Attempts to use the required number of technological sources of the coating material flow lead to an unacceptable complexity of the technological unit design [8].

The traditional cathode node of the coating material flow generator consists of a cathode itself, its fastening and sealing system, a magnetic system and a cooling system. Reducing dimensions of the cathode assembly is possible only till operation principle is maintained.

Therefore, on currently used equipment implementation of forming a multicomponent coating due to simultaneous generation of multiple streams of atoms or ions of the components which make up the future coating is seriously difficult or not possible at all.

Thus, to solve the problem, it is necessary to develop a fundamentally new methodology for the synthesis of multicomponent coatings and equipment for its implementation, which would make it possible to form coatings based on HEAs by generating particle streams of coating material with any number of components required to solve particular problem.

At the National Aerospace University "Kharkov Aviation Institute" new technological generators [9, 10] were developed, which are a sort of inverse magnetron systems [11]. On their basis pilot industrial technological units have been designed and new technologies were developed. These technologies do not require cathodes with complex composition anymore. Due to the sputtering of a large number of target cathodes made of either individual coating components or partial alloys, these technologies allow the formation of multi-component coatings of almost any component composition, including HEA and gradient coatings. Changing the electrical parameters of the sputtering system allows one to form a coating with a predetermined change in the component composition both in thickness of a coating and depending on the part location on the product surface [9, 10].

2 Research Objective Determination

In the previous publication [9], there were identified which technological parameters effect on the technological process of forming multicomponent coatings using the investigated sputtering systems. For this purpose a process simulation of material transfer from target cathodes to the substrate surface was used. It was shown that a large number of interrelated parameters effects on the technological process, for the correct determination of which it is necessary to carry out either an unacceptable number of experiments, or to create a mathematical model of the investigated sputtering systems, which would allow to calculate these parameters with the required accuracy.

To solve this task a fluid plasma model was used [12, 13]. Simulation of local plasma parameters for an inverse magnetron sputtering system with sectioned cathode nodes and axial plasma flows was carried out and the results were compared with experimental ones. It is concluded that the proposed mathematical model correctly describes the operation of the investigated sputtering systems.

Verification of the modelling method for determination the local plasma parameters was carried out for only one mode. While these sputtering systems can be used in different operation modes it is necessary to ensure that the proposed method of calculating the discharge parameters can be applied to the entire range of operation parameters.

3 Simulation of Discharge Characteristics in Magnetron Sputtering Systems with Sectioned Cathodic Units

The investigated sputtering systems must operate at different parameters to provide particular conditions for the formation of particular coatings in turn. The proposed mathematical model is supposed to describe correctly their work under all possible operating conditions. Unfortunately, it is not possible to carry out probe measurements of plasma parameters in the discharge gap of the investigated sputtering systems for many operating modes due to the high complexity of such studies.

Therefore, a comparison of experimental and calculated current-voltage characteristics of the investigated sputtering systems at different operating modes was performed. Experimental and calculated current-voltage characteristics for an inverse magnetron sputtering system with sectioned cathode nodes and radial plasma flows are presented in Figs. 1, 2, 3 and 4. To obtain experimental volt-ampere characteristics, the pressure of the plasma-forming gas in the technological unit was maintained at 0.05 Pa, the same value was for numerical simulation. Measurements and registration of operating parameters were performed using a programmable logic controller K202 and current and voltage sensors.

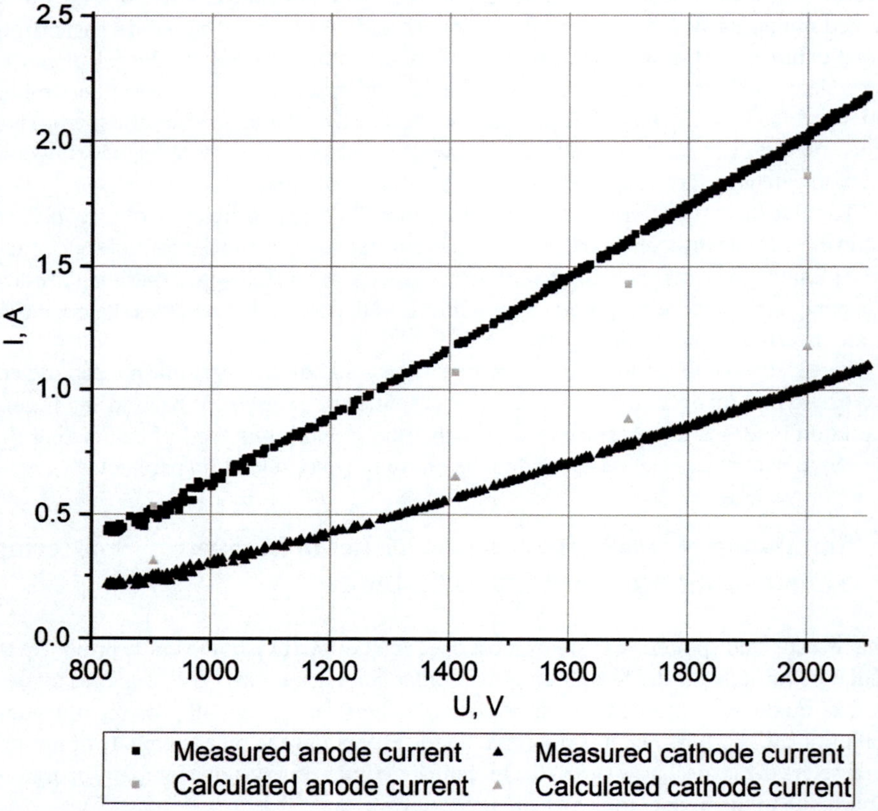

Fig. 1. Volt-ampere characteristics of the inverse magnetron sputtering system with sectioned cathode nodes and radial plasma flows for resistance value of the ballast resistor in the target cathode circuit of 0 Ω.

From the Figs. 1, 2, 3 and 4 one can see that for the inverse magnetron sputtering system with sectioned cathode nodes and radial plasma flows the behaviour of the current-voltage characteristics is identical for simulation and empirical data. For most operational modes of the sputtering system, deviation between empirical and calculated data does not exceed 15%. Only for the operating mode when resistance of the ballast resistor in the target cathode circuit is 2000 Ω, the deviation may reach 30%.

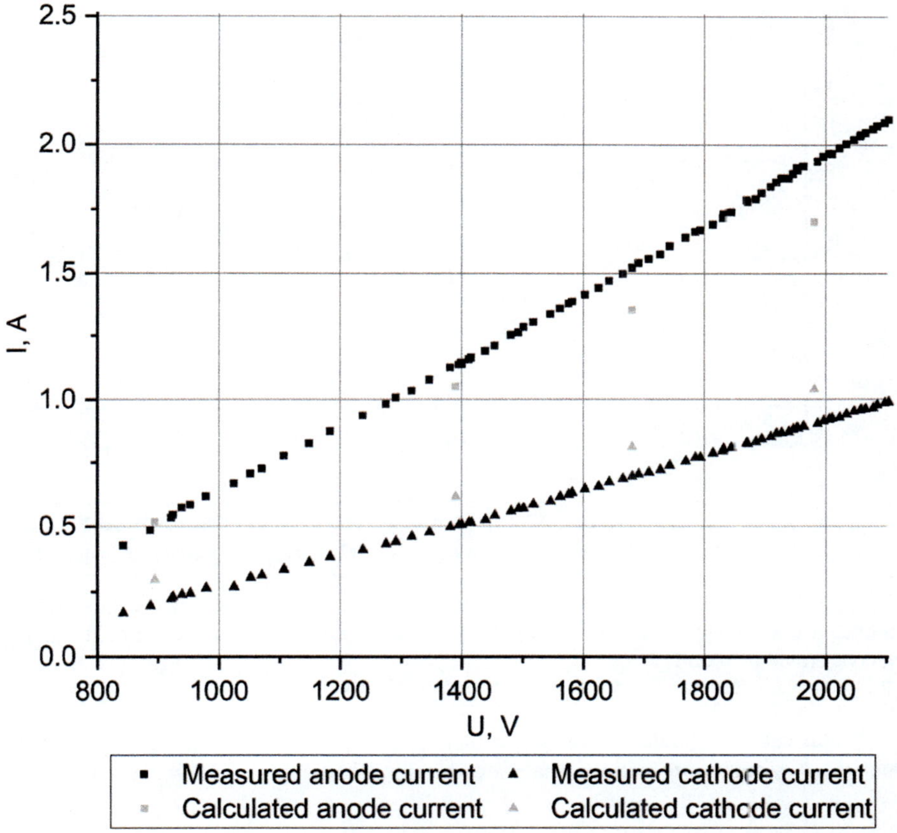

Fig. 2. Volt-ampere characteristics of the inverse magnetron sputtering system with sectioned cathode nodes and radial plasma flows for the resistance value of the ballast resistor in the target cathode circuit of 100 Ω.

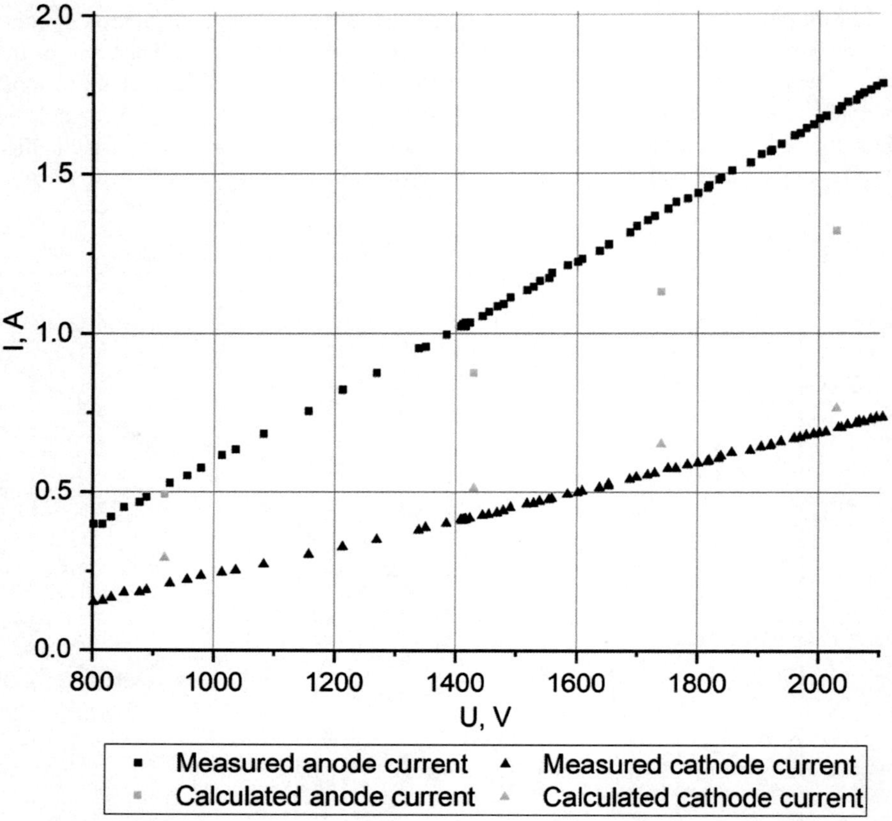

Fig. 3. Volt-ampere characteristics of the inverse magnetron sputtering system with sectioned cathode nodes and radial plasma flows for the resistance value of the ballast resistor in the target cathode circuit of 1000 Ω

Empirical and simulation current-voltage characteristics for an inverse magnetron sputtering system with sectioned cathode nodes and radial plasma flows are presented

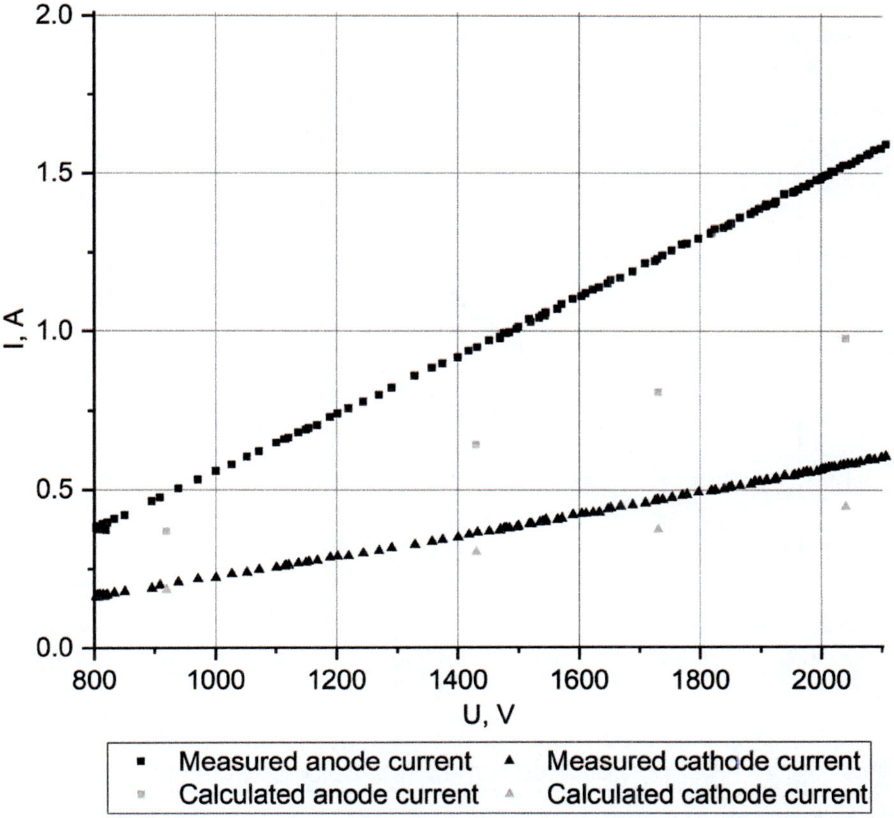

Fig. 4. Volt-ampere characteristics of the inverse magnetron sputtering system with sectioned cathode nodes and radial plasma flows for the resistance value of the ballast resistor in the target cathode circuit of 2000 Ω.

in Figs. 5, 6. During obtaining the empirical volt-ampere characteristics, the resistance of the ballast resistor in the target cathode circuit was set at 72 Ω, the same value was in the calculation.

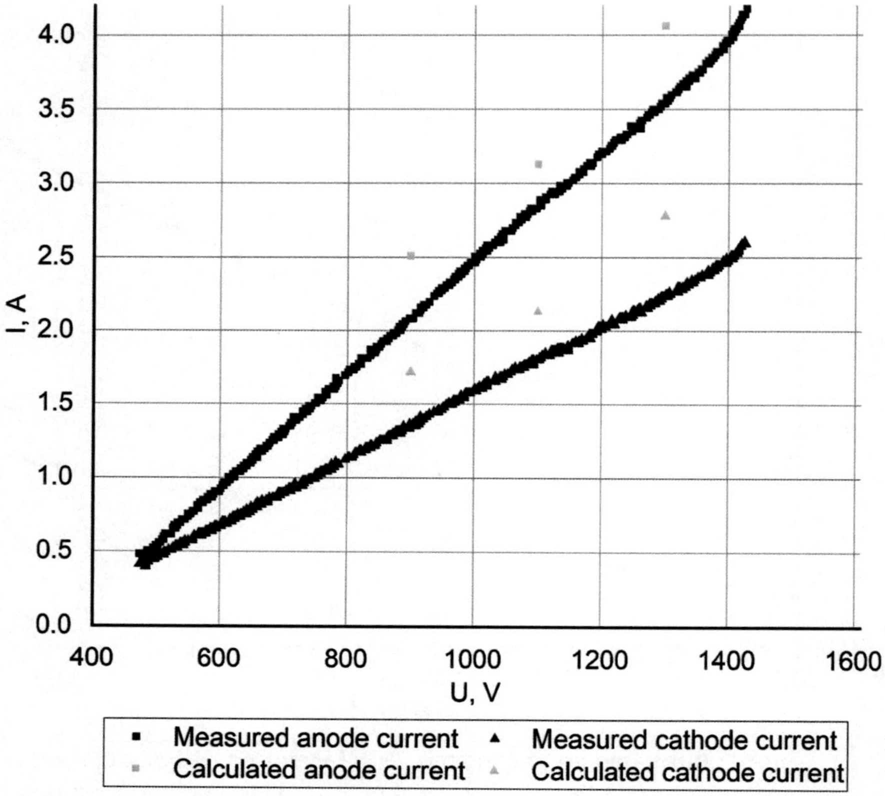

Fig. 5. Volt-ampere characteristics of the inverse magnetron sputtering system with sectioned cathode nodes and axial plasma flows at an argon pressure of 0.05 Pa.

For an inverse magnetron sputtering system with sectioned cathode nodes and axial plasma flows, one can see that behavior of the current-voltage characteristics is identical for simulation and empirical data. The difference between empirical and calculated data does not exceed 32%.

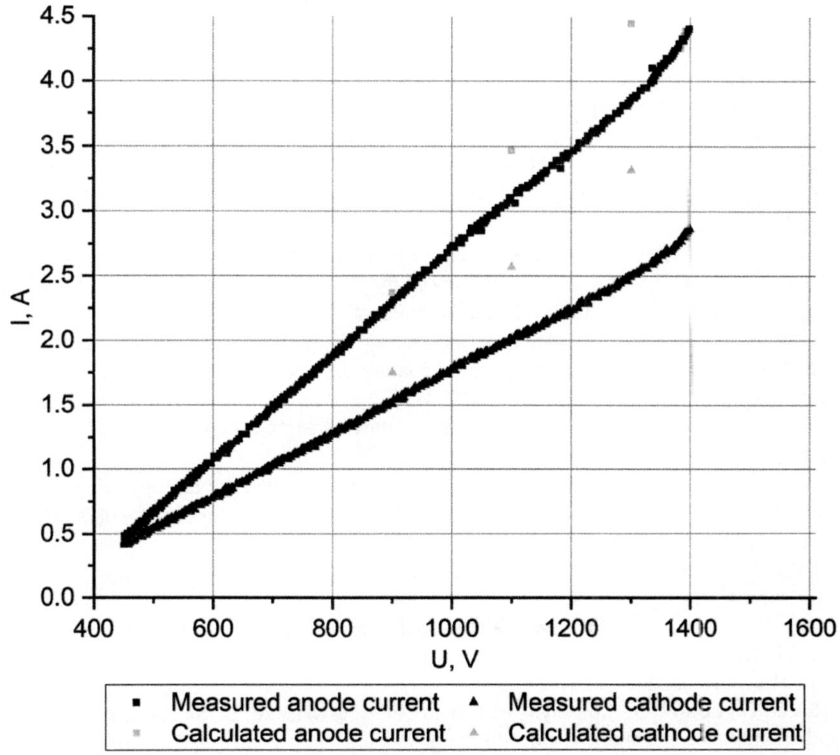

Fig. 6. Volt-ampere characteristics of the inverse magnetron sputtering system with sectioned cathode nodes and axial plasma flows at an argon pressure of 0.08 Pa.

4 Conclusion

Taking into account the comparison results of behavior of local plasma parameters' distribution in the investigated sputtering systems and both calculated and measured volt-ampere characteristics, one can conclude that the simulation technique can adequately describe the work of systems.

Such approach makes possible application of this technique during designing equipment for formation of functional coatings based on the studied sputtering systems. In this case time and costs for equipment designing will be reduced significantly, no need in a large number of laboratory specimens anymore.

At the same time applying of this modeling technique during development of technological processes for formation of functional coatings is not always available due to accuracy of the coincidence of the calculated and measured basic parameters. Thus, there is still need in search of ways increasing accuracy of calculation for basic parameters of technological process of formation of coverings for investigated sputtering systems.

References

1. Miracle, D., Senkov, O.: A critical review of high entropy alloys and related concepts. Acta Mater. **122**, 448–511 (2017). https://doi.org/10.1016/j.actamat.2016.08.081
2. Yakovchuk, K., Rudoy, Y., Mykyktchyk, A., Onoprienko, E.: Structure and properties of condensed gradient metal-ceramic thermal barrier coatings with NiAl-bond coat alloyed with Y, Hf, Dy, Electrometallurgy today, #4, pp. 8–16 (2017). https://doi.org/10.15407/sem2017.04.02
3. Shaginyan, L., Gorban', V., Krapivka, N., Firstov, S., Kopylov, I.: Properties of coatings of the Al–Cr–Fe–Co–Ni–Cu–V high entropy alloy produced by the magnetron sputtering. J. Superhard Mater. **38**, 25–33 (2016). https://doi.org/10.3103/S1063457616010044
4. Beresnev, V., Sobol', O., Andreev, A., et al. Formation of superhard state of the TiZrHfNbTaYN vacuum–arc high-entropy coating. J. Superhard Mater. **40**, 102–109 (2018). https://doi.org/10.3103/S1063457618020041
5. Yan, X., et al.: A brief review of high-entropy films. Mater. Chem. Phys. **210**, 12–19 (2017). https://doi.org/10.1016/j.matchemphys.2017.07.078
6. Li, W., Liu, P., Peter, K.: Microstructures and properties of high-entropy alloy films and coatings: a review. Mater. Res. Lett. **6**(4), 199–229 (2018). https://doi.org/10.1080/21663831.2018.1434248
7. Zhang, W., Liaw, P.K., Zhang, Y.: Sci. China Mater. **61**(1), 2–22 (2017). https://doi.org/10.1007/s40843-017-9195-8
8. Nagy, P., et al.: Processing and characterization of a multibeam sputtered nanocrystalline CoCrFeNi high-entropy alloy film. Surf. Coat. Technol. **386**, 125465 (2020). https://doi.org/10.1016/j.surfcoat.2020.125465
9. Sliusar, D., Isakov, O., Kolesnyk, V., Chugai, O., Litovchenko, L., Stepanushkin, M.: Computer simulation of abnormal glow discharge in an inverse magnetron sputtering system with axial plasma flows. In: Nechyporuk, M., Pavlikov, V., Kritskiy, D. (eds.) ICTM 2020. LNNS, vol. 188, pp. 556–564. Springer, Cham (2021). https://doi.org/10.1007/978-3-030-66717-7_47
10. Isakov, A., Kolesnik, V., Okhrimovskyy, A., Stepanushkin, N., Taran, A.: Numerical simulation of abnormal glow discharge processes in crossed electric and magnetic fields. Probl. Atomic Sci. Technol. **6**(94)(20), 171–174 (2014)
11. Thornton, J., Penfold, A.: Cylindrical Magnetron Sputtering. Thin Film Processes, pp. 76–110. Academic Press (1978)
12. Markosyan, A., Teunissen, J., Dujko, S., Ebert, U.: Comparing plasma fluid models of different order for 1D streamer ionization fronts. Plasma Sources Sci. Technol. **24**(6), 065002 (2015). https://doi.org/10.1088/0963-0252/24/6/065002
13. Gudmundsson, J., Lundin, D., Brenning, N., Raadu, M., Huo, C., Minea, T.: An ionization region model of the reactive Ar/O_2 high power impulse magnetron sputtering discharge. Plasma Sources Sci. Technol. **25**, 065004 (2016). https://doi.org/10.1088/0963-0252/25/6/065004

Modeling and Optimization of Photoelectric Device Positioning for Angular Displacement Measurements in Aircraft Computer Control and Operation Systems

Nikolay D. Koshevoy[1], Oleksandr V. Zabolotnyi[1], Vitalii P. Siroklyn[1](✉), Elena M. Kostenko[2], Tetiana G. Rozhnova[3], and Oleh L. Burlieiev[4]

[1] National Aerospace University "Kharkiv Aviation Institute", Kharkiv 61070, Ukraine
kafedraapi@ukr.net
[2] Poltava State Agrarian Academy, Poltava 36000, Ukraine
[3] Kharkiv National University of Radioelectronics, Kharkiv 61040, Ukraine
[4] Ukrainian Humanities Institute, Bucha 08292, Ukraine

Abstract. Optimization of the geometrical arrangement of photodetectors has been carried out using semi-natural modeling on a laboratory setup. The optimal geometric arrangement parameters of the photodetectors are found, which make it possible to expand the measurement range, increase the sensitivity and simplify the design of the measuring device of the computer system.

Keywords: Angular displacement · Computer system · Optimization · Automation · Photodetector · Modeling

1 Introduction

Qualitative information on angular displacements is required for various computer systems that are commonly used in robotics, agriculture, the aviation industry, and other sectors of the economy. Such systems make it possible to automate and speed up the process of obtaining, processing, and presenting information to decide for further actions in the area under study [1]. Optimization of methods and means for measuring angular displacements according to various criteria makes it possible to improve the design and operation of computer systems for the above sectors of the economy.

Problem Statement. In papers [2–7], photoelectric methods and devices for measuring angular displacements are presented, which can be successfully used to calculate accelerations [5]. In [6], mathematical models were found at each control point of the measurement range, allowing to optimize the geometric arrangement of the photodetectors concerning the radiation source in terms of sensitivity. However, for the obtained coordinates, there are no mathematical models of their dependence on the angular displacement, which makes it impossible to determine their value outside the values of the control points.

M. Nechyporuk et al. (Eds.): ICTM 2021, LNNS 367, pp. 321–330, 2022.
https://doi.org/10.1007/978-3-030-94259-5_28

The study aims to find mathematical models of the dependence of the optimal voltage on the photodetectors and their coordinates on the angular displacement from the point of view of sensitivity.

2 Materials and Methods

The goal of the study was achieved by modeling the process of measuring the angular displacement based on the photoelectric method. Its application makes it possible to carry out high-precision measurements without changing the design of the object of rotation.

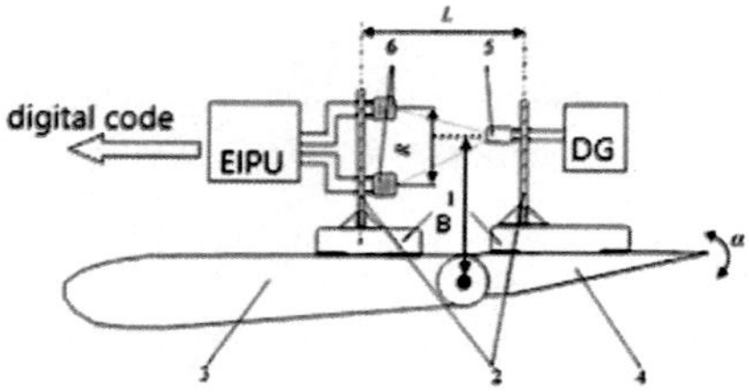

Fig. 1. Layout of the experimental setup for photoelectric measurement of angular displacement 1 - stand; 2 - holders; 3 - object of research; 4 - object of rotation; 5 - radiation source; 6 – photodetector.

In this [9], method measurements, two stands with holders are installed, one of which is attached to a stationary object, and the other to the object of rotation. In the initial position, the distance between the holders is L. The radiation source is established on the holder, the stand of which is fixed to the object of rotation at a distance B from its axis. The radiation source is also connected to the power driver. Photodetectors at a certain distance R from each other, optically connected to the radiation source, are fixed on another holder, the stand of which is installed on a stationary object. After that, the photodetectors are connected to the electronic information processing unit (EIPU). The values of the distances B and L are determined based on the geometric dimensions of the object of revolution.

Figure 1 shows the layout of the experimental setup that implements the resented method at the research object.

Figure 2 shows an experimental setup that implements the photoelectric method.

A rotary stand 2, a measuring board 3, and two stands 4 are attached to base 1, into which two guides 5 are installed and fixed in them with four nuts 6. The rotary stand 2 is equipped with a handle with a lock 7 and a graduation scale 8. G-shaped metal lever 9, the upper part of which simulates the outer side of the steering surface of the aircraft. By means of an adhesive connection, a platform 10 made of non-metallic material is mounted on lever 9, on which the holder of the transmitting part 11 with the radiation source 12 is fixed, installed at a distance B from the axis of rotation. The receiving part consists of a frame 13, the lower part of which is attached to the guides 5 using nuts 14, and the holder of the receiving part 15 with photodetectors 16 is attached to the upper part. In the initial position, the holder of the transmitting part 11 and the receiving part 15 are at a distance L from each other, and the main axis of the radiation source 12 coincides with the main axis of the central photodetector 16.

The operation principle of this experimental setup is as follows. With the help of nuts 14, frame 13 is fixed at a distance L from the holder of the transmitting part 11. The measuring board 3 is connected to a stabilized power supply +12V (not shown in the figure). Then, moving the handle with the lock 6, sequentially with a step $\Delta\alpha$ on the graduation scale 7, the angular position of the rotary stand 2 is set. Due to its rigid engagement with the L-shaped metal arm 8 and platform 9, this movement is transmitted to the holder of the transmitting part 11 with the radiation source 12. The photodetectors 16 perceive the distribution of the intensity of the light flux and convert it into a corresponding voltage, which is transmitted through the connecting wires 17 to the measuring board 3. The signal is removed from each photodetector 16 separately.

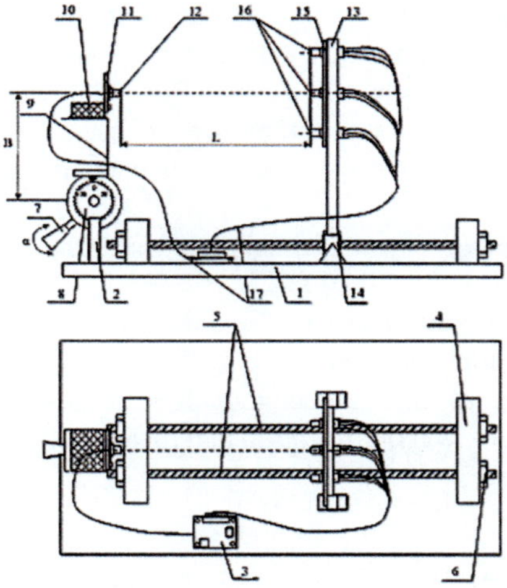

Fig. 2. Experimental setup for photoelectric measurement of angular displacement

The schematic diagram of the device is shown in Fig. 3.

The circuit includes the following elements: a light-emitting diode (VD1) in the near-infrared range with a limiting resistor R1, photodetectors of the infrared range of signals VT1…VT9 with collector resistors R2…R10, a DA1 supply voltage stabilizer with high-frequency filters in the form of capacitors C1, C2. The signal is read off with a Mastech-MS8200G digital multimeter from resistors R2…R10.

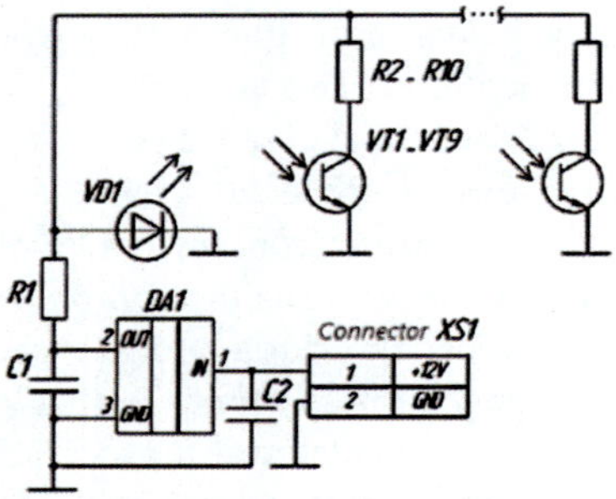

Fig. 3. Schematic diagram of the experimental setup

3 Experimental Research

As an optimization criterion, the voltage on the photodetector VT from the angular displacement α was chosen, since its value is directly proportional to the sensitivity. Range of variation of angular movement: $\alpha \in [-30°; 30°]$ with a step of $\Delta\alpha = 5°$.

The factors influencing this characteristic were chosen:

X_1 - displacement along the horizontal axis relative to the main axis of the radiation source X, mm;

X_2 - displacement along the vertical axis relative to the main axis of the radiation source Y, mm.

Table 1 shows the intervals of variation and the main levels of factors.

Since linear models allow determining only the area in the vicinity of the optimum point, it is expedient to use higher-order models to increase the accuracy of finding the optimal solution for a given optimization parameter. In this experiment, a quadratic model will be synthesized.

Table 1. Levels and intervals of factors' variation.

Factors	Factor levels			Variation intervals	Dimension
	−1	0	+1		
X_1(X)	−30	0	+30	30	mm
X_2(Y)	−70	0	70	70	mm

The placement of the photodetector elements is carried out at the points of the orthogonal compositional plan (OCCP) (Fig. 4).

The measurements were carried out at three different values of the distance between the holder of the radiation source and the holder of the photodetectors L, namely: a) $L_1 = 110$ mm; b) $L_2 = 150$ mm; c) $L_3 = 220$ mm. The distance from the axis of rotation of the measurement object to the main axis of the radiation source is B = 150 mm [9].

As a result of the experiment, the coefficients values of the mathematical models at the control points of angular displacement for L_1, L_2, L_3 were obtained [9]. This add-on uses the conjugate gradient method to find the extremum (minimum or maximum) of a nonlinear function.

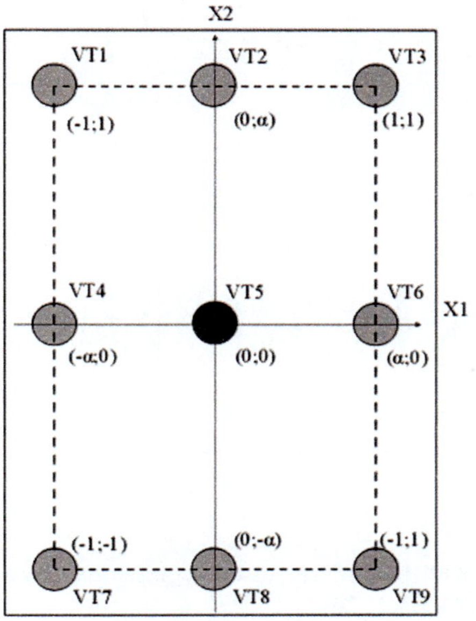

Fig. 4. Placement of the photodetector elements at the points of the OCCP

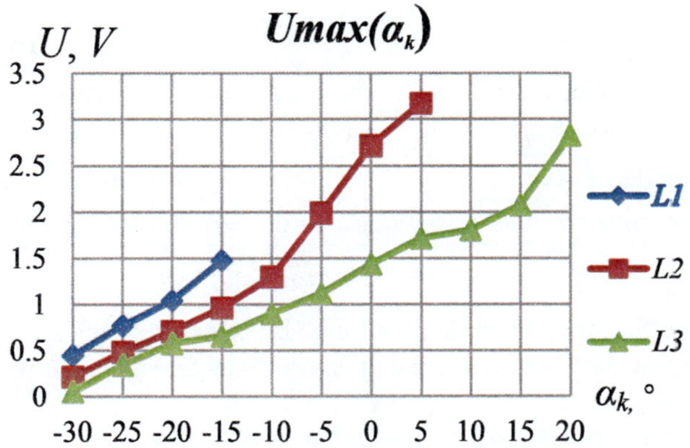

Fig. 5. Graphical representation of voltage values for L_1, L_2, L_3

Table 2 shows the maximum voltage values $U_{max}(\alpha_k)$, found by adequate mathematical models, considering the continuity of measurements for L_1, L_2, L_3.

In Fig. 5, the same data is presented graphically. Its analysis suggests that the dependence of the voltage on the photodetectors is close to linear. Therefore, it is necessary to test this hypothesis and find mathematical models of the dependence $U_{max}(\alpha_k)$ based on regression analysis.

Table 2. Voltage values Umax(α_k) for L_1, L_2, L_3

α_k, °	α_k, rad.	L_1	L_2	L_3
		$U_{max}(\alpha_k)$	$U_{max}(\alpha_k)$	$U_{max}(\alpha_k)$
−30,00	−0,52	0,45	0,21	0,05
−25,00	−0,44	0,77	0,48	0,34
−20,00	−0,35	1,04	0,70	0,58
−15,00	−0,26	1,48	0,96	0,66
−10,00	−0,17	–	1,30	0,90
−5,00	−0,09	–	1,99	1,12
0,00	0,00	–	2,71	1,44
5,00	0,09	–	3,17	1,72
10,00	0,17	–	–	1,81
15,00	0,26	–	–	2,08
20,00	0,35	–	–	2,83

Table 3 shows the statistical results of the regression analysis carried out using the Data Analysis tool in MS Excel 2016.

Table 3. Regression analysis results.

Regression analysis parameter	L_1	L_2	L_3
Multiple correlation coefficient (R)	0,995	0,976	0,984
Determination coefficient (R2)	0,991	0,952	0,967
Actual Fisher criterion (F act.)	0,567	7,746	6,727
Theoretical Fisher criterion (F theor. significance level $\alpha = 0{,}05$)	211,205	118,056	267,659

Since the coefficient of determination for all L is close to 1, and the tabular value of the Fisher criterion significantly exceeds the experimental value, the resulting models are linear and adequate.

Mathematical models of the dependence $U_{max}(\alpha_k)$ for L_1, L_2, L_3 have the following form:

$$L_1 : U(\alpha) = 2,45 + 3,8\alpha, \tag{1}$$

$$L_2 : U(\alpha) = 2,51 + 4,92\alpha, \tag{2}$$

$$L_3 : U(\alpha) = 1,48 + 2,83\alpha, \tag{3}$$

where α – angular movement, rad.

It should be noted that the models will work effectively at measurement control points. However, the voltages obtained from the models in the intervals between the control points can be obtained only when the photodetector is located relative to the radiation source at the maximum point. Therefore, it is also necessary to obtain mathematical models of the dependence of the displacements along the X and Y axes for L_1, L_2, L_3.

Table 4 shows the experimental values of the displacement along the X and Y axes for the obtained maximum voltages $U_{max}(\alpha_k)$ at each control point for L_1, L_2, L_3.

Table 4. Experimental values of the displacement along the X and Y axes for L_1, L_2, L_3

α_{k},°	α_{k}, rad	L_1		L_2		L_3	
		X, mm	Y, mm	X, mm	Y, mm	X, mm	Y, mm
−30,00	−0,52	−15	70	−5	70	−30	70
−25,00	−0,44	−15	70	−10	70	−30	70
−20,00	−0,35	−30	70	−15	70	−30	70
−15,00	−0,26	−10	70	−10	70	−30	70
−10,00	−0,17	–	–	−10	70	−30	70
0,00	0,00	–	–	−10	15	−17,5	10
−5,00	−0,09	–	–	−7,5	27,5	−30	30
0,00	0,00	–	–	−10	15	−17,5	10
5,00	0,09	–	–	−12,5	2,5	−30	−10
10,00	0,17	–	–	–	–	−30	−25
15,00	0,26	–	–	–	–	−30	−70
20,00	0,35	–	–	–	–	−30	−70

The simulation results using MS Office 2016 are shown in Table 5. In these models, the value of α is substituted in radians.

As can be seen from Table 5, the models for L1 are the most reliable, although the sector of coverage of the angular movement is only 15°.

As a result of the optimization and simulation of the geometrical arrangement of the photodetectors, mathematical models were obtained for the optimal, in terms of sensitivity, voltages on the photodetectors and the coordinates of their location from angular displacement.

Based on the results and research, the architecture of a computer system for controlling angular displacements has been developed (Fig. 5) [9]. The computer system refers to control and measuring equipment and is designed to check the technical condition of objects of rotation in various industries. Its application makes it possible to use only one photodetector for measuring angular displacements, to recreate real conditions during the control process, to change the number of control points of the specified measurement

Table 5. X and Y offset modeling.

Offset	Model, (α, rad.)	Significance coefficient, approx. (R2)
L1		
X	X = 8,33α3 – 57,5α2 + 114,17α – 80	1
Y	Y = 70	1
L2		
X	X = –0,0032α6 + 0,12α5 – 1,55α4 + 9,49α3 – 26,55α2 + 27,45α – 13,89	0,89
Y	Y = 0,5α3 – 8,82α2 + 34,55α + 38,83	0,94
L3		
X	X = 0,008α5 – 0,23α4 + 2,25α3 – 9,48α2 + 16,45α – 39,09	0,34
Y	Y = –1,66α2 + 3,88α + 72,64	0,96

range, to obtain a visual general characteristic of deviations from the norm of angular displacements at all control points.

The developed architecture of the computer system for angular displacements control is shown in Fig. 6.

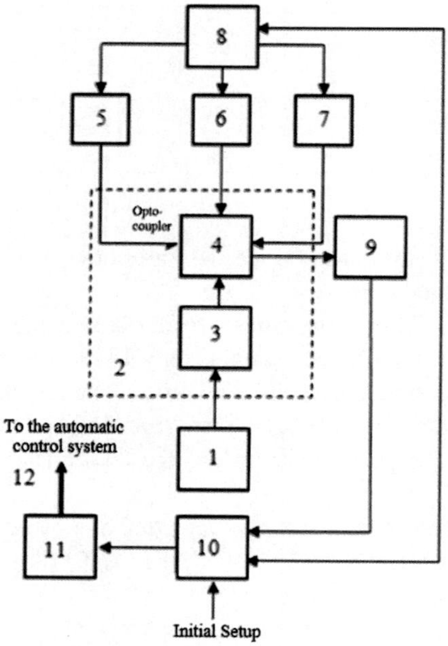

Fig. 6. Architecture of the computer system for angular displacement control

Architecture of the computer system for angular displacement control includes the following components: 1 – rotation object, 2 - optocoupler, 3 - source of infrared radiation, 4 - photodetector, 5 - displacer along the X axis, 6 - displacer along the Y axis, 7 - displacer along the Z axis, 8 - control device, 9 - converter, 10 - personal computer, 11 - connecting interface, 12 - output.

The use of a personal computer connected through the connecting interface to the output bus, as well as having a direct connection with the control device, allows to connect to the automatic control system of the rotation object.

The computer control system can operate in two modes: manual and automatic.

If the automatic control system of the rotation object of allows the connection of external devices, then the computer control system will operate in automatic mode as follows. The personal computer 10 transmits the code of the number of the control point of the angular displacement to the control device 8, which from it generates a signal for the displacers along the X axis 5, the Y axis 6, and the Z axis 7 to install the photo receiver 4 of the optocoupler 2 in corresponding provision. The control device 8 signals the completion of the installation to the personal computer 10, which then transmits the control value of the angular displacement through the connecting interface 11 to the output bus 12, from which this information is read by the automatic control system of the rotation object. When moving, the rotation object 1 changes the position of the source of infrared radiation 3 relative to the photodetector 4 of the optical unit 2, as a result of which a signal proportional to its angular displacement appears at the output of the photodetector 4. This signal enters the input of the converter 9, where a digital code is formed and transmitted for further processing in the personal computer 10. In the manual mode of operation, the signal from the computer 10 to the output bus 12 is not received.

The following data are stored in the memory of a personal computer: the number of control points, their numerical value, the corresponding code values of the coordinates of the photodetector for each control point of the corresponding type of object of rotation (determined experimentally), the function of converting the received code into the amount of angular displacement.

Thus, when using special software, the following data can be obtained: the value of the real angular displacement of the object of rotation and the error of deviation from the nominal value at each control point, a graphic image of the obtained data on the screen of a personal computer.

The practical significance of the computer control system is confirmed by the patent of Ukraine No. 96478.

4 Conclusions

Thus, the found mathematical models for a device operating on the photoelectric method of measurement make it possible to increase its average sensitivity to 0.056–0.124 V/°and to simplify the design by reducing the number of photodetectors used to 1.

References

1. Yolochkin, M.: Information technology. Oniks,Moskow (2009)

2. Oganesyan, A., Thehovsky, M., Koshevoy, N., Gordienko, V.: Investigation into optoelectronic aviation angle meter by the design – of – experiments method. Telecommun. Radio Eng. **69**(9), 841–847 (2010)
3. Koshevoy, N., Burleiev, O., Gordienko, V.: Photoelectric methods of measuring the steering angle in aviation. Telecommun. Radio Eng. **71**(8), 759–762 (2012)
4. Koshevoi, N.D., Kostenko, E.M., Oganesyan, A.S., Tsekhovskoi, M.V.: Aircraft system for measuring the angular deflections of control surfaces. Russ. Aeronaut. (Iz VUZ) **56**(4), 418–422 (2013). https://doi.org/10.3103/S1068799813040168
5. Steck, D.: Classical and Modern Optics. Oregon, University of Oregon, p. 352 (2013)
6. Korobiichuk, I., Bezvesilna, O., Kachniarz, M., Koshovyj, M., Kvasnikov, V.: Methods and ways of piezoelectric accelerometers fastening on the object of research. Acta Phys. Pol. A **133**(4), 1112–1115 (2018)
7. Koshevoy, N., Burlieiev, O., Kostenko, O.: Technologies for measuring angular displacements for computer control and management systems. In: Monograph. National Aerospace University «Kharkiv Aviation Institute», Kharkiv (2019)
8. Walkenbach, J.: Excel Formulas 2010. LLC "ID Williams", Moskow (2011)
9. Patent of Ukraine No. 96478, MPK (2015.01) G 06 N 99/00. Computer control system for angular displacement of the aircraft steering surface. National Aerospace University by name of Zhukovski "KhAI". No. u201408705. Decelerated 01.08.14. Published 10.02.15, Bulletin No. 3, 3p.

Modified Gray Codes for the Value (Time) Optimization of a Multifactor Experiment Plans

Nikolay D. Koshevoy, Volodymyr A. Dergachov, A. V. Pavlik, Vitalii P. Siroklyn(✉), Iryna I. Koshevaya, and Olena A. Hrytsai

National Aerospace University Kharkov Aviation Institute, Chkalov Street 17, Kharkov 61070, Ukraine

kafedraapi@ukr.net

Abstract. To increase the effectiveness of experimental studies, it is advisable to use the methods of experiment design. Herewith, it is necessary to optimize multifactor experiments plans in terms of value (time) costs. To solve this problem has been developed the method, based on a generation of modified Gray codes. The software has been developed to implement the proposed method. The working capacity and effectiveness of the proposed method and computer programs are proved by researching a number of technological processes and devices.

Keywords: The optimal experiment plan · Cost · Time · Modified Gray Codes · Computer programs

1 Introduction

The design of experiment methods make it possible to reduce the value (time) costs with research of a variety of objects. One way to solve this problem is to minimize a number of changes in factor levels while constructing plans of multifactor experiments. The previously proposed method [1], based on the Gray code using, solves the problem of minimizing the number of changes in factor levels, but the obtained plans are not always optimal in terms of value (time) costs for their implementation.

Therefore, there is a problem how to develop the method and tools for constructing a multifactor experiment plans with a minimum number of changes in factor levels and optimal ones in terms of value (time) costs.

2 Analysis of Literary Sources

The construction of the optimal plans of multifactor experiments can be implemented by the combinatorial optimization method [2–12]. There are also well-known methods of optimization multifactor experiment plans in terms of value (time) costs [13–15], which have their own advantages and disadvantages. So, for example, the plans, which are constructed with the method [1], based on the Gray code using, have the minimal

M. Nechyporuk et al. (Eds.): ICTM 2021, LNNS 367, pp. 331–343, 2022.
https://doi.org/10.1007/978-3-030-94259-5_29

number of changes in factor levels, but they are not always optimal in terms of value (time) costs. The method of exhaustive search with the modern development of computer technology allows to obtain optimal plans only for the number of factors $k \leq 3$. Many methods have low performance and do not give the opportunity to get the exact solution during optimization.

Therefore, the problem is to develop the method for constructing a multifactor experiment plans with a minimum number of changes in factor levels and optimal ones in terms of value (time) costs for their implementation.

3 The Main Research Materials

3.1 The Modified Gray Codes

In [15] is given a fragment of the array of multifactor experiment plans with a minimum number of changes in factor levels for the number of factors $k = 3$, which includes 144 options.

An analysis of the binary codes in this set with a minimal number of changes showed that there are codes obtained by using the Gray code with certain transformations. In this case, the value of « + 1» in the experimental plan was changed to «1», and the value of «–1» to «0». Such codes we will call the modified Gray codes.

Among the transformations, it have been distinguished E, H and G (E,H) transformations, which save the main feature of the Gray code - a minimum number of changes.

The E array of transformations E = {E1,...,Ed} is a column permutation of a k-bit Gray code. E1 corresponds to the Gray code, and the column value is denoted byB1... Bk. Table 1 shows the E transformations for the case $k = 3$.

Table 1. E transformations of the Gray code for $k = 3$.

E1			E2			E3			E4			E5			E6		
B1	B2	B3	B1	B3	B2	B2	B1	B3	B2	B3	B1	B3	B1	B2	B3	B2	B1
0	0	0	0	0	0	0	0	0	0	0	0	0	0	0	0	0	0
0	0	1	0	1	0	0	0	1	0	1	0	1	0	0	1	0	0
0	1	1	0	1	1	1	0	1	1	1	0	1	0	1	1	1	0
0	1	0	0	0	1	1	0	0	1	0	0	0	0	1	0	1	0
1	1	0	1	0	1	1	1	0	1	0	1	0	1	1	0	1	1
1	1	1	1	1	1	1	1	1	1	1	1	1	1	1	1	1	1
1	0	1	1	1	0	0	1	1	0	1	1	1	1	0	1	0	1
1	0	0	1	0	0	0	1	0	0	0	1	0	1	0	0	0	1

The H array of transformations H = {H1,..., HT} is an inversion of a k-bit Gray code. The Hi transformation is represented as a binary set, in which the value «0» means that there is no inversion of the corresponding column and it is indicated by the letter «B».

In the case of the value «1», the column is inverted and it is indicated by the letter «C». Table 2 shows the H transformations for the case k = 3.

Table 2. H transformations of the Gray code for k = 3.

H1			H2			H3			H4		
0	0	0	0	0	1	0	1	0	0	1	1
B1	B2	B3	B1	B2	C3	B1	C2	B3	B1	C2	C3
0	0	0	0	0	1	0	1	0	0	1	1
0	0	1	0	0	0	0	1	1	0	1	0
0	1	1	0	1	0	0	0	1	0	0	0
0	1	0	0	1	1	0	0	0	0	0	1
1	1	0	1	1	1	1	0	0	1	0	1
1	1	1	1	1	0	1	0	1	1	0	0
1	0	1	1	0	0	1	1	1	1	1	0
1	0	0	1	0	1	1	1	0	1	1	1
H5			H6			H7			H8		
1	0	0	1	0	1	1	1	0	1	1	1
C1	B2	B3	C1	B2	C3	C1	C2	B3	C1	C2	C3
1	0	0	1	0	1	1	1	0	1	1	1
1	0	1	1	0	0	1	1	1	1	1	0
1	1	1	1	1	0	1	0	1	1	0	0
1	1	0	1	1	1	1	0	0	1	0	1
0	1	0	0	1	1	0	0	0	0	0	1
0	1	1	0	1	0	0	0	1	0	0	0
0	0	1	0	0	0	0	1	1	0	1	0
0	0	0	0	0	1	0	1	0	0	1	1

In the case when the modified Gray code is obtained by using E and H transformations, it is denoted as G (E, H). If only one transformation group is used, then it is denoted G (E) or G (H) respectively. The table shows the modified G (E, H) codes for k = 3.

Table 3. The modified code G(E, H) for k = 3.

№	E	H	G(E,H)
1	1	1	000, 001, 011, 010, 110, 111, 101, 100
2	1	2	001, 000, 010, 011, 111, 110, 100, 101
3	1	3	011, 010, 000, 001, 101, 100, 110, 111
4	1	4	010, 011, 001, 000, 100, 101, 111, 110
5	1	5	110, 111, 101, 100, 000, 001, 011, 010
6	1	6	111, 110, 100, 101, 001, 000, 010, 011
7	1	7	101, 100, 110, 111, 011, 010, 000, 001
8	1	8	100, 101, 111, 110, 010, 011, 001, 000
9	2	1	000, 010, 011, 001, 101, 111, 110, 100
10	2	2	001, 011, 010, 000, 100, 110, 111, 101
11	2	3	011, 001, 000, 010, 110, 100, 101, 111
12	2	4	010, 000, 001, 011, 111, 101, 100, 110
13	2	5	110, 100, 101, 111, 011, 001, 000, 010
14	2	6	111, 101, 100, 110, 010, 000, 001, 011
15	2	7	101, 111, 110, 100, 000, 010, 011, 001
16	2	8	100, 110, 111, 101, 001, 011, 010, 000
17	3	1	000, 001, 101, 100, 110, 111, 011, 010
18	3	2	001, 000, 100, 101, 111, 110, 010, 011
19	3	3	011, 010, 110, 111, 101, 100, 000, 001
20	3	4	010, 011, 111, 110, 100, 101, 001, 000
21	3	5	110, 111, 011, 010, 000, 001, 101, 100
22	3	6	111, 110, 010, 011, 001, 000, 100, 101
23	3	7	101, 100, 000, 001, 011, 010, 110, 111
24	3	8	100, 101, 001, 000, 010, 011, 111, 110
25	4	1	000, 010, 110, 100, 101, 111, 011, 001
26	4	2	001, 011, 111, 101, 100, 110, 010, 000
27	4	3	011, 001, 101, 111, 110, 100, 000, 010
28	4	4	010, 000, 100, 110, 111, 101, 001, 011
29	4	5	110, 100, 000, 010, 011, 001, 101, 111
30	4	6	111, 101, 001, 011, 010, 000, 100, 110
31	4	7	101, 111, 011, 001, 000, 010, 110, 100
32	4	8	100, 110, 010, 000, 001, 011, 111, 101

(continued)

Table 3. (*continued*)

№	E	H	G(E,H)
33	5	1	0 0 0, 1 0 0, 1 0 1, 0 0 1, 0 1 1, 1 1 1, 1 1 0, 0 1 0
34	5	2	0 0 1, 1 0 1, 1 0 0, 0 0 0, 0 1 0, 1 1 0, 1 1 1, 0 1 1
35	5	3	0 1 1, 1 1 1, 1 1 0, 0 1 0, 0 0 0, 1 0 0, 1 0 1, 0 0 1
36	5	4	0 1 0, 1 1 0, 1 1 1, 0 1 1, 0 0 1, 1 0 1, 1 0 0, 0 0 0
37	5	5	1 1 0, 0 1 0, 0 1 1, 1 1 1, 1 0 1, 0 0 1, 0 0 0, 1 0 0
38	5	6	1 1 1, 0 1 1, 0 1 0, 1 1 0, 1 0 0, 0 0 0, 0 0 1, 1 0 1
39	5	7	1 0 1, 0 0 1, 0 0 0, 1 0 0, 1 1 0, 0 1 0, 0 1 1, 1 1 1
40	5	8	1 0 0, 0 0 0, 0 0 1, 1 0 1, 1 1 1, 0 1 1, 0 1 0, 1 1 0
41	6	1	0 0 0, 1 0 0, 1 1 0, 0 1 0, 0 1 1, 1 1 1, 1 0 1, 0 0 1
42	6	2	0 0 1, 1 0 1, 1 1 1, 0 1 1, 0 1 0, 1 1 0, 1 0 0, 0 0 0
43	6	3	0 1 1, 1 1 1, 1 0 1, 0 0 1, 0 0 0, 1 0 0, 1 1 0, 0 1 0
44	6	4	0 1 0, 1 1 0, 1 0 0, 0 0 0, 0 0 1, 1 0 1, 1 1 1, 0 1 1
45	6	5	1 1 0, 0 1 0, 0 0 0, 1 0 0, 1 0 1, 0 0 1, 0 1 1, 1 1 1
46	6	6	1 1 1, 0 1 1, 0 0 1, 1 0 1, 1 0 0, 0 0 0, 0 1 0, 1 1 0
47	6	7	1 0 1, 0 0 1, 0 1 1, 1 1 1, 1 1 0, 0 1 0, 0 0 0, 1 0 0
48	6	8	1 0 0, 0 0 0, 0 1 0, 1 1 0, 1 1 1, 0 1 1, 0 0 1, 1 0 1

The software (certificate of author rights registration to the work №74877, №74881, №89032; registered in the Ministry of the Economic Development and Trade of Ukraine 21.11.2017, 29.05.2019) is developed to automate the constructing process of catalogues for multifactor experiment plans with a minimum number of changes in factor levels.

The search of the optimal experiment plan in terms of value (time) costs is carried out in the set G (E, H) (see Table 3).

3.2 The Value (Time) Optimization of Experiment Plans for Researching a Technological Processes and Devices

By researching the technological process of galvanized steel drilling, the following factors were selected as affecting factors for the roughness of the treated surface: X1 – a number of rotation, X2 – an effort, X3 – a hole diameter. The initial experiment plan, optimized by the method [13], based on the genetic algorithms using, has an implementation cost of 16 conv. units.

Four plans of the experiment were obtained by using the modified Gray codes with an optimal cost of 15,5 conv. units each (Table 4).

Table 4. An optimal experiment plans for researching the technological process of drilling.

G(2,5)			G(2,6)			G(2,7)			G(2,8)		
X1	X2	X3	X1	X2	X3	X1	X2	X3	X1	X2	X3
+1	−1	−1	+1	−1	+1	+1	+1	−1	+1	+1	+1
+1	+1	−1	+1	+1	+1	+1	−1	−1	+1	−1	+1
+1	+1	+1	+1	+1	−1	+1	−1	+1	+1	−1	−1
+1	−1	+1	+1	−1	−1	+1	+1	+1	+1	+1	−1
−1	−1	+1	−1	−1	−1	−1	+1	+1	−1	+1	−1
−1	+1	+1	−1	+1	−1	−1	−1	+1	−1	−1	−1
−1	+1	−1	−1	+1	+1	−1	−1	−1	−1	−1	+1
−1	−1	−1	−1	−1	+1	−1	+1	−1	−1	+1	+1

By researching the technological process of increasing the productivity of CNC machines, the following factors were selected as a dominant: X1 – prevention time; X2 – a number of CNC machines; X3 – machine working hours throughout the day; X4 – a frequency of prevention.

As a result of the experimental design optimization by the method [13], based on the genetic algorithms using, the experiment plan is obtained with an implementation cost of 177 conv. units.

The modified Gray codes allowed us to obtain 8 optimal experiment plans with an implementation cost of 130 conv. units each (Table 5).

Table 5. An optimal experiment plans for researching the technological process of increasing the productivity of CNC machines.

G(12,2)				G(12,4)				G(12,6)				G(12,8)			
X1	X2	X3	X4	X1	X2	X3	X4	X1	X2	X3	X4	X1	X2	X3	X4
−1	−1	−1	+1	−1	−1	+1	+1	−1	+1	−1	+1	−1	+1	+1	+1
−1	+1	−1	+1	−1	+1	+1	+1	−1	−1	−1	+1	−1	−1	+1	+1
+1	+1	−1	+1	+1	+1	+1	+1	+1	−1	−1	+1	+1	−1	+1	+1
+1	−1	−1	+1	+1	−1	+1	+1	+1	+1	−1	+1	+1	+1	+1	+1
+1	−1	+1	+1	+1	−1	−1	+1	+1	+1	+1	+1	+1	+1	−1	+1
+1	+1	+1	+1	+1	+1	−1	+1	+1	−1	+1	+1	+1	−1	−1	+1
−1	+1	+1	+1	−1	+1	−1	+1	−1	−1	+1	+1	−1	−1	−1	+1
−1	−1	+1	+1	−1	−1	−1	+1	−1	+1	+1	+1	−1	+1	−1	+1
−1	−1	+1	−1	−1	−1	−1	−1	−1	+1	+1	−1	−1	+1	−1	−1

(continued)

Table 5. (*continued*)

G(12,2)				G(12,4)				G(12,6)				G(12,8)			
X1	X2	X3	X4	X1	X2	X3	X4	X1	X2	X3	X4	X1	X2	X3	X4
−1	+1	+1	−1	−1	+1	−1	−1	−1	−1	+1	−1	−1	−1	−1	−1
+1	+1	+1	−1	+1	+1	−1	−1	+1	−1	+1	−1	+1	−1	−1	−1
+1	−1	+1	−1	+1	−1	−1	−1	+1	+1	+1	−1	+1	+1	−1	−1
+1	−1	−1	−1	+1	−1	+1	−1	+1	+1	−1	−1	+1	+1	+1	−1
+1	+1	−1	−1	+1	+1	+1	−1	+1	−1	−1	−1	+1	−1	+1	−1
−1	+1	−1	−1	−1	+1	+1	−1	−1	−1	−1	−1	−1	−1	+1	−1
−1	−1	−1	−1	−1	−1	+1	−1	−1	+1	−1	−1	−1	+1	+1	−1
G(12,10)				G(12,12)				G(12,14)				G(12,16)			
X1	X2	X3	X4	X1	X2	X3	X4	X1	X2	X3	X4	X1	X2	X3	X4
+1	−1	−1	+1	+1	−1	+1	+1	+1	+1	−1	+1	+1	+1	+1	+1
+1	+1	−1	+1	+1	+1	+1	+1	+1	−1	−1	+1	+1	−1	+1	+1
−1	+1	−1	+1	−1	+1	+1	+1	−1	−1	−1	+1	−1	−1	+1	+1
−1	−1	−1	+1	−1	−1	+1	+1	−1	+1	−1	+1	−1	+1	+1	+1
−1	−1	+1	+1	−1	−1	−1	+1	−1	+1	+1	+1	−1	+1	−1	+1
−1	+1	+1	+1	−1	+1	−1	+1	−1	−1	+1	+1	−1	−1	−1	+1
+1	+1	+1	+1	+1	+1	−1	+1	+1	−1	+1	+1	+1	−1	−1	+1
+1	−1	+1	+1	+1	−1	−1	+1	+1	+1	+1	+1	+1	+1	−1	+1
+1	−1	+1	−1	+1	−1	−1	−1	+1	+1	+1	−1	+1	+1	−1	−1
+1	+1	+1	−1	+1	+1	−1	−1	+1	−1	+1	−1	+1	−1	−1	−1
−1	+1	+1	−1	−1	+1	−1	−1	−1	−1	+1	−1	−1	−1	−1	−1
−1	−1	+1	−1	−1	−1	−1	−1	−1	+1	+1	−1	−1	+1	−1	−1
−1	−1	−1	−1	−1	−1	+1	−1	−1	+1	−1	−1	−1	+1	+1	−1
−1	+1	−1	−1	−1	+1	+1	−1	−1	−1	−1	−1	−1	−1	+1	−1
+1	+1	−1	−1	+1	+1	+1	−1	+1	−1	−1	−1	+1	−1	+1	−1
+1	−1	−1	−1	+1	−1	+1	−1	+1	+1	−1	−1	+1	+1	+1	−1

In many cases, the optimal experiment plans, obtained by using the modified Gray codes and the genetic algorithms coincide in their implementation cost. However, the method, based on the genetic algorithms using, gives only one such plan, and the modified Gray codes provide several plans with a minimum implementation cost. For example, for researching the technological PCB plating process [16], 6 experiment plans were obtained by using the modified Gray codes with an implementation cost of 25,83 conv. units each: G(24,9), G(24,10), G(24,11), G(24,12), G(24,13), G(24,15) (Table 6).

Table 6. An optimal experiment plans for researching the technological PCB plating process.

G(24,9)				G(24,10)				G(24,11)			
X1	X2	X3	X4	X1	X2	X3	X4	X1	X2	X3	X4
+1	−1	−1	−1	+1	−1	−1	+1	+1	−1	+1	−1
+1	−1	+1	−1	+1	−1	+1	+1	+1	−1	−1	−1
+1	−1	+1	+1	+1	−1	+1	−1	+1	−1	−1	+1
+1	−1	−1	+1	+1	−1	−1	−1	+1	−1	+1	+1
+1	+1	−1	+1	+1	+1	−1	−1	+1	+1	+1	+1
+1	+1	+1	+1	+1	+1	+1	−1	+1	+1	−1	+1
+1	+1	+1	−1	+1	+1	+1	+1	+1	+1	−1	−1
+1	+1	−1	−1	+1	+1	−1	+1	+1	+1	+1	−1
−1	+1	−1	−1	−1	+1	−1	+1	−1	+1	+1	−1
−1	+1	+1	−1	−1	+1	+1	+1	−1	+1	−1	−1
−1	+1	+1	+1	−1	+1	+1	−1	−1	+1	−1	+1
−1	+1	−1	+1	−1	+1	−1	−1	−1	+1	+1	+1
−1	−1	−1	+1	−1	−1	−1	−1	−1	−1	+1	+1
−1	−1	+1	+1	−1	−1	+1	−1	−1	−1	−1	+1
−1	−1	+1	−1	−1	−1	+1	+1	−1	−1	−1	−1
−1	−1	−1	−1	−1	−1	−1	+1	−1	−1	+1	−1
G(24,12)				G(24,13)				G(24,15)			
X1	X2	X3	X4	X1	X2	X3	X4	X1	X2	X3	X4
+1	−1	+1	+1	+1	+1	−1	−1	+1	+1	+1	−1
+1	−1	−1	+1	+1	+1	+1	−1	+1	+1	−1	−1
+1	−1	−1	−1	+1	+1	+1	+1	+1	+1	−1	+1
+1	−1	+1	−1	+1	+1	−1	+1	+1	+1	+1	+1
+1	+1	+1	−1	+1	−1	−1	+1	+1	−1	+1	+1
+1	+1	−1	−1	+1	−1	+1	+1	+1	−1	−1	+1
+1	+1	−1	+1	+1	−1	+1	−1	+1	−1	−1	−1
+1	+1	+1	+1	+1	−1	−1	−1	+1	−1	+1	−1
−1	+1	+1	+1	−1	−1	−1	−1	−1	−1	+1	−1
−1	+1	−1	+1	−1	−1	+1	−1	−1	−1	−1	−1
−1	+1	−1	−1	−1	−1	+1	+1	−1	−1	−1	+1
−1	+1	+1	−1	−1	−1	−1	+1	−1	−1	+1	+1
−1	−1	+1	−1	−1	+1	−1	+1	−1	+1	+1	+1

(*continued*)

Table 6. (*continued*)

G(24,12)				G(24,13)				G(24,15)			
X1	X2	X3	X4	X1	X2	X3	X4	X1	X2	X3	X4
−1	−1	−1	−1	−1	+1	+1	+1	−1	+1	−1	+1
−1	−1	−1	+1	−1	+1	+1	−1	−1	+1	−1	−1
−1	−1	+1	+1	−1	+1	−1	−1	−1	+1	+1	−1

For researching an eddy current meter of dielectric coatings thickness on a metal surfaces of products [17–19], 4 experiment plans were obtained with an implementation cost of 36,7 conv. units each: G(2,5), G(2,6), G(2,13), G(2,14) (Table 7).

Table 7. An optimal experiment plans for researching an eddy current meter of dielectric coatings thickness on a metal surfaces of products.

G(2,5)				G(2,6)				G(2,13)				G(2,14)			
X1	X2	X3	X4	X1	X2	X3	X4	X1	X2	X3	X4	X1	X2	X3	X4
−1	+1	−1	−1	−1	+1	−1	+1	+1	+1	−1	−1	+1	+1	−1	+1
−1	+1	−1	+1	−1	+1	−1	−1	+1	+1	−1	+1	+1	+1	−1	−1
−1	+1	+1	+1	−1	+1	+1	−1	+1	+1	+1	+1	+1	+1	+1	−1
−1	+1	+1	−1	−1	+1	+1	+1	+1	+1	+1	−1	+1	+1	+1	+1
+1	+1	+1	−1	+1	+1	+1	+1	−1	+1	+1	−1	−1	+1	+1	+1
+1	+1	+1	+1	+1	+1	+1	−1	−1	+1	+1	+1	−1	+1	+1	−1
+1	+1	−1	+1	+1	+1	−1	−1	−1	+1	−1	+1	−1	+1	−1	−1
+1	+1	−1	−1	+1	+1	−1	+1	−1	+1	−1	−1	−1	+1	−1	+1
+1	−1	−1	−1	+1	−1	−1	+1	−1	−1	−1	−1	−1	−1	−1	+1
+1	−1	−1	+1	+1	−1	−1	−1	−1	−1	−1	+1	−1	−1	−1	−1
+1	−1	+1	+1	+1	−1	+1	−1	−1	−1	+1	+1	−1	−1	+1	−1
+1	−1	+1	−1	+1	−1	+1	+1	−1	−1	+1	−1	−1	−1	+1	+1
−1	−1	+1	−1	−1	−1	+1	+1	+1	−1	+1	−1	+1	−1	+1	+1
−1	−1	+1	+1	−1	−1	+1	−1	+1	−1	+1	+1	+1	−1	+1	−1
−1	−1	−1	+1	−1	−1	−1	−1	+1	−1	−1	+1	+1	−1	−1	−1
−1	−1	−1	−1	−1	−1	−1	+1	+1	−1	−1	−1	+1	−1	−1	+1

Eight experiment plans were obtained for researching the device for controlling the quality of dielectric materials [20, 21], with an implementation cost of 16,1 conv. units each: G(13,2), G(13,4), G(13,6), G(13,8), G(13,10), G(13,12), G(13,14), G(13,16) (Table 8).

Table 8. An optimal experiment plans for researching the device for controlling the quality of dielectric materials.

G(13,2)				G(13,4)				G(13,6)				G(13,8)			
X1	X2	X3	X4	X1	X2	X3	X4	X1	X2	X3	X4	X1	X2	X3	X4
−1	−1	−1	+1	−1	−1	+1	+1	−1	+1	−1	+1	−1	+1	+1	+1
+1	−1	−1	+1	+1	−1	+1	+1	+1	+1	−1	+1	+1	+1	+1	+1
+1	+1	−1	+1	+1	+1	+1	+1	+1	−1	−1	+1	+1	−1	+1	+1
−1	+1	−1	+1	−1	+1	+1	+1	−1	−1	−1	+1	−1	−1	+1	+1
−1	+1	+1	+1	−1	+1	−1	+1	−1	−1	+1	+1	−1	−1	−1	+1
+1	+1	+1	+1	+1	+1	−1	+1	+1	−1	+1	+1	+1	−1	−1	+1
+1	−1	+1	+1	+1	−1	−1	+1	+1	+1	+1	+1	+1	+1	−1	+1
−1	−1	+1	+1	−1	−1	−1	+1	−1	+1	+1	+1	−1	+1	−1	+1
−1	−1	+1	−1	−1	−1	−1	−1	−1	+1	+1	−1	−1	+1	−1	−1
+1	−1	+1	−1	+1	−1	−1	−1	+1	+1	+1	−1	+1	+1	−1	−1
+1	+1	+1	−1	+1	+1	−1	−1	+1	−1	+1	−1	+1	−1	−1	−1
−1	+1	+1	−1	−1	+1	−1	−1	−1	−1	+1	−1	−1	−1	−1	−1
−1	+1	−1	−1	−1	+1	+1	−1	−1	−1	−1	−1	−1	−1	+1	−1
+1	+1	−1	−1	+1	+1	+1	−1	+1	−1	−1	−1	+1	−1	+1	−1
+1	−1	−1	−1	+1	−1	+1	−1	+1	+1	−1	−1	+1	+1	+1	−1
−1	−1	−1	−1	−1	−1	+1	−1	−1	+1	−1	−1	−1	+1	+1	−1
G(13,10)				G(13,12)				G(13,14)				G(13,16)			
X1	X2	X3	X4	X1	X2	X3	X4	X1	X2	X3	X4	X1	X2	X3	X4
+1	−1	−1	+1	+1	−1	+1	+1	+1	+1	−1	+1	+1	+1	+1	+1
−1	−1	−1	+1	−1	−1	+1	+1	−1	+1	−1	+1	−1	+1	+1	+1
−1	+1	−1	+1	−1	+1	+1	+1	−1	−1	−1	+1	−1	−1	+1	+1
+1	+1	−1	+1	+1	+1	+1	+1	+1	−1	−1	+1	+1	−1	+1	+1
+1	+1	+1	+1	+1	+1	−1	+1	+1	−1	+1	+1	+1	−1	−1	+1
−1	+1	+1	+1	−1	+1	−1	+1	−1	−1	+1	+1	−1	−1	−1	+1
−1	−1	+1	+1	−1	−1	−1	+1	−1	+1	+1	+1	−1	+1	−1	+1
+1	−1	+1	+1	+1	−1	−1	+1	+1	+1	+1	+1	+1	+1	−1	+1
+1	−1	+1	−1	+1	−1	−1	−1	+1	+1	+1	−1	+1	+1	−1	−1
−1	−1	+1	−1	−1	−1	−1	−1	−1	+1	+1	−1	−1	+1	−1	−1
−1	+1	+1	−1	−1	+1	−1	−1	−1	−1	+1	−1	−1	−1	−1	−1
+1	+1	+1	−1	+1	+1	−1	−1	+1	−1	+1	−1	+1	−1	−1	−1

(*continued*)

Table 8. (*continued*)

G(13,10)				G(13,12)				G(13,14)				G(13,16)			
X1	X2	X3	X4	X1	X2	X3	X4	X1	X2	X3	X4	X1	X2	X3	X4
+1	+1	−1	−1	+1	+1	+1	−1	+1	−1	−1	−1	+1	−1	+1	−1
−1	+1	−1	−1	−1	+1	+1	−1	−1	−1	−1	−1	−1	−1	+1	−1
−1	−1	−1	−1	−1	−1	+1	−1	−1	+1	−1	−1	−1	+1	+1	−1
+1	−1	−1	−1	+1	−1	+1	−1	+1	+1	−1	−1	+1	+1	+1	−1

4 Results and Discussion

The following results were obtained by value optimization of experiment plans with using the modified Gray codes for researching the technological process of galvanized steel drilling, PCB plating process, process of increasing the productivity of CNC machines, an eddy current meter of dielectric coatings thickness on metal surfaces of products, the device for controlling the quality of dielectric materials.

The optimal experiment plans for researching the technological process of galvanized steel drilling give a realization value gain 1,23 times more, than the plan obtained by using the genetic algorithm based method. The plans for researching the process of increasing the productivity of CNC machines give a realization value gain of 1,39 times.

The optimal experiment plans for researching the technological PCB plating process, an eddy current meter of dielectric coatings thickness, the device for controlling the quality of dielectric materials, obtained by using the modified Gray codes and the genetic algorithms, coincide by their value realization.

The benefits of the modified Gray codes are as follows:

- we get an optimal plan by value realization experiment;
- the modified Gray codes make it possible to obtain several plans with a minimum value realization;
- they allow to choose the experiment plan, which is optimal not only by value realization, but also by other criteria.

5 Conclusion

Here is proposed the method for value (time) optimization of experiment plans, which are based on using the modified Gray codes. The software has been developed to implement the proposed method. The working capacity and effectiveness of the proposed method and computer programs are proved by researching a number of technological processes and devices.

It was shown, that the modified Gray codes make it possible to obtain several plans with a minimum value realization. It allows choosing an experiment plan from the set, which is optimal not only by value realization, but also by other criteria.

References

1. Koshevoy, N., Zadolitny, A., Koshevaya, I., et al.: Research of moisture-meter device for bulk and liquid materials. In: 29th International Scientific Symposium "Metrology and Metrology Assurance" MMA 2019-Proceeding, Sozopol, Bulgaria, 6–9 September 2019, pp. 1–4. Institute of Electrical and Electronics Engineers Inc. (2019). https://doi.org/10.1109/MMA.2019.8935983
2. Hoskins, D.S.: Combinatorics and statistical inferecing. In: Applied Optimal Designs, no. 4, pp. 147–179 (2007)
3. Morgan, J.P.: Association schemes: designed experiments, algebra and combinatorics. J. Am. Stat. Assoc. **100**(471), 1092–1093 (2005)
4. Bailey, R.A., Cameron, P.G.: Combinatorics of optimal designs. Surv. Combin. **365**, 19–73 (2009)
5. Montgomery, D.C.: Design and Analysis of Experiments, 9th edn., 629 p. Wiley (2017)
6. Bartos, B.J., Mc Cleary, R., McDowall, D.: Design and Analysis of Time Series Experiments, 393 p. Oxford University Press, Oxford (2017)
7. Berger, P.D., Maurer, R.E., Celli, G.B.: Experimental Design with Applications in Management, Engineering and the Sciences, 640 p. Springer, New York (2018).https://doi.org/10.1007/978-3-319-64583-4
8. Rodrigues, M.I., Iemma, A. F.: Experimental Design and Process Optimization, 336 p. CRC Press, New York (2016)
9. Wu, C.J., Hamada, M.S.: Experiments: Planning, Analysis, and Optimization, 743 p. Wiley (2015)
10. Karpenko, A.P.: Modern algorithms of search optimization. Nature-inspired optimization algorithms: the training manual, p. 446. Bauman MSTU Publ (2014)
11. Yakovlev, S.V., Pichugina, O.S.: Properties of combinatorial optimization problems over polyhedral-spherical sets. Cybern. Syst. Anal. **54**(1), 99–109 (2018). https://doi.org/10.1007/s10559-018-0011-6
12. Yakovlev, S.V.: Convex extensions in combinatorial optimization and their applications. In: Optimization and Its Applications, vol. 130, pp. 567–584. Springer, Cham (2017). https://doi.org/10.1007/978-3-319-68640-0-27
13. Koshevoy, N.D., Gordienko, V.A., Sukhobrus, Y.: Optimization for the design matrix realization value with the aim to investigate technological processes. Telecommun. Radio Eng. **73**(15), 1383–1386 (2014). https://doi.org/10.1615/TelecomRadEng.v73.i15.60
14. Koshevoy, N.D., Kostenko, E.M., Gordienko, V.A., Syroklyn, V.P.: Optimum planning of experiment in manufacturing the electronic equipment. Telecommun. Radio Eng. **70**(8), 731–734 (2011). https://doi.org/10.1615/TelecomRadEng.v70.i8.60
15. Koshevoy, N.D., Kostenko, E.M., Pavlyk, A.V., Koshevaya, I.I., Rozhnova, T.G.: Research of multiple plans in multi-factor experiments with a minimum number of transitions of levels of factors. Radio Electron. Comput. Sci. Control **2**(2), 53–59 (2019). https://doi.org/10.15588/1607-3274-2019-2-6
16. Koshevoi, N.D., Bestan, S.G., Gordienko, V.A.: Simulation and optimization of current density measuring devices. Telecommun. Radio Eng. **55**(1), 98–101 (2001). https://doi.org/10.1615/telecomradeng.v55.i1.140
17. Koshevoi, N.D., Cehovski, M.V., Gordienko, V.A.: Design of the thickness meters for dielectric coating. Telecommun. Radio Eng. **59**(7–9), 172–176 (2003). https://doi.org/10.1615/telecomradeng.v59.i7-9.140
18. Trembovetska, R.V., Halchenko, V.Ya., Tychkov, V.V.: Studying the computational resource demands of mathematical models for moving surface eddy current probes for synthesis problems. Eastern-Eur. J. Enterp. Technol. 5/5(95), 39–46 (2018). https://doi.org/10.15587/1729-4061.2018.143309, http://iournals.uran.ua/eejet/article/view/143309/144682

19. Halchenko, V.Ya., Trembovetska, R.V., Tychkov, V.V., Storchak, A.V.: Nonlinear surrogate synthesis of the surface circular eddy current probes. Przeglad elektrotechniczny **9**, 76–82 (2019). https://doi.org/10.15199/48.2019.09.15. https://pe.org.pl/articles/2019/9/15.pdf
20. Zabolotny, A.V., Koshevoi, N.D., Gordienko, V.A.: Improving efficiency of the quality control of substances with dielectric properties. Telecommun. Radio Eng. **57**(2–3), 184–196 (2002). https://doi.org/10.1615/telecom-radeng.v57.i2-3.200
21. Zabolotnyi, O.V., Koshevoi, M.D.: Conditionaliu examination of the testing algorithms for coal-water slurries moisture measurement. Naukovyi Visnyk Natsionalnoho Universytetu **1**, 51–59 (2018). https://doi.org/10.29202/nvngu/2018-1/21

Criteria and Similarity Scales in the Task of Modelling the Operation of Inflatable Air Shock Absorbers When Landing Parachute Platforms

Olexander V. Betin, Igor V. Kaluzhynov, Denys O. Betin(✉), and Olena Y. Betina

National Aerospace University «KhAI», Chkalova 17, Kharkiv 61070, Ukraine
{o.betin,o.betina}@khai.edu, kaluzhynov@gmail.com,
betindenys@gmail.com

Abstract. Research has been carried out aimed at developing the theoretical bases of physical modelling of landing processes of parachute platforms on inflatable air shock absorbers. As an object of research, the soft tissue shells filled through inlet openings with atmospheric or air injected from gas cylinders are chosen. During the landing, air is compressed in the shell, it flows out through special valves and the absorption of the kinetic energy of the blow. When performing the research, it was accepted: the shells of the air shock absorbers are impermeable; the pressure in them is distributed evenly over the entire surface of the shells; during the landing, the ideal gas law and the adiabatic process of changing its state are valid for the air in the shells. As a result, the criteria were determined and similarity scales are formalized for modeling the operation of inflatable air shock absorbers during landing on parachute platforms.

Keywords: Criteria · Similarity scales · Modeling · Inflatable air shock absorbers · Process of landing · Parachute platforms

1 Introduction

The most effective method of researches of behavior of the landing equipment is the physical modeling using the free flying dynamically similar model (FDSM) or the bench test dynamically similar model (BDSM) of the landing equipment as natural object (NO) [1–6].

Bench researches of process of landing of NO on the BDSM belong to physical modeling which differs from other types of modeling (mathematical, analog and so forth) in the fact that the model has the same physical nature, as the NO. At the same time, the physical experiment is conducted in natural conditions, and study is conducted on the basis of results of experiment, of laws of interaction of the BDSM with environment, with gravitational and other fields of Earth, and (or) impact of this environment onto aggregates and systems of the BDSM with the subsequent transfer of results of physical modeling onto NO.

M. Nechyporuk et al. (Eds.): ICTM 2021, LNNS 367, pp. 344–352, 2022.
https://doi.org/10.1007/978-3-030-94259-5_30

When modeling the landing of NO on the BDSM, not only conditions of carrying out experiments have to satisfy to certain laws, but also the BDSM has to be created taking into account these laws. The answer to the arising question about adequacy of modeling and possibility of transfer of results of model researches can be received by application of the theory of similarity allowing to formalize conditions which have to be keeped, both at all stages of creation of the BDSM, and when carrying out researches.

Transfer of results of model researches onto NO, requires execution of conditions of geometrical, kinematic and dynamic similarity. The BDSM has to have the same as the NO, form, mass and inertial parameters, both elastic and geometrical characteristics required for similarity. Velocities, accelerations, overloads, air pressure in shock-absorbers, and required parameters of the environment have to meet the accepted conditions of similarity. As a result of this the BDSM will behave in flight as the NO in the corresponding conditions.

The formalized mathematical record of conditions of similarity forms the system of connections (the system of criteria of similarity) which has to be imposed on environment parameters, parameters and characteristics of the BDSM. The satisfaction of the restrictions determined by this system at all stages of creation of the BDSM (designing, production, ground tests) and model's bench researches allows to provide necessary completeness of similarity and, therefore, to receive in safe conditions and in the absence of the landing NO reliable data about the most difficult modes of its landing.

Thus, the system of criteria of similarity is theoretical basis of the solution of applied and methodical problems of creation of the BDSM and execution of bench researches with using them, and method of forming this system jointly with the Standard atmosphere is theoretical base of modeling of landing of NO with using BDSM.

2 Basic Provisions of the Theory of Similarity

Touching upon the questions the theory of similarity, it is necessary to stop on the concepts and definitions, used further [7, 8].

So, two bodies are similar geometrically if the relation of any their analogous nominal sizes is constant, i.e.

$$\frac{\ell_n}{\ell_m} = k_\ell = \text{const}_1, \tag{1}$$

where ℓ_n, ℓ_m – randomly chosen analogous sizes (for example, characteristic sizes) of the comparable bodies (natural object and model); k_ℓ – scale of the linear sizes.

Hereinafter code "*n*" defines the relation of criterion or index to parameters or characteristics of NO; and code "*m*" – to the corresponding parameters or characteristics of BDSM.

The geometrical similarity of NO and BDSM is understood not as specified similarity of structures (though it is possible too), but as similarity of exterior surfaces of bodies of identical form.

Let's pay attention, that the requirement of geometrical similarity extends not only to shapes of streamline bodies when studying the aerodynamic phenomena, but also onto borders of flows (if those are available).

Equality of angles between corresponding lines of geometrically similar bodies is the consequence of geometrical similarity, i.e.

$$\phi_n = \phi_m \quad \text{or} \quad k_\phi = \frac{\phi_n}{\phi_m} = 1, \tag{2}$$

where k_ϕ – scale of angles.

Besides, the areas and volumes of geometrically similar bodies are in the constant ratio.

$$\frac{S_{\mathrm{H}}}{S_{\mathrm{M}}} = \frac{\ell_n^2}{\ell_m^2} = k_\ell^2 = k_s; \tag{3}$$

$$\frac{W_{\mathrm{H}}}{W_{\mathrm{M}}} = \frac{\ell_n^3}{\ell_m^3} = k_\ell^3 = k_w, \tag{4}$$

where k_s – scale of areas; k_w – scale of volumes.

Concept of geometrical similarity, at first sight, is rather simple. However, the presented ratios together with the previous formulations of geometrical similarity to a large extent reflect only theoretical ideas about such similarity of the objects under review.

As a result of execution of technology processes of production of details, modular and general assemblies, there are arising deviations of geometrical forms (roughness, waviness, gaps, ledges, and other distortions of the preset form) and discrepancies of the real surfaces to forms which are theoretically preset for NO or calculated for FDSM or BDSM are observed. It influences onto flow around any of the considered model objects, but more defines research opportunities of their FDSM [3–6, 9].

Achievement of required values of parameters of form and surface quality of FDSM depends on technological level of production to a large extent. Application of method of vacuum formation of coverings and method of usage of master-form both face-mould for ensuring interchangeability and reconciliation of technological rigging allow to solve successfully designated problem; that exempts from need of research of questions of aerodynamic similarity for aerial vehicles, whose geometrical characteristics do not meet requirements of geometrical similarity. It does not exclude wind tunnel tests of scale models (and, perhaps, of master-form of model's surface), and also FDSM after production.

As for the landing NO and its BDSM, because of rather small velocities of landing and short duration of this process, insignificant deviations of geometrical forms practically are not affecting onto results of modeling.

Two movements are geometrically and kinematically similar if geometrically similar bodies cover the similar distances into proportional intervals of time. It means that along with availability of geometrical similarity, also constancy of the relation of analogous intervals of time t_n and t_m is necessary:

$$\frac{t_n}{t_m} = k_t = \mathrm{const}_2, \tag{5}$$

where k_t – scale of times.

If V_n, V_m – linear velocities of the movement of geometrically similar bodies, then in the presence of kinematic similarity the relation of velocities V_n, V_m also presents the constant value for any pair of analogous points of the bodies:

$$\frac{V_{\mathrm{H}}}{V_{\mathrm{M}}} = \frac{\ell_n t_m}{\ell_m t_n} = \frac{k_\ell}{k_t} = k_v = \mathrm{const}_3, \tag{6}$$

where k_v – scale of linear velocities.

The similar ratio is fair also for linear accelerations:

$$\frac{a_{\mathrm{H}}}{a_{\mathrm{M}}} = \frac{\ell_n t_m^2}{\ell_m t_n^2} = \frac{k_\ell}{k_t^2} = \frac{k_v}{k_t} = k_a = \mathrm{const}_4, \tag{7}$$

where k_a – scale of linear accelerations.

The phenomena are dynamically similar if conditions of compliance of all elements creating dynamic picture are met, i.e. if polygons of forces operating on geometrically similar bodies are similar (and overloads, naturally, are equal). It means that at dynamic similarity forces of the same nature have to be equally directed and proportional:

$$\frac{R_n}{R_m} = k_R = \mathrm{const}_5, \tag{8}$$

where k_R – scale of forces.

2.1 Similarity Scale When Modeling Landing onto Inflatable Air Shock-Absorbers

According to definition, at dynamic similarity the similarity of polygons of forces operating onto NO and its BDSM is observed. In particular, it is possible to write down that

$$\frac{F_n}{F_m} = \frac{G_n}{G_m}, \tag{9}$$

where $\frac{F_n}{F_m} = \frac{m_n a_n}{m_m a_m}$ – the relation of inertial forces of NO and its BDSM when landing; a_n, a_m – accelerations of NO and BDSM when landing; $\frac{G_n}{G_m} = \frac{m_n g_n}{m_m g_m}$ – the relation of forces of weight of NO and its BDSM when landing; g_n, g_m – accelerations of gravity at heights of landing of NO and BDSM.

After substitution of the last expressions for the relations $\frac{F_n}{F_m}$, $\frac{G_n}{G_m}$ into equality (9) and reduction onto $\frac{m_n}{m_m}$, we will receive the scale of accelerations

$$k_a = \frac{a_n}{a_m} = \frac{g_n}{g_m}. \tag{10}$$

On the other hand, from the ratio (7) we have

$$k_a = \frac{k_\ell}{k_t^2}. \tag{11}$$

Equating the right parts of equalities (10) and (11), and after transformations into the scale view, we will have

$$k_t = \sqrt{k_\ell \frac{g_m}{g_n}}, \tag{12}$$

but in accordance with (6)

$$k_v = \frac{k_\ell}{k_t}, \tag{13}$$

therefore, using (12), we will receive ratio for the scale of linear velocities in form

$$k_v = \frac{k_\ell}{k_t} = \frac{k_\ell}{\sqrt{k_\ell \frac{g_m}{g_n}}} = \sqrt{k_\ell \frac{g_n}{g_m}}. \tag{14}$$

Simplification of expression (14) for range of tasks of modeling is obviously possible. In particular, when modeling landing of NO on the BDSM, the heights of experiments for both objects can be in range of 0...5000*m*. According to SA, acceleration of gravity at these heights does not change considerably. Therefore it is possible to consider that for NO and its BDSM $g_n = g_m$, and, therefore $k_t = k_v = \sqrt{k_\ell}$ and $k_a = 1$.

The central (or main) element of the shock-absorber systems under review is the air shock-absorber which can be considered as some kind of covering, with expanded or compressed check volume. Modeling of his behavior under similar loading provides modeling of process of landing of NO on the BDSM.

For obtaining necessary criteria and scales of similarity we will use the fact that because of the short duration of landing, change of condition of air in coverings of air shock-absorbers complies to adiabatic process (at which air in shock-absorbers does not exchange warmth with structural elements of the platform and the surrounding atmosphere) [1, 10–13].

We will write down the equation of adiabatic process in two equivalent forms [14, 15]:

$$\frac{p_{\text{BH}}}{\rho_{\text{BH}}^{\gamma}} = \text{const}_6 \tag{15}$$

and

$$p_{\text{BH}} W_{\text{BH}}^{\gamma} = \text{const}_7, \tag{16}$$

where $\rho_{\text{BH}} = \frac{m_{\text{BH}}}{W_{\text{BH}}}$ – air density in the shock-absorber m_{BH}, W_{BH} – the weight and volume of air in the shock-absorber; γ – adiabatic index (for air $\gamma = 1,41$).

Using (16) and concept of geometrical and kinematic similarity in the forms of (4), (5) and (12) for natural and model air shock-absorbers in analogous time-points, we will receive

$$p_{\text{BH}1.n}(t_{1.n}) W_{\text{BH}1.n}^{\gamma}(t_{1.n}) = p_{\text{BH}2.n}(t_{2.n}) W_{\text{BH}2.n}^{\gamma}(t_{2.n}); \tag{17}$$

$$\begin{aligned} & p_{\text{BH1}.m}(t_{1.m}) W^{\gamma}_{\text{BH1}.m}(t_{1.m}) \\ & = p_{\text{BH2}.m}(t_{2.m}) W^{\gamma}_{\text{BH2}.m}(t_{2.m}) \end{aligned}, \tag{18}$$

where $\frac{t_{1.n}}{t_{1.m}} = \frac{t_{2.n}}{t_{2.m}} = \sqrt{k_\ell \frac{g_m}{g_n}} = k_t$;
$\frac{W_{\text{BH1}.n}(t_{1.n})}{W_{\text{BH1}.m}(t_{1.m})} = \frac{W_{\text{BH2}.n}(t_{2.n})}{W_{\text{BH2}.m}(t_{2.m})} = k^3_\ell = k_{w.\text{BH}}$;
$k_{w.\text{BH}}$ – the scale of air volumes in shock-absorbers of NO and BDSM.

After dividing (17) onto (18), and after transformations and reductions, we will have

$$\frac{p_{\text{BH1}.n}(t_{1.n})}{p_{\text{BH1}.m}(t_{1.m})} = \frac{p_{\text{BH2}.n}(t_{2.n})}{p_{\text{BH2}.m}(t_{2.m})}. \tag{19}$$

Process of landing is considered complete when object of landing ceases to move. Usually it occurs at approach of internal pressure in shock-absorbers p_{BH} to external atmospheric pressure p_{aTM}. If landing of NO and BDSM happens onto surfaces to the same height above sea level (i.e. $H_n = H_m = H$), in the final moments of process of landing $p_{\text{BH2}.n}(t_{2.n} = t_{\text{KOH}.n}) = p_{\text{BH2}.m}(t_{2.m} = t_{\text{KOH}.m}) = p_{\text{aTM}}(H)$.

Thus, in accordance with (19) in any point of landing (that is intermediate in aspect of time) of NO and its BDSM, the condition has to be satisfied

$$\frac{p_{\text{BH}.n}(t_n)}{p_{\text{BH}.m}(t_m)} = 1 = k_{p.\text{BH}}, \tag{20}$$

where $k_{p.\text{BH}}$ – the scale of air pressures in shock-absorbers of NO and BDSM.

Using the equation of adiabatic process in the form of (15) and the received result (20); and using the similar reasoning, we will have

$$\frac{p_{\text{BH1}.n}(t_{1.n})}{\rho^{\gamma}_{\text{BH1}.n}(t_{1.n})} = \frac{p_{\text{BH2}.n}(t_{2.n})}{\rho^{\gamma}_{\text{BH2}.n}(t_{2.n})}; \tag{21}$$

$$\frac{p_{\text{BH1}.m}(t_{1.m})}{\rho^{\gamma}_{\text{BH1}.m}(t_{1.m})} = \frac{p_{\text{BH2}.m}(t_{2.m})}{\rho^{\gamma}_{\text{BH2}.m}(t_{2.m})}, \tag{22}$$

then

$$\frac{\rho^{\gamma}_{\text{BH1}.n}(t_{1.n})}{\rho^{\gamma}_{\text{BH1}.m}(t_{1.m})} = \frac{\rho^{\gamma}_{\text{BH2}.n}(t_{2.n})}{\rho^{\gamma}_{\text{BH2}.m}(t_{2.m})}; \tag{23}$$

and taking into account that

$\rho_{\text{BH2}.n}(t_{2.n} = t_{\text{KOH}.n}) = \rho_{\text{BH2}.m}(t_{2.m} = t_{\text{KOH}.m}) = \rho_{\text{aTM}}(H)$,

finally we will have

$$\frac{\rho_{\text{BH}.n}(t_n)}{\rho_{\text{BH}.m}(t_m)} = 1 = k_{\rho.\text{BH}}, \tag{24}$$

where $k_{\rho.\text{BH}}$ – the scale of density of air in shock-absorbers of NO and BDSM.

The mass flow of air in the check volume of air shock-absorbers is defined by the mass flow of air out-coming of shock-absorbers and incoming into them ($G_{\text{BH}}^{\text{BbIX}}$ and $G_{\text{BH}}^{\text{BX}}$), i.e.

$$G_{\text{BH}} = G_{\text{BH}}^{\text{BbIX}} - G_{\text{BH}}^{\text{BX}}. \tag{25}$$

Let's consider that process of filling of shock-absorbers by the air was completed until contact of platform with the ground surface; and the mass flow of air incoming into shock-absorbers $G_{\text{BH}}^{\text{BX}} = 0\,\text{kg/m}^3$.

At the same time, at adiabatic process, the mass flow of air out-coming of shock-absorbers [12, 14, 15].

$$G_{\text{вн}}^{\text{вых}} = K_{\text{кл}} S_{\text{кл}} \frac{p_{\text{вн}}}{\sqrt{RT_{\text{вн}}}} \left(\frac{p_{\text{атм}}}{p_{\text{вн}}}\right)^{\frac{1}{\gamma}} \cdot\cdot \sqrt{\frac{2\gamma}{\gamma-1}\left(1 - \left(\frac{p_{\text{атм}}}{p_{\text{вн}}}\right)^{\frac{\gamma-1}{\gamma}}\right)}, \tag{26}$$

where $K_{\text{кл}}$– the flow coefficient, which depends on outlet velocity of air, form and surface quality of valves of vent holes; $S_{\text{кл}}$– cross-sectional area of valves; R – universal gas constant (for dry air at reference conditions $R = 287, 14\,m^2/s^2K$); T_{BH} – air temperature in shock-absorbers, K.

Taking into account, that as earlier the landing NO and BDSM happens onto surfaces at the same height above sea level (i.e. $H_n = H_m = H$), and, therefore, with identical pressures and temperatures of the atmospheric air, which enters shock-absorbers (hence, at adiabatic process in analogous time-points $T_{\text{BH}.n}(t_n) = T_{\text{BH}.m}(t_m)$) and, considering ratio (20), we will receive, that for mass flows of air out of shock-absorbers of NO and BDSM the ratio is fair.

$$\frac{G_{\text{вн}.n}(t_n)}{G_{\text{вн}.m}(t_m)} = \frac{K_{\text{кл}.n}(t_n)}{K_{\text{кл}.m}(t_m)} \frac{S_{\text{кл}.n}(t_n)}{S_{\text{кл}.m}(t_m)}. \tag{27}$$

Having accepted that for the BDSM the valves are used of identical form with NO or are specially made with characteristic $K_{\text{кл}.n}(t_n) = K_{\text{кл}.m}(t_m)$., we will receive

$$\frac{G_{\text{вн}.n}(t_n)}{G_{\text{вн}.m}(t_m)} = \frac{S_{\text{кл}.n}(t_n)}{S_{\text{кл}.m}(t_m)}. \tag{28}$$

The expiration of air from shock-absorbers, among other things, is the aerodynamic phenomenon of flow around the valve of vent hole by current of air. At the same time, it is necessary to execute the condition of similarity by Froude's criterion Fr, which characterizes the relation of gravity forces to forces of inertia and means observance of similarity, taking into account gravity for the compared flows [3, 7–9].

Besides, ensuring dynamic similarity during the landing of NO and BDSM onto air shock-absorbers requires execution of geometrical similarity concerning changing of air volumes in coverings and, therefore, of the mass of this air which is under identical pressure at the time-points determined by kinematic similarity.

The change of volumes of air in coverings of shock-absorbers of NO and BDSM into analogous time-points.

$$\Delta W_{\text{BH}.n}(t_n) = \Delta S_{xz.n}(t_n) \Delta \ell_{y.n}(t_n); \tag{29}$$

$$\Delta W_{\text{BH}.m}(t_m) = \Delta S_{\text{xz}.m}(t_m)\Delta \ell_{y.m}(t_m), \tag{30}$$

where $\Delta S_{\text{xz}.n}(t_n)$, $\Delta S_{\text{xz}.m}(t_m)$, $\Delta \ell_{y.n}(t_n)$, $\Delta \ell_{y.m}(t_m)$ – change of the areas of bottoms and vertical overall dimensions of coverings of NO and BDSM.

During landing for compliance of geometrical similarity it is necessary so that

$$\frac{\Delta W_{\text{BH}.n}(t_n)}{\Delta W_{\text{BH}.m}(t_m)} = \frac{\Delta S_{\text{xz}.n}(t_n)}{\Delta S_{\text{xz}.m}(t_m)}\frac{\Delta \ell_{y.n}(t_n)}{\Delta \ell_{y.m}(t_m)} = k_\ell^2 k_\ell = k_\ell^3. \tag{31}$$

Change of mass of air in coverings of shock-absorbers of NO and BDSM in analogous time-points.

$$\Delta m_{\text{вн}.n}(t_n) = \rho_{\text{вн}.n}(t_{\text{нач}.n})\Delta W_{\text{вн}.n}(t_n); \tag{32}$$

$$\Delta m_{\text{вн}.m}(t_m) = \rho_{\text{вн}.m}(t_{\text{нач}.m})\Delta W_{\text{вн}.m}(t_m), \tag{33}$$

where $\rho_{\text{вн}.n}(t_{\text{нач}.n})$, $\rho_{\text{вн}.m}(t_{\text{нач}.m})$– initial density of air inside shock-absorbers of NO and BDSM.

Then

$$\frac{\Delta m_{\text{вн}.n}(t_n)}{\Delta m_{\text{вн}.m}(t_m)} = \frac{\rho_{\text{вн}.n}(t_{\text{нач}.n})}{\rho_{\text{вн}.m}(t_{\text{нач}.m})}\frac{\Delta W_{\text{вн}.n}(t_n)}{\Delta W_{\text{вн}.m}(t_m)} = k_{\rho.\text{вн}} k_\ell^3, \tag{34}$$

and taking into account (24)

$$\frac{\Delta m_{\text{BH}.n}(t_n)}{\Delta m_{\text{BH}.m}(t_m)} = \frac{m_{\text{BH}.n}(t_n)}{m_{\text{BH}.m}(t_m)} = k_\ell^3 = k_{m.\text{BH}}, \tag{35}$$

where $k_{m.\text{BH}}$ – the scale of mass of air in shock-absorbers of NO and BDSM.

Therefore, the relation of mass flows of air out of shock-absorbers of NO and BDSM in analogous time-points

$$\frac{G_{\text{BH}.n}(t_n)}{G_{\text{BH}.m}(t_m)} = \frac{\frac{\Delta m_{\text{BH}.n}(t_n)}{t_n}}{\frac{\Delta m_{\text{BH}.m}(t_m)}{t_m}} = \frac{\Delta m_{\text{BH}.n}(t_n)}{\Delta m_{\text{BH}.m}(t_m)}\frac{t_m}{t_n} = k_\ell^3 \frac{1}{k_t} = k_\ell^{2,5}\sqrt{\frac{g_n}{g_m}} = k_G, \tag{36}$$

where k_G – the scale of mass flows of air out of shock-absorbers.

Taking into account, that as earlier the landing NO and BDSM happens onto surfaces at the same height above sea level (i.e. $H_n = H_m = H$), and, therefore $g_n = g_m$, we will receive

$$k_G = \frac{G_{\text{BH}.n}(t_n)}{G_{\text{BH}.m}(t_m)} = k_\ell^{2,5}. \tag{37}$$

Using ratio (28) for (37), we will have

$$\frac{S_{\text{кл}.n}(t_n)}{S_{\text{кл}.m}(t_m)} = k_\ell^{2,5} = k_{s.\text{кл}}, \tag{38}$$

where $k_{s.\text{кл}}$– scale of areas of valves of shock-absorbers of NO and BDSM.

3 Conclusions

Research has been carried out aimed at developing the theoretical bases of physical modelling of landing processes of parachute platforms on inflatable air shock absorbers. As a result, the criteria were determined and similarity scales are formalized for modeling the operation of inflatable air shock absorbers during landing on parachute platforms.

References

1. Kaluzhinov, I.V.: The choice of landing system of dynamically similar flying model. In: Designing of Airplane Frameworks and Their Connections, pp. 167–178. Kharkov Aviation Institute, Kharkov (1986). (in Russian)
2. Mironov, A.D., Vladychin, G.P., Zolotukhin, R.B., Chetvergov, V.N., Hanov, I.K.: Methods of researches onto the flying models. Mechanical Engineering, 144 p. (1988). (in Russian)
3. Betin, A.V., Ryzhenko, A.I., Ryabkov, V.I., Cheranovsky, O.R.: Determination of the sizes and mass and inertial parameters of free-flying dynamically similar aircraft models: schoolbook, 101 p. Ministry of Education of Ukraine, Kharkov Aviation Institute, Kharkiv (1992). (in Russian)
4. Betin, A., Ryzhenko, A., Sadovnychiy, S., Peralta-Fabi, R.: Simulation of aircraft flight dynamics by means of dynamically similar models. In: Modeling and Simulation Technologies, pp. 64–69. American Institute of Aeronautics and Astronautics, Boston (1998)
5. Betin, A., Sadovnychiy, S., Ryshenko, A.: Simulation of aircraft emergencies by means of dynamically similar models. In: Modeling and Simulation Technologies. pp. 328–334. American Institute of Aeronautics and Astronautics, Portland (1999)
6. Betin, A., Sadovnychiy, S., Ryshenko, A.: Flight control system damage simulation using freely flying models. Aeronaut. J. **109**(1091), 45–50 (2005)
7. Gukhman, A.A.: The introduction into the theory of similarity. Mechanical Engineering, 296 p. (1973). (in Russian)
8. Sedov, L.I.: The methods of similarity and dimensionality in mechanics, 448 p. Science (1981). (in Russian)
9. Betin, A.V.: Criteria of similarity when studying flight dynamics. In: Problems of Designing of the Flying Models for Research of Critical Duties of Flight: Collected Papers of Kharkov Aviation Institute, Kh., pp. 78–89 (1989). (in Russian)
10. Bazhenov, V.I., Osin, M.I.: The landing of spacecrafts onto planets. Mechanical Engineering, 159 p. (1978). (in Russian)
11. Lobanov, N.A.: Bases of calculation and designing of parachutes. Mechanical Engineering, 364 p. (1965). (in Russian)
12. Wang, H., Hao, G.: Research on matching and evaluating for recovery system of airborne vehicle based on FEM. In: The 7th China CAE Annual Conference, Kunming, China, pp. 120–127 (2011). (in Chinese)
13. Zhiqi, D.U., Pengli, S.H.A.O.: Dynamic finite element simulation of the aluminum alloy hull at landing. Acta Armamentarii **30**(1), 1–4 (2009). (in Chinese)
14. Belyaev, N.M.: Thermodynamics of variable amount of gas, 286 p. DGU, Dnepropetrovsk (1981). (in Russian)
15. Kukhling, H.: Reference Book on Physics, 520 p. World (1982). (in Russian)

COVID-19 Epidemic Process Simulation Using ARIMA Model

Alireza Mohammadi and Dmytro Chumachenko(✉)

National Aerospace University "Kharkiv Aviation Institute", Kharkiv, Ukraine

Abstract. The substantial ascendant trend within the number of daily infected new cases with coronavirus around the world is a warning, and several other researchers are utilizing various mathematical and machine learning-based prediction models to forecast the long-term trend of the COVID-19 pandemic. During this research, the Autoregressive Integrated Moving Average or ARIMA model was implemented to forecast the COVID-19 expected daily number of cases in Ukraine. We implemented Autoregressive Integrated Moving Average for this research. The forecasting results showed that the trend in Ukraine will continue ascending and should reach up to more than 1.8 million total cases if stringent precautionary and control measures don't get implemented to prohibit the spread of COVID-19.

Keywords: Epidemic model · Machine learning · Epidemic process simulation · COVID-19 forecasting · COVID-19

1 Introduction

The coronavirus cases number is continuously escalating at an alarming rate across the globe. Among all seven coronavirus versions, the primary four caused mild symptoms and were identified within the mid-1960s, though the fifth and sixth, which are SARS-CoV and MERS-CoV that emerged in 2003 and 2012 respectively, caused severe symptoms [1]. In December 2019, the seventh version which was named SARS-CoV-2 showed up in Wuhan, China [2], and the disease was declared as an endemic in March 2020 by World Health Organization (WHO) [3]. This version of coronavirus is asymptomatic or causes mild, moderate, or severe symptoms [4]. Despite the lack of prevention and supervision of measures that have been implemented in many countries, this novel coronavirus is quite contagious and has been quickly surging worldwide [5]. The total number of infected people is over 60 million in the US by March 2021, also, regarding other countries' cases, total numbers are substantially increasing, which still adversely impact people's lives, healthcare facilities, and countries' economies. Accordingly, forecasting the long-term surge of the COVID-19 based on the available data has high importance to grasp the current situation, evaluate the pandemic severity, and help the authorities put well-tailored strategies and productive decisions to control the pandemic and constraint new cases shortly.

M. Nechyporuk et al. (Eds.): ICTM 2021, LNNS 367, pp. 353–363, 2022.
https://doi.org/10.1007/978-3-030-94259-5_31

Mathematical modeling is an effective tool for analyzing epidemic processes and the development of infectious diseases [6]. The use of modeling and information technology makes it possible to effectively solve problems related to medical diagnostics and other health problems [7–11]. It allows not only to predict the future state of the population, but also to identify the factors that most affect the development of the epidemic process [12, 13]. This allows epidemiologists and public health professionals to take effective evidence-based containment measures [14–16].

Recently, different mathematical and machine-learning-based forecasting models to estimate the breakout of the pandemic have been proposed, and evaluate its impact thoroughly [17]. In this research, the Autoregressive Integrated Moving Average or ARIMA model was implemented to predict the daily number of COVID-19 cases in Ukraine and other countries. The ARIMA model has been accustomed to forecast the process of COVID-19 disease within the most affected 15 countries [18] and showed realistic numbers that are almost similar to the present situation in those countries. It's vital to forecast the longer-term trend of COVID-19, and consequently, represent beneficial recommendations for the authorities on the way to prohibit the outbreak of the disease and implement new meticulous measures to avoid a possible suspension of those pilgrimages.

The **aim** of the paper is to develop COVID-19 epidemic process model and to provide experimental study of in based on real data of incidence.

This research is part of a comprehensive epidemic forecasting and epidemic containment decision-making model development study, which concept is presented in [19].

2 Materials and Methods

The best and the most widely applied model structure is named the Autoregressive (AR) model. In AR model, the present output *z(t)* is expressed by previous values and parameters α_{t-p}, as formulated in (1), where t represents time and p represents parameters.

$$z_t = -\alpha_1 z_{t-1} - \alpha_2 z_{t-2} - \cdots - \alpha_k z_{t-p} + \varepsilon \tag{1}$$

where α(z−1) is given by

$$\alpha\left(z^{-1}\right) = 1 - \alpha_1 z^{-1} + \cdots + \alpha_k z^{-p}. \tag{2}$$

Another model is named moving average (MA). Within the MA model, the output z(t) is expressed in terms of innovation input including filtered with the weights ßq has been formulated in (3) and (4).

$$z_t = \varepsilon + \beta_1 \varepsilon_{t-1} + \beta_2 \varepsilon_{t-2} + \cdots + \beta_t \varepsilon_{t-q} \tag{3}$$

$$\beta\left(z^{-1}\right) = 1 - \beta_1 z^{-1} + \cdots + \beta_k z^{-q} \tag{4}$$

By mixing AR and MA, an augmented structure is attained that is termed the ARMA model and defined in (5).

$$z_t = -\alpha_1 z_{t-1} - \cdots - \alpha_k z_{t-p} + \varepsilon + \beta_1 \varepsilon_{t-1} + \cdots + \beta_t \varepsilon_{t-q} \tag{5}$$

A more augmented version model is named Autoregressive Integrated Moving Average (ARIMA), including a minimum difference once and expressed in (6).

$$z_t = (1 - z_{t-d}) - \alpha_1 z_{t-1} - \cdots - \alpha_p z_{t-p} + \varepsilon + \beta_1 \varepsilon_{t-1} + \cdots + \beta_1 \varepsilon_{t-q} \tag{6}$$

This model has an appropriate performance regarding prediction practical examples in sorts of literature in many fields. To identify the ARIMA model, it is assigned the order for the three terms: p for AR, q for MA, and also the number of various steps d.

The information was tested for stationarity using Augmented Dickey-Fuller (ADF) test before estimating the parameters for the ARIMA model, whereas the null hypothesis H_0 is the statistic non-stationary. ADF test result offered that the time-series data were not stationary ($p > 0.05$) and the null hypothesis was rejected. The ARIMA model parameters were chosen considering Akaike information criterion (AIC) described as:

$$\text{AIC}_{p,q} = \frac{-2\ln(\text{maximized likehood}) + 2r}{n} \approx \ln\left(\hat{\sigma}_a^2\right) + r\frac{2}{n} \tag{7}$$

where n is the data observations number, $r = p + q + 1$, and $\hat{\sigma}_a$ is the maximum likelihood forecast.

To assess the performance of each model there are some accuracy measurement functions being used. They are described as follows:

1. The definition of Root Mean Square Error or RMSE:

$$\text{RMSE} = \sqrt{\frac{1}{N}\sum_{i=1}^{N}(z_i - \tilde{z}_i)^2} \tag{8}$$

where z_i is actual value and z-bar is predictive value

2. The definition of Mean Absolute Error or MAE:

$$\text{MAE} = \sqrt{\frac{1}{N}\sum_{i=1}^{N}|z_i - \tilde{z}_i|} \tag{9}$$

3. The definition of Coefficient of Determination or R^2:

$$R^2 = 1 - \frac{\frac{1}{N}\sum_{i=1}^{N}(z_i - \tilde{z}_i)^2}{\frac{1}{N}\sum_{i=1}^{N}(z_i - \overline{z}_i)^2} \text{ where } \overline{z}_i \text{ is defined as } \overline{z}_i = \frac{1}{N}\sum_{i=1}^{N} z_i \tag{10}$$

4. The definition of Mean Absolute Percentage Error or MAPE:

$$\text{MAPE} = \frac{1}{N}\sum_{i=1}^{N}\left|\frac{\tilde{z}_i - z_i}{\tilde{z}_i}\right| \tag{11}$$

5. The definition of Root Mean Squared Relative Error or RMSRE [20]:

$$\text{RMSRE} = \sqrt{\frac{1}{N}\sum_{i=1}^{N}\left(\frac{\tilde{z}_i - z_i}{\tilde{z}_i}\right)^2} \tag{12}$$

3 Results

The ARIMA model was utilized in this research to forecast the spread of COVID-19 in Ukraine. This model achieves the smallest RMSE, MAPE, RMSRE values, and highest R_2 value. The data used in this research refers to the daily and the total number of confirmed COVID-19 cases in Ukraine daily until 13/04/2021. Figure 1 depicts a comparison of confirmed COVID-19 cases among the countries including the highest amount until 13/04/2021.

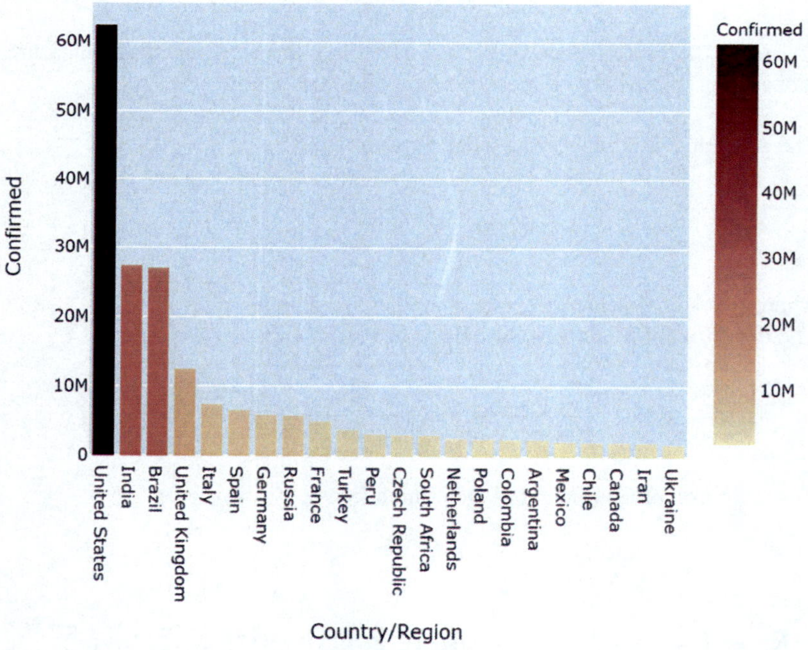

Fig. 1. Data import.

In Fig. 2 we can see the trend of COVID-19 in Ukraine until 13/04/2021. The figure shows that the total amount of confirmed cases has exceeded 1.8 million.

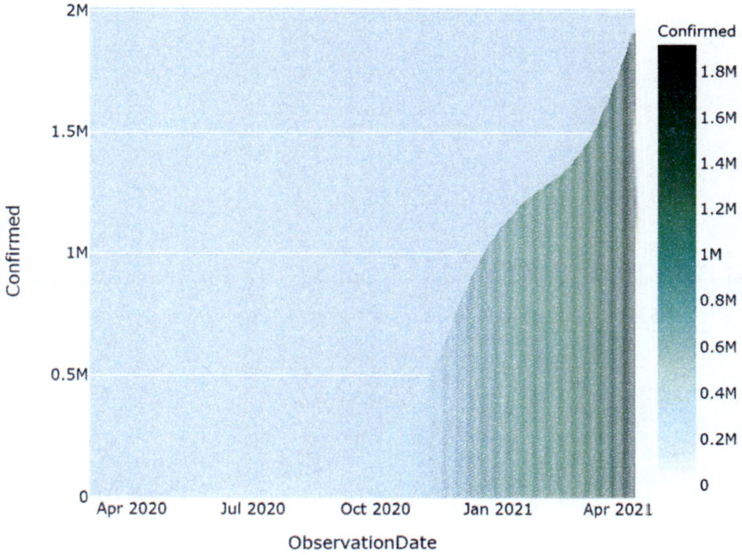

Fig. 2. Confirmed COVID-19 cases per day in Ukraine.

Figure 3, 4 and 5 are shown the evolution of the number of Confirmed, Death, and Recovered cases through time. There is an ascending trend in all three cases in Ukraine. That means although the confirmed cases of COVID-19 are increasing, the rate of Recovery cases is also enhancing and the ratio tends to be stable. On the other hand, the Death cases are uprising that is not something surprising since in some cases infection leads to death, unfortunately. Also, the comparison of Confirmed, Death and Recovered cases in five countries are shown in Fig. 6, 7 and 8. The most Confirmed cases in this comparison belong to Spain, whereas the Recovered cases in Germany are uprising. Ukraine takes place somewhere close to China in comparison with fewer cases for each parameter.

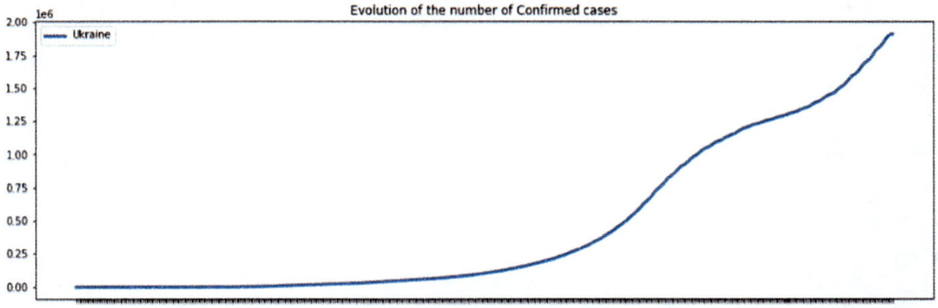

Fig. 3. Evolution of the number of Confirmed cases through time.

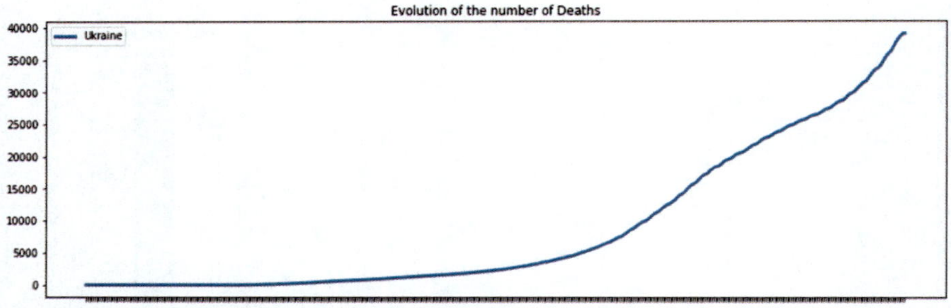

Fig. 4. Evolution of the number of Death cases through time.

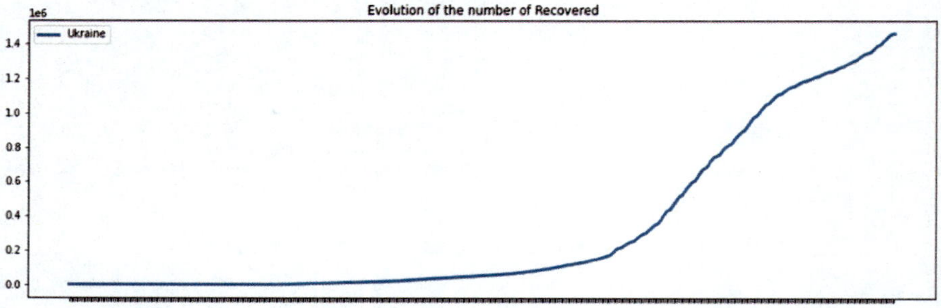

Fig. 5. Evolution of the number of Recovered cases through time.

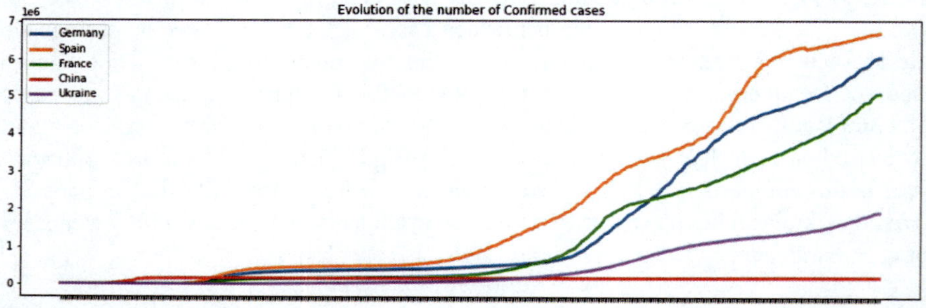

Fig. 6. Evolution of the number of Confirmed cases through time in five countries including Germany, Spain, France, China, and Ukraine.

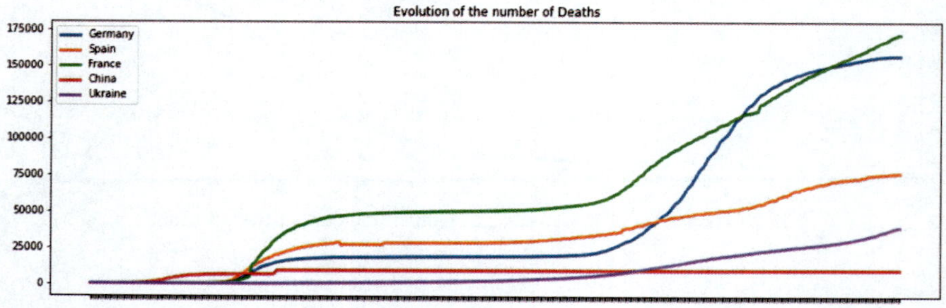

Fig. 7. Evolution of the number of Death cases through time in five countries including Germany, Spain, France, China, and Ukraine.

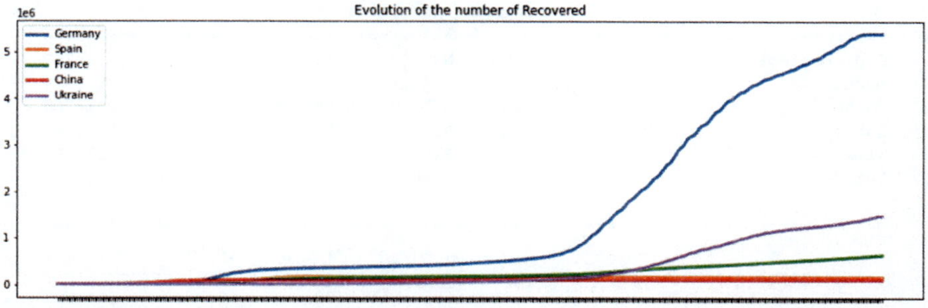

Fig. 8. Evolution of the number of Recovered cases through time in five countries including Germany, Spain, France, China, and Ukraine.

Figure 9 depicts the evaluation of case distribution in Ukraine with rolling mean and standard after implementing the functions in the model on data. Figure 10 shows the Dickey-Fuller Test results in Ukraine after implementing the function in the ARIMA model and it shows a reasonable accuracy for the data in Ukraine.

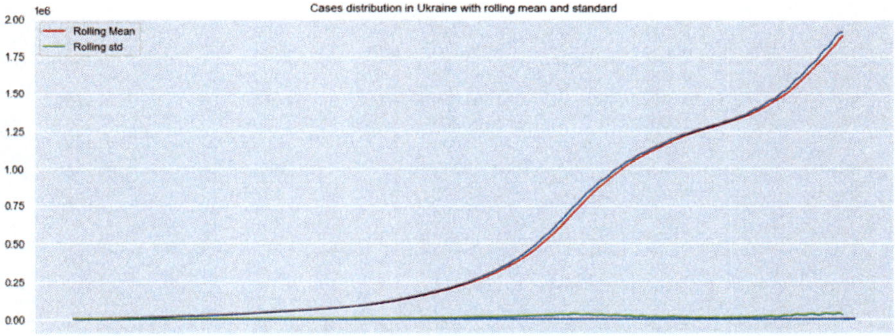

Fig. 9. Evolution of case distribution in Ukraine with rolling mean and standard.

```
Dickey-Fuller Test results for Ukraine
Test Statistic                  0.875860
p-value                         0.992767
#Lags Used                     14.000000
Number of Observations Used   375.000000
Critical Value (1%)            -3.447909
Critical Value (5%)            -2.869278
Critical Value (10%)           -2.570892
dtype: float64
```

Fig. 10. Dickey-Fuller Test results for Ukraine.

Figure 11 shows the result of implementing the ARIMA model on the data set of Confirmed cases in Ukraine. The MAPE calculated for the model is almost 13.5% which is reasonable and shows the model is able to forecast future cases well.

```
                              ARIMA Model Results
==============================================================================
Dep. Variable:                   D2.y   No. Observations:                  349
Model:                 ARIMA(4, 2, 5)   Log Likelihood               -2781.573
Method:                       css-mle   S.D. of innovations            681.208
Date:                Mon, 17 May 2021   AIC                           5585.146
Time:                        16:19:46   BIC                           5627.552
Sample:                             2   HQIC                          5602.027

==============================================================================
                 coef    std err          z      P>|z|      [0.025      0.975]
------------------------------------------------------------------------------
const        122.5179     72.591      1.688      0.091     -19.759     264.795
ar.L1.D2.y     1.0363      0.130      7.953      0.000       0.781       1.292
ar.L2.D2.y    -1.1744      0.117    -10.049      0.000      -1.403      -0.945
ar.L3.D2.y     0.3346      0.092      3.644      0.000       0.155       0.515
ar.L4.D2.y    -0.4167      0.081     -5.151      0.000      -0.575      -0.258
ma.L1.D2.y    -0.0433      0.142     -0.306      0.760      -0.321       0.234
ma.L2.D2.y     0.6409      0.081      7.893      0.000       0.482       0.800
ma.L3.D2.y     0.7663      0.062     12.427      0.000       0.645       0.887
ma.L4.D2.y    -0.4279      0.099     -4.336      0.000      -0.621      -0.235
ma.L5.D2.y     0.4899      0.094      5.233      0.000       0.306       0.673
                                    Roots
=============================================================================
                  Real          Imaginary           Modulus         Frequency
-----------------------------------------------------------------------------
AR.1            0.6242           -0.7932j            1.0093           -0.1439
AR.2            0.6242           +0.7932j            1.0093            0.1439
AR.3           -0.2228           -1.5185j            1.5347           -0.2732
AR.4           -0.2228           +1.5185j            1.5347            0.2732
MA.1           -1.0000           -0.0000j            1.0000           -0.5000
MA.2            0.1221           -0.9933j            1.0007           -0.2305
MA.3            0.1221           +0.9933j            1.0007            0.2305
MA.4            0.8146           -1.1724j            1.4276           -0.1534
MA.5            0.8146           +1.1724j            1.4276            0.1534
-----------------------------------------------------------------------------

Mean absolute percentage error: 13.503311
```

Fig. 11. Implement result of ARIMA model with the calculated MAPE.

Figure 12 shows the predicted values from 14/04/2021 to 26/06/2021 attained by using the ARIMA (0, 2, 1) and ARIMA (0, 1, 1) model with a 95% confidence interval (CI). The forecasted value (in red) is based on the fitted ARIMA model for daily confirmed cases of COVID-19 for the next 72 days and also the current confirmed cases until 13/04/2021 (in blue) are shown in Fig. 12. The results depict an increase in the trend for daily total confirmed cases in Ukraine by 26/06/2021 has a point forecast of more than 4 million (95% CI). However, Ukraine is going to control this pandemic in the middle of April.

It is a warning that the number of COVID-19 cases is continually escalating across the world. The infectious disease estimation of COVID-19 is measured by developing a hypothesis for interpreting the observed situation via time series analysis using the ARIMA model due to 'time' importance for disease management studies [21–23].

Former studies have been revealed that ARIMA is one of the most reliable models as it includes higher fitting and forecasting accuracy [24, 25]. The main purpose of this

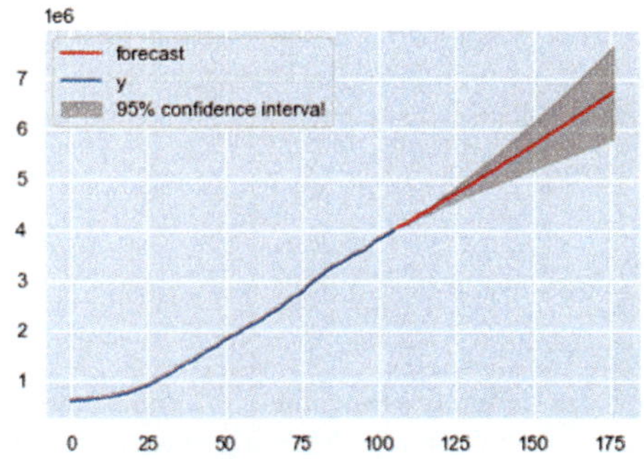

Fig. 12. Forecast and confidence intervals of total confirmed cases in Ukraine of COVID-19 using fitted model (blue line: actual data, red line: 72 days forecast, Gray zone: 95% of CI).

research was to illustrate and forecast the expected number of the new confirmed cases of COVID-19 patients in Ukraine by implementing the ARIMA model, based on the total confirmed daily cases declared by the Institute of Epidemiology, Disease Control, and Research. The forecasting indicates an ascendant trend for daily total confirmed cases. Meanwhile, total new death, and total confirmed cases are likely to become stable. The current data reveals that Ukraine is going to control the pandemic in the mid of April 2021 if the spreading pattern of the disease acts similarly. A study conducted for forecasting different countries' COVID-19 trend depicted stable conditions in China. Based on a study, for forecasting the end of COVID-19 by using the ARIMA model, it is expected that pioneer countries in COVID-19 confirmed cases will slow down by April 2021. Despite the ARIMA model's appropriate forecast, few drawbacks can be seen. Firstly, it has to be updated manually. Secondly, if more data is added to the model, the model will hand out different forecast results. Thirdly, the forecasting accuracy has a direct correlation with the observation numbers.

4 Conclusions

In this research, we showed the trend pattern of spreading of the COVID-19 in Ukraine. We realized that the ARIMA works properly as a model to forecast future cases of COVID-19 in Ukraine within 72 days. Using the ARIMA model, we could estimate the daily Confirmed cases reproduction number for the upcoming month. The total confirmed cases may exceed more than 4 million in March 2021. These findings potentially bestead in promoting policies to address the COVID-19 pandemic and put viable strategies to keep the pandemic through introducing new movement restrictions such as mass testing for detecting potential COVID-19 cases across the country. Eventually, this work has a lot of scope for future improvement in order to forecast the outbreak of the pandemic of COVID-19 more precisely by including the statistics for forecasting the daily Deaths and Recovery cases.

The disadvantage of the model is the impossibility of identifying factors influencing the development of the COVID-19 pandemic. But proposed model can be used as a verification tool to increase the accuracy of agent-based or compartment models of COVID-19.

Acknowledgement. The study was funded by the National Research Foundation of Ukraine in the framework of the research project 2020.02/0404 on the topic "Development of intelligent technologies for assessing the epidemic situation to support decision-making within the population biosafety management".

References

1. Rajnik, M., Cascella, M., Cuomo, A., Dulebohn, S.C., Di Napoli, R.: Features, Evaluation and Treatment Coronavirus (COVID-19). StatPearls Publishing (2021)
2. Meng, S., et al.: Epidemiology, causes, clinical manifestation and diagnosis, prevention and control of corona-virus disease (COVID-19) during the early outbreak period: a scoping review. Infect. Dis. Poverty **9**(1), 1–12 (2020)
3. Shih, T.P., Lai, C.C., Ko, W.C., Tang, H.J., Hsueh, P.R.: Severe acute respiratory syndrome coronavirus 2 and coronavirus disease-2019 (COVID-19): the epidemic and the challenges. Int. J. Antimicrob. Agents **55**(3), 105924 (2020)
4. Novel CPERE: The epidemiological characteristics of an outbreak of 2019 novel coronavirus diseases (COVID-19) in China. Z. Liuxingbingxue Zazhi **41**(2), 145 (2020)
5. Monllor, P., Su, Z., Gabrieli, L., Montoro, A., Taltavull de La Paz, M.: COVID-19infection process in Italy and Spain: are the data talking? Evidence from ARMA and vector autoregression models. Front Public Health **8**, 550602 (2020). https://doi.org/10.3389/fpubh.2020.550602
6. Dudkina, T., Meniailov, I., Bazilevych, K., Krivtsov, S., Tkachenko, A.: Classification and prediction of diabetes disease using decision tree method. In: CEUR Workshop Proceedings, vol. 2824, pp. 163–172 (2021)
7. Skitsan, O., Meniailov, I., Bazilevych, K., Padalko, H.: Evaluation of the informative features of cardiac studies diagnostic data using the Kullback method. In: CEUR Workshop Proceedings, vol. 2917, pp. 186–195 (2021)
8. Izonin, I., Tkachenko, R., Shakhovska, N., Lotoshynska, N.: The additive input-doubling method based on the SVR with nonlinear Kernels: small data approach. Symmetry **13**(4), 612 (2021). https://doi.org/10.3390/sym13040612
9. Gargin, V., Radutny, R., Titova, G., Bibik, D., Kirichenko, A., Bazhenov, O.: Application of the computer vision system for evaluation of pathomorphological images. In: 2020 IEEE 40th International Conference on Electronics and Nanotechnology, ELNANO 2020 - Proceedings, pp. 469–473 (2020). https://doi.org/10.1109/ELNANO50318.2020.9088898
10. Yakovlev, S.V.: The method of artificial space dilation in problems of optimal packing of geometric objects. Cybern. Syst. Anal. **53**(5), 725–731 (2017). https://doi.org/10.1007/s10559-017-9974-y
11. Davidich, N., et al.: Advanced traveller information systems to optimizing freight driver route selection. In: Proceedings - International Conference on Developments in eSystems Engineering (DeSE), pp. 111–115 (2020). https://doi.org/10.1109/DeSE51703.2020.9450763
12. Tkachenko, A.S., et al.: Semi-refined carrageenan promotes generation of reactive oxygen species in leukocytes of rats upon oral exposure but not in vitro. Wien. Med. Wochenschr. **171**(3–4), 68–78 (2020). https://doi.org/10.1007/s10354-020-00786-7
13. Meniailov, I., Krivtsov, S., Ugryumov, M., Bazilevich, K., Trofymova, I.: Application of parallel computing in robust optimization design. In: Nechyporuk, M., Pavlikov, V., Kritskiy, D. (eds.) Integrated Computer Technologies in Mechanical Engineering. AISC, vol. 1113, pp. 514–522. Springer, Cham (2020). https://doi.org/10.1007/978-3-030-37618-5_44

14. Yakovlev, S.: Convex extensions in combinatorial optimization and their applications. In: Butenko, S., Pardalos, P.M., Shylo, V. (eds.) Optimization Methods and Applications. SOIA, vol. 130, pp. 567–584. Springer, Cham (2017). https://doi.org/10.1007/978-3-319-68640-0_27
15. Sokoliuk, A., et al.: Machine learning algorithms for binary classification of liver disease. In: 2020 IEEE International Conference on Problems of Infocommunications Science and Technology, PIC S and T 2020 – Proceedings, pp. 417–421 (2021). https://doi.org/10.1109/PICST51311.2020.9468051
16. Davidich, N., et al.: Projecting of urban transport infrastructure considering the human factor. Commun. – Sci. Lett. Univ. Zilina **22**(1), 84–94 (2020)
17. Petropoulos, F., Makridakis, S.: Forecasting the novel coronavirus COVID-19. PLoS ONE **15**(3), e0231236 (2020)
18. Kumar, P., et al.: Forecasting the dynamics of COVID-19 pandemic in top 15 countries in April 2020: ARIMA model with machine learning approach. medRxiv (2020). https://doi.org/10.1101/2020.03.30.20046227
19. Yakovlev, S., et al.: The concept of developing a decision support system for the epidemic morbidity control. In: CEUR Workshop Proceedings, vol. 2753, pp. 265–274 (2020)
20. Alzahrani, S., Aljamaan, I., Al-Fakih, E.: Forecasting the spread of the COVID-19 pandemic in Saudi Arabia using ARIMA prediction model under current public health. J. Infect. Public Health **13**, 914–919 (2020)
21. Haines, L.M., Munoz, W.P., Van Gelderen, C.J.: ARIMA modelling of birth data. J. Appl. Stat. **16**, 55–67 (1989)
22. Choi, K., Thacker, S.B.: An evaluation of influenza mortality surveillance, 1962–1979: I. Time series forecasts of expected pneumonia and influenza deaths. Am. J. Epidemiol. **113**, 215–226 (1981)
23. Fedushko, S., Peráček, T., Syerov, Y., Trach, O.: Development of methods for the strategic management of web projects. Sustainability (Switzerland) **13**(2), 742 (2021)
24. Chen, P., Yuan, H., Shu, X.: Forecasting crime using the ARIMA model. In: Proceedings - 5th International Conference on Fuzzy Systems and Knowledge Discovery, FSKD 2008, vol. 5 (2008)
25. Hue, H.T.T., Pradit, S., Lim, A., Goncalo, C., Nitiratsuwan, T.: Shrimp and fish catch landing trends in Songkhla Lagoon, Thailand during 2003–2016. Appl. Ecol. Environ. Res. **16**, 3061–3078 (2018)

Takeoff and Landing Model of an Aircraft in the Aviation Simulator

Mariia Pyvovar(✉), Dmitriy Kritskiy, Tetiana Plastun, Vasylysa Kalashnikova, and Oleksii Popov

National Aerospace University Kharkiv Aviation Institute,
17, Chkalova Street, Kharkiv 61070, Ukraine

Abstract. In this paper, a mathematical model has been further developed, which makes it possible to estimate the required length of the runway, which is necessary for the landing and takeoff of the aircraft. This model has been implemented in an aviation simulator developed at the National Aerospace University Kharkiv Aviation Institute.

Keywords: Aircraft · Aviation simulator · Runway length

1 Introduction

Modern visualization systems are of two types: projection and collimation. In both types of imaging systems, the image is projected using projectors on spherical or cylindrical screens [1]. The projection of the image on the screens located in the immediate vicinity of the simulator cockpit leads to the fact that the line of sight of remote objects being projected depends on the position of the pilots' eyes [2].

The mobility system moves the simulator cockpit, which allows pilots to feel the created normal, longitudinal and lateral overload and angular acceleration on all three axes. Due to the limited running of the platform, simulation of overload is performed only for a short time, but this is considered sufficient, as the key information for the pilot is the change in congestion caused by management, and not the value of congestion [3].

During the development of the mathematical law of motion of the simulator platform on the simulated simulator, the equations of motion of the aircraft using the methods of harmonic analysis are decomposed into a series of harmonic oscillations-harmonics. The first harmonics are the harmonics of the lowest frequency that make the greatest contribution to the movement of the aircraft. At the same time the person is the least sensitive to these long-period fluctuations. If you slowly increase the overload to small values, the person in a sitting position may not even feel its change. Higher harmonics with increasing frequency make less and less contribution to the movement and they are increasingly sensitive to humans. Therefore, lower harmonics are suppressed using a high-pass filter.

In addition to short-term simulation of overload, there is also the possibility of long-term simulation of overload [4]. The simplest and most widely used method of simulating

M. Nechyporuk et al. (Eds.): ICTM 2021, LNNS 367, pp. 364–372, 2022.
https://doi.org/10.1007/978-3-030-94259-5_32

long-term overload is using the horizontal component of gravity to simulate longitudinal and lateral overload by appropriate tilt of the platform. In order to achieve this effect, during forming the law of motion of the platform, the equations of motion of the aircraft are passed through a low-pass filter, which suppresses the higher harmonics.

Another way to simulate long-term overload is installing a simulator cabin on a centrifuge. However, centrifuge simulators due to the high cost have not become widespread and are used only in developed countries for the training of fighter pilots and astronauts.

2 Main Part

To collect the entered data by the user, a developed input data input system is used, which consists of an Arduino Mega 2560 board and a set of encoders [5]. The entered data is interpreted using the Air Manager software [6], which stores a description of all the tools in the virtual panel and their data. After receiving a signal from the board to change the position of the encoder or press it, the system transmits the analog signal to digital and compares the ID of the physically engaged encoder with the virtual and determines which virtual instrument belongs to the encoder (Fig. 1).

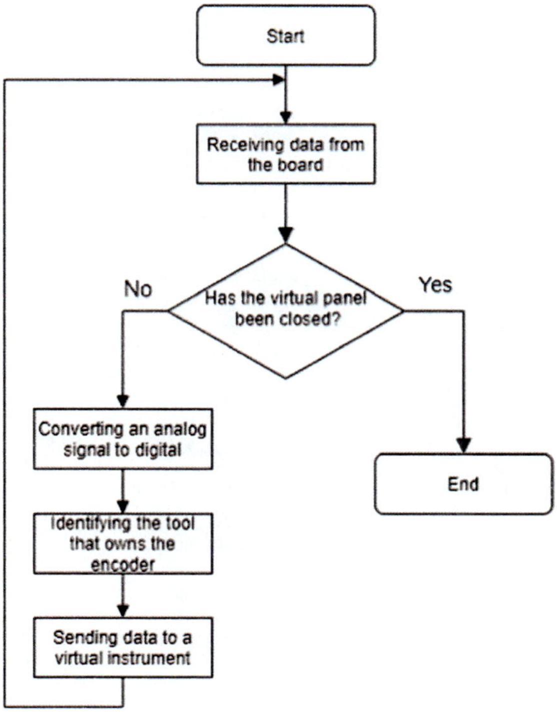

Fig. 1. Algorithm for collecting and interpreting input data

The request to the flight simulator is formed after connecting the virtual panel to it and initializing the data of the virtual instrument panel tools. Connection to the flight simulator and data initialization is performed once for the entire time of work with the panel. Based on the received data at the stage of collecting and interpreting the entered data, a request is formed to change the settings of the tool with which the user interacted (Fig. 2).

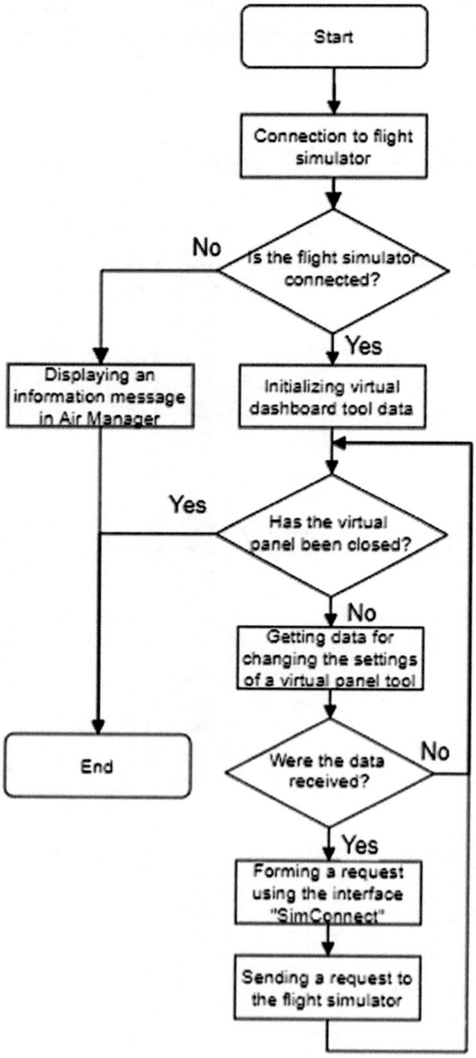

Fig. 2. Algorithm for generating a request to the flight simulator

The formation of indicators of the virtual panel occurs when working with the virtual panel, as well as when connected to the flight simulator. During the connection, the instrument indicators of the virtual instrument panel are obtained from the flight simulator in numerical format (speed, altitude, angle of rotation, etc.). The received data are formed and displayed in a user-friendly form (display of devices of the virtual panel of devices is close to real). The calculation of output data for special instruments of the dashboard is also performed.

By "special" tools are meant those tools that are not present on the instrument panel of a real aircraft, and are aimed at further informing the user. In this research work, a special tool called "brake path indicator" was developed.

The algorithm for forming the indicators of the virtual dashboard of the aircraft is shown in Fig. 3.

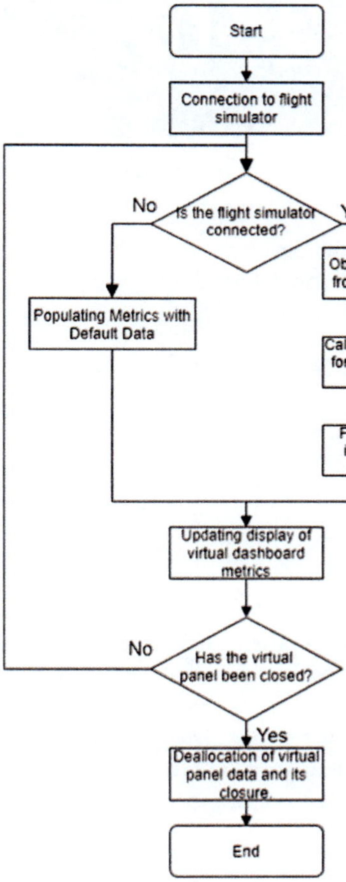

Fig. 3. Algorithm for forming indicators of the virtual panel

The safety of the entire flight depends significantly on the successful execution of takeoff and landing maneuvers, which are usually performed manually or under the control of the pilot. These modes are characterized by increased psychological stress on the pilot, which creates the preconditions for incorrect decision-making and further action.

Therefore, the creation of information support for the pilot, including the provision of objective assessments of the current or future situation, is certainly useful and relevant.

This study will use the method of predicting the braking distance on the run or in the event of an interrupted takeoff. The data found will be displayed on one of the sensors on the dashboard, which is called the brake path indicator (Fig. 4).

Fig. 4. Brake distance indicator

The methodological basis of the developments is the energy approach to aircraft motion control. The mathematical formulation of the approach is the equation of energy balance in the system of objects "LA - power plant - environment" [7] (formula 1).

$$\Delta H_E = \Delta H_E^{eng} + \Delta H_E^{D} + \Delta H_E^{w} \tag{1}$$

The extended equation of energy balance in the system of objects of units of specific energy in a compact form is shown in formula 2.

$$H_E* = E*/mg = h + \frac{V^2}{2g}, \tag{2}$$

where ΔH_E is the increase in energy height, ΔH_E^{eng} ΔH_E is the specific operation of the engine, ΔH_E^D is the energy consumption to overcome the force of drag, ΔH_E^D is the operation of the wind. Integral expressions are obtained for each member of the energy balance equation. The unit of specific energy is a meter, so it is also called energy height.

However, on the runway modes there are resistance forces from the chassis supports. Therefore, the energy balance equation (formula 3) must be generalized to ground modes by adding the term ΔH_E^{brakes}, which reflects the process of energy absorption to overcome the mechanical braking forces.

$$\Delta H_E = \Delta H_E^{eng} + \Delta H_E^{D} + \Delta H_E^{w} + \Delta H_E^{brakes} \tag{3}$$

This equation establishes quantitative relationships between the energy source and all its consumers. The equation is written in the form of increases in the specific energy of motion.

The new term ΔH_E^{brakes} is represented in formula 4.

$$\Delta H_E^{brakes} = \int_{t_1}^{t_2} V k_{brakes} dt, \tag{4}$$

where V is the speed of the aircraft, k_{brakes} is a generalized braking factor, which depends on the current coefficient of adhesion of the wheels with the runway coating and the degree of compression of the brake pads.

The predicted length of the braking distance is defined as the distance at which the energy (or speed) of motion will be repaid from the current to some small value of ε.

$$S_{pred} = S(t)|_{V(t) \leq \varepsilon} \tag{5}$$

Predicting the braking stage for the most general case, ie from the approach stage, we will take into account the flight altitude. The total mechanical energy of motion at the moment is given in formula 6.

$$E(t_0) = m\frac{V(t_0)^2}{2} + mgh(t_0) \tag{6}$$

The amount of energy in the process of motion is determined by the work of external forces on the trajectory S, given in formula 7.

$$E(t)_{pred} = m\frac{V(t_0)^2}{2} + mgh(t) + S_{pred}\sum_i F_i(t) \tag{7}$$

The number of external forces F_i includes: thrust of the engine P (including reversible), aerodynamic forces D, corresponding to the current configuration, and braking forces F_{brakes}. The latter depend on the coefficient of adhesion of the wheels to the coating, the release of brake pads and parachutes, as well as the energy consumption of the brake system.

Direct measurement or estimation of all operating forces on board is a virtually unsolvable task. The effect of the equal of all these forces is manifested in the form of acceleration. In the proposed method, the sum of the acting forces is naturally calculated through the longitudinal acceleration *a(t)* (formula 8), which on board is usually determined by the measured overload $n_x(t)$ (formula 9).

$$\sum_i F_i(t) = ma(t) \tag{8}$$

$$a(t) = gn_x(t) \tag{9}$$

Then, according to the stop criterion $V(t) \leq \varepsilon$, where ε is a small value which is close to zero, the predicted length of the braking distance will be found, which is given in formula 10.

$$D_{brakes} = S(t)_{pred} = 0.5\left(V(t)^2 - \varepsilon^2\right)\Big/ gn_x(t) \tag{10}$$

According to the calculated estimate of the maximum stopping distance of the aircraft, in the field of view of the pilot can be generated a display of data on the reserve of the distance of safe braking, the calculation of which is given in formula 11.

$$L_{reserv} = L_{runway} - x(t) - D_{brakes} \tag{11}$$

It is obvious that the informativeness and the possibility of using the forecast to increase situational informativeness is higher, the more reliable the forecast estimates of traffic parameters in relation to their real values.

The simplest and most obvious forecast of the braking distance is calculated during the movement of the aircraft on the runway, based on the current energy characteristics and forces.

All forces in the process of movement change either according to some program or due to unforeseen circumstances. Therefore, the forecast cannot coincide with the real process and always contains error or uncertainty. To increase the reliability of the forecast, this paper proposes a method of correction of forecasting algorithms.

This method assumes that the use of available technical means of braking (including the chassis brake system, traction reversal, release of interceptors) can be carried out not in accordance with regulations, but in fairly arbitrary scenarios on the run.

The correction parameters were determined on the basis of statistical processing of the results of model experiments. A large number of model experiments were performed on a variety of mileage conditions, which included the spread of landing weights, speeds and traction.

The strongest influence on the dynamics of the runway has the magnitude and direction of thrust of the engines. A typical sequence of engine modes was simulated: low gas, intermediate reverse, maximum reverse and discharge to low gas.

The purpose of the correction was to bring the calculated predicted distance closer to the end point of the run to the actual one.

The adjusted brake path length is given in formula 12.

$$D_{brakes\_cor} = Q_{cor} D_{brakes} \tag{12}$$

Formula 13 shows the correction factor.

$$Q_{cor} = Q_{cor}(J_i, k_{brakes}, V) \tag{13}$$

Where J_i is a sign of modes, V is the rolling speed.

It was found that the largest errors in the forecast occurred in areas with maximum reverse and released interceptors [8]. The identifiers of these modes are the logical variables J_{rev}, J_{int} and the correction factor also differs depending on the mode.

$$Q_{cor} = \begin{cases} Q_{rev}, \text{якщо } J_{rev} \geq 1 \\ Q_{int}, \text{якщо } J_{int} \geq 1 \end{cases} \tag{14}$$

At the beginning of the braking distance (on the reverse section) the greatest influence on the prediction errors have the coefficient of adhesion k_{adh} and rolling speed V. Thus, the coefficient of reversible thrust is given in formula 15.

$$Q_{rev} = k_{rev}(k_{adh}) \cdot k_{rev}(V) \tag{15}$$

According to the results of modeling on a set of discrete conditions, forecasting errors and corresponding scale correction factors were determined. It was found that the correction factors are weakly dependent on the mass of the aircraft [9]. Then, by the method of regression analysis for the discrete function $k_{rev}(k_{adh})$, a formulaic approximation in the form of polynomials of the 2nd, 3rd, and 4th degrees was found.

The next step in improving the plausibility of the forecast was the introduction of an additional speed correction factor in the forecasting algorithm [10]. From the many possible options, the relative scaling method was chosen. An empirical dependence on velocity was found and given in formula 16.

$$k_{rev}(V) = k_1(k_0 + (1 - k_0)V/V_{in}), \quad (16)$$

where V_{in} is the initial braking rate, k_1 is the coefficient that determines the total intensity of correction, k_0 is the coefficient that changes the degree and sign of correction as the aircraft moves along the runway. The adjustable coefficients k_0 and k_1 were determined by the search method on a set of flight conditions by the criterion of the minimum average error of the forecast on the reversible section.

3 Conclusion

The 4 variants were tested. For comparison, generalized forecasting errors were found within the reversible section and along the entire trajectory. Averaging was performed on the set of errors in planting simulations in all ranges of planting masses (from 70 to 105 tons) and adhesion coefficients (from 0.3 to 0.75).

Summary Table 1 of comparative estimates contains the results at two speeds is presented.

Table 1. Comparison of error estimates at two speeds

	Polynomial approximation	Errors on the reverse	Errors on the entire trajectory
V = 80 km/h	2nd degree	3.276711	17.3525
	4nd degree	2.928	16.849
V = 120 km/h	2nd degree	3.104756	15.97233
	4nd degree	4.312995	13.09849

Despite the differences in the approximating polynomials, the final errors differ by no more than 10%. Small values of forecasting errors indicate high correction efficiency.

References

1. Kritskiy, D., Karatanov, A., Koba, S., Druzhinin, E.: Increasing the reliability of drones due to the use of quaternions in motion. In: 2018 IEEE 9th International Conference on Dependable Systems, Services and Technologies (DESSERT), pp. 348–352 (2018). https://doi.org/10.1109/DESSERT.2018.8409157

2. CAE 7000 Series Level D full-flight simulator. https://web.archive.org/web/20100215135143/http://www.cae.com/en/sim.products/_pdf/datasheet.cae.7000.series.pdf. Accessed 15 Oct 2021
3. Kritsky, D.N., Druzhinin, E.A., Pogudina, O.K., Kritskaya, O.S.: A method for assessing the impact of technical risks on the aerospace product development projects. In: Shakhovska, N., Medykovskyy, M.O. (eds.) CSIT 2018. AISC, vol. 871, pp. 504–521. Springer, Cham (2019). https://doi.org/10.1007/978-3-030-01069-0_36
4. Kritskiy, D., Yashin, S., Koba, S.: Unmanned aerial vehicle mass model peculiarities. In: Shkarlet, S., Morozov, A., Palagin, A. (eds.) MODS 2020. AISC, vol. 1265, pp. 299–308. Springer, Cham (2021). https://doi.org/10.1007/978-3-030-58124-4_29
5. Air Manager by Sim Innovations. https://www.siminnovations.com/index.php?option=com_content&view=article&id=3. Accessed 15 Oct 2021
6. A Designer's Guide to Encoders. https://www.digikey.com/en/articles/a-designers-guide-to-encoders. Accessed 15 Oct 2021
7. Kurdjukov, A., Natchinkina, G., Shevtchenko, A.: Energy approach to flight control. In: AIAA Conference on Navigation, Guidance & Control, AAIA, pp. 543–553 (1998)
8. X-Plane 11. https://www.xplane.com/x-world/landing/. Accessed 8 Oct 2021
9. Kuznetsov, A.G., Shevchenko, A.M., Solonnikov, Yu.I.: The methods of forecasting some events during the aircraft takeoff and landing. In: Preprints of the 19th IFAC Symposium on Automatic Control in Aerospace (ACA-2013), Wuerzburg, Germany, pp. 183–187. Wuerzburg University, Würzburg (2013)
10. Shevchenko, A.M.: Some means for informational support of airliner pilot. http://lib.physcon.ru/doc?id=78f90e41e746/. Accessed 15 Oct 2021

Information Technology in the Creation of Rocket Space Systems

Bending-Twisting Deforming of Thin-Walled Composite Sections at Thermal-Mechanical Loading

Igor Taranenko[1], Anton Tsaritsynskyi[1], Tetyana Nabokina[1], and Andrii Kondratiev[2](✉)

[1] National Aerospace University "Kharkiv Aviation Institute", 17 Chkalova Street, Kharkiv 61070, Ukraine
[2] O.M. Beketov National University of Urban Economy in Kharkiv, 17 Marshal Bazhanov Street, Kharkiv 61002, Ukraine
andrii.kondratiev@kname.edu.ua

Abstract. Thin-walled composite sections with quite complicated open and closed cross-section are widely used nowadays. Internal stresses appeared at manufacturing of such sections can reach relatively high values and lead to complicated bending-twisting deformation of structural elements that has to be taken into consideration at the stage of design. The paper is devoted to studying of stress-strain state of long sections made of composites and having non-uniform cross-section. The theory of deformation of thin-walled sections with non-uniform properties over cross-section contour loaded with arbitrary system of thermal-mechanical loading based on generalized elasticity theory of thin-walled rods. System of resolving equations is obtained for determination of translations of composite contour arbitrary points at applying longitudinal stress only but distributed over countour length by arbitrary law. Quite high convergency of experimentally measured translations (difference is not more than 30%) with suggested analytical dependences allows to recommend developed approach for practical application. Conducted parametrical and experimental research allow to formulate practical recommendations for composing rational geometry and structure of sections with non-uniform parameters over contour and select proper parameters of manufacturing process. Reinforcing schemes and rational places of doublers installation are suggested, these can minimize section warping during its manufacturing.

Keywords: Residual deformations · Warping · Open and close contour

1 Introduction

Polymer composite materials (PCM) having high specific strength-to-weight and rigidity-to-weight ratio are widely used now in different branches of engineering [1, 2]. Generally monolayer of composite reinforced with system of fibers or fabric possess high physical-mechanical properties in-plane of it [3], therefore main composites advantages can be realized fully in such thin-walled structures like plates and shells. Nowadays there

M. Nechyporuk et al. (Eds.): ICTM 2021, LNNS 367, pp. 375–387, 2022.
https://doi.org/10.1007/978-3-030-94259-5_33

many papers published devoted to the theory, methods of analysis and design of plates and shells made of PCM [4, 5]. Last years composites are implemented for structural applications for which analytical model of thin-walled rods is very suitable. Analytical model of thin-walled rod allows to analyse such existing and prospective structures as composite blades of airplane propeller, blades of helicopter main rotor, blades of wind energy generator, composite driving shafts, elements of robots and manipulators actuators [6]. Also, this model can be used with definite degree of approximation for analysis of aircraft fuselage and high aspect ratio wing, PCM are used widely for production of above-mentioned structures [7]. Conventional methods of building mechanics include analysis of truss and stiffenered structures [8, 9], but nowadays thin-walled composite structures with quite complicated open and closed sections are used in national economy. For analysis and design of such sectional one has to generalize conventional theory of thin-walled structures allowing to take into consideration anisotropic and laminated structure of PCM.

2 Literature Review

Principal difference of theory thin-walled sections made of composites from conventional one developed for isotropic sections relates to anisotropy of PCM which defines relationships between axial deformation, section twisting angle and rod longitudinal sagging, i.e. displacements which can't be eliminated by selection of specific coordinate system. Application of more precise analytical models, for example, based on iteration method is spread at determination of stress-strain state of thin-walled sections made of composites. Thus, the solution of problem of elastic deforming of multi-layer cylindrical rod with orthotropic layers is shown in paper [10]. To describe structural arrangement of the rods cylindrical coordinate system is used. And mechanical properties of PCM of rod are functions of one variable only. The problem is solved by means of direct integration of total system of elasticity theory equations in selected coordinate system at condition of absolutely rigid contact of layers boundaries. Paper [11] contains iteration analytical method of the problem solving for clamped torsion of anisotropic rod with open section taking into consideration shear, lateral deformations and influence of reciprocal coefficients of linear deformation to shear one. It is shown that valid result can't be obtained at the first cycle of analysis. Iteration methods are quite universal but very bulky and complicated for practical application at higher stages of precise analysis. But at the same time precise solution for multi-layer thin-walled structures, for example [12], are restricted in term of possibility of taking into consideration different types of loads and boundary conditions. However, one can get quite precise applied solutions of problems based on such methods. Thus, paper [13] shows exact direct solution of a problem of laminated thin-walled rod bending loaded with distributed load at absolutely rigid contact between monolayers. Piecewise constant functions of elastic characteristics are used in Hook's relationships to consider non-uniform arrangement of laminated rod section. Two different approaches for determination of deformations of thin-walled rods cross-section, i.e. the method of generalized eugen vectors and generalized theory of beams, are compared in the paper [14]. Their differences, similarities and solutions obtained are estimated. The following conclusion is done – mentioned methods give precise results although they use different structural models and strategies of solution. Paper [15] suggests novel model of considering local bending of thin-walled sections. The model is based on energy method. Paper [16] analyzes asymptotic models based on neglection of relatively small members, which introduce poor portion of energy to total energy of deformation of anisotropic section with open section. Paper [17] deals with calculation of stresses in thin-walled box-shape composite sections, having variable rigidity; stress analysis is conducted based on modified analytical model. This model takes into consideration bending-twisting interaction and warping effects. In paper [18] authors suggested to use mechanics of genom structure to develop asymptotically correct analytical model of thin-walled rods with different cross-sections without application of any auxiliary conditions. Nowadays determination of stress-strain state of structural elements is conducted mostly by means of specialized software based on finite elements methods [19]. Several numerical experiments with different finite elements were conducted in the paper [20]. Polynoms of n-degree were used for rigidity matrix composing in this paper, moreover value of n was assumed to be arbitrary parameter. Paper [21] shows developed model taking into consideration shear of multi-layer composite thin-walled rods with openings. Generalized approach of the problem solving is based on finite elements method but rigidity matrix is obtained by means of virtual displacements principle. Paper [22] pays thorough attention to local deformations and to distortion of thin-walled rod cross-section. Since theory of thin-walled composite sections analysis appeared recently its realization in computational complexes stays at low level yet. Paper [23] considered method of modelling of complicated behavior of thin-walled composite rods with open and closed cross-sections. This method of modelling is more effective from the point of view of computation comparing with detailed modelling by finite elements method. In this method the total problem of 3D-analysis is splitted on 2D problem applied to rod cross-section and 1D one applied to rod length. Disadvantage of such approach of modelling is high cost of computation. Experimental studying of carbon plastic thin-walled rod at combined loading of bending and torsion is conducted in paper [24]. Experimental results have shown that elements demonstrate high ability to dissipate energy at pure torsion and compressive-bending-torsional loads but at the same time applied axial force increases load-carrying ability and rigidity. Disadvantages of such kind of experimental research is application of quite expensive materials and energy-consumptive equipment. Besides, realization of all theoretically possible structural packages of anisotropic PCM is practically impossible [25]. Application of special technologies for manufacturing of structural elements made of PCM is frequently related to appearing inside of them auxiliary stress-strain state due to influence of temperature or other factors [26, 27]. Papers [28, 29] suggest mathematical models for for studying of stress-strain state of structurally anisotropic panels made of PCM. Influence of following factors of panels manufacturing technology is considered: residual thermal stress and preliminary stretching of reinforced fibers.

Based on above-mentioned analysis one can make following conclusion – development of applied theory of thin-walled composite rods is quite actual and important problem in applied aspect.

3 Research Methodology

Let's consider thin-walled cylindrical rod with open cross-section loaded with axial forces (normal P and shear T), transversal bending moment M and torsional moment H (Fig. 1).

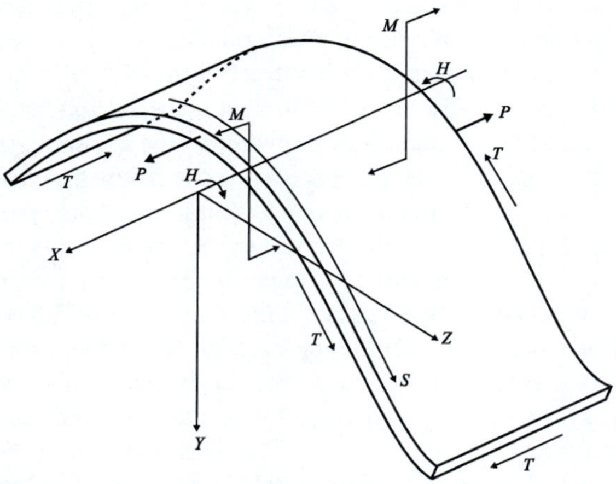

Fig. 1. Generalized loading of thin-walled rod

Analysis scheme for estimation of stress-strain state of generalized rod with non-uniform section can be composed on the bases of methodology [30]. Deformation laws for such body are grounded on following hypothesis: shear deformation of mid-surface is equal to zero and countour of rod cross-section is undeformable, i.e. the distance between two any points of cross-section at transition to deformed state stays the same.

At derivation of rod deforming equations one can use generalized formula for displacement of contour arbitrary point (Fig. 2):

$$u(x,s) = \zeta(x) - \xi^{'}(x)z(s) - \eta^{'}(x)y(s) - \theta^{'}(x)\omega(s), \quad (1)$$

where $\zeta(x)$, $\xi(x)$, $\eta(x)$, $\theta(x)$ – longitudinal, lateral and transversal (along axes z and y correspondingly) linear displacements and twisting angle of entire section; $\omega(s)$ – sectorial area of a point.

To determine functions $\zeta(x)$, $\xi(x)$, $\eta(x)$, $\theta(x)$ one has to add physical relationships and equation of rod equilibrium. For orthotropic material in selected coordinate system physical law has following view

$$\varepsilon_x = \frac{\sigma_x}{E_x} - \mu_{zx}\frac{\sigma_z}{E_z} + \alpha_x\Delta T; \quad \varepsilon_z = \frac{\sigma_z}{E_z} - \mu_{xz}\frac{\sigma_x}{E_x} + \alpha_z\Delta T, \quad (2)$$

where $\varepsilon_x, \varepsilon_z, \sigma_x, \sigma_z$ - strains and stresses of a rod; $E_x, E_z, \mu_{xz}, \mu_{zx}, \alpha_x, \alpha_z$ – elasticity moduli, Poisson's ratios and thermal linear expansion coefficients of rod material; ΔT – temperature difference.

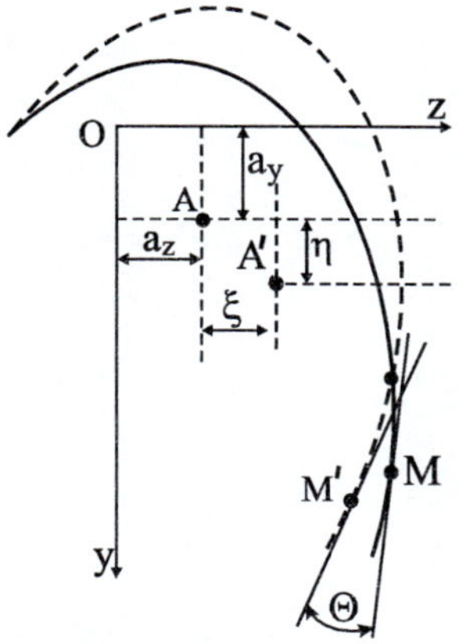

Fig. 2. Deformation of rod cross-section

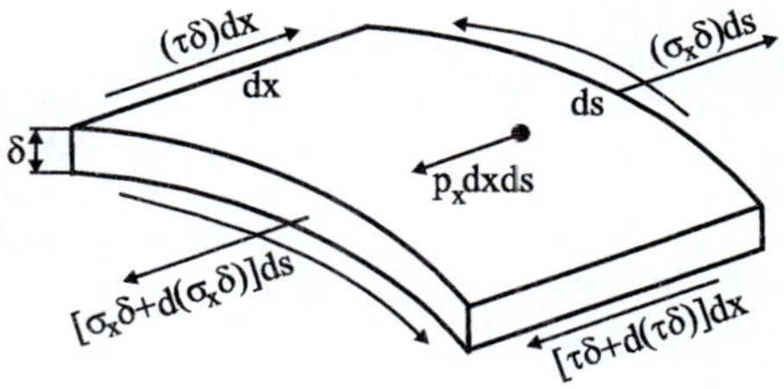

Fig. 3. Equilibrium of elementary representative element of a rod

Dependence of shear stress function one can get considering deformation conditions of elementary representative element of rod (Fig. 3).

If one assumes that rod thickness δ doesn't depend on x then we can get:

$$\tau(x,s) = \frac{1}{\delta(s)}\left[S_0(x) - \int_0^s p_x ds - \int_0^s \frac{\partial \sigma}{\partial x}\delta(s)ds\right] \tag{3}$$

where p_x – projection of external superficial loading intensity, $S_0(x)$ – arbitrary function of integration; $\delta(s)$ – section thickness function.

After inserting of deformation equation $\varepsilon(x,s) = \frac{\partial u(x,s)}{\partial x}$ to physical law (2) one can get formula for normal stress determination

$$\sigma_x(x,s) = \overline{E}_x(s)\left[\zeta'(x) - \xi''(x)z(s) - \eta''(x)y(s) - \theta''(x)\omega(s)\right], \tag{4}$$

where $\overline{E}_x = \frac{E_x}{1-\mu_{zx}\mu_{xz}}$ is reduced elasticity modulus.

After insertion of formula (4) to dependence (3) one can get function for shear stress determination

$$\tau(x,s) = \frac{1}{\delta(s)}\left[S_0(x) - \int_0^s p_x(x,s)ds - \zeta''(x)\int_0^s \overline{E}_x(s)\delta ds + \xi'''(x)\int_0^s \overline{E}_x(s)z(s)\delta ds \right.$$
$$\left. + \eta'''(x)\int_0^s \overline{E}_x(s)y(s)\delta ds + \theta'''(x)\int_0^s \overline{E}_x(s)\omega(s)\delta ds\right]. \tag{5}$$

Let's consider condition of rod static equilibrium. For this one can select elementary representative strip with width dx laying between sections with coordinates $x = const$ and $x + dx = const$ (Fig. 4).

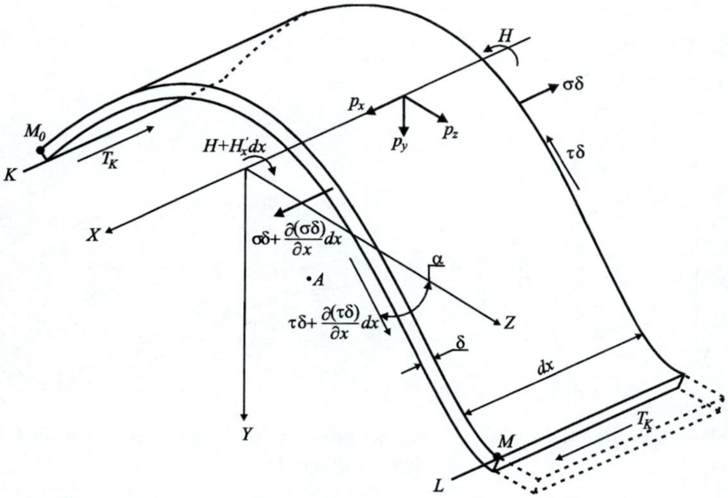

Fig. 4. To derivation of rod equilibrium equations

Let's write condition of equalto to zero of all applied forces projections to axes x, y, z and equations of all applied moments projections around arbitrary point A:

$$\begin{aligned}
&\sum F_x = 0; \int\limits_S \frac{\partial(\sigma\delta)}{\partial x} dxds + (T_L - T_K + p_x)dx = 0; \\
&\sum F_y = 0; \int\limits_S \frac{\partial(\tau\delta)}{\partial x} dxds \sin\alpha + p_y dx = 0; \\
&\sum F_z = 0; \int\limits_S \frac{\partial(\tau\delta)}{\partial x} dxds \cos\alpha + p_z dx = 0; \\
&\sum M_A = 0; \int\limits_S \frac{\partial(\tau\delta)}{\partial x} dx[(z - a_z)\sin\alpha - (y - a_y)\cos\alpha]ds + H'_x dx + mdx = 0.
\end{aligned} \tag{6}$$

Following designations are used in these equations: p_x, p_z, p_y – intensity of external loading per unit length along axes x, z and y correspondingly; $T_L(x)$, $T_K(x)$ – external shear forces applied to rod edges; H – external torsion moment per unit length; H'_x – derivative from torsion moment at its changing along rod longitudinal asis; m – total moment caused by external lateral loading.

After insertion of dependences (4) and (5) for normal and shear stress to formulas (6) and integration from the point L to the point K (Fig. 4) taking into account boundary conditions one can get the system of general equations that describe stress-strain state of rod loaded with arbitrary system of loads:

$$\begin{aligned}
&\zeta''(x)\int_F \overline{E}_x(s)dF-\xi'''(x)\int_F \overline{E}_x(s)z(s)dF-\eta'''(x)\int_F \overline{E}_x(s)y(s)dF\\
&-\theta'''(x)\int_F \overline{E}_x(s)\omega(s)dF+p_x+T_L-T_K=0;\\
&\int_F \frac{\partial p_x}{\partial x}z(s)ds+\zeta'''(x)\int_F \overline{E}_x(s)z(s)dF-\xi^{IV}(x)\int_F \overline{E}_x(s)z^2(s)dF\\
&-\eta^{IV}(x)\int_F \overline{E}_x(s)y(s)z(s)dF\\
&-\theta^{IV}(x)\int_F \overline{E}_x(s)\omega(s)z(s)dF+p_z+T_L'(x)z_L-T_K'(x)z_K=0;\\
&\int_F \frac{\partial p_x}{\partial x}y(s)ds+\zeta'''(x)\int_F \overline{E}_x(s)y(s)dF-\xi^{IV}(x)\int_F \overline{E}_x(s)z(s)y(s)dF\\
&-\eta^{IV}(x)\int_F \overline{E}_x(s)y^2(s)dF\\
&-\theta^{IV}(x)\int_F \overline{E}_x(s)\omega(s)y(s)dF+p_y+T_L'(x)y_L-T_K'(x)y_K=0;\\
&\int_F \frac{\partial p_x}{\partial x}\omega(s)ds+\zeta'''(x)\int_F \overline{E}_x(s)\omega(s)dF-\xi^{IV}(x)\int_F \overline{E}_x(s)z(s)\omega(s)dF\\
&-\eta^{IV}(x)\int_F \overline{E}_x(s)y(s)\omega(s)dF-\theta^{IV}(x)\int_F \overline{E}_x(s)\omega^2(s)dF\\
&+\frac{\alpha}{3}\theta''(x)\int_F G(s)\delta^3(s)sdF+m+T_L'(x)\omega_L-T_K'(x)\omega_K=0,
\end{aligned}\tag{7}$$

where z_K, y_K, z_L, y_L – coordinates of original and final points of section contour curve; ω_K, ω_L – sectorial areas for these points; G, δ, s, θ' – shear modulus, thickness, length and derivative by twisting angle of i-th section of contour; α_S – empirical coefficient depending on shape of rod cross-section; $dF = \delta ds$.

Angle section with rigidity doubler on one of angle section cap (Fig. 5, a) and channel section with rigidity doublers on both caps (Fig. 5, b) were selected for experimental studying of combined thin-walled composite rods deformed state.

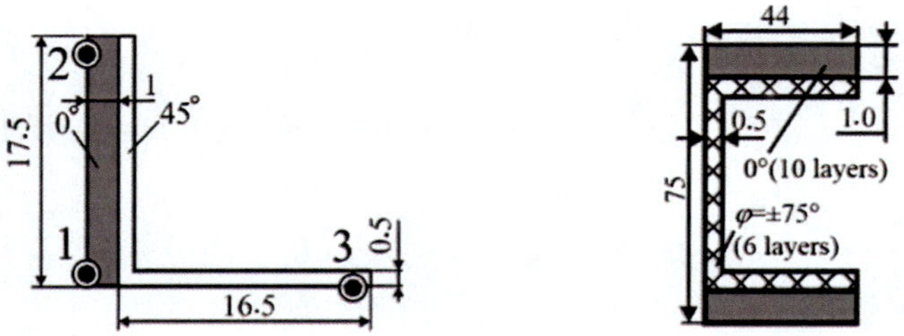

(a) (b)

Fig. 5. Arrangement of combined composite sections used for experimental testing: a – angle section; b –channel section

Thin-walled sections were produced of carbon plastic by autoclave-vacuum forming based on unidirectional carbon tape ELUR-0.08 and binder EDT 69N with characteristics: elasticity modulus in longitudinal direction $E_1 = 120$ GPa and in lateral direction $E_2 = 10$ GPa; shear modulus $G_{12} = 6$ GPa; Poisson's ratio $\mu_{12} = 0.29$; thermal linear expansion coefficient in longitudinal direction $\alpha_1 = (-1.5)\cdot 10^{-6}\cdot 1/K$ and in lateral direction – $\alpha_2 = (28.0)\cdot 10^{-6}\cdot 1/K$.

Produced sections after extraction from jig are shown on the Fig. 6.

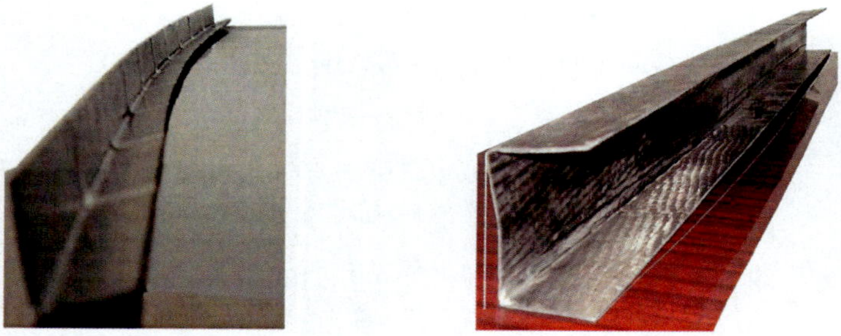

(a) (b)

Fig. 6. Angle section (a) and channel section (b) after extraction from jig

Horizontal translations ξ were measured with dial indicator, vertical translations η – with dial indicator and vernier height gauge.

4 Results

Results of theoretical (predicted) and practical (obtained experimentally) deformation components of contour were compared and analyzed.

Thus, for reference points 1, 2, 3 of angle section (Fig. 5, a) components of total displacements (η, ξ) were calculated and their correspondently deviations in space (η*, ξ*) were measured (Table 1).

Table 1. Displacements of reference points of angle section in space

x, m	0	0.1	0.2	0.3	0.4	0.5	0.6	0.7	0.8	0.9
Point 1										
η, mm	0	0.093	0.360	0.801	1.418	2.210	3.178	4.321	5.647	7.134
η^*, mm	0	0.120	0.280	0.650	1.250	1.980	2.830	3.950	5.820	7.500
Δ,%	0	22.5	28.5	23.2	13.4	11.6	12.3	9.4	3.0	4.9
ξ, mm	0	0.225	0.900	2.025	3.600	5.625	8.100	11.025	14.40	18.225
ξ^*, mm	0	0.180	0.670	1.830	3.120	4.900	8.270	11.800	16.160	19.730
Δ,%	0	25.0	34.3	10.7	15.4	14.8	2.1	6.6	10.9	7.6
Point 2										
η, mm	0	0.088	0.352	0.792	1.408	2.200	3.168	4.312	5.639	7.128
η^*, mm	0	0.110	0.280	0.630	1.280	1.970	3.300	4.520	5.940	6.990
Δ,%	0	20.0	25.7	25.7	10.0	11.7	4.0	4.6	5.1	2.0
ξ, mm	0	0.462	1.233	2.397	3.987	6.025	8.473	11.398	14.733	18.462
ξ^*, mm	0	0.380	0.970	1.850	3.690	5.940	8.100	10.870	15.230	19.200
Δ, %	0	21.6	27.1	29.6	8.0	1.4	4.6	4.9	3.3	3.8
Point 3										
η, mm	0	0.302	0.653	1.128	1.758	2.561	3.518	4.648	5.940	7.342
η^*, mm	0	0.350	0.780	1.300	1.900	2.800	3.500	4.320	5.650	6.950
Δ, %	0	13.7	16.3	13.2	7.5	8.5	0.5	7.6	5.1	5.6
ξ, mm	0	0.225	0.900	2.025	3.600	5.625	8.100	11.025	14.40	18.225
ξ^*, mm	0	0.200	0.800	1.900	3.300	5.100	8.100	11.900	15.300	19.600
Δ, %	0	12.5	12.5	6.5	9.1	10.3	0	7.4	5.9	7.0

Confidence interval of experimental values was analyzed to estimate confidence of practically measured linear displacements of section points in horizontal and vertical planes. To measure deviation of each reference point of section its displacement in space was measured five times and mathematical expectation and dispersion of a value was found. Some of experimental results are shown on the Figs. 7 and 8.

Analysis of obtained results has shown that results of linear translations prediction coincide well with experimental values. For majority of measurements the difference doesn't exceed 15...25%.

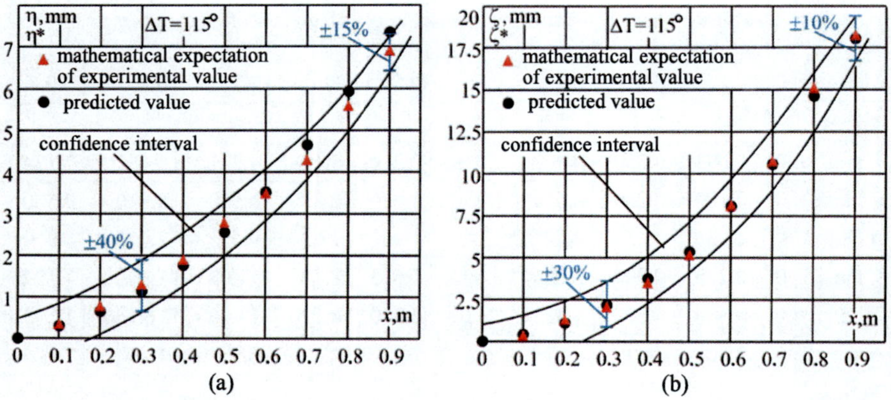

Fig. 7. Results of comparison of linear translations of reference points of angle section and confidence interval for experimental values: a – vertical translation, b – horizontal translation

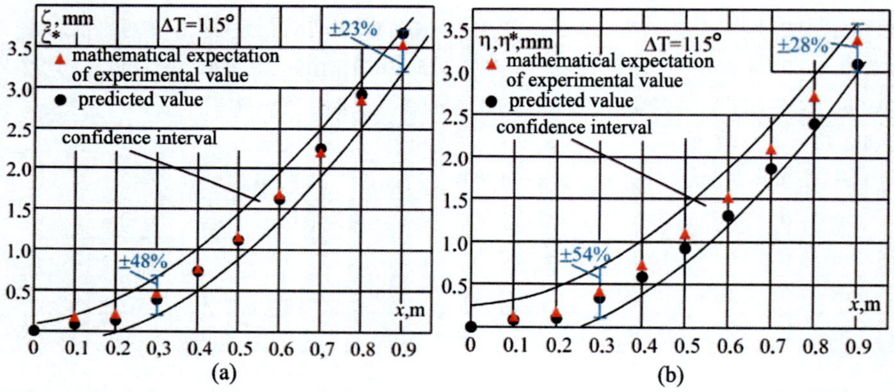

Fig. 8. Results of comparison of linear translation of channel section reference points and confidence interval for experimental values: a – vertical translation, b – horizontal translation

5 Conclusions

Following conclusions can be done based on conducted parametrical studies and comparison of obtained experimental data with results of analytical estimation of stress-strain state parameters of thin-walled combines rods made of PCM.

Generalized theory for determination of stress-strain state of thin-walled anisotropic rods at arbitrary external loading. System of resolving equations, permitting to determine displacements in space of arbitrary point of rod section countour, is obtained.

Series of typical composite sections are produced and correspondent measurement of such warping components as linear displacement in space with respect to selected coordinate system was done. Set of experimental studies allowing to check and prove of confidence of developed theory for estimation of stress-strain state is conducted.

Analysis of scientific papers devoted to studying of PCM behaviour allows quite reliably to claim that occurring internal thermal or shrinkage stress-strain state is accompanied with time by very complicated by nature relaxation processes. Repeated measurements of reference points displacements of produced sections were conducted in three months to analyze practically such phenomenon. Deviation of contour reference points up to 10…15% was found. Thus, this phenomenon requires auxiliary studying.

References

1. Hsissou, R., Seghiri, R., Benzekri, Z., Hilali, M., Rafik, M., Elharfi, A.: Polymer composite materials: a comprehensive review. Compos. Struct. **262**, 15 (2021). https://doi.org/10.1016/j.compstruct.2021.113640
2. Fomin, O., Lovska, A.: Improvements in passenger car body for higher stability of train ferry. Eng. Sci. Technol. Int. J. Jestech **23**(6), 1455–1465 (2020). https://doi.org/10.1016/j.jestch.2020.08.010
3. Kondratiev, A., Gaidachuk, V., Nabokina, T., Kovalenko, V.: Determination of the influence of deflections in the thickness of a composite material on its physical and mechanical properties with a local damage to its wholeness. East. Eur. J. Enterprise Technol. **4**(1–100), 6–13 (2019). https://doi.org/10.15587/1729-4061.2019.174025
4. Ugrimov, S.V., Shupikov, A.N.: Layered orthotropic plates. Generalized theory. Compos. Struct. **129**, 224–235 (2015). https://doi.org/10.1016/j.compstruct.2015.04.004
5. Kondratiev, A.: Improving the mass efficiency of a composite launch vehicle head fairing with a sandwich structure. East. Eur. J. Enterprise Technol. **6**(7–102), 6–18 (2019). https://doi.org/10.15587/1729-4061.2019.184551
6. Bao, Y., Wang, B., He, Z., Kang, R., Guo, J.: Recent progress in flexible supporting technology for aerospace thin-walled parts: a review. Chin. J. Aeronaut. **35**, 10–26 (2021). https://doi.org/10.1016/j.cja.2021.01.026
7. Gavva, L.M., Firsanov, V.V.: Analytical review of account methods and experimental approaches to stress-strain state investigation of structurally-anisotropic aircraft panels made from composite materials. In: IOP Conference Series: Materials Science and Engineering, vol. 927 (2020). https://doi.org/10.1088/1757-899x/927/1/012067
8. Smetankina, N., Merkulova, A., Merkulov, D., Postnyi, O.: Dynamic response of laminate composite shells with complex shape under low-velocity impact. In: Nechyporuk, M., Pavlikov, V., Kritskiy, D. (eds.) ICTM 2020. LNNS, vol. 188, pp. 267–276. Springer, Cham (2021). https://doi.org/10.1007/978-3-030-66717-7_22
9. Kovalov, A., Otrosh, Y., Rybka, E., Kovalevska, T., Togobytska, V., Rolin, I.: Treatment of determination method for strength characteristics of reinforcing steel by using thread cutting method after temperature influence. Mater. Sci. Forum **1006**, 179–184 (2020). https://doi.org/10.4028/www.scientific.net/msf.1006.179
10. Koval'chuk, S.B., Gorik, A.V., Pavlikov, A.N., Antonets, A.V.: Solution to the task of elastic axial compression–tension of the composite multilayered cylindrical beam. Strength Mater. **51**(2), 240–251 (2019). https://doi.org/10.1007/s11223-019-00070-z
11. Polinkevich, K.Y.: Opredelenie napryazhenno-deformirovannogo sostoyaniya tonkostennyh anizotropnyh sterzhnej otkrytogo profilya pri kruchenii. The thesis for candidate degree, Saint-Peterburg, p. 160 (2019)
12. Gorik, A.V., Koval'chuk, S.B.: Solution of a transverse plane bending problem of a laminated cantilever beam under the action of a normal uniform load. Strength Mater. **50**(3), 406–418 (2018). https://doi.org/10.1007/s11223-018-9984-7

13. Goryk, A.V., Koval'chuk, S.B.: Elasticity theory solution of the problem on plane bending of a narrow layered cantilever beam by loads at its free end. Mech. Compos. Mater. **54**(2), 179–190 (2018). https://doi.org/10.1007/s11029-018-9730-z
14. Garcea, G., et al.: Deformation modes of thin-walled members: a comparison between the method of generalized eigenvectors and generalized beam theory. Thin-Walled Struct. **100**, 192–212 (2016). https://doi.org/10.1016/j.tws.2015.11.013
15. Thai, H.T., Uy, B., Khan, M.: A modified stress-strain model accounting for the local buckling of thin-walled stub columns under axial compression. J. Constr. Steel Res. **111**, 57–69 (2015). https://doi.org/10.1016/j.jcsr.2015.04.002
16. Harursampath, D., Harish, A.B., Hodges, D.H.: Model reduction in thin-walled open-section composite beams using variational asymptotic method. Part I Theory Thin-Walled Struct. **117**, 356–366 (2017). https://doi.org/10.1016/j.tws.2017.03.018
17. Gunay, M.G., Timarci, T.: Stresses in thin-walled composite laminated box-beams with curvilinear fibers: antisymmetric and symmetric fiber paths. Thin-Walled Struct. **138**, 170–182 (2019). https://doi.org/10.1016/j.tws.2019.02.003
18. Deo, A., Yu, W.B.: Thin-walled composite beam cross-sectional analysis using the mechanics of structure genome. Thin-Walled Struct. **152**, 10 (2020). https://doi.org/10.1016/j.tws.2020.106663
19. Orgiyan, A., Kobelev, V., Ivanov, V., Balaniuk, A., Aymen, A.: Ensuring the bending stiffness of pre-compressed cantilever boring bars during fine boring. In: Ivanov, V., Trojanowska, J., Pavlenko, I., Zajac, J., Peraković, D. (eds.) DSMIE 2020. LNME, pp. 315–324. Springer, Cham (2020). https://doi.org/10.1007/978-3-030-50794-7_31
20. Choi, I.S., Jang, G.W., Choi, S., Shin, D., Kim, Y.Y.: Higher order analysis of thin-walled beams with axially varying quadrilateral cross sections. Comput. Struct. **179**, 127–139 (2017). https://doi.org/10.1016/j.compstruc.2016.10.025
21. Vojnic-Purcar, M., Prokic, A., Besevic, M.: A numerical model for laminated composite thin-walled members with openings considering shear lag effect. Eng. Struct. **185**, 392–399 (2019). https://doi.org/10.1016/j.engstruct.2019.01.142
22. Carrera, E., Kaleel, I., Petrolo, M.: Elastoplastic analysis of compact and thin-walled structures using classical and refined beam finite element models. Mech. Adv. Mater. Struct. **26**(3), 274–286 (2019). https://doi.org/10.1080/15376494.2017.1378780
23. Islam, A., Sheikh, A.H., Bennett, T., Thomsen, O.T.: An efficient model for laminated composite thin-walled beams of open or closed cross-section and with or without in-filled materials. Compos. Struct. **256**, 16 (2021). https://doi.org/10.1016/j.compstruct.2020.112998
24. Wang, Y.H., et al.: Combined compression-bending-torsion behaviour of CFST columns confined by CFRP for marine structures. Compos. Struct. **242**, 19 (2020). https://doi.org/10.1016/j.compstruct.2020.112181
25. Rimasauskiene, R., Jurenas, V., Radzienski, M., Rimasauskas, M., Ostachowicz, W.: Experimental analysis of active-passive vibration control on thin-walled composite beam. Compos. Struct. **223**, 8 (2019). https://doi.org/10.1016/j.compstruct.2019.110975
26. Baran, I., Cinar, K., Ersoy, N., Akkerman, R., Hattel, J.H.: A review on the mechanical modeling of composite manufacturing processes. Arch. Comput. Methods Eng. **24**(2), 365–395 (2016). https://doi.org/10.1007/s11831-016-9167-2
27. Kondratiev, A., Slivinsky, M.: Method for determining the thickness of a binder layer at its nonuniform mass transfer inside the channel of a honeycomb filler made from polymeric paper. East. Eur. J. Enterprise Technol. **6**(5–96), 42–75 (2018). https://doi.org/10.15587/1729-4061.2018.150387
28. Gavva, L.M., Firsanov, V.V.: Mathematical models and methods for calculating the stress-strain state of aircraft panels from composite materials taking into account the production technology. Mech. Solids **55**(3), 403–412 (2020). https://doi.org/10.3103/s002565442003005x

29. Boitsov, B.V., Gavva, L.M., Pugachev, Y.N.: The stress-strain state of structurally anisotropic panels from composite materials under force and process temperature exposure. Polym. Sci. Ser. D **12**(1), 85–90 (2019). https://doi.org/10.1134/S1995421219010039
30. Taranenko, I.M.: The method of analysis thermal stressed state of skin panel stringers made of composites. The thesis for candidate degree of engineering science, Kharkiv, p. 152 (2010)

Information Technology in the Design and Manufacture of Engines

The Application Features of Integrated Technologies for the Design of Elements of Systems with Gas-Turbine Aircraft Engines

Lyudmyla Rozova(✉) and Gennadii Martynenko

Department of Dynamics and Strength of Machines, National Technical University "Kharkiv Polytechnic Institute", NTU "KhPI", 2, Kyrpychova Street, Kharkiv 61002, Ukraine
luda.rozova@gmail.com, gmartynenko@ukr.net

Abstract. The work considers the issues of using integrated approach for modeling and solving engineering objects. It consists in the combined use of existing finite element software packages for general engineering solutions and specialized self-developed programs. Such integrated approach provides the possibility of obtaining a joint solution of problems of modeling the elements of gas turbine aircraft engines and allows taking into account the influence of various factors, and ensuring the reliable operation of the structure as a whole.

The paper presents the created iterative algorithm for the improved solution of the dry gas seals, used as end seals in the gas turbine engine-compressor system. This algorithm includes the joint solution of interrelated gasdynamics, heat transfer and thermoelasticity problems for dry gas seals. For the implementation of this iterative solution algorithm an integrated approach for solving individual problems of a complete solution was proposed, which includes the use of specially developed software package GasDin as well as the universal finite element software package. All stages of the iterative process and data exchange are controlled using the created software package.

Keywords: Integrated technologies · Engine-compressor system · Dry gas seal

1 Introduction

The relevance of information integrated technologies in the engineering industry has become more and more widespread in recent decades. On the one hand, this is due to sharp increase of the capabilities of computing technology, the appearance of specialized software systems for the simulation of various physical processes. The presence of special programming environments and their constant updating and development, allows implementing non-standard solutions to specialized tasks for improved modeling and solution [1, 2]. On the other hand, the high cost and shortage of natural resources in most cases reduced or excluded the possibility of carrying out the necessary amount of experimental works for obtaining various standard sizes of structures and their studies.

One of the promising directions is the use of integrated information technologies in the design and manufacture of aircraft engines. Design accuracy, which ensures the

M. Nechyporuk et al. (Eds.): ICTM 2021, LNNS 367, pp. 391–402, 2022.
https://doi.org/10.1007/978-3-030-94259-5_34

failure-free operation of the engine, is very important here. The aircraft gas turbine engines have found their application not only in aviation industry. Gas turbine engines are installed in gas pumping units in engine-compressor systems to maintain the required gas pressure in the main gas pipelines (Fig. 1) [3–6].

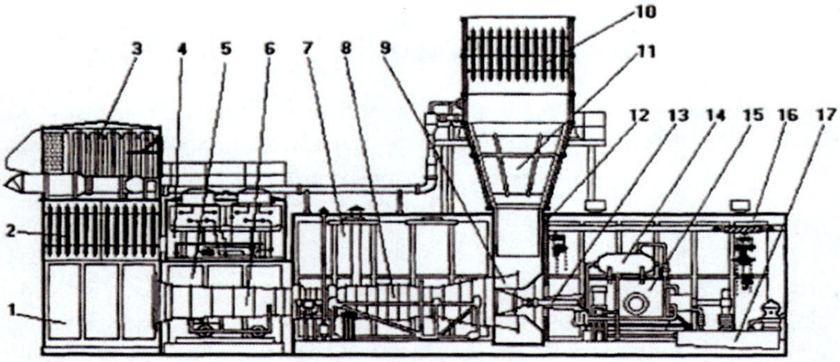

Fig. 1. Gas-pumping unit configuration: 1 – suction chamber; 2 – acoustic filter; 3 – air-cleaning facility; 4 – ventilation unit; 5 – intermediate unit; 6 – duct; 7 – engine section; 8 –gas-turbine engine NK-16ST; 9 – exhaust volute; 10 – exhaust muffler; 11 – diffuser; 12 – sealed bulkhead; 13 – subshaft; 14 – accumulator; 15 – compressor; 16 – compressor section.

The use of non-contact dry gas mechanical seals as end seals provides modern requirements for the reliability of such engine-compressor system as a whole. End seals ensure the tightness of the structure [3, 4]. Considered dry gas seals have been widely used in recent years, because they have a large number of advantages over oil lubricated seals [4, 7]. The advantages of this type of seals are the following: the ability to operate at high rotor speeds with minimal gas leakage, no need for a large and expensive oil supply system, smaller dimensions and weight. For better understanding of problems arising in the modeling of dry gas seals, it is necessary to consider in detail their operating principle (Fig. 2). It consists of two rings: rotating and axially movable.

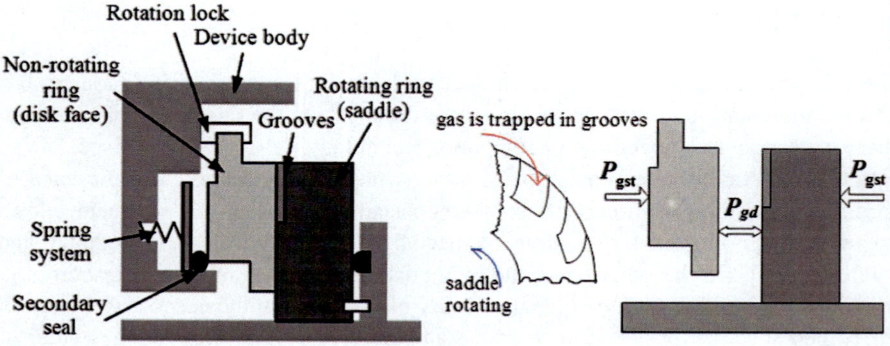

Fig. 2. The operation principle of dry gas seal.

The main operation condition for this type of seals is the presence of required gas layer in the gap between the seal rings in steady state operation. The creation and maintenance of such gas layer with a thickness of $3–4 \cdot 10^{-6}$m is provided by special grooves with a depth of $6–7 \cdot 10^{-6}$m (see Fig. 2) [8, 9]. In working condition, when rotor of engine-compressor system rotates with one of the seal rings, the gas is trapped in grooves and provides the necessary gas layer (see Fig. 2).

The use of integrated information technologies in the design and servicing of this type of seals can ensure reliable and long-term operation of the seal units, and, consequently, the reliable operation of considered engine-compressor system as a whole.

2 Literature Review

The large number of researchers and scientific schools are engaged in the issues of dry gas seals modeling and solving. The existing works are subdivided into the study of the influence of separate processes in the seals, because of the scientific intensity of analysis problem in the general setting [1].

So, in works [10–14], well-proven general technical software systems are used to analyze the gasdynamic pressure between the seal rings. These works are devoted to the selection of the optimal type of micro-grooves, for direct and reversible operation of the seal, by modeling the gas pressure distribution in the gas layer between the rings. Existing software systems also contain a well-developed graphical environment for modeling and visualization of the solution results.

In steady operating mode, when the compressor rotor reaches operating rotational speed, the rings deformed by action of gas pressure, uneven rings heating, and heat release in gas layer. Strains of the seal rings, commensurate with the size of the working gap, cause a significant change in the gap configuration and size [1].

Most of the works available in the literature on this topic, are devoted to the issues of obtaining the gas pressure distribution in the gap between the seal rings in the operating, steady state [11–14]. In this case, a gasdynamic problem, which belongs to a special section of gas lubrication theory, is solved. And in most cases, the heat release in the gas layer is not taken into account. At the same time, various types of grooves and their influence on the resulting gasdynamic force, opening the rings, are studied [8, 11–14]. The solution results in such works are in good agreement with each other. To solve these problems, general technical software systems are also used. Or, an alternative is self-developed programs based on simplified engineering solution methods. The mentioned above tasks are solved separately, due to their rather complex joint formulation.

The temperature loads on the seal rings and heat release in the gas layer are studied in much less number of works [14, 15]. However, a change in the gas temperature in the gap leads to a redistribution of the gas pressure [1, 16]. As indicated above, the presence of temperature effect on the seal rings also leads to the appearance of temperature strains, commensurate with the size of the working gap itself. This leads to a significant redistribution of gas pressure [1, 3, 16].

The dynamic phenomena, arising during the operation of the engine-compressor system, also affect dry gas seal operation. The works [10, 18–21] are devoted to the issues of dynamic stability during the seal operation. It should be noted that the high accuracy of modeling the processes occurring in dry gas seals also provides the stable operation of the units in transient processes. For the most part, for this purpose, existing general engineering software packages are used.

An alternative approach in structure design is the use of integrated technologies, which consist in the use of a combination of powerful finite element software packages and specially developed software products that allow solving a highly specialized problem. With this approach, we can take the best qualities of both methods.

3 Formal Problem Statement

The purpose of this work is to present the proposed combined approach that implements an iterative improved algorithm for the solution of dry gas seals used as end seals in the engine-compressor system as part of the gas pumping unit. The presented iterative algorithm is based on the developed software package for gasdynamic solution of dry gas seals.

4 Theoretical Basis

Let's consider the details of developed algorithm for improved solution of dry gas seals. To formulate the basic equations, we will focus on each problem separately.

The problem of gasdynamic in this case is reduced to the consideration of a viscous compressible fluid and belongs to a special section of gas dynamics – the theory of gas lubrication [16]. To derive the equation of gas lubrication, taking into account the temperature changes in gas layer under steady-state operating mode, the equilibrium equations of gas element in the gap between seal rings, the equation of gas masses equality – the equation of continuity, the state equation for gas are used. Using some conversions, the distribution of gas pressure in the gap between the rings of the mechanical seal in dimensionless form, taking into account the temperature changes and gas viscosity, has the form [16]:

$$\begin{aligned}&\frac{\partial}{\partial x}\left[\frac{1}{\mu T_{av}}\left(h^3\frac{\partial P}{\partial x}-12\frac{r_2^2\omega}{P_2\delta^2}\mu h\sqrt{P}z\right)\right]\\&+\frac{\partial}{\partial z}\left[\frac{1}{\mu T_{av}}\left(h^3\frac{\partial P}{\partial z}+12\frac{r_2^2\omega}{P_2\delta^2}\mu h\sqrt{P}x\right)\right]=0,\end{aligned} \tag{1}$$

where x and z – coordinates in the plane of the gap, related to the value of outer radius of the seal rings r_2; $T_{\text{av}} = T_{\text{av}}(x,z)$ – an average integral function of gas temperature changes over the gap thickness and $P = p^2$ – is the square of the gas pressure, related to the T_2 and P_2, the values of temperature and gas pressure at the inlet of the seals (on the outer radius r_2); ω – the angular velocity of rotating ring; h – the thickness of the gas layer, related to the value of working gap between seal rings δ; μ – gas dynamic viscosity coefficient.

The boundary conditions to Eq. (1) are the specified values of temperatures and pressure at the boundaries of the seal [1, 16]:

$$P\big|_{r=r_0} = P_0; \quad P\big|_{r=r_2} = P_2; \quad T\big|_{r=r_0} = T_0; \quad T\big|_{r=r_2} = T_2, \tag{2}$$

where P_0 – gas pressure at the inner radius (at the outlet of the seal); T_0 – gas temperature at the inner radius.

To solve a nonlinear differential Eq. (1) with boundary conditions (2), the Bubnov-Galerkin method was chosen in combination with the finite element method [1, 16]. This approach has proven its effectiveness and is widely used today to solve problems of gas dynamics.

At the second stage of creating improved methodology for dry gas seal solution, the heat transfer problem for the gas layer and the working rings is considered. The heat transfer equations for working rings for the stationary case have the form [1, 3]:

$$\nabla\left(k_{T_{1,2}}\nabla T\right) = 0, \tag{3}$$

where k_{T_1}, k_{T_2} – thermal conductivity coefficients of graphite and steel working rings of the seal, respectively. The boundary conditions for Eq. (3) are boundary conditions of the third kind, which characterize the convective heat exchange of rings with the environment. The outer surfaces of working rings are washed with gas of known temperature.

The temperature distribution in the gas layer is described by the heat equation, which was obtained from the gas energy equation, using accepted assumptions [1]:

$$\begin{aligned} &\frac{\partial}{\partial x}\left[k_T\frac{\partial T}{\partial x}\right] + \frac{\partial}{\partial y}\left[k_T\frac{\partial T}{\partial y}\right] + \frac{\partial}{\partial z}\left[k_T\frac{\partial T}{\partial z}\right] + \mu\left[\left(\frac{\partial v_x}{\partial y}\right)^2 + \left(\frac{\partial v_z}{\partial y}\right)^2\right] \\ &- \frac{PC_v}{RT}\left(v_x\frac{\partial T}{\partial x} + v_z\frac{\partial T}{\partial z}\right) = 0, \end{aligned} \tag{4}$$

where k_T – thermal conductivity of the gas film; $T = T(x,y,z)$ – gas temperature; v_x and v_z – gas velocity components along the gap; C_v – specific gas heat intensity ratio at a constant volume; R – universal gas constant.

It turns out to be difficult to determine the heat transfer coefficients on the rings sealing surfaces in the gap, therefore the joint heat transfer problem for gas layer and working rings is solved. To solve it, the finite element method in a variational formulation is used. In contrast to the problem of gasdynamics, here it is quite correct to use universal software systems.

Since heat release in the gas layer depends on the distribution of gas pressure and temperatures, which are simultaneously unknown, it is necessary to organize an iterative process, in which two tasks are involved: the gasdynamic problem, taking into account the uneven distribution of temperatures, and the heat transfer problem.

Having obtained the distribution of the pressure acting in the gap on the surfaces of seal working rings, and the temperature distribution in the rings, at the third stage, the problem of rings deformation from temperature and force effects is considered.

The classical theory of thermoelasticity assumes that the elastic body strain consists of two components: the first is caused by internal stresses, the second is change in temperature. Presented in displacements, the thermoelasticity equations for the stationary case have the form:

$$\mu \nabla^2 \mathbf{u} + (\lambda + \mu)\nabla\theta - \frac{E}{1-2\nu}\alpha\nabla T = 0, \tag{5}$$

where $\mathbf{u} = \mathbf{u}(u, v, w)$ – displacement vector; $\theta = \frac{\partial u}{\partial x} + \frac{\partial v}{\partial y} + \frac{\partial w}{\partial z}$; E – elastic modulus; ν – the Poisson's ratio; α – the linear thermal expansion coefficient; λ, μ – Lamé coefficients; ρ – density.

The static boundary conditions will be the gasdynamic pressure in the gap between the rings, the gas-static pressure on the outer surface of the rings, and the temperature distribution in the rings. The thermoelasticity problem solution can also be obtained by the standard finite element method in a variational formulation using a universal finite element software package.

5 Iterative Solution Algorithm

The proposed methodology for dry gas seal solution in a steady state operation includes the problems of gasdynamic, heat transfer, thermoelasticity, and force deformation considered above. The ideal option would be to get a joint solution to these problems. However, taking into account the nonlinearity and peculiarity of the solution of each considered problems, their joint solution is difficult.

At the same time, it is necessary to take into account the strong interrelation of formulated tasks. The temperature changes in the gas layer have a sufficient effect on the distribution of gas pressure in the gap and on the appearance of strains in working rings. They significantly change the flatness of the gap and affect the distribution of gas pressure in the gap [3, 16]. Taking into account the strong interrelation of considered above problems, an iterative algorithm for their joint solution was proposed (Fig. 3). This algorithm includes not only the main iteration loop, but also nested iterative loops to refine the solution of each problem included in the solution. At each global iteration of the developed iterative process, the field of gas pressures in the gap and gas temperatures in the gap and working rings is established by iteratively solving the gasdynamic and heat transfer problems, first with a plane-parallel gap.

After that, the problem of thermoelasticity is solved iteratively, and the gas-dynamic problem for each ring separately for a certain temperature distribution. The configuration of working gap is set first, taking into account the strains of the rotating ring, then the configuration of working gap with the strains of axially movable ring is established, taking into account the steady distribution of strains for rotating ring. Having obtained the final form of gap in iterative way with current rings strains, the gasdynamic problem is solved again and the convergence condition 1 is checked (see Fig. 3). After that the heat transfer problem is solved and the convergence condition 2 is checked. If the convergence conditions are not met, the iterative process must be continued further. The iterative process continues until the forces from the external pressure acting on the working unit (resulting gas-static force F_{gst}) and the gasdynamic pressure (F_{gd}) arising in the working gap between the seal rings are balanced with a given accuracy (Fig. 4).

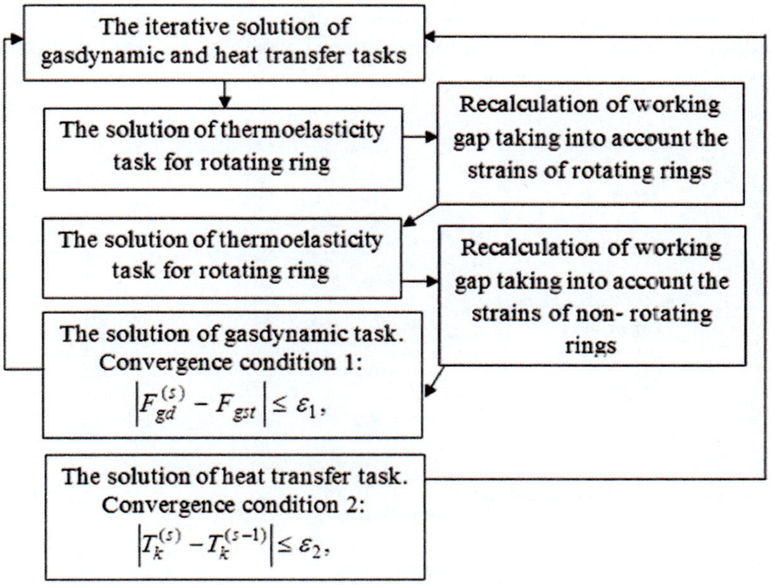

Fig. 3. An algorithm of iterative solution.

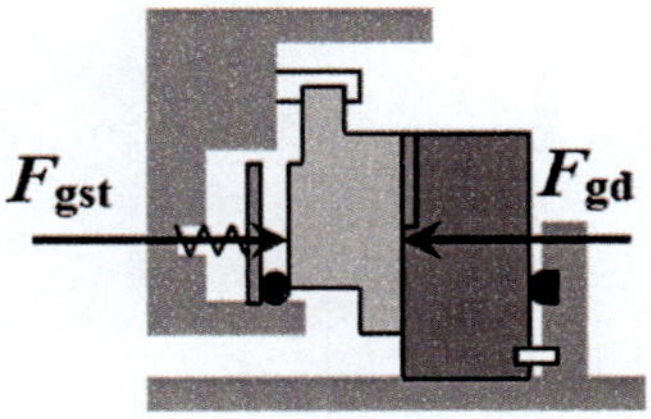

Fig. 4. Force stability of non-rotating ring

6 Program Implementation and Integrated Approach

The program implementation of proposed iterative solution algorithm was created in several stages. An integrated approach was applied for this purpose using already existing finite element software systems in combination with self-created software programs. The features of proposed integrated approach are that by taking the best characteristics of existing software systems and specialized developed programs to obtain a better solution algorithm. This approach has shown its effectiveness in implementation of developed iterative algorithm for the seal improved solution.

Taking into account the specifics of gasdynamic problem, the specialized software package GasDin was developed to solve it, using the proposed Bubnov-Galerkin algorithm in combination with the finite element method (Fig. 5) [1, 16]. This software package was developed using C++ programming language. At the same time, it was decided to use standard finite element package for the solution of heat transfer and thermoelasticity problems, which has proven to be effective for solving such problems. It should also be noted that the modules of universal finite element software package are also used for gas layer modeling, for finite elements meshing, and also for visualization the results of all types of solutions.

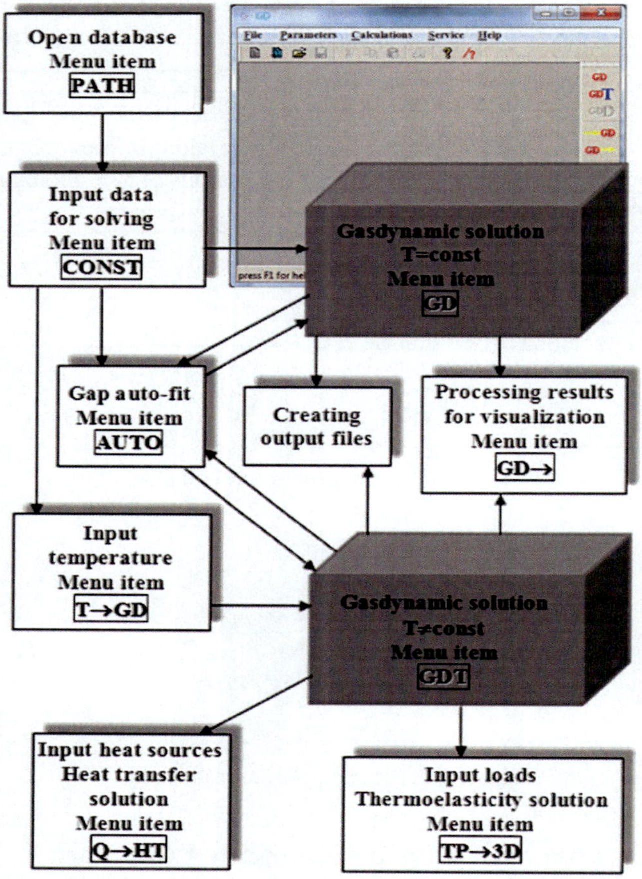

Fig. 5. The main window of specialized software package GasDin and the logical scheme of dry gas seal solution.

The developed software package also makes it possible to carry out the initial selection of working gap between the seals by aligning the gasdynamic and gas-static resulting forces (see Fig. 4). The gas-static force is calculated during the preparation of the initial data after thorough study of the working drawings, the method of sealing the non-rotating ring and setting the boundary conditions.

After the gasdynamic solution, the intensity of heat sources in gas layer is calculated in software package GasDin and entered into the database to solve the heat transfer problem. Further, the rings fields of temperatures and pressures in gas layer are entered into the corresponding databases for solving the deformation problem, taking into account the possible mismatch of the partition grids. The recalculation of the working gap, taking into account the obtained strains of seal rings, is also carried out automatically (Fig. 5). The software package GasDin can be used as a separate software package, also being integrated in the process of full solution.

7 Numerical Solutions

The testing of the proposed iterative algorithm was carried out in several stages, in the process of increasing the complexity of general solution algorithm. It was the testing of developed software package GasDin and its integration with the universal finite element software package during the implementation of all iterations of proposed algorithm (see Fig. 3). For more details of testing features are presented in papers [1, 3, 16]. The testing process included the solutions for working models of dry gas seal of engine-compressor system used at the Joint Stock Company "Sumy Machine-Building Science-and-Production Association", Sumy, Ukraine. A great advantage was the ability to compare the test results with experimental data. The experimental stand to study the seal working parameters has been created in this company. It allowed determining the pressure at the inlet and outlet of the seal, temperature, gas velocity, gas leakages through the seal, which is quite important characteristic. The results of test solutions for seal working models with different types of grooves are in good agreement with the main experimental results [1, 3, 16].

The results of solution using the presented integrated algorithm for the seal working model with spiral grooves, which are most often used, are given below. Initial data for the considered seal with spiral grooves are $r_0 = 90 \cdot 10^{-3}$ m; $r_2 = 112.5 \cdot 10^{-3}$ m; $P_0 = 1.3 \cdot 10^5$ Pa; $P_2 = 57 \cdot 10^5$ Pa; the thickness of the gas layer is $h_1 = 3 \cdot 10^{-6}$ m; in the groove $h_2 = 10 \cdot 10^{-6}$ m; $\omega = 555$ rad/s, $T_2 = 303$K, number of grooves $\eta_k = 12$; $R = 509$ N·m/(kg·K), operating environment – natural gas; $F_{gst} = 68528.9$ N. Accuracy of finding the resulting force is $\varepsilon_1 = 1 \cdot 10^{-1}$; temperature finding accuracy is $\varepsilon_2 = 1 \cdot 10^{-1}$; pressure finding accuracy $\varepsilon_3 = 1 \cdot 10^{-5}$ in relative values.

The iterative process convergence for the considered seal is achieved in 5 global iterations and 33 subiterations. The configuration of the working gap between the seal rings changes significantly from rings strains, commensurate with gap size. As a result of the iterative solution process, a steady-state schematic representation of working gap can be obtained (Fig. 6). The size of the working gap during the iterative process is estimated by its parallel component. As a result of the above solutions, the parallel component of working gap between the rings stopped at value of $3.4 \cdot 10^{-3}$m, while the gas leakages through the seal were $0.81 \cdot 10^{-3}$ kg/s. As a result of experimental studies the obtained gas leakages through this type of seal were $0.76 \cdot 10^{-3}$ kg/s (the difference is 8%). The given gas leakages exceed the norm. As a recommendation for this seal model, it is necessary to change the geometric parameters in order to reduce gas leakages through it.

The resulted distribution of gas pressure at a plane-parallel gap for the seal with spiral groove (the solution model with cyclic symmetry conditions) and at the final iteration are shown in Fig. 7.

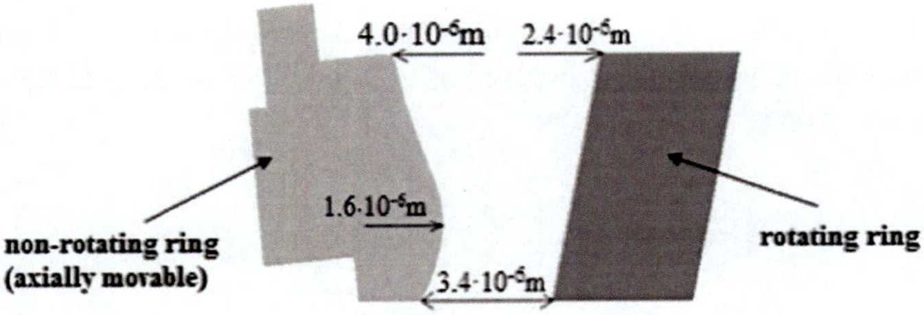

Fig. 6. Schematic configuration of final view of working gap.

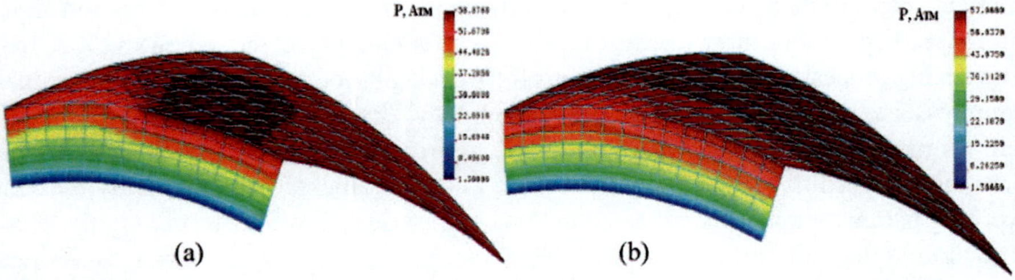

Fig. 7. The distribution of gas pressure: a) at a plane-parallel gap, b) the final distribution.

Experimental data have shown that in the absence of contact between the surfaces of seal rings, the gas heating in the gap occurs on average up to 40–50 °C. Computational studies for this model showed gas heating in the gap by 26 °C [1, 3]. Established as a result of the iterative process, the temperature distribution in the seal rings and in the gap is shown in Fig. 8.

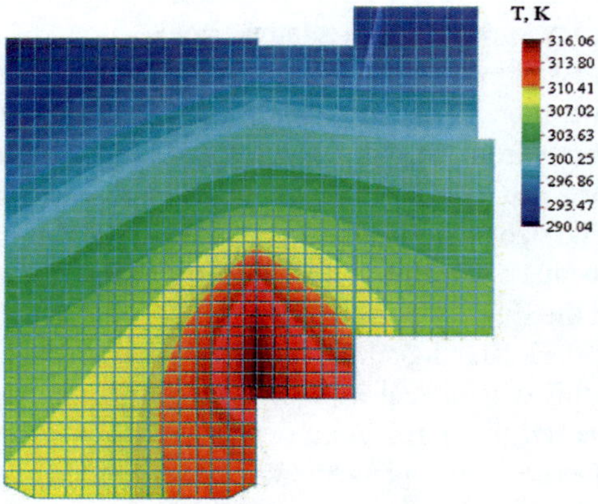

Fig. 8. The final temperature distribution in seal rings and working gap.

8 Conclusions

The use of integrated information technologies in the design and manufacture of gas turbine aircraft engines allows ensuring their long-term non-failure operation and significantly extend their service life. This goal can be achieved by applying improved computer modeling of their individual elements. The subject of research in the work is dry gas seals. They are used as end seals in the engine-compressor system, which is the main part of the gas pumping unit, which maintains the required gas pressure in the main gas pipelines. Sealing units are critical units that ensure reliable operation of the system as a whole. Studies have shown that complete mathematical model for dry gas seal solution, based on the features of their work, should include the solution of the interrelated problems of gasdynamics, heat transfer and thermoelasticity. The presented integrated approach includes the interaction of the developed specialized software package and the universal finite element software package, which makes it possible to implement the proposed iterative algorithm for the joint solution of mentioned above tasks. Improved computer modeling and solving of this type of seal will allow selecting the optimal geometric and physical seal parameters.

References

1. Rozova, L., Martynenko, G.: Information technology in the modeling of dry gas seal for centrifugal compressors. In: Subbotin, S. (ed.) The Third International Workshop on Computer Modeling and Intelligent Systems (CMIS-2020), CEUR Workshop Proceedings, vol. 2608, pp. 536–546. Aachen, Germany (2020)
2. Martynenko, G.Y., Marusenko, O.M., Ulyanov, Y.M., Rozova, L.V.: The use of information technology for the design of a prototype engine with rotor in magnetic bearings. In: Nechyporuk, M., Pavlikov, V., Kritskiy, D. (eds.) Integrated Computer Technologies in Mechanical Engineering. AISC, vol. 1113, pp. 301–309. Springer, Cham (2020). https://doi.org/10.1007/978-3-030-37618-5_26
3. Rozova, L., Martynenko, G.: The design of elements of systems with gas-turbine engines based on information technology. In: Nechyporuk, M., Pavlikov, V., Kritskiy, D. (eds.) ICTM 2020. LNNS, vol. 188, pp. 475–486. Springer, Cham (2021). https://doi.org/10.1007/978-3-030-66717-7_40
4. Boyce, M.P.: Bearings and seals. In: Gas Turbine Engineering Handbook, 4th edn., pp. 557–604. Butterworth-Heinemann, Elsevier (2012). https://doi.org/10.1016/B978-0-12-383842-1.00013-5
5. Artemova, T.G.: Gas Pumping Unit GPA-Ts-16. USTU-UPI, Yekaterinburg (2002). (in Russian)
6. Falaleev, S., Bondarchuk, P., Tisarev, A.: Development of advanced carbon face seals for aircraft engines. IOP Conf. Ser. Mater. Sci. Eng. **302**(1), 1–6 (2018). https://doi.org/10.1088/1757-899X/302/1/012004
7. Cao, S., Chen, Y.: A review of modern rotor/seal dynamics. Gongcheng Lixue/Eng. Mech. **26**(Suppl. 2), 68–79 (2009)
8. Sun, J., Liu, M., Xu, Z., Liao, T.: Research on operating parameters of T-groove cylindrical gas film seal based on computational fluid dynamics. Adv. Compos. Lett. **28**, 1–7 (2019). https://doi.org/10.1177/0963693519864373

9. Du, Q., Zhang, D.: Research on the performance of supercritical CO2 dry gas seal with different deep spiral groove. J. Therm. Sci. **28**(3), 547–558 (2019). https://doi.org/10.1007/s11630-019-1139-z
10. Blasiaka, S., Zahorulko, A.V.: A parametric and dynamic analysis of non-contacting gas face seals with modified surfaces. Tribol. Int. **94**, 126–137 (2016)
11. Chen, Y., Peng, X., Li, J., Jiang, J.: The influence of structure parameters of spiral groove on dynamic characteristics of dry gas seal. Mocaxue Xuebao/Tribology **36**(4), 397–405 (2016). https://doi.org/10.16078/j.tribology.2016.04.001
12. Xu, W., Yang, J.: Spiral-grooved gas face seal for steam turbine shroud tip leakage reduction: performance and feasibility analysis. Tribol. Int. **98**, 242–252 (2016). https://doi.org/10.1016/j.triboint.2016.02.035
13. Gao, L.: Steady simulation of T-groove and spiral groove dry gas seals. Int. J. Heat Technol. **37**(3), 839–845 (2019). https://doi.org/10.18280/ijht.370321
14. Chen, Z., Jiang, L., Li, J., Wu, B.: Numerical analysis of temperature field on the sealing rings of a dry gas seal. Sichuan Daxue Xuebao (Gongcheng Kexue Ban)/J. Sichuan Univ. (Eng. Sci. Ed.) **46**(3), 175–181 (2014)
15. Zhu, W., Li, N., Wang, H.: Finite element analysis on thermal deformation of T-shape Groove dry gas seal. In: 2010 International Conference on Measuring Technology and Mechatronics Automation, pp. 284–288. IEEE, Changsha City, China (2010). https://doi.org/10.1109/ICMTMA.2010.43
16. Rozova, L., Martynenko, G.: Mathematical modeling and program implementation of gasdynamic solution of dry gas seals for centrifugal compressors. In: Shakhovska, N., Medykovskyy, M.O. (eds.) Advances in Intelligent Systems and Computing V. CSIT 2020. AISC, vol. 1293. Springer, Cham (2021). https://doi.org/10.1007/978-3-030-63270-0_40
17. Chen, Y., et al.: Dynamic performance of dry gas seals and analysis of interactions among its influencing factors. Mocaxue Xuebao/Tribology **39**(3), 269–278 (2019). https://doi.org/10.16078/j.tribology.2018144
18. Badykov, R., Falaleev, S., Wood, H., Vinogradov, A.: Gas film vibration inside dry gas seal gap. In: Global Fluid Power Society Ph.D. Symposium (GFPS), pp. 1–6 (2018).https://doi.org/10.1109/GFPS.2018.8472383
19. Zhu, R., Zhang, Y., Ren, J., Li, H., Han, Q.: Nonlinear dynamics study of a high-temperature rotor-bearing-seal system in gas turbine. J. Vibroengineering **19**(1), 629–640 (2017). https://doi.org/10.21595/jve.2016.17491
20. Wei, Y., Liu, S.: Nonlinear dynamics analysis of rotor-brush seal system. Trans. Can. Soc. Mech. Eng. **43**(2), 209–220 (2019). https://doi.org/10.1139/tcsme-2018-0132
21. Sun, D., Li, S.Y., Zhao, H., Fei, C.W.: Numerical investigation on static and rotor-dynamic characteristics of convergent-tapered and divergent-tapered hole-pattern gas damper seals. Mater. (Basel) **12**(14), 2324 (2019). https://doi.org/10.3390/ma12142324

Mathematical and Computer Simulation of Rotor Dynamics Phenomena in Electromechanical Systems with Magnetic Bearings

Gennadii Martynenko(✉) and Volodymyr Martynenko

Department of Dynamics and Strength of Machines, National Technical University "Kharkiv Polytechnic Institute", NTU "KhPI", 2, Kyrpychova Street, Kharkiv 61002, Ukraine
gmartynenko@ukr.net

Abstract. The paper considers original computer software for analyzing of rotor dynamics in passive and active magnetic bearings. These bearings are used in motors, compressors, pumps and other rotary machines. In the design of such electromechanical systems, the use of information technologies is relevant. This implies mathematical modeling of processes and their computer simulation to find rational or optimal designs. The object of research is the dynamics of rotors in magnetic bearings. The feature of the approach is taking into account interconnected electromagnetic and mechanical controlled processes. The main aim is simulating and visualizing of the nonlinear phenomena of rotor dynamics in magnetic bearings of various types. Presentation of the results is made in graphical and textual form for determination of dynamic characteristics and evaluation of motion stability. The mathematical analytical model is based on the Lagrange-Maxwell differential equations. Implementation of computer modeling and visualization of rotor dynamics phenomena was performed in the form of a nonlinear simulation computational model of dynamics of rotors in magnetic bearings. This simulation model has an ability to construct vibrograms, spectrograms, motion paths, phase trajectories and stroboscopic Poincare sections, three-dimensional spectra, amplitude-frequency characteristics and dependences of amplitudes of sub- and superharmonics on rotation speed for selected rotor points. Verification of the simulation model was performed by comparing the results with the experimental data for existing rotor systems.

Keywords: Rotor dynamics · Passive and active magnetic bearing · Nonlinear processes and phenomena · Mathematical model · Computer simulation

1 Introduction

Rotor systems and machines with magnetic bearings (MBs) of passive and active types are nonlinear mechatronic electro-magneto-mechanical ones [1, 2]. The mathematical description of their dynamic behavior traditionally involves finite element or finite difference approaches and computational tools. However, only analytic approach allows

M. Nechyporuk et al. (Eds.): ICTM 2021, LNNS 367, pp. 403–414, 2022.
https://doi.org/10.1007/978-3-030-94259-5_35

to take into account all the features of such complex systems as "rotor in MBs". This is due to the fact that modern general-purpose software systems have universalism. But solving the problem requires an implementation of the associated electromagnetic and mechanical analysis with an application of them. Such the problem should be solved in a nonlinear formulation depending on time and taking into account a negative feedback in control system of an active MB. Implementing this analysis is practically impossible in a full formulation, ensuring an acceptable calculation time and the necessary accuracy of the result. At the same time, analytical models allow to describe such an interconnected system using systems of differential equations, taking into account the active magnetic bearing control laws. These studies are devoted to advantages and possibilities of practical implementation of this approach.

2 Literature Review

Passive and active magnetic bearings (PMBs and AMBs) are an alternative type of elastic-damper bearings of rotors, which has many advantages [3–6]. They are used in various rotary systems and machines, from miniature ones such as computer fan [7], to large ones such as engine [8] and gas turbine-generator [9]. In a mathematical description of the dynamic behavior of the "rotor in PMBs and AMBs" system, an important issue is a correct mathematical formulation of force and stiffness characteristics of the PMBs and AMBs [10–13]. To ensure an adequacy of the mathematical modelling, an important point is to take into account the relationship of dynamic electrical and magnetic processes with mechanical vibrations [14, 15].

Traditionally, the mathematical description of rotor dynamics in PMBs and AMBs utilizes methods for analyzing the dynamics of rotors in elastic-damping supports [16, 17]. They are also used for general analysis of rotor systems with various types of sliding or rolling bearings. For preliminary analysis, the elastic restoring force, regardless of the type of bearing, is linearized and its value is substituted into the calculation model. Usually, two approaches are used – analytical description or numerical modeling. The first approach is not always implemented due to the complexity of the mathematical description [18, 19], in particular with magnetic bearings [20]. The second approach involves the use of CAE software systems, which are based on numerical methods, for example, finite element (FE) or finite difference ones [21]. Commercial FE codes allow to use various computational models, for example, three-dimensional solid-state [22] and beam ones [23], and take into account some of individual features of rotor systems with magnetic fields, for example, a complex structure of shaft, AMBs and other elements [24–26]. However, these software tools do not allow to use FE models in order to find solutions that take into account the entirety of processes occurring in such a complex electromechanical system as "rotor in AMBs" [27, 28]. An analytical description of stationary and non-stationary processes in rotor systems with different types of magnetic bearings allows avoiding this drawback with complex accounting. There are known works in which modeling involves individual aspects of such task. For example, these are gyroscopic moments and the interconnectedness of contours of electromagnets of one AMB [29], AMB stator configuration [30], relationship between various parameters [31], multiple shafts [32], control system features [33, 34], active vibration suppression algorithms [35–37], including nonlinear ones [38]. Special attention should be paid to the mathematical description of the nonlinear dynamics of rotors in MBs [29, 30, 39] and application of new algorithms for dynamic analysis [40]. In any case, an experimental verification of the used approaches and tools is the necessary condition [29, 41]. The end result of a mathematical modeling is the technical implementation of created objects or machines at the prototype level [42]. The proposed in [14, 15] concept for constructing analytical mathematical models of the dynamics of rigid rotors in MBs allows one to eliminate most of the disadvantages of the existing approaches. This work is devoted to the formulation of a general approach to modeling the dynamics of rigid and flexible rotors in PMBs and AMBs based on an analytical mathematical description. A computer-based implementation of this mathematical apparatus in the form of analytical-numerical program code make it possible to obtain a simulation computational model that eliminates most of the discovered shortcomings. Such model can be used reasonably to analyze a dynamic behavior of not only the simplest rotor systems with MBs, but also to simulate the dynamics of rotors of real industrial rotary machines.

3 Problem Statement, Aim and Object of Research

The literature review indicates that modern CAE software systems have universalism. That is why their use for the analysis of the systems under consideration does not allow taking into account joint mechanical and controlled electromagnetic processes adequately. Besides, fast calculation of instantaneous values of magnetic restoring forces in PMBs and AMBs gets difficult to implement and highly costly in terms of resources and time. This practically prevents quick carrying out variant calculations in order to select the required system parameters, as well as solving the problems of optimal choice. In this case, a full account of current delay caused by coil inductance, as well as damping associated with the presence of active resistances in their circuits, is practically insoluble problem. Besides, solving rotor dynamics problems in existing CAE software packages makes impossible to fully take into account the control law which depends on the system state parameters. At the same time, the analytical model allows describing an interconnected electromechanical system using differential equation system (DES). In addition, calculating the natural frequencies and modes of vibration, critical velocities and the forms of forced vibrations of the rotors as well as assessing stability cause the problem of providing acceptable error of numerical methods. In view of the above, the choice in favor of analytical approach is obvious.

The object of research is the dynamics of rotors in magnetic bearings, taking into account interconnected electromagnetic and mechanical controlled processes. The study considers a complete magnetic suspension of a rotor in two radial and one axial magnetic bearings. Such a variant is a most common one in small and medium-sized rotor machinery. The design schematic for a rigid rotor in MBs is shown in Fig. 1.

Here, a spatial fixed orthogonal system of coordinates O^*xyz has been introduced. Its axis O^*z passes through the centers of radial MBs. Point C is centre of the mass, m is the rotor mass, J_x, J_y, and J_z are the main central moments of inertia; $e_x = e_1$ and $e_y = e_2$ ($e^2 = e_1{}^2 + e_2{}^2$) are linear whereas $\gamma_x = \gamma_1$ and $\gamma_y = \gamma_2$ are angular unbalance parameters. Points O, O_1, O_2, and O_3 are on the rotor axis of stiffness, with O being in one plane with C (perpendicular to the axis); O_1, O_2 and O_3 are centers of the radial MB1,2 and axial MB3 supporting sections. Gaps in MB1, MB2, and MB3 equal to δ_{r1}, δ_{r2}, and δ_a, respectively. The rotor rotates with a constant angular speed ω. To determine rigid rotor positions, the study suggests use of generalized coordinates $q = \{x_1, y_1, x_2, y_2, z_3\}$. Their total number being $M = 5$. Figure 2 represents a classic view of radial passive magnetic bearings and eight-pole active magnetic bearing with four electromagnets.

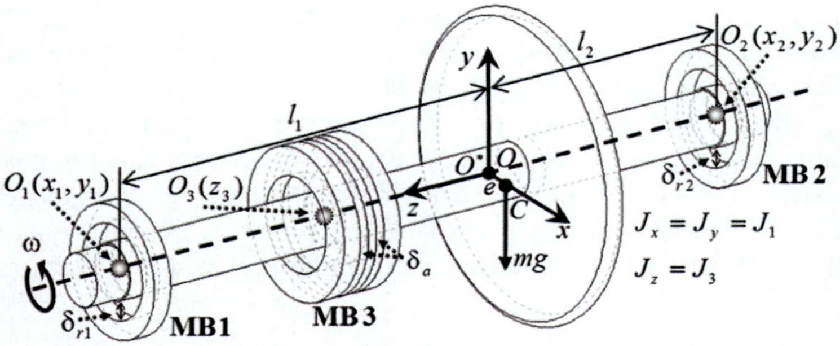

Fig. 1. Typical scheme of a complete rotor suspension in MBs

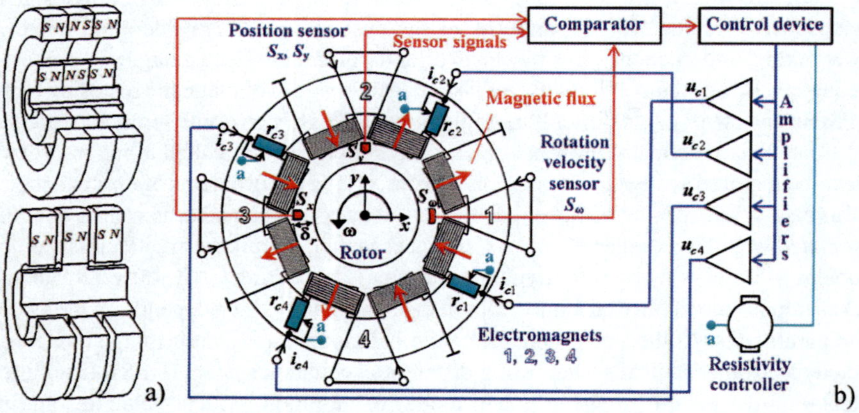

Fig. 2. Classic radial magnetic bearings designs: a) passive magnetic bearings made of permanent ring magnets; b) active magnetic bearing and structural diagram of a control system

The main aim is to analytical-numerical simulate and software implementation the phenomena of rotor dynamics in non-resonant, near-resonance, resonant and operating modes. This software product should search for dynamic parameters of rotors in various types of magnetic bearings, taking into account the nonlinearities inherent in such systems for assessing vibration state and stability of motion for rotational frequencies.

4 Mathematical Modelling of the Rotor Dynamics in MBs

For the mathematical description dynamics rotors in MBs uses an analytical model that is based on the Lagrange-Maxwell differential equations [14, 15]. For rigid rotor (Fig. 1)

the mathematical model has the form [15]:

$$\begin{cases}(ml_2^2+J_1)\ddot{x}_1/l^2+(ml_1l_2-J_1)\ddot{x}_2/l^2+\omega J_3(\dot{y}_1-\dot{y}_2)/l^2+b_{x_1}\dot{x}_1+f''_{x_1}+f'''_{x_1}= \\ \qquad =-\partial\Pi/\partial x_1-\partial W/\partial x_1+Q_{x_1}+H_{x_1}(t); \\ (ml_1^2+J_1)\ddot{x}_2/l^2+(ml_1l_2-J_1)\ddot{x}_1/l^2-\omega J_3(\dot{y}_1-\dot{y}_2)/l^2+b_{x_2}\dot{x}_2+f''_{x_2}+f'''_{x_2}= \\ \qquad =-\partial\Pi/\partial x_2-\partial W/\partial x_2+Q_{x_2}+H_{x_2}(t); \\ (ml_2^2+J_1)\ddot{y}_1/l^2+(ml_1l_2-J_1)\ddot{y}_2/l^2-\omega J_3(\dot{x}_1-\dot{x}_2)/l^2+b_{y_1}\dot{y}_1+f''_{y_1}+f'''_{y_1}= \\ \qquad =-\partial\Pi/\partial y_1-\partial W/\partial y_1+Q_{y_1}+H_{y_1}(t); \\ (ml_1^2+J_1)\ddot{y}_2/l^2+(ml_1l_2-J_1)\ddot{y}_1/l^2+\omega J_3(\dot{x}_1-\dot{x}_2)/l^2+b_{y_2}\dot{y}_2+f''_{y_2}+f'''_{y_2}= \\ \qquad =-\partial\Pi/\partial y_2-\partial W/\partial y_2+Q_{y_2}+H_{y_2}(t); \\ m\ddot{z}_3+b_{z_3}\dot{z}_3+f''_{z_3}+f'''_{z_3}=-\partial\Pi/\partial z_3-\partial W/\partial z_3+Q_{z_3}+H_{z_3}(t); \\ \partial\Psi_{c1}/\partial t+r_{c1}\,\partial W/\partial\Psi_{c1}=u_{c1}(x_1,..,z_3,\dot{x}_1,..,\dot{z}_3,i_1,..,i_N); \\ \cdot\quad\cdot\quad\cdot\quad\cdot\quad\cdot\quad\cdot\quad\cdot \\ \partial\Psi_{cN}/\partial t+r_{cN}\,\partial W/\partial\Psi_{cN}=u_{cN}(x_1,..,z_3,\dot{x}_1,..,\dot{z}_3,i_1,..,i_N).\end{cases} \tag{1}$$

This model consists of 5 nonlinear differential equations of motion and N equations describing the total current law for each kth circuit of the system (all AMB coils). Here m is a rotor overall mass; J_1 and J_3 are rotor moments of inertia; l_1, l_2 and l_3 are distances from the coordinate system center to centers of the radial and axial supporting sections respectively ($l_1 + l_2 = l$); members $-\partial\Pi/\partial q_j$ are potential forces, for example, restoring magnetic forces in the PMBs; $P_{qj} = -\partial W/\partial q_j$ are electromagnetic reactions of the AMB; $H_q(t)$ are external periodic loads caused by the dynamic rotor imbalance; Q_j are other non-potential generalized forces; $f'_{qj}(q_i), f''_{qj}(q_i)$ are nonlinear members of the equations of motion, caused by inertia forces and a potential field of the second and third order; $b_{x1,\ldots,z3}$ are viscosity coefficients; $r_{c\,1,\ldots,N}$ are active resistances in winding circuits; $u_{c\,1,\ldots,N}$ are control voltages supplied to the AMB windings, the value of which is formed in accordance with the adopted control law; $W(\Psi_{c1},\ldots,\Psi_{cN},x_1,\ldots,z_3)$ is the magnetic field energy of the all AMBs: Ψ_{ck} are the flux linkage. The expression of the magnetic energy of AMBs is recorded on the basis of the analysis of magnetic circuits [15]. In this case, the forces in the AMB consider the characteristics of electromagnetic circuits. For a flexible rotor, the potential energy in the Lagrange-Maxwell differential equations is represented as the sum of the total internal energy of the system and the energy of conservative forces (gravity, elastic forces), and kinetic energy is represented as sum of the kinetic energies of the shaft and disks. They are formed as, for example, in [43]. Such an approach allows obtaining an analytical description of the system dynamics taking into account nonlinear features of MBs (Fig. 2). This is an advantage over traditional approaches [44].

5 Simulation Computational Model (Software Implementation)

The block diagram of a nonlinear simulation computational model of dynamics of rotors in magnetic bearings (SCM-DRMB-N) is presented in Fig. 3. The software implementation was performed using C, Java and interpreted programming language.

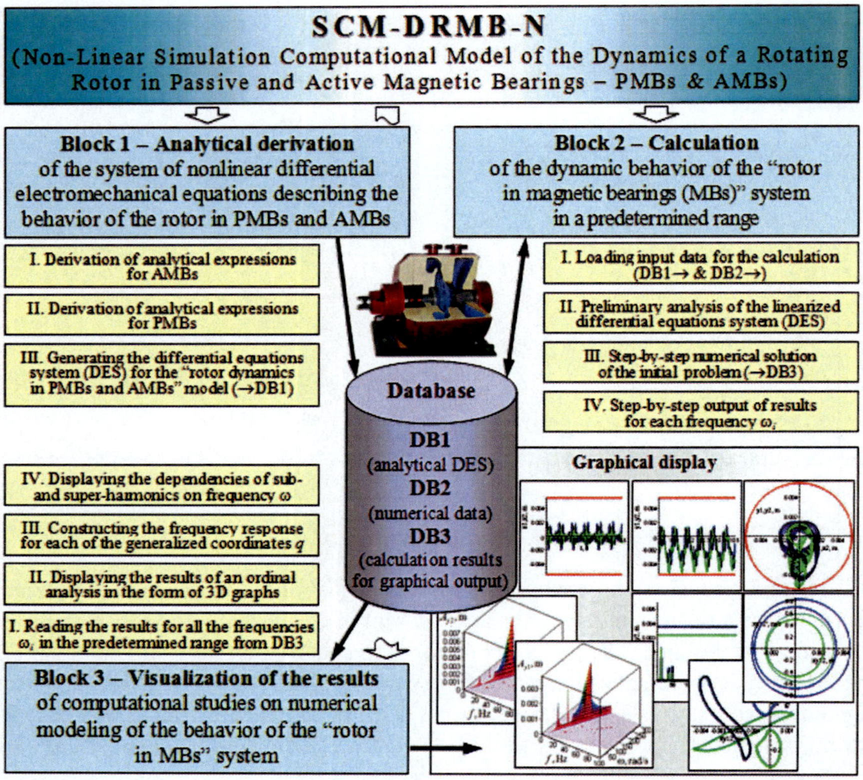

Fig. 3. The block diagram of the simulation computational model SCM-DRMB-N

The structure of the simulation model includes three program blocks (Fig. 3), communication between which is carried out through the database. The first block performs the following actions: 1. the formation of analytical expressions for height of the gap under element of AMB poles area, magnetic resistance of circuit sections, magnetic energy of AMBs; 2. finding discrete values of magnetic forces, approximation of force characteristics of PMBs; derivation of DES in regard to flux linkages Ψ_{ck} in q coordinates, derivation of differential equations of rotor motion in q coordinates, formation of the complete nonlinear DES of the magnetomechanical system and writing to the database (DB1). The second block of the program implements: 1. reading of analytical expressions of the complete DES from DB1, reading from DB2 (input data) – parameters of PMBs, AMBs & AMB control system, reading of source parameters for calculating the system dynamics from DB2; 2. linearization of the DES and reduction to the form of the eigenvalue problem, search of natural frequencies (NFs) of a nonrotating rotor in MBs (at $\omega = 0$), construction of eigen-modes of the nonrotating rotor (at zero frequency), constructing Campbell's frequency diagram and determining values of critical rotation speeds; 3. dividing the frequency range into steps (ω_i), cycle formation based on angular frequency steps ω_i, assignment of initial condition values for ω_i, solution of the nonlinear DES by the Runge-Kutta method, recording the dependencies of generalized coordinates q = $\{x_1, y_1, x_2, y_2, z_3\}$ on time, determination of steady motion parameters for ω_i, checking the uniqueness of DES solution for ω_i, recording the calculation results for ω_i into DB3; 4. display of vibrograms of generalized coordinates for ω_i, display of a motion paths of the rotor supporting points, finding of the spectra of x_1, y_1, x_2, y_2, z_3 by the FFT method, display of spectrograms of generalized coordinates for ω_i, display of phase trajectories $\dot{q}_i(q_i)$ for ω_i, display of Poincare Sections for ω_i, as well as three-dimensional spectra, amplitude-frequency characteristics and dependences of amplitudes of sub- and superharmonics on rotation speed for selected rotor points (Fig. 3). The third block is responsible for visualizing the results. Advantages of the SCM-DRMB-N include: an analytical determination of the full resolving DES of the dynamics of the "rotor in PMBs and AMB" system; an ability to perform multiple variant design calculations; an ability to change the control algorithm; a possibility of introducing a measurement error of the components of the state vector; the convenience of a graphical representation of the behavior of the "rotor in MBs" system over a calculated time interval and in a given frequency range.

6 Computational Capabilities and Validation of SCM-DRMB-N

6.1 Laboratory Rig of a Rotor in the Combined Magnetic Suspension

The laboratory rig of the rotor in a combined passive-active magnetic suspension is presented in Fig. 4 [45]. It was used to validate the SCM-DRMB-N simulation model.

Fig. 4. The model rotor in the radial PMBs and the axial AMB (Source: [45]) – the appearance of the laboratory rig, scheme and force characteristics of the PMBs

The rotor with a length of 0.37 m and a mass of 2.5 kg is suspended in a radial direction using two PMBs. They are made of two ring permanent magnets with an axial magnetization. In axial direction, the centering is provided by thrust AMB with CS.

6.2 Modeling Algorithm and Results of Numerical Analysis

The results of computational studies of forced vibrations using SCM-DRMB-N for a laboratory rig are vibrograms. These are solutions at stationary sections for generalized coordinates x_1,y_1,x_2,y_2,z_3 in the angular velocity range of 0-100π rad/s. The method for constructing a three-dimensional spectra (spectrograms) is schematically depicted in Fig. 5, where A is harmonic amplitudes obtained using the fast Fourier transform (FFT), ω_0 is an angular frequency of driving force (is related to angular velocity of the rotor as $\omega_0 = \omega$), f is spectrum frequency. Both super- and subharmonics are present on the spectra which is consistent with results of experimental studies [45].

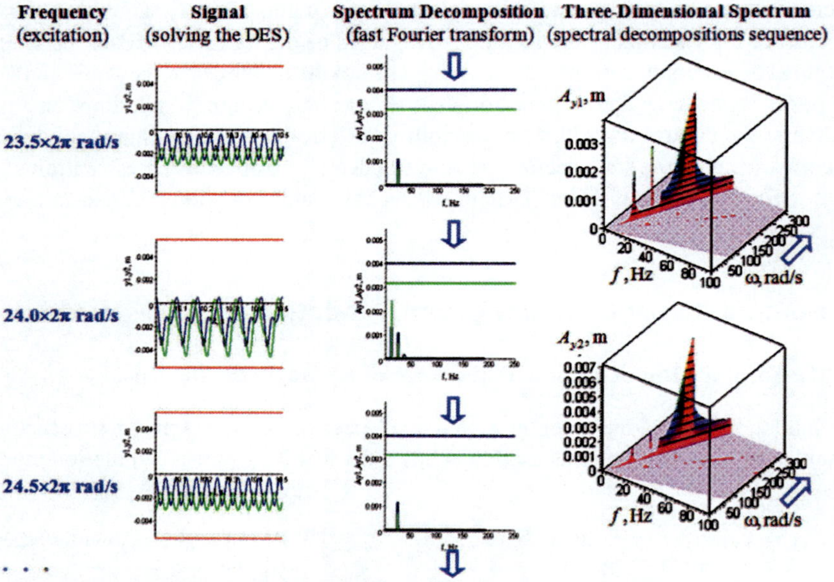

Fig. 5. Three-dimensional spectrum construction procedure

The rotor stability at each speed is estimated by vibrograms, trajectories in the gaps, phase trajectories and Poincare sections. For one of the dangerous frequencies, they are shown in Fig. 6. Singular points of the "center" type or regular attractors are observed on the phase trajectories, and the Poincare sections are stroboscopic points (correspond to a stable state) or stable limit cycles for all generalized coordinates.

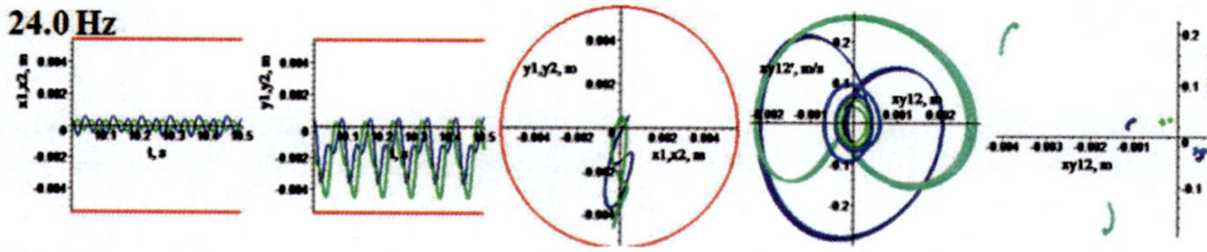

Fig. 6. Calculation results for the motion stability analysis – vibrograms of radial vibrations of rotor left and right supporting sections, motion paths, phase trajectories and Poincare sections

6.3 Comparative Analysis of Calculated and Experimental Data

Dependences of the fundamental harmonic and subharmonics on the frequency of the excitation force is presented in Fig. 7a. The characteristics are projections of three-dimensional spectra (Fig. 5) onto the $O\omega A$ coordinate planes. The dependences of super-harmonics are not given here, however, they are also present. The resonant frequencies are equal to [2π rad/s]: $\omega_{1x} = 10.5$, $\omega_{1y} = 12$, $\omega_{2x} = 22.5$ and $\omega_{2y} = 36$ (disruption).

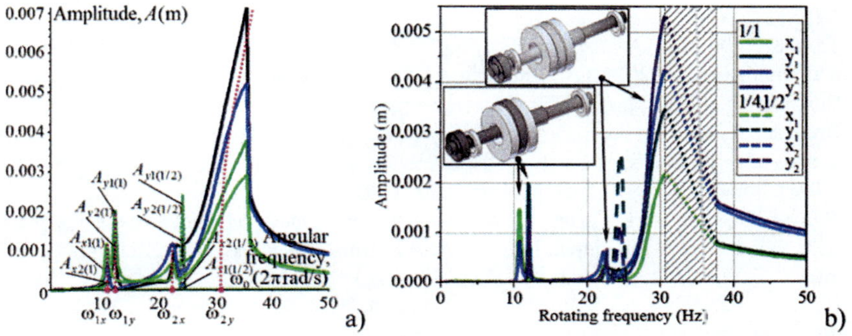

Fig. 7. Dependences of the fundamental harmonic and subharmonics on the frequency of the excitation force: a) simulation results using SCM-DRMB-N; b) experimental data (Source: [45])

The experimental data are presented in Fig. 7b. An acceleration of the rotor was carried out up to 3000 rpm (50 Hz). The region of maximum amplitudes with a breakdown of oscillations in the range of ~1850–2280 rpm (~31–38 Hz) is marked by the dashed lines in Fig. 7b. The following phenomena were found: a bifurcation both of the first (~10.5 and ~12 Hz) and the second (~22.5 and ~33 Hz) resonances; direct (<10.5 Hz) and reverse (<12 Hz) cylindrical precessions, as well as direct (<22.5 Hz) and reverse (<36 Hz) conic precessions. In terms of amplitudes, the difference does not exceed 2–5%, and according to resonance frequency values it is less than 0.2%. These allows concluding that the analysis of the rotor dynamics in MBs is reliable.

7 Conclusion

The paper proposes an approach to the formation of refined nonlinear analytical and simulation computational models describing the dynamic behavior of rotor systems in MBs. These models can be used for the refined search for dynamic parameters of rotors in magnetic bearings of different types taking into account a large set of nonlinearities inherent in such systems. They provide a unique opportunity to construct vibrograms, spectrograms, motion trajectories, phase trajectories and stroboscopic Poincaré sections, as well as three-dimensional spectra. This allows a comprehensive assessment of the vibration state and stability of motion in a given speed range.

The use of the SCM-DRMB-N improves the reliability of a calculation determination of parameters of rotors, magnetic bearings and AMB control systems in comparison with other tools.

References

1. Siva Srinivas, R., Tiwari, R., Kannababu, Ch.: Application of active magnetic bearings in flexible rotordynamic systems – a state-of-the-art review. Mech. Syst. Signal Process. **106**, 537–572 (2018)
2. Wan, Z.: A review of active magnetic bearings supported systems optimization design. Int. J. Magn. Electromagnetism **6**(1):027, 1–10 (2020)
3. Safaeian, R., Heydari, H.: Augmentation of passive magnetic bearing performance by using air or iron intervals. COMPEL Int. J. Comput. Math. Electr. Electron. Eng. **38**, 657–671 (2019)
4. Tănase, N., Morega, A.M., Chiriță, I., Ilie, C.: Passive magnetic bearing – design and numerical simulation. In: 2019 11th International Symposium on Advanced Topics in Electrical Engineering (ATEE), pp. 1–5, Bucharest, Romania (2019)
5. Breńkacz, Ł, Witanowski, Ł, Drosińska-Komor, M., Szewczuk-Krypa, N.: Research and applications of active bearings: a state-of-the-art review. Mech. Syst. Signal Process. **151**, 107423 (2021). https://doi.org/10.1016/j.ymssp.2020.107423
6. Nguyen, D., Nguyen, T., Hoang, D.: A non-linear control method for active magnetic bearings with bounded input and output. Int. J. Power Electron. Drive Syst. **11**(4), 2154–2163 (2020). https://doi.org/10.11591/ijpeds.v11.i4.pp2154-2163
7. Jungmayr, G., Marth, E., Panholzer, M., Amrhein, W., Jeske, F., Reisinger, M.: Design of a highly reliable fan with magnetic bearings. Proc. Inst. Mech. Eng. Part I J. Syst. Control Eng. **230**(4), 361–369 (2016)
8. Martynenko, V.: Analysis of strength and bearing capacity of the auxiliary mine ventilation fan connected to the rotor of its electrical drive. In: 2020 IEEE KhPI Week on Advanced Technology (KhPIWeek), pp. 19–23, 9250078, Kharkiv, Ukraine (2020). https://doi.org/10.1109/KhPIWeek51551.2020.9250078
9. Pilotto, R., Nordmann, R.: Vibration control of a gas turbine-generator rotor in a combined cycle system by means of active magnetic bearings. In: Cavalca, K.L., Weber, H.I. (eds.) IFToMM 2018. MMS, vol. 60, pp. 520–533. Springer, Cham (2019). https://doi.org/10.1007/978-3-319-99262-4_37
10. Sun, J., et al.: Dynamic stiffness analysis and measurement of radial active magnetic bearing in magnetically suspended molecular pump. Sci. Rep. **10**, 1401 (2020)
11. Santra, T., Roy, D., Choudhury, A.: Calculation of passive magnetic force in a radial magnetic bearing using general division approach. PIER M **54**, 91–102 (2017)
12. Zhao, C., Zhou, J., Xu, Y., Di, L., Ji, M.: Identification of magnetic bearing stiffness and damping based on hybrid genetic algorithm. Trans. Nanjing Univ. Aeronaut. Astronaut. **34**(2), 211–219 (2017). https://doi.org/10.16356/j.1005-1120.2017.02.211
13. Martynenko, G., Martynenko, V.: Numerical determination of active magnetic bearings force characteristics taking into account control laws based on parametric modeling. In: 2019 IEEE International Conference on Modern Electrical and Energy Systems (MEES), pp. 358–361, Kremenchuk, Ukraine (2019). https://doi.org/10.1109/MEES.2019.8896501
14. Martynenko, G., Martynenko, V.: Modeling of the dynamics of rotors of an energy gas turbine installation using an analytical method for analyzing active magnetic bearing circuits. In: 2020 IEEE KhPI Week on Advanced Technology (KhPIWeek), pp. 92–97, 9250156, Kharkiv, Ukraine (2020). https://doi.org/10.1109/KhPIWeek51551.2020.9250156
15. Martynenko, G.: Accounting for an interconnection of electrical, magnetic and mechanical processes in modeling the dynamics of turbomachines rotors in passive and controlled active magnetic bearings. In: 2018 IEEE 3rd International Conference on Intelligent Energy and Power Systems (IEPS), pp. 326–331. IEEE, Kharkiv, Ukraine (2018). https://doi.org/10.1109/IEPS.2018.8559518

16. Matsushita, O., Tanaka, M., Kanki, H., Kobayashi, M., Keogh, P.: Vibrations of Rotating Machinery: Volume 1. Basic Rotordynamics: Introduction to Practical Vibration Analysis. Springer Japan, Tokyo (2017)
17. Matsushita, O., Tanaka, M., Kanki, H., Kobayashi, M., Keogh, P.: Vibrations of Rotating Machinery, vol. 2. Advanced Rotordynamics: Applications of Analysis, Troubleshooting and Diagnosis. Springer Japan, Tokyo (2019). https://doi.org/10.1007/978-4-431-55456-1
18. Ribeiro, E.A., Alves, D.S., Cavalca, K.L., Bavastri, C.A.: Stability analysis and optimization of a hybrid rotating machinery support combining journal bearings with viscoelastic supports. Mech. Mach. Theory **156**, 104166 (2021)
19. Maldonado, D.J.G., Karev, A., Hagedorn, P., Ritto, T.G., Sampaio, R.: Analysis of a rotor-dynamic system with anisotropy and nonlinearity using the floquet theory and the method of normal forms. J. Sound Vib. **453**, 201–213 (2019)
20. Kim, H., Sikanen, E., Nerg, J., Sillanpaa, T., Sopanen, J.T.: Unbalanced magnetic pull effects on rotordynamics of a high-speed induction generator supported by active magnetic bearings-analysis and experimental verification. IEEE Access **8**, 212361–212370 (2020)
21. Martynenko, V., Hrytsenko, M., Martynenko, G.: Technique for evaluating the strength of composite blades. J. Inst. Eng. (India) Ser. C **101**(3), 451–461 (2020). https://doi.org/10.1007/s40032-020-00572-9
22. Zaccardo, V.M., Buckner, G.D.: Active magnetic dampers for controlling lateral rotor vibration in high-speed rotating shafts. Mech. Syst. Signal Process. **152**, 107445 (2021)
23. Ramos, D.J., Ferraz, A.R., Daniel, G.B., Ritto, T.G.: Dynamic analysis of rotating systems considering uncertainties in the bearings' parameters. Mech. Mach. Sci. **63**, 460–474 (2019)
24. Cao, J., Allaire, P., Dimond, T., Dousti, S.: Auxiliary bearing system optimization for AMB supported rotors based on rotor drop analysis: Part I – Rotor drop analysis method. In: ASME Turbo Expo 2016: Turbomachinery Technical Conference and Exposition, vol. 7A, Structures and Dynamics, Paper No. GT2016-56323, p. 11 (2016)
25. Kuppa, S.K., Lal, M.: Dynamic behaviour analysis of coupled rotor active magnetic bearing system in the supercritical frequency range. Mech. Mach. Theory **152**, 103915 (2020)
26. Pavlenko, I.V., Simonovskiy, V.I., Demianenko, M.M: Dynamic analysis of centrifugal machines rotors supported on ball bearings by combined application of 3D and beam finite element models. In: 15th International Scientific and Engineering Conference HERVICON and PUMPS 2017, p. 8. Sumy State University, Sumy, Ukraine (2017)
27. Yektanezhad, A., Hosseini, S.A.A., Tourajizadeh, H., Zamanian, M.: Vibration analysis of flexible shafts with active magnetic bearings. Iran. J. Sci. Technol. Trans. Mech. Eng. **44**(2), 403–414 (2018). https://doi.org/10.1007/s40997-018-0263-9
28. Martynenko, G.: Analytical method of the analysis of electromagnetic circuits of active magnetic bearings for searching energy and forces taking into account control law. In: 2020 IEEE KhPI Week on Advanced Technology (KhPIWeek), pp. 86–91, 9250138, Kharkiv, Ukraine (2020). https://doi.org/10.1109/KhPIWeek51551.2020.9250138
29. Ebrahimi, R., Ghayour, M., Khanlo, H.: Nonlinear dynamic analysis and experimental verification of a magnetically supported flexible rotor system with auxiliary bearings. Mech. Mach. Theory **121**, 545–562 (2018)
30. Wu, R., Zhang, W., Yao, M.: Nonlinear dynamics near resonances of a rotor-active magnetic bearings system with 16-pole legs and time varying stiffness. Mech. Syst. Signal Process. **100**, 113–134 (2018). https://doi.org/10.1016/j.ymssp.2017.07.033
31. Li, B.L., Zeng, L., Zhang, P.M., Zhu, Z.D.: Sliding mode active disturbance rejection decoupling control for active magnetic bearings. Dianji yu Kongzhi Xuebao/Electr. Mach. Control **25**, 129–138 (2021)
32. Ebrahimi, R., Ghayour, M., Khanlo, H.M.: Chaotic vibration analysis of a coaxial rotor system in active magnetic bearings and contact with auxiliary bearings. Comput. Nonlinear Dyn. **12**(3), 11 (2017). https://doi.org/10.1115/1.4034869

33. Hutterer, M., Schrödl, M.: Control of active magnetic bearings in turbomolecular pumps for rotors with low resonance frequencies of the blade wheel. Lubricants **5**, 26 (2017)
34. Chen, S., Song, M.: Energy-saving dynamic bias current control of active magnetic bearing positioning system using adaptive differential evolution. IEEE Trans. Syst. Man Cybern. Syst. **49**(5), 942–953 (2019)
35. Jung, D., DeSmidt, H.: A new hybrid observer based rotor imbalance vibration control via passive autobalancer and active bearing actuation. J. Sound Vib. **415**, 1–24 (2018). https://doi.org/10.1016/j.jsv.2017.11.024
36. Chen, Q., Liu, G., Han, B.: Unbalance vibration suppression for AMBs system using adaptive notch filter. Mech. Syst. Signal Process. **93**, 136–150 (2017)
37. Martynenko, G., Martynenko, V.: Rotor dynamics modeling for compressor and generator of the energy gas turbine unit with active magnetic bearings in operating modes. In: 2020 IEEE Problems of Automated Electrodrive. Theory and Practice (PAEP), pp. 1–4, 9240781. IEEE, Kremenchuk, Ukraine (2020). https://doi.org/10.1109/PAEP49887.2020.9240781
38. Shao, X., Wang, W., Jie, F., Jiang, X.: Investigation on intelligent rotor vibration control based on electromagnetic damping seal. In: ASME Turbo Expo 2017: Turbomachinery Technical Conference and Exposition, GT2017-65055, V07AT34A033, p. 10. ASME, Charlotte, North Carolina, USA (2017). https://doi.org/10.1115/gt2017-65055
39. Su, Y., Gu, Y., Keogh, P.S., Yu, S., Ren, G. Nonlinear dynamic simulation and parametric analysis of a Rotor-AMB-TDB system experiencing strong base shock excitations. Mech. Mach. Theory **155**, 104071 (2021)
40. Kuppa, S.K., Lal, M.: Dynamic behaviour analysis of coupled rotor active magnetic bearing system in the supercritical frequency range. Mech. Mach. Theory **152**, 103915 (2020). https://doi.org/10.1016/j.mechmachtheory.2020.103915
41. Xu, Y., Zhou, J., Di, L., Zhao, C.: Active magnetic bearings dynamic parameters identification from experimental rotor unbalance response. Mech. Syst. Signal Process. **83**, 228–240 (2017). https://doi.org/10.1016/j.ymssp.2016.06.009
42. Martynenko, G.Y., Marusenko, O.M., Ulyanov, Y.M., Rozova, L.V.: The use of information technology for the design of a prototype engine with rotor in magnetic bearings. In: Nechyporuk, M., Pavlikov, V., Kritskiy, D. (eds.) Integrated Computer Technologies in Mechanical Engineering. AISC, vol. 1113, pp. 301–309. Springer, Cham (2020). https://doi.org/10.1007/978-3-030-37618-5_26
43. Han, B., Huang, Z., Le, Y.: Design aspects of a large scale turbomolecular pump with active magnetic bearings. Vacuum **142**, 96–105 (2017)
44. Ishida, Y., Yamamoto, T.: Linear and Nonlinear Rotordynamics: A Modern Treatment with Applications, 2nd edn. Wiley-VCH Verlag GmbH & Co. KGaA, Weinheim (2012). https://doi.org/10.1002/9783527651894
45. Martynenko, G., Ulianov, Yu.: Combined rotor suspension in passive and active magnetic bearings as a prototype of bearing systems of energy rotary turbomachines. In: 2019 IEEE International Conference on Modern Electrical and Energy Systems (MEES), pp. 90–93, Kremenchuk, Ukraine (2019). https://doi.org/10.1109/MEES.2019.8896571

Rational Control of the Operability of the Model Unit of Electric Flywheel Motors Under Destabilizing Effects

Anatolii Kulik(✉), Konstiantyn Dergachov, Serhii Pasichnik, Vitalii Dzhulgakov, and Vasilii Petrenko

National Aerospace University "Kharkiv Aviation Institute", Kharkiv, Ukraine
{k.dergachev,s.pasichnyk,v.dzhulgakov}@khai.edu

Abstract. The results of theoretical and experimental studies of the fundamental possibility of joint use of deep diagnostics and flexible restoring procedures to ensure rational control of the operability of the model unit of electric flywheel motors are presented. Diagnostic models for 11 types of destabilizing effects are described. A dichotomous tree for searching for types of destabilizations is proposed. Algorithms for restoring the operability of a unit of electric flywheel motors are given. The results of experimental studies of one channel of the model unit of electric flywheel motors in the form of transient characteristics at step input actions are presented.

Keywords: Unit of electric flywheel motors · Rational control · Operability · Destabilizing effect · Deep diagnostics · Flexible restoring

1 Introduction

Electric flywheel motors (EFM) are used to create control torques relative to the center of mass of spacecraft (SC). The idea of using the EFM to control the angular position of the spacecraft belongs to the founder of theoretical cosmonautics E. Tsialkovsky. The practical application of EFM began in 1963 in spacecrafts «Kosmos-14», «Kosmos-23» (USSR) и «Nimbus» (USA).

The tendency to increase the duration of the active functioning of the spacecraft gives rise to a number of contradictions in the design. On the one hand, the traditional way of increasing the redundancy of the hardware to ensure the required duration of the operational state of the spacecraft control systems leads to a significant increase in the weight, size, energy and cost characteristics. On the other hand, the terms of long-term operation of the spacecraft cause an increase in the requirements for long-term, trouble-free and reliable operation of the payload. The implementation of these requirements also leads to a significant increase in such characteristics, with existing restrictions on the general characteristics of specific SC as aircraft. Such contradictions are resolved by finding acceptable compromises. Compromise solutions in most cases depend on the level of development of scientific research and spacecraft hardware.

M. Nechyporuk et al. (Eds.): ICTM 2021, LNNS 367, pp. 415–436, 2022.
https://doi.org/10.1007/978-3-030-94259-5_36

The modern level of development of on-board computing devices makes it possible to implement many functions of spacecraft control, which were previously performed in flight control centers. These functions include control over the operability of on-board systems and, in particular, automatic orientation and stabilization systems. During the execution of a space mission, these systems are affected by both external disturbing influences: aerodynamic, magnetic and gravitational forces, and internal influences: malfunctions and equipment failures. Taken together, these, in fact, destabilizing effects worsen the operation of the orientation and stabilization systems, i.e. destabilize performance and lead to emergency situations. For the rational neutralization of an emergency situation, it is required, firstly, to identify the cause of destabilization, and then, to fend off it with the available means. Identification of the cause of destabilization is associated with the process of deep diagnosis, which refers to the intellectual activities of a human. The parrying process is a process of flexible restoration of operability based on the results of the diagnosis associated with the intelligent solution of multivariate tasks according to a number of criteria in real time.

Known theoretical studies on diagnostics and fault tolerance of dynamic systems are devoted to the formalization of certain aspects of operability control [1]. A number of researchers have attempted to combine, on different grounds, the diagnostic and recovery procedures to control the operability of technical objects [2–5]. The theory of automatic control is a scientific basis for the design of attitude control and stabilization systems of a spacecraft. This theory is based on a number of provisions. Basic one is the knowledge of the features of the functioning of the control object, which is reflected in the corresponding mathematical models in the form of differential equations, transfer functions, descriptions in the state space. These mathematical models describe, as a rule, the nominal operating mode. To describe emergency, abnormal modes in accordance with the provisions of automatic control theory, appropriate mathematical models are needed that connect the causes of destabilization with the consequences. Such models are called diagnostic, and they have received appropriate development in a number of studies [6]. The use of diagnostic models made it possible to form a new principle of control by diagnosis and to develop the main provisions of rational control of operability [7, 8].

Scientific interest in the study of EFM of various structural configurations does not wane. A number of studies are known on the development of algorithms for detecting and localizing malfunctions in sensors and flywheels of a spacecraft [9], intelligent diagnostics of malfunctions of satellite flywheels bearings [10], detecting and isolating gyroscopes failures on board a satellite [11], a method for diagnosing a flywheel of a satellite system orientation [12], detection of flywheel failures using simulation in the Simulink environment [13], development and implementation of a fault-tolerant control system for electric drives [14], fault-tolerant control by reconfiguring the pyramidal structure of the EFM [15]. In the works considered, only certain aspects of ensuring the operability of the EFM are investigated, namely, the diagnosis of individual failures, and the possibilities of various means of restoring operability are not investigated according to the results of the diagnosis. Theoretical studies to ensure deep diagnostics and flexible restoration of the efficiency of EFM units for a multitude of potentially possible destabilizing effects are not found in the technical literature. The direction of the use of rational control tools to control the operability of electric flywheel motors under various destabilizing effects seems relevant and productive.

The article presents the results of theoretical and experimental studies of the fundamental possibility of joint use of deep diagnostics and flexible recovery procedures to ensure rational control of the operability of model unit of flywheel electric motors under various destabilizing effects.

The research results are presented in the following structure. In the second section "Research results" in Sect. 2.1, a description of the experimental stand for the study of the model unit of electric flywheel motors of a pyramidal structure is presented. The placement of the EFM and the external view of the experimental stand are presented. The functional diagram of the digital system for rational control of the EFM operability is described. A general view of the device for rational control of the EFM prototype unit is given.

Section 2.2 is devoted to the consideration of the main provisions of rational control. The block diagram of a rational control system by diagnosis is presented and described. The purpose of the diagnostic unit and the control unit is described in detail. The definition of diagnosability and recoverability of an object of rational control is given.

The experimental characteristics of the functional elements of the rational control object are presented in Sect. 2.3. These are the static characteristics of the EFM, power amplifier, tachogenerator and angular position sensor of the EFM unit, as well as the transient characteristics of the object. The results of processing the experimental data are presented.

Section 2.4 describes the mathematical models of the rational control object in the form of transfer functions and in the state space for both continuous and discrete time.

Diagnostic models are presented in Sect. 2.5. The set of types of destabilization of functional elements and the corresponding diagnostic models are described.

Features of diagnostics of the destabilizing effect are presented in Sect. 2.6. The substantiation of the necessity of using the production model of knowledge and two-valued predicate relations is given. Constructions of two-valued predicate equations and a dichotomous tree for searching for the types of destabilizing effects in an object of rational control are given.

Section 2.7 describes the procedures for restoring the operability of the EFM unit for all eleven types of destabilizing effects using signal tuning and reconfiguration of the equipment.

The results of experimental studies of one EFM control channel are presented in Sect. 2.8 in the form of transient characteristics for step input actions.

2 Research Results

2.1 Description of the Experimental Stand for the Study of Model Unit

EFMs belong to the class of inertial actuators of the spacecraft, which ensure the redistribution of the angular momentum between the rotating rotor and the body of the apparatus. There are several redundant EFM placement structures. Two of them are the most known.

The first one is the simplest redundant EFM configuration, where a cube is chosen as a support body, called the "NASA standard". In this configuration, four EFMs are used, three of which create control moments along the axes of the associated coordinate system, and the moment created by the fourth EFM is directed at equal angles to the axes of the coordinate system associated with the spacecraft. This configuration has a low level of redundancy. In the second configuration, a pyramid was selected as a support body. This configuration was developed by General Electric [15]. EFM is placed along the edges of the pyramid. Such placement of EFM in the actuator units made it possible to significantly increase the level of redundancy required for long-term space missions. The block diagram of the EFM pyramidal placement, consisting of four EFM (EFM1-EFM4), four flywheel angular velocity sensors (tachogenerators TG1-TG4), three angular sensors (ASx, ASy, ASz) is shown in Fig. 1.

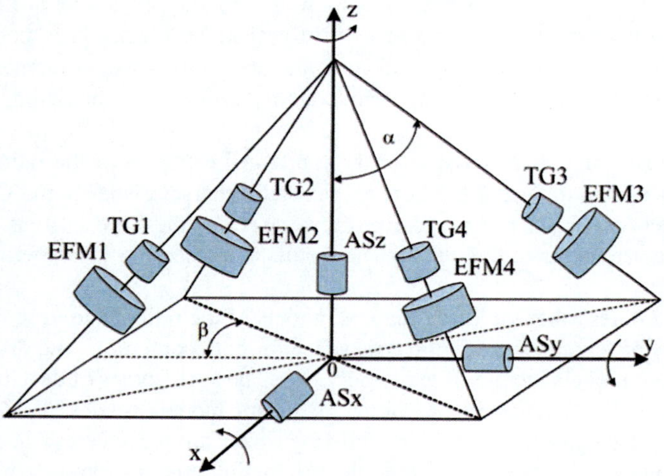

Fig. 1. Block diagram of the pyramidal placement of EFM

The external view of the experimental stand with the EFM prototype unit placed in a gimbal is shows in Fig. 2.

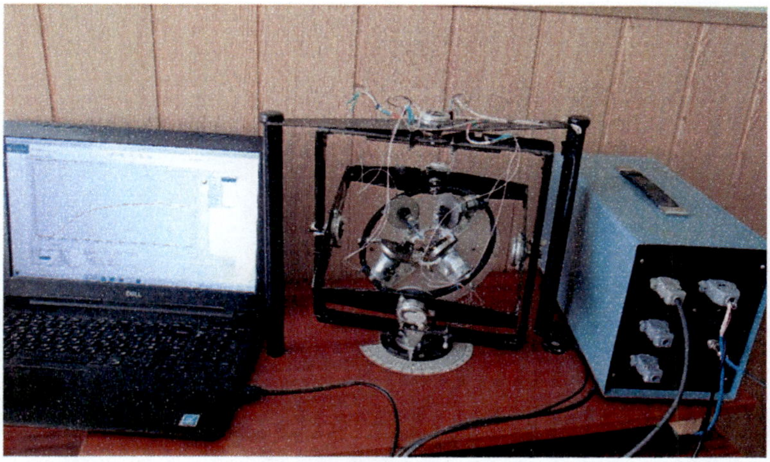

Fig. 2. Experimental stand

DC motors RC-2804RA-2865 are used in the EFM prototype unit. The flywheel is a steel disc with a radius $r = 0,0225$ m and mass $m = 0,073$ kg. The moment of inertia of such a disc $I = 0,185$ kgm^2. EFMs are placed on a square base of a support body with side $a = 0,12$ m and at an angle of inclination $\alpha = 60°$.

For each EFM of the prototype unit, a system is formed that provides both the control of the kinetic moment and the stabilization of the operability under destabilizing effects.

Functional diagram of a digital system for rational management of EFM operability for one channel is shown in Fig. 3.

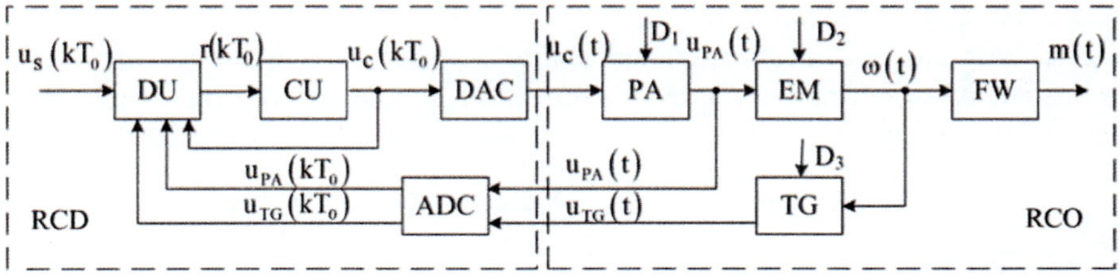

Fig. 3. Functional diagram of the digital system for the rational control of the EFM operability.

Here: DU – diagnostic unit; CU – control unit; DAC – digital-to-analog converter; PA – power amplifier; EM – electric motor; F – flywheel; TG – tachogenerator; ADC – analog-to-digital converter; RCO – rational control object; RCD – rational control device; $u_S(kT_0)$ – setting action; $r(kT_0)$ – discrete signal of DU; $u_C(kT_0)$ – discrete signal of CU; $u_C(t)$ – voltage of DAC; $u_{PA}(t)$ – input voltage of EM; $\omega(t)$ – flywheel angular velocity; $m(t)$ – kinetic moment; $u_{TG}(t)$ – voltage of tachogenerator; $u_{TG}(kT_0)$ – discrete tachogenerator signal; $u_{PA}(kT_0)$ – discrete signal of PA; D_1, D_2 и D_3 – set of types of destabilization, respectively EM and TG. Rational control of each EFM and the entire unit is carried out by the digital control device shown in Fig. 4.

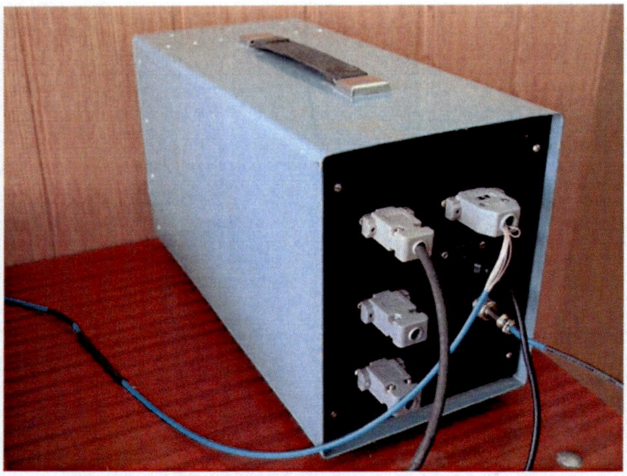

Fig. 4. Rational control device for the EFM prototype unit

The prototype unit is controlled and the results of experiments are visualized using a Dell Inspiron 3582 laptop.

2.2 Fundamentals of Rational Control

When designing satellite attitude control and stabilization systems, the classical principle of control by deviation is used. This principle does not allow to fully preserve the operability of satellite control systems under large disturbing effects and malfunctions. Accidents and disasters in aerospace technology caused by the inoperability of various control systems convince the need to consider the operability as a dynamic state that must be controlled during operation in order to stabilize it under conditions of destabilizing effects. With this consideration, the operability is presented as a new specific object, for the control of which a new principle of control is required - control by diagnosis. Figure 5 shows a block diagram of a system of rational control by diagnosis.

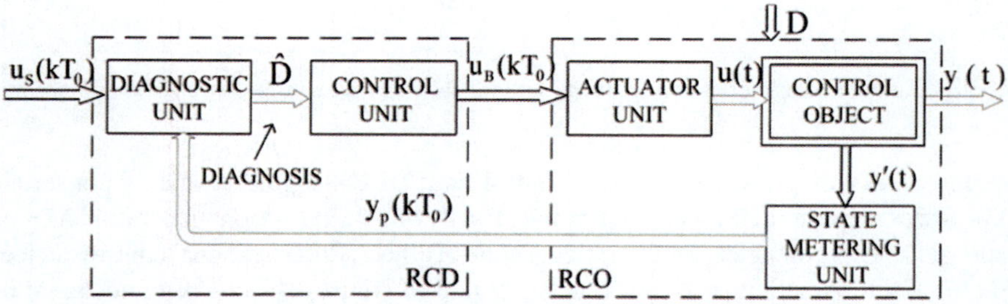

Fig. 5. Block diagram of a rational diagnosis control system

So, the block diagram of a rational control system (RCS) consists of two interconnected subsystems. The first one is a rational control object (RCO), which includes a control object (CO), an actuator unit (AU) and a state metering unit (SMU). The entire RCO subsystem is affected by a multitude of uncontrolled undefined destabilizing effects $\tilde{D}$.

The second subsystem, a rational control device (RCD), consists of a performance diagnostic unit (DU), in which, based on vector signals (setting signal $u_S(kT_0)$ and extended output signal $y_p(kT_0)$) a diagnosis of RCO operability is formed in the form of assessments of destabilizing effects $\hat{D}$. In the control unit (CU), control actions $u_B(kT_0)$ to the AU are generated to restore the operability of RCO.

The diagnostic unit (DU) is the key element in presented block diagram. The purpose of DU is to form a diagnosis of RCO. For this, it is required to detect the appearance of a destabilizing effect, localize and identify it by searching for a place, defining a class and a type.

The detection of a destabilizing effect is the identification of the fact of a deviation in the operability of RCO. The place of destabilization (concrete constructively completed part) is a fragment of RCO exposed to a destabilizing effect that caused its inoperability. A class is a group of destabilizing effects leading to the same type of disruption to the operability of a fragment of RCO. The type of destabilizing effect is its specific physical manifestation, eliminated with the help of available redundant means of restoring operability and compensating for deviations. Thus, in DU, deep diagnostics are carried out, which is a process of consistently detecting the fact of the appearance of destabilization, the place of its action, determining the class and a specific removable type.

The purpose of CO is the rational choice of such a surplus means from the available ones, the use of which makes it possible for the detected cause of destabilization $\hat{d}_i \in \hat{D}$ to restore the operation of RCO in the best possible way under the current conditions. Recovery of operability is a three-step process. At the first stage, there is a rational choice from the set of excess funds available in RCO and CU, which are most acceptable in the current emergency situation. On the second – countering the identified removable types of destabilizations using the selected tools. On the third, compensation of the deviations of the output variables of the CO from the required behavior that appeared during the diagnosis and the previous stages of restoration.

In the presented block diagram, the structure and functions of RCO differ from the traditional classical ones. Together with the necessary controllability and observability properties, RCO must have properties that ensure the diagnosability and recoverability of its operability [8].

The diagnosability of the RCO is understood as its property, which consists in the possibility of unambiguously detecting the removable cause of the destabilizing effect based on the states available for measurement in a finite time. The recoverability of RCO is understood to be such a property that, with the help of the available redundant means, it is possible to restore operability in a finite time. So, to ensure the diagnosability of RCO, the volumes of information and its nature corresponding to the depth of diagnosis are required, i.e. signals, and for recoverability – redundant funds.

Event uncertainty of destabilizing effects on RCO generates many of its possible states and necessitates rational control of its operability.

Rational control is formed as a result of a reasonable combination of analytical tools, model and bench experimental research, intuition and common sense of developers.

Rational control of objects under the action of event-driven uncertain destabilization is based on deep diagnostics of removable causes of malfunction and flexible recovery in real time. Therefore, rational control is the art of eventual intelligent control.

2.3 Experimental Characteristics of Functional Elements

For the formation of algorithms for diagnosing and restoring operability, mathematical models are required that reflect the transforming properties of the functional elements of the system. The construction of mathematical models is based on experimental static and dynamic characteristics. A series of experiments were carried out on the experimental stand and the static and dynamic characteristics presented in the Figs. 6, 7 and 8 were obtained.

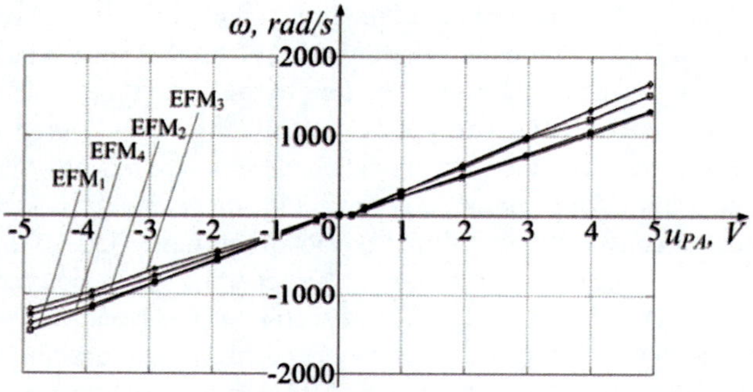

Fig. 6. Static characteristics of EFM

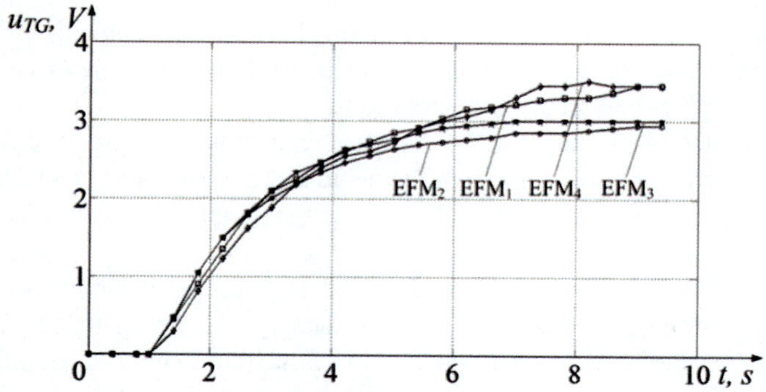

Fig. 7. Transient characteristics of EFM

The static characteristics of the EFM are symmetric with respect to the origin of the coordinates with a dead zone $[-0,3;\ +0,3]V$, due to friction. Dynamic transient characteristics are aperiodic. The static characteristics of the remaining functional elements are continuous and monotonous. Obviously, the given characteristics can be approximated by linear dependencies, and mathematical models can be represented in the class of

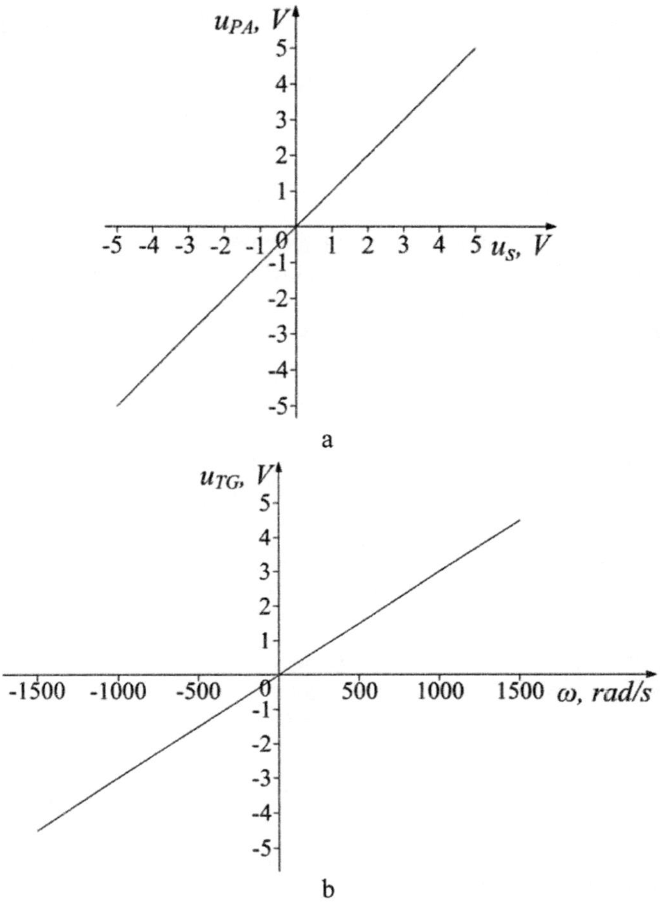

Fig. 8. Static characteristics of (a) PA and (b) TG

linear mathematical models using linear differential equations with constant coefficients and transfer functions.

As a result of processing the experimental data, mathematical models were obtained for all EFM. So, for a specific EFM1: 1) maximum linearization error $\delta_{\max} < 6\ \%$; 2) operating point [0; 0]; 3) input voltage range – $[-5;\ +5]\ V$; 4) angular velocity range – $[-1515;\ +1515]$ rad/s; 5) estimated coefficient values $\hat{\kappa}_{EM} = 303$ rad/V · s; $T_{EM} = 3$ s; transfer function $W_{EM}(s) = \frac{\Omega(s)}{U(s)} = \frac{\kappa_{EM}}{T_{EM}s+1}$; 7) time scale – $m_t = 1$.

For a specific PA1: 1) maximum linearization error – $\delta_{\max} < 0,5\ \%$; 2) operating point coordinates [0; 0]; 3) input signal range $[-5;\ +5]\ V$; 4) estimated coefficient value $\hat{\kappa}_{PA} = 2 \cdot 10^{-2}$; 5) transfer function $W_{PA}(s) = \frac{U_{PA}(s)}{\tau(s)} = \kappa_{PA}$; 6) time scale $m_t = 1$.

For the channel of the EFM prototype unit under consideration, the static characteristic of the used TG is approximated by a linear dependence with the following attributes: 1) linearization error $\delta_{\max} < 0,5\ \%$; 2) operating point coordinates [0; 0]; 3) input signal range $[-1515;\ +1515]$ rad/s; 4) output signal range $[-5;\ +5]\ V$; 5) estimated conversion factor $\hat{\kappa}_{TG} = 0,003$ V $\cdot$ s/rad; 6) transfer function $W_{TG}(s) = \frac{U_{TG}(s)}{\Omega(s)} = \kappa_{TG}$; 7) time scale $m_t = 1$.

Also, mathematical linear models were formed with all the attributes necessary for subsequent research for all functional elements of the experimental stand.

2.4 Mathematical Models

As a result of linear approximation of the experimental characteristics, linear mathematical models of the functional elements of the PA, EM and TG were obtained with the following attributes: 1) the accuracy of linearization; 2) the interval of changes in the input action; 3) the interval of the output signal change; 4) a mathematical description of the transforming properties; 5) the numerical values of the coefficients of the mathematical description; 6) the interval of variation and 7) the scale of the simulation time.

Mathematical descriptions of the transformation properties of functional elements are obtained in the form of transfer functions in the following form:

$$\text{PA:} W_{PA}(s) = \frac{U_{PA}(s)}{U_C(s)} = \kappa_{PA}; \tag{1}$$

$$\text{EM:} W_{EM}(s) = \frac{\Omega(s)}{U_{PA}(s)} = \frac{\kappa_{EM}}{T_{EM}s + 1}; \tag{2}$$

$$\text{TG:} W_{TG}(s) = \frac{U_{TG}(s)}{\Omega(s)} = \kappa_{TG}. \tag{3}$$

Here $\kappa_{PA}, \kappa_{EM}, \kappa_{TG}$ – the transfer factors of the corresponding functional elements; T_{EM} – time constant of EM.

In the vector-matrix form, the transformation of signal images in RCO can be represented in the following form:

$$\begin{bmatrix} U_{PA}(s) \\ U_{TG}(s) \end{bmatrix} = \begin{bmatrix} W_{PA}(s) \\ W_{PA}(s) W_{EM}(s) W_{TG}(s) \end{bmatrix} U_C(s). \tag{4}$$

In the space of continuous states, the transformation of temporal signals into RCO can be represented by a system of vector-matrix equations:

$$\dot{\omega}(t) = \frac{1}{T_{EM}}\omega(t) + \frac{\kappa_{PA}\kappa_{EM}}{T_{EM}}u_C(t);\ \omega(t_0) = \omega_0;$$

$$\begin{bmatrix} u_{PA}(t) \\ u_{TG}(t) \end{bmatrix} = \begin{bmatrix} \kappa_{PA} & 0 \\ 0 & \kappa_{TG} \end{bmatrix} \begin{bmatrix} u_C(t) \\ \omega(t) \end{bmatrix}. \tag{5}$$

To pass to the description in the space of discrete states, we take the following assumption. Signal conversion in functional elements of ADC and DAC is inertial-free with a transfer coefficient equal to one. The use of this assumption will simplify the subsequent presentation of the features of rational control. When forming discrete equations, the Euler formula is used:

$$\dot{\omega}(t) \approx \frac{\omega[(k+1)T_0] - \omega(kT_0)}{T_0}, \tag{6}$$

where T_0 – quantization period; $k = 0, 1, 2, ...$

After the appropriate transformations, the following system of discrete equations was obtained:

$$\omega(k+1) = \left(1 - \frac{T_0}{T_{EM}}\right)\omega(k) + \frac{\kappa_{PA}\kappa_{EM}T_0}{T_{EM}}u_C(k);\ \omega(k_0) = \omega_0;$$

$$\begin{bmatrix} u_{PA}(k) \\ u_{TG}(k) \end{bmatrix} = \begin{bmatrix} \kappa_{PA} & 0 \\ 0 & \kappa_{TG} \end{bmatrix} \begin{bmatrix} u_C(k) \\ \omega(k) \end{bmatrix}. \tag{7}$$

In these equations, to simplify notation, the parameter T_0 is omitted in the arguments of the variables.

The given mathematical descriptions reflect only the nominal operating mode of RCO.

2.5 Diagnostic Models

Diagnostic models are mathematical descriptions linking the cause of destabilization with its consequence. For the formation of diagnostic models (DM), it is necessary to identify destabilizing effects that disrupt the operability of RCO. The experimental stand has the ability to simulate 14 types of destabilization.

For a power amplifier, the set $D_1 = \{d_{11}, ..., d_{14}\}$ consists of the following types of destabilization:

1) d_{11} – reduction in transfer factor;
2) d_{12} – positive drift;
3) d_{13} – negative drift;
4) d_{14} – break at the input wire.

For an electric motor, the set $D_2 = \{d_{21}, d_{22}, d_{23}\}$ includes three types of destabilization:

1) d_{21} – reduction in transfer factor;
2) d_{22} – increasing the time constant;
3) d_{23} – break in the control circuit.

For a tachogenerator, the set D_3 includes such types of destabilization:

1) d_{31} – reduction in transfer factor;
2) d_{23} – positive drift;
3) d_{33} – negative drift;
4) d_{34} – break at the output wire.

Each of the 14 types of destabilization disrupts the operation of RCO. To detect this fact, a reference model is used in the form:

$$\hat{u}_{TG}(k+1) = \left(1 - \frac{T_0}{T_{EM}}\right)\hat{u}_{TG}(k) + \frac{T_0}{T_{EM}}\kappa_0 u_C(k);\ \hat{u}_{TG}(k) = 0, \tag{8}$$

where $\kappa_0 = \kappa_{PA}\kappa_{EM}\kappa_{TG}$. With this model, a deviation signal is generated.

$$\Delta u_{TG}(k+1) = \tilde{u}_{TG}(k+1) - \hat{u}_{TG}(k+1); \quad (9)$$

here $\tilde{u}_{TG}(k)$ – signal from the output of the destabilized RCO. To detect the fact of disruption of RCO, a predicate equation is used in the form:

$$z_0 = S_2\{|\Delta u_{TG}(k+1)| \geq \sigma_0\};\ k = \overline{0, n_0};\ \rho = 0,9, \quad (10)$$

where S_2 – two-digit predicate symbol; σ_0 – deviation tolerance; n_0 – number of measurements; ρ – confidence factor.

To determine the element under destabilization, a predicate is calculated

$$z_1 = S_2\{|\Delta u_{PA}(k)| \geq \sigma_1\};\ k = \overline{n_0, n_1};\ \rho = 0,9, \quad (11)$$

here $\Delta u_{PA}(k) = \tilde{u}_{PA}(k) - \hat{u}_{PA}(k)$.

DM for the type of destabilization d_{11} is described by the equation

$$\Delta u_{PA}(k) = \Delta\kappa_{PA}u_C(k), \quad (12)$$

where $\Delta\kappa_{PA}(k) = \tilde{\kappa}_{PA} - \kappa_{PA}$.

DM for the types of destabilization d_{12} and d_{13} will be such:

$$\Delta u_{PA}(k) = \tilde{u}_{PA0}, \quad (13)$$

where Δu_{PA0} – value of drift.

DM for the type of destabilization d_{14}:

$$\Delta u_{PA}(k) = -\hat{u}_{PA}(k), \quad (14)$$

For EM the forms of DM will be the follows:

$$d_{21}\colon \Delta u_{TG}(k+1) = \left(1 - \frac{T_0}{T_{EM}}\right)\Delta u_{TG}(k) + \frac{T_0}{T_{EM}}\kappa_{TG}\Delta\kappa_{EM}u_{PA}(k); \quad (15)$$

$$d_{22}\colon \Delta u_{TG}(k+1) = \left(1 - \frac{T_0}{T_{EM}}\right)\Delta u_{TG}(k) + \frac{T_0}{T_{EM}^2}[u_{TG}(k) - \kappa_{EM}\kappa_{TG}u_{PA}(k)]\Delta T_{EM}; \quad (16)$$

$$d_{23}\colon \Delta u_{TG}(k+1) = -\hat{u}_{TG}(k+1). \quad (17)$$

For TG four forms of DM were obtained:

$$d_{31}\colon \Delta u_{TG}(k+1) = \left(1 - \frac{T_0}{T_{EM}}\right)\Delta u_{TG}(k) + \frac{T_0}{T_{EM}}\kappa_{EM}\Delta\kappa_{TG}u_{PA}(k); \quad (18)$$

$$d_{32}, d_{33}\colon\ \Delta u_{TT}(k+1) = \left(1 - \frac{T_0}{T_{ЭД}}\right)\Delta u_{TT}(k) + \frac{T_0}{T_{ЭД}}\tilde{u}_{TT0}; \quad (19)$$

$$d_{34}\colon \Delta u_{TT}(k+1) = -\hat{u}_{TT}(k+1). \quad (20)$$

With the help of the DM, the arguments of the predicate equations are formed, from which the procedure for searching for the types of destabilization of RCO is constructed.

2.6 Diagnosis of Destabilizing Effects

Destabilizing effects disrupt the operability of the RCO. According to the main provisions of rational control, destabilizing effects are one-time undefined events. Uncertainty is due to the unknown time of their appearance, place of origin, class and specific type. In conditions of such uncertainty, diagnosis is able only with the help of the procedure for searching for a specific type of destabilizing effect according to its signs. For the formation of signs of destabilizing effects, it is advisable to form this knowledge using the corresponding languages of knowledge representation. Typical models of knowledge representation are: production; message boards; framing, semantic networks and logical [16, 17]. The most adequate for the formation of signs of destabilizing effects, as evidenced by the results of experimental studies, is the production model. In the production model, knowledge is represented by a set of rules “IF – THEN”. To form the rules, two-valued predicate equations obtained from the corresponding diagnostic models are used in the article. The construction of two-valued predicate equations and the features of their formation for predicates z_0 and z_1 were described in Sect. 2.5. Without going into the details of the formation of knowledge about the signs of destabilizing effects, only the expressions of two-valued predicate equations are presented here.

For PA, signs for the types d_{11}, ..., d_{14} are:

$$z_2 = S_2\{\sigma_2 - \Delta u_{PA}(k)\};\ k = \overline{n_1, n_2};\ \rho = 0,8, \tag{21}$$

$$z_3 = S_2\{\sigma_3 - |\Delta u_{PA}(k+1) - \Delta u_{PA}(k)|\};\ k = \overline{n_2, n_3};\ \rho = 0,8, \tag{22}$$

For combination “EM and TG”, signs of the types of destabilization d_{21}, ..., d_{23} and d_{31}, ..., d_{34} are:

$$z_4 = S_2\{\sigma_4 - \Delta u_{TG}(k)\};\ k = \overline{n_1, n_2};\ \rho = 0,8, \tag{23}$$

$$\begin{aligned} z_5 = S_2\{\sigma_5 - |[\Delta u_{TG}(k+1) - \Delta u_{TG}(k)]u_{PA}(k) \\ - [\Delta u_{TG}(k+2) - \Delta\ u_{TG}(k+1)]u_{PA}(k)|\};\ k = \overline{n_2, n_3};\ \rho = 0,8; \end{aligned} \tag{24}$$

$$\begin{aligned} z_6 = S_2\{\sigma_6 - |[\Delta u_{TG}(k+1) - \Delta u_{TG}(k)]u_{TG}(k+1) - u_{PA}(k+1) \\ - [\Delta u_{TG}(k+2) - \Delta\ u_{TG}(k+1)][u_{TG}(k) - u_{PA}(k)]|\};\ k = \overline{n_3, n_4};\ \rho = 0,8; \end{aligned} \tag{25}$$

$$z_7 = S_2\{\sigma_7 - |\Delta u_{TG}(k+2) - 2\ \Delta u_{TG}(k+1) + \Delta u_{TG}(k)|\ \};\ k = \overline{n_4, n_5};\ \rho = 0,8. \tag{26}$$

The obtained signs of destabilizing effects allow to proceed to the formation of the structure of logical inference in the form of a dichotomous tree (Fig. 9).

In the presented search procedure, the types of destabilizing effects: d_{23} – break in the control wire of EM and d_{34} – break of the output wire of TG are not diagnosed separately. Also such types are not diagnosed: d_{21} – reduction in transfer factor of EM and d_{31} – reduction in transfer factor of TG.

It makes no sense to increase the depth of diagnostics of these types of destabilizing effects due to the identity of the means used to restore the operability of RCO.

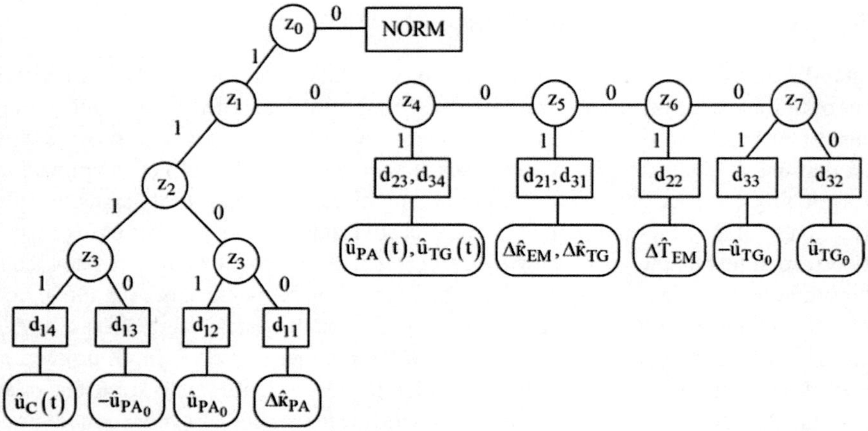

Fig. 9. Dichotomous tree for searching for types of destabilizing influences in RCO

For the formation of procedures for restoring the operability of the RCO, it is necessary to know the magnitudes of the destabilizing effects. Therefore, the dichotomous tree ends with "leaves", which contain the estimated values of the parameters that characterize the magnitude of destabilizing effects. Estimated values are obtained from the corresponding DM after arithmetic mean processing. So, for PA:

$$d_{14} \div \hat{u}_C(t) \equiv 0; \quad d_{13} \div -\hat{u}_{PA_0} = \frac{1}{m}\sum_{k=1}^{m} \Delta u_{PA}(k);$$
$$d_{12} \div u_{PA_0} = \frac{1}{m}\sum_{k=1}^{m} \Delta u_{PA}(k); \quad d_{11} \div \Delta\hat{\kappa}_{PA} = \frac{1}{m}\sum_{k=1}^{m} \frac{u_{PA}(k)}{u_C(k)}. \tag{27}$$

Both for EM and TG:

$$d_{23}, d_{34} \div \hat{u}_{PA}(t) \equiv 0; \; \hat{u}_{TG}(t) \equiv 0;$$
$$d_{21}, d_{31} \div \Delta\hat{\kappa}_{EM} = \frac{1}{mT_0\kappa_{TG}} \sum_{k=1}^{m} \frac{T_{EM}\Delta u_{TG}(k+1) - (T_{EM} - T_0)\Delta u_{TG}(k)}{u_{YM}(k)};$$
$$\Delta\hat{\kappa}_{TG} = \frac{1}{mT_0\kappa_{EM}} \sum_{k=1}^{m} \frac{T_{EM}\Delta u_{TG}(k+1) - (T_{EM} - T_0)\Delta u_{TG}(k)}{u_{PA}(k)};$$
$$d_{22} \div \Delta\hat{T}_{EM} = \frac{T_{EM}}{mT_0} \sum_{k=1}^{m} \frac{T_{EM}\,\Delta u]_{TG}(k+1) - (T_{EM} - T_0)\Delta u_{TG}(k)}{u_{TG}(k) - \kappa_{EM}\kappa_{TG}u_{PA}(k)}; \tag{28}$$
$$d_{33} \div -\hat{u}_{TG_0} = \frac{1}{mT_0} \sum_{k=1}^{m} [T_{EM}\,\Delta u_{TG}(k+1) - (T_{EM} - T_0)\Delta u_{TG}(k);$$
$$d_{32} \div \hat{u}_{TG_0} = \frac{1}{mT_0} \sum_{k=1}^{m} [T_{EM}\,\Delta u_{TG}(k+1) - (T_{EM} - T_0)\Delta u_{TG}(k).$$

As follows from the above expressions, the estimated parameter values are calculated for m discrete signal values by means of arithmetic mean averaging of the results.

The complete diagnosis obtained in this way makes it possible to proceed to the procedure for restoring the RCO operability with the help of the appropriate use of the available redundant resources.

2.7 Restoring the Operability of the Flywheel Motor Unit

Destabilizing effects from the sets D_1, D_2 and D_3 lead to disruption of the functioning of the RCO, i.e. to disrupt its performance. To restore operability, redundant tools and means of putting them into operation, algorithms and programs for compensating for the consequences of destabilizing effects are needed. The experimental stand implements the simplest version of the use of excess funds, when each type of destabilizing effect uses its own tool. Let's consider these tools in more detail for each functional element.

So, as a result of diagnosing PA, such types are identified as d_{11} – reduction of transfer factor; d_{12} – positive drift; d_{13} – negative drift; d_{14} – break of input wire.

Under the action of a destabilizing effect d_{11}, its influence is compensated by means of signal adjustment. In this case, to the control signal of the PA, $u_C(k)$ the additional signal is added

$$u_A(k) = \Delta\hat{\kappa}_{PA} u_C(k), \tag{29}$$

then such a control signal

$$u_C^{'}(k) = (1+\Delta\kappa_{PA}) u_C(k), \tag{30}$$

will compensate the decreasing of the transfer factor of PA.

With a positive drift d_{12}, its compensation is performed by a control signal of this type:

$$u_C^{'}(k) = u_C(k) - u_{PA_0}, \tag{31}$$

With a negative drift d_{13} the control signal is shifted by the value of the estimate – $\hat{u}_{YM_0}$, i.e.

$$u_C(k) = u_C(k) + \hat{u}_{PA_0}. \tag{32}$$

Under a destabilizing effect d_{14}, a complete failure of the PA occurs and therefore the channel is removed from the process of functioning by turning off the power supply of the PA, EM and TG and its input signals are blocked.

Diagnostics of the "EM and TG" fragment allows to determine the following types of destabilizing effects: d_{21} – reduction of transfer factor; d_{22} – increasing of the time constant; d_{23} – break of control wire; d_{31} – reduction of transfer factor; d_{32} – positive drift; d_{33} – negative drift; d_{34} – break of the output wire.

With a decrease in the EM transfer factor, it is possible to restore its operability by means of a static adjustment, which consists in such an additional control signal to the main control signal that the previous value of the angular velocity and, consequently, the torque $m(t)$ are restored. An additional constant signal compensating for the decrease of $\kappa_{ЭД}$is calculated in this way:

$$u_C^{'}(k) = \left(1 + \Delta\hat{\kappa}_{EM}\kappa_{TG}\right)u_C(k). \quad (33)$$

Such a new control signal provides high-quality compensation for the destabilizing effect d_{31} and restoration of the RCO functioning.

In case of a destabilizing effect d_{22} – an increase in the constant time, to restore the operability of RCO, the procedure of phase advance of the control signal is used using the following algorithm:

$$u_C^{'}(k+2) = u_C(k+1) + \left(1 - \frac{T_0}{a}\right)u_C(k), \quad (34)$$

where a – coefficient depending on the estimated value of the parameter $\Delta\hat{T}_{EM}$.

In case of a break in the control circuit – type d_{23}, the channel is completely inoperative and therefore the procedure is used to remove it from the functioning process by disconnecting the power supply of all functional elements of the RCO and blocking the input and output signals.

In case of a destabilizing effect d_{31} – reduction of transfer factor of TG, and this destabilizing effect is indistinguishable from the destabilizing effect d_{21} (Fig. 9), restoration is performed according to the following logic. If, when using the control signal (33), RCO is not completely restored, which will be shown by the repeated diagnostic procedure, then it is needed to use a control signal of this type:

$$u_C^{'}(k) = \left(1 + \kappa_{EM}\,\Delta\hat{\kappa}_{TG}\right)u_C(k), \quad (35)$$

which will provide a complete and high-quality restoration of the RCO.

When a positive drift d_{32} appears in the operation of the TG, the control signal $u_C(k)$ is shifted by the value $\hat{u}_{TG_0}(k)$. The corrective control signal is formed as follows:

$$u_C^{'}(k) = u_C(k) - \hat{u}_{TG_0}, \quad (36)$$

When a negative drift in the TG appears, the control signal will be as follows:

$$u_C^{'}(k) = u_C(k) + \hat{u}_{TG_0}. \quad (37)$$

If the destabilizing effect is a break in the output of the TG – d_{34}, then it is possible to use this channel for generating the kinetic momentum, but the accuracy of orientation and stabilization of the gimbal frame will be lower due to the lack of feedback. Increasing the positioning accuracy of the gimbal suspension frames and, accordingly, the model block of electric flywheel motors is possible with the use of angular position sensors of the frames and, thus, allows to fully restore the channel operability.

It is possible to restore the TG measurement if there is additional information about the EM operable state. In this case, it can be used the reference model of the RCO that reproduces the signal $\hat{u}_{TG}(t)$.

2.8 Experimental Characteristics of Rational Control of the Operability of One Channel of the Electric Flywheel-Motor Unit

For the prototype unit of electric flywheel motors, an algorithm and software were developed for diagnosing and restoring its operability under a given set of destabilizing effects. After appropriate debugging of algorithms and programs, a series of experimental studies was carried out on the fundamental possibility of joint use of deep diagnostics and flexible restoration of the operability of the control channels of the electric flywheel-motor unit. Using one channel as an example, let's consider the nature of transient processes in diagnosing and restoring the operability of the RCO when simulating specific destabilizing effects. The experiments were carried out for test actions in the form of a step function. In Fig. 10(a) the graph of the setting action $u_S(t)$, is presented, which is given when $t = 1$ s. In Fig. 10(b) the response of the RCO to a step signal – the transient response of the RCO is shown.

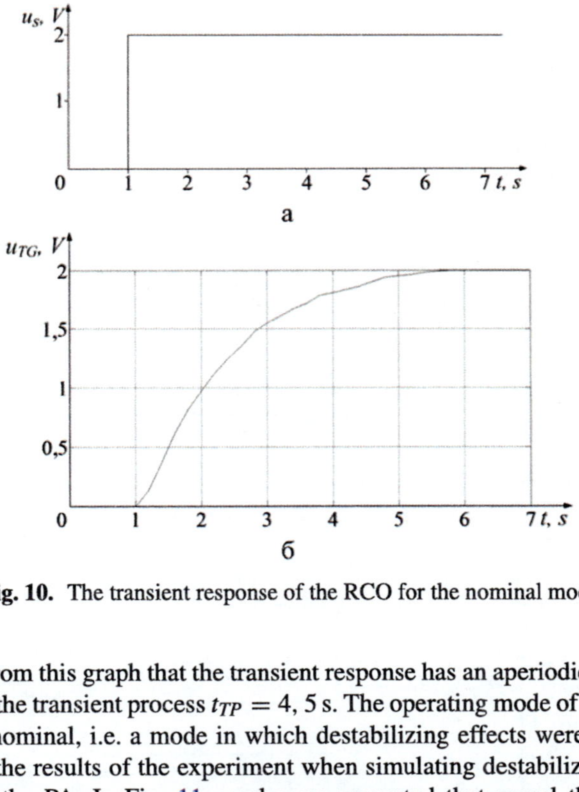

Fig. 10. The transient response of the RCO for the nominal mode

It follows from this graph that the transient response has an aperiodic character with the duration of the transient process $t_{TP} = 4, 5$ s. The operating mode of the RCO in this experiment is nominal, i.e. a mode in which destabilizing effects were not simulated. Let's consider the results of the experiment when simulating destabilization d_{11} in the functioning of the PA. In Fig. 11 graphs are presented that reveal the processes of imitation, diagnostics and restoration of the RCO operability. The step-type reference action, which is supplied to the switchgear in the first second is shown in Fig. 11(a).

Figure 11(b) shows the nature of the change in the transfer factor of the PA. So, before $t = 2$ s the transfer factor $\kappa_{PA} = 1$, and at the moment of time $t = 2$ s the transfer factor decreases by 0,2 and became $\tilde{\kappa}_{PA} = 0{,}8$. Figure 11(c) reflects the response of the

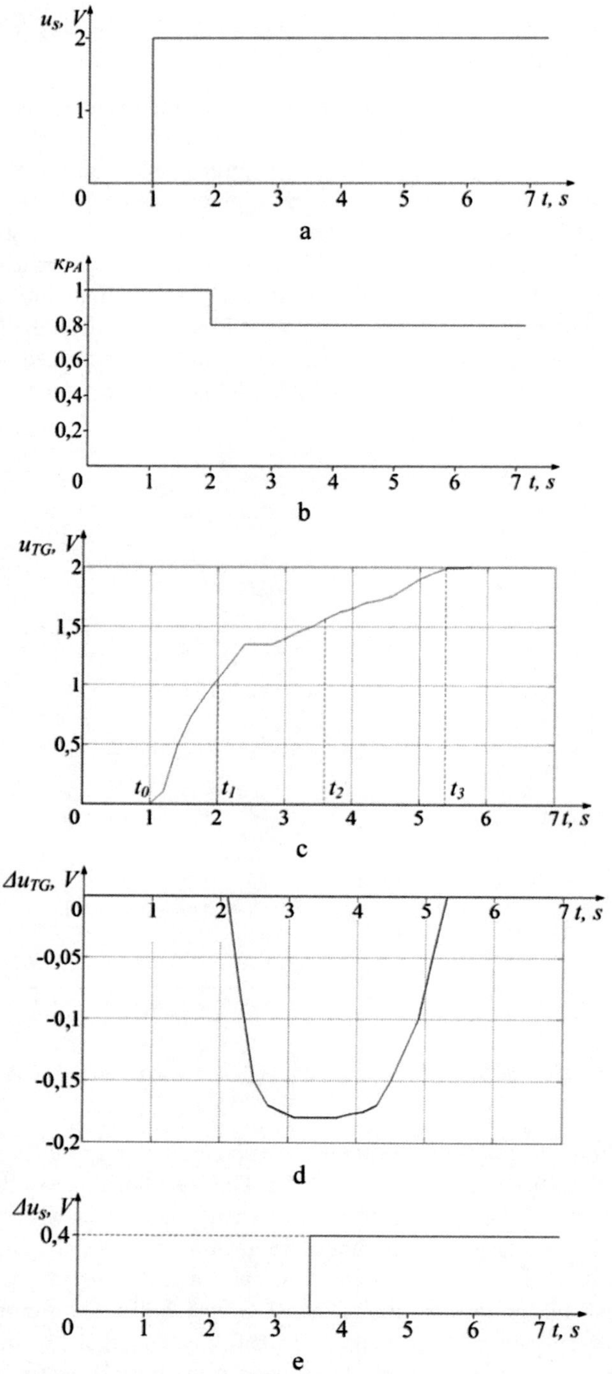

Fig. 11. Transient graphs of RCO at the type of destabilization d_{11}

RCO to a step input signal and a step reduction in the PA gain. Let's distinguish two periods in this reaction. The first is the period of diagnosis. It starts at $t_1 = 2$ s and continues until $t_2 = 3{,}5$ s. The duration of this period $\Delta t_1 = 1{,}5$ s. The second one is the recovery period. It begins after the period of diagnosis, when a complete diagnosis is formed, i.e., at $t_2 = 3{,}5$ s and continues until $t_3 = 5{,}5$ s, when the complete restoration of the RCO operability ends. The duration of this period $\Delta t_2 = 2$ s.

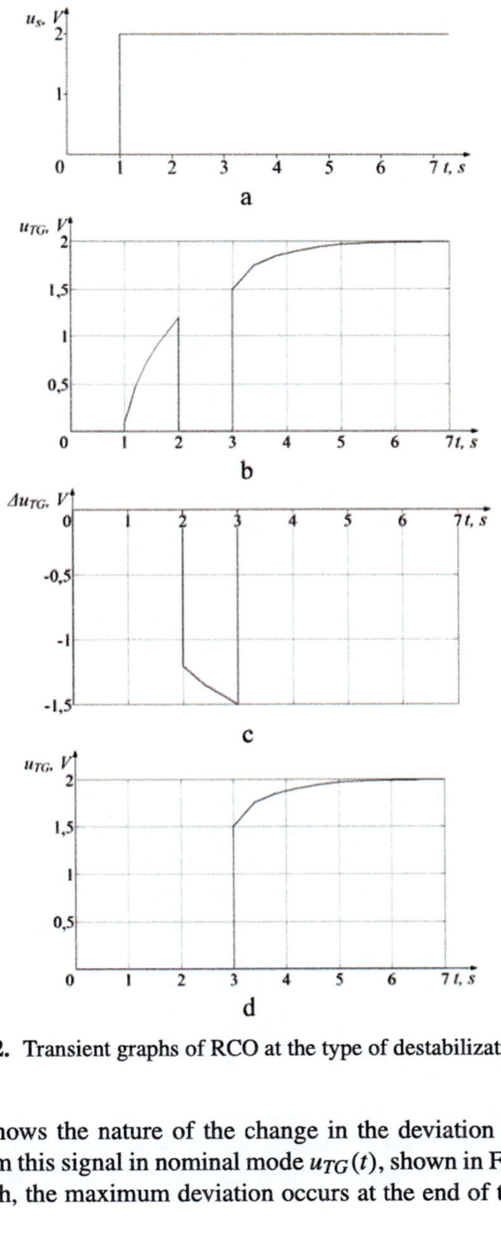

Fig. 12. Transient graphs of RCO at the type of destabilization d_{34}

Figure 11(d) shows the nature of the change in the deviation of the RCO output signal – $\tilde{u}_{TG}(t)$ from this signal in nominal mode $u_{TG}(t)$, shown in Fig. 10(b). As can be seen from the graph, the maximum deviation occurs at the end of the diagnosis and at the beginning of the recovery. Figure 11(e) shows a graph of the additional signal to the reference action. This is how a signal adjustment is implemented, which compensates for the destabilization d_{11}, which consists in a decrease in the transfer factor of the PA – $\Delta\kappa_{PA} = 0,2$. It can be seen from the graphs of transient processes that the total period of diagnostics and restoration of the RCO operability is $\Delta t = \Delta t_1 + \Delta t_2 = 1,5 + 2 = 3,5$ s. The time of the transient process, both in the nominal and in the destabilized modes, has not changed and is $t_{TP} = 4,5$ c. Thus, the time interval Δt for diagnosing and restoring the operability of a destabilized RCO is less than the time of its transient process t_{TP}. In other words, during the transition process, it was possible to carry out a complete diagnosis, i.e. identify the cause of the destabilization and calculate the estimate of its magnitude $\Delta\hat{\kappa}_{PA}$, as well as form and implement the signal adjustment of the RCO so that by the end of the transient process, the operability is restored.

The results of experimental studies of the system of rational control of the operability of one channel of the electric flywheel-motor unit are presented in Fig. 12, when simulating such a type of destabilization as a break in the output wire of the TG – d_{34}. A step action of 2 V is supplied to the RCO at $t_1 = 1$ s (Fig. 12(a)). At the moment of time $t_2 = 2$ s a break of the TG output wire is simulated and the signal disappears.

Until the moment of time $t_3 = 3$ s diagnostics takes place and the reason for the disappearance of the signal is determined – break of the output wire – d_{34} (Fig. 12(b)). The nature of the signal deviation $u_{TG}(t)$ in the nominal mode is shown in Fig. 12(c). To restore the operability of the RCO in this experiment, such a resource was used as a reference model that generates a signal $\hat{u}_{TG}(t)$. Since the point in time data from the reference model has been used (Fig. 12(d)). As one can see in the graph in Fig. 12(b), from the moment of time t_3 the RCO operability was restored, corresponding to the nominal mode of operation. The deviation signal is zero from the moment of time $t_3 = 3$ s, which testifies to the prompt and high-quality restoration of the RCO operability.

Experiments with the imitation of other types of destabilization of the RCO's operability made it possible to make sure that it is in principle possible to diagnose in real time and restore operability during the transient process for diagnosed types of destabilization.

3 Concluding Remarks

In the implementation of long-term space missions, in most cases, the main mode of spacecraft control is attitude control and stabilization. For such a mode, actuators are required that use electrical energy obtained on board with the help of solar panels. EFM represent such a class of actuators for long-term flights. The technical impossibility of providing a long-term resource of trouble-free operation of the EFM gives rise to the need to search for approaches to ensure fault-tolerant operation. Traditional approaches using structural redundancy according to the majority principle lead to a significant increase in the mass-dimensional, energy and cost characteristics of the actuators. Expanding the functionality of actuators by automating functions such as diagnosing and restoring their operability in real time is a promising direction for intellectualization of control. New theoretical results on the development of the principle of control by diagnosis made it possible to form a new class of adaptive control systems based on deep diagnostics and flexible restoration of operability. This class is called the class of rational control systems. This paper presents the main results of theoretical and experimental studies on the rational control of the operability of the EFM prototype unit of a pyramidal structure.

The results of the studies carried out confirmed the fundamental possibility of deep diagnostics of the EFM control channel during the transient process and the possibility of restoring operability before the end of the transient process when simulating a number of destabilizing effects.

The results obtained open up the opportunity to move on to solving new problems of forming a compact modular software structure for diagnosing and restoring the operability of the entire EFM unit.

The pyramidal structure of the EFM prototype unit is an excess structure and therefore it is of interest to solve the problem of using other EFM control channels when one of them fails.

It also seems promising to expand the capabilities of diagnostic procedures with the necessary means of using signals from the spacecraft angular position sensors. Such use will expand the set of diagnosed destabilizing effects and increase the efficiency of obtaining a diagnosis.

The solution of these and other problems, which will inevitably arise as a result of research and experimental work, will make it possible to develop effective actuators based on EFM for orientation and stabilization systems of modern and promising spacecraft for long-term space missions.

References

1. Isermann, R.: Fault-Diagnosis Systems—An Introduction from Fault Detection to Fault Tolerance. Springer, Heidelberg (2006). https://doi.org/10.1007/3-540-30368-5
2. Isermann, R.: Fault-Diagnosis Applications: Model-Based Condition Monitoring: Actuators, Drives, Machinery, Plants, Sensors, and Fault-Tolerant Systems. Springer, Heidelberg (2011). https://doi.org/10.1007/978-3-642-12767-0
3. Blanke, M., Kinnaert, M., Lunze, J., Staroswiecki, M.: Diagnosis and Fault-Tolerant Control, 2nd edn. Springer, Heidelberg (2006). https://doi.org/10.5555/1197019
4. Hajiyev, C., Caliskan, F.: Fault Diagnosis and Reconfiguration in Flight Control Systems. Kluwer Academic Publishers, London (2003)
5. Noura, H., Theilliol, D., Ponsart, J.-C.: Fault-tolerant Control Systems: Design and Practical Applications. Springer, Heidelberg (2009). https://doi.org/10.1007/978-1-84882-653-3
6. Kulik, A.S.: Signal'no-parametricheskoye diagnostirovaniye sistem upravleniya. Khar'kov: Gos. aerokosm. unt «KHAI», Biznes Inform. (2000) (in Russian)
7. Kulik, A.S.: Rational control of the operability of autonomous aircrafts. Part II. J. Autom. Inf. Sci. **49**(8), 1–18 (2017). https://doi.org/10.1615/JAutomatInfScien.v49.i8.10
8. Kulik, A.S.: Elementy teorii ratsional'nogo upravleniya ob"yektami. Xarkív: Nats. ayerokosm. un-t. «KHAÍ» (2016) (in Russian)
9. Baldi, P., et al.: Combined singular perturbations and nonlinear geometric approach to FDI in satellite actuators and sensors. IFAC-PapersOnLine **50**(1), 7253–7259 (2017). https://doi.org/10.1016/j.ifacol.2017.08.1375
10. Baldi, P., et al.: Fault diagnosis for satellite sensors and actuators using nonlinear geometric approach and adaptive observers. Int. J. Robust Nonlinear Control **29**(16), 5429–5455 (2018). https://doi.org/10.1002/rnc.4083

11. Muthusamy, V., Kumar, K.D.: A novel data-driven method for fault detection and isolation of control moment gyroscopes onboard satellites. Acta Astronaut. **180**, 604–621 (2021). https://doi.org/10.3390/electronics10131537
12. Long, D., et al.: A data fusion fault diagnosis method based on LSTM and DWT for satellite reaction flywheel. Math. Probl. Eng. **2020**, 2893263 (2020). https://doi.org/10.1155/2020/2893263
13. Li, G., et al.: The flywheel fault detection based on Kernel principal component analysis. In: IEEE 3rd Information Technology, Networking, Electronic and Automation Control Conference (ITNEC-2019), pp. 425–432 (2019). https://doi.org/10.1109/ITNEC.2019.8729163
14. Cheng, X., et al.: Design and implementation of a fault-tolerant magnetic bearing control system combined with a novel fault-diagnosis of actuators. IEEE Access **9**, 2454–2465 (2020). https://doi.org/10.1109/ACCESS.2020.3046668
15. Melanchenko, A.G.: Fault-tolerant control of earth observation satellite with redundant configuration of reaction wheels. Kosmichna Nauka i Tekhnolohiya **19**(1), 28–37 (2013) (in Russian). https://doi.org/10.15407/knit2013.01.028
16. Benzmüller, C., Miller, B. (eds.): CICM 2020. LNCS (LNAI), vol. 12236. Springer, Cham (2020). https://doi.org/10.1007/978-3-030-53518-6
17. Russell, S., Norvig, P.: Artificial Intelligence: A Modern Approach, 4th edn. Pearson, New York (2020)

Design and Analysis of the Bimetallic Fan Blade

Volodymyr Martynenko(✉)

Department of Dynamics and Strength of Machines, National Technical University "Kharkiv Polytechnic Institute", NTU "KhPI", 2, Kyrpychova str., Kharkiv 61002, Ukraine

Abstract. The paper considers a design of a bimetallic blade of a mine ore ventilation fan that was created to achieve a light weight combined with a high strength of its elements. The proposed concept of the bimetallic blade composes advantages of aluminum for producing an airfoil with a rigidity of steel as a material of a blade root. Chill casting of aluminum material onto a specially designed steel skeleton resulted in a close and deep interaction of two medias and required an accurate finite-element modelling of a cyclically symmetric sector of a fan rotor. Such the modelling was performed using nonlinear contact formulation that enabled correct reflection of an interaction of steel and aluminum medias and gain the stress-strain distribution along the considered bodies. Analyzing this distribution indicates that the bimetallic blade is durable enough and capable to maintain the workloads. The provided technique can be used further in aerospace applications for the creation of light and durable metallic blades.

Keywords: Bimetallic blade · Fan installation · Finite-element analysis · Structural strength · Fluid-structure interaction

1 Introduction

Fans are widely used for cooling and ventilation at commercial and industrial enterprises, coalpits and mines. The topic of evaluating fan strength is developed enough. However, there are many scientific and engineering problems related to it.

The fan airfoil is subjected to unsteady air flow which can increase static stresses and cause vibrations. The works [1, 2] presented an experimental investigation of emergence of fan flutter in a wind tunnel with the transient growth analysis. The study [3] developed a technique for simulating the flutter effect of a wide-chord axial fan airfoil. The paper [4] conducted a comprehensive study of fan flutter effect for various mass ratios and blade stiffness parameters and the work [5] investigated this effect under acoustic treatment. The papers [6, 7] described the process of static and dynamic strength evaluation of industrial axial fans.

As friction acts a significant role in holding together the bolted or prestressed assembly, it is important to take it into account when modeling the mechanical behavior of fans. The work [8] gives a comprehensive view on the problem of friction modelling for engineering purposes. The paper [9] covers the question of overcoming the nonlinearity issues caused by frictional contacts in blade joints. The article [10] investigates the

M. Nechyporuk et al. (Eds.): ICTM 2021, LNNS 367, pp. 437–448, 2022.
https://doi.org/10.1007/978-3-030-94259-5_37

localization of slip zones in blade roots for dynamic problems taking into account the hysteresis process in frictional joints.

Fans are supposed to move the air and it can be full of dirt and partial inclusions. That's why the problem of the increased wear is relevant, especially for aluminum airfoils which are common for axial fans. The work [11] proposed a numerical algorithm which allows simulating the wear process of bladed impellers while a comparison with experimental results showed its good accuracy. The authors of the work [12] performed a comprehensive experimental study of the surface quality and wear process of aluminum by analyzing its microstructure. The study [13] proposed a crack detection method based on evaluation of blade natural frequencies while the investigation [14] illustrated the research on fault diagnostics of the turbofan engine blade.

The idea of combining two materials in macro scales was already proposed by scientists and engineers. For example, the works [15, 16] describe a technique for a fabrication and calculation of strength of a bimetallic lock for a gas turbine engine impeller. In such the connection, fixation occurs by means of the locking process, that is, the elements are combined by friction.

The authors of the paper [17] state the concept of applying a thermosetting outer layer of an alloy on a light inner core of a turbine blade using the soldering process. The work [18] considers a process of forming bimetallic damps in turbine blades by an injection of a main material of molten superalloys into prepared cavities. The paper [19] discusses the possibility of creating a bimetallic blade airfoil by removing a part of the first material near its tip and then applying the second material to a resulting cavity by the surfacing process.

Eventually there is no a lot of concepts for bimetallic rotor blades, especially in application to fan systems. But the expected benefits of using such the construction are attractive, namely: the use of a steel alloy for manufacturing the blade root provides increased strength of the attachment, and an aluminum alloy used for manufacturing an airfoil results in its reduced weight and ease of profiling; in case of an abrasive wear of the airfoil, it is possible not to make a whole new blade, but to replace this airfoil by casting it onto the steel part that was used previously, which significantly reduces the cost of planned and unplanned replacements of the fan blades; the process of manufacturing the blades becomes cheaper due to a reduced use of cast aluminum in the blade design, which is more expensive than the steel elements.

The static and dynamic structural analyses of the considered main ventilation fan rotor presented in Fig. 1 (left) with an impeller of 3.2 m diameter and aluminum blades was well-studied in the previous works. A bimetallic blade concept in Fig. 1 (right) was proposed for this rotor. The initial design includes a steel rotor with cast aluminum blades fixed in their seats using two bolts with wedge-shaped parts.

Since the design of a bimetallic fan blade is novel, there is a lack of techniques and some uncertainties in its modeling. The most challenging task is to simulate the interface between the steel and aluminum domains. The only way to understand an adequate model for it is visual observations and experimental investigations. The current paper discusses the details of the physical model of the bimetallic fan blade and presents the results of nonlinear finite-element analysis of the blade. The novelty of this approach is caused by a non-developed apparatus for modeling this kind of design.

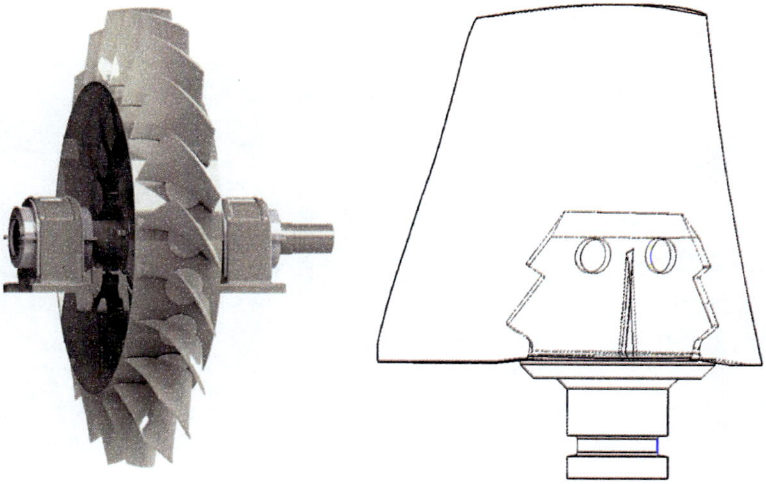

Fig. 1. Design elements of the fan (left – the rotor; right – the bimetallic blade).

2 Concept of the Bimetallic Fan Blade

As it was mentioned above, the blade root was supposed to be made of a steel hot rolled alloy. The airfoil, which should meet the requirements of the high technological and aerodynamic loads, was designed as an aluminum alloy part. A combination of two heterogeneous materials was performed by a steel core (Fig. 2) which should be welded to a steel root and contacted with an aluminum airfoil.

The shape of the steel blade root was taken similar to the one of the usual aluminum blades in order to fit their seats in the hub. The root was produced from the hot rolled steel machined in order to get the desired shape. The steel core was made by rolling the previously cut plane sheet of metal in the rolling drum. This provided a good manufacturability of the core and a high accuracy of the manufacture process.

The aluminum airfoil was produced by the casting process. The steel root welded with the steel core were put into a casting mold (Fig. 2). The fact that the steel root shape coincided with the one of the usual aluminum root allowed having a good fixation in the mold, as well as preventing the leakage of aluminum into the root zone. As a result, aluminum liquid got only to the airfoil zone which made it possible to have the aluminum airfoil deposited on the steel core after its solidification.

A shape of the core was designed considering all the requirements for reaching the best contact between steel and aluminum parts, as well as the highest bending stiffness when they are subjected to centrifugal loads. Figure 2 also illustrates the chill casting process: the aluminum part (III) is put onto the steel core and the root (I) using the chill mold (II). This results in a closed contact between steel and aluminum medias (Fig. 3). As it follows from Fig. 3 (left), there is a full penetration of aluminum into holes of the core, and Fig. 3 (right) indicates the full connection between two different parts. These observations were used for an appropriate choice of contact details for a structural analysis of the bimetallic blade.

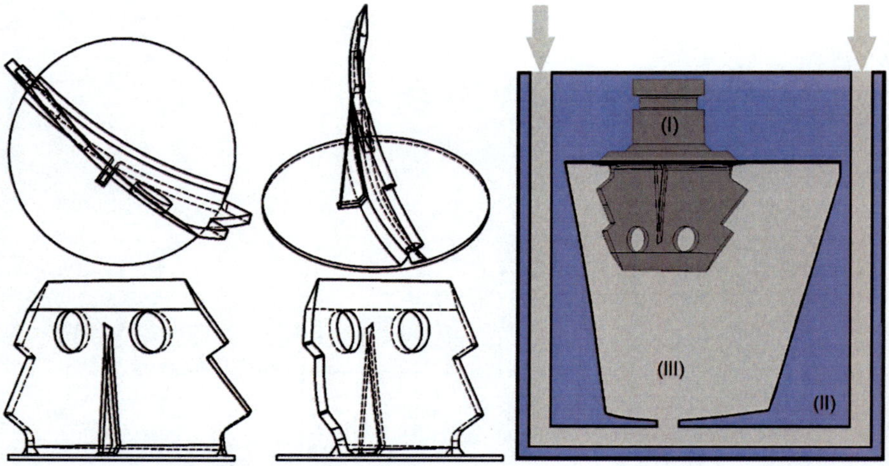

Fig. 2. Bimetallic blade – core design and casting process.

Fig. 3. The physical contact between aluminum and steel (left – casting in holes, right – cross section).

3 Structural Analysis of the Bimetallic Blade

The structural analysis of fans can be performed using modern CAE codes which provide finite-element analysis of the mechanics of solid deformable bodies. ANSYS Workbench allows modeling problems of high complexity as well as performing nonlinear static and dynamic analyses with geometrical, contact and material types of nonlinearities [20, 21]. The symmetry of the geometrical model of the rotor (Fig. 1), as well as centrifugal loads allowed building a cyclically symmetric calculation model (Fig. 4). The mesh quality was checked by executing the series of trial linear calculations with different mesh sizes and by comparing the maximal von Mises equivalent stress values for nodal and element solution plots. This pre-analysis allowed concluding that the approved finite-element mesh delivered the error less than 2% which is applicable value for both engineering and scientific purposes.

The finite-element model of the cyclically symmetric rotor with the blade included 328352 solid finite elements with the second order shape functions. Depending on regularity of the geometrical models of structural elements, the corresponding mesh consisted of either mapped hexahedron parts for the zones of interest or free hexahedron ones for the less important ones. The airfoil and the core were meshed by tetrahedrons in order to provide the best transition of finite elements in locations with the most complicated geometrical shape. This model considers a nonlinear contact between the aluminum airfoil and the steel core, as well as bonded contacts between the other contacting parts in order to reduce a calculation complexity of the analysis (Fig. 5).

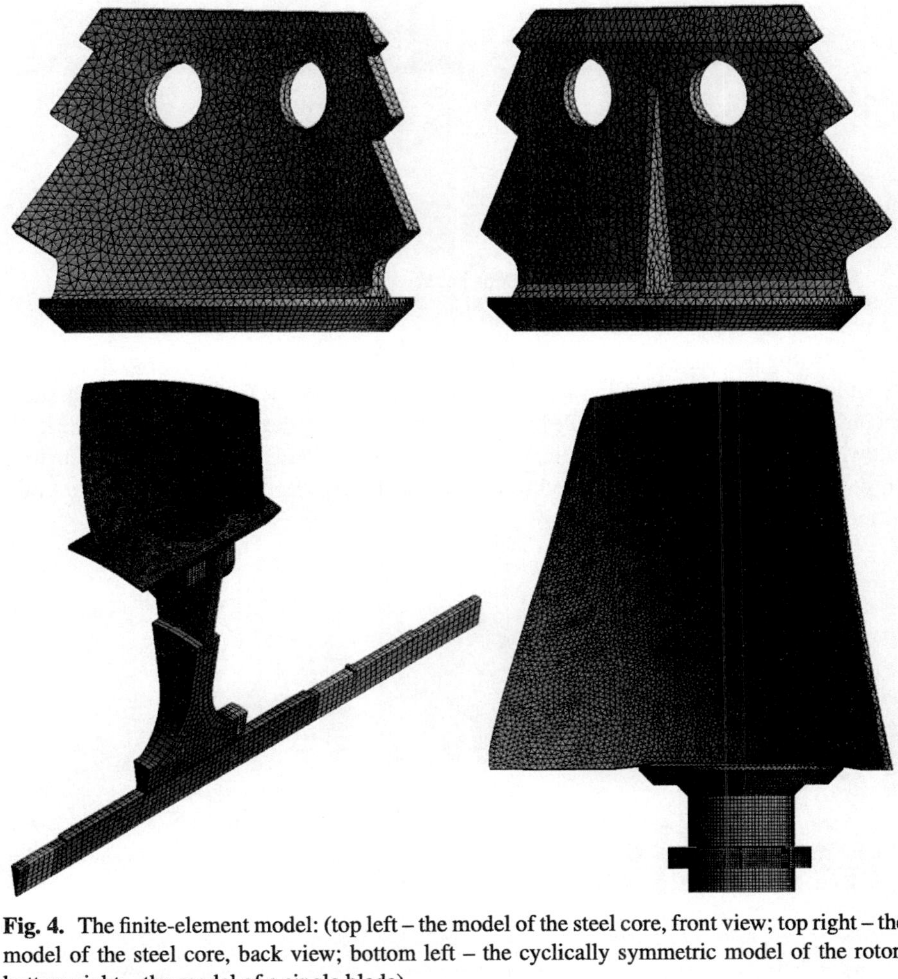

Fig. 4. The finite-element model: (top left – the model of the steel core, front view; top right – the model of the steel core, back view; bottom left – the cyclically symmetric model of the rotor; bottom right – the model of a single blade).

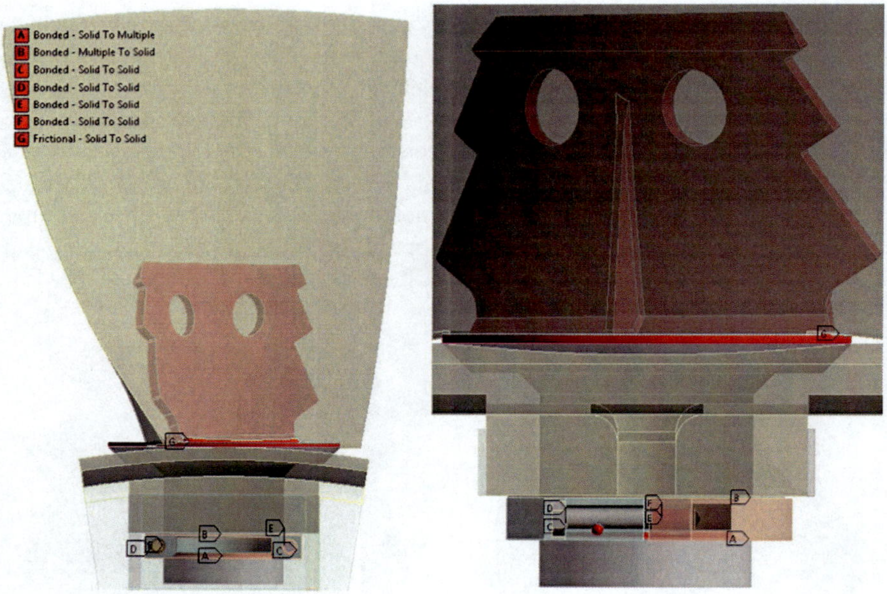

Fig. 5. Contact conditions between the parts of the blade.

Centrifugal loads corresponding to rotation at 750 rpm were applied to the model. In addition to that, performing an aerodynamic analysis of a fan installation passage, including guide vane, rotor blades and straightener (from left to right in Fig. 6, left), allowed obtaining the total pressure, which then was imported to the blade airfoil surface (Fig. 6, right). The rotor was fixed in bearing seats in order to prevent the rigid body motion of the construction.

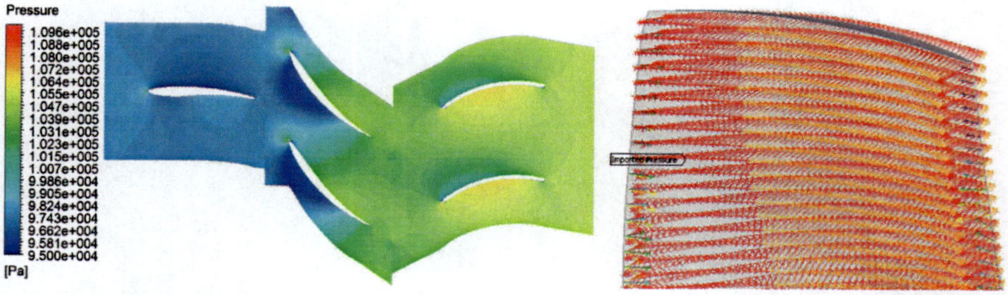

Fig. 6. Aerodynamic loads on the blade airfoil (left – distribution of the total pressure in a fan installation passage, cylindrical section; right – imported pressure).

As it follows from a contact formulation, the considered structural analysis is nonlinear. Large deflections were taken into account in addition to the nonlinearity caused by using the frictional contact.

4 Results

The nonlinear solution allowed gaining the stress-strain state of the construction that takes into account the different contact statuses in steel-aluminum connection (Fig. 7).

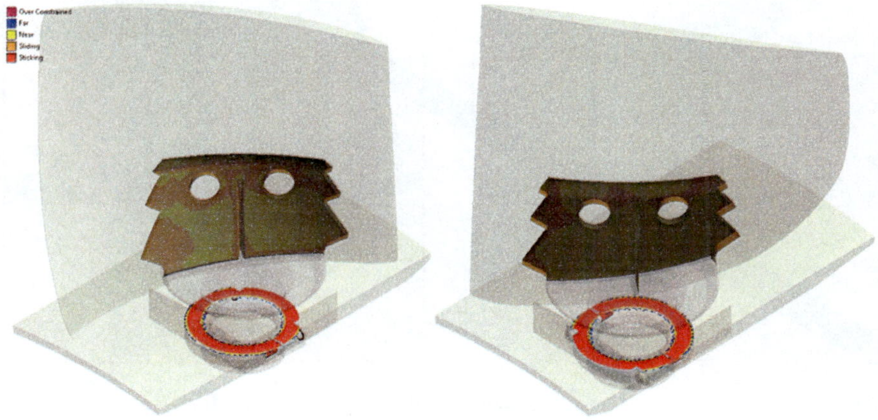

Fig. 7. Contact status at the fixation point (left – front view, right – back view).

The blades of the considered fan should have met the requirements of limited radial displacements of its shroud, so such the displacements were checked. As it follows from the figure, the maximal radial displacements occur at the blade tip and do not exceed the value of the gap between the fan blade and the casing equal to 1 mm. Figure 8 and Fig. 9 present the contour graphs of equivalent stresses in the construction.

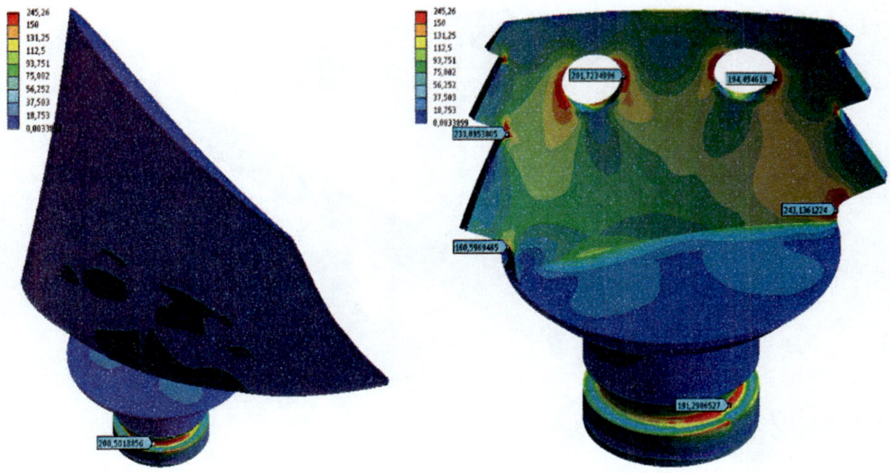

Fig. 8. Von Mises equivalent stresses in blade components, [MPa].

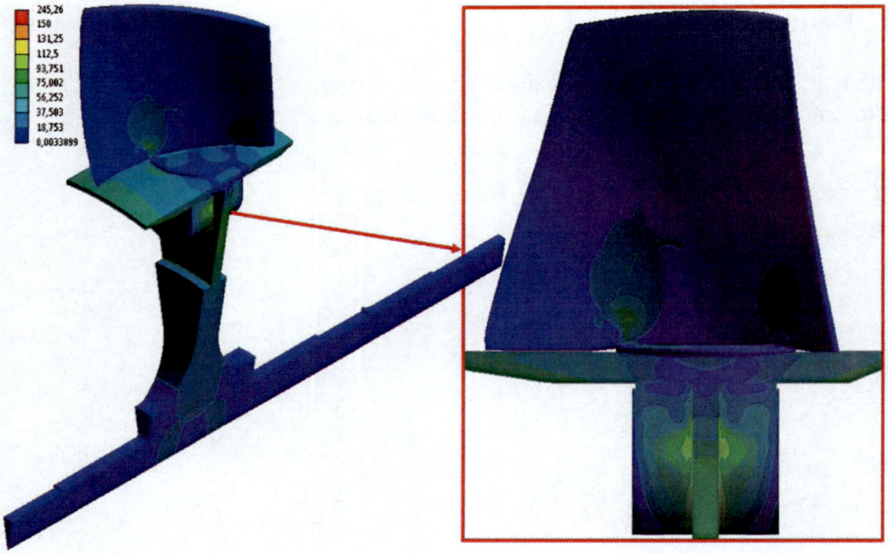

Fig. 9. Von Mises equivalent stresses in the rotor, [MPa].

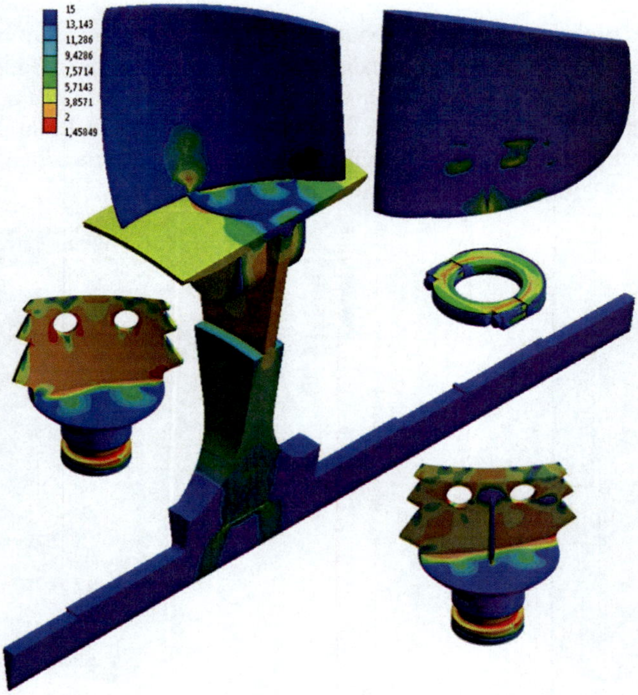

Fig. 10. The safety factor in parts of the rotor with the bimetallic blade, [−].

The Von Mises yield criterion was chosen for an equivalent stress evaluation. According to this criterion, the equivalent stress can be obtained as the RMS of principal stresses $\sigma_1, \sigma_2, \sigma_3$: $\sigma_{eqv} = \{0.5[(\sigma_1 - \sigma_2)^2 + (\sigma_2 - \sigma_3)^2 + (\sigma_3 - \sigma_1)^2]\}^{0.5}$.

If to define a failure limit [σ] by material yielding, i.e. when the maximum equivalent stress reaches or exceeds the yield limit S_y, then $[\sigma] = S_y$.

The corresponding strength condition takes a form: $\max(\sigma_{eqv}) < [\sigma]$.

Defining the safety factor as $F_s = [\sigma]/\max(\sigma_{eqv})$ allow rewriting the strength condition to the following form: $F_s > 1$. The contour graph of the safety factor in the construction and its parts is presented in Fig. 10.

As it follows from the figure, the most loaded points are located in the steel part of the blade, namely in holes of the core, in a weld between the core and the root, and in the fixing cone surfaces of the root. Despite this, the safety factor did not get the value less than 1.45, which means that the strength condition is met and the steel-aluminum joint is capable to take the workloads of the fan installation.

5 Discussion

The considered nonlinear contact of two parts showed the picture of their interaction and indicated that not all the contacting surfaces are closed. Comparing the contact picture (Fig. 7) with locations of maximal stresses in the airfoil and the core, one can notice that these locations coincide with the places of a closed contact, which means, that these places are the most loaded ones. Such an information should help in designing new bimetallic blades, especially when shaping the steel core.

Another limitation, that could have been exceeded when using the bimetallic blade design, was the maximal radial displacement of the blade shroud, since a connection of the root and the core was thin and may have been flexible enough to cause an excess of a shroud tip clearance by the loaded blade. That's why these values were also checked that indicated the maximal radial displacements to be in safe limits.

Analyzing the Von Mises equivalent stress distribution picture (Figs. 8 and 9) shows that the local maximal stresses in steel parts are located near the holes of the core and in the chamfer of the blade root. The global maximum of stresses, which is equal to 245 MPa, is in the weld between the root and the core. The maximal stresses in the aluminum airfoil are located near the weld of steel parts and are equal to 98 MPa. The safety factor picture (Fig. 10) shows, that the bimetallic blade is durable.

The comparative analysis of the natural frequencies of blade airfoils with the same shape made of different materials allow to identify the differences in their mechanical behavior. The work [22] gave the information on vibrational characteristics of the homogeneous aluminum blade and presented the results of the modal analysis of the airfoil made of fiberglass with anisotropic properties [23, 24]. Table 1 and Fig. 11 summarize the natural frequencies (NFs) of three types of blades.

Table 1. NFs of the steel-aluminum blade in comparison with aluminum and composite ones.

NF No	NF value, Hz		
	Aluminum blade	Composite blade	Steel-aluminum blade
1	142,80	120,36	95,12
2	195,85	217,69	213,55
3	300,23	285,93	283,25
4	529,92	312,91	512,44
5	607,56	334,57	554,91
6	724,89	369,10	787,72

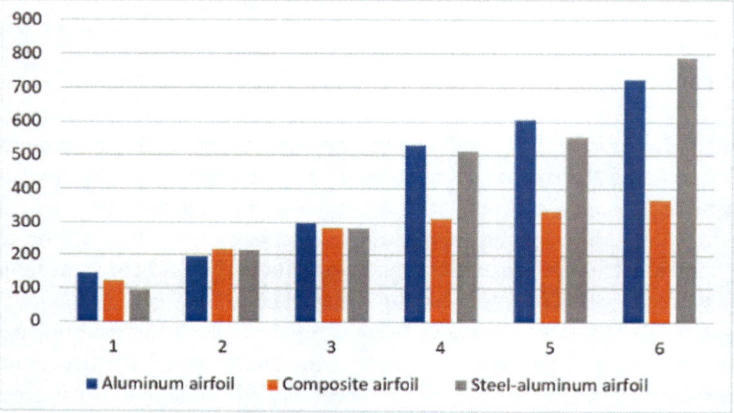

Fig. 11. The column chart of the first six natural frequencies [Hz] of different types of airfoils.

As it follows from the table and the figure, the tendency of the mechanical behavior of the steel-aluminum blade is very close to the one of the homogeneous aluminum blade for higher frequencies. This is due to the rigid fixation of either steel or aluminum blade roots in the fan hub which allows the movement of only the blade airfoil. The close natural frequencies of two blades indicate the approximately equal bending stiffness of them in the blade tip sections. At the same time, the first natural frequencies of the aluminum and steel-aluminum blades, which correspond to the natural forms with bending in root sections, differ significantly. This is connected with the fact that the bending stiffness of the airfoils in their root sections is different and conditioned by either aluminum material (for the aluminum airfoil) or steel weld (for the steel-aluminum airfoil). The bending stiffness of the welded sheet of steel is lower than the one of thick aluminum profile which leads to the first natural frequency of the steel-aluminum airfoil 1.5 times lower than the one of the aluminum airfoil. However, the first natural frequencies for all three airfoils are.

6 Conclusions

The paper proposed a technique for analyzing the structural strength of the bimetallic fan blade. The technique is quite new since not only methods for solving such problems were not proposed before, but the concept of bimetallic blade itself is original. Some amount of papers and patents proposed to use this idea for turbomachinery in order to gain new antiwear, damping and mechanical properties of the rotor blades.

The concept of a bimetallic blade in application to a main ventilation fan production gives significant economic and technological benefits when replacing such the blades during a maintenance of fan installations. In addition to that the use of steel-aluminum blade also allows making the blade root of durable, rigid, but heavy steel, while casting the blade airfoil with aluminum which is comparatively light, easy for shaping, but not as durable as steel for producing the blade root. These benefits and overall engineering and simulation approaches can be used in aerospace applications.

Observing the results of the calculation indicates that the blade design provides the appropriate stiffness to tension and bending loads, caused by centrifugal and aerodynamic ones. The obtained distribution of pressure, caused by aerodynamic loads, was imported to an airfoil surface which is important not only in terms of the root strength. Bimetallic blades were used for replacing aluminum blades of the fan rotor during a planned repair and are operating under workloads up to the present moment which proves a correctness of the proposed technique when analyzing a structural strength of such the blades. It can be used as a methodology for creating and calculating bimetallic rotor blades in different areas of turbomachinery.

References

1. Schwartz, M., Manzoor, S., Hémon, P., de Langre, E.: By-pass transition to airfoil flutter by transient growth due to gust impulse. J. Fluids Struct. **25**, 1272–1281 (2009). https://doi.org/10.1016/j.jfluidstructs.2009.08.001
2. Hémon, P., de Langre, E., Schmid, P.: Experimental evidence of transient growth of energy before airfoil flutter. J. Fluids Struct. **22**, 391–400 (2006). https://doi.org/10.1016/j.jfluidstructs.2005.11.005
3. Chew, J.W., Marshall, J.G., Vahdati, M., Imregun, M.: Part-speed flutter analysis of a wide-chord fan blade. In: Fransson, T.H. (ed.) Unsteady Aerodynamics and Aeroelasticity of Turbomachines, pp. 707–724. Springer, Dordrecht (1998). https://doi.org/10.1007/978-94-011-5040-8_46
4. Chahine, C., Verstraete, T., He, L.: A comparative study of coupled and decoupled fan flutter prediction methods under variation of mass ratio and blade stiffness. J. Fluids Struct. **85**, 110–125 (2019). https://doi.org/10.1016/j.jfluidstructs.2018.12.009
5. Sun, Y., Wang, X., Du, L., Sun, X.: Effect of acoustic treatment on fan flutter stability. J. Fluids Struct. **93**, 102877 (2020). https://doi.org/10.1016/j.jfluidstructs.2020.102877
6. Neff, T., Lahm, A.: Structural statics and dynamics on axial fan blades. 1–7 (2014)
7. Zare, A., Munisamy, K.M., Najafzadeh, A., Shahizare, B., Ahmadi, A., Jaafari, A.A.: Finite element analysis of axial fan blade with different chord lengths. Indian J. Sci. Technol. **6**, 4403–4409 (2013)
8. Tabor, D.: Friction, surface science and tribology—a personal view. Proc. Inst. Mech. Eng. Part C Mech. Eng. Sci. **205**(6), 365–378 (1991). https://doi.org/10.1243/PIME_PROC_1991_205_134_02

9. Zboinski, G.: Physical and geometrical non-linearities in contact problems of elastic turbine blade attachments. Proc. Inst. Mech. Eng. Part C J. Mech. Eng. Sci. **209**(4), 273–286 (1995). https://doi.org/10.1243/PIME_PROC_1995_209_154_02
10. Chen, J., Zang, C., Zhou, B., Petrov, E.P.: A study of friction microslip modeling for dynamic analysis of bladed discs with root joints. Proc. Inst. Mech. Eng. Part C J. Mech. Eng. Sci. **233**(8), 2599–2614 (2019). https://doi.org/10.1177/0954406218792027
11. Ahmad, K., Baker, R.C., Goulas, A.: Computation and experimental results of wear in a slurry pump impeller. Proc. Inst. Mech. Eng. Part C J. Mech. Eng. Sci. **200**(6), 439–445 (1986). https://doi.org/10.1243/PIME_PROC_1986_200_153_02
12. Gao, Q., Li, W., Chen, X.: Surface quality and tool wear in micro-milling of single-crystal aluminum. Proc. Inst. Mech. Eng. Part C J. Mech. Eng. Sci. **233**(16), 5597–5609 (2019). https://doi.org/10.1177/0954406219855409
13. Yu, M., Fu, S., Gao, Y., Zheng, H., Xu, Y.: Crack detection of fan blade based on natural frequencies. Int. J. Rotat. Machin. **2018**, 1–13 (2018). https://doi.org/10.1155/2018/2095385
14. Han, Z., Chen, J., Zhang, B., Wang, Y., Gao, Z.: Research on fault diagnosis of the fan blade of turbofan engine. In: Proceedings of 2016 Prognostics and System Health Management Conference, PHM-Chengdu 2016, pp. 1–5 (2017). https://doi.org/10.1109/PHM.2016.7819838
15. Magerramova, L., Vasilyev, B.: Bimetallic blisks with shrouded turbine blades for gas turbine engines. Sci. Educ. Bauman MSTU (2015). https://doi.org/10.7463/0615.0771150
16. Magerramova, L., Vasilyev, B., Nigmatullin, R., Kinzburskiy, V.: Presented at the Design of a Bimetallic Blisk Turbine for a Gas Turbine Engine and its Production Using Powder Metallurgy Methods (2017)
17. Lee, C.-P., Isburgh, A.M., Wilson, P.S.: Bimetallic turbine airfoil (1994)
18. Patrick, D.K., Bauer, R.C., Kelly, T.J.: Monolithic and bi-metallic turbine blade dampers and method of manufacture (2010)
19. Millet, F., Leroy, D.: Method of manufacturing a bmetallic turbine blade and use thereof (1996)
20. Martynenko, G., et al.: Numerical simulation of missile warhead operation. Adv. Eng. Softw. **123**, 93–103 (2018). https://doi.org/10.1016/j.advengsoft.2018.07.001
21. Martynenko, G., Avramov, K., Martynenko, V., Chernobryvko, M., Tonkonozhenko, A., Kozharin, V.: Numerical simulation of warhead transportation. Def. Technol. (2020). https://doi.org/10.1016/j.dt.2020.03.005
22. Martynenko, V., Hrytsenko, M., Martynenko, G.: Technique for evaluating the strength of composite blades. J. Inst. Eng. (India) Ser. C **101**(3), 451–461 (2020). https://doi.org/10.1007/s40032-020-00572-9
23. Martynenko, V.G.: An original technique for modeling of anisotropic viscoelasticity of orthotropic materials in finite element codes applied to the mechanics of plates and shells. Mech. Mech. Eng. **21**, 389–413 (2017)
24. Martynenko, V.G., Lvov, G.I., Ulianov, Y.N.: Experimental investigation of anisotropic viscoelastic properties of glass fiber-reinforced polymeric composite material. Polym. Polym. Compos. **27**, 323–336 (2019). https://doi.org/10.1177/0967391119846362

Gas Turbine Engine System Multiobjective Perfection Using the Multilevel Hierarchy Stochastic Optimization Model and Solution Searching Method

Mykhaylo Ugryumov[1], Andriy Myenyaylov[2], Ievgen Meniailov[1](✉), and Irina Trofymova[3]

[1] V.N. Karazin Kharkiv National University, Kharkiv, Ukraine
m.ugryumov@karazin.ua, j.menyailov@khai.edu
[2] Commercial Aircraft Engine Co., Ltd., Shanghai, China
myenyaylov@ukr.net
[3] National Aerospace University, Kharkiv, Ukraine
i.trofymova@khai.edu

Abstract. The object of our research is increasing the veracity of rational decision searching processes under input data uncertainly in the turbofan engine (TFE) elements perfection. The methodology of TFE system perfection is offered. This methodology bases on the interrelated multiobjective stochastic optimization (modification) problems sequential solution for subsystems and their elements by robust estimation the main variables and operating tolerances according to them. The method of multiobjective system modification problem quasisolution searching under input data uncertainly and bordering of feasible solution class is considered. Quasisolution synthesis is realized by regularization of smoothing functional minimum searching with using the A.N. Tihonov's method. The evolution method was created for modification problem numerical solution with input data uncertainly; it based on genetic algorithm using. The system aerodynamic turbofan engine compressor perfection was made by the development methodology.

Keywords: Multiobjective stochastic optimization · Incorrect problem · Regularization method · Evolution method

1 Introduction

The advances in modern computer technology and methods for modeling physical processes make it possible to move on to the formulation of more and more complex problems to optimize the design of gas turbine engines. There is a tendency to move from the "traditional" optimization of the aerodynamics of individual blades to the consideration of multistage designs and to joint problem solving for related disciplines. For example, article [1] presents a global-local approach that is applied to surrogate optimization of the design of a highly loaded transonic rotor 37 NASA, and [2] – an

M. Nechyporuk et al. (Eds.): ICTM 2021, LNNS 367, pp. 449–461, 2022.
https://doi.org/10.1007/978-3-030-94259-5_38

example of optimizing the efficiency of fan blades of a modern jet engine. The complication of the optimization problem is demonstrated in the article [3], where a coupled method for optimizing the aero-structure is used, which has efficiency as an objective function and the maximum von Mises stress set as a constraint. Article [4] is devoted to the optimization of isentropic efficiency and safety factor while fulfilling geometric, aerodynamic and mechanical constraints. Examples of multidisciplinary design for centrifugal compressors are given in [5] and [6]. The work discussed in article [7] is aimed at optimizing the holes for membranous cooling on the flat top of the turbine cascade. As a result, the optimized membranous cooling holes have improved both aerodynamic performance and heat transfer characteristics of the tip. The work [8] is devoted to the problems arising in the production of gas turbine engines and minimizing the effect of deviation from the nominal size. Paper [9] describes the optimization model in insurance system. Paper [10] describes non-linear estimation methods of robust optimal design and diagnostics. Works [11, 12] are dedicated to synthesis method of robust neural network models. This paper demonstrates an approach to representing uncertainties in a jet engine model caused by manufacturing and assembly.

This article discusses stochastic optimization models and methods for finding solutions for a complex technical system with a hierarchy (for example, a gas turbine engine). The evolutionary method is based on the classical genetic algorithm and allows solving the assigned tasks with less computer resources in comparison with classical genetic algorithms and other methods of structural-parametric optimization. This methodology is based on the sequential solution of interrelated multicriteria problems of stochastic optimization (modification) for subsystems and their elements.

The paper is organized as follows: Sect. 2 defines the modification problem. Section 3 introduces the modification problem structurization. Section 4 presents the solution method of modification problem. Section 5 introduces the regularizing algorithm of modification problem quasi-solution construction. Section 6 defines the stochastic optimization problem. Section 7 offers the decision model of stochastic optimization problem with mixed conditions. Section 8 presents the method of stochastic optimization problem solution. Section 9 presents the methods application for gas turbine engine elements modification problem solution. Section 10 gives conclusions.

2 Modification Problem Definition

The technical system modification problem description is considered.

The following data are known: researching object description, general characteristic and system's properties, functioning conditions and main demands of system's performance and economic characteristics; system's structure; modification goal and design decision quality criterion; information of analogs and prototype, feasible controls class (means and realizing facilities). Those data are formalized.

It is need to identify the optimal functional elements (FE) construction parameter values for realization desired values of modificated system quality criterions from the setting demands rational compromise condition.

Entered data view for modification object problem is formalized. Let desired values of modificated system quality criterion $W^* = \{w_i^*\}$ is defined or can be found through prototype parameters $w_i^* = \phi_i \cdot \left(w_i^\circ\right)_0$, where $\phi^\circ = \{\phi_i\}$ is vector, its elements are perfection system indexes by i-th criterions ($i = 1..I$). Let isomorphic mapping C_l : $\Phi_l^\circ \to W_l^\circ$, $W_l^\circ = \{w_{\text{ln}}^\circ\}$ exists for phase variables of every l-th subsystem (FE) $\Phi_l^\circ = \{\Phi_{\text{ln}}\}$, then values w_{ln}° estimates of phase variables $\Phi_{\text{ln}} \in D_{\Phi a}$, $l = 1...L, n = 1...N_l$. Vectors set $\phi_1^\circ = \{\phi_{\text{ln}}\}$ is entered, its elements are perfection indexes of subsystem ϕ_{ln}. Desired values of l-th modificated subsystem (FE) phase variable estimates can be defined or found through prototype parameters $w_{\text{ln}}^* = \phi_{\text{ln}} \cdot (w_{\text{ln}}^\circ)_0$.

In compliance with Nash's balance conception system modification problem can be reduce to multilevel optimization problem. Let several subsystems are present, there is system under consideration (outer system), technical system (first level subsystem), FE (second and less levels subsystems), they interact among themselves for minimization of common quality criterion. For all subsystem own quality criterions are defined:

– for outer system – $W_e^\circ(q_e^\circ, \phi^\circ)$:

$$W_e^\circ = \{w_{ei}^\circ\}, w_{ei}^\circ = \left|w_i^\circ(q_e^\circ) - w_i^*\right| = \left|w_i^\circ(q_e^\circ) - \phi_i \cdot (w_i^\circ)_0\right|, \; q_e^\circ = (\Pi_r^\circ, U_r^\circ, \Phi^\circ); \tag{1}$$

– for inner subsystems – $W_{sl}^\circ(q_l^\circ, \phi_l^\circ)$:

$$\begin{aligned} &W_{sl}^\circ = \{w_{s\text{ln}}^\circ\}, w_{s\text{ln}}^\circ = \left|w_{\text{ln}}^\circ(q_l^\circ) - w_{\text{ln}}^*\right| = \left|w_{\text{ln}}^\circ(q_l^\circ) - \phi_{\text{ln}} \cdot (w_{\text{ln}}^\circ)_0\right|, \\ &q_l^\circ = (\Pi_l^\circ, U_l^\circ, \Phi_l^\circ), \end{aligned} \tag{2}$$

where Π° – designed and mode variables vector, U° – control variables and Φ° – phase parameters vector.

For outer system and subsystem global criterion is entered, there are multiple utility functions, that will agree with (1). Thus, order to preference is define in the middle of criterions.

Unlike outer system with quality criterions (1) inner subsystems consider desired values of modificated subsystem phase parameters $\Phi_l^* = \{\Phi_{\text{ln}}^*\}$ or corresponding estimates:

$$W_l^* = C_l(\Phi_l^*), \; W_l^* = \{w_{\text{ln}}^*\}, \phi_{\text{ln}} = w_{\text{ln}}^*/(w_{\text{ln}}^\circ)_0, \tag{3}$$

where C is vector-function like initial data.

Let prototype v_0°, alternatives set $V = \left\{v_i^\circ\right\}$, estimates set of criterion desired values and preferences system G are defined. Acceptable control $\widehat{U}^\circ(\Phi^\circ)$ ($\widehat{U}^\circ \in D_U$) need to identify, its control has to convert system's initial state to other acceptable state with respect to G –

$$\hat{v}^\circ \in \hat{V} \subseteq V, \; \hat{v}^\circ R v^\circ (\forall v^\circ \in V).$$

Submitted modification problem can be referred to problem class of Chebyshev's approximation definition for incompatible system of nonlinear equations [13] and converted to nonlinear mathematical programming problem.

3 Modification Problem Structurization

Suppose all subsystems are agreement with multiple utility function (1). Then at first optimization problem for outer system of searching control vectors $\widehat{U}_r^{\circ}$ and phase parameters $\widehat{\Phi}^0 = \{\widehat{\Phi}_l^{\circ}\}$ is solving. If relations (3) use and $\Phi_{\text{ln}}^* = \hat{\Phi}_{\text{ln}}$ suppose, target values of modification subsystem (FE) phase variable estimates w_{ln}^* or perfection subsystem (FE) indexes define $\phi_1^{\circ} = \{\phi_{\text{ln}}\}$.

Next the rational values of subsystems parameters (FE) $\hat{q}_l^{\circ} = (\widehat{\Pi}_l^{\circ}, \widehat{U}_l^{\circ}, \widehat{\Phi}_l^{\circ})$ define in time of optimization problems (1, 2) solving for inner subsystems (FE) in terms of selected target perfection subsystems (FE) indexes ϕ_l° (solution for outer system).

Search $\hat{v}^{\circ} \in \widehat{V}$ process is convenient to offer as process of distributed decision making interconnected inverse and optimization problems like sequence of operation top-down from aggregated definition to more detailed object description with following available solution aggregation to general (Fig. 1).

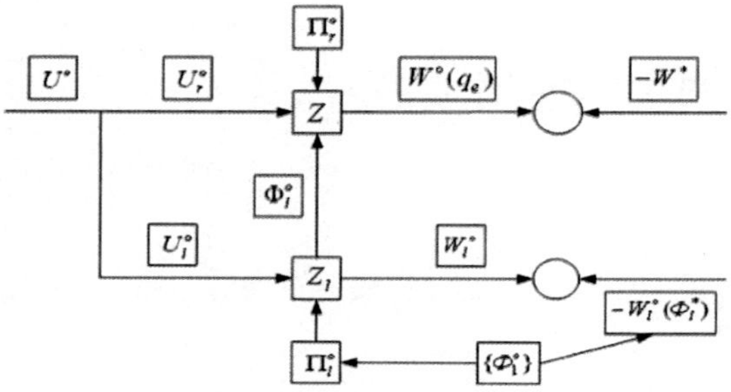

Fig. 1. Conceptual information technical system model: scheme of variable aggregation, subsystem decomposition and coordination ($\Phi_l^* = \hat{\Phi}_l^{\circ}$)

4 Modification Problem Solution Method

Modification problem solution method contains to following sequence of operations.

The best alternative for outer system is searching out:

$$\hat{v}_e^{\circ} = F_e(V_e, G_e),\, V_e = \left\{v_{ej}^{\circ}\right\}, j = 1...J, \tag{4}$$

where $v_e^{\circ} = (q_e^{\circ}, W_e^{\circ})$, $q_e^{\circ} = (\Pi_r^{\circ}, U_r^{\circ}, \Phi^{\circ})$, $\Phi^{\circ} = \{\Phi_l^{\circ}\}$, $l = 1...L+1$, $\Phi_r^{\circ} = \Phi_{L+1}^{\circ}$, $W_e^{\circ} = W_e(q_e^{\circ}, \phi^{\circ})$.

The best alternative for inner subsystem is searching out:

$$\forall l = 1...L : \hat{v}_l^{\circ} = F_l(V_l, G_l),\, V_l = \left\{v_{lj}^{\circ}\right\}, j = 1...J,$$

where $v_l^{\circ} = (q_l^{\circ}, W_{sl}^{\circ})$, $q_l^{\circ} = (\Pi_l^{\circ}, U_l^{\circ}, \Phi_l^{\circ})$, $W_{sl}^{\circ} = W_{sl}(q_l^{\circ}, \phi_l^{\circ})$, $\phi_l^{\circ} = \{\phi_{\text{ln}}\}$, $\phi_{\text{ln}} = w_{\text{ln}}^*/(w_{\text{ln}}^{\circ})_0$, $W_l^* = \{w_{\text{ln}}^*\}$, $W_l^* = C_l(\Phi_l^*)$. In this case target values of modificated subsystem phase variable are equal to computed $\Phi_l^* = \widehat{\Phi}_l^{\circ}$ up to outer system. Problem solution for inner subsystem is $\hat{v}_l^{\circ} = \left(\hat{q}_l^{\circ}, \hat{W}_{sl}^{\circ}\right)$, $\hat{q}_l^{\circ} = \left(\Pi_l^{\circ}, \hat{U}_l^{\circ}, \hat{\Phi}_l^{\circ}\right)$.

Let us consider second step in detail. For $\forall l = 1...L$ it is $\hat{v}_l^{\circ} = F_l(V_l, G_l)$. In consideration of $W_{sl}^{\circ} = W_{sl}(q_l^{\circ}, \phi_l^{\circ})$ if $\hat{W}_{sl}^{\circ} = 0$ procedure $F(V_l, G_l) : V_l \to \hat{v}_l^{\circ}$ can be presented as sequence of operations:

$$\begin{cases} \hat{q}_l^o = F\left(q_l^o, \phi_l^o, G_l\right) \\ W_{sl}^o = W_{sl}\left(q_l^o, \phi_l^o\right) \end{cases}.$$

According to entered early problems coordination for outer system and subsystem – $\phi_l^{\circ} = \{\phi_{\ln}\}$, $\phi_{\ln} = w_{\ln}^*/\left(w_{\ln}^{\circ}\right)_0$, $W_l^* = \{w_{\ln}^*\}$, $W_l^* = C_l(\Phi_l^*)$ and $\Phi_l^* = \hat{\Phi}_l^{\circ}$, then:

$$\begin{cases} \hat{q}_l^{\circ} = F_l(q_l^{\circ}, \phi_l^{\circ}(\hat{\Phi}_l^{\circ}), G_l) = F_l(q_l^{\circ}, \phi_l^{\circ}(f_l(W_e^{-1}(\hat{q}_e^{\circ}, \phi^{\circ}))), G_l), \\ W_{sl}^{\circ} = W_{sl}(q_l^{\circ}, \phi_l^{\circ}(f_l(W_e^{-1}(\hat{q}_e^{\circ}, \phi^{\circ})))), \end{cases} \tag{5}$$

where $f_l : \Phi^{\circ} \to \Phi_l^{\circ}$, $W_e^{-1}(\hat{q}_e^{\circ}, \phi^{\circ})$ is complete preimage $\hat{q}_e^{\circ}$.

Let α is hierarchy level in system with multiechelon hierarchy systemic structural model, where $\alpha = 0, 1, 2$ is system, subsystem, FE. Then in our case thereby accepted Nash's balance conception relations R^{α} of preferences system $G_{\alpha} = \{W_{\alpha}^{\circ}, R^{\alpha}, Z_{\alpha}\}$ is agreement with relations $R^{\alpha-1}$ of preferences system $G_{\alpha-1}(\alpha = 1, 2)$, because for any $v_{\alpha-1,1}^{\circ}, v_{\alpha-1,2}^{\circ} \in V_{\alpha-1}$ such as $v_{\alpha-1,1}^{\circ} R^{\alpha-1} v_{\alpha-1,2}^{\circ}$ and any $q_{\alpha,1}^{\circ} = W_{\alpha-1}^{-1}(q_{\alpha-1,1}^{\circ}, \phi_{\alpha-1,1}^{\circ})$, $q_{\alpha,2}^{\circ} = W_{\alpha-1}^{-1}(q_{\alpha-1,2}^{\circ}, \phi_{\alpha-1,2}^{\circ})$ and corresponding $v_{\alpha,1}^{\circ} = (q_{\alpha,1}^{\circ}, W_{\alpha,1}^{\circ})$, $v_{\alpha,2}^{\circ} = (q_{\alpha,2}^{\circ}, W_{\alpha,2}^{\circ})$ will find, they satisfy conditions $v_{\alpha,1}^{\circ} R^{\alpha} v_{\alpha,2}^{\circ}$. And relations R^{α} are asymmetrical. In this case statement [14] is correct: if relations R^{α} of preferences system $G_{\alpha} = \{W_{\alpha}^{\circ}, R^{\alpha}, Z_{\alpha}\}$ are asymmetrical and relations R^{α} is $W_{\alpha-1}$-agreed with $R^{\alpha-1}$ $(\alpha = 1, 2)$, then $\hat{v}^{\circ} \in \hat{V}$, where $\hat{v}^{\circ} = \bigcup_{\alpha=0}^{2} \hat{v}_{\alpha}^{\circ}$. It is supposed subset $\hat{V}$ is externally stable in preferences system G.

5 Regularizing Algorithm of Modification Problem Quasi-solution Construction

Modification problem quasi-solution (4, 5) synthesis was realized by interconnected inverse and optimization problem distributed decision on the base of smoothing functional minimum search regularization with method by A. N. Tihonov [15].

For ease the all quality criterions are reduced to dimensionless type:

$$\Delta w_n^{\circ} = \frac{2(w_n - < w_n >)}{w_{n,\max} - w_{n,\min}}, \quad \Delta w_n^* = \frac{2(w_n^* - < w_n >)}{w_{n,\max} - w_{n,\min}},$$

$$(\Delta w_n^{\circ}, \Delta w_n^*) \in [-1, 1], \quad n = 1...N_k, \tag{6}$$

where $w_n, w_{n,\min}, w_{n,\max}, w_n^*$ – are current, minimal, maximal and reference values of concerned quality criterion; $< w_n > = \frac{w_{n,\max} + w_{n,\min}}{2}$, N_k is number of quality criterions in concerned k-th subproblem.

Quality criterions for outer and inner subsystem (1, 2), which were enter early, are rewritten in the form:

$$w_{ei}^{\circ} = \left|\Delta w_i^{\circ}(q_e^{\circ}) - \Delta w_i^*\right|,\ w_{s\ln}^{\circ} = \left|\Delta w_{\ln}^{\circ}(q_l^{\circ}) - \Delta w_{\ln}^*\right|,\ w_{ei}^{\circ} \in [-1, 1],\ w_{s\ln}^{\circ} \in [-1, 1].$$

Similarly the all control variables are reduced to dimensionless type:

$$\Delta u_m^{\circ} = \frac{2(u_m - u_{m,0})}{u_m'' - u_m'},\ u_m^{\circ} \in [-1, 1],\ m = 1...M_k,$$

where $u_{m,0}$ is variable value, that is correspond to prototype, M_k is number of control variable in concerned k-th subproblem.

Let us enter a mapping $G : \{w_n^{\circ}\} \to \hat{w}^{\circ}$, which define general target function:

$$\hat{w}^{\circ}(U_k^{\circ}, \beta) = \sum_{n=1}^{N_k} f_{fit}[(w_n^{\circ})^2] + \beta \sum_{m=1}^{M_k} (\Delta u_m^{\circ})^2, \tag{7}$$

where G is preferences system of person, who makes decision, β is regularization parameter.

Assigned problem quasi-solution (normal solution) can be found by A.N. Tihonov's regularization method [15]:

$$\hat{U}_p^{\circ} = \arg \min_{U^{\circ} \in D_U} \hat{w}^{\circ}(U^{\circ}, \beta_p), \tag{8}$$

$$\mathrm{D}_U = \left\{U^{\circ} = (u_1, ...u_m, ..., u_{Mk}) : (\forall \mathrm{m} \in [1...\mathrm{M}_k]) u_m' \le u_m \le u_m''\right\}.$$

Parameter β_p was chosen ($\beta_{p+1} = \beta_p/10, p = 0, 1, 2, ...$) in according to generalized misalignment principle: $\left|w_n^{\circ}\right| \le \xi_n + \mathrm{h}\left\|\Delta u_m^{\circ}\right\|,\ n = 1...N_k$, where ξ_n is definition error Δw_n^*, h is definition error $\widehat{w}^{\circ}$. Problem solution (8) is realized with evolutional method that is presented below.

6 Stochastic Optimization Problem Statement

Stochastic optimization problem is defined as $f(x) \to min$, where f is target function, its type depends on concerned problem specification, and x is a random variable with given distribution law. In present paper x is considered with Gaussian law or uniform law. In the case of Gaussian law x is set uniquely by specifying its average $M[x] = x_c$ and variance $D[x] = \sigma_x^2$. It should be note if multiparametric optimization problem is considered then variable x and its influential values x_c and σ_x are the vectors; its dimension corresponds to current problem dimension.

In consequence of variable x random nature target function $f(x)$ also is random variable. For random target function definition, it needs to specify its average $M\left[f(x)\right] = f_c$ and variance $D\left[f(x)\right] = \sigma_f^2$. For calculation the expectancy of hitting a target function values to given interval the vector family $\overline{x} = \{x_1, x_2, x_3 \ldots x_n\}$ is generated randomly for given values of x_c and σ_x, appropriate target function values f_i $(i = 1...n)$ are calculated, and number of elements i_f satisfied to requirement $f_{\min} \leq f(x_i) \leq f_{\max}$ is found. Received value is normalized by division into n.

Often several variables enter into the target function expression. Let us assume target function has J variables. In this case the convolution of target function variables $\hat{f}$ is used. For calculation the expectancy of hitting a target function values to given interval the vector $f_i = \{f_{1,i}, f_{2,i} \ldots f_{J,i}\}$ is calculated for element family. In that case the own contingencies are defined for every variable. And instead of a requirement $f_{\min} \leq f(x_i) \leq f_{\max}$ disjunction of requirements $f_{j\ \min} \leq f_j(x_i) \leq f_{j\ \max}, j = 1...J$ are defined.

In the stochastic optimization problem class the three decision models are assigned. These models are M-problem (minimal target function average search), V-problem (minimal target function variance search) and P-problem (minimal or maximal target function achievement probability). Let us consider the every of these problems.

M-problem: it necessary to find such x_c, which ensures the minimal target function value f_c by given σ_x. For modification problem solution it is right to reduce this problem to minimization problem of function such as $|f_c - f^*|$, where f^* is desired target function value. This problem is the nearest to classic type of optimization problems. If the stochastic nature of variable x and target function is disregarded, the problem is reduced to classic problem.

V-problem: it necessary to find such x_c, which ensures the minimal spread in values σ_f by given σ_x. Conversation the optimization problem into the modification problem can be carried out similarity to M-problem conversation.

P-problem: it necessary to find such x_c in order to the probability $P(f_{\min} \leq f(x) \leq f_{\max})$ reach extremum by given σ_x.

It should be note often in stochastic optimization problem it necessary to observe the averages and variances changing of target function simultaneously and not allow exceeding them given values. For this problem solving the models with mixed conditions are considered.

7 Mathematical Decision Model for Stochastic Optimization Problem with Mixed Conditions

There is need to satisfy the several defined above conditions simultaneously for practical problem solution. For example, there is maximization the probability of hitting a target function values to given interval by minimal target function probabilistic average deviation from given value. The denoted conditions of two criterion problems can be solved in following type.

Convolution of target functions on base of power average concept by A.N. Kolmogorov –

$$\hat{f} = \frac{\Delta_R^2 + \Delta_C^2}{\frac{i_{\alpha,f}}{n_\alpha} + \gamma} + \beta \left[\sum_{m=1}^{M_k} \Delta_{x,m}^2 + \frac{1}{n_\alpha} \sum_{m=1}^{M_k} \left| \chi_{x,m}^2 - n_\alpha \right| \right], \tag{9}$$

where $\Delta_R = \frac{M[R]-R^*}{\sigma_R^*}$, $\Delta_C = \frac{M[C]-C^*}{\sigma_C^*}$; $R^*, C^*, \sigma_R^*, \sigma_C^*$ are desired target function values and its mean-square deviations; $i_{\alpha f}$ is number of points from n_α, which are in the given interval;

$\Delta_{x,m} = \frac{M_\alpha[x_m]-x_{m,0}}{\sigma_m^*}$, $\chi_{x,m}^2 = \frac{n_\alpha M_\alpha\left[(x_m-M_\alpha[x_m])^2\right]}{(\sigma_m^*)^2}$, $x_{m,0}$ are variables x_m values for prototype; σ_m^* is an upper limit of control variable definitional domain.

Convolution of target functions on base of maximum likelihood concept –

$$\hat{f} = \frac{1}{2}\left(\Delta_R^2 + \Delta_C^2\right) + \frac{\beta}{n_\alpha}\left\{\frac{1}{2}\sum_{m=1}^{M_k} q_{x,m}^2 + n_\alpha \sum_{m=1}^{M_k} \ln(\sigma_{x,m}) + \right. \qquad (10)$$
$$\left. + \frac{1}{2}\sum_{m=1}^{M_k}\left[\chi_{\sigma,m}^2 - (n_\alpha - 3)\ln\left(\chi_{\sigma,m}^2\right)\right] - n_\alpha \ln\left(\frac{i_{\alpha,f}}{n_\alpha} + \gamma\right)\right\},$$

where $q_{x,m}^2 = \frac{n_\alpha M_\alpha\left[(x_m-x_{m,0})^2\right]}{(\sigma_m^*)^2}$ and $\chi_{\sigma,m}^2 = \frac{n_\alpha(\sigma_m)^2}{(\sigma_m^*)^2}$.

8 Method of Stochastic Optimization Problems Solution

The evolutional method (EM) is defined as classic genetic algorithm (GA) modification for optimization problems solution [16]. In present paper novelty elements of available algorithm description is offered as compared with classic GA.

Usually for every individual the target function is calculated one time by computation in determine definition. If one of the described problems is solved, a mini-population with given center and variance is generated randomly (Gaussian law) instead of one individual. Then subject to defined problem the probabilistic average (M-problem) or the variance (V-problem) or probability of target function staying in given interval for mini-population (P-problem) are calculated. The mini-population size is selected on base of efficiency balance computation and the correctness ensuring of stochastic problem solution and is not less 200 individuals.

In order to protection the population from nonoptimal chromosome dominance and also prevention the early GA convergence the fitness function is used:

$$Fitness = 1 - e^{-Mf}, M > 1.$$

The evolutional method was developed on base of GA for calculation accuracy perfection and for more rational GA advantage usage.

The special feature of introduced algorithm is adaptation procedure application. The essence of this algorithm is in following. After the first running the required problem parameters with the best target function of present GA settings are given in initial definitional domain D_0 by low epochs number. Then the given individual is taken as the new definitional domain D_1 center, that has a less dimension than D_0 (if condition $D_1 \subset D_0$ is respected), and the next GA running is performed. And due to search domain reduction with usable individuals number keeping it is possible to improve the GA calculation accuracy on the regular iteration in proportion to definitional domain decrease level.

As GA settings in EM the following values were used: maximal number of epochs (iterations) was from 20 to 200, crossover probability was 0.9, mutation probability was 0.1, population size was from 50 to 100 individuals (mini-populations), algorithm stopping criterion was maximal iterations number, parent individuals' selection principle was the tape with elite selection, mutation and crossover were standard one-point.

The introduced EM allows solving the defined problems with less computer resources as compared with the classic genetic algorithms.

9 Gas Turbine Engine Elements Modification Problem Solution Example

The numerical solution of system perfection problem for three-shaft TFE for modern regional passenger aircraft is considered as example was given. The integral thermodynamic model was used for flow characteristics definition, in this model subsystem phase variable values were assigned as initial data. In calculation result the target functions were defined such as specific fuel flow rate C_S and specific impulse R_S. The convolutions of target functions (e.g. (9) or (10)) were used as effective target functions. The high-pressure compressor efficiency value η_{HPC} and high-pressure turbine efficiency value η_{HPT} with the constant air mass flow rate G_a were chosen as control variables. For prototype variables C_S and R_S values were equal to $C_S = 0.0694$ kg/(N·h), $R_S = 102.3$ N·s/kg. Its target values were chosen like that $C_S^* = 0.0687$ kg/(N·h) ($\Delta\hat{C}_S^* = -1$), $R_S^* = 103.36$ N·s/kg ($\Delta\hat{R}_S^* = 1$) with the given mean-square deviations $\sigma_C = 1.2 \cdot 10^{-4}$ kg/N·h and $\sigma_R = 0.3$ N·s/kg. The lower limit of control variable changing interval was taken equal to nominal prototype characteristics values ($\eta_{HPC} = 0.844$; $\eta_{HPT} = 0.845$), and the upper limit was taken equal to 0.9 for all control variables. The $\sigma^*_{\eta_{\mathrm{HPC}}} = 0.0030$; $\sigma^*_{\eta_{HPT}} = 0.0020$ were chosen in the capacity of mean-square deviations of required prototype phase variables.

Modification problem solution for outer system was found by evolutional method that was declared above. Calculation was done by chosen acceptable decision region for phase variable. HPC, HPT and TFE average parameters calculation results with $P \geq 0.99$ are presented in Table 1.

On the Fig. 2a and 2b there are computed solutions of average C_S, R_S and theirs mean-square deviations. In other words, the subsystem parameters operating tolerance detection problem for $P \geq P^*$, was solved on the basis of desired target function values and given tolerance of its deviation for system in whole.

On the Fig. 3 there are the level lines, which are according to the computed rational parameters mathematical expectation values and its mean-square deviations for high-pressure compressor and high-pressure turbine. The level lines correspond to the calculation probabilities $P = 0.9, 0.95, 0.99$.

Thus, the operating tolerance detection problem was solved on the basis of desired target function values and given tolerance of its deviation for system in whole.

Later HPC modification problem was decision. Found relation phase variable values $\hat{\Phi}^\circ = \left\{\hat{\eta}_{HPC}, \hat{\sigma}_{\eta_{HPC}}\right\}$ were used for desired values of upgradeable HPC phase variables estimates, that were accepted in the capacity of input data for modification problem. The convolutions of target functions (9) or (10) were used in the capacity of smoothing

Table 1. Engine system modification result

$\Delta\hat{\eta}^{\circ}_{HPC}$	$\Delta\hat{\eta}^{\circ}_{HPT}$	$\Delta\hat{C}^{\circ}_{S}$	$\Delta\hat{R}^{\circ}_{S}$
0.77	0.69	−0.84	1.34
$\hat{\sigma}^{\circ}_{\eta HPC}$	$\hat{\sigma}^{\circ}_{\eta HPT}$	σ°_{C}	σ°_{R}
0.0029	0.0016	$0.98 \cdot 10^{-4}$	0.25

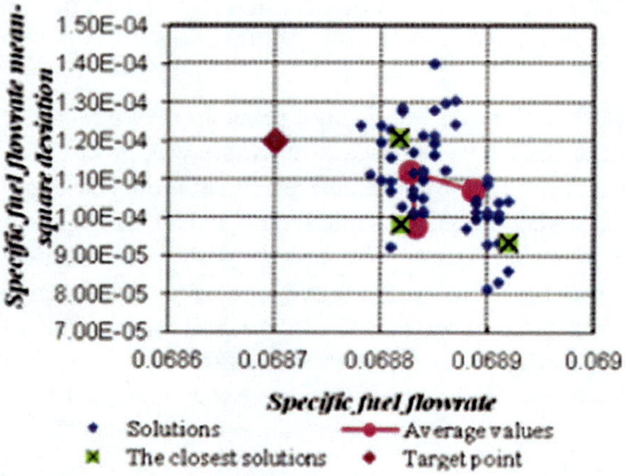

a. The computed solutions of C_S

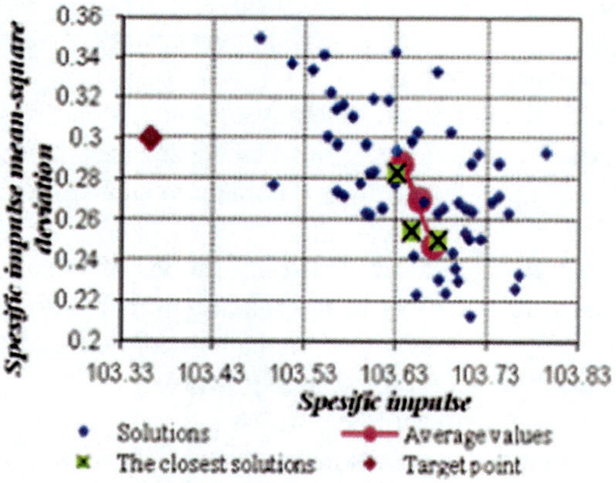

b. The computed solutions of R_S

Fig. 2. **(a)** The computed solutions of C_S. **(b)** The computed solutions of R_S

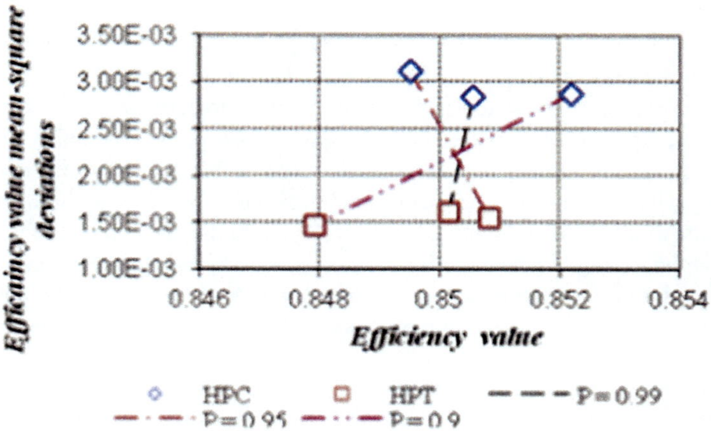

Fig. 3. The level lines of solution probability

target function for HPC modification problem. Desired HPC blade rows phase variables, that were used for concerned modification HPC variant, were chosen blades parameters $\Phi_n^\circ = \{k_{n\sigma}, k_{n\delta}, \Delta\upsilon_n\}$, where $k_\sigma = \delta_\sigma/(\delta_\sigma)_0$, $\delta_\sigma = 1 - \sigma$, $k_\delta = \delta/\delta_0$; $n = 1...2Z$, Z – number of HPC stages, σ is total pressure recovery coefficient, δ is flow deviation angle, $\Delta\upsilon$ is blade stagger angle change.

Calculations were carried out for HPC, which consist of seven stages. The $\sigma^*_{\pi\text{HPC}} = 0.12$ and $\sigma^*_{\eta\text{HPC}} = 0.0030$ were received for mean-square deviation of the required prototype phase variables $\sigma^*_\upsilon = 0.150$ degree, $\sigma^*_{k_\delta} = 0.050$, $\sigma^*_{k_\sigma} = 0.050$. The $\Delta\pi^*_{HPC} = 0.00$ and $\Delta\eta^*_{HPC} = 0.77$ with corresponding mean-square deviations $\sigma^*_{\pi\text{HPC}} = 0.10$ and $\sigma^*_{\eta\text{HPC}} = 0.0029$ were chosen in the capacity of required HPC phase variables values. İt was given, that by constraint region for required phase variables, $P \geq 0.99$, the mathematical expectations and mean-square deviations $\Delta\hat{\pi}^\circ_{HPC} = 0.00$, $\hat{\sigma}_{\pi\text{HPC}} = 0.10$, $\Delta\hat{\eta}^\circ_{HPC} = 0.77$ and $\hat{\sigma}_{\eta\text{HPC}} = 0.0023$ were attained by the variables mathematical expectations (Table 2, here $\Delta\hat{\upsilon} = \hat{\upsilon} - (\upsilon)_0$ grad., $\sigma^*_\upsilon = 0.150$ grad., $\sigma^*_{k_\delta} = 0.050$, $\sigma^*_{k_\sigma} = 0.050$).

Table 2. Parameters calculations results of $\Delta\hat{\upsilon}$, $\hat{k}_\sigma$ and $\hat{k}_\delta$ HPC blade rows

Stage	Rotor			Stator		
	$\Delta\hat{\upsilon}$	$\hat{k}_\delta$	$\hat{k}_\sigma$	$\Delta\hat{\upsilon}$	$\hat{k}_\delta$	$\hat{k}_\sigma$
1	−0.76	0.98	0.99	0.72	1.00	0.99
2	−0.49	1.03	0.97	−0.47	1.01	0.91
3	0.77	1.05	0.95	0.15	1.00	0.97
4	0.07	1.08	0.91	0.18	1.00	0.97
5	0.58	1.05	0.91	0.67	0.94	0.88
6	−1.00	0.98	0.96	0.14	1.00	0.98
7	0.91	1.06	0.93	0.01	0.95	0.93

10 Conclusion

In the issue of conducted researches, the system mathematical model and methodology of rational design decisions synthesis for multiobjective stochastic optimization (modification) problems were developed. The methodology ensures the availability of required variables permanent estimation with practically obtainable accuracy under input data uncertainty. The evolutionary method for nonlinear multiobjective stochastic optimization (modification) problems solution was developed. In this method quasisolution synthesis under bordering of feasible solution class is realized by regularization of smoothing functional minimum searching according to the A.N. Tihonov's method with using the adaptation procedure in evolutionary method parameters choice in time of decision making.

The developed evolutionary method should be considered the high-performance method because its application and the adaptation procedure from iteration to other iteration in time of relaxation parameters and fitness function parameters choice ensure the informational complexity reduction of the developed method of rational decision searching no less them 30% as compared with existing genetic algorithms.

On basis of development methodology, the system aerodynamic turbofan engine compressor improvement was made. The given data analysis reveals the significant feature of turbofan engine system aerodynamic perfection problem solution. This feature consists in the following – the problem solution results depend on chosen target function mean-square deviation values and selected probability level of their attainment.

References

1. Mondal, S., Joly, M. M., Sarkar, S.: Multi fidelity global-local optimization of a transonic compressor rotor. In: Proceedings of the ASME Turbo Expo 2021: Turbomachinery Technical Conference and Exposition, vol. 2D: Turbomachinery, ASME. GT2019-91778 (2019)
2. Lopez, D.l., Ghisu, T., Shahpar, S.: Global optimization of a transonic fan blade through al-enabled active subspaces. In: Proceedings of the ASME Turbo Expo 2021: Turbomachinery Technical Conference and Exposition, ASME. GT2021-59166 (2021)
3. Cuciumita, C., John, A., Qin, N., Shahpar, S.: Structurally constrained aerodynamic adjoint optimization of highly loaded compressor blades. In: Proceedings of the ASME Turbo Expo 2021: Turbomachinery Technical Conference and Exposition, ASME. GT2021-59717 (2021)
4. Hemmert-Pottmann, S., Voß, C., Forsthofer, N., Schneider, T.: High-dimensional multi-fidelity optimization of a 2.5 stage low pressure compressor. In: ISABE-Paper 2019-24067, 24th ISABE Conference, 22–27 September, 2019, Canberra (2019)
5. Barsi, D., Perrone, A., Ratto, L., Ricci, G., Sanguineti, M.: Centrifugal compressor aero-mechanical design: a machine learning approach. In: Proceedings of the ASME Turbo Expo 2021: Turbomachinery Technical Conference and Exposition, ASME. GT2021-59473 (2021)
6. Agnolucci, A., Marconcini, M., Arnone, A., Toni, L., Grimaldi, A., Giachi, M.: Centrifugal compressor stage efficiency and rotor stiffness augmentation via artificial neural networks. In: Proceedings of the ASME Turbo Expo 2021: Turbomachinery Technical Conference and Exposition, ASME. GT2021-59998 (2021)
7. Yang, S., Zhang, M., Liu, Y., Yang, J.: Multi-objective optimization of turbine blade tip with film. In: Proceedings of the ASME Turbo Expo 2021: Turbomachinery Technical Conference and Exposition, ASME. GT2021-60126 (2021)

8. Ludwig, N., Antinor, G., Daub, M., Duddeck, F.: Uncertainty quantification of a jet engine performance model under scarce data availability. In: Proceedings of the ASME Turbo Expo 2021: Turbomachinery Technical Conference and Exposition, ASME. GT2021-58604 (2021)
9. Bazilevych, K., et al.: Stochastic modelling of cash flow for personal insurance fund using the cloud data storage. Int. J. Comput. **17**(3), 153–162 (2018)
10. Meniailov, I., Ugryumov, M., Chumachenko, D., Bazilevych, K., Chernysh, S., Trofymova, I.: Non-linear estimation methods in multi-objective problems of robust optimal design and diagnostics of systems under uncertainties. In: Nechyporuk, M., Pavlikov, V., Kritskiy, D. (eds.) Integrated Computer Technologies in Mechanical Engineering. AISC, vol. 1113, pp. 198–207. Springer, Cham (2020). https://doi.org/10.1007/978-3-030-37618-5_18
11. Bakumenko, N., Strilets, V., Meniailov, I., Chernysh, S., Ugryumov, M., Goncharova, T.: Synthesis method of robust neural network models of systems and processes. In: Nechyporuk, M., Pavlikov, V., Kritskiy, D. (eds.) ICTM 2020. LNNS, vol. 188, pp. 3–16. Springer, Cham (2020). https://doi.org/10.1007/978-3-030-66717-7_1
12. Ugrymova, K., et al.: Synthesis of robust optimal control program for axial flow compressor turning guide vanes. Int. J. Comput. **19**(3), 347–354 (2020)
13. Zgurovsky, M.Z., Pankratova, N.D.: System Analysis: Problems, Methodology, Applications, 743p. Naukova Dumka, Kyiv (2005) (in Russian)
14. Vyazgin, V.A., Fedorov, V.V.: Automated Designing Mathematical Methods, 184p. Visshaya Shkola, Moscow (1989) (in Russian)
15. Tihonov, A.N., Arsenin, V.Ya.: Methods of Ill-posed Problems Decision, 288p. Nauka, Moscow (1986) (in Russian)
16. Rutkovskaya, D.: Neural network, genetic algorithms and fuzzy systems. In: Rutkovskaya, D., Pilinskij, M., Rutkovskij, L. (eds.) M.: Gorjachaya liniya, Telekom, 452p (2004) (in Russian)

Peculiarities of the Modelling of the Bird Dynamic Impact on Fan Blades of an Aircraft Turbojet Engine at Operating Modes

Vyacheslav Merculov[1], Mykola Kostin[1], Gennadii Martynenko[2,3], Natalia Smetankina[3(✉)], and Volodymyr Martynenko[2]

[1] Zaporozhye Machine-Building Design Bureau Progress State Enterprise named after Academician A.G. Ivchenko, 2, Ivanova Street, Zaporizhzhia 69068, Ukraine
{03530,kostinnn}@ivchenko-progress.com

[2] National Technical University "Kharkiv Polytechnic Institute", 2, Kyrpychova Street, Kharkiv 61002, Ukraine
gmartynenko@ukr.net

[3] A. Pidhorny Institute for Mechanical Engineering Problems of the National Academy of Sciences of Ukraine, 2/10, Pozharskogo Street, Kharkiv 61046, Ukraine
nsmetankina@ukr.net

Abstract. The paper considers the problem of adequate mathematical modelling of the impact of flying birds on the fan blades of an aircraft turbojet engine at its normal operating mode. Bird strike with the blades is modelled to evaluate the post-strike deformation and strength state of the blades. Such an evaluation is required for analysing the further serviceability of an aircraft turbojet engine. The suggested approach enables performing such an evaluation using implicit dynamic analysis methods by replacing the flying bird with a special equivalent dynamic load distributed across the strike spot. In contrast to the conventional use of explicit methods for simulating the process of impact of rotating fan wheel blades with a bird flying with a certain speed, such implementation of simulation modelling enables using sufficiently detailed finite-element models. This allows determining the parameters of the stress-strain state of the blades after impact with high accuracy. In this case, the displacements and the deformations of blade points and the stresses therein are suitable for evaluating the wheel serviceability according to airworthiness standards. The problem solution algorithm consists in building computational models for dynamic analysis with detailed account of geometric, physical, mechanical and operating characteristics of the structure. Simulation modelling of the most hazardous variant of bird strike with the fan blades being considered yields the patterns of displacement of blade points, deformations and stresses. The computational model is verified using several error estimation criteria, whereas the approach and results were validated by comparison with test bench experimental data.

Keywords: Aircraft gas turbine engine · Fan blades · Bird strike · Mathematical simulation · Equivalent dynamic loads

M. Nechyporuk et al. (Eds.): ICTM 2021, LNNS 367, pp. 462–473, 2022.
https://doi.org/10.1007/978-3-030-94259-5_39

1 Introduction

The most common problem during the operation of turbojet engines in modem aircraft, without account of engineering and technological challenges and the human factor, is the flying bird strike. Such impact is essentially accidental and it is practically impossible to predict and avoid it. Such impact often has detrimental consequences [1].

Therefore, the development of methods and approaches to refined modelling of such a process is always relevant and necessary from the viewpoint of checking existing aircraft engines and those being developed for compliance to airworthiness standards [2]. The first design component of an aircraft turbojet engine that meets the incoming air, and hence, the birds is the fan bladed wheel. Hence, the fan blades (FB) are the structural elements requiring particular attention from the strength viewpoint after a possible bird strike at operating modes. In this case, simulation modelling of the results of such impact is a method for replacing full-scale experiments with variant calculation studies. The paper is dedicated to developing new approaches to solving this problem.

2 Literature Review

Bird strike is an extremely critical event for air transport safety [3]. Modern aircraft structures have increasingly intricate components, which require improving their investigation methods. Some complex problems, such as different kinds of birds, velocities and angle of impact are challenging to analyse using analytical approaches [4, 5]. On the other hand, experimental testing using prototypes can be expensive and time consuming [6]. With the advent of highly effective FEM-based software packages, further consideration is given to the problem of joint deformation of the bird and target. The numerical evaluation of numerous bird strike scenarios provides for improved design efficiency and ensures safety of engines [7].

Therefore, numerical bird strike simulation has been adopted by the aircraft industry and has proved beneficial in the design validation, testing and certification of manufactured products.

3 Problem Statement and Object of Research

3.1 Problem Formulation

A review of present-day periodical scientific information sources points to the relevance of solving the problem of strength of the component of aircraft turbojet engines (TJE) in refined statements. In particular, such problems are those of analysing the deformation and strength of fan wheel blades under unexpectedly occurring dynamic loads at normal engine operating modes. Such loads occur when flying birds are ingested into the engine, and their impact can be critical from the viewpoint of further serviceability. Special care is taken to check this when considering aircraft TJEs of practically all modifications.

The dynamic strength design of the blading of different rotary equipment and machines is a requisite condition for development and upgrading or for determining the causes of contingencies [8, 9].

An analysis of information sources has shown that different analytical and numerical approaches are used for modelling this process [10, 11]. Analytical approaches enable to model the bird in detail and represent the blades in simplified form. In this case, the blades are modelled as beam or plate elements without account of their geometry. This leads to an inaccurate determining of the parameters of the stress-strain state of the blades and an erroneous evaluation of the probability of their failure and further serviceability. Numerical methods, for example, the finite-element method admits a detailed geometric modelling of intricate blade shapes. Numerical calculations of the strength of TJE fan blades involve a variety of approaches – from replacing the bird's impact on the blade with a quasistatic equivalent loading using implicit analysis methods [12] and to modelling the impact process itself using explicit methods with discretization of bird model volumes and the following approaches: Lagrangian [13], Smooth Particles Hydrodynamics (SPH) [14, 15], Arbitrary Lagrangian-Eulerian (ALE) [16, 17] or Coupled Eulerian-Lagrangian (CEL) [18, 19].

Each of these approaches has its advantages and drawbacks. For instance, replacing the bird dynamic impact on the FB with an equivalent quasistatic system of forces distributed across the blade aerofoil surface enables using a sufficiently detailed mesh and implicit solution methods [12]. This yields a minimal error of determining the stress-strain state parameters. In turn, explicit analysis methods allow modelling the process of bird collision with the blades, but fail to ensure a sufficient accuracy when determining displacements, deformations and stresses [20, 21]. This solves the posed problem. However, a comprehensive accounting of all factors and processes makes it impossible to use detailed geometric and finite-element models due to the problem's big computational burden [22, 23]. In this case, the issue of ensuring accuracy when modelling the process dynamics after FB impact with a bird remains open [24].

The paper offers a kind of middle approach. It allows using sufficiently detailed finite-element meshes to determine the parameters of the stress-strain state of a FB as a function of time.

3.2 Object and Aim of Research

The object of research in the paper is typical fan blades of a gas turbine engine (GTE). The titanium blades have a wide-chord aerofoil without antivibration blade shrouds. Their circular assembly makes up the fan wheel. The blades are dovetailed to the disk. The fan wheel disc is bolted to the flanges of the shaft that provides a mechanical linkage of the fan wheel with the turbine. Accounting for the disc and shaft in the computational model has little effect on the bird resistance of the blades. Therefore, these elements are not considered in the suggested approach. The geometrical model of the fan blading of the GTE fan wheel are shown in Fig. 1.

The aim of research is to develop a technique to simulate a bird strike with the FB at operating modes with an equivalent system of dynamic forces to check the FB strength state and further serviceability. The technique is based on using the finite element method and dynamic strength analysis of GTE fan blades. Here, in contrast to using explicit analysis, modelling the dynamic process of interaction of a flying bird with the rotating wheel is not performed. Implicit finite element analysis methods are used. This ensures a mesh discretization level with minimal errors of numerical determination of

Fig. 1. Geometric model of the fan wheel as a circular FB assembly.

displacements, deformations and stresses as a function of time. That is, the result is the finding of the dynamic response of the blade to the bird's impact, with the response characteristics being fit for evaluating the dynamic strength and further serviceability.

The mesh discretization error can be estimated while performing relevant implicit method analyses using standard procedures [12]. The approach and results are validated by acceptable experimental data on the FB state after impact with a single big bird with a mass of about 2 kg, flying with a speed of slightly more than 100 m/s, with a reduced body diameter of about 0.1 m, with its reduced length of about 0.25 m, and the fan wheel rotational speed corresponding to the GTE takeoff mode (5,000–5,500 rpm). It implies testing an engine with bird body ingestion into the flow part by means of compressed-air guns. The tests were conducted according to airworthiness standards. They give a full picture of the character and degree of FB deformation after bird strike, and therefore they are suitable for comparative analysis [12].

4 Modelling the Dynamic Behaviour of GTE Fan Blades During Bird Strike

Modelling the dynamic behaviour of the GTE blading presupposes that a single blade is considered. This is because the bird's biggest impact is taken up by a single blade with which initial contact occurs. This enables to build a sufficiently accurate finite element model with a minimal discretization-related error. In doing so, it is offered to replace the impact of flying birds with an equivalent dynamic impulse force load. The suggested approach, without performing involved and computation-intensive nonlinear dynamic analysis with explicit methods, enables estimating the localisation of hazardous zones for both stresses and deformations. The hazardous zones primarily are the sites of maximum deflections and stresses.

4.1 Technique of Simulation Modelling of Bird Impact on the Fan Blades

To achieve the goal, a numerical study program is offered. It comprises several stages, such as: building a physical model of the object of research with selection of a configuration of parameters and substantiation of simplifications of geometry, materials, loading and fastening; creating a computational finite element scheme of an ordered structure for one FB using the tools and elements of implicit dynamic analysis; conducting preliminary variant static analyses in a linear statement to estimate the error using different methods; performing a simulation computational study with impulse action of the force load and accounting for the quasistatic centrifugal force and nonlinear material characteristics; verifying equivalent forces by the physical content of the results and their compliance with experimental data. The computation studies themselves deal with studying the behaviour of the structure under dynamic impulse action of loads that model the impact of a medium-size or big bird on the FB.

4.2 Finite Element Model of the Fan Blades

The finite element model, as a regular mesh of implicit analysis finite elements, is shown in Fig. 2. The blade aerofoil has an ordered mesh of 20-node finite elements, and the shrouds, the blade leading edge tip and the blade root are divided into a free mesh comprising 10, 15 and 13-node finite elements. The degrees of freedom of these elements are displacements in the direction of the axes in the global Cartesian system of coordinates. The blade aerofoil (in the middle) has a zone for subsequent application of a bird strike load. The area of this zone corresponds to the cross-section area of a big bird. Such a configuration yields the least error of all possible ones [12].

The action of static centrifugal forces is accounted for as preloading. From the viewpoint of the greatest hazard, two computational cases were distinguished. They model the strike of a single big bird with the aforementioned parameters on the designated blade surface and its leading edge. The boundary conditions are the restraint of the root bottom end over all degrees of freedom. Hence, the disk is excluded from consideration. This is possible because its stiffness exceeds that of the FB aerofoil by an order.

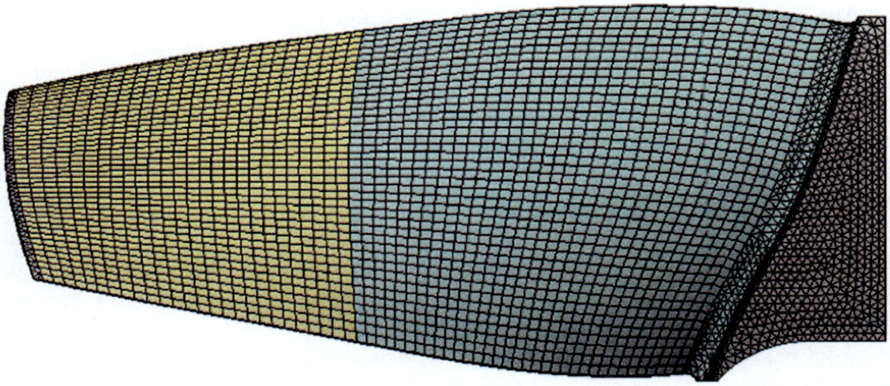

Fig. 2. Finite element model of the fan blade (13,540 finite elements; 13,7082 degrees of freedom).

5 Numerical Experiments in Modelling Bird Strike on the Rotating Wheel Blades

Computational investigations are dedicated to studying the structure response with account of nonlinear material behaviour at a static action of joint loading by centrifugal forces and impulse loads that model the bird's impact on the FB.

5.1 Computational Cases and the Model

A substantiated choice of the most hazardous cases of the collision a flock of medium-size birds or a single big bird with a blade is presented in [12]. It was made by considering the forms of GTE FB deformation during bench tests. The most hazardous case, from the viewpoint of disturbance of strength properties, was found to be that of a strike of a single big bird with a mass of about 2 kg, a body diameter of about 0.1 m and a length of about 0.25 m, and which is flying with a speed of slightly over 100 m/s. In this case, the bird strike locations on the FB are relatively random, though some of them are real and present a particular hazard. They were those that were considered when choosing computational cases and corresponding computational models.

The most critical bird strike location, from the viewpoint of FB strength and further serviceability, is the surface on its aerofoil. Its area is equal to that of the bird body cross-section. The computational scheme for this case is shown in Fig. 3a. In the scheme, the load distributed across the surface is directed along the wheel axis. Its value is found based on the equivalence of force and body impulses $F\Delta t = mv$, where F is force, Δt is its impact time, m is bird mass, v is bird speed, with Δt being equal to the time interval during which the bird covers a distance equal to its length with the speed v. The total force value of the load is 97,180 N, and the wheel rotational speed corresponds to the takeoff mode. The suggested approach to representation with an equivalent load enables replacing the dynamic impact of the bird on the FB with a quasistatic loading. In doing so, a fairly high correspondence to mechanical action authenticity is retained.

The second location of a big bird strike [12] is that of a part of the FB edge. In this setting, the edge cutting into the body of the bird injected into the interblade space was modelled. In this case, the force load is separated into axial and lateral components. Hence, this computational case is less hazardous because a part of the load acts in the blade plane, with the blade bending stiffness in this direction being much higher than that in the perpendicular one. This is confirmed by the results of quasistatic strength analyses [12].

Static analyses (Fig. 3b) used a stress-strain diagram of the titanium alloy and the model of the elastoplastic behaviour of the material to account for the nonlinear plastic behaviour of the FB material. This model includes the yield point condition, the flow rule and the hardening rule. The model used was that of polygonal isotropic hardening with the von Mises yield criterion without account for the Bauschinger effect. The Newton-Raphson method was used for nonlinear analyses.

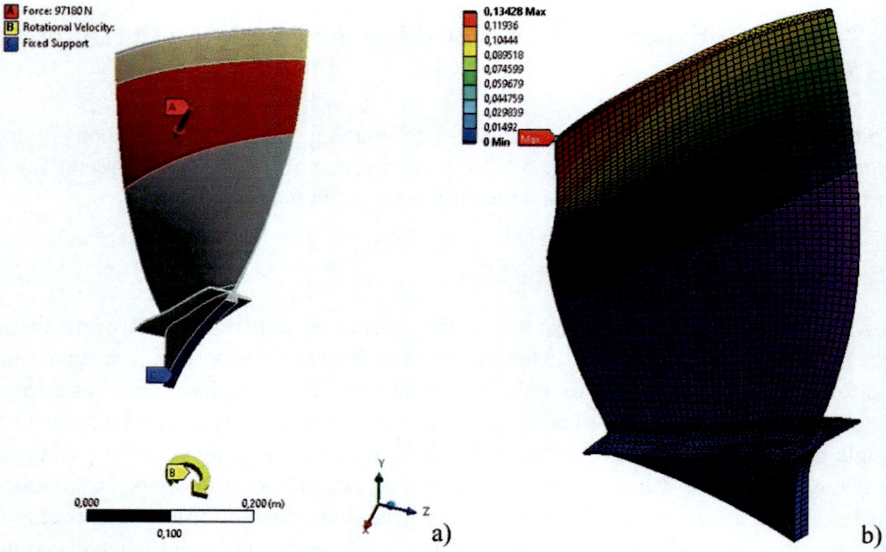

Fig. 3. Computational characteristic: a) FB computational scheme with a load equivalent to the impact of a big bird on the designated part of the blade surface; b) module of the total displacement vector from static load action, m.

5.2 Statement of the Dynamic Strength Problem

The subject of the main part of the investigation is studying FB behaviour with account of the system dynamics with the same load as that in the described computational case (Fig. 3b), though applied as an impulse during a period of $\Delta t = 0.0022$ s. This is the time interval during which the bird flies over a distance equal to its length with speed v of about 100 m/s.

To obtain the system's dynamic response in the computational scheme (Fig. 3a), impulse dynamic loading was introduced. In this case, analyses of transient process were performed using the Newmark numerical method for solving a system of differential equations. The Newton-Raphson method was used to account for nonlinear effects.

5.3 Results of Numerical Experiments

The results of analysis of transient dynamic processes are shown in Fig. 4 and Fig. 5 as graphs of the dependencies of displacements and stresses, respectively. Figure 4 shows the graphs of total and component displacements in a point located on the leading edge on the blade aerofoil tip. This is the point of maximum values of total displacements. The vertical dashed lines designate the time interval of load impulse action. Analysis shows that the transient process of free oscillations is quasiperiodic with damping caused by both the action of resistance forces and energy expenditure on deformation with build-up of residual deformations. Here, one can observe a polyharmonic response caused by the nonlinear behaviour of the system.

Figure 5 shows the stress curves with a display of equivalent stresses and stress tensor components. The stress response is of the same character as that of displacements. The pattern of total displacements and equivalent von Mises criterion stresses at the point of the maximum values of the latter (shown on the graph) are shown in Fig. 6 and Fig. 7, respectively.

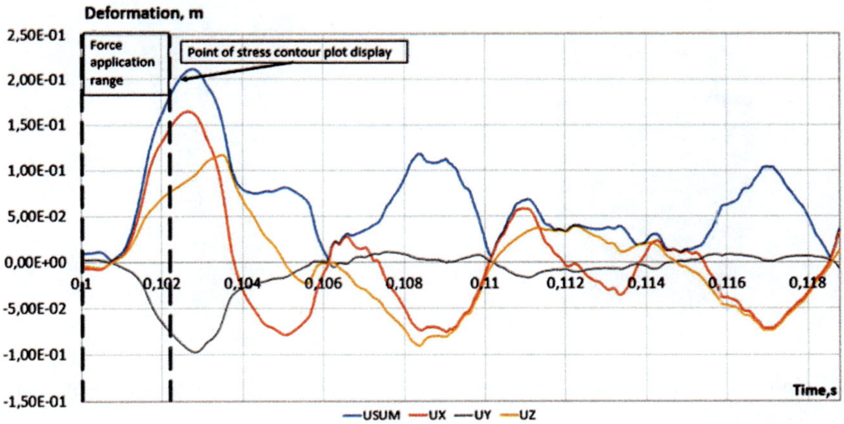

Fig. 4. Displacements vs time in the point of maximum FB displacements located on the leading edge of the aerofoil tip.

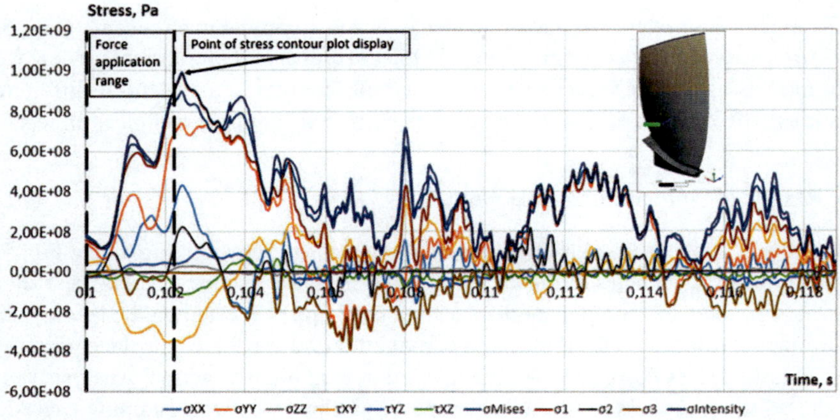

Fig. 5. Different components and equivalent stresses vs time in the point of maximum FB stresses located on the aerofoil leading surface.

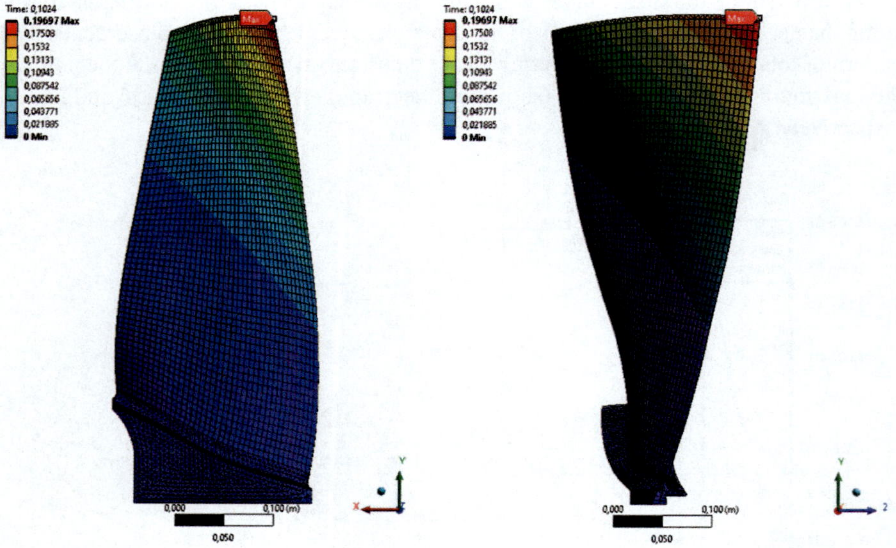

Fig. 6. Module of total displacement vector by the action of an impulse load on the designated part of the aerofoil surface, m.

5.4 Discussion of Results

The distribution of stresses and deformations, as well as the configuration of level isolines correspond well to the expected actual behaviour of the structure under the applied loading. The results are fairly meaningful physically. An evaluation of the modelling results by the qualitative behaviour of the structure has shown an acceptable match when compared with experimental data. In this case, quantitative parameters can show a certain

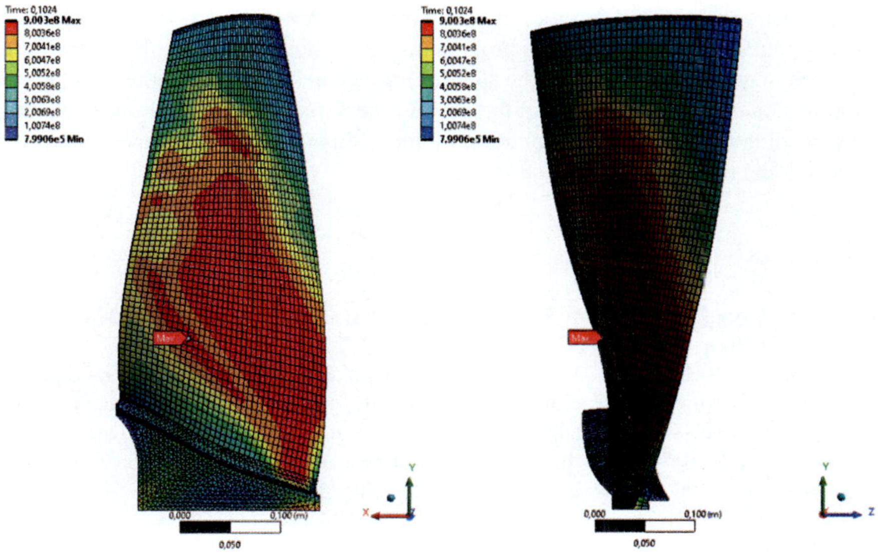

Fig. 7. Equivalent von Mises criterion stresses at a point in time corresponding to their maximum values, Pa.

discrepancy because it must be taken into account that, under actual conditions, not the entire bird's load is "rigidly" transmitted to the blade because of its "soft" structure.

Analysis of results shows that the distribution of displacements at the specified time point coincides with the distribution of displacements at static application of the same load (Fig. 3b) when using the same finite-element model. However, the maximum deviations are 0.197 m (Fig. 6) and differ from the static solution of 0.134 m (Fig. 3b) by about 32%. This is indicative of that replacing a dynamic action with a static analog to achieve a desired accuracy of simulation modelling of dynamic processes, including impact ones occurring during a bird strike on FB is impossible.

As concerns equivalent stresses, their values at a critical time point are 900 MPa (Fig. 7). This practically coincides with the values of the same stresses under static loading that models a big bird strike on the blade surface, with the stresses being equal to 901 MPa [12]. Although the distribution pattern is not completely identical, it is similar and the maximum stress localisation is found in a very close neighbourhood of the aerofoil.

6 Conclusion

The simulation study enables drawing a conclusion that such correlation of results of static and dynamic analyses confirms the correctness of the suggested calculation scheme, and it enables to finally verify the finite-element model as one with an acceptable discretization and numerical solution error level.

Of course, even the solution of the initial problem with process modelling in time by using implicit methods, specifically solving the motion equation with account of inertia, elastic and resistance forces by direct integration with regard for nonlinearities, can be no substitute for solving the problem using explicit analysis tools. However, as compared to quasistatic analyses, the solution in such a statement is more close to real conditions, and helps to understand the peculiarities of structure behaviour depending on the loading configuration. In follow-up studies, this will help to choose appropriate computational models.

References

1. El-Sayed, A.F.: Bird Strike in Aviation: Statistics, Analysis and Management. Wiley, New Jersey (2019)
2. Shahimi, S.S., Abdullah, N.A., Hrairi, M., Ahmad, M.I.M.: Numerical investigation on the damage of whirling engine blades subjected to bird strike impact. J. Aeronaut. Astronaut. Aviat. **53**(2), 193–200 (2021). https://doi.org/10.6125/JoAAA.202106_53(2).11
3. Smetankina, N., Kravchenko, I., Merculov, V., Ivchenko, D., Malykhina, A.: Modelling of bird strike on an aircraft glazing. In: Nechyporuk, M., Pavlikov, V., Kritskiy, D. (eds.) Integrated Computer Technologies in Mechanical Engineering. AISC, vol. 1113, pp. 289–297. Springer, Cham (2020). https://doi.org/10.1007/978-3-030-37618-5_25
4. Riccio, A., Cristiano, R., Saputo, S.: A brief introduction to the bird strike numerical simulation. Am. J. Eng. Appl. Sci. **9**(4), 946–950 (2016). https://doi.org/10.3844/ajeassp.2016.946.950
5. Abrate, S.: Soft impacts on aerospace structures. Prog. Aerosp. Sci. **81**(2), 1–17 (2016). https://doi.org/10.1016/j.paerosci.2015.11.005
6. Yuecheng, J.: A review of research on bird impacting on jet engines. IOP Conf. Ser. Mater. Sci. Eng. **326**, 012014 (2018). https://doi.org/10.1088/1757-899X/326/1/012014
7. Zhang, D., Fei, Q.: Effect of bird geometry and impact orientation in bird striking on a rotary jet-engine fan analysis using SPH method. Aerosp. Sci. Technol. **54**, 320–329 (2016). https://doi.org/10.1016/j.ast.2016.05.003
8. Rusanov, A., Martynenko, G., Avramov, K., Martynenko, V.: Detection of accident causes on turbine-generator sets by means of numerical simulations. In: 2018 IEEE 3rd International Conference on Intelligent Energy and Power Systems (IEPS 2018), pp. 51–54. IEEE, Kharkiv (2018). https://doi.org/10.1109/IEPS.2018.8559546
9. Martynenko, V., Hrytsenko, M., Martynenko, G.: Technique for evaluating the strength of composite blades. J. Inst. Eng. (India) Ser. C **101**(3), 451–461 (2020). https://doi.org/10.1007/s40032-020-00572-9
10. Hou, N., Li, Y., Liu, J.: Numerical simulation of bird impact on hollow blades of titanium fan assembly. J. Aerosp. Eng. **32**(4), 04019044 (2019). https://doi.org/10.1061/(asce)as.1943-5525.0001024
11. Liu, J., et al.: Numerical simulation and test on damage of rotary engine blades impacted by bird. Int. J. Crashworthiness **24**(1), 106–120 (2019). https://doi.org/10.1080/13588265.2018.1452548
12. Merculov, V., Kostin, M., Martynenko, G., Smetankina, N., Martynenko, V.: Force simulation of bird strike issues of aircraft turbojet engine fan blades. In: Cioboată, D.D. (ed.) ICoRSE 2021. LNNS, vol. 305, pp. 129–141. Springer, Cham (2022). https://doi.org/10.1007/978-3-030-83368-8_13
13. Kumar, R.V.: Failure analysis of rotorcraft composite end plate structure under high-velocity bird impact. J. Fail. Anal. Prev. **16**(5), 864–873 (2016). https://doi.org/10.1007/s11668-016-0165-3

14. Guo, P., Liu, Z., Zhang, G., Reza, H., Zhang, J.: Study on effect of bird impact position and attitude on fan blade damage. J. Vib. Shock **40**(12), 124–131 (2021). https://doi.org/10.13465/j.cnki.jvs.2021.12.016
15. Tatlier, M.S.: A numerical investigation of a bird strike on the structure of an aircraft wing leading edge. Eur. Mech. Sci. **4**, 37–40 (2020). https://doi.org/10.26701/ems.622830
16. Chuan, Z., Xiang-Hua, J., Xiang-Hai, C., Tong-Cheng, S.: TC4 hollow fan blade structural optimization based on bird-strike analysis. Procedia Eng. **99**, 1385–1394 (2015). https://doi.org/10.1016/j.proeng.2014.12.674
17. Hedayati, R., Sadighi, M.: Finite element bird-strike modeling. In: Bird strike. an experimental, theoretical and numerical investigation, Chapter 6, pp 113–159. Elsevier (2016). https://doi.org/10.1016/b978-0-08-100093-9.00006-6
18. Saeed, M.H., Muhammad, N.: Investigate the damage of kinetic energy high-velocity bird impact on three different model spar by using the coupled Eulerian-Lagrangian (CEL) approach. Mechan. Mach. Sci. **99**, 215–223 (2021). https://doi.org/10.22541/au.160684062.24928465/v1
19. Saputo, S., Sellitto, A., Riccio, A., Di Caprio, F.: Crashworthiness of a composite wing section: numerical investigation of the bird strike phenomenon by using a coupled Eulerian–Lagrangian approach. J. Mater. Eng. Perform. **28**(6), 3228–3238 (2019). https://doi.org/10.1007/s11665-019-03944-0
20. Li, K.-Y., Jiang, X.-H., Wei, D.-S.: Numerical simulation of bird strike on a S-shaped stealth inlet. In: Zhang, X. (ed.) APISAT 2018. LNEE, vol. 459, pp. 1696–1706. Springer, Singapore (2019). https://doi.org/10.1007/978-981-13-3305-7_136
21. Zhang, Z., Li, L., Zhang, D.: Effect of arbitrary yaw/pitch angle in bird strike numerical simulation using SPH method. Aerosp. Sci. Technol. **81**, 284–293 (2018). https://doi.org/10.1016/j.ast.2018.08.010
22. Marinkovic, D., Zehn, M.: Survey of finite element method-based real-time simulations. Appl. Sci. **9**(14), 2775 (2019). https://doi.org/10.3390/app9142775
23. Touzeau, C., Magnain, B., Serra, Q., Florentin, E.: Accuracy and robustness analysis of geometric finite element model updating approach for material parameters identification in transient dynamic. Int. J. Comput. Methods **16**(1), 1850084 (2019). https://doi.org/10.1142/S0219876218500846
24. Zhou, Y., Sun, Y., Huang, T.: Impact responses of slender composite plates for bird-strike testing of fan blades. Latin Am. J. Solids Struct. **16**(9), 1–12 (2019). https://doi.org/10.1590/1679-78255809

Increasing Ecological and Energy Efficiency of Combustion Engines by Using a Thermopressor

Dmytro Konovalov(✉), Mykola Radchenko, Halina Kobalava, Maxim Pyrysunko, and Andrii Andreev

Admiral Makarov National University of Shipbuilding, 9, Heroes of Ukraine Avenue, Mykolayiv 54025, Ukraine

Abstract. The development of environmentally efficient technologies for modern heat engines has a negative impact on their energy efficiency. Emission reduction measures require additional external energy costs. Reducing the formation of nitrogen oxides (NOx) requires a reduction in the maximum combustion temperature of the fuel. However, lowering this temperature decreases the fuel and energy efficiency of the internal combustion engine. Environmentally and energy-efficient technologies which would reduce emissions into the environment and at the same time eliminate the negative impact on the fuel and energy efficiency of power plants are promising today. The efficiency analysis of thermopressor technologies application in gases recirculation systems of ship low-speed main diesels is considered in the paper. The operation of thermopressors is based on the effect of thermo-gas-dynamic compression, which consists in increasing the air flow pressure with contact air cooling. Cooling of exhaust gases by the thermopressor in recirculation systems allows to reduce the back pressure in the gas exhaust system and, as a consequence, to exclude the negative impact of reducing the maximum temperature in the ICE cycle by reducing energy costs. The thermopressor application allows to reduce or fully compensate for hydraulic losses, which in turn decreases the fan load (or electric compressor load) of the recirculation system, reduces the load on the scrubber and the heat removal system from the recirculation gases, and accordingly reduces the specific fuel consumption by 1.5–1.7%.

Keywords: Scrubber · Compressor · Internal combustion engine · Recirculation system

1 Introduction

The consequence of modern power plants using is an excessive increase in fuel consumption and emissions of harmful substances into the environment. The largest share belongs to power plants based on internal combustion engines due to pollution by toxic substances contained in combustion products.

The formation of harmful gases (carbon dioxide CO_2, nitrogen oxides NOx, carbon monoxide CO, sulfur oxides SOx) depends on the organization of work processes in the

M. Nechyporuk et al. (Eds.): ICTM 2021, LNNS 367, pp. 474–483, 2022.
https://doi.org/10.1007/978-3-030-94259-5_40

internal combustion engine. Therefore, the solution should be sought in the development of technologies related to the ecologization of engines. An effective way in this direction is the artificial neutralization of harmful substances in the exhaust gases, for example, gas recirculation (EGR technology), that can be used on ships.

2 Literature Review

Ecologization of heat engines leads to a decrease in their energy efficiency, as emission reduction measures require additional external energy costs. Reducing the formation of nitrogen oxides NOx requires a reduction in the maximum combustion temperature of the fuel, and on the other hand, lowering this temperature decreases the fuel and energy efficiency of the internal combustion engine [1, 2].

Hence, it is important to create such environmentally and energy efficient technologies that would reduce emissions into the environment and at the same time would not have a negative impact on the fuel and energy efficiency of internal combustion engines [3, 4].

A promising direction today is to use the jet thermopressor technologies, which are based on the effect of thermos-gas-dynamic compression [5, 6], that in turn provide cooling and simultaneous increase of gas flow pressure in circuit design solutions for neutralization of harmful substances in fuel combustion products. The operation of thermopressors is based on the effect of thermo-gas-dynamic compression, which consists in increasing the air flow pressure with contact air cooling.

A lot of researches are devoted to the improvement of the exhaust gas recirculation system [7, 8]. The efficiency of recirculation system application for diesel engines in the operation was substantiated experimentally and theoretically, which allows to reduce NOx with a recirculation fraction of Kr = 9% by 35%, CO by 90% and soot particles by 50%. The specific fuel consumption increases by 4% (at a capacity of 75% of nominal), when a comprehensive system using to reduce the toxicity of exhaust gases. The efficiency of recirculation gas cooling is also investigated. For example, a decrease in the temperature at the diesel inlet by 60 °C (from 200 °C to 140 °C) leads to an additional reduction of NOx by 3% and a simultaneous reduction in the fuel consumption by 2% [8].

The combustion temperature decreases slightly because of CO_2 and water molecules have a higher heat capacity. According to the data by Chub T.V. [9], the increased mass consumption leads to the exhaust gases temperature reducing up to 93%, while the increase in specific heat leads to ones up to 7%. Cooling of recirculation gases leads to a decrease in emissions of NOx and particles at comparable degrees of recirculation Kr.

Studies [9] have shown that at an engine load of 50%, recirculation reduces nitrogen oxides by 22.5% with an increase in fuel consumption by 4%, and together with evaporative cooling, nitrogen oxides are reduced by 28.5%.

The authors of the study [10] conducted an experimental research to observe the effect of exhaust gas recirculation on the temperature and opacity of the exhaust gases. It has been shown that increasing the degree of recirculation reduces the temperature of the exhaust gases, for example, with recirculating coefficient Kr = 21% decrease in the exhaust gases temperature was up to 10 °C [10].

The results of the exhaust gas recirculation application for the MAN 12K90ME9 engine, that was installed as the main engine on the container ship, are shown in the research by N. Bent [11]. At the degree of recirculation at rated and partial engine loads which was Kr = 30–40%, NOx emissions were reduced up to 80%, SOx emissions were reduced by 25%, particulate matter were reduced (PM) by 50%. However, it led to an increase specific fuel consumption by 5–6 g/(kW·h). Increasing the specific fuel consumption on low-speed engines to 5 g/(kW·h) was also confirmed by another authors [12].

A typical recirculation system for a ship's main low-speed engine includes: a scrubber; a cooler; a water mist catcher; a fan; and a NaOH solution support system with a pump and tank. It should be noted that the components of the system are quite large (Fig. 1), the pumps of the cleaning system, cooling system and fan (or electric compressor) require the consumption of electricity [13, 14].

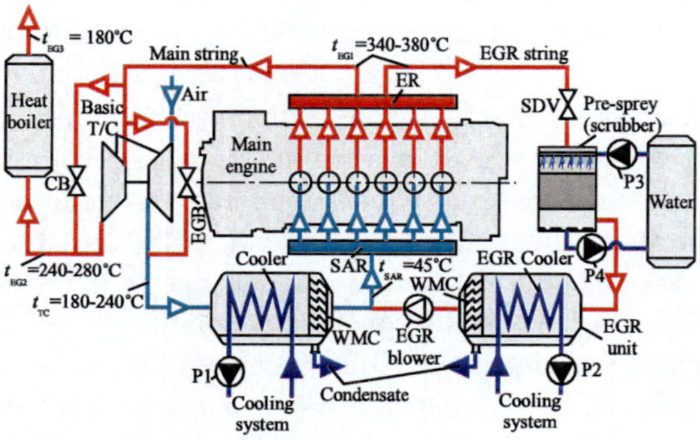

Fig. 1. Scheme of EGR-technology with bypass matching for the MAN marine diesel engine: P1, P2, P3, P4 – Pump; TC – turbocharger; EGB – exhaust gas bypass valve; CBV – cylinder bypass valve; ER – exhaust receiver; SAR – scavenge air receiver; SOV – EGR shut-off valve; WMC – water mist catcher

It can be used two main schemes of EGR-technology in marine two-stroke diesels [15, 16]:

1. bypass recirculation system, with one supercharged turbocharger (used for engines with a cylinder diameter up to 700 mm);
2. recirculation system with two or more turbochargers (used for engines with a cylinder diameter of more than 700 mm).

The scrubber of the recirculation system is an important element of the gas cleaning system. The direction of the inlet flow of exhaust gases to the scrubber of the recirculation system is usually radial, and the outlet – vertically from above [16]. The scrubber must be equipped with a moisture collector. The purified recirculation exhaust gases pass through the droplet separator before the inlet. The scrubber always must be installed vertically. Horizontal arrangement is not possible because effective purification of the exhaust gases requires countercurrent interaction between the gases and the spray water.

In the process of cleaning the water is pumped out with the moisture collector through the heat exchanger to the top of the scrubber. Water is injected into the flow of exhaust gases from the spray nozzles in the scrubber, and is also fed into the middle part of the scrubber to improve the efficiency of SOx removal. Water absorbs sulfur oxides SOx and other components and cools the exhaust gases. The disadvantages of these devices include large size and weight, as well as the need to use in the construction of highly corrosion-resistant materials to prevent the danger of low-temperature corrosion.

Another disadvantage of the gas recirculation system is the increase in the back pressure of the exhaust gases due to additional pressure losses in the scrubber and the heat exchanger-cooler, which, in turn, contributes to the deterioration of fuel efficiency of the engine.

It should be noted that there is another very important element in the exhaust gas recirculation system – the gas heat exchanger-cooler. Creating a reliable and efficient heat exchanger for cooling recirculating gases is a difficult task due to deposits and contaminants emitted by these gases. The presence of a large number of particles in the exhaust gases of diesel engines makes it more difficult to ensure the reliability of the recirculation system due to pollution and deposits. Solid particles entering the cylinder with recirculation gases increase its wear. All this reduces emissions of nitrogen oxides, but increases the formation of smoke and there are problems with cleaning parts [13]. Based on numerous studies, it can be concluded that it is essential to develop an efficient heat exchanger (cooler) of fuel combustion products in the recirculation system [17, 18].

In view of the above, the following hypothesis was formulated: cooling the exhaust gases of the internal combustion engine with an thermopressor in recirculation systems allows to reduce the back pressure in the gas exhaust system and, as a result, eliminate the negative impact of reducing the maximum temperature in the internal combustion engine cycle.

This hypothesis determines the following direction of research - ensuring efficient energy conversion in the internal combustion engine using an thermopressor for evaporative contact cooling (recirculation) of exhaust gases, as a consequence, reducing back pressure in the exhaust system and increasing efficiency and reducing fuel consumption by power plant.

3 Research Methodology

The study used a comparative calculation method to determine the thermodynamic and energy efficiency of the thermopressor cooling system in the power plant. To apply this method, own software package was developed for:

- modeling all working processes in the thermopressor;
- calculating the main structural elements of the thermopressor, taking into account the developed methodology of rational design;
- calculating the energy efficiency and the main indicators of gas turbine engines and internal combustion engines when using thermopressor cooling systems, taking into account changes in climatic and hydrometeorological conditions, as well as partial modes of operation of the power plant;

– calculating the performance of the recirculation gas system of the internal combustion engine, taking into account the exhaust gases cooling in the thermopressor.

The software package was developed on the basis of the Visual Basic programming language, as well as the Excel program.

The calculation of the thermopressor characteristics (air pressure P_{air}, air temperature T_{air}, air speed w_{air}, water speed w_w, etc.) and its main geometric parameters of the flow part under given operating conditions of the engine of the power plant was carried out according to methods [19–22]. The calculation took into account the pressure loss in the nozzle (confuser), working chamber, diffuser, as well as the frontal resistance of injected water droplets.

The calculation of the main internal combustion engine parameters was performed using software complexes Diesel-RK [23] and CEAS (MAN B&W) [24] taking into account changes in the corresponding parameters of the air at the inlet to the engine cylinders, as well as taking into account partial modes. To use these software, an additional control program was developed, which determined the parameters of moist gas (air) at the inlet and outlet of the turbocharger, which is cooled in an thermopressor, as input data for Diesel-RK and CEAS.

4 Results

One of the disadvantages of the exhaust gas recirculation system is an increase in back pressure due to additional pressure losses in the scrubber and heat exchanger-cooler, and this, in turn, contributes to the deterioration of fuel efficiency of the engine.

Promising heat exchangers-coolers that could be used in the exhaust gas recirculation system include contact heat exchangers of jet type is called thermopressor.

The comparison was made on the basis of the basic scheme with exhaust gas recirculation (Fig. 1) when analyzing the effectiveness of the proposed solution [25]. These schemes are used to reduce harmful emissions (CO2, NOx, SOx) for low-speed engines from MAN (according to Tier III). This involves bypassing part of the exhaust gases with the following cooling and simultaneous condensation (cleaning) of environmentally unsafe gases in the scrubber and cooling in the heat exchanger-cooler gas to a temperature equal to the air temperature in the receiver. The system includes a scrubber, cooler, water mist trap, fan, and NaOH solution support system with a pump and tank. The water supply to the pre-spray in the scrubber in front of the cooler is part of the water treatment system.

The bypass recirculation system (Fig. 1) works as follows: exhaust gases in the amount of up to 30–40% are fed through a valve from the exhaust gas receiver to the scrubber, where they are partially cooled and cleaned by spraying water with special nozzles. The gases are then cooled in a heat exchanger-gas cooler, the condensate is discharged into a condensate trap, and the purified and cooled gas is fed in a fan or electricity compressor to an air receiver, where it mixes with fresh air.

Schematic solutions are considered for the ship low-speed diesel two-stroke engine MAN B&W brand 6G70ME-C9.5. The CEAS software package presented by MAN was used for the analysis [12]. The calculation was made for the following initial data:

rated power and engine speed – $N_n = 21840$ kW, $n_n = 83$ min^{-1}; specific consumption of conventional fuel ($Q^r{}_n = 42700$ kJ/kg) – $g_n = 172$ g/(kW·h) under ISO conditions; turbocharger – ABB280L; EGR recirculation circuit – bypass with scrubber and cooling gas; operational characteristics - load on the engine – 90%; power – $N_e = 19656$ kW; speed – $n_e = 80.1$ rpm; specific consumption of conventional fuel – $g_e = 169.8$ g/(kW·h) at ISO.

The calculation of engine characteristics was performed both in nominal and operational (90% of the engine load) modes and in the entire possible range of partial loads (30–100%), as well as for the following conditions:

- ISO – air temperature at the inlet of the turbocharger $t_{air1} = 25$ °C, the temperature of the cooling water in the supercharged air cooler $t_{w1} = 25$ °C, the relative humidity - $\varphi_1 = 30\%$;
- tropical zone – $t_{air1} = 45$ °C, $t_{w1} = 36$ °C, $\varphi_1 = 60\%$;
- cold zone – $t_{air1} = 10$ °C, $t_{w1} = 10$ °C, $\varphi_1 = 60\%$.

From the point of view of ensuring efficient cooling of exhaust gases in the recirculation system, it is possible to offer a circuit solution with the installation of the thermopressor in front of the scrubber (Fig. 2). In this case, the thermopressor provides a certain increase in gas pressure, which, in turn, will reduce the back pressure of the gas path of the engine and, accordingly, reduce the load on the additional fan (or electric compressor). In addition, no less important, this device will provide cooling of the exhaust gases to a temperature slightly above the dew point (based on the hazard conditions of low-temperature sulfuric acid corrosion). This solution will also reduce the heat load on the scrubber and the surface gas cooler.

Injection should be carried out with more water than necessary for evaporation (not less than 10%). This solution is necessary to provide a greater effect from the increase in pressure as a result of thermogas-dynamic compression (by reducing friction losses), on the one hand, and to provide conditions for the injection of liquid into the scrubber. This will reduce the load on water consumption for injection in the scrubber, as well as partially or completely compensate for pressure losses.

The thermopressor application in front of the scrubber of the exhaust gas recirculation system allows to increase the pressure of the exhaust gases bypassed P_{tp} through the engine turbocharger by $0.2 - 0.4 \cdot 10^5$ Pa (6–12%) (Fig. 3). At the same time, the decrease in temperature as a result of thermo-gas-dynamic compression effect is ΔT_{tp} = 150–300 K (relative temperature is $T_{tp1}/T_{tp2} = 1.3$–1.7, Fig. 4).

The amount of water required for evaporation in the thermopressor is $G_{w.tp} = 0.35$–0.50 kg/s (Fig. 5), $g_{w.tp} = 0.05$–0.08 or 5–8% of the exhaust gas consumption, at the recirculation coefficient Kr = 0.12–0.20 (12–20%) with water consumption per scrubber $g_{w.egr} = 0.02$–0.04 (2–4%) (Fig. 6).

Increasing the pressure in the thermopressor allows to reduce the back pressure in the gas exhaust system, and thus reduce the fan load of the exhaust gas recirculation system. Thus, the fan power decreased from $N_f = 120$–160 kW to $N_{f.tp} = 20$–110 kW (Fig. 7), and in the cold zone when the engine is loaded above 75%, the use of the fan is not required at all, because the gas pressure after the thermopressor is sufficient to ensure that is injected into the air receiver without additional compensation.

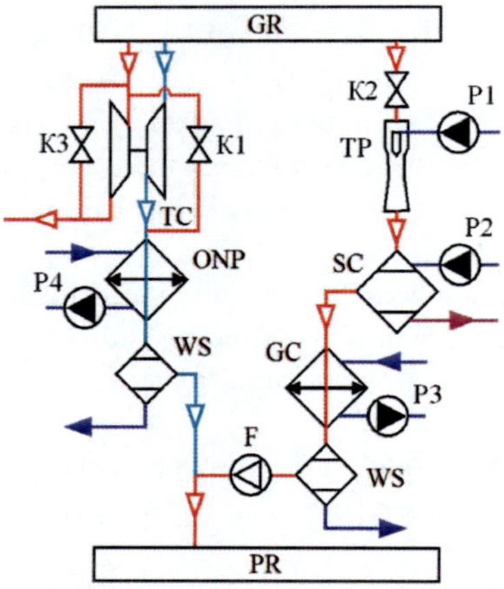

Fig. 2. An exhaust gas recirculation scheme by using the thermopressor: P1, P1, P3, P4 – pumps; TC – turbocharger; K1 – EGR bypass valve; K2 – EGR control valve; K3 – bypass valve on the exhaust line; GR – gas receiver; PR – charge air receiver; WS – water separator; TP – thermopressor; GC – gas cooler; ONP – supercharged air cooler; F – fan; SC – scrubber

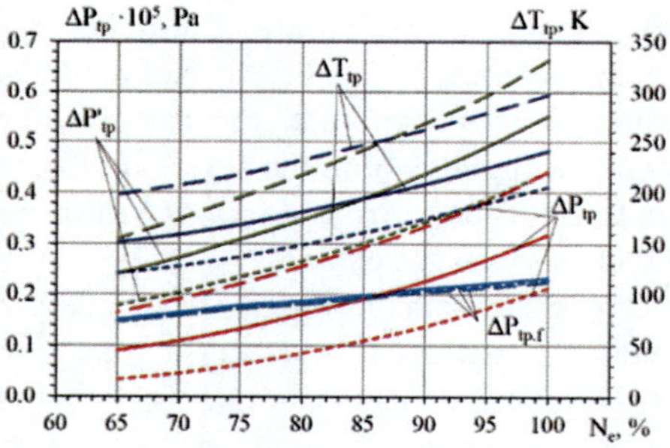

Fig. 3. Dependences of pressure increase in TP without taking into account friction $\Delta P'_{tp}$, taking into account friction P_{tp}, friction losses $\Delta P_{tp.f}$, decrease in gas temperature in TP ΔT_{tp} on the engine load N_e: ______ – ISO conditions (t_{air1} = 25 °C; t_{w1} = 25 °C); __ __ __ – tropical conditions (t_{air1} = 45 °C; t_{w1} = 36 °C); _ _ _ _ _ – cold zone (t_{air1} = 10 °C; t_{w1} = 10 °C)

Reducing the fan load of the recirculation system, as well as reducing the back pressure in the system reduces the specific fuel consumption of the engine from $g_{e.egr}$ = 166.5–174.0 g/(kW·h) to $g_{e.tp}$ = 164.0–171.0 g/(kW·h), Δg_e = 2.5–3.0 g/(kW·h) (1.5–1.7%) with a total specific fuel consumption without waste recirculation gases – g_e = 158.0–170.0 g/(kW·h) (Fig. 7). It is possible to understand that at loading on the engine close to operational (90–95% of engine power) fuel expenses at application of recirculation system with the thermopressor and without recirculation in general are equal that does such circuit decision rational from the point of view of economy. In this case, it is possible to combine high environmental and economic benefits from the application of such a solution.

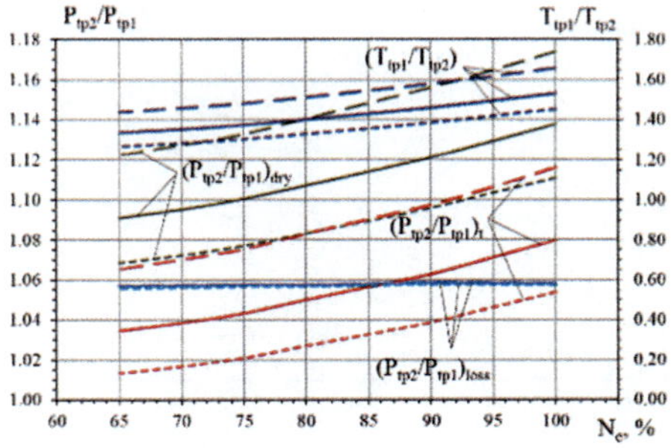

Fig. 4. Dependences of relative pressure increase in TP without taking into account friction (P_{tp2}/P_{tp1}), taking into account friction P_{tp2}/P_{tp1}, relative friction losses $(P_{tp2}/P_{tp1})_t$, relative decrease of gas temperature (T_{tp1}/T_{tp2}) on the engine load N_e: _______ – ISO conditions (t_{air1} = 25 °C; t_{w1} = 25 °C); — — – tropical conditions (t_{air1} = 45 °C; t_{w1} = 36 °C); _ _ _ _ _ – cold zone (t_{air1} = 10 °C; t_{w1} = 10 °C)

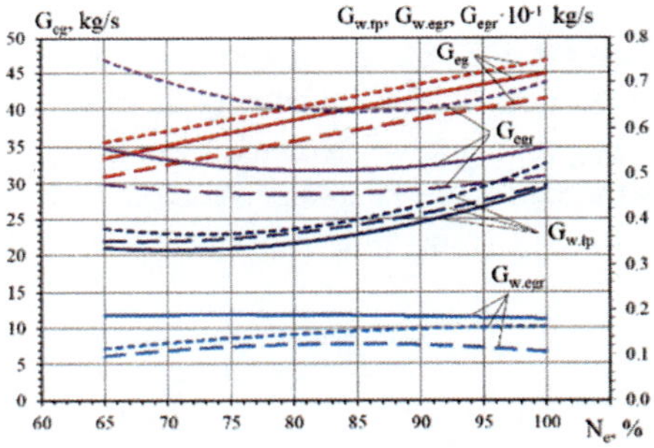

Fig. 5. Dependences of exhaust gas flow rate G_{eg}, recirculation gas consumption G_{egr}, water consumption for injection in TP $G_{w.tp}$, water consumption for injection in scrubber $G_{w.egr}$ on the engine load N_e: _______ – ISO conditions (t_{air1} = 25 °C; t_{w1} = 25 °C); — — – tropical conditions (t_{air1} = 45 °C; t_{w1} = 36 °C); _ _ _ _ _ – cold zone (t_{air1} = 10 °C; t_{w1} = 10 °C)

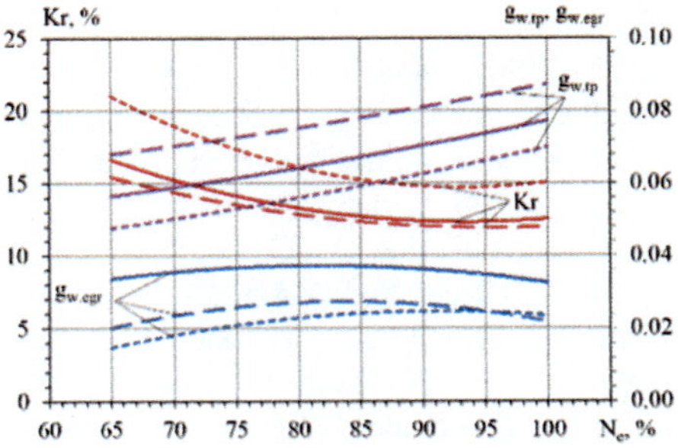

Fig. 6. Dependences of the recirculation coefficient Kr, relative water flow for injection in TP $g_{w.tp}$ and in the scrubber $g_{w.egr}$ on the engine load N_e: _______ – ISO conditions (t_{air1} = 25 °C; t_{w1} = 25 °C); — — – tropical conditions (t_{air1} = 45 °C; t_{w1} = 36 °C); _ _ _ _ _ – cold zone (t_{air1} = 10 °C; t_{w1} = 10 °C)

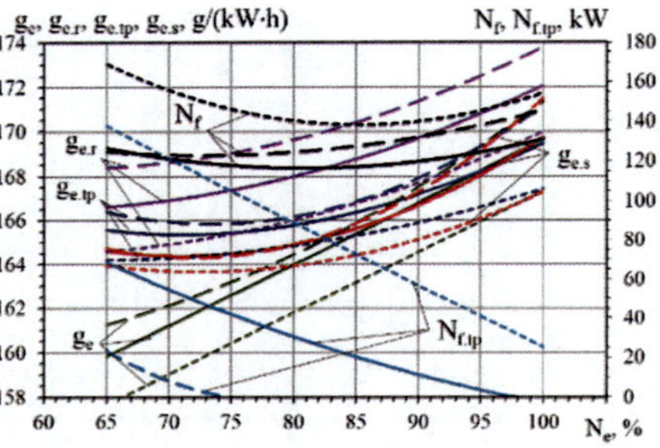

Fig. 7. Dependences of specific fuel consumption: without recirculation g_e; with recirculation $g_{e.egr}$; with recirculation and TP $g_{e.tp}$; specific fuel consumption with recirculation taking into account the reduction of the load on the fan $g_{e.s}$, the fan power of the recirculation system N_f, the fan power of the recirculation system with TP $N_{f.tp}$ on the engine load N_e

5 Conclusions

The principle of increasing the efficiency of ecological recirculation of combustion products of internal combustion engines by cooling them with thermopressors, which reduces the load on the gas recirculation system: reduce or fully compensate for hydraulic losses, respectively, and the load on the fan (or electric compressor) recirculation system, reduce heat from recirculating gases with seawater, which reduces the specific fuel consumption by 1.5–1.7%.

References

1. Trushliakov, E., Radchenko, A., Forduy, S., Zubarev, A., Hrych, A.: Increasing the operation efficiency of air conditioning system for integrated power plant on the base of its monitoring. In: Nechyporuk, M., Pavlikov, V., Kritskiy, D. (eds.) Integrated Computer Technologies in Mechanical Engineering. AISC, vol. 1113, pp. 351–360. Springer, Cham (2020). https://doi.org/10.1007/978-3-030-37618-5_30
2. Radchenko, A., Scurtu, I.-C., Radchenko, M., Forduy, S., Zubarev, A.: Monitoring the efficiency of cooling air at the inlet of gas engine in integrated energy system. Therm. Sci. (2020). https://doi.org/10.2298/TSCI200711344R
3. Radchenko, M., Radchenko, R., Tkachenko, V., Kantor, S., Smolyanoy, E.: Increasing the operation efficiency of railway air conditioning system on the base of its simulation along the route line. In: Nechyporuk, M., Pavlikov, V., Kritskiy, D. (eds.) Integrated Computer Technologies in Mechanical Engineering. AISC, vol. 1113, pp. 461–467. Springer, Cham (2020). https://doi.org/10.1007/978-3-030-37618-5_39
4. Trushliakov, E., Radchenko, A., Radchenko, M., Kantor, S., Zielikov, O.: The efficiency of refrigeration capacity regulation in the ambient air conditioning systems. In: Ivanov, V., Pavlenko, I., Liaposhchenko, O., Machado, J., Edl, M. (eds.) DSMIE 2020. LNME, pp. 343–353. Springer, Cham (2020). https://doi.org/10.1007/978-3-030-50491-5_33
5. Konovalov, D., Kobalava, H., Radchenko, M., Scurtu, I.C., Radchenko, R.: Determination of hydraulic resistance of the aerothermopressor for gas turbine cyclic air cooling. In: TE-RE-RD 2020, 14 p. E3S Web of Conferences, vol. 180, p. 01012 (2020). https://doi.org/10.1051/e3sconf/202018001012

6. Radchenko, M., Radchenko, A., Radchenko, R., Kantor, S., Konovalov, D., Kornienko, V.: Rational loads of turbine inlet air absorption-ejector cooling systems. Proc. IMechE Part A J. Power Energy, pp. 1–13 (2021). https://doi.org/10.1177/09576509211045455
7. Samoilov, N.P., Igonin, V.I., Kashevarov, O.A., Samoilov, D.N.: Toxicity of autotractor engines and ways to reduce it. Kazan, KGU (1997)
8. Kirpichenkov, S.V.: Controlled exhaust gas recirculation in the system of complex reduction of toxic emissions of medium-speed diesel engine of a river vessel. Moscow (2002)
9. Chub, T.V.: Recirculation of exhaust gases of a marine diesel generator as a means of reducing emissions of nitrogen oxides. Moscow (2000)
10. Agrawal, K., Singh, S.K., Sinha, S., Shukla, M.K.: Effect of EGR on the exhaust gas temperature and exhaust opacity in compression ignition engines. Acad. Proc. Eng. Sci. **29**(3), 275–284 (2004)
11. Bent, N.: 8500 TEU container ship green ship of the future concept study. Odense Steel Shipyard Ltd. (2009)
12. MAN B&W. Two-stroke Marine Engines. Emission Project Guide, 9th edn. MAN Diesel & Turbo, Copenhagen (2018)
13. Kornienko, V., Radchenko, R., Mikielewicz, D., Pyrysunko, M., Andreev, A.: Improvement of characteristics of water-fuel rotary cup atomizer in a boiler. In: Tonkonogyi, V., et al. (eds.) InterPartner 2020. LNME, pp. 664–674. Springer, Cham (2021). https://doi.org/10.1007/978-3-030-68014-5_64
14. Radchenko, R., Pyrysunko, M., Radchenko, A., Andreev, A., Kornienko, V.: Ship engine intake air cooling by ejector chiller using recirculation gas heat. In: Tonkonogyi, V., et al. (eds.) InterPartner 2020. LNME, pp. 734–743. Springer, Cham (2021). https://doi.org/10.1007/978-3-030-68014-5_71
15. Baert, R.S., Beckman, D.E., Veen, A.: Efficient EGR Technology for Future HD Diesel Engine Emission Targets. SAE 1999-01-0837 (1999)
16. Wärtsilä Environmental Product Guide. Wärtsilä, Marine Solutions. Vaasa (2017)
17. Lance, M.J., Mills, Z.G., Seylar, J.C., Storey, J., Sluder, C.S.: The effect of engine operating conditions on exhaust gas recirculation cooler fouling. Int. J. Heat Mass Transf. **126**, 509–520 (2018)
18. Hoseini, S.S., Najafi, G., Ghobadian, B.: Thermal and fluid simulation of a new diesel engine cooling exhaust gas recirculation system to reduce exhaust gas emissions. J. Adv. Res. Fluid Mech. Thermal Sci. **51**(2), 197–208 (2018)
19. Radchenko, A., Trushliakov, E., Kosowski, K., Mikielewicz, D., Radchenko, M.: Innovative turbine intake air cooling systems and their rational designing. Energies **13**(23), 6201 (2020). https://doi.org/10.3390/en13236201
20. Shapiro, H., Wadleigh, K.R., Gavril, B.: The aerothermopressor – a device for improving the performance of a gas-turbine power plant. Trans. ASME **78**(7), 617–653 (1956)
21. Erickson, J.: A Theoretical and Experimental Investigation of the Aerothermopressor Process. Massachusetts Institute of Technology, Massachusetts (1958)
22. Wang, S.K. (Shan Kuo): Handbook of Air Conditioning and Refrigeration, 2nd edn. McGraw-Hill (2000)
23. Diesel-RK. http://www.diesel-rk.bmstu.ru/Eng/index.php. Accessed 10 Nov 2019
24. CEAS Engine Calculations. MAN Diesel Turbo. https://marine.man-es.com/two-stroke/ceas. Accessed 10 Nov 2019
25. MAN, Humid Air Motor Technology for Green Profits, p. 7

Modelling the Process of Interaction of a Pulsed Jet with a Workpiece by Electrohydraulic Forming

Mykhaylo Taranenko(✉) and Oleksandr Naryzhniy(✉)

National Aerospace University "Kharkiv Aviation Institute", 17, Chkalov st., Kharkiv 61070, Ukraine
{m.taranenko,o.naryzhniy}@khai.edu

Abstract. Necessity of thin-walled bottoms local forming are proved and conclusion about demanding of local loading of liquid jet modelling is made in the paper. Analytical model and scheme for creation of impulse jets by group and electrohydraulic discharge are described briefly. The character of an workpiece deformation is studied and it is shown that it has quite complicated waved oscillating character. Deformation rates are changed in the range of several orders during mentioned process. Possibility of stress-strain state controlling is grounded. Application of developed mathematical model is suggested for optimization processes of forming with different objective functions. The novelty of the results of the article lies in the simulation of multi-pulse loading of a workpiece.

Keywords: Electrohydraulic discharge · Discharge cavity · Blank · Impact waves · Hydrostream · Porous-gas cavity · Velocity · Plastic deformation intensity · Deformation work

1 The Problem

Thin-walled large-dimensional articles (TLA) are widely used in in up-to-date machine-building. These group of articles include bottoms of different shape – elliptical and semi-spherical in cross-section and round in the plan, car body articles for different vehicles, space communication antenna mirrors; rigidity panels for airplanes and helicopters, heat-exchangers and articles of chemical apparatus. Different requirements to dimensional precision, deformation distribution or residual stress apply to mentioned articles.

Main manufacturing processes for such articles for producing them on widely used sheet-forming equipment in majority are well-proven. But in case of necessity of manufacturing TLA in low-volume production some difficulties appear related to technical-economic parameters of existent manufacturing processes.

There are such processes for TLA manufacturing by impulse methods: explosive forming with brizant explosive and electrohydraulic forming (EHF) [1, 2]. Presses with large saved energy (up to 500 kJ) and spatial-temporary loading control are developed and used for trial production for the second above-mentioned method of forming [2]. Such presses can be implemented for consequent local forming of TLA.

M. Nechyporuk et al. (Eds.): ICTM 2021, LNNS 367, pp. 484–496, 2022.
https://doi.org/10.1007/978-3-030-94259-5_41

In view of this the problem for determination rational consequence of loading of deformable blank with different goals appears. These goals can be: maximum energy efficiency, required distribution of deformations to minimize blank sagging or residual stress and also to satisfy other technical requirements. Results of a part of prepared natural experiments and results of trial developing of manufacturing process are shown in papers [2–5]. These results are shown high potential possibility of loading control for increasing forming processes efficiency but also high labor intensity and cost of natural studies. Reduction of these parameters can realized by means of numerical modeling of forming processes.

2 Main Part

Results of Preliminary Experiments. Preliminary (prearranged) experiments were conducted on trial multi-electrode installation developed by KhAI. On this installation following experimental data were obtained: about shape of locally created blank zone at different consequence of loading and distribution deformation of thinning along radius of loading zone (see below). Manufacturing processes of TLA of different shapes forming were conducted on natural equipment. Trial batches of articles with different shape were formed on trial-industrial press ПЭГ-500 [2].

Some forming results of thin-walled bottoms are shown on the Fig. 1.

Generalized preliminary conclusions are the following:

- at central loading of flat blank one can observe intensive and progressive wrinkles appearing on flange and dome part of a blank (Fig. 1, *a*). Wrinkles (corrugations)move over female die rib and pass to dome part at increasing degree of flange drawing. During this process blank tearing is observed (Fig. 1, *b*);
- this processes become more intensive with increasing blank flexibility (ratio of blank diameter to its thickness);
- flange non-uniform drawing degree along perimeter of matrix space is observed that can be explained by non-uniform conditions of drawing. To increase uniformity of drawing it is necessary to shift loading zone from the center to zones with less drawing degree;
- higher drawing degree can be realized at definite consequence of local loading (Fig. 1, *b*) but in this case one has to control less thinning on loaded blank zones because elevation of thinning above allowable values can lead to local tearing (Fig. 1, *c*).

Results of determination of thinning distribution on formed article (Fig. 1, *d*) permit to make conclusion that at definite consequence of local impulse loading it is possible to control both the shape of article in wide range and distribution of thinning deformations along article surface. Examples of positive results of consequent local forming of TLA can be found in [2]. For rational application of such technology one has to use computer modelling with application of interrelated mathematical models of processes of load creation and blank deforming.

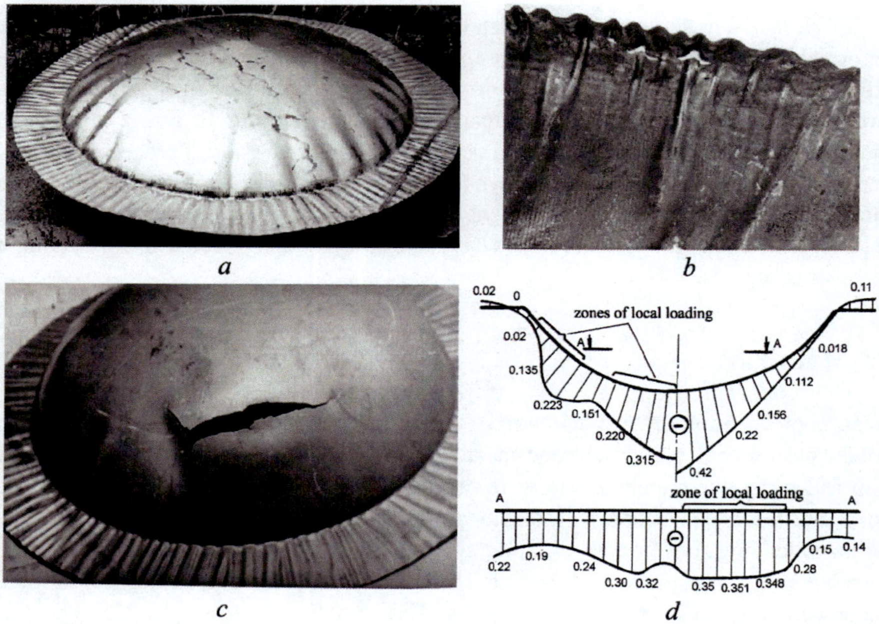

Fig. 1. Typical examples of defects at forming of thin-walled sheet large-dimensional bottoms [2]: *a* – intermediate shape of a blank after central loading, diameter of drawing rib of female-die is 1050 mm, diameter of semi-finished article is 1480 mm, material – steel X18H9T with thickness 1.6 mm; *b* – progressive wrinkle-creation after consequent forming stages: wrinkles can be seen which transfer ty dome part and torn zones at the tops of wrinkles; *c* – blank discontinuities at the zone of bending local loading; *d* – distribution of logarithmic deformations of thinning at local consequent loading over entire surface (at the left) and central loading (at the right) of a bottom with diameter 1070 mm, depth is 415 mm, material – steel 08кп, thickness is 1.6 mm. Diametral section is above, latitudinal section is below. Bold line – article outline, dashed line – distribution of ε_δ in latitude direction at central loading

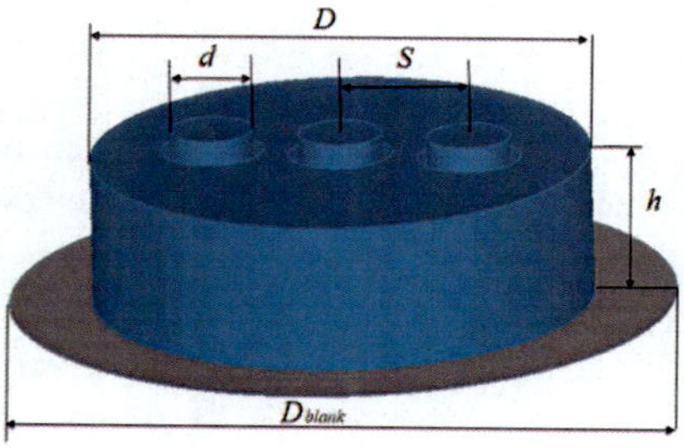

Fig. 2. Triple-planed analysis scheme of numerical modeling: *D* – chamber diameter; *d* – discharge cavity diameter; *S* – spacing between axes of immersed chambers; *h* – discharge spacing; D_{blank} – blank diameter

Conditions of Modelling. The system shown on Fig. 2 is considered. It consists of three discharge cavities immersed to liquid. Deforming blank is disposed below, it is installed on drawing die with diameter 252 mm. Diameter of a blank is 320 mm and it is clamped to the die. Such conditions exclude material drawing from the flange zone to die cavity.

Zone of energy releasing in discharge cavity was located asymmetrically with respect to axis of cavity and at definite spacing from its bottom. This corresponds to real structure. Geometrical dimensions of model are following:

- diameter of discharge cavity is 40 мм, depth of zone of energy releasing from the edge of cavity – 30 mm;
- spacing of discharge cavities location – 60 mm;
- distance from the edge of discharge cavity to blank was selected in four variants: 40, 32, 20 and 12 mm. This corresponds to relative distances of discharge *d/h* equals to 1.0; 0.75; 0.5 and 0.25. The last value is the most close to experimental conditions.

Consequence of numerical experiment was selected as following.

1. Mechanism of blank deforming was considered at single discharge at central discharge cavity on selected distances;
2. Mechanism of impulse loading appearing was considered at discharges in two cavities (central and side) shifted in time;
3. Interaction of three discharges was considered: in two side cavities simultaneously and in central cavity – shifted in time.

Analysis of energy releasing process at high-voltage underwater discharge was considered at condition that definite quantity of heat energy (9.375 kJ) releases during 32 μs in the volume of channel with continuous conductivity. In this case energy losses on impact waves creation, electrode material evaporation and heat losses were not consider at all. Thus power function of heat releasing is assumed to be a triangle with height 400 MW and basement 32 μs. These values come in an agreement with analogous parameters given in the paper [5].

Blank had diameter 320 mm and thickness 1.0 mm. Its material is considered to be elastic-plastic aluminum alloy with deformation strengthening describes as following power law.

$$\sigma_T = Ae^n,$$

where $A = 604.9$ MPa; $n = 0.275$.

Original value of yielding strength is 100 MPa, density – 2100 kg/m^3, elasticity modulus – 7×10^{10} Pa, Poisson's ratio 0.33. These mechanical properties correspond approximately to widely used aluminum alloys like AMr6 or Д16AT.

Modeling was conducted in medium LS-DYNA by means of ALE [3, 7] method application.

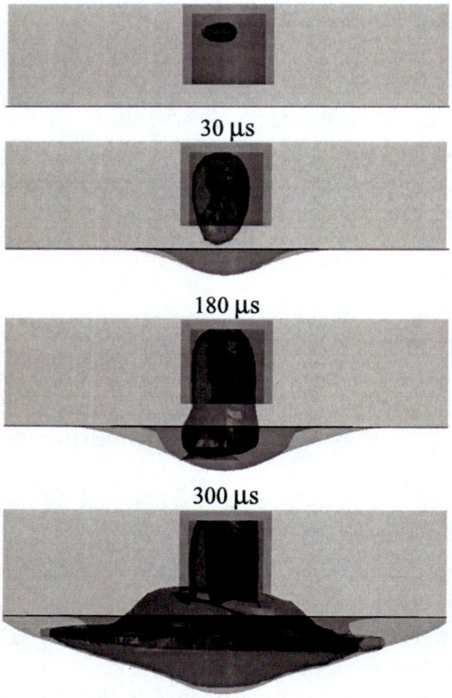

Fig. 3. Consequence of porous-gas bubble development at single discharge and bubble interaction with deformable blank at spacing 12 mm (0.3 *d*). Arrow shows changing of convective shape of a blank

For numerical experiments following parameters of analysis scheme was selected:

- high-voltage discharges are conducted in central discharge cavity (DC) and simultaneously in central and two side DC with the same level of released energy;
- discharges spacing with respect to a blank *h* were selected as a consequence: 40 (1.0*d*), 32 (0.75*d*). 20 (0.5*d*) and 12 (0.3*d*) mm that corresponds to conditions of natural experiments.

Results of Researches. At realization of the first stage, i. e. modelling of process of interaction single impulse jet with blank following situation was met – developing of steam gas bubble (SGB) inside DC, SGB leaving out and distribution in space of transferring medium between side of DC and blank, interaction of jet (directed by head part of SGB of liquid stream) with blank (Fig. 3). During first 30 μs after beginning of energy releasing (32 μs) impact waves leave discharge zone (not shown on Fig. 3), then waves distribute at first in DC and then pass through the side of DC. Velocity of waves movement from side to direction of blank lays in the range 1600…1900 m/s. They lead to beginning of blank movement and create directed to blank the field of liquid velocities movement. After that boundaries of SGB begin to move at first in frames of DC rigid boundaries and then in space between DC side and blank. SGB become stretched egg-like shape and its head part pushes ahead definite volume of liquid which supports blank has begun to move. Next stage is the following: axial translation of SGB boundaries starts to change its direction (frame 300 μs), i. e. horizontal components of SGB boundaries movement appear. This leads to non-productive kinetic energy consumption of moving liquid.

On the frame 1110 μs one can see definite zone of selected analysis scheme – low width of rigid wall of DC, which is began to be enveloped by SGB at its expansion upper (out of blank. In future this drawback was eliminated).

At the beginning of deformation blank takes convex axis-symmetrical shape. Then this monotonic convex shape begins to distort – zones with non-monotonic shape as movable wave appear in transversal section. This distortion of shape is shown with arrow on the frame 1110 μs.

Modelling of deforming process of flat blank shows that its stress-strain state changes non monotonic but has wave-like (oscillating) character (Fig. 5) through blank thickness in a plane projection. Ring-like waves of thinning deformation changing can be seen.

More thoroughly mechanism of waves distribution is analyzed in the paper [6]. In epicenter of impulse distortion source (Fig. 4) curved line impact wave 2 moves to blank 4 and gives to it definite movement velocity downwards. Small by amplitude compressive wave 5 moves in opposite direction and reflects from blank. The zone of cavitation 6 appears at interaction of compressive wave 5 with hydrostream which is moved by action of SGB 3. Upper boundary of region 7 begins to move to opposite direction and steadily envelopes cavitation zone under liquid jet influence which takes acceleration from expanding SGB.

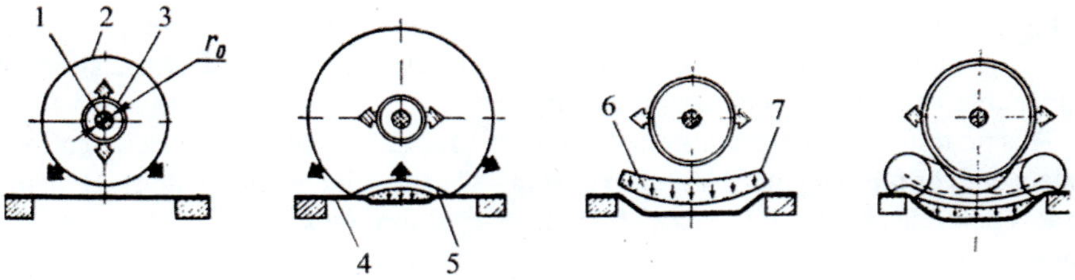

Fig. 4. Schematic imaging of interaction of impulse loading with thin-walled blank at charge explosion in water [6]: 1 – charge; 2 – impact wave front; 3 – SGB boundaries; 4 – blank; 5 – reflected wave of compression; 6 – cavitation zone; 7 – upper boundary of cavitation zone

Deforming impact-waved zone of blank has the shape of truncated cone basement boundaries of which move to boundaries of blank clamping.

Analysis of plate movement at such conditions shows that two types of waves distribute in blank: longitudinal and transversal. They move with different velocity which depend on stress-strain state parameters. Changing of this state under influence of impulse jet loading has complicated waved character (Fig. 5). Deformations of thinning (third principal deformation) appear inside epicenter of loading and my means of longitudinal waves transfer to peripheral zones of deformation. This considerations are adjusted well between each other.

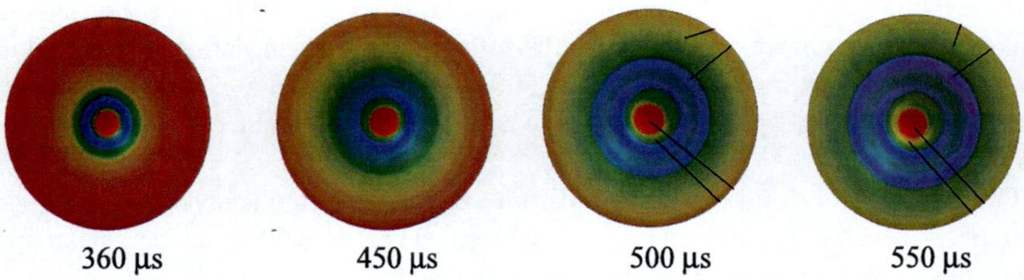

Fig. 5. Character of changing of the third principal strain at impulse loading at the distance 12 mm (0.3 *d/h*)

To estimate adequacy of used mathematical model to real process it was decoded to conduct comparison of results got with experimental results given in the paper [2]. Figure 6 shows frames from high-rate photo registering the process of leaving of liquid impulse jet from horizontally disposed DC and interaction of jet with perpendicularly installed rigid plate (Fig. 6, *b*) and the same plate but which applies angle with respect to cavity axis (Fig. 6, *c*). It can be seen consequently on frames that at first the cloud of small energy particles are removed from DC cut and then move in axial direction. This cloud is braked up by impact wave (foggy zone). Then certain morphosis (black color) having cylindrical shape appears and leaves at the cut. This is immersed liquid jet moves away from the chamber side, begins to expand in sides and then close up with rigid plate.

At interaction of sloped plate with jet the last expands in radial direction non-uniformly: at less gap between chamber edge and plate radial expansion is less than at large gap. One can see a sort of jet changes direction to the side of more gap.

Frame-by-frame increment of both time and velosity of jet side translation are comparable in numerical and natural experiments.

This permits to make conclusion about quite adequate numerical model to reality in respect with of mechanism of jet leaving from DC and jet interaction with rigid blank.

It is possible to compare shapes of obtained local formings and distribution of logarithmic deformations of thinning by means of results of natural experiments and numerical modelling (Fig. 7). One can see that at comparison of dependencies predefined shape of local forming climbs above value of natural local forming depth in center on 30%. It can be explained by certain non-compatibility of mechanical properties of natural blank (aluminum alloy AMцAM) and blank thickness: natural blank has thickness 1.5 mm but modelled – 1.0 mm.

Calculated value of thinning in center of local forming equals to 0.13, that twicw less that natural blank has. This mismatching also can be explained by the same reasons.

Based on above-mentioned considerations one can make conclusion that used in numerical analysis mathematical model is adequate to reality from the point of view of SGB developing mechanism and SGB interaction with deformable blank but not very precise at calculation of thinning deformations and shape of local forming.

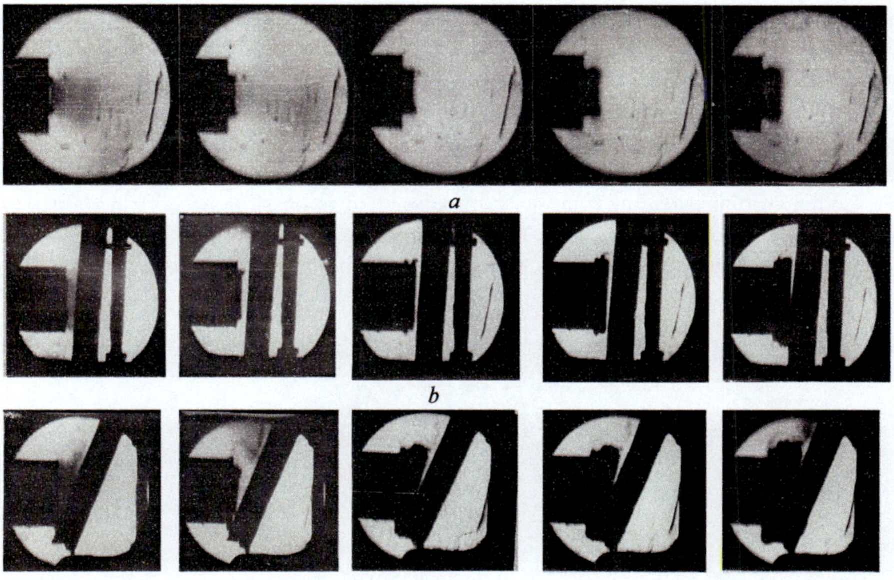

Fig. 6. High-rate photo registering-gramma of impulse jet appearing process at electrohydraulic discharge in low volume chamber with cylindrical inner cavity: *a* – original stage of SGB leaving from DC – leaving of small particles of liquid under impact wave influence (frame 1) and cylindrical SGB leaving (frame 3, 4 and 5), time between frames is 64 μs; *b* – interaction of cylindrical shape SGB with rigid barrier, installed perpendicular to DC, time between frames is 96 μs; *c* – interaction of SGB with sloped rigid barrier, time between frames is 96 μs [6]

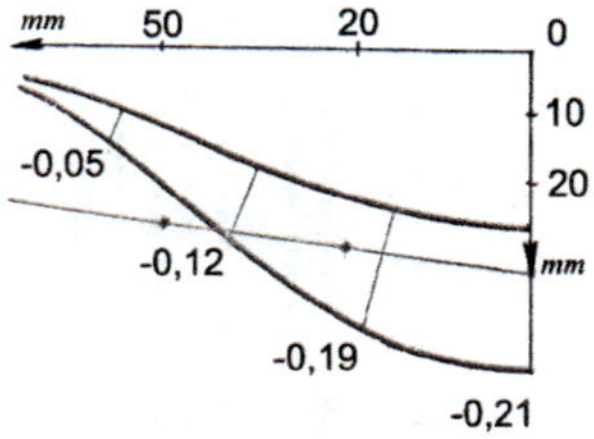

Fig. 7. Shape of local forming (half of diametral section) – bold line with values; distribution of logarithmic deformations of thinning – bold line; estimated shape of local forming – thin line with dots [2]

Compressing of impulse jet from side, i.e. increased resistance of jet to spreading in horizontal direction has to stipulate more effective usage of released energy at electrohydraulic discharge. This is possible to realize by generation of pressure increasing in liquid by sides of main jet.

For numerical modelling of such process triple-cavity analysis scheme was used (Fig. 2). Saved energy releases simultaneously or with definite shifting in time in three DC. Illustration of final stages of such process is shown on Fig. 8. One can see as side

SGB compress central one and as consequence concentrate its energy in direction to blank. At different time delay values degree of compression of central (main) SGB is different. Optimal delay time of discharges in side SGB equals to 300 μs in considered variants of structural-manufacturing parameters values of given analysis scheme. But value of this optimal time depends on distance of blank disposing from discharges zone, released energy in side cavities and spacing between cavities.

Center of blank loading zone with impulse jet it is possible to shift in frames of definite range at double discharge. This mechanism of jet controlling rationally to use at calibration stage of zone of conjunction of sides with bottom of box-like article.

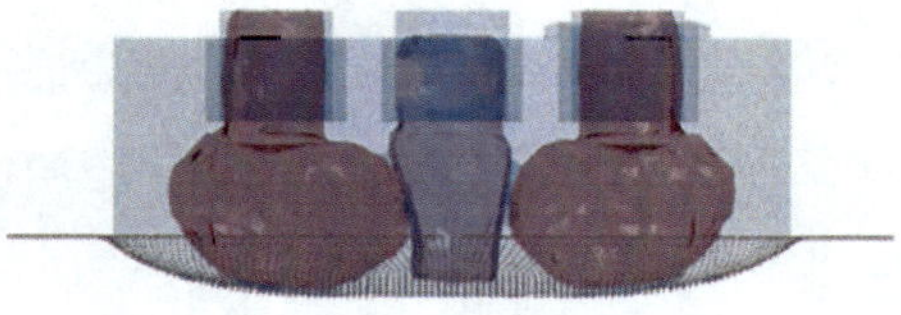

0,0027 sec

Fig. 8. Longitudinal section of triple SGB interacting with deformable blank at equal value of energy released in each DC and delay of electrohydraulic discharges in side LC on 300 μs and spacing 40 mm

Kinetics of blank behavior at delaying of side discharges with respect to central one can be demonstrated by temporal dependences of velocity and translations of reference points (Fig. 9 and 10).

From the process beginning central zone of a blank (points D, E, F) start intensive movement which continues during 190 μs, i. e. moment of energy releasing in central DC. It is possible under impact waves influence only. Then uniform accelerated movement of this zone stops but movement of peripheral zones of blank starts. After that dependence of velocity on time has monotonic growing character. It can be called as oscillatory and velocity of central point reaches values up to 25 m/s. After reaching maximum (duration is about 0.5…1.0 μs) decelerated motion starts which also has oscillatory character. After total running out negative (reversed) movement with oscillations can be observed. Suspectedly these oscillations can be explained by elastic aftereffect.

Dependencies of blank points translations in time have monotonic character (Fig. 10). Velocities oscillations are not shown at this figure because velocity oscillation period is quite low (up to 30 μs) and comparing with scale of temporary dependencies is not seen. One has to pay attention that maximum deflection is shown not in the blank center (point E) but at epicenters of side discharge cavities (points F and G) that is quite understandable. Blank plastic deformation appears in DC epicenter (point A, Fig. 11) in which first by time discharge is conducted. Three focal points of plastic deformation in corresponding epicenter appear in case of energy releasing simultaneously in three DC. Deformations spread in wavy mode as ring zones from blank center to peripheral zones after plastic focal point appearing. Duration of this process of deformation close to 1.0 μs that corresponds to duration of maximum sagging of blank central zone (Fig. 10).

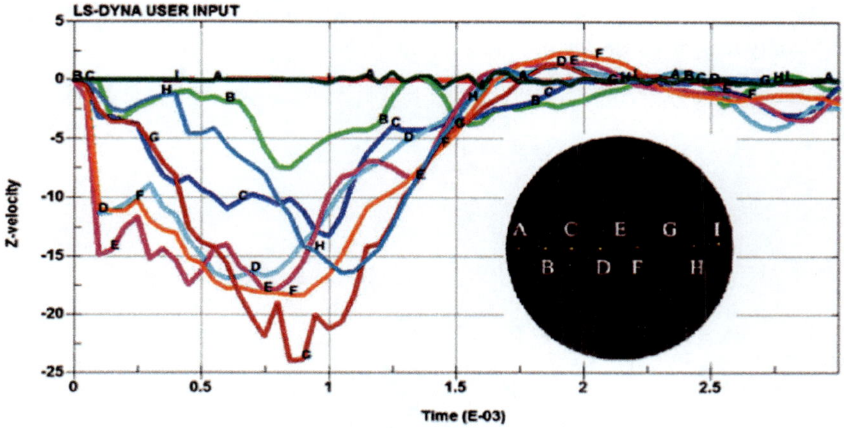

Fig. 9. Dependence in time of blank reference points shifting rates at triple discharge with delay 300 μs

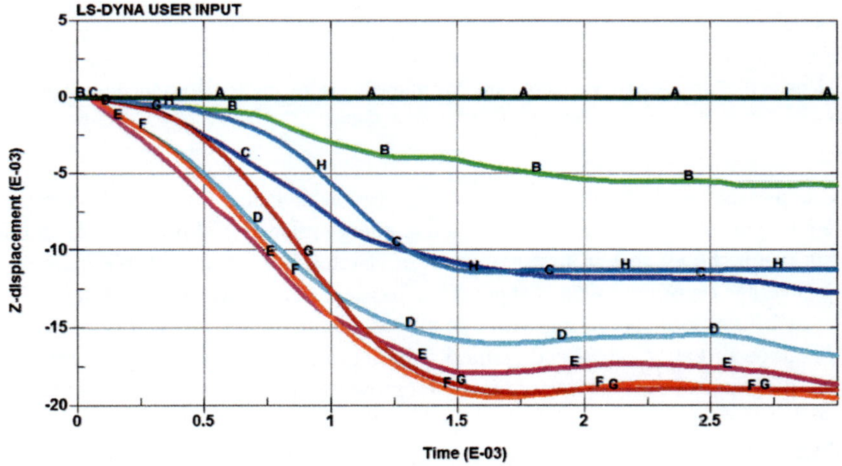

Fig. 10. Dependence in time of blank reference points translations at triple discharge with delay 300 μs

It is easier to estimate quantitative indexes of intensive deformation changing as function of time by graph shown on Fig. 11. It can be seen that developing and accumulation of parameter $\dot{\varepsilon}_i^{plast}$ differs for different points. For point A close to blank center and point E close to die rib accumulation of plastic deformations happens monotonically with average deformation rates, for point A it equals to 14×10^2 s^{-1} and point E – 49 s^{-1}. Accumulation of plastic deformations in points B, C, and D occurs non-monotonically, in stepped way. It is result of waved deformation accumulation. Average deformation rate in these points close to 20…30 s^{-1}. Velocity of plastic deformation wave distribution from point A to point B equals to 240 m/s, but in point B and point

C – 54 m/s. These values of waves velocities of deformation transferring to a blank significantly less that velocity of sound in aluminum alloy in 20 times (sound wave is an elastic distortion).

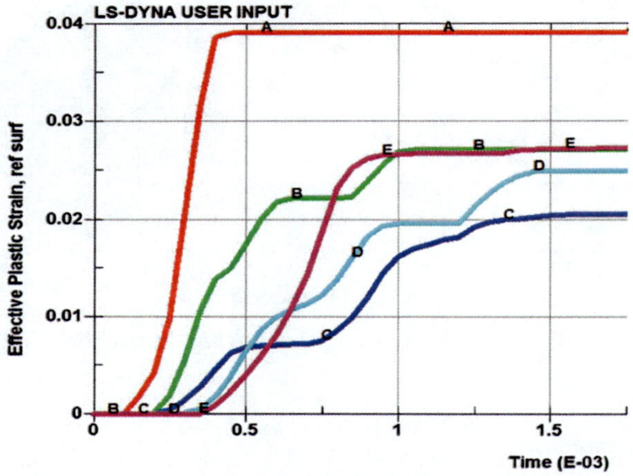

Fig. 11. Temporary dependence of plastic deformation intensity ($\dot{\varepsilon}_i^{plast}$) for correspondent points of a blank at double power developed. Duration of energy releasing is 16 μs. Increment of points is near 27 mm, point A is disposed at the center of a blank

Comparison of deformation velocities at the blank center for loading variant in triple-cavity discharge system with duration of energy releasing 32 μs and variant under consideration shows, that in first variant even at releasing factually triple quantity of energy average deformation rates at point located near female die rib are equal to 14 s^{-1} that is in 4 times less comparing with variant considered. In another words one can say that blank deformation rates become more with increasing rate of energy releasing and this growing prevail over the rate of parameter growing as function on quantity of energy released.

As confirmation of possibility to control sheet blank deformation rate in quite wide range one can consider selection of optimal combination of electrical parameters of discharge circuit which is very important for creation of necessary regime of vibro-impulse loading to reduce deformation warping of such kind of articles [4, 5].

One has to mention that intensive oscillations of blank zones disposed out of direct influence of immersed jet. Maximum of displacements shift periodically over poor deformable field of such zones (Fig. 12) that corresponds to obertons of oscillations. This evidences about complicated oscillational process of deformable blank.

One can suggest that fixation of obertons is stipulated by influence of transversal waves reflected from rigidly clamped perimeter of blank, velocity of these waves distribution is near several dozens meters per second, the intensity of the waves reduces sharply close to blank center.

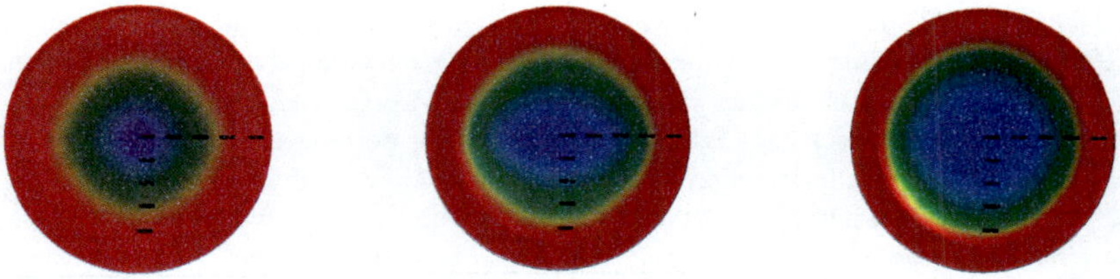

Fig. 12. Demonstration of oscillations in vertical directions of blank zones translations, located out of direct loading zone

More deep studying of such oscillation process is very important since one can determine conditions of resonance at which elevation of plastic properties of blank material happens and also conditions for higher degree of residual stress (deformational) relaxation occurs.

3 Conclusion

1. Conducted numerical modelling for process of interaction of impulse jet with deformable blank at electrohydraulic forming have shown qualitative adequacy of used model both for the stage of jet deformation, displacement in operational volume and at the stage of blank deformation. Numerical analysis has revealed possibility of controlling ang concentration of energy released in space and time.
2. At the same time comparison of numerical results with natural experiment has shown necessity of correction of used model by the parameters of blank material mechanical properties and parameters of analysis scheme.
3. Numerical experiments allowed to make more precise and effective the structure of energy-releasing cell, i.e. the group of seven operational cavities united in one block in which discharges are realized with definite delay in time. Exact energy-releasing cell has to be moved by definite trajectory over blank surface.
4. Blank deformation by impulse jet has waved oscillation character. In this case the changing rate of plastic deformation intensity varies in wide range from 1400 s^{-1} to 14 s^{-1} for different zones of a blank. Velocities of plastic deformation wave distribution vary from 240 m/s to zero value. In first turn these parameters depend on duration of energy releasing and then on its quantity.
5. Further researches with application of developed mathematical model allow to optimize processes of consequent local forming of large-dimensional sheet articles.

References

1. Pihtovnikov, R.V., Zav'jalova, V.I.: Shtampovka listovogo metalla vzryvom [Stamping of sheet-metal by an explosion], p. 175. Mashinostroenie. Publ, Moscow (1964)

2. Taranenko, M.E. Jelektrogidravlicheskaja, S.: Teorija, oborudovanie teh-processy [The Electro-hydraulic stamping : is a theory, equipment of technological processes] : monografija v 2 ch. Kharkiv : Nac. ajerokosm. un-t im. N. E. Zhukovskogo HAI, p. 272 (2011)
3. Naryzhnyj, A.G., Taranenko M.E.: Issledovanie processa posledovatel'nogo deformirovanija listovoj zagotovki pri jelektrogidravlicheskom nagruzhenii [Research of process of successive deformation of sheet purveyance at an electro-hydraulic ladening]. Visnik NTU «HPI». Serija «Innovacionnye tehnologii i oborudovanie dlja obrabotki materialov v mashinostroenii i metallurgii». Kharkiv : NTU HPI, no. 26 (1351), pp. 21–27 (2019)
4. Naryzhnyj, A.G., Taranenko M.E.: Mehanicheskie processy v sisteme s pogruzhnoj jelektrorazrjadnoj kameroj [Mechanicprocesss in the system with a immerse electro-bit chamber]. Obrabotka metallov davlenim : sb. nauch. tr. DGMA. Kramatorsk : DGMA, no. 2 (49), pp. 136–142 (2019)
5. Gulyj, G.A.: Nauchnye osnovy razrjadno-impul'snyh tehnologij. [Scientific bases of bit-impulsive technologies]. Otv. red. B. Ja. Mazurovskij. Kiev : Naukova dumka . Publ., p. 208 (1990)
6. Shtampovka, V.. Osnovy, T.: [Stamping by an explosion. Bases of theory]. Pod red. M. A. Anuchina. Moscow : Mashinostroenie. Publ. 152 (1972)
7. Golovashhenko, S.F., Mamutov A.V., Mamutov V.S.: Metodika komp'juternogo modelirovanija jelektrogidroim-pul'snogo formoizmenenija tonkolistovyh metallov s primeneniem programmnogo kompleksa LS-DYNA 971 [Methodology of computer design electrogidroimpulsive firmng of thin-sheet metals with the use of programmatic complex LS - DYNA 971] Metalloobrabotka : Nauchno-proizvodstvennyj zhurnal. **3**(69), 23–29 (2012)

Reducing the Harmful Emissions While Water-Fuel Emulsions Combustion

Victoria Kornienko[1](✉), Roman Radchenko[2], Serhiy Forduy[3], Andrii Andreev[1], and Victor Khaldobin[2]

[1] Kherson Branch of Admiral Makarov National University of Shipbuilding, 44 Ushakova Avenue, Kherson 73000, Ukraine
[2] Admiral Makarov National University of Shipbuilding, 9 Heroes of Ukraine Avenue, Mykolayiv 54025, Ukraine
[3] PepsiCo, Inc., Kyiv, Ukraine

Abstract. As consequence of deteriorated fuel quality, soot deposits on the EGB tubes, had increased and, in some cases, had resulted in soot fires and in extreme cases into a high temperature iron fire in which the boiler itself burned. The best burning out of fuel combustible components due to applying a water-fuel emulsion (WFE) provides decreasing a concentration of solids and soot in the exhaust gases and hence their toxicity. The kinetics of low-temperature pollution on exhaust gas boiler (EGB) condensing surfaces with WFE combustion are investigated to obtain approximation equations for predicting processes development. When WFE combustion with a water content of 30%, the low-temperature corrosion and pollution intensity decreases, which makes it possible to install condensing heating surfaces in the EGB. The installation of a condensing heating surface in the EGB reduces the content of NOx in gases by 55%, SO_2 by 50%, and the content of solid particles by 3 times. The use of a complex system ensures that gases are purified from toxic ingredients and heat emissions to the level recommended by IMO.

Keywords: Condensing heating surface · Low-temperature corrosion · Exhaust gas boiler · Toxic ingredients

1 Introduction

Receiving additional energy due to deep utilization of heat losses of an internal combustion engine (ICE) saves fuel consumed for the operation of a ship's power plant. This, accordingly, leads to a decrease in emissions of harmful substances into the atmosphere, contributes to the satisfaction of the more stringent standards of the International Maritime Organization (IMO), which regulate the limits of these emissions.

According to the MAN specialists [1], the IMO requirements (III level from SO_2, NOx emissions) can be fulfilled using the following technologies: Water-fuel emulsion combustion (WFE) - WIF (Water in Fuel Emulsion); Scavenge Air Monistening (SAM); Exhaust Gas Recirculation (EGR); Selective Catalytic Reduction (SCR).

M. Nechyporuk et al. (Eds.): ICTM 2021, LNNS 367, pp. 497–508, 2022.
https://doi.org/10.1007/978-3-030-94259-5_42

The use of a combined SAM & WIF scheme to reduce NOx emissions is promising: water vapor in the combustion chamber increases the heat output and reduces the O_2 content. According to MAN, an increase in thermal power and a reduction of O_2 in the charge air provides a decrease of the combustion temperature, which leads to a decrease in NOx emissions. In addition, it should be noted that with a decrease of the combustion temperature, the soot concentration increases, as well as the amount of CO [2, 3].

The use of WIF technology leads to an increase of fuel consumption up to 1.2% (if the condensing surfaces of isn't used when gases cooling below the dew point temperature of H_2SO_4 and H_2O vapors), SAM technology – 2.3%, EGR technology – 4.6%, SCR technology – 7.5_15% (taking into account the price of urea). But WIF technology provides a reduction of NOx to 30%, SAM technology – 45%, the existing EGR system – 70%; SCR technology – by 80% (at the required level of IMO requirements (III level) – 80% reduction of NOx). In addition, the SCR system must be accompanied by scrubber technology for the removal of SO_2 [4, 5].

2 Literature Review

In [6, 7] as consequence of deteriorated fuel quality, soot deposits on the EGB tubes, had increased and, in some cases, had resulted in soot fires. In extreme cases, the soot fire had developed into a high temperature iron fire in which the boiler itself burned [8]. According to data of [9, 10], when water-fuel emulsion (WFE) is combustion, the deposits become loose, or fully absent due to decrease of soot and coke generation. During combustion of WFE drops micro-explosions are observed, which intensify burning process, which is also observed in combustion chambers of combustion engines [11]. However, in these literary sources there is no quantitative data of pollution intensity. In [12] the results of experimental studies on effect of standard diesel fuel (DF) combustion and environmental diesel fuel (EDF), with water content of 12…31% by mass on indicators of toxicity and smoke of exhaust gases are presented. According to [13], indisputable advantage of using EDF in diesel engine is its effect on exhaust gases smoking, which in maximum load is decreased in 1.3…2.7 times. According to the data [13, 14] using of WFE does not require constructive alterations of diesel engine and can significantly improve environmental characteristics of engine.

Majority of the studies reported that soot and particulate matter (PM) were reduced with increase in water concentration [15, 16]. Numerical and experimental study showed reduction in the soot emission of 68% and 75% that when 10% and 15% respectively of water content by volume in the WFE were used [17, 18]. From the various experimental results reported, the majority of the studies confirmed that soot and PM are reduced by using the WFE emulsion fuel [19, 20]. WFE is used as alternative fuel and decreases the emissions of NOx and PM in diesel engines [21]. Based on the obtained results [22], it can be stated that an operation of the engine with diesel fuel and WFE reduces emissions of PM in 2.5…3.5 times, that can be explained by the presence of water vapour in combustion chamber as a catalyst that helps burning carbon particles and other components difficult to oxidize. In [23] the influence of parameters of combustion process in existing low capacity boiler plants on the level of formation of nitrogen oxides, carbon monoxide and soot was studied.

For estimating the efficiency of such greening and fuel saving [24, 25] technologies during the operation in actual climatic conditions various methods of modelling [26, 27], statistical treatment of data of influence of ambient air parameters on the operating characteristics of waste heat recovery cooling systems [28, 29], monitoring [30–32] and experiments [33–35] are applied.

The aim of the research is to develop a system for the integrated purification of ICE exhaust gases. Research tasks:

- to carry out experimental research of pollution processes of EGB, that influence a heat transfer intensity, and EGB working reliability;
- to obtain dependences of pollution rate from wall temperature and values of wall temperature range with minimum pollution;
- to obtain emission rates of toxic ingredients before and after using system for the integrated purification of exhaust gases.

3 Methodology

Experimental researches of pollution intensity at wall temperature values below dew point temperature of sulfuric acid vapors were carried out in an experimental setup (Fig. 1, a) with combustion of fuel oil and WFE based on them.

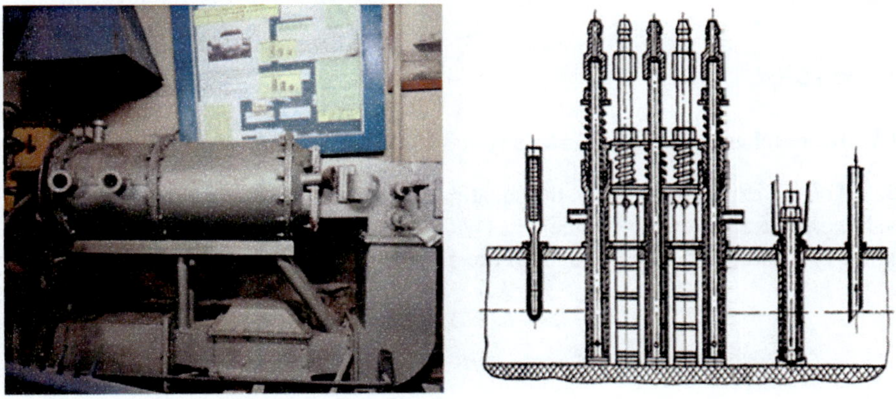

a b

Fig. 1. General view of experimental setup (a) and tube-samples of low-temperature condensing economizer (b).

The experimental installation includes the following elements: combustion chamber, burner, fuel preparation system, gas duct. The form of combustion chamber provides a good filling of torch. The dimensions of the chamber: length – 800 mm, internal diameter – 300 mm. A rotary nozzle is used as a burner [20, 21]. Much attention during development and commissioning of experimental setup was paid to fuel system, which is designed to burn from 1 to 3 kg/h of fuel. Preparation of WFE for combustion in furnace of experimental setup was carried out in a separate installation. Air is supplied to the burner by a fan through the air heater. The temperature of hot air was 150…180 °C. The flue gases were removed from the installation by exhaust fan.

There were working areas (Fig. 1, b) with sample tubes with an outer diameter of 25 mm provided in installation gas duct to study pollution process. Cooling of samples to study LTHS pollution was carried out with air from a receiver or from four thermostats with water and oil.

The assemblies of three tube samples were installed in gas duct of experimental setup, where gas velocity was 7…8 m/s and its temperature about 200 °C.

Experiments were carried out measurements of the geometric characteristics of each tube-sample, which made it possible to determine the outer area of the sample F to the test. After that we weighed each sample (before the experiment) - mass m_1. At the end of experiment, packs of tube samples were pulled out from gas pipeline. Samples with corrosion products, acid and soot deposits were weighed against by analytical weights (mass of m_2).

The pollution speed of metal surface K_p at a certain temperature of the tube wall was determined by formula

$$K_{\mathrm{p}} = (\mathrm{m}_1 - \mathrm{m}_2)/F \cdot \tau \tag{1}$$

where: K_p – pollution speed of metal surface [g/(m^2·h)], m_1 – mass of sample before experiment [g] m_2 – mass of sample after experiment [g], F – average surface of the outer surface of the sample to the experiment [m^3], τ – duration of experiment [h].

4 Results

4.1 Researches of Pollution Intensity

Based on the experimental data, the equation of the pollution rate K_p depending on the wall temperature t_w during the fuel oils ($W^r = 2\%$) combustion (1 mode) was obtained by the approximation method. In this case, the polynomial equation was selected:

$$K_p = 3082.92 - 117.228 \times t_{\mathrm{w}} + 1.6613 \times t_{\mathrm{w}}^2 - 1.0344 \times 10^{-2} \times t_{\mathrm{w}}^3 + 2.3881 \times 10^{-5} \times t_{\mathrm{w}}^4 \tag{2}$$

This equation (regression coefficient R = 0.9839; $\mathrm{R}^2 = 0.9677$ is obtained for the following characteristics of the pollution intensity: $t_w = 85…130$ °C, $W^r = 2\%$. Figure 2 shows the calculated (predicted) values for K_p using the fitted model. In addition to the best predictions, the figure shows: 95% prediction intervals for new observations and 95% confidence intervals for the mean of many observations. The prediction and confidence intervals correspond to the inner and outer bounds on the graph of the fitted model.

Comparison of the calculated values of the pollution rate K_{pC} (Eq. (2)) from those obtained during the experimental study K_{pE} is $\delta_{\mathrm{K}} = \pm 5\%$ (Fig. 3).

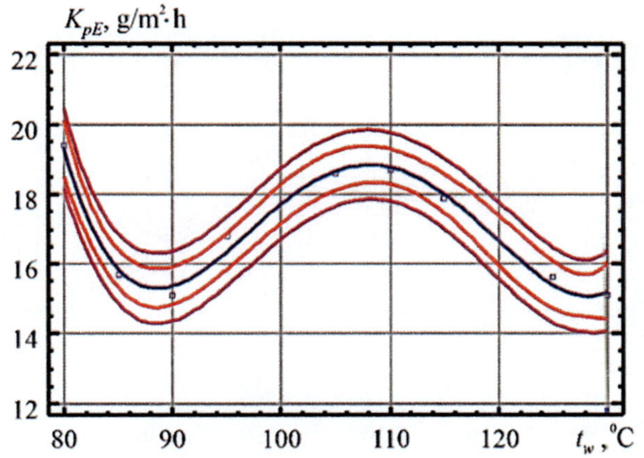

Fig. 2. Experimental dependences of pollution rate K_{pE} from wall temperature t_w with confidence and prediction curves during the fuel oils combustion.

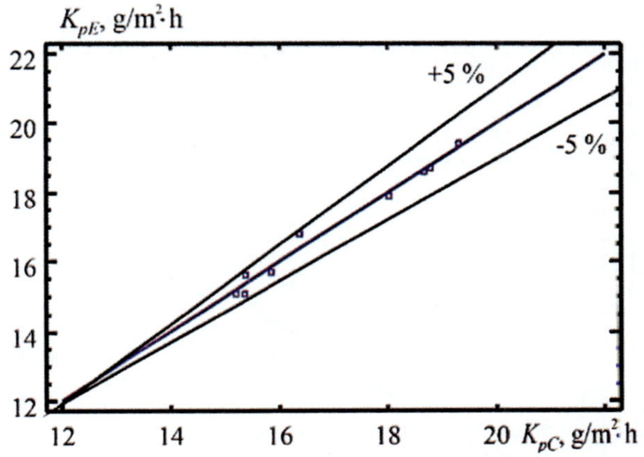

Fig. 3. Comparison of calculated pollution rate K_{pC} values with experimental K_{pE} during the fuel oils combustion.

The polynomial equation of the pollution rate K_p depending on the wall temperature t_w during the WFE ($W^r = 30\%$) combustion (2 mode) based on the experimental data, was selected:

$$K_p = 624.931 - 23.3676 \times t_w + 0.3306 \times t_w^2 - 2.0583 \times 10^{-2} \times t_w^3 + 4.7526 \times 10^{-6} \times t_w^4 \quad (3)$$

This equation (regression coefficient R = 0.9968; $R^2 = 0.9937$ is obtained for the following characteristics of the pollution intensity: $t_w = 80...130$ °C, $W^r = 30\%$. Figure 4 shows the calculated (predicted) values for K_p with prediction and confidence intervals.

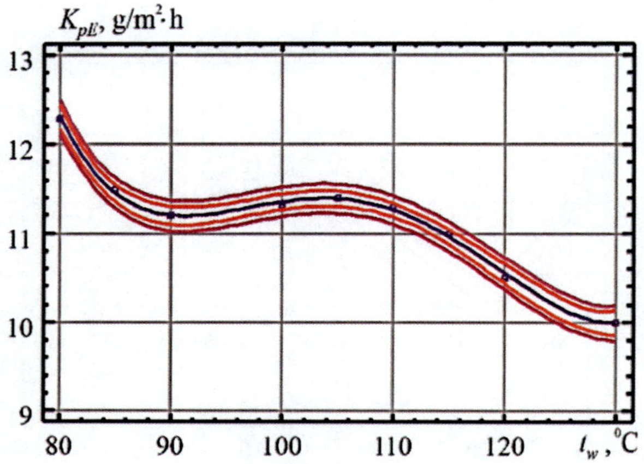

Fig. 4. Experimental dependences of pollution rate K_{pE} from wall temperature t_w with confidence and prediction curves during the WFE combustion.

Comparison of the calculated values of the pollution rate K_{pC} (Eq. (3)) from those obtained during the experimental study K_{pE} is $\delta_K = \pm 5\%$ (Fig. 5).

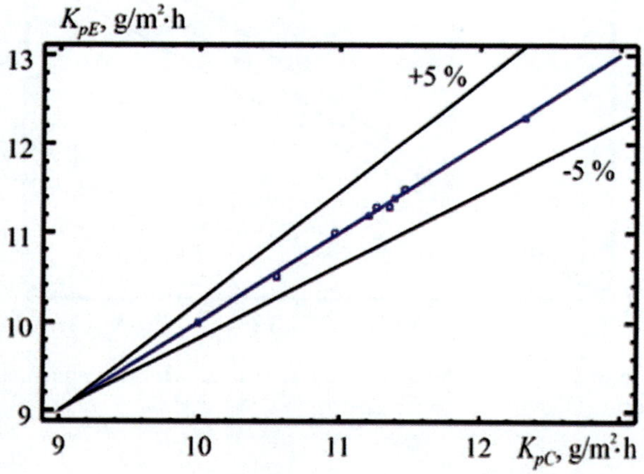

Fig. 5. Comparison of calculated pollution rate K_{pC} values with experimental K_{pE} during the WFE combustion.

4.2 System for Complex Gas Cleaning Method

Based on the experimental and theoretical studies (Fig. 1), a scheme of system for complex gas cleaning method of an ICE was developed (Fig. 6).

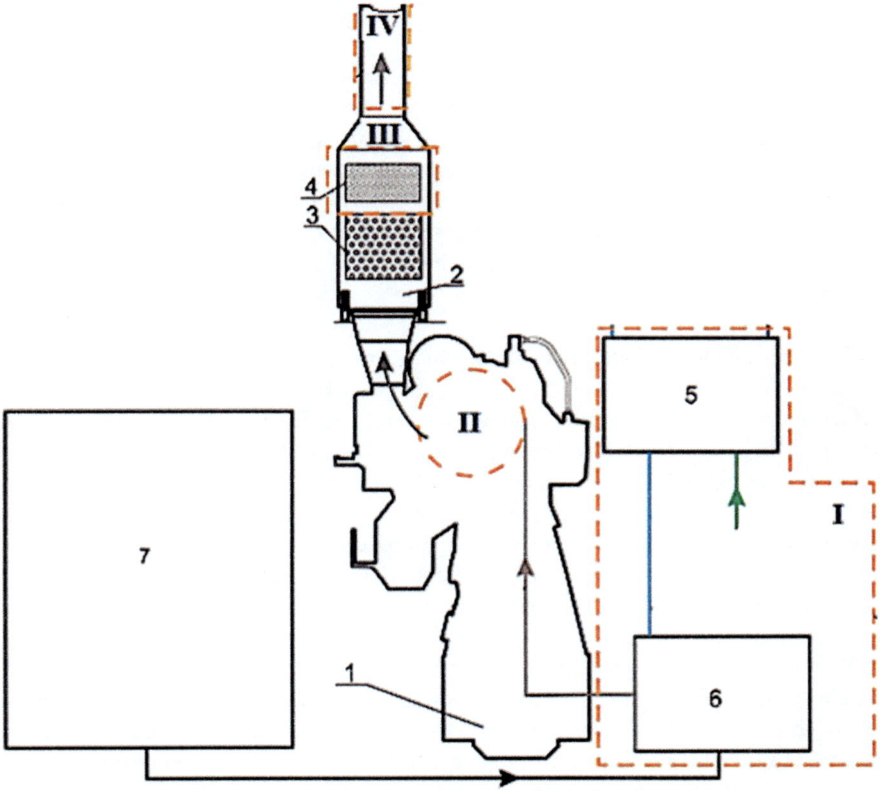

Fig. 6. A scheme of system for complex gas cleaning method and stages of cleaning: I – WFE preparation with a water content of about 30%; II – reducing concentration of toxic substances and solids in gases when WFE is burnt with water content 30%; III – adsorption processes occurring on condensing surfaces of exhaust gas boilers; IV – processes occurring on condensing surfaces of gas flues; 1 – ICE: 2 – EGB; 3 – dry convective surface; 4 – condensing heating surface; 5 – water preparation unit; 6 – water-fuel emulsion preparation unit; 7 – fuel tank.

Possibility of solving complex problems in proposed technology is ensured by combustion of WFE with specifically recommended value of water content ($W^r = 30\%$). Such WFE composition substantially affects not only running of thermal and physico-chemical processes along the entire path of fuel combustion aggregates (starting from combustion zone and to cut of flue), but also directs them in the required direction. For performing tasks in technology of proposed method, providing solutions to problems of improving economic efficiency, improvement of environmental indicators and reliability, it is envisaged 4 stages of technological process:

1) WFE preparation with a water content of about 30%;
2) WFE combustion with a water content of about $W^r = 30\%$ leads to formation of equimolar ratio NO_2: NO in exhaust gases at outlet of combustion zone (as

confirmed by patent and providing low-temperature corrosion (LTC) reduction), as well as reducing NOx, SO_2, PM emissions;
3) installation of condensing surfaces, on which conditions are created for passivation of metal and a sharp decrease LTC intensity, as well as conditions from side of gases and in condensate to intensify NOx, SO_2, PM absorption;
4) continuation of absorption intensification on condensing surfaces of gas flues (providing conditions for reliable operation of their metal) or maintaining temperature of metal of these gas flues above the dew point temperature of sulfuric acid vapour H_2SO_4 without NOx, SO_2 absorption, but ensuring reliability of work (at low LTC level).

The main elements of the power plant, which provides for the combustion of specially prepared WFE with a water content of 30%, are the ICE and the EGB. A dry convective surface and a condensing surface must be installed In the EGB to perform tasks. It is also mandatory to install a water treatment unit and WFE (the first stage of gas purification). Specially prepared WFE is supplied to the ICE injectors.

As a result of the combustion of activated WFE at the engine outlet we obtain exhaust gases of the corresponding composition with a reduced amount of toxic ingredients up to 35% or more, and most importantly, the equimolar ratio of NO_2: NO in NOx (which is confirmed by our experimental and literature data). This is the second stage of exhaust gas cleaning, which allows to reduce, for example, the concentration of NOx by 30–50%.

Further, the exhaust gases enter to the EGB, in which a dry convective surface is installed at the inlet (superheater, vapor-generating surface), and a condensing convective surface in the form of an economizer and (or) a hot water supply section with a metal temperature of 70–130 °C at the outlet, which leads to condensation of sulfuric acid vapors in the exhaust gases of ICE.

In the acid condensate under the indicated conditions, an average concentration of about 57% is established. The result is a sharp increase of SO_2 and NOx absorption. The presence in them of an equimolar ratio of NO_2: NO provides passivation of the condensing surface made of carbon steel. This provides a sharp decrease of the LTC intensity, an increase of the operation reliability of these condensing surfaces and the possibility of a sharp increase of the depth of exhaust gases utilization to ~ 80–90 °C instead of 160 °C (when standard fuels combustion). Thus, the third stage of gas cleaning is carried out.

Further, the gases after the EGB enter to the gas duct. With such a low gas temperature and ensuring the temperature of the gas duct metal after the ICE at a level of 70–80 °C, that is, in the presence of sulfuric acid condensate on the inner surface of the gas ducts, the process of absorption of toxic substances will continue with reliable operation of the gas duct metal, and the intensity of the mass flow will additionally decrease H_2SO_4 and LTC (this is the fourth stage of gas purification).

This is due to the fact that in gases there is an equimolar ratio of NO_2: NO, which means that the passivation of the metal surface and a decrease of the LTC will be ensured with a minimum temperature difference between the gases and the metal of the gas ducts.

Thus, the implementation of these stages of gas purification provides a decrease of the content of toxic ingredients in gases by almost 50% (compared to existing technologies, which provide a decrease of the exhaust gas temperature of the EGB to 160 °C) and partial removal of solid particles, contained in gases (when WFE combustion, solid and soot particles are 80% less than standard fuels combustion) (Fig. 7).

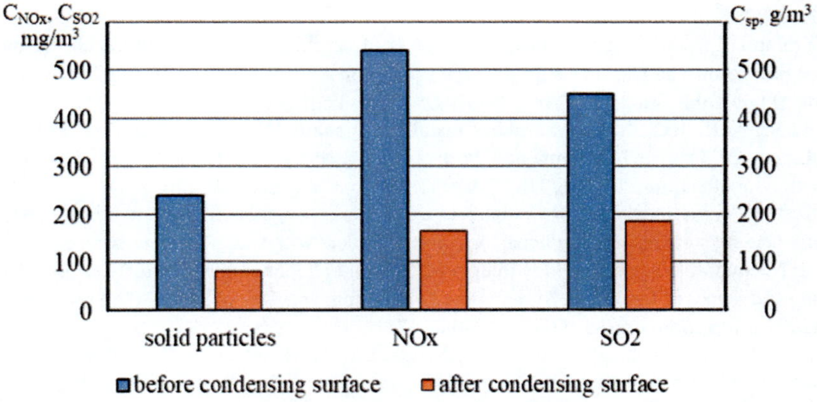

Fig. 7. Emission rates of toxic ingredients.

It is necessary to use scrubber technologies as the fifth stage of cleaning in the presented technology in order to ensure the final cleaning of exhaust gases from NOx and SO_2 to the level corresponding to the IMO regulations.

5 Conclusions

The best burning out of fuel combustible components due to applying a WFE provides decreasing a concentration of solids and soot in the exhaust gases and hence their toxicity.

The kinetics of low-temperature pollution on EGB condensation surfaces with WFE combustion are investigated to obtain approximation equations for predicting processes development.

When WFE combustion with a water content of 30%, the LTC and pollution intensity decreases, which makes it possible to install condensing heating surfaces in the EGB. The installation of a condensing heating surface in the EGB reduces the content of NOx in gases by 55%, SO_2 by 50%, and the content of solid particles by 3 times.

It is necessary to use scrubber technologies as the last stage of cleaning in the presented technology in order to ensure the final cleaning of exhaust gases from NOx and SO_2 to the level corresponding to the IMO regulations.

References

1. MAN Diesel Turbo. https://marine.man-es.com/two-stroke/ceas. Accessed 22 July 2021
2. Ecospec $CSNO_x^{TM}$ Broshure. https://www.environmental-expert.com/downloads/ecospec-csnox-3-in-1-emission-abatement-system-brochure-785875. Accessed 22 July 2021

3. Landet, R.D.: PM emissions and NO_x – reduction due to water in fuel emulsions in marine diesel engines. Norwegian University of Science and Technology, Department of Marine Technology, Student thesis, p. 73 (2010). https://ntnuopen.ntnu.no/ntnu-xmlui/bitstream/handle/11250/237793/375078_FULLTEXT01.pdf?isAllowed=y&sequence=1. Accessed 22 July 2021
4. Kjolholt, J.: Assessment of possible impacts of scrubber water discharges on the marine environment. The Danish Environmental Protection Agency (2012) p. 93. https://www2.mst.dk/Udgiv/publications/2012/06/978-87-92903-30-3.pdf. Accessed 22 July 2021
5. Hansen, J.P.: Exhaust Gas Scrubber Installed Onboard MV Ficaria Seaways. Public Test Report the Danish Environmental Protection Agency p.30 (2012). https://www2.mst.dk/Udgiv/publications/2012/06/978-87-92903-28-0.pdf. Accessed 22 July 2021
6. Kornienko, V., Radchenko, R., Bohdal, Ł., Kukiełka, L., Legutko, S.: Investigation of condensing heating surfaces with reduced corrosion of boilers with water-fuel emulsion combustion. In: Nechyporuk, M., et al. (eds.) Integrated Computer Technologies in Mechanical Engineering. Advances in Intelligent Systems and Computing, vol. 188, pp. 300–309. Springer, Cham (2021). https://doi.org/10.1007/978-3-030-37618-5
7. Wojs, M.K., Orliński, P., Kamela, W., Kruczyński, P.: Research on the influence of ozone dissolved in the fuel-water emulsion on the parameters of the CI engine. In: IOP Conference Series: Materials Science and Engineering vol. 148, pp. 1–8 (2016)
8. Soot deposits and fires in exhaust gas boilers. MAN Diesel & Turbo (2014)
9. Kornienko, V., Radchenko, R., Bohdal, T., Radchenko, M., Andreev A.: Thermal characteristics of the wet pollution layer on condensing heating surfaces of exhaust gas boilers. In: Ivanov V., et al. (eds.) Advances in Design, Simulation and Manufacturing IV. Lecture Notes in Mechanical Engineering, pp. 339–348. Springer, Cham (2021). https://doi.org/10.1007/978-3-030-77823-1
10. Baskar, P., Senthil Kumar, A.: Experimental investigation on performance characteristics of a diesel engine using diesel-water emulsion with oxygen enriched air. Alex. Eng. J. **56**(1), 137–146 (2017)
11. Miao, Y.C., Yu, C.L., Wang, B.H., Chen, K.: The applied research of emulsified heavy fuel oil used for the marine diesel engine. Adv. Mater. Res. **779**, 469–476 (2013)
12. Parsadanov, I.V., Solodovnikov, V.V.: Application of water fuel emulsion in autotractor diesel. Ecological efficiency (part 1). Inter.Combust. Eng. **2**, 118–120 (2011)
13. Zang, T., Okada, H., Tsukamoto, T., Ohe, K.: Experimental study on water particles in the combustion of marine four-stoke diesel engine operated with emulsified fuels. Paper 193. Vienna, CIMAK (2007)
14. Vellaiyan, S., Amirthagadeswaran, K.S.: The role of water-in-diesel emulsion and its additives on diesel engine performance and emission levels: a retrospective review. Alex. Eng. J. **55**, 2463–2472 (2016)
15. Kornienko, V., Radchenko, M., Radchenko, R., Konovalov, D., Andreev, A., Pyrysunko, M.: Improving the efficiency of heat recovery circuits of cogeneration plants with combustion of water-fuel emulsions. Therm. Sci. **25**(1B), 791–800 (2021)
16. Gupta, R.K., Sankeerth, K.A., Sharma, T.K., Rao, G., Murthy, K.M.: Effects of water-diesel emulsion on the emission characteristics of single cylinder direct injection diesel engine - a review. Appl. Mech. Mater. **592**, 1526–1533 (2014)
17. Radchenko, R., Pyrysunko, M., Kornienko, V., Scurtu, I.-C., Patyk, R.: Improving the ecological and energy efficiency of internal combustion engines by ejector chiller using recirculation gas heat. In: Nechyporuk, M., et al. (eds.) Integrated Computer Technologies in Mechanical Engineering. Advances in Intelligent Systems and Computing, vol. 188, pp. 531–541. Springer, Cham (2021). https://doi.org/10.1007/978-3-030-37618-5
18. Patel, K.R., Dhiman, V.: Research study of water- diesel emulsion as alternative fuel in diesel engine – An overview. Int. J. Lat. Eng. Res. Appl. **2**(9), 37–41 (2017)

19. Radchenko, R., Pyrysunko, M., Radchenko, A., Andreev, A., Kornienko, V.: Ship engine intake air cooling by ejector chiller using recirculation gas. In: Tonkonogyi V. et al. (eds.) Advanced Manufacturing Processes II. InterPartner-2020. Lecture Notes in Mechanical Engineering, pp. 734–743. Springer, Cham (2021). https://doi.org/10.1007/978-3-030-68014-5
20. Patel, N., Modi, M., Patel, T.: Investigation of diesel engine with water emulsifier - a review. Int. Res. J. Eng. Technol. **4**(2), 879–883 (2017)
21. Kruczyński, P., Orliński, P., Kamela, W., Ślęzak, M.: Analysis of selected toxic components in the exhaust gases of a CI engine supplied with water-fuel emulsion. Pol. J. Environ. Stud. **27**(1), 129–136 (2018)
22. Katin, V., Kosygin, V., Akhtiamov, M., Vol'khin, I.: Mathematical models of the output of major pollutants in the process of burning water fuel oil emulsions in boiler plants. In: Murgul, V., Popovic, Z. (eds.) EMMFT 2017. AISC, vol. 692, pp. 987–997. Springer, Cham (2018). https://doi.org/10.1007/978-3-319-70987-1_107
23. Sugeng, D.A., Ithnin, A.M., Amri, N.S.M.S., Ahmad, M.A., Yahya, W.J.: Water content determination of steam generated water-in-diesel emulsion. J. Adv. Res. Fluid Mech. Thermal Sci. **49**(1), 62–68 (2018)
24. Trushliakov, E., Radchenko, A., Radchenko, M., Kantor, S., Zielikov, O.: The efficiency of refrigeration capacity regulation in the ambient air conditioning systems. In: Ivanov V., et al. (eds.) Advances in Design, Simulation and Manufacturing III (DSMIE 2020). Lecture Notes in Mechanical Engineering, pp. 343–353. Springer, Cham (2020). https://doi.org/10.1007/978-3-030-50491-5
25. Trushliakov, E., Radchenko, A., Forduy, S., Zubarev, A., Hrych, A.: Increasing the operation efficiency of air conditioning system for integrated power plant on the base of its monitoring. In: Nechyporuk, M., et al. (eds.) Integrated Computer Technologies in Mechanical Engineering (ICTM 2019). Advances in Intelligent Systems and Computing, vol. 1113, pp. 351–360. Springer, Cham (2020). https://doi.org/10.1007/978-3-030-37618-5
26. Volintiru, O. N., Scurtu, I. C., Stefănescu1, T. M.: Modeling and optimization of HVAC system for special ships. J. Phys. Conf. Ser. **1122**, art no 012004 (2018)
27. Bohdal, L., Kukiełka, L., Legutko, S., Patyk, R., Radchenko, A.M.: Modeling and experimental analysis of shear-slitting of AA6111-T4 aluminum alloy sheet. Materials **13**(14), 3175 (2020)
28. Radchenko, A., Trushliakov, E., Kosowski, K., Mikielewicz, D., Radchenko, M.: Innovative turbine intake air cooling systems and their rational designing. Energies **13**(23), 6201 (2020). https://doi.org/10.3390/en13236201
29. Cherednichenko, O., Tkach, M., Dotsenko, S.: The usage of a waste heat recovery metal-hydride unit of continuous operation in the maritime energy. In: IEEE International Conference on Modern Electrical and Energy Systems (MEES), pp. 510–513 (2019)
30. Bohdal, L., Kukielka, L., Świłło, S., Radchenko, A.M., Kułakowska, A.: Modelling and experimental analysis of shear-slitting process of light metal alloys using FEM, SPH and vision-based methods. In: AIP Conference Proceedings vol. 2078, p. 020060 (2019)
31. Shukla, A.K., Singh, O.: Thermodynamic investigation of parameters affecting the execution of steam injected cooled gas turbine based combined cycle power plant with vapor absorption inlet air cooling. Appl. Therm. Eng. **122**, 380–388 (2017)
32. Radchenko, A., Trushliakov, E., Tkachenko, V., Portnoi, B., Prjadko, O.: Improvement of the refrigeration capacity utilizing for the ambient air conditioning system. In: Tonkonogyi, V. et al. (eds.) Advanced Manufacturing Processes II. InterPartner 2020. Lecture Notes in Mechanical Engineering, pp. 714–723. Springer, Cham (2021). https://doi.org/10.1007/978-3-030-68014-5

33. Cherednichenko, O., Serbin, S., Dzida, M.: Investigation of the combustion processes in the gas turbine module of an FPSO operating on associated gas conversion products. Polish Maritime Res. **26**(4), 149–156 (2020)
34. Radchenko, N., Radchenko, A., Tsoy, A., Mikielewicz, D., Kantor, S., Tkachenko, V. Improving the efficiency of railway conditioners in actual climatic conditions of operation. In: AIP Conference Proceedings, vol. 2285, p. 030072 (2020)
35. Kuznetsov, V., Dymo, B., Kuznetsova, S., Bondarenko, M., Voloshyn, A.: Improvement of the cargo fleet vessels power plants ecological indexes by development of the exhaust gas systems. Polish Maritime Res. **28**(1), 97–104 (2021)

Effect of Utilization Exhaust and Recirculation Gases of Ship Diesel Engine in Absorption Chiller

Roman Radchenko[1](✉), Maxim Pyrysunko[2], Victoria Kornienko[2], Viktor Gorbov[1], and Ivan Kalinichenko[2]

[1] Admiral Makarov National University of Shipbuilding, 9 Heroes of Ukraine Avenue, Mykolayiv 54025, Ukraine

[2] Kherson Branch of Admiral Makarov National University of Shipbuilding, 44 Ushakova Avenue, Kherson 73000, Ukraine

Abstract. Technical regulation is carried out based on the real possibility of ensuring the required level of environmental friendliness of harmful emissions. Requirements for the environmental safety of ships are developed by the International Marine Organization (IMO). The requirements of IMO standards, applications and protocols of the MARPOL 73/78 convention regulate technical, organizational and legal issues of environmental protection at sea. Their implementation often leads to reduction of the economic and energy performance of marine diesel engines. However, the main difficulties in ensuring the environmental safety of internal combustion engines are associated with the selectivity of the components of the exhaust gases system. The innovative scheme of Exhaust Gas Recirculation (EGR) system with using the heat of recirculation gas by absorption chiller (ACh) for cooling the air at the intake of main ship engine is proposed. The effect of using the heat of recirculation gas for cooling engine intake air is analyzed taking into account the changing climatic conditions on the vessel's route line. The results of calculations have decrease in the specific fuel consumption of engines of the 6G50ME-C9.6-TIII type to 163.0–167.3 g/(kW·h). When the air at the engine inlet is cooled to ACh using the heat of exhaust and recirculated (ecological) gases, the annual fuel economy increases, respectively: to 200.9 t.

Keywords: Absorption chiller · Harmful emissions · Ecology · Exhaust gas recirculation

1 Introduction

Reducing the toxicity and the amount of harmful emissions with the exhaust gases of the engines of ship power plants is a complex scientific and technical problem. The need to solve it is dictated by regulatory and legislative requirements, which are becoming more stringent. Most often, their implementation is accompanied by a decrease in the economic and energy performance of the engines [1].

M. Nechyporuk et al. (Eds.): ICTM 2021, LNNS 367, pp. 509–519, 2022.
https://doi.org/10.1007/978-3-030-94259-5_43

One of the promising directions in the greening of marine engines is the neutralization of harmful substances in exhaust gases by recirculation with cleaning in scrubbers and heat removal by seawater and the supply of purified gases together with inflated air after the turbocharger back into the engine cylinders (EGR technology) [2].

However, the use of such technologies is associated with additional energy costs for the circulation of gases and heat losses from the sea water, which leads to an increase in fuel consumption and a decrease in engine power. At the same time, cooling the intake air of the engines would provide a reduction in fuel consumption.

As a result, it is justified to use the technologies of utilization of recirculation exhaust gas heat for cooling engine intake air to improve the environmental and fuel efficiency of the marine diesel engine using EGR systems [3, 4]. These technologies allow to increase diesel fuel economy by cooling the combustion air with waste heat using chillers (WHUCh) [5].

2 Literature Review

One of the constituent components that ensure human life is air. The person consumes about 10 m^3 of air per day, hence it is clear how much importance should be attached to its purity. Clean air is a multi-component mixture consisting of gases, water vapor and solid (aerosol) particles. Naturally, the dominant role belongs to gas, which includes constant and variable gases. Permanent gases include nitrogen (78%), oxygen (21%), carbon dioxide (0.03%) and inert gases. Variable gases include: methane, carbon monoxide, ozone, sulfur oxides and nitrogen oxides [6, 7].

However, nowadays it is possible to speak about clean air only conditionally, especially in industrial zones. In most cases the ambient air is polluted. Presenting in the atmospheric air and polluting it, harmful substances have the most negative effect on human life. Consequently, the problem of air protection in areas of large industrial enterprises and traffic congestion is multifaceted. Firstly, it requires a comprehensive study of sources of pollution and harmful substances emitted by them. Secondly, it is needed to complete the assessment of the impact of these substances on human health. The distribution of harmful emissions by transport mode can be seen from the Fig. 1. For a number of years the standards on emissions from ships have been constantly tightened. This is largely due to the fact that the amount of emissions from ships is constantly growing [8, 9].

Table 1 shows the time of formation of nitrogen oxide depending on the temperature of the fuel flame.

Gas recirculation slows down the fuel combustion process and, ultimately, leads to a sharp decrease in NO_X emissions. Along with a decrease in nitrogen oxides, the yield of incomplete combustion products (in particular, soot) increases and the fuel efficiency of a diesel engine deteriorates. These negative effects accompanying recirculation are especially noticeable at full loads. For this reason, bypassing gases is economically justified only in the fractional modes of the internal combustion engine and with a recirculation rate of $12 \div 20\%$. Typically, the recirculation coefficient K_r value is optimized by controlling the wastegate according to load, speed, fuel efficiency, NO_X and C.

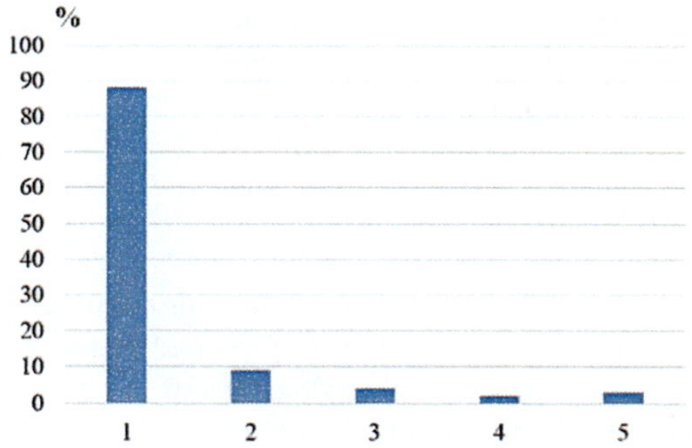

Fig. 1. Distribution of harmful emissions by mode of transport: 1 - automobile transport; 2 - railway transport; 3 - road complex; 4 - air transport; 5 - river and marine transport

Table 1. The time of formation of nitrogen oxide depending on the temperature of the fuel flame [32]

T, K	1750	2000	2250	3000	4000
τ, sec	140	1	$5.3 \cdot 10^{-3}$	$7.8 \cdot 10^{-5}$	$7.2 \cdot 10^{-7}$

The IMO standard establishes the maximum permissible value of specific NO_x emissions depending on the nominal speed of the crankshaft [10]. This approach to standardizing NO_x emissions is justified, as it allows taking into account the technical capabilities of diesel engines of various sizes and purposes. In addition to nitrogen oxides, IMO regulations also limit emissions of sulfur oxides. As noted earlier, SO_x emissions (as opposed to NO_x) are not affected by combustion by itself. All sulfur enters the combustion chamber along with the fuel; therefore, it is only possible to reduce SO_x emissions by using low-sulfur fuel. In ECA zones, diesels must operate on fuels with a sulfur content of no more than 1.0%, and from 2016 the limit for sulfur in fuel is limited to 0.1%.

Today, it is promising to use technologies that would ensure an increase in fuel and energy efficiency of combustion engines with EGR systems [11], i.e. would combine high environmental efficiency [12] with engine fuel efficiency due to deep waste heat utilization [13, 14] and engine cyclic air cooling by applying refrigeration [15, 16], trigeneration or integrated energy systems [17], and jet technologies, and statistical treatment of data of influence of actual climatic conditions [18, 19] on performance characteristics of cooling systems [20, 21]. To evaluate the effectiveness of technologies for reducing harmful emissions into the atmosphere and saving fuel, methods and modeling [22, 23] the influence of environmental parameters on the performance of cooling systems with waste heat [24, 25] recovery are used, monitoring and experiments are carried out. Such technologies provide engine cyclic air cooling by waste heat [26, 27] using chillers (WHUCh): ejector chillers (ECh) as the simplest in design and absorption chillers (ACh) as highly efficient heat transformer with coefficient of performance ζ = 0.6–0.7.

The aim of the scientific research is to increase the ecological and energetic efficiency of a ship power plant by cooling the cycle air with an ACh, which utilizes the heat of exhaust and recirculated (ecological) gases.

3 Research Methodology

Provided schematic solution with the use of the heat-using circuit of the ACh for the two-stroke engine 6G50ME-C9.6 [28, 29], responsible for Tier III environmental conditions. To analyze the system parameters and characteristics of the diesel engine, the CEAS software package was used. The next initial data were used for the calculation: power – $N_e = 9288$ kW; engine load – $N_{MCR} = 90\%$; SFC – $g_e = 166.0$ g/(kWh); speed – $n_e = 96.5$ rpm; EGR system.

Determining the effect and development of a mathematical model of cyclic air cooling processes for engines of marine power plants with exhaust gas recirculation and heat utilization, fuel and environmental efficiency indicators were calculated according to the programs of developers of marine engines with exhaust gas recirculation, "MAN Energy Solution") [30]. The temperature of the seawater t_w along the vessel's voyage line was determined using the electronic resource NOAA's (National Oceanic and atmospheric Administration) of the American National Environmental Information Center (NCEI) "nodc.noaa.gov", available on-line [31]. The parameters of the ambient air (temperature t_a and relative humidity φ_a) were taken from the electronic resource "meteomanz.com" [32]. In the Fig. 2 there are changes of the climatic conditions during the voyage.

The calculation of the operating parameters of the heat-recovery contour based on the absorption chillers was carried out according to the equations that are used in the software package developed at the Department of Conditioning and Refrigeration of admiral Makarov National University of Shipbuilding (Ukraine).

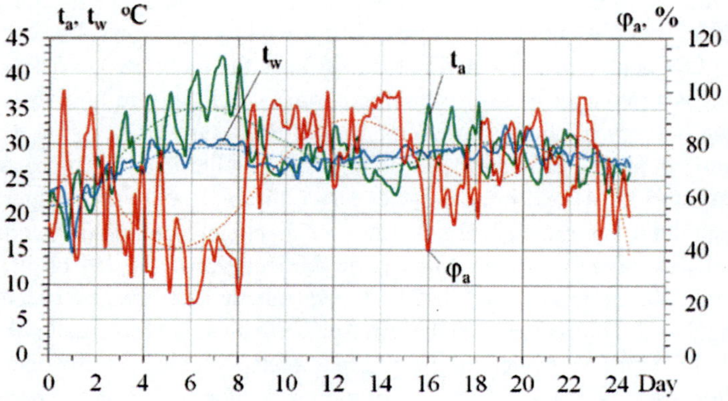

Fig. 2. The variation of climatic conditions (air temperature t_a, temperature of sea water t_w, and relative humidity φ_a) during the vessel voyage

The heat of the exhaust gases required to heat the condensate and generate steam is calculated taking into account the mass flow rate of the exhaust gases G_{eg}, their heat capacity c_{eg} and the temperature of the exhaust gases at the inlet t_{eg1} and at the outlet t_{eg2}:

$$Q_{g \cdot eg} = G_{eg} \cdot c_{eg} \cdot \Delta t_{eg} = G_{eg} \cdot c_{eg} \cdot (t_{eg1} - t_{eg2}) \quad (1)$$

The heat of recirculated (ecological) gases is calculated taking into account the mass flow rate of recirculated (ecological) gases G_{rg}, their heat capacity c_{rg} and the temperature of the exhaust gases at the inlet t_{rg1} and at the outlet t_{rg2} from the generator:

$$Q_{g \cdot rg} = G_{rg} \cdot c_{rg} \cdot \Delta t_{rg} = G_{rg} \cdot c_{rg} \cdot (t_{rg1} - t_{rg2}) \quad (2)$$

The amount of recirculated (ecological) gases G_{rg} is calculated taking into account the amount of exhaust gases G_{eg} and the recirculation coefficient K_r and is calculated:

$$G_{rg} = G_{eg} \cdot K_r \quad (3)$$

According to the total heat of exhaust and recirculation (ecological) gases required for the calculation is:

$$\sum Q_g = Q_{g \cdot eg} + Q_{g \cdot rg} \quad (4)$$

During the voyage, the available cooling capacity $Q_{0rg(0.7)}$ exceeds its value $Q_{0.15}$ required to cool the air at the engine inlet to the temperature $t_{a2} = 15$ °C, as the heat of exhaust gases $Q_{h.rg}$ is much higher than its required value $Q_{h.15(0.7)}$ for cooling the air at the engine inlet to $t_{a2} = 15$ °C.

4 Results

The solution with using the ACh was developed and analysed (Fig. 3). During the usage of the EGR system, some of the exhaust gases (10–40%) go into the scrubber recycling system (SRS). The greening of diesel engines of marine power plants through the recirculation of exhaust gases in EGR systems with their resupply to the engine cylinders leads to the significant reduction in fuel economy, which is due primarily to fuel processes. at the expense of energy costs which are directed on purification of the fulfilled gases with the subsequent removal of heat by seawater.

The results of calculating the air cooling system at the engine inlet in ACh ($\zeta = 0.6$; 0.7) using the heat of recirculated (ecological) gases $Q_{g.rg} = 985...1075$ kW show that the obtained cooling capacity is (Fig. 4): $Q_{0.rg(0.6)} = 590...640$ kW ($\zeta = 0.6$); $Q_{0.rg(0.7)} = 690...750$ kW ($\zeta = 0.7$). At the same time, the cooling capacity $Q_{0.15}$ required for air cooling at $t_{a2} = 15$ °C practically does not exceed $Q_{g.rg} = 700$ kW. In the figure, these values are presented during the Odessa-Shanghai route line (1.07... 27.07.2018).

Utilization of heat of recirculated gases in ACh with a high value $\zeta = 0.6$; 0.7 provides the amount of air cooling $\Delta t_{a.rg(0.6)} = 13...25$ °C and $\Delta t_{a.rg(0.7)} = 14...28$ °C against $\Delta t_{15} = 20...30$ °C while cooling to 15 °C (Fig. 5).

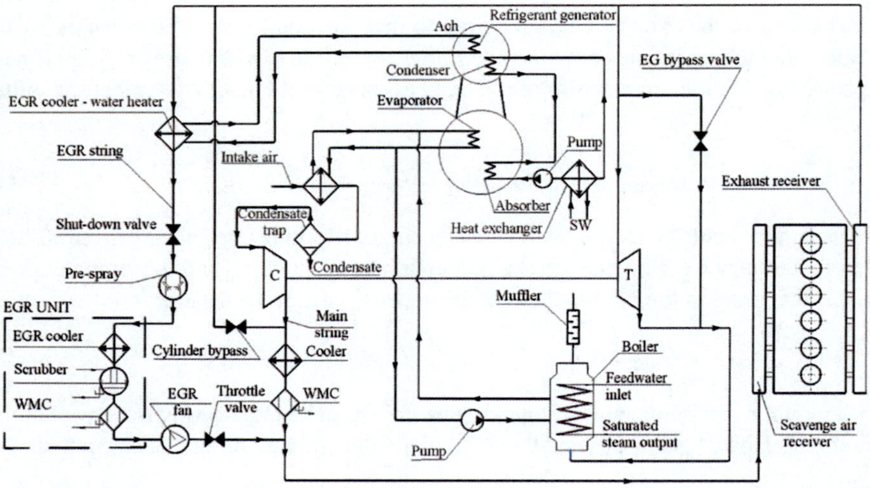

Fig. 3. Scheme of EGR-technology for the two-stroke diesel engine 6G50ME-C9.6 with ACh.

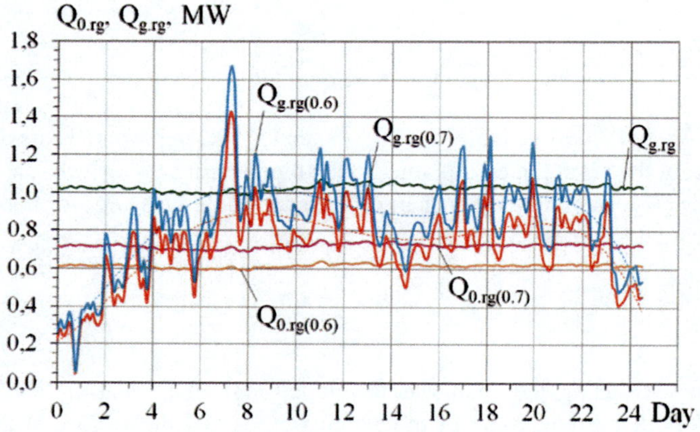

Fig. 4. Current values of the heat of recirculation gases $Q_{g.rg}$, refrigerating capacity $Q_{0.rg(0.6)}$ and $Q_{0.rg(0.7)}$, obtained by its transformation into ACh, and heat $Q_{g.rg(0.6)}$ and $Q_{g.rg(0.7)}$, required to cool the air at the engine inlet to $t_{a2} = 15$ °C with thermal coefficients $\zeta = 0.6$; 0.7

Cooling the temperature of the air at the engine inlet by utilizing the heat of exhaust and recirculation gases in ACh allows to reduce the value of the specific fuel equivalent consumption depending on the value of the thermal coefficient ACh. In this case, the specific fuel consumption is (Fig. 6): $\Delta g_{erg+eg(0.6)} = 163.6\ldots167.9$ g/(kW·h) ($\zeta = 0.6$); $\Delta g_{e.rg+eg(0.7)} = 163.9\ldots167.6$ g/(kW h) ($\zeta = 0.7$).

In this case, the fuel economy for the voyage due to the air cooling at the inlet to the motor by heat recovery and exhaust gas recirculation to ACh, is (Fig. 7): $\Sigma B_{f.rg+eg(0.6)} = 14.8$ tons ($\zeta = 0.6$); $\Sigma B_{f.rg+eg(0.7)} = 17.3$ tons ($\zeta = 0.7$).

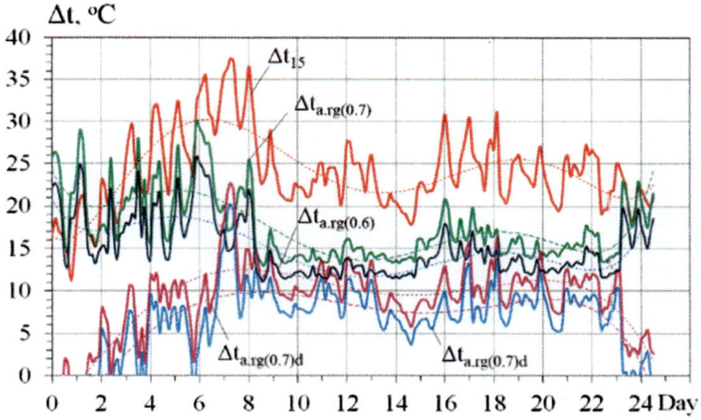

Fig. 5. Current values of air temperature decrease $\Delta t_{15} = 15$ °C during cooling to $t_{a2} = 15$ °C, real air temperature decrease in ACh $\Delta t_{a.rg(0.6)}$ and $\Delta t_{a.rg(0.7)}$ and its undercooling (deficit) in ACh: $\Delta t_{a.rg(0.7)d} = \Delta t_{15} - \Delta t_{a.rg(0.7)}$; $\Delta t_{a.rg(0.6)d} = \Delta t_{15} - \Delta t_{a.rg(0.6)}$

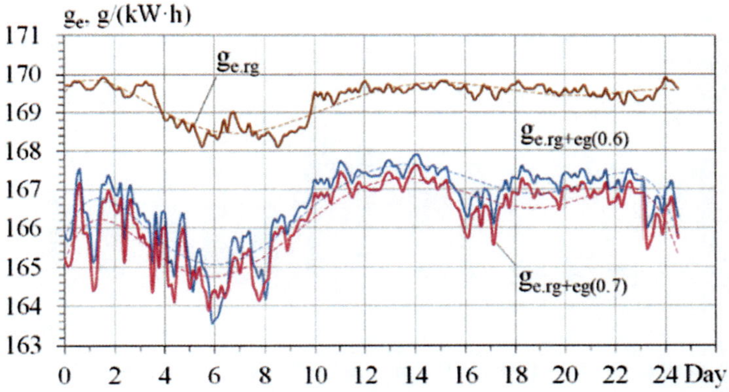

Fig. 6. Current values of the specific fuel consumption of the base engine with gas recirculation $g_{e.rg}$, the specific fuel consumption of the engine $g_{e.rg+eg(0.6)}$ and $g_{e.rg+eg(0.7)}$ with air cooling at the inlet to ACh ($\zeta = 0.6$; 0.7) during the voyage

According to [33], the price for heavy fuel of the IFO 380 brand for the 6G50MEC-9.5 engine is 390 $/t USD.

When the air at the engine inlet is cooled to ACh (Fig. 8, 9) using the heat of exhaust and recirculated (ecological) gases, the annual fuel economy increases, respectively: $\Delta B_{f(0.6)} = 172.2$ t, $\Delta B_{f(0.7)} = 200.9$ t.

When using ACh with high values of the thermal coefficient, the reduction in fuel costs is: at $\zeta = 0.6$ to 67.2 ths. USD; at $\zeta = 0.7$ to 78.4 ths. USD.

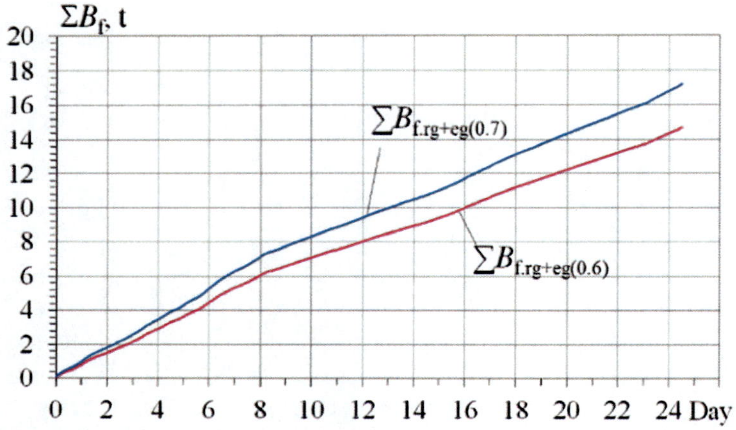

Fig. 7. Total fuel economy ΣB_f for the voyage. $\Sigma_{Bf.\ rg+eg(0.6)}$ and $\Sigma_{Bf.\ rg+eg(0.7)}$ due to cooling of the air at the engine inlet to ACh ($\zeta = 0.6$; 0.7) using the heat of the recirculation and exhaust gases during the voyage of the vessel

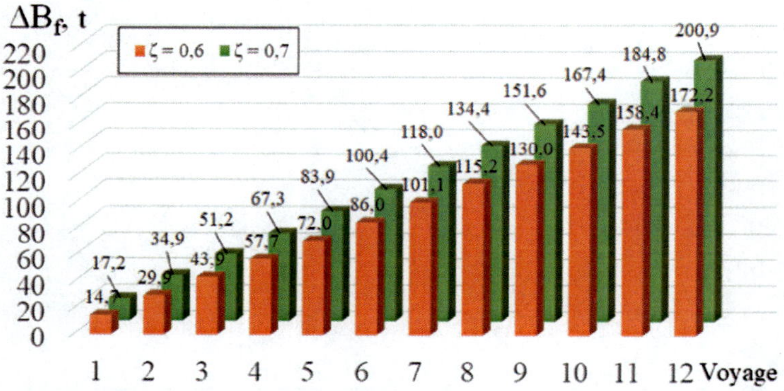

Fig. 8. The total value of the fuel consumption reduction per 1… 12 Odessa low-Shanghai (12 voyages per year) while cooling air inl

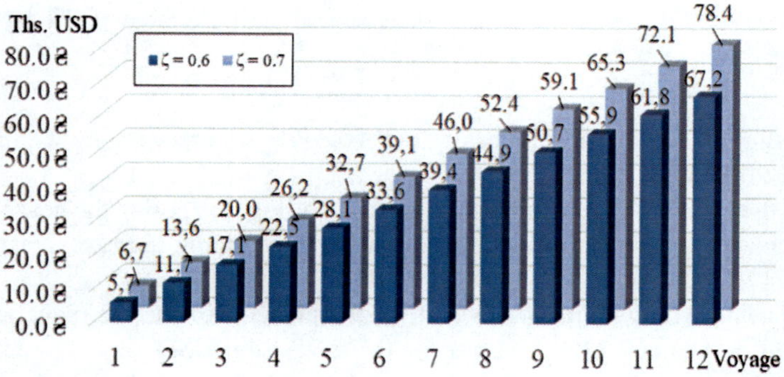

Fig. 9. The total value of the reduction in fuel costs for 1… 12 voyages Odessa-Shanghai (12 voyages per year) when cooling the air at the engine inlet to ACh ($\zeta = 0.6$; 0.7) using the heat of exhaust and recirculation gases

According to the leading manufacturer WHUCh "York" [34, 35], the price of a heat-using absorption refrigeration machine is about 100,000 ths. USD. Therefore, taking into account the calculations with ACh with the heat coefficient of $\zeta = 0.7$, as the annual savings are about 80 ths. USD., The developed scheme can pay for itself in a year and a half.

5 Conclusions

The variant of the scheme-designed solution and calculations are proposed and analyzed using the heat of recirculation (ecological) and exhaust gases of the engine of the ship power plant for cooling the air at the engine inlet in ACh.

The most energetically efficient system for cooling the air is at the engine inlet by utilizing the heat of recirculation and exhaust gases in ACh with the thermal coefficient $\zeta = 0.7$, which ensures decrease in the specific fuel consumption of engines of the 6G50ME-C9.6-TIII type to $g_{e.rg + eg(0.7)} = 163.0\ldots 167.3$ g/(kW·h) compared to $g_{e.rg} = 168\ldots 170$ g/(kW·h) for an engine with recirculation of ecological exhaust gases. In this case, the annual fuel economy is $\Delta B_{f(0.7)} = 200.9$ t, and the reduction in fuel costs at $\zeta = 0.7$ to 78.4 ths. USD.

References

1. Kumar, J., Thirumala, S., et al.: Effect of reformed EGR on the performance and emissions of a diesel engine: a numerical study. Alexandria Eng. J. **57**, 517–525 (2018)
2. Ghosh, S., Dutta, D.: The effects of EGR on the performance and exhaust emissions of a diesel engine operated on diesel oil and pongamia pinata methyl ester (PPME). Int. J. Eng. Inven. **12**(1), 39–44 (2020)
3. Saravanana, P., Kumar, N., Ettappan, M., Dhanagopale, R., Vishnupriyan, J.: Effect of exhaust gas recirculation on performance, emission and combustion characteristics of ethanol-fueled diesel engine. Case Stud. Thermal Eng. 20, 100643 (2020). https://doi.org/10.1016/j.csite.2020.100643
4. Radchenko, A., Radchenko, N., Tsoy, A., Portnoi, B., Kantor, S.: Increasing the efficiency of gas turbine inlet air cooling in actual climatic conditions of Kazakhstan and Ukraine. AIP Conf. Proc. **2285**, 030071 (2020)
5. Pure SOx Exhaust gas cleaning (EMD00281EN 1208). http://www.alfalaval.com/industries/marine/oiltreatment/Documents/PureSOx%20product%20brochure.pdf. Accessed 04 Mar 2013
6. MARPOL Consolidated edition 2020/International Maritime Organization. London. p. 459 (2011). http://www.idgca.org/doc/app5_290115.pdf
7. Trushliakov, E., Radchenko, A., Forduy, S., Zubarev, A., Hrych, A.: Increasing the operation efficiency of air conditioning system for integrated power plant on the base of Its monitoring. In: Nechyporuk M., et al. (eds.) Integrated Computer Technologies in Mechanical Engineering (ICTM 2019). Advances in Intelligent Systems and Computing, vol. 1113, pp. 351–360. Springer, Cham (2020). https://doi.org/10.1007/978-3-030-37618
8. Bohdal, L., Kukiełka, L., Legutko, S., Patyk, R., Radchenko, A.M.: Modeling and experimental analysis of shear-slitting of AA6111-T4 aluminum alloy sheet. Materials **13**(14), 3175 (2020)

9. Hiraoka. N., Miyanagi, A., Kuroda, K., Ito, K., Nakagawa, T., Ueda, T.: The world's first onboard verification test of UE engine with low pressure EGR complied with IMO s NOx tier III regulations. Mitsubishi Heavy Ind. Tech. Rev. **53**(2) (2016)
10. Bohdal, L., Kukielka, L., Radchenko, A. M., Patyk, R., Kułakowski, M., Chodór, J. Modelling of guillotining process of grain oriented silicon steel using FEM AIP Conf. Proc. **2078**, 020080 (2019)
11. Kumar, J., Thirumala, S.: Effect of reformed EGR on the performance and emissions of a diesel engine: A numerical study. Alexandria Eng. J. **57**, 517–525 (2018)
12. Radchenko, R., Pyrysunko, M., Kornienko, V., Scurtu, I.-C., Patyk, R.: Improving the ecological and energy efficiency of internal combustion engines by ejector chiller using recirculation gas heat. In: Nechyporuk, M., Pavlikov, V., Kritskiy, D. (eds.) ICTM 2020. LNNS, vol. 188, pp. 531–541. Springer, Cham (2021). https://doi.org/10.1007/978-3-030-66717-7_45
13. Hiraoka, N., Miyanagi, A., Kuroda, K., Ito, K., Nakagawa, T., Ueda, T.: The world's first onboard verification test of UE engine with low pressure EGR complied with IMO s NOx tier III regulations. Mitsubishi Heavy Ind. Tech. Rev. **53**(2) (2016)
14. Radchenko, A., Trushliakov, E., Tkachenko, V., Portnoi, B., Prjadko, O.: Improvement of the refrigeration capacity utilizing for the ambient air conditioning system. In: Tonkonogyi, V. et al. (eds.) Advanced Manufacturing Processes II. InterPartner 2020. Lecture Notes in Mechanical Engineering, pp. 714–723. Springer, Cham (2021)https://doi.org/10.1007/978-3-030-68014-5_69
15. Trushliakov, E., Radchenko, A., Radchenko, M., Kantor, S., Zielikov, O.: The efficiency of refrigeration capacity regulation in the ambient air conditioning systems. In: Ivanov V., et al. (eds.) Advances in Design, Simulation and Manufacturing III (DSMIE 2020). Lecture Notes in Mechanical Engineering, pp. 343-353. Springer, Cham (2020)https://doi.org/10.1007/978-3-030-50491-5_33
16. Roh, H.G., Park, S., Lee, C.S.: Effect of exhaust gas recirculation on the combustion and emissions of dimethyl ether in a passenger vehicle diesel engine. J. Energy Eng. **144**(5), 04018061 (2018)
17. Radchenko, N., Radchenko, A., Tsoy, A., Mikielewicz, D., Kantor, S., Tkachenko, V.: Improving the efficiency of railway conditioners in actual climatic conditions of operation. AIP Conf. Proc. **2285**, 030072 (2020)
18. Shukla, A.K., Singh, O.: Thermodynamic investigation of parameters affecting the execution of steam injected cooled gas turbine based combined cycle power plant with vapor absorption inlet air cooling. Appl. Therm. Eng. **122**, 380–388 (2017)
19. Radchenko, A., Stachel, A., Forduy, S., Portnoi, B., Rizun, O.: Analysis of the efficiency of engine inlet air chilling unit with cooling towers. In: Ivanov V., et al. (eds.) Advances in Design, Simulation and Manufacturing III (DSMIE 2020). Lecture Notes in Mechanical Engineering, pp. 322-331. Springer, Cham (2020)https://doi.org/10.1007/978-3-030-50491-5_31
20. Nag, S., Sharma, P., Gupta, A., Dhar, A.: Experimental study of engine performance and emissions for hydrogen diesel dual fuel engine with exhaust gas recirculation. Int. J. Hydrogen Energy **44**(23), 12163–12175 (2019)
21. Radchenko, N., Trushliakov, E., Radchenko, A., Tsoy, A., Shchesiuk, O.: Methods to determine a design cooling capacity of ambient air conditioning systems in climatic conditions of Ukraine and Kazakhstan. AIP Conf. Proc. **2285**, 030074 (2020)
22. Radchenko, M., Mikielewicz, D., Tkachenko, V., Klugmann, M., Andreev, A.: Enhancement of the operation efficiency of the transport air conditioning system. In: Ivanov V., et al. (eds.) Advances in Design, Simulation and Manufacturing III (DSMIE 2020). Lecture Notes in Mechanical Engineering, pp. 332-342. Springer, Cham (2020)https://doi.org/10.1007/978-3-030-50491-5_32

23. Bohdal, L., Kukielka, L., Świłło, S., Radchenko, A., Kułakowska, A.: Modelling and experimental analysis of shear-slitting process of light metal alloys using FEM, SPH and vision-based methods. In: AIP Conference Proceedings vol. 2078, 020060 (2019)
24. Kornienko, V., Radchenko, R., Konovalov, D., Andreev, A., Pyrysunko, M.: Characteristics of the rotary cup atomizer used as afterburning installation in exhaust gas boiler flue. In: Ivanov V. et al. (eds.) Advances in Design, Simulation and Manufacturing III (DSMIE 2020). Lecture Notes in Mechanical Engineering, pp. 302-311. Springer, Cham (2020)https://doi.org/10.1007/978-3-030-50491-5_29
25. Kornienko, V., Radchenko, M., Radchenko, R., Konovalov, D., Andreev, A., Pyrysunko, M.: Improving the efficiency of heat recovery circuits of cogeneration plants with combustion of water-fuel emulsions. Therm. Sci. **25**(1B), 791–800 (2021)
26. Radchenko, R., Pyrysunko, M., Radchenko, A., Andreev, A., Kornienko, V.: Ship engine intake air cooling by ejector chiller using recirculation gas. In: Tonkonogyi, V. et al. (eds.) Advanced Manufacturing Processes II. InterPartner-2020. Lecture Notes in Mechanical Engineering, pp. 734–743. Springer, Cham (2021). https://doi.org/10.1007/978-3-030-68014-5
27. Kornienko, V. Radchenko, R., Bohdal, Ł., Kukiełka, L., Legutko, S.: Investigation of condensing heating surfaces with reduced corrosion of boilers with water-fuel emulsion combustion. In: Nechyporuk M., et al. (eds.) Integrated Computer Technologies in Mechanical Engineering. Advances in Intelligent Systems and Computing, vol.188, pp. 300–309. Springer, Cham (2021). https://doi.org/10.1007/978-3-030-37618-5
28. MAN Diesel Turbo.: CEAS Engine Calculations. https://marine.man-es.com/two-stroke/ceas. Accessed 22 June 2019
29. Ship & Bunker.: https://shipandbunker.com/prices#VLSFO. Accessed 10 Apr 2020
30. METEOMANZ: http://meteomanz.com/
31. NOAA's: https://www.nodc.noaa.gov/OC5/WOD/datawodgeo.html
32. York YHAP-C Absorption Heat Pumps. York a Johnson Controls Company. p. 12 (2016). http://www.Johnsoncontrols.com
33. Model YPC Two Stage Steam Fired Absorption Chiller Style D/York a Johnson Controls Company. p. 36 (2017). http://www.Johnsoncontrols.com
34. MAN Diesel & Turbo: MAN B&W Two-stroke Marine Engines. Emission Project Guide (2018). https://marine.man-es.com/applications/projectguides/2stroke/content/special_pg/7020-0145-09_uk.pdf
35. Influence of ambient temperature conditions. Main engine operation of MAN B&W two-stroke engines. MAN Diesel & Turbo. Copenhagen, Denmark, p. 17 (2010). http://www.mandieselturbo.com.Solstice®zeRefrigerantHFO-1234ze

Analyzing the Efficiency of Using an Installed Refrigeration Capacity of Ambient Air Conditioning System

Andrii Radchenko[1(✉)], Serhiy Forduy[2], Serhiy Kantor[3], Veniamin Tkachenko[3], and Felix Tsaran[1]

[1] Admiral Makarov National University of Shipbuilding, 9 Heroes of Ukraine Avenue, Mykolayiv 54025, Ukraine
[2] PepsiCo, Inc., Kyiv, Ukraine
[3] PJSC "Zavod" "Ekvator", Mykolaiv, Ukraine

Abstract. One of the most reasonable reserves to improve the efficiency of ambient air conditioning systems is to enable the operation of refrigeration compressors in close to nominal modes by selecting a rational design refrigeration capacity and its distribution in response to the current thermal load according to the actual variable climatic conditions to provide a closed to the maximum annual refrigeration production and matching current air conditioning duties at the same time. The approach to improve the efficiency of using the installed air conditioning systems refrigeration capacity is based on comparing the excess of potentially possible refrigeration production (based on the installed refrigerating capacity) for a certain period with its value needed for precooling the ambient air to a certain threshold (intermediate) temperature, that provides a relatively stable thermal load within subsequent deep air subcooling to a target temperature. The range of a stable thermal load can be provided with operation of a conventional compressor in a mode closed to a nominal value, whereas precooling the ambient air with significant fluctuations in thermal load requires regulation of the refrigeration capacity by application of a variable speed compressor or by using excessive cooling capacity accumulated at reduced thermal load.

Keywords: Ambient air conditioning · Design refrigeration capacity · Current thermal load

1 Introduction

Energy consumption for the processing of ambient air in air conditioning systems (ACS) depends on the ambient air temperature t_a and relative humidity φ_a, which changes significantly during the day. It is obvious that the production of refrigeration by refrigeration machine (RM) according to air conditioning duties over a certain period of time, for example a year (annual refrigeration production) $\sum(Q_0 \cdot \tau)$, [kWh], where: Q_0 – is the current refrigeration capacity of RM (thermal load on the ACS), [kW]; τ – is the duration of the ACS operation in hours, is chosen as a criterion to estimate the efficiency of

M. Nechyporuk et al. (Eds.): ICTM 2021, LNNS 367, pp. 520–531, 2022.
https://doi.org/10.1007/978-3-030-94259-5_44

using the installed refrigeration capacities of the RM. Since the amount of refrigeration capacity spent on ambient air processing in the ACS depends on the current ambient air parameters and the duration of the ACS operation, the efficiency of using the installed refrigeration capacity of ACS can be estimated by comparing its value spent for air conditioning over a certain period of the operation and potentially possible refrigeration production due to operation of RM on full, nominal, load.

2 Literature Review

Many publications are devoted to improving air processing in ACS by intensification of heat transfer processes in air coolers [1, 2], evaporators [3, 4] and condensers [5], application of various refrigerant circulation contours [6, 7], alternate safe refrigerants [8, 9], waste heat recovery technics, including combined cooling, heating and power generation [10, 11], modelling [12, 13], optimization [14, 15], experimental and monitoring [16, 17] methods to match current cooling demands. Some of principal technical innovations and methodological approaches in waste heat recovery refrigeration [2, 18] might be successfully applied for air conditioning [19], in particular, evaporative cooling [20, 21], two-stage air cooling [22, 23].

Numerous researchers have studied the energy efficiency of the Variable Refrigerant Flow (VRF) systems [24, 25] and proposed some practical recommendations. Most of the studies have been conducted on solutions of efficient operation of the VRF system in buildings and control strategies of the systems [26]. A control algorithm of the supply air temperature as a threshold temperature in the outdoor air processing (OAP) unit to run the VRF-OAP system more efficiently for buildings was developed [27]. The control algorithm was conducted to adjust the refrigerant flow supply to the OAP and the indoor unit appropriately through supporting the supply air temperature. Results [28] show that ACS have a great potential for energy saving and the adjustability of VRF.

The VRF system with heat recuperative ventilation [29] and a dedicated outdoor ACS was introduced [30]. The evaluation of indoor thermal-humidity environments and energy consumption of the VRF system [31] with a heat pump was conducted [32].

The author [33] proposes the method to calculate thermal load of building. The VRF systems operate with high part-load efficiency [34], that results into high daily and seasonal energy efficiency, so as ACS typically spend most of their operating hours within the range of 40% to 80% of maximum cooling capacity [35].

The majority of ACS designing methods issue from the assumption to cover the maximum current cooling needs or with some limitation [34–36] or annual maximum cooling consumption [37–40]. Such approach inevitable leads to considerable overestimation of design cooling capacities and systems oversizing. So, the problem to determine a rational design cooling capacity without ACS oversizing needs further solution.

The aim of the study is to develop an approach to analyzing the efficiency of using the installed refrigeration capacity of ACS and method to determine a rational design (installed) refrigeration capacity value and its distribution according to the current climatic conditions to avoid oversizing a refrigeration machine and its enlarged cost.

3 Methodology

The efficiency of ACS and their RM performance depends on their loading and a duration of their yearly operation. Therefore, the annual refrigeration production in response to air conditioning duties is considered as a primary criterion for the choice of a rational design overall thermal load of ACS. For this the current refrigeration capacities, generated by RM at any time period in response to the air conditioning duties for ambient air conditioning down to the target leaving air temperature, have been summarized over the year to determine the rational design overall refrigeration capacity of ACS.

In order to conduct this procedure the authors develop a method of rational designing based on the yearly loading characteristic curve of annual summarized refrigeration production dependence on the design specific refrigeration capacity of the RM to choose its value, that provides closed to maximum annual production of refrigeration.

The specific annual production of refrigeration:

$$\sum(q_0 \cdot \tau) = \sum(\xi \cdot c_{ma} \cdot (t_{amb} - t_{a2}) \cdot \tau) \tag{1}$$

where: $\sum(q_0 \cdot \tau)$ – specific annual production of refrigeration [kg·h/kW]; ξ – specific heat ratio; t_a – ambient air temperature [°C]; t_{a2} – air temperature at the air cooler outlet [°C]; c_{ma} –specific heat of moist air [kJ/(kg·K)]; τ – time interval [h].

The specific refrigeration capacity is calculated as:

$$q_0 = \xi \cdot c_{ma} \cdot (t_a - t_{a2}) \tag{2}$$

A rational specific refrigeration capacity $q_{0.rat}$ is determined to exclude unproductive expenses of refrigeration capacity q_0 caused by oversizing RM without obtaining a noticeable effect in increasing the annual production of refrigeration $\sum(q_0 \cdot \tau)$, that was limited by the value of $(0.95\ldots0.97)\sum(q_0 \cdot \tau)$ to choose corresponding rational specific refrigeration capacity $q_{0.rat}$.

Proceeding from a different behavior of current thermal loads, the ambient air treatment in the ACS is considered as a two-stage processing and includes a range of hermal load fluctuation as the first (high-temperature) stage and a range of comparatively stable thermal load as the second (low-temperature) stage. The threshold air temperature is determined to provide a rational distribution of design overall cooling capacity of ACS between two stages with different thermal load behaviors.

Taking into account a relatively stable behavior of the specific thermal (cooling) load on the air cooler (AC) of the ACS within the range $q_{0.10–15} = q_{0.10} - q_{0.15}$ (or $q_{0.10–17} = q_{0.10} - q_{0.17}$) when the air is cooled from the temperature t_{a2} = 15 (or 17)°C to t_{a2} = 10 °C compared to the ambient air cooling from $t_a$to t_{a2} = 15 (or 17)°C the first thermal load range is taken as a basic design (installed) for deep cooling of air from t_{a2} = 15 (or 17)°C to t_{a2} = 10 °C. Accordingly, a design refrigeration capacity for precooling the ambient air from the current temperature $t_a$to t_{a2} = 15 (or 17)°C, as booster component, is determined by the residual principle as the difference between the design specific refrigeration capacity $q_{0.10rat}$ for the entire process of cooling the ambient air from the current temperature $t_a$to t_{a2} = 10 °C and its stable component $q_{010–15}$: $q_{0.b10–15} = q_{0.10rat} - q_{010–15}$ (or $q_{0.b10–17} = q_{0.10rat} - q_{010–17}$).

Since the fluctuations of the current refrigeration capacity spent for cooling the ambient air from t_a to t_{a2} = 10 °C are caused mainly by its booster part $q_{0.15}$ (or $q_{0.17}$), which corresponds to precooling the ambient air from t_a to t_{a2} = 15 (or 17)°C, at elevated current thermal loads $q_{0.15}$ there is some deficit of the installed booster component $q_{0.b10\text{–}15}$ ($q_{0.b10\text{–}17}$) of refrigeration capacity $q_{0.b10\text{–}15}$ ($q_{0.b10\text{–}17}$), calculated by the residual principle, whereas at reduced current thermal loads $q_{0.15}$ (or $q_{0.17}$), on the contrary, its excess $q_{0.bex10\text{–}15} = q_{0.10rat} - q_{0.b10\text{–}15} = q_{0.10rat} - q_{0.15}$ compared to the current specific thermal loads $q_{0.15}$ and in according $q_{0.bex10\text{–}17}$.

How efficiently the installed booster component of the refrigeration capacity of the ACS is spent on precooling the ambient air from its current temperature t_a to t_{a2} = 15 °C (or 17 °C) with the change of current thermal loads according to the actual climatic conditions can be judged by comparing the total monthly summarized specific refrigeration consumption to cover the current thermal loads $\sum(q_{0.15}\cdot\tau)$ (or $\sum(q_{0.17}\cdot\tau)$) for cooling the ambient air from the current t_a to t_{a2} = 15 °C (or 17 °C) with potentially possible refrigeration production by a booster stage with design specific refrigeration capacity calculated on the residual basis, $\sum(q_{0.b10\text{–}15}\cdot\tau)$ (or $\sum(q_{0.b10\text{–}17}\cdot\tau)$), on the other hand.

4 Results

In order to generalize the results and simplify calculations for any total refrigeration capacities Q_0, it is convenient to present the refrigeration capacity of the RM ACS not in absolute Q_0, but in relative (specific) values per unit air mass flow rate (G_a = 1 kg/s) – in the form of specific refrigeration (cooling) capacity, $q_0 = Q_0 / G_a$, kW/(kg/s), or kJ/kg, where Q_0 is the total refrigeration capacity when cooling the air with the flow rate G_a: $Q_0 = (c_a\cdot\xi\cdot\Delta t_a)G_a$, where $\Delta t_a = t_{amb} - t_{a2}$ – decrease in air temperature.

To justify the approach to the analysis of the efficiency of using the installed (design) refrigeration capacities of ACS chillers taking into account the change in thermal loads according to the current climatic conditions, the current values of specific refrigeration capacity q_0 of RM ACS when cooling the ambient air from the current temperature t_a to t_{a2} = 10, 15 and 17 °C, respectively $q_{0.10}$, $q_{0.15}$ and $q_{0.17}$ for July of 2017 year, Nikolaev region, Ukraine (Fig. 1) have been considered.

As can be seen, when the ambient air is cooled from its current temperatures t_a to t_{a2}, the thermal load fluctuations of $q_{0.10}$, $q_{0.15}$ and $q_{0.17}$ are very significant. The almost equidistant trend lines of the specific thermal load $q_{0.10}$, $q_{0.15}$ and $q_{0.17}$ indicate that these fluctuations are due primarily to changes in the specific thermal load $q_{0.15}$ and $q_{0.17}$ for precooling the ambient air to the temperatures t_{a2} = 15, 17 °C, within which there is practically damping of the fluctuations of the current thermal load.

At the same time, with the further cooling of the air from the intermediate temperature t_{a2} = 15 °C to t_{a2} = 10 °C the fluctuations of the specific thermal loads on the ACS $q_{0.10\text{–}15} = q_{0.10} - q_{0.15}$ are relatively small, without taking into account 3–5 short-term bursts-drops, caused by a decrease in the current values of the ambient air temperature below 15 °C (Fig. 2, *a*).

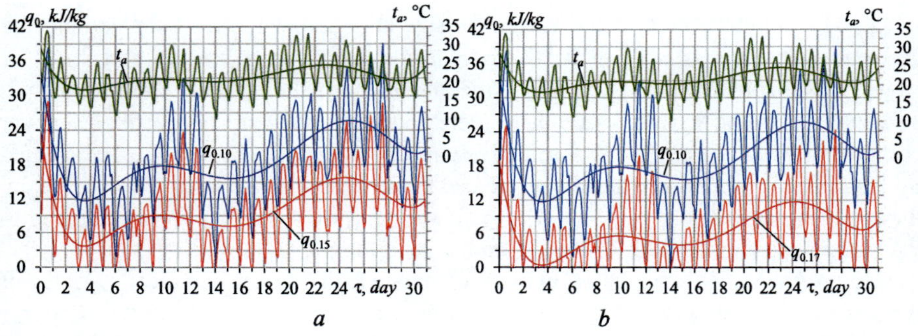

Fig. 1. Current values of ambient air temperature t_a, specific refrigeration capacity $q_{0.10}$, needed for cooling ambient air from t_a to $t_{a2} = 10$ °C, specific refrigeration capacity $q_{0.15}$ and $q_{0.17}$, needed for cooling ambient air from t_a to various intermediate temperatures $t_{a2} = 15$ °C and $t_{a2} = 17$ °C: *a* – $q_{0.15}$ for $t_{a2} = 15$ °C; *b* – $q_{0.17}$ for $t_{a2} = 17$ °C

Obviously, the higher the value of the precooling temperature of the ambient air (intermediate temperature t_{a2}), i. e., the narrower the range of fluctuations of thermal (cooling) loads, the longer the operating life of the refrigeration machine for booster precooling ambient air during the year at current thermal loads.

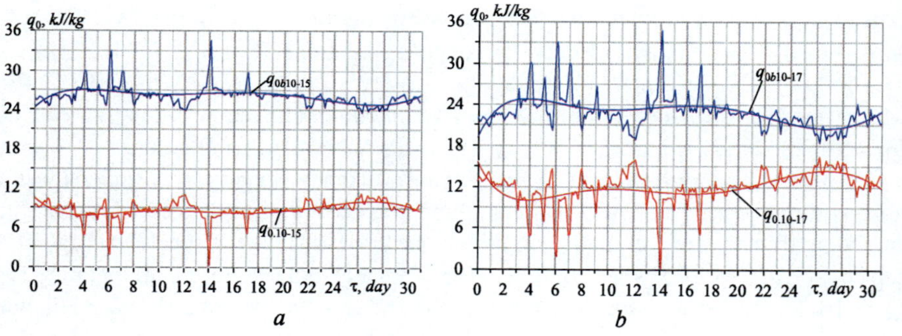

Fig. 2. Current values of booster specific refrigeration capacity $q_{0.b10\text{–}15} = q_{0.10rat} - q_{0.10\text{–}15}$ with basic specific refrigeration capacity $q_{0.10\text{–}15} = q_{0.10} - q_{0.15}$, needed for subcooling air from 15 °C to $t_{a2} = 10$ °C (*a*) and booster specific refrigeration capacity $q_{0.b10\text{–}17} = q_{0.10rat} - q_{0.10\text{–}17}$ with basic specific refrigeration capacity $q_{0.10\text{–}17} = q_{0.10} - q_{0.17}$, needed for subcooling air from 17 °C to $t_{a2} = 10$ °C (*b*) with rational design overall (for two-stage ambient air cooling) specific refrigeration capacity $q_{0.10rat} = 35$ kJ/kg

In order to determine the upper threshold temperature for precooling ambient air, the calculations of the processes of cooling the ambient air from current t_a to a higher intermediate temperature $t_{a2} = 17$ °C were made and the corresponding specific refrigeration capacity for the subsequent subcooling the air $q_{0.10\text{–}17} = q_{0.10} - q_{0.17}$ from $t_{a2} = 17$ °C to $t_{a2} = 10$ °C was determined (Fig. 2, *b*).

As can be seen, the specific refrigeration capacity $q_{0.10–17} = q_{0.10} - q_{0.17}$ for subcooling the air from the intermediate temperature t_{a2} = 17 °C to t_{a2} = 10 °C becomes very unstable compared to the lower intermediate temperature t_{a2} = 15 °C (Fig. 2, *a*). This is caused by an earlier (at t_{a2} = 17 °C) fall to zero of the specific refrigeration capacity $q_{0.17}$ for precooling the ambient air due to the narrowing of the temperature range $\Delta t_{17} = t_a$ – 17 °C of cooling the ambient air at an elevated intermediate temperature t_{a2} = 17 °C (compared to $\Delta t_{15} = t_a$ – 15 °C at t_{a2} = 15 °C), that results in replacing the fluctuations of specific refrigeration capacity, previously damped when cooling the ambient air to the temperature t_{a2} = 15 °C, with corresponding increase in specific refrigeration capacity $q_{0.10–17} = q_{0.10} - q_{0.17}$ for further subcooling air from the temperature t_{a2} = 17 °C to t_{a2} = 10 °C.

Taking into account a relatively stable behavior of the specific thermal load on the air cooler of the ACS within the range $q_{0.10–15} = q_{0.10} - q_{0.15}$ when the air is subcooled from t_{a2} = 15 °C to t_{a2} = 10 °C (Fig. 2,*a*) compared to large fluctuations of refrigeration capacity $q_{0.15}$ within the range of ambient air precooling from t_a to t_{a2} = 15 °C (Fig. 1,*a*), the first thermal load $q_{0.10–15}$ is taken as a design stable component of refrigeration capacity for deep cooling of air from t_{a2} = 15 °C to t_{a2} = 10 °C.

Accordingly, a design refrigeration capacity $q_{0.b10–15}$ for booster precooling the ambient air from the current temperatures t_a to t_{a2} = 15 °C is determined by a residual principle as the difference between a design specific refrigeration capacity $q_{0.10rat}$ for the entire process of cooling the ambient air from the current temperatures t_a to t_{a2} = 10 °C and its stable component $q_{010–15}$: $q_{0.b10–15} = q_{0.10rat} - q_{010–15}$.

With this a rational design refrigeration capacity $q_{0.10rat}$ = 35 kJ/kg for the entire process of cooling the ambient air from the current temperatures t_a to 10 °C is assumed to provide closed to a maximum annual refrigeration production $\sum(q_0 \cdot \tau)$ (Fig. 3).

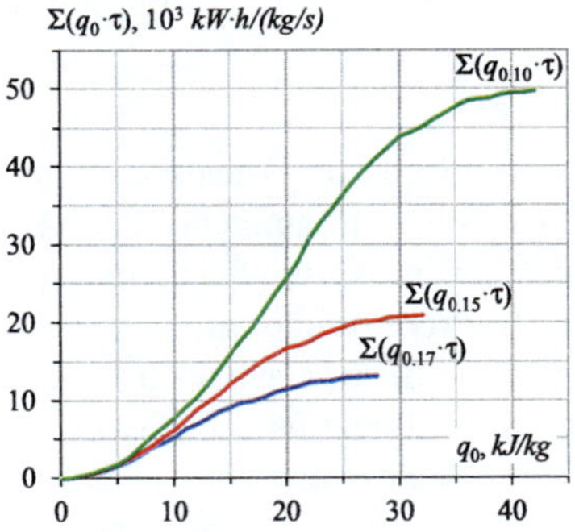

Fig. 3. Specific annual refrigeration production $\sum(q_0 \cdot \tau)$ required for cooling the ambient air to the temperatures t_{a2} = 10, 15 and 17 °C against a design specific refrigeration capacity q_0

As can be seen, for the considered climatic conditions when the air is cooled to the temperature of t_{a2} = 10 °C in the ACS with installed (design) specific (at G_a = 1 kg/s) refrigeration capacity of RM $q_{0.10rat}$ = 35 kJ/kg, which provides close to the maximum annual refrigeration production $\sum(q_0 \cdot \tau) \approx 48 \cdot 10^3$ kWh/(kg/s), while maintaining its increment with a noticeable high rate. Similarly, for cooling ambient air to the temperature of t_{a2} = 15 °C the rational value of specific refrigeration capacity of RM $q_{0.15rat}$ = 25 kJ/kg, and to t_{a2} = 17 °C – the value of $q_{0.17rat}$ = 22·kJ/kg.

Current values of booster specific refrigeration capacity $q_{0.b10–15} = q_{0.10rat} - q_{0.10–15}$ and $q_{0.b10–17} = q_{0.10rat} - q_{0.10–17}$ and booster specific refrigeration capacity excess $q_{0.bex10–15} = q_{0.b10–15} - q_{0.15}$ and $q_{0.bex10–17} = q_{0.b10–17} - q_{0.17}$ with subcooling air from t_{a2} = 15 °C and t_{a2} = 17 °C accordingly to t_{a2} = 10 °C with rational design overall (for two-stage ambient air cooling) specific refrigeration capacity $q_{0.10rat}$ = 35 kJ/kg (according to Fig. 3) are presented in Fig. 4.

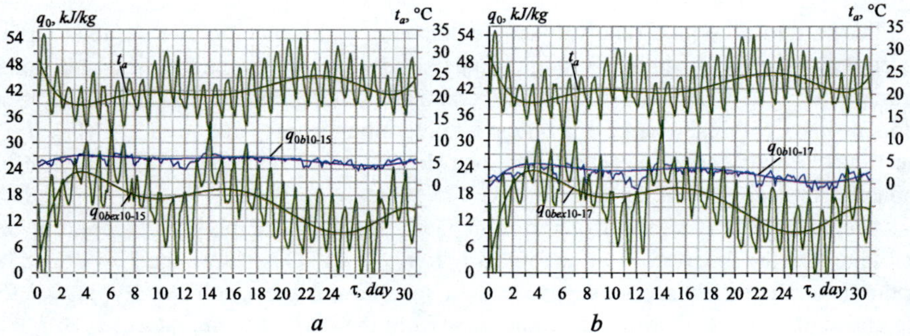

Fig. 4. Current values of ambient air temperature t_a, booster specific refrigeration capacity $q_{0.b10–15} = q_{0.10rat} - q_{0.10–15}$ and booster specific refrigeration capacity excess $q_{0.bex10–15} = q_{0.b10–15} - q_{0.15}$ with subcooling air from t_{a2} = 15 °C to t_{a2} = 10 °C (*a*) and booster specific refrigeration capacity $q_{0.b10–17} = q_{0.10rat} - q_{0.10–17}$ and booster specific refrigeration capacity excess $q_{0.bex10–17} = q_{0.b10–17} - q_{0.17}$ with subcooling air from t_{a2} = 17 °C to t_{a2} = 10 °C (*b*) with rational design overall specific refrigeration capacity $q_{0.10rat}$ = 35 kJ/kg according to Fig. 3

As Fig. 4 shows, the excess of the current values of the specific refrigeration capacity $q_{0.b10–15} = 35 - q_{010–15}$ for precooling the ambient air from its current temperature t_a to t_{a2} = 15 °C above its design value drops from the current specific refrigeration capacity $q_{0.15}$ to zero, and the fact that the excess of the current values of the specific refrigeration capacity $q_{0.b10–15}$ in some days (4, 6, 7, 14 and 17.07.2017) exceeds the design value, is explained by the reduction of refrigeration capacity $q_{0.10–15}$ for deep cooling the air while reducing the current ambient air temperature t_a below 15 °C.

Potentially possible monthly summarised refrigeration production for precooling the ambient air to the intermediate temperature t_{a2} = 15 °C in a booster stage of two-stage air cooler available in accordance with its design specific refrigeration calculated on a residual basis, $\sum(q_{0.b10–15}\cdot\tau)$, its potential excess $\sum(q_{0.bexc10–15}\cdot\tau) = \sum(q_{0.b10–15}\cdot\tau) - \sum(q_{0.15}\cdot\tau)$ over the current values of the specific refrigeration spent for precooling the ambient air to t_{a2} = 15 °C in a booster stage (refrigeration consumption) monthly summarised $\sum(q_{0.15}\cdot\tau)$ and monthly summarised refrigeration consumption $\sum(q_{0.10–15}\cdot\tau)$ for deep subcooling the air from t_{a2} = 15 °C to t_{a2} = 10 °C is plotted in Fig. 5, *a*, as well as their values for precooling the ambient air to an increased intermediate temperature $t_{a2} = 17$ °C are presented in Fig. 5, *b*.

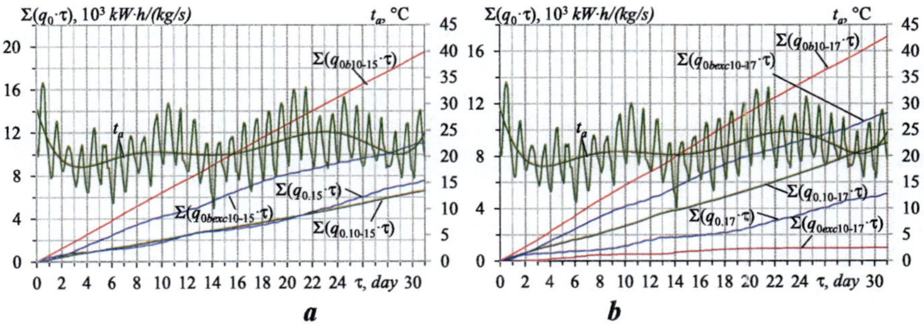

Fig. 5. Current values of total summarized per month specific refrigeration consumption $\sum(q_{0.15}{\cdot}\tau)$ for precooling the ambient air to 15 °C, refrigeration spent for subcooling air from 15 °C to 10 °C, $\sum(q_{0.10\text{–}17}{\cdot}\tau)$, potentially possible refrigeration production for precooling ambient air $\sum(q_{0.b10\text{–}15}{\cdot}\tau)$ according to a design refrigeration capacity of booster stage, potentially possible excess of refrigeration production for precooling ambient air in a booster stage $\sum(q_{0.bexc10\text{–}15}{\cdot}\tau) = \sum(q_{0.b10\text{–}15}{\cdot}\tau) - \sum(q_{0.15}{\cdot}\tau)$ (*a*) and total monthly summarized specific refrigeration consumption $\sum(q_{0.17}{\cdot}\tau)$ for precooling the ambient air to the intermediate temperature 17 °C, refrigeration spent for subcooling air from 17 °C to 10 °C, $\sum(q_{0.10\text{–}17}{\cdot}\tau)$, potentially possible refrigeration production for precooling ambient air $\sum(q_{0.b10\text{–}17}{\cdot}\tau)$ according to a design refrigeration capacity of booster stage, potentially possible excess of refrigeration production for precooling ambient air in a booster stage $\sum(q_{0.bexc10\text{–}17}{\cdot}\tau) = \sum(q_{0.b10\text{–}17}{\cdot}\tau) - \sum(q_{0.17}{\cdot}\tau)$ and $\sum(q_{0exc10\text{–}17}{\cdot}\tau) = \sum(q_{0.10\text{rat}}{\cdot}\tau) - \sum(q_{0.17\text{rat}}{\cdot}\tau) - \sum(q_{0.10\text{–}17}{\cdot}\tau)$

An elevated intermediate temperature $t_{a2} = 17$ °C of ambient air precooling does not provide stabilization of thermal loading of subsequent subcooling the air from increased intermediate temperature $t_{a2} = 17$ °C down to the target temperature $t_{a2} = 10$ °C that leads to arising an excess of refrigeration production $\sum(q_{0.bexc10\text{–}17}{\cdot}\tau) = \sum(q_{0.b10\text{–}17}{\cdot}\tau) - \sum(q_{0.17}{\cdot}\tau)$ compared to its consumption $\sum(q_{0.10\text{–}17}{\cdot}\tau)$ $q_{0.10\text{–}17} = q_{0.10\text{–}17} = q_{0.10} - q_{0.17}$ when cooling the air from $t_{a2} = 17$ °C to $t_{a2} = 10$ °C (Fig. 5, *b*).

As can be seen, the specific refrigeration capacity $q_{0.10\text{–}17} = q_{0.10} - q_{0.17}$ for subcooling the air from the increased intermediate temperature $t_{a2} = 17$ °C to the target $t_{a2} = 10$ °C becomes very unstable compared to the lower intermediate temperature $t_{a2} = 15$ °C (Fig. 2, *a*). This is caused by an earlier (at $t_{a2} = 17$ °C) fall to zero of the refrigeration capacity $q_{0.17}$ for precooling the ambient air due to the narrowing of the temperature range $\Delta t_{17} = t_a - 17$ °C of cooling the ambient air at an elevated intermediate temperature $t_{a2} = 17$ °C (compared to $\Delta t_{15} = t_a - 15$ °C at $t_{a2} = 15$ °C), that results in replacing the fluctuations of specific refrigeration capacity, previously damped when cooling the ambient air to the temperature $t_{a2} = 15$ °C, with corresponding increase in specific refrigeration capacity $q_{0.10\text{–}17} = q_{0.10} - q_{0.17}$ for further subcooling air from the temperature $t_{a2} = 17$ °C to $t_{a2} = 10$ °C.

Comparison of refrigeration capacity of air subcooling $q_{0.10\text{–}15} = q_{0.10} - q_{0.15}$ and $q_{0.10\text{–}17} = q_{0.10} - q_{0.17}$ shows that the stabilization of thermal load occurs as a result of precooling the ambient air to the lower intermediate temperature $t_{a2} = 15$ °C compared to $t_{a2} = 17$ °C. Therefore, the intermediate temperature $t_{a2} = 15$ °C is assumed as a threshold value and specific refrigeration capacity $q_{0.10\text{–}15} = q_{0.10} - q_{0.15}$ is taken as the base, which ensures the effective operation of the RM in the mode close to the nominal.

As can be seen, the total consumption of the specific refrigeration generation per month $\sum(q_{0.15}\cdot\tau)$ for cooling the ambient air to $t_{a2} = 15$ °C is $\sum(q_{0.15}\cdot\tau) \approx$ 7 kW·h/(kg/s), which is less than the excess of the potential refrigeration production for a precooling of the ambient air (to $t_{a2} = 15$ °C) $\sum(q_{0.bexc10\text{–}15}\cdot\tau) \approx 11$ kW·h/(kg/s) and is less 40% of the potential refrigeration output of booster stage $\sum(q_{0.b10\text{–}15}\cdot\tau) \approx$ 19 kW·h/(kg/s).

In the first approximation, we can assume that the total per month consumption of the specific refrigeration is $\sum(q_{0.15}\cdot\tau) \approx 7$ kW·h/(kg/s), which is 40% of the potentially possible refrigeration production of booster stage $\sum(q_{0b10\text{–}15}\cdot\tau) \approx 19$ kW·h/(kg/s), corresponds to the required refrigeration capacity range from 100% to 40% of nominal. Based on this hypothesis, the use of refrigeration compressors with a frequency regulation to 40% could be considered as a way of energy-efficient operation for precooling ambient air to $t_{a2} = 15$ °C.

At the same time, the total per month consumption of specific refrigeration $\sum(q_{0.17}\cdot\tau)$ for cooling the ambient air to $t_{a2} = 17$ °C is about $\sum(q_{0.17}\cdot\tau) \approx 5$ kW·h/(kg/s), which is much less than the excess of the potential refrigeration production according to a design refrigeration capacity of a booster stage for precooling the ambient air (to $t_{a2} = 17$ °C) $\sum(q_{0bexc10\text{–}17}\cdot\tau) \approx 11{,}5$ kW·h/(kg/s) over the actual refrigeration needed for precooling the ambient air to $t_{a2} = 17$ °C.

Insufficient damping of the drop in the current thermal load within precooling the ambient air to $t_{a2} = 17$ °C (compared to $t_{a2} = 15$ °C) leads to arising the excess of design refrigeration production $\sum(q_{0exc10\text{–}17}\cdot\tau) \approx 1$ kW·h/(kg/s), though insignificant in comparison with its consumption $\sum(q_{010\text{–}17}\cdot\tau) \approx 9$ kW·h/(kg/s) within further subcooling the air from the temperature $t_{a2} = 17$ °C to $t_{a2} = 10$ °C. Obviously, the use of compressors with frequency regulation of refrigeration capacity can be effective in the range of thermal loads $q_{010\text{–}17}$ for subcooling the air to $t_{a2} = 10$ °C, whereas the damping of booster refrigeration capacity $q_{0b10\text{–}17}$ for precooling the ambient air to $t_{a2} = 17$ °C their application is possible only in conjunction with the accumulation of excessive refrigeration.

Thus, the proposed approach to analyzing the efficiency of using the installed refrigeration capacity of the ACS can be applied both to choose its rational distribution depending on the character of the change in actual thermal load, and to determine the ranges of thermal loads for the effective application of different methods of managing the ambient air processing in the ACS.

5 Conclusions

A novel method to determine a rational design refrigeration capacity to provide a closed to maximum annual refrigeration production according to conditioning duties at reduced by about 20% design refrigeration capacity compared with traditional designing assumption to cover the maximum current cooling needs has been developed.

Proceeding from a different behavior of current thermal loads, the ambient air processing in ACS is proposed to consider as a two-stage process that includes a range of thermal load fluctuation as the first (high-temperature) stage and a range of comparatively stable thermal load as the second (low-temperature) stage.

The method of rational design refrigeration capacity and its distribution is quite useful for a comfort air conditioning, as well as for cooling air at the inlet of combustion engines to enhance their power output (electricity production) in trigeneration systems. It provides the effective application of energy saving methods, in particular, to cover increased thermal loads by accumulated excessive refrigeration and by application of compressors with variable refrigeration capacity to cover changeable loads.

References

1. Marques, R.P., Hacon, D., Tessarollo, A., Parise, J.A.R.: Thermodynamic analysis of trigeneration systems taking into account refrigeration, heating and electricity load demands. Energy Build. **42**, 2323–2330 (2010)
2. Eidan, A.A., Alwan, K.J.: Enhancement of the performance characteristics for air-conditioning system by using direct evaporative cooling in hot climates. Energy Proc. **142**, 3998–4003 (2017)
3. Kornienko, V., Radchenko, M., Radchenko, R., Konovalov, D., Andreev, A., Pyrysunko, M.: Improving the efficiency of heat recovery circuits of cogeneration plants with combustion of water-fuel emulsions. Therm. Sci. **25**(1B), 791–800 (2021)
4. Alahmer, A., Alsaqoor, S.: Simulation and optimization of multi-split variable refrigerant flow systems. Ain Shams Eng. J. **9**(4), 1705–1715 (2017)
5. Radchenko, A., Trushliakov, E., Kosowski, K., Mikielewicz, D., Radchenko, M.: Innovative turbine intake air cooling systems and their rational designing. Energies **13**(23), 6201 (2020). https://doi.org/10.3390/en13236201
6. Butrymowicz, D., Gagan, J., Śmierciew, K., Łukaszuk, M., Dudar, A., Pawluczuk, A., Łapiński, A., Kuryłowicz, A.: Investigations of prototype ejection refrigeration system driven by low grade heat. In: Proceedings, HTRSE-2018, E3S Web of Conferences, vol. 70, 03002 (2018). https://doi.org/10.1051/e3sconf/20187003002
7. Konovalov, D., Kobalava, H., Radchenko, M., Scurtu, I.C., Radchenko, R.: Determination of hydraulic resistance of the aerothermopressor for gas turbine cyclic air cooling. In: TE-RE-RD 2020, E3S Web of Conferences, 18074343
8. Kuznetsov, V., Dymo, B., Kuznetsova, S., Bondarenko, M., Voloshyn, A.: Improvement of the cargo fleet vessels power plants ecological indexes by development of the exhaust gas systems. Pol. Maritime Res. **28**(1), 97–104 (2021)
9. Radchenko, R., Pyrysunko, M., Radchenko, A., Andreev, A., Kornienko, V.: Ship engine intake air cooling by ejector chiller using recirculation gas. In: Tonkonogyi, V. et al. (eds.) Advanced Manufacturing Processes II. InterPartner-2020. Lecture Notes in Mechanical Engineering, pp. 734–743. Springer, Cham (2021). https://doi.org/10.1007/978-3-030-68014-5
10. Volintiru, O.N., Scurtu, I.C., Stefănescu1, T.M.: Modeling and optimization of HVAC system for special ships. J. Phys. Conf. Ser. 1122, art no 012004 (2018)
11. Trushliakov, E., Radchenko, A., Radchenko, M., Kantor, S., Zielikov, O.: The efficiency of refrigeration capacity regulation in the ambient air conditioning systems. In: Ivanov V., et al. (eds.) Advances in Design, Simulation and Manufacturing III (DSMIE 2020). Lecture Notes in Mechanical Engineering, pp. 343–353. Springer, Cham (2020). https://doi.org/10.1007/978-3-030-50491-5

12. Cherednichenko, O., Serbin, S., Dzida, M.: Investigation of the combustion processes in the gas turbine module of an fpso operating on associated gas conversion products. Pol. Maritime Res. **26**(4), 149–156 (2020)
13. Radchenko, M., Radchenko, R., Tkachenko, V., Kantor, S., Smolyanoy, E.: Increasing the operation efficiency of railway air conditioning system on the base of its simulation along the route line. In: Nechyporuk, M., et al. (eds.) Integrated Computer Technologies in Mechanical Engineering (ICTM 2019). Advances in Intelligent Systems and Computing (2020), vol. 1113, pp. 461–467. Springer, Cham (2020). https://doi.org/10.1007/978-3-030-37618-5
14. Bohdal, L., Kukiełka, L., Legutko, S., Patyk, R., Radchenko, A.M.: Modeling and experimental analysis of shear-slitting of AA6111-T4 aluminum alloy sheet. Materials **13**(14), 3175 (2020)
15. Radchenko, R., Pyrysunko, M., Kornienko, V., Scurtu, I.-C., Patyk, R.: Improving the ecological and energy efficiency of internal combustion engines by ejector chiller using recirculation gas heat. In: Nechyporuk, M., et al. (eds.) Integrated Computer Technologies in Mechanical Engineering. Advances in Intelligent Systems and Computing, vol. 188, pp. 531–541. Springer, Cham (2021). https://doi.org/10.1007/978-3-030-37618-5
16. Shukla, A.K., Singh, O.: Thermodynamic investigation of parameters affecting the execution of steam injected cooled gas turbine based combined cycle power plant with vapor absorption inlet air cooling. Appl. Therm. Eng. **122**, 380–388 (2017)
17. Sur, A., Das, R.K., Sah, R.P.: Influence of initial bed temperature on bed performance of an adsorption refrigeration system. Therm. Sci. **22**(6A), 2583–2595 (2018)
18. Trushliakov, E., Radchenko, A., Forduy, S., Zubarev, A., Hrych, A.: Increasing the operation efficiency of air conditioning system for integrated power plant on the base of its monitoring. In: Nechyporuk, M., et al. (eds.) Integrated Computer Technologies in Mechanical Engineering (ICTM 2019). Advances in Intelligent Systems and Computing, vol. 1113, pp. 351–360. Springer, Cham (2020). https://doi.org/10.1007/978-3-030-37618-5
19. Radchenko, M., Mikielewicz, D., Tkachenko, V., Klugmann, M., Andreev, A.: Enhancement of the operation efficiency of the transport air conditioning system. In: Ivanov V., et al. (eds.) Advances in Design, Simulation and Manufacturing III (DSMIE 2020). Lecture Notes in Mechanical Engineering, pp. 332–342. Springer, Cham (2020). https://doi.org/10.1007/978-3-030-50491-5
20. Im, P., Mini, M., Munk, J. D., Lee, J.: Cooling season full and part load performance evaluation of variable refrigerant flow (VRF) system using an occupancy simulated research building. In: International Refrigeration and Air Conditioning Conference. Paper 1788 (2016). http://docs.lib.purdue.edu/iracc/1788
21. Konovalov, D., Kobalava, H., Radchenko, M., Scurtu, I.C., Radchenko, R.: Determination of hydraulic resistance of the aerothermopressor for gas turbine cyclic air cooling. In: TE-RE-RD 2020, E3S Web of Conferences 180, 01012, p. 14 (2020)
22. Khatri, R., Joshi, A.: Energy performance comparison of inverter based variable refrigerant flow unitary AC with constant volume unitary AC. Energy Procedia **109**, 18–26 (2017)
23. Konovalov, D., Kobalava, H., Maksymov, V., Radchenko, R., Avdeev, M.: Experimental research of the excessive water injection effect on resistances in the flow part of a low-flow aerothermopressor. In: Ivanov V., et al. (eds.) Advances in Design, Simulation and Manufacturing III (DSMIE 2020). Lecture Notes in Mechanical Engineering, pp. 292–301. Springer, Cham (2020). https://doi.org/10.1007/978-3-030-50491-5
24. Zhang, L., Wang, Y., Meng, X.: Qualitative analysis of the cooling load in the typical room under continuous and intermittent runnings of air-conditioning. Procedia Eng. **205**, 405–409 (2017)
25. Lee, J.H., Yoon, H.J., Im, P., Song, Y.–H.: Verification of energy reduction effect through control optimization of supply air temperature in VRF-OAP system. Energies 11(1) (2018)

26. Kornienko, V., Radchenko, R., Bohdal, Ł, Kukiełka, L., Legutko, S.: Investigation of condensing heating surfaces with reduced corrosion of boilers with water-fuel emulsion combustion. In: Nechyporuk, M., et al. (eds.) Integrated Computer Technologies in Mechanical Engineering. Advances in Intelligent Systems and Computing, vol. 188, pp. 300–309. Springer, Cham (2021). https://doi.org/10.1007/978-3-030-37618-5
27. Liu, C., Zhao, T., Zhang, J.: Operational electricity consumption analyze of VRF air conditioning system and centralized air conditioning system based on building energy monitoring and management system. Procedia Eng. **121**, 1856–1863 (2015)
28. Fengxia, H., Zhongbin, Z., Hu, H., Zemin, C.: Experimental study on the all-fresh-air handling unit with exhaust air energy recovery. Energy Procedia **152**, 431–437 (2018)
29. Park, D.Y., Yun, G., Kim, K.S.: Experimental evaluation and simulation of a variable refrigerant-flow (VRF) air-conditioning system with outdoor air processing unit. Energy Build. **146**, 122–140 (2017)
30. Kornienko, V., Radchenko, R., Mikielewicz, D., Pyrysunko, M., Andreev, A.: Improvement of characteristics of water-fuel rotary cup atomizer in a boiler. In: Tonkonogyi, V. et al. (eds.) Advanced Manufacturing Processes II. InterPartner 2020. Lecture Notes in Mechanical Engineering, pp. 664–674. Springer, Cham (2021). https://doi.org/10.1007/978-3-030-68014-5
31. Aynur, T.N., Hwang, Y., Radermacher, R.: Integration of variable refrigerant flow and heat pump desiccant systems for the cooling season. Appl. Therm. Eng. **30**, 917–927 (2010)
32. Aynur, T.N., Hwang, Y., Radermacher, R.: Field performance measurements of a heat hump desiccant unit in dehumidification mode. Energy Build. **40**, 2141–2147 (2008)
33. Sait, H.H.: Estimated thermal load and selecting of suitable air-conditioning systems for a three story educational building. Procedia Comput. Sci. **19**, 636–645 (2013)
34. Radchenko, M., Radchenko, R., Tkachenko, V., Kantor, S., Smolyanoy, E.: Increasing the operation efficiency of railway air conditioning system on the base of its simulation along the route line. In: Nechyporuk M. at al. (eds) Integrated Computer Technologies in Mechanical Engineering (ICTM 2019). Advances in Intelligent Systems and Computing (2020), vol. 1113, pp. 461–467. Springer, Cham (2020). https://doi.org/10.1007/978-3-030-37618-5
35. Kalhori, S.B., Rabiei, H., Mansoori, Z.: Mashad trigeneration potential–An opportunity for CO2 abatement in Iran. Energy Conv. Manag. **60**, 106–114 (2012)
36. Cardona, E., Piacentino, A.: A methodology for sizing a trigeneration plant in mediterranean areas. Appl. Thermal Eng. **23**, 15 (2003)
37. Rodriguez-Aumente, P.A., Rodriguez-Hidalgo, M.C., Nogueira, J.I., Lecuona, A., Venegas, M.C.: District heating and cooling for business buildings in Madrid. Appl. Therm. Eng. **50**, 1496–1503 (2013)
38. Freschi, F., Giaccone, L., Lazzeroni, P., Repetto, M.: Economic and environmental analysis of a trigeneration system for food-industry: A case study. Appl. Energy **107**, 157–172 (2013)
39. Oktay, Z., Coskun, C., Dincer, I.: A new approach for predicting cooling degree-hours and energy requirements in buildings. Energy **36**(8), 4855–4863 (2011)
40. Forsyth, J.L.: Gas turbine inlet air chilling for LNG. In: IGT International Liquefied Natural Gas Conference Proceedings, vol. 3, pp. 1763–1778 (2013)

Efficiency of Gas Turbine Cyclic Air Cooling by Thermopressor

Halina Kobalava, Dmytro Konovalov(✉), Mykola Radchenko, Serhiy Kantor, and Olena Girzheva

Admiral Makarov National University of Shipbuilding, 9, Heroes of Ukraine Avenue, Mykolayiv 54025, Ukraine

Abstract. The use of thermo-gas-dynamic compression (thermopression) in the cyclic air-cooling systems of power plants based on gas turbine engines and internal combustion engines makes it possible to combine two processes: contact air cooling and pressure increase, which reduce the cost of compressor power. A thermopressor is used to implement these processes. The water spraying systems are analyzed and a scheme for improving the charge air cooling systems for ship engines is developed, which consists in the use of overheated water for injection into the thermopressor. This technology makes it possible to increase the relative increase in air pressure at the thermopressor outlet by 5–8%, with a corresponding decrease in the engine turbocharger power.

Keywords: Superheated steam · Injection · Dispersed flow · Intercooling · Cyclic air

1 Introduction

In order to search for reserves to increase the efficiency of heat utilization of power plants by heat-using circuits based on both thermal use of refrigeration machines and methods and methods of cooling cyclic, charge air or air at the inlet, one can turn to the effect of thermo-gas-dynamic compression (thermopression). The effect of thermo-gas-dynamic compression is to increase the pressure while decreasing the temperature in the process of evaporation of a finely dispersed liquid, injected into a stream of vapor (gas) moving at a speed of about sound.

2 Literature Review

In technology, processes are widely used in which the movement of gas through channels occurs under various external influences. These include a change in the cross-sectional area of the channel, the exchange of energy with the environment in the form of a transfer of mechanical energy or heat, friction into the channel walls, a change in the gas flow rate as a result of the supply of liquid into the flow, the process of mechanical and thermal interaction of liquid droplets with a gas flow, etc.

M. Nechyporuk et al. (Eds.): ICTM 2021, LNNS 367, pp. 532–540, 2022.
https://doi.org/10.1007/978-3-030-94259-5_45

Intensive heat supply causes an increase in aerodynamic resistance, and removal - its decrease [1, 2]. With intensive heat removal and appropriate organization of the working process, it is possible not only to significantly reduce the resistance, but also to increase the total pressure in the gas flow. In this case, due to the overwhelming thermal effect (heat removal), the gas flow is compressed [3].

Heat removal can be carried out by evaporative cooling of the coolant injected into the gas stream. The possibility of the process with an increase in the total pressure of the flow during evaporative cooling was first theoretically shown in [4].

The device, in which due to the removal of heat from the gas flow, an increase in the total gas pressure occurs, was called a thermopressor (aerothermopressor) [1–3]. Heat removal can be carried out by contact heat exchange through the channel walls and by evaporative cooling of the cooling water injected into the gas stream.

Questions of the theory of the working process, design and experimental research of a thermopressor are widely presented in works [5–7].

In a real thermopressor, the gas expansion process takes place in a well-profiled snot in an almost adiabatic manner. In the narrow part of the nozzle, where the gas flow moves at a speed of (0.5–0.9) M, a flow of cooling water is supplied, mechanically finely atomized. In the area of evaporation, the interaction of these flows takes place, it is expressed in the acceleration and fragmentation of droplets and some deceleration of the gas flow, heating and evaporation of droplets, and cooling of the gas. A two-phase flow is formed, moving at a high speed, in which the processes of heat and mass transfer, changes in the composition of the vapor-gas-liquid mixture and all flow parameters take place. These processes, to one degree or another, continue in the diffuser, where there is a general deceleration of the flow and an increase in static pressure.

Consideration of the operation of a real thermopressor, carried out by a number of authors [1–3], shows that when the vapor-gas-liquid mixture moves in the evaporation section, three characteristic modes can be distinguished.

1. Mode I: the influence of the resistance of liquid droplets prevails over the positive effect of evaporation and determines the behavior of the gas flow. There is an increase in speed, a decrease in static pressure, and the temperature of the liquid rises, approaching the saturation temperature.
2. Mode II: the decisive process is the evaporation of the liquid. The flow rate decreases and the total and static pressures increase. In this mode, the retarding action of the gas into the liquid affects, since in a certain area the liquid can have a velocity greater than the velocity of the gas. Evaporation rate gradually decreases. This is due to the fact that the surface area of the droplets decreases and the difference in the velocity and temperature of the phases decreases.
3. Mode III: surface friction (in modes I and II was relatively insignificant) becomes the predominant factor.

It is advisable to organize the working process of a real aerothermopressor in such a way as to minimize energy losses in mode I and turn off mode III completely. The flow must enter the diffuser immediately after the end of mode II, or mode II must also apply to the diffuser. In the diffuser, the velocity of the vapor-gas mixture decreases and the static pressure rises. It is important that this process takes place with minimal aerodynamic drag – for this it can be proposed to inject more liquids than is necessary for evaporation, thus ensuring incomplete evaporation.

To date, quite a lot of experience has been accumulated in the use of a thermopressor (aerothermopressor) and technologies based on it for cooling exhaust gases and air of power plants based on internal combustion engines and gas turbines [6, 7].

3 Research Methodology

The droplet diameter has a great influence on the length of the evaporation section and the speed regime: with a decrease in the initial droplet size, the length of the evaporation section and friction losses decrease significantly.

The necessary effective atomization can be achieved by increasing the Mach number M at the entrance to the evaporation chamber or by using special nozzles at sufficiently high pressures of the liquid injected. In [2], a combined spraying system was tested and showed a positive effect, in which the main part of the leads was introduced into the flow at a very low speed, and the other (5–8%) through the nozzle.

Water injection increases the coefficient of friction by 10–20%, and in order to avoid a subsequent increase in losses, it is necessary to reduce the aerodynamic resistance of the structural elements of the injection system, which are located inside the channel (flow path). For this purpose, it is necessary to place a device for injecting liquid into a stream with a low gas velocity (in front of the nozzle – a confuser) and their more optical form.

An improved design for liquid atomization is difficult, especially at high gas flow rates. Today, injection devices used in thermopressors are simple designs for pneumatic atomization of a liquid supplied to a gas flow at a low speed, or low-pressure centrifugal nozzles with a relatively coarse (ineffective) cut [2, 3].

The calculation of the characteristics of the thermopressor (air pressure P_{tp}, air temperature T_{tp}, air speed w_{air}, water speed w_w, etc.) and its basic geometric parameters of the flow path under the given climatic conditions of the ship's navigation, as well as the operational characteristics of the main ship engine was carried out according to the methods [6, 7]. The calculation took into account the pressure loss in the nozzle (confuser), working chamber, diffuser, as well as injecting water with the frontal resistance of drops.

The study uses a computational model developed by the authors for the use of thermopressors in gas turbine cycles, which takes into account the features of the calculation of gas turbine cycles and schemes, the features of thermo-gas-dynamic compression of gases (cyclic air) and the calculation of the processes of air thermal content during contact heat exchange. On the basis of a computational model, a software package has been developed that allows one to determine the parameters of air and working fluid (gas) at characteristic points of the cycle, the parameters of the fuel-energy and thermodynamic efficiency of a gas turbine engine [8, 9].

The calculation of the cycles of gas turbine operations was carried out according to well-known methods [10–12]. Calculation of heat and humidity processes (cooling, air humidification and heating of cyclic air) was carried out by classical dependences taking into account contact heat transfer in gas turbine and gas-steam turbine plants [13, 14].

4 Results

As mentioned earlier, the diameter of the liquid droplet rather strongly affects the length of the evaporation section and the speed regime: with a decrease in the initial droplet size, the length of the evaporation section and friction losses decrease significantly.

Water injection increases the coefficient of friction by 10–20%, therefore, in order to avoid a further increase in losses, it is necessary to reduce the aerodynamic resistance of the structural elements of the system. The use of improved systems for supplying and spraying water that meet these requirements will significantly increase the effect of thermo-gas-dynamic compression at a relatively low cost of water.

One of the promising modern ways to improve the efficiency of spraying is the use of overheated water, for example, when implementing water injection at the compressor inlet of a gas-steam turbine plant, it can increase the efficiency of the plant by 1–2% [15, 16]. In thermopressor systems, the spraying of a liquid superheated relative to the saturation temperature was previously considered.

For modern marine medium-speed engines (MSE), as a rule, a multi-circuit cooling system is used. At the same time, two cooling sections are used in the charge air cooler (CAC): high-temperature, in which heat is removed from the air to the water of the engine cooling system, and low-temperature, with heat removal to the fresh water circuit of the central cooler.

The thermo-pressor is installed by the turbocharger in front of the CAC. This solution makes it possible to exclude the high-temperature section from the CAC, leaving only the low-temperature section for aftercooling the air in front of the receiver to a predetermined temperature of 45 °C. In order to improve the quality of water spraying in the aerothermal pressure system, it is proposed to preheat water in the PV heat exchanger to a temperature of 85–90 °C due to the heat of the water leaving the engine cooling system. The next overheating of water to a temperature of 100–120 °C is carried out due to the utilization of the heat of waste gases in a recovery boiler (RB). In addition to reducing energy costs for organizing the spraying process itself, this allows part of the low-grade heat to be returned to the power plant cycle together with water injections, which, in turn, makes it possible to increase the heat utilization factor.

Features of the process of contact cooling in the charge air thermopressor with superheated water are as follows. During contact cooling of an air stream with superheated water, a number of processes take place, which mutually influence each other. The processes of heat and moisture exchange between air and water are decisive. Depending on the relationship between the vapor content of saturated air near a water droplet and the vapor content in the volume of air, either evaporation or condensation occurs [17, 18]. In the process of contact of air with drops of stubborn superheated water, a layer of saturated air with the temperature of water is formed at the surface of the drop. As a result, if the air is unsaturated, a driving force of moisture exchange appears, in which the vapor that has formed around the surface of the drop passes into the nearby air layers and then under the influence of diffusion spreads in the total volume [19, 20].

A special software package was developed on the basis of known methods for calculating thermopressor devices, as well as taking into account the peculiarities of the process of contact cooling of charge air with superheated water. The calculation of the thermopressor system was carried out for the main ship's cleaning and cleaning system of the "Wartsila" company, brand 6L20 ($N_e = 1200$ kW, $n = 1000$ rpm).

The injection of superheated water (temperature 120 °C) allows to minimize or completely exclude the mode and from the working process of the thermopressor (the drag of the drop in the air flow during acceleration or deceleration prevails) (Fig. 1). This is due to the fact that water droplets immediately begin to intensively evaporate (there is no heating to the saturation temperature). As you can see, the process of increasing pressure is carried out from section I, and not II, reduces the pressure loss due to the influence of the drag of liquid droplets. The pressure increase in the thermopressor without injection of superheated water is $\Delta P_{tp} = 0.031$ MPa (9%). Reducing the mode and in the working process of the thermopressor makes it possible to additionally increase the pressure by another 0.025 MPa (6%), the total pressure increase will be 0.056 MPa (15%).

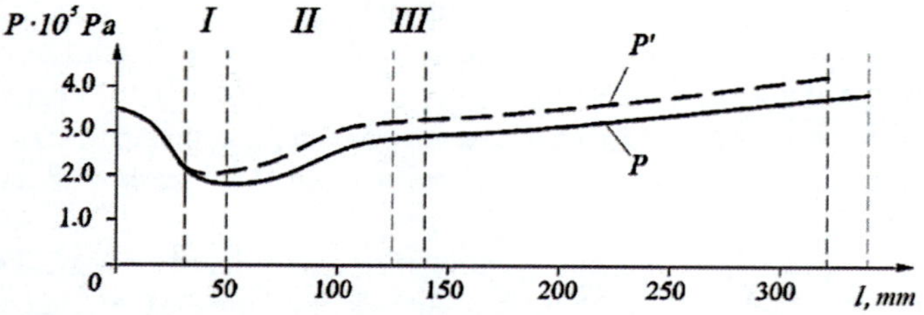

Fig. 1. Dependences of the pressure change P relative to the length of the thermopressor flow path l at different temperatures of water injection $t_{w1} = 50$ °C; 120 °C ($\pi_c = 3.5$; $t_{a1} = 185$ °C; $G_a = 2.39$ kg/s; $G_w = 0.1$ kg/s)________ – at $t_{w1} = 50$ °C; _ _ _ _ – at $t_{w1} = 120$ °C.

The analysis of the research results shows (Figs. 2–5) that with an increase in the temperature of the water for injection into the thermopressor from 25 °C to 120 °C, that is, to a temperature above the saturation temperature at the pressure in the working chamber, the pressure loss in the I mode decreases, and, as a consequence, the overall pressure increase ΔP_{tp} increases. So for $\pi_c = 2.5$ at an air velocity at the entrance to the working chamber $M = 0.35$, the decrease in pressure losses was $\Delta P_{fr}'' = 0.3 \cdot 10^4$ Pa (1%), with a general increase in pressure $\Delta P_{tp} = 0.4 \cdot 10^4$ Pa (2%), which is 0.8% higher than when water is injected without overheating. For the air velocity at the inlet to the working chamber $M = 0.95$: $\Delta P_{fr}'' = 2.6 \cdot 10^4$ Pa (10%), $\Delta P_{tp} = 2.6 \cdot 10^4$ Pa (10%)–7% higher than when water is injected without overheating.

For $\pi_c = 3.5$ at $M = 0.35$: the decrease in pressure losses was $\Delta P_{fr}'' = 0.4 \cdot 10^4$ Pa (1%), $\Delta P_{tp} = 0.8 \cdot 10^4$ Pa (2%) – by 0, 8% higher than when water is injected without overheating. At an air velocity at the inlet to the working chamber $M = 0.95$: ΔP_{fr}" = $3.7 \cdot 10^4$ Pa (10%), $\Delta P_{tp} = 5.6 \cdot 10^4$ Pa (15%)–8% higher than with water injection without overheating.

The air temperature at the outlet of the thermopressor was $t_{a2} = 52$ °C (relative temperature decreases $T_1/T_2 = 1.26$), moisture content $d_{a2} = 31$ g/kg – for $M = 0.95$ and $\pi_c = 2.5$. For $M = 0.95$ and $\pi_c = 3.5 - t_{a2} = 66$ °C (relative temperature decreases $T_1/T_2 = 1.35$), moisture content $d_{a2} = 41$ g/kg. The obtained parameters of the air at the

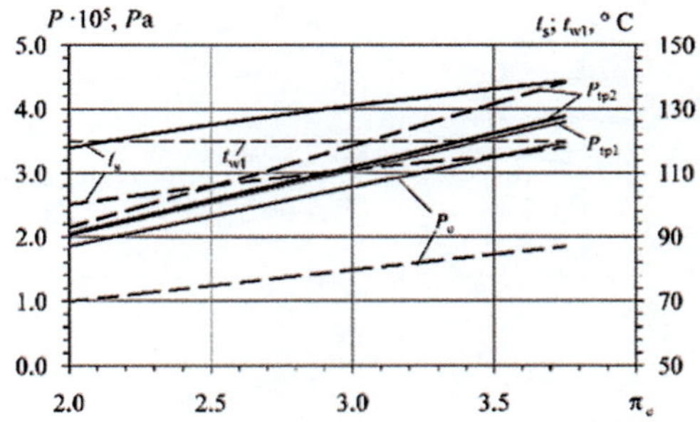

Fig. 2. Dependences of the pressure at the thermopressor inlet P_{tp1}, at the outlet P_{tp2}, after the nozzle P_c, the saturation temperature of the water vapor t_s on the degree of pressure increase in the turbocompressor π_c and various air velocities at the inlet to the working chamber of the thermopressor ______ – at M = 0.35; _ _ _ _ – at M = 0.95

outlet of the thermopressor correspond to those recommended for this type of engine by the manufacturer, according to which the temperature of the charge air after the CAC should not exceed 50–70 °C).

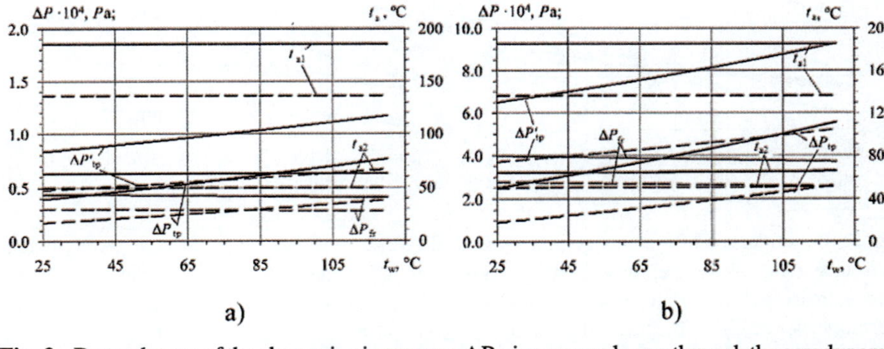

Fig. 3. Dependences of the change in air pressure ΔP, air pressure losses through the aerodynamic resistance ΔP_{fr}, air temperature t_a on the water temperature t_w at various degrees of pressure increase in the turbocharger $\pi_c = 2.5$; 3.5 and air velocities at the working chamber inlet of the thermopressor: a) M = 0.35; b) M = 0.95______ – at $\pi_c = 2.5$; _ _ _ _ _ – at $\pi_c = 3.5$

The use of injection of water overheated relative to the saturation temperature (Fig. 5) allows to additionally increase the pressure with the thermopressor by $(\Delta P_{tp}''-\Delta P_{tp}) = 0.05\text{–}0.3 \cdot 10^4$ Pa (1–8%). The amount of water, what is injected into the thermopressor (Fig. 5a) is $G_w = 0.06\text{–}0.11$ kg/s (2–5% of the charge air flow rate).

Overheating of water is proposed to be carried out by utilizing the heat of waste gases in a utilization boiler. This allows part of the low-grade heat to be returned to the power plant cycle together with water injection, which, in turn, will increase the heat utilization coefficient.

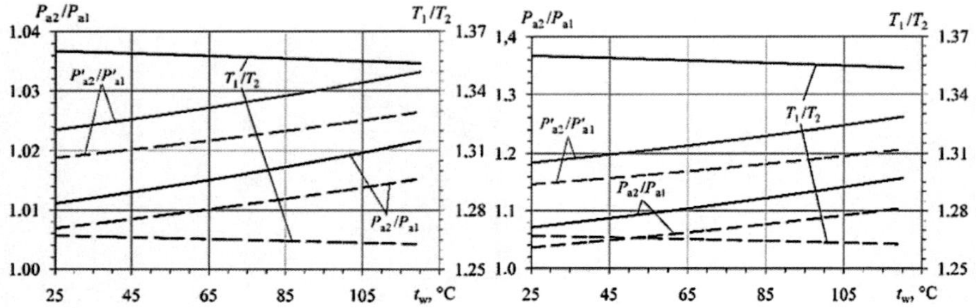

a) b)

Fig. 4. Dependences of the relative change in the air pressure P_{a2}/P_{a1} and the relative temperature decrease T_1/T_2 on the water temperature t_w at various degrees of pressure increase in the turbocharger $\pi_c = 2.5$; 3.5 and air velocities at the working chamber inlet of the thermopressor: a) M = 0.35; b) M = 0.95. ________ – at $\pi_c = 2.5$; _ _ _ _ _ – at $\pi_c = 3.5$

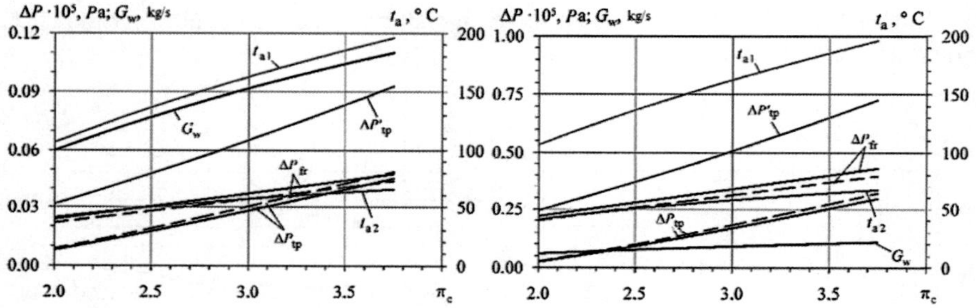

a) b)

Fig. 5. Dependences of the change in air pressure ΔP, air pressure losses through aerodynamic resistance ΔP_{fr}, air temperature t_a, water consumption for injection G_w on the degree of pressure increase in the turbocharger π_c, and air velocities at the working chamber inlet of the thermopressor: a) M = 0, 35; b) M = 0.95; ________ – at $\pi_c = 2.5$; _ _ _ _ _ – at $\pi_c = 3.5$

5 Conclusions

The use of superheated water for injection into the thermopressor of the marine engines charge air cooling system is a promising way to increase the efficiency of the spraying process, which can significantly increase the effect of thermo-gas-dynamic compression. An additional increase in pressure in the thermopressor is 5–8%.

The injection of superheated water into the thermopressor allows to minimize or completely exclude from the working process the mode associated with the influence of the drag of liquid droplets, which in turn reduces the pressure loss for aerodynamic resistance in this area by 8%.

Overheating of water is proposed to be carried out by utilizing the heat of waste gases in a utilization boiler. This will allow part of the low-grade heat to be returned to the power plant cycle together with water injection, which, in turn, will increase the heat utilization factor.

References

1. Fowle, A.: An experimental investigation of an aerothermopressor having a gas flow capacity of 25 pounds per second. Massachusetts Institute of Technology, USA (1972)
2. Erickson, J.: A theoretical and experimental investigation of the aerothermopressor process. Massachusetts Institute of Technology, USA (1958)
3. Shapiro, H., Wadleigh, R., Gavril, D., Fowle, A.: The aerothermopressor – device for improving the performance of a gas turbine power plant. Massachusetts Institute of Technology, USA (1955)
4. Vulis, L.A.: Gas Flow Thermodynamics. Gosenergoizdat, Moscow (1950)
5. Pflaum, W., Hause, W.: Leistungserhohung der Gasturbine durch Absenkung des Entspannungsenddrucres. MTZ **1**, 21–29 (1968)
6. Konovalov, D., Kobalava, H., Radchenko, M., Sviridov, V., Scurtu, I.C.: Optimal sizing of the evaporation chamber in the low-flow aerothermopressor for a combustion engine. In: Tonkonogyi, V., et al. (eds.) InterPartner 2020. LNME, pp. 654–663. Springer, Cham (2021). https://doi.org/10.1007/978-3-030-68014-5_63
7. Konovalov, D., Kobalava, H., Radchenko, M., Scurtu, I.C., Radchenko, R.: Determination of hydraulic resistance of the aerothermopressor for gas turbine cyclic air cooling. In: TE-RE-RD 2020, 14 p. E3S Web of Conferences vol. 180, p. 01012 (2020). https://doi.org/10.1051/e3sconf/202018001012
8. Trushliakov, E., Radchenko, A., Forduy, S., Zubarev, A., Hrych, A.: Increasing the operation efficiency of air conditioning system for integrated power plant on the base of its monitoring. In: Nechyporuk, M., Pavlikov, V., Kritskiy, D. (eds.) Integrated Computer Technologies in Mechanical Engineering. AISC, vol. 1113, pp. 351–360. Springer, Cham (2020). https://doi.org/10.1007/978-3-030-37618-5_30
9. Radchenko, A., Trushliakov, E., Kosowski, K., Mikielewicz, D., Radchenko, M.: Innovative turbine intake air cooling systems and their rational designing. Energies **13**(23), 6201 (2020). https://doi.org/10.3390/en13236201
10. Sexton, W.R., Sexton, M.R.: The effects of wet compression on gas turbine engine operating performance. In: Proceedings of GT2003 ASME Turbo Expo: Power for Land, Sea and Air. Atlanta, Georgia, USA, pp. 1–7 (2009)
11. Chaker, M.: Key parameters for the performance of impaction-pin nozzles used in inlet fogging of gas turbine engines. In: Proceedings of GT2005 ASME Turbo Expo: Power for Land, Sea and Air. Reno-Tahoe, Nevada, USA, pp. 1–7 (2005)
12. Bhargava, R., Bianchi, M., Peretto, A., Spina, P.R., Feasibility, A.: Study of existing gas turbines for recuperated, intercooled, and reheat cycle. J. Eng. Gas Turbines Power **126**(3), 531–544 (2004)
13. Bhargava, R.K., Bianchi, M., De Pascale, A., Negri di Montenegro, G., Peretto, A.: Gas turbine based power cycles - a state-of-the-art review. In: International Conference on Power Engineering-2007. Hangzhou, China, pp. 23–27 (2007)
14. Jonsson, M., Yan, J.: Humidified gas turbines – a review of proposed and implemented cycles. Energy **30**, 1013–1078 (2005)
15. Farouk, N., Sheng, L., Hayat, Q.: Effect of ambient temperature on the performance of gas turbines power plant. Int. J. Comput. Sci. **10**(3), 439–442 (2013)
16. Kowalski, M., Badyda, K.: Performance analysis of a gas turbine air heat recovery unit using GateCycle® software. J. Power Technol. **92**(1), 48–54 (2012)

17. Sun, J., Hou, H., Zuo, Z.: Numerical study on wet compression in a supercritical air centrifugal compressor. Proc. IMechE Part A J. Power Energy **234**(3), 384–397 (2020)
18. Paepe, W., Delattin, F., Bram, S.: Water injection in a micro gas turbine – assessment of the performance using. Appl. Energy **112**, 1291–1302 (2013)
19. Oh, H.W.: Advanced fluid dynamics. Rijeka, Croatia (2012)
20. Bergman, T.L., et al.: Fundamentals of heat and mass transfer, 7th edn. John Wiley and Sons, New Jersey (2011)

Improving the Efficiency of Operation of Gas Turbine Plants Based on Turbo-Gear Units

Oleg Savenkov[1], Andrii Radchenko[1](✉), Borys Bileka[2], Ionut Cristian Scurtu[3], and Ivan Kalinichenko[2]

[1] Admiral Makarov National University of Shipbuilding, 9 Heroes of Ukraine Avenue, Mykolayiv 54025, Ukraine
[2] National Academy of Sciences in Ukraine, Kiev, Ukraine
[3] Academia Navala "Mircea Cel Batran", Constanta, Romania

Abstract. During the operation of power plants based on gas turbine installations of ships, imbalances in the axes of the connected shafts of ship units inevitably occur, which have a negative impact on the reliability and performance of the main turbo-gear units. To compensate for these distortions, the design of a gear coupling with a combined longitudinal modification of the external and internal teeth is proposed, which due to the equal. The distribution of forces between all paired pairs of teeth is characterized by an increased load capacity of 2.19 times in comparison with traditional designs. When using these gear couplings, there is no need for re-entding of the increased rates of uncenters of the axes of the connected shafts of the main turbo-gear units with distortions of the axes of the connected shafts up to $5 \cdot 10^{-3}$ rads. Elimination of axle re-entding leads to an increase in the reliability of the vessel's operation and a decrease in the likelihood of emergency situations.

Keywords: Geared-turbine set · Misalignment of axes · Gear coupling · Load capacity · Realignment

1 Introduction

The practice of operating ships shows that after 4–5 years from the beginning of their operation, the initial values of the misalignments of the axes of the connected shafts of the main turbo-gear units of power plants increase significantly, which leads to a deterioration in their performance and technical and economic indicators. These distortions exceed the established norms by 2–6 times, significantly overload the output and input shaft of the engine and gearbox, respectively, the intermediate shaft, as well as bearings, which in general negatively affects the reliability, durability and repair suitability of the main turbo-gear units, which can lead to unscheduled shutdowns of ships for repair, as a result of which the reliability of the vessel operation decreases, the likelihood of emergency situations increases and repair costs increase.

M. Nechyporuk et al. (Eds.): ICTM 2021, LNNS 367, pp. 541–552, 2022.
https://doi.org/10.1007/978-3-030-94259-5_46

2 Literature Review

To date, the problem of misalignment of the axes of the connected shafts in the field of shipbuilding and ship repair is solved in two ways, either by re-centering the axes or by using compensating gear couplings [1, 2]. Re-centering leads to downtime of ships, requires significant material costs for docking and repair work, which generally reduces their carrying capacity [3–5].

During the operation of traditional designs of ship gear couplings due to the instant application of the load on the conjugate pairs of teeth during axle misalignments, which also varies from zero or minimum value to the maximum value, there is a shock load of the teeth, there is an extremely uneven distribution of forces between the conjugate pairs of teeth, deterioration of the lubrication conditions of the teeth, an increase in friction power losses and a decrease in load capacity [5, 6]. Thus, the development of new designs of gear couplings with high load capacity in the conditions of uncoupling of the axes of the connected shafts of the units is an urgent task, the successful solution of which predetermines the increase in efficiency, reliability, service life and efficiency, both main turbo-gear units and the vessel as a whole [7, 8].

The most complete issues of cardinal increase of efficiency of gear couplings and, as a consequence, the main turbo-gear units as a whole through development of new technical solutions and methods of calculation are set out in [9–11]. Three directions are used as the main directions of increasing the load capacity of gear couplings. **The first** direction is to profile the sweeps of the dividing cylinders of the hinges with the outer and the wallpaper with the inner teeth of the gear couplings [12, 13]. **The second** direction provides for the manufacture of three-crown bushings and clips with the first and third rims of the bushings and clips rotated in the circumferential direction at certain angles relative to the second rims of the bushings and clips [14, 15]. **The third** direction is based on the establishment of the relationship between the parameters of the longitudinal combined modification of the outer and inner teeth at the specified sizes of the gear couplings, the parameters of the gearing of the teeth and the angles of the misalignment of the axes, in which the teeth in the conditions of rearrangements of the axes are characterized by a uniform distribution of forces between them [16, 17].

Ways to improve the characteristics and technical and economic indicators of power plants by eliminating the negative impact of misalignments of the axes of the connected shafts by using gear couplings of new designs are most fully described in [18–20].

3 Methodology

Functions of distribution of lateral gaps in highly effective gear couplings are received on the basis of a method of definition of laws of distribution of normal lateral gaps between teeth. The equation of total deformation, consisting of contact and bending deformations, as well as deformations elastically embedded in the body of the sleeve and the holder of the outer and inner teeth and the dependence of the force distribution are obtained by calculating the load capacity of gear couplings. Determination of friction coefficients in ship toothed couplings was based on the method of calculation of efficiency coefficients in ship toothed couplings, based on a mathematical model of friction.

In this work, the object of research is the gear coupling, in which the external and internal teeth are longitudinally modified. In this case, the outer teeth of the sleeve *1* are barrel-shaped with a radius of curvature *R,* and the inner teeth of the frame *2* at the end sections are symmetrically beveled at an angle oϕ Ψ_0, and in the middle part - side-shaped with a radius of curvature equal to also *R* (Fig. 1, ***a, b***).

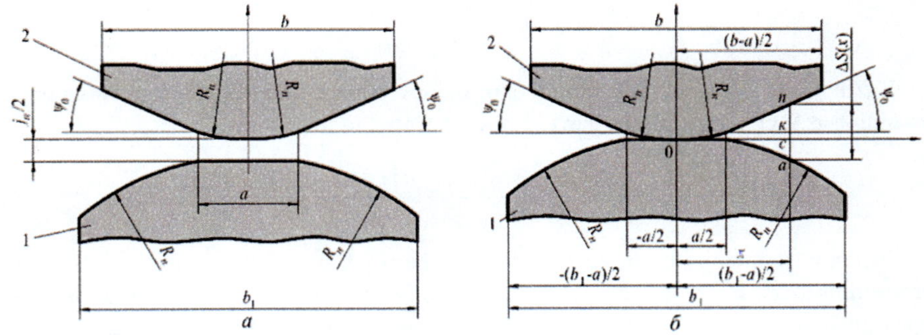

Fig. 1. Arrangement of teeth 1 and 2 in the absence of (*a*) and the presence (*b*) between them of contact

The basis for the study is the function of the distribution of normal lateral gaps between the teeth with misalignment of the axes by the angle Ψ, which in the general case has the form [21–23]

$$j_n(x, \varphi) = j_n(x) - x\psi \cos\varphi + \frac{mz\psi^2}{8}\sin 2\varphi + \frac{mz\psi^2\alpha_w}{4}\cos\varphi, \tag{1}$$

where j_n (x) is the maximum lateral gap between the teeth at point *x; x* –current abscissa; φ – angle of rotation arbitrarily taken to consider the conjugate pair of teeth; *m* – engagement module; *z* – number of; α_w – gear angle of teeth.

In accordance with Fig. 2, *and* the maximum lateral gap between the teeth will look like this.

$$j_n(x) = \frac{j_n}{2} + \Delta S(x), \tag{2}$$

where j_n is half the size of the normal side clearance between the teeth before the skew of the axes; $\Delta S(x) = ac + cto + \text{kn}$ - the gap between the teeth due *to* the longitudinal modification of the outer 1 and inner 2 teeth.

The dimensions *ac, ck* and *kn,* characterizing the gap Δ*S*(*x*), are determined in accordance with Fig. 1, *b* as follows:

$$ac = \frac{(R\psi_\tau)^2}{2R_n} = \frac{R\psi^2}{2\cos\alpha_w}\cos^2\varphi; \quad c\kappa = \frac{a^2}{8R_n} = \frac{a^2\cos\alpha_w}{8R};$$

$$\kappa n = R_n\psi_\tau\psi_0 = \frac{R\psi\psi_0}{\cos\alpha_w}\cos\varphi,$$

where $\Psi\tau = \Psi\cos\varphi$ is the axis misalignment angle corresponding to the angle of rotation φ of the gear coupling; a – the length of the middle section of the inner tooth 2, characterized by rectilinear forming; $R_n = R/\cos\alpha_w$ is the radius of barrel-like teeth in normal cross-section.

The variable *value* x, included in Eq. (1), is determined (Fig. 2, *b)* based on the expression

$$x = \frac{a}{2} + R_n\psi_\tau = \frac{a}{2} + \frac{R}{\cos\alpha_w}\cos\varphi.$$

Taking into account the found expressions *ac*, *cto*, *to n* and the variable x, assuming $\Delta S(x) = \Delta S(\varphi)$, the gap (2) is represented in final form

$$\Delta S(x) = \Delta S(\varphi) = \frac{R\psi^2\cos^2\varphi}{2\cos\alpha_w} + \frac{a^2\cos\alpha_w}{8R} + \frac{R\psi\psi_0}{\cos\alpha_w}\cos\varphi \tag{3}$$

When introducing into the expression (1) the right part of the dependence (3) and proceeding from the replacement of the notation of the function $j_n(x, \varphi)$ with the designation $j_n(\varphi)$, we present the law of distribution of normal lateral gaps between the teeth.

$$j_n(\varphi) = \frac{j_n}{2} - \left(\frac{R\psi^2}{2\cos\alpha_w} - \frac{mz\psi^2\alpha_w}{4}\right)\cos^2\varphi + \left(\frac{R\psi\psi_0}{\cos\alpha_w} - \frac{a\psi}{2}\right)\cos\varphi + \frac{a^2\cos\alpha_w}{8R} + \frac{mz\psi^2}{8}\sin 2\varphi. \tag{4}$$

Equation (4) characterizes the distribution of normal lateral gaps between the teeth in the absence of exposure to the gear clutch of torque. If the gear coupling is loaded with the specified moment, then the pair of teeth in which the value of the normal lateral clearance is minimal, that is, $j_n(\varphi) = j_{n\min}$, which is possible at an angle $\varphi = 0^\circ$ in Eq. (4) will come into contact first. According to the above, that is, at $\varphi = 0^\circ$, the expression $j_{n\min}$, based on Eq. (4), will take the form.

$$j_{n\min} = \frac{j_n}{2} - \frac{R\psi^2}{2\cos\alpha_w} + \frac{mz\psi^2\alpha_w}{4} + \frac{R\psi\psi_0}{\cos\alpha_w} - \frac{a\psi}{2} + \frac{a^2\cos\alpha_w}{8R}. \tag{5}$$

Then the movement of the contact line in question of length h_p, which we denote through W, of the loaded pair of teeth in the circumferential direction, taking into account the expression (5) will be equal to

$$W = W_{\max} + j_{n\min} = W_{\max} + \frac{j_n}{2} - \frac{R\psi^2}{2\cos\alpha_w} + \frac{R\psi\psi_0}{\cos\alpha_w} + \frac{mz\psi^2\alpha_w}{4} - \frac{a\psi}{2} + \frac{a^2\cos\alpha_w}{8R}, \tag{6}$$

where $W_{\max}$ is the maximum total deformation of the most loaded conjugate pair of teeth.

Proceeding from the expressions (4) and (6) the deformation equation of an arbitrarily considered conjugate pair of teeth, characterized by the angle of rotation φ, we write as a function

$$W(\varphi) = W - j_n(\varphi) = W_{\max} - \left(\frac{R\psi^2}{2\cos\alpha_w} - \frac{mz\psi^2\alpha_w}{4}\right)\left(1 - \cos^2\varphi\right)$$

$$+\left(\frac{R\psi\psi_0}{\cos\alpha_w}-\frac{a\psi}{2}\right)(1-\cos\varphi)-\frac{mz\psi^2}{8}\sin 2\varphi \quad (7)$$

Let us denote the angle of coverage of the gear coupling, within which all paired pairs of teeth participate in the transmission of torque, that is, are loaded, through γss and the angle $\gamma \leq \pi/2$ (Fig. 2). In the future, the angle γ will be called the parametric angle. Thus, at some value of the angle $\varphi = \gamma$, the deformation value $W(\gamma)$ will be zero, that is, $W(\gamma) = 0$, and the deformation of the most loaded conjugate pair, taking into account the function of the teeth (7) in this case will take the form of

$$W_{\max}=\left(\frac{R\psi^2}{2\cos\alpha_w}-\frac{mz\psi^2\alpha_w}{4}\right)\left(1-\cos^2\gamma\right)-\left(\frac{R\psi\psi_0}{\cos\alpha_w}-\frac{a\psi}{2}\right)(1-\cos\gamma)-\frac{mz\psi^2}{8}\sin 2\gamma\cdot \quad (8)$$

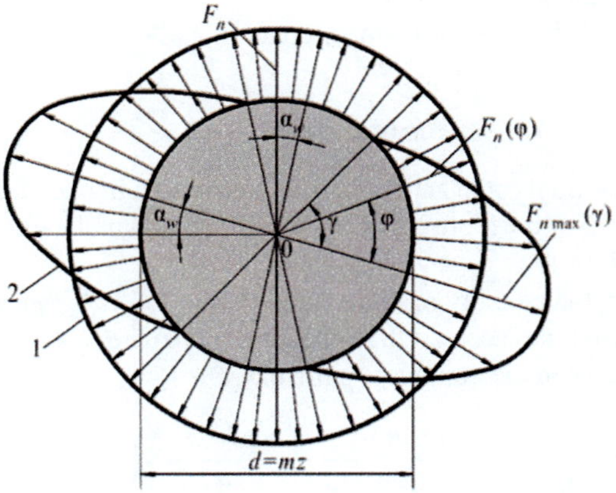

Fig. 2. Load distribution between the teeth in the proposed (curve 1) and traditional (curve 2) gear couplings

After substituting the right side of the expression (8) in Eq. (7), we write down the function of changing the deformations of the conjugate pairs of teeth of the tooth coupling, namely:

$$W(\varphi)=\left(\frac{R\psi^2}{2\cos\alpha_w}-\frac{mz\psi^2\alpha_w}{4}\right)\left(\cos^2\varphi-\cos^2\gamma\right)-\left(\frac{R\psi\psi_0}{\cos\alpha_w}-\frac{a\psi}{2}\right)(\cos\varphi-\cos\gamma)$$
$$-\frac{mz\psi^2}{8}(\sin 2\varphi-\sin 2\gamma). \quad (9)$$

At the angle $\varphi = \pi/2$, the last paired pair of teeth will choose a normal lateral clearance, so it will come into contact but not yet be loaded. In this case, the deformation of this pair of teeth $W(\varphi) = 0$, while the deformation of the most loaded pair will reach the maximum value in accordance with the function (8) takes the form.

$$W_{\max}=\left(\frac{R}{\cos\alpha_w}-\frac{mz\alpha_w}{2}\right)\frac{\psi^2}{2}+\left(\frac{R\psi_0}{\cos\alpha_w}-\frac{a}{2}\right)\psi.$$

If, when the axes are skewed, all conjugate pairs of teeth are in engagement, then the deformation of the maximum loaded conjugate pair of teeth is indicated through $W^*_{\max}$, and in this case it will be equal to $W^*_{\max} = W_{\max} + W_F$, where W_F is the deformation of the minimally loaded conjugate pair of teeth at angle $\varphi = \pi/2$.

Using the superposition principle for all conjugate pairs of teeth within one quarter of the circle (Fig. 2), write the expression (9) in general form

$$W(\varphi) = W_F + \left(\frac{R\psi^2}{2\cos\alpha_w} - \frac{mz\psi^2\alpha_w}{4}\right)\left(\cos^2\varphi - \cos^2\gamma\right) - \left(\frac{R\psi\psi_0}{\cos\alpha_w} - \frac{a\psi}{2}\right)(\cos\varphi - \cos\gamma) - \frac{mz\psi^2}{8}(\sin 2\varphi - \sin 2\gamma) \tag{10}$$

The deformation *of* $W(\varphi)$ should be understood as the total deformation of the conjugate pair of teeth in question, due to the contact interaction and bending of the teeth, as well as elastic movements adjacent to the outer and inner teeth of the sections of the rims of the bushing and the frame [24, 25].

The total value of these deformations, denoted as $W_{\sum}(\varphi)$, is determined from the expression

$$W_{\sum}(\varphi) = \delta_{\sum} F_n(\varphi), \tag{11}$$

where $F_n(\varphi)$ is the normal force acting on the pair of teeth in question, characterized by the angle of rotation φ of the tooth coupling; $\delta_{\sum}$ is the total malleability of the conjugate pair of teeth, mm/H.

Proceeding from the equality of deformations $W(\varphi) = W\sum(\varphi)$, we obtain from the expression (11), taking into account the dependence (10) the function of the distribution of normal forces between the conjugate pairs of teeth

$$F_n(\varphi) = \frac{1}{\delta_\Sigma}\left[W_F + \left(\frac{R\psi^2}{2\cos\alpha_w} - \frac{mz\psi^2\alpha_w}{4}\right)\left(\cos^2\varphi - \cos^2\gamma\right) - \left(\frac{R\psi\psi_0}{\cos\alpha_w} - \frac{a\psi}{2}\right)(\cos\varphi - \cos\gamma) - \frac{mz\psi^2}{8}(\sin 2\varphi - \sin 2\gamma)\right] \tag{12}$$

The expression (12) contains unknown parameters W_F and γ, the definition of which will be considered separately. Based on the presence in the engagement of part of the conjugate pairs of teeth, the deformation $W_F = 0$, and therefore this expression will take the form of

$$F_n(\varphi) = \frac{1}{\delta_\Sigma}\left[\left(\frac{R\psi^2}{2\cos\alpha_w} - \frac{mz\psi^2\alpha_w}{4}\right)\left(\cos^2\varphi - \cos^2\gamma\right) - \left(\frac{R\psi\psi_0}{\cos\alpha_w} - \frac{a\psi}{2}\right)(\cos\varphi - \cos\gamma) - \frac{mz\psi^2}{8}(\sin 2\varphi - \sin 2\gamma)\right]. \tag{13}$$

The unknown parametric angle γ is determined based on the condition of equality of the transmitted torque of the gear coupling both when the axes are skewed and in the absence of misalignment of these axes, which is possible when

$$T = rzF_t = 4\int_0^\gamma r dF_n(\varphi)\cos\alpha_w = \frac{4zr\cos\alpha_w}{r\cdot 2\pi}\int_0^\gamma \omega_n(\varphi)d\varphi, \tag{14}$$

where $dF_n(\varphi) = \omega_n(\varphi)d\varphi$; $\omega_n(\varphi) = zF_n(\varphi)/2\pi$ - normal force intensity, N/rad; F_t is the circumferential force.

As a result of substitution in Eq. (14) of expression (13) we get the dependence of the circumferential force

$$F_t = \frac{2\cos\alpha_w}{\pi\delta_\Sigma}\int\limits_0^{\gamma}\left[\left(\frac{R\psi^2}{2\cos\alpha_w} - \frac{mz\psi^2\alpha_w}{4}\right)\left(\cos^2\varphi - \cos^2\gamma\right)\right.$$
$$\left. -\left(\frac{R\psi\psi_0}{\cos\alpha_w} - \frac{a\psi}{2}\right)(\cos\varphi - \cos\gamma) - \frac{mz\psi^2}{8}(\sin 2\varphi - \sin 2\gamma)\right]d\varphi \quad (15)$$
$$= \frac{2\cos\alpha_w}{\pi\delta_\Sigma}\left[\left(\frac{R\psi^2}{2\cos\alpha_w} - \frac{mz\psi^2\alpha_w}{4}\right)\cdot A(\gamma) - \left(\frac{R\psi\psi_0}{\cos\alpha_w} - \frac{a\psi}{2}\right)\cdot B(\gamma) - \frac{mz\psi^2}{8}\cdot C(\gamma)\right],$$

where $A(\gamma)$, $B(\gamma)$, $C(\gamma)$ are the functions of the parametric angle γ, defined from the expressions:

$$A(\gamma) = \frac{\gamma}{2} + \frac{\sin 2\gamma}{4} - \gamma\cos^2\gamma; \quad B(\gamma) = \sin\gamma - \gamma\cos\gamma; \quad C(\gamma) = \sin^2\gamma - \gamma\sin 2\gamma$$

The parametric angle γ is determined by solving the transcendental Eq. (15) at given values F_t, δ_Σ, R, m, z, α_w, Ψ, Ψ_0.

The unknown value of the deformation W_F assuming the angle γ is known, we will find from Eq. (12), taking into account the substitution of its right side into the expression (14), namely:

$$W_F = \frac{\pi F_t\delta_\Sigma}{2\gamma\cos\alpha_w} - \left(\frac{R\psi^2}{2\cos\alpha_w} - \frac{mz\psi^2\alpha_w}{4}\right)\cdot\frac{A(\gamma)}{\gamma}$$
$$+\left(\frac{R\psi\psi_0}{\cos\alpha_w} - \frac{a\psi}{2}\right)\cdot\frac{B(\gamma)}{\gamma} + \frac{mz\psi^2}{8}\cdot\frac{C(\gamma)}{\gamma} \quad (16)$$

As a result of replacement in Eq. (12) of deformation W_F by the expression of the right part of the dependence (16), we obtain in the final form the function of the distribution of normal forces between the conjugate pairs of teeth, taking into account the misalignment of the axes of the connected shafts of the aggregates

$$F_n(\varphi) = \frac{\pi F_t}{2\gamma\cos\alpha_w} - \frac{\psi^2}{2\delta_\Sigma}\left(\frac{R}{\cos\alpha_w} - \frac{mz\alpha_w}{2}\right)\left(\frac{1}{2} + \frac{\sin 2\gamma}{4\gamma} - \cos^2\right)$$
$$+\frac{\psi}{\delta_\Sigma}\left(\frac{R\psi_0}{\cos\alpha_w} - \frac{a}{2}\right)\left(\frac{\sin\gamma}{\gamma} - \cos\varphi\right) + \frac{mz\psi^2}{8\delta_\Sigma}\left(\frac{\sin^2\gamma}{\gamma} - \sin 2\varphi\right) \quad (17)$$

The expression (17) is characteristic of the case of exit from the engagement of part of the conjugate pairs of teeth during the transmission of torque by the clutch in conditions of misalignment of the axes of the connected shafts of the units. If during the operation of the gear coupling all the conjugate pairs of teeth are involved in the transfer of the load, then in Eq. (17) it is necessary to take the angle $\gamma = \pi/2$, and therefore this equation is converted to the form

$$F_n(\varphi) = \frac{F_t}{\cos\alpha_w} - \frac{\psi^2}{2\delta_\Sigma}\left(\frac{R}{\cos\alpha_w} - \frac{mz\alpha_w}{2}\right)\left(\frac{1}{2} - \cos^2\varphi\right) + \frac{\psi}{\delta_\Sigma}\left(\frac{R\psi_0}{\cos\alpha_w} - \frac{a}{2}\right)\left(\frac{2}{\pi} - \cos\varphi\right) + \frac{mz\psi^2}{8\delta_\Sigma}\left(\frac{2}{\pi} - \sin 2\varphi\right) \tag{18}$$

At the skew angle $\Psi = 0$, Eqs. (17) and (18) are characterized by a uniform distribution of forces between all conjugate pairs of teeth, on the basis of which $F_n(\mathrm{j}) = F_t/\cos\alpha_w = F_n$.

If in Eqs. (17) and (18) we take the angle of rotation of the teeth $\varphi = 0°$, we get the dependencies of the maximum normal forces acting on the most loaded conjugate pair of teeth:

$$F_{n\max}(\gamma) = \frac{\pi F_t}{2\gamma\cos\alpha_w} + \frac{\psi^2}{4\delta_\Sigma}\left(\frac{R}{\cos\alpha_w} - \frac{mz\alpha_w}{2}\right)\left(1 - \frac{\sin 2\gamma}{2\gamma}\right) - \frac{\psi}{\delta_\Sigma}\left(\frac{R\psi_0}{\cos\alpha_w} - \frac{a}{2}\right)\left(1 - \frac{\sin\gamma}{\gamma}\right) + \frac{mz\psi^2}{8\delta_\Sigma}\cdot\frac{\sin^2\gamma}{\gamma} \tag{19}$$

$$F_{n\max} = \frac{F_t}{\cos\alpha_w} + \frac{\psi^2}{4\delta_\Sigma}\left(\frac{R}{\cos\alpha_w} - \frac{mz\alpha_w}{2}\right) - \frac{(\pi-2)\psi}{2\delta_\Sigma}\left(\frac{R\psi_0}{\cos\alpha_w} - \frac{a}{2}\right) + \frac{mz\psi^2}{4\pi\delta_\Sigma} \tag{20}$$

4 Results

Equation (20) indicates that with the participation of all conjugate pairs of teeth in the engagement, the maximum force $F_{n\max}$ is always greater in magnitude than the magnitude of the force $F_n$acting on any conjugate pair of teeth in the absence of misalignment of axes, that is, at $\Psi = 0$. However, as follows from Eq. (20), it is possible to evenly distribute the load between all conjugate pairs of teeth, provided that the sum of the second, third and fourth terms standing on the right side of the specified equation is zero. In this regard, proceeding from the equality $F_{n\max} = F_t/\cos\alpha_w$ and equating the specified sum of terms to zero, we get the equality

$$\frac{(\pi-2)R\psi_0}{\cos\alpha_w} = \frac{R\psi}{2\cos\alpha_w} + \frac{(\pi-2)a}{2} + \frac{mz\psi}{2}\left(\frac{1}{\pi} - \frac{\alpha_w}{4}\right) \tag{21}$$

Assuming that all the parameters included in the equality (21) are known, except for the angle Ψ_0, for the definition of which the specified equality is solved with respect to the angle Ψ_0 in connection with which

$$\psi_0 = \frac{\psi}{2(\pi-2)}\left[1 + \frac{(4-\pi\alpha_w)mz\cos\alpha_w}{4\pi R}\right] + \frac{a\cos\alpha_w}{2R}. \tag{22}$$

Thus, at given values m, z, α_w, a, R and Ψ, it is always possible to find such a value of the bevel angle Ψ_0 of the teeth, at which the load during the skew is distributed evenly between all the conjugate pairs of teeth, and therefore the specified tooth coupling can be considered practically insensitive to the misalignment of the axes.

As an example, consider a toothed coupling, with the following parameters: $m = 5$ mm; $z = 60$; $\alpha_w = 20°$; $a = 5$ mm; $R = 3000$ mm; $\psi = 5\cdot10^{-3}$ rad; $\delta_\Sigma = 5{,}5\cdot10^{-6}$ mm/H; $F_t = 3000$ N. In accordance with the specified data, according to the formula (22), we determine the value of the angle $\psi_0 = 3.338\ 10^{-3}$ rads, at which all the teeth of the gear coupling are loaded with the same forces $F_n = F_t/\cos\alpha_w = 3000/\cos20°$ $= 3191$ N under operating conditions with distortions of the axes of the connected shafts of the units.

Let's determine the maximum force acting on the most loaded conjugate pair of teeth of a traditional tooth coupling, in which the outer teeth are barrel-shaped with a radius of curvature $R = 3000$ mm, and the inner teeth are straight. For this purpose, for this coupling, we determine the value of the function $A(\gamma) = \pi F_t\delta_\Sigma/R\psi^2\cos\alpha = 0{,}735$, which corresponds to the angle $\gamma = 77°$ [24].

By formula

$$F_{n\max}(\gamma) = \frac{\pi F_t \delta_\Sigma}{2\gamma\cos\alpha_w} + \frac{R\psi^2}{4\delta_\Sigma}\left(1 - \frac{\sin 2\gamma}{2\gamma}\right)$$

define the maximum force $F_{n\max}(\gamma) = 6581$ N.

From the comparison of the forces $F_{n\max}(\gamma) = 6581$ N and $F_n = 3191$ N, we find that the load capacity of the proposed gear coupling design exceeds the load capacity of the traditional gear coupling in 6581/3191 = 2.19 times.

Thus, the proposed design of the gear coupling with a combined longitudinal modification of the outer and inner teeth is characterized by an increased load capacity more than twice in comparison with traditional designs and a uniform distribution of forces between the teeth at the angles of misalignment of the axes of the connected shafts of the main turbo-gear units up to $\psi = 5\cdot10^{-3}$ rads.

5 Conclusions

The best burning out of fuel combustible components due to applying a WFE provides decreasing a concentration of solids and soot in the exhaust gases and hence their toxicity.

1. It is shown that an increase in the efficiency of operation of gas turbine plants based on turbo-gear units by eliminating the negative effect of misalignments of the axes of the connected shafts is possible due to the use of toothed couplings with a combined longitudinal modification of the outer and inner teeth, insensitive to the indicated misalignments of the axes.
2. With regard to the proposed gear couplings operating with distortions of the axes of the connected shafts, the equations of distribution of normal lateral gaps, total deformations and forces between longitudinally modified inner and outer teeth are obtained.
3. On the basis of the equations of distribution of forces between the teeth, expressions are obtained that establish the relationship between the parameters of longitudinal modification of the internal and external teeth, taking into account the size of the gear couplings and the angles of misalignment.

4. It is shown that in comparison with traditional, the proposed couplings are characterized by an increased load capacity of 2.19 times and a uniform distribution of forces between the teeth at the angles of misalignment of the axes of the connected shafts of the main turbo-gear units $\psi = 5 \cdot 10^{-3}$ rad.
5. From the results of theoretical studies, it follows that when using the gear couplings in question, there is no need for re-retches of the increased rates of uncoupling of the axes of the connected shafts of the main turbo-gear units, in connection with which the downtime of ships is excluded and their carrying capacity is preserved in full.

References

1. Nobuo, K.: Reliability estimation for failures on main engine system by ship reliability database system. Statistical study on reliability of ship equipment and safety management. Technical papers. Bull. JIME **29**(2), 64–70 (2000)
2. Podgurenko, V.S.: Improving the performance of gear couplings operating at imbalances in the axes of the connected shafts. Bull. Nat. Tech. Univ. "KhPI". Collect. Sci. Works Ser. Mech. Drive Probl. **41**(1014), 94–98 (2013)
3. Yang, Z., Wen, H., Yang, X., Gorbov, V., Mitienkova, V., Serbin, S.: Marine Power Plant (Shanghai Scientific and Technical Publishers 2021), p. 335. Springer, Singapore (2021) https://doi.org/10.1007/978-981-33-4935-3
4. Kornienko, V., Radchenko, M., Radchenko, R., Konovalov, D., Andreev, A., Pyrysunko, M.: Improving the efficiency of heat recovery circuits of cogeneration plants with combustion of water-fuel emulsions. Thermal Sci. **25**(1B), 791–800 (2021)
5. Radchenko, M., Radchenko, R., Tkachenko, V., Kantor, S., Smolyanoy, E.: Increasing the operation efficiency of railway air conditioning system on the base of its simulation along the route line. In: Nechyporuk, M., et al. (eds.) Integrated Computer Technologies in Mechanical Engineering (ICTM 2019). Advances in Intelligent Systems and Computing, vol. 1113, pp. 461–467. Springer, Cham (2020)
6. Slobodianskii, M.G., Antsupov, A.V., Lukinskih, S.V.: Failure model for gear couplings under the criterion of working surface endurance. In: Radionov, A.A., Kravchenko, O.A., Guzeev, V.I., Rozhdestvenskiy, Y.V. (eds.) ICIE 2019. LNME, pp. 65–73. Springer, Cham (2020). https://doi.org/10.1007/978-3-030-22041-9_8
7. Bachschmid, N., Curami, A., Petrone, F.: Vibrational behaviour of rotors with gear couplings in case of insufficient coupling lubrication. In: Goodwin, M.J. (eds.) Rotordynamics 1992. Springer, London (1992). https://doi.org/10.1007/978-1-4471-1979-1_29
8. Cherednichenko, O., Tkach, M., Mitienkova, V.: Improving of energy efficiency of cruise ships by applying of thermochemical recuperation. In: Nechyporuk, M., Pavlikov, V., Kritskiy, D. (eds.) Integrated Computer Technologies in Mechanical Engineering - 2020. ICTM 2020. LNCS, vol. 188, pp. 320–331. Springer, Cham (2021). https://doi.org/10.1007/978-3-030-66717-7_27
9. Radchenko, M., Mikielewicz, D., Tkachenko, V., Klugmann, M., Andreev, A.: Enhancement of the operation efficiency of the transport air conditioning system. In: Ivanov, V., Pavlenko, I., Liaposhchenko, O., Machado, J., Edl, M. (eds.) DSMIE 2020. LNME, pp. 332–342. Springer, Cham (2020). https://doi.org/10.1007/978-3-030-50491-5_32
10. Konovalov, D., Kobalava, H., Maksymov, V., Radchenko, R., Avdeev, M.: Experimental research of the excessive water injection effect on resistances in the flow part of a low-flow aerothermopressor. In: Ivanov, V., Pavlenko, I., Liaposhchenko, O., Machado, J., Edl, M.

(eds.) DSMIE 2020. LNME, pp. 292–301. Springer, Cham (2020). https://doi.org/10.1007/978-3-030-50491-5_28
11. Shevchenko, S., Mukhovaty, A., Krol, O.: Gear clutch with modified tooth profiles. Procedia Eng. **206**, 979–984 (2017). http://dspace.snu.edu.ua:8080/jspui/handle/123456789/1783
12. Khanmamedov, S.A.: Experimental determination of bending moments at static loading of gear couplings. In: Bulletin of the National Technical University "KhPI". Collection of Scientific Works. Thematic Issue: Machine Llearning and CAD, vol. 22, pp. 175–180 (2012)
13. Zhao, G., Liu, Z.-S., Ye, J.-H., Chen, F.: Meshing force model of misaligned gear coupling and its influence on a rotor system. Harbin Gongcheng Daxue Xuebao/J. Harbin Eng. Univ. **30**(1), 33–39 (2009)
14. Li, M., Yu, L.: Analysis of the coupled lateral torsional vibration of a rotor-bearing system with a misaligned gear coupling. J. Sound Vibr. **243**(2), 283–300 (2001). https://doi.org/10.1006/jsvi.2000.3412
15. Antsupov, A.V., Antsupov, A.V., Antsupov, V.P.: Analytical method for project resource estimation of metallurgical machinery parts. Izvestiya Vysshikh Uchebnykh Zavedenij. Chernaya Metallurgiya **60**, 62–66 (2017). https://doi.org/10.17073/0368-0797-2017-1-30-35
16. Kolesnikov, V.I., Koropets, P.A., Sinyutin, E.S.: Vibration-based diagnostics of helicopter transmission gear couplings. Russ. Aeronaut. **63**, 7–13 (2020). https://doi.org/10.3103/S106879982001002X
17. Elkholy, A.H., Alfares, M.A.: Misalignment loads in splined gear couplings. Int. J. Comput. Appl. Technol. **15**(1–3), 128–137 (2010)
18. Mikhailov, A.N.: Methodology and principles of improving the technological quality assurance and efficiency of manufacturing spatially modified tooth couplings. In: Progressive Technologies and Systems of Machinery and Equipment: International Zb. Science Parts. DonNTU, Donetsk, vol. 35, pp. 136–151 (2008)
19. Shevchenko, S.V., Mukhovaty, A.A., Krol, O.S.: Modification of a gear coupling. J. Adv. Res. Eng. Sci. **7**(1), 18–23 (2017). North Charleston, USA: SRC MS, CreateSpace
20. Radchenko, R., Pyrysunko, M., Radchenko, A., Andreev, A., Kornienko, V.: Ship engine intake air cooling by ejector chiller using recirculation gas heat. In: Tonkonogyi, V., et al. (eds.) InterPartner 2020. LNME, pp. 734–743. Springer, Cham (2021). https://doi.org/10.1007/978-3-030-68014-5_71
21. Konovalov, D., Kobalava, H., Radchenko, M., Scurtu, I.C., Radchenko, R.: Determination of hydraulic resistance of the aerothermopressor for gas turbine cyclic air cooling. In: TE-RE-RD 2020, E3S Web of Conferences, vol. 180, p. 14, Article no. 01012 (2020)
22. Belkov, V.A., Maksimov, V.A., Shitikov, I.A.: Numerical study of characteristics of gear couplings of high-speed turbomachines. Bull. Kazan Technol. Univ. **10**, 517–520 (2010)
23. Podgurenko, V.S.: Determination of bending moments in real conditions of operation of gear couplings. In: Bulletin of the National Technical University "KhPI". Collection of Scientific Works. Thematic Issue: Machine Learning and CAD. NTU "KhPI", Kharkiv, vol. 22, pp. 102–106 (2012)
24. Podgurenko, V.S.: Load capacity of gear couplings taking into account errors in the manufacture of teeth. Bull. Agrarian Prychornomor'ya **1**(77), 197–203 (2014)

25. Trushliakov, E., Radchenko, A., Radchenko, M., Kantor, S., Zielikov, O.: The efficiency of refrigeration capacity regulation in the ambient air conditioning systems. In: Ivanov, V., Pavlenko, I., Liaposhchenko, O., Machado, J., Edl, M. (eds.) Advances in Design, Simulation and Manufacturing III DSMIE 2020 Proceedings of the 3rd International Conference on Design, Simulation, Manufacturing: The Innovation Exchange, DSMIE-2020, Kharkiv, Ukraine. Mechanical and Chemical Engineering. LNME, vol. 2, pp. 343–353. Springer, Cham (2020).https://doi.org/10.1007/978-3-030-50491-5_33
26. Bohdal, L., Kukiełka, L., Legutko, S., Patyk, R., Radchenko, A.M.: Modeling and experimental analysis of shear-slitting of AA6111-T4 aluminum alloy sheet. Materials **13**(14), 3175 (2020)

Nano-modeling

Linking Dynamics of Growth of Copper Oxide Nanostructures in Air

Andrii Breus, Sergey Abashin, Oleksii Serdiuk, and Oleg Baranov(✉)

National Aerospace University, Kharkiv 61070, Ukraine
O.Baranov@khai.edu

Abstract. The main purpose of the research is to expand the range of the theoretical and experimental knowledge about the synthesis methods of copper oxide nanowires. For the purpose, a set of experiments was carried out in the conditions of ambient air at the temperatures of 600 to 1000 °C, and then the experimental results were simulated by use of the recently developed theoretical model. The results confirmed that the model can be applied in the oxygen pressure range of about 10^4–10^5 Pa, and the set of energies introduced in the model, stays almost constant for these pressures and in the temperature range of 600–800 °C. In addition to the traditional modes of the copper nanowire synthesis, a high temperature mode at 1000 °C was found where the thin nanowires are grown not on the flat layer of oxide evenly distributed along the sample surface, but on the bulky 3D microstructures of copper oxide. However, the optimal temperature for the growth of 1D copper oxide nanostructures was confirmed to be close to 600–700 °C, where the longest nanowires with the most abundant yield are obtained. At the same time, the geometrical characteristics of the nanowires are almost independent on the tenfold decrease in the oxygen pressure for the optimal temperatures. The obtained results imply the necessity of further development of the model to explain the nanowire nucleation.

Keywords: Copper oxide · Nanostructures · Thermal growth · Ambient air

1 Introduction

One-dimensional nanostructures of metals oxidized at elevated temperatures, have gained an attention both in research and industrial communities, due to their outstanding properties conditioned by their morphology [1]. Generally, the properties are described in terms of the large surface-to-volume properties (specified as a ratio of the surface of the material spent for the nanostructure formation to its volume), high aspect ratio (that is a ratio of the nanostructure length to its diameter), sharp tips (measured as a diameter of the highest part of the nanostructure), and general sizes (the length of the nanostructures is usually between 1 to 100 μm while a diameter is of 10 to 100 nm) [2]. At that, the difference in morphology is accompanied by a large variety of physical and chemical characteristics, such as distinguished mechanical strength and electrical conductivity, to mention but a few [3]. 1D copper oxide nanostructures are studied over the decades and widely recognized as a typical, which are considered as 'reference' when talking

M. Nechyporuk et al. (Eds.): ICTM 2021, LNNS 367, pp. 555–564, 2022.
https://doi.org/10.1007/978-3-030-94259-5_47

about 1D nanostructures called 'nanowires' [4]. That is why they are considered as a perspective candidate for energy storage [5], catalytic [6] and gas sensing application [7], field emission [8] and photovoltaic devices [9].

There are different technologies of synthesis of the copper oxide nanowires based on thermal [10], chemical [11], and plasma-enhanced methods [12], with different advantages and drawbacks inherent for each of them. Thus, thermal methods are the simplest and are recognized for their equipment cost-efficiency yet the time of nanowire synthesis is the longest for them [13]; use of chemical regents results in shortening the treatment time and expands the diversity of the nanostructures yet accompanied with possible environmental pollution [14]; plasma-based methods are considered as the most promising due to high reactivity of components [15] and ability to use the electric [16] and magnetic control [17] of the media in addition to the temperature and pressure gradients yet the cost-efficiency and total production yield (number of nanowires per unit area) is not so high.

In spite of the fact that CuO nanowires are studied for rather long time, their origins as well as the mechanisms of growth are not fully understood that is considered as the biggest obstacle for large-scale production and industrial application. Up-to-now, there are two experimentally observed morphologies of CuO nanowires: bi-crystalline [18] and single-crystalline [19] with two corresponding mechanisms of growth through internal boundary between the crystals [3, 20] (bi-crystalline growth) and along the external surface of the single-crystalline nanowire [21, 22]. Recently, a theoretical model describing the growth of both types of nanowires has been proposed [23] and tested by use of the data obtained from the literature, as well as by use of the experimental results found by the authors [12]. However, the relevance analysis allows establishing some discrepancies in the operation modes to reach the specified parameters of the nanowires. Thus, the most accepted temperature range to grow the longest nanowires with the largest yield is considered to be located between 500 and 700 °C with a maximum in the middle of the range; however, this maximum can be shifted towards ±100 °C [24], and both high and low temperature synthesizes for the temperatures of about 400 and 800 °C are reported [25].

That is why the purpose of this paper is to expand the experimental knowledge on the synthesis conditions of copper oxide nanowires as well as to use the experimental data to test the developed theoretical model. Since the main attention in the cited above paper was paid to the mode with oxygen pressure of 10^5 Pa, now it was reduced to the ambient air condition, i.e. the pressure was maintained at about $2 \cdot 10^4$ Pa. At the same time, the temperature range was expanded from 600 to 1000 °C, and such parameters as thicknesses of Cu_2O and CuO oxide layers, length and diameter of CuO nanowires were experimentally measured.

2 Materials and Methods

To conduct the process of copper oxide synthesis, MTI Corporation Dual Zone Split Tube furnace OTF-1200X was used as a reactor. Copper samples of 15 mm in diameter and 2 mm in thickness were put in the cylindrical quartz tube on the ceramic substrates. The samples were stayed at ambient atmospheric conditions for 5 h, and the experiments were conducted at the temperatures of 600 °C, 700 °C, 800 °C, and 1000 °C. After the annealing, the samples were taken out of the furnace and stayed at ambient temperature for 24 h. A schematic of the setup is shown in Fig. 1. The surfaces of oxidized samples were observed by field emission scanning electron microscope (SEM) to measure the average length and diameter the nanowires, as well as thicknesses of Cu_2O and CuO layers.

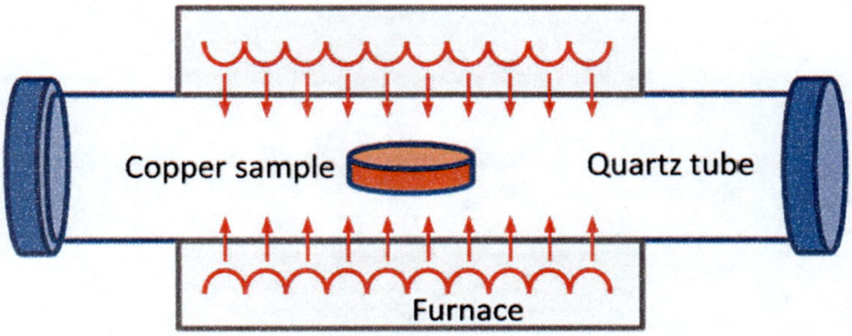

Fig. 1. Schematic of the experimental setup.

3 Experimental Results

After the experiment, all samples exhibited evidence of oxidation with black layer of oxide formed on their surface. However, the appearance of the oxide was different at the dependence on the oxidation mode, and it correlated with the density of the nanowires observed by use of SEM. Unlike the results obtained for the pressure of 10^5 Pa of oxygen [12, 23], the five-fold decrease up to $2 \cdot 10^4$ Pa of a partial oxygen pressure leads to shifting the maximum of the nanowire production yield from 600 °C to 700 °C, as it can be concluded after comparing Fig. 2, a and Fig. 2, c. For 800 °C, more short and bulky nanowires were found (Fig. 2, e) yet the diameters of the nanowires stayed almost constant (300–400 nm) through the modes (Fig. 2, b and Fig. 2, d). At 1000 °C (Fig. 2, f) the appearance of the oxide structures changed drastically. The relatively uniform distribution of the nanowires grown along the even structure of the oxide layers was transformed to 3D aggregates covered by the small and thin nanowires with average diameters and length of 70 nm and 1 μm, respectively. After the SEM analysis, the less reflective oxide on the surfaces of the samples was associated with the layer with the densest array of nanowires.

Thus, the main experimental findings can be summarized as follows. Lack of oxygen molecules (5 times lesser) resulted in generation of 12 μm of total thickness of oxide layer at ambient air conditions instead of 30 μm for 10^5 Pa of oxygen pressure at 600 °C. Supposedly, it leads to the decrease of surface stress that is considered as a driving source of the nanowire nucleation process [26, 27]. To compensate this factor, the increase in temperature is necessary – in the experiment the temperature rise up to 700 °C resulted in generation of the oxide layer with a total thickness of about 75 μm yet the thickness of CuO layer is three times lesser than was reported for the pressure of 10^5 Pa. Thus, one may conclude about 6–10 μm/h as the optimal oxidation rate to grow dense arrays of CuO nanowires.

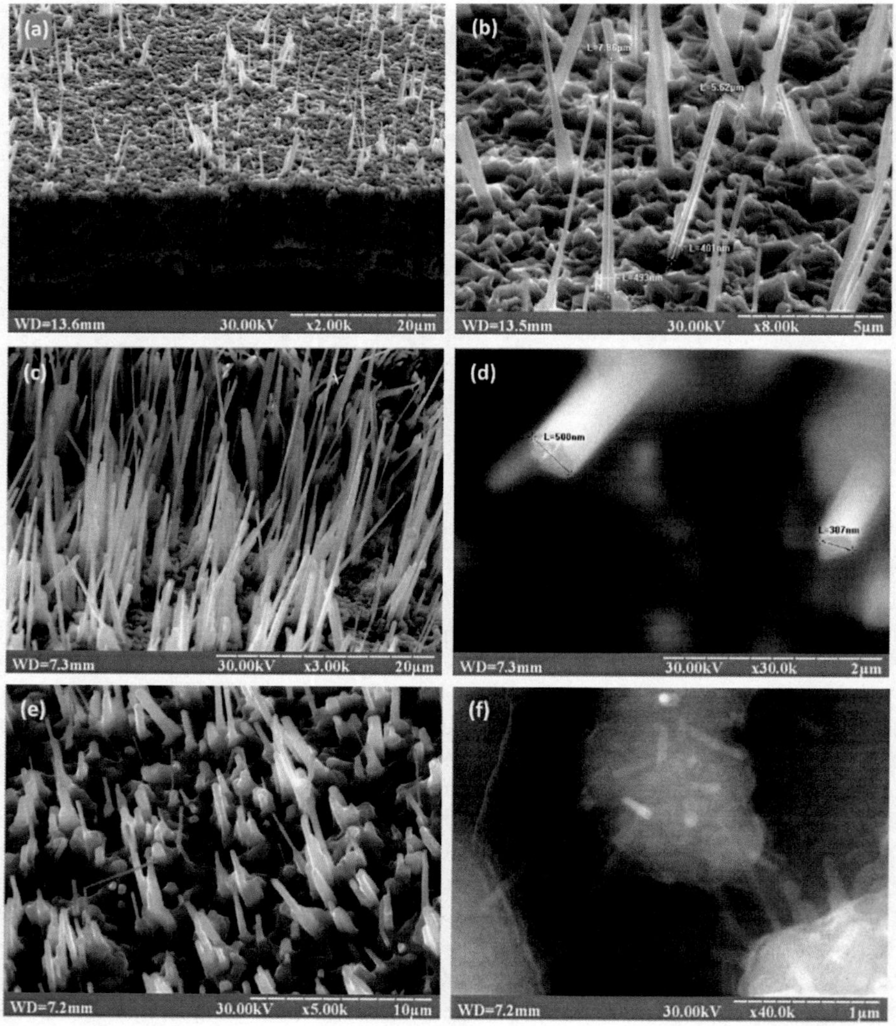

Fig. 2. Copper oxide nanostructures synthesized under ambient air conditions at the different temperatures: a, b – 600 °C; c, d – 700 °C; e – 800 °C; f – 1000 °C.

4 Theoretical Model

A theoretical model was developed to reveal the mechanism of 1D nanostructure growth, and the schematic explaining the linking dynamics of formation of Cu_2O, CuO oxide layers, as well as 1D nanostructures, can be found elsewhere [23]. The model does not consider the formation of a nanowire nuclei (seed) and proposes a nanowire growth mechanism under the condition of the seed generation. In the model, the growth of the nanowire due to the diffusion of copper along the side surface of the nanowire or through the bi-crystalline plane is considered to develop a set of equations, and it describes as the single crystal as the bi-crystal growth. After the evaluations, the following set of equations is obtained to find the time dependent thicknesses of Cu_2O layer (L_{ox1}), CuO layer (L_{ox2}), total thickness of the oxides (L_{ox}), length and diameter of the nanowires (L_{nw} and $2R_{nw}$, respectively); and the initial conditions are specified for these equations: $L_1(0) = L_{10}$, $L_2(0) = L_{20}$, $L_{ox}(0) = L_{10} + L_{20}$.

$$\frac{dL_{ox1}}{dt} = \begin{cases} \left(\dfrac{P_{b1}}{1 + a_{b1} L_{ox1\,\min}} + \dfrac{P_{l1}}{1 + a_{l1} L_{ox1\,\min}}\right) A_c, \\ (L_{ox1}(t) \le L_{ox1\,\min}) \wedge (L_{ox2}(t) \le L_{ox2\,\min}); \\ \left(\dfrac{P_{b1}}{1 + a_{b1} L_{ox1}} + \dfrac{P_{l1}}{1 + a_{l1} L_{ox1}}\right) A_c - \left(\dfrac{2B_c}{1 + bL_{ox2\,\min}} + \dfrac{2C_c}{1 + cL_{ox2\,\min}}\right) \\ -D_{nw} \dfrac{a_{ox1}^3}{a_{ox2}^3} D_c(L_{ox2\,\min}), \\ (L_{ox1}(t) > L_{ox1\,\min}) \wedge (L_{ox2}(t) \le L_{ox2\,\min}); \\ \left(\dfrac{P_{b1}}{1 + a_{b1} L_{ox1}} + \dfrac{P_{l1}}{1 + a_{l1} L_{ox1}}\right) A_c - \left(\dfrac{2B_c}{1 + bL_{ox2}} + \dfrac{2C_c}{1 + cL_{ox2}}\right) \\ -D_{nw} \dfrac{a_{ox1}^3}{a_{ox2}^3} D_c(L_{ox2}), \\ (L_{ox1}(t) > L_{ox1\,\min}) \wedge (L_{ox2}(t) > L_{ox2\,\min}); \\ \left(\dfrac{P_{b1}}{1 + a_{b1} L_{ox1\,\min}} + \dfrac{P_{l1}}{1 + a_{l1} L_{ox1\,\min}}\right) 2A_c, \\ (L_{ox1}(t) \le L_{ox1\,\min}) \wedge (L_{ox2}(t) > L_{ox2\,\min}); \\ 0, \text{ otherwise.} \end{cases} \tag{1}$$

$$\frac{dL_{ox2}}{dt} = \begin{cases} \left(\dfrac{2B_c}{1 + bL_{ox2\,\min}} + \dfrac{2C_c}{1 + cL_{ox2\,\min}}\right), \\ (L_{ox1}(t) > L_{ox1\,\min}) \wedge (L_{ox2}(t) \le L_{ox2\,\min}); \\ -\left(\dfrac{P_{b1}}{1 + a_{b1} L_{ox1}} + \dfrac{P_{l1}}{1 + a_{l1} L_{ox1}}\right) A_c + \left(\dfrac{2B_c}{1 + bL_{ox2}} + \dfrac{2C_c}{1 + cL_{ox2}}\right), \\ (L_{ox1}(t) > L_{ox1\,\min}) \wedge (L_{ox2}(t) \le L_{ox2\,\min}); \\ -\left(\dfrac{P_{b1}}{1 + a_{b1} L_{ox\,\min 1}} + \dfrac{P_{l1}}{1 + a_{l1} L_{ox1\,\min}}\right) A_c, \\ (L_{ox1}(t) \le L_{ox1\,\min}) \wedge (L_{ox2}(t) > L_{ox2\,\min}); \\ 0, \text{ otherwise.} \end{cases} \tag{2}$$

$$\frac{dL_{ox}}{dt} = \begin{cases} \left(\frac{P_{b1}}{1+a_{b1}L_{ox1\min}} + \frac{P_{l1}}{1+a_{l1}L_{ox1\min}}\right)A_c, \\ (L_{ox1}(t) \le L_{ox1\min}) \wedge (L_{ox2}(t) \le L_{ox2\min}); \\ \left(\frac{P_{b1}}{1+a_{b1}L_{ox1}} + \frac{P_{l1}}{1+a_{l1}L_{ox1}}\right)A_c - D_{nw}\frac{a_{ox1}^3}{a_{ox2}^3}D_c(L_{ox2\min}), \\ (L_{ox1}(t) > L_{ox1\min}) \wedge (L_{ox2}(t) \le L_{ox2\min}); \\ \left(\frac{P_{b1}}{1+a_{b1}L_{ox1}} + \frac{P_{l1}}{1+a_{l1}L_{ox1}}\right)A_c - D_{nw}\frac{a_{ox1}^3}{a_{ox2}^3}D_c(L_{ox2}), \\ (L_{ox1}(t) > L_{ox1\min}) \wedge (L_{ox2}(t) > L_{ox2\min}); \\ \left(\frac{P_{b1}}{1+a_{b1}L_{ox1\min}} + \frac{P_{l1}}{1+a_{l1}L_{ox1\min}}\right)A_c, \\ (L_{ox1}(t) \le L_{ox1\min}) \wedge (L_{ox2}(t) > L_{ox2\min}); \\ 0, \text{ otherwise.} \end{cases} \tag{3}$$

$$\frac{dL_{nw}}{dt} = \begin{cases} \alpha_L(t) n_{Cu20nw}(L_{ox2\min}) \cos(\beta_L L_{nw}), \\ (L_{ox1}(t) > L_{ox1\min}) \wedge (L_{ox2}(t) \le L_{ox2\min}); \\ \alpha_L(t) n_{Cu20nw}(L_{ox2}) \cos(\beta_L L_{nw}), \\ (L_{ox1}(t) > L_{ox1\min}) \wedge (L_{ox2}(t) > L_{ox2\min}); \\ 0, \text{ otherwise.} \end{cases} \tag{4}$$

$$\frac{dR_{nw}}{dt} = \begin{cases} \alpha_L(t)\frac{n_{x2s}(t)}{n_{x2t}(t)} n_{Cu20nw}(L_{ox2\min}), \\ (L_{ox1}(t) > L_{ox1\min}) \wedge (L_{ox2}(t) \le L_{ox2\min}); \\ \alpha_L(t)\frac{n_{x2s}(t)}{n_{x2t}(t)} n_{Cu20nw}(L_{ox2}), \\ (L_{ox1}(t) > L_{ox1\min}) \wedge (L_{ox2}(t) > L_{ox2\min}); \\ 0, \text{ otherwise.} \end{cases} \tag{5}$$

5 Results and Discussion

The developed model was used to calculate the nanowire growth under the different substrate temperatures as the main growth parameters. The experimental dependence of the sample temperature is approximated for the calculations by the expression: $T_s(t) = (T_{max} - T_0)(1 - exp[-t/\tau]) + 293$ (K), where T_{max} was 600, 700, and 800 °C; $T_0 = 20$ °C, $\tau = 2000$ s. This dependence was used in the system of Eqs. (1–5) to calculate the thickness of the cuprous oxide (Cu_2O) layer $L_{ox1}(t)$, cupric oxide (CuO) layer $L_{ox2}(t)$, and total oxide thickness $L_{ox}(t)$, nanowire length $L_{nw}(t)$ and diameter $2R_{nw}(t)$ on time t. In the simulation, the following values are used [23]: adsorption energies $\varepsilon_{aO2} = 1.15$ eV; $\varepsilon_{aO2nws} = 0.27$ eV; $\varepsilon_{aO2nwt} = 0.5$ eV; dissociation energy of O_2 on the Cu_2O surface $\varepsilon_{x\text{-}dis} = 0.65$ eV; energy of oxygen diffusion $\varepsilon_{x2} = 0.8$ eV. The list of other energies used in the calculations is shown in Table 1.

Table 1. Energies used in the calculations.

T, °C	Energies, eV					
	ε_{c1b}	ε_{c1l}	ε_{c2}	ε_{c2s}	$\varepsilon_{i\text{-}O2s}$	$\varepsilon_{i\text{-}O2t}$
600	1.35	1.94	0.7	0.2	0.94	0.64
700	1.27	1.86	0.5	0.2	0.94	0.64
800	1.39	2.02	0.45	0.2	1.2	0.8

The results of the calculations are shown in Fig. 3. From the comparison of the results for the air conditions, when the oxygen pressure is $2\cdot10^4$ Pa, with the calculations performed for the pressure of 10^5 Pa, the following conclusions can be made. The energies in the table are close to the values used for the high pressure. The difference of about 0.1 eV in the diffusion energies can be attributed to the difference in the surface stress generated under the different growth conditions. At the same time, the rates of the growth of Cu_2O layer for air conditions at 700 °C correspond to the rates obtained for 10^5 Pa at the same temperature in spite of the five-fold decrease of oxygen concentration, while the rate of the growth of CuO layer in the present experiment is three times lesser that follows the trend of the oxygen concentration decrease.

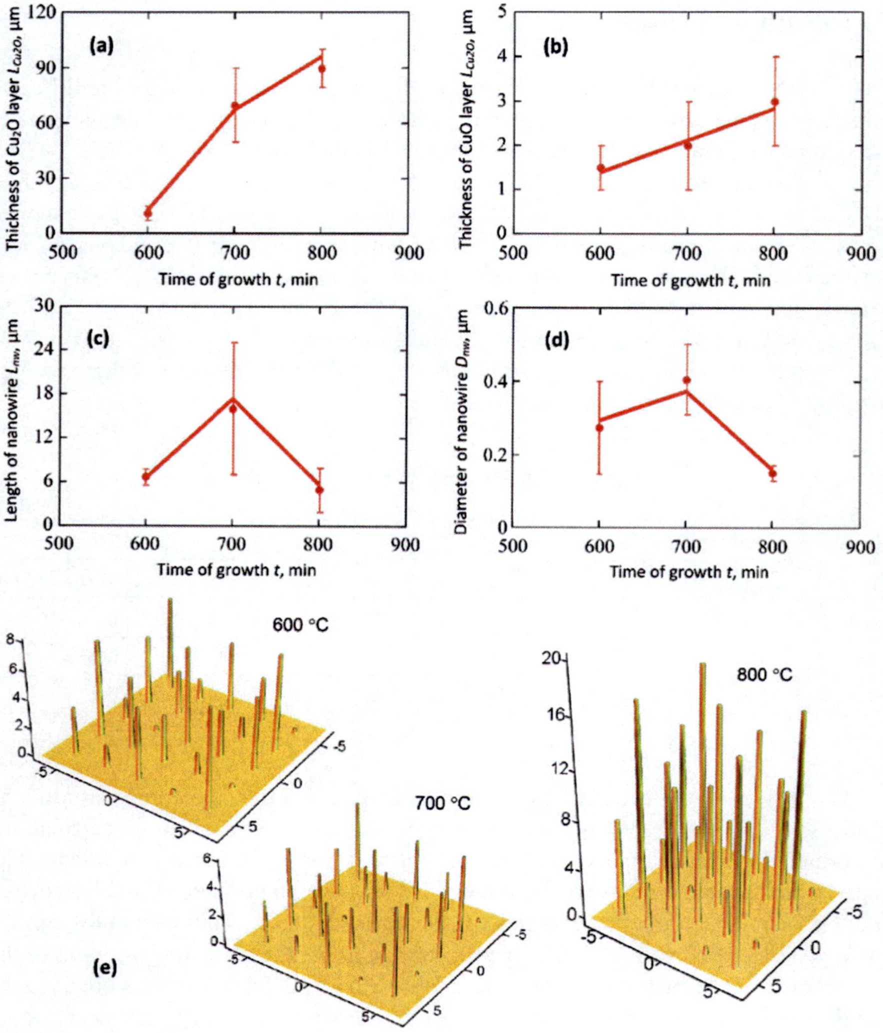

Fig. 3. Results of the experiments and calculations: a – c – dependencies on time of growth: a – thickness of Cu_2O layer; b – thickness of CuO layer; c – length of oxide nanowires; d – diameter of oxide nanowires; e – 3D view after 5 h of growth (simulations).

6 Conclusions

The results of the experiments and calculations allowed expanding the range of application of the developed theoretical model. It was confirmed that the dynamics of the oxide layers and nanostructures can be simulated by use of the same energy values as for the five times higher oxygen pressure. It was found that the thickness of Cu_2O layer is almost non-dependent on the oxygen pressure in the range of $2{\cdot}10^4$ to 10^5 Pa for 700 °C, while the thickness of CuO oxide decreases with the pressure decrease. The length of the nanowires also does not change with the pressure decrease yet the large measurements error for 700 °C in the present experiment should be mentioned. At the same time, the increase of the temperature to 800 °C leads to the decrease of the nanostructure size and density, which correspond to the trend reported in the literature. Further temperature rise results in forming 3D aggregates with thin nanowires.

Acknowledgements. The authors acknowledge the support from the project funded by National Research Foundation of Ukraine, under grant agreement No. 2020.02/0119.

References

1. Mahmoodi, A., Solaymani, S., Amini, M., Nezafat, N.B., Ghoranneviss, M.: Structural, morphological and antibacterial characterization of CuO nanowires. SILICON **10**(4), 1427–1431 (2017). https://doi.org/10.1007/s12633-017-9621-2
2. Zúñiga, A., Fonseca, L., Souza, J.A., Rivaldo-Gomez, C., Pomar, C.D., Criado, D.: Anomalous ferromagnetic behavior and size effects in CuO nanowires. J. Magn. Magn. Mater. **471**, 77–81 (2019)
3. Rackauskas, S., Nasibulin, A.G.: Nanowire growth without catalysts: applications and mechanisms at the atomic scale. ACS Appl. Nano Mater. **3**(8), 7314–7324 (2020)
4. Majumdar, D., Ghosh, S.: Recent advancements of copper oxide-based nanomaterials for supercapacitor applications. J. Energy Storage **34**, 101995 (2021)
5. Su, Y., Liu, T., Zhang, P., Zheng, P.: CuO nanowire arrays synthesized at room temperature as a high-performance anode material for Li/Na-ion batteries. Thin Solid Films **690**, 137522 (2019)
6. Nkhaili, L., et al.: A simple method to control the growth of copper oxide nanowires for solar cells and catalytic applications. Adv. Conden. Matter Phys. **2020**, 1–8 (2020)
7. Yang, F., et al.: Design and understanding of a high-performance gas sensing material based on copper oxide nanowires exfoliated from a copper mesh substrate. J. Mater. Chemis. A **3**, 20477–20481 (2015)
8. Feng, L., et al.: Excellent field emission properties of vertically oriented CuO nanowire films. AIP Adv. **8**(4), 045109 (2018)
9. Jafari, A., Terohid, S., Kokabi, A., Moradiani, A.: Electrical, structural, and photocatalytic properties of copper oxide nanowire. J. Chem. Res. **44**(7–8), 471–474 (2020)
10. Guo, B., et al.: Single-crystalline metal oxide nanostructures synthesized by plasma-enhanced thermal oxidation. Nanomaterials **9**(10), 1405 (2019)
11. Kim, M.J., Alvarez, S., Yan, T., Tadepalli, V., Fichthorn, K.A., Wiley, B.J.: Modulating the growth rate, aspect ratio, and yield of copper nanowires with Alkylamines. Chem. Mater. **30**(8), 2809–2818 (2018)
12. Baranov, O., Filipič, G., Cvelbar, U.: Towards a highly-controllable synthesis of copper oxide nanowires in radio-frequency reactive plasma: fast saturation at the targeted size. Plasma Sources Sci. Technol. **28**, 084002 (2019)
13. Sondors, R., et al.: Size distribution, mechanical and electrical properties of CuO nanowires grown by modified thermal oxidation methods. Nanomaterials **10**(6), 1051 (2020)
14. Filipič, G., Cvelbar, U.: Copper oxide nanowires: a review of growth. Nanotechnology **23**, 194001 (2012)
15. Baranov, O., Romanov, M.: Current distribution on the substrate in a vacuum arc deposition setup. Plasma Process. Polym. **5**, 256 (2008)

16. Levchenko, I., Romanov, M., Baranov, O., Keidar, M.: Ion deposition in a crossed E×B field system with vacuum arc plasma sources. Vacuum **72**, 335 (2004)
17. Baranov, O., Romanov, M., Wolter, M., Kumar, S., Zhong, X., Ostrikov, K.: Low-pressure planar magnetron discharge for surface deposition and nanofabrication. Phys. Plasmas **17**, 053509 (2010)
18. Zhu, Y., Mimura, K., Isshiki, M.: Influence of oxide grain morphology on formation of the CuO scale during oxidation of copper at 600–1000 °C. Corros. Sci. **47**(2), 537–544 (2005)
19. Zhang, Q., et al.: CuO nanostructures: synthesis, characterization, growth mechanisms, fundamental properties, and applications. Prog. Mater. Sci. **60**, 208–337 (2014)
20. Tu, C.H., et al.: Resistive memory devices with high switching endurance through single filaments in Bi-crystal CuO Nanowires. J. Alloy. Compd. **615**(5), 754–760 (2014)
21. Filipič, G., Baranov, O., Mozetič, M., Cvelbar, U.: Growth dynamics of copper oxide nanowires in plasma at low pressures. J. Appl. Phys. **117**, 043304 (2015)
22. Shi, J., et al.: Synergistic effects on thermal growth of CuO nanowires. J. Alloys Compounds **815**, 152355 (2020)
23. Baranov, O., Košicek, M., Filipic, G., Cvelbar, U.: A deterministic approach to the thermal synthesis and growth of 1D metal oxide nanostructures. Appl. Surface Sci. **566**, 150619 (2021). https://doi.org/10.1016/j.apsusc.2021.150619
24. Mohamed, S.H., Al-Mokhtar, K.M.: Characterization of Cu_2O/CuO nanowire arrays synthesized by thermal method at various temperatures. Appl. Phys. A **124**(7), 493 (2018)
25. Tran, T.H., et al.: Effect of annealing temperature on morphology and structure of CuO nanowires grown by thermal oxidation method. J. Cryst. Growth **505**, 33–37 (2019)
26. Chason, E., Engwall, A.M., Rao, Z., Nishimura, T.: Kinetic model for thin film stress including the effect of grain growth. J. Appl. Phys. **123**, 185305 (2018)
27. Jagtap, P., Chason, E.: A unified kinetic model for stress relaxation and recovery during and after growth interruptions in polycrystalline thin films. Acta Mater. **193**, 202–209 (2020)

Deposition of Titanium-Containing Coatings by Low-Pressure Cold Spraying

Andrey Volkov[1(✉)], Oleksandr Shorinov[1], and Sergii Polyviany[2]

[1] National Aerospace University "Kharkiv Aviation Institute", 17 Chkalova Street, Kharkiv 61070, Ukraine
a.volkov@khai.edu

[2] Motor Sich JSC, 15 Motorobudivnykiv Avenue, Zaporizhzhia 69068, Ukraine

Abstract. The paper carried out the study of the possibility of obtaining titanium-containing coatings by low-pressure cold spraying. Using the method of numerical simulation, the energy parameters of particles of sprayed mixture of Ti6Al4V + Al_2O_3 powders were determined and analyzed upon collision with the substrate with different weight percentages of Al_2O_3 powder in the mixture, which makes it possible to predict and more accurately select the process parameters. A practical study was carried out with the deposition of a mixture of Ti6Al4V and A-20-11 powders for metallographic analysis of resulting coatings. Low-pressure cold-sprayed coatings for metallographic analysis were prepared in accordance with ASTM E3-01 followed by scanning electron microscopy. The research results showed that the obtained coatings have a rather dense structure when using mixtures with more plastic materials and corundum, and the analysis of influence of energy parameters of spraying mode on microstructure of coatings obtained enable the control of the quality of coatings when changing spraying modes in the considered range. The carried out experimental and theoretical studies of the spraying process will make it possible to find scientifically grounded technological recommendations for performing work on restoration and repair of parts from titanium alloys, as well as obtaining protective coatings with a predetermined gradient of thickness properties.

Keywords: Cold spraying · Protective and restorative coatings · Titanium alloys · Physical and mechanical characteristics · Technological recommendations

1 Introduction

Taking into account the high cost of materials used and the general trend towards less material-intensive production, additive technologies have been increasingly introduced into industry in recent decades, which include the methods of spraying functional coatings described in many papers [1–3].

The ability of cold spraying method to create protective and restorative coatings with properties close to those of the base material and/or functional properties superior to the base material not only allows extension of the life of new parts, but also provides a second

M. Nechyporuk et al. (Eds.): ICTM 2021, LNNS 367, pp. 565–574, 2022.
https://doi.org/10.1007/978-3-030-94259-5_48

life to damaged parts that would be otherwise discarded both at the manufacturing stage and during operation.

The unique features of cold spraying method include obtaining of coatings without phase transformations, oxidation and changes in properties in coating and substrate materials due to use of low process temperatures, which has already made this method very popular in the industry, where the quality and reliability of products are paramount, and versatility and relative cheapness of the method makes it possible to use it in many industries.

To date, the use of materials from Ti and Ti alloys in cold spraying is mainly focused on commercially pure Ti, with limited use of Ti6Al4V alloy [4–6].

Studies of obtaining coatings by cold spraying of titanium-containing materials are mainly focused on the use of high-pressure equipment and expensive gases He and N_2 [7, 8].

Thick coatings of titanium-containing materials (several millimeters) can be obtained using modern cold spraying technology; although it is still difficult to achieve a pore-free microstructure. The mechanical properties of cold-sprayed titanium coatings are critical for industrial applications [9–11].

Taking this into account, the purpose of this article is to determine the possibilities of obtaining titanium-containing coatings by low-pressure cold spraying.

2 Numerical Study

2.1 Particle Impact: Features of the Process

The processes of high-velocity impact of solid particles with substrate are of significant interest for understanding the mechanism of formation of coatings during cold spraying, however, the coating formation is yet to be fully studied [12] and the mechanism of bond formation between a particle and a substrate during high-velocity impact is the most discussed topic of cold spraying.

The process of coating forming during cold spraying can be conditionally divided into 2 stages:

1) impact of particles with the surface of substrate, its activation followed by adhesion;
2) interaction of subsequent particles with previously fixed ones and shock compression of particle layers.

It is generally accepted that during cold spraying the material should have plasticity features which ensure pressing, shearing of surfaces and fixing. According to the theory of dynamics, the pressure shock wave and the resulting plastic shock wave lead to high deformations of the particle. When the particle collides with the substrate, the pressure and elastic deformation increase up to the dynamic yield point followed by plastic deformation with significant deformation of the material structure and its heating.

A critical velocity by particles of the sprayed material is required to achieve the above properties and states. To create favorable conditions for fixing and additional deformation of particles, a commercially available powder of titanium alloy Ti6Al4V was mixed with a composite powder on a metal matrix $Al + Zn + Al_2O_3$ (see Table 1).

Table 1. Composition of a composite powder on a metal matrix $Al + Zn + Al_2O_3$.

Material content and size	Zinc content and size	Corundum content and size
60% wt. Al; 22 μm, SD6	25% wt. Zn; 10 μm, SD5	15% wt. Al_2O_3; 22 μm, SD6

2.2 Selection of Materials and Setting Initial Conditions

The effect of particles on the substrate was simulated using the ANSYS AUTODYN software. This study used a combined Lagrange-Euler method to avoid the problems associated with the deformation of computational mesh. Simulation of various spraying conditions was carried out in a two-dimensional deformation plane with certain initial conditions and selection of necessary models for materials.

Materials were selected from the AUTODYN material library; substrate material is titanium with the impact interaction equation of state and Steinberg-Guinan strength model. The titanium alloy Ti6Al4V with the impact interaction equation of state and Steinberg-Guinan strength model was adopted as the material of metal particles. The solid particles are aluminum oxide with the impact interaction equation of state and Von Mises strength model. The initial conditions were taken as follows: initial speed of Ti6Al4V – 525 m/s, substrate – 0 m/s; different initial conditions for particles of aluminum oxide were selected for each individual experiment. The size of computational mesh was taken to be 2 μm to improve the computation time; and as noted in [13–15], with a decrease in the mesh size, the velocity for the onset of shear instability decreases almost linearly and the degree of particle elongation during collision remains almost the same.

It was found in [16] that the weight percentage (wt.%) of aluminum oxide powder in the mixture is the determining factor in the efficiency of coating. For this purpose, different weight percentages of Al_2O_3 powder were used in the mixture (see Table 2).

Table 2. Particle size of the powder used in simulation.

Experiment No.	Ti6Al4V particle size, μm	Al_2O_3 particle size, μm	Al_2O_3 weight percentage in the mixture	Al_2O_3 particle velocity, m/s
1	30	38	54	515
2	30	35	49	530
3	30	30	42	560
4	30	25	33	600
5	30	20	24	650
6	30	15	15	720
7	30	10	7	830

(continued)

Table 2. (*continued*)

Experiment No.	Ti6Al4V particle size, μm	Al_2O_3 particle size, μm	Al_2O_3 weight percentage in the mixture	Al_2O_3 particle velocity, m/s
8	30	5	2	1070
9	30	–	–	–

Figure 1 shows a general 2D model of experiments with Ti6Al4V particles and four Al_2O_3 particles with a distance of 5 μm between the particles.

Fig. 1. Finite element model

Ti6Al4V particle diameter was taken to be 30 μm in all experiments; Al_2O_3 diameter of particles changed with each experiment from 38 μm and decreased with decreasing weight percentage of Al_2O_3 particles in the mixture. Since the acceleration of particles during cold spraying depends on their size, the speed of Al_2O_3 particles was calculated for each experiment. The calculation of particle velocity was carried out according to equation [17]:

$$V_p = V\sqrt{\frac{C_D A_p \rho x}{m}}, \tag{1}$$

where V_p is the particle velocity; V is the gas flow rate; C_D is the coefficient of front drag; A_p is the cross-sectional area of the particle; ρ is the gas density; x is the axial position; m is the particle mass.

2.3 Simulation Results

Figure 2 shows simulation results of interaction of particles of a powder mixture with different weight content of Al_2O_3 upon impact on the substrate.

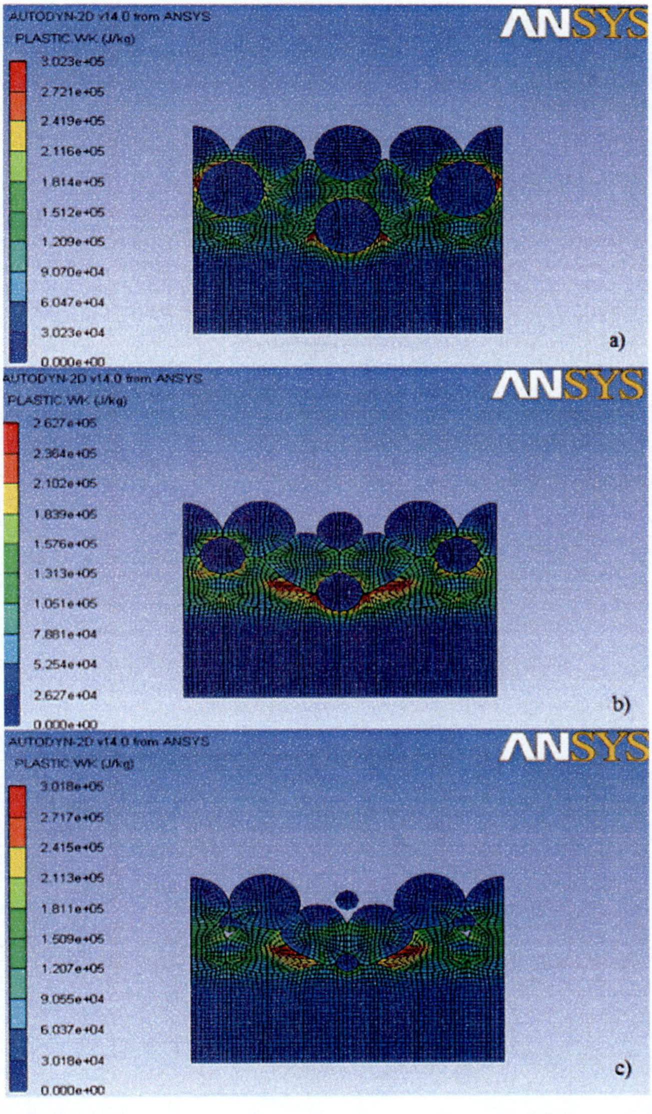

Fig. 2. Simulation results: change in plastic response at different weight content of Al_2O_3 particles in the powder mixture: a) Ti6Al4V + 42wt.% Al_2O_3; b) Ti6Al4V + 24wt.% Al_2O_3; c) Ti6Al4V + 7wt.% Al_2O_3

The outlines show the change in the plastic response of the coating. It is clearly seen that only Ti6Al4V particles were deformed, while solid Al_2O_3 particles remain in a non-deformed state. It was found that the highest strain rates are in the surrounding areas of Al_2O_3 particles, which deform Ti6Al4V particles. It can also be concluded that higher substrate deformation rates occur with larger particles.

Part of mechanical energy used during the plastic deformation process is converted into heat, while the remainder of the energy accumulates in the material. The fraction of plastic response velocity is dissipated in the form of heat and is often taken as a constant parameter equal to 0.9 [18].

Figure 3 shows the energy changes during the collision of particles of the powder mixture Ti6Al4V + 33wt.% Al_2O_3.

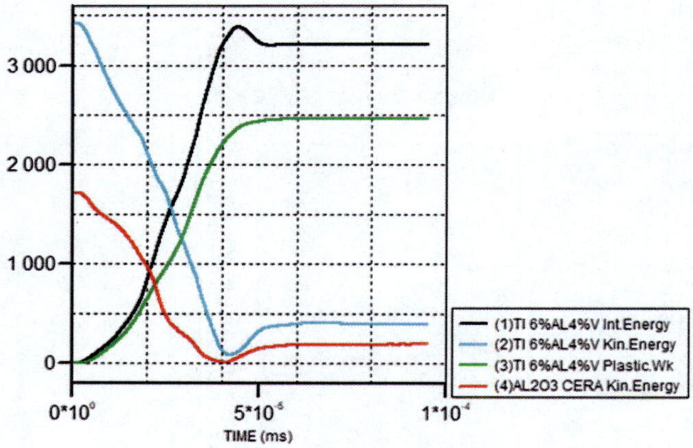

a

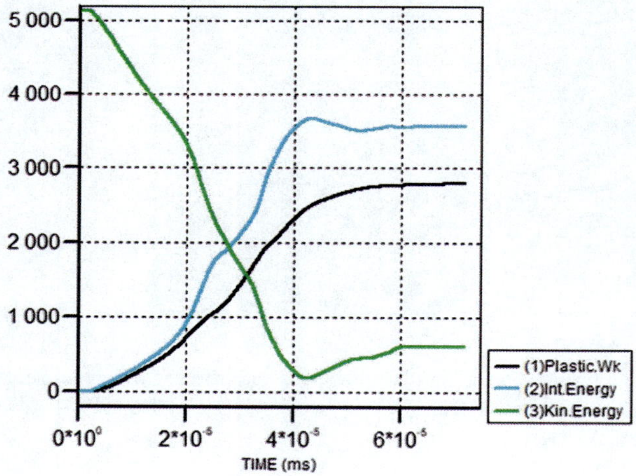

b

Fig. 3. Energy change during collision: a) for Ti6Al4V + 33wt.% Al_2O_3; b) for pure Ti6Al4V

As can be seen from Fig. 3, the amount of kinetic energy transferred by Ti6Al4V particles in the Ti6Al4V + Al_2O_3 mixture is theoretically completely retained in the metal; conversely, the amount of energy contained in the Al_2O_3 particles is sufficient to carry out the plastic response and is also dissipated as heat. For the case of Ti6Al4V spraying without Al_2O_3 particles, the energy is transferred to material particles, and the residual energy is subsequently dissipated as heat during the plastic response. In both cases, part of the total energy of the system is used to perform the plastic response of the substrate, which increased with increasing weight percentage of Al_2O_3 particles in the mixture, the amount of energy is also stored in Ti6Al4V, and the plastic response remains the same in both cases.

With an increase in the weight percentage of Al_2O_3 particles by 5%, the plastic response of Ti6Al4V decreases; when a higher Al_2O_3 content is added, the plastic response begins to increase, reaching an optimal point of about 30%; this optimal value corresponds to the results of experiments, which determined that the optimal value of the content of Al_2O_3 particles in the mixture amounts to 30% [16, 19]. After this point, plastic response begins to decrease by up to 50%; then it increases to saturation of the Al_2O_3 particles, in which the excess of solid particles deforms the residual metal particles. On the side of the substrate, this plastic response exponentially increases with increasing weight percentage of Al_2O_3 particles.

It can be concluded that a certain amount of kinetic energy of Al_2O_3 particles plays an important role in the deformation of metal particles during the deposition of a Ti6Al4V $_+$ Al_2O_3 mixture. Thus, having calculated the required kinetic energy for Al_2O_3 particles in the mixture and selecting Al_2O_3 particles of a certain diameter with a certain weight content in the mixture, the formation of a coating is possible. Computational simulation can be used in order to understand an effect of low-pressure cold spray process parameters on particle kinetic energy as well as to verify calculated results of Eq. 1 [20].

3 Numerical Study: Metallographic Analysis of Coatings

Intense plastic deformation of particles during cold spraying leads to strengthening of the resulting coating. This suggests that the extensive plastic flow of material particles is the main process that determines the structure of the coating formation. It is well known that metals are among the materials that in most cases have excellent ductility and low dynamic yield strength. Metal particles with high kinetic energy build up a dense coating through deformation on impact. Thus, the macroscopic mechanical properties of composites are strongly influenced by the bulk properties of the constituents; they also affect the mechanical characteristics of the matrices/reinforcement interfaces.

Low-pressure cold spray coatings for metallographic analysis were prepared in accordance with ASTM E3-01 followed by scanning electron microscopy.

Figure 4 shows images of a scanning electron microscope at 150× magnification for samples deposited at initial temperatures of 500 °C (*a*), 600 °*C* (*b*), and 700 °*C* (*c*) and the percentage composition of 70/30 mixture of Ti6Al4V titanium alloy powder with Al + Zn + Al_2O_3. In the microstructure, black particles are particles of non-deformed corundum incorporated into the coating. The coating thickness amounted to 200 μm. The visual inspection of coatings shows a dense structure of softer aluminum and zinc with small particles of titanium alloy and practically non-deformed large particles of titanium alloy. In the microstructure, the boundaries of powder particles are clearly visible. The flattened shape of small particles showed a high degree of deformation.

Despite high coating density, some open boundaries near the coating surface and some oxidized boundaries in the coating structure can be noted. The oxide layer can be explained by the electrolytic method of powder production [21, 22]; this is typical for all powders of a dendritic structure, which have a large surface area compared to the spherical shape of particles obtained by spraying. In addition, the source of oxidation can be the use of air as a working gas [23, 24].

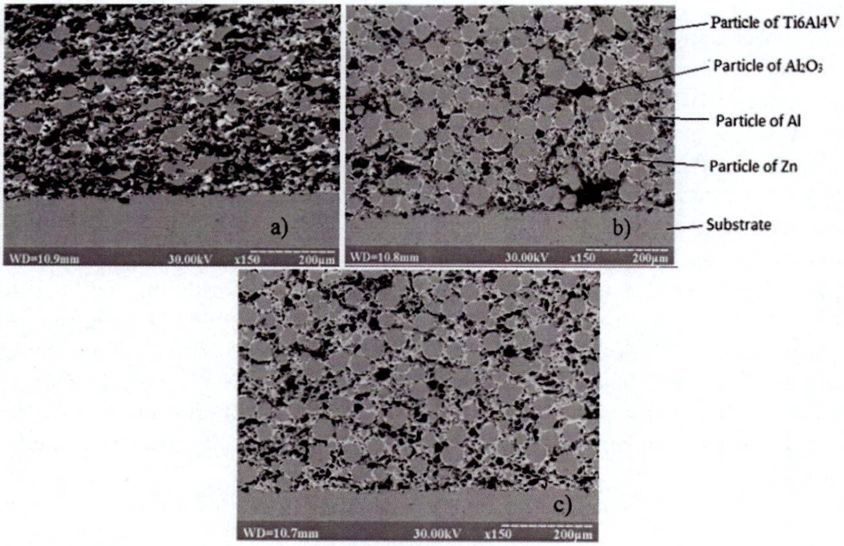

Fig. 4. SEM image at 150× magnification of a sample coated with a mixture of Ti6Al4V titanium alloy powder with Al + Zn + Al_2O_3 in 70/30 ratio, respectively, at initial temperatures of 500 °C (*a*), 600 °*C* (*b*) and 700 °*C* (*c*)

4 Conclusions

A numerical experiment was carried out to study the effect of powder particles of various compositions on a substrate for a better understanding and assessment of the coating process by cold spraying using the ALE (Arbitrary Lagrangian Eulerian) method available in the ANSYS AUTODYN software.

A full-scale experiment was carried out to study the effect of powder particles of various compositions on a substrate for a better understanding and assessment of the coating process by the method of low-pressure cold spraying.

ASTM standards were used for the analysis of obtained experimental data. The analysis of the microstructure is carried out.

The research results enable establishment of the possibility to obtain titanium-containing coatings by the method of low-pressure cold spraying and the influence of energy parameters of spraying mode on the microstructure of the resulting coatings, which makes it possible to control the quality of coatings when changing the spraying modes in the considered range.

References

1. Bagherifard, S., Guagliano, M.: Fatigue performance of cold spray deposits: coating, repair and additive manufacturing cases. Int. J. Fatigue **139**, 105744 (2020)
2. Huang, J., et al.: Pure copper components fabricated by cold spray (CS) and selective laser melting (SLM) technology. Surf. Coat. Technol. **395**, 125936 (2020)
3. Wu, H., Liu, S., Zhang, Y., Liao, H., Raoelison, R.-N., Deng, S.: New process implementation to enhance cold spray-based additive manufacturing. J. Therm. Spray Technol. **30**(5), 1284–1293 (2021). https://doi.org/10.1007/s11666-021-01205-y
4. Tan, K., Markovych, S., Hu, W., Wang, Y., Shorinov, O., Wang, Y.: On the characteristics of cold spray technology and its application in aerospace industries. IOP Conf. Ser. Earth Environ. Sci. **719**, 032023 (2021)
5. MacDonald, D., Fernández, R., Delloro, F., Jodoin, B.: Cold spraying of armstrong process titanium powder for additive manufacturing. J. Therm. Spray Technol. **26**(4), 598–609 (2016). https://doi.org/10.1007/s11666-016-0489-2
6. Petrovskiy, P., Khomutov, M., Cheverikin, V., Travyanov, A., Sova, A., Smurov. I.: Influence of hot isostatic pressing on the properties of 316L stainless steel, Al-Mg-Sc-Zr alloy, titanium and Ti6Al4V cold spray deposits. Surf. Coatings Technol. **405**, 126736 (2021)
7. Tan, A.W.-Y., et al.: Influence of particle velocity when propelled using N_2 or N_2-He mixed gas on the properties of cold-sprayed Ti6Al4V coatings. Coatings **8**(9), 327 (2018)
8. Garrido, M.A., Sirvent, P., Poza, P.: Evaluation of mechanical properties of Ti6Al4V cold sprayed coatings. Surf. Eng. **34**, 399–406 (2018)
9. Tan, A.W.-Y., et al.: Microstructure, mechanical and tribological properties of cold sprayed Ti6Al4V–CoCr composite coatings. Compos. Part B Eng. **202**, 108280 (2020)
10. Li, W., Assadi, H., Gaertner, F., Yin, S.: A review of advanced composite and nanostructured coatings by solid-state cold spraying process. Crit. Rev. Solid State Mater. Sci. **8436**, 1–48 (2018)
11. Tortuero , S., Garrido, M. A., Poza, P., Rodríguez, J.: Evaluating the erosion resistance of Ti6Al4V coatings deposited by cold spray. Wear **454–455**, 203337 (2020)
12. Assadi, H., Gärtner, F., Klassen, T., Kreye, H.: Comment on "Adiabatic shear instability is not necessary for adhesion in cold spray." Scripta Mater. **162**, 512–514 (2019)
13. Hassani-Gangaraj, M., Veysset, D., Champagne, V.K., Nelson, K.A., Schuh, C.A.: Response to Comment on "Adiabatic shear instability is not necessary for adhesion in cold spray." Scripta Mater. **162**, 515–519 (2019)
14. Dolmatov, A.I., Bilchuk, O.V.: Modelling of gas flow with solid particles in a short nozzle. Metallofiz. Noveishie Tekhnol. **40**(9), 1257 (2018)
15. Dolmatov, A.I., Bilchuk, O.V.: Simulation of process of solid particles sputtering with dimet nozzle. Metallofiz. Noveishie Tekhnol. **41**(7), 927 (2019)
16. Zhang, Z., Liu, F., Han, E.-H., Xu, L., Uzoma, P.C.: Effects of Al_2O_3 on the microstructures and corrosion behavior of low-pressure cold gas sprayed Al 2024-Al_2O_3 composite coatings on AA 2024–T3 substrate. Surf. Coatings Technol. **370**, 53–68 (2019)
17. Dolmatov, A.I., Sergeev, S.V., Kurin, M.O., Voronko, V.V., Loza, T.V.: Kinematics of the solid particle accelerated by a flow of gas in a supersonic nozzle and work hardening of the processed surface. Metallofiz. Noveishie Tekhnol. **37**(7), 871–885 (2015)

18. Guruswami, R., Rosakis, A.J., Hodowany, J., Rosakis, P.: On the conversion of plastic work into heat during high strain-rate deformation. AIP Conf. Proc. **620**, 557 (2002)
19. Shockley, J.M., Descartes, S., Vo, P., Irissou, E., Chromik, R.R.: The influence of Al_2O_3 particle morphology on the coating formation and dry sliding wear behavior of cold sprayed Al-Al_2O_3 composites. Surf. Coatings Technol. **270**, 324–333 (2015)
20. Dolmatov, A.I., Polyviany, S.A.: Interaction of solid particles from a gas stream with the surface of a flat nozzle. Metallofiz. Noveishie Tekhnol. **43**(3), 319–328 (2021)
21. ASTM E3–01. Standard Guide for Preparation of Metallographic Specimens (2007)
22. Lagerbom, J., Mäkinen, H., Vuoristo, P.: Effect of heat treatment on properties of cold sprayed coatings. In: Thermal Spray 2005: EXPLORE its Surfacing Potential, pp. 240–244. ASM International, Basel (2005)
23. Koivuluoto, H., Kulmala, M., Vuoristo, P.: Structural properties of high-pressure cold- sprayed and low-pressure cold-sprayed coatings. In: Surface Modification Technologies, vol. XXII, pp. 65–72 (2008)
24. Zou, Y., Qin, W., Irissou, E., Legoux, J.-G., Yue, S., Szpunara, J.A.: Dynamic recrystallization in the particle/particle interfacial region of cold-sprayed nickel coating: electron backscatter diffraction characterization. Scripta Mater. **61**, 899–902 (2009)

Nanocomposites Implementation for Oil Storage Systems Electrostatic Protection

Olena Sierikova[1(✉)], Elena Strelnikova[2], Vasil Gnitko[2], Anatoliy Tonkonozhenko[3], and Leonid Pisnia[4]

[1] National University of Civil Defence of Ukraine, Kharkiv, Ukraine
sierikova_olena@ukr.net
[2] A. M. Pidhorny Institute for Mechanical Engineering Problems NAS of Ukraine, Kharkiv, Ukraine
[3] M. K. Yangel Yuzhnoye State Design Office, Dnipro, Ukraine
[4] Ukrainian Scientific Research Institute of Ecological Problems, Kharkiv, Ukraine

Abstract. Storage tanks are responsible engineering constructions designed to get, store, release and account the oil and petroleum products. Tanks elements in operational conditions sustain significant changes of temperature, high pressure, vibration, seismic loads, uneven subsidence, corrosion. The crucial issue today is to ensure the reliability of operation and trouble-free operation of oil storage systems. One of the main causes of explosions and fires in oil storage tanks (reservoirs) is the static electricity generated in the pipeline during the oil transportation. The aim has been to increase the environmental safety of oil storage systems by effectively neutralizing static electricity using composites and nanocomposites with disordered and ordered systems of fibrous inclusions. Therefore, the innovative oil storage construction with improved characteristics has been proposed in the paper. It has been established the use of filler in the carbon fibers form demonstrates the growth of material strength characteristics in the fibers direction. Complete structured ordering of nanofibers further increases the strength properties in the main direction of the fibers location, but does not significantly reduce the strength characteristics in the other two directions. The results of research prove the use of carbon nanofibers to create the quality nanocomposite is appropriate and could be used to neutralize static electric charges that occur in oil storage and could lead to emergencies.

Keywords: Emergency · Environmental safety · Technogenic object · Oil reservoirs · Nanomaterial · Carbon fiber inclusions

1 Problem Statement

Storage tanks are responsible engineering constructions designed to get, store, release and account the oil and petroleum products. Tanks elements in operational conditions sustain significant changes of temperature, high pressure, vibration, seismic loads, uneven subsidence, corrosion. The crucial issue today is to ensure the reliability of operation and trouble-free operation of oil storage systems. One of the main causes of explosions

M. Nechyporuk et al. (Eds.): ICTM 2021, LNNS 367, pp. 575–583, 2022.
https://doi.org/10.1007/978-3-030-94259-5_49

and fires in oil storage tanks (reservoirs) is the static electricity generated in the pipeline during the oil transportation.

As a result, electrostatic charges introduced into the tank together with the oil, creating the electric field and, respectively, the conditions for the occurrence of the spark breakdown of the gas space above the oil surface. Issues of static electricity require prompt research and justification for the introduction of new structures and materials in the transportation and petroleum products storage. The formation of electric charges in petroleum products during their movement calls electrification. Electric charges in the volume or on the surface of petroleum products call electrostatic. The strongest electrification manifested at high speeds of transportation through pipes. The importance of the processes of electrostatic charges accumulation is especially great, as electrified materials are flammable liquids. Discharges of static electricity in flammable environments could cause explosions and fires [1].

There is the great danger of static electricity discharges and damage in the oil and oil refining industries. Therefore, the development of measures to prevent and eliminate dangerous manifestations of static electricity is very important issue. Taking into account the increasing requirements for protection against static electricity in the chemical, petrochemical and refining industries, as well as the growing volume of transportation and storage of oil and petroleum products, there is the necessity for systematic research and development of methods to reduce the intensity of electrification of oil and petroleum products.

According to statistics and numerous research papers, the accumulation of dangerous amounts of electrostatic charges in tanks is possible, which leads to electric field strengths exceeding the critical value of 3 kV/mm, resulting in conditions for sparking and explosion and fire, respectively. Therefore, the development of fundamentally innovative and more effective ways to neutralize electrostatic charges in the oil flow is the important scientific and technical issue.

In the assessing process the scale of possible technogenic hazards at oil depots and storage facilities, the main scenarios of accidents have been identified:

- the most dangerous - the explosion of the ground oil tank with subsequent ignition, fire and complete destruction of the tank;
- the most probable - partial destruction of the pump unit; local leaks from technological equipment, pipelines.

In the accidents development by these scenarios, it is possible to ignite the oil (petroleum product) and the strait fire. The main striking factors of these accidents are thermal radiation, the shock wave impact, exposure to open flames, damage by fragments. In accidents with the most severe consequences, the affected area (destruction) could reach several hundred meters. Few tens of meters from the edge of the strait will limit the size of the danger zones in the event of the fire spill. In case of realization of the considered emergencies, destruction of the equipment adjacent on the industrial site could be possible. This fact contributes to the affected area increasing.

The occurrence of static electricity could lead to the accumulation of the very large charge and eventually to the discharge in the spark form, which in the flammable and especially in the explosive environment could cause the fire (explosion), as to ignite the explosive mixture of flammable substance vapors with air electrostatic charge of 300–500 V [2].

The aim has been to increase the environmental safety of oil storage systems by effectively neutralizing static electricity using composites and nanocomposites with disordered and ordered systems of fibrous inclusions.

2 Analysis of Recent Research and Publications

Hyperion Catalyst Company (Cambridge, USA) has improved the method of obtaining nanotubes and today 60% of American cars equipped with gas pipelines containing carbon nanotubes. Due to their high electrical conductivity, they neutralize static charges that could generate the spark in the fuel line. Today, nanotubes actively introduced into containers - for example, in boxes for transporting chips. They protect electronic devices from mechanical damage, remove electrostatic charges that could damage chips [3].

The nanotubes introduction as a filler gives some composite materials electrical conductivity and other important properties. Conductive plastics could be implemented in aerospace engineering, for example as signal wires, devices of protection against discharges of static electricity, for creation of various drives and flexible electric schemes. The so-called multifunctional polymeric materials could be used in the manufacture of rocket engines, dynamic in space structures, as well as in the manufacture of many standard items of military equipment (for example tanks, tires, etc.) [4].

In the recent years, advanced polymer-based composites or nanocomposites are preferred structural components for aircraft due to their light weight and comparable mechanical properties, but such composite systems do not provide low resistance to charge flux and are subsequently vulnerable to lightning and precipitation [5].

The emergence of nanotechnology in the field of polymer composites has been demonstrated the ability to bypass the prevalence of such static charge accumulation and at the same time provides the synergistic feature of the polymer and nanocomposite [6]. Although polymers have natural isolation origin, but nanofiller variants such as carbon black, carbon fiber, 1-n tetradecyl-3-methylimidazolium bromide, metal particles and nanofibers such as polyaniline [7], have demonstrated the significant increase in conductivity exceeding the insulation threshold. Although detailed evaluation of composite or nanocomposite materials for aerospace research has been widely used [8, 9], but the use of such materials to circumvent the phenomenon of static charge in structural and space applications has been the least studied.

In the [10] have demonstrated the use of different nanoparticles and its research for various aeronautical applications. In the context of static charge reduction in avionic composite, embedded fillers in the polymer sometimes referred to as antistatic agents and are generally classified based on their model composition and chemical structure [11]. Pionteck and Wypych have presented the antistatic agents in the mode of their application could be alienated to internal and external antistatic agents, where the external deposited fragments are surfactant ionic or nonionic, while intrinsically active agents used to develop conductive pathways in polymeric materials by reinforcement [12].

Carbon antistatic agents such as graphene, carbon nanotubes (multi-walled carbon nanotubes and single-walled carbon nanotubes), carbon nanofibers and graphene oxide widely used to reduce the accumulation of static charge on the polymer surface. Soot - the traditional nanofilling non-hypoxide resin as the electrostatic or antistatic agent in aerospace applications, formed by incomplete combustion of aromatic hydrocarbons at high temperatures [13]. From this point of view, the dissociation of the C-H bond of aromatic hydrocarbons leads to the reaction between carbon atoms and aromatic radicals, which leads to the formation of the crystallographic layered structure consisting of the hexagonal carbon ring [14].

Authors [15–17] have demonstrated that the gap distance between the particles plays the significant role in the final bulk conductivity of the sample. It is reported the concentration increasing of the filler reduces the distance between the particles of the filler, which, in turn, increases the crosslinking density, which leads to the increase in composite conductivity.

Known the implementation of electrically conductive coatings as shielding materials on polymers or other dielectrics. However, they could peel off or crack under load. This problem could be solved by using nanocomposites of polymers with electrically conductive fillers [16–18]. However, in the case of micron-sized fillers (metal and carbon fibers and particles, soot, etc.) to achieve sufficient electrical conductivity of the composite requires the use of high fillings up to 16 vol. %. As a result, the mechanical properties of the material deteriorate. To solve this problem, nanosized electrically conductive fillers, such as carbon nanotubes, provide high conductivity and strength of the material even with very low content of carbon nanotubes.

It is important to create the new hybrid nanocomposite that effectively shields from electromagnetic fields in the wide frequency range of 10 MHz–20 GHz, has the high electrical conductivity and is mechanically strong. Therefore, the properties of composites and nanocomposites with disordered and ordered systems of carbon fiber inclusions, which could be used to neutralize static electric charges that occur in oil storage and could lead to emergencies, have been investigated [19].

3 Materials and Methods

3.1 Finite Element Method in the Study of Composites and Nanocomposites with Disordered and Ordered Inclusion Systems

The mathematical and computational models developed in [18] on the basis of FEM has been allowed to carry out the analysis of nanomaterials with spherical continuous or hollow inclusions, and also with short nanofibers and nanotubes.

The series of nanocomposite materials based on epoxy matrix with carbon fillers of different shapes has been treated. Solid spherical and fibrous spherical inclusions have been used. Due to the software complex, the dependence of the strength characteristics of nanocomposite material on the size, shape, means of placement of carbon nanofillers, as well as the level of filling the matrix with inclusions have been found and investigated.

It has been assumed that the matrix of the nanocomposite of cured epoxy resin has the following anisotropic properties: density $\rho_M = 1160$ kg/m^3, elasticity modulus $E_M = 3{,}78$ Gpa, Poisson's ratio $\nu_M = 0{,}35$.

As the inclusion material, carbon fillers with orthotropic properties have been considered: density 1800 kg/m^3, elasticity modulus in the direction OX, $E_1 = 290$ GPa, elasticity modulus in the direction OY, $E_2 = 23$ GPa, elasticity modulus in the direction OZ, E_3 =23 GPa, Poisson's ratio XY, $\nu_{12} = 0{,}2$, Poisson's ratio YZ, $\nu_{23} = 0{,}4$, Poisson's ratio XZ, $\nu_{13} = 0{,}2$, shear module XY, $G_1 = 9$ GPa, shear module YZ, $G_2 = 214$ GPa, shear module XZ, $G_3 = 9$ GPa.

The calculated characteristics of the new nanocomposite material have been clarified as the result of the performed research.

3.2 Partially Ordered Inclusions

Similar to the study of matrices with cylindrical inclusions, the nanocomposites analysis with filler in the partially ordered longitudinal carbon fibers form has been performed [18–20]. Carbon nanofibers located along the entire length of the matrix, have the diameter of 1 nm, and 10° of the maximum allowable deviation from the main axis. The volume fraction of inclusions from 0,025 to 0,25 has been analyzed. The calculated models of the nanocomposite with the volume fraction of inclusions equal to 0,2 have been presented on Fig. 1.

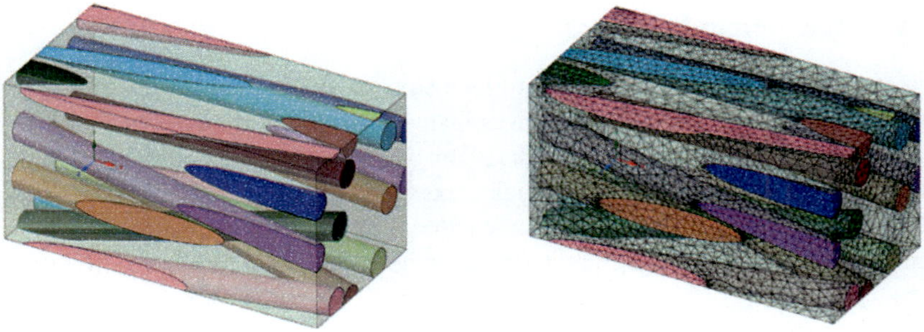

a) b)

Fig. 1. Representative cell (a) and finite-element (b) model of nanomaterial with fibrous partially ordered inclusions

The results of the calculations have been presented in Table 1, the fibrous carbon inclusions of the same size and orthotropic properties of the material have been studied.

It has been clarified the use of inclusions in the carbon nanofibers form demonstrates the increase in the strength characteristics of the material in the fibers direction. For the same inclusions volume fraction, fiber reinforcement was more than 10 times more effective than cylindrical or spherical reinforcement of the same material. In addition to significant hardening, which characterized by the module E1, there was also hardening in other directions, close in level compared to other types of inclusions.

Table 1. Mechanical properties of nanocomposite with fibrous inclusions

Volume fraction	E_1, MPa	E_2, MPa	E_3, MPa	G_{12}, MPa	G_{23}, MPa	G_{31}, MPa	ν_{12}	ν_{13}	ν_{23}	Density
0.0259	7815,8	4205,3	4090,2	1483,7	1447,2	1686,4	0,3220	0,4032	0,4333	1176
0.0504	14788	4419,4	4397,9	1524,9	1490,8	1576,8	0,3382	0,3623	0,4775	1192
0.0764	22541	4621	4589	1675	1544	1601	0,3242	0,3451	0,4885	1211
0.1002	30985	4780,1	4791,8	1636,7	1584,5	1654,8	0,3384	0,3377	0,4979	1224
0.1261	35518	4934,2	5027,8	1870,3	1629,8	1794,2	0,3900	0,3218	0,4876	1240
0.1510	39187	5106,1	5160,1	1815,5	1689,3	1838,1	0,3596	0,3333	0,4909	1256
0.1791	40591	5378,5	5398,8	1915	1761,7	2246,9	0,3160	0,4448	0,4783	1274
0.2039	40471	5568,8	5511,3	1959	1814,7	2477,5	0,2549	0,5065	0,4785	1290
0.2294	50946	5753,8	5723,5	2200,6	1885,8	2340,9	0,3242	0,4638	0,4779	1306
0.2525	58330	6301,5	6012,1	2385,2	1911,8	2113,5	0,3707	0,3199	0,4785	1321

3.3 Ordered Fibrous Inclusions

The matrices with the filler in the longitudinal and ordered form of carbon fibers according to the diamond scheme have been investigated. Carbon nanofibers have located along the entire length of the matrix, have the 1 nm diameter, the fibers direction coincides with the direction of one of the coordinate axes. The inclusions volume fraction from 0,025 to 0,25 has been analyzed. The calculated models of the nanocomposite with the volume fraction of inclusions equal to 0,2 have been presented on Fig. 2 [18].

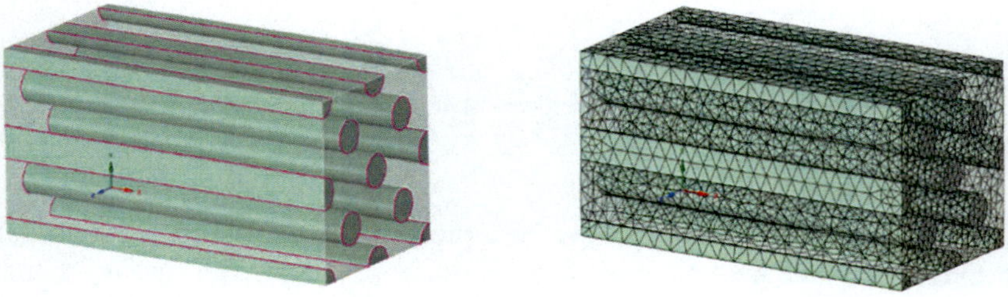

a) b)

Fig. 2. Representative cell (a) and finite-element (b) model of nanomaterial with ordered fibrous inclusions

The results of the treatment have been presented in Table 2, the calculations have been performed for fibrous carbon inclusions of the same size, with orthotropic properties of the material.

As in the previous case, with fibrous partially ordered inclusions, the use of carbon fibers form as the filler demonstrates the increase in the strength characteristics of the

Table 2. Mechanical properties of nanocomposite with fibrous inclusions

Volume fraction	E_1, MPa	E_2, MPa	E_3, MPa	G_{12}, MPa	G_{23}, MPa	G_{31}, MPa	ν_{12}	ν_{13}	ν_{23}	Density
0,025	10935	4228,8	4228,8	1452,3	1444,9	1452,3	0,3454	0,3454	0,4649	1,176
0,05	18095	4429,3	4429,3	1506,5	1492,8	1506,5	0,341	0,341	0,490	1,192
0,075	25252	4589,9	4589,9	1562,8	1544,4	1562,8	0,3365	0,3365	0,5004	1,208
0,1	32404	4739,1	4739,1	1621,4	1600,6	1621,4	0,3328	0,3321	0,5057	1,224
0,125	39561	4883,3	4883,3	1682,3	1660,5	1682,3	0,3278	0,3278	0,5089	1,24
0,15	46719	5026,3	5026,3	1745,6	1724,6	1745,6	0,3255	0,3235	0,5110	1,256
0,175	53875	5169,8	5169,8	1811,5	1793	1811,5	0,3193	0,3193	0,512	1,272
0,2	61033	5316,2	5316,2	1880,3	1866,6	1880,3	0,3151	0,3150	0,513	1,288
0,225	68190	5466,5	5466,5	1952,2	1945,9	1952,2	0,3109	0,3109	0,5146	1,304
0,25	75347	5621,4	5621,4	2027,5	2030,7	2027,5	0,3068	0,3068	0,515	1,32

material in the direction of the fibers. Compared with partially ordered fibers, the complete structured ordering further increased the strength properties in the main direction of the fibers, the difference was more than 29%, but underestimated the strength characteristics in the other two directions by 6–12%. Thus, the carbon nanofibers using allows to create the high-quality nanocomposite with pronounced orthotropic properties [21, 22].

4 Outcomes

It has been established the use of filler in the carbon fibers form demonstrates the growth of material strength characteristics in the fibers direction. Complete structured ordering of nanofibers further increases the strength properties in the main direction of the fibers location, but does not significantly reduce the strength characteristics in the other two directions. The results of research prove the use of carbon nanofibers to create the quality nanocomposite is appropriate and could be used to neutralize static electric charges that occur in oil storage and could lead to emergencies. Thus, the innovative oil storage construction with improved characteristics has been proposed in the paper. In future it will be treated the different kinds of inclusions, their orientation in the representative cell to achieve the antistatic effect, increase the material durability and reliability.

References

1. Naqash, W., Majid, K.: Fabrication of a novel PANI/[Co(NH3)4(C3H4N2)2]Cl3 nanocomposite with enhanced dielectric constant and ac-conductivity. J. Mater. Sci. Mater. Electron. **28**, 14217–21422 (2017). https://doi.org/10.1007/s10854-017-7279-6
2. Li, L., Dai, L.: Review on fire explosion research of crude oil storage tank. E3S Web Conf. **236**, 01022 (2021). ICERSD 2020

3. Guo, B., et al.: Highly flexible, thermally stable, and static dissipative nanocomposite with reduced functionalized graphene oxide processed through 3D printing. Mater. Sci. Compos. Part B Eng. **208**, 108598 (2021). https://doi.org/10.1016/j.compositesb.2020.108598
4. Zhang, M., Zhang, C., Du, Z.J., Li, H.Q., Zou, W.: Preparation of antistatic polystyrene superfine powder with polystyrene modified carbon nanotubes as antistatic agent. Compos. Sci. Technol. **138**, 1–7 (2017). https://doi.org/10.1016/j.compscitech.2016.11.010
5. Yadav, R, Tirumali, M., Wang, X., Naebe, M., Kandasubramanian, B.: Polymer composite for antistatic application in aerospace. Defence Technol. **16**(1), 107–118 (2020). https://doi.org/10.1016/j.dt.2019.04.008
6. Braga, N.F., Zaggo, H.M., Montagna, L.S., Passador, F.R.: Effect of carbon nanotubes (CNT) functionalization and maleic anhydride-grafted poly (trimethylene terephthalate) (PTT-g-MA) on the preparation of antistatic packages of PTT/CNT nanocomposites. J. Compos. Sci. **4**, 1–18 (2020). https://doi.org/10.3390/jcs4020044
7. Mora, A., Verma, P., Kumar, S.: Electrical conductivity of CNT/polymer composites: 3D printing, measurements and modeling. Compos. Part B. **183**, 107600/1–107600/11 (2020). https://doi.org/10.1016/j.compositesb.2019.107600
8. Silva, T.F., Menezes, F., Montagna, L.S., Lemes, A.P., Passador, F.R.: Preparation and characterization of antistatic packaging for electronic components based on poly (lactic acid)/carbon black composites. J. Appl. Polym. Sci. **136**(13), 47273 (2019). https://doi.org/10.1002/app.47273
9. Jawaid, M., Thariq, M.: Sustainable Composites for Aerospace Applications. Woodhead Publishing, Cambridge (2018)
10. Gonçalves, C., Gonçalves, I.C., Magalhães, F.D., Pinto, A.M.: Poly (lactic acid) composites containing carbon-based nanomaterials: a review. Polymers **9**(7), 269/1–269/37 (2017). https://doi.org/10.3390/polym9070269
11. Satish, G., Prasad, V.V.S., Ramji, K.: Manufacturing and characterization of CNT based polymer composites. Math. Models Eng. **3**(2), 89–97 (2017). https://doi.org/10.21595/mme.2017.19121
12. Wypych, G., Pionteck, J.: Handbook of Antistatics, 2nd edn. ChemTec Publishing, Toronto (2017)
13. Chow, W.S., Lim, Y.T.: Antistatic and thermal properties of Poly (Lactic Acid)/Polypropylene/Carbon nanotube nanocomposites. J. Eng. Sci. **16**(2), 57–69 (2020)
14. Monti, M., Zaccone, M., Frache, A., Torre, L., Armentano, I.: Dielectric spectroscopy of PP/MWCNT nanocomposites: relationship with crystalline structure and injection molding condition. Nanomaterials **11**, 550 (2021). https://doi.org/10.3390/nano11020550
15. Taj, M., Manohara, S.R., Hanagodimath, S.M., Gerward, L.: Novel conducting poly(3,4-ethylenedioxythiophene) – graphene nanocomposites with gigantic dielectric properties and narrow optical energy band gap. Polym. Testing **90**, 106650 (2020). https://doi.org/10.1016/j.polymertesting.2020.106650
16. Fadil, Y., Agarwal, V., Jasinski, F., Thickett, S.C., Minami, H., Zetterlund, P.B.: Electrically conductive polymer/rGO nanocomposite films at ambient temperature via miniemulsion polymerization using GO as surfactant. Nanoscale **11**(14), 6566–6570 (2019). https://doi.org/10.1039/C9NR00816K
17. Cho, J., et al.: Enhanced electrical conductivity of polymer nanocomposite based on edge-selectively functionalized graphene nanoplatelets. Compos. Sci. Technol. **189**, 108001 (2020)
18. Sierikova, O., Koloskov, V., Degtyarev, K., Strelnikova, O.: The deformable and strength characteristics of nanocomposites improving. In: Materials Science Forum, vol. 1038, pp. 144–153. Trans Tech Publications Ltd., Switzerland (2021)
19. Surianinov, M., Andronov, V., Otrosh, Y., Makovkina, T., Vasiukov, S.: Concrete and fiber concrete impact strength. Mater. Sci. Forum **1006**, 101–106 (2020)

20. Tulskyi, H.H., Liashok, L.V., Shevchenko, H.S., Vasilchenko, A.V., Stelmakh, O.A.: Synthesis of functional nanocomposites based on aluminum oxides. Functional Mater. **26**(4), 718–722 (2019)
21. Kovalov, A., Otrosh, Y., Rybka, E., Kovalevska, T., Togobytska, V., Rolin, I.: Treatment of determination method for strength characteristics of reinforcing steel by using thread cutting method after temperature influence. In: Materials Science Forum, vol. 1006, pp. 179–184. Trans Tech Publications Ltd. (2020)
22. Smetankina, N., Merkulova, A., Merkulov, D., Postnyi, O.: Dynamic response of laminate composite shells with complex shape under low-velocity impact. In: Nechyporuk, M., Pavlikov, V., Kritskiy, D. (eds.) Integrated Computer Technologies in Mechanical Engineering-2020, vol. 188, pp. 267–276. Springer, Cham (2021). https://doi.org/10.1007/978-3-030-66717-7_22

Project Management and Business Informatics

Violation of Flight Rules and Airspace Use: Current Criminal Situation

Andrii Lysenko[1(✉)], Iryna Lysenko[2], Ivan Hrabazii[1], and Anton Polianskyi[3]

[1] Department of Operational Investigative Activities, Kharkiv National University of Internal Affairs, KhNUIA, Kharkiv, Ukraine

[2] Department of Law, National Technical University «Kharkiv Polytechnic Institute», NTU «KhPI», Kharkiv, Ukraine

irina8119@ukr.net

[3] Laboratory of R&D and Publishing at National Scientific Centre «Hon. Prof. M. S. Bokarius Forensic Science Institute», Kharkiv, Ukraine

anton_polianskyi@ukr.net

Abstract. It is determined on the basis of analysis of official statistics of the Office of the Prosecutor General on criminal offenses under Articles 281 and 282 of the Criminal Code of Ukraine, professional readings, prevalence of criminal offenses currently under investigation, trends, causes and conditions conducive to their commission. The research used a system of general scientific and special methods providing an objective subject analysis, in particular, sociological, comparative law, statistical analysis, normative, formal, logical and content analysis of documents. Research results are presented by means of calculated data on quantitative indicators of detection dynamics of criminal offenses under Articles 281 and 282 of the Criminal Code of Ukraine in 2014–2020, as well as a list of key determinants of these offenses and related criminogenic situation.

Scientific research paper novelty is that it offers a number of new scientific provisions and proposals, namely: identifies the current state of the criminogenic situation associated with violations of air traffic rules and the Ukrainian airspace use, trends, its main causes and conditions that contribute to their commission. Practical significance lies in the fact that the scientific provisions formulated and presented in the paper contain specific data that have both general theoretical and practical significance are the basis for improving the efficiency of law enforcement and can be used in organizing detection and investigation of air traffic violations and airspace use.

Keywords: Aviation industry · Criminogenic situation · Criminal offenses · Violation of air flight rules · Violation of airspace use

1 Introduction

At present, aviation industry plays an important role in ensuring state interests, national security, meeting the needs of society and economy in air transportation and aviation. Rapid development of aviation in the world in the XX-XXI century has led to its global

M. Nechyporuk et al. (Eds.): ICTM 2021, LNNS 367, pp. 587–598, 2022.
https://doi.org/10.1007/978-3-030-94259-5_50

spread. Today it is impossible to imagine civilized states without aircraft. At the same time, increase in the number of air transportation, aviation works and frequency of their use, and in Ukraine and around the world, leads to an increase in potential risks of negative consequences in the areas of flight safety, aviation safety, environmental protection, economic and information security.

Ukraine pays considerable attention to ensuring aviation safety through state regulation of aviation activities and the use of Ukraine airspace. To this end, in 2011 the leading legislative act in the research area was adopted – the Air Code of Ukraine. Ukraine has also acceded to the Convention on International Civil Aviation. In addition, Articles 281 and 282 of the Criminal Code of Ukraine establish criminal liability for violation of air flight rules and violation of airspace use rules, respectively.

2 Literature Review

In professional readings scientists have paid considerable attention to the study of criminal-legal characteristics of criminal offenses against traffic safety and operation of transport including violations of air flight rules and airspace use. In research papers of such scientists as S. V. Babanin, V. I. Borysov, S. V. Hizimchuk, V. O. Hlushkov, H. V. Zhylinskii, M.I. Korzhanskyi, Ya. V. Matviychuk, V. A. Myslyvyi, A. B. Razumov et al., First of all, the topic, composition signs of these criminal offenses, forms of illegal manifestation, problems of their criminal and legal qualification, criminal liability and criminal and legal classification, etc. were considered in detail.

However, the issue of determining current criminogenic situation state associated with commission of such criminal offenses as violations of air traffic rules and airspace use has remained out of scientist attention. This fact led us to choose this research topic.

3 Main Content Presentation

One of the factors that leads to the risk of damage to life, health or property is violation of air traffic rules and airspace use. Stabilization of the criminogenic situation in this area is gaining additional importance due to the fact that Ukrainian airspace is used by aircraft of other countries. Thus, Article 2 of the Air Code of Ukraine states that Ukraine has full and exclusive sovereignty over the airspace of Ukraine that is part of the territory of Ukraine [1]. In addition, in accordance with Art. 2 Convention on the Law of the Sea (UNCLOS), sovereignty of a coastal State also extends to the airspace over the territorial sea [2]. At the same time, in accordance with the provisions of paragraph 1 of part 1 of Article 8 of the Air Code of Ukraine, the state bodies of Ukraine should provide conditions for activities of national and international users of Ukrainian airspace.

Ukraine, as an acceding state to the Convention on International Civil Aviation [3] is responsible for fulfilling international obligations arising from this Convention and for guaranteeing and creating conditions for society security, protection of interests while conducting activities in the field of civil aviation and Ukrainian airspace use. Thus, in order to maintain Ukrainian global image as a legal, democratic and civilized country, one of its responsibilities is to ensure air safety not only for domestic but for foreign aircraft.

Despite relatively low prevalence in modern Ukraine of criminal offenses under Articles 281 and 282 of the Criminal Code of Ukraine, stabilization of criminogenic situation associated with their commission is a very important task. This statement is due to increased public danger of violating air traffic rules and using airspace that can lead to death, great material damage, other serious consequences and, as noted above, undermine the image of Ukraine in the world.

One of the ways to increase the level of counteraction by law enforcement agencies to criminal offenses under Articles 281 and 282 of the Criminal Code of Ukraine is to improve its theoretical foundations through comprehensive analysis and assessment of the current state of the criminogenic situation in this area. Providing a criminogenic characterization of violations of air flight rules and the use of airspace allows to determine the trends of their commission, reasons and conditions contributing to their commission. Accordingly, law enforcement agencies carrying out operational and investigative and criminal proceedings having these data can focus their efforts in the right direction, to take measures to improve management of their activities.

Considering the current state of criminogenic situation related to the commission of criminal offenses under Articles 281 and 282 of the Criminal Code of Ukraine, it should be emphasized that relevant analysis basis is to study the state and dynamics of their commission. Note that on the one hand, information obtained indicates prevalence of criminal offenses and their public danger and on the other hand to identify trends in their commission, in order to determine forms and measures for their combating by law enforcement agencies. At the same time, it should be borne in mind that statistics cannot always objectively reflect the true state of crime situation and in some cases can obscure and complicate real problems rather than solve them.

In this regard British criminologists are quite apt to point out that one should always be careful when drawing conclusions from crime statistics. Only reported crimes are recorded not those committed. The rise in crime can reflect the fact that police are catching more criminals than the fact that more crimes are being committed. The opposite is true [4].

The status and dynamics of accounting for detected criminal offenses under Articles 281 and 282 of the Criminal Code of Ukraine can be traced to statistics published on official website of the Office of the Prosecutor General of Ukraine reflecting the main indicators of pre-trial investigation by the number of detected crimes. Proceedings sent to court with an indictment [5].

Thus, in 2014 in Ukraine, 5 unified criminal offenses provided for in Article 281 of the Criminal Code of Ukraine were recorded in the Unified Register of Pre-trial Investigations. At the same time, only in 2 criminal proceedings the persons were served with a notice of suspicion and 1 criminal proceeding was sent to court with an indictment. In 2015, respectively, 4 criminal offenses were recorded, in 1 criminal proceeding a person was served with a notice of suspicion and no criminal proceeding was sent to court with an indictment. In 2016, we see a decrease in the number of registered criminal offenses – 1, in no criminal proceedings was the person notified of the suspicion and an indictment was drawn up. Instead, in 2017 there was a significant increase in registered criminal offenses – 9, in 2 criminal proceedings persons were served with a notice of suspicion and 2 criminal proceedings were sent to court with indictments. In 2018, 7 criminal offenses were recorded, in 2 criminal proceedings persons were served with a notice of suspicion and 2 criminal proceedings were sent to court with indictments. In 2019, 3 criminal offenses were recorded, but in no criminal case was a notice of suspicion served and an indictment was drawn up. In 2020, statistics show a sharp increase in the number of reported criminal offenses – 12, in 5 criminal proceedings, individuals were served with a notice of suspicion and 5 criminal proceedings were sent to court with indictments.

Thus, violations of air traffic rules in Ukraine for the period 2014–2020 indicates rather unstable dynamics. Trends in changes in the quantitative indicators of detection of criminal offenses in percentage terms shown in the diagram in Fig. 1. While calculating these changes over the years, the indicators of 2014 were defined by us in this research as basic ones (starting), at the level of 100%.

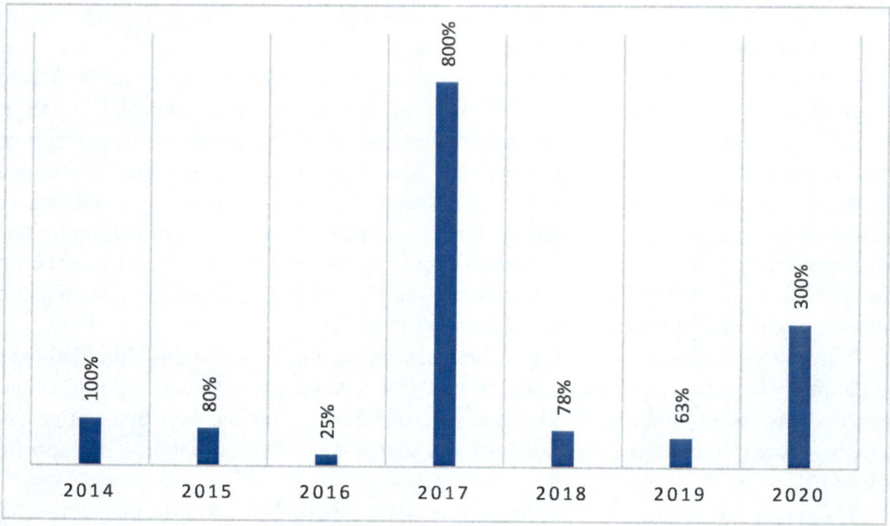

Fig. 1. Detection dynamics of criminal offenses under Art. 281 of the Criminal Code of Ukraine: *Violation of air flight rules*

As for detection dynamics of second composition of criminal offenses i.e. violation of the rules of airspace use, in Ukraine in 2014 was recorded in the Unified Register of pre-trial investigations 2 criminal offenses under Article 282 of the Criminal Code of Ukraine. At the same time, in none of criminal proceedings was the person notified of the suspicion and an indictment was drawn up. In 2015, respectively, 3 criminal offenses were recorded, only in 1 criminal proceeding a person was served a notice of suspicion and no criminal proceeding was sent to court with an indictment. In 2016, we see a decrease in the number of recorded criminal offenses to the level of 2014 – 2, in one criminal proceeding a person was served a notice of suspicion and no criminal proceedings were sent to court with an indictment. In the following 2017, 2 criminal offenses were also recorded but in none of the criminal proceedings was the person served with a notice of suspicion and an indictment was drawn up. In 2018, 1 criminal offense was recorded, and as in the previous year, in no criminal proceedings, a person was served with a suspicion notice and indictment was drawn up. In 2019 according to quantitative indicators, situation of 2017 is completely repeated, i.e. 2 criminal proceedings were taken into account, suspicions were not handed over, indictments were not drawn up. Instead, in 2020, statistics show a sharp increase in the number of reported criminal offenses – 4, in default of criminal proceedings when a person was served with a suspicion and an indictment was drawn up.

Thus, violations of airspace use rules in Ukraine for the period 2014–2019 show relative stability, with a significant increase in 2020. Trends in the change of quantitative indicators of detection of criminal offenses in percentage terms are shown in the diagram in Fig. 2. Calculation of these changes over the years, indicators of 2014 were defined by us in this research as basic (starting), at the level of 100%.

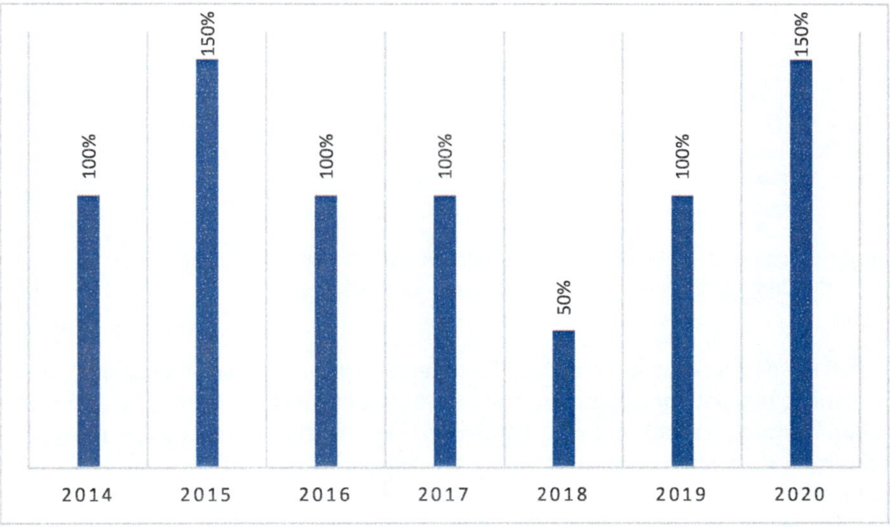

Fig. 2. Detection dynamics of criminal offenses provided by Art. 282 of the Criminal Code of Ukraine: *Violation of the rules of airspace use*

While research on criminogenic situation related to commission of criminal offenses under Articles 281 and 282 of the Criminal Code of Ukraine, it is of great interest to study statistics published on the official website of the Office of the Prosecutor General [5] on the completion of criminal proceedings and refer them to court with an indictment. This approach allows us to determine effectiveness of operational and investigative units of law enforcement agencies in combating the crimes under investigation.

Thus, in 2014, percentage of criminal proceedings initiated for committing criminal offenses under Article 281 of the Criminal Code of Ukraine which successfully completed the pre-trial investigation with the drafting of an indictment and sending them to court – 20%. In the following years of the period we studied, these indicators were respectively: in 2015 – 0%, in 2016 – 0%, in 2017 – 20.2%, in 2018 – 28.6%, in 2019 – 0% and in 2020 – 41.7% (see the diagram in Fig. 3). Thus, based on the analysis of these statistics, we can conclude that in 2014–2020, average percentage of criminal proceedings in which the pre-trial investigation was successfully completed was – 15.8%.

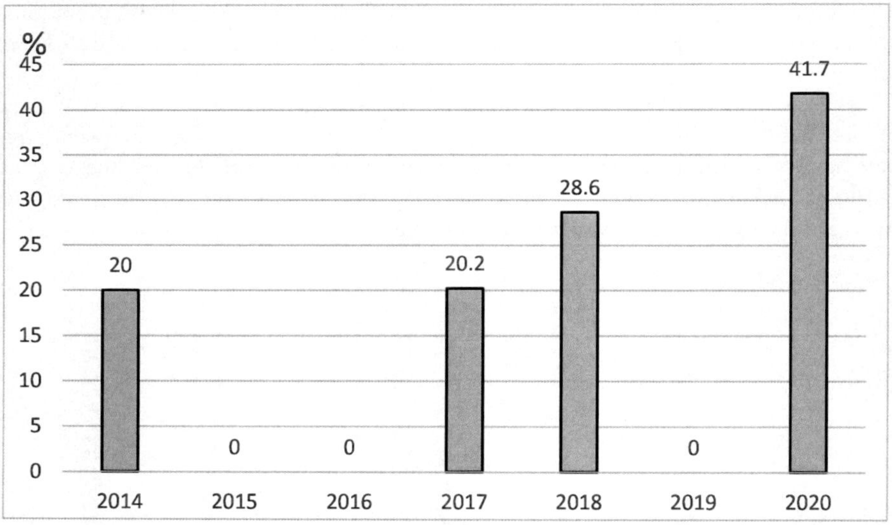

Fig. 3. Dynamics of criminal proceedings initiated for committing criminal offenses under Art. 281 of the Criminal Code of Ukraine sent to court with indictment

Let us consider data on dynamics of referral to court of criminal proceedings with an indictment, initiated for committing criminal offenses under Article 282 of the Criminal Code of Ukraine. It should be noted that analysis of statistical data led to a very interesting result from a scientific point of view. Thus, in the period of 2014–2020, 16 considered criminal offenses were recorded in the Unified Register of Criminal Proceedings. During the same period, only two suspects were handed over to the persons within the framework of these initiated criminal proceedings. At the same time, no criminal proceedings were sent to court with indictment. Thus, based on the analysis of these statistics, it can be stated that in 2014–2020, percentage of criminal proceedings when pre-trial investigation was successfully completed is 0%.

It should be noted that such low percentages can indicate on the one hand, as a low level or certain shortcomings in management and implementation of operational and investigative, criminal proceedings by law enforcement agencies in combating criminal offenses under Articles 281 and 282 of the Criminal Code of Ukraine and on the other hand, imperfection of the current Criminal Procedure Code of Ukraine associated with premature initiation of criminal proceedings in the case of obtaining information about a criminal offense without sufficient verification of its authenticity.

Adoption in 2012 of the new Criminal Procedure Code of Ukraine has led to significant changes in criminal procedure. Thus, in accordance with Part 1 of Article 214 of the Criminal Procedure Code of Ukraine regulating the procedure for initiating pre-trial investigation, the investigator, coroner, prosecutor immediately but not later than

24 h after filing an application, notification of a criminal offense or which source of circumstances that can indicate a criminal offense commission is obliged to enter the relevant information in the Unified Register of pre-trial investigations and initiate an investigation [6].

Thus, in fact, the pre-trial investigation begins from the moment the information is entered into the Unified Register of Pre-trial Investigations. This fact virtually eliminates the possibility of a full preliminary verification of obtained information. Analysis and generalization of experience of practical activities of bodies of inquiry and pre-trial investigation indicates that such a procedural procedure leads to the further termination of criminal proceedings.

It should be noted that before introduction of the above legal norm, before initiation of a criminal case, a thorough verification of the received operational information was carried out. In case of its confirmation, in availability of the bases and necessity, further operative development within the established operative case was carried out. While verification of received information and operative development, in accordance with Article 1 of the Law of Ukraine: *On Operative Investigation Activity* search and fixation of factual data on illegal actions of individuals and groups which responsibility is provided by the Criminal Code of Ukraine [7].

In other words, prior to initiation of a criminal case, not only a preliminary check of information about the crime was carried out, but documentation of criminal activity of the defendants. This approach minimized the facts of unfounded criminal proceedings. Instead, in modern legal conditions before beginning of criminal proceedings this possibility is missing. This provision accordingly leads to unjustified initiation of criminal proceedings and their subsequent termination in this connection.

It should be noted that actual state of prevalence of criminal offenses under Article 282 of the Criminal Code of Ukraine differs in large quantitative indicators of their implementation, as the analysis of statistics should take into account the difficulties that arise while their qualification. This situation is due to construction of disposition under consideration of the criminal law. It belongs to the mixed type and includes features of descriptive and blanket dispositions that leads to difficulties in act qualification.

Thus, the disposition of Article 282 of the Criminal Code of Ukraine states that violation of the rules of airspace use is a violation of rules of launching missiles, conducting all types of firing, blasting or other actions in airspace, if it posed a threat to air safety [8]. Thus, from an objective point of view, this criminal offense can be committed in such a form as "committing other acts in airspace, if it posed a threat to air safety". This element of the disposition is blanket in nature and refers us to other laws and regulations.

Such legislative and regulatory acts include the Air Code of Ukraine and the Regulation on the use of airspace of Ukraine, approved by the Resolution of the Cabinet of Ministers of Ukraine of December 6, 2017 № 954 [9]. The norms of Article 23 of the Air Code of Ukraine regulate management of Ukrainian airspace use and responsibility for violating the procedure for its use. Clause 3 of this legal norm stipulates that violations of the procedure for the use of Ukrainian airspace are considered to be actions or omissions of legal entities or individuals that have violated the requirements of legislative and other regulations governing the use of Ukrainian airspace. Paragraph 4 states that the list of violations of the procedure for the Ukrainian airspace use is determined by the Regulations on Ukrainian airspace use [1].

In paragraph 46 of the Regulations on Ukrainian airspace use to violations of Ukrainian airspace use include:

- airspace use without submission of applications, obtaining a permit and/or conditions of its use, except for the cases provided by this Regulation;
- non-compliance with conditions of Ukrainian airspace use or flight project (deviation from the route, specified flight level, take-off or landing of aircraft from an unscheduled (unplanned) or closed aerodrome, permanent runway/heliport, etc.) without permission or approval of the air service traffic (air traffic control body), except for emergencies;
- flight of aircraft group which number exceeds specified ones in application;
- violation of requirements for procedure of crossing the state border and performing a flight in a zone with a special regime of the Ukrainian airspace use;
- flight performance in the restricted zone or in the restricted areas and temporarily reserved airspace of Ukraine while their use without the aircraft permission;
- flight of an aircraft that does not respond to the request of radar system of state recognition, except for aircraft on which the equipment of radar system of state recognition is not provided or which are returned with faulty equipment of the radar system of state recognition from airfields, runways, heliports repair of such equipment is impossible;
- non-compliance with the requirements for informing control bodies over observance of the procedure for Ukrainian airspace use on performance of flights and other activities related to the Ukrainian airspace use [9].

Thus, the Regulation on the of Ukrainian airspace use provides a detailed list of procedure violations for the use of airspace of Ukraine provided that the threat to air safety should be classified as criminal offenses under Art. 282 of the Criminal code of Ukraine.

However, the investigated Regulation is a by-law and was adopted in accordance with the Air Code of Ukraine, other laws of Ukraine cannot contradict them. At the same time, in Article 23 of the Air Code, legislator made a mistake which was then duplicated by the Cabinet of Ministers of Ukraine in approved Regulations on the use of airspace of Ukraine. This error is that Article 23 of the Air Code and paragraph 46 of the Regulation on the use of airspace of Ukraine states such a term as "procedure violation of airspace use" that does not correspond to the title of Article 282 of the Criminal Code.

In our opinion, this directly affected not only possibility of correctly classifying the act as a criminal offense and its qualification while opening criminal proceedings under this article but the subsequent closure of criminal proceedings under paragraph 2 of part 1 of Article 284 of the Criminal Code. This legal norm stipulates that criminal proceedings are closed if there is no lack in the act of a criminal offense [6].

This fact is confirmed by the above analysis of statistical data of the Prosecutor General's Office of Ukraine on accounting in 2014–2020 in the Unified Register of Criminal Proceedings under Article 282 of the Criminal Code of Ukraine and their further movement. At the same time, as already mentioned during this period, no criminal proceedings were sent to court with an indictment.

This circumstance becomes an obstacle to objective reflection in the statistics of the results of combating criminal offenses under Article 282 of the Criminal Code of Ukraine and monitoring the real state of criminogenic situation and the actual level of crime. Incorrect assessment of the criminogenic situation negatively affects adoption by law enforcement agencies of adequate measures to combat investigated criminal offense.

Given the above, we consider it appropriate to amend Article 23 of the Air Code of Ukraine, paragraph 46 of the Regulations on the Ukrainian airspace use of in terms of bringing them into line with Article 282 of the Criminal Code of Ukraine by replacing the word “order” with “rules”.

In order to stabilize criminogenic situation associated with air traffic violations and airspace use, it is necessary, first of all, to establish the reasons and conditions that contribute to their commission. The rule of law should make appropriate efforts at regulatory and organizational level in order to create an effective system for eliminating or at least reducing determinants of criminal offenses. This approach is an important step in preventing the crimes under investigation.

The causes and conditions of crime act as its determinants. In criminology, the determinants of crime (derived from Latin: *determinare*), scientists understand the complexes of social phenomena which joint impact on the person leads to commission of criminal offenses. This influence is explained by the fact that cause of crime creates socially dangerous consequences only under certain conditions.

Scientists define the causes of crime as factors that give rise to crime as a consequence [10, p. 95]. However, the cause of crime in itself does not cause a socially dangerous act. As criminologists aptly point out, for the onset of the consequence (realization of the cause in the dynamics) it is necessary to combine it with the condition [11, p. 269]. The causes of crime play a major role in a person’s decision to commit a criminal offense. It should be noted that decision is usually influenced not by one but at the same time a combination of such reasons.

As for the conditions of crime, they do not directly generate it, but “open the way” for the action of causes, facilitate onset of socially dangerous consequences [10, p. 95]. Under the crime conditions, scientists understand a set of negative phenomena and processes that contribute to formation and manifestation of crime causes [11, p. 274]. In other words, conditions of crime are social phenomena that do not independently generate crime but contribute to its determination.

Thus, we share the view of those scholars who believe that the causes and conditions of crime is a collective concept, where the causes are negative social phenomenon of public life that generates and supports crime, causes its growth or decline; and conditions are social, natural and anthropogenic factors that do not generate crime, but contribute to crime development [12, p. 8].

It should be noted that causes and conditions of crime do not always have an exclusively negative social meaning. Thus, along with negative determinants, positive social phenomena can also lead to criminal offense commission. For example, this applies to criminal liability liberalization and so on.

Given limited scope of our research paper, we will briefly consider the most common determinants of criminal offenses under Articles 281 and 282 of the Criminal Code of Ukraine. In our opinion, these include general causes and conditions that contribute to the spread of crime in general and special which directly contribute to commission of the criminal offenses we investigate. Both groups of determinants are closely related so we will not separate them into separate groups.

While analyzing commission determinants of criminal offenses under research, we should take into account the views of scientists who point out that crime determinants, as well as crime itself are not permanent social phenomena. Crime causes are dynamic and change over time [13]. We believe that as determinants of such criminal offenses as violation of air traffic rules and use of airspace, first of all, it is necessary to highlight the following:

1. Deficiencies in legal regulation of criminal proceedings, caused by the imperfection of current Criminal Procedure Code of Ukraine (they have already been mentioned above) affecting the results of combating the criminal offense in question. Thus, failure to prosecute perpetrators leads to a lack of preventive influence on perpetrators and others that certainly contributes to the spread of crime.
2. Imperfection of criminal and legal regulation of criminal offenses provided by Article 282 of the Criminal Code of Ukraine. This is due to a legislator error that led to inconsistency of the norms of this article: the norms of the Air Code of Ukraine and the Regulations on Ukrainian airspace use, which norms it refers. As mentioned above this circumstance directly affected qualification of these socially dangerous acts and accordingly the level of their counteraction.
3. Moral degradation and spiritual society crisis manifested in blurred moral guidelines, spiritual devastation, high prevalence of alcoholism, drug addiction in Ukraine that causes a low level of responsibility for their actions and leads to criminal arrogance (e.g., transportation of flammable substances, unauthorized opening the aircraft door during the flight, etc.).
4. Protest against accepted values and norms in society that can lead to disregard for certain rules of conduct and lead to criminal negligence (for example, smoking on board an aircraft which led to a fire).
5. Ignoring safety rules during high-hazard work, such as by taking action (for example, launching missiles in the airspace area, conducting training firing with inappropriate ammunition, launching balloons in places or at heights where air flights are carried out; carrying out explosives) work near runways before or after permitted time, holding competitions or other actions with birds that can cause them to collide with aircraft, etc.) and by inaction (for example, not turning on signal lights on TV towers or high-rise buildings, disagreement installation work carried out on the runway, with air traffic services).
6. Availability of corruption facts in judiciary, regulatory, law enforcement and other state bodies, which is reflected in the results of the fight against crime and contributes to its spread. According to the Transparency International Corruption Perceptions Index, Ukraine ranked 117th out of 180 countries in terms of corruption in the public sector in 2020, ranking second in Europe after Russia [14].
7. The low level of combating crime in Ukraine, inter alia, due to circumstances set out in paragraphs 1–3.

It should be noted that list of reasons and conditions that contribute to commission of such criminal offenses as violation of air traffic rules and use of airspace is not exhaustive but they have a dominant influence on the commission of these socially dangerous acts and together create favorable for of this basis.

4 Conclusions

Thus, on the basis of the above, it can be concluded that currently criminal offenses under Articles 281 and 282 of the Criminal Code of Ukraine are relatively rare in Ukraine. At the same time, we have identified certain shortcomings in activities of law enforcement agencies to combat this criminal offense and regulatory regulation of this activity (imperfection of criminal procedural and criminal legislation). These circumstances affected the results of law enforcement activities. Thus, in the period from 2014 to 2020, percentage of criminal proceedings initiated for committing criminal offenses under Articles 281 and 282 of the Criminal Code of Ukraine sent to court with indictments averaged only 15.8% and 0%, respectively.

This situation is unacceptable, given that criminal offenses under investigation have a significant public danger and can lead to serious irreversible consequences. The latter factor increases relevance of such a form of crime prevention by law enforcement agencies as the prevention of these criminal offenses. An important prerequisite for successful implementation of this form of combating crime is a set of measures to eliminate or minimize the impact of the causes and conditions we have identified on violations of air traffic rules and airspace use.

References

1. Verkhovna Rada of Ukraine. Legislation of Ukraine: Air Code of Ukraine: Law of Ukraine dated 19.05.2011 No. 3393-VI (n.d.). https://zakon.rada.gov.ua/laws/show/3393-17?lang=en#Text
2. United Nations: United Nations Convention on the Law of the Sea: International legal act dated 10.12.1982 No. 995_057 (n.d.). https://www.un.org/depts/los/convention_agreements/texts/unclos/closindx.htm
3. International Civil Aviation Organization: Convention on International Civil Aviation: International legal act dated 7.12.1944 No. 7300/9 (n.d.). https://www.icao.int/publications/Documents/7300_cons.pdf
4. BBC Bitesize: Nature of Crime: National 5 modern studies revision (2021). https://www.bbc.co.uk/bitesize/guides/zqb2pv4/revision/1
5. Prosecutor General's Office of Ukraine: Official website: Statistics – Prosecutor General's Office (n.d.). https://www.gp.gov.ua/ua/1stat
6. Verkhovna Rada of Ukraine. Legislation of Ukraine: The Criminal Procedural Code of Ukraine: Law of Ukraine dated 13.04.2012 No. 4651-VI (n.d.). https://zakon.rada.gov.ua/laws/show/4651-17#Text
7. Verkhovna Rada of Ukraine. Legislation of Ukraine: On Operative Investigation Activity: Law of Ukraine dated 18.02.1992 No. 2135-XII (n.d.). https://zakon.rada.gov.ua/laws/show/2135-12#Text

8. Verkhovna Rada of Ukraine. Legislation of Ukraine: Criminal Code of Ukraine: Law of Ukraine dated 05.04.2001 No. 2341-III (n.d.). https://zakon.rada.gov.ua/laws/show/2341-14#Text
9. Cabinet of Ministers of Ukraine: On approval of the Regulations on the use of airspace of Ukraine: resolution of the Cabinet of Ministers of Ukraine dated 06.12.2017 No. 954-2017-p (n.d.). https://zakon.rada.gov.ua/laws/show/en/954-2017-п?lang=uk#Text
10. Bogatyrev, I.G., Topchiy, V.V. (eds.): Criminology. VD Dakor, Kyiv (2018)
11. Golovkin, B.M.: Causality in system determinations to criminality. Theory Pract. Jurisprudence **1**(5), 268–275 (2014). http://nbuv.gov.ua/UJRN/tipp_2014_1_24
12. Grin, M.V., Kondratov, M.A.: Basic Concepts in Criminology: Terminological Dictionary. KubGAU, Krasnodar (2017)
13. Negi, R.: Causes of crime. Law Times J. (2019). https://lawtimesjournal.in/causes-of-crime/
14. Transparency International: Corruptions Perceptions Index 2020 for Ukraine (2020). https://www.transparency.org/en/cpi/2020/index/ukr

Transformation of Legal Labor Regulation Under Influence of Artificial Intelligence

Svitlana Gutsu[1](✉) and Andriy Bublikov[2]

[1] Department of Law, National Aerospace University «Kharkiv Aviation Institute», Kharkiv, Ukraine
s.gutsu@khai.edu

[2] National Scientific Center, «Hon. Prof. M. S. Bokarius Forensic Science Institute», Kharkiv, Ukraine

Abstract. The article is devoted to research on prospects artificial intelligence implementation in the field of labor relations. The tendencies inherent in digitalization processes influencing labor field are marked, and also the changes which will occur in the field of labor law are forecasted. It is proved that artificial intelligence implementation leads to rapid changes not only in technologies used in workflow but to changes in labor nature, employment, management of relations, labor protection and labor guarantees. Necessity of modernization of the system of professional training, education and advanced training of staff in conditions of digitalization is determined. Issues of transformation of employee status in labor legal relations are investigated. It is determined that with development of artificial intelligence, the range of subjects of labor law will be expanded. With acquisition of the legal status of *Electronic personhood* in civil law, "smart" jobs will be able to obtain the status of a subject in the employment relationship. Emphasis is placed on the emergence of workers with neural prostheses that restore or enhance natural human abilities. As a result, it will be necessary to determine the legal status of at least three categories of employees: employees with disabilities who need neural prostheses for medical reasons; employees who have expressed a desire to install neural prosthesis without medical indications: employees who do not use biotechnological means. It is concluded that labor legislation requires reformulation of labor law principles in accordance with the new conditions for implementation of labor relations, as well as the development of new standards and guarantees for labor protection.

Keywords: Artificial intelligence · Electronic personhood · Robots · Subject of labor law · Principle of equality and non-discrimination in labor field · Occupational safety · Vocational training · Digital skills · Labor relations

1 Introduction

Technology growth is the main driving force of economic growth of peoples and improving living standards of people for a long time. It increases overall productivity, thereby increasing per capita income and consumption. Growing ability of artificial intelligence

M. Nechyporuk et al. (Eds.): ICTM 2021, LNNS 367, pp. 599–617, 2022.
https://doi.org/10.1007/978-3-030-94259-5_51

systems (AISs) to solve complex problems autonomously can radically change our economy and society, for example by developing new transport types, unmanned aerial vehicles, cargo transportation, space programs or health care. Technology does not always destroy professions. They change certain labor tasks. Often, new technologies create jobs and require employees to acquire new skills. Throughout history, innovations have increased employee productivity, created new products and markets, thus creating new jobs in economy.

Just as these new technologies open up enormous prospects for human civilization development, they are considered as a threat potentially destroying traditional labor markets, increasing inequality and discrimination. The great public fear is that robots and AISs will replace human jobs on a large scale. In conditions of mass unemployment or underemployment, mass population impoverishment and economic society delamination will increase. The ever-increasing inequality between technology owners and employees can lead to protracted social conflict.

Rapid development of artificial intelligence raises many issues in terms of determining its legal status, capabilities and ethical constraints. All this is especially relevant for labor world, because society has deep concerns about potential dangers and risks associated with implementation of artificial intelligence in life. Attracting robot's changes labor relations, fills them with new content and forms. Software using artificial intelligence elements are now used to automatically perform those tasks that were solved only a few years ago by employees. Labor legislation formed in general in the era of industrial society cannot respond effectively to changes taking place; a set of new risks and new problems related to development of digital technologies is not taken into account [1].

Research Purpose: analyze the changes taking place in society in implementation of artificial intelligence in labor relations. Highlight those that have the greatest impact on the field of hired labor; taking into account positions of Ukrainian and foreign experts in this field; identify the main blocks in legal labor regulation making changes in will be made in the coming years; formulate some necessary amendments to Ukrainian labor legislation.

2 Artificial Intelligence Influence on Labor Relations

Implementation of artificial intelligence in economic and social processes certainly affects relations between employee and employer. These relations depend on the economy type and the level of development of economic relations in society. Currently, the question is not so much whether artificial intelligence affects change in the type and content of labor relations but how exactly they change and what modernization of labor legislation they need.

Despite the overall low economic level of Ukraine, information technology field in the country is developing rapidly and has great potential for further growth. In early 2019, the Deep Knowledge Analytics agency of Deep Knowledge Ventures London investment fund specialized in AISs, blockchain and technology trends, published a research on artificial intelligence market in Eastern Europe. As it turned out, Ukraine is in the top three in the field of artificial intelligence and related technologies. The report states that countries such as Ukraine and Poland are leaders in the number of artificial intelligence providers even compared to Western European countries. (Fig. 1). 28 Ukrainian companies supply solutions with artificial intelligence compared to 226 suppliers worldwide [2].

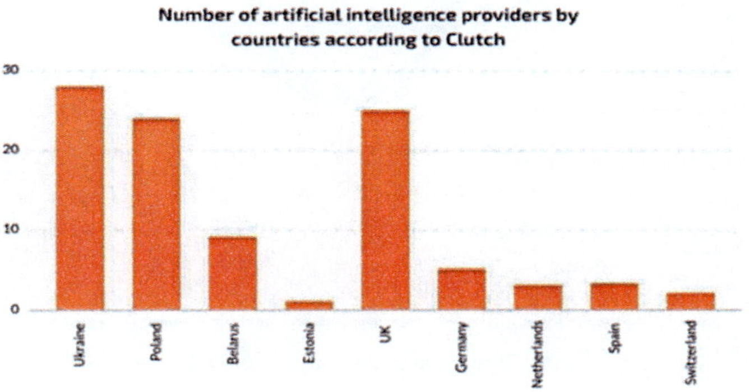

Fig. 1. Number of AI companies in Europe

Ukrainian scientific and staff pull is huge: Ukrainian share in digital technology market is more than 10%. (Fig. 2). Currently, this situation persists [3], as evidenced by the *Interactive Mind Map AI in Eastern Europe Industry Landscape* [4]. However, the downside is that nowadays Ukraine is developing more digital economy of other countries using our professional capabilities on outsourcing basis. The task of state policy for the coming years should be to develop a legal concept of attracting human resources for development of the national digital economy requiring labor legislation transformation.

Analysis of professional readings and statistical research suggests that most experts: lawyers, economists, sociologists support the view that nowadays in the field of labor relations are undergoing major transformations in connection with society digitalization. Researches by A. Korinek, M. Webb indicate how introduction of artificial intelligence affects reduction of jobs, advantages of robots over humans in the field of efficiency to reduce the cost of production [5, 6]. Research papers by I. A Filipova who believes that in the context of digitalization leading to emergence of new professions, new types of communications updating occupational safety standards and displacing people from workplaces by artificial intelligence systems, law will have to be adjusted to maintain legal regulation effectiveness excite interest [7].

Among Ukrainian researchers of artificial intelligence legal definition research papers by Baranova O. A, Kharitonova E., Kharitonova O., O. E. Radutny attract attention but impact of implementation of robots and technology on labor relations in domestic professional readings remain unnoticed by most scientists.

Every year globalization intensifies in the world. Technology development has turned the Earth into a single global information area. Certainly, Ukraine is its part.

Transformational processes in economy, production, education, labor cannot but affect our state. This is confirmed, by Government Priority Action Project for 2020, where in paragraph 18 it was planned "Expanding cooperation between Ukraine and the

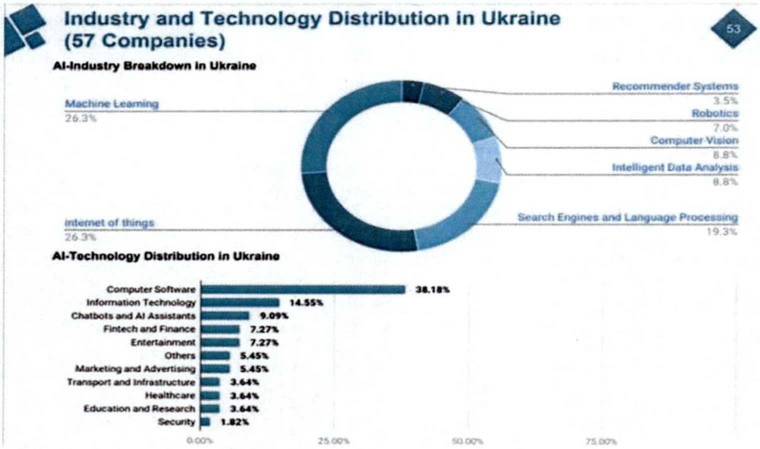

Fig. 2. Digital technology market in Ukraine

EU in the digital economy in order to integrate Ukraine into the EU Digital Single Market. As a result, certain advantages of the EU Digital Single Market should be extended to Ukraine. Another important point of the same Plan is item 55: "Accession to the work of Digital Economy Committee and Task Force on the Digital Economy (TFDE) of the Organization for Economic Co-operation and Development. The aim is to ensure the presentation of Ukraine position in the Organization for Economic Cooperation and Development; creating preconditions for Ukraine accession to standards of the Digital Economy Committee of Organization for Economic Co-operation and Development" [8].

In pursuance of the above document, in connection with the need to develop a coherent state policy to regulate the field of artificial intelligence and eliminate Ukraine lag in relevant field, Cabinet of Ministers of Ukraine on December 2, 2020 approved the Concept of Artificial Intelligence in Ukraine. This Concept defines the purpose, principles and objectives for development of artificial intelligence technologies in Ukraine as one of priority areas in the field of scientific and technological research. Important document innovations were the legal definition of the *artificial intelligence* term, as well as the definition of artificial intelligence as a field of activity in the field of information technology providing creation, implementation and use of artificial intelligence technologies [9]. In addition, the *risk of rising unemployment due to the use of artificial intelligence technologies* has been identified as one of the problems that needs to be addressed. For this purpose, the Concept proposes the following:

- development of specialized educational programs for professional development and adult education in the field of artificial intelligence, as well as the provision of scholarships to persons studying in such programs including specialized for certain population groups, including people with disabilities, internally displaced persons;
- providing social protection for professionals obtaining additional education in the artificial intelligence field;

- development and implementation of a nationwide information campaign on general aspects, risks and threats of use and further dissemination of artificial intelligence technologies in Ukraine.

Government believes that artificial intelligence technologies should contribute to the economy transformation, labor market, government institutions and society as a whole. For this purpose, priority areas for implementation of the Concept among others, are:

- raising the level of professional training to provide the field of artificial intelligence technologies with qualified staff;
- bringing legislation in the field of artificial intelligence technologies in line with international regulations;
- development of a Roadmap for people retraining whose work can be automated in the next five to ten years and as a result, significant increase in the number of skilled professionals in the artificial intelligence field, including scientific and scientific and pedagogical staff and dissemination of skills competent use of artificial intelligence.

Thus, the issue of artificial intelligence impact on labor market and labor relations is an indisputable fact. Currently, in domestic legislation it is the only document that defines the mentioned problem. Therefore, it is necessary to refer to international community experience in this matter for more effective implementation of their results in national legislation.

Let us refer to the EU law which Ukraine is so eager to integrate with. Legal regulation of the use of new technologies, use of robots and use of artificial intelligence is one of the priority issues for the European Parliament. Thus, based on performed research results, a number of normative legal acts have been developed, *Robotics 2020* roadmap for development of robotics in Europe has been created, and the *Horizon 2020* program, framework program for research and technological development of the European Union. Note that most of the rules concerning legal status of robots and artificial intelligence lie in the plane of civil law. However, this is the norm, the basis for further extension to criminal, administrative and labor relations.

On February 16, 2017, the European Parliament approved a Resolution on the rules of civil law on robotics and proposed to recognize the special legal status of Electronic personhood for complex robots making independent decisions [10]. On December 7, 2018, the Agreed Project for Development and Use of Artificial Intelligence prepared by the European Commission for the European Parliament and the European Council was published. The Project includes a set of measures to increase investment and to study the impact of artificial intelligence on labor relations. In April 2019, the European Commission website published the final report of expert group studying the impact of digital transformation on labor markets that indicated the need to adapt labor relations to new realities in order to prevent risks to occupational safety and health [11].

Process of society transition from an industrial economy to a digital one (Industry 4.0) significantly changes labor relations. These changes are both quantitative and qualitative ones. Implementation of digital technologies in production and service is gradually increasing the level of their automation: the use of complex software and robotic systems is growing that leads to changes in hired labor management. Software with artificial intelligence elements are now used to automatically perform those tasks that have always been solved by human. Among new conditions which the current labor legislation is not designed for: mass replacement of employees with robotics at workplaces; significant change in employee functionality in connection with the automation of production; spread in working environment of virtual and augmented reality technologies, etc.

In February 2021, PwC commissioned a survey of 32,517 members of general public. Respondents included employees, business owners, contractors, students, unemployed looking for job and those on leave or temporarily laid off. According to the survey, the following results were obtained:

- 60% of respondents are concerned that automation threatens many jobs.
- 48% believe there will be no traditional employment in the future and that we will sell our skills on a short-term basis to those who need them.
- 56% think that not all people will have stable long-term jobs in the future;
- 61% believe that their government should act to protect jobs and this feeling is more acute among people 18–34 years old (66%) than among people over 55 years old (51%).
- 39% believe that their job will become obsolete within five years.
- 80% believe that they can adapt to new technologies in their workplace,
- 46% of people with higher education say that their employer offers them opportunity to improve their digital skills but only 28% of people with a graduate degree say the same [12] (Fig. 3).

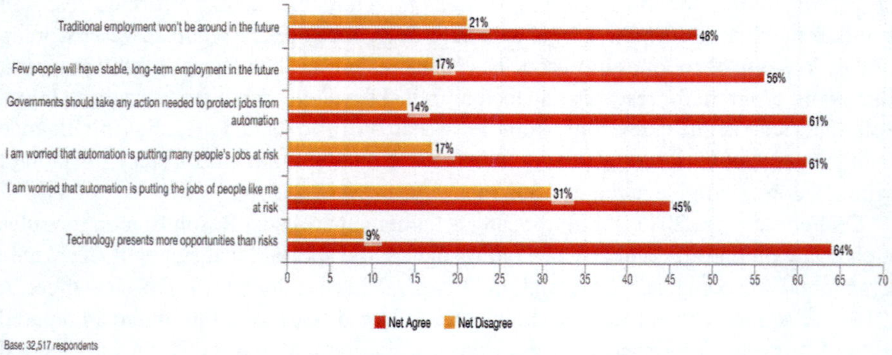

Fig. 3. The impact of automation on jobs

Processes related to digitalization of economy and society as a whole, significantly affect relations between employees and employers, so among researchers in labor law, economics and sociology of labor in recent years, more and more issues regarding ways to change regulation of relations in the labor field. It is a popular conclusion that automation reduces labor costs and reduces the overall demand for labor, so in the era of rapid automation increases the likelihood of deterioration of employee conditions [13]. It is difficult to disagree with this position.

According to a PwC analysis, artificial intelligence, robotics and other forms of intelligent automation can bring huge economic benefits that will bring up to $ 15 trillion in world GDP by 2030. Expecting such profits will also create demand for many jobs, but there are also fears they could displace many existing jobs. By the mid-2030s, up to 30% of jobs can be automated with slightly more men affected in the long run, as autonomous vehicles and other machines will replace many manual jobs where their share of employment is higher. However, while first and second waves, women could be at greater risk of automation due to their higher representation in clerical and other administrative functions (Fig. 4) [14].

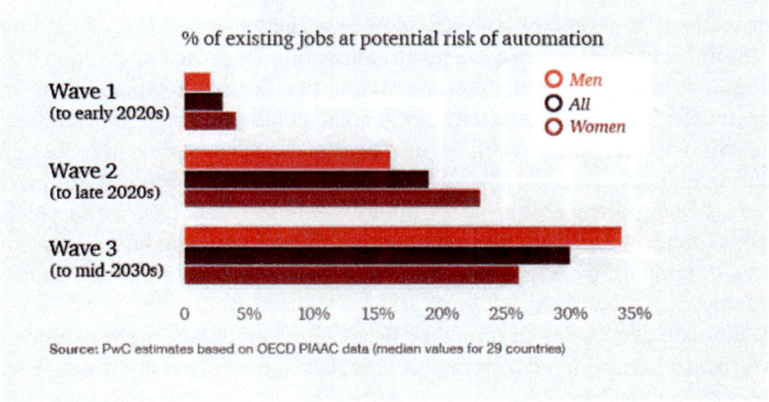

Fig. 4. Risk of job loss due to production automation by gender

In January 2020, the European Foundation for the Improvement of Living and Working Conditions published a report entitled: *Technologies that change games: transforming production and employment in Europe* examined the impact of eight factors (additive manufacturing, advanced robotics, the Internet of Things, industrial biotechnology, electric cars, autonomous vehicles, blockchain and virtual and augmented reality) in different economic sectors. The report forecasts changes in industry, services, labor market and employment under influence of each of the eight factors [15].

Thus, labor law will be transformed, as it should respond to change and create effective tools for regulating labor relations in a digital society. In our opinion, scientists have to pay special attention to transformation of vocational training and employee retraining, development of new standards for occupational safety and health, as well as the expansion of labor relation forms.

3 Training of Employees as a Determining Factor of Competitiveness in Conditions of Artificial Intelligence Implementation

The need to build a new system of vocational training and improve digital employee skills is not a new idea. These issues have been repeatedly raised both in studies by independent organizations and at the level of regional and international institutions. Some countries and organizations have positive experience in implementing employee training programs. We have studied and analyzed this experience in our research papers [16; 17].

As companies accelerate their implementation of artificial intelligence projects and many jobs remain remote, employees of all economic sectors need to acquire new skills that will allow them to think and work differently.

On August 29, 2017, the UN Department of Economic and Social Affairs published a research on the technology impact on labor market ("Frontier Issues: The impact of the technological revolution on labor markets and income distribution") [18]. This research in the artificial intelligence field indicates that technology will have a "huge impact" on labor markets. If necessary laws are adopted at the local, national and international levels, this impact can be directed in a positive direction. In particular, it is noted that new technologies do not replace entire professions but specific tasks. Technological change is changing the demand for skills in the labor market and in order to remain competitive in jobs that are complemented by artificial intelligence, workers must acquire the necessary skills during their education or during vocational training. Providing appropriate skills to workers is an important area of public policy. Labor supply at each skill level depends on the education and vocational training system. Greater access to higher education and vocational training for workers will provide them with access to jobs that require higher qualifications.

In addition to acquisition of formal skills, technology growth speed requires greater employee flexibility and rapid learning. Education and training systems need to prepare employees to be flexible and develop new skills in response to the rapid changes brought about by new technologies. States can provide incentives and resources for employees to obtain additional training and education not only in the technical field, but also in creativity, management, social and communication skills. Availability of low-cost online courses has greatly expanded the opportunities for lifelong learning as employees learn new skills and invest in their own human capital. However, government programs in this area can have a much greater effect. Thus, the Priority Action Project of the Government of Ukraine for 2021 provides for paragraph 54 "Improving digital skills, ability to use public digital gadgets and the Internet for various categories of citizens, as well as increasing the digital competence of civil servants, doctors and teachers. As part of these activities, access to the National Digital Literacy Platform "Diia: Digital Education" to 12 new educational series on digital literacy, 1000 qualified trainers were trained; employs 1,500 trainers who teach digital literacy to various population segments in person and via the Internet; information campaigns were conducted to motivate Ukrainians of all ages to acquire digital skills. The result of these actions should be creation of opportunities for citizens to master digital skills in a convenient and accessible form by watching educational series on digital literacy and training with the help of qualified trainers…". In addition to the above, the document provides for measures for:

- infrastructure creation for development of population digital skills (clause 56);
- creation of a partner network for digital literacy training of Ukrainian citizens who do not have PC or smartphone at home and providing access to computers, digital gadgets and Internet, especially for elderly and vulnerable groups (clause 57);

- digital literacy training for different population segments via creating systematic educational content on digital literacy in Ukrainian language and conducting staff training in all regions (clause 58).

As a result, 500,000 people have already been trained in digital skills; 100 memoranda were signed with private companies to provide their own technical resources for training citizens; The "Trainers for Trainers" training program has been created for people who will teach digital literacy.

Another example of the government work in this direction is the approval of the Resolution № 39 of the Cabinet of Ministers of Ukraine dated on January 20, 2021 *on the implementation in Kyiv of an experiment on vocational training of registered unemployed for the needs of employers in the field of information technology* [19]. Experiment participants are employers in the field of information technology, educational entities that provide services in non-formal education system and registered unemployed who have the right to receive social services for vocational training, retraining and advanced training in accordance with the Laws of Ukraine: *On Mandatory State Social Unemployment Insurance* and *On Employment of Population* – unemployed. The list of educational programs for which the experiment is conducted is formed by the Ministry of Digital Transformation of Ukraine on the basis of applications from employers and will be published on the official websites of the Ministry and the Kyiv City Employment Center. Vocational training is funded by the Compulsory State Social Insurance Fund in the event of unemployment within the approved estimates for such purposes in 2021.

Thus, importance of digital learning is recognized at the state level and implemented in general education and social protection programs. However, in our opinion, state regulation of the field of professional training and advanced staff training involved in working with artificial intelligence programs is a necessary condition for transformations in labor relations.

Currently in Ukraine, usual periodicity of mandatory training for individual professionals is 5 years. In the public sector, it is at the employer's expense, with provision of certain labor guarantees to such employees for the training period. Professional training in the workplace is carried out in accordance with regulations in the education field, labor and others governing public relations in the field of providing employers with qualified and competitive workers in the labor market. Thus, the Regulations on Vocational Employee Training while Manufacturing stipulate that: "Vocational employee training is continuous and is carried out while their work in order to gradually expand and deepen their knowledge, skills and abilities in accordance with production requirements" [20]. Law of Ukraine: *On Professional Development of Employees* among the main activities of employers in the field of professional development of employees establishes that: [21]. However, the law stipulates specific terms and obligation to improve professional skills only for employees of certain professions and positions. For example, scientific and pedagogical employees of higher and postgraduate education institutions improve their skills in accordance with the Procedure for professional development of pedagogical and scientific employees at least once every five years [22].

According to Article 29 of the Law of Ukraine: *On Civil Service*, civil servants improve their skills constantly including through training in relevant educational institutions, as a rule, at least once every five years. Learning outcomes and training are one of the reasons for promotion.

The attestation system and advanced training of physicians and pharmacists has been updated. The Regulation: *On Carrying out Attestation of Physicians* stipulates that persons who have been assigned a qualification category are attested to confirm it at least once every five years. At physician request, attestation for qualification category assignment can be carried out in a shorter period but not earlier than one year from the date of previous attestation.

Frequency of in-service training of civil servants is defined in the Regulation: On the system of professional training of civil servants, heads of local state administrations, their first deputies and deputies, local government officials and deputies of local councils and is usually held at least once every three years [23].

As we can see, decision regarding the issue of professional development of most categories of employees is left to the employer discretion. The result is too slow changes in the system of digital learning and training. Advanced training is understood as modern companies as providing employees with opportunity to gain knowledge, tools and skills needed to use advanced and ever-changing technologies in workplace and in everyday life. Not everyone needs to learn to code but people in many professions already need to understand and manage artificial intelligence, data analytics, autonomous vehicles and other technologies that are impossible to even predict, those that are emerging now and those that will be created in the future.

In 2021 research by PwC Global found that:

- 40% of employees successfully improved their digital skills while pandemic;
- 77% are ready to learn new skills or completely retrain;
- 74% consider training as a matter of personal responsibility.

Young people are twice as likely to get opportunities for training than older people and urban residents are 1.5 times more likely than people in cities [12].

By 2025, full robotization is possible in certain economy segments, all this will provoke abandonment of the system of lifelong employment and a rapid change in qualification requirements for employees. In such a situation, the state should act as a regulator to ensure the progressive society development as a whole and normal living and working conditions for each citizen. In our opinion, condition of employee training, retraining or certification in connection with the introduction of new technologies should be included in employment contracts. Further development of labor law is not possible without a review of the regulation of lifelong learning. This should be a prerequisite for concluding employment relationships and distinguishing them from civil law contracts.

4 Changing Labor Law Principles Under Influence of Artificial Intelligence Implementation

Labor law of Ukraine as an integral part of general theory of national law is an independent law branch has certain principles of legal labor regulation. The law principles are the result of centuries of human development, result of understanding the patterns of society development as a whole, embodiment of democratic and humanistic traditions. First of all, they have a regulatory value, influencing legal behavior of labor subjects and other closely related relations and act as a starting point for the further development of legislative and law enforcement activities [24]. However, stability of principles should not mean their inviolability. In connection with society development and emergence of new important components, the principles can evolve and be filled with new content. Legal principles "arise, mature and, ultimately, can disappear" in connection with filling of the principle scope with a different meaning. We should agree with the position of scientists that formation of new approaches to the interpretation and application of law principles, consequence of changing living conditions in society. Existence and content of principles of labor law depends on development of labor relations and they depend on development of society, so the normative "content" of principles inevitably changes over time.

In connection with a significant change in conditions of hired labor, in addition to the creation of new standards in labor, there is an objective need to reformulate or clarify the content of certain principles of labor law.

In the theory of law of Ukraine, it is accepted to allocate the basic (general), interbranch and branch principles. The basic principles of labor law include:

- equality in labor and employment, prohibition of discrimination at while job performing (Article 24 of the Constitution of Ukraine, Articles 2, 2 and Labor Code of Ukraine);
- freedom of labor and employment, prohibition of forced labor (Articles 43 of the Constitution of Ukraine, Articles 5, 21, 49–4 of the Labor Code of Ukraine);
- employment and moneyed assistance for unemployed (Article 43 of the Constitution of Ukraine, Articles 2, 5, Chapters III and III-A of Labor Code of Ukraine).

In labor law, as in any other branch of law, intersectoral principles are not fully manifested but only in relation to features governed by labor law of public relations (principle of property protection; principle of association freedom) and sectoral ones express general essential properties of rules of this law branch only. Most of these principles, they are specified and manifested in individual institutions and norms depending on their content and purpose (right to safe working conditions; right to remuneration, etc.).

Implementation of robots and artificial intelligence in the field of labor changes its nature and form and thus inevitably lead to the transformation of both basic and sectoral principles which labor relations are built on.

In our opinion, principle of equality and non-discrimination at work, will primarily fall under the transformational processes associated with involvement of artificial intelligence in labor relations. This is due to emergence in coming years of new subjects of legal relations: the so-called "digital" or "Electronic personhood" work with artificial intelligence properties and secondly "human enhancement" with person with elements of biotech implants.

Observing the speed with which works are being implemented in all production fields and social life, idea of giving artificial intelligence legal personality no longer seems such a distant fantasy. Gradually, national laws of individual countries prepare the ground for appropriate changes. The Council of Europe has set up a temporary committee on artificial intelligence to study development and legal regulation of relations with its participation. The problem of the legal status of robots is under active discussion both in Ukraine and in the world as a whole. Changes in national labor legislation will depend on the solution found.

In February 2017, the European Parliament in a resolution for the European Commission proposed to recognize the special legal status of electronic personhood for complex robots that make independent decisions. According to H. Eidenmüller, Professor of Commercial Law in the Faculty of Law at the University of Oxford [25], "smart" robots should be considered by such people in the near future. This means that they will be endowed with legal personality. Among Ukrainian scientists, the supporter of this concept is Radutny O. E., who does not rule out the possibility of including artificial intelligence in the circle of criminal law subjects [26].

For labor law, the first question is what will be the status of robots that have received subjective rights and replaced workers in workplace? Current labor legislation, as national as foreign one does not yet provide answers to this question. Machine is no longer a working tool but it is the law object. However, it becomes a performer of certain labor functions along with an employee, in some cases instead of. The need to regulate new relations in the working process requires systemic changes in legislation. The legal employee position as a subject of legal relations changes. Will the job be granted the employee status or will it be a separate subject of the employment relations? The issue of recognizing robots as law subjects that would have a certain set of rights and responsibilities is actively discussed by researchers and practitioners. For example, it is proposed to give artificial intelligence systems the right to inviolability of their "body" and code.

Recognition of the legal personality of "smart" robots is a matter of civil law. Interest for labor law is the discussion of the status of artificial intelligence system in connection with the need to regulate the rights to the results of intellectual and economic activities created by this system in labor process.

No less important is the question of the future determination of the legal status of employees – carriers of neural prostheses. Currently, supporters of transhumanist movement demand forced measures that involve direct introduction of technology into the human body, their fusion with each other. The result should be what is defined as the improvement of human nature, i.e. "human enhancement", a person "formatted" with achievements of medical and biotechnological sciences. The broad definition of *enhancement* refers to biotechnological interventions designed to improve human nature, and often have little to do with the goal of maintaining or restoring human health.

Neural prostheses are electronic implants that can restore motor, sensory and other functions lost due to injury or illness. The global market for medical neural implants is expected to reach $ 54.28 billion by 2025.

In a technologically advanced society, quantity and quality of work will depend on the presence (absence) of neural prostheses combined with other technological devices at workplace. It can turn out that employees, namely: people will also be divided into certain categories. Here lies the contradiction between the interests:

- disabled employees in need of neural prostheses for medical reasons, they are protected by the Convention on the Rights of Persons with Disabilities 2006 № 61/106164 according to which States parties undertake to ensure and promote the full realization of all human rights and fundamental freedoms by all persons with disabilities without discrimination on disability symptom;
- employees who have expressed a desire to install a neural prosthesis without medical indications, for example, sensory organs, superior in their parameters to biological prototypes or use tools designed to enhance cognitive abilities taking into account specialization and personal characteristics (neural control interface, powered exoskeleton);
- other employees who do not use the biotechnological means specified in previous paragraphs.

Neural prostheses "raise" capabilities of people with disabilities to ordinary people and even higher. In the future thanks to development of technology these capabilities will increase. It is believed that some people who do not have medical indications will want to "improve" by taking advantage of neural prostheses. If they are banned, it will be discrimination against them in comparison with disabled people and to deprive disabled people of this means to deprive them of the right to equalize life quality with non-disabled people.

On the one hand, it is necessary to use possibilities of neural prostheses, on the other hand, it is necessary to regulate the status of such employees. These employees will have a number of abilities that are not inherent in other employees. Some people without medical indications for neural prostheses want to "improve" their capabilities, how to deal with their status? From what level of complexity of the neural prostheses can we talk about the special status of such employees? [27, p. 62]. Unexplored possibilities of artificial intelligence increase the risks of working with them for employees in neighboring workplaces and for customers of employers. Among the risks: threat of hacker attacks, to the artificial intelligence system of neural prostheses (for example, with the help of "Malicious brain" can be malicious manipulation of brain implants); problem of protecting any confidential information that can be recorded by such systems, etc. With proliferation of people with neural prostheses, legal regulation of the legal status of disabled employee becomes insufficient.

New technologies of human improvement that will lead to overcoming aging and death are associated with both positive and negative trends in human development and create not only new opportunities but also risks. Because equal access to biotechnology is impossible, over time it will change the conditional natural equality between people that forms the basis of what is considered society. The problem of inequality in society, or rather formation of new inequality forms, as a result, availability or lack of access to "improvement" technology.

The newly formulated principle should take into account the possibility of using a neural prosthesis as a part of the human body, that is actually artificial intelligence. It will be more profitable for employers to hire people with neural prostheses for some jobs, as possibility of using them in the working process will allow you to get the best result from an employee with a ready-made device with artificial intelligence. Employee with a neural prosthesis can become more efficient than a normal worker (due to exceeding the natural indicators of functionality) and then ordinary employee will lose. Subsequently, protection against discrimination can be needed for other employees who will not be hired due to the lack of such devices or fired, replaced by more efficient workers with neural prostheses [27, p. 68].

All this allows us to formulate tasks for professionals in the field of labor law: to be prepared for the fact that in the next 5–10 years in countries that are leaders in digitalization, the problem of job regulating of people with neural prostheses (especially the problem of formulating standards other employees) will be on the agenda. Taking into account rapid technology spread in a relatively short period of time, this problem will reach the world level and will have to be considered within national legal systems.

Thus, labor law will follow the path of expanding the range of subjects of labor relations and the very meaning of the *employee* term. Obviously, *Electronic personhood* with acquisition of legal personality will be part of participants in the employment relations. Either as a separate individual or as an employee. In turn, the classical *employee* definition as an individual, endowed with legal capacity and legal capacity is also changing. With the advent of *human enhancement* there is a need to study and separately determine the legal status of each category of employees to ensure the principle of equality and non-discrimination at work.

Occupational safety also needs to be rethought with introduction of artificial intelligence in labor relations. The main provisions on implementation of constitutional right of citizens to protection of their life and health in working process are determined by the Law of Ukraine: *On Labor Protection* dated on October 14, 1992. The document contains norms on implementation of the constitutional right of citizens to proper, healthy and safe working conditions at workplace defined by law reflecting the principle of occupational safety with participation of relevant state bodies regulates relations between the owner of the enterprise employee on safety, occupational health and working environment and establishes a single procedure for labor protection management in Ukraine. However, with the advent of robots and *human enhancement* along with the jobs of ordinary employees there are new production risks not provided by current labor legislation. The problem of labor safety regulation will become most relevant with the advent of such cooperation. When robots share a work space with people the need to develop new occupational safety and health standards should be legally developed and approved. One of main professional problems is the need for legal regulation in the field of occupational safety when the robots will share workspace with people. Security standards should be developed and approved including at international level.

It should be noted that several research in this area have already been performed at the level of European and international institutions. For example, the report of the European Agency for Safety and Health at Work: "Forecast of new and emerging risks to safety and health related to digitalization until 2025" is interesting to study. The document analyzes four possible scenarios for introduction of digital technologies taking into account various social, political and economic factors. Each scenario presents different challenges and opportunities for occupational safety. Changing digital technologies and how they are used at workplace probably depend on various factors including:

- level of economic growth and investment in ICT, AISs and skills;
- demand and acceptance of AISs by the public and employees;

– as management, management and investment decisions support innovations in ICT, AISs.

Each scenario provides conclusions on how technology itself can affect occupational safety, physical and mental health. Certainly, changes can have both positive and negative consequences. For example, positive changes include the fact that automation, robotics and autonomous vehicles or drones can reduce the need for employees to work in dangerous and difficult conditions, with toxic or harmful substances. They also offer ability to transfer routine or repetitive tasks to fast, accurate and tireless machines.

However, the same technology can be a source of damage due to capture, entanglement, shock, noise and vibration. The number of accidents is a direct result of machine to human collisions that will increase over time due to the growing number of robots in the work environment. Sensors at work should help ensure the physical safety of employees; however, if the sensors fail, become contaminated or are subject to cyberattacks, security systems may fail and there is a risk of contact with human. Safety risks will also be caused by equipment used by the work that can pose a danger for employees nearby (job includes lasers, radiation sources, welding electrodes and mechanical equipment, etc.). Increased risk of chemical exposure to materials used for 3D and 4D printing/production at home and in local manufacturing centers where micro-enterprises cannot have necessary skills or knowledge to manage risks.

Involving artificial intelligence in the work environment also has an impact on the emotional area of a person. The amount of monitoring and the feeling of lack of personal control, information overload, "blurring" of working time is increasing. The growing use of robotics reduces personal contact and increases human isolation and reduced communication leads to social tensions and lack of informal training. Less variation of work leads to increased monotony. Many people will be employed through online platforms and many of their work tasks are sedentary. Most employees experience high cognitive and visual loads. Others suffer from voice health problems due to increased use of voice-controlled Internet connected devices at work [28]. Stress in its various forms is one of the main psychosocial risks due to the digitalization of society. Council of Europe Resolution № 2267 on Stress at Work dated on 1 March 2019 emphasizes that impact of artificial intelligence and its introduction into production on the growth of employee stress [29].

Cars will never be perfect, the error risk, accident or making wrong decision remains. Thus, governments face the challenge of revising occupational safety and health standards. It is necessary to develop a state policy on the safe integration of robots at workplace and this is a new area for health and safety professionals. The principle of occupational safety should be filled with new content reflecting development of digitalization of society expanding the range of subjects of labor relations through robots and *human enhancement* to ensure physical and psychological safety of employees.

Another principle of labor law that is undergoing significant changes with implementation of artificial intelligence technologies: the principle of stability of labor relations. Employees enter into an employment contract in order to have a permanent source of livelihood for themselves and their families. Therefore, they are interested in ensuring that their relations with the enterprise, institution, organization which they have concluded with an employment contract is long-term and stable. Duration and stability of labor relations is a guarantee of employment.

The report of the European Agency for Safety and Health at Work "Forecast of new and emerging risks to safety and health related to digitalization until 2025" predicts possibility of changing the form and employment period. Thus, one scenario assumes that rapid progress in ICT has a profound impact on performance. Economy is dominated by freelance growth, zero hours and short-term contracts (so-called gig economy). Many people work for at least five employers are registered on several online platforms at any time and often change jobs. Ukrainian labor law from 2020 also allows an employee to enter into employment contracts at one or more companies at the same time.

Online business places, as a rule are not currently required to provide regular work but provide ability to call an employee on request. In some cases, the operation of the online workplace offers desired flexibility in terms of performance but in many cases this is due to forced flexibility. Online business places can allow employees to change frequently jobs and job types, as such platforms provide access to a wide range of job types and cannot have mechanisms in place to verify that employees have appropriate skills for each job. It also leads to a short relationship duration. Features of online business place, such as the three-sidedness of the parties involved, temporality, informality, autonomy and mobility, make it difficult to establish employment relationships. Platform owners are not inclined to consider themselves employers (as well as users on the part of performers) and treat employees as self-employed. Working online platforms work differently and exercise different levels of control. Some platforms set rates and exercise significant behavioral control, for example, through fines for employees who do not meet certain standard that some believe is equivalent to the actions of a party as an employer. The gradual departure from the traditional form of employment contract requires a new definition of the meaning of this term and its features. Ukrainian legislation in 2020 has undergone significant changes in this regard. On March 30, 2020, the Law of Ukraine "On Amendments to Certain Legislative Acts Aimed at Providing Additional Social and Economic Guarantees in Connection with the Spread of Coronavirus Disease (COVID-2019)" [30] was adopted that changed the definition of the *employment contract* term. Very important phrases were removed from the text: "defined by this agreement, subject to internal labor regulations." In addition, the remote job concept is a form of job management when work is performed by an employee at his place of residence or elsewhere at his choice, including with the help of information and communication technologies, but outside the employer premises. At remote job, employees distribute working time at their discretion, they are not subject to the rules of internal labor regulations. In other words, now the employee is not obliged to follow the internal regulations and perform work that is clearly defined in the employment agreement. However, it is on these grounds that judicial practice and the science of labor law distinguish between labor and civil law relations between the employee and the employer. Now the problem of vague definition of the role and responsibility of employer in relation to the employee increases risks of transfer of labor relations between the parties in the plane of civil law, thereby violating the principle of stability of labor relations.

Thus, from the point of view of labor law, use of digital technologies and the introduction of artificial intelligence leads to rapid changes not only in the technologies used while working but also to changes in the nature of work, employment, relationships, labor protection and labor guarantees.

5 Conclusions

In the age of digital globalization introducing artificial intelligence into the workspace is proceeding at a rapid pace. This primarily applies to "knowledge-intensive" industries, such as: production of electrical and electronic equipment, aviation, missile, space industries, instrument making, microbiological industry, computer science industry and others.

Artificial intelligence allows you to perform tasks that were previously economically unprofitable or too dangerous. In the near future increasing the requirements for the qualification level of employees, such as those who "teach" and control AISs or whose working tasks change with implementation of digital technologies.

Among the main tasks facing labor law in the context of digitalization:

1. Regulating labor relations in those workplaces where works, ordinary employees and people with neural prostheses interact;
2. Conditions for acquisition of digital skills and continuous training of employees should be enshrined in employment contracts.
3. Determination of the legal status of new subjects of labor relations: "smart" robots (*Electronic personhood*) and *human enhancement*.
4. Reformulation of the principles of labor law in accordance with new conditions for realization of labor relations.
5. Development of new standards for labor protection and guarantees for realization of principle of the employee's right to safe and healthy working conditions.

In order to adequately respond to digital calls, it is important to systematically address existing problems. The changes will primarily affect the institutions of employment contract, labor protection, vocational training, guarantees and compensation. There will be rules governing use of high-tech means of production and biotechnology by employers, obligation of employers to retrain employees when replacing them with AISs and so on.

References

1. Intelligence artificielle et travail. Rapport à la ministre du travail et au secrétaire d'État auprès du Premier ministre, chargé du numérique (2018) [in English]. https://www.strategie.gouv.fr/sites/strategie.gouv.fr/files/atoms/files/fs-rapport-intelligence-artificielle-28-mars-2018.pdf
2. AI in Eastern Europe. Deep Knowledge Analytics (2018) [in English]. https://www.dka.global/ai-in-eastern-europe
3. DeepTech in Europe Industry Landscape Overview (2020) [in English]. https://mindmaps.dka.global/reports/ai-in-eastern-europe
4. Interactive Mind Map AI in Eastern Europe Industry Landscape [in English]. https://mindmaps.dka.global/ai-in-eastern-europe

5. Korinek, A.: Labor in the age of automation and artificial intelligence (2019) [in English]. https://econfip.org/wp-content/uploads/2019/02/6.Labor-in-the-Age-of-Automation-andArtificial-Intelligence.pdf
6. Webb, M.: The impact of artificial intelligence on the labor market (2020) [in English]. https://web.stanford.edu/~mww/webb_jmp.pdf
7. Filipova, I.A.: Labour law: challenges of digital society. pravo. Zhurnal Vysshey shkoly ekonomiki **2**, 162–182 (2020). [in Russian]
8. Order of the Cabinet of Ministers of Ukraine On approval of the plan of priority actions of the Government for 2020 dated 9 September 2020, 1133 p. [in Ukrainian]. https://www.kmu.gov.ua/npas/pro-zatverdzhennya-planu-prioritetni-a1133r
9. The concept of artificial intelligence development in Ukraine. Order of the Cabinet of Ministers of Ukraine of 2 December 2020 № 1556-p. [in Ukrainian]. https://zakon.rada.gov.ua/laws/show/1556-2020-%D1%80Text
10. European Parliament Resolution of 16 February 2017 with recommendations to the Commission on Civil Law Rules on Robotics [in English]. http://www.europarl.europa.eu/sides/getDoc.do?pubRef=-//EP//TEXT+TA+P8-TA2017–0051+0+DOC+XML+V0//EN
11. Final report of the High-Level Expert Group on the Impact of the Digital Transformation on EU Labour Markets [in English]. https://ec.europa.eu/digital-single-market/en/news/policy-andinvestment-recommendations-trustworthy-artificial-intelligence
12. Hopes and fears (2021) [in English]. https://www.pwc.com/gx/en/issues/upskilling/hopes-and-fears.html#content-free-1-1dad
13. Acemoglu, D., Restrepo, P.: The wrong kind of AI? Artificial intelligence and the future of labor demand (2019) [in English]. https://www.nber.org/papers/w25682.pdf
14. PwC, How will automation impact jobs? (2018) [in English]. https://www.pwc.co.uk/services/economics/insights/the-impact-of-automation-on-jobs.html?utm_campaign=sbpwc&utm_medium=site&utm_source=articletext#data-explorer
15. Game-changing technologies: Transforming production and employment in Europe. Eurofound, 2020, Publications Office of the European Union, Luxembourg [in English]. https://www.eurofound.europa.eu/de/publications/report/2020/game-changing-technologies-transforming-production-and-employment-in-europe#tab-01
16. Gutsu, S., Mkrtchyan, M., Strielkina, A.: Social and legal aspects of the transition to industry 4.0. In: Nechyporuk, M., Pavlikov, V., Kritskiy, D. (eds.) ICTM 2020. LNNS, vol. 188, pp. 726–737. Springer, Cham (2021). https://doi.org/10.1007/978-3-030-66717-7_62
17. Gutsu, S.F.: The impact of industry 4.0 on labor relations and the labor market: international experience for Ukraine. Sci. Bull. Uzhgorod Natl. Univ. RIGHT Ser. **1**(61), 101–106 (2020) [in Ukrainian]. http://visnyk-juris-uzhnu-uz.com/61-tom-1
18. Frontier Issues: The impact of the technological revolution on labour markets and income distribution. Department of Economic & Social Affairs (2017) [in English]. https://www.un.org/development/desa/dpad/publication/frontier-issues-artificial-intelligence-and-other-technologies-will-define-the-future-of-jobs-and-incomes/
19. Resolution of the Cabinet of Ministers of Ukraine On the implementation in Kyiv of an experiment on vocational training of registered unemployed for the needs of employers in the field of information technology of 20 January 2021 № 39 [in Ukrainian]. https://www.kmu.gov.ua/npas/pro-realizaciyu-v-m-kiyevi-eksperimentu-z-profesijnogo-navchannya-zareyestrovanih-bezrobitnih-dlya-potreb-robotodavciv-sferi-informacijnih-tehnologij-i200121-39
20. Order of the Ministry of Labor and Social Policy of Ukraine On approval of the Regulations on vocational training of workers at work from 26.03.2001 № 127/151 [in Ukrainian]. https://zakon.rada.gov.ua/laws/show/z0315-01Text
21. On the professional development of employees. Law of Ukraine. Information of the Verkhovna Rada of Ukraine, 2012, № 39, article 462 [in Ukrainian]. https://zakon.rada.gov.ua/laws/show/4312-17#Text

22. The order of advanced training of pedagogical and scientific-pedagogical workers. Resolution of the Cabinet of Ministers of Ukraine of 21 August 2019 № 800 [in Ukrainian]. https://zakon.rada.gov.ua/laws/show/800-2019-%D0%BF#Text
23. Regulations on the system of professional training of civil servants, heads of local state administrations, their first deputies and deputies, local government officials and deputies of local councils of 6 February 2019 № 106 [in Ukrainian]. https://zakon.rada.gov.ua/laws/show/106-2019-%D0%BF#n14.
24. Vasylko, I.: On the problem of determining the principles of ensuring the labor rights of employees working for employers - individuals. Entrepreneurship, economy and law. № 7, pp. 82–86 (2018) [in Ukrainian]. http://pgp-journal.kiev.ua/archive/2018/7/14.pdf
25. Eidenmueller, H.: The rise of robots and the law of humans. Oxford Legal Studies Research Paper. No. 27 (2017) [in English]. https://ssrn.com/abstract=2941001
26. Radutny, O.E.: Artificial intelligence as a subject of crime. Inf. Law **4**(23), 106–115 (2017). [in Ukrainian]
27. Filipova, I.A.: Transformation of legal regulation of labor in a digital society. Artif. Intell. Labor Law Sci. Edt. Nizhny Novgorod: Nizhny Novgorod State University. N. I. Lobachevsky. **89** (2019) [in Russian]. http://telecomlaw.ru/club/filippova.pdf
28. Stacey, N., Ellwood, P., Bradbrook, S., Reynolds, J., Williams, H., Lye, D.: Foresight on new and emerging occupational safety and health risks associated with digitalisation by 2025. Luxembourg: Publications Office of the European Union. 159 (2018) [in English]. https://osha.europa.eu/en/publications/foresight-new-and-emerging-occupational-safety-and-health-risks-associated/view
29. Resolution 2267 Stress at work (2019) [in English]. http://assembly.coe.int/nw/xml/XRef/Xref-XML2HTML-EN.asp?fileid=26466&lang=en
30. On amendments to some legislative acts aimed at providing additional social and economic guarantees in connection with the spread of coronavirus disease (COVID-2019). Law of Ukraine. Information of the Verkhovna Rada of Ukraine. 16, 100 (2020) [in Ukrainian]. https://zakon.rada.gov.ua/laws/show/530-20Text

Some Issues of Corruption Risk Assessment in the Activities of State Bodies, Legal Entities (in Case Study of the Aviation Industry of Ukraine)

Olena Syniavska[1(✉)] and Olha Pushko[2]

[1] Department of Law-Enforcement Activity and Policeistics, Kharkiv National University of Internal Affairs, Kharkiv, Ukraine

[2] Department of Economic Cybernetics, Sumy State University, Kharkiv, Ukraine

Abstract. The paper is devoted to the issues on identification and corruption risks assessment as one of the necessary prerequisites for preventing corruption in the aviation industry of Ukraine. The process of the corruption risks assessment in the activities of state bodies is considered in case study of the State Aviation Administration of Ukraine, and in the activities of legal entities – in case study of *Boryspil International Airport* State Enterprise.

Keywords: Corruption risk · Identification of corruption risk · Corruption risk assessment · Anti-corruption program · National Agency on Corruption Prevention · State Aviation Administration of Ukraine · *Boryspil International Airport* State Enterprise

1 Introduction

In the contemporary world, unfortunately, there are no countries that are completely free from corruption. However, for Ukraine, corruption became not only a subject of unfailing interest but also a factor of real threat to the national security and the constitutional order. Over the past 2020, the Ukrainian's indicators in Corruption Perceptions Index (CPI) increased by 3 points. With 33 points out of 100 possible, Ukraine took the 117th place out of 180 countries in the CPI list. Along with our country in the ranking, there are Egypt, African Eswatini (Swaziland), Nepal, Sierra Leone, and Zambia — all these countries were also scored 33 points in CPI-2020 [1].

This disgraceful phenomenon concerned the area, controlled by the Ministry of Infrastructure of Ukraine, namely the aviation industry. Thus, according to the information from September 7, 2020, NABU and SAPO exposed the corruption scheme of illegal disposal of property of *Boryspil International Airport* State Enterprise that caused more than UAH 16.5 million damage to the State Enterprise.

As the investigation found, the officials of *Boryspil International Airport* State Enterprise from 2014 to 2019, on the basis of conducted preliminary lease agreements, illegally leased out premises to the business entities at a reduced price, and since the property

M. Nechyporuk et al. (Eds.): ICTM 2021, LNNS 367, pp. 618–634, 2022.
https://doi.org/10.1007/978-3-030-94259-5_52

belongs to the Ministry of Infrastructure of Ukraine, only the State Property Fund of Ukraine has the right to lease it under the relevant tender procedure [2].

At the same time, it is clear that it is impossible to control corruption without knowing and understanding the conditions and circumstances in which it arises and develops. It is recognized all over the world that rational planning of anti-corruption measures should always include the identification of risks, their influence, and on this basis, it is necessary to implement countermeasures planning. The so-called *vulnerability to corruption assessment* has become very popular because its conducting is absolutely necessary before the implementation of any initiative in the sphere of the anti-corruption policy. This paper is devoted to namely certain issues of the corruption risk assessment in the activities of state bodies, legal entities (in case study of the aviation industry of Ukraine).

2 Legal Framework of the Corruption Risk Assessment in the Activities of State Bodies and Legal Entities

After single efforts in the sphere of corruption assessment (such as, e.g., the Decree of the President of Ukraine No. 175/2004 of 9 February 2004 *On the system of measures to eliminate the causes and conditions that contribute to criminal acts and corruption* [3], in which the term: *corruption risk* is even not used), Law of Ukraine of 14 October 2014 *On the Principles of State Anti-Corruption Policy in Ukraine (Anti-Corruption strategy) for 2014–2017* [4] introduces a legal basis for such assessments.

In accordance with the Anti-Corruption strategy, the central executive bodies and state-owned enterprises, based on the results of conducted analysis of the corruption risks in the field of activity of the relevant body or enterprise, have to create anti-corruption programs, which should include:

- methodology of the corruption risk assessment in the activities of the body or state-owned enterprise, the corruption risks, reasons that give rise to them, and conditions that lead to them;
- measures to eliminate the corruption risks, persons responsible for their implementation, deadlines, and necessary resources.

Law of Ukraine of 14 October 2014 *On Prevention of Corruption* [5] also contains certain commitments in the field of corruption risk assessment. So,

- according to Article 61 (§ 2) of this Law *Head, founders (participants) of a legal entity provide regular corruption risk assessment in its activities and carry out appropriate anti-corruption measures. In order to reveal and eliminate the corruption risks in the activities of the legal entity, independent experts may be involved, in particular for conducting audits;*
- according to Article 63 (Part 1 § 2) legal entities have to include *the procedure for conducting periodic assessment of the corruption risks in the activities of the legal entity* in their Anti-Corruption Programs.

In accordance with the aforementioned Law, the National Agency on Corruption Prevention is the body whose powers also include coordination and providing methodological assistance in identifying the corruption risks in their activities by state bodies, authorities of the Autonomous Republic of Crimea and local self-government bodies, and realization of measures to eliminate them, including preparation and implementation of Anti-Corruption Programs (Article 11 Part 1 § 7), and to the rights – to prove the Methodology of the corruption risk assessment in the activities of state bodies and analyze the Anti-Corruption Programs of authorities and provide mandatory proposals for such programs (Article 12 Part 1 § 11).

Accordingly, the National Agency on Corruption Prevention, by its decision No. 126 of 2 December 2016, approves the Methodology of the corruption risk assessment in the activities of authorities [6].

This Methodology defined the complexity of rules and procedures on the corruption risk assessment in the activities of authorities where in accordance with the Law of Ukraine *On Prevention of Corruption*, the Anti-Corruption Programs are accepted, as the corruption risk assessment is carried out in preparation of the Anti-Corruption Program of the authority or during its periodic review.

In the mentioned Methodology, the list of terms facilitating the work regarding the corruption risk assessment has been created. Thus, in particular, the concept of the corruption risk was defined as the probability that *there will be the event of corruption offence or corruption-related offense that will have negative impact on the achievement of definite goals and tasks by the authority,* as well as the process of the corruption risk identification as *an identification of corruption risk factors in the activity/inactivity of government officials that may contribute to the commission of a corruption offense or an offense related to corruption,* and the corruption risk assessment as *the process of identification, analysis and direct assessment of corruption risk* [6, § 2]. Also important for the process of corruption risk assessment are the following terms, defined in the Methodology of corruption risk assessment in the activities of the authorities:

1. An object of the corruption risk assessment—functions and tasks of the authority, its officials defined by legislative acts, and the ways of their implementation;
2. The corruption risk factors—conditions and reasons that promote, arouse, or allow the official to commit acts that may lead to the commission of a corruption offense or an offense related to corruption while performing the functions of the state or local self-government;
3. Internal environment—organizational and managerial processes in the authority aimed to provide with the implementation of the statutory functions and tasks by the authority;
4. External environment—a number of environmental factors (legislation, other regulations, except acts adopted directly by the authority, natural phenomena, etc.) that affect the exercise of authorities by the authority, while such an authority has no direct influence on these factors [ibidem].

Other terms in this Methodology are used in the meanings given in the Law of Ukraine *On Prevention of Corruption.*

The Corruption Risk Assessment Commission of the authority may conduct the corruption risk assessment which consists of employees of structural units of the authority, and in concert, public representatives, experts with knowledge of the internal and external environment of the authority and has experience in this area.

The aforementioned Methodology defines the only approach to the authorities in the work organization of the corruption risk assessment that arises during the implementation of their powers, a set of rules and procedures for assessing corruption risks in their activities. In addition, it should be noted that some aspects of the corruption risk assessment, raised issues on their practical implementation, needed clarification. As the corruption risk assessment in the activities of the authority is one of the preparation stages of the Anti-Corruption Program of the authority, so such more detailed explanations can be found in the Methodological recommendations on the preparation of the Anti-Corruption Programs of the authorities, proved by the decision of the National Agency on Corruption Prevention No. 31 of 19 January 2017 [7], which have such an informational and explanatory nature and do not establish legal norms.

Let us consider the process of the corruption risk assessment of the authority in case study of the State Aviation Administration of Ukraine as a central executive body, the activities of which is directed and coordinated by the Cabinet of Ministers of Ukraine through the Ministry of Infrastructure who implements the state policy in the field of civil aviation and the use of airspace of Ukraine and is the authorized body for civil aviation [8].

3 The Process of the Corruption Risk Assessment in the Activities of the Authority (in Case Study of the Aviation Industry of Ukraine)

The process of the corruption risk assessment in the activities of the authority consists of the following main stages:

- organizational and preparatory measures;
- Identification (Detection) of Corruption Risks;
- Corruption Risk Assessment;
- Compiling a Report on the Results of the Corruption Risk Assessment [6].

Let's consider these steps in more detail.

3.1 Organizational and Preparatory Measures for the Corruption Risk Assessment in the Authority

According to the Methodology of the corruption risk assessment in the activities of the authority [6] preparation for the corruption risk assessment includes the following stages:

- making a decision on the corruption risk assessment by an authorized subject to approve the Anti-Corruption Program, according to Article 19 Part 1 of Law of Ukraine

On Prevention of Corruption (hereinafter referred to as the subject of approval of the Anti-Corruption Program), *a decision to conduct a corruption risk assessment.* Thus, in conformity with the Order of the State Aviation Administration No. 1563 of 15.11.2019 *On Corruption Risk Assessment in the State Aviation Administration*, the work on the corruption risk assessment was begun [9];

- posting the decision on the subject of approval of the Anti-Corruption Program on the corruption risk assessment on the official website of the authority. Information about the beginning of the work on the corruption risk assessment was posted on the official site of the State Aviation Administration;
- establishment of the Commission (in case of the corruption risk assessment by the Commission). Unfortunately, there isn't any information about the composition of the Commission on the corruption risk assessment in this authority and its work plan on the official site of the State Aviation Administration. Some information can be taken only from the Report, based on the corruption risk assessment in the activities of the State Aviation Administration of Ukraine 2020–2021 [9], according to which such a Commission was established by the Order of the State Aviation Administration No. 1699 of 05.12.2019 *On establishment of the Corruption Risk Assessment Commission in the State Aviation Administration*;
- making a work plan of the corruption risk assessment (hereinafter referred to as the work plan).

The subject of approval of the Anti-Corruption Program (Head of the State Aviation Administration of Ukraine) makes decisions on the corruption risk assessment which, in order to involve members of the public and experts in the composition of the Commission, are posted on the official website of the authority no later than seven working days before the moment of its approval. He appoints the Head of the Commission and its personal composition, the term of the corruption risk assessment, as well as approves the regulations on it.

It is recommended to create the Commissions with a number of 10 – 20 members, depending on the number of employees and functions performed by the authority. For example, for institutions with up to 50 employees, the Commission may consist of 3–5 members. The composition of the Commission obligatory involves management personnel of structural units that ensure the performance of the general functions of the authority: personnel management services, accounting departments and legal services, the structural unit of internal audit, the authorized unit (person) for the prevention and detection of corruption (in case of their formation).

In addition to that, it is meaningful to engage leadership (employees) responsible for performing special functions of the authority, e.g., a control-and-auditing unit, a subdivision of licensing, the unit that coordinates the provision of administrative services, etc., as well as employees of territorial authorities (in case of their formation).

At different stages of the corruption risk assessment, other employees of the authority (by consent) may be involved in the work of the Commission in accordance with the procedure established by law, who are not members of the Commission but may provide information for objective and qualitative assessment of the corruption risks.

The Head of the Commission can be appointed the Head of the authorized unit (if any) or another person of the management authorized by the subject of approval of the Anti-Corruption Program. The subject of approval of the Anti-Corruption Program provides the Commission with the appropriate powers (approving its provisions) and the material resources necessary for the corruption risk assessment.

From the information, contained in the Report, on the results of the corruption risk assessment in the activities of the State Aviation Administration of Ukraine 2020–2021 [9], it follows that the representatives of all independent structural subdivisions of the State Aviation Administration of Ukraine and a representative of the State Aviation Administration' Public Council, in total 20 persons, were included in the Commission.

The Commission as any other subsidiary body of the authority keeps minutes of the meeting. The Commission is recommended to meticulously document the conclusions and decisions made during the corruption risk assessment. The information about the periodicity of the meetings of the Commission, the powers of the Head, rights, and responsibilities of the Commission's members are defined in the regulations on it.

At the same time, subjects, methods and techniques of the corruption risk assessment, sources of information for its conduct, persons responsible for the corruption risk assessment, and terms of the assessment for each object are determined directly by the Commission.

The Commission, in compliance with the objects of the corruption risk assessment, makes the work plan that is approved by the Head of the Commission.

The work plan reflects the information about:

- objects of the corruption risk assessment;
- sources of information for the corruption risk assessment;
- methods and techniques of the corruption risk assessment;
- persons responsible for the corruption risk assessment for each object;
- terms of the corruption risk assessment for each object.

Thus, the work plan of the Corruption Risk Assessment Commission in the State Aviation Administration was accepted at the first meeting of the Commission and approved by the Head of the Commission January 16, 2020.

If it is necessary, the work plan may be adjusted by the Commission at the different stages of the corruption risk assessment.

3.2 Identification (Detection) of the Corruption Risks

The corruption risks assessment is preceded by their identification.

When identifying the corruption risks, the authority determines its areas of activity, vulnerable to risks, certain functions and tasks, specific areas of the activity of the structural divisions during the performance of their functions, and tasks.

Identification of the corruption risks is conducted by studying (analyzing) the external and internal environment of the authority in order to identify factors of corruption risks in *regulations and organizational* and *managerial activities* of the authority.

Regulations, which govern the activity of the authority, are analyzed for the purpose of identifying norms on the exercise of discretionary powers and norms by the authority that promote the commission of the corruption offense or the corruption-related offense (vaguely worded definitions and language constructions, legal collisions, etc.).

The identification of the corruption risks in the organizational and managerial activities of the authority involves the study and analysis of the following issues:

- an organizational structure of the authority;
- a system of internal control;
- staff management;
- conducting procurement procedures;
- provision of administrative services;
- implementation of control and supervisory functions;
- adherence to the requirements, restrictions, prohibitions established by Law of Ukraine *On Prevention of Corruption*;
- other issues arising from the activities of the authority.

Information sources for the identification of the corruption risks are:

- an interview (sociological survey), questionnaire survey (including anonymous ones) of government officials, as well as members of the public who have interacted (interact) with the authority, in particular, with the use of an e-mail, Internet, phone, etc.;
- normative and legal and organizational-administrative acts regulating the activity of the authority;
- analysis of the implemented measures to eliminate the identified the corruption risks for previous periods;
- the results of inspections, audits, conducted by regulatory authorities, as well as internal inspections and internal investigations in the authority;
- mass media and social networks posting;
- information from open registers;
- appeals received by the authority from individuals and legal entities.

The identification of the corruption risks involves the determination and classification of the corruption risks by categories and types.

By categories the corruption risks can be:

- external (the probability of occurrence of the corruption risks is not related to the performance of the relevant functions and tasks by the authority);
- internal (the probability of occurrence of the corruption risks is directly related to organizational-administrative activities in the authority in accordance with the functions and tasks assigned to it).

By types the corruption risks can be:

- regulations (absence, contradiction, or inaccurate regulation in the legislation of provisions on the performance of functions and tasks by the authority);
- organizational (uncertainty or unclear definition of implementation procedures of the functions and tasks by the authority);
- personnel;
- financial and economic;

- administrative;
- control and supervision, etc.

The list of the corruption risk types is not complete. Each authority can determine the list of the types of the corruption risks, taking into consideration the specifics of its tasks and functions.

Based on the results of identification of corruption risks, their formal determination is carried out, which includes the analyzed functions, tasks of the authority or specific areas of activities of structural units, existing control measures, and description of the corruption risk [6].

Addition 1 to the Report, by the results of the corruption risk assessment in the activities of the State Aviation Administration of Ukraine 2020–2021, preliminary contains a description of certified corruption risks in the activities of the State Aviation Administration of Ukraine. Thus, such risks include, for example:

- discriminatory rules on admission to the market of ground handling services at airports;
- absence of an established order of approval or exemption from the provisions of Technical Instructions Doc 9284 AN/905 ICAO;
- implementation of control and supervisory functions in the part of aircrafts of the amateur structural system without provision of regulations;
- absence of the established order of reporting on the volume of transportation of dangerous goods;
- lack of requirements for regulating the operation of unmanned aircrafts;
- lack of qualification requirements for managers of aviation entities, representatives of the management of such entities who are responsible for aviation security, and criteria for certification of such employees, as well as lack of procedure by the State Aviation Service to conduct inspections of deputy Heads of aviation security entities in law enforcement agencies, as provided by the tenth part of Article 10 of the Air Code of Ukraine, eighth part § V of the State Program of Civil Aviation Security, approved by the Law of Ukraine No. 1965-VIII of 21 March 2017, and others [9].

3.3 The Corruption Risks Assessment

The identified corruption risks are subject to further assessment to identify the factors of occurrence and existence of these corruption risks and the possible consequences of corruption or corruption-related offenses that may occur as a result of the existence of identifying risks.

Corruption risks assessment is conducted according to the criteria of *the likelihood of occurrence* of identifying corruption risks and *the consequences* of the corruption offense or the offense related to corruption.

The assessment of the likelihood of occurrence of corruption risk is determined according to the frequency of cases of corruption or corruption-related offenses, taking into account the time limits.

In accordance with *the criteria for the likelihood of occurrence* of the corruption risk, it is assigned a low, medium, or high level.

The frequency of cases of corruption or corruption-related offenses is determined by the following levels:

- probably or almost certainly, the corruption or the offense related to corruption can be committed in the short term: the next few months (up to one year), and can be repeated;
- rarely – the corruption or the offense related to corruption has been committed only once for the recent three years and it is likely to be committed within three years;
- never – the corruption offense or the offense related to corruption has not been committed and the likelihood of its commission is almost impossible.

The likelihood of the corruption risk is assessed three-point scale:

- low – 1 point;
- medium – 2 points;
- high – 3 points [6].

An example of the results of the likelihood assessment of the corruption risk is shown in Table 1.

Table 1. RESULTS of the likelihood assessment of the corruption risk to occur

	Likelihood		
Identified risks	Low	Medium	High
Risk 1	1		
Risk 2		2	
Risk 3			3
Risk 4	1		
Risk 5	1		

According to the criteria of the consequences of the corruption or the offense related to corruption, the corruption risk is assigned a low, medium, or high level.

The consequences of the corruption or the corruption-related offense are assessed in accordance with the number of losses of the authority, which the commission establishes itself [6].

An example of the levels of the likelihood of the consequences of the corruption or the corruption-related offense, and the criteria for their definition is shown in Table 2.

The consequences of the corruption or the corruption-related offense are assessed on the three-point scale:

- low – 1 point;
- medium – 2 points;
- high – 3 points [6].

Table 2. EXAMPLE of the likelihood of the consequences of the corruption or the corruption-related offense, and the criteria for their definition

Likelihood level	Criteria for determining the likelihood level		
	Financial losses of the authority	Level of liability	Reputational losses of the authority
High	Significant financial losses	Trials against the authority or its officials are expected; committing the corruption or the offense related to corruption involves criminal liability	Loss of reputation within the broader population
Medium	Financial losses are present, but insignificant	The commission of the offense related to corruption involves administrative liability	Loss of reputation among parties of the direct contacts
Low	No financial losses are expected	The commission of the corruption or the offense related to corruption involves disciplinary liability	Loss of reputation among the employees of the authority

The example of the results of the assessment of the consequences of the corruption or the corruption-related offense is shown in Table 3.

Table 3. RESULTS of the assessment of the consequences of the corruption or the corruption-related offense

	Consequences		
Identified risks	Low	Medium	High
Risk 1		2	
Risk 2			3
Risk 3			3
Risk 4	1		
Risk 5			3

The priority (degree) of corruption risks is established by their quantitative level, which is determined by the product of the likelihood level of the corruption risk on the level of the consequence of the corruption or the corruption-related offense [6].

The example of the corruption risks assessment by quantitative level is shown in Table 4.

Table 4. RESULTS of the corruption risks assessment by quantitative level

Identified risks	Probability of occurrence of a corruption offense or an offense related to corruption (points)	Consequences of a corruption offense or an offense related to corruption (points)	Priority of the corruption risks *	Assessment
Risk 1	1	2	2 (1 × 2)	Low
Risk 2	2	3	6 (2 × 3)	High
Risk 3	3	3	9 (3 × 3)	High
Risk 4	1	1	1 (1 × 1)	Low
Risk 5	1	3	3 (1 × 3)	Medium

* Total numeric value.

The priority of the corruption risks is assessed on-scale:

- low – from 1 to 2 points;
- medium – from 3 to 4 points;
- high – from 6 to 9 points [6].

The example of determining the priority of the corruption risks is given in Table 5.

Table 5. PRIORITY of corruption risks

Consequence	Likelihood		
	Low	Medium	High
High	3 (1 × 3)	6 (2 × 3)	9 (3 × 3)
Middle	2 (1 × 2)	4 (2 × 2)	6 (3 × 2)
Low	1 (1 × 1)	2 (2 × 1)	3 (3 × 1)

3.4 Report on the Results of the Corruption Risk Assessment

Based on the results of the assessment of corruption risks in the activities of the authority, the commission prepares a report [6].

The report on the results of the corruption risk assessment contains:

a description of the identified corruption risks, corruption risk factors and possible consequences of the corruption or the corruption-related offense;

proposals for measures to eliminate (reduce) the level of identifying corruption risks.

Measures to eliminate identified corruption risks are to identify possible mechanisms for counteracting and preventing corruption risks, contain proposals for ways to implement them, and aimed at eliminating or minimizing the conditions (causes) of corruption risks.

The table of the assessed corruption risks in the activities of the State Aviation Administration and measures for their elimination is given in Annex 2 to the Report on the results of corruption risk assessment in the State Aviation Administration 2020–2021 [9].

The Report on the results of the corruption risk assessment in the activities of the authority is approved by the subject of approval of the Anti-Corruption Program and is included in the Anti-Corruption Program.

In case of detection, based on the corruption risk assessment, external corruption risks, which consist in the imperfection of normative legal acts, the subject of which is not the authority, the identified corruption risks and proposals to eliminate (minimize) are sent to the National Agency for Prevention of Corruption.

The National Agency for Prevention of Corruption notes that the Anti-Corruption Program analysis is primarily aimed at verifying the results of the government's assessment of corruption risks in its activities in terms of the adequacy and sufficiency of work to identify risks and the specificity and proportionality of the proposed measures to eliminate them.

Therefore, before approving the Report, it is necessary to make sure that it is about corruption risks (the probability that an offense will occur), and not about the objects of evaluation (for example, public procurement), not about the corruption itself or the corruption-related offenses (failure to report conflicts of interest, abuse of official position, disclosure of information, etc.) or failure to violate the requirements of special legislation (failure to provide information on requests for information, violation of budget legislation, etc.).

In addition, consideration should be given to if the measures to eliminate corruption risks are aimed at reducing the factors of such risks (e.g., whether a risk is identified as an unregulated procedure, such a measure to eliminate it as *carrying out explanatory work* is insufficient, as its implementation will not eliminate the factor of such risk — the absence or imperfection of a certain normative legal or organizational act that regulates the relevant procedure — after the adoption of which it is possible to carry out explanatory work on its application. The same goes for measures *to ensure control, to ensure employees to be informed, to execute powers in compliance with legislative requirements,* as aforementioned is existing control measures (i.e., that is, what is already carried out by the body in a particular area), and measures to minimize corruption risks can be only identified exceptionally something new, something that is not enough to eliminate the likelihood of corruption or the corruption-related offenses.

Thus, in conformity with the Report on implementation in 2020 of measures to eliminate corruption risks, identified by the Anti-Corruption Program of the State Aviation

Administration for 2020–2021, which is posted on the official website of the State Aviation Administration of Ukraine (https://avia.gov.ua/pro-nas/zapobigannya-koruptsiyi/), such corruption risk as *Implementation of control and supervisory functions in the part of amateur aircraft without normative legal acts* was identified. The measure to eliminate such risk was: *To develop and adopt Decree: On approval of the Aviation Rules of Ukraine, Technical Requirements and Administrative Procedures for the Issuance of Permits for Flights of Amateur Aircraft.* The expected result from the implementation of such a measure here is *the establishment of clear requirements for maintaining the airworthiness of aircraft of amateur design.* According to the Report, the state of implementation of this measure to eliminate corruption risk is as follows: *The draft Aviation Rules of Ukraine: Technical requirements for the assessment of amateur aircraft, is being finalized for further internal approval by the State Aviation Service.*

4 Corruption Risk Assessment in the Activities of Legal Entities (Case Study of *Boryspil International Airport* State Enterprise)

Regarding the corruption risk assessment in the activities of legal entities, we emphasize that on March 2, 2017, the National Agency on Corruption Prevention by its decision No. 75 approves the Standard Anti-Corruption Program of the legal entity [10], where, in particular, it is stated that the legal entity ensures the development and implementation of measures that are necessary and sufficient to prevent, detect and combat corruption in its activities. In addition to that, the list of anti-corruption measures in the activities of a legal entity includes periodic assessment of corruption risks in the activities of the legal entity.

Section II of the Anti-Corruption Program of the legal entity *Scope and Circle of People Responsible for the Implementation of the Anti-Corruption Program* includes the subsection *Periodic Assessment of Corruption Risks in the Activities of the Legal Entity*, which states that the legal entity *at least once a year carries out an internal assessment of corruption risks* in their activities.

In the given document, the corruption risk is determined as a substantiated likelihood of occurrence of corruption or corruption-related offense, or violation of the Anti-Corruption Program.

The corruption risk assessment in the legal entity is performed by the Commission on the corruption risk assessment (hereinafter referred to as the Commission).

The operating procedure and composition of the Commission are approved by the Head of a legal entity.

The Commission includes an official of the legal entity responsible for the implementation of the Anti-Corruption Program (Chairman of the Commission), Heads of structural units of the legal entity, as well as other employees designated by the Head in agreement with the Commissioner.

During the corruption risk assessment at the instance of the Commissioner, other employees of the Legal Entity, as well as independent experts or specialists may be involved in the work of the commission without including it in its membership.

The Commissioner, in order to avoid conflicts of interest or bias in the work of the commission, when allocating functions among the members of the commission, takes into account the scope of their duties in the legal entity.

The purpose of the Commission is to prevent, identify and eliminate corruption risks in the activities of the founders (participants), Head and employees of the legal entity.

The corruption risks in the activities of the legal entity are divided into internal and external. The internal corruption risks are identified in organizational and managerial, financial and economic, personnel, legal procedures of the legal entity. External corruption risks are identified in the activities of business partners, including public authorities, local governments, with which the legal entity is in a business relationship.

On the results of corruption risk identification by the Commission, in accordance with it Based on the results of the identification of corruption risks by the Commission, in accordance with the procedure of its activity, their definition and description, classification by categories and types are carried out. On the results of the corruption risk assessment in the activities of the Legal entity, the Commission prepares a written report, which is signed by the members of the Commission.

The Report is made according to the form and structure, determined in the order of the Commission's activity.

The Report, based on the results of the corruption risk assessment, is submitted to the Head, founders (participants) of the legal entity and should include:

- identified corruption risks, as well as the reasons that give rise to them and the conditions that contribute to them;
- assessment of identified corruption risks;
- proposals on measures to prevent, eliminate (reduce) the level of identified corruption risks.

The text of the report is provided to the employees of the legal entity, and may also be published on the website of the legal entity (if any).

If during the measures to assess corruption risks the Commissioner finds a violation of the Anti-Corruption Program, committing a corruption or corruption-related offense, he initiates to the Head the issue on an internal investigation in the manner, prescribed by Section XV of the Anti-Corruption Program.

In addition, the Legal entity *has to undergo an external assessment of corruption risks at least once every three years*, which is provoked by an organization, that provides audit, legal or consulting services, or independent experts.

Based on the results of the report on internal and/or external corruption risk assessment, the Head, founders (participants) take the necessary measures to prevent, detect and combat corruption in the activities of the Legal entity, including by changing existing anti-corruption standards and procedures.

Thus, the Anti-Corruption Program of *Boryspil International Airport State Enterprise* was approved by the Order of the General Director No. 01-07.5-81 [11] of 8 November 2019 and published on the official website of this legal entity. In its content, its second section meets the aforementioned requirements, specified in the Standard Anti-Corruption Program of the legal entity. The Corruption Risk Assessment Commission in *Boryspil IA SE* was created by the Order of the General Director No. 01-07-274 of 31 October 2017 as amended in accordance with the Order No. 01-07.3-31 of 06 September 2019. As can be seen from the minutes of the meeting of the Corruption Risk Assessment Commission in *Boryspil International Airport* dated 18.09.2020, posted on the official website of the airport [12], the Chairman of the Commission was appointed the Commissioner on Prevention and Detection of Corruption in *Boryspil IA SE*—Head of the Department on Prevention and Detection of Corruption. The following issues were considered during the meeting:

1. Review, discussion and assessment of the main corruption risks in *Boryspil IA SE*.
2. Approval and signing of the draft report on the corruption risk assessment and measures to eliminate them.

The Report of the corruption risk assessment and measures to eliminate them in *Boryspil International Airport State Enterprise* was approved by the Order of the General Director No. 41-11-1 of 28 September 2020 [13]. It should be noted here that some positions, mentioned in the Report, revealed as identifying risks, are in fact corruption itself or corruption-related offenses (for example, § 6 *Violation of established rules of ethical conduct. Dishonesty of officials of the enterprise in the performance of official duties*, or § 9 *Occurrence of conflict of interest during collegial decision-making*, etc.), or violation of the Law (for example, § 10: Improper compliance with the Law of Ukraine: On Prevention of Corruption). In this regard, it should be noted that it might be appropriate for the Anti-Corruption Programs of legal entities to be subject to approval by the National Agency on Corruption Prevention. Because the formally established Anti-Corruption Program, that does not identify genuine corruption risks and, as a result, does not identify measures to address them, is unlikely to be an effective *set of rules, standards and procedures for detecting, combating and preventing corruption in the activities of legal entities* [10]. In addition, in our opinion, it would be appropriate for the National Agency on Corruption Prevention to compile and constantly update the list of corruption risks and measures to eliminate them on the basis of reports on corruption risk assessment of the authority, as well as existing and new reports of international and foreign organizations and legal entities.

5 Conclusion

It is impossible to prevent and counteract the existence and spread of corruption without understanding its preconditions. Therefore, the identification and corruption risks assessment in the activities of state bodies, legal entities and identification of measures to eliminate them is extremely important for the prevention of corruption both in the aviation industry and in Ukraine.

In conclusion, I have a mind to quote Kos Drago, the Head of Working group of the Organization for Economic Cooperation and Development (OECD) on combating bribery in international commercial transactions, author of the technical manual *Management of Corruption Risks in the Civil Service of Ukraine* (Procedures for Assessing Corruption Risks in the Activities of Officials),

"*An active approach to corruption risk management is one of the most effective steps, which can be used by the leadership of the institution to reduce the level of corruption activity. Corruption is not an issue that any institution wants to find problematic for itself, but the reality is that most of them are vulnerable to a certain level of corruption.*

Although it is not possible to eliminate all corruption risks, institutions can take positive and constructive actions to reduce their vulnerability. The combination of effective corruption risk management, comprehensive corruption risk assessment, strong internal control, as well as coordination and timely implementation of risk reduction measures can significantly reduce the level of corruption risks.

It is necessary to stress that combating corruption is a constructive activity. A strong and effective institution exists as the leadership takes preventive measures to manage the corruption risks before they arise. The leadership of the institution, that creates an ethical culture, has an impact on employees who are satisfied with their work in such an institution, they work better, and eventually, the work of the institution as a whole becomes more effective (May 6, 2015)".

References

1. Corruption Perceptions Index – 2020. Transparency International Ukraine [in Ukrainian]. http://cpi.ti-ukraine.org/#/
2. NABU has terminated a five-year corruption scheme at Boryspil International Airport [in Ukrainian]. https://nabu.gov.ua/novyny/nabu-prypynylo-pyatyrichnu-korupciynu-shemu-na-dp-ma-boryspil?fbclid=IwAR3CnYcBwhFJS8qo1ZWM7RfRZ5lZG_B_5tTxLIp21
3. On the system of measures to eliminate the causes and conditions that contribute to criminal acts and corruption: Decree of the President of Ukraine № 175/2004 of 9 February 2004 [in Ukrainian]. https://zakon.rada.gov.ua/laws/show/175/2004#Text
4. On the principles of state anti-corruption policy in Ukraine (Anti-Corruption Strategy) for 2014–2017: Law of Ukraine of 14 October 2014 № 1699-VII [in Ukrainian]. https://zakon.rada.gov.ua/laws/show/1699-18#Text
5. On prevention of corruption. Law of Ukraine of 14 October 2014 № 1700-VII [in Ukrainian]. https://zakon.rada.gov.ua/laws/show/1700-18
6. Methodology for assessing corruption risks in the activities of public authorities: approved by the Decision of the National Agency for Prevention of Corruption of 2 December 2016 № 126 [in Ukrainian]. https://zakon.rada.gov.ua/laws/show/z1718-16#Text
7. Methodical recommendations on preparation of anti-corruption programs of government bodies: approved by the decision of the National Agency for Prevention of Corruption of 19 January 2017 № 31 [in Ukrainian]. https://nazk.gov.ua/wp-content/uploads/2019/04/Metodychni -recommendations-1.pdf
8. Regulations on the State Aviation Service of Ukraine: approved by the resolution of the Cabinet of Ministers of Ukraine of 8 October 2014 № 520 [in Ukrainian]. https://zakon.rada.gov.ua/laws/show/520-2014-%D0%BF#Text
9. Report on the results of corruption risk assessment in the activities of the State Aviation Service of Ukraine 2020–2021 [in Ukrainian]. https://avia.gov.ua/wp-content/uploads/2020/08/ilovepdf_merged-2.pdf
10. Typical anti-corruption program of a legal entity: approved by the decision of the National Agency for Prevention of Corruption of 2 March 2017 № 75 [in Ukrainian]. https://zakon.rada.gov.ua/laws/show/z0326-17#Text
11. Anti-corruption program of the State Enterprise Boryspil International Airport: approved by the order of the General Director of the State Enterprise Boryspil International Airport dated 8 November 2019 № 01-07.5-81. https://kbp.aero/wp-content/uploads/2019/11/Nova-Antykoruptsijna-programa.pdf

12. Minutes of the meeting of the commission for corruption risk assessment in the State Enterprise IA Boryspil dated 18.09.2020 [in Ukrainian]. https://kbp.aero/wp-content/uploads/2021/03/Protokol_zasidannya_komisiyi_z-otsinky-koruptsijnyh-ryzykiv-2020.pdf
13. Report on the assessment of corruption risks and measures to eliminate them at the State Enterprise Boryspil International Airport: approved by the order of the General Director of the State Enterprise Boryspil International Airport dated 28 September 2020 № 41-11-1 [in Ukrainian]. https://kbp.aero/wp-content/uploads/2021/03/Zvit-otsinky-koruptsijnyh-ryzykiv-ta-zahodiv-shhodo-yih-usunennya-2020.pdf

Role and Significance of Situational Analysis in Crime Counteraction at Critical Infrastructure and Aerospace Industry

Oleksii Lytvynov[1(✉)], Hanna Spitsyna[2], Yurii Danylchenko[3], and Serafyma Hasparian[4]

[1] Department of Criminal Law and Criminology, Kharkiv National University of Internal Affairs, Kharkiv, Ukraine

[2] Humanities and Law Faculty, National Aerospace University "Kharkiv Aviation Institute", NAU "KhAI", Kharkiv, Ukraine
spitsyna_hanna@ukr.net

[3] Academician Stashis Scientific Research Institute for the Study of Crime Problems, National Academy of Law Sciences of Ukraine, Kharkiv, Ukraine
ket_khrom@ukr.net

[4] Department of International Cooperation and Work with Foreigners of National Scientific Centre, Bokarius Forensic Science Institute, Kharkiv, Ukraine
femida2015@ukr.net

Abstract. Modern means and methods of counteracting crime are considered. Based on the analysis of domestic and foreign experience, main trends in improving national and international legal mechanisms of the state in combating criminal offenses and crimes are presented. Main organizational and legal mechanisms (models) of counteraction to criminal offenses and crimes are given.

It is proved that main purpose of situational analysis in combating crime at critical infrastructure and aerospace industry is to ensure meaningfulness, certainty and adequacy of the actions of control component (both at the level of the crime subject and at state level as a whole) in rapidly changing and uncertain conditions. From the point of view of situational approach, crime management on objects of critical infrastructure and aerospace is first of all art of the head to understand a situation, to define its characteristics and to choose the corresponding management and only then to follow the scientific criminological recommendations having general and universal character.

Keywords: Crime · Situational analysis · Activity approach · Criminogenic situation · Aerospace industry · Critical infrastructure facilities

1 Introduction

Over the past half century, the world has seen an increase in the level of complexity and mobility of social processes, there is an acceleration of socio-economic transit. As a result, effectiveness of the functioning and progressive development of the state is largely

M. Nechyporuk et al. (Eds.): ICTM 2021, LNNS 367, pp. 635–648, 2022.
https://doi.org/10.1007/978-3-030-94259-5_53

determined by flexibility, dynamism and adaptability of its management to requirements of external and internal environment. At the moment, criminological science and practice in this segment are at the formation stage.

Situational Crime Prevention is a method that allows to reduce a specific situation to appropriate standard for which a certain response mechanism has already been developed while studied event is analyzed in relationship and dynamics of all components. With regard to combating criminal offenses and crimes in critical infrastructure and the aerospace industry, essence of the situational approach is that offenses are committed "due" to a situation where the situation is a set of determinants that shape this situation. It has not only applied but also methodological value as it allows you to scientifically develop specific guidelines, pointing to the specific features of the investigation of criminal offenses and crimes.

2 Literature Review

Some aspects of consideration of the issues of general counteraction to criminal offenses and crimes were investigated in the works of such domestic and foreign scientists as H. A. Avanesov, O. I. Alekseev, Yu. M. Antonyan, O. M. Bandurka, Yu. D. Blushtein, M. H. Verbenskyi, T. S. Gavrish, V. V. Golina, I. M. Danshin, O. M. Dzhuzha, A. I. Dolgova, A. P. Zakalyuk, A. F. Zelinskyi, O. H. Kalman, M. H. Kolodyazhnyi, I. V. Leshukova, O. M. Litvinov, P. V. Melnik, P. H. Nazarenko, M. I. Pashkovskyi, N. Ye. Filipenko, I. Ya. Friedman, Sh. M. Khaziev, Yu. M. Chornous, V. Yu. Shepitko, I. I. Shepitko and others. The range of scientific developments on this issue is quite wide but not all problematic issues of combating criminal offenses and crimes are covered, some of them have only been identified, and some still remain controversial. However, these studies mostly focus on the field of legal regulation and activities of law enforcement officers. Therefore, without diminishing the scientific and practical value of published research papers regarding this issue, we note that the issue of situational analysis in combating crime in critical infrastructure and aerospace requires comprehensive coverage, taking into account current legislation and problems of practice.

3 Main Content Presentation

One of the reasons for numerous difficulties experienced by state and society is the lack of clear theoretical guidelines and values in the management in favor of the state. In these conditions the task of search of methodology of research of administrative criminological processes becomes actual. An example of such a methodology, relevant to modern social, economic and political conditions can be a situational approach [1].

Specific reasons for formation of a situational approach to combating criminal offenses and crimes at critical infrastructure and aerospace industry can be considered:

- accelerating the rate of social, economic and technological development;
- strengthening diversity of types of organizational structures;
- inconsistency of *traditional theories* with the needs of complicated crime management practice;

- insufficient practical efficiency and isolation *from the real world* of existing theories of crime counteraction management;
- inability of a particular theory of management to serve as a specific tool in practice of crime counteraction;
- transformation of *ordinary* crime into high-tech that leads to its transnationality;
- increasing the threat of terrorist acts and complicating geopolitical situation, especially in the context of a permanent military conflict in the east of our state;
- increasing threats to life and normal functioning at critical infrastructure and aerospace facilities, etc.

The main purpose of situational analysis is to ensure meaningfulness, certainty and adequacy of the control component of critical infrastructure and aerospace (both at the level of the crime subject and at the level of the state as a whole) in rapidly changing and uncertain conditions. From the point of view of the situational approach, crime counteraction management is first of all the art of manager of a strategically important object to understand the situation, determine its characteristics and choose the appropriate management and only then follow scientific criminological recommendations that are general and universal. Experts emphasize that situation analysis is basically the process of critically evaluating the internal and external conditions that affecting organization which is done prior to a new initiative or project. It provides the knowledge to identify current opportunities and challenges to your organization [2].

Situational analysis, on the one hand is one of the best tools for consistent, comprehensive, systematic analysis of important management decisions, on the other hand situational approach allows to identify the main trends that determine the dynamics of the decision-making situation, as well as the main factors on situation.

Uniqueness of situational approach to combating criminal offenses and crimes and its management is precisely the desire to combine two opposing ideas of management: idea of exclusivity or uniqueness based on the fact that each situation is unique and the idea of universality, according to which there is one best solution [3, p. 8].

Thus, situational approach is not a summary of specific principles, tools and management procedures but is a general methodology, way of thinking in the field of organizational problems and ways to solve them. It is used not in itself but as the basis of certain concepts of crime counteraction management. To one degree or another, the situational approach affects almost all basic criminological concepts, teachings and directions, and in the long run has every opportunity to become the dominant trend in the development of theory.

Reforming the crime counteraction system in critical infrastructure and aerospace industry, which has unfolded in the country in recent years against the background of legal, administrative, municipal, judicial and other reforms, requires active work to find ways, mechanisms and, of course, methodology management in a state more adequate to the new social and political conditions. Speaking about the grounds for such a reform, it should be emphasized that the assessment of the situation implies something more than just an analysis of the current situation. We need to find out the reasons for what is happening and find new opportunities. Only on this basis will it be possible to solve the problem of innovative development of combating criminal offenses and crimes in critical infrastructure and aerospace and to enter the trajectory of sustainable growth.

In a broad sense, the situation (derived from the Latin *situs* and *situtio*: situation) is interpreted as a set, a combination of conditions and circumstances that create a certain situation and state [4, p. 208]. Analyzing the concept of the situation, we must name the following essential features. The situation is a fragment of being, where, firstly, there is a change of provisions, i.e. a set of conditions and circumstances that change other conditions and circumstances, and secondly, it is possible to predict, control any part of being [5, with. 56–59].

An important situation feature should be considered as existence of three circumstances: where any event occurs, when it occurs and with whom (or with what) something happens. So, at intersection of these conditions of existence there is a point, a quantum, a moment, which is one element of the space-time system or one link in the chain of the system, which can be called situational. Each moment or link of being is formed into a chain of system, which forms a situational, i.e. duration, a series of events. If we look at the object of combating crime as a certain quality (system), procedure of formation of this quality may have different levels or hierarchies of situational conditions. The set of elements, subsystems and systems forms a situational system [6]. In general, the world appears as a situation of situations.

Another, no less important feature of the situation is that it has a "genetic" continuity, "heredity" of existence, because a certain situation and situation that constitute every moment (situational conditions) arise due to existence of the previous moment. This is how temporal genetic determination occurs. Everything that exists is in a certain situation, state and position due to the fact that before that there were a number of previous events of different complexity, different orientation and different intensity of various factors.

Given the above, we can provide the following situation definition of *Situational Crime Prevention* in critical infrastructure and aerospace: fragment of the system which arose as a result of external or internal factors and reflects the conditions and circumstances of formation and further transit of events that were the basis of these factors.

Situation requires any reason, driving force is a factor or factors due to which there is a change in one state of the object (objects) of combating crime to another. It is from the standpoint of the situational approach that theory essence of factors becomes clear, because the factor or set of factors determine social situation nature and determine the further development of events occurring in society. The external environment (external factor) which influences the internal situation (internal factor), plays an important role in the formation and functioning of counteraction to criminal offenses and crimes. As a result, at intersection of external and internal factors there is a situation that is a fragment of the system of combating criminal offenses and crimes in critical infrastructure and aerospace, and the set of situations consists of the situational dynamics of counteraction and its history. At the same time, in a changing environment, dominant role of certain factors changes situationally. Methodology of situational approach allows to explain corresponding movements of social matter [7].

Situational analysis algorithm of crime prevention on objects of critical infrastructure and aerospace branch depends on types of situations and the purposes of application of situational analysis. All types of crime prevention situations can be divided according to the level of stability: stable, uncertain, conflict, crisis.

Stable situation can be used to successfully combat crime when the result is systematically achieved and all factors of instability are compensated.

Uncertainty of situation arises with sharp increase in social and information interactions. This is a kind of lack of awareness about quality of crime counteraction. Uncertainty, as a sign of crime prevention system management can arise in the case of unclear correlation nature between events and their factors due to a time disorder or an increase in the number of simultaneous events, for example.

Conflict situation of crime counteraction arises in connection with the unstable situation in the country, liquidation or re-profiling of the subject of counteraction to crime, deterioration of quality of its activity, loss of controllability.

Crisis situation arises when the normal procedure of crime counteraction is stopped and when unforeseen events arise that call into question the logic of combating crime, as well as its mission. The management crisis of crime counteraction is emergence of destructive procedures in its mechanism. This is due to excessive growth, or vice versa, obvious stagnation. The crisis manifests itself in the following forms: lack of funding, emergence of interpersonal conflicts, problems of division of functions.

Determining the type of situation in which crime counteraction is nothing more than a basic basis, but on the other hand, it is an intermediate result of situational analysis. The end result of this analysis will be the development of recommendations to improve the fight against crime.

The task of situational analysis at critical infrastructure and aerospace facilities is to create a conditional report on where most threatening situation has arisen in terms of security factors, how to scale it and solve it with the available forces, means and methods. The set of analyzes will help to understand where to invest resources in the first place and what to pay attention to in the first place.

After analyzing the opinions of scientists and practitioners, we allow ourselves to draw parallels in the activities of critical infrastructure and aerospace with the work of a commercial enterprise (firm), as methodology for assessing emerging risks is described by almost identical models of their response [8].

Situational analysis of critical infrastructure and the aerospace industry can include several subtypes of analysis, all of which ultimately help to create a common report to develop a strategy for combating criminal offenses and crimes.

Here are some key methods:

1. *SWOT analysis* is assessment of strengths and weaknesses of critical infrastructure and aerospace. SWOT is a method used to study the features of the life of strategic objects and external influences on them. For this purpose, it is necessary to analyze the following factors:

a. *Strengths.* All factors that contribute to safety of life in critical infrastructure and aerospace (for example, geographical advantages, availability of human, material and technical resources, developed infrastructure, etc.).
b. *Weaknesses.* Factors that negatively affect viability of critical infrastructure and the aerospace industry (for example, need for high fixed costs to maintain security, special staffing requirements and consequently complexity of hiring).

c. *Opportunities (opportunities offered by external environment).* These are long-term trends in industry, legislative changes, government support, rapid development of technical means of protection and discovery of new technologies, social trends and more.
d. *Threats (threats to environment).* Factors that could potentially be detrimental, such as special political decisions, complications on the world stage (escalation of fighting in eastern Ukraine), economic development crisis, need to be taken into account, etc.

It has generally a good idea to work with several people to identify the greatest number of ideas for the SWOT matrix, which is why completing a SWOT analysis is much like a brainstorming session [9].

Terminology of the textbook of the Department of Political Analysis and Forecasting states that *brainstorming* is a collective assessment regulated by special rules based on stimulating creative activity of experts in this field of analysis, research [10, p. 18]. H. O. Kovalchuk characterizes brainstorming as a method of solving urgent problems in a very limited time. Essence of the method is that "it is necessary to express the largest number of ideas in a small amount of time, discuss and select them. This method is used to develop creativity or to solve complex problems. Method of *brainstorming* can be used in various forms of activity: in work with small groups, teams, large groups (*play with audience*), individual work *face to face* [11, p. 36].

2. *PEST analysis is a study of the external environment that affects the life of critical infrastructure and aerospace.* PEST analysis is a method of studying the external/macro environment. It is used to assess the development trends of strategic objects and compile a list of threats that actually affect their livelihoods. Sometimes PEST analysis is an integral part of SWOT analysis.

For PEST analysis, all influencing factors are grouped into four blocks:

a. *P – political factors.* This block contains factors of a political and legal nature. For example, an analysis of the legal framework that is planned to be developed and adopted in the coming years. This will enable the development of strategic plans for development of critical infrastructure and the aerospace industry. This block also includes factors and trends in foreign policy (strategic forecasts of relations between countries, labor market, migration, etc.).
b. *E – economic factors.* Critical infrastructure and aerospace executives need to make a forecast on several parameters that characterize the state of the economy and its impact on the security and viability of the whole structure: study dynamics of economic development (whether it is in decline or growth, because it significantly effects on the level of industry financing, forecast of fluctuations in exchange rates, because it depends on equipment and the number of means of protection that must be purchased from foreign partners, etc.).
c. *S – social and demographic factors.* This block analyzes the demographic situation and level of education of the population, the level of qualification of the future workforce (for example, assessment of the professional level of graduates of specialized educational institutions and calculation of qualified staff lack, etc.).

d. *T – technological factors*. This block examines innovations that will help ensure the safety of critical infrastructure and aerospace industry. Experts understand the technologies that are just beginning to be widely used in the industry. It is necessary to study the latest trends in the development of IT and engineering sector related to the industry. Understand what new discoveries and startups will affect the development of the industry in the near future: what will be automated that services will change key technologies and production mechanics, etc.
   There is an extended type of PEST analysis when two additional blocks of questions are considered. The legal block analyzes the legal environment that affects the activities of strategic objects (for example, studies the regulations that legislators at various levels plan to adopt in the coming years). The ecological block specifies the degree of object impact on the environment and potential ways to overcome the ecological crisis.

3. *Resource-based View (RBV)*. The resource-based view (RBV) argues that focusing on an organization's strengths is essential to achieve a sustained competitive advantage. RBV supporters look for ways to use internal resources (assets) to take advantage of external opportunities. To understand RBV, resources are classified into two groups: tangible and intangible.

Tangible assets are physical things such as land, equipment and machines, and real estate.

Intangible assets are anything an organization can own that is not physical. Examples include intellectual property, and the organization's reputation and goodwill [12].

Situational Crime Prevention describes factors that influence management of combating criminal offenses and crimes in critical infrastructure and aerospace, as well as its communication with other systems of security and defense sector. This information helps to understand the situation more objectively and create an effective plan that will work in the right direction. In order to assess the current situation, it is necessary to conduct a comparative analysis with previous data. This approach to situational analysis takes into account all the changes that are likely to have occurred in external and internal circuits of the system. Detection of changes allows you to identify factors that can affect the situation and its change.

While Situational Crime Prevention at critical infrastructure and aerospace industry, the following characteristics are distinguished.

Structure is relations and correlation of direct participants while their interaction, as well as the channels of interaction of external and internal driving forces of situation. These are her protagonists in their connections and relationships. Situation structure sets organizational forms and resource base for implementation of internal driving forces of its development. The main issues of analysis of the situation: nature of its internal structure and the composition of its immediate participants.

Content – main contradictions, for and around which solution of the interaction of the participants in the situation is built and which are the internal driving forces of its development. The content of the situation consists of such contradictions which solution leads to its irreversible, purposeful and natural changes, i.e. to development. The situation mechanism is based on the collision and struggle of different, often opposite, tendencies to resolve the main contradictions that made up the situation, forces and participants behind. The main question of situation content analysis: what are the contradictions that served as the impetus for its launch and implementation, contradictions for and around the solution of which the interaction of participants; contradictions, which, transforming into individual problems of all participants in the situation, serve as a motivating motive for each of them to interact. Finally, how (compromise or conflict) these contradictions are resolved as a situation result.

Meaning is a way of understanding the situation by its immediate participants, as well as the forces behind them and around them; current structure and state of consciousness of the participants in the situation, the ways in which they ask themselves the question "what is happening" and "what to do", as well as the ways in which they answer these questions. Analysis of meaning of the situation is reduced to identifying the impact on its dynamics of various human factors: intellectual, motivational, volitional, mental, moral and ethical and other "ideal" characteristics of the participants in the situation. The main question is: how, why and what model of the situation does each of its participants build? The answer to it involves, first of all, the analysis of their worldview, "world picture" which is formed by the basic premises and bases of judgments and actions, including ways to build a model of the situation. The connection between the ideal and the real in the situation is interactive: one or another understanding of meaning of the situation becomes ontological, becomes an integral part of the situation, actively directs its development. Understanding the meaning of a situation is not some mechanical "reading", it is always a creative act, generation of individual consciousness.

Significance is situation consequences for its direct participants and for the interacting systems of basic activities that shaped the situation and acted as external driving forces of its development. The result of the situation will be the same for everyone, but the consequences, impact and results, i.e. values are different. Significance idea of situation is always based on partially incomplete, unreliable and insufficient knowledge, in conditions of uncertainty which is exacerbated by instability of the situations themselves and the increasing pace and scale of changes in the comprehensive systems of basic activities. However, the need to make decisions and act "here and now" forces the participants to act as if they have necessary and sufficient information about outcome of the situation. The first group of questions, object-aspect, concerns, for example, the political, economic, financial and other consequences of the outcome of the situation or its individual events; the second, subjective, specifies the consequences of a particular outcome of the situation for each of its participants, as well as the forces behind.

Situational Crime Prevention at critical infrastructure and aerospace is also used when there is a need to make management decisions in real time in a complex situation in relation to controlled objects [13, p. 46]. Situational analysis provides on the basis of event monitoring of the situation identification of situations (threats), monitoring the situation, forecast scenarios of their development, as well as support the process of developing control effects to achieve the target situation (normalization) [14, p. 200].

There are four levels in management processes in large systems: strategic, operational, tactical and basic [15, p. 88], we also identify four levels of decision support for crime.

At the strategic and operational level of crime counteraction management at critical infrastructure and aerospace facilities, tasks related to the analysis of mass flows of events in the area of responsibility of the prevention subject, in the most important areas and areas of responsibility of individual units. To analyze the situation and support decision-making procedure at these levels, mathematical methods of statistical analysis of flow of events are used in order to create dynamic models of processes (for example, a significant number of weapons illegally in population, etc.). Based on them, the analysis of the current situation, comparative analysis with the past period of time, identification of trends and anomalies in the flow of events, forecasting development of situations and threats.

At tactical and basic levels of crime counteraction management at critical infrastructure and aerospace facilities, firstly, decision-making processes are planned to plan the use of forces and means for a period of time and in accordance with instructions of higher authorities and secondly, special analysis of factual data on specific elements of the criminogenic situation, formalized by means of a semantic network which allows to identify non-obvious connections between elements of the situation, to determine similar spatial event situations, to identify associative connections and patterns.

Situational approach also assumes that unique situations require development of unique solutions based on a systematic consideration of the situation, taking into account many details based on their analysis. This approach is used in cases of difficult to formalize the management object (crime prevention) its uniqueness, incomplete description when it is not possible to build an accurate model, nevertheless, object can be verbally described by experts [16, p. 18].

The idea of Situational Crime Prevention at critical infrastructure and aerospace in the above conditions is based on the hypothesis that all information about the object of management and ways to manage it can be expressed in natural language. Therefore, when creating a system of situational management of crime, all possible descriptions of the state of the object are classified and presented by means of a multi-cluster model superimposed on description of the current situation. As a result, certain management steps are generated, resulting in a description of the new state of the object and so on indefinitely.

If generalization of states and their classification is not possible, then the need to describe all possible states of the object, as well as possible control effects makes this approach difficult to implement for large systems, such as crime counteraction. It requires large material and time costs for experts to create a database of situations and their corresponding management influences. In essence, it is vital to formalize databases and knowledge of experience of the forensic expert (experts) in managing a complex organization.

Described approach is a conceptual basis of situational management of counteraction to criminal offenses and crimes on objects of critical infrastructure and aerospace branch which realization can be various. In particular, the development of computer and information technology has led to the emergence of a new technology of situational management - the creation of situational centers [17]. They combine decision-making processes in complex strategic, tactical or emergency situations provided with services (databases, repositories of unstructured information), mathematical modeling, forecasting, visualization, communication, etc. This symbiosis of expert knowledge and capabilities of modern information technology allows you to make decisions in complex and confusing situations quickly and efficiently.

Depending on the field of application, the name "Situation Center" can be transformed into "command and control center", "crisis center", "emergency center", "meeting room" (corporate boardroom), conference room). For this purpose, the center means not only a specially equipped room, but also the appropriate information, telecommunications, software and methodological tools that provide the process of delivery, aggregation of information in order to develop an appropriate management decision [18, p. 65–69].

In Ukraine and abroad, since the middle of the last century, significant experience has been gained in creating situation centers. The disadvantage of all top-level situational management centers is the analytical functionality. At the same time, fascination with the formalization of problem situations with the use of even the most modern intelligent information technologies in any case leads to the detachment of models from reality, loss of integrity of discourse and awareness (awareness) of decision makers.

At creation of any situational centers the essential role is given to issues of information maintenance, modeling and regulation. However, the latent, transcendent and emotional aspects of decision-making are left out of consideration. As a result, in most cases, the creation of situational centers is multidisciplinary complexity, insufficient attention to the analytical functionality is the reason for rupture (fragmentation) of the problem situations in critical infrastructure and aerospace. As a result, decisions made in situational centers do not always meet the stringent requirements of efficiency and do not provide the necessary synergies. The principle of integrity can be the basic principle of creating a situation center on critical infrastructure and aerospace. The main element of its implementation is semantic communication and hence the image, language and text of decision-makers in the situation center and information space that surrounds it. However, there is no theoretical model that covers all, or at least the main, aspects of the process of semantic communicative interaction, although research in this area has been conducted for a long time.

At the same time, the creation of an intersectoral situation center, working on the basis of interaction and coordination between law enforcement agencies and specialists in aerospace industry and critical infrastructure, makes any problem situation extremely accessible for consideration and awareness of decision makers as holistic as possible.

The intersectoral situation center will provide:

- making decisions on degree of regulatory impact (whether to continue monitoring, take additional measures and forecast, influence the situation in other means);
- accounting for overt and covert crisis factors determining effectiveness of management influences;
- rapid development of conceptual models ready for change that take into account the risks of emergencies in critical infrastructure and aerospace;
- obtaining clear, accessible and informative results.
- For this purpose, intersectoral situation center should provide a so-called operational understanding of situation that includes the following essential components:

- holistic perception of heterogeneous elements (factors, phenomena) of the environment in a single spacetime representation;
- awareness of importance of elements and connections between;
- projection of states of elements and connections between them in the future;
- state of operational understanding of situation by an individual or group of people as a result of cognitive analysis and assessment of the situation;
- creating mechanisms and conditions for timely formation, dissemination and use of a common operational understanding of situation for all levels and all teams, departments and organizations.

  International experience in formation of communication and information support, involvement of network experts allows us to identify three main shortcomings of the existing traditional approach to maintaining "operational understanding of the situation" based on verbal or documentary exchange of information in a hierarchical system:

  - it takes time for a situation report prepared by one of participants to reach the addressee in other groups, especially those above the hierarchy;
  - situation report provides its static reflection and usually becomes obsolete by the time it is received by the addressee (especially at the operational or political level);
  - not everyone who needs the information contained in the message can access it.

  Even if there is a high degree of integration of different departments, exchange of information based on the transmission of voice messages (radio or video channels) or printed documents (courier, fax, e-mail) does not meet modern requirements for shared situational awareness. Analysis of emergencies involving the commission of criminal offenses or crimes in critical infrastructure and aerospace industry indicates that functionality to support "operational understanding of the situation" is a critical foundation, prerequisite for making the right decision in an emergency.

In this regard, it should be noted that competent and timely decisions that can change the situation of combating crime based on the achievement of hierarchically organized goals, there are a number of features. Based on the correct identification of the main factors and trends in dynamics of transformation of the criminogenic situation, they clearly record:

"Corridor of opportunities" for development: a set of immutable data-constraints, within any of selected trajectories of development will inevitably remain, striving to get the most out of what is actually achievable;

Both positive and negative tendencies of development of crime counteraction, that is not only for what and in which sequence it moves but from what and in which sequence refuses;

Strict connection of formulated goals with the mechanisms of their achievement, identifying and presenting these mechanisms, paying special attention to realization of set tasks;

"Natural" trends in the situation dynamics that can act as drivers of achieving targets for combating crime.

At the same time, situational approach to the analysis of combating criminal offenses and crimes in critical infrastructure and aerospace has a number of methodological shortcomings that limit its effective use:

1. Situations are considered as one of the factors of situational approach and not as the main, primary, initial, system-forming factor.
2. Situational factors are understood purely subjectively, arbitrarily, without connection with situation itself.
3. Issues of the form of situation presentation (complex of its factors) for criminological analysis of counteraction to criminal offenses and crimes at the objects of critical infrastructure and aerospace industry have not been worked out.
4. The methodology of analysis and use of situations is mainly developed in relation to unified or typical situations used in learning process and poorly substantiated in the field of real situations used to make decisions for crime counteraction.

And one more. The concept of situation is functional not attributive; it is a set of properties and features that relate to the situation, are its necessary and sufficient signals, but they are not absolute, once and for all data. On the contrary, this set is always determined by position, tasks and methodological capabilities of the "observer". Its settings and capabilities will determine the grounds and criteria for assigning objects, including events, to the situation, principles and operations for their individualization and differentiation (consideration of their differences and similarities and selection of sets on this basis), spatial, temporal and functional boundaries (distribution, duration, impact, change and consequences, speed, pace and rhythm of implementation). Therefore, under the same circumstances, different "observers" can see different situations.

The scope of situational approach can be significantly expanded in modern conditions of rapidly changing factors of the external and internal environment and require further methodological development.

4 Conclusions

Summarizing the above, we want to note that the role and importance of situational analysis in combating crime at critical infrastructure and aerospace is constantly growing, because it is possible to develop operational and strategic plans to combat criminal offenses and crimes at these sites. To adequately assess the situation when the decision is made, first, you need an in-depth analysis of the current object state consisting in a comprehensive analysis and consideration of the following crisis factors:

- decision has a clearly defined goal and is multiple in nature (i.e. takes into account the prospect of its application in the future);
- has restrictions on resource provision, as well as on the timing of implementation;
- includes elements of complexity and action in the combined uncertainty of future risks to the life of critical infrastructure and aerospace;
- has a detailed architecture and sequence (i.e., each problem to be solved should be structured from public to private; the problem should be solved in prescribed sequence or in phases, such as planning, development, implementation, etc.;

- includes the development of a risk assessment model that allows to identify risks and assess their impact on the activities of critical infrastructure and aerospace at different stages of life cycle, develop management influence to reduce overall risk and improve countering criminal offenses and crimes;
- includes tools for finding the optimal solution to combat criminal offenses and crimes in critical infrastructure and aerospace.

  Secondly, it is necessary to create an intersectoral situation center, working on the basis of interaction and coordination between law enforcement agencies and aerospace professionals and critical infrastructure that will be a hub of intellectual, institutional and technological model of high-quality aerospace cluster management. At a high professional level, complex processes that are implemented in areas related to critical infrastructure and aerospace can be regulated. Urgency of creating such a situational center is also due to the fact that domestic science and practice have faced complex problems of scientific, technical, economic, social and institutional nature, and some of them arose as a consequence of the general state of the global economy. Range of issues that can be addressed in it is very wide, namely: to raise the effectiveness of the leadership of interacting parties and increase the effectiveness of such cooperation; to ensure management effectiveness of the of the implementation of projects for crime counteraction in critical infrastructure and aerospace by improving the efficiency and quality of management decisions; reduce labor costs in the collection, processing, retrieval and use of data, as well as the preparation of consolidated reporting documentation; continue to improve the information support of the management system through the introduction of decision support subsystem at all stages of management; to provide an opportunity to optimize decisions made by their expert assessment and simulation of situation in real time; to ensure the active development of interregional and interstate cooperation in various areas of preventive activities at critical infrastructure and aerospace industries.

Thus, the situational analysis allows, based on a deeper understanding of the situation and the dynamics of its development, to make more informed management decisions, as well as to anticipate possible crises and take timely measures to combat crime in critical infrastructure and aerospace.

References

1. Surmin, A.Y, et al.: Situational analysis or case-method anatomy. Kyiv: Center for Innovation and Development, 286 (2002). [in Ukrainian]
2. The Easy Guide to Performing an Effective Situation Analysis [in English]. https://creately.com/blog/diagrams/what-is-a-situation-analysis/#:~:text=Situation%20analysis%20is%20basically%20the,your%20organization%2C%20service%20or%20product
3. Pospelov, D.A.: Situational Management: Theory and Practice. Nauka (1986).[in Russian]
4. Dictionary of the Ukrainian language: in 11 volumes. **9** (1978). [in Ukrainian]
5. Osipov, G.S.: From situational management to applied semiotics. Artif. Intell. News. **6**, 56–59 (2002). [in Ukrainian]
6. Klykov, Y.I.: Situational control of large systems. Energy (1974). [in Russian]
7. Lysov, O.E.: A situational approach to knowledge management. GUAP (2009). [in Russian]

8. Proceedings of the 2016 2nd International Conference on Social Science and Higher Education. Atlantis-Press [in English]. https://www.atlantis-press.com/proceedings/icsshe-16/25859135
9. How to Complete a Situational Analysis (With Definition and Elements). 23 February 2021 [in English]. https://www.indeed.com/career-advice/career-development/situational-analysis
10. Teleshun, S.: Political forecasting as a type of management activity: practice. rivers/S. Teleshun, O., Titarenko, I., Reiterovich. K.: NADU, 2008 (2008). [in Ukrainian]
11. Kovalchuk, G.: Activation of education in economic education: textbook. way. / G. Kovalchuk; Kyiv. nat. econ. un-t. 2nd ed., Supplement. K.: КНЕУ (2003). [in Ukrainian]
12. Common Frameworks for Situational Analysis. Module 4: Environments and Strategic Management. Principles of Management [in English]. https://courses.lumenlearning.com/wm-principlesofmanagement/chapter/common-frameworks-for-situational-analysis
13. Litvak, B.G.: Development of an administrative decision: textbook. M: Business (2000). [in Ukrainian]
14. Smirnov, E.A.: Management decisions. RIOR (2016). [in Russian]
15. Axelrod, R.: The Structure of Decision: Cognitive Maps of Political Elites. Princeton, University Press (1976).[in Russian]
16. Filippovich, A.Y.: Integration of situational, simulation and expert modeling systems. Elix + LLC (2003). [in Russian]
17. Situational center of the National Police of Ukraine [in Ukrainian]. https://112.ua/statji/situacionnyy-centr-nacpolicii-analiz-prestupnostioperativnoy-obstanovki-i-koordinaciya-podrazdeleniy-431302.html
18. Danchulo, A.N.: Information-analytical technologies and situational centers. Publ. Serv. 4, 65–69 (2004). [in Russian]

Responsibility for Administrative and Criminal Offenses in the Field of Use of Airspace of Ukraine

Anatolii Komziuk[1], Olena Salmanova[2](✉), and Oleksii Salmanov[3]

[1] Department of Administrative Law and Procedure, Kharkiv National University of Internal Affairs, Kharkiv, Ukraine

[2] Administrative Law and Procedure, Kharkiv National University of Internal Affairs, Kharkiv, Ukraine

[3] Criminal Procedure and Pre-Trial Investigation, Kharkiv National University of Internal Affairs, Kharkiv, Ukraine

Abstract. Concept analysis of Ukrainian airspace and its use is carried out, normative bases of regulation relations in the field of use of airspace, value of legal responsibility in maintenance of safety of such use, parity of administrative and criminal responsibility are defined.

Notion of administrative misdemeanors in the field of airspace use and their signs including social harm, illegality, guilt and administrative punishment are defined. Emphasis is placed on practical significance of administrative offense composition. The most characteristic features of administrative responsibility, including offenses in the field of airspace of Ukraine are highlighted. The types and signs of administrative penalties applied for committing these offenses are established. The types of corpus delicti of these offenses are determined, the peculiarities of proceedings in cases about are singled out.

Concepts and types of criminal offenses in the field of airspace use, peculiarities of their legal structures, as well as criminal liability for their commission are defined.

Keywords: Airspace · Use · Administrative offenses · Administrative liability · Criminal offenses · Criminal misdemeanors · Crimes · Criminal liability

1 Introduction

Airspace is one of the basic categories in characterization of a sovereign state. According to Art. 2 of Constitution of Ukraine, its sovereignty extends to the entire territory that within existing border is integral and inviolable one [1]. Article 1 of the Air Code of Ukraine defines airspace as a part of the air area located above land and water territory of Ukraine including its territorial waters (territorial sea) and limited by a vertical surface passing along the state border of Ukraine. Basic principles of using airspace of Ukraine are practically defined in Art. 4 of this Code. Firstly, Ukraine, as a state that has acceded to the Convention on International Civil Aviation is responsible for fulfilling

M. Nechyporuk et al. (Eds.): ICTM 2021, LNNS 367, pp. 649–663, 2022.
https://doi.org/10.1007/978-3-030-94259-5_54

international obligations arising from this Convention and for guaranteeing and creating conditions of security for society, protection of interests regarding airspace use of Ukraine; secondly, state regulation of activities in this area is formation of state policy and development strategy, defining tasks, functions, conditions of activity regarding airspace of Ukraine, application of aviation security measures, adoption of mandatory aviation rules of Ukraine, state control over their implementation and establishing responsibility for their violation [2].

Thus, responsibility, this means legal responsibility, for violation of the rules of airspace use of Ukraine is legally recognized as one of the means of ensuring its proper functioning. It is well known that most effective types of legal liability in ensuring compliance with certain mandatory rules are two types: administrative and criminal one devoted to this intelligence.

Sufficient attention is paid to airspace use in scientific publications and they relate to all aspects of such use including airspace systems [3], trends in air traffic in airspace [4], aircraft monitoring systems and potential consequences for air safety [5], issues of using unmanned aerial vehicles and helicopters [6], analysis of airspace violations [7], state regulation in the field of air transport [8], etc. At the same time, issue of liability for offenses regarding airspace use in Ukraine needs in-depth research.

Research Purpose

The purpose of this article is to define concept, grounds and peculiarities of liability for administrative and criminal offenses regarding airspace of Ukraine.

2 Main Content Presentation

Experts from the National Bureau of Air Accidents Investigation of Ukraine (NBAAI) note that violations of airspace use adversely affect the level of flight safety and in some cases can lead to dangerous approach of aircraft or even their collision in the air. According to statistics, violations of the use order of Ukrainian airspace account for more than 50% of the total number of events that occur during aircraft operation and their number tends to increase. The number of general aviation aircraft is increasing every year and in recent years the airspace use has been significantly affected by unmanned aerial vehicles (UAVs). Violations of airspace use contain a high potential threat not only to flight safety but to the national security of Ukraine as a whole [9]. According to information of the mentioned Bureau, during the operation of civil aircraft of Ukraine in 2020 there were: 5 accidents (1 during transport, 3 during private flights and 1 during a test flight); 3 accidents; 1 serious incident; 28 incidents; 1 emergency; 15 violations of the use order of airspace [10].

Thus, existence of real and potential threats to the safe and efficient use of airspace requires a research on the means to ensure implementation of such use, especially the means of legal liability. Article 126 of the Air Code of Ukraine stipulates that for illegal actions legal entities and individuals whose activities are related to the use of Ukrainian airspace, development, manufacture, repair and operation of aircraft, economic activities in civil aviation, air traffic services, aviation safety are responsible in accordance with the law [2].

Violations of airspace use are quite diverse in nature, liability for their commission can be provided by various regulations of Ukraine including the Code of Administrative Offenses [11] and the Criminal Code [12]. It should be emphasized that administrative and criminal responsibilities regulated by these codes have quite different essential characteristics. They differ from each other on the grounds: administrative offense for administrative liability and a criminal offense for criminal; content: each type of liability consists in application of specific influence measures for administrative offence; they are administrative penalties and for criminal liability criminal penalties; procedure for implementation: administrative liability is implemented during proceedings on administrative offenses that is a type of administrative procedure, procedure for criminal prosecution is regulated by a separate Code, namely: Criminal Procedural Code of Ukraine [13] and execution of criminal penalties is regulated by another Code: Criminal Executive Code of Ukraine [14].

At the same time, expediency of joint liability consideration for administrative and criminal offenses in the field of use airspace Ukraine is due to availability of common features. It means similarity of objects of encroachment on these offenses: ensuring aviation security [15], security of use of the airspace of Ukraine [16] that should be discussed in more detail.

The main normative regulatory document that regulates use of Ukrainian airspace and its security is the Air Code of Ukraine. In its article 1, first of all the mentioned basic concepts are defined: use of airspace of Ukraine is carrying out of activity connected with flights of aircraft, with movement (stay) of material objects in airspace of Ukraine and with explosive works, launches of rockets, all types of firing including in order to influence the hydrometeorological processes in atmosphere that pose a threat to the safety of flights of aircraft and other aircraft [2]. It is worth noting that the Air Code uses the terms: *aviation security*, *flight safety* and *aviation security* that are directly related to the safety of airspace use, as this space is the place of flight and aviation operation.

According to Art. 23 of the Air Code of Ukraine, use organization of its airspace is a set of measures taken to ensure safe, economical and regular air traffic, as well as any other activities related to the airspace use of Ukraine. Cabinet of Ministers of Ukraine approves the relevant Regulation on the procedure for airspace use [17]. According to these acts, violation of the procedure for the airspace use of Ukraine is considered to be actions or inaction of legal entities or individuals that have led to a violation of requirements of laws and other regulations governing this procedure. It is clear that they can be administrative or criminal offenses which specifics will be discussed later.

Administrative offense is a type of offenses committed in the field of public administration. Article 9 of the Code of Administrative Offenses defines it as an illegal, culpable (intentional or negligent one) action or inaction that encroaches on public order, property, rights and freedoms of citizens, established order of management and which the law provides for administrative liability. The same article enshrines the so-called principle of "priority of criminal liability" which meaning is that administrative liability for such violations occurs if they do not entail in accordance with the law of criminal liability. V. K. Kolpakov emphasizes that this legislator approach is due to existence of cases when the separation of crime from administrative misconduct is complicated [18, p. 34].

We would like to emphasize once again that administrative offenses are closely connected with the field of public administration with ensuring observance of the rules established in this field. It is worth agreeing with opinion that society has a large number of different legally binding rules, standards, requirements that are established or sanctioned by the state. Such rules are aimed at regulating public relations, standardizing certain procedures, actions, ensuring law and order and proper discipline in certain areas of activity. The rules of conduct established by the state directly affect the interests of all or most citizens, enterprises, institutions and the state as a whole [19, p. 122]. One of properties of these rules is that they are protected by administrative sanctions and their violation is recognized as an administrative offense [20]. Such mandatory rules include the rules for use of Ukrainian airspace, in particular.

Earlier we drew attention to the fact that an administrative offense is first of all an act, behavior, deed of a person, action or inaction, it is an act of external detection of a person's attitude to reality, other people, the state, society. The law is not subject to the beliefs, opinions of people, if they have not found external manifestation. Such an act is recognized as an administrative offense in the availability of certain signs that traditionally include social harm (sometimes talking about danger), illegality, guilt and administrative punishment [21, p. 14–15]. These features are inherent in all acts that are recognized as administrative offenses including violations of the rules of use of airspace of Ukraine. They form material concept of an administrative offense, i.e. characterize it as a phenomenon of material reality.

A similar opinion is held by other authors who believe that legislative definition of an administrative offense makes it possible to understand the main functional components of content of this type of illegal behavior by establishing its characteristics. The main of them is an illegal act (fact, phenomenon, event) occurred or took place in reality. After all, an administrative misdemeanor is considered committed only when a person allows illegal, active, conscious and willful behavior that causes or can create a real opportunity to harm the object that is protected by administrative law. It is necessary to pay special attention to the fact that in most cases administrative misconduct is committed through active illegal actions. At the same time, a significant number of administrative offenses listed in a special part of the Code of Administrative Offenses are enshrined through passive inaction, when a person in order to prevent illegal behavior should take positive action but for various reasons does not [22, p. 601].

The person committing offense violates his constitutional duty, as Part 1 of Art. 68 of the Constitution of Ukraine establishes that everyone is obliged to strictly abide by the Constitution and laws of Ukraine. In turn, fulfillment of this obligation necessitates observance of all rules adopted in the manner prescribed by law. Assessment of violation of the requirements of laws and bylaws as an illegal act necessitates the application of appropriate coercive measures and bringing the perpetrator to justice, to administrative responsibility, in particular [22, p. 256].

It is necessary to distinguish from the material concept of an administrative offense its legal characteristics that means determining characteristics of a particular action, which the law associates with, possibility of administrative liability. The material concept of the offense is specified in the legislation on administrative offenses through of specific offenses, when law provides necessary and specific signs of a particular action.

Substantive notion of an offense answers the question of what all misdemeanors have in common (for example, between violating flight safety rules and violating the rules of conduct on an aircraft), and composition notion of misdemeanor answers the question of what distinguishes some misdemeanors from others. Administrative misdemeanors differ in specific signs that characterize their objective and subjective aspects and form the legal structure of an administrative offense [21, p. 16–17].

V. K. Kolpakov quite rightly notes that doctrine of the offense composition occupies one of central places in administrative and legal science and has great practical significance. First, it helps to identify the most significant signs of anti-social acts, their delimitation and establishment of fair sanctions; secondly, it helps legal bodies to correctly classify offenses and to choose measures of influence adequate to; thirdly, it allows to understand the law helps to train lawyers and legal education of citizens [23, p. 218].

We will speak about specific components of administrative offenses in the field of airspace use in Ukraine a little later. At present, it is necessary to dwell briefly on the main essential characteristics of administrative liability that is applied for their commission, because an administrative offense is recognized only as an illegal offense, which the law provides for this specific type of state coercion. However, realization of its measures is not always accompanied by administrative misconduct but possibility of their application is its mandatory property.

It should be borne in mind that in recent years in Ukraine a new doctrine of administrative law is being developed, one of which main tasks is formation of its modern system. Within the framework of this research paper, transformation of certain administrative and legal institutions in the sub-branch of administrative law is carried out. This also affected institution of administrative responsibility. In scientific literature of recent years, attention is no longer paid to the institution but to the sub-branch of administrative tort law. However, there are disputes about the place of this sub-branch in the system of administrative law, i.e. where it should be in General or Special Administrative Law. Speaking about the content of administrative tort law, it should be borne in mind that its subject are relations that arises in connection with commission of administrative offenses and the implementation, as a consequence, measures of administrative responsibility. This means that this sub-sector, as well as its predecessor, namely: institution of administrative responsibility, needs to be reformed as soon as possible, as we have previously expressed our opinion [24].

As D. M. Lukyanets rightly points out, administrative responsibility, more precisely its conceptual principles and the most essential characteristics of the normative construction have a long history. Having undergone many transformations due to peculiarities of functioning in existence conditions of various forms of organization state of society, administrative responsibility has survived to our time. However modern realities of the Ukrainian state, desire to join the European values require a rethinking of the place of this institution in the domestic law, transition from views on it as a repressive-fiscal instrument of the state, understanding it as a means of law and order in those areas of public relations. objectively necessary is state regulation and public administration, means of ensuring rights and human and citizen freedoms [25, p. 5].

Most characteristic signs or properties of administrative liability, including offenses in the field of use of airspace of Ukraine, it occurs for a specific type of offense, administrative one. Secondly, administrative liability is imposition of administrative penalties on violators that are specific in content and different from other measures of legal influence. Third, administrative penalties are imposed administratively, i.e. without going to court. The list of bodies and officials to whom this right is granted is given in a specific section of the Code of Administrative Offenses. However, nowadays courts also consider cases of administrative offenses and their powers are constantly expanding, although this state of affairs cannot be considered normal. Fourth, there are no official relations between the bodies (officials) imposing administrative penalties and offenders. Fifthly, the procedure for bringing to administrative responsibility is special, it differs significantly from other types of legal proceedings. Sixthly, administrative liability is regulated by the rules of administrative law which together constitute its normative basis. Therefore, administrative liability as a special type of legal liability is the application to persons who have committed administrative offenses (misdemeanors), administrative penalties that entail for these persons aggravating consequences of property, moral, personal or other nature and imposed by authorized bodies or officials on grounds and in the manner prescribed by the rules of administrative law [21, p. 6].

It is also necessary to say a few words about the principles which administrative responsibility is based on, the main of which are legality, responsibility only for committing an illegal act, responsibility only in in case availability of guilt, inevitability of responsibility, presumption of innocence, individualization of responsibility, humanism. These principles are a kind of theoretical guidelines that establish the principles of normative definition and actual application of measures of this responsibility including for administrative offenses in the field of use of airspace of Ukraine.

With regard to responsibility content, it should be borne in mind that, as previously noted, it is the application of administrative penalties, the types of which are set out in Art. 24 of Code on Administrative Offenses. For committing administrative offenses regarding airspace use of Ukraine provides for the use of only some of them, namely, warning, penalty, paid seizure of dangerous substances or objects and confiscation of such substances or objects.

Warning is the mildest administrative penalty, it consists in official condemnation of the committed misdemeanor and the warning of the offender against its repetition. For committing administrative offenses in the field of airspace use, it is made in writing. The main administrative penalty has always been and remains a penalty which amount depends on the size of the tax-free minimum income. The maximum penalty that can be applied for committing offenses in the field of airspace use is five hundred non-taxable minimums, i.e. 8,500 UAH. Paid seizure of dangerous substances or objects consists in their forced seizure and subsequent sale with the transfer of the proceeds to the former owner less the cost of selling the seized object, in contrast to their confiscation forced transfer of property to the state.

With regard to the types of administrative offenses in the field of airspace use, it should be noted that in the current Code of Administrative Offenses, none of them directly contains instructions on the airspace use. They relate to flight safety and rules for the use of air transport, i.e. relationship between the use of airspace is indirectly encroached upon. These warehouses are included in Chapter 10: *Administrative Offenses in Transport, Roads and Communications*. In general, these compositions are provided by five articles, namely: Art. 111: *Violation of flight safety rules*), Art. 112: (*Violation of the rules of conduct on the aircraft*, Article 113: *Violation of the rules of international flights*, Article 120: *Violation of fire safety rules on rail, sea, river and air transport and Article 133 using transport*.

Administrative liability is provided by these articles for the following actions that constitute the objective offense side:

- allocation in aerodrome area of any signs and devices similar to markings and devices adopted for recognition of aerodromes, or burning of pyrotechnic articles without permission of the airport administration, or the arrangement of facilities that promote mass crowds of birds dangerous to flight aircraft;
- non-compliance with the rules on allocation of night and day markings or devices on buildings and structures;
- damage to aerodrome equipment, aerodrome signs, aircraft and their equipment;
- passage without proper permission on the territory of airports (except airports), airfields, radio and light facilities of flights;
- performance of flights in violation of legal regulations governing aviation activities;
- non-compliance with the orders of aircraft commander by persons on board the aircraft;
- violation of photography rules, filming and use of radio communications on board the aircraft;
- violation of rules of international flights;
- violation of fire safety rules established on air transport;
- violation of rules of transportation of dangerous substances or objects by air.

As can be understood from the above, objective side of administrative offenses is mainly determined by outlining the specific actions (inaction) of the person. Only the last four syllables are blanket, i.e. they constitute a violation of certain rules established by other regulations.

The subjects of these offenses can be individuals: citizens or officials. Literature suggests that such subjects in some cases are flight crew (violation of the rules of international flights (Article 113), the performance of flights in violation of regulations (Article 111) [8, p. 432]), however Code on Administrative Offenses does not contain any clarifications in this regard.

Here it is necessary to pay attention to responsibility of legal entities for violating the order of use of the airspace of Ukraine. As mentioned earlier, these violations are considered to be actions or inactions not only by individuals but also of legal entities. The only problem is that the Code of Administrative Offenses applies only to individuals, it does not recognize legal entities as subjects of administrative offenses. However, this does not mean that legal entities are not subject to administrative liability, it is simply regulated by other regulations. In particular, for violation of use order of the airspace of Ukraine, their responsibility is established by the Air Code of Ukraine [2], Art. 127 that provides for financial sanctions in the form of fines applied to legal entities: aviation entities. Even in the definition of these sanctions there is the *penalty* term indicating the administrative nature of these sanctions.

Therefore, it is reasonable to think that having gone from denying the very possibility of bringing legal entities to administrative responsibility to recognizing them as subjects of this responsibility in some individual laws, legislation on administrative offenses in its development should go to the next stage to define general rules and procedures of administrative responsibility legal entities, without which it is impossible to ensure proper protection of these persons from wrongful prosecution [26].

Proceedings in cases of administrative offenses in the field of use of the airspace of Ukraine also have certain peculiarities, during the implementation of which administrative liability is realized. Proceedings in cases of administrative offenses are understood as a clearly defined sequence of procedural actions carried out by authorized entities on the grounds and in the manner prescribed by law, in order to timely, comprehensive, complete and objective clarification of the circumstances of each case, resolution in strict accordance with the law, ensuring the implementation of the decision [19, p. 292]. These proceedings have the task of timely, comprehensive, complete and objective clarification of the circumstances of each case, resolving it in strict accordance with the law, ensuring the implementation of the decision, as well as identifying the causes and conditions conducive to administrative offenses, crime prevention, education citizens in the spirit of observance of laws, strengthening of legality (Art. 245 Code on Administrative Offenses).

These proceedings are based on a number of principles, the main of which include: legality; equality of citizens before the law; publicity; ensuring the right to protection; national language; material truth; presumption of innocence; publicity; humanism.

Officials of the State Aviation Service of Ukraine as the central executive body for civil aviation are empowered to consider and resolve cases on most administrative offenses in the field of airspace use. Cases of two misdemeanors: violation of the rules of photography, filming and use of radio communications on board the aircraft and violation of the rules of transportation of dangerous substances or objects by air are subject to district, district in cities, city or city courts (judges).

Protocols on these offenses are authorized to be drawn up mainly by the officials of the State Aviation Administration of Ukraine. Only on violations of the rules of transportation of dangerous substances or objects by air protocols are drawn up by officials of the National Police and on violations of fire safety rules established by air employees of departmental, local fire protection and members of voluntary fire protection.

Institution of criminal responsibility is one of the main institutes of criminal law. Such well-known domestic and foreign scientists as Yu. V. Baulin, A.V. Vasylevskyi, O. D. Hrin, O. I. Donchenko, O. O. Dudorov, S. O. Zahorodniuk, S. I. Zeldov, V. O. Navrotskyi, Yu. Kolomiiets, M. I. Korzhanskyi, I. V. Krasnytskyi, L. L. Kruhlikov, N. F. Kuznetsova, M. I. Panov, A. I. Santalov, Yu. M. Tkachevskyi, S. M. Shkola and many others.

As for essence and significance of the *criminal liability* category M. I. Panov argues that performs an important integrative function in criminal law and legislation, combining (uniting) their General and Special parts, turning their rules and institutions into organically interconnected and complementary structural elements of a single and holistic system of criminal law as an independent branches of law [27]. However, despite the repeated mention of criminal liability in the articles of the current Criminal Code of Ukraine, its statutory definition does not exist. According to Yu. A. Ponomarenko: "this state of affairs not only does not contribute to the certainty of criminal law regulation, but also has a significant impact on the content and results of research in the field of criminal law" [28, p. 100].

Variety of concept definitions of criminal liability, on the one hand, provides opportunity to improve knowledge about its essence, content, grounds and legal regulation. However, some scientific papers analyze and propose the use of alternative formulations, such as *criminal liability*, *anti-criminal liability*, *human criminal liability* and so on.

It is worth agreeing with the opinion of Yu. A. Ponomarenko, who conducted a detailed analysis of a large number of definitions and formulations of the essence and meaning of the *criminal liability* term and agreed with a group of scholars who argue that criminal liability should be understood as "relating to criminal offenses, their fighting and preventing." Given the above, there are no grounds for refusing to use in further research the traditional term: *criminal liability* [28, p. 102].

The vast majority of scientists to the main features of criminal liability include:

- first, the public legal nature of criminal liability, which, according to V. O. Navrotskyi, consists of several points (entity authorized to establish, apply and enforce it is exclusively public authority in person of the state, has a coercive procedure for its application by the state, is applied by the state regardless of the desire or consent of the subject or any other subject, etc.) [29];
- secondly, has a special order of application;
- third, it is a legitimate restriction on the exercise of human rights and freedoms;
- fourthly, it points to a legal fact that is recognized as the basis for such a restriction, so the only type of offense for criminal liability can be applied, that is exclusively a criminal offense;
- fifth, it can be applied only to the person who has committed a criminal offense;
- sixth, provided exclusively by law; seventh, it aims to punish a person for a criminal offense and it is through the achievement of this goal that criminal liability is able to achieve other goals set before criminal law in general.

Generalization of these features allowed Yu. A. Ponomarenko define criminal liability as "prescribed by criminal law restrictions on exercise of human rights and freedoms that are applied compulsorily by a court decision to the person who committed a criminal offense, in order to punish him" [28, p. 161].

In general, supporting a detailed analysis of the set of signs of criminal liability conducted by Yu. A Ponomarenko we will try to provide an author's definition of criminal liability in our research topic.

In our opinion, criminal liability for violation of the procedure for the use of airspace is the application to individuals who have committed criminal offenses, criminal penalties that cause these persons negative consequences of moral, material, personal or other nature and legitimately restrict their rights and freedoms imposed by the court are implemented in criminal proceedings and are regulated by the Criminal, Criminal Procedural Code of Ukraine and Criminal Executive Code of Ukraine.

As noted above, the only type of offense which criminal liability can be applied for is a purely criminal offense.

Institution of criminal offense, along with the institute of criminal responsibility, is the second main institute of criminal law. According to P. N. Panchenko, "around these two institutions are built all the provisions of both General and Special parts of criminal law, moreover - all other institutions of criminal law are, in fact, sub-institutions of criminal offense or criminal liability" [30]. A separate issue of determining the essence of a criminal offense was devoted to the study of O. A. Banchuk, V.I. Borysov, D. O. Balobanov, N.L. Berezovska, M.M. Dmytruk, V. M. Kuts, M. P. Karaiev, M. I. Kovalov, N. M. Miroshnychenkщ, V.O. Navrotskyi, I. S. Samoshchenko, F. V. Taranovskyi, V. Ya. Tatsii, V. O. Tuliakov, M. I. Khavroniuk and other scientists.

Let us dwell in more detail on understanding of the concept and types of criminal offenses in general and in the field of ensuring the use of airspace in particular. According to Part 1 of Article 11 of the Criminal Code, "Criminal offense is a socially dangerous criminal act (action or omission) committed by the subject of a criminal offense provided by the Criminal Code of Ukraine" [12].

In 2008, the Decree of the President of Ukraine approved the Concept of Reforming the Criminal Justice of Ukraine, and this can be considered beginning application of *criminal offense* concept of at the legislative level, although the Concept itself uses the *criminally punishable acts* concept [31]. The real legislative consolidation of the *criminal offense* concept in the Criminal Code of Ukraine took place only in 2020 [32].

According to V. M Kuts, "a criminal offense is an illegal act that caused a conflict over the infliction of significant damage by one of its participants to another or the creation of a threat of such damage, the legal consequence of which are statutory measures of a criminal nature." N. M. Miroshnychenko defines a criminal offense as "socially dangerous, illegal, culpable act (action or omission) committed by the subject of a criminal offense" [34] and this concept is as close as possible to legislative definition.

All acts without exception that are recognized as criminal offenses are listed in a single legal act: the Criminal Code of Ukraine. Socially dangerous criminal act, in turn, is recognized as a legally significant, voluntary prohibited by law. In turn, Acts are divided into action or inaction.

The basis for criminal liability in accordance with Art. 2 of the Criminal Code is commission of a socially dangerous act including criminal offense. Elements of criminal offenses include: object, objective side, subjective side of subject.

According to Article 12 of the Criminal Code, criminal offenses are divided into criminal offenses and crimes. Thus, according to G. V. Fedotova, the *criminal offense* concept actually took the place of the generic (daughter) concept to denote only specific concepts of "crime" and "criminal offense" [35, p. 220].

According to M. M. Dmytruk, "internal and external systemic preconditions for the introduction of a criminal offense have been identified." He called danger inconsistency of the act with consequences of criminal prosecution (criminal record and imprisonment or threat of its application) internal preconditions for introduction of a misdemeanor into the criminal legislation of Ukraine; failure of the legislator to take into account the perception in the domestic legal consciousness of the crime as an act associated with serving such a punishment as *imprisonment*; social and legal unconditionally of certain provisions of the Criminal Code of Ukraine. External prerequisites M. M. Dmitruk determined the division of criminal activity in the states of the Romano-Germanic legal system into several types and recommendations of interregional institutions on decriminalization of some mass crimes that do not pose a high degree of public danger. [36, p. 9].

According to the Criminal Code of Ukraine, a criminal offense is an offence (action or inaction) provided by the Criminal Code, for which the main punishment is a fine of not more than three thousand non-taxable minimum incomes or other punishment not related to imprisonment.

Thus, criminal offenses have the least public danger due to the object of encroachment, nature and severity of damage, manner of committing an act, form and degree of guilt, motives and purpose, as well as other objective and subjective features, so under criminal misdemeanor is proposed to understand the type of criminal offense, consisting in socially dangerous behavior that encroaches on objects of criminal law protection that causes or can cause minor damage, and consequences of such damage cannot be irreversible [35, p. 223].

In the field of airspace use, criminal offenses include acts established by Part 1 of Art. 281 of the Criminal Code, indicating "violation of the rules of flight safety by persons who are not employees of air transport, if it created a danger to human life or other serious consequences is punishable by a fine of one thousand to three thousand tax-free minimum incomes or arrest for up to six months, or restriction of liberty for up to three years and Part 1 of Article 282 of the Criminal Code that criminalizes violations of the rules of airspace use, and provides that "violation of the rules of launching missiles, all firing, explosives work or other actions in the airspace, if it posed a threat to air safety shall be punishable by a fine of one thousand to three thousand non-taxable minimum incomes or correctional labor for up to two years, or restriction of liberty for up to five years.

Crime is an act (action or inaction) that has the following mandatory features:

- committed by the crime subject;
- is guilty;
- is socially dangerous;
- is illegal (i.e. relevant act is provided by the current Criminal Code);
- is criminally punishable (i.e. for committing such an act there is a criminal penalty). Missing at least one of these signs indicates crime lack.

Crimes are divided into minor, serious and especially serious. A minor crime is part 4 of Art. 12 of the Criminal Code, an act (action or inaction) for which the main punishment is a penalty of not more than ten thousand non-taxable minimum incomes or imprisonment for a term not exceeding five years. Felony under Part 5 of Art. 12 of the Criminal Code, an act (action or inaction), for which the main punishment is a penalty of not more than twenty-five thousand non-taxable minimum incomes or imprisonment for a term not exceeding ten years. Particularly serious crime is under Part 6 of Art. 12 of the Criminal Code, an act (action or inaction) for which the main punishment is a fine of more than twenty-five thousand non-taxable minimum incomes, imprisonment for more than ten years or life imprisonment.

The crime severity for which the main punishment through penalty and imprisonment is provided at the same time is determined based on the term of punishment through imprisonment provided for relevant crime.

In the field of airspace use, crimes include acts provided for:

1. Part 2 of Art. 281 of the Criminal Code: violation of flight safety rules by persons who are not employees of air transport, if it created a danger to human life or other serious consequences (Part 1 of Article 281 of Criminal Code), if they caused moderate injuries or caused great material damage (punishable by imprisonment for a term of two to five years);
2. Part 3 of Art. 281 of the Criminal Code: violation of the aircraft safety rules by persons who are not employees of air transport, if it created a danger to human life or occurrence of other serious consequences (Part 1 of Article 281 of Criminal Code) or violation of safety rules of aircraft by persons are employees of air transport, if it created a danger to human life or other serious consequences, if they caused moderate injuries to victim or caused great material damage (Part 2 of Article 281 of Criminal Code) and if they caused death or other serious injuries. consequences (punishable by imprisonment for a term of five to twelve years);
3. Part 2 of Art. 282 of the Criminal Code: violation of the rules of launching missiles, conducting all types of firing, blasting or other actions in airspace, if it posed a threat to air safety (Part 1 of Article 282 of the Criminal Code), if they caused victim moderate or serious injuries or caused great material damage (punishable by imprisonment for a term from two to five years);
4. Part 3 of Art. 282 of the Criminal Code: violation of the rules of launching missiles, conducting all types of firing, blasting or other actions in the airspace, if it posed a threat to air safety (Part 1 of Article 282 of the Criminal Code) or violation of the rules of launching missiles, conducting all types of firing, blasting or other actions in airspace, if it posed a threat to air safety, if they caused the victim moderate or serious injuries or caused great material damage (Part 2 of Article 282 of Criminal Code) and if they caused death (shall be punishable by imprisonment for a term of five to twelve years).

The elements of these crimes are enshrined in Section XI: *Criminal offenses against traffic safety and transport operation.* As can be seen from the above, these compositions are material, unambiguous (Article 281 of Criminal Code) or alternative (Article 282 of the Criminal Code), blanket, i.e. they are a violation of certain rules established by other regulations governing performance of flights (Article 281 of the Criminal Code) or regulate the rules of anti-aircraft and other types of shooting; rules for launching various missiles; rules of blasting; rules for the use of helicopters in the construction of complex facilities, etc. (Article 282 of the Criminal Code) (statutes, codes, rules, instructions, etc.).

Objective side of these crimes is characterized by three features:

- action (action or inaction);

– onset or possibility of socially dangerous consequences;
– causal link between the act and consequences.

Subjective side of these crimes can be described as mental attitude of guilty person to the fact of violation of certain rules and these violation consequences. The act may be committed with direct intent or criminal negligence. Guilt in relation to the consequences under Article 281 of the Criminal Code can be negligent (criminal arrogance or negligence). Given the consequences of the crimes provided for in parts 2 and 3 of Article 282 of the Criminal Code, only a negligent form of guilt (criminal arrogance or negligence) is possible for. In general, these crimes are considered as negligent ones.

The subject of crimes is specific one. It can be a pilot or a member of the aircraft crew who are not employees of air transport (Article 281 of Criminal Code) or a person responsible for performing these works or for compliance with safety rules in a particular area of work (Article 282 of Criminal Code).

Criminal punishment for crimes under Articles 281 and 282 of Criminal Code of Ukraine is defined as imprisonment and consists in isolating convict and allocation for a certain period in a closed penitentiary institution. Punishment specified in the sanctions of these articles is non-alternative and is imposed by the court in accordance with general principles, rules and deadlines.

3 Conclusions

In view of the above, we can conclude that to ensure the proper use of Ukrainian airspace it is important to take legal action for violating the rules established in this area, the leading place among which belongs to its two types, administrative and criminal one.

Administrative liability in the field of airspace use of Ukraine is established for offenses that encroach on flight safety and rules of air transport use, i.e. they indirectly encroach on the relations of airspace use. These articles are included in the five articles of Chapter 10: *Administrative Offenses in Transport, Roads and Communications.* A significant problem of administrative liability in the field of airspace use of Ukraine is the improper regulation of liability of legal entities which the Code of Administrative Offenses does not recognize as subjects of administrative offenses. Therefore, with adoption of the new Code of Administrative Offenses, general rules and procedures of administrative liability of legal entities should be defined which without it is impossible to ensure proper protection of these individuals from wrongful prosecution.

Criminal liability in the field of airspace use in Ukraine is established for criminal offenses that encroach on the safety of aircraft or pose a threat to air safety. These components are enshrined in two articles of Section XI: *Criminal offenses against traffic safety and transport operation.* Actions and consequences described in the first parts of these articles form the objective side of criminal offenses, and enshrined in the second and third parts of the articles, objective side of the crimes. Sanctions for criminal offenses in the use of Ukrainian airspace provide for penalties through penalty, correctional labor, arrest or restriction of liberty. Issues are caused by the non-alternative punishment by the means of imprisonment for crimes that is enshrined in the sanctions of mentioned articles of the Criminal Code of Ukraine.

References

1. The Constitution of Ukraine [in Ukrainian]. http://www.prostopravo.com.ua/klub_yuristov/zakonodatelstvo/konstitutsiya
2. Air Code of Ukraine [in Ukrainian]. https://zakon.rada.gov.ua/laws/show/3393-17n214
3. National Airspace System. https://www.cfinotebook.net/notebook/national-airspace-system/national-airspace-system [in English]
4. Majumdar, A., Ochieng, W.Y.: A trend analysis of air traffic occurrences in the UK airspace. J Navig. **56**(2), 211–229 (2003). [in English]
5. Aircraft health monitoring systems and potential effects on airspace safety [in English]. https://www.captechu.edu/blog/aircraft-health-monitoring-systems-and-potential-effects-airspace-safety
6. Chris Hill 2020 Drone and helicopter pilots: act now to improve airspace safety, National Drone Safety Awareness Week, Nov., 16–22 [in English]
7. Zuschlag, M., John, A.: Volpe Violations of Temporary Flight Restrictions and Air Defense Identification Zones: An Analysis of Airspace Violations and Pilot Report Data [in English]
8. Brusakova, O.V.: Administrative and legal principles of state regulation in the field of air transport in Ukraine: a monograph. Kharkiv, 588 p (2020). [in Ukrainian]
9. Analysis of violations of the order of use of the airspace of Ukraine in 2013–2016 [in Ukrainian]. http://nbaai.gov.ua/wp-content/uploads/2020/05/ppvpp_2013-2016.pdf
10. Analysis of the state of flight safety based on the results of the investigation of aviation incidents and incidents with civil aircraft of Ukraine and foreign-registered vessels that occurred in 2020 [in Ukrainian]. http://nbaai.gov.ua/wp-content/uploads/2021/03 /analysis2020full.pdf
11. Code of Ukraine on Administrative Offenses [in Ukrainian]. https://zakon.rada.gov.ua/laws/show/80731-10#Text
12. Criminal Code of Ukraine. https://zakon.rada.gov.ua/laws/show/2341-14#Text
13. Criminal Procedure Code of Ukraine [in Ukrainian]. https://zakon.rada.gov.ua/laws/show/4651-17Text
14. Criminal Enforcement Code of Ukraine [in Ukrainian]. https://zakon.rada.gov.ua/laws/show/1129-15Text
15. Code of Ukraine on Administrative Offenses. Scientific and practical commentary [in Ukrainian]. http://uazakon.ru/ukr/kupap/
16. Commentary to the Criminal Code of Ukraine [in Ukrainian]. http://yurist-online.com/ukr/uslugi/yuristam/kodeks/024/279.php
17. Regulations on the use of airspace of Ukraine: approved by the resolution of the Cabinet of Ministers of Ukraine of 6 December 2017. № 954 [in Ukrainian]. https://zakon.rada.gov.ua/laws/ show/954–2017-%D0%BF#n12
18. Kolpakov, V.K.: Course of administrative law of Ukraine. Kyiv: Yurinkom Inter, 872 (2013). [in Ukrainian]
19. Panasyuk, O.V.: Administrative-tort law of Ukraine: theoretical issues of development and reform: a monograph. Kharkiv: LLC In business. 508 (2020) . [in Ukrainian]
20. Golosnichenko, I.P.: Gaps in the legal regulation of the initiation of proceedings in the case of an administrative offense. Law Ukraine **7**, 73–77 (2003). [in Ukrainian]
21. Administrative responsibility in Ukraine: textbook; edited by A. T. Komzyuk. 2nd edition. Revised and supplemented. Kharkiv: University of Internal Affairs. Affairs, 112 (2000). [in Ukrainian]
22. Code of Ukraine on administrative offenses: Scientific and practical commentary / R.A. Kalyuzhny, A.T. Komzyuk, O.O. Pogribny, etc.; 2nd type. Kyiv: All-Ukrainian. Legal Unity Publishers Association, 655 p (2008). [in Ukrainian]

23. Kolpakov, V.K.: Administrative-tort legal phenomenon: monograph. Kyiv: Yurinkom Inter. 528 p. (2004)
24. Komzyuk, A.T., Salmanova, O.Y.: Current issues of reforming administrative tort law. Kharkiv National University of Internal Affairs: 20 years in the status of national: materials intern. scientific-practical conference (Kharkiv, March 2, 2021)/editors: Sokurenko, V.V., (chairman), Shvets, D.V., (deputy chairman), Bandurka, O.M., and others ; order. Grechenko, V.A.; Ministry of Internal Affairs of Ukraine, Kharkiv. nat. University of Internal Affairs affairs. Kharkiv: KhNUVS, pp. 92–95 (2021). [in Ukrainian]
25. Lukyanets, D. M.: Development of the institute of administrative responsibility: conceptual principles and problems of law enforcement: dis. ... Dr. jurid. Sciences / National Academy of Sciences of Ukraine; Institute of State and Law. V. M. Koretsky. Kyiv. 410 p (2007). [in Ukrainian]
26. Komzyuk, A.T., Lipiy, E.A.: Administrative liability of legal entities: some problems of legal regulation. Law Secur. **2**(65), 34–38 (2017). [in Ukrainian]
27. Panov, M.I.: Criminal liability and its basis. Bull. Natl. Acad. Prosecutor's Off. Ukraine. **4**, 45 (2010). [in Ukrainian]
28. Ponomarenko, Y.A.: General theory of determining the punishment of criminal offenses: a monograph. Kharkiv: Pravo, 2020. 720 p. http://library.nlu.edu.ua/POLN_TEXT/POSIBNIKI_2021/Ponomarenko-2020.pdf (2020). [in Ukrainian]
29. Navrotsky, V.O.: Criminal law as a branch of public law. Fundamental problems of criminal responsibility: materials of sciences. Polylogue. Kharkiv, 7 September 2018 Kharkiv: Law (2018). [in Ukrainian]
30. Panchenko, P.N.: Legal constructions in the system of criminal legislation and their importance for improving this system and teaching criminal law disciplines in law schools. Legal technique. No. 7 (part 2) 2013. [in Ukrainian]
31. The concept of reforming the criminal justice of Ukraine, approved by the Decree of the President of Ukraine of 8 October 2008 № 311/2008 [in Ukrainian]. https://zakon.rada.gov.ua/laws/show/311/2008#Text
32. On amendments to some legislative acts of Ukraine to simplify the pre-trial investigation of certain categories of criminal offenses: Law of Ukraine of 20.12.2019 № 2617-VIII [in Ukrainian]. https://zakon.rada.gov.ua/laws/show/2617-19?lang=ruText
33. Kuts, V.M.: Problems of criminal liability: a textbook (2013) [in Ukrainian]. https://textbook.com.ua/pravo/1474612665
34. Miroshnichenko, N.M.: On the definition of criminal offense. Law Soc. № Part 6 2 / 2015, 53–157 (2015). http://pravoisuspilstvo.org.ua/archive/2015/6_2015/ part_2/29.pdf [in Ukrainian]
35. Fedotova, G.V.: Criminal offense as a type of criminal offense. Sci. Law Enforcement **4** (38), 219–225 (2017) [in Ukrainian]. http://naukaipravoohorona.com/journal/ukr/2017_4/32.pdf
36. Dmitruk, M.M.: The category of misdemeanor in criminal law doctrine. Abstract of the dissertation for the degree of Candidate of Law in the specialty 12.00.08 - criminal law and criminology; criminal executive law. National University Odessa Law Academy. Odessa (2012) [in Ukrainian]. https://core.ac.uk/download/pdf/50593791.pdf

Offenses Related to Financial and Credit Operations in Aerospace Industry: Causes of Occurrence and Consequences

Iryna Petrova[1(✉)] and Katerina Hovorova[2]

[1] R&D and Publishing Laboratory, National Scientific Center «Hon. Prof. M. S. Bokarius Forensic Science Institute», Kharkiv, Ukraine

[2] National Scientific Center «Hon. Prof. M. S. Bokarius Forensic Science Institute», Kharkiv, Ukraine

Abstract. Authors cover the causes and consequences of offenses related to financial and credit operations in aerospace industry. It is established that offenses related to financial and credit operations in aerospace industry are unprofitable, destructive, illegal, punishable acts committed by violating the order of financial and credit operations and encroach on economic interests not only of aerospace enterprises but state and society as a whole. Mechanism of committing these illegal acts involves the use of various methods, and often their complexes consisting in unreasonable granting (receipt) of loans, breach of loan agreement, misuse, late and incomplete loan repayment, other breaches of obligations arising as a result concluding transactions for financial and credit operations in aerospace industry.

It is argued that knowledge of offense causes related to financial and credit operations in aerospace industry will prevent their commission, as well as timely identify and establish all circumstances of their commission. It is proved that offense causes related to financial and credit operations in aerospace industry include both moral and psychological offender qualities and conditions affecting their formation, level of legal regulation of financial and credit operations in particular. It is noted that one of the causes of offenses related to financial and credit operations in aerospace industry is the anti-social attitude of offenders characterized by such moral and psychological traits as selfishness, desire for quick and easy profit, greed, often envy, adventurism irresponsibility, disregard for generally accepted rules of conduct, including laws, etc. Consequences of offenses related to financial and credit operations in aerospace industry are identified and characterized as an element of the objective offense side and as a result of illegal activities in the field of financial and credit operations in general.

Keywords: Financial and credit operations · Aerospace industry · Crimes

1 Introduction

Offenses related to financial and credit transactions are characterized by special public danger, complex mechanism of commission, as well as difficulty of their detection and

M. Nechyporuk et al. (Eds.): ICTM 2021, LNNS 367, pp. 664–689, 2022.
https://doi.org/10.1007/978-3-030-94259-5_55

pre-trial investigation or resolution in court. All this is due primarily to specifics which field these offenses are committed in, especially when it comes to aerospace industry. Therefore, in order to ensure effectiveness of establishment and resolution of such illegal acts, in particular, its organization and possibility of using specific expertise, it is necessary first to clarify the nature of offenses related to financial and credit transactions; determinants and consequences of their commission; features of ways of committing these acts and their typical traces. In this case, it will be possible to determine the main activities of law enforcement agencies and the court to establish offenses related to financial and credit operations in aerospace industry to identify methods and means of gathering evidence including through the use of forensic science possibilities including forensic economic documents financial and credit operations [1, p. 56].

Therefore, we consider it necessary to outline theoretical and general methodological provisions of forensic economic analysis of documents of financial and credit operations on the basis of determining theoretical and methodological foundations of research on offenses related to financial and credit transactions, namely to clarify the causes and consequences of such offenses in aerospace industry, as well as issues of documentary validity of financial and credit transactions and their reflection in accounting of enterprises in this industry as one of distinguishing features of the mechanism of studied category offenses.

As rightly noted by M. O. Dunas, in order to develop and effectively implement measures to prevent crime in our society ensuring its reduction and in historical perspective and the elimination of it as a social phenomenon, it is necessary to know clearly what generates it and to feed. This knowledge is achieved by causally explaining existence of crime and analyzing its dynamics [2, p. 92]. It is justified and logical to find out the causes of offenses related to financial and credit operations in aerospace industry. In addition, it is impossible to determine the causes of offenses related to financial and credit transactions without clarifying nature and content of the latter.

2 Causes of Offenses Related to Financial and Credit Operations in Aerospace Industry

First of all, it should be noted that depending on nature and degree of public danger of illegal acts related to financial and credit operations in aerospace industry, distinction should be made between administrative, economic, criminal and civil offenses committed through financial and credit operations. Therefore, clarifying nature of offenses related to financial and credit operations in aerospace industry, it is necessary to analyze not only scientific approaches of legal scholars specializing in various areas of law, namely: administrative, economic, criminal and civil, but and the provisions of administrative, commercial, criminal and civil law.

The greatest attention is paid to the issues of committing offenses related to financial and credit operations and accordingly, to the issues of their qualification, definition, detection and pre-trial investigation in research papers on criminological and criminal law sciences. This is due to the fact that according to the statistics of crimes in the financial and credit system, they account for 1.5% of all crimes which 15% belong to economic offenses [3]. In addition, as rightly pointed out by V. V. Kornienko and O. V. Pchelin, the banking sector accounts for more than 75% of all losses associated with economic crime. Such losses are most often related to technologies of illicit enrichment in such areas as: tax evasion; embezzlement of bank customer funds; legalization of illegally obtained income, etc. [4].

It should be noted that authors attributed lending to common banking operations that are the most vulnerable in terms of criminology. They stressed that entrepreneurs and citizens in financial market have recently been growing interest in loans. This is most likely due to the fact that banks are too risky to expand their loan portfolios willing to attract new customers while neglecting researches on solvency of potential customers. This situation has led to mass non-payment of debts, abuses and fraud in the defined area [4, p.4; 5, p. 370].

In addition, in criminal law cycle sciences to denote criminal offenses which mechanism is associated with implementation of financial and credit transactions, use various terms among which should be distinguished such as:

- crimes (offenses) in banking sector;
- crimes (offenses) in financial and credit sector;
- crimes (offenses) committed through credit and financial transactions;
- crimes (offenses) in the field of banking;
- crimes (offenses) committed with use of banking services and technology;
- credit and banking crimes (offenses);
- crimes (offenses) in the field of banking;
- crimes (offenses) committed in the field of economic relations of credit and financial activities, etc.

As rightly emphasized by O. V. Gerasimov, such terminological uncertainty does not contribute to the effective use of these terms either in research or in practice [6, p. 37].

Regarding the content of criminal offenses committed through financial and credit transactions, there is also no unanimity among scientists. In particular, T. Poshivaniuk notes that "crimes in the field of banking are selfish encroachments on financial bank resources other credit and financial institutions committed with the use of certain credit and banking operations by authorized business entities or other persons." In this case, banks can act as objects of criminal encroachment and tools to achieve a criminal goal [7, p. 69; 3; 8, p. 287].

S. S. Chernyavskyi suggests that crimes in the field of banking should be understood as provided by Criminal Code of Ukraine socially dangerous acts that encroach on funds (property) of banks or are committed using banking operations by authorized officials of financial institutions and other business entities or other persons who due to certain circumstances gained access to their conducting [9, p. 195]. According to O. V. Gerasimov, crimes in the banking sector should be understood as those criminal offenses that encroach on the property, property rights of banks or their customers, established order of operation of banks and the banking system, banking secrecy or insider information of banks [10, p. 7].

I. V. Masliy defining crime in the financial and economic field, identifies such features of investigated crimes as their homogeneity; illegality; social danger; selfishness; causing significant material damage as a result of encroachments on the established order of management of economic processes, economic development of the state, state of financial security [11, p. 558]. A. M. Klochko, V. V. Nezhevelo under crimes in the field of banking understand a set of intentional socially dangerous acts aimed at seizing financial resources of banking institutions or their customers that can be committed while banking operations or with their use by the subjects of crimes in the banking field [12, p. 203].

Along with defining concept of the studied category of offenses, some scholars focus on its correlation with other related offenses. As an example, the position of O. V. Gerasimov indicating that comparative characteristics of concepts of crimes in the banking sector and crimes in the financial and economic field are correlated as part and whole is noting that crimes in banking sector is an integral part of the concept of crimes in the financial and economic field. This is due to the fact that crime in the financial and economic field covers a larger range of crimes than the first concept. Under current law, financial institutions include banks, credit unions, pawnshops, leasing companies, trust companies, insurance ones, funded pension institutions, investment funds and companies and other legal entities whose sole activity is provision of financial services and in cases where directly defined by law, other services (transactions) related to provision of financial services. The same applies to relationship between the concepts of crimes in banking sector and crimes in the credit and financial field. In addition to the banks, subjects of credit and financial field include credit unions that in accordance with current legislation also provide lending and financial services through joint cash contributions of credit union members [10, p. 39–40].

A. M. Klochko, V. V. Nezhevelo emphasize that crimes in the field of banking are limited to crimes committed in banking system. In their opinion, the concept of crimes in the banking system is broader than crimes in the field of banking, as it covers various types of socially dangerous acts that are not due to specifics of banking [12, p. 203].

Therefore, based on the above, we can conclude that criminal offenses related to financial and credit operations are primarily socially dangerous, illegal, criminally punishable acts committed in the banking system by violating the terms and conditions of financial and credit operations as borrowers and creditors. At the same time, offenses in the field of financial and credit operations in aerospace industry are widespread and less socially dangerous. In particular, depending on nature and degree of social harm, it is also necessary to distinguish between administrative, economic and civil offenses related to financial and credit operations in aerospace industry.

It should be noted that in scientific sources, issues related to the definition of such types of offenses have not been investigated. In general, most scholars-administrators pay attention to general provisions on administrative offenses, as well as peculiarities of qualification of administrative offenses in the field of finance and business and the issue of bringing to administrative responsibility [13, p. 21–30; 14, p. 65–67].

In particular, the studied category of offenses is a kind of administrative offenses in the field of banking that according to M. Chernov can be committed by banking structures based on performance of their powers and functions and other financial institutions [15, p. 104]. Thus the scientist correctly states that ways of administrative offenses in the field of banking activity differ in that:

- firstly, most of them are impossible without appropriate prior training, study and assessment of situation in the banking sector, creation of conditions that will contribute to the offense, adoption of security measures;
- secondly, commission of these offenses is associated with a violation of the procedure for conducting banking activities or specific banking operations and other agreements established by banking legislation and NBU regulations.

The distinctive and specific feature of this type of administrative offenses is that they are committed under the guise of legal banking operations, as a result of illegal nature of violator actions is latent [15, p. 106].

In addition to administrative and criminal offenses related to financial and credit transactions, violations of economic and civil law in this area are also common. In particular, economic offenses related to financial and credit operations in aerospace industry are violations of terms and conditions of financial activities of economic entities including financial intermediation in through banking operations. Civil offenses in this area are non-performance or performance in violation of specified conditions in the content of obligations arising from conclusion of transactions for financial and credit operations in aerospace industry.

Thus, analyzing scientific literature and legislation that determining specifics of financial activities and procedure for bringing perpetrators to justice for violations in this area, we can conclude that offenses related to financial and credit operations in aerospace industry are unprofitable, illegal, punishable acts committed by violating the order of financial and credit operations in aerospace industry and encroach not only on this industry stability but on economic interests of the state and society as a whole.

The mechanism of committing these illegal acts involves the use of various methods and often their complexes consisting in unreasonable granting (obtaining) loans, breach of loan agreement, misuse, late or incomplete loan repayment, other breaches of obligations arising as a result concluding transactions for financial and credit operations in aerospace industry.

In order to prevent commission of offenses related to financial and credit operations in aerospace industry, as well as to identify them in a timely manner and establish all circumstances of their commission, it is necessary to know the reasons that give rise to these illegal acts. Moreover, the industry affected by investigated offenses is strategic for development and existence of any state. This underlines once again importance of ensuring and organizing effective activities to identify offenses related to financial and credit operations in aerospace industry. That is why the state pays special attention to proper regulation, as well as control over the banking system, namely: a set of different types of national banks and credit institutions operating under general monetary mechanism [7, p. 68].

This is exacerbated by the fact that banking system plays an important role in the economic development of both the industry and the state as a whole, as it operates to ensure the stability of pricing policy, support the national currency, organization and functioning of payment system. That is why, as noted by A. M. Klochko, one of the important tasks for effective work management of credit institutions and banks is the need to their protecting from illegal encroachments [7, p. 68]. One of the areas of such security is to identify the determinants of offenses related to financial and credit operations in the aerospace industry, to identify ways to their eliminating, as well as developing effective means of detecting, investigating or resolving disputes related to these violations in court that include improving theoretical and general methodological provisions of forensic economic analysis of documents of financial and credit transactions.

Offenses related to financial and credit operations in aerospace industry are a type of economic offenses that is why the causes of economic crime and other offenses committed in this area, in most cases will be inherent in the investigated category of illegal acts. However, before identifying the causes of offenses related to financial and credit transactions, it is necessary to clarify nature and meaning of the cause term in general and in relation to the offenses.

In etymological sense the term: cause is used to denote a phenomenon that causes or generates another phenomenon; grounds, reason for any actions, deeds; motivation, reasoning, motives that cause any actions; contingencies, difficult circumstances; etc. [16, p. 1140]. In the philosophical understanding of the world, as well as in science, causality plays an important role. It is nothing but a certain type of correlation between objects, in which some (causes) give rise to others (consequence). Sometimes causality doctrine is called determinism. If we project this doctrine on social life, we can see that basis of determination are motives [17, p. 279–282, 585].

As V. L. Petrushenko notes, in a common sense the cause is interpreted as something that causes a phenomenon or forms a result. In this sense, the cause is often defined as one that precedes something else but the rigid connection of the cause with time often does not justify itself or is the basis for gossip emergence, prejudice. Therefore, gradually the notion of cause began to be associated with the transfer of action from one phenomenon (objects) to another. In this understanding, the cause should be considered together with the conditions, circumstances, factors of certain phenomena and processes. One of acute problems arising against the cause cognition is the basis for actions that manifest free will, because if everything has a cause, then arbitrary actions can be considered as a misunderstanding of phenomena arising from incomplete knowledge of all their circumstances [18, p. 168].

The above indicates that cause term is used to denote those circumstances and conditions that are the basis of a particular act, phenomenon, event; generate something; act as incentive committing something. That is why the cause is considered in close and inseparable connection with the consequences, because the cause stimulates commission of certain actions that lead to the corresponding consequences. This scheme is fully acceptable in relation to offenses, namely their causes, consequences and their correlation. If we speak about the causes of offenses, it should be noted that their study is devoted to the research of scientists in criminology, psychology, legal psychology, criminalistics and so on. It should be emphasized that in these research papers the main attention is paid to determination of criminal offenses.

In particular, T. V. Kovaleva, L. P. Kovriga, B. V. Korobeynikov, G. M. Minkovskyi, I. V. Pisarchuk and others indicate that cause is called the phenomena (system) that generate, produce other phenomena (systems) – consequences. In other words, there is a genetic, producing connection between cause and effect [19, p. 496; 20, p. 547], because the causes are nothing but those social phenomena that in a binary correlation generate and reproduce criminality and crimes as a natural consequence [21, p. 116].

In this regard, some scientists (for example, I. V. Pisarchuk, O. V. Pchelina) consider the reasons are subjective (in other words, are internal) factors about offender identity forming his criminal motivation by influencing his psychological qualities and they define the conditions as external circumstances creating a favorable situation, encourage the crime commission [22, p. 142; 20, p. 548]. In other words, crime causes are the factors determining and characterizing person views, his motivation, form his anti-social attitude [23, p. 36].

Note that anti-social attitude as the person views is formed in his mind for a long time, whereupon it can manifest itself in a specific offense. The formation source of particular anti-social attitude should be recognized factors of:

- biological nature associated with inherited psychophysiological person characteristics, his temperament, etc.;
- social nature depending on specific life circumstances, environment and upbringing.

Namely, under influence of such factors, a person can develop selfish views, individualistic habits and skills [20, p. 549; 22, p. 142].

V. O. Konovalova points out that main reason for illegal actions is expressed in the anti-social person attitude (behavior model). According to her, any activity is generated and at the same time depends on motives of anti-social behavior, individual needs and aspirations, his attitude to accepted social norms in his environment and society itself. According to the scientist, the reason for any anti-social behavior is contradiction between individual desires, needs and the real, specific situation when he should come to a decision. At the same time, the level of legal awareness of population plays a special role in choosing behavior and formation of anti-social attitude accordingly. The latter is nothing but necessary prerequisite for legal culture [24, p. 41–45].

V. V Pankratov emphasizes the psychological component of offense causes, he rightly notes that problem of the causes of misdemeanors is a problem of social and psychological conditionality of certain behavior [25, p. 21]. This position is supported by O. M. Dzhuzha, V. V. Vasilevich, O. G. Kolb and other criminologists, who suggest that crime causes should be understood as "socio-psychological circumstances that directly generate and reproduce crime and crimes as a natural consequence" [26, p. 45]. In other words, offense causes are those psychological properties of the individual that are formed under certain circumstances and determine its priorities, values and motives, incentives for its activities (inaction).

Along with this, speaking about offense causes in general and related to financial and credit operations in aerospace industry in particular, it is appropriate to study not only psychological offender traits but the conditions which such properties were formed under. This means these phenomena are not isolated, do not occur and cannot exist separately. They are closely related because they complement each other and in some cases create a favorable environment for their formation. Therefore, it is not for nothing that in philosophical and legal literature, scholars consider the causes and conditions of actions, namely, illegal, inextricably linked, as a single phenomenon.

In particular, L. P. Skalozub, V. I. Vasilinchuk analyzing scientific approaches to determining the crime causes found that common among scientists is the position that external circumstances in themselves cannot give rise to crime and therefore cannot be its reason. They can only form the cause or contribute to commission of crimes. This opinion is valid for the causes of a particular crime, because a crime cannot be committed without the will of the person himself. Therefore, it can be reasonably assumed that cause of criminal behavior is formed not simultaneously and not by one group of conditions but by their whole complex and, as a rule, for a long time. In addition, along with the causes of illegal acts are caused by conditions that are phenomena that can be necessary for this event occurrence but do not directly cause unlike the first [27, p. 307].

For example, O. V. Shevchenko investigating the causes and corruption crime conditions notes that the latter are a complex set of criminogenic circumstances for varying degrees in interaction with the offender identity and other factors determine the commission of corruption crimes [28, p. 131]. O. M. Litvinov suggested that causes and conditions contributing to crime commission (an act prohibited by criminal law), to understand the set of necessary, legally significant phenomena that naturally and inevitably lead to the crime commission [29, p. 324].

L. P. Skalozub and V. I. Vasylynchuk emphasize that it is anti-social attitude in the mind of a person was formed under influence of various negative phenomena and processes that becomes the cause of offenses. Such an attitude makes her want to commit an offense or can lead to its negligence. Emphasis is placed on the fact that phenomena and processes will be the only reasons for formation of negative personality traits among which should be singled out anti-social views, skills and habits that are responsible for anti-social attitude in his mind. Under such conditions, a certain situation will motivate a person to act in accordance with the previously formed in his mind his own attitude. Therefore, finding out the cause of anti-social attitudes is of great practical importance. This attitude, formed in the mind of a person is nothing more than possibility of committing an offense and for implementation of criminal intent requires certain conditions under which possibility can become reality [27, p. 309].

Along with the reasons for the offense, conditions that contribute to its commission are inextricably linked. In particular, they are phenomena, processes that are necessary for the causes of this type of offense, activate these causes. These include shortcomings: in regulatory legislation; in state institution activities, economic organizations, enterprises, and individual citizens. They are the ones who objectively contribute to offense commission or occurrence of a criminal result [30]. That is why it is necessary to distinguish between the broad and narrow meaning of the concept of crime cause. In the first case, we are speaking about a set of interrelated mandatory conditions and reasons in narrow sense [27, p. 309], i.e. a set of anti-social person attitude and conditions contributed to its formation. In our opinion, this interpretation most fully reveals content of factors that cause criminal offenses and determine their impact on the crime mechanism and ways for their elimination.

In context of the above, it is true and noteworthy that causes and conditions of a particular offense are a multifaceted concept that is repeatedly pointed out by scientists, highlighting various aspects of this category. In particular, V. M. Kudryavtsev uses the term circumstances that contribute to the crime commission to denote the outlined category. According to the scientist, the latter include: specific external circumstances that prompted a person to commit a crime; conditions of moral personality formation; circumstances that contribute to achievement of a criminal result [31, p. 133–134].

G. M. Minkovskyi identifies the following aspects of determining causes and conditions of the crime:

- "environment that forms a disharmony or deformation of the needs, interests, values of a particular person which become the basis of criminogenic motivation;
- criminogenic motivation itself;
- situations when a person is in formation, activity and directly in committing a crime that contribute to emergence and implementation of criminogenic motivation in behavior (conditions that contribute to a particular crime);
- psychophysiological and psychological features of a person that increase his sensitivity to criminogenic influences from the outside and stimulating their transformation into an internal position" [32, p. 136].

Therefore, based on the above, we believe that causes of offenses related to financial and credit transactions in aerospace industry, include both moral and psychological offender qualities and conditions that affect their formation:

- state of Ukrainian economy in general and the banking system in particular;
- level of regulatory regulation of the procedure for financial and credit transactions;
- procedure for training and selection of employees of financial institutions, etc.

Note that it is necessary to begin to find out what circumstances are the causes of investigated offenses by establishing the external and internal factors that cause and contribute to commission of illegal acts. After all, as O. V. Zemlyanska rightly points out that study of the consciousness of the individual is one of the prerequisites for establishing the crime determinants, in particular, we are speaking about the legal beliefs of the individual. The scientist also emphasizes that crisis in the social and economic field of Ukraine, the weakening of the role of the state in managing socio-political processes, legal nihilism, destruction of old moral principles and the lack of clear new guidelines and other contributed to crime and a number of phenomena. Thus, under influence of these factors, the idea of everyday consciousness of the individual is formed, perception of others and behavior in situations of interaction is determined. In particular, the scientist found that "the causes of crime are a complex category including such three main blocks as:

- internal or subjective reasons:

  - individual features: greed, selfishness (30.8%); aggression, malice, contempt (23.1%); envy (15.4%), others);
  - defects of morality and self-control: lack of moral principles; desire to achieve anything for self-affirmation, for glory, for enjoyment and for pleasure; unwillingness to restrain – 20.9%;
  - mental and somatic diseases or disorders, psychological traumas (16.5%);
  - defects of cognitive area: lack of fear of punishment and misunderstanding of its inevitability, misunderstanding of the essence of crimes – 8.8%;
  - drug dependence, alcoholism (4.4%);

- external ones:

  - dissatisfaction with life due to social injustice, low living standards in the country (48.6%);
  - destructive environment: bad family, school, difficult childhood – 18.9%;
  - media influence: information about crime routine disseminated by television and Internet; positive criminal image –16.2%;
  - bad legislation, impunity for crimes (8.2%);
  - environment that does not allow to develop: education and recreation inaccessibility– 8.1%;

- reasons justifying the crime:

  - no other way out, extreme necessity, blackmail (50.0%);
  - under influence of others: religion, hypnosis, bigotry– 27.8%;
  - by inadvertence (22.2%) [33, p. 129–131].

We note that offenses related to financial and credit operations in aerospace industry are usually committed by persons who receive loans or persons working in financial institutions. These illegal acts can be committed simultaneously by both of the above groups of persons by conspiracy. Accordingly, speaking about the reasons for such actions, we should first of all speak about the anti-social attitude of these groups. They have such features as: selfishness; desire for quick and easy profit; avarice; often envy; adventurism; irresponsibility; disregard for generally accepted behavior rules; including laws, etc. In general, it can be argued that these moral and socio-psychological traits are mostly characteristic of a large number of economic offenses.

In particular, the position of O. V. Pchelina is considered appropriate in this context, that the psychological and social qualities of persons who commit economic offenses are inextricably linked with their mentality uniting set of mental, intellectual, ideological, religious, aesthetic and other thinking features of Ukrainian population. She argues that offense root cause including economic should be determined by mentality of a person who under influence of negative external factors and conditions (which the author attributes the growth of property stratification, deteriorating crime situation in a crisis in the economy, decline former industrial relations, etc.) forms in a person other negative psychological and social qualities: selfishness, greed, desire for a comfortable life, negative or contemptuous attitude to law, contempt for interests of others. These formed negative person qualities are defined by the scientist by the direct reasons of a specific offense [22, p. 143].

A similar position is taken by contributors of the Criminal Prevention of Economic Crimes guidance manual that states that anti-social views of the criminal are formed by the psychological atmosphere in the state environment (competition, desire for power and enrichment). At the same time, main properties that motivate criminals to commit criminal offenses, scientists include:

- "progressing material needs, the desire to be a wealthy person who has the money not only to meet every day needs but to create a certain "image" belonging to the highest circles of society;
- belief that money is the highest value in life and constant need to "make money" becomes the main and sometimes the only goal;
- legal and moral nihilism, i.e. disregard for laws and norms of morality and the desire to achieve a certain goal by any illegal and sometimes openly criminal methods, in particular;
- neglect of the interests of society as a whole and individuals, contempt for people who are below their financial capabilities;
- propensity to take risks in official activities and in life in general, growing into adventurism (disregard in decision-making of clearly existing dangerous circumstances);
- dynamics of behavior depending on life circumstances (possibility of their change to the opposite) [34, p. 145–148].

The mentioned features of violator's personality that are one of reasons for committing offenses related to financial and credit operations in aerospace industry are formed, modified and established under influence and action of external factors. In other words, we are speaking about formation conditions of anti-social attitudes which are conditions that contribute to the offense commission and therefore are one of the constituent causes of such illegal acts in their broadest sense. In particular, in professional readings literature there are proposals to group conditions that contribute to the commission of crimes by identifying social (economic, political, social relations and ideological factors), legal factors and shortcomings of organizational activities. Also common in scientific sources is generalization and systematization of causes and conditions of occurrence and spread of offenses by classifying them into socio-economic, political, organizational and institutional (bureaucratic), legal, moral and psychological factors that have quite conditional boundaries [27, p. 310–311].

In the context of conditions for committing offenses related to financial and credit operations in aerospace industry, it is necessary to analyze the conditions that are characteristic of economic offenses including crimes in this area. As an example, O. V. Pchelina notes that "specific conditions characteristic of economic crimes are:

- lack of proper control by government agencies (55%);
- victim (reckless) behavior of victims (10.8%);
- shortcomings in management of economic security service of enterprises and banks (37.8%);
- insufficiently thorough verification of candidates for employment (41.4%);
- shortcomings of regulatory regulation of business activities of this type (6.3%);
- improper performance of their duties by officials (0.9%);
- inaccurate reflection of information in documentation of financial statements, improper execution of contractual relations (0.9%)" [22, p. 144–145].

From the above it can be seen that main conditions for committing economic offenses in general and related to financial and credit operations in aerospace industry, in particular are primarily:

- improper functioning of public authorities in determining the implementation mechanism of financial and credit operations and control over;
- imperfect regulatory and legal regulation of the activities of financial institutions in terms of financial and credit operations and control over, as well as law enforcement, regulatory and judicial authorities in terms of prevention, termination and counteraction to offenses in this area;
- low level of economic state development (high unemployment rate, low wages, low level of liquidity in banking sector, financial instability, etc.);
- poor training and recruitment by financial institutions, etc.

It is considered appropriate to dwell in more detail on some of the defined conditions that are part of offense causes related to financial and credit operations in aerospace industry.

First of all, the state of crime and other offenses in the field of financial and credit operations in aerospace industry, as well as the mechanism of such illegal acts are directly affected by risks in financial sector which together determine the state of financial stability in Ukraine. In particular, we are speaking about conditions for banks, fullness of the funding base, level of lending to individuals, profitability, capitalization and high liquidity of banking sector. In other words, the modern life of the average citizen, functioning of enterprises, institutions and organizations, any individuals or legal entities and general economic state system is ensured by the use of banking technologies and is associated with the stable functioning of banking system. Therefore, stability of the banking system is characterized by reliability, balance and proportional functioning of its structural elements, ability to maintain a stable balance and reliability for a long time [12, p. 204].

Note that level of economic development of any state is no exception and Ukraine, directly depends on the state of its banking system. As a confirmation of this opinion, the following axiomatic statement should be made: banking system is one of the main areas of the national economy of any developed country. Creating effective banking infrastructure is one of the main tasks of economic stability not only in countries with developed economies but in countries that are at the stage of formation and development of market economy [35, p. 10].

Banking system is represented by a branched but internally organized, interconnected, united common goal [36], a set of banks, banking institutions, financial institutions operating within a single financial and credit mechanism headed by central bank and subordinated to. [37], through accumulation of funds and the provision of various services to customers to accept deposits and loans [38]. Therefore, in a modern democracy it is difficult to imagine development of a market economy without a banking system.

Complex process of market transformation that has taken place in recent years in our country exacerbates efficiency of the banking system in Ukraine. Given that effective operation of banking institutions significantly increases competitiveness of enterprises in aerospace industry and the economy as a whole, it also contributes to achieving positive results in its reform [39, p. 200]. In particular, as contributors of the Banking Operations textbook edited by O. V. Dzyublyuk quite rightly points out in this regard, "the purpose of effective operation of the banking system is to ensure possibility of effective monetary regulation of the economy, credit and settlement services of economic turnover, as well as stable activities of banking institutions" [40, p. 24].

According to the above, stable functioning of banking system is an integral condition for the viability of national economy and at the same time the whole state [41, p. 19]. Moreover, D. O. Getmantsev, O. P. Orlyuk, N. G. Shuklina rightly emphasize that "banking system is a component of credit system of the state participating in performance of the main functions of the financial system, in particular, ways to provide:

- national currency stability;
- ways to move financial resources over time, across borders and between industries, etc.;
- mechanism of pooling financial resources and their distribution between individual entities;
- smooth operation of payment systems, in particular by improving clearing methods and settlements that facilitate trade;
- market saturation with price information that allows to coordinate decentralized decision-making in certain economy sectors;
- by developing and providing ways for risk management" [42; 43, p. 20].

Unfortunately, despite importance of economic development state in general and banking system in particular to ensure population welfare of Ukraine, these areas of life in our country are characterized by unsatisfactory condition, insufficient, openly low, development level that is one of prerequisites and reasons committing offenses related to the financial and credit field of aerospace industry.

Currently, Ukrainian economy state today can be characterized only as unsatisfactory, as the vast majority of its industries demonstrate negative indicators. In particular, hryvnia has depreciated, raising prices for gasoline, food and basic necessities. This affected financial condition of population and led to an increase in demand for consumer lending. However, management of banks did not realize main crisis lessons. Although borrower creditworthiness is assessed at a higher level, however, it is insufficient to avoid mass loan non-repayment, as in previous years [44, p. 64–65].

At the same time according to the financial stability reporting for the first half of 2019, due to high operational efficiency and low cost of credit risk in the first quarter, sector rate of return (ROE) exceeded 30%. Profitability risks are now low. One of systemic risk sources is high dollarization of bank balance sheets and high share of non-performing loans on bank balance sheets (over 50%) [45, p. 4–5].

The need to pay attention to peculiarities of banks forming preconditions for committing crimes (and other offenses) in banking including in the field of bank lending, highlighted by A. Tolstosheeva, who highlights:

- lack of favorable conditions for effective counteraction to illegal encroachment on property and activity of banks;
- insufficiency or lack of preventive security measures for banks aimed at detecting, localizing, preventing offenses committed by bank employees;
- unfair remuneration of bank employees;

- lack of promotion of safe behavior in banks, appropriate ideology that would form the appropriate mentality of employees regarding their attitude to their own safety, bank security and others [48, p. 107].

Examining the issue of banking crime and the crime source in the field of credit activities of banks, it should be noted that recently there has been a tendency to worsen the situation with abuse of office and dishonest behavior of employees. This circumstance is one of the factors causing commission of offenses related to financial and credit transactions. Experts explain this situation by staff professionalism level, because with its increase, the level of threat from them to banks increases. In other words, threats from their own employees in the lending activities of banks exist constantly and the importance of solving this problem remains urgent, as the professional level of skills of employees is constantly growing. It should be emphasized that implementation of such threats can be carried out only under conditions conducive to criminal behavior of employees. Possibility of committing offenses under adverse conditions is significantly reduced and only one quarter of employees can look for opportunities to commit offenses under unfavorable conditions [46; 47; 48, p. 109].

Unsatisfactory condition of the financial sector of Ukraine, as well as its impact on the level of offenses related to financial and credit operations, is indicated by the analysis of judicial practice in the cases of the studied category. In particular, A. Vronska, Judge of the Supreme Court emphasizes that in recent years' liquidity in banking sector has deteriorated, profits of enterprises including aerospace industry and population incomes have significantly decreased that has led to a significant increase in litigation and bank deposit agreements [49].

In addition, the analysis of case law in civil proceedings arising from credit relations, "allows in addition to the above to identify such negative factors that have led to a large number of violations of credit agreements, bank deposit agreements, securing contractual obligations, recovery on mortgaged property, recognition of loan agreements, sureties, pledges, mortgages, etc., as:

- unsatisfactory state of affairs with liquidity;
- rapid growth of foreign exchange rates;
- significant reduction in household income;
- impossibility of timely fulfillment of contractual obligations to repay the loan and the deposit due to changes in financial condition of the parties to such legal relations;
- reluctance of one of the parties on various grounds to return the received under the contract (subjective factor)" [50].

Thus, we conclude that offenses related to financial and credit operations in aerospace industry as human behavior manifestation are committed according to the principle of determinism which means that they are caused by certain factors and lead to specific consequences. To identify and characterize the causes of these illegal acts should use the cause interpretation in a broad sense, as a set of anti-social person attitude and conditions that contributed to its formation. The main causes of offenses related to financial and credit operations in aerospace industry are such moral and socio-psychological offender traits forming his anti-social attitude, such as:

- selfishness, desire for quick and easy profit, greed, often envy, adventurism, irresponsibility, disregard for generally accepted rules of conduct, including laws, etc.;
- improper functioning of public authorities in determining implementation mechanism of financial and credit operations and control over;
- imperfect regulatory and legal regulation of the activities of financial institutions in terms of financial and credit operations and control over, as well as law enforcement, regulatory and judicial authorities in terms of prevention, termination and counteraction to offenses in this area;
- low level of economic state development (high unemployment rate, low wages, low level of liquidity in banking sector, financial instability, etc.);
- poor training and recruitment by financial institutions; etc.

3 Causes of Offenses Related to Financial and Credit Operations in Aerospace Industry

As already mentioned, any actions are generated by certain factors (causes) and lead to a certain result (consequences). Therefore, in order to better understand the mechanism of committing offenses related to financial and credit operations in aerospace industry, as well as ways to detect, investigate and prosecute them, it is necessary to find out consequences of these illegal acts and the consequences they lead to.

Coverage of the outlined issues should begin with clarifying the etymological interpretation of the consequence term as well as the meaning of this concept in relation to offenses. Thus, the consequence term in the modern Ukrainian language is used in the following meanings:

- coming out follows from something; result;
- for indicating the reason that gave the results named in the sentence [16, p. 736].

In other words, we can see that concept meaning of consequence is expressed by such a word as "result". In this regard, we consider it necessary to consider interpretation of the latter. "The result in etymology is understood as the final, final result of any occupation, activity, development, etc.; consequence of any action, some phenomenon, etc.; the amount received after certain mathematical operations; indicator of something" [16, p. 1209].

In view of the above, it can be argued that concepts of consequence and result are synonymous and denote what certain actions lead to their end result. However, it should be noted that a specific action can both lead to a certain result which is a direct, close consequence that is inextricably linked to the action (inaction) and at the same time lead to a delayed consequence that is the result of such activities in in general. Therefore, speaking about consequences of offenses related to financial and credit operations, it is necessary to distinguish consequences as an element of the objective offense side and as a result of illegal activities in the field of financial and credit operations in general. In other words, consequences of investigated offenses should be covered from the two above-mentioned aspects [51, p. 45].

It is a well-known and unquestionable fact that any action inherently leads to certain consequences. The latter by their nature can be both positive and negative ones. If we speak about offenses, negative and illegal nature of such acts immediately determines nature of consequences. They are marked by their social harm or danger. That is why in criminal cycle sciences can be found the socially dangerous consequences term suggesting the harm (damage) caused by criminal acts to public relations that are protected by criminal law, or as a real danger (threat) of such harm [52, p. 124].

Given the above, we note that in legal science, consequence is an element of the objective offense side – a system of signs of illegal behavior, necessary and sufficient to bring a person to justice [53, p. 8–9]. At the same time, nature of consequences of illegal actions related to financial and credit operations in aerospace industry allows for sectoral identification of specific offense (determine sectoral affiliation: public or private offense) and correctly classify the act.

Thus, offenses related to financial and credit operations in aerospace industry being a type of economic wrongdoing are in most cases material in nature. This means that composition of such offenses involves occurrence of negative consequences specified in the rule of law. At the same time qualifying investigated acts, it is necessary to emphasize nature of consequences, because it will determine the industry affiliation of a particular offense and to properly organize further activities of law enforcement or court.

In this context, the issue of harmfulness and social danger of an illegal act becomes important, because the act nature determines nature of the consequences caused by. That is why while speaking about offenses related to financial and credit operations in aerospace industry, we are talking about administrative, economic, criminal and civil offenses in this area. These actions differ primarily in the size and nature of damage, as well as the mechanism of regulation of social relations arising, change and cease as a result of the offense. In other words, in the present research, offenses related to financial and credit transactions term is used as a collective term to denote any offenses, regardless of their industry, committed in the banking system through financial and credit transactions.

However, in the context of detecting and investigating/prosecuting offenses related to financial and credit transactions and aerospace industry, the issue of their sectoral affiliation is important as it determines specifics of such activities. In this case, special attention should be paid to law enforcement agencies and/or the court to establish such a circumstance to be proved as a consequence/consequences of illegal act and its conditionality by such an act (identify the causal link between).

In the context of the above, in order to determine consequences of offenses related to financial and credit operations in aerospace industry, as an element of the objective offense side, we consider it necessary to clarify interpretation of the consequence/consequences term in administrative, commercial, criminal and civil rights.

According to Art. 218 of the Commercial Code of Ukraine: "basis of economic and legal liability of a participant in economic relations is an offense committed by him in the field of management" [54]. In this case, composition of economic offense includes the following mandatory elements:

- illegal behavior of business entities;
- damage existence;

– causal link between the offender's wrongful behavior and consequences;
– offender's guilt" [55, p. 160].

In general, violations of economic obligation consequences in the field of lending or established requirements for financial and credit operations are a violation of established legal order in the field of management and the rights and legitimate citizen interests, institutions and the state. The latter in economic law is understood to mean "costs incurred by the entitled party, loss or damage to its property, as well as income not received by it, which the entitled party would receive in the event of proper performance of obligations or compliance with the rules of economic activity by the other party" [54].

Civil law enshrines the right of every person to defend his civil right in case of its violation, non-recognition or challenge. In this case, if a person has suffered damages as a result of violation of his civil law, he has the right to compensation [56]. Accordingly, the basis for civil liability is the act commission containing civil offense (a set of such elements: wrongful act, damage, causal link between the wrongful act and harmful consequences, guilt) [57, p. 68].

In other words, from the above it is clear that one of the mandatory elements of civil offenses is harm existence that is a consequence of illegal behavior. At the same time, civil offenses related to financial and credit operations in aerospace industry can cause both material damage in the form of damage and moral damage by means of undermining financial institution business reputation and trust in by current and prospective customers. Mentioned illegal acts in accordance with Part 2 of Art. 22 of the Civil Code of Ukraine can cause losses as a result of non-performance or performance in violation of conditions of financial and credit operations through real losses and/or lost profits [56].

In general, consequences of economic and civil law violations related to financial and credit operations in aerospace industry, determine the harmful nature of such acts. Moreover, existence of such a consequence as damage (direct and/or indirect) that can be expressed in monetary terms is one of the characteristic features of offenses under research [58, p. 101–102].

In addition to the above, offenses related to financial and credit operations in aerospace industry can be public law, i.e. administrative and criminal ones. Therefore, the issue of consequences of such offenses as an integral element of the objective side of their composition is important. As has been repeatedly noted, nature and extent of the damage caused by an illegal act indicates the level (degree) of its public danger and allows to distinguish between administrative and criminal offenses, as well as to correctly classify them. In particular, economic and civil offenses are harmful because the violation, respectively, of economic and civil rights of the individual leads to damages, usually by the means of pecuniary damage.

At the same time, administrative and criminal offenses are characterized by such features as social harm and public danger. As M. V. Okereshko rightly points out in this regard, any public offense, both a crime and an administrative misdemeanor, carries a potential threat of causing real damage to public relations and are not only socially dangerous but also socially harmful, if consequence of such an act was causing harm [59, p. 80].

At the same time, as noted by V. P. Zhushman, P. Yu. Shvedenko, the offense in credit relations is a conscious, willful act of socially dangerous illegal behavior of an individual or legal entity [60, p. 56]. Under such conditions, subjects of the investigated offenses understand the consequences that will lead (may lead) to their actions and, nevertheless, knowingly commit offenses.

However, it should be noted that terminology used to describe the consequences as a criminal phenomenon is not the same. Such a designation, as rightly noted by T. V. Kolesnikov, traditionally in criminal law is carried out in two directions: within characteristics of only those concepts that relate to the basis of criminal liability; "Outside" such a ground but in relation to any object protected by criminal law [61, p. 539]. In other words, there are some types of consequences of the crime, in particular, consequences in general that do not affect the offense qualification and consequences as a crime element (its objective side) [62, p. 38].

The same can be said about terminology used in administrative law. In other words, administrative and criminal offenses related to financial and credit operations in aerospace industry, as a result, lead to a violation of legally protected public relations in the banking system in general, as well as to violations of specific conditions of financial and credit operations in particular. As an example, there are a number of offenses, including:

– administrative offenses:

- violation of the legislation on financial issues (Article 164^{2} of the Code of Ukraine on Administrative Offenses);
- violation of budget legislation (Article 164^{12} of the Code of Administrative Offenses);
- "violation of banking legislation; legislation in the field of state regulation of non-banking financial services markets; currency legislation; legislation governing transfer of funds in Ukraine; regulations of the National Bank of Ukraine or implementation of risky transactions that threaten the interests of depositors or other creditors of the bank" (Article 1665 of the Code of Administrative Offenses);
- procedure violation for engaging in financial services (Article 1668 of the Code of Administrative Offenses);
- legislation violation in the field of guaranteeing deposits of individuals (Article 166^{19} of the Code of Administrative Offenses) [63];

– criminal offenses:

- "misuse of budget funds, budget expenditures or the provision of loans from the budget without established budget allocations or with their excess" (Article 210 of the Criminal Code of Ukraine);
- bringing bank to insolvency (Article 218^{1} of the Criminal Code of Ukraine);
- causing bankruptcy (Article 219 of the Criminal Code of Ukraine);
- "procedure violation for maintaining a database of depositors or procedure for compiling reports" (Article 220^{1} of the Criminal Code of Ukraine);
- "falsification of financial documents and reports of financial institution, concealment of insolvency of a financial institution or grounds for revocation (cancellation) of financial institution license" (Article 2202 of the Criminal Code of Ukraine);

- fraud with financial resources (Article 222 of the Criminal Code of Ukraine), etc. [64].

Separately, we note that being public law, these offenses not only cause personal harm to victims but violate established by state grounds and procedures for financial and credit operations in aerospace industry, thereby encroaching on public relations in banking system. This is manifestation of social danger of these illegal acts. The degree of such danger expressed through including nature and extent of the consequences caused by these acts, allows to distinguish and differentiate, to qualify them as administrative or criminal offenses.

These illegal actions violate the established conditions of banks and other financial institutions in provision of relevant financial services related to financial and credit operations, in particular. This consequence directly determines their social danger.

However, taking into account mechanism of committing a specific administrative or criminal offense related to financial and credit operations in aerospace industry, consequences of such illegal acts will vary and be detailed depending on the specific components of such offenses. In particular, illegal actions such as:

- budget legislation violation (Article 164^{12} of the Code of Administrative Offenses);
- fraud with financial resources (Article 222 of the Criminal Code of Ukraine).
- "misuse of budget funds, budget expenditures or provision of loans from the budget without established budget allocations or with their excess" (Article 210 of the Criminal Code of Ukraine); etc.

The same offenses as violation of legislation on financial issues (Article 164^2 of the Code of Administrative Offenses); "Falsification of financial documents and reports of a financial organization, concealment of insolvency of a financial institution or grounds for revocation (cancellation) of financial institution license" (Article 2202 of the Criminal Code of Ukraine), etc. lead to such consequences as:

- income concealment;
- procedure violation of keeping accounting records, namely lack of accounting or maintaining it in violation of established procedure, introduction of false data in financial statements;
- failure to submit financial statements;
- untimely or low-quality inventory of cash and tangible assets, etc.

The above effects in aerospace industry can ultimately lead to misappropriation, misappropriation of property, non-receipt of taxes to relevant budgets, insolvency of the entity, and so on.

O. B. Poisner points out some consequences of offenses related to financial and credit operations, noting that as a result of committing certain crimes committed with use of banking services and technologies, the bank is insolvent in front of its customers [8, p. 288].

As mentioned earlier, offenses related to financial and credit operations in the aerospace industry have negative consequences for banking system as a whole and affect its condition. In other words, consequences of such illegal acts are not only an integral part of the relevant offenses but a general negative social phenomenon that hinders the proper functioning of such an important sector of state economy as the banking system.

It should be noted that unsatisfactory state of banking system, in particular its branch such as lending, as a result of offenses related to financial and credit operations is also one of the causes of the same offenses. In other words, it is a prerequisite for the formation of anti-social attitude of the offender.

Thus, it can be seen that consequences of the investigated offenses in aerospace industry as a general negative phenomenon that does not affect their differentiation and qualification, at the same time is one of the reasons for committing such illegal acts in the future.

Note that these categories form a kind of vicious circle, a system in which cause and effect are interdependent and interdependent. In this context, it is appropriate to say L. P. Skalozub, V. I. Vasilinchuk, who argue that "correlation of cause and effect is realized not directly (cause – effect) but through conditions that trigger the cause in a given situation. In other words, the cause preceding consequence occurrence can manifest its action only in existence of certain conditions that are also necessary for consequence occurrence" [27, p. 307]. In this case, consequence under certain conditions can also act as a cause of another action.

Thus, a special role in the banking system is played by one of the main functions of the bank: lending to businesses, including aerospace industry and citizens. Increased public danger of offenses in the field of bank lending is determined, in particular, by the fact that these actions cause significant damage not only to commercial banks and depositors but also negatively affect functioning of entire economic system. Most often, the funds received by offenders are taken out of Ukraine and there are stored in illegally opened accounts. Thus, along with direct losses caused by non-return of credit resources to banks, public interests, it is necessary to determine the task of huge damage such as: accelerating inflation, devaluation of the national currency that can gradually lead to "bleeding" credit and financial system [48, p. 106].

O. B. Poizner also points to the negative impact of offenses committed with use of banking services and technologies on the state of Ukrainian economy. He rightly notes that these illegal acts can be considered one of the most dangerous in the field of financial, economic and economic activities in aerospace industry. This is due to the fact that their consequence is not only to harm interests of individual economic entities in aerospace industry but also to destabilize the economic development of the state as whole and distrust citizens in ability of public authorities to protect their constitutional rights. In addition, due to illegal activities in financial, credit and banking fields, Ukrainian economy has lost and continues to lose a significant amount of funds that are taken out of the country. Currency values illegally transferred outside Ukraine are estimated at billions of dollars.

These offenses, especially criminal ones, cause damage to banking system which proper management and functioning is the key to economic development and therefore consequence of these crimes is undermining development of Ukrainian economy. This is due to the fact that banking sector is a strategic segment of market economy and plays a leading role in ensuring stable economic relations.

It should be noted that in a market economy it is possible to meet the needs of individuals and legal entities to receive quality banking services from proper payments, loans, securities transactions to timely return of deposits, etc. The level of customer confidence in specific banking institutions and the state of control over credit and financial system in general is a factor that ensures effectiveness of economic policy [8, p. 286]. As a result of committing criminal offenses related to financial and credit operations in aerospace industry, such a sector of Ukrainian economy as banking is losing the trust of population and investors.

Along with the above, economic offenses negatively affect the economy, inhibit development of market relations in Ukraine, generate inflation, distort public consciousness, violate principles of social justice, undermine power and law authority, state institutions, lead to a restriction of the rights and legitimate interests of economic entities and citizens. As noted by N. Yu. Cantor, crime is formation basis for shadow economy sector, as a result of which a significant amount of material resources is removed from fiscal control and are often used as sources of crime. At the same time, shadow economy social danger lies in existence of uncontrolled production, distribution, exchange and consumption of goods and services that are hidden from public administration for selfish purposes [65, p. 40].

Consequences of offenses related to financial and credit operations in aerospace industry are that banking system becomes unstable and does not create comfortable conditions to meet population needs. As a result, the risks of transferring risks to responsible borrowers, who pay for both themselves and debtors are becoming more frequent, thus ensuring the break-even point of the ban lending activities. This scheme is used by banks throughout the period of the banking system of Ukraine and retail lending [44, p. 65]. This undermines once again the undermining of confidence in financial institutions that is already at a very low level and sharply reduces attractiveness of our state for investment.

Thus, offenses related to financial and credit operations in aerospace industry lead both to a certain result that is a direct close consequence which is inextricably linked with action (inaction) and to deferred consequence that is result of such activities in general. Therefore, it is necessary to distinguish consequences of offenses related to financial and credit operations in aerospace industry, as an element of the objective offense side and as a result of illegal activities in the field of financial and credit operations in general.

4 Conclusions

It is established that offenses related to financial and credit operations in aerospace industry are unprofitable, destructive, illegal, punishable acts committed by violating the order of financial and credit operations and encroach on the economic interests not only of aerospace enterprises but state and society as a whole. Mechanism of committing these illegal acts involves the use of various methods and often their complexes consisting in unreasonable granting (receipt) of loans, breach of loan agreement, misuse, late and incomplete loan repayment, other breaches of obligations arising as a result concluding transactions for financial and credit operations in aerospace industry.

It is argued that knowledge of offense causes related to financial and credit transactions in the aerospace industry will prevent their commission, as well as timely identify and establish all circumstances of their commission. It was found that causes of offenses related to financial and credit transactions include both moral and psychological offender qualities and such conditions affecting formation, as the state of Ukrainian economy in general and the banking system in particular, level of regulatory and legal regulation of procedure for conducting financial and credit operations, procedure for training and selection of employees of financial institutions, etc.

It is noted that one of the causes of offenses related to financial and credit operations in aerospace industry is the anti-social attitude of offenders characterized by such moral and socio-psychological traits as selfishness, desire for quick and easy profit, greed, often envy, adventurism irresponsibility, disregard for generally accepted rules of conduct, including laws, etc. The mentioned features of the violator's personality are formed, modified and established under influence and action of external factors and factors in relation to the person:

- improper functioning of public authorities in determining implementation mechanism of financial and credit operations and control over;
- imperfect regulatory and legal regulation of the activities of financial institutions in terms of financial and credit operations and control over, as well as law enforcement, regulatory and judicial authorities in terms of prevention, termination and counteraction to offenses in this area;
- low level of economic state development (high unemployment rate, low wages, low level of liquidity in banking sector, financial instability, etc.);
- poor training and recruitment by financial institutions.

It is proved that offenses related to financial and credit operations in aerospace industry lead to a certain result that is a direct, close consequence that is inextricably linked to the action (inaction) and to a delayed consequence which is the result of such activities in general. Consequences of offenses related to financial and credit operations in aerospace industry are identified and characterized as an element of objective offense side and as a result of illegal activities in the field of financial and credit operations in general. It is noted that nature of consequences of illegal actions related to financial and credit operations in aerospace industry allows for sectoral identification of a specific offense (determine the sectoral affiliation – composition of a public or private offense) and correctly classify the act.

References

1. Govorova, K.I.: Causes of offenses related to financial and credit transactions. Law as an effective social regulator: materials of the international scientific-practical conference, Lviv, 15–16 February 2019. Lviv: Western Ukrainian organization Center for Legal Initiatives, pp. 56–59 (2019). [in Ukrainian]
2. Dunas, M.O.: Criminological measures to prevent crimes in the sphere of economic activity, which are committed by officials: dis. ... cand. jurid. Science: 12.00.08. Lviv. 220 p. (2019). [in Ukrainian]

3. Taras Poshivaniuk acquainted the LHS students with the specifics of criminal proceedings with the participation of banks. Legal High School. Educational platform [in Ukrainian]. https://lhs.net.ua/ua-zi-spetsyfikoiu-kryminalnykh-provadzhen-za-uchastiu-bankiv-slukhachiv-lhs-oznaiomyv-taras-poshyvaniuk-ru-so-spetsyfykoi-uholovnykh-proyzvodstv uchastyy-bankov-slushatelei-lhs-oznakomyl-taras-po
4. Pchelina, O.V., Kornienko, V.V.: Features of investigation of crimes committed by credit and financial operations: guidelines. Kharkiv. Hark. nat. University of Internal Affairs Affairs. 36 p. (2011). [in Ukrainian]
5. Didkivska, G.V., Chukhray, V.V.: Features of the investigation of crimes committed through credit and financial transactions. Legal Sci. Electron. J. **6**, 370–372 (2019) [in Ukrainian]. https://doi.org/10.32782/2524-0374/2019-6/88
6. Gerasimov, O.V.: Counteraction to crime in the banking sector: dis. ... cand. jurid. science: 12.00.08. Kharkiv. 235 p. (2019). [in Ukrainian]
7. Klochko, A.M.: Crimes in the field of banking. Legal Bull. Ukrainian Acad. Bank. **1**(10), 68–71 (2014). [in Ukrainian]
8. Poisner, O.B.: General characteristics of crimes committed using banking services and technology. Legal life of modern Ukraine: materials International. scientific-practical conf.: 16–17 trav. T. 2. Kyiv: «Phoenix», pp. 286–289 (2013). [in Ukrainian]
9. Chernyavsky, S.S.: Crimes in the field of banking as an object of criminological research. Bull. Criminological Assoc. Ukraine. **4**,192–198 (2013). [in Ukrainian]
10. Gerasimov, O.V.: Countering crime in the banking sector: author. dis. ... cand. jurid. science: special. 12.00.08. Kharkiv. 18 p. (2019). [in Ukrainian]
11. Masliy, I.V.: On the question of defining the concept and structure of crime in the financial and economic sphere as an element of its criminological characteristics. Scientific works of the National University Odessa Law Academy. T. 13, pp. 553–561 (2013). [in Ukrainian]
12. Klochko, A.M., Nezhevelo, V.V.: Actualization of the need to determine the generic object of crimes in the field of banking: problematic aspects. Leg. Sci. Electron. J. **3**, 201–204 (2019) [in Ukrainian]. http://www.lsej.org.ua/3_2019/53.pdf
13. Kurylo, V., Chernetskaya, O.: Administrative offenses in the field of finance and entrepreneurship. Leg. Ukraine. 6, 21–30 (2014). [in Ukrainian]. http://nbuv.gov.ua/UJRN/urykr_2014_6_6
14. Borodin, I.L., Bortnik, V.A.: Administrative responsibility: textbook. way. Kiev. DP Type. House Staff. 136 p. (2011). [in Ukrainian]
15. Chernov, M.: General characteristics of the legal structures of offenses in the field of banking. Natl. Leg. J. Theor. Pract. 6,104–107 (2014). [in Ukrainian]
16. Large explanatory dictionary of the modern Ukrainian language (with appendix, supplement and CD) . way. and heads. ed. W.T. Busel. Kyiv. Irpen: VTF Perun, 1736 p. (2009). [in Ukrainian]
17. Trailer, E.M., Chernov, A.M., Chekal, L.A.: Philosophy: a textbook. Kyiv: Akademvydav. 592 p. (2007). [in Ukrainian]
18. Petrushenko, V. L.: Philosophical dictionary: terms, personalities, sentences. Lviv: Magnolia. 352 p. (2011). [in Ukrainian]
19. Kovaleva, T.V., Kovriga, L.P.: Explanatory dictionary of the Ukrainian language / incl. Kharkiv. Sintex, 672 p. (2002). [in Ukrainian]
20. Pysarchuk, I.V.: Causes and conditions that contribute to the misuse of budget funds by officials. bulletin of the national university Lviv Polytechnic. Leg. Sci. **845**, 546–552 (2016). [in Ukrainian]
21. Criminology: textbook. / ed. Korobeynikov, B.V., Kuznetseva, N.F., Minkovsky, G. M.: Moscow. BEC, 556 p. (1998). [in Russian]
22. Pchelina, O.V.: Causes and conditions that contribute to the commission of economic crimes. Law Secury. **5**, 141–146 (2009). [in Ukrainian]

23. Tararukhin, S.A.: Criminal behavior. Social and psychological traits. Moscow. Jurid. lit. 224 p. (1974). [in Russian]
24. Konovalova, V.E.: Legal psychology: textbook. manual for jurid. universities and faculties studying in special. Jurisprudence. Kharkiv. Osnova, 198 p. (1990). [in Ukrainian]
25. Pankratov, V.V.: Methodology and methods of forensic research. Moscow. Jurid. lit. 136 p. (1972). [in Russian]
26. Dzhuzha, O.M., and others. Criminology: textbook. manual / for general ed. Dr. jurid. Sciences, Prof. OM Dwarf. Kyiv. Attica. 312 p. (2009). [in Ukrainian]
27. Skalozub, L.P., Vasilinchuk, V.I.: Causes and conditions that contribute to the commission of crimes in the budget sphere, and ways to eliminate them. Scientific Bulletin of Lviv State University of Internal Affairs. The series is legal. Vip. 3. Pp. 306–316 (2011). [in Ukrainian].
28. Shevchenko, O.V.: Causes and conditions of corruption crime. Bull. High Counc. Justice **3**(7), 122–132 (2011). [in Ukrainian]
29. Litvinov, O.M.: The concept and classification of causes and conditions that contribute to the crime. Law Forum. 2, 323–327 (2008) [in Ukrainian]. http://www.nbuv.gov.ua/e-journals/FP/2008-2/08lomsnz.pdf
30. Skalozub, L.P., and others 2010 Glossary of terms of operational and investigative activities (Section - Fight against economic crime): practice. manual. Kyiv. State Enterprise Printing House of the Ministry of Internal Affairs of Ukraine. 143 p. [in Ukrainian]
31. Kudryavtsev, V.N.: Genesis of crime. Experience of criminological modeling: textbook. allowance. Moscow. FORUM-INFRA-M. 216 p. (1998). [in Russian]
32. Criminology/ed. Kuznetsova, N.F., Minkovsky, G.M., Moscow. MSU Publishing House, 566 p. (1994). [in Russian]
33. Zemlyanskaya, O.V.: Legal ideas of the individual on the determination of the crime. Problems of extreme and crisis psychology. Vip. **7**, 129–138 (2010). [in Ukrainian]
34. Velikanov, S.V., and others 2006 Forensic prevention of economic crimes: scientific-practical. way. / for ed. V. A. Crane. Kharkiv. Kharkiv is legal. 236 p. [in Ukrainian]
35. Melnyk, P.V., Tarangul, L.L., Gordey, O.D.: Banking systems of foreign countries: a textbook. Kyiv. Alert, Center for Educational Literature. 586 p. (2010). [in Ukrainian]
36. Zhukov, A.M., and others 2000 Banking law of Ukraine: textbook / for general. ed. A.O. Selivanov. Kyiv. In Jure Publishing House. 384 p. [in Ukrainian]
37. Kostyuchenko, O.A.: Banking law of Ukraine: a textbook. 4th type. Kyiv. A.S.K. 448 p. (2006). [in Ukrainian]
38. Music-Stefanchuk, O.A.: Financial law: a textbook. 3rd ed., Add. and processing. Kyiv. Attica. 264 p. (2007). [in Ukrainian]
39. Voloshchuk, M.G., Karabin, T.O., Mendzhul, M.V.: Financial law: textbook. way. Kind. 3rd, add. and processing. Uzhhorod: Oleksandra Garkusha Publishing House, 244 p. (2017). [in Ukrainian]
40. Banking: a textbook / ed. Doctor of Economics, Prof. OV Dzyublyuk. Ternopil: TNEU Publishing House Economic Thought, 696 p. (2009)
41. Starynsky, M.V.: Banks as participants in administrative-tort relations in the field of banking: author's ref. dis. … Cand. jurid. Science: special. 12.00.07. Kyiv. 19 p. (2009). [in Ukrainian]
42. Orlyuk, O.P.: Legal problems of organization and activity of the banking system of Ukraine: dis. Dr. Jurid. Science: 12.00.07. Kyiv. 474 p. (2003). [in Ukrainian]
43. Getmantsev, D.O., Shuklina, N.G.: Banking law of Ukraine. Kyiv. Center for Educational Literature. 344 p. (2007). [in Ukrainian]
44. Marchenko, K.V.: The influence of the economic environment on the features of bank consumer lending. Development of financial services in the context of the new financial architecture: All-Ukrainian. Science Forum (Kyiv, May 14–15, 2015): thesis add. / resp. ed. N.P. Shulga. Kyiv. Kyiv'. nat. trade and economy Univ., 2015. pp. 64–65 (2015). [in Ukrainian]

45. Financial Stability Report approved by the Financial Stability Committee of the National Bank of Ukraine on 14 June 2019 [in Ukrainian]. https://bank.gov.ua/admin_uploads/article/FSR_2019-R1.pdf?V= 4
46. Samoukina, N.V.: Indispensable employee and personnel security. Moscow. Vershina. 176 p. (2008). [in Russian]
47. Zubok, M.I., Yaremenko, S.M.: Ensuring the security of the enterprise, the bank in working with staff. Part one. Bus. Secur. 5, 44–46 (2010). [in Russian]
48. Tolstosheeva, A.: The impact of the criminogenic situation in Ukraine on the lending activities of banks. Econ. Anal. **12**(2), 106–110 (2013). [in Ukrainian]
49. Vronskaya, A.: Execution of credit agreements: a review of the case law of the Supreme Court. Judicial power of Ukraine 2012 [in Ukrainian]. https://supreme.court.gov.ua/supreme/pres-centr/zmi/488409/
50. Luspenyk, D.D., Melnyk, Z.P.: Judicial practice of civil cases arising from credit relations. The Supreme Court of Ukraine: official website (2021) [in Ukrainian]. http://www.viaduk.net/clients/vsu/vsu.nsf/ (documents) / 1E6E164907ADC7BBC2257B7C004A74D2
51. Govorova, K.I.: Consequences of offenses related to financial and credit transactions. Law as an effective social regulator: materials of the international scientific-practical conference, Lviv, 14–15 February 2020. Lviv: Western Ukrainian organization Center for Legal Initiatives, pp. 45–47 (2020). [in Ukrainian]
52. Baulin, Y.V., and others Criminal law of Ukraine: General part: textbook/ed. prof. Stashisa, V.V., Tatsia, V.Y., 4th ed., Reworked. and add. Kharkiv. Law. 456 p. (2010). [in Ukrainian]
53. Komlyk, V.V.: The composition of the offense: theoretical approaches and practical measurements: author. dis. ... Cand. jurid. Science: special. 12.00.01. Kyiv. 18 p. (2017). [in Ukrainian]
54. Economic Code of Ukraine: Law of Ukraine of 16.01.2003 № 436 – IV. (with changes and additions). Information of the Verkhovna Rada of Ukraine. № 18. Article 144 (2003). [in Ukrainian]
55. Baganets, V.O.: Economic and legal responsibility of public authorities (on the example of the prosecutor's office). Eur. Perspect. **6**, 157–161 (2014). [in Ukrainian]
56. Civil Code of Ukraine: Law of Ukraine of January 16, 2003 № 435 – IV. Information of the Verkhovna Rada of Ukraine. № 40. Article 356 (2003). [in Ukrainian]
57. Ivashchenko, V.V.: Civil liability as a type of legal liability. Leg. Sci. Electron. J. 2, 66–69 (2019). [in Ukrainian]. http://www.lsej.org.ua/2_2019/18.pdf
58. Kikot, O.O.: Grounds for economic and legal responsibility: theoretical and legal issues. Sci. Notes NaUKMA. Leg. Sci. **103**, 100–102 (2010). [in Ukrainian]
59. Okereshko, M.V.: The concept and composition of an administrative offense committed by a doctor in the field of public health. Young Sci. **5.1**(45.1), 80–84 (2017) . [in Ukrainian]
60. Zhushman, V.P., Shvedenko, P.Y.: The concept of offenses arising in the field of lending to agricultural producers, their characteristics, conditions and causes. Sci. Bull. Natl. Univ. Life Environ. Sci. Ukraine. Vip. 197. Ch 7 2, pp. 55–58 (2014). [in Ukrainian]
61. Kolesnikova, T.V.: Public danger and consequences of crime: the question of the relationship and criminal nature. Current issues of state and law. Vip. **64**, 538–545 (2012). [in Ukrainian]
62. Kuznetsova, N.F.: Significance of criminal consequences for criminal liability. Moscow. Gosuryizdat. 219 p. (1958). [in Russian]
63. Code of Ukraine on Administrative Offenses: Law of Ukraine of 07.12.1984 № 8073 – X. (with changes and additions). Information of the Verkhovna Rada of the USSR. 1984. № 51. Article 1122. [in Ukrainian]

64. Criminal Code of Ukraine: Law of Ukraine of 5 April 2001 № 2341 – III. (with changes and additions). Information of the Verkhovna Rada of Ukraine. № 25. Article 131 (2001). [in Ukrainian]
65. Kantor, N.Y.: Administrative and legal mechanism of crime prevention in the economic sphere as a basis for ensuring the economic stability of the state. Bull. Natl. Univ. Lviv Polytech. Leg. Sci. 824, 39–44 (2015) [in Ukrainian]. http://nbuv.gov.ua/UJRN/vnulpurn_2015_824_9

Legal Support of Aerospace Environmental Monitoring

Olena Savchuk(✉)

Law Department of National Aerospace University – "Kharkiv Aviation Institute", Kharkiv, Ukraine

Abstract. The paper considers the problematic issues that arise in the legal field of aerospace environmental monitoring. The importance and relevance of ensuring the collection, processing, storage and analysis of information about the state of the environment with the introduction of modern innovative technologies is proved. The purpose of aerospace environmental monitoring is recognized as the preservation and prevention of anthropogenic climate change, rapid response to problems arising from environmental pollution, as well as natural phenomena, man-made accidents and disasters and maintaining a favorable environment for society as a whole.

An analysis of the state environmental policy aimed at achieving good environmental status by introducing an ecosystem approach to all areas of socio-economic development of Ukraine to ensure the constitutional right of every citizen of Ukraine to a clean and safe environment, the introduction of sustainable use and preservation and restoration of natural ecosystems.

Keywords: Ecological innovations · Environmental monitoring · Aerospace monitoring · Ecological policy · Climate · Environment · Ecologization

1 Introduction

A person, his life and health, honor and dignity, inviolability and security are the highest social value in Ukraine. Implementation of this norm of the Constitution of Ukraine is the main goal of the state national security policy. One of the priorities of Ukraine's national interests and national security is to ensure environmental security, create safe living conditions, in particular in areas affected by hostilities, and build an effective civil defense system.

However, today the irrational use of natural resources, degradation of forests, water basins, agricultural lands continues. The system of household and industrial waste management is inefficient. Opportunities to adapt the economy, livelihoods and civil protection to climate change are insufficient. The negative impact of ongoing hostilities in some districts of Donetsk and Luhansk regions and the consequences of the Chornobyl disaster on the environment and human health persists.

M. Nechyporuk et al. (Eds.): ICTM 2021, LNNS 367, pp. 690–703, 2022.
https://doi.org/10.1007/978-3-030-94259-5_56

Environmental safety, in particular the safety of the human environment, is one of the highest priorities. We must create conditions for maintaining ecological balance on the territory of Ukraine, modernization of communal infrastructure, strengthen environmental protection, introduce the latest waste management systems and reduce industrial emissions, ensure efficient use of natural resources, protect forests and reservoirs, develop protected areas, prevent consequences of natural and man-made emergencies and eliminate them [1].

The formation of an environmental monitoring system in Ukraine is one of the priority measures to stabilize and improve the quality of the environmental situation in the country [2].

2 Literature Review

This issue has been studied to one degree or another by such representatives of domestic environmental law as V.I. Andreytsev, G.V. Anisimova, G.I. Balyuk, M.M. Brynchuk, A.P. Hetman, N.V. Gulak, V. M. Zavgorodnya, V.A. Zuev, M.I. Erofeev, I. O. Ikonytska, M.V. Krasnova, I.I. Karakash, N.R. Kobetska, V.M. Komarnytsky, V.V. Kostytsky, N.V. Kruchinina, P.F. Kulinych, V.E. Lyzgaro, N.R. Malysheva, M.V. Mozalyova, M.I. Malyshko, V.V. Petrov, E.V. Plznyak, I.O. Soloshich, Yu. O. Tsibulsky, Yu. S. Shemshuchenko, V. M. Shmandiy, M.V. Shulga O.A. Shompol, S.V. Sharapova and others.

3 Main Content Presentation

In accordance with the Law of Ukraine "On Environmental Protection" in order to collect, process, store and analyze information about the environment, forecast its changes and develop scientifically sound recommendations for effective management decisions in Ukraine creates a system of state environmental monitoring. Monitoring of the state of the environment, the level of its pollution is carried out by the central executive body implementing the state policy in the field of environmental protection, other specially authorized state bodies, as well as enterprises, institutions and organizations whose activities lead or may lead to deterioration environment. These enterprises, institutions and organizations are obliged to transfer the analytical materials of their observations to the relevant state bodies free of charge. The procedure for state monitoring of the natural environment is determined by the Cabinet of Ministers of Ukraine. State bodies together with relevant scientific institutions ensure the organization of short-term and long-term forecasting of environmental changes, which should be taken into account in the development and implementation of programs and measures for economic and social development of Ukraine, including environmental protection, use and reproduction of natural resources and ensuring environmental safety (Article 22).

Organization of environmental monitoring, creation and operation of a network of national ecological automated information and analytical system to ensure access to environmental information, the provisions of which are approved by the Cabinet of Ministers of Ukraine (paragraph "c" of Article 20).

Resolution of the Cabinet of Ministers of Ukraine of March 30, 1998 № 391 "On Approval of the Regulations on the State Environmental Monitoring System" regulates the procedure for the establishment and operation of the monitoring system. The resolution defines the monitoring system as: "an open information system, the priorities of which are the protection of vital environmental interests of man and society; preservation of natural ecosystems; prevention of crisis changes in the ecological state of the environment and prevention of ecological emergencies.

Creation and functioning of the monitoring system for the purpose of integration of ecological information systems covering certain territories is based on the principles: coherence of normative-legal and organizational-methodical providing, compatibility of technical, information and software of its constituent parts; systematic observations of the state of the environment and man-made objects that affect it; timeliness of receiving, complexity of processing and use of information on the state of the environment (ecological information) received and stored in the monitoring system; objectivity of primary, analytical and forecast information on the state of the environment (environmental information) and efficiency of its delivery to public authorities, local governments, public organizations, media, the population of Ukraine, interested international institutions and the world community [3].

The system of ecological monitoring promotes the implementation of the state ecological policy of Ukraine, which provides for the ecological and rational use of the natural and socio-economic potential of the state, the preservation of a favorable living environment of society; socio-economic and economically rational solution of problems that arise as a result of environmental pollution, dangerous natural phenomena, man-made accidents and catastrophes; development of international cooperation on nature biodiversity conservation, protection of the ozone layer of the atmosphere, prevention of anthropogenic climate change, etc. [4, p. 129].

The monitoring system is aimed at: increasing the level of study and knowledge about the ecological state of the environment; improving the efficiency and quality of information services to users at all levels; improving the quality of justification of environmental measures and the effectiveness of their implementation; promoting the development of international cooperation in the field of environmental protection, rational use of natural resources and environmental security.

The main tasks of the subjects of the monitoring system are: long-term systematic monitoring of the state of the environment; analysis of the ecological state of the environment and forecasting of its changes; information and analytical support for decision-making in the field of environmental protection, rational use of natural resources and environmental safety; information services for public authorities, local governments, as well as providing information on the state of the environment (environmental information) of the population and international organizations [3].

Resolution of the Cabinet of Ministers of Ukraine of December 5, 2007 № 1376 "On approval of the State target environmental program for environmental monitoring" The purpose of the Program is to ensure the development of a single state environmental monitoring system aimed at: improving its efficiency; providing executive authorities, local governments and the population with timely and reliable information about the state of the environment, raising the level of environmental knowledge of citizens; determination of common priorities when planning the actions of the subjects of the monitoring system; creation of a single network of observations; technical re-equipment and improvement of regulatory and methodological support of the monitoring system; coordination of elements of information technologies used by the subjects of the monitoring system [5].

Ukraine is a party to more than 70 international bilateral and multilateral agreements and conventions, the implementation of which requires the use of information on the state of the environment and forecasting its changes. Therefore, the development of the monitoring system should be carried out taking into account the common European requirements. Over the years of monitoring, the subjects of the monitoring system have accumulated some experience and a significant amount of information on the state of the environment and its sources of pollution. However, as the information is collected on paper and placed in databases that do not meet the requirements and recommendations of the United Nations Economic Commission for Europe on the establishment of the European Network of Observations and Information on the State of the Environment, it is not possible to make full use of available data. comprehensive environmental assessment. The solution of these problems is achieved by combining the efforts of the monitoring system and coordination of their actions, ensuring the continuity of observations of the components of the environment, improving regulatory, methodological, technical and organizational support to create a single network of observations, introduction of advanced information technologies and banks data on his condition. A comprehensive monitoring of the state of atmospheric air, water, land, forests, geological environment, biological diversity, waste management, physical factors affecting the environment is provided. To ensure monitoring, it is necessary, in particular, to identify information needs, create a system for assessing the state of the environment, develop monitoring programs at various levels, observe, collect, process and analyze monitoring data and evaluate information, prepare conclusions and proposals [5].

In order to ensure the integration of information resources and interaction of the subjects of the monitoring system, a single automated subsystem for data collection, processing, analysis and storage will be created and operated, for which: unification of hardware and software is envisaged; integration of information subsystems of separate subjects of the monitoring system for complex assessment of information; introduction of advanced information and analytical technologies and unified forms of data presentation and storage; ensuring the openness of information systems for their expansion and implementation of new directions. The creation and further development of banks of information resources in the monitoring system will be carried out taking into account the level principle of division of tasks and will provide information content of the relevant databases and top-level banks. Multilevel is introduced in order to increase the functional and economic efficiency of information support [5].

The aim of the state environmental policy is to achieve good environmental status by introducing an ecosystem approach to all areas of socio-economic development of Ukraine in order to ensure the constitutional right of every citizen of Ukraine to a clean and safe environment, introduction of sustainable use and preservation and restoration of natural ecosystems. One of the principles of state environmental policy is the prevention of emergencies of natural and man-made nature, which involves analysis and forecasting of environmental risks based on the results of strategic environmental assessment, environmental impact assessment and comprehensive monitoring of the environment. The main tools for implementing state environmental policy are comprehensive environmental monitoring and supervision (control) in the field of environmental protection, rational use, reproduction and protection of natural resources – will ensure the transition to crime prevention and environmental monitoring, reducing pressure on business environment, broad public involvement in environmental control through the construction of an effective system of monitoring compliance with environmental legislation, taking into account the best practices of organizing the functioning of similar institutions in the member states of the European Union [6].

Concept of implementation of state policy in the field of space activities for the period up to 2032 "№48-r of 25.01.2012 the results of the implementation of this Concept considers the provision of systematic data from the domestic space system of Earth observation and geophysical monitoring "Sich", as well as foreign space Earth remote sensing devices (within the framework of joint international programs and agreements), as well as to create a national system of geographic information support and emergency monitoring as part of the European (GMES) and global (GEOSS) systems and ensure the operation of its information services to interested users [7].

Improving the efficiency of space activities, solving urgent problems of national security and defense and realizing the geopolitical interests of the state, promoting socio-economic, scientific, educational and ecological-cultural development of society, increasing the prestige of Ukraine as a space state on the world stage, transforming Ukraine's space industry into a high-tech multiplier development of the domestic economy. Preservation of Ukraine's status as a space state, gradual acquisition of the status of a member of the European Space Agency, active participation in European and international space programs and projects, including space debris control program, implementation of European space standards in Ukraine, and meeting public needs for space information services systems, including remote sensing data and services for geographic information systems for various purposes, satellite navigation and telecommunication systems using orbital means of national and foreign space system operators, geophysical monitoring data of the Earth are the goal and main direction of Ukraine's space activities until 2022 [8].

It should be noted that in ecological and legal science the term "environmental monitoring" does not have a generally accepted definition, and its concept is still not fully formed [9, p. 74].

In the specialized literature, monitoring is generally understood as (from the Latin "control", "supervise") the process of continuous observation of objects – air, water, plants, animals, land. This definition was proposed in 1972 at the UN Stockholm Conference on the Environment.

The term "monitoring" is used as a long-term, standardized measurement, observation, environmental assessment and reporting to determine its status and trends [10].

According to N.R. Malysheva, monitoring is a system of observation, collection, processing, transmission and analysis of information about the state of the environment, forecasting its changes and developing scientifically sound recommendations for management decisions [11, p. 419].

V.V. Petrov, it understands environmental monitoring as observation and control of the environment (atmospheric air, reservoirs, oceans, soils) and changes in its economic development, the system of collecting, summarizing, evaluating and transmitting information about real or expected harmful effects [12, p. 548].

S.V. Sharapova under ecological monitoring considers the scientific information-analytical system regulated by the legislation, which allows to define and forecast changes in the state of the environment, to control and manage them [2, p. 2].

In addition, the term "monitoring" is used as a long-term, standardized measurement, observation, environmental assessment and reporting to determine its status and trends [13, p. 2].

Analyzing the monitoring function of the management of faunal legal relations, V.V. Shekhovtsov emphasizes that the basis for building an environmental monitoring system should be based on the principle of management based on a systematic analysis of information about the quality of the environment. According to the scientist, the monitoring system does not cover environmental quality management activities, but is a source of information needed to make environmentally significant decisions. Instead, the main tasks of the environmental monitoring system can be reduced to three: 1) observation of factors affecting the environment; 2) assessment of the actual state of the environment; 3) assessment of the projected state of the environment [13, p. 116].

A similar opinion is held by V.M. Bogolyubov, M.O. Klimenko, V.B. Mokin, highlighting along with the already mentioned tasks another – providing scientific and information support for management decisions [14, p. 5].

Thus, undoubtedly, the main purpose of environmental monitoring is to meet the need for information for an integrated ecosystem approach to environmental management [15, 16].

According to O.V. Ilyina, the main tasks of environmental monitoring today are:

- the organization of a single state system of control over the components of the natural environment;
- setting up an automated system for collecting, processing, summarizing and storing information on the quantity and condition of natural resources (data bank);
- assessment of natural resource potential and possible level of resource use;
- inventory of pollution sources and study of the degree of anthropogenic impact on the components of the natural environment;
- modeling and forecasting of changes in the ecological situation and the level of environmental health;
- development of management decisions aimed at ensuring the rational use of nature and sustainable development of the region.

According to the author, the importance of ecological monitoring deserves special attention, which is that it is an open information system, the priorities of which are the protection of vital ecological interests of man and society, preservation of natural ecosystems, reproduction of crisis changes and prevention of emergencies. situations. In the process of conducting ecological monitoring of the use of natural resources, information on the natural state of natural objects is obtained and this information is provided for processing and analysis; use of the received information on a condition of natural resources in the course of their operation; implementation of management functions in relation to natural resources that are exploited. In addition, the data obtained through environmental monitoring are used to determine the limits of use of natural resources, as well as to provide natural resources for use (including on lease) or ownership [17, p. 201].

We agree with the above tasks of environmental monitoring proposed by scientists, we consider it appropriate to emphasize the issue of its classification. Traditionally, in the specialized scientific literature, the most common option is to divide environmental monitoring by its objects. Legislation includes atmospheric air, water, biodiversity, forests, land, waste management, physical factors, and geological environments. The latter cannot negate the fact that this monitoring is comprehensive, i.e., in addition to assessing the state of individual natural objects, must provide a holistic assessment of the state of the environment, taking into account the relationships, interdependence in ecological systems. This circumstance determines the organizational structure of environmental monitoring, which should ensure the functioning of the subjects of this monitoring as a whole, facilitate the exchange of information between them, provide an opportunity to summarize this information, issue objective assessments of individual natural resources and the environment as a whole [18, p. 176].

Consider a few of them. Air quality is monitored by the State Hydrometeorological Service. Atmospheric air pollution is monitored in 53 cities of Ukraine at 162 stationary, two route observation posts and two transboundary transfer stations. The chemical composition of precipitation and the acidity of precipitation are monitored. The mandatory air quality monitoring program includes seven pollutants: dust, nitrogen dioxide (NO_2), sulfur dioxide (SO_2), carbon monoxide, formaldehyde (H_2CO), lead and benzene (a) pyrene. Some stations monitor for additional pollutants. The analysis of the presence of pollutants in precipitation and snow cover is carried out.

The State Ecological Inspectorate carries out selective sampling at emission sources. More than 65 parameters are measured. The Sanitary and Epidemiological Service monitors air quality in residential and recreational areas, in particular near major roads, sanitary protection zones and residential buildings, in schools, preschools and medical facilities in cities and in the work area. In addition, the analysis of air quality in the residential area is carried out according to the complaints of residents [19].

Monitoring of land waters The State Hydrometeorological Service monitors the hydrochemical status of waters at 151 water bodies, as well as carries out hydrobiological observations at 45 water bodies. Data on 46 parameters are obtained, which make it possible to assess the chemical composition of water, nutrients, the presence of suspended particles and organic matter, major pollutants, heavy metals and pesticides. Chronic water toxicity is monitored at 8 water bodies. Indicators of radioactive contamination of surface waters are determined. The State Ecological Inspectorate (Ministry of Environment) takes water samples and receives data on 60 measured parameters. The State Committee for Water Management monitors rivers, reservoirs, canals, irrigation systems and reservoirs within integrated water management systems, water supply systems, transboundary watercourses and reservoirs in areas affected by nuclear power plants. Water quality control by physical and chemical indicators is carried out on 72 reservoirs, 164 rivers, 14 irrigation systems, 1 estuary and 5 complex canals. In addition, within the framework of radiation monitoring of waters, water management organizations control the content of radionuclides in surface waters. The Sanitary and Epidemiological Service monitors the sources of centralized and decentralized drinking water supply, as well as recreation areas along rivers and reservoirs. Enterprises of the State Geological Survey (Ministry of Environment) monitor the state of groundwater. At the monitoring sites, the level of groundwater (presence) and their natural geochemical composition are assessed. 22 parameters are determined, including the concentration of heavy metals and pesticides. The Sanitary and Epidemiological Service performs chemical analysis of groundwater intended for drinking consumption. Coastal water monitoring The State Hydrometeorological Service operates a coastal water monitoring network consisting of monitoring stations at wastewater discharge sites and research stations located in the coastal areas of the Black and Azov Seas. At existing stations, measurements of 16 to 26 hydrochemical parameters of water and bottom sediments are performed. The State Inspectorates for the Protection of the Black and Azov Seas (Ministry of Environment) have their own surveillance systems. Their responsibilities include monthly sampling and analysis of the impact of pollution sources located on the coast; monitoring of ship discharges; pollution from exploration and production of oil, gas and construction materials on the sea shelf; supervision of the use of living marine resources. The State Sanitary and Epidemiological Service monitors the quality of sea water in areas of recreational and health water use [19].

Soil condition monitoring. The State Hydrometeorological Service monitors the contamination of agricultural lands with pesticides and heavy metals in settlements. Samples are taken every five years, samples for heavy metals in the cities of Konstantinovka and Mariupol are taken annually. The State Ecological Inspectorate (Ministry of Environment) carries out sampling at industrial sites within the country. Total number of parameters measured 27. Institutions of the Ministry of Health monitor the condition of soils in the territories of their possible negative impact on public health. The most covered areas are agricultural products, areas where pesticides are used, soils in the area of residential areas, playgrounds and institutions. Soil samples in the places of storage of toxic waste on the territory of enterprises and outside the territory of enterprises in the places of their storage or disposal are investigated. The Ministry of Agrarian Policy monitors the soils of agricultural use. Radiological, agrochemical and toxicological determinations, residual amount of pesticides, agrochemicals and heavy metals are carried out [19].

Monitoring of biodiversity indicators Due to limited budget funding, monitoring is carried out only for species of industrial interest (trees, fish, game). The enterprises of the State Forestry Committee monitor forest vegetation in 24 regions of the country. Biomass is assessed, its damage by biotic and abiotic factors; hunting fauna, biodiversity; radiological determinations. Some research is carried out through international assistance, or through international programs. Radiation monitoring The State Hydrometeorological Service monitors radioactive contamination of the atmosphere by daily measurements of gamma radiation exposure doses, deposition of radioactive particles from the atmosphere and the content of radioactive aerosol in the air. Measurements of radioactive contamination of surface waters on 8 water objects are carried out. Near nuclear power plants, the State Hydrometeorological Service measures radioactive contamination of surface water with cesium-137 y and soil contamination. The monitoring laboratories of the Ministry of Agrarian Policy control the places of concentration of radioactive substances in soils and food products. The Ministry for Emergencies monitors GRE doses at 10 automated points near nuclear power plants. Within the 30-km zone around the Chornobyl NPP (Exclusion Zone), the Ministry for Emergencies monitors the concentration of radionuclides; radionuclides in precipitation, as well as the concentration of "hot" particles in the air. The International Radioecological Laboratory of the Chornobyl Center for Atomic Safety, Radioactive Waste and Radioecology in Slavutych monitors the impact of radiation in the exclusion zone [19].

As for aerospace environmental monitoring, the following should be noted. As practice shows, the best results are achieved under the condition of complex, synchronous conducting of space and ground researches when results of ground measurements are extrapolated to the maps received on the basis of space and aviation pictures. Such research should be implemented on the basis of the use of aerospace monitoring system, the purpose of which is to introduce modern methods of aerospace observation in Ukraine and on the basis of the results of such observation to predict the occurrence of natural and man-made threats [20, p. 95].

Assessment of environmental consequences on the example of the accident at the Fukushima nuclear power plant showed the feasibility and effectiveness of the use of unmanned aerial vehicles for radiation monitoring of the environment and video surveillance, and others [21].

The main tasks solved by the aerospace monitoring system are: continuous space monitoring of the territory of Ukraine in order to identify early signs of dangerous natural and man-made processes that may occur; observation, assessment, forecast and control of dangerous natural and man-made processes and phenomena, the environment; collection, processing, analysis and generalization of data on energy-intensive processes on Earth and in near-Earth space in order to identify anomalies that are precursors of dangerous natural phenomena; observation, assessment and forecast of danger of cross-border and transregional transfers of dangerous substances, other negative influences; operational collection, processing and submission to public authorities and local governments of information on potential sources of natural and man-made disasters, the creation and maintenance of a database of emergencies and their sources, forecasting the occurrence of dangerous natural and man-made phenomena and their consequences; control over the dynamics of processes at individual industrial, agricultural and other facilities; timely detection, identification and forecasting of accidents and man-made disasters, as well as harmful and affecting factors, in particular: levels of physical fields, concentrations of radioactive, chemical, biological substances; assessment of the degree of danger of a complex environmental situation during the normal functioning of dangerous objects, as well as emergencies that occur during accidents and disasters on them [20, p. 96].

The introduction of innovative environmental technologies is based on obtaining high-precision (in terms of spatial and temporal resolution) aerospace images, which requires the use of improved and adequate methods of processing, able to calculate and analyze a sufficient number of indicators of mining areas for a limited period of time [22, p. 29].

The use of space surveillance systems for environmental monitoring is one of the main sources of remote sensing methods to determine the state and nature of the environment. This makes it possible to identify anthropogenic impact on the environment, in particular to determine its nature (e.g., emissions into the atmosphere) and its areas with the necessary operational and spatial completeness of observation, as well as comprehensively assess: impact properties, size, concentration, forecasting effects, content of interaction and observe the anthropogenic zone. The use of space images, as well as their complex processing makes it possible to more fully reflect the nature of the impact of harmful processes on the environment. In addition, due to the analysis of space images it is possible to conduct a comprehensive assessment of the ecological state of the Earth's surface and the nature of harmful substances, but also the levels of their concentration and impact zone. Therefore, the report reveals the possibilities of environmental monitoring using modern space and geoinformation technologies based on space image processing [23, p. 4].

Thanks to the cooperation of the Ministry of Environment with the State Space Agency of Ukraine, from now on illegal logging, mining, natural landfills, as well as the construction of reservoirs will be monitored by aerospace surveying. In accordance with the "Memorandum between the Ministry of Environment and the SCA on cooperation in science, technology and education", specialists of the National Center for Space Management and Testing provide the Ministry of Environment space images, the results of their thematic processing and relevant analysis. Thus, during 2020, more than 900 information and analytical documents based on space data were transferred to the Ministry of Environment. At the moment, it is possible to take images from space in retrospect and collect reliable information about illegal logging or quarrying. Now, based on the inspection, the inspectors will clearly know the coordinates of the places to pay attention to first, and will identify the culprits faster. This new tool for approaching environmental control is being tested. It will help simplify the procedure of conducting the inspections themselves, speed up the calculation of damages and become a substantial piece of evidence in the courts. Based on space images, inspectors are already conducting inspections of forestries in a test mode. The next step will be to monitor violations in the field of subsoil use. Natural landfills will also be identified, as well as the condition of coastal protection strips will be assessed. For example, received a request from citizens that SSE "Northern Forest" which carries out forest use without the prescribed approvals and limits, without obtaining an opinion on the environmental impact assessment, as well as conducts continuous felling on more than 1 ha of forest. This verification was carried out using satellite images that confirmed the facts of deforestation.

The central executive body that formulates and implements state policy in the field of space activities, within its competence, monitors and maintains a database of geophysical observations and ensures its interaction with the National Data Center of the system of seismic observations and improving the safety of population in seismically dangerous regions [24].

Over the last decade, powerful tools for data collection and processing have been developed, and a new level of interaction between information infrastructure, surveillance tools and research has been provided. The widespread introduction of Earth remote sensing technologies is associated with the restructuring of geoinformation support of the economy. This adjustment is based on a significant improvement in environmental assessments and threat identification based on a new forecasting methodology that uses large arrays of observational aerospace and ground data. To accomplish this task, it is necessary to create a qualitatively new information system that will meet the needs of scientists, politicians and the public in the interests of joint solutions to environmental problems. Today, large amounts of information are collected in various archives that do not have relevant catalogs and data in which are not available to researchers and users on a systematic basis. Therefore, innovative research and development of new technologies that will ensure the synergistic effect of different sources of information are essential [25].

In recent years, the question of the introduction of 3D-cadastres has become more frequent, the introduction of which should be accompanied by the creation of three-dimensional maps and terrain plans based on constantly updated information from satellite communication systems [26].

For aerospace monitoring, the water surface is a natural integrator of information, which allows to detect processes occurring not only on the sea surface, but also in the water column and bottom space [27, p. 68].

The organization of monitoring of anthropogenic impact within the protected areas will solve the problem in the periodic receipt of relevant information. Monitoring of anthropogenic impact by terrestrial methods will ensure the accounting of most types of violations of the protected regime, but requires periodic field observations to identify and geodetic reference of objects of anthropogenic origin. At the same time, protected areas are subject to additional workload by registrars. In turn, aerospace monitoring of anthropogenic impact will solve the problem of non-contact research in determining the actual state of anthropogenic load on specially protected natural areas. The composition of certain objects of anthropogenic origin, identified by remote sensing of the Earth (remote sensing), depends on the spatial resolution of aerospace information used [28, p. 490].

Principles of building a system of aerospace surveillance and forecasting of emergencies: unconditional compliance with the norms and principles of international space and air law, as well as compliance with unilateral and multilateral government obligations in the field of aerospace activities; the widest possible use of the results of profile research and development conducted in the framework of international space aerospace monitoring programs; sequence and phased creation of system components taking into account the priority of forecasting tasks of global natural and man-made processes, the latest achievements in the field of aerospace monitoring and the resources involved; development of the terrestrial segment of the system on the basis of full-scale testing (practical testing) of basic technologies and software and hardware of aerospace monitoring; wide information and organizational-technological interaction of the orbital segment with ground and aviation monitoring means, as well as with functioning space systems; comprehensive monitoring of the occurrence and development of dangerous events occurring on Earth and in outer space, acceptable reliable prediction of their occurrence in order to take adequate measures to prevent or mitigate the devastating effects, timely evacuation of people, the maximum possible salvation of material resources and cultural values with extensive use of navigation, telecommunications and information resources of the world community to solve the full range of modern humanitarian problems of mankind [20, p. 96].

As a result, the inspectors found a number of violations of environmental legislation and calculated losses amounting to UAH 15 million. The materials were handed over to law enforcement officers. The practice of using space images ensures fast results in recording environmental offenses and identifying the culprits. A retrospective analysis helps to understand in which areas you need to conduct enhanced control [29].

4 Conclusions

Therefore, having examined the issue of legal support for aerospace environmental monitoring, the following conclusions can be drawn.

The current situation regarding the state of the environment requires immediate decisive steps aimed at reducing the risks of anthropogenic pollution, improving and increasing and/or preventing the reduction of quantitative and species status of biodiversity, public awareness or insufficient public awareness of negative environmental changes. and threats.

Irrational and depleting use of natural resources, pollution and littering of the environment lead to global environmental problems that threaten the stability and balance of the system in the system man – nature, have irreversible consequences – leads to ecological crisis.

Sustainable development of Ukraine includes three main components: political, economic and environmental, and accordingly the solution of global problems of rational and inexhaustible use and special protection of natural resources needs to be solved at all levels. It is necessary to have comprehensive objective information about the state of natural objects, as well as anthropogenic impacts on it. After all, the lack of such information hinders the development of appropriate decisions in the field of nature management, environmental protection and environmental safety. To overcome the consequences of the ecological crisis, it is necessary to be informed about its causes, in order to assess the emerging processes and determine the necessary actions to eliminate them. The basis for the development of any environmental assessment systems is environmental monitoring. Therefore, an important step in determining the quality of the environment is the system of collecting information about its condition.

The existing system of environmental monitoring is not ideal, and therefore cannot effectively perform the overall function of assessing the state and level of resource use to effectively forecast changes and develop recommendations for management decisions to optimize economic activity and nature management in individual regions. To this end, we propose further implementation and development of aerospace environmental monitoring. After all, aerospace activities include fundamental and applied space research in order to exchange innovative technologies and bring them to world markets, which is a very important issue for our country today.

References

1. On the decision of the National Security and Defense Council of Ukraine of September 14, 2020 “On the National Security Strategy of Ukraine”: Decree of the President of Ukraine of September 14, 2020 № 392/2020. Official Gazette of Ukraine dated 25.09.2020, № 75, Art. 2377 [in Ukrainian]
2. Sharapova, S.V.: Legal support of environmental monitoring in Ukraine: author’s ref. dis. cand. jurid. science: 12.00.06. Kharkiv, 19 p. (2002) [in Ukrainian]
3. On approval of the Regulations on the state system of environmental monitoring: Resolution of the Cabinet of Ministers of Ukraine of March 30, 1998 № 391. Official Gazette of Ukraine official publication of April 16, 1998, № 13, art. 91 [in Ukrainian]
4. Petlyuk, Y.S.: Legal basis for environmental monitoring. Bull. Ministry Justice Ukraine **12**, 126–121 (2013) [in Ukrainian]
5. On approval of the State target ecological program for environmental monitoring: Resolution of the Cabinet of Ministers of Ukraine of December 5, 2007 № 1376. Official Gazette of Ukraine official publication of 17.12.2007, № 93, Art. 3403 [in Ukrainian]
6. On the Basic Principles (Strategy) of the State Environmental Policy of Ukraine for the period up to 2030: Law of Ukraine of February 28, 2019 № 2697-VIII. Information of the Verkhovna Rada of Ukraine dated April 19, 2019, № 16, Art. 70 [in Ukrainian]
7. On approval of the action plan for the implementation of the Concept of implementation of state policy in the field of space activities for the period up to 2032: Order of the Cabinet of Ministers of Ukraine of January 25, 2012 № 48-r. Government courier dated February 29, 2012, № 39. https://zakon.rada.gov.ua/laws/show/48-2012-%D1%80#Text [in Ukrainian]
8. Strategy of space activity of Ukraine for the period till 2022: Order of the State space agency of Ukraine 21.05.2015 № 100. https://zakon.rada.gov.ua/rada/show/v0100814-15#Text [in Ukrainian]
9. Environmental law: a textbook/AP Hetman and others; for order. A.P. Hetman. Kharkiv: Pravo, 552 p. (2019) [in Ukrainian]
10. Aighewi, I.T., Ekundayo, E.O.: Monitoring of the Environment as a Whole. Encyclopedia of life support systems. Environmental Monitoring, vol. 1, 449 p. (2009) [in English]
11. Malysheva, N.R.: Basic management functions in the field of environmental protection. Environmental law of Ukraine. Academic course: textbook/for general. ed. Y.S. Shemshuchenko; Institute of State and Law. V.M. Koretsky NAS of Ukraine, Kyiv University of Law. 2nd type. 720 p. (2008) [in Ukrainian]
12. Petrov, V.V.: Ecological Law of Russia: Textbook for Universities. MBEK Publishing House, 557 p. (1995) [in Russian]
13. Shekhovtsov, V.V.: Legal regulation of the implementation of the monitoring function of management in the field of faunal legal relations and ways to improve it. Legal Sci. J. **4**(2020), 115–118 (2020). [in Ukrainian]
14. Environmental monitoring: textbook. V.M. Bogolyubov, M.O. Клименко, B.B. Mokin and others]; ed. V.M. Bogomolov [2nd ed., Reworked. and ext.]. Vinnytsia: VNTU, 232 p. (2010) [in Ukrainian]
15. Convention on the Protection and Use of Transboundary Watercourses and International Lakes. Małgorzata LandsbergUczciwek, Martin Adriaanse, Rainer Enderlein//Management of Transboundary Waters in Europe. Convention on the Protection and Use of Transboundary Watercourses and International Lakes. 22–25 September 1997 Mrzezyno, Poland. Economic Information Bureau, 536 p. (1998) [in English]
16. Nutrient Balances for Danube Countries. Project EU/AR/102A/91. Danube Applied Research Program [Text]/Final Report. 07/07/1997 [in English]

17. Ilyina, O.V.: Problems of monitoring in environmental law at the present stage. Bull. Kharkiv Natl. Univ. Internal Affairs **4**(63), 199–206 (2013) [in Ukrainian]
18. Shershun, S.M.: Management of the environmental monitoring system in Ukraine: legal and organizational issues. Law Soc. **1**, 175–181 (2015). [in Ukrainian]
19. Ecological Monitoring of the Environment. Official portal of the Ministry of Environmental Protection and Natural Resources of Ukraine. https://mepr.gov.ua/content/ekologichniy-monitoring-dovkillya.html [in Ukrainian]
20. Litovchenko, A.A.: The concept of aerospace emergency monitoring system. Sci. Bull. Civ. Prot. Fire Saf. **2**(4), 95–100 (2017) [in Ukrainian]
21. Babak, V.P., Kanchenko, V.A., Klyuchnikov, A.A., Krasnov, V.A., Chepur, N.L.: Unmanned aerial vehicles as a means of radiation monitoring of nuclear power plants and the environment. Safety Issues of Nuclear Power Plants and Chernobyl, Vip. 19, pp. 60–69 (2012) [in Russian]
22. Sergeeva, K.L.: Cognitive approach in the problems of aerospace monitoring of mining regions. Inf. Process. Syst. **1**, 27–32 (2017). [in Ukrainian]
23. Pashkov, D.P.: Features of the use of co funny pictures during ecological monitoring of territories. Modern aerospace technologies in environmental monitoring. In: Proceedings of the Scientific-Practical Seminar (November 27, 2017). Kyiv, 16 p. (2017) [in Ukrainian]
24. On space activities: Law of Ukraine of 15.11.1996 № 502/96-VR. Information of the Verkhovna Rada of Ukraine dated January 21, № 1, Art. 2 (1997) [in Ukrainian]
25. Fedorov, O.P.: On the target complex program of NAS of Ukraine «Aerospace environmental surveillance for sustainable development and security as a national segment of Horizon-2020 ERA-PLANET project (ERA-PLANET/UA). Visnik Nacional'noi' academii' nauk Ukrai'ni **12**, 35–41 (2017). https://doi.org/10.15407/visn2017.12.035
26. Kondratenko, D., Savchuk, O., Liubchych, A.: Foreign practices in land cadaster systems: an experience for Ukraine. Eur. J. Sustainable Dev. **9**(2), 325–337 (2020). https://doi.org/10.14207/ejsd.2020.v9n2p325
27. Shchypsov, O.A., Fedorovsky, O.D., Khyzhniak, A.V.: Aerospace monitoring of the Ukrainian near-shelf areas of the Black Sea as a dual purpose methodology. Visnik Nacional'noi' academii' nauk Ukrai'ni **4**, 68–75 (2018). https://doi.org/10.15407/visn2018.04.068
28. Taranenko, O.S.: Fundamentals of organization of aerospace monitoring of anthropogenic impact on the territory of the Dnieper-Oryol Nature Reserve. Urban Plan. Spatial Plan. **44**, 490–499 (2012). [in Ukrainian]
29. We are introducing aerospace monitoring to record crimes against nature. Official portal of the Ministry of Environmental Protection and Natural Resources of Ukraine. https://mepr.gov.ua/news/36824.html?fbclid=IwAR2d0uc3RW9Pi3IC-GAGMzeanu-IaLjG6pF9jtRYjbOerpyLbQUPAy_X61M [in Ukrainian]

Legal Regulation of Public Security and Tasks Related to Its Provision in the Aerospace Industry

Iryna Petrova[1(✉)] and Olha Hetmanets[2]

[1] R&D and Publishing Laboratory of National Scientific Center «Hon. Prof. M. S. Bokarius Forensic Science Institute», Kharkiv, Ukraine

[2] Department of Legal Support of Economic Activity of the Faculty № 6 at Kharkiv National University of Internal Affairs, Kharkiv, Ukraine

Abstract. The authors found a public security as a component of national security that draws attention to the need for its improvement of society and scientists, bringing the issue of its provision to the center of scientific researches. Taking into account the fact that among the components of the national security, public security has a special place, covering the protection of public and citizens interests; rights and freedoms of each person, makes it a priority task of the State to protect against those socially dangerous internal threats, which are characterized by the level of crime and violation. Consideration of scientific developments allowed to formulate the definition of public security as a state of protection of vital public interests, concentrated in its material and spiritual values, from sources of the danger of natural or artificial nature at which prevention of threats of harm by such sources of danger is provided.

It is noted that public security, including in the aerospace industry, in modern conditions, should be considered in a broad sense of the term as the security of society, taking into account the systemic approach and the relationship with all types of security and existing threats to the security. The main types of threats to public security, including in the aerospace industry, were determined: socially dangerous acts and emergencies of social, natural, man-made, and biological nature. A threat to the public security has been defined as a set of conditions and factors that create a real and potential danger to public security objects. Supporting the position of scientists, it is emphasized that legal forms of the public security can be systematized by content (law-making and law-enforcement, as regulatory and law enforcement), purposefulness (by field of protection), method of expression, etc. In addition, it has been established that non-legal forms include forms that express organizational and material-and-technical actions aimed at ensuring the public security, including in the aerospace industry, and that non-legal forms, as well as legal ones, are related to the competence of security subjects and their legal status.

Keywords: Legal regulation · Public security · Aerospace industry · Tasks

M. Nechyporuk et al. (Eds.): ICTM 2021, LNNS 367, pp. 704–719, 2022.
https://doi.org/10.1007/978-3-030-94259-5_57

1 Introduction

The development of Ukraine as a democratic State is accompanied by the orientation of society to the principles of humanism and justice, and the search for ways to build it actualizes the filling of real democratic content the processes of using basic principles of normative and legal regulation of the whole national security system. One must point out that this system is the main component of the mechanism of ensuring public order and determines its content in essence. This necessitates the clarification of the principles on the basis of which the system is formed and exists, namely the content of the principles of legal regulation of its organization and activities.

Taking into consideration the fact that the protection of national security is not only a state, but also a sociopolitical matter, normative and legal regulation of the national security will not achieve the desired results without a balanced state policy aimed at protecting the foundations of democracy and, accordingly, national and public security [1]. The Constitution of Ukraine (Article 3) stipulates that assertion and providing of the rights and freedoms of man is the main duty of the State [2].

2 Literature Review

It should be noted that the public security is a component of the national security that draws the attention of society and scientists to the need to ensure it, bringing them to the center of their scientific researches. Scientific works by V.B. Averianova, O. M. Bandurka, Yu. P. Bytiaka, S. M. Husarova, S. Yu. Zhyla, A. T. Komziuka, B. P. Kondrashova, O. Ye. Korystina, M. V. Korniienko, Ya. L. Kurbanov, O. Yu. Syniavska, V. H. Fatkhudinov, and others are devoted to the study of the public security.

As an example, M. V. Korniienko clarified the issues on protection of public order and ensuring the public order as functions of law enforcement activities, outlining the content and the main functions of law enforcement activities of the State and emphasized the direction of protection of the public order and ensuring the public security, pointing to the need for a clear understanding of the meaning of the public order [3].

Considering the correlation of the concepts: *public security* and *public order,* Ya. L. Kurbanov outlined the basic requirements for the system of security assurance; identified elements of the public security: the content of the public security; system of subjects of its guarantee; their tasks and principles of activity; a list of threats that pose a threat to the public security; means of guaranteeing it. He determined socially dangerous acts and emergencies of social, natural, man-made, and biological nature as threats to the public security [4].

Having studied the issue on legal security of the public security V. H. Fatkhudinov found that a secure legal environment and a secure legal space are their relationships or the given opportunity to assimilate that security information through appropriate government mechanisms [5]. Along with this, V. V. Novikov analyzed the legal definition of forms of the public security in Ukraine, for which a number of leading tasks were solved, among which was the formulation of methodological issues related to determining the essence of forms of the public security in Ukraine [6].

S. Yu. Zhyla outlined certain aspects of ensuring the public security by the National Police of Ukraine as a result of studying the content and essence of the concept of national security and its relation with the public security, as well as analyzed the police powers, the system, and content of police measures, their significance and a place in the mechanism ensuring the public security [7].

All of them together came up with a common position on the content of this concept, however, comprehensive consideration of the forms of the public security of Ukraine, their legal support, and areas of implementation in the aerospace industry was not conducted.

3 Main Content Presentation

In the modern world, the issues on ensuring the safe environment are raised very acutely due to economic, environmental, social, and other factors, and to ensure human existence and material development need to be solved in full. The National Security Strategy of Ukraine *Human Safety – National Safety,* approved by the Decree of the President of Ukraine No. 392/2020 of 14.09.2020 (hereinafter referred to as Strategy) determines the fundamental provisions of the national security system to ensure …. *high level of readiness of society and the state to respond to a wide range of threats* [8]. According to the Strategy, the implementation of the provisions on the establishment of the national security system provides for:

- risk assessment, timely identification of threats and determination of vulnerabilities;
- effective strategy planning and crisis management, in particular, implementation of universal protocols for crisis responses and restoring, taking into account NATO recommendations;
- acting coordination to clear interaction of security and defense sector bodies, other state bodies, hromadas, business, public society and population in prevention and responding to threats and overcoming the consequences of emergencies;
- expansion of necessary knowledge and skills in this area;
- establishing and maintaining reliable lines of communication between state government bodies and the population throughout Ukraine [8].

Given the aforementioned, one must be noted that the government proposes to establish the powerful public security system and connects this activity with the establishment of a civil protection system which provides for the creation of legal opportunities for the public, on the one hand, to exercise rights, guaranteed by the Constitution of Ukraine, and on the other hand, to protect their own interests.

Among types of the national security, the public security takes a peculiar place, as it covers protection of the interests of citizens, rights, and freedoms of every single person that is the priority of the State. The Declarative Norm of the Law of Ukraine of 17.09.2020 *On National Security of Ukraine* determines *National security and defense – a protection of vital interests, human and civil rights and freedoms for society and the individual, ensuring which is a priority of the Security Services of Ukraine, other state bodies, local governments, their officials and the public, which take concerted measures to implement and protect national interests from the impact of threats* [9].

Given definition emphasizes that in contrast to the national security, the essence of which is the protection of state sovereignty, territorial integrity and democratic constitutional order and other vital national interests of the country from real and potential threats of non-military nature, the public security determines the security of citizens state to guarantee the protection of vital interests, human and civil rights and freedoms for society and the individual, the prevention of pressing threats to the lives of citizens.

In a legal dictionary, the public security is determined as *the system of social relationships and legal norms that regulate these relationships with the objective of ensuring the public peace, inviolability of life and health, normal work and rest of citizens, normal activities of state and public organizations, institutions and enterprises* [10, p. 180].

In science the public security is defined as one of the important components of the national security, which determines the level of protection of citizens and, as a consequence, society and the state, mainly from socially dangerous internal threats, which are characterized by level of crimes and offences in society. However, the essence of the public security remains controversial among scientists.

In this regard, Ya. L. Kurbanov notes that *there are some difficulties in defining the concept: public security, which is from the point of a general methodological approach has all the necessary features that allow it to be classified as a complex legal phenomenon. Today, there are cases when scientists criticize the fact that the State, as the leading subject in ensuring public security, protects something vague. This is due to insufficient scientific development of this concept, because currently there is a lack of comprehensive researches that would be devoted to the problems of ensuring public security, both under normal conditions and in the event of emergencies of various kinds* [4, p. 90].

The definition of the concept: *public security* in detail was given by V. P. Tykhyi, who noted that the *public security, in the narrow sense of the word, as an object of crime, is a certain system of public relations, which ensures the prevention and elimination of the general danger of violent harm to law enforcement interests in general, thus guaranteeing their stability and reliability* [11, p. 24–25].

Kondrashov B. P. proposes that the public security must be understood as *the system of social relationships regulated with legal norms with the objective of ensuring human safety, public peace, comfortable facilities for work and rest of citizens, normal activity of state bodies, public associations, enterprises, institutions and organizations, from the threat posed by criminal and other illegal acts, violation of the order of use of sources of increased danger, objects and substances removed from civil turnover, phenomena of negative man-made and natural character, as well as other special circumstances* [12, p. 8]. It should be stressed that the scientist in this definition not only listed the sources and types of dangers, but also quite thoroughly identified socially dangerous acts (not only crimes) and other harmful natural and man-made phenomena.

Considering a socio-political concept of the term: *public security* through the prism of the constituent concepts, O. S. Dotsenko noted that *this category can be determined as dynamic and stable state of adverse effect, which are constantly monitored by government agencies, which should create conditions for skills development and implementation of socially significant human needs, their legal rights and freedoms, ensuring the development and preservation of material and spiritual values of society, as well as territorial integrity, sovereignty and constitutional order of the State* [13, p. 28]. It is necessary to agree with the category's components, considered by the scientists in the given definition, among which:

- objects, namely a person, society, State;
- subject, it is a question of the bodies formed for this purpose;
- circumstances of the threat as adverse effects;
- ways of providing to which the author referred: activities aimed at maintaining a dynamic and stable state.

There is a view of the public security as one of basic categories in the methodology of law, which is characterized by ambiguity, multifacetedness, the presence of a significant volume and a certain content. The public security is considered as a state in which nothing threatens anyone [14, p. 12].

A separate group of scientists in the content of the concept: *public security* unities all objects of its protection, namely, *a public security – a set of bodies and means that prevent and eliminate threats to human life and health, property and environment* [15, p. 21].

In the legislation of foreign states and scientific works we sometimes face the lack of a definition of the *public security*. Researches of certain scientists have noted that*…in some developed western countries the term: public security, does not have a clear definition and differs in its determinations depending on national and historical characteristics. This is due to the fact that in democratic societies, public security is considered primarily as a process of providing it with special structures (police, gendarmerie, national guards, etc.), who themselves must choose the best option and form of action to achieve the best results in order to maintain public order in each certain case* [16].

A review of foreign legislation provides grounds to assert that the essence of public security is related to the activities of the police to prevent and eliminate socially dangerous offenses and crimes. In Great Britain, the features of providing of the public security practices are governed by *the United Kingdom Police Act 1996*, which obliges the police to maintain public order. At the same time, the work of the British police in the protection of public order and security is not related to administrative activities, as it is carried out by all units and police services. A specific feature of the British police's public security activities is that the interaction between its various forces (services, departments) is not clearly regulated and is implemented on a contractual basis between senior officers [17].

A German jurist R. Samper also believes that the *public security* is an indefinite legal concept and it is impossible to fix by the law what in a certain case (specific time and place) should be understood as the public security and order and what the police should do for this. On his opinion, specific actions should be determined in each certain case, time, place, taking into account all available circumstances and opportunities. In accordance with the decision of Bavarian constitutional court (Bayerisches verfassungsgericht in German) dated on October 13, 1951, the term: *public security* was defined as *the integrity of health, dignity, freedom and property of citizens, as well as the inviolability of the rule of law and basic state institutions* [18, 19].

In turn, a broader definition of the public security is offered by the World Police Encyclopedia, which defines the public security as a system of public relations, governed by legal norms in order to ensure personal safety, public peace, favorable working and leisure conditions of citizens, normal functioning of public authorities, public services. associations, enterprises, institutions and organizations from the threat, posed by criminal and other illegal encroachments, violation of the procedure for using a source of increased danger, objects and substances withdrawn from free circulation, phenomena of a negative man-made and natural nature, as well as special circumstances.

The list of definitions of the category *public security* can be continued, however, the results of their analysis reveal the dual position of scientists on the determination of the essence of this concept, offering to understand it:

- firstly, as a system of relations in the field of law enforcement to protect citizens from offenses and crimes, i.e. related to the activities of law enforcement agencies, and the subjects of this activity are directly individual citizens;
- secondly, as relations that are realized in society, as general dangerous phenomena, that cover all citizens at the same time, i.e., technical or natural disasters, pandemics, etc., and are associated with various spheres of social activity, in particular, with the aerospace environment.

Thus, the *public security* should be understood as the state of protection of vital interests of society, concentrated in its material and spiritual values, from sources of danger of natural or artificial nature, which ensures the prevention of threats of harm by such sources of danger. Creating a legal environment for the organization of public security requires a legislative definition on the one hand of threats, that have a public direction and coincide in importance with threats and challenges to national interests, and on the other – the rights and responsibilities of citizens, public entities, public associations on their prevention and elimination.

So, the public security, in modern conditions, should be considered in a broad context of the concept, namely, as the security of society, taking into account a systematic approach and the relationship with all types of security and existing threats to the security. In this case, the public security is closely linked to the security of each individual as a member of that society and is an integral part of national, collective, international and global security.

It should be noted separately that in the legal literature and official documents the concept: the *public security* is used to describe the state in which the interests of the state of different nature are protected, the likelihood of occurrence of dangerous real and potential internal and external threats is minimal, as well as the likelihood of occurrence of effective development of society in general and the individual, in particular, is maximally guaranteed. This general theoretical definition fully reflects the administrative and legal (administrative and informational) aspects, and the guarantee of public security is associated with the development, adoption and implementation of regulations as guarantees of general protection of society and each person as an individual, guaranteeing their safe living, realization of their rights and freedoms [4, p. 90–91].

Professor O. E. Korystin outlined the main requirements for the security guarantee system, namely:

- operation within the limits regulated by law;
- ensuring reliable protection of national interests;

– forecasting and timely prevention of security threats;
– effective operation in both normal and emergency conditions;
– presence of a clear structural and functional delimitation of the legislative, executive and judicial branches of government;
– openness and transparency within the necessary limits regarding the activities of the relevant bodies, included in the specified system [20, p. 90].

Guaranteeing the public security not only in the normal conditions of society and the state, but also in the event of emergencies by pursuing a unified state policy in the field of guaranteeing the internal security of the state. That is why public security is associated with the implementation of measures of economic, political, organizational and other nature, aimed primarily at preventing threats to the vital interests of society and the individual. To create and maintain the required level of protection of public security facilities, the necessary things are:

– development of a system of legal norms that regulate public relations in this area;
– determination of main directions of state authorities activities on issues of guaranteeing public security and prevention of emergencies of different genesis;
– formation of special bodies for realization of directions of the state activities in the field of prevention of emergencies and guarantee of public security under this conditions;
– determination of control mechanism and supervision over their activity [21, p. 27].

It is necessary to agree with the mentioned elements, and note that they can be combined into a certain system, which in science is called *public security system.*

Analysis of the current legislation proves, that the state of threat to the public security may arise when violation of rules is allowed:

– operation of air, rail, sea, river, and road transport;
– road, sea and traffic;
– building and other works;
– operation of fire-hazardous objects and devices;
– handling of weapons, ammunition, explosives, potent poisons, radioactive isotopes and other dangerous objects and substances, etc.

Socially dangerous violations of these rules are classified in accordance with criminal law as crimes against public security (Section IX of the Criminal Code of Ukraine) [22]. Simultaneously, such offenses encroach on the personal safety of citizens. Offenses, that encroach on the public security, are also described in a number of chapters of the Code of Ukraine *On Administrative Offenses* (chapters 8, 9) [23].

Thus, the main types of threats to the public security are socially dangerous acts and emergencies of social, natural, man-made, and biological nature. The threat to the public security may be defined as a set of conditions and factors that create a real and potential danger of the public security objects.

Public security objects include: rights and freedoms, life and health of citizens, property, environment, as well as state and public institutions, the functioning of which provides normal living conditions for citizens, society and state. It is these objects of the public security that are vital interests of the individual, society and the state and, in fact, represent the highest level of interests. In turn, these interests can be attributed to the most important categories of individual and social life, the deviation of the parameters of which from the established level negatively affects the integrity, system and progress in the development of the individual, society and state. Therefore, there is an objective need for their protection by the state and society. That is, vital interests are a set of needs, the satisfaction of which reliably ensures the existence and possibility of progressive development of the individual, society and state [4, p. 92].

It should be emphasized that the security environment, according to V. H. Fatkhudinov, is *a set of characteristics of law that operates in the spatial and temporal dimensions in accordance with the action of the security space of society. The correlation of security space and security environment is based on the assimilation of security information, the ability to perceive it in certain coordinates of time and space by the individual and society through appropriate government mechanisms, which gives them a sign of normalization* [5, c. 50].

In the process of studying the legal definition of the public security forms of Ukraine, V. V. Novikov proved that *the form should be understood as an expression of the essence of the security subjects' actions, which was objectified. Since the security functions are practically implemented with the help of forms, the effectiveness of public security activities largely depends on the use of certain forms. The forms are designed to promote the appropriate implementation of security functions, the achievement of security goals with the least expenditure of effort, resources and time. It is important not only what actions are taken to ensure security, but also in what form they are implemented* [6, p. 382].

According to scientists, legal forms of public security can be systematized by content, purpose, mode of expression, etc. In terms of content, the legal form of public security, in turn, is divided into law-making and law enforcement. Scientists have determined that law-making activity consists in the creation of legal norms, their improvement, change and cancellation, that is, in the issuance of regulations governing relations in the field of public security [24, p. 187].

Law enforcement is carried out by all subjects of public security. Public authorities and their officials, local governments and their officials use all types of realization of the right: application, execution, use, observance, etc. Within the limits of their powers, established by the state, the same types of law enforcement may be used by public associations and other non-governmental organizations. As for individuals, they implement law enforcement in the field of public security in the form of compliance, enforcement and use [25, p. 98].

In Ukraine, such enforcement of the public security and civil order is entrusted, inter alia, to the security and defense Sector of Ukraine, which in accordance with the Law of Ukraine *On National Security of Ukraine* consists of four interrelated components: security forces; defense forces; defense-industrial complex; citizens and public associations that voluntarily participate in ensuring national security, whose powers are determined by law.

Note that the Ministry of Internal Affairs of Ukraine has responsibilities for the formation and implementation of state policy in the areas of:

- ensuring the protection of human rights and freedoms, the interests of society and the state, combating crime, maintaining public security and law and order, and providing police services;
- protection of the state border of Ukraine and the sovereign rights of Ukraine in its maritime economic zone;
- civil protection, protection of the population and territories from emergencies and prevention of their occurrence, liquidation of emergencies, rescue, fire extinguishing, fire and technogenic safety, activities of emergency rescue services, as well as hydrometeorological activities;
- migration (immigration and emigration), in particular, combating illegal migration, citizenship, registration of individuals, in particular, refugees and other statutory categories of migrants (Article 18 of the Law of Ukraine *On National Security in Ukraine*) [9].

The current Resolution of the Ministry of Internal Affairs of Ukraine Procedure for reviewing the public security and civil protection of Ukraine states that public security – protection of vital interests of society and the individual, human and civil rights and freedoms from unlawful encroachments, emergencies that pose a threat causing damage to life, health and property of a significant number of people [26].

It should be noted that law enforcement, depending on the content and properties of legal norms, can be combined into the following forms:

- regulatory, which is used to address individual specific cases and issues related to public safety, to realize the rights, freedoms and legitimate interests of man and citizen, government agencies, public associations, enterprises, institutions and organizations and other entities security rights;
- law enforcement, which is implemented in order to protect the legal relations of public relations in the field of security, ensuring their inviolability.

Systematizing the legal forms of activity of security subjects on the basis of purposefulness distinguish between internal and external forms. Separately, we note that domestic legal forms are used for:

- solving internal management and organizational and staffing issues;
- records management in the organization or institution, that is the subject of security;
- management of structural units and employees within the security entity (body, organization, etc.);
- management of subordinate bodies, services that are a part of the system of that or another security entity.

In turn, external legal forms are used to carry out the tasks and functions assigned to the security entity, which constitute the content of activities to ensure the public security.

Note that both internal and external legal forms of security entities can be both law-making and law-enforcement. According to the method of expression, the legal forms of activity of security subjects are divided into verbal (written and oral) and implicit, and the choice and use of these forms of security entities are determined by the legal properties of this activity.

Given the scope of the environment, covering the objectives and a large number of regulations governing these relations, the question arises as to the appropriateness of their systematization and unification of the components of national security, in particular in the field of public security in aerospace industry. As an example of establishment of legal basis of the organization of one of the components of national security – economic, is the decision of the Cabinet of Ministers of Ukraine dated on 10.04.2021 to support the draft Decree of the President of Ukraine *On Decision of NSDC of Ukraine on Economic Strategy of Ukraine for the Period up to 2025* [8]. The document identifies challenges and threats to economic security; measures to prevent and combat these threats; targets and critical limits of economic security indicators, etc. Normative legal acts on public security, which are currently determined by Articles of laws of different legal definitions by normative legal acts, including the Constitution of Ukraine, also need to be studied accordingly.

The Constitution of Ukraine establishes the basic provisions for the security of human life and the public. The second section of the Constitution of Ukraine contains a list of basic rights, freedoms and responsibilities of man and citizen, which is difficult to call exhaustive. In particular, the Fundamental Law stipulates that:

1. All human beings are free and equal in dignity and rights. Rights and freedoms are inalienable and inviolable.
2. Constitutional rights and freedoms are guaranteed and cannot be revoked. When adopting new laws or amending existing laws, it is not allowed to narrow the content and scope of existing rights and freedoms.
3. Everyone has the right to the free development of his personality, provided that the rights and freedoms of others are not violated, and has responsibilities to a society in which the free and comprehensive development of his personality is ensured [2].

In a separate section of the Constitution of Ukraine, a number of Articles are directly related to ensuring the security of human life and citizenship, enshrines the rights of the individual in this area. In particular, in accordance with Article 27 *Everyone has the inalienable right to life. No one can be arbitrarily deprived of life. The duty of the State is to protect human life. Everyone has the right to protect his life and health, the lives and health of others from unlawful encroachment* [2]. Thus, the components of the right to life are its inalienability, the prohibition of arbitrary deprivation of life, the right to protect one's life and the lives of others from unlawful encroachments. Article 27 of the Constitution of Ukraine emphasizes the inalienability of the right to life, which belongs to everyone, regardless of race, color, political beliefs, citizenship, etc. The Constitution not only enshrines the right to life, but also imposes on the state the obligation to protect human life and, consequently, to ensure this right.

On the basis of the Constitution of Ukraine, a number of laws and bylaws have been adopted, which define the rights and responsibilities of employees, professionals whose activities, in various spheres of society, ensure a normal living environment and public security in the country. Among which are:

- *codified regulations*, for example, the Civil Protection Code of Ukraine regulates relations related to the protection of the population, territories, environment, and property from emergencies, response to them, functioning of a single state system of the civil protection, and defines the powers of state bodies, the Council of Ministers of the Autonomous Republic of Crimea, local governments, rights and duties of the Ukrainian citizens, foreigners, and stateless persons, enterprises, establishments and organizations regardless of ownership of the Air Code of Ukraine, which states that *aviation security – protection of civil aviation from acts of unlawful interference, which is provided a set of measures involving material* [27];
- *incorporative legal acts,* for example – the basics of Ukrainian legislation on health care, which declares that everyone has a natural inalienable and inviolable right to health care, as well as defines the general principles of activity in the field of health care. Society and the state are responsible to present and future generations for the level of health and preservation of the gene pool of the population of Ukraine, ensure the priority of health care in the state, improving working conditions, educations, living and recreation, solving environmental problems, improving health care and the introduction of a healthy lifestyle;
- *laws of Ukraine,* which determine the legal basis for the activities of employees in various areas of life and apply to all citizens. For example, legislation on environmental protection, which establishes the legal framework for regulating relations in the field of protection, use and reproduction of natural resources, environmental safety; prevention and elimination of negative impact of economic and other activities on the environment; conservation of natural resources, genetic fund of wildlife, landscapes and other natural complexes, unique territories and natural objects related to historical and cultural heritage (Article 1 of the Law of Ukraine *On Environmental Protection*) [28]; or the Law of Ukraine of 24.02.1994 *On Ensuring Sanitary and Epidemic Safety of the Population* which *regulates public relations arising in the field of sanitary and epidemiological well-being, determines the relevant rights and responsibilities of state bodies, enterprises, institutions, organizations and citizens, establishes the procedure for organizing the state sanitary-epidemiological service and state sanitary-epidemiological supervision in Ukraine* [29]; or the Law of Ukraine of 14.10.1992 *On Labor Protection*, which defines *the basic provisions for the implementation of the constitutional right of workers to protection of their lives and health in the course of work for proper, safe and healthy working conditions, regulates with the participation of public authorities relations between the employer and the employee on safety, occupational health and the working environment and establishes a single procedure for the organization of labor protection in Ukraine* [30];
- *bylaws*, regulating relations on safety of activity of employees in separate areas and issued by the state authorities for this activity. For example, the Resolution of the Ministry of Internal Affairs of Ukraine *On approval of the Procedure for conducting a review of public security and civil protection of the Ministry of Internal Affairs of Ukraine,* which defines the organizational, methodological principles of preparation and conduct of the MIA review of the state of protection of human and civil interests, society and the state from socially dangerous acts, negative impact of the criminogenic situation and state of protection of the population and territories from emergencies [26], or the Order of the Ministry of Infrastructure of Ukraine *On Approval of the Rules*

of Emergency Rescue and Fire Safety of Flights in Civil Aviation of Ukraine, which determine the organization and conduct of emergency rescue and fire support of flights, rescue operations at airports (aerodromes) and in the area of responsibility aerodromes for conducting search and rescue operations in civil aviation of Ukraine [31], or Orders of the State Aviation Safety Supervision Service of Ukraine, which determine the technical and operational requirements for permanent and temporary runways, used for take-off and landing of light aircraft and helicopters with a maximum take-off mass not exceeding 5700 kg. The rules also set requirements for the preparation of documentation on the procedure for admission and operation of these runways by legal entities and individuals – their owners or operators. The legislation stipulates that these rules are designed to ensure the safety of light aircrafts, as well as the safety of the population of the areas adjacent to the place of flight of these aircrafts, the safety of other aircrafts in the airspace of Ukraine [32].

In addition, in the practice of security entities apply various types of administrative agreements, including:

- agreements on cooperation between security entities;
- contracts for service in security entities;
- contracts on cooperation of security subjects with non-governmental organizations, etc.

We would like to emphasize that non-legal forms include forms that express organizational and logistical actions aimed at ensuring public security, including in the aerospace industry, which are characterized. Non-legal forms, as well as legal ones, are related to the competence of security subjects and their legal status. It should be emphasized that non-legal forms do not directly entail legal consequences, so they are used by all security subjects, including citizens and non-governmental organizations.

Organizational forms of activity are not related to the emergence, change or termination of specific legal relations. They are used in the course of current day-to-day work to ensure public safety in the aerospace industry. The organizational forms of activities to ensure public safety in the aerospace industry include:

- collection, analysis, generalization and implementation of positive experience of public security;
- training of security officers, their internships and systematic training,
- providing practical assistance to field executors;
- development of scientifically substantiated recommendations and formation of measures on introduction and use of achievements of science and technology on questions of maintenance of public safety, etc.

Material and technical forms of activity are used in ancillary actions, aimed at creating the necessary conditions for the successful implementation of the basic functions of public security. They are designed to ensure the effective operation of security actors, increase productivity and culture of their management. The material and technical forms include:

- analysis of the process and results of activities;
- preparation of reports on the activities of the institution;
- preparation of certificates on certain processes of activity;
- registration of various documents (financial, personnel, technical, etc.);
- record keeping of a specific institution of the aerospace industry;
- registration of facts, etc.

Given the complexity of public security management processes in the aerospace industry, we are convinced that the increase in the volume and scope of activities aimed at ensuring public security, the role and importance of logistical actions and their forms of expression should be expected to increase gradually.

We emphasize that among these regulations the fundamental documents have a fundamental importance that determine the legal order of the organization of the authorized entity in a particular area of society. Thus, the Strategy of Public Security and Civil Protection of Ukraine is developed by the Ministry of Internal Affairs of Ukraine by the decision of the Minister of Internal Affairs of Ukraine after approval of the National Security Strategy of Ukraine and submitted to the Cabinet of Ministers of Ukraine (Article 29 of the Law of Ukraine *On National Security of Ukraine*). However, there is no relevant strategic document on public security in the aerospace industry.

The current Decree of the President of Ukraine of 09.12.2010 *On the State Service of Ukraine for Aviation Safety Supervision,* which states that *in order to improve state regulation in the area of airspace and air traffic services, to create conditions for the implementation of measures, provided by the Air Code of Ukraine, On Ensuring Aircraft Safety, Aviation Safety, Quality of Work and Services in this Area, it is advisable to create the State Aviation Safety Supervision Service of Ukraine on the basis of the State Department of Aviation Transport of Ukraine, which is being liquidated* [33].

Note that the tasks of ensuring the safety of activities in various sectors of the national economy and other areas of social activity are entrusted to authorized entities. Tasks on aviation security are assigned to the Aviation Security System Department (hereinafter referred to as the Department) is a structural unit of the State Aviation Service of Ukraine (hereinafter referred to as the State Aviation Service), which is directly subordinated to the Head of the State Aviation Service. The main tasks of the Department are such as:

- implementation of a unified state policy in the field of aviation security, protection of civil aviation from acts of illegal interference and cyber threats;
- development and implementation of rules, practices and procedures for the protection of civil aviation from acts of unlawful interference, taking into account aspects of safety, regularity and efficiency of flights;
- preparation of draft laws and regulations on the implementation of state policy and planning in the field of aviation security;
- analysis and risk of threats to civil aviation and others [33].

However, the task of ensuring public order at airports is carried out by the relevant law enforcement agencies (Article 79 § 2 of the Air Code of Ukraine) [27].

It is important to form and approve sectoral Strategies and Concepts by the relevant ministries on the basis of the Strategy of Public Security and Civil Protection of Ukraine in order to ensure the security of the country's life and the realization of national interests. In particular, to develop and implement a Strategy for the safety of activities in the aerospace industry which is able to ensure the protection and safeguarding of activities, related to the safe use of Ukrainian airspace and critical infrastructure of the aerospace industry, as well as for the development of operational plans and force plans in crisis situations in this area.

The Strategy of Public Safety and Civil Protection of Ukraine, as a fundamental document, will determine the general basis of public security in various areas of society, and sectoral Strategies, including in the aerospace industry, specify activities to ensure public security of employees in a particular area and, thus, ensure sustainable, systematic and effective development of the national economy in a safe environment.

4 Conclusions

Taking into account the scientific developments has established that *public security* should be understood as the state of protection of vital interests of society, concentrated in its material and spiritual values, from sources of danger of natural or artificial nature, which ensures the prevention of threats from such sources. It was proved that ensuring public security, as a social phenomenon, activities to ensure a safe environment, including in the aerospace industry, is realized by:

- determination of threats and means of overcoming them, ability to perform tasks as intended, infrastructure, directions of development, other indicators necessary for planning activities of the Ministry of Internal Affairs of Ukraine, the National Guard of Ukraine, the National Police of Ukraine, the State Border Guard Service of Ukraine, the State Migration Service of Ukraine, the Civil Service of Ukraine on emergency situations;
- establishing the need for budget funding, sufficient to achieve the defined goals and objectives, and planning the main areas of use of financial resources;
- preparation of regulations, defining strategy indicators in the relevant sector of society, in particular in the aerospace sector, that to be aimed at implementing the Strategy of Public Security and Civil Protection of Ukraine, their objectives, responsible for developing documents, monitoring their implementation and evaluation, etc.

References

1. The concept and principles of legal regulation of national security. https://studopedia.com.ua/1_27579_ponyattya-ta-printsipi-normativno-pravovogo-regulyuvannya-zabezpechennya-natsionalnoi-bezpeki.html [in Ukrainian]
2. The Constitution of Ukraine: Law of Ukraine of June 28, 1996. № 254k / 96 of the Verkhovna Rada (as amended) Information of the Verkhovna Rada of Ukraine (1996). № 30. Art. 141 [in Ukrainian]
3. Kornienko, M.V.: Ensuring public order and public safety as a function of law enforcement. Scientific Bulletin of Uzhhorod National University, Helvetica Publishing House, Uzhhorod **2**(22 Part 1), pp. 160–163 (2013) [in Ukrainian]

4. Kurbanov, J.L.: Ensuring public order and public safety in modern conditions: questions of theory. South Ukrainian Law J. **3–4**, 89–93 (2016). [in Ukrainian]
5. Fatkhudinov, V.G.: Legal content of public safety. Scientific Bulletin of the International Humanities University, **31**, 48–51 (2018) [in Ukrainian]
6. Novikov, V.V.: Forms of public safety. Law Forum **1**, 382–386 (2014). [in Ukrainian]
7. Zhyla, S.: Ensuring public safety by the national police: some issues. South Ukrainian Law J. **2**, 137–140 (2015). [in Ukrainian]
8. On the National Security Strategy of Ukraine: Decree of the President of Ukraine of May 26, 2015 № 287/2015. Database "Legislation of Ukraine". Verkhovna Rada of Ukraine. https://zakon.rada.gov.ua/laws/show/287/2015#n7 [in Ukrainian]
9. On the National Security of Ukraine: Law of Ukraine № 2469-VIII of August 3, 2018 (as amended). Information of the Verkhovna Rada of Ukraine (VVRU), № 31, p. 5, Article 241 (2018) [in Ukrainian]
10. Dictionary of terms on the theory of state and law: textbook. allowance. Hand. aut. count NI Panov. Kharkiv. Basis. 180 p. (1997) [in Russian]
11. Tykhy, V.P.: Criminal legal protection of public safety. Kharkiv. Higher School. 172 p. (1981) [in Russian]
12. Kondrashov, B.P.: Public safety and administrative and legal means of its maintenance: dis. Dr. Jurid. Science: 12.00.02. Moscow, 303 p. (1998) [in Russian]
13. Dotsenko, S.O.: Organization of public security police management in modern conditions: dissertation. jurid. science; 12.00.07. Kyiv, 206 p. (2003) [in Ukrainian]
14. Panov, M.I.: Security as a fundamental category in the methodology of jurisprudence (before the problem). Bull. Acad. Legal Sci. Ukraine **3**(22), 10–16 (2000) [in Ukrainian]
15. Tumanov, G.A., Frizko, V.I.: Public safety and its provision in extreme conditions. Owls. State Right **8**, 20–23 (1989). [in Russian]
16. Semin, V.S.: Conceptual approaches to public safety: foreign experience and conclusions for Ukraine. http://old2.niss.gov.ua/content/articles/files/1_Siomin-8f832.pdf (2018) [in Ukrainian]
17. The Police Act 1996 (equipment) regulations 2011 and the Police Act 1996 (services) regulations 2011: impact assessment. https://www.gov.uk/government/publications/the-police-act-1996-equipment-regulations-2011-and-the-police-act-1996-services-regulations-2011-impact-assessment [in English]
18. Samper, R.: Commentary on the Bayerischen Police Department, Munich (1965) [in German]
19. International commission of jurists. The review. N 36 (1986) [in English]
20. Korystin, O.E.: Methodological and organizational principles of economic security of Ukraine. Law Soc. **4**, 88–93 (2010). [in Ukrainian]
21. Sapozhnikov, A.I.: Administrative and legal regime of public safety: dis. cand. jurid. Science: 12.00.14. Moscow. 198 c. (2005) [in Russian]
22. Criminal Code of Ukraine: Law of Ukraine of April 5, 2001. № 2341-III (with changes and additions). Information of the Verkhovna Rada of Ukraine (VVR), № 25–26, Art. 131 (2001) [in Ukrainian]
23. Code of Ukraine on Administrative Offenses: Law of Ukraine of 07.12.1984 № 8073-X. (with changes and additions). Information of the Verkhovna Rada of the Ukrainian SSR (VVR), supplement to № 51, Art. 1122 (1984) [in Ukrainian]
24. Tsvik, M.V., Tkachenko, V.D., Rogachev, L.L.: General theory of state and law: a textbook. for ed. M.V. Tsvik and others. Kharkiv. Law. 428 p. (2002) [in Ukrainian]
25. Popov, L.L., Kornev, A.P.: Administrative Law and Administrative Activities of Internal Affairs Bodies. Acad. Ministry of Internal Affairs of the USSR, Moscow, 223 p. (1990) [in Russian]

26. Procedure for conducting a review of public security and civil protection by the Ministry of Internal Affairs of Ukraine: approved. Resolution of the Cabinet of Ministers of Ukraine of May 22, 2019. № 507. Official Gazette of Ukraine dated June 27, № 49, Art. 1656 (2019) [in Ukrainian]
27. Air Code of Ukraine: Law of Ukraine of 19.05.2011 № 3393-VI (as amended). Information of the Verkhovna Rada of Ukraine (VVR), № 48–49, p. 536 (2011) [in Ukrainian]
28. On environmental protection: the law of Ukraine of 25.06.91 № 1264-XII (with changes and additions). Information of the Verkhovna Rada of Ukraine (VVR), № 41 (1991) [in Ukrainian]
29. On ensuring the sanitary and epidemic well-being of the population: the law of Ukraine of 24.02.1994 № 4004-XII (as amended). Information of the Verkhovna Rada of Ukraine (VVR), № 27, p. 218 (1994) [in Ukrainian]
30. On labor protection: the law of Ukraine of 14.10.1992 №2694-XII (as amended and supplemented). Information of the Verkhovna Rada of Ukraine (VVR), № 49, Art. 668 (1992) [in Ukrainian]
31. Rules of emergency rescue and fire safety of flights in civil aviation of Ukraine: approved. by the order of the Ministry of Infrastructure of Ukraine dated 07.05.2013. № 286. Official Gazette of Ukraine, № 41.yu St. 1490 (2013) [in Ukrainian]
32. Rules of admission to operation of runways for flights of light aircraft: approved. by the order of the State Service of Ukraine for Aviation Safety Supervision dated 01.12.2004 № 205. Official Gazette of Ukraine, № 52, Art. 3471 (2004) [in Ukrainian]
33. About the State Service for Supervision of Aviation Safety. Decree of the President of Ukraine № 1085/2018 of 09.12.2010 № 1085/2010. Official Gazette of Ukraine, № 28, vol. 1, p. 74, Art. 1869 (2004) [in Ukrainian]

Legal and Conceptual Aspects of Securing Space Activity Entities-Facilities in Ukraine

Ihor Shynkarenko[1], Vasyl Zakharov[2], Iryna Shynkarenko[3](✉), Oleksandra Zakharova[4], and Olha Semenovykh[5]

[1] Law Department of National Aerospace University "Kharkiv Aviation Institute", Kharkiv, Ukraine
{sir,sir2009}@ukr.net

[2] Department of Operative Investigative Activity of the Lviv State University of Internal Affairs, Lviv, Ukraine
zakharov_vp@ukr.net

[3] Department of Operative Investigative Activity and Crimes Investigation of the Kharkiv National University of Internal Affairs, Kharkiv, Ukraine
i_shinkarenko@ukr.net

[4] Department of Criminal Producer and Forensics of the Lviv State University of Internal Affairs, Lviv, Ukraine
alexandra_lv@ukr.net

[5] Law Department of National Aerospace University "Kharkiv Aviation Institute", Kharkiv, Ukraine
semyonovih@ukr.net

Abstract. The place of the aerospace field in the system of critical infrastructure in Ukraine has been identified. Current problems of legal regulation of securing aerospace activity facilities as a part of Ukraine critical infrastructure have been outlined. Legal and theoretical-practical problems of securing aerospace infrastructure facilities have been considered. Organization and legal aspects of counteraction to criminal offenses and threats in the aerospace field have been disclosed. The emphasis is made on existing legal and organizational problems in functioning of non-state security organizations of economic entities in the field of critical infrastructure in directions of counteraction to different types of threats in their activities. It has been suggested to formulate the model of securing economic entities in the aerospace field regarding fight against threats to security and counteraction to criminal offenses, including of terrorist nature. Directions in the development of organizational and legal model of aerospace field security and the use of the experience of international community in the scientific, managerial and educational field have been proposed.

Keywords: Legal regulation · Security · National security · Aviation security · Space security · Air transportation · Space activity · Threats to security

1 Introduction

In conditions of the present, the world community has formulated a single opinion that state policy in the field of national security should comprise measures to protect national critical infrastructure facilities from various types of threats, including terrorist threats.

M. Nechyporuk et al. (Eds.): ICTM 2021, LNNS 367, pp. 720–735, 2022.
https://doi.org/10.1007/978-3-030-94259-5_58

In accordance with the world experience and current views of scientists in the field of national security, critical infrastructure includes military and military-industrial facilities, nuclear and hydroelectric generating stations, oil and gas pipelines, oil refinery and chemical plants, that is, facilities being particularly important for the normal support of state's life, and an accident or terrorist act in relation to them can cause a significant damage to life and health of citizens of a country where an accident happened: Biryukov D. S, Kondratov S. I (2012) [1, p. 32–53; 2, p. 15].

Over the last few years, a scientific thought in the field of theoretical and legal framework of securing critical infrastructure has included aviation and space activity to its components: Zubko G.Yu. (2019) [3, p. 144–148].

Meanwhile, modern Ukrainian scientific law community and as a result, national legislation do not maintain full focus on aerospace infrastructure in Ukraine as an integral part of the national critical infrastructure: Shynkarenko I.R. (2021) [4, p. 281–285].

Complex accounting and safety risk management should embrace the security of a state in multiple directions (especially the organization of information and technological, environmental, space and personal security of citizens). At the same time, current scientific developments touch upon only problems of state organizations activity ensuring the security of critical infrastructure facilities. Analysis of national and international acts in the field of civil aviation and space activities demonstrates that they are underdeveloped:

- organizational and legal mechanism and not defined place of security services of economic entities in the aerospace field (air transportation and space activity facilities);
- theoretical and legal model of activity coordination of security organizations in the aerospace field and system of interaction between law enforcement agencies as to neutralization of security threats as well as crime counteraction are underdeveloped.

Given the above, it could be argued that today there is a lack of research on:

1. Theoretical and legal framework in the activity of critical infrastructure organization elements singling out the organizational and legal model of security in the aerospace field of Ukraine, as a component of critical infrastructure.
2. Legal and organizational issues of creating non-state security organizations of economic entities in the field of critical infrastructure.
3. Formation of the system of coordination and interaction between national and international entities of securing aerospace field economic entities.
4. Legal issues of using nanotechnologies and biometric identification methods.

2 Literature Review

Scientific problems that are the subject of our scientific paper have been outlined in scientific papers on security science and legal principles of the use of nano and forensic technologies in the area of facilities security in the aerospace field. Research on the achievements of the National School of Security Science shows that scientific papers by Thomas Hobbes, Immanuel Kant, George Hegel, etc. form their basis. Their views have been further developed in scientific papers by O. Vlasiuk, V. Horbulin, Ye. Marchuk,

V. Palamarchuk; also in recent years in researches by O. P. Dzoban, O. H. Sydorchuk who have identified the theoretical content and ratio of the concepts: *National-State Social Security – Aviation Security – Space Security* and other fundamental principles of securing various areas of national security: Vlasyuk O. S. (2016), Gorbulin V. (1997), Sidorchuk O. G. (2018), Dzioban O. P. (2004) [5; 6, p. 76–88; 7; 8, p. 106–113].

Furthermore, at the national and international level, a theoretical representation of space security is formed which derives from the content of national security, as well as legal principles of control over space activities are developed in the last decade [9].

Stelmakh O. S. (2012) [10, p. 522–530] formed conceptual issues in the field of space security and its genesis, patterns of formation through the prism of fundamental concepts and categories of global collective security system in his scientific papers.

In his scientific papers, Omelchenko A. V. (2019) has analyzed basic trends and problems of space industry development on the basis of conformity in technological, industrial, institutional, personnel and regulatory support of projects [11].

As a result, H. Iu. Zubko focuses on the problems of classifying space industry as critical infrastructure: «Threats to the national security of Ukraine in the area of critical infrastructure are identified. It is emphasized that the efficient functioning of space infrastructure will be a guarantee of the effectiveness of the infrastructure system in general in all dimensions: land, air, water, cyberspace, and space. The conclusion is substantiated that Ukraine should develop space infrastructure, launch its satellites into orbit, and create a powerful space infrastructure. Noted that in order to achieve the set goal and tasks, it is necessary to develop appropriate directions of the state policy in this field and adequate directions of their administrative and legal support» [3, p. 144–148].

We support the opinion that for the formation of a modern airspace field security model as a component of critical infrastructure of Ukraine has world significance: Yermenchuk O. P. (2018) [12].

The analysis of mentioned and other scientific papers demonstrates the absence of scientific papers which, based on systematic and complex approaches to critical infrastructure, form theoretical and applied bases of the organizational-legal model of the aerospace field in Ukraine as a component of critical infrastructure.

Taking into account a strong link both at production and efficient level between space and aviation fields, security principles should be considered through the prism of problems with the security of air transport facilities.

At the present stage, security research in this area is carried out in the following directions:

- theoretical research as to identification and study of relevant problems of aviation transport functioning regulation, algorithms of conceptual models development in formation of transport security system: Beschastny V. M., Sobakar A. A. (2010) [13, p. 2–5];
- identification of theoretical and applied problems in security of rail, aviation and other types of transport;
- definition of the content of aviation security, peculiarities of control organization and supervision over aviation and flights security (V. P. Babak (2004), M. V. Karpenko (2004), H. I. Kasperovych (2004), V. O. Maksymov (2004), V. I. Tokarev (2004), A. V. Filipov (2010) and others) [14, p. 554; 15; 16, p. 10];

– research on theoretical and applied problems of the formation of Public policy in the field of aviation transport infrastructure security and counteraction to illegal interference in civil aviation (T. Ye. Vysotska (2012), O. Dubynskyi (2018), R. A. Kaliuzhnyi (2013), D. P. Lotiuk (2014), S. I. Moskalenko (2017), I. M. Sopilko (2013), V. V. Ustymenko (2014) and others) [17; 18, p. 131–135; 19, p. 249–251; 20, p. 10; 21, p. 149].

At the same time, in mentioned scientific papers, attention is drawn to the formation of directions in the activity of security services and structure of critical infrastructure facilities to solve tasks related to identification and neutralization of threats for their normal performance, including identification of organizational and legal direction in formation of security services for aerospace field economic entities. Especially it concerns cyberterrorism counteraction. Mentioned issues are outlined in scientific papers by M. V. Nechiporuk (2020), H. O. Spitsyna (2020), N. Ye. Filipenko (2020), I. R. Shynkarenko (2020), I. I. Shynkarenko (2020) and others [22, p. 263–262; 23, p. 206–217; 24, p. 278–281; 25, p. 100–102].

A separate direction of theoretical and applied researches is the scientific exploration of preventive activity problems on detection of potential threats and counteraction to criminal activity based on the analysis of people's behavior patterns (profile) present at facilities and adjoining territory of the air transport infrastructure.

Initially, the *profiling* term was used in the context of making a search psychological portrait (profile) of an unknown person by traces at the scene. Pioneers in the field of profiling use while combating security threats were representatives of Western School: Holmes R. M. (2009), Fallon, L., Snook, B. and others [26; 27, p. 7–29].

At the end of the XX – early XXI century, research on the use of biometrical technologies in the field of crime counteraction and neutralization of security threats at various facilities was performed by Zakharov V. P., Rudeshko V. I. (2015) [28, p. 312].

Analysis of scientific exploration results on the mentioned and other theoretical developments indicates the need for a systematic research in securing aerospace field starting with theoretical- legal and structural problems concerning neutralization of various types of threats and ending in counteraction to different types of criminal offenses, including of a terrorist nature.

3 Main Content Presentation

In the concept of the National Targeted Scientific and Technical Program of Ukraine for 2021–2025, it is outlined that the purpose of the program is to ensure the efficient use of space potential and increase of its influence on the solution of current tasks of socio-economic, environmental, information, scientific and educational development of society, ensuring realization of state interests in the field of national security and defense due to:

– "gradual establishment of the national space observation system based on domestic and foreign orbital facilities and information technologies as an innovative instrument of economy digitalization, ensuring of sustainable development, control over

emergency situations of natural and man-made nature, cooperation with the GEOSS international system in solving global and regional issues;
- creation of space observation satellite constellation on the basis of domestic platforms and medium and high distinction scanners to meet national needs and ensure joint work with the European system: COPERNICUS [29].

The grounds for the formation of such an opinion is growth at 2.2% of the world space economy in 2018, which is estimated at $ 414.75 billion.

Commercial income increased to $ 336.89 billion in 2019, grew by 6.3% comparing to $ 328.86 billion in 2018.

Space products and services, sector of commercial income from space, amounted to 217,72 billion dollars, which is 1.7% less compared with 214.18 billion dollars in 2018.

Space infrastructure and supporting industries, another main commercial sector, have grown to 119.17 billion dollars: it is 16.1% more than $ 102.66 billion in 2018; and for the first time since 2016, this sector has exceeded 100 billion dollars [30]. The indicated tendency is confirmed in 2020–2021 (Fig. 1). Such an increase in the characteristic significance of space economy determines the likely possibility of profound research on its structures as objects of terrorist threats. It is needed to accept the opinion of the participants of the International scientific and practical conference on combating environmental, space and cyber terrorism:

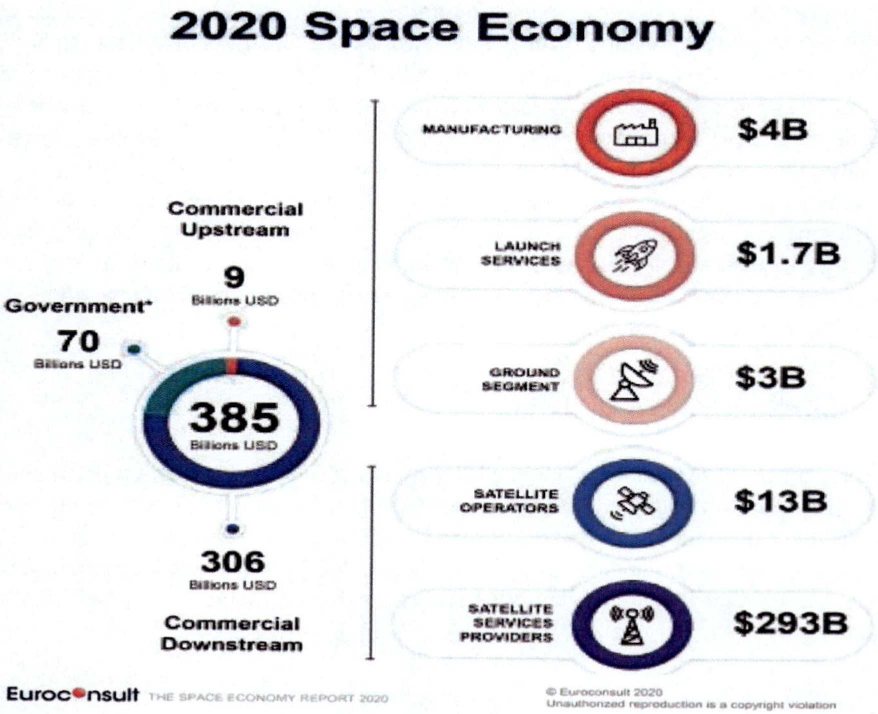

Fig. 1. Space economy from 2020.

It is about intentional use or threat of use by individuals, terrorist groups or organizations of spacecrafts with weapons placed on them to inflict significant human and material losses in a particular country or countries.

It is manifested in illegal seizure of aircrafts operating in space, violation of their operation quality to destroy them both on earth and in space [31].

Considering the previously mentioned views of scientists and the wording of the draft Law of Ukraine: *On Critical Infrastructure* dated 2021, space activities, space technologies and services are classified as vital functions which violation leads to negative consequences for Ukrainian national security [32]. In turn, to identify the organizational and legal model of security of space activity entities, space technologies and services regardless of ownership, the requirements of the Concept of the National Targeted Scientific and Technical Program of Ukraine for 2021–2025 [33] should be taken into account.

The concept stipulates a comprehensive update of space legislation: the formation of the system of laws and bylaws on a new market-oriented basis in accordance with international law and European Union law. This will contribute to the formation of the mechanism of state regulation and self-regulation, creation and development of the internal market of space technologies, information and services, introduction of advanced space technologies in the civil sector and civil technologies in the space industry and formation of security branches in the private sector involved in space activity.

Article 1 of the Law of Ukraine: *On Space Activities* stipulates that entities of space activities include enterprises, institutions and organizations of any form of ownership and organizational-legal form, including international and foreign conducting space activities [34].

At the same time, taking into account the experience of securing air transport to create an organizational and tactical model of aerospace industry security in general and each of its constituent elements in particular, it is crucial to define types of threats and corresponding subsystems of its security component.

To ensure the security of aerospace field facilities belonging to various entities of aerospace activity, preventive measures are taken to identify individuals and organizations whose activities may create conditions for illegal interference in their activity [4, p. 282]. To identify existing problems and directions for their overcoming, it is necessary to specify the content of main theoretical and applied security categories, to compile the list of airspace infrastructure facilities and entities of airspace activities. In accordance with the established procedure, economic entities are allowed to carry out the following types of space activity:

- test (excluding computer modeling) of carrier rockets, including their units, components and spacecrafts;
- launch of carrier rockets or spacecrafts;
- control of spacecrafts in low Earth orbit or in outer space;
- return of spacecrafts and / or their components from low Earth orbit or outer space to Earth [35].

As the experience of world countries demonstrates, such diversity of economic entities and space activity infrastructure facilities requires the creation of an integrated system of their security. We are of the view that its main content is the formation of such a state of the industry where the risk of threats of causing losses to people or property decreases to a manageable level. Therefore, we should agree with the view of Dr. Kestutis Paulauskas (2020): "Space threats can take multiple forms because the space system is a large target, comprising a space-based segment, a ground control terminal, data links and the actual user" [36].

According to Cain, J. R. (2016), another factor that determines the need to form an optimal airspace field security model is the presence of terrorist threats to entities and facilities of space activities and humanity in general [37, p. 93–110].

Considering the mentioned and other opinions and having compared them with the analysis of current legislation in the field of national security, counteraction to various types of crime, air transport, civil aviation and space activity, it could be deduced that their security is the result of an ongoing process in the areas of space activity security, environmental protection, economic security, information security of economic entities and neutralization of terrorist threats.

In turn, in accordance with the provided definition and opinions of scientists, including in related areas of activity, security consists of:

- ensuring the condition of the aerospace field in which the risk of losses to people or property is reduced to an acceptable level of security (airspace field security): Ajey Lele, Gunjan Singh (2008) [38];
- general neutralization of threats based on prevention and counteraction to acts of illegal interference with the activities of entities and space activity facilities (space security): Stelmakh O.S. (2016) [34, 40];
- enforcing rule of law on the territory of space activity entities and providing protection of vital interests for society and a person, rights and freedoms of a man/citizen, which support is a priority task of security forces activity, other international and national state bodies, local government bodies, their officials and public, which implement agreed measures regrading realization and protection of national interests under the influence of threats (public security): Stelmakh S. O. (2016) [40, p. 39–52];
- personal safety provision for enterprise employees: entities of space activity and those operating in their territory (personal safety): Shyra T. B. (2015) [41];
- ensuring of a sustainable economic and financial development for an entity of space activity, strong protection of property and personnel of a company from the influence of internal and external threats, prevention of undesired events resulting in negative consequences (economic and financial security): S. D. Geley, M. I. Krupka, M. D. Lesechko, J. J. Malik, E. J. Mayovets (2002) [42];
- security of entrepreneurial activity: ensuring protection of enterprises from risks: Grishchenko V. F., Grishchenko I. V., Samofalova O. A. (2015) [45];
- protection against possible losses due to theft of real estate, equipment, products, inventions, programs, etc. as well as protection against possible losses due to accidents, catastrophes resulted from the use of out-of-date equipment, low qualification and irresponsibility of employees (material and technological security (MTB): Lapin V. M. (2010) [44].
- support of normal functioning of collecting, processing, storage and use of information at space activity facilities or a set of measures aimed at ensuring the protection of

information about a person, society and state from unauthorized access, use, disclosure, destruction, amendments, familiarization, record check or destruction (in this regard, the *information protection*: *cybersecurity* term is used more often: Kovalenko Yu. O. (2010) [45, p. 123–129].

Taking into account definitions for various types of security in the aerospace field and critical infrastructure, main features can be distinguished, namely the provision of:

- prevention of illegal interference in the activity of space activity entities-facilities;
- protection of society vital interests;
- personal safety and interests of employees at space activity enterprises;
- sustainable economic and financial development of space activity entities;
- reliable protection of property and personnel of space activity organizations from the influence of internal and external threats;
- smooth functioning of the systems of collection, processing, storage and use of information on entity-facility-field, etc.

Thus, it could be assumed that the main characteristic of security is the condition of an entity and facility of space activity in which the risk of internal and external threats decreases to the minimum.

At the same time, we agree with the view regarding that all definitions of aviation security lack:» first, all available threats; second, measures and methods of counteracting them»: Brusakova, O., Iemets, L., Shulha, A., Kolotik, A. (2020) [46, p. 549].

Given the above, one of the main conditions for effective security is the formation of theoretical idea about the content of the concepts of *security* identifying their content and place in the formation of state policy and strategy in the field of national security of Ukraine:

- national security;
- space security;
- security of critical infrastructure facilities;
- security of entities;
- facilities of infrastructure space activity;
- security of economic entities at facilities of critical infrastructure, including space activities.

As a result, the formation of a sustainable idea of the security of aviation and space industry in general, and in particular entities and facilities of space activity, it is possible to determine all threats and measures that are used to detect and neutralize them.

According to the above and analysis of scientists' opinions on the latest paradigm of national security, we can deduce that the content of entities security: space infrastructure facilities and business entities located in their territory can be viewed from several points of view:

Firstly, as conditions faced by the complex system when the impact of external and internal factors does not lead to emergence of threats and negative objective and subjective processes.

Secondly, as a state of space activity entity and space infrastructure facility where the risk of losses to people or property is reduced to a manageable level.

Thirdly, both *as the type of activity* of authorized structural units on the creation of regulatory environment for entities functioning: facilities of space activity.

The above relates to the concept of the subject security in general: a facility of critical infrastructure and space in particular and its specific content based on the type of threats against a facility and individual directions of the entity activity.

Given the general theoretical approaches to the distribution of security threats by levels and sources of emergence, they can be differentiated in accordance with the following:

By *source of origin* all threats can be divided into internal and external:

- *internal threats:* factors associated with disadvantages and miscalculations in the activity of airspace activity entities-facilities which can lead to negative consequences, as well as with the effectiveness of measures taken to neutralize them;
- *external threats:* factors which origin is outside space activity entities-facilities.

In accordance with *the degree of severity* of their effects: threats with low, medium and high degree of severity:

- threats with a low degree of negative consequences: do not significantly interfere with activities of space activity entities-facilities and do not threaten the environment, human life and health;
- threats of the medium degree of negative consequences: lead to negative consequences, which overcoming does not require significant efforts;
- threats of a high degree of negative consequences: their implementation creates threat to the environment, human life and health and can lead to destruction of space infrastructure.

In relation to entities-facilities of space activity, depending on directions of their activities, types of threats, forces and means used while their neutralization, security is differentiated by:

- air security industry (ASI);
- space security (SS);
- public security (PS);
- economic security (ES);
- financial security (FS);
- personnel security (PS);
- personal security (PS);
- physical security of a facility (PSF);
- information security (IS);
- information and technological security (ITS).

If consider the space activity entity in accordance with the mentioned types of security, its normal functioning as a structural element of critical infrastructure is determined by the degree of protection from the whole complex of threats. Thus, there is a systemic problem of forming a complex concept of its security, especially in the field of cybersecurity: M. Merabti, M. Kennedy, W. Hurst (2011) [47].

Efficiency of activity regarding maintenance of the corresponding level of security of space activity entities-facilities must start with the identification of possible security threats. Therefore, in accordance with the National Security Strategy of Ukraine approved by the Decree of the President of Ukraine No. 287/2015 dated on May 26, 2015, the security of critical infrastructure (CI) is exposed to the following threats:

- critical depreciation of fixed assets of Ukrainian infrastructure facilities and insufficient level of their physical security;
- insufficient level of critical infrastructure protection from terrorist attacks and sabotage;
- inefficient security management of critical infrastructure and life support systems.
- critical depreciation of Ukrainian infrastructure fixed assets and insufficient level of their physical security;
- insufficient level of critical infrastructure protection from terrorist attacks and sabotage;
- inefficient security management of critical infrastructure and life support systems.

Threats to critical infrastructure in the event of their implementation will manifest themselves in the form of cessation of services and goods provision that are crucial for population, economy and public administration.

Identified threats are not exhaustive for entities and facilities of aerospace activity, especially in conditions of modernization of its infrastructure in compliance with the standards of critical infrastructure security.

Generalizing the above and using a systematic approach, we can make the following conclusions about levels of security threats regarding space activity entities-facilities:

- firstly, multifactoriality of a risk, security threats to entities-facilities: the aerospace field facilities;
- secondly, non-system gaps in the operational management of an entity (a facility of space activity) are not a threat to security and does not disrupt normal technological activity and security standards;
- thirdly, threats include actions aimed at destabilizing proper operation of space activity entities-facilities and obtaining a certain benefit from it or achieving another goal;
- fourthly, threats include systemic illegal violations of the current legislation regulating various areas of activity of entities-facilities of aerospace field infrastructure.

Emergence of security problems in the field of space activity is observed against the background of reorientation of the scientific field due to concentration of resources in priority directions of scientific and technical activity, introduction of scientific developments in production [48].

For implementation of provisions in mentioned procedure, innovative aerospace clusters are created:

1. Innovative Aerospace Cluster "Mechatronics" (Kharkiv): association of Ukrainian enterprises and a scientific and educational complex (National Aerospace University – Kharkiv Aviation Institute, etc.), in the area of creating modern aviation and space systems: all which work is based on synergetic integration of mechanics, electronics and computer technologies [49].
2. Scientific and Practical Cluster "Regional Center for Earth Land Monitoring "Slobozhanshchyna": formation of theoretical framework and methodology for solving tasks of territories ecological monitoring, detection of ecosystems contamination, definition of landholdings quality, identification of man-made disasters, fire conditions control, mathematical modeling while emergencies, monitoring of the state of afforestation to prevent illegal logging, unauthorized extraction of mineral deposits and determination of losses for the state [50].
3. Dnipro Space Cluster: founders of the Cluster will be up to 10 space industry enterprises, educational and research institutions, including *Southern Machine-Building Plant* State Enterprise, *Space Valley* Public Organization, Oles Honchar Dnipro National University, etc.

Such expansion of space activity entities creates conditions for the emergence of new security threats in the aerospace field and raises the problem of boosting preventive activities at the strategic, organizational and tactical levels. Given the experience of other countries, we can draw the following conclusions and suggestions [51].

4 Conclusions

1. Therefore, in XXI century, it is needed to create such a system of airspace activity that would ensure proper ongoing development of space activities of the aviation-space field production area and neutralize all types of security threats.
2. Based on the indicated state of issues related to facilities security: entities of the aerospace field, and existence of different types of security, basic categories as to formation of the strategy and tactics of security should become categories:

   - airspace security (AS);
   - space security (SS);
   - public security (PS);
   - economic security (ES);
   - financial security (FS);
   - personnel security (PS);
   - personal security (PS);
   - physical security of facilities (FSF);
   - information security (IS);
   - logistics security (LS).

Systematic categories are the aerospace industry security (AIS):

- space security (SS);

– public security (PS), which determine the content of facilities security: entities of aerospace activities and the aerospace field in general.

These types of security can be viewed from a theoretical point of view and meaningful content:

As a theoretical category, aerospace field security (AFS) is the system of internal and external conditions for the entity functioning: the facility of space infrastructure in which the risk of causing losses to people or property is reduced to an acceptable level and when the effect of external and internal factors does not lead to the emergence of threats and negative objective and subjective processes.

As an effective category, aerospace field security (AFS) is a type of authorized structural units activity regarding creation of regulatory framework of entities functioning: space activity facilities on the basis of prevention of illegal interference in their activities, creation of conditions for protection of personal security and interests, protection of personnel property of space activity entities from the influence of internal and external threats, etc.

It can be stated that aerospace activities infrastructure is a structural component of Ukrainian critical infrastructure which addresses the pressing task of ensuring its protection from external and internal threats.

Possible directions to solve the problems of forming an effective system of entities-facilities security in Ukrainian space activity:

1. Theoretical aspects:

- *clarification of the theoretical content* and ratio of categories of field security and its types, as well as measures to protect entities-facilities of the aerospace field:

  – aerospace field security (AFS);
  – space security (SS);
  – public security (PS);
  – economic security (ES);
  – financial security (FS);
  – personnel security (PS);
  – personal security (PS);
  – physical security of a facility (PSF);
  – information security (IS);
  – logistics security (LS);

- *determining the content* of causes and conditions of criminal and other offenses that are the basis for the formation of threats to the security of the whole air transport infrastructure in Ukraine.
- *development of tactics and* methods of implementation of measures and use of technical means to protect facilities of the I–IV type of air transport critical infrastructure and neutralization of security external and internal threats.
- *scientific substantiation of methods and tactics* of counteraction to criminal offenses concerning facilities and entities of aerospace activity;

- to develop scientific bases for introduction of nanotechnologies, biometric, multibiometric, multimodal and multifactor methods of implementation of persons monitoring at the sites of aerospace industry infrastructure facilities.

2. Legal aspects:

- analysis of other countries experience and views of scientists enables to determine factors that must be taken into account during the legislative process in the field of securing critical infrastructure and space activities, namely creation of the list of legislative acts requiring corresponding changes, amendments, as well as those that need to be developed, establishment of the regulatory significance of such acts and the priority of their preparation, determination of the structure and content of each of these acts, determination of the range of drafters of legislative acts and time for their preparation;
- bringing the legislation of Ukraine in line with modern requirements of world experience in securing critical infrastructure and space activity, namely the adoption of the Law of Ukraine*: On Critical Infrastructure* with the provision of norms on attribution of space activity infrastructure to its elements; specification of all types of transport, including the air, as building blocks; determination of criteria for internal and external evaluation of security units efficiency, regardless of the ownership form; formation of a mechanism to ensure legality while implementation of measures on neutralization of security threats;
- to protect cybersecurity of air transport infrastructure and aerospace field, to adopt regulations aimed at forming the legal, technological and organizational framework for information space protection;

3. Organizational aspects:

- to protect cybersecurity of aerospace field infrastructure entities-facilities, to take actions aimed at forming the legal, technological and organizational framework for the protection of information space and neutralization of illegal interference in space activity;
- introduce joint preventive measures by the security services of institutions, firms and organizations of air transport infrastructure facilities and the national police to identify suspects by their location;
- to introduce the establishment of an integrated database of "unreliable passengers" similar to the US and to create a database of such persons in the field of threats to aerospace infrastructure along with the National Police, SBGSU, SSU and Ukraine Aviation Administration considering information about persons committing criminal offenses of theoretical orientation.

4. Aspects of ensuring training for space security entities personnel in the field of security science:

- to provide training for qualified personnel within the framework of specialization 081 *Law* of the first and second level of higher education for security units for air transport and aerospace facilities;
- to create a research laboratory on legal and organizational issues of aerospace field security at National Aerospace University "Kharkov Aviation Institute" and so on.

References

1. Biryukov, D.S., Kondratov, S.I.: Critical Infrastructure Protection: Problems and Prospects for Implementation in Ukraine: An Analytical Report. NISS, Kyiv, 96 p. (2012). https://niss.gov.ua/sites/default/files/2012-09/zah_ynfrastr-b98c0.pdf [in Ukrainian]
2. Biriukov, D., Kondratov, S., Sukhodolia, O.: Green Paper on Critical Infrastructure Protection in Ukraine: Proceedings of International Expert Meetings. NISS, Kyiv, 176 p. (2015) [in Ukrainian]
3. Zubko, G.Y.: Security of space infrastructure as a course of the state infrastructure policy of Ukraine. Actual Probl. Native Jurisprudence **4**, 144–148 (2019). https://doi.org/10.15421/391931
4. Shynkarenko, I.R.: Theoretical and legal problems of securing the aviation and space industry of Ukraine. International and National Security: Theoretical and Applied Aspects: Mater. V International Scientific-Practical Conf. (Dnipro, March 12, 2021). Dnipro State University of Internal Affairs, Dnipro, 418 p. https://niss.gov.ua/sites/default/files/2012-09/zah_ynfrastr-b98c0.pdf (2021) [in Ukrainian]
5. Vlasyuk, O.S.: National Security of Ukraine: The Evolution of Domestic Policy Problems: Select. Science. Works. Kyiv. NISS. 528 p. https://niss.gov.ua/sites/default/files/2012-09/zah_ynfrastr-b98c0.pdf (2016) [in Ukrainian]
6. Gorbulin, V.: National security of Ukraine and international security. Polit. Thought **1**, 76–88 (1997). [in Ukrainian]
7. Sidorchuk, O.G.: Social security: state regulation and organizational and economic support: a monograph. LRIDU NADU, Lviv, 492 p. (2018) [in Ukrainian]
8. Dzioban, O.P.: On the question of the expediency of modeling the national security system using the Bartlett model. Scientific notes of Kharkiv Military University. Social philosophy, psychology. KhMU, Kharkiv **2**(20), 106–113 (2004) [in Ukrainian]
9. Space Law: Emerging Trends and Policy Hardcover, 224 p. https://www.amazon.com/Space-Law-Emerging-Trends-Policy/dp/9382823972 (2018) [in English]
10. Stelmakh, O.S.: International legal regime of space security: to the theory of the question. State and Law. Issue 56 (2012) [in Ukrainian]
11. Omelchenko, A.V.: Security component of space industry development management. Electronic scientific publication “Public Administration and National Security” (2019) [in Ukrainian]
12. Yermenchuk, O.P.: Basic approaches to the organization of protection of critical infrastructure in European countries: experience for Ukraine: monograph. Dnipro State University of Internal Affairs, Dnipro, 180 p. ISBN: 978-617-7665-36-5 (2018) [in Ukrainian]
13. Beschastny, V.M., Sobakar, A.A.: State policy of transport safety of Ukraine: current issues of implementation. Chamber **4**, 2–7 (2010). [in Ukrainian]
14. Babak, V.P., Kharchenko, V.P., Maksimov, V.O., others 2004 Aviation Security Ed. VP Babak. Technology, Kyiv, 584 p. (2004) [in Ukrainian]
15. Tokarev, V.I., Kasperovich, G.Y., Karpenko, M.V.: Flight Safety: Textbook. Manual. National Aviation University, Kyiv, 175 p. (2004) [in Ukrainian]
16. Filippov, A.V.: Administrative and legal security of civil aviation in Ukraine: author’s ref. dis. PhD. Science: 12.00.07. Irpen: Nat. Univ. tax. services of Ukraine. 17 p. (2010) [in Ukrainian]
17. Vysotska, T.E.: Directions for improving the state policy to ensure the safety of civil aircraft. Public administration: improvement and development, No. 11. http://www.dy.nayka.com.ua/?op=1&z=510 (2012) [in Ukrainian]
18. Dubynsky, O.: Subjects of state control over compliance with transport legislation in Ukraine. Enterprise Econ. Law **10**(2018), 131–135 (2018). [in Ukrainian]

19. Moskalenko, S.I.: Problems of aviation security in Ukraine. Comp. Anal. Law **6**, 249–251 (2017). [in Ukrainian]
20. Kalyuzhny, R.A., Sopilko, I.M.: Organizational and legal support of flight safety on air transport. Legal Bull. **3**(28), 10–15 (2013). [in Ukrainian]
21. Ustymenko, V.V., Lotyuk, D.P.: Basic aspects of control and supervision of safety in air transport of Ukraine. Legal Sci. Electron. J. **5**, 149–152 (2014). [in Ukrainian]
22. Shynkarenko, I., Spitsyna, H.: Security of aircraft infrastructure objects: theoretical aspects. Naukovyy Visnyk Dnipropetrovs'kogo Derzhavnogo Universytetu Vnutrishnikh Sprav **2**(2), 253–262 (2020). https://doi.org/10.31733/2078-3566-2020-2-253-262
23. Nechyporuk, M., Pavlikov, V., Filipenko, N., Spitsyna, H., Shynkarenko, I.: Cyberterrorism attacks on critical infrastructure and aviation: criminal and legal policy of countering. In: Nechyporuk, M., Pavlikov, V., Kritskiy, D. (eds.) Integrated Computer Technologies in Mechanical Engineering – 2020: Synergetic Engineering, pp. 206–217. Springer International Publishing, Cham (2021). https://doi.org/10.1007/978-3-030-66717-7_17
24. Shynkarenko, I.R., Spytsyna, G.O.: Actual problems of securing air transport facilities in today's conditions. Actual problems of ensuring public order and security in modern conditions: domestic and international experience: mater. II International scientific-practical conf. (Dnipro, October 25, 2019). DSUIA, Dnipro, pp. 278–281 (2019) [in Ukrainian]
25. Shynkarekno, I.R., Spytsyna, G.O.: Directions of formation of a scientific basis of safety of objects of infrastructure of air transport. International and national security: theoretical and applied aspects: mater. IV International scientific-practical conf. (Dnipro, March 13, 2020). Dnipro State University of Internal Affairs, Dnipro, pp. 100–102 (2020) [in Ukrainian]
26. Holmes, R.M.: Profiling Violent Crimes: An Investigative Tool. Sage Publications Inc., Thousand Oaks, 319 p. https://www.amazon.com/Profiling-Violent-Crimes-Investigative-Tool/dp/1412959985. ISBN-13:978-1412959988 (2009) [in English]
27. Fallon, L., Snook, B.: Criminal profiling. In: Brewer, N., Douglass, A.B. (eds.) Psychological Science and the Law. The Guilford Press, pp. 7–29 (2019). https://psycnet.apa.org/record/2018-34834-001 [in English]
28. Zakharov, V.P., Rudeshko, V.I.: Biometric Technologies in the XXI Century and Their Use by Law Enforcement Agencies: A Guide, 2nd ed., Lviv State University of Internal Affair, Ext. Lviv, 492 p. (2015) [in Ukrainian]
29. On approval of the Concept of the National Targeted Scientific and Technical Space Program of Ukraine for 2021–2025: Order of the Cabinet of Ministers of Ukraine of January 13, 2021. № 15-p. https://zakon.rada.gov.ua/laws/show/15-2021-%D1%80#Text [in Ukrainian]
30. Global turnover of the space economy from 2009 to 2019. Statista (2020). https://www.statista.com/statistics/946341/space-economy-global-turnover/ [in English]
31. Global challenges in building European security. Countering environmental, space and cyberterrorism: International scientific-practical conference. https://www.ukrinform.ua/rubric-presshall/2542682-globalni-vikliki-evropejskij-bezpeci-protidia-ekologicnomu-kosmicnomu-i-kiberterorizmu.html [in English]
32. On Critical Infrastructure: Draft Law of Ukraine 5219 of 09.03.2021. http://w1.c1.rada.gov.ua/pls/zweb2/webproc4_1?pf3511=71355 [in Ukrainian]
33. On approval of the Concept of the National Targeted Scientific and Technical Space Program of Ukraine for 2021–2025: Order of the Cabinet of Ministers of Ukraine of January 13, 2021. № 15-p. [in Ukrainian]
34. Law of Ukraine on space activities. Information of the Verkhovna Rada of Ukraine (VVR). 1997. № 1, p. 2. {With changes made in accordance with the Laws № 912-IX dated 17.09.2020 [in Ukrainian]
35. Shinkarenko, I.R.: Theoretical and legal problems of securing the aviation and space industry of Ukraine. International and national security: theoretical and applied aspects. Theses of the

V International scientific-practical conference (Dnipro, March 12, 2021). Dnepropetrovsk State University of Internal Affairs, Dnipro, 418 p. (2021) [in Ukrainian]
36. Procedure for issuance (refusal to issue, revocation) of permits for certain types of space activities: Resolution of the Cabinet of Ministers of Ukraine №197, dated February 26 2020. https://zakon.rada.gov.ua/laws/show/197-2020-%D0%BF#Text [in Ukraine]
37. Dr. Kestutis Paulauskas Space: NATO's latest frontier. NATO Review. 13 March 2020. https://www.nato.int/docu/review/articles/2020/03/13/space-natos-latest-frontier/index.html [in English]
38. Cain, J.R.: Space terrorism – a new environment; new causes. In: Cockell, C.S. (ed.) Dissent, Revolution and Liberty Beyond Earth, pp. 93–109. Springer International Publishing, Cham (2016). https://doi.org/10.1007/978-3-319-29349-3_7
39. Lele, A., Singh, G.: Space Security and Global Cooperation. Academic Foundation, 264 p. (2008). https://www.amazon.com/Space-Security-Global-Cooperation-Ajey/dp/8171887414/. ISBN-13: 978-8171887415 [in English]
40. Rathgeber, W., Remuss, N.-L., Schrogl, K.-U.: The Need to Counter Space Terrorism – A European Perspective ESPI Perspectives. No 17, January 2009 (2009). https://www.peacepalacelibrary.nl/ebooks/files/UNIDIR_pdf-art2909.pdf [in English]
41. Stelmakh, O.S.: International legal regime of security research and use of outer space for peaceful purposes: dis. For science. Degree Cand. Law of Sciences: specialty 12.00.11 – international law. Kyiv: National Academy of Sciences of Ukraine, Institute of State and Law. VM Koretsky/240 p. (2016) [in Ukrainian]
42. Shyra, T.B.: Personal safety of the employee as a basis for guaranteeing the economic security of the enterprise. Efficient Economy **10** (2015). http://www.economy.nayka.com.ua/?op=1&z=4408 [in Ukrainian]
43. Economic security of Ukraine: internal and external factors: Textbook. way. for higher textbook lock/S.D. Geley, M.I. Krupka, M.D. Lesechko, J.J .Malik, E.J. Mayovets; ed.: J.J. Malik L.: Type. center of LNU named after I. Franko, 255 p. (2002) [in Ukrainian]
44. Grishchenko, V.F., Grishchenko, I.V.: Samofalova Business security as a component of national security (2015). http://www.dy.nayka.com.ua/pdf/12_2015/5.pdf [in Ukrainian]
45. Lapin, V.M.: Life Safety: A Textbook (2007). https://pidruchniki.com/19310710/bzhd/vstup_bezpeka_zhittyediyalnosti_lyudini [in Ukrainian]
46. Kovalenko, Y.O.: Ensuring information security at the enterprise. Ind. Econ. **3**, 123–129 (2010). https://uk.wikipedia.org/wiki/ Процитовано 2021-04-28 [in Ukrainian]
47. Brusakova, O.V., Iemets, L.O., Shulha, A.M., Kolotik, A.S.: Legal and organizational-technical assurances of aviation security. Revista Amazonia Investiga **9**(29), 542–549 (2020). https://doi.org/10.34069/AI/2020.29.05.59
48. Merabti, M., Kennedy, W.: Hurst critical infrastructure protection: a 21st century challenge. 2011 International Conference on Communications and Information Technology (ICCIT)/29–31 March 2011. Aqaba, Jordan (2011). https://doi.org/10.1109/ICCITECHNOL.2011.5762681 [in English]
49. Reforming the Scientific Sphere as a Factor in the Transition of Ukraine's Economy to an Innovative Model of Development: An Analytical Note. National Institute for Strategic Studies, Kyiv (2018). http://old2.niss.gov.ua/articles/2118/ [in Ukraine]
50. Innovative aerospace cluster "Mechatronics". http://www.fed.com.ua/ua/mehatronika.html [in Ukrainian]
51. In Kharkiv on the basis of the National Aerospace University named after ME Zhukovsky (KHAI) opened the Regional Center for Space Monitoring of the Earth "Slobozhanshchina". https://kharkivoda.gov.ua/news/105963 [in Ukrainian]
52. National Research Council: Toward Infrastructure Improvement: An Agenda for Research. National Academies Press, Washington, DC (1994). https://doi.org/10.17226/4753

Certain Aspects of Staffing and Training of Forensic Experts as a Preventive Factor of Safety Ensuring Aerospace Industry Facilities

Oleh Uhrovetskyi[1] and Oleg Kurdes[2(✉)]

[1] National Scientific Centre «Hon. Prof. M. S. Bokarius Forensic Science Institute», Honored Lawyer of Ukraine, Kharkiv, Ukraine
oleg0023@i.ua

[2] National Scientific Center «Hon. Prof. M. S. Bokarius Forensic Science Institute», Kharkiv, Ukraine
oleg700615@ukr.net

Abstract. Some aspects of staffing of forensic science institutions and professional training of forensic experts are considered. Emphasis is placed on the growing expert role in ensuring preventive security measures in various areas of public life including aerospace industry. It is emphasized that the quality of drawing up an expert opinion depends on the possibility of bringing the perpetrators to justice. It is established that technical and other reasons that led to emergencies at critical infrastructure facilities and identified by an expert are the basis for appropriate work to eliminate this in the future and, ultimately, strengthen the safety of these facilities. Ways of improving staffing of specialized forensic institutions and professional training of experts were scientifically substantiated. Attention is focused on the need to improve the legal regulation of internships for experts abroad that will have positive consequences for further interstate integration in the field of forensic activities to ensure safety in aerospace industry.

Keywords: Aerospace industry · Security · Forensic science · Forensic expert · Staffing · HR management · Selection of candidates · Training · Advanced training · Internship · Mentoring

1 Introduction

Currently, world community, in addition to other global current problems faces such threats as terrorism, extremism, local and hybrid wars. The most common reasons for their emergence are the desire of key players in the political arena to prevent the loss of global leadership and others to strengthen their own role in world politics. Recently, the interstate confrontation has intensified sharply, which is most often manifested in non-military methods: political, economic, information. Some international actors use criminal methods of struggle in the form of local and hybrid wars, as well as sponsorship of terrorist and extremist organizations. All of this poses a real threat to the functioning of critical infrastructure, which includes a number of aerospace facilities. Only joint

M. Nechyporuk et al. (Eds.): ICTM 2021, LNNS 367, pp. 736–753, 2022.
https://doi.org/10.1007/978-3-030-94259-5_59

interstate efforts can stop global threats to the world order. Ensuring the security of critical infrastructure encourages governments in all countries to look for effective ways to combat criminal encroachment.

Forensic science activity plays an important role in implementation of preventive measures to ensure safety of aerospace facilities. Given the fact that terrorism, extremism, local and hybrid wars have gone beyond national borders and become a global threat, currently the issue of accelerating integration processes, including in the field of forensic science, has become acute. Exchange of scientific achievements between forensic experts, methods of conducting forensic examinations, joint training of expert personnel will be useful for any state and will lead to further interstate recognition of the forensic expert conclusion, regardless of the country of origin. This will simplify the process of prosecuting international criminals. The causes of emergencies, both technical and other, identified by forensic expert while forensic examination can well be the basis for the application of preventive measures by eliminating shortcomings or shortcomings that contributed to these events and prevent similar in the future.

Thus, the issue of improving the selection of candidates for the forensic expert positions, who by their moral and business qualities are able to perform the tasks set before them, their professional training, becomes relevant, requires comprehensive research and development of specific ways to improve. Among the areas of improving the staffing of specialized expert institutions and professional training of forensic experts should be research on such HR management components as:

- planning the need for staff for each forensic expert area of specialization;
- selection and employment;
- disposition of employees within the organizational and staffing structure of the institution;
- staff adaptation;
- staff training;
- forensic expert assessment for compliance with the job positions held;
- formation of applicant pool for managerial positions;
- creation of staff motivation system and introduction of own corporate culture.

2 Literature Review

Certain aspects of staffing of specialized forensic institutions and training of expert staff were paid attention to by such scientists as O. M. Kliuiev, O. P. Uhrovetskyi, E. B. Simakova-Yefremian, I. A. Petrova, O. P. Snigerov, N. M. Tkachenko, S. He. Hasparian, K. S. Dmitriev, O. V. Agapova, N. D. Komissarov, M. E. Bondar, O. A. Lopata and others. The scientific achievements of scientists make possible to analyze the state of regulatory framework and legal support of work with expert staff in these institutions. A significant contribution to the theory of personnel management of legal entities was made by: V. P. Chemekov, P. A. Komirchyi, P. I. Repeshko, K. B. Kozak, Yu. O. Ruban, A. M. Klochko, M. E. Morozova, I. I. Gazda and others. At the same time, it must be acknowledged that the staffing of specialized forensic institutions and the training of experts in the context of judicial reform needs to be improved through the introduction of modern methods and techniques.

Legal managerial and methodical support of work with expert personnel needs further scientific substantiation and improvement, therefore it is relevant now. Interstate integration processes in the field of forensic science including in the scientific field and professional training of experts will have positive consequences in counteracting global threats: terrorism, extremism and hybrid wars. Therefore, in time there is a need to improve the legal regulation of internships by experts abroad and mutual training of forensic experts.

3 Main Content Presentation

The end of the twentieth century and the beginning of the twenty-first marked the end of the confrontation between the two global political systems, the capitalist and the communist. Communism became closed within individual countries and, in general, ceased to influence global world politics. Political confrontation between the two political systems has been replaced by other global threats to the world order: extremism and terrorism, hybrid wars and local military conflicts. Politicians, businessmen, ordinary people, high-risk facilities and critical infrastructure including air transport, are threatened by the new challenges. Organized crime has gone beyond the borders of countries has taken on an international character and has a tendency to grow rapidly. All this complicates the dangerous situation in every country in the world, regardless of living standards. Hybrid wars actively waged by the Russian Federation in Ukraine and Syria pose a threat of escalation into the Third World War. Undermining stability in any country of the world through terror and extremism becomes an instrument of influencing the leadership of any state.

Damage to critical infrastructure, on the other hand, weakens a country ability to withstand an external aggressor and is part of a hybrid war. Today it is important to improve all components of security in each country, to develop new mechanisms for countering threats, both internal and external. Only with the coordinated efforts of the world community can the aggressor, terrorism and extremism be defeated.

O. M. Cukhodolia singles out the following cases of purposeful physical impact on the object of critical infrastructure:

- physical fascination that does not change object functionality;
- shutdown (physical seizure, redemption of assets, impact on suppliers of raw materials, creation of artificial barriers to the market, etc.);
- partial dismantling;
- complete destruction;
- creating obstacles to restore the object [1, p. 65–66].

Forensic science provides the country law enforcement agencies with independent and objective evidence in court i.e. forensic expert conclusion. Possibility of bringing the perpetrators to justice depends on the quality of these conclusions. The impossibility of punishing crimes leads to the self-confidence of criminals in the impunity of their actions. Professional forensic examination can reveal technical deficiencies in the operation of the mechanisms and other factors that led to emergency. Therefore, directly or indirectly, the role of the expert in providing preventive security measures in various areas of public life is growing.

Legal and organizational bases of forensic activity in Ukraine are defined by the Law of Ukraine: *On Judicial Examination* dated on February 25, 1994 № 4038-XII [2]. According to Article 7 of this Law, forensic activities related to forensic examinations are carried out exclusively by state specialized institutions. This article contains a list of state specialized institutions that can conduct engineering and technical types of examinations, namely:

- forensic science institutions of the Ministry of Justice of Ukraine;
- Scientific Research Forensic Centers (SRFC) of the Ministry of Internal Affairs of Ukraine, Ministry of Defense of Ukraine, Security Service of Ukraine, State Border Guard Service of Ukraine [2].

According to Article 10 of the Law of Ukraine: *On Judicial Examination*, forensic experts may be persons who have the necessary knowledge to provide an opinion on the issues under investigation. At the same time, the Law defines the requirements for a forensic expert of a state specialized institution: the availability of appropriate higher education and educational qualification level not lower than a specialist, appropriate training and qualification of a forensic expert in a particular expert specialty. Article 12 of this Law obliges a forensic expert to conduct a full investigation with the subsequent provision of a reasoned and objective written forensic expert conclusion [2].

Section IV of the Law of Ukraine: *On Judicial* Examination is devoted to international cooperation in the field of forensic science. Thus, in accordance with Article 24 of this section of the Law, state specialized institutions that perform forensic examinations have the right to establish international scientific relations with forensic and forensic institutions of other states. They have the right to organize joint scientific conferences, symposia and seminars, as well as to exchange trainees and scientific information, to carry out joint publications in the field of forensic science and criminalistics [2].

The Order of the Ministry of Justice of Ukraine (hereinafter referred to as Minjust) dated on October 8, 1998 № 53/5 (registered with the Ministry of Justice dated on November 3, 1998 under № 705/3145) approved the Instruction on the appointment and conduct of forensic examinations and expert research and Scientific methodical recommendations on the preparation and appointment of forensic examinations and expert studies (hereinafter referred to as Instruction) [3]. In accordance with paragraph 1.2.1 of the Instruction, the forensic examination carried out in connection with encroachments on critical infrastructure (including aerospace industry) includes explosive examination and in accordance with paragraph 1.2.2 of the Instruction on engineering types examinations that are carried out in connection with encroachments on critical infrastructure, include: fire and technical, life safety and aviation and technical.

Clause 1.7 of the Instruction on examination and research conducted by certified forensic experts who are not employees of state specialized institutions are limited by Article 7 of the Law of Ukraine: *On Judicial Examination.* Thus, according to Annex 6 to the Regulation on the Central Expert Qualification Commission at the Ministry of Justice of Ukraine and certification of forensic experts, approved by the order of the Ministry of Justice dated on 03.03.2015 № 301/5 [4], certified forensic experts who are not employees of state specialized institutions, following types of examinations cannot be carried out:

- forensic aviation and technical examination majoring in 10.25 *Research on aviation events and incidents*;
- explosive technical examination majoring in: 5.1 *Research on explosives, explosion products and shots*, 5.2 *Research on explosive devices, traces and circumstances of the explosion* and 5.4 *Research on circumstances and mechanism of anthropogenic explosions.*

Currently, forensic expert has more technical capacity to conduct examinations than before. Emergence of new devices and scientific developments has significantly affected the development of the theoretical foundations of forensic science. Therefore, a forensic expert must not only have basic knowledge of the specialty, which he receives while studying in higher education, but also must undergo appropriate training, internships and advanced training in specialized forensic institutions. He must also have deep knowledge in the field of jurisprudence, without which it is impossible to perform a forensic examination, the procedure for which is governed by applicable law. All this necessitates the staffing of specialized forensic institutions with qualified and responsible persons who, by their moral and business qualities, are able to perform the tasks assigned to them.

Scientists noted that the formation of highly qualified personnel of forensic institutions begins with a properly developed personnel policy and taking the necessary measures to implement in the work of forensic institutions international and domestic experience in forming a professional core of employees capable of providing the law enforcement system with quality expert opinions [5, with 312]. Therefore, today the role of the personnel department of the forensic institution in the formation of a team of experts-professionals capable of performing at a high level the tasks set before them is significantly increasing. The harmonious functioning of the entire forensic institution depends on the professionalism HR department of the institution.

We agree with the opinion of A. M. Klochko on the staff planning role as a directed activity of training, ensuring staff development, determining its professionally qualified structure and needs for employees, as well as control over their official use [6, p. 177]. In addition, it should be noted that the formation of personnel policy of the forensic institution is due to both strategic and tactical planning of personnel activities including a list of specific tasks of working with staff as a whole and with each employee.

According to V. P. Chemekov, the main tasks of HR management are:

- analysis of HR management;
- planning the need for employees;
- selection and employment;
- payroll management;
- development of a system of employee motivation;
- assessment of knowledge and skills of employees;
- performance appraisal;
- management of educational process and staff development;
- career development management of employees [7, p. 11].

P. A. Komirchyi emphasizes normative regulation of all components of public service staffing and connects it with the need for proper staffing with high-quality staff. The scientist concludes that it is necessary to regulate the provisions of administrative legislation to work with the staff of public law entities and offers the following forms of staffing:

- organizational and staff work;
- professional training;
- employee selection;
- employment;
- public service;
- employment termination.

We are convinced that we should agree with the opinion of P. A. Komirchyi that staffing forms of public service function as a single component and are generally aimed at transforming regulatory requirements into practice [8, p. 140–142].

P. I. Repeshko in scientific work proves that use of such tools of personnel management as: effective management, fair remuneration for work performed, creating appropriate working conditions, establishing a psychological climate in team, monitoring employee duties, encouraging and punishing employees for results of work, promoting the professional growth of employees should become important means of improving work efficiency [9, p. 139].

According to K. B. Kozak and Y. O. Ruban, new functions of personnel departments include:

- HR monitoring (study of the following components: dynamics of staff movement, employee motivation and the level of their job satisfaction, staff development, labor discipline, etc.);
- HR controlling (maintaining an information database on personnel, studying effectiveness of the impact of staffing on team results and effectiveness of management methods);
- personnel marketing (study of the labor market and search for specialists);
- HR consulting (study of the need for different categories of employees, analysis of ways to improve staff skills);
- HR audit (assessment of compliance of employees with their job responsibilities, study of the structure of the legal entity and corporate culture);
- work with intermediary entities on the personnel selection issue;
- use of human resources management system (HRMS) [10, p. 26].

Thus, summarizing the scientific views of scientists, we can formulate the main components of work with staff of forensic institutions:

- planning the need for specialists, selection, hiring of employees;
- staffing in accordance with the staff list;
- adaptation of staff through the institute of mentoring;
- professional training of employees;

- employee's expert assessment for compliance with job positions held;
- formation of applicant pool for managerial positions;
- development of an effective system of encouragement and punishment;
- introduction of staff motivation system;
- formation of corporate culture.

Currently it is important to review the existing legal framework for staffing of forensic institutions. Development and approval of a number of regulations will improve the work with the staff of these institutions. For this purpose, it is considered appropriate to develop and adopt:

1. Typical staffs of state forensic institutions.
2. Instruction on the organization of personnel accounting, staffing and work with staff of state forensic institutions.
3. Regulations on the procedure for forming and working with applicant pool for managerial positions of state forensic institutions.
4. Code of professional ethics of a forensic expert.
5. Instructions for the organization of professional training of staff of state forensic institutions.

In resolving a significant number of issues related to the personnel, planning and accounting activities of forensic institutions, the presence of the staff list of the institution is important. This is a local legal act that determines the composition of employees of the institution, which indicates the number of positions in the structural units with an indication of salaries. Having a properly designed staff list is also important to avoid problems with regulatory authorities.

In order to build a management hierarchy of forensic institutions, it is important to form a reserve of staff for management positions and work with this reserve, which is a significant motivation for professional growth of employees of the institution. The procedure for selecting employees for the reserve of personnel for management positions and their training should be regulated. When forming a reserve, it is necessary to follow the principle of consistent career growth of the employee. A necessary condition for quality training of reservists is the division of the educational process into theoretical and practical parts. To this end, it is necessary to develop standard training programs. The practical part of the training should be carried out by performing the duties of a temporarily absent manager, for which the reserve is applying. Working with the reserve of personnel for promotion is a constant process and aims to qualitatively update the management of the institution.

Development and adoption of the Code of Professional Ethics of a forensic expert, normatively approved procedure for prosecuting for violation of its requirements will promote formation of moral and business qualities of employees, increase their level of responsibility for the results of their work. A forensic expert is a bearer of truth and justice, so formation of his professional ethical qualities is important. The behavior of a forensic expert should comply with the norms of morality and values in force in society.

Currently, the world's leading companies use the institute of mentoring as the main method of staff adaptation that allows to solve the problem of accelerating the infusion of a new employee into the team and his professional development as a specialist. In general, mentoring can be defined as a way to transfer skills and knowledge from less experienced professionals. Development and approval of the legal support of the institute of mentoring to work with employees of forensic institutions will accelerate procedure of adaptation of employees to the new place of work and their training.

The Dictionary of Ukrainian Language (1970–1980) defines the word *mentor* as "one who gives advice, teaches; counselor, teacher" [11, p. 198]. In the same Dictionary, the word "mentoring" is defined as "the activity of assisting with advice, training, etc." [12, p. 697].

E. M. Fomin proposes to understand mentoring as a personally oriented pedagogical process that promotes the mastery and understanding of everyone who learns their profession, and creates conditions for mastering the professional competence of individual experience in solving real problems. The scientist draws attention to the negative impact of ineffective adaptation on the work of the hired employee that can negatively affect his mental state and contribute to the creation of tension in the team, will reduce productivity. Therefore, the scientist emphasizes the need to form a positive microclimate in the team and the idea of social partnership [13, p. 7].

Thus, the analysis of the works of scientists proves that functioning of the institution of mentoring will contribute to:

- formation of positive motivation to work in sponsors;
- improving the quality of professional training of employees;
- reduction of terms of adaptation of new employees to specific working conditions of forensic expert;
- development of corporate culture;
- reducing staff turnover;
- increase efficiency of the team as a whole.

Separately, it should be noted that main areas of introduction of a new employee should be familiarization: with functional responsibilities, workplace and equipment with specifics of work, with the structure of institution and location of the premises with the document management system. The duration of this stage, the mentor can significantly reduce for his subordinate. Noteworthy is the opinion of M. E. Morozova that mentor should be not only a highly qualified employee but have the desire to work with a new employee, as well as have such qualities as sensitivity, discipline, diligence and sociability [14, p. 5].

In order to improve the functioning of the institute of mentoring, it is considered appropriate to propose the establishment of an advisory body in each forensic institution: The Council of Mentors that will coordinate the work of mentors, provide them with practical assistance, offer mentors, make proposals to encourage or punish and sponsored.

The positive effect of a properly organized mentoring system is that the new employee adapts more quickly to working conditions, increases his professional level, reduces the likelihood of conflict with employees due to insufficient level of training. At the same time, the mentor increases his authority and adds status to the team, as well as realizes his own potential through the successful implementation of new functions. The institution benefits from a reduction in the time required to adapt to a new employee and receives well-trained and trained staff. This increases the overall cultural level of the team, focused on teamwork.

Mentoring should be a component of professional training of forensic experts, to help in mastering professional knowledge and skills, to promote education of employees, to form discipline and necessary business and moral and psychological qualities in beginners, to adapt them to specific conditions of forensic activity, to be an effective learning tool. It should be noted that the appointment of a mentor should be carried out not only for the hired employee, but also for the employee who has been transferred to another position that requires new knowledge and skills. An important factor in the success of mentoring is the period for which the mentor is assigned to the sponsor. We are convinced that this term should coincide with the term of training and advanced training of the sponsor.

Important requirements for a mentor should be not only his qualifications, authority, discipline, demanding, experience in the institution, as well as the presence of his pedagogical tact. The psychological compatibility of the mentor with the sponsor plays an important role. Pedagogical tact is the ability in each case to find the right course of action, but while maintaining the demands of the sponsored. Factors such as lack of interest in performing the functions of the mentor due to lack of financial incentives, his workload and lack of skills in educational work, as a rule, lead to formalism in the work and do not have positive consequences. Based on the above, we note that among the ways to improve the functioning of the institution of mentoring should be:

- material and intangible incentives for mentors;
- taking appropriate measures to improve the image of the mentor;
- appointment of mentors from more experienced and qualified employees;
- holding competitions for the title of *Best Mentor of the Institution* with awarding the winners with cash prizes, valuable gifts, etc.;
- conducting classes with mentors on basics of pedagogical skills.

Thus, the introduction of the institute of mentoring in the system of professional training of employees will significantly improve the quality of training and accelerate the adaptation to the specific working conditions of a forensic expert. We are convinced that achieving a balance in the system: "requirements for the mentor → the results of his work → encouragement of the mentor" will lead to the implementation of mentoring tasks. Failure in any component of this link will lead to formalism in this activity and will have negative consequences in the process of training and adaptation of the sponsored. Important in the proper functioning of the institute of mentoring is the joint coordinated activities of the units of the forensic institution: staffing and training, as well as the interest of the management of the structural units in which the sponsors work.

According to N. M. Tkachenko, the training of forensic experts is one of the most important means of expert support of justice in the country and emphasizes that only a trained expert is able to conduct a full study and provide an objective opinion on the issues before him [15, p. 203]. In turn, K. S. Dmitrieva proposes to consider the training of forensic experts as one of the components of strategic areas of international cooperation of specialized forensic institutions of different states. The scientist comes to the conclusion about the need for fluency of experts in some foreign languages and their training in the field of international cooperation in forensic science with knowledge of international humanitarian law [16, p. 291].

In connection with dynamic development of forensic activity and the introduction of new technologies in the practice of forensic examinations, the position of S. He. Hasparian on the need to improve the training of experts is noteworthy. The scientist comes to the conclusion about the need to train highly qualified scientific personnel for specialized research institutions [17, p. 281] and in order to improve the professional training of forensic experts proposes to resolve this issue in the administrative and legal plane [18, p. 185].

E. B. Simakova-Yefremian and his colleagues believe that law enforcement system of the state is interested in availability of highly qualified forensic experts and draw attention to the use of domestic and foreign experience in the training of forensic experts. Scholars insist on the introduction of such an effective form of international cooperation as mutual internships of experts and scientists in forensic institutions of other countries [19, p. 175]. Convinced that current legal framework for experts training in other countries needs to be revised and improved for this purpose; introduction of normative regulation of the coordination issue of training at the level of central executive bodies that would have a positive impact on the interstate exchange of experience in conducting forensic examinations with further improvement of methods of their conducting.

Among the elements of stimulating the work of a forensic expert, I. A. Petrova, in addition to material component singles out the intangible one determined by the employee's desire for professional development through acquisition of new knowledge and skills during advanced training. Interesting is her position on the need for a forensic institution "investing in human capital" that should significantly improve professionalism of employees and lead in the future to the stable development of this institution [20, p. 216]. Noteworthy are her statements about the need to create a centralized system of training and retraining of forensic experts by creating appropriate units within specialized research forensic institutions. The scientist proposes the mandatory training and retraining for all categories of forensic experts, regardless of the employment form [21, p. 19] and provided a list of necessary requirements for a forensic expert as a "specialist of high scientific qualification who can develop new creative expert methods" [22, p. 79].

While studying the views of scholars on administrative and legal regulation in the field of education, O. V. Agapova concludes that public relations are regulated by means of administrative and legal regulation. In addition, she believes that education field is covered by public relations which are governed by administrative law and proposes to consider regulations as a basis for administrative and legal regulation of training of highly qualified scientific staff for forensic science institutions of [23, p. 58–59].

In this regard, S. He. Hasparian emphasizes the need for perfect organizational and methodological changes in the methodology of training forensic experts and considers the main condition for the success of this training regulatory framework and structural elements of educational activities, development of methodological and technological education standards, control over assimilation of theoretical knowledge and practical skills by individuals [17, p. 286].

Interesting is the opinion of N. L. Komissarov that such traditional forms of educational process as lectures and seminars, during the training of forensic experts are ineffective because they "do not provide a combination of all levels of perception of information: visual, auditory and tactile". The scientist offers a method of professionally oriented games for practical classes [24, p. 297].

In addition, in order to improve the training of experts M. E. Bondar sees the need to expand the range of their specific expertise and believes that the learning process should be carried out comprehensively in the following areas:

- performing scientific workshops and practical classes on research methods;
- independent training by studying professional readings by experts;
- consultations by specialists in the field of forensic science [25, p. 291].

Currently, the world is actively developing integration processes in all areas of life. Forensic activities that provide the state's law enforcement system with objective expert opinions cannot exist within a single country. Therefore, international cooperation of forensic institutions of different countries in order to exchange methods of forensic examinations and professional training of experts becomes relevant. We are convinced that integration in forensic activity will eventually lead to legal unification of this activity and bringing national standards to a single international one that will further develop a legal mechanism for recognizing a forensic expert conclusion not only in the country where the examination was conducted.

Analysis of the work of scientific best practices allows us to state that forms of international cooperation in training of experts should be offered such as:

- conclusion of interstate agreements on cooperation;
- conclusion of agreements on cooperation between forensic science institutions of different states;
- exchange of experience in professional training of forensic expert staff by forensic science institutions of various states;
- creation of unified international methodological approaches to forensic expert training [26, p. 101].

Interesting are the views of O. M. Kliuiev who proposed the principle of mandatory states to cooperate with each other in the field of forensic science and provides forms of this cooperation, including:

- exchange of information and various types of scientific, educational, methodical and reference publications;
- organization of scientific workshops and exhibitions in the field of forensic science;
- joint development of methods of forensic examination;
- joint training, retraining and internship of forensic experts;
- mutual review of the expert's opinion;
- mutual scientific and methodological assistance on forensic activities [27, p. 156].

In Ukraine, training of Ministry of Justice forensic experts, as well as forensic experts who are not employees of state forensic science institutions, are regulated by the following regulations:

1. Law of Ukraine: On Judicial Examination.
2. Regulations on the Central Expert Qualification Commission under the Ministry of Justice of Ukraine and certification of forensic experts approved by the order № 301/5; of the Ministry of Justice of Ukraine dated on 03.03.2015.
3. Procedure for assigning qualification classes of forensic experts to employees of forensic science institutions of the Ministry of Justice of Ukraine approved by the order № 3660/5of the Ministry of Justice of Ukraine dated on 30.12.2011 [28].

Professional activity of a forensic expert is related to stressful situations that can arise in a person while forensic examination of scene objects. Therefore, HR department should first pay attention to the comprehensive study of candidates for the position of expert, including their psychological readiness for the specific working conditions inherent in forensic science. Poor quality work on the study of mental characteristics of candidates will have further negative consequences in the form of staff turnover and unsatisfactory moral and psychological climate in the team. The application of modern scientifically based methods of testing the psychological readiness of employees to work in critical situations will positively affect their ability to perform their duties in stressful conditions. The experience of negative emotional states by the employee, as a rule, leads to stress reducing his ability to work and worsens the end results. In addition, the destructive mental state of the expert leads to violations of discipline and legality and subsequently to professional deformation. The development and conduct of professional and psychological training of experts will have a positive practical significance for the formation of employees' psychological readiness for self-regulation of emotional states and behavior in stressful situations and self-confidence. I. I. Gazda considers forms of psychological support:

- professional and psychological selection (while employment);
- psychological support for relationships in the team;
- professional and psychological training of employees;
- psychological support during the performance of official duties in implementation of official activities [29, p. 16].

We state that main educational time of psychological training of forensic experts should be used through practical classes, while use innovative methods, technologies, interactive forms of learning and apply business games, trainings, simulation of situations. Employees of structural subdivisions of psychological research of forensic science institutions, specialists of higher education institutions and specialized institutions should be involved in psychological training classes. We consider it expedient to put the organization of psychological training on the units for the organization of professional training of experts. Successful psychological training will be facilitated by development of psychological training programs for experts, which will be of practical importance for formation of their psychological readiness for the performance of official duties. Initiation of psychological training is a component of the personnel policy of forensic institutions and will affect the creation of a corporate culture in the team in which there will be an atmosphere of unity and mutual interest in working results.

Quality of forensic examination directly depends on professional forensic expert training. In this regard, it is important to properly conduct this training with. Emergence of new technologies and equipment requires from the expert not only basic knowledge in the area of specialization, but he should also be sufficiently knowledgeable in other areas of knowledge.

Given that a forensic expert is a procedural person, this requires a basic knowledge of jurisprudence. Technological progress in various fields of science obliges expert to constantly update his knowledge. Therefore, it is safe to say that forensic expert training should be an ongoing process. Given the above, it can be argued that the professional forensic expert training should be divided into the following components:

- initial training in the area of specialization (performed while initial forensic expert qualification in a particular forensic expert area of specialization);
- advanced training (performed while the forensic expert qualification confirmation in a particular his area of specialization after a specified period of time);
- internship (practical training in a specific area of specialization under guidance of a responsible person after theoretical training);
- independent training (according to a separate curriculum under the control of the responsible person);
- in-service training (conducted during the year to study the theory of forensic science and by studying new regulations).

It is proposed to divide initial training in area of specialization and forensic expert advanced training into two stages:

- first stage: preparation on theoretical and procedural forensic science issues and forensic science activity;
- second stage: practical training (theoretical part of specialization and internship).

After each stage of training it is necessary to assess knowledge of employees by conducting differentiated tests. In case of positive results, employee passes the final test. In order to accept these final tests, it is necessary to introduce an automated system of anonymous evaluation of results that should be constantly updated, in accordance with changes in legislation and methods of expert research. In our opinion, for the proper functioning of this system within the central executive body, it is advisable to create a separate unit to service this system and conduct final certification for assignment (confirmation) of qualifications in expert specialties of all forensic experts in the country, regardless of their subordination and form of employment relations. In the case of a positive certification, the expert should obtain a certificate of assignment (confirmation) of the qualification of a forensic expert of a single state standard.

Development of methodological recommendations as for the professional training management of forensic experts will have a positive impact on the process of their training. These guidelines should have a standardized procedure for conducting all components of training. Approval of methodological recommendations at the level of the central executive body unifies the training of forensic experts that in our opinion will also have a positive effect. Given the specifics of expert research, it is necessary to develop separate guidelines for the training of forensic experts in each expert specialty. Given the above, we consider it necessary to add guidelines for the development of forensic experts to add, as a separate topic of research, to the research plan of the Ministry of Justice of Ukraine.

In addition to the main types of training, the following additional forms of the educational process will have a positive effect on quality of forensic expert training:

- organization of scientific conferences, workshop and practical classes by specialists in the field of specific expertise;
- performing lectures that cover issues related to forensic science by foreign experts;
- independent study of professional readings by experts;
- obtaining consultations from specialists in the field of specific expertise in an area of specialization.

Given the current technical capabilities, holding scientific conferences, workshops, lectures by foreign experts and receiving consultations, including online, which will ensure their efficiency, save on business trips and save time without compromising educational process quality. In addition, conducting lectures online by foreign experts will facilitate further integration processes in interstate relations in the field of forensic science.

Possibility of conducting professionally oriented games and trainings, solving typical situations within practical classes requires a separate study. For this purpose, it is necessary to develop separate methodological recommendations on these issues in which the appendix should include a list of typical situations and ways to their solving. Forms of these classes can be both workshops and practical classes in the workplace or in the laboratory.

We emphasize the need to implement the principle of an integrated approach to forensic expert staff training. Only by combining main types of educational process with additional, as well as forensic expert self-training, you can get a qualified employee who can make a quality and objective forensic expert conclusion that will provide law enforcement an important evidence base for prosecuting criminals.

Personal interest of forensic expert in improving their professional level is important in educational process. Combining material incentives (promotions, salary increases, additional bonuses, valuable gifts, etc.) and intangible (diplomas, inclusion of employees in various types of advisory bodies, etc.) will have a significant impact on the motivation of the expert to improve their skills.

An example of successful professional training of persons who intend to obtain (confirm, extend) qualification of a forensic expert is creation of the National Scientific Center «Hon. Prof. M. S. Bokarius Forensic Science Institute» (hereinafter referred to as NSC «Hon. Prof. M. S. Bokarius FSI») of the advanced training department (hereinafter referred to as Department) whose main tasks are training (internship) and advanced training of forensic experts of both state specialized institutions and forensic experts who are not employees of these institutions.

While educational process the Department practices the assignment of highly qualified experts by orders of the director to employees who for the first time intend to obtain the qualification of a forensic expert which in fact bears the hallmarks of mentoring. Examining motivation of interns to the classes quality with sponsors, we conclude that it is necessary to develop a system of material incentives. In this regard, it is considered appropriate to make appropriate additions to the terms of remuneration of employees of state specialized forensic science institutions. NSC «Hon. Prof. M. S. Bokarius FSI» has introduced internal examinations based on the results of theoretical and practical training courses with subsequent passing of a person who intends to obtain forensic expert qualification, exam in the qualification chamber of the Central Expert Qualification Commission at the Ministry of Justice of Ukraine.

In addition, NSC «Hon. Prof. M. S. Bokarius FSI» constantly conducts internships for persons from other specialized forensic institutions in order to obtain (confirm) the qualification of a forensic expert, both from Ukraine and from other countries. Currently it is important to develop and approve an interagency legal act that will regulate the procedure for such an internship.

We believe that the following measures will significantly improve forensic expert professional training:

- introduction of institute of mentoring for training persons who intend to obtain the forensic expert qualification and normative consolidation of the procedure for conducting this area of activity;
- development of methodical recommendations for conducting internships in each forensic expert area of specialization;
- procedure approval for management and supervision over the independent training of forensic expert staff;
- development and publication of a manual on the course: *Theoretical, organizational and procedural issues of forensic examination*;
- use of visual information (illustrations, diagrams, tables, illustrations, etc.) through the introduction of presentations on each educational topic;
- division of the practical part of training into theoretical and internship and the main time of this part should be spent on training directly at the workplace in laboratory;
- passing practical part of training by forensic experts in institutions that have a modern material and technical base for a certain area of specialization;
- normative regulation of all components of professional training of forensic experts.

The need to develop scientific and methodological literature on both theoretical and practical issues deserves special attention while organization of professional training of forensic experts. Given constant changes in procedural legislation and the improvement of methods of conducting forensic examinations, updating of existing and publication of new educational literature should be permanent.

We are convinced that the assignment (confirmation or extension) of a forensic expert qualification should be decided by a single state interdepartmental expert-qualification commission, whose powers should extend to all forensic experts in the country, regardless of means of employment and subordination. For this purpose, it is necessary to develop uniform requirements for the professional qualities of an expert and a list of questions on both theoretical and procedural knowledge of forensic science, as well as specific expertise in forensic expert areas of specialization. Work of this commission should be based on the principles of legality, independence and objectivity. It is proposed to form the commission from representatives of all ministries subordinated to forensic experts, as well as representatives of forensic experts who are not employees of state specialized institutions.

4 Conclusions

Summarizing the above, we note that, due to growing role of forensic expert in ensuring preventive security measures for critical infrastructure, including in aerospace industry, requirements for him as a specialist who by his professional and moral and business qualities able to conduct a full, objective and independent expert study, draw up a qualified forensic expert conclusion to meet the needs of the judiciary. Circumstances identified by forensic expert while research will be an impetus to eliminate the shortcomings served to cause the emergency. The following measures will be taken to improve staffing of specialized forensic science institutions and forensic expert professional training:

- regulatory and legal regulation of all components of work with staff and training of experts;
- approval of organizational and staffing structure of a specialized forensic science institution that should have a clear hierarchy of positions and no duplication of responsibilities among management, dynamic in nature, ability to meet the needs of law enforcement agencies in quality forensic expert conclusions, meet staff uniformity;
- analysis of the need for specialists, study of moral and business qualities of candidates for forensic expert positions, selection management and hiring of employees;
- arrangement of institution staff in accordance with the staff list and expert workload;
- organization of adaptation of employees to the specific working expert conditions, both for the first time hired and those who intend to obtain an additional area of specialization. For this purpose, it is proposed to use mentoring institute;
- development and approval of the system of evaluation of experts for compliance with the positions held and, in case of adoption of a conclusion on non-compliance with the position held, normative consolidation of the list of organizational consequences for employee;
- formation of applicant pool for managerial positions and its training;
- development of a system of motivation based on encouragement and expediency of punishing expert staff;
- formation of corporate culture of the institution that will promote unity of employees in implementation of tasks and create a positive working ambience;
- division of professional training into the following components: initial training, advanced training, independent training and in-service training;
- deepening of international cooperation in the field of forensic activities, including during the training of forensic experts and exchange of scientific and methodological literature.

With the use of all components of staffing and their training in working with staff of specialized forensic institutions, you can eventually get a qualified expert who can conduct a full, independent, objective forensic expert research and create an effective preventive factor in ensuring the safety of aerospace industry.

References

1. Sukhodolya, O.M.: Protection of critical infrastructure in a hybrid war: problems and priorities of state policy of Ukraine. Strategic priorities. National Institute for Strategic Studies, Kyiv **3**(40), 62–76 (2016) [in Ukrainian]
2. On forensic examination: Law of Ukraine of 25.02.1994 № 4038-XII (with changes and additions). Information of the Verkhovna Rada of Ukraine, № 28 (1994). https://zakon.rada.gov.ua/laws/show/4038-12 (20.04.2021) [in Ukrainian]
3. On approval of the Instruction on appointment and conduct of forensic examinations and expert research and Scientific and methodological recommendations on preparation and appointment of forensic examinations and expert research: Order of the Ministry of Justice of Ukraine dated 08.10.1998 № 53/5 (with changes and additions). https://zakon.rada.gov.ua/laws/show/z0705-98 (20.04.2021). [in Ukrainian]
4. On approval of the Regulation on the Central Expert Qualification Commission at the Ministry of Justice of Ukraine and certification of forensic experts: order of the Ministry of Justice of Ukraine dated 03.03.2015 № 301/5 (with changes and additions). https://zakon.rada.gov.ua/laws/card/z0249-15 (20.04.2021) [in Ukrainian]
5. Kurdes, O.V.: Regarding the personnel policy of forensic institutions of Ukraine. Problems and prospects for the development of forensic science and criminology: Mat-ly International. scientific-practical conf. (Odessa, 16.10.2020). Odessa: Odessa Research Institute of Forensic Science of the Ministry of Justice of Ukraine, pp. 308–312 (2020) [in Ukrainian]
6. Klochko, A.M.: Staffing of internal affairs bodies of Ukraine: order and basic elements. Law Forum. Kharkiv National University of Internal Affairs, Kharkiv **4**, pp. 175–180 (2014) [in Ukrainian]
7. Chemekov, V.P.: Grading: Technology of Building a Personnel Management System. Vershina, Moscow, 208 p. (2007) [in Russian]
8. Komirchy, P.A.: Forms of staffing the public service in the law enforcement sphere of Ukraine. South Ukrainian Law Journal. Odessa State University of Internal Affairs, Odessa **3**, pp. 139–142 (2019) [in Ukrainian]
9. Repeshko, P.I.: Mechanisms for staffing forensic science in Ukraine. Economy State **8**, 138–140 (2011). [in Ukrainian]
10. Kozak, K.B., Ruban, Y.O.: Improving the work of personnel service at the enterprise. Econ. Food Ind. **2**(18), 24–28 (2013) [in Ukrainian]
11. Dictionary of the Ukrainian language: in 11 volumes/Academy of Sciences of the Ukrainian SSR, Inst. of Linguistics. O.O. Potebnya; redcol.: I.K. Bilodid and others. Kyiv: Nauk. dumka, 1970–1980. Vol. 5: V.O. Vinnik, L.A. Yurchuk, 840 p. (1974) [in Ukrainian]
12. Dictionary of the Ukrainian language: in 11 volumes / Academy of Sciences of the Ukrainian SSR, Inst. of Linguistics. O.O. Potebnya; redcol.: I.K. Bilodid and others. Kyiv: Nauk. dumka, 1970–1980. Vol. 11: S.I. Golovashchuk, 699 p. (1980) [in Ukrainian]
13. Fomin, E.N.: Diversification of the institute of mentoring as a potential for successful adaptation of a young specialist. *Secondary vocational education*: monthly. theoretical and scientific method magazine. Moscow, № 7, pp. 6–8 (2012) [in Russian]
14. Morozova, M.E.: Mentoring as a process of forming the personality of a young specialist. Scientific Bulletin of University of Education Management "Economics and Management". Kyiv: University of Education Management, Issue 1, pp. 1–12 (2016) [in Ukrainian]

15. Tkachenko, N.M.: Improving the professional training of forensic experts – an important factor in the development of expert didactics. Theory and practice of forensic science and criminology: collection of scientific works. Kharkiv: Pravo, Issue 18, pp. 203–214 (2018) [in Ukrainian]
16. Dmitrieva, K.S.: Current trends in international forensic cooperation. Legal Sci. Electron. J. **1**, 288–291 (2020) [in Ukrainian]
17. Gasparyan, S.G.: Ways to improve the quality of training for forensic institutions of the Ministry of Justice of Ukraine. Theory and practice of forensic science and criminology: collection of scientific works. Kharkiv: Pravo, Issue 18, pp. 281–287 (2018) [in Ukrainian]
18. Gasparyan, S.G.: The place of administrative and legal norms in the legal regulation of staffing of forensic institutions of Ukraine. Theory and practice of forensic science and criminology: collection of scientific works. Kharkiv: Pravo, Issue 20, pp. 175–188 (2019) [in Ukrainian]
19. Simakova-Efremian, E.B., Derecha, L.M., Myasoedov, V.V.: Actual problems of cooperation with organization of foreign states in the sphere of forensic expert activity. Theory Pract. Forensic Sci. Criminalist. **18**, 175–184 (2018). https://doi.org/10.32353/khrife.2018.19
20. Petrova, I.A.: Motivation ways for employees of forensic science institutes of Ukraine. Theory Pract. Forensic Sci. Criminalist. **18**, 214–221 (2018). https://doi.org/10.32353/khrife.2018.23
21. Petrova, I.A.: Ways to improve the system of training forensic experts in Ukraine. Current issues of forensic science and criminology: collection of materials from the international scientific-practical conference dedicated to the 95th anniversary of the Kharkiv Research Institute of Forensic Science prof. M. S. Bocarius. Kharkiv: KhNDISE, pp. 17–19 (2018) [in Ukrainian]
22. Petrova, I.A.: Scientific component in the formation of a forensic expert. Current issues of forensic science and criminology: collection of materials from the international scientific-practical conference dedicated to the 100th anniversary of the birth of Doctor of Law, Professor, Honored Worker of Science and Technology of Ukraine MV Saltevsky. Kharkiv: KhNDISE, pp. 78–79 (2017) [in Ukrainian]
23. Agapova, O.V.: Educational activities of research institutions in the field of expert support of justice: current issues of administrative and legal regulation. Bull. KhNUVS **2**(85), 56–65 (2019) [in Ukrainian]
24. Komissarov, N.L.: Features of training forensic experts for the expert service of the Ministry of Internal Affairs of Ukraine. Expert support of justice: materials from the international scientific-practical conference. Simferopol: KrymNDISE, pp. 296—299 (2006) [in Ukrainian]
25. Bondar, M.E.: Improving the professional training of a handwriting expert, taking into account the expansion of his range of special knowledge. Expert Support of Justice: Materials from the International Scientific-Practical Conference. Simferopol: KrymNDISE, pp. 286–291 (2006). [in Ukrainian]
26. Lopata, O.A.: Content, tasks and forms of international cooperation in the field of forensic science. Visegrad J. Hum. Rights **3**, 97–102 (2016). [in Slovak]
27. Kliuiev, OM: About international cooperation in the field of forensic science. Theory Pract. Forensic Sci. Criminalist. **18**, 154–162 (2018). https://doi.org/10.32353/khrife.2018.17
28. On approval of the Procedure for assigning qualification classes of forensic experts to employees of research institutions of forensic examinations of the Ministry of Justice of Ukraine: order of the Ministry of Justice of Ukraine dated 30.12.2011 № 3660/5 (with changes and additions). https://zakon.rada.gov.ua/laws/show/z0002-12 (20.04.2021) [in Ukrainian]
29. Gazda, I.I.: Analysis of psychological components of professional reliability of law enforcement officers. Sci. Bull. Lviv Natl. Univ. Internal Affairs **2**, 15–24 (2014). [in Ukrainian]

Organizational and Legal Support of Economic Security of Aerospace Industry of Ukraine

Nikolay Shelukhin[1], Ivan Kharaberiush[1], Oleksandr Shelukhin[2](✉), Vitalii Tsymbalistyi[3], and Volodymyr Selevko[3]

[1] Law and Public Administration Department, Mariupol State University, Mariupol, Ukraine
[2] Chair of Navigation and Handling Ship, Azov Maritime Institute of the National University "Odessa Maritime Academy", Mariupol, Ukraine
[3] Law Department, National Aerospace University "Kharkiv Aviation Institute" NAU "KhAI", Kharkiv, Ukraine

v.selevko@khai.edu

Abstract. Some directions and conditions for ensuring the economic security of the aerospace industry are considered at the conceptual level in this research paper. Attention is paid to formation of the concept of economic security of the aerospace industry of Ukraine in historical aspect. It is emphasized that economic security is one of important components of national security and is directly dependent on safe activities of lower levels of its economy: individual enterprises in general and the aerospace industry as well. Attention is focused on the fact that at present the main threat to the economic security of enterprises is a group of internal and external factors, such as: insufficient debugging of state and legal support, low level of fair competition among business entities, criminalization of society in economic field and other factors that pose a potential threat to national interests. From the standpoint of identifying and substantiating issues, objects and subjects of ensuring the economic security of aerospace industry of Ukraine are considered. The most significant issues affecting the low level of economic security in the industry include: instability of organizational and structural management means, miscalculations in personnel policy, uncertainty of the state strategy in industry development, low efficiency in the use of finance in the presence of a high corruption component. The article reveals peculiarity of subjects ensuring economic security of enterprises and institutions in aerospace industry that include state and non-state bodies and certain persons who take part in ensuring sustainable and stable functioning of enterprises and institutions in aerospace industry. Necessity of control is demonstrated as the most important element of public administration which disciplines the workflow of enterprises in aerospace industry and significantly affects the level of its economic security.

Keywords: State · Control · Security · Economic · Enterprise · Activity · Law · Ukraine · Industry · Economic security · Aerospace · Aerospace industries

1 Introduction

One of important conditions for sustainable development of enterprises in aerospace industry while market transformations in Ukraine, formation of high end results of

M. Nechyporuk et al. (Eds.): ICTM 2021, LNNS 367, pp. 754–779, 2022.
https://doi.org/10.1007/978-3-030-94259-5_60

production and economic activity is existence of an effective system for ensuring their economic security.

Against missing background of a coherent, conceptually verified security strategy, a weak legislative base, unworked effective mechanisms for protecting enterprises in aerospace industry, issues of their economic security [1] in market conditions are becoming especially relevant.

In modern professional readings the *economic security* concept is quite often derived directly from a more general category: *security* [2]. The Issue of ensuring security was formed simultaneously with state formation. Ancient thinkers, realizing this need, dedicated their scientific treatises to it: for example, Jean-Jacques Rousseau, who wrote two hundred years ago that concern for self-preservation and security is the most important of all concerns of the state. Safety in the initial and in the general sense of the word is "a state in which there is no danger, there is protection from dangers". However, it is impossible for an enterprise to function without dangers. Hazards, problems and complexities can be helpful to some extent, playing a mobilizing role for enterprise [3]. It is only by going through the tests of economic competition that businesses achieve their goals in the marketplace. They are forced to constantly protect their economic security. Thanks to its own security, the level of economic security of the state rises. In the modern world, nations and states experience their own vulnerability to the actions of not only other countries or their associations. Activities of individual transnational corporations and banks are classified as an independent type of threat to economic security that can undermine economic sovereignty of individual states [4]; and the economic security of individual domestic enterprises in aerospace industry and their associations can be even more vulnerable to the activities of MNCs and monopoly banks.

Economic security is one of important components of national security reflecting the causal relationship between the economic level of a country's development, its military-economic potential and national security [5]. In this context, economic security of the state is formed from safe activities of the lower levels of its economy, individual enterprises.

The issue of ensuring economic security was practically not considered in the studies of the Soviet period, although protection of state ownership of production means was determined at the legislative level. Researchers of economic security such as L. I. Abalkin, N. P. Shmelev considered the problem from a national standpoint within the framework of socialist system [6]. The issue of ensuring the economic security of business entities from the point of view of law is still insufficiently studied.

Currently, the issue of economic security of enterprises in the aerospace industry has a high degree of relevance in Ukraine and around the world [7]. Indeed, in recent years, many reforms have been carried out in the country which have changed essence and principles of the entire economy of the state. Despite this, the level of ensuring the safety of these enterprises and the principles of its implementation lag behind the economic development of the state.

Currently, the main threat to economic security of enterprises is a group of internal and external factors, such as: inadequate state and legal support, low level of fair competition among business entities, criminalization of society in the economic field and other factors. These problems pose a potential threat to national interests.

We believe that solution to these issues lies in the constant monitoring and assessment of the influence degree of the above factors on the country's economy. In turn, the state should develop and implement effective measures to neutralize these factors.

In world economic science, concept of economic security is not used. In Ukraine, noticeable attention has been paid to the issue of economic security in recent years. Among other things, the concept and criteria of legal support of economic security were also studied. Different authors have the very concept of economic security; its criteria differ significantly. Therefore, definition of criteria for economic security is in the process of scientific research.

A. I. Illarionov considers economic policy to be one of the most important factors in ensuring economic security [8]. Russian scientist reduces interstate differences in the conduct of economic policy to different scales of state participation in economic life. The main directions of the state economic policy are: state entrepreneurship; budgetary policy; monetary policy; monetary policy; foreign trade policy. As we can see, there is a search for theoretical ways to protect economy.

In order to provide high quality protection of economic security of enterprises in aerospace industry, in addition to identifying threats and the correct choice of indicators of their manifestation, it is necessary to pay attention to the subjects of ensuring economic security that will determine compliance of economic security of an enterprise existing in production with real conditions and a set of necessary measures to prevent and repel threats.

Unlike other types of security, economic security itself acts as a generic concept in relation to its species: industrial, financial, food based ones, etc. Therefore, the subject of ensuring enterprise security should be specific and at the same time capable of protecting vital economic interests from external and internal threats [9].

Problems of their own economic security arise before each enterprise not only during periods of crisis but while working in a stable economic environment, complex of target tasks solved in this case has a significant difference.

As *economic security of enterprises in aerospace industry, we mean protection of its scientific and technical, technological, production, personnel and other potential from direct or indirect threats aimed at destabilizing enterprise activities.* No enterprise can feel economically secure if its products are not in demand on the market; no enterprise producing means of production can feel safe if there is a prolonged decline in the technological development of a country [10]. Security is the state of protection of the individual, society, state from external and internal dangers and threats. The methods of protecting economic security depend on: the type of market, provision scope, time of implementation and provision, method of assessment, the level of achievement [11]. Determination of factors as criteria for choosing the types of protection of economic security of enterprises in the aerospace industry allows the latter in practice to effectively maneuver forces and means starting with the choice of a professional security structure or their own security service.

1. Objects of ensuring economic security of aerospace industry of Ukraine

The objects of ensuring the economic security of the aerospace industry of Ukraine include:

- enterprises of the *Ukroboronprom* (Ukrainian Defense Industry) concern;
- enterprises manufacturing dual-use goods or their components;
- airlines and airports;
- educational institutions for professional training for aerospace industry.

Let us consider in more detail objects of ensuring economic security of aerospace industry of Ukraine and highlight the specific features that affect provision of their economic security.

Enterprises of the *Ukroboronprom* concern. Aerospace industry is strategic for the military-industrial complex. It plays one of the main roles in the defense capability of the state. Ukraine is rightfully one of the dozen of the most developed countries in the world with a full high-tech cycle of creating aviation technology. The key enterprises of the Ukrainian aviation industry are the companies that are part of the aviation cluster of the *Ukroboronprom* concern: Antonov State Enterprise, that is industry leader, civil aviation plant 410, aircraft repair plants, as well as private companies, represented primarily by the *Motor Sich* PJSC that is one of leading manufacturers of helicopters and aircraft engines.

By the time Ukraine gained independence, 3,594 enterprises were operating on the country territory producing military-purpose and dual-use products with a total number of personnel of about 3 million people. Almost 700 enterprises were involved in the production of directly military products, including 205 production associations and 139 research and production associations, with a total headcount of 1.45 million people.

Aerospace enterprises related to the defense capability of Ukraine have been operating since 2010 as part of the *Ukroboronprom* State Concern. *Ukroboronprom* was created in accordance with the Decrees of the President of Ukraine and the Resolutions of the Cabinet of Ministers of Ukraine [12] in 2010 as an association of the military-industrial complex (hereinafter referred to as MIC) of strategic importance. The concern unites more than 100 state-owned enterprises and factories in the main branches of the defense industry – aircraft construction and aircraft repair, engine and turbine construction, shipbuilding, rocketry, ammunition and armored vehicles. Ukroboronprom has more than 10 design bureaus and six special exporting companies. The company employs about 67 thousand people.

From 2010 to the beginning of 2014, the concern included 134 enterprises of the defense industry of Ukraine (125 state and state-owned, as well as 9 joint-stock companies, the management of the corporate rights of the state in respect of which was transferred to the Concern) The Concern consisted of five structural divisions: aircraft construction and aircraft repair; high-precision weapons and ammunition, armored, automotive, engineering and special equipment; shipbuilding and marine engineering; radar, radio communication; Antiaircraft Defense Systems.

The main tasks of the Concern were defined in the relevant regulatory documents which include:

- implementation of centralized regulation, control and coordination of activities of the Concern members;
- assistance in the use and development of export, production, scientific and technical potential of the Concern members;

- creation and maintenance of a system for the sale of high-tech industrial products, generalization of the results of intellectual activity and working performance and service providing;
- assistance to the Concern members in conducting applied research in promising areas of science and technology development, introducing advanced technologies into production;
- attracting investments in military-industrial complex to create competitive samples of high-tech industrial products;
- development of relevant documents and implementation of marketing strategies;
- development of international markets for weapons, military and specialized vehicles.

As of the beginning of 2018, in the production of military products and components, out of 447 Ukrainian enterprises, aircraft manufacturing and aircraft repair enterprises, 27 ones [13].

Due to the fact that workflow on some of enterprises of aerospace industry of Ukraine was stopped, and other enterprises did not participate in the state defense order for a long time, they were put up for privatization by the State Property Fund in 2021. These are, first of all: Vinnytsia Aviation Plant "VIAZ", Research Institute "Buran", etc. [14].

Currently in Ukraine there are only 147 operating state-owned enterprises of the *Ukroboronprom* Concern with a total staff of no more than 100 thousand people. It should be noted that in the military-industrial complex of Ukraine, about 250 private business entities, created over the past twenty years, have appeared and are functioning according to a number of indicators are quite effective in production and mobile in decision-making. The share of private enterprises in the fulfillment of the defense order is growing every year and today their number in the state defense order is almost twice the number of executors from public sector enterprises [15].

In accordance with the Decree No. 59 by President of Ukraine dated on February 27, 2020, the *Ukroboronprom* State Concern is currently undergoing reorganization [16]. On its basis, it is planned to form a central executive body responsible for the formation and implementation of the state military-industrial policy and performing the functions of managing state property objects in the military industrial complex i.e. a new defense holding: *Defense Systems of Ukraine* and from the former enterprises of "*Ukroboronprom*" 6 holding companies were created under the auspices of the management company *Defense Systems of Ukraine* under the conditional names: *Radar Systems*, *Precision Weapons*, *Marine Systems*, *Special Chemistry and Ammunition*, *Armored Equipment*, *Aviaremont*, as well as the holding *Aerospace Systems of Ukraine* (under the auspices of "*Defense Systems of Ukraine*" and the State Space Agency of Ukraine), which will include the leading Ukrainian enterprises for the production of rocket and space technology: " *Production Association Yuzhny Machine-Building Plant*, *Antonov serial production plant*, etc. [17].

Enterprises producing dual-use products or components. These enterprises include economic entities at whose facilities the production of special metals and alloys, engines and other power plants, radar and communications equipment, electronic computers, weapons is established [18]. For example, the Ukrainian coastal missile system Neptune with RK-360 cruise missiles was developed by the Luch design bureau, Motor Sich Joint Stock Company in Zaporizhzhia supplies the MS-400 turbojet engine, developer of the seeker is the *Radionix* LLC, and the Kharkov Aviation Plant supplies some components. The *Vizar* state enterprise which has recently undergone modernization, will be engaged in the serial production of missiles of this type [19].

Airlines and airports. As of 01.07.2020, 16 national airlines were officially registered in Ukraine: Azur Air Ukraine, Kharkiv Airlines, Anda Air, Atlasjet Ukraine, AeroSvit Airlines, Bravo Airways, WizzAir Ukraine, Dniproavia, Donbassaero, Ukraine International Airlines, Motor Sich Airlines, Windrose (WINDROSE), Ukrainian-Mediterranean Airlines (UM Air), Air Onix (Air Onix), UTair Ukraine (UTair Ukraine), YanAir (YanAir) and 26 airports: Berdyansk (Berdyansk), Vinnitsa (Gavrishovka), Dnipro (Dnepropetrovsk), Donetsk (Donetsk), Zhitomir (Zhitomir), Zaporozhye (Wet), Ivano-Frankovsk (Ivano-Frankovsk), Kiev (Boryspil), Kiev (Igor Sikorsky Kyiv International Airport (Zhuliany), Kiev (Antonov), Kryvyi Rih (Lozovatka), Kropyvnytskyi (Kirovohrad), Luhansk (Luhansk), Lviv (Lviv), Mariupol (Mariupol), Mykolaiv (Mykolaiv), Odessa (Odessa), Poltava (Suprunovka), Rivne (Rivne), Sumy (Sumy), Ternopil (Ternopil), Uzhgorod (Uzhgorod), Kharkiv (Kharkiv), Kherson (Kherson), Khmelnitskyi (Khmelnitskyi), Chernivtsi (Chernivtsi)[1].

According to the air navigation services provider Ukraerorukh, in 2020 the number of flights in the Ukrainian sky among airlines was distributed as follows:

- Turkish Airlines: 14 623 flights (10.3%);
- Ukraine International Airlines: 14 406 flights (10,1%);
- Belavia Belarusian Airlines: 10 644 flights (7.5%);
- Wizz Air Hungary: 9 151 flights (6,4%);
- WINDROSE airlines: 8 511 flights (6%);
- SkyUp: 8 434 flights (5.9%);
- LOT: 6 260 flights (4,4%);
- Ryanair: 5 965 flights (4,2%);
- Azur Air Ukraine: 4 663 flights (3,3%);
- Pegasus Airlines: 3 446 flights (2,4%).

In total, *Ukraerorukh* dispatchers escorted 142 047 flights in 2020, which is 57.6% less than in 2019. Ukrainian airlines made 49 733 flights (−54.7%), foreign air carriers performed 92 314 flights (−59.1%).

Decrease in the number of aircraft in the Ukrainian sky is associated with the COVID-19 pandemic and travel restrictions that were introduced around the world [20].

Educational institutions. Staff training for aerospace industry of Ukraine is carried out by two specialized higher educational institutions:

- National Aerospace University “Kharkiv Aviation Institute” NAU “KhAI”
- National Aviation University (Kiev) with its separate structural divisions – Aviation and Space Lyceum. Sikorsky, Higher Vocational School of NAU, College of Engineering and Management, Slavic College, Kiev College of Computer Technologies

[1] However, it should be emphasized that for various reasons (major repairs of airports, the military conflict in Donbas, restrictions related to COVID-19, etc.) more than half of airlines and airports do not implement economic activities related to air transportation.

and Economics of NAU, Vasylkiv College, Kryvyi Rih College, Kropyvnytskyi Flight Academy.

Specifics of aerospace industry enterprises in Ukraine affecting the provision of economic security. Collapse of the military-industrial complex of the former Soviet Union led to the destruction of traditional system-forming relationships between enterprises and industries in the aerospace field that have not yet been restored or reformatted. Strategic objectives for:

- formation of a public administration system in the military-industrial field, its organizational structure;
- creation of a regulatory and legal framework for military–industrial complex (MIC);
- development and practical implementation of the state military technical and defense-industrial policy;
- ensuring the presence of the domestic MIC accessibility in the world arms market as a mandatory attribute of the state international policy has not been fulfilled [21].

In connection with this, specificity of the enterprises of aerospace industry of Ukraine affecting the provision of economic security is destructive and has such features as:

1. Instability of the organizational and structural form of management. In 1991, Ukraine got a part of the aviation industry enterprises, certain elements of production of special chemicals, production capacity of rocket and space industry, production and development of aircraft engines, turbines, and element base. All enterprises of the Ukrainian military-industrial complex had a state ownership, so they could work according to unified plans under state administration. Transformation and privatization of enterprises in the aerospace industry in Ukraine led to replacement of the overwhelming majority of enterprises with a form of ownership, from state to private. This in led to a weakening of state control over quality of products, a significant rise in prices and the generation of corruption schemes [22]. In the history of independent Ukraine, the defense-industrial complex management system is constantly being reformed:

    - 1990–1992: State Committee of Ukraine for Defense Industry and Mechanical Engineering
    - 1992–1997: Ministry of Mechanical Engineering, Military-Industrial Complex and Conversion;
    - 1997–2011: Agency for the Defense Industry Complex, Ministry of Industrial Policy;
    - 2011–2020: *Ukroboronprom* State Concern;
    - since 2020: *Defense Systems of Ukraine* State Holding.

2. Miscalculations in personnel policy have led to: lower requirements for management of enterprises in the industry; problems in the formation of personnel policy, taking into account the main strategic directions of development of the aerospace industry in Ukraine; lack of a system for training highly qualified personnel for important and

promising areas of the aerospace industry; constant drain of scientific personnel; loss of technologies and technical documentation of the latest Ukrainian developments. Frequent changes in the leadership of the industry (appointment to top management positions according to political quotas), due to which there is a lack of continuity in economic, organizational and management activities.

3. Uncertainty the state strategy in the industry development, hence the lack of workflow planning of enterprises in the industry, placement of ad hoc orders, focus on foreign products and difficulties with import substitution [23].

Aerospace industry of Ukraine was tasked with creating its own high-precision missile weapons, not inferior in performance to the best world models [24]. At the same time, the state did not fully finance these works and significant results were not achieved.

Since 2015, Antonov State Enterprise has ceased production of aircraft (Antonov State Enterprise and more than 25 Ukrainian aircraft manufacturing enterprises specialized in military transport aircraft). Instead of new promising developments in aircraft technology, the work of the aviation and space industry of Ukraine is focused on the repair of existing equipment. For example, in 2017, aircraft building and aircraft repair enterprises transferred to the Air Force of Ukraine about 40 units of restored and modernized aircraft and helicopter equipment including up to 10 fighter aircraft [25].

4. Low efficiency in the use of finance involving a high corruption component. Over 85% of enterprises are financially unstable, more than 45% of enterprises are teetering on the brink of bankruptcy, due to the lack of working capital, they are unable to fulfill large-scale orders [26]. At the same time, a number of corruption scandals related to embezzlement of funds and their misuse were reflected in the media (Svinarchuk case, purchase of spare parts in Russia, theft at the Lviv Armored Plant, etc.).
2. Subjects of ensuring economic security of aerospace industry of Ukraine.

Many entities directly and indirectly participate in ensuring management of economic security of enterprises in aerospace industry of foreign countries [27] and Ukraine.

Subjects of economic security of enterprises can be classified, depending on their affiliation, into two groups:

- state bodies belong to legislative, executive and judicial branches of government, designed to ensure the safety of all law-abiding participants in business relations without exception, and the activities of these bodies cannot be controlled by entrepreneurs themselves. These bodies form legislative basis for functioning and protection of business activities in the aerospace industry in its various aspects. The set of actions of state bodies to ensure the economic security of enterprises constitutes state regulation in the field of ensuring the protection of the economic security of objects in the aerospace industry. State bodies include: directly the state itself, represented by the President, the Verkhovna Rada, the Cabinet of Ministers; bodies exercising control powers in the field of the aerospace industry: ministries, state committees and other sectoral executive bodies; law enforcement agencies as subjects of ensuring economic

security in the field of entrepreneurial activity: the police, the Ministry of Emergency Situations, the Security Service of Ukraine, the Bureau of Economic Security, etc.; bodies of the prosecutor's office and the judicial system that carry out prosecutorial and judicial supervision in the field of ensuring economic security in the field of entrepreneurial activity;
- non-state legal entities and individuals directly carrying out activities to protect the economic security of a particular enterprise. Such subjects can be: employees of the company own security service and invited employees from other specialized enterprises that provide services for the protection of entrepreneurial activities.

Let us consider each of these groups in more detail.

State bodies, mainly take part in ensuring economic security at all its levels (state, regional, security of enterprises). Therefore, considering their role in the issue under study, we will start from the fact that all levels of economic security are interdependent. Depending on the scope of competence, the functions they perform, assigned tasks and powers we divide them into three subgroups:

1. State bodies of general competence take part in ensuring economic security of business entities along with the implementation of other functions inherent in such bodies. This subgroup includes: The Verkhovna Rada of Ukraine, President of Ukraine, National Security and Defense Council, Cabinet of Ministers of Ukraine, the Constitutional Court of Ukraine, courts of general jurisdiction, local government bodies and local self-government.
2. Ministries and departments that in accordance with the powers assigned to them, solve the tasks of ensuring economic security in the field of economic activity, which they head, or which they belong to.
3. Law enforcement, regulatory and other bodies directly entrusted with the task of preventing, detecting and solving crimes (offenses) in the financial and economic fields, carrying out control measures to identify violations of the current financial and economic legislation and their elimination.

Law enforcement agencies form an independent group of state bodies, isolated on the basis of professional activity, with clearly defined tasks and performing certain law enforcement functions. The purpose, tasks and functions of law enforcement agencies in general are to ensure the rule of law and law and order in all field of life in our country. For ensuring economic security, activities of law enforcement agencies are mainly reduced to combating economic crime, namely: preventing and exposing offenses in all sectors of the Ukrainian economy [28].

Fight against crime in general and economic crime in particular, as the main activity of law enforcement agencies is in three main areas: general management; crime prevention, actually law enforcement.

Law enforcement activity to ensure the economic security of the state is a type of state activity carried out by specially authorized state bodies and aimed at ensuring legality and law and order in the economic field of the state, protecting the rights and interests of both an individual and collectives, organizations and institutions, as well as the state in the whole.

Aforementioned activity is carried out through the application of measures of state coercion and public influence in strict accordance with the law and in compliance with the procedure (order) established for with the bringing of the perpetrators to legal responsibility.

Functions of a separate subdivision of a law enforcement agency engaged in the fight against economic crime are considered in regulatory legal acts. These functions include: administrative, preventive, investigative measures, criminal procedure, executive and security.

Administrative function consists in the executive and administrative leadership activities, services and subdivisions of internal affairs bodies for implementation and management of public order protection and ensuring public safety within the powers determined by law.

Preventive function is crime and offence hindering, identify and eliminate the causes and conditions that contribute to.

Investigative function is carried out in accordance with the Law of Ukraine: *On Operative Investigation Activity* and consists in the identification of information by the relevant departments of law enforcement agencies about committed or impending crimes, as well as about persons hiding from the investigation or serving a sentence for their arresting.

Criminal procedural function is to consider applications and reports on crimes committed, conduct an inquiry and pre-trial investigation in criminal cases, the investigation of which is entrusted by law to law enforcement agencies.

Executive function is to carry out administrative penalties within its competence.

Security function is to protect property of individuals and legal entities of all types of property on a contractual basis.

Control and supervisory function is to exercise, in accordance with legislation, administrative supervision over persons released from places of confinement, other control and supervisory powers.

System of regulatory bodies involved in ensuring the economic security of enterprises in the aerospace industry is quite numerous and diverse. It is customary to divide all control bodies into bodies of general competence that perform control functions along with other organizational and managerial functions and specialized control bodies.

Specialized state regulatory bodies exercise control over professional areas of activity (tax, budget, currency, banking, customs, environmental, veterinary control, fire, sanitary supervision, etc.). As a rule, the legal status of each of these bodies is determined by a separate normative act.

Activities of non-state actors of economic security of enterprises and institutions are often understood as non-state law enforcement activities.

Procedure for the implementation of non-state law enforcement activities in Ukraine is regulated by both the norms of public and private law. In Ukraine, this type of activity is still not normatively fixed, and the very concept of *non-state law enforcement activity* is missing in legislation.

The legal basis for implementation of non-state law enforcement activities in Ukraine is the provision of Art. 3 of the Constitution of Ukraine which established that a person, his life and health, honor and dignity, inviolability and safety are recognized as the highest social value, the provision of Part 4 of Art. 55 of the Constitution, which enshrined the right of everyone to protect their rights and freedoms from violations and unlawful encroachments by any means not prohibited by law and the provision of Part 1 of Art. 59 of the Constitution of Ukraine which established that everyone has the right to legal assistance.

In conditions of modern Ukraine, a system of subjects of non-state law enforcement activity has already been formed, but the process of legal status legislative registration of certain types of non-state law enforcement activity subjects remained incomplete. Currently in our state, such activities are carried out by a non-state system, represented by private security, security-technical and detective enterprises, commercial security services, enterprises of various forms of ownership, information bureaus, bank security services, training centers, specialized funds, research structures, specialized faculties institutions of higher education. It should be noted that a significant part of these faculties are professionals, former employees of the Security Service, internal affairs bodies, Armed Forces and special forces units.

Tasks solved by the non-state system are aimed at protecting legal rights and interests of entrepreneurs and contribute to the overall development of entrepreneurship in Ukraine. Therefore, the non-state system for ensuring the security of entrepreneurship and the individual acts as an objective ally of the state system, and this determines the possibility and necessity of interaction between state services and security agencies with representatives of the non-state security system.

Emergence of non-state law enforcement activities in Ukraine in the early 90s of 20th century associated with fundamental transformations that have taken place in the political, economic and legal field of society. It was in the context of the increased crime situation in the country, corruption growth and imperfection of legal mechanisms for protecting human rights that the inability of state law enforcement agencies to provide mobile security in all social fields of the society life turned out to be mobile and completely on their own. This, on the one hand, led to emergence of the non-state sector of law enforcement that is designed to protect persons engaged in professional activities associated with an increased risk to life and safety (entrepreneurs, journalists, etc.), as well as to meet the needs of society for security services. On the other hand, the implementation of non-state law enforcement activities is conditioned by the fact that in certain cases it acts as an entrepreneurial activity which purpose is to make a profit.

Distinctive features that differentiate non-state and state law enforcement:

1. While implementation of non-state law enforcement activities, non-state (private) interests are subject to guarding and protection. Protection of public, state interests in this situation is primarily a derivative goal.
2. Non-state law enforcement actors (carriers of non-state law enforcement activities) do not act on behalf of the state are not its agents or authorized official representatives, but act exclusively on behalf of their clients who have hired on their behalf.
3. Non-state law enforcement actors in the course of non-state law enforcement activities have no right to apply measures of legal responsibility, including sanctions, to violators of the law.
4. The implementation of non-state law enforcement activities is primarily entrusted to specially created enterprises that have passed state registration and received a license

for the right to carry out a certain type of non-state law enforcement activity. These enterprises are completed with trained professionals mainly persons with knowledge in the field of jurisprudence, forensic science and other disciplines, who have undergone specific training and received a state license. Management and activities of such enterprises are detailed and comprehensively regulated by law including by departmental by-laws of various ministries and departments that establish separate procedural rules for solving the most important issues.

Realization subjects of non-state law enforcement activities are participants in public relations arising in connection with ensuring security of the private field of public life, protecting the rights and legitimate interests of non-state entities and individual citizens.

In contrast to the state, non-state law enforcement activity is predominantly a set of contractual relations where the boundaries and time of the emergence and termination of these relations are clearly defined. The state law enforcement activity has no restrictions in terms of volume, time and circle of persons.

As we have noted, species of non-state law enforcement activities include: commercial and non-commercial; security-detective and human rights (advocacy) activities, educational activities in the field of non-state law enforcement; activities carried out in favor of certain persons (clients) and activities carried out in favor of an unlimited number of persons.

It should be emphasized that security and detective activity (private investigation) is autonomous in nature which covers security of entrepreneurship and does not interfere with other types of law enforcement [29].

Depending on the purpose of activity of non-state law enforcement entities, they are divided into those for whom this activity is the main activity (private security guards) and those for whom this activity is not the main one (educational institutions that train security professionals), as well as commercial and non-profit ones. In a state governed by the rule of law, an entrepreneur may be the subject of non-state law enforcement activity.

Depending on the activity field, these subjects are divided into those who perform their functions exclusively within the framework of public law relations (public assistants of internal affairs bodies, voluntary public associations created to protect the state border, lawyers specializing in criminal law) and those who perform their functions in the field of both public and private law (private detectives or security guards, non-state security services, private notaries).

Main functions of non-state law enforcement activities: human rights; preventive; security; detective; function of criminal or administrative justice.

Procedure regulation for realizing non-state security activities as a special species of entrepreneurial activity is provided for by the Law of Ukraine: *On Security Activities* by departmental normative acts of the Ministry of Internal Affairs of Ukraine.

All of the above confirms that non-state law enforcement actors are primarily involved in realization of the main types of law enforcement activities, contributing to their effective conduct.

Thus, the subjects ensuring the economic security of enterprises and institutions are state and non-state bodies, as well as certain persons who according to their duties, competencies, tasks, in accordance with legislation or for their own reasons, take part in ensuring sustainable and stable functioning of enterprises and institutions [30].

3. Control in the field of enterprise activities of aerospace industry of Ukraine as a way to ensure their economic security

Control is the main way to ensure the rule of law and discipline in public administration. It is one of the most important elements of government, an important function of the state. Without organizing and exercising control, the work of the state apparatus and other state structures is impossible. Any decision, any action in the field of public administration ends with control over their implementation, otherwise they will not make sense.

Control is one of the main factors that disciplines the behavior of enterprises in the aerospace industry [31], this is what makes the activities of the state "transparent" for society and for the state, social relations, the state of work of individual state and other entities, etc.

There is a point of view according to which control should be considered as an independent branch of government. For the first time, back in the early twentieth century, such an opinion was expressed by Sun Yat-sent [32], outstanding Chinese scientist. At the same time, he formulated *Constitution of the branches of government* where he distinguished, in addition to the classical (*legislative, executive, judicial*) and independent *examination* and *control* authorities.

Considering control in connection with its performance in a reflective, cognitive role allows you to capture differences by comparing results with goals. The very fact of accurate and timely recording of deviations is of paramount importance. After all, the reasons for deviations are not always considered through the prism of control.

Essence of control is to check result compliance with the specified parameters, its goals, and the content is that the authorized bodies and persons, using organizational and legal methods and means, within the framework of their competence, find out whether the activities of the controlled objects comply with regulatory requirements and regulatory legal acts the tasks assigned to; analyze impact results of control subjects on controlled objects, degree of deviation from the set goals, planned methods and results and directly take measures to eliminate them and bring the perpetrators to justice.

Thus, control is a form of legal activity in which authorized bodies and persons within the framework of control proceedings, in order to obtain legally significant results and provide (ensure) regulatory impact, collect and verify information on the actual fulfillment of regulatory requirements, compliance with regulatory requirements at controlled facilities and legal acts and directly take necessary measures.

In terms of the form of control, control can be: state, non-state, internal and public one.

State control as a guarantee of the right of enterprises of the aerospace industry of Ukraine to economic security. State control includes the controlling actions of all state structures without exception and control in the field of public administration. Its main purpose is to study the relationship between subjects of public administration in order to maintain the rule of law and discipline in public administration. The goals of state control can be divided into strategic and tactical (or applied) one. Strategic purpose of State control is to ensure the rule of law and discipline in society in general and in public administration in particular. Tactical purpose of State control consists in "observing and analyzing the compliance of the activities of all subjects of social relations with the parameters established by the state, as well as a certain" correction "of deviations from these parameters" [33]. In public administration, control extends to all areas: economic, socio-cultural, administrative and political activities and intersectoral public administration.

According to Yu. M. Kozlov, State control includes the following components that reveal its content: monitoring activities of corresponding management object; obtaining necessary objective and reliable information about the level of legality and discipline in public administration; taking measures to prevent and eliminate violations of law and discipline; identification of the reasons and conditions conducive to the commission of an offense; taking measures to bring the offender to legal responsibility [34]. Approximately the same idea is expressed by V. P. Belyaev: "control function in its content includes three elements: obtaining the necessary information; its analysis and assessment; response to identified deviations from the established rules and requirements" [35]. In addition, the content of control activities includes obtaining and analyzing relevant information; identification of trends and reasons for non-fulfillment (improper fulfillment) of tasks, instructions; the adoption in a number of cases of measures to suppress revealed violations of the law and to bring the perpetrators to justice.

State control is a complex phenomenon with many sides and forms of its manifestation. State control as a realization form of public administration provides execution verification of laws and other regulations in order to prevent deviations from the norms and rules established by law. The control function is an indispensable attribute of every state at any stage of its historical development. State control is characterized by all essential features that are characteristic of the state as a whole. Like the state as a whole, state control is:

- political management configuration that is at the same time the most important instrument of the state for the implementation of its goals and objectives;
- apparatus that unites people professionally engaged in control activities to verify the execution of decisions of state authorities;
- control is carried out on behalf of state bodies, regardless of their tasks and the type of performed activity;
- control acts as a state-power activity; control bodies give mandatory instructions to controlled objects on elimination of the identified deficiencies;
- control bodies can raise the issue of bringing to responsibility those responsible for the violations found and directly apply measures of state coercion in a number of cases [36].

In the state-legal sense, control is viewed through the prism of state activities, as a means of verifying compliance with the instructions of state power. While State formation control is carried out within the framework of general State activities without being allocated to the appropriate system of bodies and officials. While further development and improvement of the state, it acquires clarity, consistency and strict legal regulation.

Principles of State control:

- controllability. This principle application fully extends to control activities carried out while State control over timely and high-quality fulfillment by authorities, public associations, various commercial and non-commercial structures of the requirements of regulatory legal acts, decisions taken;
- legality. Legality principle is a universal legal principle which has found its normative embodiment in numerous articles of Constitution of Ukraine. In the aspect we are considering, this principle means that the state control apparatus and supervised bodies must exercise their functions in strict accordance with the laws and regulations in force on the territory of Ukraine;
- consistency. This principle presupposes existence of a functional system of state control over the activities of controlled bodies. For example, an analysis of the legal and organizational support of the units of the Administration of the President of Ukraine exercising presidential control indicates that there can be several system-forming bodies (units) of control. A functional system as such does not imply an obligatory hierarchical structure of subordination;
- centralization and independence of subjects exercising State control;
- independence. Bodies of state control and controllers should have organizational, functional, material and other independence from controlled bodies of state power, legal entities and individuals, as well as persons associated with;
- responsibility. Responsibility of managers and controllers exercising state control means that each official must be disciplined for improper performance of the tasks assigned to;
- publicity (publicity or openness). Transparency of State control is a necessary attribute, especially in a democratic society, since it is necessary to inform the public about the results of control. The principle of publicity assumes possible (permissible) openness of activities of state bodies including those carrying out law enforcement activities and access of citizens to information on control results;
- expediency. Means the compliance of set control tasks with certain goals, rationality, rationality, practical and theoretical justification of the need for control settings (tasks) facing both a particular subject of state control and a controlled body to determine the appropriateness of a decision made related to time, place, period, with the involved resource for monitoring;
- efficiency that provides: optimal ratio between the costs of monitoring (control measure) and the beneficial effect of its result; no unnecessary (unproductive) costs (frugality); achieving the goals and objectives of control with minimal costs or achieving the best maximum) result when exercising control over the results of the work of executive authorities when attracting a strictly defined amount of resources allocated for the audit;
- citizen honor and dignity protection.

For economic policy realization, the implementation of targeted economic and other programs and programs of economic and social development, the state uses a variety of means and mechanisms to regulate the activities of enterprises in the aerospace industry. The main means of the State regulatory influence on activities of enterprises in the aerospace industry are:

- government procurement;
- licensing, patenting and quotas;
- certification and standardization;
- application of standards and limits;
- regulation of prices and tariffs;
- provision of investment, tax and other benefits;
- provision of grants, compensations, targeted innovations and subsidies.

State control purpose is to identify the actual state of affairs in various fields of public life, it has the goal of providing objective information about the state of public relations, execution of made decisions, compliance of activities of civil servants with the assigned task and certain powers, compliance with the rule of law. Control allows you to get a significant information scope which analysis helps to find the most appropriate and rational solutions for further development of various areas. The control bodies actually have all the current state bodies while carrying out inspections have the right to directly intervene in the operational activities of controlled structures (limit, suspend or terminate their activities) and independently bring offenders (legal entities and individuals) to legal responsibility. Among approaches to determining the types of forms of state control that can be single out:

- publication of regulatory acts of management;
- publication of individual (non-normative) acts;
- carrying out direct organizational measures;
- implementation of material and technical operations.

Currently, our State is going through an important stage on the path of the progressive development of civil society and the rule of law, such important goals as fight against corruption, support for small and medium sized businesses and development of social and economic fields come to the fore. Under these conditions, the most important management mission is performed by State control.

Modern State control depending on subject composition is subdivided into presidential, parliamentary one, Ombudsman's control, executive authority control and judicial control. In turn, it is proposed to distinguish: judicial constitutional control, judicial control in the sphere of executive power (including within the framework of administrative justice), judicial control in criminal proceedings and intradepartmental (hierarchical) judicial control. Court while deciding the issue of reinstating an unlawfully released person at work, actually makes a managerial decision on the extension of labor legal relations, although the issue of staffing is attributed to competence of the enterprise administration institution.

Direct state control over enterprise activities in aerospace industry is carried out in the areas of state influence on the activities of business entities. Control over the activities of business entities is carried out by applying: technical regulations; standards; codes; established practice; classifiers; technical conditions, etc.

Application of standards or their individual provisions is mandatory for:

- business entities, if there is a reference to standards in technical regulations;

- parties to an agreement (contract) for development, manufacture or supply of products, if it contains a reference to certain standards;
- manufacturer or supplier of products, if he has drawn up a declaration of conformity of products to certain standards or applied the designation of these standards in its labeling.

The state exercises control in order to protect the rights of enterprises in aerospace industry and consumers. Every business entity and consumer has the right to protect their rights and legitimate interests. The rights and legitimate interests of these subjects are protected by:

- recognition of availability or lack of rights;
- recognition of acts of state authorities and local self-government bodies, acts of other subjects contradicting the legislation, infringing on the rights and legitimate interests of enterprises of the aerospace industry or consumers, as fully or partially invalid;
- invalidation of business agreements on the grounds provided for by law;
- restoration of situation existed before the violation of the rights and legitimate interests of enterprises of the aerospace industry;
- suppression of actions that violate the right or create a threat of its violation;
- adjudication to performance of an obligation in kind;
- compensation for losses;
- penalty application;
- application of contractual sanctions;
- establishment, change and termination of entrepreneurial legal relations;
- in other ways provided by the current legislation.

Non-state control. Non-state control includes: international control; control of financial and credit authorities; audit (independent) control; control exercised by local governments.

1. International control is carried out by international institutions and representatives of foreign states on implementation of international treaties or obligations arising from those treaties. Creation of numerous international bodies, institutions and their subdivisions, endowed with control powers, testifies to the critical importance of control in the management system. They operate on the basis of international agreements. For example, the UN secretariat operates within the United Nations; within the European Union: Council of Europe. The UN Secretariat monitors the observance by the UN member states of international acts adopted by the UN General Assembly. The bodies of the Council of Europe control the observance of the international acts of the European Union.
2. Control of financial and credit authorities: monetary control over financial and economic activities carried out by banks while lending operations, financing and settlements. Financial control methods: preliminary and current control while conducting monetary transactions through verification of submitted documents, establishes the right to receive a loan, make payments of loan accounts. When preparing documents, financial and credit institution checks whether the company has its own circulating

assets and gets acquainted with the most important indicators of economic activity. When issuing a loan, a financial and credit institution checks its collateral, that is, whether the company has inventory. Financial control while lending is combined with control exercised when organizing settlements. Accepting an order from a business entity to transfer money to another company, financial and credit institution checks the legality of this transaction;

3. Audit (independent) control. Auditing is a business activity for independent verification of accounting and financial (accounting) statements of enterprises in the aerospace industry. The purpose of the audit is to verify the reliability of the financial (accounting) statements of enterprises and compliance of accounting procedure with the current legislation;
4. Control of local government bodies. Local self-government is a system of organization and activity of citizens, ensuring the independent solution of local issues by population, management of municipal property, proceeding from the interests of all residents of a given territory. The broader concept of local self-government implies the right and real ability of local self-government bodies to regulate and manage a significant part of public affairs, acting within framework of the law, under their responsibility and in the interests of local population.

Internal control. Internal control is an audit of production and economic enterprise activities as a whole and of its individual structural units, carried out by the accounting department, the financial department and some other economic services of the enterprises of aerospace industry.

The types of internal control include: accounting control, management control, as well as integrated systems for the functioning of enterprises in the aerospace industry, based on permanent self-control and mutual control, strict adherence to established corporate procedures and ethical standards: compliance control, controlling, the *Know your customer* system.

Accounting control is a set of measures to ensure correctness of accounting in an organization, in particular, control of analytical and synthetic accounting data, compliance with accounting policies, drawing up accounting registers, filling out internal and external accounting forms. The chief accountant is a key link in the internal control system. Powers of chief accountant are as follows: together with the head of the enterprise, he signs all documents that serve as the basis for the acceptance and issuance of inventory and cash, as well as settlement, credit, financial obligations and business contracts; does not accept for execution and execution documents on transactions that contradict the current legislation and violate contractual and financial discipline; carries out preliminary financial control over the correctness and legality of spending funds.

Management Control. Control carried out while activities of enterprise management bodies. It is a procedure of tracking progress of implementation of the adopted strategic decisions and evaluating the results achieved while their implementation. Management control is carried out in the areas of the development strategy of an economic entity: supply and sales policy, innovation policy, pricing policy, financial policy, investment policy, human resource policies.

Compliance control is monitoring compliance of activities with legal requirements, professional practice and ethics, in order to minimize or prevent the occurrence of risks associated with application of sanctions, both by government bodies and the company clients, as a result of non-compliance with legal requirements, internal documents, procedures or other standards.

This term came into Ukrainian practice with the arrival of large foreign companies on the market where this function is an integral activity part [37].

As a rule, basic document of compliance control is the Code of conduct governing standards of conduct of institution staff while interacting with customers, other employees, counterparties, suppliers, supervisors and third parties that an employee encounters while fulfillment of their professional duties and applies to such business systems: business integrity, company assets, and the like. In addition to the Code of Conduct, an institution can develop such documents as:

- Code of Ethics. Code of Ethics (code of corporate conduct) – as a rule, it is a fairly general document that affects almost all aspects of institution activities. It talks about moral and ethical principles, standards of conduct, management priorities and employee responsibilities;
- Gift policy. Its function is to distinguish between the concepts of "gift" and "bribe"/"kickback" or, in other words, to mark the border after which the gift becomes an offering in order to obtain the opportunity to manipulate the official in their interests. Existence of such a policy is especially relevant in countries where national traditions dictate the need for this kind of interaction with regulators, partners and contractors. As a rule, policy does not prohibit donations, but limits their value and introduces procedures for implementation of appropriate controls;
- Whistleblowing policy. The policy of reporting violations of ethical standards exists in most Western institutions and regulates the procedure and methods for reporting violations by employees of business entities (right to anonymity), as well as the procedure for further investigation and documentation of these violations. It should be noted that the high-quality implementation of this function is one of the most effective ways to combat violations within the company;
- Financial Crime/Fraud Prevention can be carried out independently and in cooperation with regulatory authorities. The issue of effective and correct interaction with regulatory bodies is very relevant today, since even very law-abiding institutions face a number of practical difficulties in such a situation;
- Fight against Money Laundering and Terrorist Financing policy. This policy is implemented in one way or another in all financial and many non-financial organizations in developed and developing countries. It prevents penetration of criminally acquired income into the legal sector of the economy and prevents terrorism financing. Currently, it is one of the main tools for combating the legalization of shadow income and is based on the norms of international law and the recommendations of international groups;
- Privacy Policy. The privacy policy regulates the non-disclosure of data about customers and their transactions. It implies not only the formation of a general culture of handling customer data, but the organization of storage and compliance with certain standards while personal data processing;

- Conflict of interest policy. This policy regulates a conflict of interest, sets ethical standards of behavior for employees in case of a conflict of interest, namely in cases where: an employee's interests may conflict with the interests of the company; the interests of one client can conflict with the interests of another client, etc. In particular, an obligation is imposed on employees to help identify and prevent cases of conflicts of interest and it is also declared that interests of the company should always be superior regarding personal interests of its individual employees;
- *Chinese Walls* policy. The *Chinese Walls* policy is necessary to delimit information field in institution activities regarding financial sector, in order to prevent conflicts of interest and create conditions for fair competition. This policy is implemented in almost all leading investment companies where such a distinction is especially important, since the possession of non-public information about someone financial conditions, investment plans, additional emissions can lead to its use by employees of another division in order to obtain additional profit. Carrying up this information barrier allows the organization not only to prevent emergence of a conflict of interest but to serve all clients without restrictions.

Currently, compliance functions are not only about setting workflow standards but overseeing the proper implementation of policies and procedures. At the same time, each institution, due to the tasks set by senior management, shareholders and creditors can implement additional processes in this area while creating its own unique compliance control structure.

Controlling is a comprehensive management support system for an organization aimed at coordinating the interaction of management systems and monitoring their effectiveness. Controlling can provide information and analytical support for decision-making processes in enterprise management and can be the basis for making certain decisions in framework of management activities.

Modern controlling includes risk management, information supply system for a business entity, a notification system by managing a system of key indicators, managing a system for implementing strategic, tactical and operational planning and a quality management system.

The practice of controlling was borrowed from England as part of common business practice in accordance with the requirements of *Company Acts* rules dating back to the 1700s. The *controlling* term itself originated in the United States.

Controlling is viewed, on the one hand, as a philosophy, a way of thinking for managers, focused on the efficient use of resources and development of a business entity in the long term and as an integrated system of information, analytical and methodological support for managers in the process of planning, control, analysis and adoption, oriented towards achieving goals, management decisions in all functional areas of the enterprise.

While building controlling systems, it always requires comparing its effect (both in the short and long term) with complication of the enterprise management system (increasing the labor intensity of processes).

Controlling system includes the following aspects:

- determination of activity goals;

- reflection of these goals in the system of effective and balanced key performance indicators (KPI)
- regular control (measurement) of actual indicator values;
- analysis and identification of reasons for deviations of actual indicator values from the planned ones;
- on this basis, making management decisions to minimize deviations.

Target task of controlling is to build an effective system for making, implementing, monitoring and analyzing management decisions at the enterprise.

Public control at the enterprises of aerospace industry in Ukraine. The state, as a form of public political power, tends to preserve social and political institutions and, as a consequence, to a slow reaction to changes in socio-political realities. The authorities are often intolerant of internal criticism and mutual responsibility among officials is encouraged. Sometimes, authorities, instead of the interests of society, represent interests of large financial and industrial groups, subordinating their entire apparatus to. Citizens often find themselves defenseless against abuses by the authorities, including corruption. All this leads to destructive processes for the social system.

As world practice indicates, in most cases, the apparatus of public authorities is not able to effectively deal with its destructive factors, using only internal power-apparatus mechanisms, just as a patient cannot successfully independently perform a complex surgical operation for himself.

Public control is carried out by non-state entities, citizens and permeates all fields of human activity including public administration (control of trade unions over the activities of enterprise administration) and the fields of everyday and personal relations (control by parents over children), self-control, etc.

The *public control* term is often understood to mean the participation of citizens in local self-government, in resolving issues, most often local, domestic purposes, that is, the very initial form of public control. This happens because the authorities are inclined to transfer public control to a lower level, direct it to solving small issues and distance it from the big ones that are decisive. In our opinion, public control should be considered in a much broader sense, as a social phenomenon in which civil society determines the main directions of the state domestic and foreign policy, controls the process of its implementation, and takes part in solving all socially significant issues at all levels.

Civil society using its internal potential and the laws of social development is capable of activating at critical moments and directing state policy on the path of the most effective development. At the same time, the more closed from civil society the state apparatus becomes and the stronger entropy manifests itself in the social system, the faster and more efficiently the consolidation of society takes place that allows civil society to act simultaneously and harmoniously.

There is a relative regularity: the more tendencies are expressed in the apparatus of government bodies to replace state interests with particular ones, the more the intellectual resources of society remain unclaimed for. These resources are transferred to civilian control field. There is also another pattern: the more vicious the apparatus of authorities, greater the intellectual resource of society is mobilized to fight the destructive state factors. It is in these phenomena that the procedure of self-organization of the social system is most clearly manifested.

Effectiveness of public control is largely due to the fact that it has a flexible structure and has the properties of an open system. The criterion for the significance of any subject of public control is public authority that consists not only of words but of specific deeds of public importance. Unlike elected government bodies, subjects of public control are not elected for any specific term. Public authority emerges as a result of a complex and usually long-term process that is permanent. Any subject of public control, having abused his public power, immediately loses this power without any formal procedures, excludes the usurpation of such power, since it is impossible to usurp power that is supported solely by the power of public authority. Thus, elements of civil control are also present in the very system of public control. Such structural dialectic contains stability guarantee of this system [38].

Progressive ideas for transforming social system most often come to mind only to individual people or small social groups, that make these ideas of the entire society property.

Public control has provided and continues to provide a certain of public life organization, adequate behavior of society members. It is because of this control that feedback principle is implemented in management of social processes or systems. He is inextricably linked with society. While development and complication of social relations, structure of social control is also rapidly developing, its mechanism is being improved and its importance in functioning of all social systems increases. Such flexibility of social control mechanism with the emergence of new conditions and tasks in the process of social development is an important prerequisite for social progress.

Forms, methods, subjects of public control and foundations of their legal status are reflected in current legislation of Ukraine.

Conclusions. Summing up, we can assert that change in the economy under influence of market mechanisms has singled out the system of economic security of enterprises in the aerospace industry into a special object of management. The issues of economic security of enterprises in aerospace industry turned out to be the least developed in legal terms: both in theoretical and practical aspects.

Under the economic security of enterprises in aerospace industry, we mean protection of its scientific and technical, technological, production, personnel and other potential from direct or indirect threats aimed at destabilizing sustainable enterprise activities.

Economic security implies: market demand for products; technological development of the country and industry; state of protection of the individual, society, state from external and internal dangers and threats. The methods of protecting economic security depend on: the market type, provision scope, the time of implementation and provision, assessment method, achievement level.

The factors influencing the economic security of enterprises and the aerospace industry in Ukraine are highlighted. These include: instability of the organizational and structural form of management; failed personnel policy; lack of a state strategy in industry development; high corruption.

The subjects ensuring the economic security of enterprises and organizations of aerospace industry of Ukraine are state and non-state bodies, as well as certain persons who ensure their effective work.

The most important role in formation and provision of economic security of the aerospace industry is played by control (as one of the main management elements) that in its form can be state, non-state, internal and public.

2 Conclusion

Considering and analyzing objects of ensuring economic security of aerospace industry of Ukraine, it was found that it is inherent in factors that negatively affect the level of economic security. This is due to instability of the organizational and structural form of management. For thirty years, the transformation and privatization of enterprises in the aerospace industry of Ukraine has taken place, which has led to the replacement of the form of ownership of the overwhelming majority of enterprises. This, as a natural consequence, became the reason for the weakening of state control over the quality of products, its significant rise in prices, and the generation of corruption schemes. Frequent changes in the management system of the military and industrial complex negatively affect formation of the economic security system of aerospace industry.

Secondly, failed personnel policy, or rather its absence. Currently there is no system for training highly qualified personnel for important and promising areas of the aerospace industry which could form a system of economic security of enterprises and the industry as a whole. Frequent and unprofessional changes in the leadership of industry have a negative impact on economic security system formation: from January 2011 to December 2020, eleven of them were appointed.

Thirdly, lack of a state strategy in the development of the aerospace industry that is directly related to its insufficient funding and as a consequence, decrease in the level of economic industry security.

Fourthly, high corruption level that leads to ineffective use of available funding.

The subjects of ensuring the economic security of aerospace industry of Ukraine are classified, depending on their affiliation, into two groups: state bodies that belong to legislative, executive and judicial branches of government and non-state bodies.

State bodies take part in ensuring economic security at the state level, regional level and the level of security of enterprises in the aerospace industry. Depending on the scope of their competence, they are divided into three subgroups: state bodies of general competence (Verkhovna Rada of Ukraine, President of Ukraine, Council of National Security and Defense, Cabinet of Ministers of Ukraine, Constitutional Court of Ukraine, courts of general competence, local government bodies, local self-government); ministries and departments; law enforcement, regulatory and other bodies.

The main features characteristic of state and non-state law enforcement are:

1. Non-state law enforcement activities can be carried out only through the application of legal measures of influence determined by specific law. These include measures of prevention, suppression and administrative procedural support.
2. The application of measures of legal influence should strictly comply with law requirements or other legal act (for example, an agreement). Only a law or other legal act can serve as the basis for the application of a specific measure of legal influence and clearly define its content.

3. Implementation of legal influence measures is carried out in the manner prescribed by law in compliance with certain procedures.
4. Implementation of non-state law enforcement activity is entrusted to specially created enterprises that have passed state registration and received a license for the right to carry out a certain type of non-state law enforcement activity.

Activities of non-state subjects of economic enterprise security and are understood as non-state law enforcement activities. Now in our state, such activities are carried out by a non-state system, represented by private security, security and technical and detective enterprises, commercial security services, enterprises of various forms of ownership, information bureaus, bank security services, training centers, specialized funds, research structures, specialized faculties of higher educational institutions.

Control in the field of activities of enterprises of aerospace industry of Ukraine is a way to ensure their economic security that disciplines activities of enterprises in the industry, makes it "transparent" for society and the state.

Control is a form of legal activity in which authorized bodies and persons within the framework of control proceedings, in order to obtain legally significant results and provide (ensure) regulatory impact, collect and verify information on the actual fulfillment of regulatory requirements, compliance with regulatory requirements at controlled facilities and legal acts and directly take necessary measures.

In terms of the form of control, control can be: state, non-state, internal and public one.

References

1. Gillen, D., Morrison, W.G.: Aviation security: costing, pricing, finance and performance. J Air Transp. Manage. **48**, 1–12 (2015)
2. Cohendet, P.: Evaluating the industrial indirect effects of technology programmes: the case of the European Space Agency (ESA) programmes. In: Proceedings of the OECD Conference on policy evaluation in innovation and technology, pp. 189–223. Paris (1997)
3. Cletus, C., Cohen, J.P., Khan, S.R.: Aviation security and terrorism: a review of the economic issues, 37 p. Canadian Parliamentary Review Economics Published (2002)
4. Shelekhov, A.O., Nikolsko-Rzhevskyi, A.S.: On the concept of economic security of the enterprise. South Ukr. Law J. **4**, 212–214 (2010)
5. Belousov, E.N.: Institutional issues of economy. Econ. Law **2**, 119–122 (2009)
6. Tretiak, V.V., Hordiienko, T.M.: Economic security: the essence and conditions of formation. Econ. State **1**, 6–8 (2010)
7. Shmelev, N.P., Popov, V.V.: At the Turning Point: Economic Adjustment in the USSR. M.: *APN* Publishing House, 398 p. (1989)
8. Abalkyn, L.: Economic security of Russia: threats and their reflection. Issues Econ. **12**, 4–13 (1994)
9. International Civil Aviation Organization: Safety report 2015 edition. Retrieved https://www.icao.int/safety/documents/icao_safety_report_2015_web.pdf (2015)
10. Illarionov, A.: Criteria of economic security. Issues Econ. (10), pp. 35–58 (1998)
11. Berlach, A.I.: Business Security: A Training Manual. K.: *Ukraine* University, 280 p. (2007)
12. Kyrychenko, O.A., Denysenko, M.P., Sidiak, V.S., Laptiev, S.M.: Problems of economic security management by economic entities: a monograph. K .: «KROK» University, 412 p., p. 23 (2010)

13. Ismailova, S.S.: Economic security: the essence and its main types. Form. Market Relat. Ukr. **10**(113). P. 126–129, p. 128 (2010)
14. On the optimization of the system of central executive bodies Decree of the President of Ukraine dated on December 9, 2010 № 1085/2010. https://zakon.rada.gov.ua/laws/show/1085/2010#Text
15. On the measures for increasing efficiency of activity of Ukrainian military-industrial complex. Decree of the President of Ukraine dated on December 28, 2010 № 1245/2010. https://zakon.rada.gov.ua/laws/show/1245/2010#Text
16. On the establishment of the *Ukroboronprom* State Concern. Resolution of the Cabinet of Ministers of Ukraine dated on December 29, 2010 № 1221. https://zakon.rada.gov.ua/laws/show/1221-2010-%D0%BF#Text
17. Certain issues of the *Ukroboronprom* State Concern dated on August 31, 2011 №. 993. https://zakon.rada.gov.ua/laws/show/993-2011-%D0%BF#Text
18. Development of the military-industrial complex https://www.kmu.gov.ua/diyalnist/reformi/ekonomichne-zrostannya/rozvitok-oboronno-promislovogo-kompleksu
19. The privatized enterprises of Ukroboronprom suspended activities until 2015, the *Ukrainian Truth* Internet edition dated on 22.04.2021. https://www.epravda.com.ua/rus/news/2021/04/22/673268/
20. Mendeleev, D.: Military-industrial complex which Ukraine has lost. The Mirror Weekly dated on August 23, 2020. https://zn.ua/ukraina-1991-2020/opk-kotoryj-ukraina-poterjala.html
21. On the decision of the National Security and Defense Council of Ukraine dated on February 18, 2020: *On the main indicators of the state defense order for 2020 and 2021, 2022.* Decree of the President of Ukraine dated on February 27, 2020 № 59/2020. https://zakon.rada.gov.ua/laws/show/59/2020#Text
22. *Ukroboronprom* is expected to be liquidated by the end of June. 11.02.2021 Aviation EXplorer The site of the aviation experts community. https://www.aex.ru/news/2021/2/11/223274/
23. Handbook of Research on Artificial Intelligence Applications in the Aviation and Aerospace Industries. IGI Global, 517 p. (2019)
24. Putin and Biden will have to come to an agreement on the Donbas subject: Russian media about Ukraine. The site: *Ostrov*. Wednesday, May 5, 2021 17:27. https://www.ostro.org/general/world/articles/611176/
25. The TOP-10 airlines in the Ukrainian sky in 2020 are named. Website www.avianews.com https://www.avianews.com/ukraine/2021/01/13/top10_airlines_in_ukrainian_sky_in2020/
26. Kinakh A.K.: Complex reforming and development of the military-industrial complex of Ukraine. In: Problematic issues of Armament and Military Technology Development. Theses at the VII Scientific and Technical Conference, 588 p. K.: Central Scientific Research Institute of Armament and Military Equipment of the Armed Forces of Ukraine, Kyiv (2019)
27. A controlling interest of Motor Sich JSC was sold to an offshore for over $ 700 million: Arakhamia. Materials of IA *LIGABusinessInform* Press Center. https://biz.liga.net/all/all/novosti/kontrolnyy-paket-motor-sichi-prodali-v-offshorah-za-700-mln-arahamiya
28. Rockets in Kyiv. Will Ukraine be able to create high-precision weapons? RIA News. August 22, 2020. https://vpk.name/news/230536_raketyi_pokievski_smozhet_li_ukraina_sozdat_vyisokotochnoe_oruzhie.html
29. Shield and sword of Ukraine: main achievements of the military-industrial complex in 2017. Part II, Website Informnapalm. https://informnapalm.org/42200-shhit-i-mech-ukrainy-part-2/
30. Mendeleev, D.: Military-industrial complex which Ukraine has lost. The Mirror Weekly dated on August 23, 2020. https://zn.ua/ukraina-1991-2020/opk-kotoryj-ukraina-poterjala.html
31. Jonas, A., Sinkevicius, A., Flannery, S., et al.: Space: Investment Implications of the Final Frontier. Morgan Stanley Research, 82 p. (2017)

32. Campa-Planas, F., Kalemba, N.: Safety as a management concept in the air transport sector: a systematic literature review. Intangible Capital **13**(1), 71 (2017)
33. Liao, M.Y.: Safety culture in commercial aviation: differences in perspective between Chinese and Western pilots. Saf. Sci. **79**, 193–205 (2015)
34. Madsen, P.M.: Perils and profits: a reexamination of the link between profitability and safety in U.S. Aviation. J. Manage. **39**(3), 763–791 (2013)
35. Korystin, O.I.: Methodological and organizational principles of ensuring the economic security of Ukraine. Law Soc. **4**, 88–93 (2010)
36. Korystin, O.I., Baranovskyi, O.I., Herasymenko, L.V., Dolia, L.M., Kaliuk, O.M.: Economic security: a training manual. In: Dzhuzhi, O.M. (ed). K.: Alerta, 368 p. Center for Educational Literature, KNT (2010)
37. Ortynskyi, V.L., Kernytskyi, I.S., Zhyvko, Z.B. et al.: Economic security of enterprises, organizations and institutions: a training manual [for students of higher education institutions], 544 p. Legal Unity (2009)
38. Prospects for Space Exploration. Forecasts to 2027. 1st edn. A Euroconsult Executive Report. Brochure 2018. http://euroconsult-ec.com/research/space-exploration-2018-brochure.pdf. Aaccessed 10 May 2021
39. Spence, J.: The Search for Modern China, p. 339. W. W. Norton & Company, USA (1991)
40. Andriiko, O.F.: Organizational and legal issues of state control in the area of executive power: author's abstract of Doctor of Law: 12.00.07. In: Andriiko, O.F. (ed.) NAS of Ukraine, the Institute of State and Law, 38 p. V.M. Koretsky National Academy of Sciences of Ukraine (1999)
41. Alekhin, A.P., Karmolitskiy, A.A., Kozlov, Y. M.: Administrative law of the Russian Federation: a textbook, p. 297. Zercalo-M (2001)
42. Konyn, N.M.: Russian administrative law. General part. Series of Lectures, P. 327. Saratov State Academy of Law, Saratov (2001)
43. Beliaev, V.P.: Control and supervision as a means of implementing legal policy. Legal Policy Legal Life **4**, 38–42 (2003)
44. Markov, K.V.: Ways to strengthen the constitutional framework of state control in the modern period. Legislation Comments Probl. **4**(70), 13–16 (2006)
45. Schumpeter, J.: Harvard University Press Theory of economic Development: An Inquiry into Profits, Capital, Credit, Interest and the Business Cycle, 255 p. Cambridge Mass (1934)
46. Goncharov, A.A.: Civil control as a mechanism of social system self-management. Management of social and economic systems, vol. 1 (2007). http://umc.gu-unpk.ru/umc/arhiv/2007/1/Goncharov.doc

Legal Regulation of Ukrainian Space Industry: Perspective Trends and International Cooperation

Ganna Goncharenko[1](✉), Olena Agapova[2], Yurii Ryabukha[3], and Oleksandr Shevchuk[4]

[1] Department of Scientific and research work of the Institute of Legal Personnel Training for the Security Service, Ukraine Yaroslav Mudryi National Law University, Kharkiv, Ukraine
anna_sovetnik@ukr.net
[2] IR Department at National Scientific Center, «Hon. Prof. M. S. Bokarius Forensic Science Institute», Kharkiv, Ukraine
agapova-lena-@ukr.net
[3] Interim Head Institute of Legal Personnel Training for the Security Service of Ukraine, Yaroslav Mudryi National Law University, Kharkiv, Ukraine
[4] Juridical Sciences, Department of Administrative Law and Administratively Activities, Yaroslav Mudryi National Law University, Kharkiv, Ukraine
Oleksandr-shevchuk@ukr.net

Abstract. Analysis of the current state of legal support of the space industry in Ukraine, ranging from legally established or internationally ratified by Ukraine to bylaws on organization of activities and development of the space industry of the country of promising and strategic level. The international legal direction of Ukraine cooperation in space is considered, where the main attention is focused on the vector of cooperation with the European Union. Public-private relations in the space sphere are studied. The directions of development and improvement of legal bases of realization of space activity in Ukraine are singled out, including it concerns introduction of branch educational component as one of directions of activization of researches of branch of space law in Ukraine. The importance of the issues of adaptation of the legislation of Ukraine to the legislation of the European Union in the space industry is emphasized. Importance of determining the ways of development of space industry of Ukraine indicates importance of the State Space Agency of Ukraine, which became the developer of the Strategy of space activities of Ukraine. Necessity of launching one's own satellite as one of the perspective directions of comic industry development in Ukraine is substantiated.

Keywords: Legal regulation of space industry · Space law · Space industry of Ukraine · International cooperation in space · Space activities · Satellite of Ukraine · Space security

1 Introduction

Security situation in the world as a whole at the beginning of 2021 is difficult. Against the backdrop of a global pandemic, new threats to international security and the stability

M. Nechyporuk et al. (Eds.): ICTM 2021, LNNS 367, pp. 780–794, 2022.
https://doi.org/10.1007/978-3-030-94259-5_61

of society are emerging. At the same time, we consider establishment of transatlantic and international cooperation and the unification of efforts to meet numerous challenges to be the only rational solution to overcome modern security threats, which fully applies to the space industry as well.

The need to present this topic as a real problem that can affect the security of states is emphasized annually by the participants of the Munich Security Conference/Münchner Sicherheits konferenz/Munich Security Conference. The final document, the Munich Security Report 2020, which analyzes current security policy developments in China, Europe, Russia and the United States, reveals an understanding importance of space and climate security, and identifies the threats posed by new technologies in general. Intensification of research in the field of legal regulation of the aerospace industry will allow Ukraine to join the legal requirements and standards of the EU and join the course of space policy of EU member states.

Space activities have indeed led to many revolutionary steps. Ukraine is no exception, there is a significant demand for space services in our country. According to data for 2020, only agriculture consumes space observation data and forecasts based on them in the amount of about 30 million dollars. for a year. Currently these services are purchased in the US [1]. Data on amber development and deforestation, the state of infrastructure in the temporarily occupied territories are in great demand, in particular, in the law enforcement system. A significant number of space services are provided by the National Space Management and Testing Center, which has access to a number of modern foreign-made satellites. The domestic market could increase significantly if the culture of space services consumption changes in Ukraine. Therefore, creation of a closed cycle of space services in Ukraine which own funds will be financed by own demand, has, in addition to undoubtedly important security factors, a significant economic potential that can become a driver of growth.

2 Literature Review

Sufficient attention is paid to formation and development of "*space law*", *space management*, *space security* in legal literature. Some issues and problems of development of international and national space law have been covered in research papers of such scientists as A. H. Abashidze, I. P. Andrushko, L. P. Anufriev, O. V Begly, K. A. Bekyashev, Yu. S. Shemshuchenko, N. R. Malysheva, V. V. Semenyaka, N. D. Krasilich, A. M Gurova, Yu. M. Utko-Maslyanik and others.

We support scientific position of such researchers as N. R. Malysheva and A. M. Gurova, who emphasize that existence of an independent space industry is the prestige and international recognition of our country, because one of the criteria determining the place in the world space ranking is the possibility of independent implementation of relevant activities. At the same time, this status imposes on the state and many obligations to the world community, which is expressly provided by international space law [2].

Legal aspects of cooperation between the member states of the European Union in the space field are considered in the research papers of Yu. M. Utko-Maslyanik. Thus, according to the researcher, space activities, as a special type of joint EU competence, consists in implementation of EU measures in space in such a way that the implementation of such competence does not prevent Member States from using their competence, as well as limited application of EU harmonization measures. space legislation of the Member States [3]. We fully share the author's opinion and note that the adaptation of Ukrainian legislation to the legislation of the European Union should comply with Community acquis in the field of space activities.

Given importance of international agreements on space activities, we also agree with the scientific justification of Yu. M. Utko-Maslyanik that for current stage of development of international space cooperation are such international agreements on space activities, adopted under the auspices of the UN, as the Treaty on Principles the activities of States in the exploration and use of outer space including the Moon and other celestial bodies, 1967, the Convention on International Liability for Damage Caused by Space Objects, 1972, the Convention on the Registration of Objects Launched into Outer Space, 1975 should be implemented in national legislation [3].

The approach of Dr. G.S. Sachdeva deserves attention, moots upon the universal applicability of space treaties. In his paper, he argues that the Outer Space Treaty does not bind all countries of the world nor new states get obliged to provisions of Space Law unless obligations of the Member States are specifically acknowledged. He suggests that since International Law has its jus cogens as peremptory norms binding on all states, it is desirable to select and evolve some precepts of Outer Space Treaty as Jus Cogens of Space Law. To attain this status, the selected principles need to be "lobby-ed" for opinion juris of Space Law scholars. He concludes that in this way Jus Cogens of Space Law would act as peremptory norms with supervisory status to govern space activities and regulate interstate relations and human conduct in Outer Space so as to command universal obedience and strict compliance [4].

Space treaties and other norms of international law, including customary norms and general principles of international law, provide a basis for the conduct of States while performing space activities. Space law principles, like the benefit of all countries, international cooperation, the freedom to explore and use outer space, non-appropriation of outer space, the peaceful uses of outer space, and the rational, efficient and economic use of the Geostationary Ring, form, among others, the body of primary responsibilities of States [5].

Detailed analysis of research papers revealed that European scientific opinion in the field of space law as well as Ukrainian, has similar approaches to space activities and application of international legal documents.

3 Main Content Presentation

Space exploration is always linked to geopolitics and national interests. Space is seen as a "common goal" which use is responsibility of all states and which requires cooperation for safe use. However, the emergence of numerous state and non-state players gaining access to outer space is a matter of concern. Space powers, such as the United States, claim that "space is no longer a shrine. Now this is a war zone" ("Space is no longer a sanctuary. It is now a warfighting domain"). The United States invests three times as much in military space research as the rest of the world [6].

NATO has declared space the fifth domain of operations, along with land, sea, air and cyberspace. Subsequently, NATO Defense Ministers agreed to establish a NATO

SPACE CENTER under NATO Allied Aviation Command during a meeting in October 2020 [7]. The newly established center will support NATO operations, missions and activities to raise awareness of space by coordinating data, products and services with NATO allies.

Effective control of the space industry and a "code of conduct" can prevent some of the risks of abuses, but new international treaties have not been initiated for a long time. The United States, the Russian Federation, and China have for years been at odds with the basics of space arms control. In an increasingly crowded orbit and threats, competition for supremacy, diversity of institutions and initiatives to promote space use can cause real damage to outer space, contrary to the "letter" of the 1967 Outer Space Treaty [9] and the "spirit" of the Declaration of Legal Principles. activities of states on the exploration and use of outer space [10].

Ukraine has the status of a space state due to scientific, engineering, production potential and a number of measures taken in 1992–2020. The use of outer space in Ukraine is carried out in accordance with acts of general and strategic direction with a detailed analysis which we propose to get acquainted in the next part of research paper.

3.1 Legal Bases of Space Activity in Ukraine

It should be noted that the general legal basis for space activities in Ukraine and under the jurisdiction of Ukraine outside its borders is determined by the Law of Ukraine № 502/96-VR *On Space Activity* dated on 15.11.1996 [11]. According to Art. 3 space activities are aimed at:

- promoting the socio-economic and scientific progress of the state, increasing the welfare of citizens;
- participation in solving common problems of mankind;
- development of space science and technology, space services and technologies that determine stable development of the national economy;
- creation of a strong export potential of the space industry;
- providing access to space, conducting scientific research of the Earth and outer space;
- creation and support by modern means of modern information space of the state;
- ensuring the long-term interests of the state in the field of national security and defense capabilities;
- promoting the improvement of education;
- participation in monitoring the implementation of agreements to which Ukraine is a party relating to international security [11].

Space activities in Ukraine in modern period are carried out on the basis of the National Targeted Scientific and Technical Space Program of Ukraine, which is developed and approved by the Verkhovna Rada of Ukraine on the proposal of the Cabinet of Ministers of Ukraine. According to the order of the Cabinet of Ministers of Ukraine № 15-r *On approval of the Concept of the National target scientific and technical space program of Ukraine for 2021–2025* dated on 13.01.2021 space activity provides realization of national interests in the field of national security and defense, the branches defining economic development, general scientific, technical and technological level, ensuring sustainable development, as well as improving the quality of life of citizens of Ukraine. The main purpose of the space program is to ensure the effective use of space potential and increase its impact on solving urgent problems of social and economic, environmental, informational, scientific and educational development of society, ensuring the interests of the state in national security and defense [12].

It is worth emphasizing the main ways of realizing the national interests of Ukraine in the field of space activities, among which we highlight:

- measures for introduction of space technologies, services and information in government and industry programs; training of qualified users;
- state support of scientific and technical projects aimed at ensuring development and modernization of infrastructure as a factor in attracting foreign investment; combination of budget support, public and private partnership and investment attraction;
- comprehensive updating of space legislation: formation of a system of laws and bylaws on new market-oriented principles in accordance with international law and the law of the European Union;
- stimulation and expansion of international cooperation, implementation of joint projects and developments, in particular within the framework of the Horizon-Europe EU program and within the framework of cooperation with the European Space Agency, NASA and other leading space agencies;
- implementation of full scale cooperation with private economy sector [12].

3.2 Regulatory and Legal Aspects of the Development of the Space Industry of Ukraine of a Promising and Strategic Level

Some issues of the impact of space industry on national security are covered in the Decree of the President of Ukraine № 392/2020 On the decision of the National Security and Defense Council of Ukraine On the National Security Strategy of Ukraine dated on 14.09.2020. Among the current and projected threats to national security and national interests of Ukraine, taking into account foreign policy and domestic conditions, the National Security and Defense Council of Ukraine includes the development of a weapons system based on new physical principles, using quantum, information, space, hypersonic, biotechnology and artificial intelligence, creation of new materials, robotics and autonomous unmanned aerial vehicles. At the same time, this strategic document states that along with other measures, it is necessary to promote the development of the aviation and space industries as having significant potential and opportunities for the production of high-tech products for civil and defense purposes [13].

Changes in political and economic situation in the country, new challenges posed to the space industry in this regard have necessitated the development of a new strategy for development of space field in Ukraine.

Given the accumulation of problems in the space sector, the need for a thorough understanding and finding ways to solve them, the developer of the new Strategy was identified by the State Space Agency of Ukraine with which we fully agree.

*Strategy of space activities of Ukraine until 2022*Adopted and approved by the order № 100 of the State Space Agency of Ukraine dated on 21.05.2015 identified the existing problems that need to be addressed, purpose, directions and ways of implementation and contains a set of priority measures to ensure development of the space industry.

Special attention should be paid to the urgent issues identified by experts in the space field that need to be solved. Their importance lies in the fact that each of them is a direction for the study, research and development of directions of overcoming which will give impetus to the development of the comic field in Ukraine. So we detail these issues:

1. It is a mismatch between opportunities provided by modern space technologies including information systems and the state of their implementation in Ukraine for the needs of national security and defense, education and science, civil society, good governance and high-tech economic development;
2. It is the lack of legislative motivational grounds for attracting investment in research and production on principles of public and private partnership and private enterprise;
3. It is limited in economic relations with the former traditional markets for goods and services involved in space activities;
4. It is low efficiency of economic activity of enterprises due to the imperfect structure of the space industry;
5. It is the lag of the technical and technological level of space industry, the slow introduction of modern technologies of design, production, use of modern materials and components;
6. Vast majority of senior workers at leading enterprises and institutions, non-compliance with modern social standards for space field workers, the need to transfer experience and knowledge from senior workers to young people [14].

Among the main directions and ways of implementation of this Strategy, the first thing that is undoubtedly true is the preservation of the status of a space state in Ukraine. It is difficult to overestimate the importance of the fact that Ukraine, having the status of a space state, enjoys this as an advantage in the international arena, it is part of our image of the country as part of the "club of space countries".

In his speech on April 12, 2021, the President of Ukraine stressed that "the task of the state in the space industry now consists of two points: we must preserve everything that has been done by the older generations. The same time, make every effort to ensure that the successes of the past years were not the only ones. Our generation must make a significant contribution to the development of the rocket and space industry. In conditions of military aggression, this direction should be considered a key element of national security and defense. This is another incentive to restore leadership and the status of a leading aerospace state" [15].

Among the main directions and ways of implementing the Strategy of space activities of Ukraine until 2022 is the gradual acquisition of the status of a member of the European Space Agency; active participation in European and international space programs and projects, including the space debris control program; introduction of European standards of space activity in Ukraine; introduction of space technologies and space information systems to the sphere of national security and defense, the real sector of the national economy, scientific and educational sphere and systems of state and regional government; meeting public needs in the services of space information systems; implementation of youth projects in the space industry related to education, practical participation of students and graduate students in space projects; use of space information systems and technologies in related fields, increasing social attractiveness for space field workers [14].

Strategy is implemented under the auspices of the State Space Agency of Ukraine.

In addition, in April 2021, at the level of the President of Ukraine, it was announced the creation of the State Space Program that aims to streamline intentions and provide understanding of what and how to do to achieve the large scale goal of industry establishing.

3.3 International Legal Direction of Cooperation of Ukraine in the Space Field

Given the strong measures to ensure space security in EU cranes, Ukraine is trying to comply with the common European space policy.

On the positive side, Ukrainian contribution to the work of United Nations Committee on the Peaceful Uses of Outer Space is worth noting. In accordance with General Assembly resolution 75/92, on 23 April 2021, the Scientific and Technical Subcommittee considered agenda clause 11: *Objects approaching the Earth.* From this agenda item, the following presentations were made:

- Achieving the Hayabusa-2 mission: discovering the world of asteroids using interplanetary flight technology with a return to Earth (representative of Japan);
- Activities of Ukraine in the field of observation of near-Earth space in 2020 (representative of Ukraine). The Scientific and Technical Subcommittee noted a number of national initiatives and activities aimed at developing capacity in the field of detection and monitoring of potentially dangerous objects approaching the Earth, early warning of them and reduction of the threat posed by them, and stressed the importance of strengthening international cooperation. and the exchange of information in this regard, so that all countries, especially countries with disabilities, are able to predict and mitigate the effects of the fall of such facilities, and are informed of potential threats [16].

Equally important from the point of view of the application of law and security in space is the report of the United Nations Workshop on Space Law on *the contribution of space law and space policy to space management and security in the 21st century* (10/10/2016). General Assembly, in its resolution on international cooperation in the peaceful uses of outer space, reaffirms annually the importance of international cooperation in establishing the rule of law, the application of space law, which play an important role in international cooperation, research and peaceful use of outer space. accession to international agreements. Within this procedure it is recommended to strengthen the relationship between international space law and practical space activities [17].

Workshop provides an overview of legal regime governing peaceful uses of outer space, examines various aspects of application of the concept of security in space in the context of global space management, including measures to ensure transparency and confidence in space, and addresses specific space issues. The workshop aims to achieve the following objectives: a) to promote the understanding, recognition and implementation of United Nations treaties and principles relating to outer space; (b) space management in the enhanced space security context, including codes of conduct and space policy development; (c) Consideration of space law and space policy in the context of the space economy, space society, space accessibility and space diplomacy; d) study of tendencies and problems of progressive development of space law; e) analysis of further capacity building, assistance and awareness raising needs in space law and space policy [17].

IV. Introduction of the educational component as a direction of development of space law branch in Ukraine. Participants in the above mentioned United Nations Space Law Workshop noted with satisfaction that many governmental and non-governmental institutions are taking initiatives to increase space law competence, namely: to apply to universities to develop space law curricula; establish scholarship programs in the field of space law for graduate students and graduates of higher education institutions; contribute to the development of a national regulatory framework governing space activities. In addition, the importance of creating an innovative approach that builds capacity in the field of space law and space policy taking into account the needs of developing countries [17].

In our deep conviction, successful implementation of space activities in Ukraine, rule-making techniques and practice of application of international legal documents in this area, depends on the level of training and knowledge gained by future professionals in the field of space law. Availability of professionals capable of resolving complex legal issues and providing legal advice depends on provision of sufficient conditions and opportunities for education in the field of space law and space policy.

In Ukraine, academic disciplines are devoted to space law and space activities, are at the stage of scientific development and are taught only in some institutions of higher education in law. We propose to single out institutions and educational establishments that have relevant disciplines in their curricula and are engaged in research in the field of space law, namely: International V. M. Koretsky Institute of state and law of National Academy of Sciences of Ukraine, Space Law International Centre of Ukraine, Department of Economic, Air and Space Law of National Aviation University, Yaroslav the Yaroslav Mudryi National Law University, National Aerospace University – "Kharkiv Aviation Institute", NAU "KhAI" etc.

Given the global management of space activities, Ukraine desire to comply with international space policy, it should be noted that international recommendations on the need for further study of trends and problems of space law should intensify domestic institutions to promote research in space law and space activities.

V. Public and private relations in space. An important event for the development of the private sector in space was adoption by Parliament of the Law № 143-IX on Amendments to Certain Laws of Ukraine on State Regulation of Space Activities dated on October 2, 2019. This law creates conditions for the development of the domestic space industry, increasing its investment attractiveness, the formation of a competitive environment for space entities of various forms of ownership. The document stipulates that the subjects of space activities can be enterprises, institutions and organizations of any form of ownership and organizational and legal form [18].

It is promising to use opportunities that have arisen as a result of the removal of the state monopoly on space activities. In particular, the Southern Machine-Building

Plant has high hopes for a partnership with the Ukrainian-American private company Firefly Aerospace, which develops lightweight space carriers. Prior to the abolition of the monopoly, a significant part of its activities Firefly Aerospace took out of Ukraine [1].

The removal of the state monopoly raises the issue of corporatization, i.e. transformation of state owned enterprises into joint-stock companies with the simultaneous removal of at least some of them from the list of enterprises that are not subject to privatization. It does not require costs and simplifies the search for investors, including strategic ones.

Legal consolidation of right of private institutions to participate in development of outer space will promote promotion of scientific and applied research in space in Ukraine carried out by private companies, associations of young scientists, projects and startups with the direct participation of Ukrainian scientists: Firefly Aerospace, Space 1000, Spacebit, Council of Young Representatives of the Space Industry, Ukrainian Youth Aerospace Association "Suziria", etc.

These and other projects are reflected in the new Concept of the National Targeted Scientific and Technical Space Program of Ukraine, which was developed by the State Space Agency and approved by the order of the Cabinet of Ministers of Ukraine on January 13, 2021 under № 15-r. In this Concept, among the ways to realize the national interests of Ukraine in the field of space activities is the state support of scientific and technical projects aimed at developing and modernizing infrastructure as a factor in attracting foreign investment combining budget support, public and private partnerships and investment.

VI. Vector cooperation with the EU. Tragic events of 2014 became a driving factor in changing the priorities of our country's cooperation, including in the space industry, respectively, the termination of cooperation with Russia, finding new and improving existing relations with foreign countries in the space industry. At the same time, taking into account the previous long-term cooperation with Russia and focusing on its needs and capacities, the issue of not gradual withdrawal but termination of Ukraine cooperation with Russia put the national space industry in a state of rapid reorientation and cooperation with other foreign space powers.

Thus, analytical report of the National Institute for Strategic Studies in 2015 demonstrated that for political reasons and given previous experience, the reorientation of activities in the space industry of Ukraine should focus on deepening cooperation with the European Space Agency, the United States and China. from possible directions of such joint activity and it is noted that the success of cooperation will depend on the interest of foreign companies in domestic projects and the conditions of motivation to ensure joint productive activities [19]. Therefore, given the priority of European integration for Ukraine, we will focus on the EU in more detail.

Since 2008, Ukraine has been cooperating with the European Space Agency (ESA) to stimulate cooperation in space exploration and use. Admission to the ESA takes place in three stages. At the first stage, the country signs a Cooperation Agreement. At the same time, if a country wants to cooperate more effectively with ESA, a State Agreement on European Cooperation is signed for five years, after which the country can start negotiations on full membership.

Cooperation of our country with ESA was established by the *Agreement between the Government of Ukraine and the European Space Agency on cooperation in the use of outer space for peaceful purposes* dated on 25.01.2008 (№ 982_001). This Agreement, signed on January 25, 2008 in Paris, was ratified by the relevant Law of Ukraine on October 30, 2008 under № 633-VI. It was important for Ukraine to expand and consolidate in this Agreement certain areas of cooperation that have the potential, agreements between the parties to inform each other about all relevant measures and programs and the course of their implementation and regular consultations in accordance with the procedures established by the Agreement.

Among the areas considered by ESA and Ukraine as having the potential for cooperation under this Agreement, the following stand out in particular:

- Space sciences, namely astronomy and astrophysics, solar system research and solar and terrestrial physics;
- Earth exploration and application programs, namely environmental monitoring, meteorology, aeronomy and geodesy, disaster prevention;
- Telecommunications, namely the provision of services and satellite navigation;
- Microgravity research, namely space biology and medicine, data processing;
- Ground segment development and use;
- Launchers [21].

This Agreement remained in force for a period of 5 years from the date of its entry into force in 2009.

While this period of bilateral cooperation, implementation mechanisms also covered:

- establishment of scholarships to enable nominees from each Party to receive training or participate in scientific or technical activities in institutions proposed by the other party establishing the scholarship;
- holding joint conferences and symposia;
- joint promotion of products and services developed under ESA programs;
- promoting educational activities in space science and technology;
- providing expertise and assistance in space project management [21].

At the end of the 5 years specified in the Agreement, taking into account the positive experience of bilateral cooperation under this international agreement, in February 2014 the *Protocol amending the Agreement between the Government of Ukraine and the European Space Agency on cooperation in the peaceful uses of outer space* (document 982_002) desiring to extend the cooperation period under this Agreement beyond the initial period of 5 years.

Pursuant to the above mentioned Protocol, Ukrainian side designates the State Space Agency of Ukraine to implement this Agreement and authorizes the State Space Agency of Ukraine to conclude implementation agreements. The new Agreement will remain in force for a period of 10 (ten) years from the date of its entry into force, i.e. until February 14, 2024.

Over time, the domestic industry has gained some experience of cooperation with ESA in the field of science, participating in European projects, including FP7, Twinning, Twinning-2 and Horizon 2020. РД-868Р propulsion engine for the fourth stage was created for the Vega launch vehicle developed by the European and Italian space agencies of the Yuzhnoye State Design Office. In another words, Ukraine has gained some experience of cooperation with ESA. However, the results of the analysis of the industry potential show that the scope of cooperation with ESA may be increased if Ukraine becomes a member of ESA. To join ESA, Ukraine must meet one of the prerequisites for membership in the European Space Agency: adhere to the principle of free circulation of funds. According to this principle, each state receives such a volume of orders from the ESA, which corresponds to its financial contribution to the budget of the agency [22].

It should be noted that each country within the ESA specializes in the areas of activity in which it provides the largest contribution to development of rocket and space technology, spacecraft or space systems. For example, France specializes in the development and manufacture of major European medium and heavy Arian launch vehicles, space satellites and space systems. Italy is building satellites, interplanetary stations, light launch vehicles, and manufacturing ESA and NASA sealed modules for the International Space Station. Germany develops satellites, interplanetary stations, automatic cargo ships, maintains major European space control centers and trains European astronauts. Therefore, to expand Ukraine's cooperation with ESA, it is advisable to justify the main activities of the industry within ESA, prepare a draft decision of the Government of Ukraine on possible financial contributions to the ESA budget and prepare a draft Plan for Ukraine accession to the European Space Agency [22].

As of 2021, about 25 countries (full and associate members) are members of this agency. The State Space Agency of Ukraine together with ESA is currently working on a roadmap to develop a specific action plan for the stages for Ukrainian full membership in ESA [23].

VII. Own satellite as a direction of space industry development of Ukraine. Currently, more than 2,000 satellites operating in about 75 countries orbit our planet, providing basic aspects of modern life: communication, navigation, science – regardless of national borders [24]. And the US Global Positioning System (GPS), for example, has become the basis for a multibillion dollar civilian industry.

Looking for ways to address security issues in the information field and the effective provision of the European digital system, EU member states have found a way to potentially address this issue. Thus, innovation with the setting "bringing Europe various stakeholders together around a common, concrete and ambitious goal of cooperation" can, on a large scale, be a potential solution. Such large, ambitious missions of the Future can foster public-private partnerships to stimulate innovation. The security, defense and outer space sector in Europe offers real and concrete opportunities to set up such missions today. As an example, the construction of the *Digital Galileo*, the EU global satellite navigation system for the European digital system, is a purposeful sovereignty and independence that could potentially unleash such unifying efforts. If Europe took this opportunity, it would also demonstrate that European cooperation can bring tangible victories in a rapidly evolving technological change [25].

Let us consider why the EU member states place such hopes for security both at the level of a human citizen and at the level of state security, construction of *Digital Galileo.*

Galileo is a global satellite navigation system (GNSS) in Europe that provides improved positioning and timing information and has significant positive effects for many European services and users. For example:

- Galileo allows users to know their exact position with greater accuracy than what other available systems offer;
- critical, emergency response services receive reliability from Galileo system;
- products that people use on a daily basis, from the navigation device in their car to the mobile phone, are endowed with the increased accuracy provided by Galileo;
- Galileo services will make European roads and railways safer and more efficient;
- it stimulates European innovation by contributing to the creation of many new products and services, job creation, etc. [26].

The launch of the Galileo system was entirely related to the need to ensure security issues, and even some sources stated that "main goal of the Galileo program is European independence".

Until now, GNSS users had to depend on non-civilian American GPS or Russian GLONASS signals. With Galileo, users now have a new reliable alternative that, unlike these other programs, remains under civilian control. In addition, Galileo provides European citizens with independence and sovereignty, a number of environmental benefits and several new services specific to the Galileo program (Open Service, Commercial Service, Search and Rescue) [26].

Given experience of EU member states, the need for Ukraine to launch its own satellite is justified not only by its own security, although this would be enough, but by the fact that the presence of such spacecraft will help solve a wide range of problems in the interests of both economy and law enforcement. structures, as well as save budget funds that are currently spent on the purchase of services provided by similar devices abroad.

Currently, the Government of our country focuses on accelerating the implementation of the task of the President of Ukraine to launch into Earth orbit spacecraft remote sensing of the earth "SICH 2-30 (2-1)" while emphasizing the need to take all necessary measures to achieve this by the end 2021, the year of the 30th anniversary of Ukraine's independence [27].

Since February of this year, the Cabinet of Ministers of Ukraine has intensified the long-standing problem of creating Ukrainian spacecraft for remote sensing of the Earth, the work continues. The goal is for Ukraine to launch its satellite into orbit by the end of the year. At the same time, according to experts, the launch of the satellite to Ukraine this year can be carried out under the conditions of the most effective coordinated work that is already underway by the Government and Parliament (Committee on Economic Development) and the State Space Agency.

Agreement reached with SpaceX, which agreed to launch the Ukrainian satellite SICH 2-1, can be considered a positive intermediate result. The terms of the contract have been agreed in advance and Ukraine now has the opportunity to sign a contract for start-up services with the legendary American company. The executor of the project of

Yuzhnoye State Design Office was given the opportunity to sign a contract and launch in December 2021 [28].

Parliament of Ukraine supports these initiatives and Verkhovna Rada Committee on Economic Development, in particular has expressed readiness to further create the necessary climate for the development of the space industry approved a number of practical steps to achieve this goal.

4 Conclusions

Complete and comprehensive analysis of selected topic allowed to draw the following conclusions.

1. Space industry is a specific area of state activity that leads to construction of a unique system of legislation which includes different legal force and nature of regulations. It should be emphasized that the national legal system is represented by acts of general and strategic nature. Ukraine chosen course of rapprochement with the EU also influences the adoption of relevant national acts taking into account general European policy in the space field.
2. Significant role in the development of space industry in Ukraine is played by international legal direction of Ukraine cooperation, namely: active participation of Ukraine in the work of the United Nations Committee on the Peaceful Uses of Outer Space, cooperation with the European Space Agency, participation in European projects (FP7, Twinning, Twinning-2 and Horizon 2020), etc.
3. On the positive side, Ukrainian efforts to launch the SICH 2-30 (2-1) remote sensing spacecraft into orbit by the end of 2021, both to ensure its own security and to address a wide range of issues in the interests of both the economy and the economy and law enforcement agencies.

Given the above, Ukrainian space industry has significant potential for active cooperation with European partners. Given the consistent and purposeful state policy of the European Union in space industry and Ukraine European integration aspirations in this area, we propose to consistently implement a set of measures to further establish and establish Ukraine as a space state, the need for such steps is justified:

1. Adapt legislation of Ukraine to the legislation of the European Union by bringing the laws of Ukraine and other normative legal acts in line with the Community acquis in the field of space activities.
2. Deepen cooperation with the European Space Agency for the joint and efficient use of outer space.
3. Bring Ukraine satellite into Earth orbit by the end of 2021.
4. Exchange information and experience between Ukrainian and European scientists and specialists in the field of space activities for development of joint (interdisciplinary) research, new technological solutions at Ukrainian public and private enterprises.
5. Attract foreign investors within the framework of public and private partnership for financing from external sources of the space industry of Ukraine.

6. Provide legal support for space activities by professionals.
7. Carry out a comprehensive reset of higher education institutions of legal orientation which will allow to develop curricula, lecture courses, etc., dedicated to space law and introduce them into initial and scientific work.
8. Popularize space topics among young people, including by: holding scientific conferences; initiating youth events dedicated to the contribution of outstanding astronauts of Ukraine; starting startups, etc.

References

1. NISS expert took part in a field meeting of the Verkhovna Rada of Ukraine Committee in Dnipro https://niss.gov.ua/news/novini-nisd/ekspert-nisd-vzyav-uchast-u-viiznomu-zasidanni-komitetu-verkhovnoi-radi-ukraini-v [in Ukrainian]
2. Malysheva, N.R., Hurova, A.M.: Novations in permissive procedure of space activities: an objective necessity or a tribute to timE? Kosmìčna nauka ì tehnologìâ **22**(6), 37–44 (2016)
3. Utko-Maslianyk, Y.M.: The legal aspects of cooperation between European Union member states in the field of space. In: Dissertation for obtaining of the Candidate of Legal Sciences degree, specialty 12.00.11 "International law". Yaroslav Mudriy National Law University, Ministry of Education and Science of Ukraine, Kharkov (2018). https://dspace.nlu.edu.ua/bitstream/123456789/15411/1/Utko-Maslianyk_2018.pdf [in Ukrainian]
4. Venkata Rao, R., Gopalkrishnan, V., Kumar, A. (eds.): Recent Developments in Space Law Opportunities & Challenges. Springer Singapore, Singapore (2017). https://doi.org/10.1007/978-981-10-4926-2_2
5. Mejía-Kaiser, M.: The Geostationary Ring: Practice and Law. Basic Concepts of Space Law Series: Studies in Space Law. Brill, Nijhoff (2020). ISBN 978-90-04-41102-9 (e-book) [in English] The Economist, "Attacking Satellites Is Increasingly Attractive – and Dangerous," 18 July 2019. https://www.economist.com /briefing/2019/07/18/attacking-satellites-is-increasingly-attractive-and-dangerous [in English]
6. The Economist: Attacking Satellites Is Increasingly Attractive – and Dangerous, 18 July 2019. https://www.economist.com/briefing/2019/07/18/attacking-satellites-is-increasingly-attractive-and-dangerous [in English]
7. NATO Space Centre: https://shape.nato.int/about/aco-capabilities2/nato-space-centre [in English]
8. Patrick, S., Evanoff, K.L.: The Right Way to Achieve Security in Space. Foreign Affairs, 17 September 2018. https://www.foreignaffairs.com/articles/space/2018-09-17/right-way-achieve-security-space [in English]
9. Treaty on the Principles Governing the Activities of States in the Exploration and Use of Outer Space, including the Moon and Other UN Celestial Bodies (ukr/ros): Treaty, International Document dated January 27, 1967. https://zakon.rada.gov.ua/laws/show/995_480#Text [in Ukrainian / Russian]
10. Declaration of Legal Principles of States for the Exploration and Use of UN Space: Declaration, International document of 13.12.1963 https://zakon.rada.gov.ua/laws/show/995_a31#Text [in Ukrainian]
11. On space activities Law of Ukraine of 15.11.1996 № 502/96-VR. https://zakon.rada.gov.ua/laws/show/502/96-%D0%B2%D1%80#Text [in Ukrainian]
12. On approval of the Concept of the National Targeted Scientific and Technical Space Program of Ukraine for 2021–2025 Order of the Cabinet of Ministers of Ukraine; Concept from 13.01.2021 № 15-р https://zakon.rada.gov.ua/laws/show/15-2021-%D1%80#Text [in Ukrainian]

13. On the decision of the National Security and Defense Council of Ukraine of September 14, 2020 "On the National Security Strategy of Ukraine". Decree of the President of Ukraine; Strategy from 14.09.2020 № 392/2020. https://zakon.rada.gov.ua/laws/show/392/2020#Text [in Ukrainian]
14. Strategy of space activity of Ukraine for the period till 2022 SCA of Ukraine: Order, Strategy dated 21.05.2015 № 100 https://zakon.rada.gov.ua/rada/show/v0100814-15#Text [in Ukrainian]
15. Ukraine is a member of the "club of space powers" and occupies a worthy place among its 13 members. President https://uatv.ua/ukrayina-vhodyt-do-klubu-kosmichnyh-derzhav-i-posidaye-gidne-mistse-sered-13-jogo-uchasnykiv-prezydent [in Ukrainian]
16. Resolution adopted by the General Assembly on 10 December 2020 on the Continuity of the work of the Committee on the Peaceful Uses of Outer Space and its subsidiary bodies A/RES/75/92 https://documents-dds-ny.un.org/doc/UNDOC/LTD/V21/027/41/pdf/V2102741.pdf?OpenElement [in English]
17. Report on the United nations workshop on space law on the theme "contribution of space law and policy to space governance and space security in the twenty-first century" A/AC.105/1131. 10/10/2016 URL: https://documents-dds-ny.un.org/doc/UNDOC/GEN/V16/087/12/pdf/V1608712.pdf?OpenElement [in English]
18. On Amendments to Certain Laws of Ukraine Concerning State Regulation of Space Activities Law of Ukraine of October 2, 2019 № 143-IX. https://zakon.rada.gov.ua/laws/show/143-20#Text [in Ukrainian]
19. Problematic issues of reorientation of the space industry in the new conditions. Analytical note https://niss.gov.ua/doslidzhennya/mizhnarodni-vidnosini/problemni-pitannya-pereorientacii-diyalnosti-kosmichnoi-galuzi [in Ukrainian]
20. The European Space Agency (ESA): https://www.esa.int [in English]
21. Agreement between the Government of Ukraine and the European Space Agency on cooperation in the peaceful uses of outer space "dated 25.01.2008 (number 982_001) (expired) https://zakon.rada.gov.ua/laws/show/982_001 Text [in Ukrainian]
22. Problematic issues of reorientation of the space industry in the new conditions: Analytical note https://niss.gov.ua/doslidzhennya/mizhnarodni-vidnosini/problemni-pitannya-pereorientacii-diyalnosti-kosmichnoi-galuzi [in Ukrainian]
23. Seven satellites and the creation of a Ukrainian spaceport is a five-year space program: www.althoughda.com.ua/news/2021/05/9/673695/ [in Ukrainian]
24. Union of Concerned Scientists: UCS Satellite Database, 16 December 2019. https://ucsusa.org/resources/satellite-database [in Ukrainian]
25. Munich Security Report 2020. https://securityconference.org/publikationen/munich-security-report-2020/ [in German]
26. Galileo is the European global satellite-based navigation system. https://www.gsa.europa.eu/european-gnss/galileo/galileo-european-global-satellitebased-navigation-system [in English]
27. The head of state instructed to launch a Ukrainian satellite - Urusky - into orbit by the end of the year. uatv.ua/glava-derzhavy-doruchyv-do-kintsya-roku-vyvesty-na-orbitu-ukrayinskyj-suputnyk-uruskyj/ [in Ukrainian]
28. SpaceX has agreed to launch the Ukrainian satellite Sich 2-1. https://www.ukrinform.ua/rubric-technology/3202589-spacex-pogodilasa-zapustiti-ukrainskij-suputnik.html [in Ukrainian]

Current Issues of Legal Regulation of Intellectual Property in Space Field

Alla Hordeiuk[1(✉)], Andrii Ostropilets[2], and Ihor Bohdaniuk[3]

[1] Law Department of National Aerospace University "Kharkiv Aviation Institute" NAU "KhAI", Kharkiv, Ukraine
a.hordeiuk@khai.edu
[2] Kharkiv, Ukraine
[3] Forensic Examinations of National Scientific Centre, "Hon. Prof. M. S. Bokarius Forensic Science Institute", Kharkiv, Ukraine

Abstract. Topical issues of legal regulation of intellectual property relations in the field of space activities at the international and national level are considered. International and domestic normative acts are analyzed and peculiarities of legal protection provision of intellectual rights of intellectual property creators of objects used in space industry, in particular on International Space Station and under jurisdiction of various partner states are determined. The Correlation between regulation of property rights and intellectual property rights in the specified space object is determined. Analytical conclusions have been made regarding expediency of intensifying activities rule-making, international and domestic control bodies on space affairs. Commercialization procedure taking place in the space field and necessitates adaptation of intellectual property legislation to new realities is noted. Position on the need to improve international and national legislation on both space law and intellectual property law at the present civilization stage in development of human relations and space is substantiated.

Keywords: Intellectual property · Intellectual property rights · Objects of intellectual property rights · Intellectual rights · Property rights · Space activities · Space field · Space industry · Outer space · Commercialization of space activities · International space law · National space law

1 Introduction

In current realities, the issue of legal protection of intellectual property is relevant as in Ukraine as in other countries, so the international community pays much attention to improving legal regulation in this area. The World Intellectual Property Organization (hereinafter referred to as WIPO) is especially conducive to intellectual property protection at the international level by ensuring cooperation between states and monitoring compliance with provisions of multilateral treaties governing legal and administrative aspects of intellectual property. With the active participation of member countries, WIPO carries out activities aimed at shaping global policy, harmonizing national laws and procedures in the field of intellectual property. However, existing international system of

M. Nechyporuk et al. (Eds.): ICTM 2021, LNNS 367, pp. 795–813, 2022.
https://doi.org/10.1007/978-3-030-94259-5_62

intellectual property protection cannot fully meet the needs of the development of civilizational processes in the world. The use of technologies by modern man as objects of intellectual rights, in the field of communications, satellite navigation, natural disaster prediction, mineral exploration and other opportunities provided by human space activities is especially relevant nowadays. It is an important factor influencing the dynamics of innovative development of country economy, indicator of scientific and technological potential, effective tool for ensuring its interests in areas that are vital at this civilization stage. However, these processes are not always provided by an effective mechanism of legal regulation both at the national and international levels. Therefore, one of priorities of international community lawmaking and the national legislator today is to create a strong legal protection of the results of intellectual creativity in space industry belonging to the field of high technology that is most saturated with innovation.

It follows from the above that issues of legal regulation intellectual property in this area of relations at both international and domestic levels are important and relevant for researches by modern scientists. In this context, it is advisable to pay attention to the peculiarities of legal regulation of property rights in space activities, to determine the connection and impact of its legal provisions on peculiarities of protection of intellectual property rights in this area.

2 Literature Review

Many domestic and foreign scientists have devoted their research papers to research on issues of improving the regulation of intellectual property but this issue in space activities is not sufficiently studied nowadays. The issues of protection of intellectual property in space are partially covered in research papers of scientists analyzing issues of international space law, but there are some research papers of domestic scientists who purposefully choose as a field of research, topic on intellectual property protection in space. Thus, among the researchers who pay attention to the problems of improving legal regulation of intellectual property: O. K. Dakhno, O. V. Dzera, R. E. Ennan, R. A. Maidannik, O. M. Melnik, O. P. Orlyuk, O. P. Svitlychnyi, E. O. Kharitonov, R. B. Shishka, I. E. Yakubovskyi and others. V. M. Borychevskyi, O. V. Biehlyi, V. H. Hornyk, N. E. Zubach, V. V. Luts, O. S. Kurilo, N. R. Malysheva, I. M. Sopilko, A. V. Khridochkin and others devoted their research to the study of intellectual property or its individual aspects in the field of space activities. In foreign scientific literature, this issue has been studied by such scientists as: R. Avveduto, A. M. Balsano, R. Berkley, S. Bhatt, D. M. Cameron, K. G. Hammerle, D. Irimies, V. Kumar, L. B. Malagar, M. Molineaux S. Mosteshar, T. Myamoto, Chukeat Noichim, R. Oosterlinck, B. L. Smith, J. A. L. Sterling, H. Van Traa, Dr. Ian Zhao.

Scholars believe that the theoretical underdevelopment of a number of issues of legal protection of intellectual property, in particular in space, significant amount of regulations and law enforcement practices, certain difficulties associated with interpretation of legislation, necessitate a comprehensive research on legal regulation intellectual property relations in the field of space activities at the international and domestic levels [1, p. 25; 2, p. 217; 3].

3 Main Content Presentation

The *intellectual property* term in the modern world is actively used in international and domestic regulations as well as in national and foreign doctrine and practice. In defining the *intellectual property* concept professionals usually rely on the provisions of Art. 2 of Convention Establishing the World Intellectual Property Organization (hereinafter referred to as WIPO) dated on 14 July 1967 defining rights that are the result of intellectual activity in industrial, scientific and literary activities and art [4].

Part 2 of Art. 41 of the Constitution of Ukraine prescribes provisions on the right of everyone to own, use and dispose of the results of their intellectual and creative activities. Part 1 of Art. 54 of the Basic Law also establishes that citizens are guaranteed freedom of creativity including scientific and technical ones [5].

Currently, many civilized countries are on the way of restoring respect for intellectual property rights. The concept of intellectual property arose in the process of long-term practice of legal consolidation of certain human rights to the results of intellectual activity in the field of science, production and art. Intellectual property is a set of rights to the results of intellectual creativity, as well as means of individualization of participants in civil turnover, their goods and services that are provided with legal protection as objects of intellectual property rights in accordance with international and domestic laws [6, p. 7]. Professionals believe that objects of intellectual property rights are defined in the law as results of intellectual creativity that are new and original one. It is noted that creating some objects is not necessarily creative, in particular commercial names, trademarks, geographical indications of origin, etc. [7, p. 89, p. 91].

The WIPO Convention defines a list of intellectual property rights which the relevant legislation of the participating countries, including Ukraine is based on. Thus, in Art. 2 of this international normative act it was established that intellectual property includes rights belonging to: fiction writing and researches; artistic activities of artists, sound recording, radio and television programs; inventions in all fields of human activity; scientific discoveries; industrial designs, trademarks, brand names and commercial designations; protection against unfair competition, as well as all other rights related to intellectual activity in industrial, scientific, literary and artistic fields [4].

Art. 420 of the Civil Code of Ukraine determines the following list of objects of intellectual property rights: fiction writing; software; (databases), performance, video recording, audio recording, broadcasting; scientific discoveries; inventions, utility models, industrial designs; layout (topography) of integrated circuits; efficiency suggestions; plant varieties, animal breeds; commercial (brand) names; trademarks (marks for goods and services), geographical indications; trade secrets. In other words, Ukrainian legislator in determining the objects of intellectual property rights subject to legal protection in Ukraine, was guided by international law, including the provisions of the Convention Establishing the World Intellectual Property Organization and the Berne Convention for the Protection of Literary and Artistic Works adopted in 1886 (entered into force for Ukraine on 25.10.1995), the Paris Convention for the Protection of Industrial Property adopted in 20.03.1883 (entered into force for Ukraine on December 25, 1991, etc. (hereinafter referred to as The Paris Convention) [8–10].

In accordance with the above Article 2 of the WIPO Convention, all intellectual property can be divided into two groups: copyright and related rights; objects of industrial property [4]. Criterion for dividing objects into groups is the scope of their application in civil (economic) turnover. Most objects of copyright and related rights are the result of intellectual creativity in the spiritual field and objects of industrial property are the result of intellectual creativity the field of scientific and technical creativity.

In the Agreement on Trade-Related Aspects of Intellectual Property Rights (TRIPS) adopted in 1994 classification of intellectual property includes the following groups: copyright and related rights; trademarks; geographical indications; industrial designs; patents; topography of integrated circuits; protection of confidential information; control over anticompetitive practices in contractual licenses [11]. This classification obviously provides an economic approach to the definition of intellectual property from the point of view that they are considered as a commodity. It should be noted that this approach is relevant nowadays, taking into account commercialization of intellectual property in various sectors of economy including in the field of space activities [3]. Thus, V. M. Borycheskyi believes that with development of science-intensive technologies in space field and in connection with the commercialization of this industry, new and diversified existing subjects and objects of intellectual property rights, there are new ways for their using. In addition, the risks of unauthorized encroachment on the rights and legitimate interests of owners, including those related to the regulation lack of new legal relations in the field of peaceful uses and space exploration [12, p. 25].

In context of this topic we should pay attention to the objects of intellectual property rights that are the result of intellectual scientific and technical creativity, i.e. those related to industrial property (including patent law and a group of non-traditional objects in the legal system. industrial property) and are used for the development of high-tech areas, in particular in space field, although experience of space activities has led to the conclusion that use of copyright and related rights is also relevant [2, p. 218; 13, p. 345–418].

Development of space intensified sharply at the end of the twentieth century, which marked the transition of mankind to a global new level of civilization. Normative regulation of space activities at the international level began simultaneously with the launch into space in 1957 of the first artificial body. As a result, international cooperation began to gain momentum for each state, and in modern realities also the rapid commercialization of space activities, as mentioned above. There is no doubt that space exploration and use is currently not possible without proper legal regulation at the level of international space law. Today, international space law is a combination of customs and treaties aimed at ensuring free, unhindered and non-discriminatory human access to space. Thus, in 1961, the General Assembly of the United Nations (hereinafter referred to as UN General Assembly) adopted a Resolution: *International Cooperation in the Peaceful Uses of Outer Space* and enshrined in this document two fundamental principles of space law: international law, including the UN Charter, applied to outer space; outer space and celestial bodies may be freely explored and used by all States in accordance with the norms of international law and shall not be subject to any of them. In 1963, the UN General Assembly adopted the Declaration of Legal Principles Governing the Activities of States in the Exploration and Use of Outer Space. The next stage in the development of international space law should be determined by the adoption of Treaty on Principles Governing the Activities of States in the Exploration and use of an Outer Space, including the Moon and other Celestial Bodies adopted in 27 January 1967 (hereinafter referred to as Outer Space Treaty). It entered into force for Ukraine on October 31, 1967. This Outer Space Treaty is considered a priority treaty that regulates space exploration and is recognized as a codification of customary international law which results in the definition of the basic principles of space activities. Thus, international law has established that the guiding norm for the development of space is the definition that outer space is recognized as an international common territory outside the national jurisdiction of individual states [14–16].

In addition to these sources, the legal support of international space law are: Agreement on the Rescue of Astronauts, Return of Astronauts and the Return of Objects launched into an Outer Space adopted in 22 April 1968 (Rescue Agreement)); Convention on International Liability for Damage Caused by Space Objects of March 23, 1972 Convention on International Liability for Damage Caused by Space Objects of 29 March 1972 (Liability Convention); Agreement on the Activities of States on the Moon and Other Celestial Bodies of December 18, 1979 Agreement Governing the Activities of States on the Moon and other Celestial Bodies of 18 December 1979 (Moon Treaty); Convention on the Registration of Objects Launched into Outer Space of 15.09.1976 date of entry into force for Ukraine: 14.09.1977 Convention of Registration of Objects Launched into Outer Space of 15 September adopted in 1976 (Registration Convention) [17–20].

It is important to understand that when it comes to the legal regulation of certain relations, in addition to legal support requires targeted activities of specialized bodies and organizations. Thus, the United Nations Office for Outer Space Affairs (UNOOSA) is the main international forum for the coordination of member country space activities for peaceful purposes for the benefit of all nations and plays a key role in the United Nations Open Space Exploration System. The executive body of the United Nations Office for Outer Space Affairs is the United Nations Committee for the Peaceful Uses of Outer Space (UNCOPUS) established in 1959 that focuses on the development of international cooperation with the United Nations. in order to promote the orderly expansion of space activities, the gradual economic growth and sustainable development of all countries, the identification of new areas of application of space technologies, the rational use of natural resources and the prevention of natural disasters. The structure of the UNOOSA includes its working bodies: The Scientific and Technical Subcommittee (SCC) and the Legal Subcommittee (SCC), which, in particular, examines general trends in space law. In its reports, the UNOOSA points to the incompleteness of existing international agreements in the field of space law. For example, current issues that need to be improved today are: establishing the role of the private sector in space; revision of the regime of international responsibility to protect not only astronauts but also space tourists in the event of accidents on space vehicles; problems of space debris control, as well as cleaning of outer space from pollution; protection of space assets; legal regime of anti-satellite weapons. We believe that all the problematic issues related to the field of intellectual property, i.e. the problem of lawful use of intellectual property in the relevant sectors for a specific purpose, as well as the protection of intellectual property rights of creators of intellectual property used in space activities. However, the activities of the КВК ООН are not always positively assessed by member countries. Some of them believe that the committee is not doing enough real work in this area, so it may lose its competitiveness among research institutions and priority functions for global space management. We consider this trend undesirable, because there should be a single coordination center for space management which can influence the codification of international law that currently regulates the legal protection of intellectual property used in space activities, as well as rule-making processes associated with the improvement of legal regulation in this area, taking into account new trends in commercialization in the study of outer space [21, 22].

Thus, one of the main issues of international space law is the problem of ensuring the protection and safeguarding of intellectual property rights that have been created in the process of space projects. However, it should be noted that this issue is related to peculiarities of the legal regulation of property rights in the space industry, so in this context, it makes sense to first pay attention to this aspect.

With regard to the legal regulation of property rights in outer space, it should be noted that under the Outer Space Treaty, no space object is subject to national appropriation, either by declaring sovereignty, use or occupation, or in any other way. Outer space is open for exploration and use by all States without any discrimination on the basis of equality and in accordance with international law, with free access to all areas of celestial bodies. Outer space is recognized as an international common territory outside the national jurisdiction of individual states, as noted above. (Articles 1, 2). [16]. Another international legal instrument, the Moon Treaty, explicitly prohibits not only the establishment of ownership of the Moon and its surface but any use of it (other than scientific) and the appropriation of the results of such activities. Under this Agreement, the surface, subsoil or natural resources of the Moon and other celestial bodies may not be the property of any State, international, national organization or individual. States are allowed to collect samples of substances on the Moon and other celestial bodies and to take them out during scientific research. These samples remain at the disposal of the respective states. Placement on the surface of the Moon or in its bowels of personnel, spacecraft, equipment, installations, stations and structures is not a basis for establishing ownership of the surface, bowels of the Moon or their parts. However, the 1979 Moon Treaty was ratified by only 14 states, so its provisions are not customary law rules to be adopted and adhered to by the entire international community. But from the above we can conclude that outer space, the Moon and any other celestial bodies cannot belong to the ownership of any country in the world. As for objects launched into space, Art. Article 8 of the *Outer Space Treaty* stipulates that objects, including spacecraft, which are launched and, accordingly, entered in the register of the relevant state, belong to the jurisdiction of the state in which the register of a certain object is entered. Such countries control their spacecraft and their crew while in space, including in the celestial body. Ownership of space objects launched into space, as well as objects delivered or constructed on a celestial body, and their constituent parts, remain intact during their stay in outer space or on the celestial body, or after returning to Earth. It follows from the norm that all space objects on Earth or in space are registered in the prescribed manner, are owned by the persons who built and launched them. The procedure for registering space objects is established by the *Registration Convention*. This Convention defines the need to register a space object at the national level with its entry in the register, which must be in each state engaged in space activities. And at the international level, registration is carried out by Secretary-General of the United Nations [16, 19, 20].

Considering the peculiarities of the legal regulation of property rights in the field of space activities, we cannot ignore the regulation of ownership of the International Space Station (hereinafter referred to as ISS). The ISS operates using a three-tier legal mechanism. The first level is Arrangement Between the United States of America and other Governments originally called the IGA adopted on January 29, 1998, (hereinafter referred to as the ISA Agreement) signed by fifteen governments Participants: the United States of America, Canada, Japan, the Russian Federation and ten member states of the European Space Agency (hereinafter referred to as ESA) (Belgium, Denmark, France, Germany, Italy, the Netherlands, Norway, Spain, Sweden, Switzerland), Great Britain, Hungary, Luxembourg, Slovenia. According to the ISA Agreement (IGA), each project participant owns its segments on which certain objects (modules) are located. Project partners retain jurisdiction and control over the facilities they register and the personnel on the ISS. This means that the above countries are co-owners of the ISS and are legally responsible for the facilities used on the ISS. The countries of the European Union are considered as one entity, which is called the European partner in the space station. The second tier, which is the legal basis for the functioning of the ISS, is based on four memoranda of cooperation between a key project participant (MOUs) – the US National Aeronautics and Space Administration (NASA) and relevant agencies of other partner countries. The purpose of these documents is to develop an effective project management model, define in detail the roles of each participant and the division of responsibilities for the implementation of individual project programs between national agencies. The third level consists of various bilateral agreements between the governmental organizations of the participating countries, which contain guidelines and principles aimed at implementing the agreements set out in the ISA Agreement (IGA) and memoranda. But any European country can pass national laws that will apply to facilities, equipment and personnel that will be sent to the ISS. That is, the main purpose of the ISA Agreement (IGA) is to ensure the national jurisdiction of each party to the ISS modules (for example, Japan has a Kibo module, Columbus module, solar panels and other modules owned by NASA) [12, p. 26–27; 23, 24].

Based on the peculiarities of the legal regulation of property rights in space, the question arises as to how the relationship of intellectual property rights to the use and disposal of intellectual property rights to certain objects as a result of intellectual creativity used in space. Researchers believe that international acts in force in the field of use and exploration of outer space do not provide a mechanism for the transfer of intellectual property rights to space objects (VM Borychevskyi, N. E. Zubach, C. Noichim, and others) [3, 12, p. 25; 25, p. 45]. This issue is addressed at the national level or at the level of international agreements concluded for each space project separately, as well as international conventions in force in the field of intellectual property. The legal regime of property rights in the field of space activities established by international acts significantly affects the relations of intellectual property. In particular, in practice, agreements between states in which their parties determine the legal regime of protection created in process of realization of the space project of a certain object of intellectual property and solve questions of distribution of intellectual rights to it. For example, in the process of conducting space activities related to the creation and maintenance of the ISS, certain obligations are established between the countries party to the ISS Agreement (IGA) on intellectual property rights.

In particular, in paragraph 2 of Art. 21 of the ISA Agreement (IGA) defines the procedure for the protection of intellectual property and stipulates that any activity carried out in the middle or in the orbital element of the space station is considered to be carried out only in the territory of the partner state that registered this element, except that, for elements registered in ESA, any European Partner State may consider such activities as activities taking place in its territory. In Clause 3. of Art. Article 21 prohibits a partner State from using national legislation on the secrecy of inventions in relation to inventions obtained in the middle of an orbital station or in a separate element by a person who is not a national of that country or does not reside in its territory, in order to prevent patent applications. Invention in any other partner country in which the protection of the confidentiality of patent applications containing confidential information is ensured by legal means. Clauses 4 and 5 of the ISS Agreement (IGA) also address the legal status of intellectual property that will receive legal protection in more than one European Partner country. According to these clauses, an organization or any other person for the same violation of their intellectual property rights can be compensated only in one of these countries. Scientists believe that the launch of the ISS is associated with extensive research on intellectual property issues in space[12, p. 25; 23].

It follows from above provisions of the IGA Agreement that intellectual property issues play an important role in establishing jurisdiction and control over station elements, as well as determining the rights to intellectual property that will be exercised aboard a permanent space laboratory. In other words, in this case, the connection of ownership of space objects taking into account principle of territoriality with the objects of intellectual property rights located in, used in space activities. According to scientists, for example, A. Vahrenwald, V. M. Borychevskyi and other researchers on this topic, the use of intellectual property rights created on board the ISS is insufficiently regulated, in particular, in the case where a particular object is common as a result of intellectual creative activity of several participants and ESA individually disposes of intellectual property rights to it [12, p. 25; 24].

In addition, both domestic and foreign scientists (R. Oosterlinck, C. Noichim, S. Mosteshar, D. M. Cameron, V. M. Borychevskyi, O. V. Biehlyi and others) believe that there is a need to define in more detail the position on improving the legal protection of inventions used in the field of space activities taking into account the opinion of experts that such inventions should be divided into two types: space inventions related to activities on Earth; inventions that are created in outer space. In that regard, C. Noichim states: "At first glance, the concept of legal regime governing an outer space is quite contrary to that of patent law. Inventions made in space will or may, indeed, get serious problems protection in intellectual property rights against infringements. Although there is still an argument on this issue, it is generally accepted that the Space Treaty does not commercial space activities…Therefore, problems will occur when an invention is used or infringed in outer space. Since there is no provision of the patent law applicable to the extra-territorial area of an outer space". S. Bhatt also devoted his work to the analysis of the need to revise and improve the legal regime of inventions in outer space. A. V.

Khridochkin, A. Vahrenwald note that the issue of the legal regime of inventions created in outer space is unresolved, as the latter does not fall under the jurisdiction of any state. Taking into account the views of scientists, we can identify a gap in the legal regulation of intellectual property rights for inventions created on objects located in space and propose in international and national space law and relevant intellectual property law to provide legal support for "space inventions" [3; 26; 27; 2, p. 219, 220; 24].

Protection of intellectual property rights of those entities that are not directly involved in the project is important, but the results of their work are used on the ISS. For example, Ukraine is not officially a party to the ISS Agreement (IGA), but such domestic enterprises as the Kyiv Radio Plant (spacecraft convergence system), State Enterprise Production Association Kyivprylad (command and measurement system) and others took an active part in the ISS research. Thus, the ISA Agreement (IGA) provides for the same legal regime for both such persons and project participants [12, p. 25].

It is important to note that since the inception of the ISS, all discussions on the protection of intellectual property rights have taken place in relation to patent law, but over time, after extensive research, it has become clear that the protection of intellectual property rights should extend beyond scientific and technical results. creativity, but also copyright and related rights [2, p. 218].

Thus, it follows from the above that currently at the international level a separate international normative act that should regulate the emergence, exercise and protection of intellectual property rights to space objects has not yet been adopted and the legal regulation of intellectual property relations in space find their decisions either in national regulations or in international bilateral and multilateral agreements on the implementation of certain space projects. We believe that it is appropriate to codify the rules governing legal protection of intellectual property at the international level, because such law-making activities can regulate relations in this area between all stakeholders and help resolve legal conflicts to protect their intellectual property rights. Thus, V. M. Borychevskyi notes that currently the "magnitude" of the legal force of intellectual property law is almost not agreed at the international level [12, p. 26].

Special attention in the context of this research should be paid to the issue of legal regulation of intellectual property rights in outer space at the national level. Until recently, no normative legal document on space activities in Ukraine provided for fundamental provisions on the protection of intellectual property rights. However, in the field of international cooperation, Ukraine has tried to use the experience of international practice in signing annexes on intellectual property to framework agreements on space exploration and use. In particular, Ukraine concluded a relevant agreement with the United States of America, namely, *Framework Agreement between the Government of Ukraine and the Government of the United States of America on Cooperation in the Exploration and Peaceful Uses of Outer Space*ratified on 17.12.2008 and entered into force for Ukraine on 22.01.2009 This Framework Agreement in Art. 1 establishes the main legal provisions and conditions of cooperation between the National Space Agency of Ukraine (existed with this name until December 9, 2010, currently it State Space Agency of Ukraine (SSAU) that is a specially authorized central executive body in Ukraine, which ensures the implementation of state policy in the field of space activities and is responsible for the state of development of the industry) and NASA or any other designated Office of each Party in the study and use of outer space for peaceful purposes in areas of mutual interest and equality and mutual benefit [29]. On May 14, 2015, the Cabinet of Ministers of Ukraine adopted Resolution №. 281 "*On Approval of the Regulations on the State Space Agency of Ukraine*". According to item 4 of this Provision, STAU according to the tasks assigned to it, in particular: generalizes practice of application of the legislation on the questions belonging to the scope of SSAU activity, develops offers on its improvement, prepares projects of legislative acts, acts of the President of Ukraine and submits them to the Deputy Prime Minister for Strategic Industries in the prescribed manner; approves draft laws, other acts of legislation approved by other central executive bodies, prepares, within the powers provided by law, conclusions and proposals for draft laws, other acts of legislation submitted to the Cabinet of Ministers of Ukraine, and draft laws, submitted to the Verkhovna Rada of Ukraine by other subjects of the right of legislative initiative; carries out international cooperation and implementation of foreign policy within its competence in the field of space activities; participates in the implementation of measures to adapt the legislation of Ukraine to the legislation of the European Union in the field of space activities; ensures effective use of scientific, technical and production potential of enterprises, institutions and organizations belonging to the sphere of its management, and creates conditions for the introduction of space technologies in the production of competitive products for the needs of the domestic and foreign markets; maintains the State Register of unique objects of space activity, carries out state supervision over their condition and use, takes measures to support them; organizes scientific, scientific and technical, investment, information, publishing activities, promotion of achievements and best practices, promotes the creation and implementation of modern information technologies and computer networks in the field of space activities [30]. From these defined tasks of the State Tax Administration it follows that the competence of this body includes the possibility of improving the legal regulation of intellectual property in space activities, taking into account the powers of law enforcement agencies, support and organization of space technology to meet the needs of the modern market.

It is important to note that the development of the legal system of the European Union (hereinafter referred to as EU) has also recently intensified, in particular in the field of space activities and legal protection of intellectual property. Experts believe that there is a so-called "Europeanization" of private law of EU member states, due to the purpose of creating and functioning of a common market [31], p. 1]. Thus, on June 27, 2014, the economic part of the Association Agreement was signed between Ukraine, on the one hand, and the European Union, the European Atomic Energy Community and their Member States, on the other hand. In accordance with Chapter 8 of Chapter V of this Agreement, its Parties shall promote the development of mutually beneficial cooperation in the field of civil space exploration and the use of outer space, in particular in the following areas: global navigation satellite systems; Earth observation and global monitoring; space science and research; applied space technologies, in particular launch technologies and rocket engine technologies. Cooperation may also include the exchange of experience in the management of space research and scientific institutions, as well as the creation of favorable conditions for research and the introduction of new technologies and the proper protection of relevant intellectual, industrial and commercial property rights. The expected results of cooperation are: active participation of Ukraine in European space projects; increasing the volume of production by Ukrainian enterprises in the space industry; increase in exports of Ukrainian space products to the EU; expanding the scope of use of satellite navigation systems in various sectors of the economy; expansion of remote sensing data of the Earth [32].

In Art. 92 of the Constitution of Ukraine stipulates that only the laws of Ukraine determine the principles of space exploration, in particular, relations in the field of space are regulated by the Law of Ukraine: *On Space Activity* which was adopted on 15.11.1996 This normative act stipulates that its effect extends to all activities related to the study and use of outer space, including the development of space science and technology, space services and technologies that determine the stable development of the national economy, providing access to space, conducting research on Earth and outer space. However, some non-specific indications regarding the regulation of intellectual property relations in space activities are not enough. There is, for example, a provision on the rules of space activities, which states that they include regulations governing activities in the space industry and its security, as well as compliance with the requirements for the protection of intellectual property rights, state, military and trade secrets. (Article 8 of the Law of Ukraine: *On Space Activity*) [4, 33].

According to Art. 7 of the Law of Ukraine: *On Space Activity* this activity in Ukraine is carried out on the basis of the National Targeted Scientific and Technical Space Program of Ukraine, which is developed for five years and approved by the Verkhovna Rada of Ukraine at the request of the Cabinet of Ministers of Ukraine. Thus, on January 13, 2012, the Cabinet of Ministers of Ukraine issued an order: *On approval of the Concept of the National Targeted Scientific and Technical Space Program of Ukraine for 2012–2025*. The document highlighted the problem to be solved by this program, taking into account that space activities ensure the realization of national interests in the field of national security and defense of industries that determine economic development, general scientific, technical and technological level, ensuring sustainable development. and also helps to improve the quality of life of citizens. The modern space economy is characterized by the emergence of new commercial applications, the involvement of private initiative, the introduction of new organizational forms and increased competition. Ukraine has extensive experience in the field of space technology development, so it should capitalize on it in accordance with modern requirements and world trends [33, 34].

The current global trend in space activities is to increase the participation of private actors. The study of this process in space has attracted the attention of many experts, for example, H. Van Traa in *Commercialization of Space Activities: Patent Rights for inventions of Outer Space* analyzed the features of commercialization in space and protection of inventions in space. Initially, space exploration and use activities were the domain of public entities, and state space programs concerned the defense sphere. Changes related to space activities can be traced by analyzing the work of the United Nations in this area, in particular during its conference. After the Second United Nations Conference on the Exploration and Peaceful Uses of Outer Space (UNISPACE 82) (1982), it became clear that after the end of the Cold War, the process of eliminating strategic weapons began, so budget contributions in this area were significantly reduced. At the next conference on the exploration and use of outer space for peaceful purposes (UNISPACE-99) (1999), it was noted that there is a tendency to involve international private companies in space activities. But at that time, international space law was not yet ready for the rapid commercialization of space activities. For the first time, NASA decided to use the help of private commercial companies to send astronauts to the moon. In 1984, the United States passed a federal law on commercial space launches. Thus began the history of New Space - a movement for space without a state monopoly, as a result of which American startups SpaceX, Rocketlab, Blue Origin appeared. Thus, today the commercialization of space activities is one of the priority areas in this area. The experience of the leading space agencies (NASA, EKA) in current market conditions indicates that due to lack of budgetary resources, work to improve the efficiency of space activities (often called this area is the commercialization of space technology). For example, NASA has established a corresponding Office (Technology Utilization Office) to coordinate the work on the commercialization of space activities and technologies, in particular, in the United States established ten centers for the commercialization of space technology. The agency receives about $ 500 million annually through the commercialization of space technology. At NASA, even anyone who wants to get commercial support is given access to watch the next launch of American media on a commercial basis. A similar system was created in ESA, thanks to the commercialization of European technologies, the agency received about two hundred million euros [35, p. 72–73; 36, p. 85–91; 37].

Thus, the commercialization of space, including the ISS project, is the path chosen by NASA, the first important step in this direction is related to the launch of the Crew Dragon spacecraft of Elon SpaceX company founded by Reeve Musk to the ISS in 2020, together with Axiom Space, the company plans to deliver tourists to the ISS [37].

It should be noted that space activities are regulated not only by the rules of public law, i.e. international acts that apply only to states. Thus, the International Institute for the Unification of Private Law (UNIDROIT) developed the Convention on the International Property Rights of Movable Property, which Ukraine ratified in 2012. The document defines the procedure for acquiring and exercising international property rights to uniquely identified real estate (including aviation). projects), as well as ensuring compliance with these rights by registering them in the International Register [38].

Every year, more and more countries and companies unite under the slogan: *Space for All*. Ukraine is no exception. Since 2017, the phrase "Demonopolization of space industry" in one context or another is increasingly heard in specialized publications. Relevant draft laws *On Amendments to Some Laws of Ukraine on Promoting Space Development and Attracting Investments in the Space Industry of Ukraine* dated on October 18, 2018, *On Amendments to Some Laws of Ukraine on Demonopolization and Development of Space Activity* began to be submitted to the Parliament of Ukraine from 02/27/2019, but they were rejected. The third bill in this area was adopted and as a result, the Law: *On Amendments to Certain Laws on State Regulation of Space Activities* dated on October 29, 2019 (hereinafter referred to as Law) came into force in Ukraine today. This Law, introducing certain changes, creates conditions for the development of the domestic space industry, increasing its investment attractiveness, the formation of a competitive environment for space entities, in particular, the regulation provides that the subjects of space activities can be enterprises, institutions and organizations of all forms ownership and organizational and legal form. So in Art. 4 of the Law of Ukraine: *On Space Activity* in accordance with the changes defined in the Law stipulates that the principles of space activities, in particular are: the discovery of space for citizens and legal entities and the ability to freely explore and use space, including the use of space and resources with restrictions only when necessary to ensure compliance with the national security interests of the state of Ukraine, as well as the permission and supervision of non-governmental space events to ensure that such actions are carried out in accordance with Ukraine international obligations in space; effective use of the scientific and technical potential of Ukraine, opportunities provided by space activities in the interests of the national economy, science, state security and for commercial purposes. This approach of the legislator can contribute to the improvement of the mechanism of legal protection of intellectual property rights in the field of space activities, because the provision in Art. 4 of the Law of Ukraine: *On Space Activity* on "effective use of scientific and technical potential of Ukraine" can be implemented only under the condition of creating an effective system of intellectual property protection in the space industry [33, 39].

After the adoption of the Law adopted in 29.10.2019 and the introduction of relevant changes in the legislation of Ukraine, so much time has passed, but at the present stage work is underway to prepare and implement the restructuring of the space industry to ensure further development of enterprises in market conditions. private companies involved in the commercialization of the space sector. For example, in the so-called "Space Race", NASA has selected nine companies that will compete for contracts for $ 2.6 billion to develop technologies for further study of the Moon, Mars, comets, asteroids for peaceful purposes in program Artemis Program, SSAU of Ukraine joined the *Agreement on the Exploration of the Moon and Mars* dated on November 13, 2020 and Ukraine became the ninth country to sign this Agreement, along with Canada, Australia, Japan, Great Britain and others. In this regard, James Frederick Bridenstine, NASA said that Ukraine was part of the broadest international coalition in the history of the study of the Moon and welcomed it as a new member who signed the agreement, and noted that many processes, technologies and principles will be tested program, which is designed to not only master the Moon, but subsequent Mars colonization [40, 41].

Thus, the Ukrainian space industry has entered the new millennium, introducing a system of market transformations that helps increase the investment attractiveness of the industry and attract investors to mutually beneficial cooperation. All these processes directly affect the field of intellectual property, so there is a need to intensify rule-making processes to improve the legal regulation of intellectual property rights at both international and domestic levels and harmonize national law with international law. In particular, the Ukrainian legislator should pay attention to this issue and in the Law of Ukraine: *On Space Activity*, define certain provisions on the protection of intellectual property rights of private companies, which in connection with the commercialization of space activities have a chance to be direct participants. In particular, according to VM Borychevskyi, it is obvious that in order to achieve stable economic growth in Ukraine, increase the competitiveness of the national economy, it is necessary to rapidly introduce innovative technologies into production processes and develop effective methods of their commercial application. One of the main implementation areas of this goal is the commercial use of space technology, including dual-use. This approach will create conditions not only for the preservation but also further development of space industry of Ukraine in the direction of commercialization of space technologies [1, p. 26].

In the context of this research, attention should be paid to the legislation of Ukraine on intellectual property. In addition to the Fourth Book of Civil Code of Ukraine is devoted to intellectual property law as a subsector of civil law, legal support in this area at the national level are the Law of Ukraine: *On Protection of Rights to Inventions and Utility Models* dated on: 15.12.1993, Law of Ukraine: *On Protection of Rights to Industrial Designs* dated 15.12.1993, Law of Ukraine: *On Protection of Rights to Integrated Circuit Designs* dated on 05.11.1997, Law of Ukraine: *On Copyright and Related Rights* dated on 23.12.1993," Law of Ukraine: *On State Secret* dated on 21.01.1994 etc. Recently in Ukraine there is a reform of legislation on intellectual property rights, including patent law. In this regard, the Law *On Amendments to Certain Legislative Acts of Ukraine Concerning the Reform of Patent Legislation* was adopted in July 27, 2020, so the Law of Ukraine: *On Protection of Rights to Inventions and Utility Models* works today changes in the new edition of 14.10.2020 [42].

The legislation of Ukraine on intellectual property does not prescribe a special legal regime for intellectual property objects used in space activities. But in particular, in Art. 4 of the Law of Ukraine "On Protection of Rights to Inventions and Utility Models" determines the priority in the application of the provisions of the international treaty of Ukraine, the binding nature of which was approved by the Verkhovna Rada of Ukraine before the norms established by Ukrainian legislation on inventions (utility models). That is, the norms of international conventions to which the state-Ukraine is a party are currently extended to the relevant intellectual property objects. It should be noted that at the international level, in addition to the Paris Convention, the Berne Convention and other international treaties, valid for Ukraine in the field of patent relations since 25.12.1991 is Patent Cooperation Treaty – PCT, concluded in 1970 According to Art. 1 of this Law, an invention (utility model) is the result of intellectual and creative activity of a person in any field of technology. Based on this concept, it can be concluded that the validity of this regulation extends its effect to intellectual property used in space activities, which is a field of high technology. Certain provisions of the Law of Ukraine: *On Protection of Rights to Inventions and Utility Models* define the legal regime of secret inventions (utility models), which are objects containing information classified as state secrets and rights certified by a patent for a secret invention (secret utility). model). Given the importance of research that can be used in space and support the position of Ukraine as a space power, the results of such research can be stored as a secret invention (utility model). In Part 2 of Art. 28 stipulates that the use of a secret invention (utility model) by the patent owner must be carried out in compliance with the requirements of the Law of Ukraine: *On State Secret* and in agreement with a state expert. In addition, taking into account the position of scientists studying the problems of patent protection in space, the need and feasibility of dividing inventions into two types, as mentioned above, the national patent law should define rules that take into account the specifics of legal protection of intellectual property rights. for inventions that can be created in outer space on a space object. It makes sense to study, analyze and use foreign experience in resolving a similar issue. For example, U.S. law provides a separate rule for determining the legal regime of inventions in outer space which states that any invention made, used or sold in outer space on a space object or its component under US jurisdiction or control is considered. as made, used or sold in the territory of the United States, except for any space object or its component, which is separately specified or otherwise provided in an international agreement to which the United States is a party, or any space object, or its component entered in the register of a foreign state in accordance with the Convention on the Registration of Objects Launched into Outer Space [43; 44, p. 168–169].

Thus, the norms of patent legislation of Ukraine extend to all achievements in the field of technology including space, so it is advisable to provide for the legal regime of inventions in space at the level of national law and determine the legal regime of intellectual property rights to results of intellectual and technical creativity, in particular those created on objects that are in outer space.

4 Conclusions

Summarizing the above, we note that at the new stage of development of space relations, special attention should be paid to intellectual property, because it is the achievements of industrial property creators, copyright and related rights, provide new impetus for further development of space activities. In this regard, it is necessary:

1. At the international level to adopt a normative act that can ensure global unification of legal regulation in the field of protection of intellectual property rights to objects used in space activities, while specifying the legal regime for the protection of intellectual property rights created on Earth and used in space activities, as well as objects created in outer space (including the ISS), which is characterized by extraterritoriality.
2. To improve the work of international bodies that control the use of outer space for measured purposes, in particular to intensify the activities of the UNOOSA, which can promote lawmaking procedures aimed at codifying the rules of international law designed to regulate the intellectual property rights of intellectual property rights. on objects used in space activities. It is expedient for the UNOOSA to initiate rule-making activities to improve the protection of intellectual property in space, as it is not a regional body like NASA and ESA, so its approach should be based on the balance of interests of each country without any discrimination in this matter in the rights of developing countries.
3. At the national level, Ukrainian legislator should pay special attention to the legal regulation of intellectual property rights to objects used in space activities and define the relevant provisions in regulations that constitute legal support in the field of intellectual property, in particular in Law of Ukraine: *On Protection of Rights to Inventions and Utility Models*, Law of Ukraine: *On Copyright and Related Rights* Law of Ukraine: *On Protection of Rights to Integrated Circuit Designs* etc. It is also advisable to regulate intellectual property issues in space legislation, primarily in the Law of Ukraine: *On Space Activity* and prescribe guidelines on legal protection of intellectual property rights in intellectual property in the field of copyright and related rights, industrial property rights used in space activities, taking into account their creation in different locations (i.e., either on Earth or in outer space).

4. To improve the work of SSAU that is a specially authorized central executive body of Ukraine ensuring realization of state policy in the field of space activities, guided by the approved by the Cabinet of Ministers of Ukraine: *Regulations on the State Space Agency of Ukraine* and is responsible for space development. We believe that intellectual creativity is an important component of space activities, so the SSAU as a responsible entity, guided by the tasks defined for it by the said Regulation, should comprehensively contribute to improving the legal regulation of intellectual property in space, taking into account the powers of the SSAU and the organization of the introduction of space technologies to meet the needs of the modern market, use of information technology in space field, etc.
5. Taking into account commercialization intensification of space activities, fully take into account the interests of private entities in the field of space activities. This process directly affects the field of intellectual property, so there is a need to improve the legal regulation of intellectual property rights not only states as owners of certain objects within space activities but private law entities both internationally and nationally, with to adhere to the direction of harmonization of national law with international law in this matter. In particular, the Ukrainian legislator, in our opinion, it is expedient in the Law of Ukraine: *On Space Activity* and in other regulations, in particular on intellectual property rights, to define certain provisions on protection of intellectual rights of private companies in connection with commercialization of space activities got a chance to be its direct participants.

References

1. Borychevs′kyy, V. M. (2007) Napryamy vdoskonalennya zakonodavstva Ukrayiny shchodo zakhystu prav intelektual′noyi vlasnosti v sferi kosmichnoyi diyal′nosti (Eng. Directions for improving the legislation of Ukraine on the protection of intellectual property rights in the field of space activities). https://law.nau.edu.ua/images/Nauka/Naukovijjurnal/2007/statgi_n2_3_2007/Borichevsryj_26.pdf [in Ukrainian]
2. Khridochkin, A.V.: Aktual′ni problemy pravovoyi okhorony intelektual′noyi vlasnosti normamy mizhnarodnoho ta natsional′noho kosmichnoho prava (Eng. Actual problems of legal protection of intellectual property by the norms of international and national space law). Pravo i suspil′stvo **5**, 217–222 (2018). [in Ukrainian]
3. Noichim, C.: The Protection of Intellectual Property Rights in Outer Space of the European Union and Thailand. https://asialaw.tripod.com>articles. [in English]
4. Convention Establishing the World Intellectual Property Organization. https://www.Wipolex.wipo.int/en/text/283854 [in English]
5. Konstytutsiya Ukrayiny (Eng. The Constitution of Ukraine): Vidomosti Verkhovnoyi Rady Ukrayiny (1996). № 30. St. 141 (iz zmin i dopovn.) [in Ukrainian]
6. Svitlychnyy, O.P.: Pravo intelektual′noyi vlasnosti: Pidruchnyk (Eng. Intellectual property law: Textbook). NUBiP Ukrayiny, Kyiv 355 p. (2017) [in Ukrainian]
7. Intelektual′ne pravo Ukrayiny (Eng. Intellectual law of Ukraine)/za zah. red. prof. O.S. Yavors′koyi. Pidruchnyky i posibnyky, Ternopil′, 608 p. (2016) [in Ukrainian]
8. Tsyvil′nyy kodeks Ukrayiny (Eng. Civil Code of Ukraine): Zakon Ukrayiny vid 16 sichnya 2003 r. Vidomosti Verkhovnoyi Rady (2003). № 40–44. St. 356 [in Ukrainian]
9. The Berne Convention for the Protection of Literary and Artistic Works in 1886. https://www.wipo.int/treaties/en/ip/berne/ [in English]

10. The Paris Convention for the Protection of Industrial Property of March 20, 1883. https://unido.org/sites/default/files/2014-04_Convention_)pdf [in English]
11. Agreement on Trade-Related Aspects of Intellectual Property Rights (TRIPS) in 1994. https://www.wto.org/English/docs_e/legal_e/27-trips.pdf [in English]
12. Borychevs′kyy, V.M.: Osoblyvosti okhorony ob″yektiv intelektual′noyi vlasnosti, stvorenykh na Mizhnarodniy kosmichniy stantsiyi (Eng. Protection of intellectual property created on the International Space Station). Yurydychnyy visnyk. № 4 (9). pp. 25–28 (2008). [in Ukrainian]
13. Sterling, J.A.L.: Space copyright law: the new dimension: a preliminary survey and dimension. J. Copyright Soc. USA **54**, 345–418 (2007). [in English]
14. Mizhnarodne spivrobitnytstvo u vykorystanni kosmichnoho prostoru v myrnykh tsilyakh: Rezolyutsiya Heneral′noyi Asambleyi OON vid 20 hrudnya 1961 r. № 1721 (XVI). (Eng. International cooperation in the peaceful uses of outer space: UN General Assembly Resolution of 20 December 1961 № 1721 (XVI). https://www.books.google.com.ua.books?isbn=5041136548. [in Ukrainian]
15. Deklaratsiya pravovykh pryntsypiv diyal′nosti derzhav z doslidzhennya ta vykorystannya kosmichnoho prostoru: pryynyata Heneral′noyu Asambleyeyu OON vid 13 hrudnya 1963 r. (Eng. Declaration of Legal Principles of States for the Exploration and Use of Outer Space: adopted by the UN General Assembly on December 13, 1963). https://zacon.rada.gov.ua/laws/show/995_a31#Text [in Ukrainian]
16. Treaty on Principles Governing the Activities of States in the Exploration and use of an Outor Space, including the Moon and other Celestial Bodies of 27 January 1967. https://unoosa.org/oosa/en/ourwork/spacelaw/treaties/intuoouterspacetreaty.html [in English]
17. Agreement on the Rescue of Astronauts, Return of Astronauts and the Return of Objects launched into an Outer Space of 22 April 1968. https://unoosa.org/oosa/ourwork/spacelaw/treaties/rescueagreement.html [in English]
18. Convention of International Liability for Damage Caused by Space Objects of 29 March 1972. https://www.faa.gov/about/office_org/headouarters_offices/ast/media/Conv_international_Liab_Damage.pdf [in English]
19. Agreement Governing the Activities of States on the Moon and other Celestial Bodies of 18 December 1979. https://www.megill.ca/iasl/centre/research/space-law/moon-treaty [in English]
20. Convention of Registration of Objects Launched into Outer Space of 15 September 1976. https://www.faa.gov/about/office_org/headquarters_offices/ast/media/Conv_Regi_Obgects_Launched.pdf [in English]
21. United Nations Committee for the Peaceful Uses of Outer Space. https://www.unoosa.org/oosa/en/ourwopk/copus/index.html [in English]
22. Upravlinnya OON z vykorystannya kosmichnoho prostoru v myrnykh tsilyakh (Eng. United Nations Office for the Peaceful Uses of Outer Space). https://Vienna.mfa.gov.ua/ustanovi-oon-u-vidni/2726-kvk-oon [in Ukrainian]
23. Arrangement between the United States of America and other Governments. Signed at Washington January 29, 1998. https://www.state.gov/wp-content/aploads/2019/02/12927-Multilateral-Space-Space-Station-1.29.1998.pdf [in English]
24. Vahrenwald, A.: Intellectual Property on the Space Station Freedom. https://Vahrenwald.eu/wp-content/uploads/2017/01/12.IntPropSpaceStation.EIPR_1993.318.pdf (2017) [in English]
25. Zubach, N.Y.: Problemy zakhystu prav intelektual′noyi vlasnosti v sferi kosmichnoyi diyal′nosti (Eng. Problems of protection of intellectual property rights in the field of space activities). https://law.nau.edu.ua/imajes/Nauka/Naukovijjurnal/2006/statji_n_1_2006/Zubach_45pdf. (2006) [in Ukrainian]

26. Borychevs′kyy, V.M.: Zakhyst prav intelektual′noyi vlasnosti Yevropeys′koho partnera Mizhnarodnoyi kosmichnoyi stantsiyi (Eng. Protection of intellectual property rights of the European partner of the International Space Station). Yurydychnyy visnyk. Seriya «Povitryane i kosmichne pravo». № 4. pp. 9–11 (2009). [in Ukrainian]
27. Byehlyy, O.V.: Tretya konferentsiya OON z kosmosu i problemy rozvytku prav intelektual′noyi vlasnosti (Eng. Third UN Conference on Space and the development of intellectual property rights). Intelektual′na vlasnist′. № 4. pp. 3–9 (2000). [in Ukrainian]
28. Bhatt, S.: Invention in outer space: need for reconsideration of the patent regime. J. Space Law, **36**(1) (2010). [in English]
29. Ramkova uhoda mizh Ukrayinoyu ta Uryadom Spoluchenykh Shtativ Ameryky pro spivrobitnytstvo v doslidzhenni ta vykorystanni kosmichnoho prostoru v myrnykh tsilyakh (Eng. Framework Agreement between Ukraine and the Government of the United States of America on Cooperation in the Exploration and Peaceful Uses of Outer Space): ratyfikovano Zakonom № 681-VI vid 17.12.2008 r. https://rada.gov.ua/laws/show/840_137#Text [in Ukrainian]
30. Postanova Kabinetu Ministriv Ukrayiny «Pro zatverdzhennya Polozhennya pro Derzhavne kosmichne ahent·stvo Ukrayiny (Eng. Resolution of the Cabinet of Ministers of Ukraine "On approval of the Regulations on the State Space Agency of Ukraine). https://zacon.rada.gov.ua/laws/show/281-2015-n#Text [in Ukrainian]
31. Ennan, R.Y.: Pravove rehulyuvannya vidnosyn intelektual′noyi vlasnosti v Yevropeys′komu Soyuzi (Eng. Legal regulation of intellectual property relations in the European Union). Avtoref. dys. …kand. yuryd. nauk. Odesa. 22 p. (2010). [in Ukrainian]
32. Uhoda pro asotsiatsiyu mizh Ukrayinoyu, z odniyeyi storony, ta Yevropeys′kym Soyuzom, Yevropeys′kym spivtovarystvom z atomnoyi enerhiyi i yikhnimy derzhavamy-chlenamy, z inshoyi storony vid 27.06.2014 r. Ekonomichna chastyna (Eng. Association Agreement between Ukraine, of the one part, and the European Union, the European Atomic Energy Community and their Member States, of the other part of 27.06.2014. Economic Part). https://zacon.rada.gov.ua/laws/show/984_011#Text [in Ukrainian]
33. Pro kosmichnu diyal′nist′ (Eng. On space activities): Zakon Ukrayiny. Vidomosti Verkhovnoyi Rady Ukrayiny.1997. № 1. St. 2 (iz zmin. i dopovn.) [in Ukrainian]
34. Pro skhvalennya Kontseptsiyi Zahal′noderzhavnoyi tsil′ovoyi naukovo-tekhnichnoyi kosmichnoyi prohramy Ukrayiny na 2021–2025 roky: Rozporyadzhennya Kabinetu Ministriv Ukrayiny vid 11 sichnya 2021 r. № 15-r. (Eng. On approval of the Concept of the National Targeted Scientific and Technical Space Program of Ukraine for 2021–2025: Order of the Cabinet of Ministers of Ukraine of January 11, 2021 № 15). https://zacon.rada.gov.ua/laws/show/25-2021-p#Text [in Ukrainian]
35. Hornyk, V., Simak, S.: Commercialization of space technologies is a perspective direction of increasing the competitiveness of UKRAINE. Sci. Notes of Taurida National V.I. Vernadsky Univ. Ser.: Econ. Manage. (2020). https://doi.org/10.32838/2523-4803/70-2-12
36. van Traa-Engelman, H.: Commercialization of space activities: patent rights for inventions in outer space. J. Luchtrecht, Special edn. № 4.2009, pp. 85–91 (2009) [in English]
37. MKS – na pensiyu? (Eng. ISS – retired?). https://hromadsre.ua/posts/mks-na-pensiyu-chomy-mizhnarodnu-kosmichnu-stanciyu-yaka-isnye-ponad-20-rokiv-nevdovzi-mozhut-zakriti. [in Ukrainian]
38. Konventsiya pro mizhnarodni maynovi prava na rukhome obladnannya: pryynyata u m. Keyptaun 16.11.2001 r. Keyptauns′ka konventsiya), ratyfikovana v Ukrayini vid 06.06.2012. Nabrala chynnosti dlya Ukrayiny 01.11.12. (Eng. Convention on the International Property Rights of Movable Property: adopted in Cape Town on 16.11.2001 (Cape Town Convention), ratified in Ukraine on 06.06.2012. Entered into force for Ukraine) 01.11.12.). https://zacon.rada.gov.ua/iaws/show/995_k62#Text [in Ukrainian]

39. Pro vnesennya zmin do deyakykh zakoniv shchodo derzhavnoho rehulyuvannya kosmichnoyi diyal′nosti (Eng. On amendments to some laws on state regulation of space activities): Zakon Ukrayiny vid 29.10.2019. Vidomosti Verkhovnoyi Rady Ukrayiny, (2019). № 46. St. 300. [in Ukrainian]
40. Ukrayina doluchylasya do prohramy NASA «Artemida» z doslidzhennya Misyatsya i Marsu (Eng. Ukraine has joined NASA's Artemis program to study the Moon and Mars). https://www.ukrinform.ua/rubric-technology/3135920-ukraina-dolucilasa-do-programi-nasa-artemida-2-doslidzenna-misaca-i-marca.html [in Ukrainian]
41. Pro vnesennya zmin do deyakykh zakonodavchykh aktiv Ukrayiny shchodo reformy patentnoho zakonodavstva (Eng. On Amendments to Certain Legislative Acts of Ukraine Concerning the Reform of Patent Legislation) : Zakon Ukrayiny vid 27.07.2020. Vidomosti Verkhovnoyi Rady Ukrayiny, (2020). № 52. St. 496 [in Ukrainian]
42. Pro okhoronu prav na vynakhody i korysni modeli (Eng. On protection of rights to inventions and utility models): Zakon Ukrayiny vid 15.12.1993 r. Vidomosti Verkhovnoyi Rady Ukrayiny, (1994). № 7. St. 32 [in Ukrainian]
43. Sopilko, I.M.: Povitryane i kosmichne pravo; dyskusiyni pytannya (Eng. Air and space law; discussion questions). Yurydychnyy visnyk. Povitryane i kosmichne pravo, Seriya № 4. pp. 161–171 (2015). [in Ukrainian]

Current Directions of Development of International Cooperation of Forensic Science Institutions of Ukraine in The Fields Historical Forensics, Forensic Archaeology and Aerospace Forensics

Ella Simakova-Yefremian[1] and Kateryna Sylenok[2](✉)

[1] National Scientific Center «Hon. Prof. M. S. Bokarius Forensic Science Institute», Science and Technology of Ukraine, Kharkiv, Ukraine
simakova@hniise.gov.ua

[2] National Scientific Center «Hon. Prof. M. S. Bokarius Forensic Science Institute», Kharkiv, Ukraine

Abstract. Current directions and levels of international cooperation of forensic institutions are considered, taking into account experience of National Scientific Center «Hon. Prof. M. S. Bokarius Forensic Science Institute». Attention is drawn to the need for cooperation at all levels of cooperation: bilateral cooperation, regional and global one. Role of standardization of forensic activities in Ukraine in accordance with international experience to ensure a high level of research quality is emphasized. Attention is paid to educational activities in the field of cooperation: conferences of forensic experts which are held in cities of different countries and serve the purpose of improving the methods of forensic research. The scale of international cooperation of Ukraine in these fields is highlighted on example of historical, archeological and aviation forensic science examinations and perspective of their development is emphasized.

Keywords: International cooperation · Forensic science institutions · Forensic historical and archaeological science · Aerospace forensics

1 Introduction

One of conditions for harmonious development of forensic science in Ukraine is strengthening of international relations between specialized forensic science institutions in Ukraine and abroad. International cooperation is an effective means of exchanging international experience, harmonizing and unifying approaches to resolving forensic science issues. European path of Ukraine is accompanied by implementation procedure of international norms and standards into national legislation. In turn, any legal regulation should be structured have an effective form of implementation of norms for their effectiveness and should not have conflicts. The reference point for cooperation of Ukrainian

M. Nechyporuk et al. (Eds.): ICTM 2021, LNNS 367, pp. 814–824, 2022.
https://doi.org/10.1007/978-3-030-94259-5_63

specialized forensic science institutions with foreign colleagues opens new opportunities for improving quality of forensic examinations and forensic research, development of practical experience and theoretical developments, provides new opportunities for professional development of scientists and forensic experts.

A number of legal documents, in particular procedural legislation, the Law of Ukraine: On Judicial Examination, as well as draft legislation (draft law "On Forensic Science Activity in Ukraine"), etc., provide for international cooperation of forensic science institutions.

2 Literature Review

The issues of improving forensic activity in Ukraine and international cooperation in the field of expert support of justice were considered by such domestic researchers as: V. Galunko, L. Golovchenko, I. Gora, L. Derecha, V. Zhuravel, N. Klymenko, O. Kliuiev, O. Lopata, A. Rusetskyi, I. Petrova, I. Pyrih, E. Simakova-Yefremian, N. Tkachenko, O. Uhrovetskyi, N. Filipenko, G. Strelets, V. Shepitko, M. Shepitko, M. Shcherbakovskyi [E.g.,: 1, 2, 3, 4, etc.].

In their research papers, these scholars considered general issues of international cooperation of forensic science institutions. Given the above, there is a need to explore issues in certain areas of forensic science, in particular in the fields of historical, archaeological and aerospace forensics.

3 Main Content Presentation

Scholars note that national forensics cannot exist in isolation, i.e. only within a single state. One of functions of forensic science is to ensure law enforcement. Without integration with international community, this function will not be fully implemented, or will not be implemented at all. Therefore, international cooperation of expert institutions is necessary in order to exchange experience taking into account modern advances in science and technology, avoid duplication of scientific and methodological support of regional distribution of tasks between forensic science institutions, creating specialized forensic science areas, etc. [1, p. 131].

The main reasons for the need to intensify international cooperation of forensic experts today are: the processes of globalization of all spheres of human activity, the rapid development of means of communication, international scientific relations, accelerating migration and related intensification of international transnational organized crime. and international terrorism [2, p. 154]. O. M. Kliuiev draws attention to the United Nations Congress on Crime Prevention and Criminal Justice that in its resolution: *International cooperation in the field of forensic science* noted that in the fight against crime, terrorism and human rights is constantly growing role of forensic science [5].

In the field of international cooperation, the main tasks of forensic science institutions include:

- experience exchange;
- taking into account modern achievements of science and technology;

- elimination of duplication of scientific and methodological support of the regional distribution of tasks between forensic science institutions;
- creation of profile forensic science directions, etc. [1, p. 131].

Thanks to international cooperation, forensic science institutions can achieve certain goals:

- harmonization and improvement of national forensic science legislation and theoretical foundations of forensic science;
- development of uniform guidelines on different forensic science species;
- obtaining international recognition of forensic expert conclusions in general;
- improvement of forensic expert activity
- improving the professional skills of employees of forensic science institutions, etc. [6, p. 66].

One of the broadest and most important areas of international cooperation in the field of forensic science nowadays is interaction of domestic forensic science institutions with international networks of forensic science institutes, forensic science institutions.

Scholars distinguish the following levels of international cooperation:

- bilateral cooperation (between forensic science institutions);
- regional (participation in regional international forensic science institutions, such as ENFSI);
- world (participation in international forensic science institutions at the global level).

At the first level of bilateral cooperation: contractual relations in the field of cooperation of forensic science institutions provide for exchange of information on R&D, new educational guidelines and reference publications, scientific conferences, workshops, exhibitions of modern forensic equipment, direct exchange of methodological, scientific and content and documentation.

Within the framework of the concluded agreements, measures are provided for approbation of research methods, technical means developed by forensic science institutions, mutual review of scientific developments and forensic expert conclusions, etc. [7, p. 31].

National Scientific Center «Hon. Prof. M. S. Bokarius Forensic Science Institute» (hereinafter referred to as NSC «Hon. Prof. M. S. Bokarius FSI») actively cooperates at the international level. During 2016–2020, the Center signed 15 international agreements on cooperation with foreign expert institutions. Thanks to such interaction with foreign colleagues, the Center's experts have the opportunity to share experience in conducting examinations, learn about new technologies used in the research and conduct joint research. Expansion of international contacts testifies to the successful activity of NSC «Hon. Prof. M. S. Bokarius FSI» and his high authority.

Cooperation at the regional level is facilitated by international forensic science institutions at appropriate level. Here are at least six of the most powerful networks that bring together forensic science institutions from different countries:

- European Network of Forensic Science Institutes (ENFSI) since 1995. This is the most developed and strong network; which member is the NSC «Hon. Prof. M. S. Bokarius FSI»;
- The Association of Firearm and Tool Mark Examiners (AFTE) was founded in 1969 and includes representatives of NSC «Hon. Prof. M. S. Bokarius FSI»;
- Southern Africa Regional Forensic Science (SARFS) Network since 2008, covering the entire African region;
- Asian Forensic Sciences Network (AFSN), existed since 2008;
- The International Network of Environmental Forensics (INEF) existed since 2008 to assist forensic environmental experts in investigation environmental offences;
- TWFN existed since 2006, is an NGO to assist in conducting forensic examinations in cases of crimes against wildlife and to preserve the biological diversity of flora and fauna [1, p. 131].

Let us pay attention, to European Network of Forensic Science Institutes (ENFSI). Scientists note that membership in ENFSI is an important strategic direction of development for all domestic forensic science institutions, because it provides high international standards of forensic science, interaction on a wide range of issues with European partners, opens new promising areas of development [7, p. 31].

As already mentioned, NSC «Hon. Prof. M. S. Bokarius FSI» since 2018 is a member of the European Network of Forensic Science Institutes. As a member of this representative organization, the Center is an institution that initiates implementation of European experience in the work of forensic institutions of Ukraine, an institution with a high level of the most complex forensic researches and latest scientific developments that have theoretical and practical significance and are essential for a high level of forensic science in the state.

Cooperation at global level is strategically important. In general, we can speak about international cooperation carried out to address strategic issues in various fields of public life (Ukraine has been a member of the UN since 1945), international cooperation in law enforcement, implemented with the involvement of Interpol (USSR membership since 1990, Ukraine, since 1992) and others. Today it is quite promising to establish international cooperation with international forensic organizations [7, p. 31].

European practice, on the basis of which legal standards are implemented, as noted by M. H. Haustov, indicates existence of mandatory standards and standards that are of a recommendatory nature. Mandatory standards are enshrined in legal norms that require proper conduct. It is this feature of the standards largely determines the general direction of development of the legal system, because by their legal nature such norms belong to the norms-principles, i.e. have a constitutive nature [8, p. 44–45].

Optional standards of a recommendatory nature are the product of international law of the twentieth century, when the "speed" of their development did not coincide with the capabilities and desires of individual states. Division of European legal standards into these varieties has led to the functioning of law in the ordinary sense, i.e. that is binding, and the so-called "soft" law.

International standards exist in any area of legal regulation but they are mostly identified with standards governing the technical field, as they are the most common.

However, it is difficult to imagine any sphere of public life in which there are no generally accepted international standards. European legal standards are formed within the framework of the two largest regional international associations – the Council of Europe and the European Union. The Council of Europe sets first and foremost standards in the humanitarian sphere: the protection of human rights, the protection of the environment, and constitutional law, which is determined by the objectives of its operation. The European Union – through directives, regulations and other regulations – sets the standards for most areas of life in the EU.

It should be noted that while of standardization development in European law is a kind of practice of public relations in the EU [9, p. 159].

Laboratories of NSC «Hon. Prof. M. S. Bokarius FSI» meet international requirements for their equipment, research methods, qualifications of experts. The Center received an Accreditation Certificate for compliance with the requirements of the international standard ISO/IEC 17025 and an Accreditation Certificate for compliance with the requirements of the international standard ISO/IEC 17020 ensuring technical competence, independence and impartiality of forensic examinations and forensic researches.

O. M. Kliuiev notes that activity of ensuring compliance with international standards, unification of forensic science methods in accordance with positive experience of foreign countries is one of important tasks of international cooperation in the field of forensic science that is solved at different levels: from bilateral to global one [9, p. 10].

Current legislation provides for various forms of information exchange in the field of theory and practice of forensic science and criminalistics including exchange of trainees, joint scientific conferences, exchange of printed publications, joint publications, etc.

Implementation of international legal standards in national practice of forensic science in Ukraine is not only by improving national law, but due to the fact that there are many crimes of an international nature which investigation requires forensic examination by many professionals in various fields of forensic science.

In the field of investigation of such international offenses, interaction and cooperation of states is regulated both by the norms of national legislation and by international treaties and agreements. This also applies to forensic science, as one of the forms of application of specific expertise in criminal proceedings [11, p. 74].

Growing number of serious international crimes requires closer cooperation between states in their investigation. For this reason, it is very important that each country has access to methods of forensic researches used in other countries.

International cooperation provides clear benefits to partner institutions. Within the framework of cooperation agreements, partners can deal with specific areas of expert science, develop methods of handwriting research, analytical methods of comparing recorded voices, as well as research of other evidence obtained while crime investigation and etc. Forensic science covers a large number of forensic areas of specialization and many current issues of forensic science are difficult to solve using the resources of one laboratory in one state [1, p. 133].

Cooperation on a contractual basis of forensic science institutions and their associations, educational institutions and research institutions of Ukraine and Europe performing applied researches, will help increase competitiveness and improve the investigation of various crimes through implementation of research results, methodological developments on forensic science. Conferences of forensic experts held in different countries of the world also serve the purpose of improving the methods of expert researches.

In order to take into account positive experience of forensic institutions of foreign countries, National Scientific Center «Hon. Prof. M. S. Bokarius Forensic Science Institute» introduced a number of forms of scientific cooperation with involvement of foreign partners. Such forms include holding scientific conferences and round tables, workshops on development of new methods and techniques of forensic science, exchange of information on new scientific publications and publications in professional journals, joint development of forensic science and interlaboratory research, training of scientific staff, etc.

Thus, on April 15–16, 2021 on the basis of National Scientific Center «Hon. Prof. M. S. Bokarius Forensic Science Institute» the International Research and Practice Conference-Polylogue on *Topical Issues of Forensic Science and Criminalistics* took place online. The conference was attended by scientists and forensic experts not only from Ukraine but also from Azerbaijan, Belarus, Great Britain, Armenia, Greece, Georgia, Israel, Kazakhstan, Latvia, Lithuania, Macedonia, Moldova, Poland, Russia, Uzbekistan, Montenegro and other countries. In total, the organizing committee accepted for publication 187 abstracts of conference proceedings and communications from 257 authors. This testifies to the unconditional interest of the scientific community and practitioners around the world to the current problems of forensic science and criminology. This landmark international scientific and practical event in the context of the COVID-19 pandemic made it possible to unite scientists and practitioners from different countries to perform common tasks to establish law and order, implement measures of forensic science in Ukraine in accordance with the best international standards.

Implementation of forensic integration will result in harmonization and improvement of national expert legislation and theoretical foundations of forensic science, development of common methodological recommendations for different species of forensic science, international recognition of forensic expert conclusions in general, improvement of forensic expert activity, professional development of forensic experts [1, p. 131].

Historical and archeological forensic researches can be considered as a direction of a new and perspective field of cooperation. This area of forensic science is important for preservation of Ukraine archaeological heritage. Each state must ensure the preservation of historical monuments for future generations, and Ukraine is currently actively developing this area. Pursuant to the Order of the Ministry of Justice № 243/5 *On Approval of Amendments to Certain Legal Acts on Judicial Expertise* dated on January 20, 2021 amendments were made to the List of Types of Forensic Examinations and Expert Specialties research institutions of forensic examinations of the Ministry of Justice of Ukraine [12], namely, the list is supplemented by a new specialty – historical and archaeological research on Land lots. This addition was initiated by National Scientific Center «Hon. Prof. M. S. Bokarius Forensic Science Institute». Since 2021, the Center has been conducting forensic examinations in the area of historical and archaeological research on land plots. Forensic historical and archaeological expertise is an important element in the mechanism of protection and preservation of archaeological sites. Objects are land plots with material remnants of human existence in past epochs, regardless of their state of preservation (mounds, cemeteries, necropolises, settlements, settlements, individual burials, dwellings, farm buildings, their remains, etc.; anthropogenic layers and movable archaeological objects found in) and accounting and research documentation for historical (located below Earth surface) and/or archaeological cultural heritage sites.

The main tasks of historical forensics, forensic archaeology regarding land plots:

- determination of availability/lack of archeological and historical heritage objects within the land plot;
- determination of availability/lack of the protection subject object of archaeological heritage within the land plot;
- assignment of archeological heritage object to a certain chronological group, determination of its cultural value that has historical significance, level of rarity and preservation, attribution of archeological objects and calculation of costs for works on archeological researches of monument (decided during object discovery) archaeological heritage within the land plot and/or multidisciplinary appraisal, construction, historical, archaeological and art research in calculating the value of the monument and/or establishing the amount of damage caused to the state due to destruction, destruction or damage of archaeological heritage sites);
- determination of availability/lack of in the accounting documentation for historical (located below earth surface) and/ or archaeological objects of cultural heritage materials on research, factual data on the authenticity and subject of protection and other tasks [13].

Cooperation in the field of historical and archaeological research is provided by national legislation (for example, in the Law of Ukraine: *On Protection of Archaeological Heritage* (Article 21 [14]), current international agreements of Ukraine with foreign countries in the field of scientific cooperation.

According to the provisions of the Constitution of Ukraine and the Law of Ukraine: *On International Agreements of Ukraine* [15], existing international treaties, the binding nature of which was approved by the Verkhovna Rada of Ukraine, are part of the national legislation of Ukraine. International cooperation in the field of archeology is provided by a number of international legal documents: European Convention on the Protection of the Archaeological Heritage [16]; Convention on the Protection of the Underwater Cultural Heritage [17]; Convention on the Protection of the Architectural Heritage of Europe [18] and others.

However, the UNESCO Convention Concerning the Protection of the World Cultural and Natural Heritage plays a key role in the preservation of cultural and natural heritage in all regions of the world. It was adopted at the 17th session of the General Conference of UNESCO in Paris in 1972. [19] Under the Convention Concerning the Protection of the World Cultural and Natural Heritage, adopted in 1972, its member states recognize that the protection of the World Heritage is the responsibility of the entire international community. The level of culture of the society is reflected in the preservation of cultural and natural heritage. The protection of monuments is not limited to restoration work – it is a phenomenon in which the political, economic, legal, moral and ethical processes that the world, every country, civil society live in are intertwined. Monuments are inseparable from society, they arise in it, they are evaluated, preserved or destroyed [20, p. 213–214].

As for forensic historical and archaeological examination, since this area is newly introduced and currently there are no comprehensive developments in this area, it can be argued that one of the promising areas of international cooperation of forensic institutions may be the field of historical and archaeological research. Moreover, since the archaeological heritage is an inexhaustible resource for studying the history and culture of the past that is being destroyed, stolen and needs protection from criminal encroachments, such cooperation will bring positive results not only for forensic research, but also another step on the way to preserving the archaeological heritage of Ukraine, its protection and restoration.

One should not overlook another new area of forensic researches. Aviation technical expertise is crucial in the investigation of aviation incidents. In many cases, the victims are citizens of different countries, and therefore states are interested in the quality, speed and openness of investigations and forensic investigations including aerospace forensics.

Aerospace forensics is a forensic research to establish mechanism and circumstances of an aviation event (incident), technical condition of aircraft, aeronautical instruments, ground aeronautical equipment, the correctness of measures for preparation and organization of flights, and other circumstances that contributed or could contribute to aviation event (incident). Objects of aerospace forensics are aircraft, ground aerodrome equipment, separate parts (assemblies of parts) of aircraft, devices and technical means of flight safety, flight documentation, records of negotiations between the crew and the flight director, records of flight recorders, investigation materials and others documents relating to the studied objects, phenomena and processes. The main tasks of forensic aviation technical examination are to establish the causes and circumstances of aviation events (incidents), to establish the technical condition of aircraft, aeronautical instruments, ground aeronautical equipment, to assess the correctness of measures for the preparation and organization of flights [21].

Aviation incidents are a special group of events that are complicated by conditions of occurrence (technical, weather, criminal) and the situation (airspace, large area of detection of traces of crime, destruction and damage of evidence and objects of study), which determines the feasibility of international cooperation professional assistance in establishing the causes of the accident and conducting an effective investigation. International cooperation also contributes to ensuring the rights of victims of citizens of different states.

Issues of civil aviation security and cooperation to ensure it are regulated by international conventions. Among them are the Convention on International Civil Aviation of December 7, 1944, to which Ukraine acceded on August 10, 1992 and which has been in force for Ukraine since September 9, 1992; Convention on Crimes and Certain Other Acts Committed on Board Aircraft (Tokyo, September 14, 1963) (Ukrainian SSR acceded on December 21, 1987); Convention for the Suppression of Unlawful Seizure of Aircraft (The Hague, December 16, 1970) (ratified by the Decree of the Presidium of the Verkhovna Rada of the Ukrainian SSR of December 27, 1971, entered into force on March 23, 1973); Convention for the Suppression of Unlawful Acts against the Safety of Civil Aviation (Montreal, September 23, 1971) (ratified by the Decree of the Presidium of the Verkhovna Rada of the Ukrainian SSR of January 16, 1973); Convention on the Marking of Plastic Explosives for the Purpose of Detection (Montreal, 1 March 1991) and others [22, p. 242].

Forensic agencies have a prominent place in the investigation of aviation incidents. Forensic expert professionalism, his training, practical skills and abilities ensure the quality and speed of the examination. It is from the results of the examination in many cases depends on the establishment of the truth in a particular criminal proceeding. Usually, an expert must not only have specialized knowledge in his/her forensic area of specialization, but also should be guided by international standards, so that forensic research is recognized as valid by the world community.

Aviation disasters have negative consequences for legal entities, individuals and states. Legal entities suffer huge material losses to compensate the victims and relatives of the dead, individuals suffer damage to their own health or lose their lives altogether, and states risk their status in the international arena. The task of each state in whose territory the aviation incident took place is to ensure the transparency and efficiency of the investigation. To do this, the state must maintain international relations, implement standards used by countries around the world. In particular, this applies to forensic research.

On February 2, 2013, Aviation Accident Investigation Manual was put into effect and Guidelines for research on Role of Human Factor in the Investigation of Aviation Accidents and Incidents were developed. Regulation on Investigation of Aviation Accidents and Incidents of March 26, 2010 is currently in force [22, p. 245; 23].

As can be seen from the above, Ukraine has broad international legal obligations in the field of aviation. This is confirmed by Ukraine's membership in many international organizations, such as the International Civil Aviation Organization, European Civil Aviation Conference (ECAC), European Organization for the Safety of Air Navigation, (Eurocontrol), Interstate Council on Aviation and Airspace Use (IAIS), Interstate Aviation Committee (IAC) etc. [24, p. 484].

All the above emphasizes the important role of our state in the international arena and the active movement of Ukrainian policy in implementation direction of international regulations in Ukrainian legislation, approximation to European standards, in particular, in in the field of expert support of justice.

4 Conclusions

Given the above, it should be noted that active development of international cooperation in the field of forensic research contributes to development of a modern system of expert support of justice in Ukraine. Involvement of Ukrainian expert forensic science institutions in international forensic science networks and cooperation with international forensic expert institutions expands opportunities for exchange of expertise in various fields and at various levels of international cooperation, thus improving quality of knowledge of professionals. Accreditation according to international standards of quality management systems, harmonization of expert legislation of Ukraine in accordance with international requirements, as well as various forms of information exchange and best practices in the field of theory and practice of forensic science are important and necessary elements of cooperation of forensic science institutions in various fields. Promising areas of international cooperation include the areas of historical forensics, forensic archaeology and aerospace forensics.

References

1. Klimenko, N.I., Kuprievich, O.A.: International cooperation of forensic institutions. Bull. Criminal Proc. **4**(2015), 130–134 (2015). [in Ukrainian]
2. Kliuiev, OM: About international cooperation in the field of forensic science. Theory Pract. Foren. Sci. Crim. **18**, 154–162 (2018). https://doi.org/10.32353/khrife.2018.17
3. Kliuiev, O.M.: International scientific cooperation is an effective mechanism for improving forensic science. Current issues of forensic science and criminology: Coll. materials international. Scientific-practical conference-polylogue (Kharkiv, April 15–16, 2021). Law, Kharkiv, 320 p. (2021). [in Ukrainian]
4. Simakova-Efremian, E.B.: To the question of interdependence of the European integration processes and integration tendencies of special knowledge in Ukraine. Theory Pract. Foren. Sci. Crim. **17**, 152–158 (2017). https://doi.org/10.32353/khrife.2017.18
5. International cooperation in the Forensic Field. Resolution 19/5 of the Commission on Crime Prevention and Criminal Justice UNODC. UN Document E/CN.15/2010/L.13/ Rev.1. https://www.unodc.org/documents/commissions/CCPCJ/Crime_Resolutions/2010-2019/2010/CCPCJ/Resolution_19-5.pdf [in English]
6. Kontymyrova, V.V.: Forensic institutions of Ukraine in the system of subjects of international cooperation. Administrative law and process: current problems and solutions. Kharkiv, pp. 66–68 (2020). [in Ukrainian]
7. Chornous, Y.M., Lopata, O.A.: Directions, levels and forms of international cooperation in the field of forensic science. Foren. Foren. **62**, 28–37 (2020). [in Ukrainian]
8. Haustova, M.G.: International standards in the context of European integration processes in Ukraine. Bull. Nat. Acad. Right. Sci. Ukraine. **2**(89), 43–53 (2017). [in Ukrainian]
9. Rusetskyi, A.A.: About the organization of forensic-expert activity: the international aspect. Theory Pract. Foren. Sci. Crim. **17**, 158–162 (2017). https://doi.org/10.32353/khrife.2017.19
10. Kliuiev,O.M.: Reforming the basic legislation of Ukraine on expert support of justice is an urgent need of the time. Problems of reforming the basic legislation of Ukraine on expert support of justice: materials of the round table (Kharkiv, February 6–7, 2020), pp. 6–11. KhNDISE, Kharkiv (2020). [in Ukrainian]
11. Matveevsky, O.V., Babin, B.V.: International legal standards of forensic examination and their national implementation. Prykarpattya Legal Bulletin. Issue 1 (22). vol. 4. pp. 73–79 (2018). [in Ukrainian]
12. List of types of forensic examinations and expert specialties for which the qualification of forensic expert is assigned to specialists of research institutions of forensic examinations of the Ministry of Justice of Ukraine. https://zakon.rada.gov.ua/laws/show/z0249-15Text [in Ukrainian]
13. National Scientific Centre "Hon. Prof. M.S. Bokarius Forensic Science Institute". Historical and archaeological examination. https://www.hniise.gov.ua/14683-storiko-arxeologchna-ekspertiza.html [in Ukrainian]
14. On the protection of archaeological heritage. Law of Ukraine of December 12, 2012 № 1626-IV. https://zakon.rada.gov.ua/laws/show/1626-15Text [in Ukrainian]
15. On international agreements of Ukraine. Law of Ukraine of June 29, 2004 № 1906-IV. https://zakon.rada.gov.ua/laws/show/1906-15Text [in Ukrainian]

16. European Convention on the Protection of the Archaeological Heritage: signed on 16.11.1992, ratified by the Law of Ukraine of 10.12.2003. № 1369VIV. https://zakon.rada.gov.ua/laws/show/994_150#Text [in Ukrainian]
17. Convention on the Protection of the Underwater Cultural Heritage: adopted by UNESCO on November 2, 2001, 2006. № 164-V. https://zakon.rada.gov.ua/laws/show/995_c52Text [in Ukrainian]
18. Convention for the Protection of the Architectural Heritage of Europe: adopted on October 3, 1985, ratified by the Law of Ukraine of September 20, 2006. № 165-V. https://zakon.rada.gov.ua/laws/show/994_226Text [in Ukrainian]
19. Convention Concerning the Protection of the World Cultural and Natural Heritage: N 6673-XI. https://zakon.rada.gov.ua/laws/show/995_089Text [in Ukrainian]
20. Prokaeva, V.V.: The role of UNESCO in preserving the cultural heritage of mankind. Current issues of international relations. Issue 103 (Part I), pp. 209–217 (2011). [in Ukrainian]
21. Kyiv Scientific Research Institute of Forensic. https://kndise.gov.ua/activity/expertise-view/c-aviacijno-tehnicna [in Ukrainian]
22. Akhtyrska, N.M.: International cooperation in the investigation of aviation events: problematic issues. Juridical Sci. Electron. J. **1**, 242–246 (2020). https://doi.org/10.32782/2524-0374/2020-1/58
23. On approval of the Regulations on the investigation of aviation incidents and incidents. Order of the Ministry of Transport and Communications of Ukraine dated March 26, 2010 № 159. https://zakon.rada.gov.ua/laws/show/z0590-10 [in Ukrainian]
24. Plahotniuk, N.V., Savitskaya, K.A.: International cooperation of Ukraine in the field of aviation security. Comp. Anal. Law **4**(2018), 482–485 (2018). [in Ukrainian]

Development Strategy of International Cooperation of Forensic Science Institutions of Ukraine with Foreign Experts in Prevention of Terrorist Attacks on Aerospace Industry and Critical Infrastructure

Mykola Nechyporuk[1], Oleksandr Kliuiev[2], Aleksandar Ivanović[3], and Nataliia Filipenko[4](✉)

[1] National Aerospace University "Kharkiv Aviation Institute", Kharkiv, Ukraine
[2] National Scientific Center «Hon. Prof. M. S. Bokarius Forensic Science Institute», Kharkiv, Ukraine
[3] Forensic Center Montenegro, Danilovgrad, Montenegro
[4] Law Department of National Aerospace University "Kharkiv Aviation Institute", Kharkiv, Ukraine
n.filipenko@khai.edu

Abstract. The issue of using modern foreign experience of preventive activity in criminological work of forensic science institutions of Ukraine is considered. Peculiarities of the main organizational forms of forensic science activity are analyzed through specialized (forensic science) institutions and through specific specialists, namely: forensic experts (for example, the practice of the institute of sworn experts: specialists who took the oath or received a license for forensic examination). The analysis of international standards used in forensic science activity is carried out. Necessity position of legislative introduction of international standards in process of forensic examination is revealed and substantiated. The main emphasis is on highlighting problems of cooperation of forensic institutions of Ukraine with foreign experts in preventing terrorist attacks on aerospace and critical infrastructure.

Keywords: Forensic science · Forensic expert · International standards · Criminological activity · Prevention · Forensic science institutions · Terrorist acts · Cooperation · Aerospace industry · Critical infrastructure facilities

1 Introduction

Issues of international cooperation and in-depth cooperation in the field of forensic science have always been and will be relevant for both domestic justice and organization of forensic science management. The great role of forensic science in crime and terrorism counteraction, protection of justice and human rights is highlighted by many international organizations, including the UN Commission on Crime Prevention and Criminal Justice

M. Nechyporuk et al. (Eds.): ICTM 2021, LNNS 367, pp. 825–848, 2022.
https://doi.org/10.1007/978-3-030-94259-5_64

in the *International cooperation in the field of forensic science* resolution [1]. This issue has received due attention in other international documents, in particular in the *Global Counter-Terrorism Strategy* (A/RES/60/288) adopted by the General Assembly in 2006, United Nations Member States recognized that capacity-building in all States is a core element of the global counter-terrorism effort [2].

The main reasons for cooperation intensification in the field of international forensic science are the processes of internationalization and globalization of all fields of human activity and rapid development of communication means, international scientific relations, significant growth of population migration, intensification of international transnational organized crime and international terrorism.

According to O. M. Kliuiev, international cooperation in the field of forensic science is a system that is developing quite strongly and is influenced by a variety of factors. Strengthening such cooperation, according to the researcher, in any country is always in interests of forensic experts, forensic institutions, judiciary, law enforcement agencies, the individual and civil society as a whole. Among the main reasons for the need to intensify international cooperation of forensic experts, the scientist attributes the processes of globalization of all fields of human activity, rapid development of means of communication, international scientific relations, accelerating migration and related intensification of international transnational organized crime and international terrorism [3, p.156].

This is especially true in responding to the criminological situation complication, as well as in emergencies, including the prevention of terrorist attacks on aerospace and critical infrastructure. This is reflected in the official position of the state, expressed in a number of recently issued legal acts. In particular, they emphasize that priority in achieving the goals of crime counteraction and eliminating external and internal threats to Ukrainian national security belongs to all state bodies that should have appropriate forces and means capable of performing specific tasks [4].

Given all this, it can be argued that today there is a lack of research on criminological activities of forensic science institutions of Ukraine to prevent terrorist attacks on aerospace and critical infrastructure, cooperation of forensic science institutions of Ukraine with foreign experts in these issues increasing the relevance chosen direction and opens prospects in determining effective practices for improving this activity.

2 Literature Review

Some aspects of foreign practice of counteracting crime by special subjects were studied in their works by such domestic and foreign scientists as H. A. Avanesov, O. I. Alekseiev, Yu. M. Antonian, O. M. Bandurka, Yu. D. Bluvshtein, M. H. Verbenskyi, T. S. Havrysh, V. V. Holina, I. M. Danshyn, O. M. Dzhuzha, A. I. Dolhova, A. P. Zakaliuk, A. F. Zelinskyi, O. H. Kalman, O. M. Kliuiev, M. H. Kolodiazhnyi, I. V. Lieshukova, O. M. Lytvynov, O. A. Lopata, P. V. Melnyk, P. H. Nazarenko, M. V. Nechyporuk, M. I. Pashkovskyi, E. B. Simakova-Yefremian, N. M. Tkachenko, N. Ie. Filipenko, I. Ia. Fridman, Sh. M. Khaziiev, Yu. M. Chornous, V. Yu. Shepitko, I. I. Shepitko and others. The issue of the need for criminological activities of experts and expert institutions was covered in their works by criminologists from Bulgaria, Montenegro, Latvia, Poland and Germany, in particular Z. Bernachek, Ya. Dimov, R. Zelvianskyi, A. Ivanović, V. Moshinskii, A. Solash, K. Rote, K. Shpindler and others.

Scientists point out that forensic experts should note circumstances identified while examination that facilitated the commission of offenses and can develop preventive recommendations to prevent certain crimes including terrorist attacks on aerospace and critical infrastructure [5, p. 61].

3 Main Content Presentation

Important part of Ukrainian modern criminal law policy is countering various terrorist threats. The most dangerous and devastating consequences are terrorist acts related to encroachment on aerospace facilities and critical infrastructure, as they pose a real threat to the stable operation of such facilities that threatens human life and health, violates work of enterprises, institutions and organizations, industrial and economic facilities, etc. [6, p. 188–191].

According to the provisions of the Law of Ukraine: On Fight against Terrorism, terrorism is a socially dangerous activity that consists in the deliberate, deliberate use of violence by taking hostages, arson, murder, torture, intimidation of the population and the authorities or other encroachments on life or health. innocent people or threats of committing criminal acts in order to achieve criminal goals [7].

As a criminal phenomenon, terrorism is an illegal, criminally punishable act expressed in the use of weapons, explosion, arson or other acts that endanger human life or health or cause significant property damage or other serious consequences, if any actions were committed for the purpose of violating public safety, intimidating population, provoking a military conflict, international complication, or in order to influence decisions or acts or omissions of public authorities or local governments, officials of these bodies, associations of citizens, legal entities persons, international organizations, or drawing public attention to certain political, religious or other views of perpetrator (terrorist), as well as the threat of committing these acts for the same purpose (Article 258) [8].

Terrorism includes ideology of violence and terrorist activities in various forms. Terrorist activities include planning creation and (or) creation of terrorist structures, involvement in terrorist activities, financing and other assistance to these activities, propaganda of violent methods to achieve social and political goals, as well as the actual commission of terrorist acts.

Terrorism is a multi-object crime, which main purpose is public safety, as well as encroachment on:

- citizen lives and health;
- critical infrastructure facilities;
- air transport facilities; natural environment;
- information environment;
- public administration bodies;
- statesmen and public figures, etc.

Experts in crime counteraction identify about 200 types of modern terrorist activity. The main ones are:

- political terrorism,
- nationalist terrorism,
- religious terrorism,
- technological terrorism, etc.

The most dangerous is technological terrorism that is the use or threat of use of weapons of mass destruction, including nuclear, chemical and bacteriological, radioactive and highly toxic chemicals, biological substances, as well as the threat of seizure of aerospace and critical infrastructure. increased danger to human life and health.

There is a growing danger of cyberterrorism i.e. actions to disrupt the automated information systems of aerospace facilities and critical infrastructure, which create a risk of death, significant material damage or other socially dangerous consequences. The main form of cyberterrorism is an attack on PC information, computer systems, data transmission equipment, other components of information structure that allows to penetrate attacked system, intercept control or suppress the means of network information exchange, perform other destructive actions [9].

Currently, the so-called "aerospace terrorism" is developing rapidly that has acquired an international, global nature and is a real threat to national and international security. This is evidenced by the total number of terrorist acts and the number of countries involved in terrorist activities [10]. The current nature of aerospace terrorism, increase in its scale in the world, is of great concern in the world. According to the National Center for Space Management and Testing of the State Space Agency of Ukraine, there were 6,758 space objects in Earth orbits. Of these, artificial satellites of the Earth - 1948 (497 in force), and 53% had a military purpose. There were already 8089 space objects in Earth orbits. Of these, 2370 artificial satellites of the Earth (668 in force). These are mainly military and dual-use spacecraft. Equipment for remote sensing of the Earth, as well as for scientific research, is regularly launched into space. The number of countries participating in space exploration is growing. Currently, the United States, Russia, China, France, Japan, Great Britain, and Italy have access to space. Each of these states is able to independently implement large space programs. In addition to these countries, countries such as Ukraine, Sweden, Norway, Canada, Brazil, Argentina, Indonesia have the opportunity to launch high-altitude, geodetic, meteorological and small spacecraft [11].

"Aerospace terrorism" can be understood as terrorist acts committed using ground-based or orbital-based space systems. The current level of scientific and technological progress allows us to talk about the following forms of manifestation of space terrorism:

- seizure (destruction) of space objects, elements of ground control systems, centers for receiving scientific information, control of outer space and other elements of space agency structures;
- intimidation of certain significant organizations, districts, regions, countries by choosing specific targeted crash sites of captured spacecraft;
- intimidation by terror (actions) from space against sea, land, air objects, etc.

The scale and consequences of implementation of space threats can be equated to the use of weapons of mass destruction and therefore pose a real threat to civilization. At the same time, it should be taken into account that boundary between normal combat operations in space and protection (counter-terrorist operations) from terrorists is being leveled.

So far, no cases of "aerospace terrorism" have been identified. At the same time, this type of terrorism requires a detailed scientific analysis. Professional readings and developments in this area are virtually missing. And it will be difficult to return to this issue after commission of terrorist acts.

The analysis of domestic and foreign sources makes it possible to attribute the following to the main current trends in terrorism evolution [12]:

– consolidation of local terrorist groups and their rapid internationalization;
– strengthening the mutual influence of various internal and external social, political, economic and other factors that contribute to emergence and spread of terrorism;
– increasing the level of organization of terrorist activities, creation of large terrorist groups with developed infrastructure;
– strengthening the link between terrorism and organized crime;
– growth of financial and logistical support of terrorist structures;
– desire of terrorism subjects to master the means of mass destruction of people;
– attempts to use terrorism as a tool for interfering in internal affairs of states;
– development and improvement of new forms and methods of terrorism aimed at expanding the consequences of terrorist acts and increasing the number of victims, etc.

The degree of danger of threats of terrorist acts is determined by the improvement level of forms, methods, forces and means of terrorist activity, tactics of its implementation, as well as the effectiveness of anti-terrorist measures of national and international systems of terrorism counteraction.

The purpose of countering terrorism in Ukraine is to protect the individual, society and the state from terrorist threats and to prevent such manifestations.

The main tasks in achieving these goals are:

– identification and elimination of factors that contribute to emergence and terrorism spread;
– detection, prevention and cessation of actions of persons and organizations aimed at preparation and commission of terrorist crimes and (or) assistance in such activities;
– bringing to justice the subjects of terrorist activity in accordance with the current legislation of our state and international community;
– cessation of attempts to transfer the activities of international terrorist organizations to the territory of Ukraine, involvement in this process of the potential of the international anti-terrorist coalition;
– constant improvement of the national system of terrorism counteraction, maintenance in a state of readiness for the use of forces and means designed to detect, prevent, stop terrorist acts and minimize their consequences;
– ensuring effective anti-terrorist protection of aerospace facilities and critical infrastructure, livelihoods and places of mass stay of people;

- counteraction to the spread of terrorism ideology, implementation of active information and propaganda measures of anti-terrorist orientation.

The national system of terrorism counteraction is a set of organizational structures (subjects of terrorism counteraction), which within the powers established by laws and regulations issued on their basis, carry out activities to combat terrorist threats, develop and implement a set of measures to prevent terrorist threats, detection and cessation of terrorist activities, minimization and elimination of possible consequences of terrorist acts.

The subjects of terrorism counteraction are authorized bodies of state power and local self-government responsible for conducting anti-terrorist measures, non-governmental organizations and associations, as well as individual citizens who provide assistance in the implementation of measures in this area. One of the central authorities that protects not only life, health, safety, but justice in general is the state forensic science institutions of Ukraine.

That is why consideration of the strategy of international cooperation development of forensic science institutions of Ukraine with foreign experts in prevention of terrorist attacks on aerospace and critical infrastructure is of great practical and scientific importance. The strategy is based on the principles of international law, the main international documents that determine the conditions and procedure for cooperation between states in the field of international forensic science cooperation. The basis of strategic planning should be science, because only on the basis of reliable scientific knowledge can assess the real state of affairs and existing opportunities to determine strategic and tactical goals and the trajectory of their achievement.

The main purpose of state forensic science institutions is to protect the interests of the state, rights and freedoms of citizens and the rights of legal entities by conducting objective, scientifically sound forensic examinations and forensic research. International cooperation of forensic science institutions is important for implementation of the rule of law, improving forensic activities and improving quality of forensic science as one of the main forms of using specific expertise in modern justice, as well as the formation of preventive recommendations for law enforcement agencies. Active activity in the field of international integration of our country into the world legal space puts before the state forensic institutions a wide range of tasks to establish international cooperation and expand cooperation with foreign specialized forensic science institutions.

One of the main directions of work of all public authorities in Ukraine on its integration into the European and world community is preparation of proposals on international legal relations and Ukrainian accession to international treaties and conventions, signing agreements on legal cooperation with relevant foreign bodies and international organizations, interaction with them within their powers, etc. That is why the current trend is to expand the participation of forensic institutions in such international cooperation, increasing their role in the development of theory and practice of forensic science and criminalistics [13, p. 288–291]. International cooperation of expert institutions is necessary for exchange of experience, advanced training of specialists of forensic institutions, taking into account modern achievements of science and technology, creation of standardized forensic methods, profile forensic directions [14, p. 113].

As N. I. Klimenko rightly points out, national forensic examination cannot exist in isolation, i.e. within a separate state [15, p. 165–168]. It will not actively carry out its law enforcement functions without integration with the international community. Therefore, international cooperation of expert institutions is necessary in order to exchange experience taking into account modern achievements of science and technology, prevention of terrorist attacks on aerospace and critical infrastructure, elimination of duplication of scientific and methodological support of regional distribution of tasks between forensic science institutions and creation of specialized forensic directions.

At the same time, the results of Sh. M. Khaziiev's analysis of the state and dynamics of international forensic science cooperation allowed him to identify the following four groups of main trends in this area:

- main trends in management of international forensic cooperation (strengthening of regional cooperation in the field of forensic science, creation of international forensic teams, improvement of means and methods of communication between forensic science institutions, creation of joint databases and data banks);
- main trends in development of international cooperation in the field of scientific support of forensic science (intensification of exchange of scientific achievements in forensic science, joint efforts of the international forensic community in developing current issues of forensic science, increasing interest in general theory of forensic science, development of human rights issues while forensic examination);
- main trends of international cooperation in the field of methodological support of forensic activities;
- main trends of international cooperation in the field of technical support of forensic activities [16, p. 73].

However, the area related to strengthening the preventive activities of forensic institutions is not mentioned or highlighted as a priority that is an incomplete reflection of criminological potential of these entities.

It should be noted that in recent years the creation of international forensic networks has intensified. Currently there are five such networks that unite scientific reserves of forensic science institutions of different countries:

1. European Network of Forensic Science Institutes existed since 1995 is the most developed and strong network;
2. South African Regional Network of Forensic Science since 2008 covering the entire African region;
3. Asian Network of Forensic Science existed since 2008;
4. International Forensic and Environmental Expert Network (existed since 2008) to assist forensic experts in the field of ecology and environmental offenses;
5. Trace evidence forensic environmental network existed since 2006 (is a non-governmental organization to assist in conducting forensic examinations in cases of crimes against wildlife and to preserve the biological diversity of flora and fauna) [17, p. 200–204; 18, p. 130–134].

For example, within the European Network of European Network of Forensic Science Institutes (ENFSI), one of the most important conditions for the effective operation of forensic institutions is validation (assessment of suitability) of forensic methods. Validation of methods is an important system element of ensuring and quality control of forensic research results, and its practice is widespread in forensic science institutions around the world.

As innovative mechanisms for improving the quality of forensic activities in recent years, validation and certification of methodological materials for forensic examination should be included in the cooperation plans of forensic science institutions. In addition, standardization (unification) of forensic methods, common terminological standards, unified requirements for forensic experts and professional forensic expert education programs will solve another problem: conducting forensic examinations on the basis of a single scientific and methodological approach [19].

European Network of European Network of Forensic Science Institutes has existed since 1995 [20]. It consists of 54 forensic science institutions, 41 of which are located in EU Member States. It is the world < s largest international organization of forensic experts with international recognition. Among the countries whose institutions are part of the European network are Armenia, Austria, Azerbaijan, Belgium, Bulgaria, Great Britain, Hungary, Greece, Georgia, Denmark, Spain, Ireland, Italy, Lithuania, Latvia, the Netherlands, Norway, Poland, Romania, Russia, Serbia, Slovakia, Slovenia, Turkey, Ukraine, Sweden, Switzerland, etc. The most famous ENFSI institutes conducting expert research are the Dutch, Polish, Hungarian and Prague forensic science institutes [21, p. 236–241].

The purpose of ENFSI is enshrined in its Constitution: be at the forefront of the world to ensure the quality of the development and conduct of forensics throughout Europe. The main activity of ENFSI is to strive to achieve a high reputation of the organization in the field of forensic science in Europe and the world by developing the quality of forensic services at all stages of the proceedings, from the scene to the court, provided by the following principles:

- ENFSI membership combines production, scientific and methodological capabilities of forensic science institutions;
- ENFSI membership expansion strengthens the trust of this organization by law enforcement agencies and judges;
- business relations with other organizations related to forensic science and criminalistics are established and maintained in an organized manner;
- activities of all ENFSI member institutions are actively encouraged that actively implement modern research methods and international standards in forensic practice and ensure high competence of experts in specific types of forensic science [20].

Currently ENFSI Members from Ukraine are three institutions: The State Scientific Research Forensic Center of the Ministry of Internal Affairs of Ukraine (since 2002)[1],

[1] The expert service of the Ministry of Internal Affairs consists of the State Research Forensic Center of the Ministry of Internal Affairs of Ukraine and territorial subdivisions – research forensic centers.

Kyiv Scientific Research Institute of Forensic Expertise of Ministry of Justice of Ukraine and National Scientific Center «Hon. Prof. M. S. Bokarius Forensic Science Institute» (from 2018) [22].

The 31st Annual ENFSI Meeting, organized by the Italian Forensic Police Service, took place on May 29–31, 2018 in Rome, Italy; it was attended by representatives of 55 forensic institutes from 31 European countries. Attention was paid to the exchange of experience and new developments in the following areas: DNA analysis, drug analysis, document verification, instrument identification, ballistics and forensic imaging (face recognition, photogrammetry, police portraiture). At this forum, Ukraine and Kharkiv were worthily represented by the KhRIFE delegation headed by O. M. Kliuiev, institution head.

Within international cooperation, this is important for combating crime, in particular for combating illegal migration, human trafficking, arms and drug trafficking, smuggling, spread of transnational and cross-border crime, search for criminals wanted internationally, search for missing persons, etc. [22].

It should be noted that the European Union recognizes the leading role in provision of judicial assistance by ENFSI, in particular in December 2011 at a meeting of the Council of Europe on Justice and Home Affairs adopted a Decision on the Strategy for European Forensic Science until 2020 providing for a European Forensic Science Area and development of forensic infrastructure in Europe [22].

What is the vision of the European Forensic Science Development Strategy 2020? Thus, in order to improve police cooperation with forensic institutions throughout the European Union and create a European forensic space by 2020, member states and the National Forensic Commission (NCFS) [22], seeking to ensure the effective administration of justice and security of citizens, decided achieve success in relevant areas, among which, given the focus of this study, the following should be highlighted:

- accreditation of forensic institutes and laboratories;
- approval and use of best practices and their use in the work of forensic science institutions;
- determination of optimal ways to create, update and use forensic databases;
- use of forensic science achievements terrorism counteraction, organized crime and other types of crimes [23, p. 206–220];
- research and implementation of projects for further development of forensic infrastructure, etc.

As we can see, this document only indirectly traces the tendency to strengthen the preventive capabilities of forensic science institutions of the European Union.

At the same time, another international document emphasizes strengthening of international cooperation between forensic science institutions of foreign countries for administration of criminal justice. Thus, in 2010, at the 19th session of the Commission on Crime Prevention and Criminal Justice of the UN Economic and Social Council, Resolution 19/5 *International cooperation in the field of forensic science* was discussed and adopted. This document notes that the role of forensic science in the administration of criminal justice and the need to further develop international relations of forensic science institutions within the existing regional networks of forensic institutions to ensure a global exchange of forensic knowledge, information and data [24].

Further study of professional readings made it possible to find out that today in the world, in addition to ENFSI, there are a number of other international organizations involved in development and improvement of forensic science, international cooperation in this field:

1. International Association for Identification (IAI) is one of the largest and most authoritative professional forensic associations. Currently, MAI has become one of the most prestigious professional associations in the world, has more than 5,000 members from 69 countries. It is chaired by a 15-member board of directors chaired by the president. The IAI has 29 standing committees, including on science and practice, scene research, fingerprint identification, trace evidence, forensic composite drawing, polygraph, forensic identification standards, international relations, etc. In addition, authoritative commissions for certification of experts in seven area of specialization have been established. The association has 45 branches in various US states and other countries.
2. International Association of Forensic Toxicologists. Founded in 1963 in London, the association has more than 1,400 members from all over the world working in forensic toxicology or related fields. Association members are police officers, medical and laboratory experts, pharmacology and toxicology experts.
3. International Association of Arson Investigators. The association has branches in 45 US states and 30 countries. It currently has about 9,000 members. The association includes a committee of forensic science developing methods and tools for collecting, researching and evaluating evidence that is found in the process of fire investigation and reconstruction of the fire scene.
4. International Crime Scene Investigators Association was established to assist individuals and organizations specializing in forensic investigation. Scene investigation is comprehensive and requires knowledge and skills in almost all forensic disciplines.
5. International Association of Forensic Nursing. The only international professional organization registered and formed solely for the purpose of developing and disseminating information on forensic medicine and forensic science (national and international). Activity areas of the association experts are assessment and research of injuries, assessment of the quality of treatment and care in cases of suspicious death, psychological rehabilitation of victims of violent crimes, examination of care for the elderly, children, patients and more. Association was established in 1992 and has more than 2,700 members.
6. International Association for Forensic Phonetics and Acoustics was founded in 1991 in the United Kingdom. Its purpose is to promote research in the field of forensic phonetics ensuring the exchange of knowledge and experience. Together with the International Association of Forensic Linguistics, Association publishes *The International Journal of Speech, Language and the Law.*
7. The International Association of Forensic Science, established in 1957, brought together scientists and practitioners from various fields of forensic science in various

fields of forensic science: forensic medicine, toxicology, fingerprint identification, forensic biology, drug research, ballistics, trace evidence, reconstruction of catastrophes and accidents, forensic psychiatry, etc. The main tasks of the Association are the development of forensic science, the exchange of scientific and technical information between forensic experts and scientists and management of training seminars.

8. International Association of Bomb Technicians and Investigators is an independent non-profit professional association formed to counter the criminal use of explosives and explosives. The Association facilitates the exchange of experience between representatives of law enforcement agencies, the armed forces, fire brigades and other services involved in emergencies and terrorist acts, as well as forensic experts. Founded in 1974, the association has more than 4,000 members from more than 60 countries. It consists of seven regional structures, including six located in the United States. The main tasks of the Association are: promoting the exchange of information in the field of explosion investigation and technical research of explosive devices, development of research in the field of combating criminal explosions, financial support of the most promising research, collection and processing of statistics in the field of combating criminal explosions. specializing in anti-explosion, providing appropriate forensic examinations and providing assistance while expert training. Association publishes *The Detonator* journal.
9. International Association of Forensic and Security Metrology is a non-profit professional organization of users, suppliers and manufacturers of equipment and technology for judiciary was established in July 2006 in the United States. The association brings together professionals who manufacture, supply and use high-precision equipment, including laser scanners, measuring instruments, reverse engineering systems, 3D printers and computer systems. The purpose of the Association is to develop and promote high-precision equipment for documentation and research to solve problems of forensic science and other areas.
10. Association of Firearm and Tool Mark Examiners was established in 1969 as a non-profit organization and now brings together forensic experts from the United States and other countries in the field of instrument and firearms research. Currently, the Association has more than 850 members from more than 40 countries.
11. The for Crime Scene Reconstruction began to function as an association of a group of professionals in Oklahoma and Texas (USA). These experts realized the need for a comprehensive investigation of the crime scene as a single whole object, which allows you to reproduce many elements of a criminal event, to identify and preserve physical evidence. The association members are law enforcement officials responsible for investigating crimes, forensic experts and teachers. The association currently has more than 550 members. The Association objectives, among others are as follows: to encourage the exchange of information and procedures (methods) useful for the reconstruction of crime scenes; stimulating research into existing and developing new improved methods of investigating crime scenes for reconstruction; promotion of educational programs designed to improve the skills of practicing experts in the field of crime scene reconstruction; providing members of the association with the opportunity to consult with colleagues in specific cases; publication of an information bulletin; development of cooperation and relations

between agencies and representatives of various forensic science institutions; providing association members with access to experts of various expert specializations within the association and dissemination of information about themselves.

12. Association of Forensic Quality Assurance Managers was formed at the initiative of the Southern California Group of Forensic Laboratories Quality Control Managers. The first conference of the association was held in Texas in October 2002. Association activities are aimed at promoting and supporting quality control of the forensic community, providing guarantees for a high level of forensic services to criminal justice institutions and protection of forensic quality control standards. The association has eight standing committees: Strategic Planning, Nomination, Statutory, Membership, Marketing and Communications, Resources and Training, Programs and Auditing.
13. International Society for Forensic Genetics was established in 1989 in Germany to promote achievements of genetics in forensic science and justice. The Society has taken on the role of a leading organization in the field of forensic genetics and coordinates research of genetic markers for the purposes of justice. It was assumed that opening of European borders within the EU could lead to an increase in international crime, and forensic geneticists would need standard research methods and results. Currently, the Society includes the main laboratories of forensic genetics of leading forensic and forensic science institutions in European countries such as Austria (Institute of Forensic Medicine), Belgium (National Institute of Forensics), Denmark (Department of Forensic Genetics at the Institute of Forensic Medicine), Great Britain. Expert Scientific Service, Institute of Cellular and Molecular Sciences), Finland (Forensic Laboratory of the National Bureau of Investigation), France (Institute of Forensic Medicine, Lyon Scientific Police Laboratory), Germany (Institute of Forensic Medicine, Federal Forensic Administration), Greece (Office of Forensic Science of Greek Police), Ireland (Police Guard Forensic Science Laboratory), Italy (Catholic University Institute of Forensic Medicine), Netherlands (Institute of Forensics), Norway (Institute of Forensic Science), Portugal (Institute of Forensic Medicine), Scotland (Department of Chemistry of University), Spain (Institute of Forensic Medicine), Sweden (National Laboratory of Forensic science) and Switzerland (Institute of Forensic Medicine). This group is carrying out a very important project: creation of a DNA database of different populations.
14. International Society of Computer Forensic Examiners is a private organization. The basic rules for its members are as follows: everyone should be included in the register of certified forensic experts; Anyone who has passed a specially designed certification program, regardless of race, gender, beliefs and nationality, can become a member of the society. If the Board of the Society or any member learns of the unethical or unprofessional behavior of any of members, internal investigation shall be conducted by the Ethics and Professional Responsibility Committee. Serious misdemeanors include perjury, falsification of diplomas, commission of an intentional crime, falsification or distortion of expert research results, falsification of training reports, violation of the Code of Conduct for Members of the Society and Professional Responsibility, refusal to cooperate in investigations, damage to the organization reputation and chess. Commission of one or more of these offenses

may result in re-certification, expulsion from the Society, suspension of membership or certificate revocation. The company has a committee on ethical standards consisting of four most reputable forensic experts, a certification board of seven people and a development committee.

15. Fingerprint Society was established in 1974 at the initiative of several fingerprint experts in the United Kingdom. The company publishes a quarterly *Fingerprint Word* journal and publishes an electronic newsletter (E-Newsletter). Once a year a conference of the Society is held, at which its members, invited scientists and practitioners make reports on various aspects of fingerprinting. During its existence, the Society has organized research on several important issues of fingerprinting, namely: papillary patterns of twins, papillary patterns of primates, fingerprints on archaeological finds and medical aspects of fingerprinting.
16. International Forensic Science Society brings together professionals from more than 60 countries. Founded in October 1959 in Nottingham, the company now has more than 2,500 members and is one of the world's oldest and most reputable forensic associations. The Society has its own Articles of Association and Code of Conduct and disciplinary rules, publishes a quarterly journal: "*Science & Justice* and an *Interfaces* bulletin. Most of the members of the Society are British scientists, as well as police experts and crime scene experts. The Society organizes scientific conferences both in the UK and in other countries, participates in certification of forensic experts and accreditation of forensic laboratories, as well as organizes training of certified professionals in crime scene investigation, forensic examination of documents, ballistics, fire and computer and technical examination.
17. International Forensic Strategic Alliance [25] unites the American Society of Crime Laboratory Directors, the European Network of Forensic Science Institutes (ENFSI), the Senior Managers of Australian and New Zealand Forensic Laboratories (SMANZFL) and The Academia Iberoamericana de Criminalistica y Estudios Forenses (AICEF). The alliance purpose is to create opportunities for strategic cooperation within the global forensic community, as well as to provide forensic assistance to other international institutions. In 2007 while a meeting at the Interpol International Symposium on Forensic Science in Lyon, the heads of joint institutions signed a memorandum of cooperation. This document consolidates global international cooperation in fight against crime by combining resources in the field of forensic science of participating countries, information exchange, exchange of new methods, technologies and technical means. The institutions also agreed to pool their resources to assist developing countries in establishing and developing their forensic agencies, believing that this would be a key factor in ensuring human rights. The Alliance is not a legal entity, although it represents a cooperation of four regional networks of forensic science institutions [26]. The Alliance efforts to strengthen the global role of forensic science in drug control and crime prevention have received active support from various United Nations bodies that have recognized the Alliance leading role in promoting internationally recognized quality standards in forensic science and development of international forensic cooperation on the basis of innovative interaction methods [24].
18. The International Forensic Summit is a forum of the most representative regional and international organizations and experts in the field of forensic science, created

to provide opportunities for communication between practitioners and scientists from different countries. Institution partners are Interpol, the US National Institute of Justice, the US National Center for Forensic Technology and US National Institute of Forensic Science. The core of the structure is the steering committee, which is responsible for the strategy and coordination of the organizations involved in the project. The main activities of the summit are forensic practice, activities of professional institutions, training of forensic experts and forensic medicine. While regular meetings, the summit members identify the most pressing problems of forensic science and seek ways to solve them through the efforts of the entire forensic community. In addition, the summit provides support to forensic institutions in the new states.

19. In 2015, Forensic Science Ireland adopted a strategic program for 2015–2018. In addition to plans to improve the forensic support of criminal justice authorities, creation of a DNA database and a number of organizational measures, it planned to further strengthen international cooperation in the field of forensic science [27].
20. In March 2016, in the United Kingdom, the Secretary of State introduced a document to Parliament entitled *Forensics Strategy: A National Approach to Forensic Criminal Justice*. It lists and describes the main measures aimed at improving the efficiency of forensic practice, ensuring high quality standards and proper ethical control both for forensic experts and in relation to law enforcement officers. Particular emphasis is placed on the need to accelerate the development of forensic computer science and methods of molecular genetic identification of criminals. The document obliges forensic laboratories of the country to use in their activity international standards for quality assurance of forensic examinations and to seek international accreditation [28].
21. In Washington State (USA) in 2009, the Forensic Council adopted a two-year strategic forensic development plan. The plan provided for improvement of forensic science services including training and retraining of staff, improvement of laboratory service and identification procedures [29].
22. The American Academy of Forensic Sciences [30] founded in 1948 to promote forensic science and their implementation in forensic science practice. The purpose of the Academy is to promote professionalism, integrity, competence, education, practice improvement (Journal of Forensic Sciences) [31] and cooperation in the field of forensic science. The Academy promotes the highest level of support in the field of professional development, exchange of knowledge and experience, judicial accreditation of its members in the United States and 71 other countries including through world-renowned journal of forensic science and annual scientific meetings. This year (2021) the 73rd meeting took place.

Thus, it should be noted that the international community is constantly interested in the field of forensic science and the use of its capabilities in combating crime and ensuring the administration of justice. In order to exchange information, improve forensic expert methods, form common standards of forensic science, solve practical problems of law enforcement, train and improve staff skills and solve other important tasks, currently more than thirty international organizations are successfully operating.

In addition, the study of preventive activities of forensic science institutions in Ukraine and abroad reveals a problem related to the legal regulation of general crime prevention and terrorist attacks on aerospace and critical infrastructure, in particular.

Thus, P. V. Melnyk, N. O. Dankovych and I. Ya. Friedman, exploring the issues of foreign experience of forensic science institutions in the general prevention of crimes, especially related to prevention of terrorist attacks on aerospace and critical infrastructure, note that obligation to establish the causes and conditions that contributed to a crime commission, as well as to make recommendations aimed at preventing or making it impossible to commit them, should, of course, rest not only on investigators and judges, but also on experts, because to establish such causes and conditions often require specific expertise [5, p. 101–114]. However, implementation of this idea requires appropriate changes to the current legislation governing forensic science. In domestic science, scientists offer different ways to solve this problem. It should be noted that ways of solving this problem indicated by scientists are reasonable and can be fully supported.

The next thing to pay attention to is the fact that both in Ukraine and in foreign countries it is important to introduce the latest information technologies in the preventive activities of forensic institutions (due to digitalization processes). Use in forensic activities of the latest information technologies (hereinafter referred to as IT) and computer technology, achievements of various sciences - mathematics, semiotics, modeling theory, algorithm theory, use in expert research of integrated automated database (hereinafter referred to as AD), intelligent interactive (dialogue) information systems and the formation of automated information systems (hereinafter referred to as AIS) of the new generation allow timely receipt and processing of relevant information, record illegal activities of individuals and organized criminal groups, prevent illegal activities of "authorities" and "leaders" of criminality, to organize work on active search of criminals and other persons, to identify and eliminate the causes and conditions that contribute to commission of crimes, thereby effectively implementing requirements of current legislation to protect the rights and interests of people, society and the state as a whole from criminal encroachment [33, p. 351–356; 34].

Creation of international and regional reference and information databases of international forensic institutions, foreign forensic science institutions is an urgent area of international cooperation of forensic science institutions of Ukraine with foreign professionals in preventing terrorist attacks on aerospace facilities and critical infrastructure. [35]. These problems could be tackled and the full potential of point-of-care and mobile forensic analyses could be realized if measurement devices could be operated in an integral forensic network. Through the network, the necessary calibration and quality control measures could be taken that would enable deployable forensic instrumentation to yield robust findings that can directly be used as evidence. The network would allow forensic experts to assess data generated outside the forensic laboratory and to provide direct assistance to the operators on location. From these activities, it also becomes apparent for which samples a more detailed follow-up investigation is required at the forensic laboratory. The forensic expert capacity is thus used more effectively and findings can be fed into the platform creating a continuous cycle of platform and data development. This approach would combine central data gathering allowing forensic intelligence and knowledge management with rapid and efficient decentralized forensic analysis. This novel concept, although technologically challenging, could lead to a step change in the efficiency and efficacy of the forensic information gathering process. It could also cause a paradigm shift in the role of forensic institutes and forensic experts in the criminal justice system: a shift towards a new role for forensic institutes and laboratories as custodians of the forensic platforms and point-of-care and portable equipment and methods. It would also allow forensic institutes to develop powerful forensic intelligence tools to reveal potential case and evidence connections, to better understand criminal activities, to monitor and optimize policing, to improve the efficiency of forensic investigations and to assist in crime prevention and disruption [36, 37, p. 38–43; 38, p. 10–16; 39, p. 181–190; 40, p. 8–15].

As for forensic science databases, the said Agreement is very useful. In order to combat terrorism as successfully as possible, seven countries of the European Union (Austria, Belgium, France, Germany, Luxembourg, the Netherlands and Spain) signed an agreement on 27 May 2005 in Prüm, Germany that allows, inter alia, exchange of DNA and fingerprints. The Council of the European Union adopted this initiative by adopting the Decision of the Council of the European Union on the promotion of cross-border cooperation, especially in the fight against terrorism and cross-border crime 208/615/JNA of 23 June 2008 [41]. The agreement provides for interstate exchange of forensic data of the Member States of the European Union in the field of DNA profiles, fingerprints and vehicle registration data.

Understanding this document importance cite certain provisions of the Agreement, especially those relating to forensic databases. For example, Sect. 2: *DNA profiles, fingerprinting material and other data*, Art. 2 *Creation of national DNA analysis files* emphasizes: The Contracting Parties undertake to open and maintain national DNA analysis files for the investigation of crimes. The processing of data stored in these files in accordance with this Convention shall be governed by other provisions in accordance with the national law applicable to the processing of the data. For the purpose of applying the Convention, the Parties shall make available reference data from national DNA analysis files. Reference materials should only contain DNA profiles established from non-coding DNA. Reference data should not contain any data which subject can be directly identified from. Reference data that cannot be linked to an identifiable person (cannot be traced) should be identified as such.

Article 3, *Automatic DNA profile search*:

1. For the purpose of criminal investigations, Contracting Parties shall allow national contact points of other States Parties to access reference data in DNA analysis files with possibility of performing an automatic search by comparing DNA profiles. A search permit can be exercised only in certain cases and in accordance with the domestic law of the Contracting Party conducting the search.
2. If automatic search indicates that submitted DNA profile matches the DNA profile entered in the searched files, contact person performing the search will receive an automatic notification of the hit and the link. If no matches are found, automatic notification will be issued.

Article 4, *Automatic comparison of DNA profiles*:

1. For investigation of criminal offenses, the Contracting Parties shall, by mutual agreement, through their national contact points, compare their DNA profiles of reference data that cannot be linked to the person being identified with all DNA profiles from other national reference data. DNA analysis files. Profiles need to be submitted and compared automatically. DNA profiles of reference data that cannot be linked to an identified person shall be provided for comparison only where provided for by the national law of the requesting Contracting Party.
2. If a Contracting Party determines that some of the submitted DNA profiles match those contained in the DNA analysis file, it shall immediately submit to national contact point of the other Contracting Party the reference data with which it was found to correspond.

Article 7, *Collection of cellular material and delivery of DNA profiles*:

When while investigation or criminal proceeding, there is no DNA profile for a particular person on territory of the requested Contracting Party, the requested Contracting Party shall provide legal assistance by collecting and examining that person cellular material and submitting the resulting DNA profile if: 1) determines the purpose for which it is necessary; 2) The requesting Contracting Party shall submit an order to initiate an investigation or a statement issued by the competent authority as required by the legislation of that Contracting Party stating that the requirements for the collection and testing of cellular material will be met if the person concerned is present. 3) The contracting parties request that requirements for collection and testing of cellular material and submission of the obtained DNA profile be met [41].

We would like to emphasize that this Convention is open for accession by any member state of the European Union. Upon accession, acceding states will also be required to implement it into national law. As the text of the Agreement indicates, only Member States have access to forensic databases (in the field of DNA profiles and fingerprints). However, candidate countries for accession to the European Union should be intensively preparing to join forensic database with the help of the Prussian Agreement. This primarily concerns the accreditation of forensic laboratories in the field of DNA profiles and fingerprint analysis in accordance with the international standard ISO 17025 [42].

The use of the latest devices and equipment, such as unmanned aerial vehicles (UAVs), has a powerful potential in countering terrorist attacks. With their help, in particular, it is possible to carry out various preventive measures: round-the-clock monitoring of the territory adjacent to the objects of critical infrastructure and aerospace; monitoring activities of ground safety equipment and protecting the data they integrate and transmit. According to researchers, there is far more technology developed and ready for the war on terror. Unmanned aerial vehicles - ranging from the small quad copters to the large military UAVs - have surveillance potential that should be exploited to its fullest. Aerospace and defense systems integrators always should be thinking about how they could be using today's off-the-shelf technologies in cutting-edge new ways to detect, track, and deter potential terrorists [43].

Unfortunately, there are still many unresolved issues in the Ukrainian legislation on the use of such devices (for example, the issues of using information obtained with the help of UAVs still remain unresolved). There are also no uniform safety requirements for certain classes of drones. This provision requires the scientific development and implementation of methods for using UAVs to counter terrorist attacks on critical infrastructure and air transport; training of professional staff of operators and users of unmanned aerial vehicles, etc.

The possibility of involving foreign specialists as forensic experts in the criminal proceedings of Ukraine is expressly provided for in Art. 9 of the Convention on Legal Assistance and Legal Relations in Civil, Family and Criminal Matters, the binding nature of which has been approved by the Verkhovna Rada of Ukraine in accordance with the established procedure. It should also be borne in mind Art. Art. 7–12 of the European Convention on Mutual Assistance in Criminal Matters containing similar provisions that is also part of the legislation of Ukraine [44, p. 35].

Section IV *International Cooperation in the Field of Forensic Science* of the Law of Ukraine: *On Judicial Examination* is devoted to the international cooperation of state forensic institutions of Ukraine in the field of forensic science. Articles 22, 23 and 24 of this section state that in the case of a forensic examination on behalf of the relevant body or person of another state with which Ukraine has an agreement on mutual legal assistance and cooperation, legislation of Ukraine shall apply, unless otherwise provided. Payment of the cost of forensic examination is made by agreement between the customer and the executor of forensic examination. Heads of state specialized institutions conducting forensic examinations, if necessary, have the right, with the consent of the body or person who appointed the forensic examination, to include in the expert commissions of leading specialists of other states. Such joint expert commissions carry out forensic examinations in accordance with the norms of the procedural legislation of Ukraine. Payment to foreign professionals for participation in forensic examination and reimbursement of other expenses related to its conduct shall be made by agreement of the parties. State specialized institutions performing forensic examinations have the right to establish international scientific relations with institutions of forensic science, criminology, etc. of other states, to hold joint scientific conferences, symposia, seminars, exchange trainees, scientific information and printed publications and conduct joint publications in the field. forensic science and criminology [45]. These issues are also considered in the Draft Law *On Forensic Activity in Ukraine* [46]. Unfortunately, it also does not take into account issues of international cooperation in the field of crime prevention.

Given significant changes in criminal procedure legislation, the provisions of the Law of Ukraine: *On Judicial Examination* and in light of the discussion of the draft new law: *On Forensic science activities*, we consider it necessary to specify procedural regulations of expert prevention in forensic expert conclusion, as well as provide procedural status preventive recommendations made by a professional. And, perhaps most importantly, to develop a *State program for the long-term development of forensic science institutions of Ukraine* and *Strategy for the development of forensic science institutions of Ukraine until 2030*.

It is also considered necessary to intensify cooperation with the relevant UNDP organization assisting national forensic systems in conducting truly independent forensic examinations and ensures access to justice for all segments of the particular country population. Successful examples of such cooperation exist, for example with the Republic of Moldova (2011). The UNDP project "Strengthening Efficiency and Access to Justice in Moldova" (A2J) is a multi-year institutional development project designed to contribute to an increased efficiency of justice services and to improved access to justice of men and women in Moldova, in particular from vulnerable and marginalized groups, through enhanced capacities of forensic institutions to provide qualitative justice services, strengthened capacities of the justice sector actors in the selected pilot areas to provide coordinated response to men's and women's justice needs and strengthened civil society able to claim the respect of rights and engage in a constructive dialogue with the justice chain actors. Project interventions will offer and encourage equal opportunity for the participation of men and women [47].

Effective crime counteraction can be only under the condition of further development of special scientific knowledge, development of effective methods and scientific and technical means of obtaining, processing and analysis of evidence.

Current stage of forensic science development is characterized by high dynamism, active influence on its development of scientific and technological progress, purposeful and active search for effective ways to improve forensic expert practice on a fundamental theoretical basis. The main features of forensic research at the present stage include its high science intensity, use of cybernetics integrating into forensic science and criminalistics are a catalyst for further development of its traditional tools and methods based on achievements of natural, technical and human sciences.

4 Conclusions

Summarizing the above, we note that for development and effectiveness of international cooperation of forensic science institutions of Ukraine with foreign experts in preventing terrorist attacks on aerospace industry and critical infrastructure, it is necessary to do the following:

Firstly, development and adoption of the *State program of long-term development of forensic institutions of Ukraine* that provide the following

- strengthening international cooperation of state forensic science institutions of Ukraine;
- support for establishment of international and regional forensic science networks;
- passing by the state FSI accreditation for compliance with international standards with subsequent receipt of international certificates in all expert specializations;
- increasing the authority of forensic institutions of Ukraine in the international arena, strengthening the position of domestic FSI in international associations of forensic experts including through participation of institutions and individual forensic experts in working groups and state representation in the Presidiums of international forensic expert organizations and unions;
- monitoring implementation of international standards in activities of state forensic science institutions, as well as the creation of a unified system of quality management of examinations and forensic researches;

- creation of a common information space in the field of international forensic cooperation (international and regional databases of current forensic methods; reference and information databases of international forensic organizations; reference and information databases of foreign forensic science organizations, foreign forensic science institutions, foreign universities that provide training and retraining for forensic experts; reference and information databases of international and foreign periodicals on forensic science, etc.).

Secondly, to develop the Strategy for the development of forensic institutions of Ukraine until 2030" that provide:

- improving efficiency of forensic activities and quality of forensic examinations in Ukraine, creating favorable conditions for the international exchange of scientific achievements and practical experience in this field;
- creation of an effective model of interaction of state forensic science institutions of Ukraine and domestic forensic experts with international associations of forensic experts and foreign forensic institutions and professionals;
- promoting development of integration processes in the field of forensic science activities in the European Union and other international and regional organizations with participation of Ukrainian forensic experts;
- creation of scientific, methodical and organizational bases of forensic expert support of prevention and investigation of terrorism with use of domestic and international experience. The main types of examinations required in the investigation of acts of terrorism are forensic explosives, ballistics, forensic aeronautical, DNA testing, computer forensics, forensic examinations of radioactive materials, forensic speech and audio analysis;
- use of international cooperation results in creating scientific, methodological and organizational bases of forensic support for the crime investigation in the field of information technology including the illegal use of outsider personal data. In addition to computer forensics in investigation of such crimes, it is important to conduct other engineering and technical examinations, as well as forensic economics, forensic linguistic, forensic intellectual property examination;
- use of international cooperation results in creating scientific, methodological and organizational bases of forensic support for investigation of transnational and international financial crimes. The investigation of such crimes should make full use of results of forensic economics, computer forensics;
- management of practical implementation of long-term strategic projects for development of international forensic cooperation, as well as monitoring their implementation;
- training of highly qualified forensic experts who are fluent in foreign languages and have received additional training in the field of international forensic cooperation and necessary basic knowledge in the field of international law and other international legal subject matters.

It is also important to harmonize the Strategy with the documents of the UN, the European Union, Interpol and Europol, which define the goals and objectives of international cooperation in the field of forensic science.

Thirdly, a separate and very promising area of international cooperation of forensic science institutions of Ukraine with foreign professionals and international forensic organizations is exchange of achievements in the field of general forensic science theory (forensic expertology). To date, there is no international scientific organization or association of experts that specializes in general theory of forensic science or theoretical foundations of criminology. Certainly, creation of such an international scientific society or association in Ukraine will objectively reflect the high level of domestic research in the field of forensic science. It should also be emphasized there are no international forums in the world devoted entirely to the problems of forensic science theory. These issues are periodically discussed at some international forensic conferences, but they are discussed alongside the narrower and more practical tasks of forensic science and criminalistics. The *forensic expertology* term began to be actively used abroad (Bulgaria, Poland, Lithuania), but its meaning is not always identical to theoretical foundations of domestic expertology.

Fourthly, creation of a new forensic association (such as the International Forensic Association for Combating Aerospace Terrorism) which will deal exclusively with cooperation in the crime investigation in the aerospace sector. Members of the Association can be domestic and foreign forensic organizations; forensic science institutions, higher education institutions that provide training and retraining of forensic experts; specialized higher education institutions that train professionals in aerospace industry; editors of professional periodicals, etc.

The co-founders of the Association can be two powerful intellectual centers: *National Aerospace University – "Kharkiv Aviation Institute", NAU "KhAI"* of the Ministry of Education of Ukraine and National Scientific Center «Hon. Prof. M. S. Bokarius Forensic Science Institute» of the Ministry of Justice of Ukraine.

The main theoretical and applied tasks, which will be dealt with at a high professional level by representatives of this association can be presented as follows:

- development of scientific and methodological recommendations to eliminate the causes and conditions that contribute to emergence and spread of terrorist attacks on objects of aerospace industry and critical infrastructure;
- drawing up short-term and long-term forecasts on occurrence and spread of terrorist threats, informing domestic and foreign specialized bodies about them in order to take measures to their neutralization;
- development and improvement of regulatory framework of activity of anti-terrorist bodies;
- development and introduction into activity of forensic science institutions of latest achievements of science and technology, equipment including those produced by domestic industry;
- development of a list of anti-terrorist measures at aerospace facilities and critical infrastructure;
- development and implementation of scientific and methodological recommendations on anti-terrorist protection of potential aerospace industry and critical infrastructure;
- conducting cultural, educational and informational and advocacy activities for the formation of socially significant values in Ukrainian society;

– improving measures of interaction and coordination of forensic science institutions of Ukraine with foreign professionals in prevention of terrorist attacks on aerospace and critical facilities in order to develop a common strategy for international cooperation in the field of terrorism counteraction.

References

1. International cooperation in the Forensic Field. Resolution 19/5 of the Commission on Crime Prevention and Criminal Justice UNODC. UN Document E /CN.15/2010/L.13/Rev.1. URL: www.unodc.org/documents/commissions/CCPCJ/Crime_Resolutions/2010-2019/2010/CCPCJ/-Resolution_19-5.pdf [in English]
2. Transport-related (civil aviation and maritime). Terrorism Offences. UNITED NATIONS OFFICE ON DRUGS AND CRIME. Vienna. UNITED NATIONS. New York. URL: https://www.unodc.org/documents/terrorism/Publications/Module_on_Transport/13-89032_Ebook_from_DM_9-9-2014.pdf [in English] (2014)
3. Kliuiev, O.M.: On the issue of international cooperation in the field of forensic science. theory and practice of forensic science and criminology. Vip. **18**, 154–162 (2016). [in Ukrainian]
4. Decree of the President of Ukraine № 121/2021 On the decision of the National Security and Defense Council of Ukraine of March 25. "On the Strategy of Military Security of Ukraine". URL: https://www.president.gov.ua/documents/1212021-37661 [in Ukrainian] (2021)
5. Melnyk, P.V., Dankovich, N.O., Ya, I.: Prevention of offenses (criminological and forensic aspects): monograph. Friedman and Others, pp. 170. National University of the State Tax Service of Ukraine, Irpin, [in Ukrainian] (2011)
6. Spitsyna, H.O., Filipenko, N.E.: Terrorist activity: criminal policy of counteraction. criminal and criminological means of counteraction to crimes against public safety and public order: collection of report theses of international scientific and practical conference dedicated to the 25th anniversary of KhNUIA (April 18, 2019, Kharkiv). ministry of internal affairs of ukraine, kharkiv national university of internal affairs; criminological association of ukraine. Kharkiv. KhNUIA **2019**, 188–191 (2019). [in Ukrainian]
7. On the fight against terrorism: the Law of Ukraine. URL: https://zakon.rada.gov.ua/laws/show/638-15#Text. [in Ukrainian]
8. Criminal Code of Ukraine: Law of Ukraine dated April 5, No. 2341-III. URL: http://zakon.rada.gov.ua/laws/show/2341-14 [in Ukrainian] (2001)
9. The use of the Internet for terrorist purposes. United Nations, New York. URL: https://www.unodc.org/documents/frontpage/Use_of_Internet_for_Terrorist_Purposes.pdf [in English] (2012)
10. Military Aviation: Issues and Options for Combating Terrorism and Counterinsurgency January 27. URL: https://www.everycrsreport.com/files/20060127_RL32737_15be14f27327ab1ddcb14fb8aaf56704832f2f76.pdf [in English] (2006)
11. Satellites in orbit - how many and whose? BBC News Ukraine. URL: http://www.bbc.com/ukrainian/news/2011/04/110411_satellites_orbit_it#:~:text=Перший%20штучний%20супутник%20Землі%20запустили,новому%20поглянути%20на%20нашу%20планету [in Ukrainian]
12. Foreign Terrorist Fighters Manual for Judicial Training Institutes South-Eastern Europe Updated Edition. Vienna. URL: https://www.unodc.org/pdf/terrorism/Foreign_Terrorist_Fighters_Handbook/EN_Foreign_Terrorist_Fighters_Ebook.pdf [in English] (2019)
13. Dmitrieva, K.S.: Current trends in international forensic cooperation. Legal Scientific Electronic Journal. (1), 288–291. URL: http://lsej.org.ua/1_2020/73.pdf. [in Ukrainian] (2020)

14. Zakovyrko, O.M.: European Network of Forensic Institutions (ENFSI): the history of creation and prospects for development. Current issues of standardization of forensic justice in Ukraine. Development prospects: mater. International scientific-practical conference, dedicated 105th anniversary of forensic examination in Ukraine and 95th anniversary of the birth of Academician. In: Segai, M.Y. (ed.) (July 4–5, 2018). KNDISE Ministry of Justice of Ukraine, Kyiv, 2018. [in Ukrainian] (2018)
15. Klimenko, N.I.: International significance of forensic activity. In: Kobylyansky, O.L. Antonyuk, P.E., Svoboda, E.Y. (ed.) Forensic Activity: Current Status and Prospects: A Collection of Round Table Materials, pp. 165–168, [in Ukrainian]. NAVS, Kyiv. (2015)
16. Khaziev, S.N.: International forensic organizations (part 1). Theory and Practice of Forensic Science: Scientific and Practical Journal. 1(13), 156–177. [in Russian] (2009)
17. Jan de Kinder: European network of forensic science institutions (ENFSI). Forensic theory and practice. No **4**(24), 200–204 (2011). [in Russian]
18. Klymenko, N.I.: International significance of forensic activity. Forensic activity: current status and prospects: a collection of materials of the round table (Kyiv, April 23, 2015), pp. 165–168. NAVS, Kyiv, [in Ukrainian] (2015)
19. Nadizhko, M.M.: Some issues of scientific and methodological support of forensic activities in the system of the Security Service of Ukraine (on the example of computer and technical research). URL: http://www.academy.ssu.gov.ua/ua/page/page_1581425700.htm [in Ukrainian]
20. Official site ENFSI. URL: https://enfsi.eu. [in English]
21. Linnyk, O.V., Omelchuk, L.V.: Relevance of Ukraine's accession to international forensic networks. International Legal Bulletin: current issues (theory and practice). Vip. 2–3 (6–7), pp. 236–241. [in Ukrainian] (2017)
22. ENFSI held the 10th conference of the working group of experts on documents (19.11.2018)/Kyiv Research Institute of Forensic Science. URL: https://kndise.gov.ua/news/news-view/c-evropejska-mereza-sudovo-ekspertnih-ustanov-enfsi-provela-10-tu-konferenciu-robocoi-grupi-ekspertiv-iz-dokumentiv. [in Ukrainian]
23. Nechyporuk, M., Pavlikov, V., Filipenko, N., Spitsyna, H., Shynkarenko, I.: 2020 Cyberterrorism Attacks on Critical Infrastructure and Aviation: Criminal and Legal Policy of Countering. Integrated Computer Technologies in Mechanical Engineering – 2020. Synergetic Engineering, pp. 206–220. ISBN 978–3–030–66716–0 ISBN 978–3–030–66717–7 (eBook). URL: https://doi.org/10.1007/978-3-030-66717-7 [in English]
24. International cooperation in the field of forensic science. Report of the Executive Director of the Commission on Crime Prevention and Criminal Justice, Vienna. URL: https://documents-dds-ny.un.org/doc/UNODC/GEN/V12/507/72/PDF?-OpenElement [in Ukrainian] (2012)
25. International Forensic Strategic Alliance. URL: https://www.ifsa-forensics.org [in English]
26. History of IFSA. URL: www.enfsi.eu/ifsa/letter-understanding; UNODC hosts International Forensic StrategicAlliancemeeting.URL: www.unodc.org/unodc/en/frontpage/2012/February/unodc-hosts-international-forensic-strategic-alliance-meeting.html [in English]
27. FSI Strategic Plan 2015–2018. URL: www.forensicscience.ie/News/FSI-Strategy-Statement-2015-2015.pdf [in English]
28. Forensic Science Strategy. A national approach to forensic science delivery in the criminal Justice System. March 2016. URL: www.gov.uk/uploads/system/uploads/-attachment_data/file/ 506652/54493_Cm_9217_Forensic_Science_Strategy_Accessible.pdf [in English]
29. Washington State ForensicInvestigation Council Strategic Plan. URL: www.ofm.wa.gov/budget/manage/strategic/0911/167strategicplan.pdf [in English]
30. The American Academy of Forensic Sciences. URL: https://aafs.org [in English]
31. Journal of Forensic Sciences. URL: https://aafs.org/aafs/Journal-of-Forensic-Sciences/AAFS/Affiliates/JFS/JFS.aspx?hkey=7f1b9581-68ab-47eb-b798-234f9eeff389 [in English]

32. Annual-Scientific-Meeting. URL: https://aafs.org/aafs/Meetings/2021-AAFS-Annual-Scientific-Meeting [in English]
33. Filipenko, N.E.: Forensic institutions of Ukraine as subjects of preventive activity. Bulletin of the Criminological Association of Ukraine. **2**(21), 224–233 [in Ukrainian] (2019)
34. Filipenko, N., Andrenko, S., Bublikov, A.: Participation of forensic science institutions in international cooperation on crime prevention. In: The 7th International Scientific and Practical Conference "World Science: Problems, Prospects and Innovations" (March 24–26, 2021), pp. 56–60. Perfect Publishing, Toronto, Canada. ISBN 978–1–4879–3793–5. [in English] (2021)
35. Filipenko, N., Bublikov, A., Obolientseva-Krasyvska, O.: Research methodology of criminological activity of forensic science institutions. In: The 3rd International Scientific and Practical Conference — Achievements and Prospects of Modern Scientific Research|| (February 7–9, 2021) Editorial EDULCP. Buenos Aires, Argentina, pp. 351–356. ISBN 978–987–859–237–4. [in English] (2021)
36. Kloosterman, A., et al.: The interface between forensic science and technology: how technology could cause a paradigm shift in the role of forensic institutes in the criminal justice system. URL: https://www.ncbi.nlm.nih.gov/pmc/articles/PMC4581008/ [in English] (2020)
37. Bell, C.: Concepts and possibilities in forensic intelligence. Forensic Sci. Int. **162**, 38–43 (2006). [in English]
38. Ribaux, O., et al.: Intelligence-led crime scene processing. Part I: forensic intelligence. Forensic Sci. Int. **195**, 10–16 (2010). [in English]
39. Morelato, M., et al.: Forensic intelligence framework-Part I: introduction of a transversal model by comparing illicit drugs and false identity documents monitoring. Forensic Sci. Int. **236**, 181–190 (2014). [in English]
40. Ross, A.: Elements of a forensic intelligence model. Aus. J. Forensic Sci. **47**, 8–15 (2015). [in English]
41. Agreement 208/615/JNA URL: https://eur-lex.europa.eu/LexUriServ/LexUriServ.do?uri=OJ:L:2008:210:0012:0072:EN:PDF [in English]
42. Ivanović, A., Rump, M.: Accreditation process forensic center of Montenegro to the mentorship of the european union (Projects EMFA-2). In: 10th Symposium of Forensic Sciencies. Bratislava. Symposium Journal. [in English] (2011)
43. The growing role of technology in the global war on terrorism. THE MIL & AERO BLOG, 8 Dec. URL: https://www.militaryaerospace.com/computers/article/16714122/the-growing-role-of-technology-in-the-global-war-on-terrorism. [in English] (2015)
44. Forensic examination: legal regulation and scientific comments: a textbook. Odyssey, Kharkiv, [in Ukrainian] (2004)
45. On forensic examination: Law of Ukraine. URL: https://zakon.rada.gov.ua/laws/show/4038-12#Text
46. On forensic activity: draft Law of Ukraine. URL: https://minjust.gov.ua/news/ministry/ministerstvo-yustitsii-ukraini-vinosit-na-publichne-gromadske-obgovorennya-proekt-zakonu-ukraini-pro-sudovo-ekspertnu-diyalnist [in Ukrainian]
47. International consultant to enhance the collaboration and uniform approaches within the system of forensic institutions of the Republic of Moldova. URL: https://jobs.undp.org/cj_view_job.cfm?cur_job_id=97848 [in English]

Legal Regulation of the Transport and Logistics System

Matvieieva Anastasiia[1(✉)], Zakharov Vasyl[2], Ostapenko Olena[3], Larisa Yerofyeyenko[3], and Olena Perunova[1]

[1] National Aerospace University «Kharkiv Aviation Institute», Kharkiv, Ukraine
a.v.matvieieva@khai.edu
[2] Lviv State University of Internal Affairs, Lviv, Ukraine
[3] Simon Kuznets Kharkiv National University of Economics, Kharkiv, Ukraine

Abstract. The paper considers the legal regulation of the transport and logistics system of Ukraine and the European Union countries in order to gain experience in this area with further adaptation to current legislation. The main priorities and problems that need to be solved by the transport and logistics system are identified. Areas of Ukraine's participation in transport and logistics integration are those we concentrate on.

Keywords: Transport · Transport system · Transport and logistics system · Transport sector

1 Introduction

Formulation of the Problem. The main advantages of our country in the international arena are an incredibly convenient geographical location at the intersection of major freight flows from Europe to Asia, the presence of non-freezing Black Sea ports, a strong and developed network of railways and highways, extensive pipeline system, gradual development of transport. However, according to official data, the transit potential of the state is used only by a quarter.

The transport system of Ukraine consists of more than 22 thousand km of railways, which are coordinated with the railways of 7 neighboring countries. There are 40 interstate railway crossings, about 170 thousand km of roads, 30 airports with 130 runways, 18 state sea trade ports with a total technological capacity of over 170 million tons, 4 fishing ports, 13 private port points and berths, more than 42 thousand km of oil and gas pipelines. The traffic statistics are as follows: about 2.2 million tons of cargo and more than 19 million passengers [1].

However, despite such excellent baseline indicators, the efficiency of the state transport system is very low.

M. Nechyporuk et al. (Eds.): ICTM 2021, LNNS 367, pp. 849–860, 2022.
https://doi.org/10.1007/978-3-030-94259-5_65

In addition, the situation is exacerbated by the fact that transport legislation is quite confusing, contains many conflicts and is imperfect. There is also a lack of comprehensive research on the development and implementation of effective transport mechanisms as a system.

Analysis of Recent Research and Publications. The main feature of the article is to determine the optimal model of legal regulation of the transport system on the basis of individual studies of such issues as: the concept and content of the transport and logistics system [2, 4, 7]; integration of Ukraine's transport system into the European transport space [9, 10].

The Aim of the Study is to study the mechanism of legal regulation of the transport and logistics system of Ukraine and ways to modernize the regulatory framework in this area.

2 Forms of Legal Regulation of Transport and Logistics Activities

It is worth noting that the transport and logistics system plays a major social, economic and environmental role. This is due to its direct impact on the functioning of society and the environment and rising living standards. Creating legal mechanisms for the development of the transport and logistics system is one of the factors strengthening the economy and improving the welfare of the population. Transport is an integral part of the logistics chain. According to Art. 1 of the Law of Ukraine "On Transport" transport is one of the most important branches of social production and is designed to meet the population and social production in transport needs. State target programs are aimed at ensuring sustainable development and improvement of the transport system. The emphasis is on taking into account its priority, on the basis of the latest achievements of scientific and technological progress, and is provided by the state [2].

Transport is an integral part of production and trade. That is why the transport component is involved in solving many logistical problems. However, there is a fairly independent transport sphere of logistics, in which multifaceted coherence between the participants in the transport process can be considered outside the direct links with the associated production and warehousing areas of material flow. Transport as a branch of material production carries passengers, luggage and cargo, and in the structure of social production belongs to the sphere of material services. The term "transport" covers not only vehicles of different types and kinds, but also transport infrastructure, specially trained personnel and information system [3].

The transport sector is one of the main sectors of the economy, has an extensive railway, road network, seaports and river terminals, airports and a wide network of air connections, cargo customs terminals, which creates the necessary conditions to meet the needs of transport users in providing transport services and business development. in general [4]. The Law of Ukraine "On Transport" outlines the purpose and objectives of public administration in the field of transport. In particular, public administration in the field of transport should ensure: timely, complete and high-quality satisfaction of the population and social production needs in transportation and defense needs of Ukraine;

protection of citizens' rights during their transport services; safe operation of transport; observance of the necessary rates and proportions of development of the national transport system; protection of economic interests of Ukraine and legitimate interests of enterprises and organizations of transport and consumers of transport services; creation of equal conditions for the development of economic activity of transport enterprises; restriction of monopoly and development of competition; coordination of work of different types of transport; licensing of certain activities in the field of transport; protection of the environment from the harmful effects of transport [2].

Today, the volume of cargo transportation compared to the previous year shows a negative trend to a significant decrease (Fig. 1).

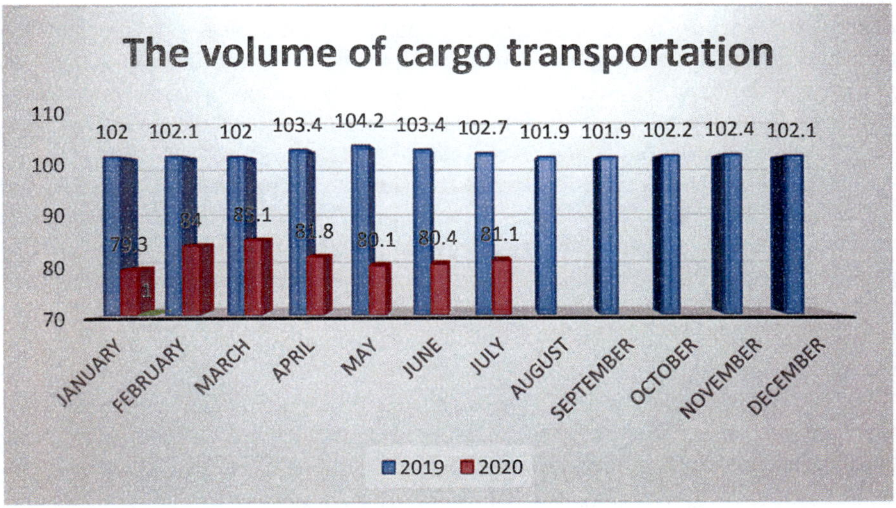

Fig. 1. The volume of cargo transportation

When studying the legal regulation of the transport and logistics system, we consider it appropriate to clarify the legal nature of this concept. The following definitions prevail in the scientific literature:

1. a set of functional and supporting subsystems, consisting of numerous interacting and interdependent elements that are parts of the state transport and distribution network, integrated with the flow of goods, general business objectives, consistent with the strategy of socio-economic development of the region and Ukraine in general [5];
2. multilevel and multifunctional system consisting of micro-, meso- and macro-levels, which not only meets the needs of transport and logistics services, but also increases the competitiveness of the whole region. Although this definition contains the concept of the system and the purpose of its operation, but not all elements of the transport and logistics system are considered comprehensively [6];
3. integrated set of subjects of transport and logistics activities and objects of transport and logistics infrastructure, which interact with each other in order to optimize

> the movement of freight flows "door to door" while minimizing and applying the most favorable conditions. This system, like any other system, consists of specific subsystems and elements, to which, depending on the specific functional phase of the logistics service of the cargo flow in the process of movement. It is advisable to include such subsystems as: transport and delivery, transport and warehousing, transport and logistics services and management, integrated information and transport and logistics solutions, as well as transport and sales system [7].

Obviously, the latter definition is the most complete for defining the concept of transport and logistics system. Therefore, we also support the positions of these scientists.

As for the actual tools of public administration of the transport sector, they include: rules of law that regulate the relationship between the transport industry and ensure the effective interaction of different modes of transport on the principles of logistics and joint competition; national and regional programs for the development of both the transport system as a whole and various modes of transport and their interaction; transport activity taxation system; development of national and regional transport balances related to plans for the development of productive forces, etc.

The priority of state regulation was and remains the improvement of regulatory and legal support for the development of both the transport system and the transport services market, including the creation of a regulatory framework governing the quality of transport services, mobilization training of transport organizations and their development of public-private partnership mechanisms; distribution of rights, responsibilities and risks between the state and the investor, as well as the definition of priority areas of these mechanisms in transport [7, p. 110].

In Ukraine, the transport and logistics system consists of public transport (road and air, sea, river, rail, as well as urban electric transport and subway); industrial railway transport; departmental transport; pipeline transport; public roads (Fig. 2).

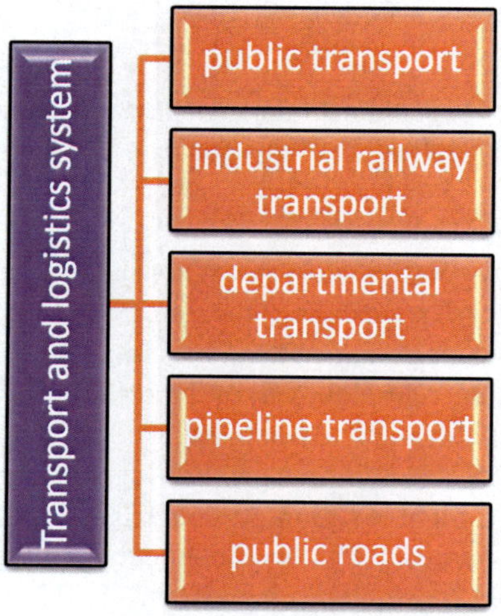

Fig. 2. The transport and logistics system

At the legislative level, this issue is regulated by such legal acts as: Customs Convention on the International Carriage of Goods under the Carnet TIR of 1975, version 29.05.2008 No 995_012, Air Code, Code of Merchant Shipping of Ukraine, Charter of the Railways of Ukraine, the Order of the Cabinet of Ministers of Ukraine "On approval of the National Transport Strategy of Ukraine for the period up to 2030" from 30.05.2018 No 430-r, Laws of Ukraine "On Transport", "On Road Transport", "On Railway Transport", "On transit of goods", "On pipeline transport", "On transport and forwarding activities", "On foreign economic activity", "On the functioning of the unified transport system of Ukraine in a special period" from 05.01.2013 No 194 - XIV, etc.

It is necessary to focus on the main problems that stand in the way of the development of transport legislation of Ukraine: the existence of outdated legislation of the USSR, which contradicts the new market conditions; a large number of regulations, the provisions of which contradict each other; non-compliance of existing transport standards with international (and, above all, European) standards, which causes great difficulties in the implementation of international freight.

According to the Commercial Code of Ukraine and the Law of Ukraine "On Transport" in Ukraine there is a single transport system, which consists of rail, road, water, air and pipeline transport. The most important problem not only in Ukraine but also in the world is the coordination of all elements of the transport system. For its successful functioning, it is necessary, first of all, to bring to a "single denominator" the legislative array that regulates them. In our opinion, there are several ways to achieve this goal:

1. Consolidation of basic and common for each institute provisions in the Civil Code of Ukraine.
2. Adoption of the Transport Code of Ukraine [8, p. 45].

The second way is the most effective and rational - the adoption of the Transport Code of Ukraine, which would serve as the Basic Law in the field of transport. All other legal acts would expand the provisions of this Code and be adopted to fulfill its general principles.

Also quite problematic is the inconsistency of transport and other areas of legislation, as well as the inconsistency of some provisions of transport legislation with international standards, in particular, in the field of international road and rail transport.

If we turn to the most common problems of transport system development, the following groups can be distinguished among them (Table 1).

At the same time, in accordance with the provisions of the National Transport Strategy of Ukraine, the need to introduce new principles of formation and coordination of state policy in the transport sector is emphasized, creating conditions for quality control over the performance of relevant executive bodies. For the most complete and effective realization of the transport potential of Ukraine, first of all, as a transit country, it is necessary to create a customer-oriented system of transport services. It is also necessary to take measures to ensure the optimal organization of the transport and road complex of the country and to obtain a synergy effect from the effective combination of potential and capabilities of all modes of transport based on the principles of partner competition during transportation [3].

Table 1. Common problems of transport system development.

No	Group	Characteristics
1	Economic	Significant depreciation of fixed assets Rising prices for fuel and material and technical resources, a significant percentage of the import component Insufficient efficiency of financial and economic mechanisms that stimulate the allocation of investment in the development of the transport system High level of tax burden, lack of preferences Low level of competitiveness of domestic enterprises in the domestic and international markets of transport services Decline in production in cargo-forming sectors of the economy Low level of innovative and high-tech components of the transport sector, tariffs subject to state regulation, imperfection of vehicle leasing mechanism on the terms of concessions and public-private partnerships
2	Organizational and managerial	Low level of intersectoral and interregional coordination in the development of the transport services market. Weak degree of use of geographical location benefits and opportunities of transport communications for delivery of transit cargo Irrational use of resources and low efficiency of vehicles Low level of management preparation for work in fierce competition The functions of the state on management of transport subjects in the field of formation of the market of transport services are lost Low level of informatization of transport process and information interaction of transport with other branches of economy Imperfection of the legislation of Ukraine in the field of customs, tax, investment regulation

(*continued*)

Table 1. (*continued*)

No	Group	Characteristics
3	Technical and technological	Slow improvement of transport technologies and their lack of consistency with production, trade, warehousing technologies Significant excess of international environmental standards of negative impact of transport on the environment The technical level of transport infrastructure and production and technical base of transport does not meet modern market requirements

Common problems that need to be addressed are:

1. the lack of an effective system of collection and boxes of administrative data in the transport sector, which can not objectively assess its condition and future development;
2. lack of a systematic approach to coordinating the development and long-term planning of all modes of transport, taking into account the socio-economic needs of the population, business, defense and geopolitical interests of Ukraine;
3. the fragmentary nature of administrative reform, which is manifested primarily in the process of delimitation of functions of state regulation and control, including the formation of relevant government agencies, operational activities and economic functions of transport enterprises;
4. low effective systems of control over the effectiveness of management decisions, insufficient transparency of reporting and behavior of public authorities and economic entities of the public sector of the economy in the transport sector;
5. outdated means of communication and feedback between transport authorities, transport companies and users of transport services, which reduces the efficiency of transport management and the quality of transport and logistics services;
6. imperfection of the system of criteria and indicators for assessing the quality of transport services;
7. too low level of development of intermodal, multimodal transportations, transport logistics;
8. the tendency to reduce the efficiency of capacity utilization of the national transport system (insufficient level of competition in the market of transport services and non-compliance with European requirements for access to the market of transport services);
9. imperfect tariff policy in the field of transport services;
10. relatively low speed of delivery of goods "door to door" and timeliness;
11. the presence of "narrow cities" of transport infrastructure;
12. low economic attractiveness of shipowners using the Ukrainian flag;

13. imperfect system of prevention of destruction of public roads, in particular control of the sizes and weight of vehicles and the corresponding level of responsibility for excess of the allowed parameters, admissible values of axial loadings of vehicles [3].

In connection with Ukraine's desire to join the EU in the near future, we consider it crucial to turn to foreign experience. The development of transport and logistics systems in the world is also successful due to the integration and creation of macro-logistics structures, such as the Benelux countries (economic union of Belgium, the Netherlands and Luxembourg), the EU, USA and Canada, Southeast Asia.

The regional factor reinforces the trend of globalization due to such aspects as the similarity of political systems and economic levels of countries; proximity of social way of life, traditions, historical roots; population migration; the only sources of energy, raw materials; unified transport system and road infrastructure; unified telecommunication communication; absence of customs and trade barriers, etc.

At the present stage of its development, the transport and logistics system of Ukraine is developing in the direction of European integration (due to the desire of the state to become part of the European Union). Also a priority is joining the world transport and logistics system due to Ukraine's inclusion in world socio-economic processes. From the point of view of European integration, Ukraine must rapidly develop the transport and logistics system as a branch of infrastructure in order to promote the rapid economic and social development of the country and its participation in the international division of labor. Ukraine's accession to international transport organizations and structures, ratification of a number of international agreements, conventions and other documents on the organization of transport systems and transportation were the first steps that began the integration of the Ukrainian transport and road complex into the European transport system. An important direction in the development of the transport and logistics system of Ukraine is to reduce the environmental hazards posed by transport to the environment. This will not only improve the environmental situation in the regions, but will also contribute to the formation of a positive image of transport enterprises [9, p. 57].

There are the following areas of Ukraine's participation in European transport and logistics integration: 1) Pan-European Transport Corridors (PETC); 2) Transcontinental Transport Corridors (TTCC); 3) Pan-European Transport Zones (PETZ); 4) International Logistics Terminals (ILT).

The integration of our country into the European transport and logistics system is a strategic direction of Ukraine's development. It is carried out on the basis of the "Concept of development of the transport and road complex of Ukraine in the medium term and until 2020" and "The program of forming a network of logistics centers in the international transport corridors of Ukraine". According to these documents, the purpose of creating a network of PETC and TTCC, which provides for the development of their territorial schemes in Ukraine and proposals for the establishment of 45 logistics centers in the system of international transport corridors of Ukraine. In order to create a single transport and logistics infrastructure of the EU, which would meet the latest requirements at a new stage of development of logistics in a united Europe, new guidelines were identified, which are the development of a Common Europe - space for the East, including

Ukraine as an integral part of Eurologistics through a network of international transport corridors and logistics centers will be connected to the transport and logistics systems of Asia and other continents [10, p. 137].

According to the Euro-vector of Ukraine's development, cooperation with the EU has common goals: to promote the restructuring and renewal of the Ukrainian transport sector; gradual harmonization of existing standards and policies with ones in the European Union (acquis communautaire); improving the movement of goods and passengers by eliminating administrative, technical and other obstacles; cooperation, including measures to improve border crossings. Particular priorities include cooperation to expand trans-European transport networks (TEN-T), transport axises, within which Ukraine will identify and allocate resources to finance transcontinental transport corridors (including the Europe-Caucasus-Asia transport corridor (ECATC), sea lanes and the Black Sea Transport Zone (Black Sea PETrA)) in accordance with procedures for ensuring clear implementation and effective implementation of international transport agreements and conventions.

Cooperation also includes interoperability and support for intermodality of transport networks, cooperation in the use of space transport systems under the "Agreement on Cooperation in Civil Global Navigation Satellite Systems (GNSS) between the European Community, its Member States and Ukraine", signed in 2005. In addition, Ukraine will take an active part in the work of the World Transport Forum, the United Nations Economic Commission for Europe (UNECE), the International Civil Aviation Organization (ICAO), the International Air Transport Association (IATA), the International Maritime Organization (IMO), and Pan-European Aviation Organizations: European Civil Aviation Conference, European Aviation Safety Agency (EASA), Intergovernmental Organization for International Carriage by Rail (OTIF) [10, p. 137].

Today, Ukraine receives financial and technical assistance from the EU to support the implementation of the principles of national transport development under three programs: "Support to the Implementation of Ukraine's Transport Strategy" (65 million euros), "Support to the Implementation of Ukraine's National Environmental Policy Strategy" (35 million euros) and "Twinning and technical assistance to support the European Neighborhood Policy" (16 million euros) [13, p. 127].

The latter is one of the key means of supporting the mechanism of adaptation of Ukrainian legislation in the field of transport services to EU norms and standards through the spread of twinning (support for the development and improvement of training of civil servants in Ukraine). To some extent, the implementation of other EU technical assistance programs in Ukraine also contributes to the development of the transport sector, in particular: INOGATE (Interstate Oiland Gas Transport to Europe), the Interstate Oil and Gas Supply Program to Europe; IBPP (Institution Building Partnership Program), Tacis Program for Partnership in Institutional Development; CBC (Cross-Border Cooperation Program). In fact, EU support is aimed at reforming Ukraine's transport sector, in particular, rail, improving road, air and shipping safety, increasing traffic flows through Ukraine and bringing transport legislation to an international level [10, p. 137].

For its part, Ukraine actively participates in the activities of European organizations that determine the development of international economic cooperation and mechanisms of cooperation between countries in the field of transport: the European Conference of

Ministers of Transport (ECMT), which operates within the Organization for Economic Cooperation and Development (OECD), and as well as the Committees of the United Nations Economic Commission for Europe and the Committee on Transport and the Subcommittee on Inland Transport, Intermodal Transport and Logistics, and the Trade Committee of the Center for Trade Facilitation and Modernization (CEFACT), European Railway Agency (ERA), Eurasian Transport Conference (MCAT), etc.

Over the past few years, experts of these international organizations and Ukrainian specialists have developed a "road map" of the integration of Ukraine's transport and logistics system into the European subregional space: 1) technical and technological modernization of border crossing infrastructure and reasonable reduction of control procedures, simplification of control mechanisms at checkpoints across the state border of Ukraine; 2) accelerated development of container service in accordance with ICD-transit technology; 3) introduction of a multifunctional complex system "Electronic Customs": customs information electronic document management of transportation infrastructure (including transit) and its integration into international customs databases; 4) improvement of the system of collection and dissemination of customs statistics, which will create a system of monitoring and forecasting of international transit of goods [10, p. 137].

The EU's transport and logistics strategy is called "Eurologistics", it is considered as an infrastructural basis and an important part of the European integration process. The course of creating a single transport system is enshrined in all major EU documents since the Treaty of Rome (1957) [11, p. 56].

Thus, the development of the transport and logistics system of Ukraine in modern conditions should be based on the following principles: promote the economic development of the country as a whole and its regions, which will increase the welfare of the population and the competitiveness of the country; rely on innovative methods of organization, management, marketing, which ensure the implementation of its main purpose - to meet consumers' needs on the basis of coordinated work of all stakeholders; ensure the ecological quality of life of the population; promote high social living standards, increase employment. Implementation of these principles will create conditions for effective competition of transport and logistics companies, the development of transport infrastructure of the country and its integration with transport and logistics systems of other countries [9, p. 59].

Today the situation with the export of transport services is unstable and unsatisfactory for the Ukrainian market of transport services. Thus, in particular, the structure of exports of services in the first half of 2020 indicates a decrease in the level of just transport services, exports of which in 2019 amounted to 58.3% according to the Ministry of Economic Development, Trade and Agriculture of Ukraine (Fig. 3). In particular, due to hostilities in eastern Ukraine, as well as the annexation of Crimea, the volume of traffic decreased both as a whole and by individual modes of transport. If we talk about 2020, the volume of traffic in general has decreased due to the introduction of a number of restrictions related to quarantine measures.

The current state of the transport and road management system needs to be improved, which should be carried out by public authorities at the regional level in the following areas: monitoring compliance with current legislation in the transport sector of the region,

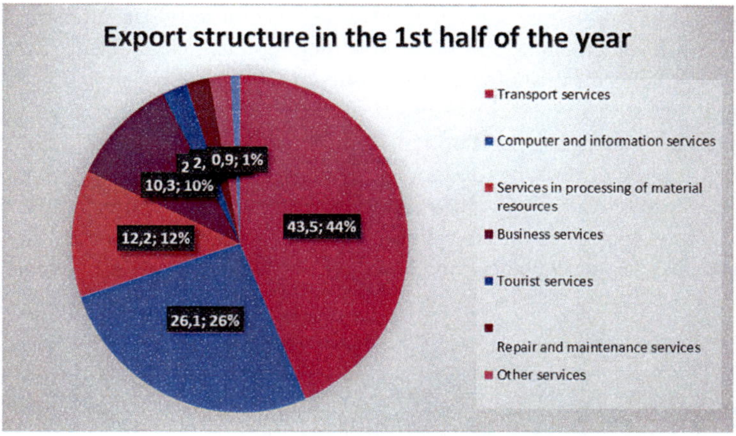

Fig. 3. Export structure in the first half of the year

including safe use of transport, environmental protection; assistance in attracting domestic and foreign investments in the transport and road complex of the region, financing of transport services provided to socially vulnerable categories of the population, as well as ensuring the special needs of state organizations in transport services; improving the regulation of regional markets for transport services, avoiding the implementation of discriminatory measures against both carriers and recipients of this type of service, ensuring tariff regulation of the activities of transport organizations-monopolists [7, p. 110].

Today in Ukraine there is a single logistics system, the main task of which is to maintain the level of compliance with the requirements of social production, with national security requirements, create an expanded infrastructure, ensuring foreign economic relations of Ukraine.

Analysis of the activities and main trends in the freight market has shown that the need for a highly developed transport and logistics system is further enhanced by Ukraine's integration into the European and world economy and can become the basis for Ukraine's effective entry into the world community. To design a new or modernize the existing organizational and production structure of a particular international transport and logistics center, a tree of goals is formed, stable links and contacts between individual logistics structures are established, as well as with service companies and organizations, suppliers and various modes of transport. As a result of the departmentalization processes, the organizational structure of the center and the management system of its operation are developed, what implements the main production functions of logistics management [7, p. 112].

Creation of multimodal terminal complexes and logistics transport and distribution centers will allow effective use of advanced logistics technologies and ensure the integration of material, information, service and financial flows of transit traffic between Europe, Asia and the Middle East [7, p. 113].

Today there is a number of problems of state regulation that hinder the effective development of the transport sector of Ukraine. In particular, it is an imbalance in the development of a single transport system; the presence of territorial and structural disparities in the development of transport infrastructure; insufficient level of accessibility of transport services for the population, mobility of labor resources; insufficient quality of transport services; low level of exports of transport services, including the use of transit potential; insufficient level of transport safety [7, p. 114].

Further development of the transport and logistics system of Ukraine should follow the path of innovations that will allow adapting to modern market conditions. Based on the experience of foreign countries, we can choose the right path for our country to make appropriate changes to regulations, with mandatory adaptation to current legislation to ensure (improve existing) resources and the ability to take into account the interests of all parties.

References

1. Preiher, D., Zhalilo, Y., Sobkevych, O., Yemelianova, O.: Implementation transit potential of Ukraine as a factor strengthening the economic links between Europe and Asia. Econ. Ukraine **4**, 47–59 (2012)
2. On transport: Law of Ukraine. The Verkhovna Rada Bulletin. 1994. No 51. Art.446 (1994)
3. Stokolyas, V.S.: Efficiency of transport logistics as a component of the logistics system. Eff. Econ. (7) (2014). http://nbuv.gov.ua/UJRN/efek_2014_7_50
4. On approval of the National Transport Strategy of Ukraine for the period up to 2030: Order of the Cabinet of Ministers of Ukraine of May 30, 2018 No 430-r. Official Gazette of July 10. No 52. Art. 1848 (2018)
5. Dorokhovsky, O.M.: Formation and functioning of regional transport and logistics systems: theory, methodology, practice: extended abstract of the thesis... Dr. Econ. Science: 08.00.05. Kherson. 369 p. (2013)
6. Syzdykbaeva, B.V.: Methodology of formation and development of the integrated transport and logistics system of Kazakhstan: extended abstract of the thesis... Dr. Econ. Science: 08.00.05. Turkestan (2010). http://dislib.ru/ekonomika/246331-metodologiya-formirovaniya-razvitiya-integrirovannoytransportno-logisticheskoy-sistemi-kazahstana.php
7. Braginsky, V.V.: Development of transport and logistics system as a form of realization of transit potential of Ukraine. Public administration: theory and practice. Electronic scientific professional publication. Nac. Acad. state manage under the President of Ukraine (2011). http://academy.gov.ua/ej/ej14/txts/Braginskiy.pdf
8. Matvieieva, A.V.: Transport corridor: legal nature, content, types and participants (comprehensive study). Kharkiv, FINN (2011)
9. Fedorchuk, E.I.: Features of state regulation of transport and logistics of Ukraine. Invest. Pract. Exp. (22), 109–114 (2017)
10. Popova, N.V.: Modern trends in transport and logistics systems. Bull. Trans. Econ. Ind. **53**, 54–60 (2016)
11. Dashkuev, M.A.: The current state of the process of integration of the transport and logistics system of Ukraine in the European subregional space. Bus. Inform. (5), 133–140 (2015)
12. Malovychko, A.S.: Integration of Ukraine into the transport and logistics system of the European Union. Black Sea Econ. Stud. **7**, 55–58 (2016)

Robotics and UAV

Topological Design for Robotic Filament Winding of Composite Beam Structures

Zheng Hu[1(✉)] and Oleksii Vambol[2]

[1] Nanchang Hangkong University, 696 Fenghe South Avenue, Nanchang 330063, China
[2] National Aerospace University "Kharkiv Aviation Institute", 17 Chkalova Street, Kharkiv 61070, Ukraine
olexii.vambol@khai.edu

Abstract. Robotic filament winding process is to place the impregnated fiber along the predefined winding trajectory to manufacture fiber-reinforced composite parts with complex geometry. The structure performance of composite parts is maximized when the fiber direction is consistent with the major principal stress direction of the parts. Hence, this work develops a novel method combining topological design and filament winding trajectory based on load transfer path for the complex beam structures with continuous fiber- reinforced composite. First, the density-based topology method is applied to obtain the optimal topology for maximizing structural stiffness. Second, based on the direction of major principal stress within the optimized topology, the continuous fiber placement paths are defined for robotic filament winding. In order to ensure the manufacturability of the topological beam structures, the conceptual design of a novel robotic filament winding equipment is developed. Finally, the FEM simulation studies of composite beam structures are performed by adopting Workbench, which verifies the feasibility and effectiveness of the proposed optimization method.

Keywords: Robotic filament winding · Topology optimization · Composite beam structures · Winding trajectory

1 Introduction

Fiber-reinforced composite materials are increasingly utilized in aerospace, marine engineering, advanced rail transit, new energy vehicles, biomedicine, electronic and electrical fields due to their superior mechanical properties, designability, corrosion resistance, fatigue resistance, and many other advantages [1]. In particular, composite materials have become the first-choice structural materials for large passenger aircraft, and the percentage of their amount to the weight of the aircraft has become an important indicator of the advancement of civil aircraft. Take Airbus as an example, the amount of composite materials has grown from 5% of A300 to 53% of A350; and the amount of composite materials on Boeing's most advanced aircraft B787 has reached 50% of the

M. Nechyporuk et al. (Eds.): ICTM 2021, LNNS 367, pp. 863–873, 2022.
https://doi.org/10.1007/978-3-030-94259-5_66

total weight of the structure [2]. Composite beam structures are commonly used as connectors and support parts, made from aluminum alloy in conventional structural design. Compared with beam structures made from aluminum alloy, fiber-reinforced composite beam structures have the advantage of high strength and stiffness-to-weight ratio.

Topology optimization is a finite-element-based computational tool commonly used to obtain the material layout corresponding to the most effective loading transmission path within a predefined design domain [3]. In structural mechanics, it is usually to minimize the structural compliance, thereby maximizing the stiffness, while limiting the volume of materials utilized in the design. In recent years, with the widespread use of composite materials in various fields, topology optimization has been attempted to be applied in the design of composite structures to make full use of the advantages of composites and new manufacturing technologies [4–6].

There are various forming processes for fiber-reinforced composite materials, including hand lay-up process, filament winding process, RTM (resin transfer molding) process, AFP (automated fiber placement) technique, 3D- printing, etc. The applications and characteristics of these manufacturing techniques are different. The criteria for choosing a manufacturing technique depend on the cost, quality, accuracy, production efficiency, structural strength and the shape and size requirements of the composite parts [7]. Among those techniques, hand lay-up, filament winding, and 3D-printing are suitable for the manufacture of topology beam structures in terms of manufacturing feasibility. In the hand lay-up process, prepregs are cut, laid down in the desired fiber angle on a mold, and then vacuum-bagged and cured. However, the topological structures have large gaps in the material. If the traditional lay-up technology is used to manufacture them, many manufacturing defects will be expected, such as the surface quality will be adversely affected and unsupported surfaces will sag.

In recent years, the rapid development of 3D printing technology is expected to realize the effective manufacturing of composite structures with complex geometry. Some researches to date have integrated topology optimization with additive manufacturing using continuous fiber-reinforced composite materials. For example, the authors of [8] proposed a topology optimization method for the continuous fiber angle optimization (CFAO) to calculate the optimal layout and direction of fiber for 3D-printing structures. However, the fiber orientations are directly regarded as design variables, which brings the optimization easily fall into a local minimum. Moreover, the obtained fiber angles are discrete and hard to generate continuous fiber printing paths. In order to solve the issues, the work [9] proposed a level-set based method for the simultaneous optimization of topology and the fiber printing paths. In view of the lack of manufacturing applications and experimental researches, the paper [10] developed a multidisciplinary design and analysis approach, which combines topology optimization, fiber reinforcement techniques and 3D-printing of continuous fiber-reinforced composite structures. However, in the current 3D-printing technology, printing continuous fiber-reinforced composite materials with complicated shapes and high mechanical properties is still a great challenge [11]. The 3D-printing of continuous fiber-reinforced composite parts still faces the prominent problems of low fiber volume content, low interface bonding strength and high porosity, which will seriously limit the effective application of 3D-printing composites in engineering structures [12]. Furthermore, due to the long polymerization and crosslinking time of thermosetting resin, it is difficult to realize real-time in-situ curing [13]. Hence, the vast majority of resin materials used for 3D-printing are thermoplastic resin [14]. Compared with thermosetting composites, thermoplastic composites have lower working temperature and are not suitable for some harsh environments of aeronautics and astronautics applications.

Filament winding is a technology in which impregnated fibers are wound on a mold in a designed shape to produce composite parts [7]. Generally, the starting materials of filament winding are continuous fiber and liquid thermosetting resin. Products manufactured by filament winding process have an extremely high strength-to-weight ratio and are suitable for high-temperature and high-pressure environments, such as composite grid structure for launch vehicles [15], fuel tanks of fighters, and fuselage of aircraft [16]. The traditional filament winding process is mainly applied to produce axisymmetric geometry parts. Afterwards, due to the development of automation systems and advanced processing equipment, it became possible to manufacture composite parts with complex geometries; this improved process is called robotic filament winding. For example, the paper [17] used the robotic filament winding technology to manufacture a fork part. The results verify the effectiveness of robot filament winding technology in realizing three-dimensional complex trajectories. Unfortunately, to the best knowledge of authors, studies on the combining topology optimization method and filament winding technology for composite structures are still lacking. As is known to all, the strength or stiffness of filament winding structures is primarily determined by the fiber direction, and the maximum strength or stiffness can be obtained when the fiber direction matches the direction of the major principal stress acting on the structure [18]. Therefore, the design of the winding trajectory should match the direction of the major principal stress as much as possible.

This work aims to develop a multidisciplinary design method that combines topology optimization approach and filament winding technique for composite beam structures. To achieve this goal, the following tasks have been accomplished:

- The density-based topology method is applied to obtain the optimal topology for maximizing structural stiffness. Based on the direction of major principal stress within the optimized topology, the continuous fiber placement paths are defined for robotic filament winding. Through the studies of FEM simulation, the feasibility and effectiveness of the proposed method are verified.
- In order to demonstrate the manufacturability of the topological beam structures, the conceptual design of novel robotic filament winding equipment is developed. So as to provide valuable insights for practical production in the future.

2 Topological Design

For simplicity, structural compliance minimization (stiffness maximization) is often taken as the design goal. Based on the density-based approach and solid isotropic material with the penalization (SIMP) interpolation model [3], the value of element pseudo density varying from 0 (void) to 1 (solid) for determining the optimal material distribution is considered as the design variable. Assuming the material is isotropic, the topology optimization model can be constructed as

$$\begin{aligned} \min: \quad & C(\boldsymbol{\rho}) = \boldsymbol{U}^T \boldsymbol{K} \boldsymbol{U} \\ \text{s.t.}: \quad & \sum_{e=1}^{n} \rho_e v_e - g V_0 \le 0 \\ & \rho_e \in [\rho_{\min}, 1] \end{aligned} \tag{1}$$

where C represents the structural compliance; $\boldsymbol{K}$ and $\boldsymbol{U}$ are the global stiffness matrix and the displacement vector, respectively; ρ_e is the density variable of the eth element; $\rho_{\min}$ is usually set to 0.001 in order to avoid stiffness matrix singularity; V_0, v_e and g denote the initial design volume, the volume of the eth element and predefined volume fraction, respectively.

The cantilever beam is often used as a standard test case to present the characteristics of the proposed approach. The geometry and boundary/load conditions are shown in Fig. 1, a. Two predefined holes on the left side of the structure are used for installation, and one predefined hole on the right is suffered concentrate force $F = 1000$ N. The volume fraction $g = 0.25$, and the material properties applied in the topology optimization process are: $E = 70$ GPa, $\upsilon = 0.3$.

The topology optimization process is performed by the Altair-Inspire software. The manufacturing constraint of "extrusion" is applied alone the thickness direction, so that the shape of the optimized topology maintains a constant cross-section along the thickness direction and the hollowing out phenomenon is avoided. As shown in Fig. 1, b, the optimized structure is composed of eight ribs, and the width of the ribs is very close. In order to facilitate manufacturing using filament winding technology, all ribs are set to the same width. The reconstructed geometric model and the corresponding principle stress tensor distribution are depicted in Fig. 2. The stress analysis is performed by the ANSYS software. It can be seen that the direction of compressive and tensile stresses is almost-parallel to the contour of the ribs. Therefore, the higher mechanical properties can be obtained by placing the fibers along the principle stresses direction of the part.

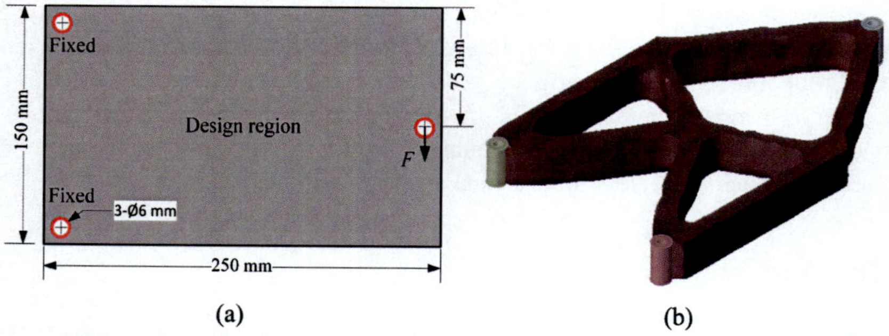

Fig. 1. Cantilever beam: (a) geometry and boundary/load conditions; (b) optimized topology.

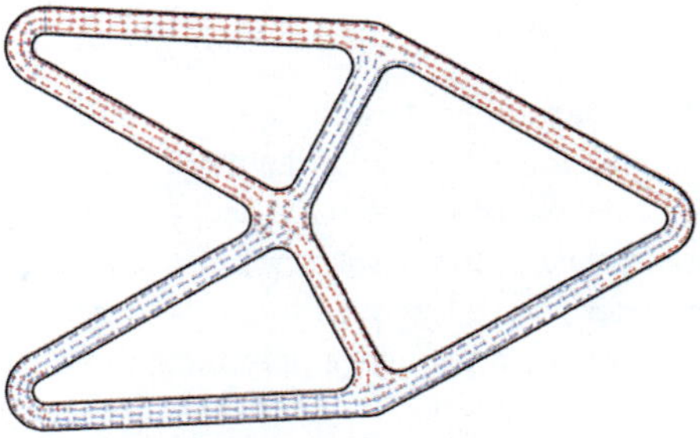

Fig. 2. Principal stress tensor visualization of the cantilever beam after reconstructing. The tension and compression stresses are denoted by the red and blue color arrows, respectively. The direction and length of the arrow indicate the direction and magnitude of the principal stress in each subregion.

3 Conceptual Design of Robotic Filament Winding System

In this section, the conceptual design of a novel robotic filament winding system is developed for the manufacturing of the optimized structure in Fig. 2. The Epoxy Carbon UD Prepreg is adopted, and the thickness of a single layer fiber band is assumed to be 0.18 mm. The basic material properties are: E_x = 121 GPa, E_y = 8.6 GPa, E_z = 8.6 GPa, G_{xy} = 4.7 GPa, G_{yz} = 3.1 GPa, G_{xz} = 4.7 GPa, v_{xy} = 0.27, v_{xz} = 0.4, v_{yz} = 0.27. According to the measurement of the optimized cantilever beam structure size, the width and thickness of the ribs are 23 mm and 10 mm, respectively. Therefore, 128 layers of 10 mm wide fiber band are necessary to manufacture the optimized cantilever beam structure.

Considering compactness, stiffness, structural lightness, and functionality principles, the conceptual design of the robotic filament winding equipment is shown in Fig. 3, which composed of four subsystems, including integrated deposition system, forming tooling, robot, and support bracket. The forming tooling is fixed and supported by the support bracket. The novel integrated deposition system with a tension system is installed on the end-effector of the robot, as detailed in paper of [19].

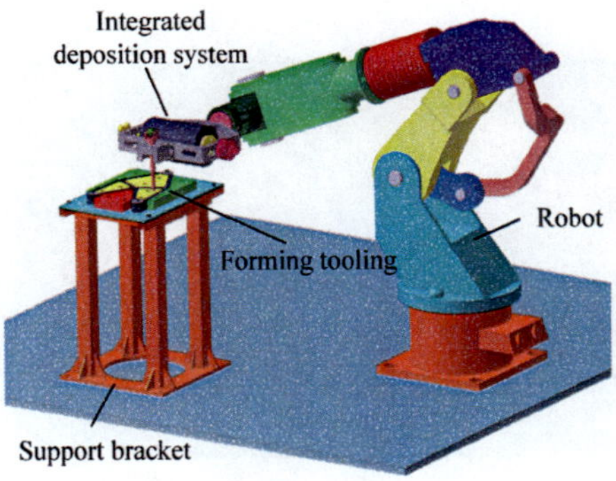

Fig. 3. Conceptual design of robotic filament winding equipment.

3.1 Design of Forming Tooling

As shown in Fig. 4, the forming tooling of the composite cantilever beam structure can be subdivided into three main elements: i) Mold: the mold is dismountable and composed of many mold parts. It is noteworthy that the mold parts (1, 3, 5) in Fig. 4 need to be removed for the convenience of the robot to perform the fiber deposition. Once the winding process is completed, these removed mold parts need to be assembled to the original position for the subsequent autoclave polymerization. All mold parts and reels can be fixed with the baseplate by bolts; ii) Reel: in order to ensure the continuity of the fiber, reels are needed at the end of the segments, so that it allows continuous turn around and uninterrupted winding operation without cutting fibers. Moreover, the reels can be used as the transition connector and can aid the composite structures to support loads; iii) Baseplate: the function is to support and fix the mold parts and reels.

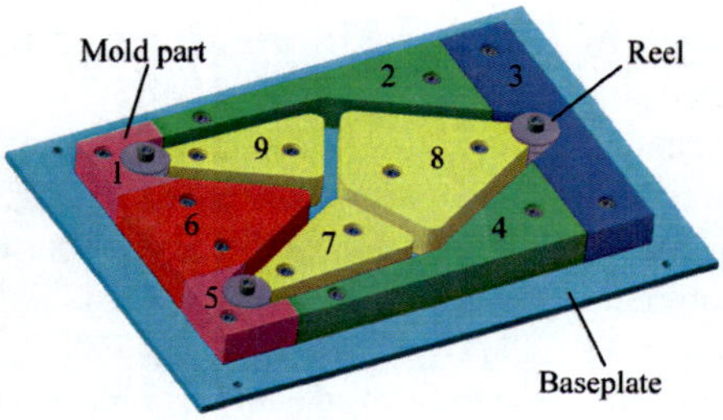

Fig. 4. Design of forming tooling for the cantilever beam.

3.2 Design of Winding Trajectory

The scheme of the winding trajectories for the composite cantilever beam structure in one cycle is shown in Fig. 5, and the filament winding steps in one cycle are presented in Table 1. The point "1" is the starting point of filament winding, and the winding direction is always oriented toward anti-clockwise. In order to keep the fibers continuous in the winding process, two layers are placed evenly in each cycle.

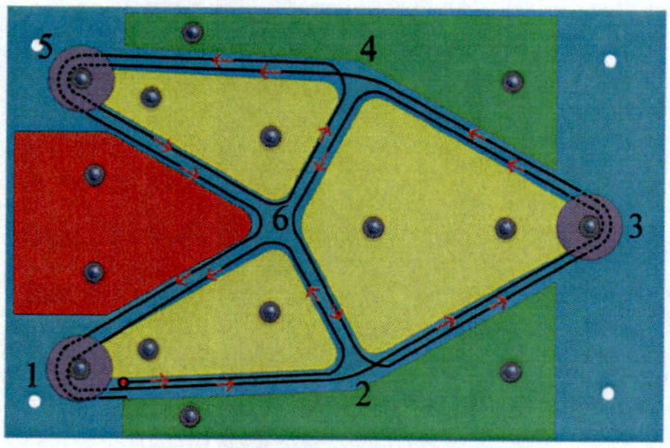

Fig. 5. The schemes of the winding trajectories for the cantilever beam.

The whole operation is cyclical and it will be repeated as many times as the number of layers to be implemented. It can be seen from Fig. 5 that most of the planned winding trajectories well match the direction of major principle stress, so as to enhance the performance of the beam structure to the largest extent.

Table 1. Filament winding steps of the cantilever beam in one cycle

Step	Trajectory	Step	Trajectory	Step	Trajectory	Step	Trajectory
1	1–2	5	2–3	9	2–3	13	6–4
2	2–6	6	3–4	10	3–4	14	4–5
3	6–1	7	4–6	11	4–5	15	5–6
4	1–2	8	6–2	12	5–6	16	6–1

4 FEM Simulation

In order to prove the effectiveness of the proposed method, the studies of FEM simulation are performed, where the anti-deforming capability is compared between optimized and un-optimized structures. The un-optimized structures with horizontal fiber and vertical fiber are shown in Fig. 6b and Fig. 6c, respectively. Since the actual volume fraction is 25% after reconstructing the topological structure, the number of layers of the un-optimized structure is 32 to meet the use of the same material weight. These un-optimized structures with fixed fiber directions can be manufactured by hand lay-up process or other available manufacturing technologies. The deformation of the three structures under the same loads and constraints are analyzed by applying the ACP module in Workbench software. The results of the FEM simulation are demonstrated in Fig. 7. It is seen that the maximum deformation value of the un-optimized structures with horizontal fiber and vertical fiber are 0.11 mm and 0.42 mm, respectively, while the optimized structure is 0.069 mm. The optimized structure of the cantilever beam can decrease 37% and 84% deformations compared with the un-optimized structure with vertical fiber and horizontal fiber, respectively. Therefore, the optimized structure has better load-bearing capability.

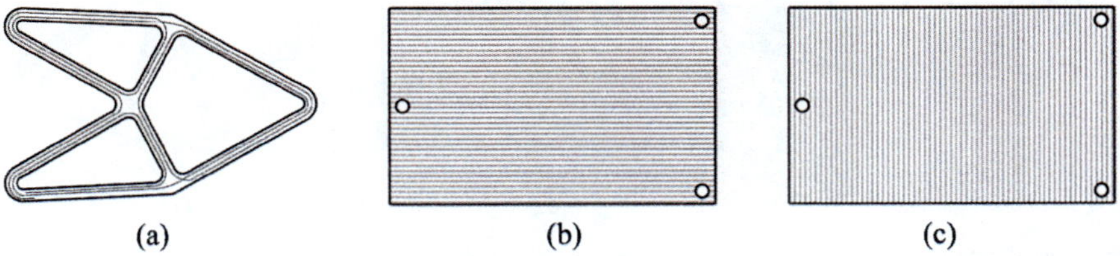

Fig. 6. The composite cantilever beam with different fiber infill patterns: (a) winding trajectory; (b) $[0°]_{32}$; (c) $[90°]_{32}$.

(a) (b) (c)

Fig. 7. The FEM simulation results of the cantilever beam, and scale factor is 50: (a) optimized structure; (b) un-optimized structures with horizontal fiber; (c) un-optimized structures with vertical fiber.

5 L-shaped beam

Without loss of generality, an L-shaped beam is also evaluated and discussed in this section. As shown in Fig. 8a, one predefined hole on the right-side is suffered concentrate force $F = 1000$ N and the upper-side edge of the beam is completely fixed. The thickness of the L-shaped beam is 23 mm and the volume fraction is set to be 20%. The design region is marked in gray color.

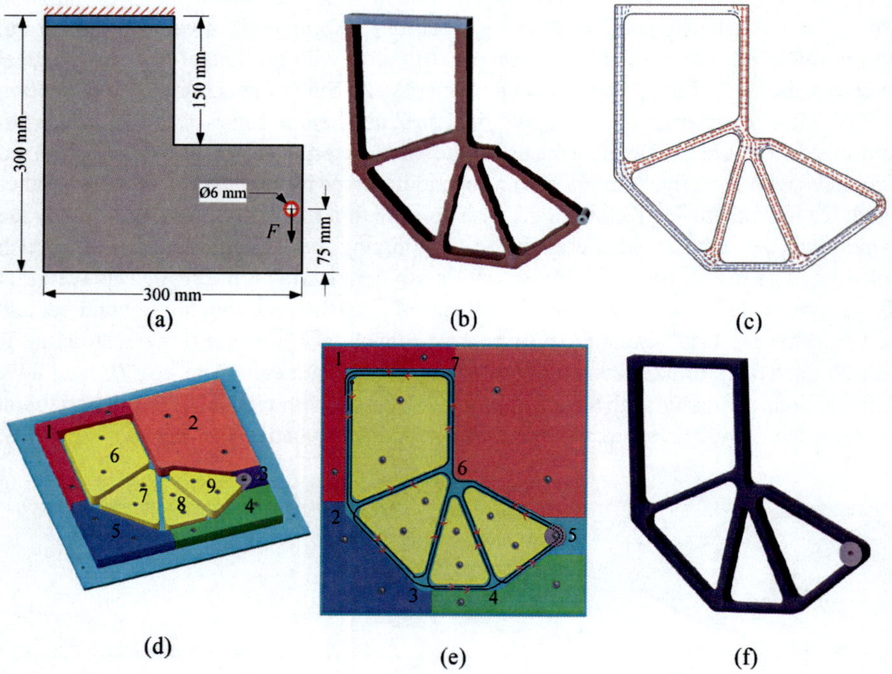

Fig. 8. L-shaped beam: (a) dimensions and boundary/load conditions; (b) optimized topology; (c) principal stress tensor; (d) design of forming tooling; (e) planning of winding trajectory; (f) 3D model of final production.

Using the similar approach as previous example, the obtained optimal topology is shown in Fig. 8b. It can be seen that the optimized structure is composed of ten ribs, and the width of the ribs is very close. Therefore, in order to manufacture the part by filament winding process, it is assumed that the width of all ribs is the same. The reconstructed geometric model and the corresponding principle stress tensor distribution are given in Fig. 8c. The direction of principle stress almost-parallel to the contour of the ribs. The design scheme of the forming tooling for the L-shaped beam is presented in Fig. 8d. During the fiber deposition process, the mold part (3) needs to be removed. The planning of winding trajectory in one cycle is depicted in Fig. 8e, and winding steps in one cycle are shown in Table 2. The whole operation is cyclical and it will be repeated as many times as the half number of layers to be implemented. After the end of fiber deposition process, the forming tooling with the composite part is placed in a vacuum bag and co-cured in an autoclave. Figure 8f shows the 3D model of final production.

Table 2. Winding steps of the L-shaped beam in one cycle

Step	Trajectory	Step	Trajectory	Step	Trajectory	Step	Trajectory
1	1–2	6	2–3	11	4–6	16	6–4
2	2–6	7	3–6	12	6–3	17	4–5
3	6–7	8	6–2	13	3–4	18	5–6
4	7–1	9	2–3	14	4–5	19	6–7
5	1–2	10	3–4	15	5–6	20	7–1

Since the actual volume fraction is 22% after reconstructing the topological structure, the number of layers of the un-optimized structure is 28 to meet the using of same material weight. The results of FEM simulation are shown in Fig. 9, and the maximum deformation value of the un-optimized structures with horizontal fiber and vertical fiber are 1.1 mm and 0.37 mm, respectively, while the optimized structure is 0.15 mm.

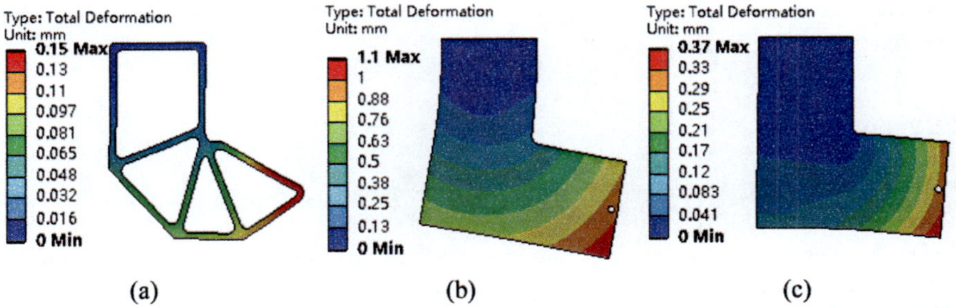

Fig. 9. The FEM simulation results of the L-shaped beam, and scale factor is 50: (a) optimized structure; (b) un-optimized structures with horizontal fiber; (c) un-optimized structures with vertical fiber.

The deformation of the optimized structure is reduced by 86% and 59%, compared with the un-optimized structure with vertical fiber and horizontal fiber, respectively. Therefore, it is found that our method still has excellent applicability in other composite beam structures.

6 Conclusion

In this work, a novel structural design method of high load-bearing composite beams based on the combination of topology optimization and filament winding trajectory was proposed. The key results are summarized as follows:

1. Topology optimization using volume-controlled based on the density method was used to obtain the optimal topological structures for isotropic materials. Based on the major principal stress direction within the optimal topological structures, the continuous fiber placement paths were defined for robotic filament winding. The FEM simulation studies of composite beam structures were performed by adopting Workbench, and the anti-deforming capability was compared between optimized and un-optimized structures. Under the condition of using the same material weight, the optimized structure of the cantilever beam can decrease 37% and 84% deformations compared with the un-optimized structure with horizontal fiber and vertical fiber, respectively, while the decrements are 86% and 59%, respectively, for L-shaped beam. The simulation results also verify the feasibility and effectiveness of the proposed optimization method.
2. The conceptual design of novel robotic filament winding equipment was developed. Moreover, the forming tooling and winding trajectory were carefully defined, so as to provide valuable references for practical production in the future. The manufacturability of the topological structures demonstrates that the proposed method can be an effective tool in the design of fiber-reinforced composite beam structures.

As a new design method, the anisotropic behavior of continuous fiber-reinforced composites has not been addressed in the topology optimization process. In order to further reduce the weight, this anisotropic behavior can be considered in further research. Moreover, the authors also intend to extend the proposed optimization method to solve 3D problems in the future.

References

1. Irving, P.E., Soutis, C.: Polymer Composites in the Aerospace Industry, 2nd edn. Woodhead Publishing, UK (2019)
2. Du, S., Guan, Z.: Strategic considerations for development of advanced composite technology for large commercial aircraft in China. Acta Mater. Compos. Sin. **25**, 1–10 (2008). https://doi.org/10.3321/j.issn:1000-3851.2008.01.001
3. Bendsoe, M.P., Sigmund, O.: Topology Optimization: Theory, Methods, and Applications. 2nd edn. Springer Science & Business Media (2013)
4. Hu, Z., Vambol, O.: Topological designing and analysis of the composite wing rib. Aerosp. Technic. Technol. (6), 4–14 (2020) https://doi.org/10.32620/aktt.2020.6.01

5. Zheng, H., Vambol, O., Sun, S.: A hybrid multilevel method for simultaneous optimization design of topology and discrete fiber orientation. Compos. Struct. **266**, 113791 (2021). https://doi.org/10.1016/j.compstruct.2021.113791
6. Hu, Z.: A review on the topology optimization of the fiber-reinforced composite structures. Aerosp. Technic. Technol. **3**(171), 54–72 (2021). https://doi.org/10.32620/aktt.2021.3.07
7. Mazumdar, S.: Composites Manufacturing: Materials, Product, and Process Engineering. CRC Press (2001). https://doi.org/10.1201/9781420041989
8. Jiang, D., Hoglund, R., Smith, D.: Continuous fiber angle topology optimization for polymer composite deposition additive manufacturing applications. Fibers **7**(2), 14 (2019). https://doi.org/10.3390/fib7020014
9. Papapetrou, V.S., Patel, C., Tamijani, A.Y.: Stiffness-based optimization framework for the topology and fiber paths of continuous fiber composites. Compos. B. Eng. **183**, 107681 (2020). https://doi.org/10.1016/j.compositesb.2019.107681
10. Chen, Y., Ye, L.: Topological design for 3D-printing of carbon fibre reinforced composite structural parts. Compos. Sci. Technol. **204**, 108644 (2021). https://doi.org/10.1016/j.compscitech.2020.108644
11. Kabir, S.M.F., Mathur, K., Seyam, A.-F.M.: A critical review on 3D printed continuous fiber-reinforced composites: history, mechanism, materials and properties. Compos. Struct. **232**, 111476 (2020). https://doi.org/10.1016/j.compstruct.2019.111476
12. Parandoush, P., Lin, D.: A review on additive manufacturing of polymer-fiber composites. Compos. Struct. **182**, 36–53 (2017). https://doi.org/10.1016/j.compstruct.2017.08.088
13. Ming, Y., Duan, Y., Wang, B., Xiao, H., Zhang, X.: 3D printing for high performance fiber reinforced polymer composites. Aeronautical Manuf. Technol. (in Chinese). **62**, 34–8 (2019)
14. Chen, X., Yao, L., Guo, L., Sun, Y.: 3D printed continuous fiber-reinforced composites: State of art and perspective. Acta Aeronaut. Astronaut. Sin. (in Chinese) (2020). https://doi.org/10.7527/S1000-6893.2020.24787
15. Giusto, G., Totaro, G., Spena, P., De Nicola, F., Di Caprio, F., Zallo, A., et al.: Composite grid structure technology for space applications. Mater. Today: Proc. **34**, 332–340 (2021). https://doi.org/10.1016/j.matpr.2020.05.754
16. Gagauz, F., Kryvenda, S., Shevtsova, M., Smovziuk, L., Taranenko, I.: Manufacturing and testing of composite wafer components with dual-purpose integrated semi-loop joints. In: Seville: 16th European Conference on Composite Materials (2014)
17. Sorrentino, L., Anamateros, E., Bellini, C., Carrino, L., Corcione, G., Leone, A., Paris, G.: Robotic filament winding: an innovative technology to manufacture complex shape structural parts. Compos. Struct. **220**, 699–707 (2019). https://doi.org/10.1016/j.compstruct.2019.04.055
18. Jianhui, F., Yun, J., Jung, Y., Lee, D.: Generation of filament-winding paths for complex axisymmetric shapes based on the principal stress field. Compos. Struct. **161**, 330–339 (2017). https://doi.org/10.1016/j.compstruct.2016.11.022
19. Carrino, L., Polini, W., Sorrentino, L.: Modular structure of a new feed-deposition head for a robotized filament winding cell. Compos Sci. Technol. **63**(15), 2255–2263 (2003). https://doi.org/10.1016/S0266-3538(03)00174-X

Tuning Methodology for Multi-circuit Digital Regulators of Robot Drives with Adjustable Dynamic Characteristics

Olena Tachinina[1(✉)], Oleksandr Lysenko[2], Kateryna Nesterenko[1], Serhii Zybin[1], and Iryna Alekseeva[2]

[1] National Aviation University, Kyiv, Ukraine
zysv@ukr.net
[2] National Technical University of Ukraine «Igor Sikorsky Kyiv Polytechnic Institute», Kyiv, Ukraine
alexirl@ukr.net

Abstract. The article is devoted to the presentation of the tuning methodology for multi-circuit digital controllers of drives of anthropomorphic robots with adjustable dynamic characteristics.

The novelty of the methodology is a comprehensive application of several known methods. The correction method allows us to build a correction algorithm that improves the dynamic properties of the drive. The Ziegler–Nichols method allows obtaining the first approximation to the optimal value of the parameters of digital proportional integral derivative regulators (DPID-regulators) of the internal and external circuits of the digital automatic control system. This first approximation is used later in the numerical optimization method. The numerical method is used for parametric optimization of DPID controllers according to the selected criterion. The method of construction of the drive condition observer allows to construct the algorithmic measuring instrument of parameters for the condition drive vector. This state observer is used to calculate the values of the mechanical component of the drive state vector for the measured electrical component. The drive condition observer avoids the need for special hardware, such as a mechanical component meter. The drive status observer provides the necessary information for correction and digital control algorithms.

In the proposed method of tuning the parameters of DPID-regulators by the Ziegler-Nichols method and the numerical method is performed using a computer mathematical model (MATLAB + Simulink). This model includes computer mathematical models (MATLAB + Simulink) of the drive and the state observer. The Ziegler-Nichols methodology is considered as a prototype methodology that is proposed to be improved.

Keywords: Numerical methods · Optimization · Digital PID-regulator · Mathematical modelling · Digital control

M. Nechyporuk et al. (Eds.): ICTM 2021, LNNS 367, pp. 874–886, 2022.
https://doi.org/10.1007/978-3-030-94259-5_67

1 Introduction

The generalized digital system of automatic control movements for the developed anthropomorphic robot (by this term in the future we will understand not only the humanoid robot, but also the humanoid exoskeleton) is hierarchical and consists of several levels [1–5].

At the 5th, highest level (level of organization) the general organization of management processes is performed. The process is carried out to control heterogeneous groups of manipulators and groups of moving devices (pediculators). That is, the organization (harmonization) of groups for heterogeneous objects of control (OC) is performed. The activities of the groups are balanced by the consumption of raw materials, energy and information resources.

At the 4th level (level of coordination) the interaction of several homogeneous or heterogeneous OC within a separate group is coordinated. The coordination of functioning processes for OC is carried out. The coordination is aimed at maintaining the nominal (estimated) performance of OC groups within a particular group.

At the 3rd level (optimization level) the optimization of the program for controlling the movements of individual OCs is performed. As a result of this optimization, the consumption of resources and energy is minimized and the efficiency of a single OC is maximized.

At the 2nd level (control level) the operation of control over the work of a separate OC is implemented. The state of OC is observed and predicted. Violation of the value of the allowed limits for changing the OC parameters is also recorded. The service personnel are provided with information about the possibility of an emergency. A decision is made to overcome the emergency situation by switching to emergency or crisis management.

At the 1st level (control level) direct control movements of drives for separate links of manipulators and pediculators is carried out. The object of control for the first level is these separate links. The control is in accordance with the information about the state vector. The state vector refers to these individual links. The control also takes place in accordance with the information about the input control action coming from the higher levels of control.

Providing anthropomorphic movements for the terminal links of an anthropomorphic robot allows us to perceive these movements as a copy of human movements.

The number of drives for a modern anthropomorphic robot (drives for joints of manipulators and pediculators, articulation and facial expressions) exceeds several hundred. The effectiveness of the methodology for tuning digital control systems for motion drives by degrees of mobility is estimated by the time interval. It is spent on tuning, provided that the specified quality of the transient process at the output of the digital control system is achieved. The digital controller controls the movement of an individual link for an anthropomorphic robot. When tuning a digital controller, the nominal load is usually considered. It is assumed that all "reasonable (rational)" deviations from this design load will be compensated for in the normal operation of the robot by a quasi-invariant (quasi-adaptive) digital system for automatic control of the movement of this link. Quasi-invariance (quasi-adaptability) is provided by the structure of the control law. Usually, a two-circuit digital automatic control system (DACS) is used to control an individual link of the robot according to one degree of mobility. Each of the loops of this system includes digital PID-regulators.

It is necessary to classify the typical degrees of mobility of typical links and provide approximately the same dynamic characteristics of these links for these degrees of mobility. In this case, we can significantly (at least twice) reduce the number of tunable digital regulators. Here we are talking about some standardized anthropomorphic robot. This is understandable for reasons of symmetry of the anthropomorphic robot. However, if we want to copy individuality (individual gait or posture and gestures, etc.), then it is necessary to give each link its own personal dynamics for each degree of mobility.

It should be especially emphasized that the state vector of a drive for any type (from electric motors to artificial muscles) consists of at least two components that have different physical nature. One component is electrical current or voltage. Another component is mechanical movement or speed or force. By measuring the electrical component of the state vector and using a mathematical model of the drive, we can calculate with a sufficient degree of accuracy the quantitative value (perform an algorithmic measurement) of the mechanical component of the state vector. We can use the measured and calculated signals in the dynamic property correction algorithm for the link drive and in the DACS internal loop for the robot link in terms of the degree of mobility. It is proposed to use either a Luenberger observer or an optimal Kalman filter type observer as such an algorithmic meter. The use of an algorithmic speed meter or coordinates of mechanical movement for the moving part of the drive allows us to get rid of a special device. It would be designed to measure this movement and thereby increase the reliability and energy efficiency of the drive control system.

The structure and parameters of the correction algorithm of dynamic properties for the drive of a link are calculated on known nominal characteristics of this drive (i.e., on its nominal LTI – mathematical model).

There are three parameters (transmission coefficients for proportional, integral and differential signals) of the PID control algorithm for the inner circuit and three similar coefficients for the outer circuit. The authors intend to adjust these parameters according to the methodology described below. Thus, the tuning of the robot link according to the degree of mobility is subject to six parameters in each DACS. Taking into account all DACS parts of the robot by the degree of mobility, the total number of configurable parameters will be equal to six multiplied by the total number of degrees of mobility, i.e., will exceed several thousand.

A simple express methodology is needed that minimizes the time to set the DACS parameters by the robot link according to the degree of mobility while ensuring the required quality of digital control. The reason is the large number of parameters that require adjustment.

The article is devoted to the presentation of the tuning methodology for multi-circuit digital controllers of drives of anthropomorphic robots with adjustable dynamic characteristics.

2 Analysis of Research and Publications

According to the classical theory of automatic control, parametric optimization (tuning) for stationary linear control systems can be performed using frequency, time and integral criteria [1, 4].

Given the large number of drives, it is desirable to be able to quickly and almost optimally adjust and reconfigure the regulators. Today, the Ziegler-Nichols method [3, 4], which is based on time criteria for the quality of operation for DACS, is most widely used to adjust DPID regulators. However, when tuning the parameters of DPID controllers of multi-circuit systems, the quality of adjustment using only this method significantly deteriorates [6–14].

The presence of a DPID regulator in each of the circuits allows us to give this circuit the properties of quasi-invariance (quasi-adaptability). This is a fundamentally important point for anthropomorphic robots, which usually operate in conditions of significant parametric and non-parametric perturbations [8–14]. A formal application of only numerical methods of multidimensional optimization usually takes a long time and significantly depends on a successful first approximation [15, 16]. For example, the application of NCD-technology of the computer mathematics system MATLAB + Simulink [15] with many parametric optimization is almost an art.

Therefore, an applied (scientific and technical) task related to the development of an effective method of preliminary (during preparation for practical use) and operational (real-time) reconfiguration of multi-circuit control systems for anthropomorphic works with digital PID-regulators is relevant.

3 Problem Statement

The challenge is to develop a methodology that is convenient for practical use in parametric optimization by scalar integrated criterion of multi-circuit control systems with digital PID-regulators.

4 Tuning Methodology

Overall, the general methodology of tuning the DACS link robot on the degree of mobility consists of the methodology of its pre-tuning and the methodology of the final tuning. The methodology of pre-tuning is the tuning on the mathematical model of the drive and the link according to the degree of mobility. The final tuning methodology is the tuning ("adjustment") on a real object. This article discusses the methodology of pre-tuning. This means that the source data for this methodology is LTI. That is, the initial data are mathematical models of drives and links. These models must be pre-structurally and parametrically identified.

As an example of the initial data, consider LTI – the mathematical model (see Fig. 1), which describes the dynamics of the drive and the links of the working body of the anthropomorphic robot. This robot is used in search and rescue operations [9–14, 16].

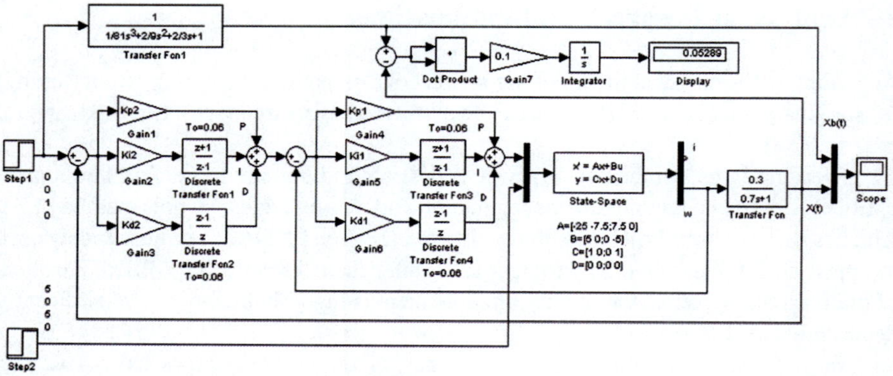

Fig. 1. The computer mathematical model of the two-circuit digital control system for the movement of the working body of an anthropomorphic robot: Kp1, Ki1, Kd1 and Kp2, Ki2, Kd2 – transmission coefficients of proportional, integral, differential signals of PID-regulators of internal (1) and external circuits (2), respectively; P, I and D – transmission channels of proportional, integral and differential signals, respectively; State Space and Transfer Fcn, respectively, computer LTI continuous mathematical models of the MIMO type for the drive and the SISO type for the link of the working body according to the degree of mobility of the anthropomorphic robot; i and w – respectively, the electric current and the speed of mechanical movement of the drive mechanism according to the degree of mobility; Step1 and Step2 – respectively, the control signal for the drive and the disturbance acting on it; Transfer Fcn1 – "standard form of Butterworth" of the third order; sampling period in time To = 0.06 s.; duration of simulation time T = 10 s.

The parameters that need to be adjusted are the transmission coefficients of DPID regulators k_{pj}, k_{ij}, k_{dj} for proportional, integral and differential signals for the first (j = 1) and second (j = 2) regulators, respectively.

As an optimality criterion to choose the average interval of observation T integral of squared deviations $\Delta_x(t)$ for the output mathematical model of the circuit working body anthropomorphic robot $x(t)$ from the reference (desired) change in time of the signal $x_b(t)$:

$$W(k_{p1}, k_{i1}, k_{d1}; k_{p2}, k_{i2}, k_{d2}) = \frac{1}{T}\int_0^T (\Delta_x(t))^2 dt, \tag{1}$$

where $\Delta_x(t) = x_b(t) - x(t)$.

As a mathematical model of the reference (desired) motion of the working body of the anthropomorphic robot, it is possible to use any of the known models [1, 3–5]. Choose, for example, the model called the "standard form of Butterworth" of the third order.

As a prototype-methodology that is proposed to improve, consider the methodology (method) Ziegler-Nichols [3, 4]. As a result of using the prototype-methodology to adjust the DPID-regulators of the internal and external circuits DACS speed for the link of the working body of the anthropomorphic robot manipulator, we obtain the following values of transmission coefficients, respectively, proportional, integral and differential signals:

Kpl = 15; Kil = 3.5; Kd1 = 7.81 (for the inner circuit) and Kp2 = 9; Ki2 = 1.8; Kd2 = 5.6 (for the outer circuit). The results of modelling the transition process, presented in Figs. 1 and 2, have showed the following.

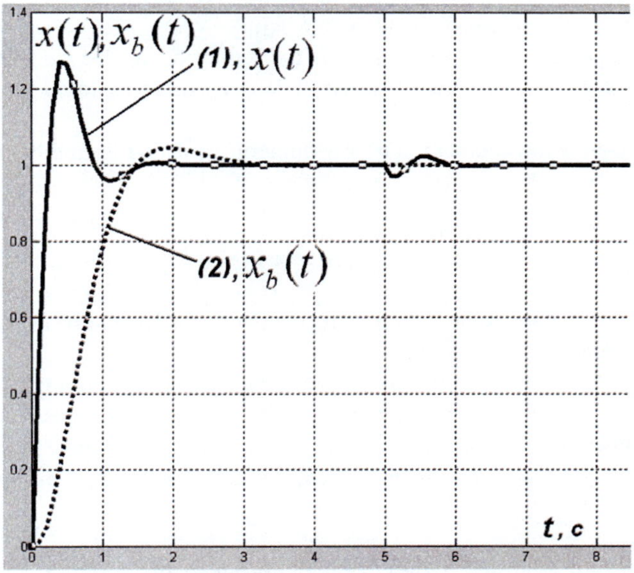

Fig. 2. The result of applying the Ziegler-Nichols method to the computer mathematical model presented in Fig. 1: the transient process and perturbation reaction (Step2 in Fig. 1 on the 5th sec.) at the DACS output of the link speed for the working body of the anthropomorphic robot manipulator by the degree of mobility (1) and its desired form (2).

The average square error *W*, which is presented on the display (Fig. 1), is 0.05289 conventional units (c.u). The transient process (Fig. 2) is oscillating with overregulation, which is almost 5 times higher than the desired value. The positive is the fact that the DACS is almost unresponsive to stepwise quasi-invariance at the drive input from the fifth second. We emphasize that DACS is quasi-invariant (quasi-adaptive) due to the use of a DPID controller in each of the two circuits.

Let us consider the methodology of tuning for multi-circuit digital regulators of robot drives with adjustable dynamic characteristics.

The methodology consists of the following stages:

- the 1st stage is the correction of the dynamic properties of the drive;
- the 2nd stage is the construction of the algorithmic meter of the drive state vector (state observer);
- the 3rd stage is the selection of the reference (desired) model of change in the time of the transient process at the output of the digital drive control system;
- the 4th stage is the application of the Ziegler-Nichols method to find the first approximation to the optimal value of the parameters for DPID-regulators of the first and

second circuits $k_{p10}, k_{i10}, k_{d10}, k_{p20}, k_{i20}, k_{d20}$. We use these values at the next stage of the methodology as initial conditions in the numerical optimization method.

The fourth stage consists of two sub-stages:

The 4th stage and the 1st sub-stage is the tuning of the first DPID only for the internal circuit and storing the values found $k_{p10}, k_{i10}, k_{d10}$.

The 4th stage and the 2nd sub-stage is the tuning of the second DPID regulator, provided that the first DPID regulator uses previously found and stored parameter values $k_{p10}, k_{i10}, k_{d10}$.

The 5th stage is the application of any numerical method to find the minimum for criterion (1):

$$W \to \min_{k_{p1}, k_{i1}, k_{d1}; k_{p2}, k_{i2}, k_{d2}}, \tag{2}$$

where, the optimal value of the criterion and control variables will be denoted as $\hat{W}$ and $\hat{k}_{p1}, \hat{k}_{i1}, \hat{k}_{d1}, \hat{k}_{p2}, \hat{k}_{i2}, \hat{k}_{d2}$.

We emphasize, first: the initial conditions for the numerical optimization algorithm are determined at the 4th stage of the methodology (the most successful first approximation and guarantees rapid convergence of the numerical optimization algorithm); secondly: already outside this methodology, additional studies of the impact on the work of the DACS are performed:

- The level quantization step studies in analog-to-digital and digital-to-analog converters are performed.
- The sensitivity of the found optimal solution to the change of the parameters of the drive and the link is being studied. These changes can be both deterministic (occur at certain points in time and at known quantities) and random.

Let us consider the result of application for the offered methodology on the example of the model (Fig. 3).

The 1st Stage. According to the known continuous MIMO LTI – the mathematical model of the drive for the working body of the anthropomorphic robot by the degree of mobility (Fig. 1) we build the discrete MIMO LTI – the mathematical model. The sampling period for time To = 0.06 s. is chosen according to the recommendations [1–4] and the c2d function of the computer mathematics system MATLAB + Simulink is applied to the continuous model [15]. The sampling period for time To = 0.06 s. is chosen according to the recommendations [1–4] and the c2d function of the computer mathematics system MATLAB + Simulink is applied to the continuous model [15].

As the result, we get

$$x(n+1) = A \cdot x(n) + Bu(n), \quad y(n) = C \cdot x(n) + D \cdot u(n),$$

where, A = [0.1841 –.2256;0.2256 0.9359]; B = [0.1504 0.04274;0.04274 –0.2928]; C = [1 0;0 1]; D = [0 0;0 0]; To = 0.06 s.

Correction of the dynamic properties of the drive is performed algorithmically. The matrix coefficient of corrective feedback is calculated and the digital feedback signal is fed to the signal channel controlling the drive (Fig. 3).

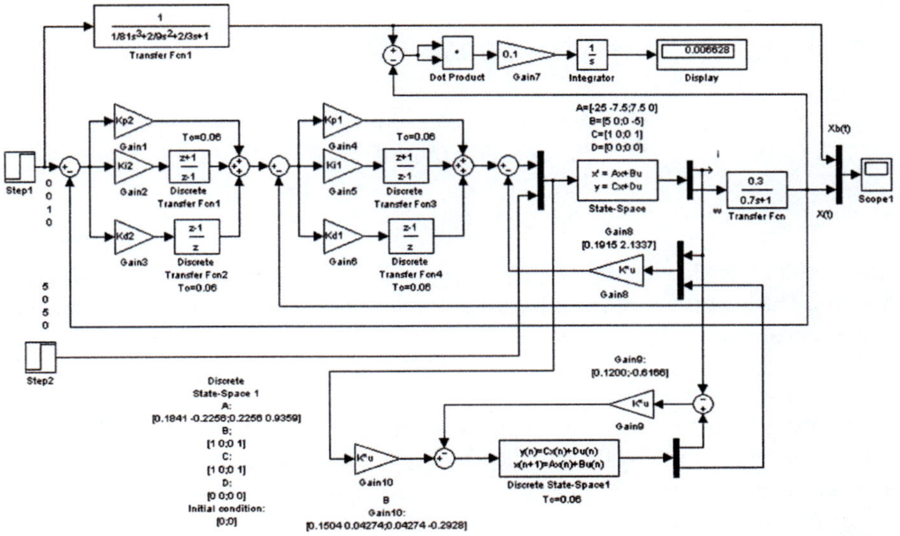

Fig. 3. The computer mathematical model of the two-circuit digital motion control system of the working body of an anthropomorphic robot with digital correction of dynamic properties (Gain8) and digital observer of the drive state vector (Gain9, Gain10, Discrete State Space1).

Changing the dynamic properties of the drive as desired corrective action is due to feedback. The dynamic properties of the drive are given mathematically by the roots of the characteristic equation for the mathematical model of the drive. Assume that both roots of the characteristic equation for the mathematical model of the adjusted drive should be equal to 0.5.

Let us calculate the matrix of corrective feedback coefficients in the macro calculator mode using MATLAB functions [17]:

$$\mathrm{B1} = [0.1504; 0.04274]; \mathrm{K05} = \mathrm{acker}(\mathrm{A}, \mathrm{B1}, [0.5\;0.5]), \mathrm{K05} = 0.1915\;2.1337,$$

where K05 – is the matrix of corrective feedback coefficients, which is implemented using a matrix amplifier Gain8 in a computer mathematical model (Fig. 3).

The 2nd Stage. Assume that the Luenberger observer is chosen as the algorithmic meter of the drive state vector (drive state observer) [1, 3, 4].

The input for the Luenberger digital observer is the digital signal from the output of a digital current sensor. Mathematically, this means that the matrix C in the original mathematical model of the drive takes the form C = [1 0].

Let's calculate the matrix of Luenberger coefficients, which ensure the convergence of the calculation results to the true values with the improvement of the result twice in each iteration (i.e. set the roots of the characteristic equation of the Luenberger observer to be equal to 0.5). The calculation is performed in macro calculator mode using MATLAB functions [15]:

$$
\begin{aligned}
&\mathrm{A} = [0.1841 - 0.2256; 0.2256\, 0.9359];\\
&\mathrm{C} = [1\, 0];\\
&\mathrm{HT} = \mathrm{acker}(A', C', [0.5\, 0.5])\\
&\mathrm{HT} = 0.1200 - 0.6166\mathrm{H05} = HT'\\
&\mathrm{H05} = HT'\\
&\mathrm{H05} = 0.1200\ - 0.6166,
\end{aligned}
$$

where H05 – is the matrix of gain used in the construction of the Luenberger observer. This matrix is implemented using the Gain9 matrix amplifier in the computer mathematical model (Fig. 3).

The 3rd Stage
As the model of the reference (desired) change in time of the signal at the DACS output by the speed of the working body of the anthropomorphic robot according to the degree of mobility, we choose the signal acting at the output of the model of the standard Butterworth form of the third order.

The 4th Stage
The 1st sub-stage.
As the result of applying the Ziegler-Nichols method [3, 4], the following values of the parameters of the first DPID-regulator for the internal circuit were obtained (Fig. 3): Kp1 = 12; Ki1 = 2.8425; Kd1 = 6.3325, that is $k_{p10} = 12, k_{i10} = 2.8425, k_{d10} = 6.3325$.

The 2nd sub-stage.

As the result of applying the Ziegler-Nichols method [3, 4], the following values of the parameters of the second DPID-regulator for the external circuit were obtained (Fig. 3): Kp2 = 6; Ki2 = 0.2; Kd2 = 10, that is $k_{p20} = 6, k_{i20} = 0.2, k_{d20} = 10$.

Let us substitute the calculated values of parameters for the first and second DPID-regulators to the simulation scheme (Fig. 3). As the result of computer simulation we get the transition process (see Fig. 4) and the following values of the criterion $W_0 = W(k_{p10}, k_{i10}, k_{d10}; k_{p20}, k_{i20}, k_{d20}) = 0.007846\ c.u.$

The 5th Stage. To finally tuning (optimize) the parameters of DPID-regulators (Fig. 3) it is necessary to solve the task (2) using the numerical method of zero order. This task is called the Hooke-Jeeves method [8, 15, 16]. Recall that as the initial conditions in the Hooke-Jeeves method, we use the values of the parameters of DPID-regulators, which were calculated during the fourth stage of the methodology. As the result, we obtain the optimal parameters: the transient process at the CSAC output (see Fig. 5); the point in the six-dimensional space of the parameters that were configured $\hat{k}_{p1} = 5, \hat{k}_{i1} = 1.8425, \hat{k}_{d1} = 4.3325; \hat{k}_{p2} = 5, \hat{k}_{i2} = 0.18, \hat{k}_{d2} = 3$; and the optimal value of the criterion at this point $\hat{W}(\hat{k}_{p1}, \hat{k}_{i1}, \hat{k}_{d1}; \hat{k}_{p2}, \hat{k}_{i2}, \hat{k}_{d2}) = 0.006628$ (see Fig. 3).

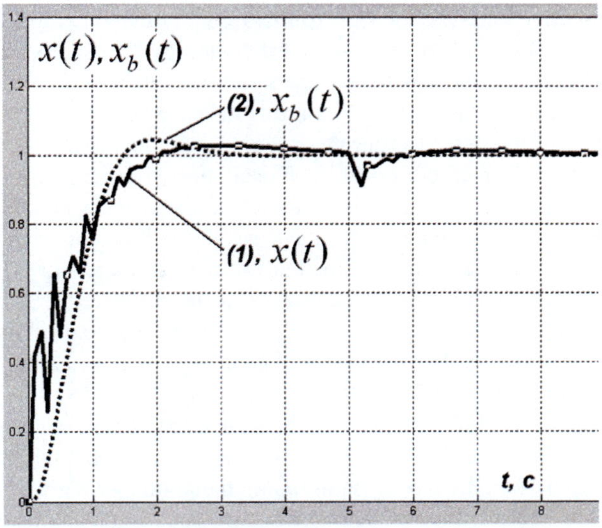

Fig. 4. The result of applying the Ziegler-Nichols method to the computer mathematical model presented in Fig. 3: the transient process and reaction to perturbation (Step2 in Fig. 3 on the 5th sec.) at the DACS output of the link speed for the working body of the anthropomorphic robot manipulator by the degree of mobility (1) and its desired form (2).

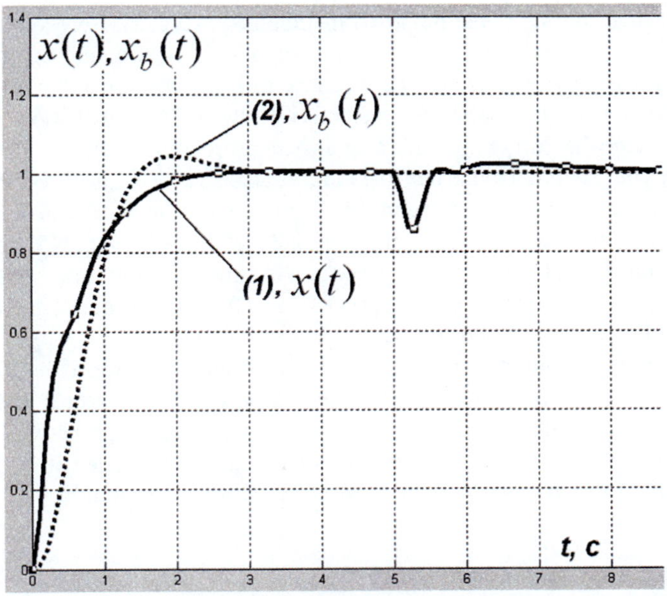

Fig. 5. The application result of the proposed method of tuning multi-circuit digital controllers for robot drives with adjustable dynamic characteristics: the transient process and perturbation reaction (Step2 in Fig. 3 on the 5th sec.) at the DACS output of the link speed for the working body of the anthropomorphic robot manipulator by the degree of mobility (1) and its desired form (2).

The comparative analysis of the application results for the Ziegler-Nichols method (methodology) and the proposed methodology. The methodology proposed in the article makes it possible to obtain an aperiodic transient process close to the desired one at the DACS output (Fig. 5). DACS controls the speed of movement of the link for the working body of the anthropomorphic robot by one degree of mobility. Applying only the Ziegler-Nichols method results in an oscillatory transient with large overshoot (Fig. 2). The improvement of the transient process has been achieved due to algorithmic correction of dynamic characteristics for the drive and refinement of values (optimization by criterion (1) using a numerical method) of the parameters of DPID-regulators. The simulation experiment has been performed for the comparative assessment of the properties of quasi-invariance (quasi-adaptability) of digital controllers tuned by the Ziegler-Nichols method and by the proposed method. The essence of the experiment has been as follows. When simulating in various computer models (Fig. 1 and Fig. 3) at the input of the drive at the 5th second, the same type of step disturbance has been applied (Step2 in Fig. 1 and Fig. 3). Quasi-invariance (quasi-adaptability) is achieved when applying each of the DPID-regulator tuning methodology. This is understandable for the reason that the property of quasi-invariance (quasi-adaptability) is due to the structure of the regulator (the use of DPID-regulators in both circuits). However, the quantitative indicators of the effect of quasi-invariance (quasi-adaptability) are different. The amplitude of the deviation from the initial stationary state and the time interval for returning to the stationary state, which has been tuned only by the Ziegler-Nichols method (Fig. 2), is approximately two times less in DACS, compared to DACS, which has been tuned according to the proposed methodology (Fig. 5). However, the integral indicator (1) with averaging over the time interval of 10 s. (taking into account both the nature of the transient process and the response to a stepwise disturbance) for DACS with tuning according to the proposed method is $W/\hat{W} = 0.05289/0.006628 = 7.9753$ times better than when using only the Ziegler-Nichols method.

We draw your attention to the fact that the combined use of the method for correcting the dynamic properties of the drive and the Ziegler-Nichols method leads to a decrease in the root mean square error by $W/W_0 = 0.05289/0.007841 = 6.7453$ times. However, at the same time, in the transient process, a trimmer effect occurs (high-frequency oscillations, Fig. 4). It is possible to get rid of these fluctuations due to the application of the method of numerical optimization of the criterion (1).

The use of an algorithmic observer of the drive state vector (observer Luenberger) allows calculating the mechanical coordinate of the drive state vector and using the calculated value both in the algorithm for correcting the dynamic properties of the drive and in the DPID-regulator of the internal circuit DACS with the speed of the link for the working body of the anthropomorphic robot in one degree of mobility.

5 Conclusions

The novelty of the methodology is a comprehensive application of following methods: correction of the dynamic properties of the drive; analytical measurement for the mechanical component of the drive state vector using the Luenberger observer algorithm; Ziegler-Nichols and zero-order numerical optimization. The Ziegler-Nichols method makes it possible to obtain initial conditions that provide a rapid approximation to the optimum point with the subsequent application of numerical zero-order methods. If the initial conditions are set randomly, it leads, in most cases, to the instability of the DACS and the inability to algorithmically calculate the value of the optimization criterion.

The calculations have confirmed the high efficiency of the proposed method compared to the case of using only the Ziegler-Nichols method. In the example that has been studied, the mean square error of the deviation has decreased by almost 8 times for the transient at the output of the digital control system. The system controls the speed of movement of the link for the working body of the anthropomorphic robot by one degree of mobility (efficiency indicator (criterion), according to which the quality of tuning has been evaluated).

The timing of all actions related to the application of the proposed methodology has showed that the tuning time of both DPID controllers increases by no more than 5–10% compared to the tuning time only by the Ziegler-Nichols method.

The application of the proposed methodology allows to achieve smooth control of movements according to the degrees of mobility of the robot, which ensure its anthropomorphism.

References

1. Repnikova, N.: Teoriya avtomatychnoho keruvannya: klasyka i suchasnist [Theory of automatic control: classics and modernity]. NTUU "KPI", Kiev (2011)
2. Kargin, A.: Vvedeniye v intellektual'nyye mashiny. Kniga 1. Intellektual'nyye regulyatory [An introduction to intelligent machines. Book 1. Intelligent Regulators]. Nord Press, Donetsk (2010)
3. Dorf, R., Bishop, R.: Sovremennyye sistemy upravleniya [Modern Control Systems]. Laboratory of Basic Knowledge Moscow (2002)
4. Goodwin, G., Grebe, S., Salgado, M.: Proyektirovaniye sistem upravleniya [Design Control Systems]. Laboratory of Knowledge, Moscow (2004)
5. Yakhno, O., Uzunov, O., Luhovskyi, A.: Vvedeniye v mekhatroniku [Introduction to mechatronics]. NTUU "KPI", Kiev (2011)
6. Lysenko, O., Tachinina, O.: Method of path constructing of information robot on the basis of unmanned aerial vehicle. Proceedings of the National Aviation University **4**(73), 60–68 (2017)
7. Lysenko, O., Alekseeva, I.: Method of application of numerical methods of zero order in problems of parametric optimization of digital PID controllers of drones. Mathematics in Modern Technical University **2**, 64–72 (2019)
8. Briskin, E., Kalinin, Ya., Artemyev, K.: Ob ustoychivosti ploskogo dvizheniya mobil'nykh robotov s shagayushchimi dvizhitelyami, rabotayushchimi v "tyanushchem" rezhime [On the stability of the plane movement of mobile robots with walking propulsion devices working in "pulling" mode], Mekhatronika, avtomatizatsiya, upravlenie **22**(1), 28–34 (2021)
9. Lavrovsky, E., Pismennaya, E.: Control of regular walking for an exoskeleton with the electric drive. Mekhatronika, Avtomatizatsiya, Upravlenie **19**(3), 160–168 (2018)
10. Rud, V.: Robotychne khapannya z unyknennyam pereshkod za dopomohoyu derev oktantiv [Robotic grasping with the avoidance of obstacles with the help of octane trees]. Artif. Intell. **25**(3), 7–12 (2020)
11. Mashchenko, S., Zadnipryanny, O.: Walking robot system. Artif. Intell. **17**(1), 243–249 (2012)
12. Koubaa, A.: Robot operating system. Springer International Publishing Switzerland (2019)

13. Fairchild, C., Harman, T.: ROS Robotics By Example - Second Edition: Learning. to control wheeled, limbed, and flying robots using ROS Kinetic Kame. Packt Publishing (2017)
14. Larionov, Y., Levikin, V., Khazhmuradov, M.: Doslidzhennya operatsiy v informatsiynykh systemakh [Research of operations in information systems]
15. Dyakonov, V., Kruglov, V.: MATLAB. Analysis, identification and modeling of systems. SPb, Peter (2002)
16. Romanchenko, I., Lysenko, O., Chumachenko, S., et al.: Models of application of information and telecommunication technologies on the basis of unmanned aerial vehicles in emergencies. NAU, Kiev (2016)

Determination of the Parameters of a Conceptual Quadcopter Model for a Group Outdoor Flights

Olha Pohudina[1] (✉), Tetiana Plastun[1], Ihor Biletskyi[2], and Igor Vasiliev[1]

[1] National Aerospace University "Kharkiv Aviation Institute", 17 Chkalova Street, Kharkiv 61070, Ukraine
olga.pogudina@khai.edu

[2] O. M. Beketov National University of Urban Economy in Kharkiv, 17 Marshala Bazhanova Street, Kharkiv 61002, Ukraine

Abstract. The article analyzes the Ukrainian companies which sale, manufacture, design of UAVs, and develop software for group flight of quadrocopters. Demonstrated the classification of UAVs intended for group flights based on existing and popular models according to five criteria. A special prototype was created based on the budget model. Reviewed a service which facilitates the hardware design process, due to the possibility of choosing the electronics of the quadrocopter, weight, frame size, propellers and additional equipment. The prototype of the quadrocopter is designed with the help of the service, selecting individually all the parameters and equipment given the purpose and climate for the operation of the quadrocopter. The calculated values, received configuration and flight characteristics are presented, a graph was constructed with optimum values: flight time, speed and distance.

Keywords: UAV · Group flight · Flight characteristics · UAV configuration · Hardware

1 Introduction

There are several clusters in the world focused on the development of unmanned aerial vehicles (UAVs) and related technologies: MassRobotics [1, 2]; Pittsburgh Robotics Network [3]; Odence Robotics [4]; Robotdalen [5]; COBOTEAM [6], Xiaoshan Robot Town [7]. The Massachusetts Robotics Cluster includes 122 commercial companies operating at various stages of product development. The main element of the cluster is powerful institutions of higher education, proposing concepts and conducting experimental prototyping of UAVs.

Recently Ukraine has also been active in the development and operation of UAVs. More than 200 enterprises are engaged in the development of UAVs (Fig. 1). At the XII International Aviation and Space Salon "AviaWorld-XXI" a record number for Ukraine of UAVs of various types and purposes was presented [8].

M. Nechyporuk et al. (Eds.): ICTM 2021, LNNS 367, pp. 887–893, 2022.
https://doi.org/10.1007/978-3-030-94259-5_68

Fig. 1. The main companies that manufacture and operate UAVs in Ukraine.

Representatives of universities, research institutes, enterprises, manufacturers, and users usually have cooperation agreements, but there is no single cluster. Nevertheless, the represented organizations occupy their positions both at various stages of the UAVs development life cycle and in the corresponding industry solutions.

The paper will consider the industry of performing group flights to organize UAVs shows. In Ukraine today there are two organizations that operate foreign UAVs: dronarium [9], skypix_droneshows [10]. Ukrainian manufacturers of UAVs for performing group flights are not represented and there are also no software developers for such tasks on the market. Therefore, it is relevant to calculate the conceptual model of UAV for the group flight, compare its parameters with the presented analogs and assess the possibility of establishing the production of such UAVs in Ukraine, taking into account the use of existing software systems.

2 Classification of UAVs Used in Group Flights on the Street

UAV formation has been studied and applied worldwide. Intel held a performance of 500 copters formations in November 2016, breaking the Guinness World Record at the time. On April 29, 2018, EHang performed a show in Xi'an consisting of 1,374 UAVs, breaking Intel's record. On September 19, 2019, this record was broken by Gao Ju innovation, with the number of UAVs up to 2,100 [11].

The UAV show can be held in venues with sufficient space. The minimum defined space is 150 × 100 m for 100 UAVs, no people can be in this space. The length of a UAV show is normally up to 10 min of flight. Therefore, for the successful execution of a group autonomous flight by a group of multicopters it is necessary to take into account: the density of their location, the accuracy of positioning of each UAV, the possibility and quality of communication with the control panel. If the flights will be performed outdoors, the UAVs' resistance to wind gusts, turbulence and other factors are important. Table 1 discusses the characteristics of known prototype quadcopters used for group flight. The cheapest model is the CLOVER from COEX. Since a large number of quadcopters are needed to create a group, this option will be considered.

Table 1. Prototypes of quadcopters used for group flight.

Model name	Weight, grams	Brightness	Longest battery life, min	Positioning technology	Networking communication (WiFi)
IFO [13, 14]	786 (without battery)	huge LED light	25	GPS RTK	2.4 GHz and 5.8 GHz range
EMO [12]	540	12 lamp beads	35	GPS RTK	5.8 GHz
CLOVER [15]	1000	LED strip	15	GPS	2.4 GHz

3 Determination of UAV Parameters for a Group Flight on the Street

When developing the quadcopter prototype based on the COEX CLOVER model, we took into account the data available in the public domain on the representative's website. Based on the obtained data and using the xcopterCalc service [16], it becomes possible to perform basic calculations necessary for designing your own copter, and get close to real results regarding flight time, compatibility, and correctness of selected components, thereby varying the parameters to obtain optimal values for the selected design.

The service has a wide range of settings and a large base of electronic parts. The main settings you need to specify for the calculation (Fig. 2):

- weight of the model (you can specify without the propulsion group (PPG) When selecting the PPG, the calculator itself takes its weight into account. Taking into account the maximum takeoff weight of 1000 g indicated in the prototype, let's consider a model with a weight of 1200 g, taking the PPG into account;
- the number of motors – for our model 4, as in the prototype, for clarity, the service immediately draws the resulting configuration;
- frame dimensions – the diagonal distance between the axes of the motors, the size is 450 mm;
- battery cell – from the list of available ones in the database, select the required, in our case, the most similar battery "LiPo 4500 mAh 50C" and the state of charge "Nominal", in the "P" field write the number of parallel-connected batteries (in the case of using a circuit with several batteries);
- electronic speed controller (ESC) – we choose from the list a speed regulator with approximate parameters – max 60A;

- in the attachments, we write the total consumption and weight of all additional equipment, which is planned to be installed, for example: camera, lantern, etc. (attachments are not taken into account);
- motor – we choose the motor SunnySky V4006–380;
- propeller – we choose the type of propeller from the list by brand, take DJI, they have a good performance, 254 mm diameter, with three blades and a pitch of 127 mm.

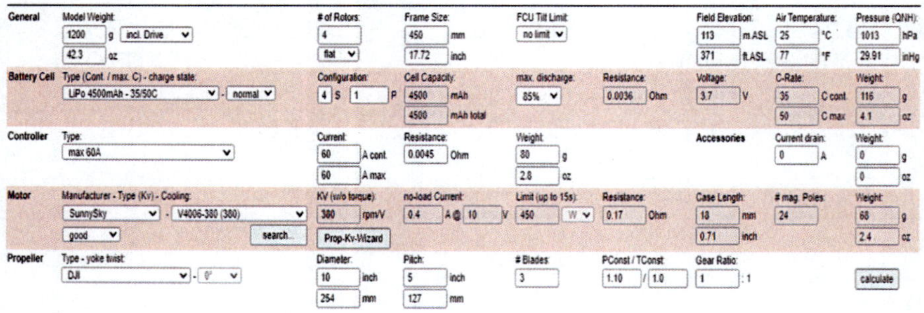

Fig. 2. Value setting.

In Fig. 3. the results and configuration of the multicopter are displayed. It is important to understand that the calculation is approximate and does not give guaranteed results, but at the same time, it well reflects how the flight characteristics change when the components and other parameters change. In Fig. 3 shows – the efficiency of the motor at optimal, maximum operating modes and in the hovering mode, the efficiency of the PPG in hovering and at maximum operation; payload, maximum tilt, rate of climb, and probable time with flight range.

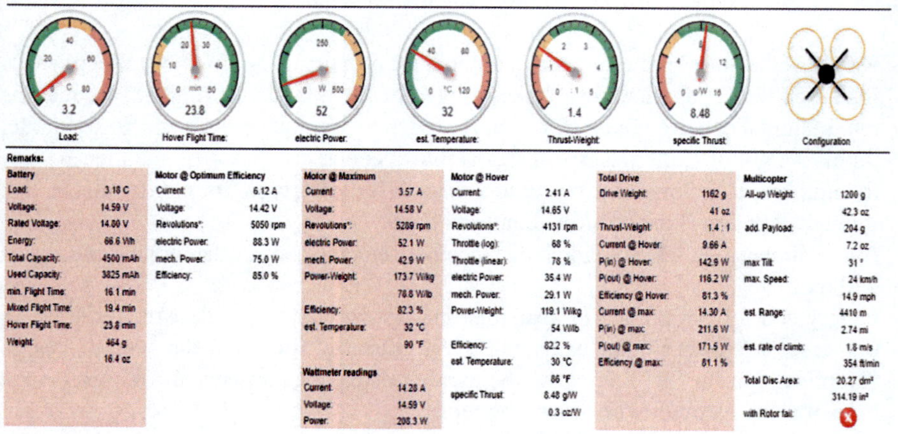

Fig. 3. Calculation results.

Consider the results of the range calculation using the graph shown in Fig. 4. The left side of the graph shows the flight time (in minutes), the bottom – the speed of the copter during the flight (in kilometers per hour), the right one – the range (in meters), the best range of the obtained flight time and range data is shown in green. The time and range are presented on the graph in two versions depending on climatic conditions, namely the air resistance. One of the most important simulated parameters is the range of distance over time depending on the speed of the UAV and air resistance. On the graph we see that the optimal speed is from 12.5 km/h to 22.5 km/h. When flying without air resistance in 25 min at a flight speed of 22.5 km/h, the UAV covers a distance of about 8 km. For the objectivity of the results, consider the flight with air resistance. At a speed of 12.5 km/h, the battery is discharged in 20 min after a distance of about 6.4 km, and at a speed of 65 km/h the battery is discharged in less than 5 min, overcoming a distance of 1.5 km.

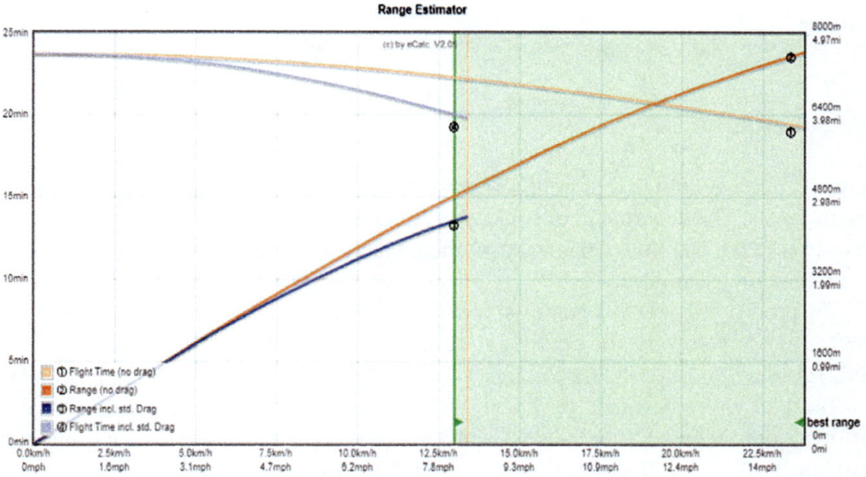

Fig. 4. Simulated flight time and distance graph for the survey copter build.

The dependence of the engine-propeller combination characteristics on the motor current is shown in Fig. 5. The total power is marked with a circle on each characteristic curve. Based on the graph, we see that the electric power limit of the motor can be even higher. Maximum flight time, calculated in Fig. 3 (23.8 min), calculated for the planned situation, when the battery is discharged by 85%. In real life, the flight time to avoid complete discharge of the battery will be no more than 19 min, ie discharge of a maximum of 80%. That is enough for a group flight of UAVs.

Given the fact that the service accuracy is ±15% and the UAV doesn't maintain a constant speed, the acquired data would differ a little bit from the real life scenario.

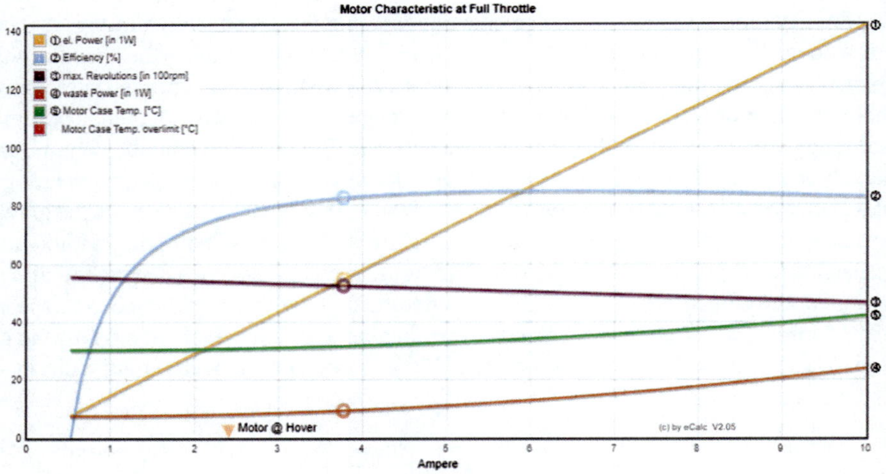

Fig. 5. Motor characteristics at full throttle

4 Conclusion

In this paper, we offered a thorough analysis of Ukrainian companies that produce and design UAVs. The department of KhAI also performs development of UAVs of different types [17, 18]. It was noted that at the moment there are no developers engaged in UAVs for organizing group flights.

The prototypes of UAVs that exist in the copter show marketwere considered. Using the xcopterCalc service, the configuration and basic flight characteristics were generated. Although most of the performance values are approximate, they clearly show the feasibility of UAVs designed for the group flight. In the future, we will further analyze various additional aspects of UAVs design of this type and the possibility of using modern known software [19] to form a group flight for the selected structural decomposition of the UAVs.

References

1. Theobald, D., Heger, F.: The safety-to-autonomy curve: An incremental approach to introducing automation to the workforce. In: Bostelman, R., Messina, E. (eds.) STP1594-EB Autonomous Industrial Vehicles: From the Laboratory to the Factory Floor, pp. 82–90. ASTM International, Conshohocken (2016). https://doi.org/10.1520/STP159420150053
2. Eller, L., et al.: Advanced autonomy on a low-cost educational drone platform. In: 2019 IEEE/RSJ International Conference on Intelligent Robots and Systems (IROS), pp. 1032–1039. IEEE, Macau (2019). https://doi.org/10.1109/IROS40897.2019.8967595
3. Kolling, A., Walker, P., Chakraborty, N., et al.: Human interaction with robot swarms: a survey. IEEE Trans. Hum. Mach. Sys. **46**(1), 9–26 (2015). https://doi.org/10.1109/THMS.2015.2480801
4. Lamberty, J., Nevers, J.: The entrepreneurial state in action: The Danish robotics cluster and the role of the public sector. Enterp. Soc. 1–31 (2020). https://doi.org/10.1017/eso.2020.36

5. Höglund, L., Linton, G.: Smart specialization in regional innovation systems: a quadruple helix perspective. R&D Management **48**(1), 60–72 (2018). https://doi.org/10.1111/radm.12306
6. COBOTEAM Auvergne-Rhône-Alpes. https://www.coboteam.fr/. Accessed 5 Oct 2021
7. Karahan, Ç., Canberk, B.: An intelligent 3D placement methodology for drone networks. In: 2021 IEEE 18th Annual Consumer Communications & Networking Conference (CCNC), pp. 1–6. IEEE, Las Vegas (2021). https://doi.org/10.1109/CCNC49032.2021.9369483
8. IEC: Report on the results of XII International Aviation and Space Salon Aviasvit-XXI. https://www.iec-expo.com.ua/en/aviaen-2021.html. Accessed 5 Oct 2021
9. DRONARIUM. https://www.dronarium.com.ua/. Accessed 5 Oct 2021
10. Skypix drone shows. https://www.facebook.com/Skypixdroneshows/. Accessed 5 Oct 2021
11. Sun, H., Qi, J., Wu, C., Wang, M.: Path planning for dense drone formation based on modified artificial potential fields. In: 2020 39th Chinese Control Conference (CCC), pp. 4658–4664. IEEE, Shenyang (2020). https://doi.org/10.23919/CCC50068.2020.9189345
12. High Great: EMO. https://www.lightshow-hg.com/customer-level-product/emo-light-show-drone.html. Accessed 5 Oct 2021
13. UVi-fy: IFO – Swarm light show drone. https://www.uvify.com/ifo/. Accessed 5 Oct 2021
14. Drone Rush: DJI Mavic Air 2 vs UVify IFO-S. https://dronerush.com/dji-mavic-air-2-vs-uvify-ifo-s-20863/. Accessed 5 Oct 2021
15. Bogatov, N., Kostin, A., Maiorov, N.: Control and analysis of quadcopter flight when setting a complex trajectory of motion. J. Phys. Conf. Ser. 1925(1), 012043 (2021). https://doi.org/10.1088/1742-6596/1925/1/012043
16. XcopterCalc: Multicopter Calculator. https://www.ecalc.ch/index.htm. Accessed 5 Oct 2021
17. Kritsky, D.N., Ovsiannik, V.M., Pogudina, O.K., Shevel, V.V., Druzhinin, E.A.: Model for Intercepting Targets by the Unmanned Aerial Vehicle. In: Palagin, A., Anisimov, A., Morozov, A., Shkarlet, S. (eds.) MODS 2019. AISC, vol. 1019, pp. 197–206. Springer, Cham (2020). https://doi.org/10.1007/978-3-030-25741-5_20
18. Yeremiyev, M.B., Karatanov, O.V., Kritsky, D.N., Pogudina, O.K.: Automated system of the variable-pitch propeller rotation control. In: 2019 IEEE 14th International Conference on Computer Sciences and Information Technologies (CSIT), Vol. 2, pp. 186–191. IEEE, Lviv (2019). https://doi.org/10.1109/STC-CSIT.2019.8929757
19. Drone Show Software. https://droneshowsoftware.com/. Accessed 5 Oct 2021

The Method of Flight Mission Formation for a Group Autonomous Flight of Unmanned Aerial Vehicles

Olha Pohudina(✉), Andrii Bykov, Dmitriy Kritskiy, and Mykhailo Kovalevskyi

National Aerospace University "Kharkiv Aviation Institute", 17 Chkalova Street, Kharkiv 61070, Ukraine
olga.pogudina@khai.edu

Abstract. The object of the research is the process of planning a group flight of unmanned aerial vehicles (UAVs). The aim is to reduce the time needed to form the flight trajectories of the UAV group and increase the quality of the mission by control the keeping of the distance and flight time on a given hardware platform. The following tools were used to organize an autonomous group flight: flight programming libraries for indoor UAVs, flight visualization libraries for outdoor UAVs. The existing tools for modeling and flight control are considered. The results obtained are: classification of UAV group flight based on the existing classification of UAV autonomy levels; a method of preparing a flight mission for a group autonomous synchronized flight and a flight along a predetermined trajectory of each member of the group. Scientific novelty: for the first time, a method for planning a group flight based on existing modeling tools has been proposed, which allows choosing a sufficient set of software for the group's tasks.

Keywords: Unmanned aerial vehicle · Outdoor group flight · Indoor group flight · Animation · SDK · Add-ons

1 Introduction

Today, manufacturers of unmanned aerial vehicles (UAVs) offer affordable solutions for the implementation of the following group flight missions: collecting high-resolution images [1], exploring hard-to-reach areas or exploring remote areas [2], delivering goods [3], organizing shows [4], etc. In addition, most of the UAVs on the market perform a number of tasks autonomously [5], for example: return to the launch point, flight stabilization, flight along a given trajectory, etc.

For the organization of an autonomous group flight, the following are offered:

- libraries for UAV programming. For example, Tello EDU contains a built-in application programming interface (API) for flight programming [6];
- software products for animation and modeling at the stage of pre-flight preparation. Cinema4D, Blender3d [7], Unity are used for animation, SITL Simulator is used for flight simulation [8];

M. Nechyporuk et al. (Eds.): ICTM 2021, LNNS 367, pp. 894–901, 2022.
https://doi.org/10.1007/978-3-030-94259-5_69

- software products for UAV control during the flight. One of the most common products is Drone Show Software [9] from SPH Engineering.

The available and open methods of formalizing the UAV flight in the group are not presented, therefore, it is relevant to consider the known instruments and develop methods for the autonomous group flight.

2 Classification of UAV Autonomy Levels and Types of UAV Group Flight

There are several approaches to the classification of UAV autonomy. One of the best ways to understand the autonomy of UAVs is the spectrum (Fig. 1), developed by Drone Industry Insights [10]. If autonomy is a measure of independence from external influence and the level of self-government, then different platforms may be autonomous, but at a different point in the spectrum of autonomy. The key thing is an understanding of these different levels and how they look in practice in the un-manned industry. Consider the levels of autonomy.

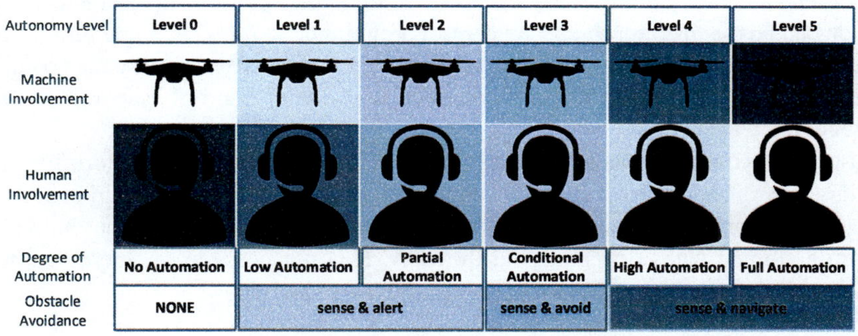

Fig. 1. The 6 Levels of Drone Autonomy

Level 0: no automation. The operator fully controls every movement of the UAV. Practical use: UAVs for racing and model airplanes.

Level 1: low automation. The pilot remains in control of the overall operation and safety of the UAV.

Level 2: partial automation. The pilot is still responsible for the safe operation of the UAV and must be prepared to take control of the UAV if something happens. However, under certain conditions, the UAV can take control regarding course, altitude and speed. Currently, most manufacturers create UAVs at this level.

Level 3: conditional automation. As at level two, the UAV can fly by itself, but the operator must still pay attention and be ready to take control at any time. The UAV notifies the operator of the need for intervention.

Level 4: high automation. The UAV can be operated by an operator, but this is not always required. The UAV is expected to have backup systems. Its behavior depends on a fixed built-in functionality or on a fixed set of rules that dictate the behavior of the system.

Level 5: full automation. The UAV controls itself under any circumstances, without waiting for human intervention. This includes full automation of all flight tasks under any conditions. There are no examples of such UAVs yet, and they are not currently being produced. However, they are expected to be able to use AI tools to plan their flights.

Starting from the second level of automation, it is possible to use UAVs for group flights. A group flight is a flight involving several UAVs, which act as a single UAV regarding navigation and location notifications, as well as obtaining permits from air traffic control authorities. The standard UAV group performs a common task in which each element of the group is located at a distance of not more than 1.85 km horizontally and 30 m vertically from the leading UAV. A non-standard group is a group that flies outside the standard group. UAV group flight can be performed outdoor and indoor. In addition, we consider the classification of group flight levels in accordance with the above classification of UAV autonomy.

Level 0 – controlled flight of each unit. Each UAV is operated by a separate pilot, but there is a joint mission and location. Examples of such groups are races and shows.

Level 1 – synchronous flight. All UAVs in the group perform the same task. One person can be the operator. A special control program or script is developed for a small group of UAVs, which will simultaneously send the commands necessary for execution to all UAVs in the group. Current methods of use: organization of indoor shows and presentations.

Level 2 – conditional asynchronous flight. Each UAV in the group performs its own function, i.e. flies on its own trajectory. Practical use: a special control program or script is developed for a small group of UAVs, which transmits information about individual pre-programmed paths depending on the UAV. It is not possible for the operator to interfere during the flight. Current methods of use – indoor shows.

Level 3 – organized automated flight. This is the quintessence of the first two levels, but the number of UAVs in the group is much higher than in previous cases. Practical use: the flight is prepared in the already existing special software, which organizes the flight of the UAV group on the specified paths. There is a modeling and automatic control of the distance between the UAVs in the group at each time of the flight. The operator may be alone, but the software has the ability to interfere with the mission of any UAV at any time during the flight. Current methods of use: outdoor shows, area monitoring.

Level 4 – flight of a swarm formation. An approach to group UAV flight that draws inspiration from the self-organized behavior of social animals. The operator determines the rules of conduct of a particular UAV, but at the stage of flight planning and modeling. When completing a flight mission, each UAV has full autonomy. The operator can only decide to end the group flight.

Consider examples of preparing a mission for a group flight of the first and third levels.

3 Flight Mission Preparation Method for Group Autonomous Synchronized Flight

Consider the software for the preparation of the flight task on the example of Tello EDU. Tello SDK allows you to create applications for a variety of operating systems and platforms.

First, in order for the UAV and the application to be able to communicate with each other, all Tello SDK commands must be wrapped in programming language functions. After that, special commands for UAV control become available and can be used with any algorithmic constructions (cycles, branches, etc.) and data structures (lists, dictionaries, etc.) of the programming language.

Secondly, in case of communication of the Tello EDU UAV group with the computer, it is necessary to use an auxiliary link that will connect all UAVs and the program as it is impossible to connect all UAVs to the PC directly (Fig. 2). Such a link is a router. The program will send commands to the router, which, in turn, will send these commands to all UAVs in the swarm. The answers will be sent in the same way, only in reverse order – from the UAV to the computer via a router.

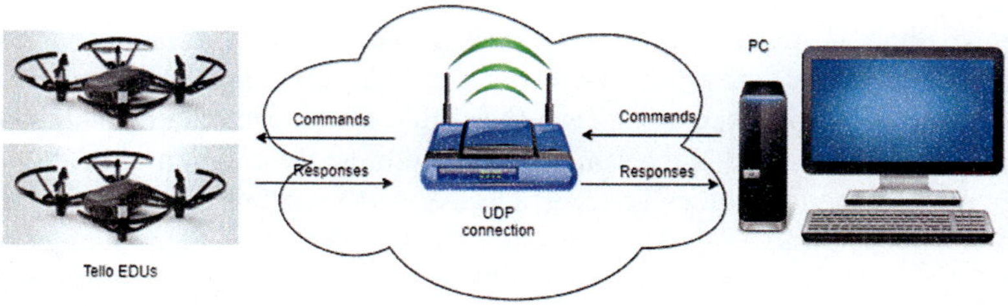

Fig. 2. General scheme of communication of the Tello EDU UAV group with a PC

Thirdly, in addition to the standard tools provided by Tello EDU, you can use additional libraries, frameworks, etc. For example, to give the UAV more autonomy and to expand its functions, you can connect and use the OpenCV computer vision library. It will allow you to analyze in more detail the images obtained from the UAV camera.

Tello EDU control automation software, such as DroneAI [11], is already available. It combines control functions with computer vision functions, but this is almost the only prototype of such software and there is currently no stable commercial version of it.

However, it is not necessary to write your own program, since the Tello SDK connects to the UAV via a UDP port over Wi-Fi, which allows users to control the UAV using text commands. You can use the usual packet manager, which will send the necessary commands in the form of packets. A good example of such software is Packet Sender [12]. This is an open source project, it can be downloaded and installed for free from the developer's website.

An experiment using the flight task preparation method by the Tello EDU UAV group was considered in [13].

4 The Method of Preparing a Flight Mission for a Group Autonomous Flight Along a Predetermined Trajectory of Each Participant

One of the important and crucial stages in the preparation of a flight mission for a group autonomous flight along a predetermined trajectory is the creation of animation, namely, the positioning of a number of UAVs in the sky. When working with animation, you need to take into account many parameters that can later affect the entire flight. It is required to maintain a path that does not intersect, maintain a safe distance between drones and limit their speed. Also, it is important to take into account the distance to the audience and the terrain on which the show will be implemented. All the necessary animation with effects can be done in 3D editors such as Cinema4D, Blender3d, or even in the Unity game engine [14]. Blender3d can extend its functionality by add-ons, which already contain in their settings the flight parameters that were mentioned earlier. Using Blender3d, work on preparing a flight mission for a group autonomous flight is greatly simplified. When creating animation, there are practically no restrictions on the formation of any shape in three dimensions. Let's consider several ways to create a flight mission for a UAV using add-ons in the open source software for creating three-dimensional computer graphics Blender3d.

Drone Show Animation Toolbox [15] is a set of tools for preparing a group autonomous flight mission in Blender. This set includes checking the distance between drones, speed and export operators. Using this add-on, you can set a bounding box for the flight of drones located in the same space, determine the duration of the flight mission in minutes, the location and number of UAVs, and configure the number of main waypoints. Drone positioning is visualized on the screen by naming each object that is responsible for the position of the object in the sky, as well as displaying materials that are responsible for lighting effects. When checking the final animation, the drone proximity zone is displayed. In case of intersection of trajectories, warnings are displayed on the screen. The result of all the animation is an export file that allows you to test it virtually. For flight testing, a so-called virtual environment is used that simulates the real behavior of the drone.

Vimdrones Designer is a paid add-on to Blender that allows you to design UAV flight paths in 3D space [16]. To use Vimdrones Designer to its fullest, your PC must be connected to a network. Thanks to this tool, preparation of a flight mission is easier. The add-on contains settings for the parameters of lighting effects and the creation of three-dimensional shapes for flight. To start working with Vimdrones Designer, you need to download it and add it to Blender as an add-on in the standard way. It is advisable to remove all objects from the scene so as not to interfere with the further construction of the flight mission for a group autonomous flight. By pressing the N key, all the designer tabs will be displayed containing the following settings: creating a UAV group; LED setup (drone light show); formation; flight check.

To construct a flight mission for a group autonomous flight, we enter the total number of UAVs (for example, 32) and their location along the X and Y axes (for example, 4 by 4). The recommended distance between UAVs is set to two meters, but it can be changed if necessary. As a result, we get a group of UAVs, consisting of 32 drones and built with the help of two subgroups where each group consists of 16 drones (Fig. 3).

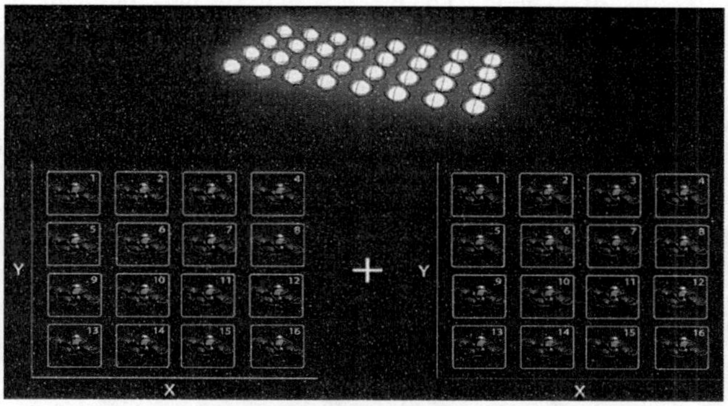

Fig. 3. Creation of a UAV group in Vimdrones Designer

After creating a group of drones, the following collections are formed: UAVs, distance check and effects. It is advisable not to change the names of objects and created collections for more stable animation work.

In the LED settings tab, set the value of the light intensity and the time of their update. The formation tab allows you to work with collections, change and add new properties inherent in the UAV group.

By opening the Vimdrones Designer console, it is possible to set the distance limit, speed limit, horizontal speed limit and vertical speed limit. Running the final animation, we will see the visualization of the above limits in real time. If any value of at least two drones reaches the limits, then it will be displayed in the window. Thanks to this, you can correct the position of the drones in the animation. After final processing, you can output the export file.

Clever-show also created their add-on for creating animation in Blender and converting it into flight paths taking into account the drones' illumination [17]. The names of objects and materials must correspond to the instructions, only in this case the color of the object's material will be exported as the color of the LED. In the export options panel, you can set a name filter for objects. If you do not use this parameter, the paths of all visible objects will be exported. If the minimum and maximum boundaries of speeds and distances are incorrect, warnings will be displayed in the animation. After setting the necessary parameters of the flight mission for a group autonomous flight, animations of the specified objects from the Blender3d project in the.csv format are exported to the specified folder.

Basically, all of the above add-ons are closed and do not have precise instructions for preparing a flight mission for a group autonomous flight in 3D editors. Therefore, based on the Drone Show Software, a trajectory planning algorithm was proposed to form a group of drones. In addition, this approach allowed us to synchronously control dozens of drones at the same time.

To design a group flight in Blender3d, we created 100 drones, and also set a bounding plane, the distance between the drones to 2.5 m and limited the speed of their movement. The animation for creating the planned label has a size of 70 m in width and 27 m in height. The LEDs on drones should change their light from the moment the flight starts. (Fig. 4).

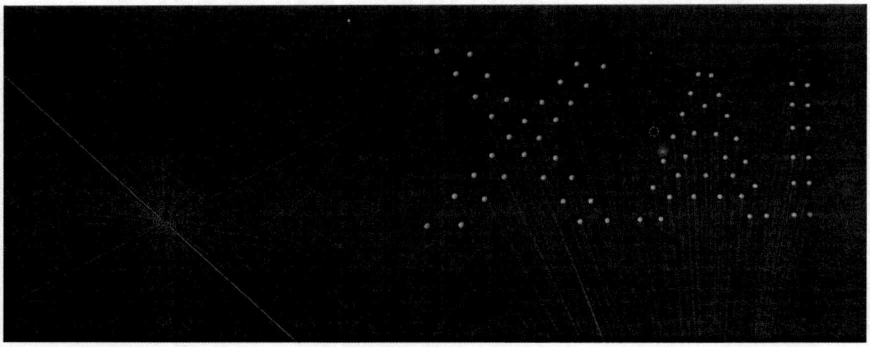

Fig. 4. Preparing a flight mission for a group autonomous flight in Blender3d

After designing the animation, it is necessary to test the animation virtually to make sure that there is no danger of the drones colliding with each other and that they can develop the required speed. The software-in-the-loop (SITL) software from Drone Show Software [18] made it possible to simulate the behavior of a group of drones. When checked, all animation formations corresponded to reality, and all takeoff and landing positions remained unchanged.

5 Conclusion

The paper considers the features of planning a group UAV flight. An overview of current information technologies used for visualization, modeling and control of a group flight is made. 6 levels of UAV autonomy are considered and 5 levels of group flight control are proposed, taking into account their outdoor and/or indoor placement.

A method for preparing a flight mission for a group autonomous synchronous flight is proposed. An architecture for transmitting control commands has been developed, a mock-up of 5 Tello EDU UAVs has been formed and tested. The existing alternatives for organizing and planning a group flight within this flight mission preparation method are considered.

A flight mission preparation method for a group autonomous flight along a predetermined trajectory of each UAV is proposed. Alternative visualization tools are proposed, namely: Drone Show Animation Toolbox, Vimdrones Designer, Clever-show. On the basis of Drone Show Software, a trajectory planning algorithm was proposed, which was tested for 100 UAVs. The distance between the UAVs was set to at least 2.5 m. The animation of the creation of the "XAI" label in the size of 70 by 27 m has been simulated.

References

1. Schilling, F., Lecoeur, J., Schiano, F., Floreano, D.: Learning vision-based flight in drone swarms by imitation. IEEE Robotics and Automation Letters **4**(4), 4523–4530 (2019). https://doi.org/10.1109/LRA.2019.2935377
2. Petrlík, M., Báča, T., Heřt, D., Vrba, M., Krajník, T., Saska, M.: A robust uav system for operations in a constrained environment. IEEE Robotics and Automation Letters **5**(2), 2169–2176 (2020). https://doi.org/10.1109/LRA.2020.2970980
3. Voronko, V., Zaitsev, V., Voronko, I.: Analiz pytannia dostavky poshty bezpilotnymy litalnymy aparatamy. Zbirnyk naukovykh prats ΛΌΓΟΣ (Analysis Of The Issue Of Mail Delivery By Unmanned Aircraft. In Ukranian), 57–59 (2020). https://doi.org/10.36074/24.04.2020.v2.14
4. Sun, H., Qi, J., Wu, C., Wang, M.: [IEEE 2020 39th Chinese Control Conference (CCC) - Shenyang, China (2020.7.27–2020.7.29)]. 2020 39th Chinese Control Conference (CCC) – Path Planning for Dense Drone Formation Based on Modified Artificial Potential Fields, pp. 4658–4664 (2020). https://doi.org/10.23919/CCC50068.2020.9189345
5. Alotaibi, E.T., Alqefari, S.S., Koubaa, A.: Lsar: multi-uav collaboration for search and rescue missions. IEEE Access **7**, 55817–55832 (2019). https://doi.org/10.1109/ACCESS.2019.2912306
6. Tello SDK 2.0 User Guide. (2018). https://dl-cdn.ryzerobotics.com/downloads/Tello/Tello%20SDK%202.0%20User%20Guide.pdf Accessed 06 Oct 2021
7. Kung, C.M., Yang, W.S., Wei, T.Y., Chao, S.T.: The fast flight trajectory verification algorithm for Drone Dance System. In: 2020 IEEE International Conference on Industry 4.0, Artificial Intelligence, and Communications Technology (IAICT), pp. 97–101. IEEE. (2020 July). https://doi.org/10.1109/IAICT50021.2020.9172016
8. SITL Simulator (Software in the Loop). https://ardupilot.org/dev/docs/sitl-simulator-software-in-the-loop.html Accessed 5 Oct 2021
9. Drone Show Software. https://github.com/ugcs/ddc Accessed 5 Oct 2021
10. Levels of Drone Autonomy. https://droneii.com/project/drone-autonomy-levels Accessed 5 Oct 2021
11. Drone AI: Software. https://www.aidronesoftware.com/ Drone AI Software. Accessed 5 Oct 2021
12. Packet sender. https://packetsender.com/ Accessed 5 Oct 2021
13. Pohudina, O.K., Kovalevskyi, M.I, Pyvovar, M.K.: Group flight automation using Tello EDU unmanned aerial vehicle. In: 2021 IEEE 16th International Conference on Computer Sciences and Information Technologies (CSIT), vol. 2, pp. 186–191. IEEE (2021 September)
14. Karatanov, O., Bykov, A., Serginko, M., Miroshnichenko, D.: Implementation of augmented reality technologies in the training process with the design of aircraft equipment. Radioelectronic and Computer Systems **1**, 110–118 (2021). https://doi.org/10.32620/REKS.2021.1.10
15. Drone Show Animation Toolbox. https://github.com/martinsh/drone_show_toolbox Accessed 5 Oct 2021
16. Vimdrones Drone Light Show. https://www.vimdrones.com/ Accessed 5 Oct 2021
17. Clever-show. https://github.com/CopterExpress/clever-show/blob/master/docs/ru/blender-addon.md Accessed 5 Oct 2021
18. Software in the loop (SITL). https://droneshowsoftware.com/ Accessed 5 Oct 2021

Simulation of Flight Control of Two UAVs Based on the "Master-Slave" Model

Mariia Pyvovar(✉), Olha Pohudina, Andrii Pohudin, and Olha Kritskaya

National Aerospace University Kharkiv Aviation Institute, 17, Chkalova Street, Kharkiv 61070, Ukraine

Abstract. The object of consideration is the process of planning a group flight of unmanned aerial vehicles (UAVs). The goal is to improve the quality of the mission by two UAVs by monitoring the observance of the distance and flight time on a given hardware platform. The levels of control of an individual UAV were classified and the levels of organization of a group UAV flight were considered. The following tools were used to organize the "master - slave" group flight such as the "master - slave" structural model, the "follow the leader" algorithm, and simulation methods. The advantages of the modern modeling tool CoppeliaSimEDU are considered. Such results as classification of UAV group flight based on the existing classification of UAV autonomy levels; a model that simulates the flight of two UAVs of the flying wing type based on the "master - slave" model were obtained. The scientific novelty of the work lies in the fact that the "master - slave" model was further developed based on the application for flight simulation in the existing simulation tool CoppeliaSimEDU. This will allow varying the parameters of the group work before flying on real equipment. A patent "Visualization of the flight of unmanned aerial vehicles according to the "master - slave" model" was obtained for the developed program.

Keywords: UAV · Unmanned aerial vehicle · Swarm · "master – slave" model · Simulation the flight

1 Introduction

According to many experts, Unmanned Aerial Vehicles (UAVs) are expected to play a key role in the future, by providing support to various services to the population [1, 2]. This is explained not only by the fact that their mass production and use is cheaper and easier than using manned aircraft, but also by the fact that some types of UAVs, due to their small size, are able to solve problems that are inaccessible to manned aircraft. UAVs although originally were developed for defense and military purposes but already today, their use helps to solve a huge range of scientific and applied problems.

Several advanced capabilities have been implemented with multiple networked UAVs, including military applications such as communications networking, reconnaissance, threat engagement, surveillance and electronic attack [3]. Such systems must have the flexibility to remain effective against known short-term disruptions, as well as the adaptability to remain effective.

M. Nechyporuk et al. (Eds.): ICTM 2021, LNNS 367, pp. 902–907, 2022.
https://doi.org/10.1007/978-3-030-94259-5_70

In this work, we present a logic-based approach oriented to the "leader follower" formation control of a swarm of UAVs that allows the swarm to avoid collision between UAVs and keep the formation based on the leader movements.

2 Principles of Control of an Individual UAV and a Group

UAV control can be divided into nine levels (Fig. 1). At the first level, interaction with the environment occurs through control and changes in the operation of the executive mechanisms. The second level adds the ability to control the rotation speeds of the propellers. Today, the simplest flight controllers include a microcontroller with a minimum set of sensors for assessing position, orientation and altitude (accelerometers, gyroscopes, barometers), which allow to estimate position and automatically influence the rotation speed of propellers, for example, to stabilize position in the air and keeping altitude.

At the third level, the orientation of the UAV is controlled. A GPS sensor is inserted into the UAV controller for self-orientation in space. At the fourth level, linear speed control is automated. At the fifth level, movement in space is controlled. For example, the operator sets the coordinates of a virtual point in space, the autopilot calculates the direction relative to the current coordinates of the UAV and generates a velocity vector that is sent to the speed controller. The speed controller generates the desired angle of the UAV, which is sent to the controller responsible for the angle, etc.

At the sixth level, intelligent planning systems are built into the UAV control system. The UAV must be equipped with algorithms for flying around obstacles or self-reconnaissance of difficult terrain. For this, an on-board computer is additionally used with such devices: cameras, lidars, radars and others. At the seventh level, automatic execution of the application appears. The tasks of the UAV include independent planning of the route of movement: for example, follow an object or automatically explore and map the space. The eighth control level implies the presence of UAV group interaction. The products offered for UAV group interaction are implemented to provide shows or conduct training and research. The ninth level assumes built-in air traffic control systems described in [4].

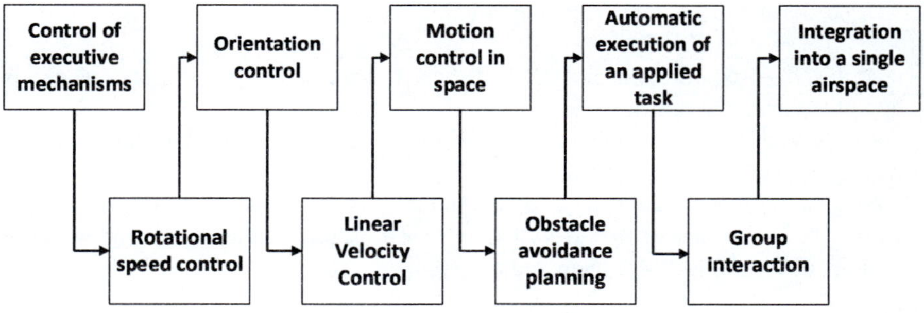

Fig. 1. UAV motion control levels

Let's consider in more detail the 8th stage - the group interaction of the UAV.

Sometimes it is necessary to use several groups of UAVs in order to cover a larger area or to obtain several points of view. Such unmanned aircraft work together and report their position and other useful information, taking into account predetermined time intervals. A number of problems that complicate the use of small-sized UAVs disappear with group use.

UAV groups can be classified based on the types of interactions between UAV (Fig. 2). Direct physical interaction: UAVs are connected directly to each other and their movement is limited by forces that depend on the movement of other UAVs (from the point of view of path planning and collision avoidance, all agents can be considered as a whole).

Formations: UAVs are not directly connected to each other, but their relative positions are strictly defined to maintain their shape (path planning can be viewed as a whole. Collision avoidance between vehicles can be built into shape control algorithms). Swarm: a team of many UAVs, interaction algorithms between which provide collective behavior (the resulting movement of the group does not necessarily represent a formation). Cooperation: UAVs from the group plan their movement in accordance with the individual tasks that must be assigned to carry out a mission of a higher order in the control system hierarchy (these trajectories are usually not geometrically interconnected as in the case of formations).

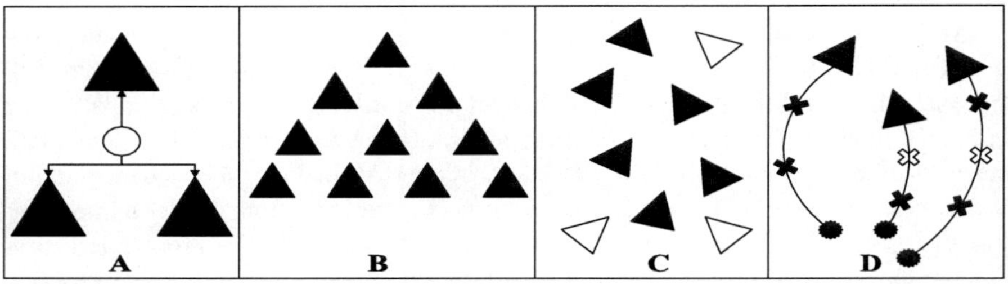

Fig. 2. Types of interactions between UAV: a – direct physical interaction, b – formations, c – swarm, d – cooperation

Three possible management strategies can be considered:

- centralized – remote control with a dedicated base station, the swarm leader is appointed from the central node;
- decentralized – the swarm leader is determined on the basis of any algorithm and does not depend on the central control station;
- mixed – combines the advantages of centralized and decentralized strategies by allocating the swarm leader if it is necessary based on one of the algorithms with the transfer of control rights to the operator [5].

Most of the models used in the developed projects consist of devices of the same type that perform one common task and do not have any centralized behavior control system. Therefore, it was decided to pay attention to the UAV according to the type of control, when the control commands come from the "appointed" leader.

3 Formal Methods and Simulation of the "Follow the Leader" Algorithm

One of the most widespread approaches to formation control in the literature is the "master – slave" approach [6, 7]. The "slave" UAV build their relative positions only according the "master" (Fig. 3). We surmised that, rather than the follower responding to more 'local' level information, the inter-agent coupling took place on a more 'global' level. Local information may be derived from the parameters of controller, whereas global information refers to the perceived object as a whole in relation to the environment and, of course, these two categories of information sources are not necessarily mutually exclusive. In the model was used a simple speed-matching model. Therefore, the current study further explores the hypothesis that follow-the-leader coupling can be regulated by global information alone in the absence of local information.

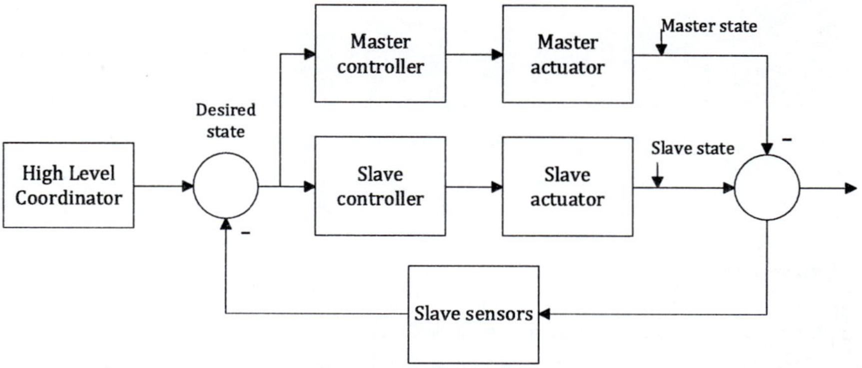

Fig. 3. Block diagram of the "master – slave" control method

A simulation model was developed that coordinates the swarm in accordance with the parameters of the leader and follower controller, taking into account the maintenance of the distance between the UAVs.

The CoppeliaSimEDU [8] is a 3D robot simulation software, with integrated development environment, that allows you to model, edit, program and simulate any robot or robotic system (e.g. sensors, mechanisms, etc.). It offers a multitude of functionality that can be easily integrated and combined through an exhaustive API and script functionality. CoppeliaSimEDU supports handling of the inverse/forward kinematics of any type of mechanism, proximity sensor simulation, camera sensor simulation with many built-in image processing filters, path planning and minimum mesh-mesh distance calculation. Using the CoppeliaSimEDU information system, a program was built that displays the scene and behavior model of two UAVs, where one aircraft obeys the other, which is the leader of the swarm.

The UAV model in the simulation is a cruise rocket [9] which was loaded from the CAD model and its properties were set manually. There are 2 programmer codes in the UAV model: one for practicing the kinematics of motion and another one is for the realistic movement. Two cameras were set onboard, one front and one floor camera to monitor the flight process in time. Figure 4 shows the UAVs target in visible layer and the simulation environment.

Fig. 4. Model of flight visualization of unmanned aerial vehicles according to the "master - slave" model

4 Conclusion

The present work shows the use of "master - slave" model for the generation of leader follower behavior in a robot swarm. When the leader moves from one place to another, in this behavior the slave UAV follows the leader towards the new location. Once the leader arrives at the new location, slave UAV localize near the leader. UAV move according to the leader's location and at the same time, they avoid colliding with the master UAV.

The following results were obtained: schema of interaction of two UAVs in the "teacher" or "mentor" modes; developed software product for visualization of UAV flight according to the "master – slave" model using the CoppeliaSimEDU information system. Conclusions. The scientific novelty of the results obtained is as follows: for the first time a simulation model of the behavior of an unmanned aerial vehicle in a group was obtained by automating the schema of interaction between two UAVs according to the "master – slave" model; the method of visualizing the dynamics of the movement of complex objects in a three-dimensional scene was further developed, which also includes objects of different geometric structures (trees, buildings, etc.). This makes it possible to obtain a rational flight scenario on a predetermined terrain, taking into account natural and artificial obstacles on the way.

A patent "Visualization of the flight of unmanned aerial vehicles according to the "master - slave" model" (№ 100840, registration date – 21.12.2020) was obtained for the developed program.

References

1. Baca, T., et al.: The MRS UAV system: pushing the frontiers of reproducible research, real-world deployment, and education with autonomous unmanned aerial vehicles. J. Intell. Rob. Syst. **102**(1), 1–28 (2021). https://doi.org/10.1007/s10846-021-01383-5
2. Kritskiy, D., Yashin, S., Koba, S.: Unmanned aerial vehicle mass model peculiarities. In: Shkarlet, S., Morozov, A., Palagin, A. (eds.) MODS 2020. AISC, vol. 1265, pp. 299–308. Springer, Cham (2021). https://doi.org/10.1007/978-3-030-58124-4_29
3. Kerr, C., Jaradat, R., Hossain, N.: Battlefield mapping by an unmanned aerial vehicle swarm: applied systems engineering processes and architectural considerations from system of systems. IEEE Access **8**, 20892–20903 (2020). https://doi.org/10.1109/ACCESS.2020.2968348
4. Pohudina, O., Kritskiy, D., Koba, S., Pohudin, A.: Assessing unmanned traffic bandwidth. In: Nechyporuk, M., Pavlikov, V., Kritskiy, D. (eds.) Integrated Computer Technologies in Mechanical Engineering. AISC, vol. 1113, pp. 447–458. Springer, Cham (2020). https://doi.org/10.1007/978-3-030-37618-5_38
5. Liping, H., Xiaolong, L., Jiaqiang, Zh., Dan, H.: Analysis of the current situation of aviation swarm operations. In: 2020 5th International Conference on Computer and Communication Systems (ICCCS), pp. 956–961. IEEE (2020). https://doi.org/10.1109/ICCCS49078.2020.9118481
6. Quesada, W.O., et al.: Leader-follower formation for UAV robot swarm based on fuzzy logic theory. In: Rutkowski, L., Scherer, R., Korytkowski, M., Pedrycz, W., Tadeusiewicz, R., Zurada, J.M. (eds.) ICAISC 2018. LNCS (LNAI), vol. 10842, pp. 740–751. Springer, Cham (2018). https://doi.org/10.1007/978-3-319-91262-2_65
7. Dano, E.B.: Resilient system engineering in a multi-UAV system of systems (SoS). In: 2019 International Symposium on Systems Engineering (ISSE), pp. 1–8. IEEE (2019). https://doi.org/10.1109/ISSE46696.2019.8984509
8. CoppeliaSim. https://www.coppeliarobotics.com/. Accessed 5 Oct 2021
9. Ukraine is developing a new cruise missile. https://naked-science.ru/article/tech/ukraina-razrabatyvaet-novuyu-k. Accessed 5 Oct 2021

Signal and Image Processing

Check for updates

Analysis of Noisy Image Lossy Compression by BPG

Victoria Naumenko(✉), Vladimir Lukin, and Sergey Krivenko

National Aerospace University, 17 Chkalov Street, Kharkiv 61070, Ukraine
v.naumenko@khai.edu, lukin@ai.kharkov.com, krivenkos@ieee.org

Abstract. Acquired remote sensing images are often large and distorted by noise. On the one hand, this leads to the need to compress them. On the other hand, the presence of noise should be taken into account at the image compression stage and while evaluating the compression efficiency.

Our work investigates the noise influence on the BPG method of image compression with losses supposing that data are distorted by additive Gaussian noise of different intensity. Within the framework of this study, the PSNR metric (peak signal to noise ratio, measured in dB) was chosen as a criterion for evaluating the quality of compression.

It is shown that three typical dependences of the metric calculated between compressed and noise-free images are possible. Depending on the situation, there are different recommendations for setting a parameter that controls compression (PCC). In particular, if there is an optimal operating point (OOP), the compression should be done in its neighborhood. The examples of dependences of PSNR on PCC for several test images are given. It is shown that existence of OOP and its position depend on image complexity and noise intensity.

Keywords: Lossy image compression · Additive noise · Optimal operation point · Better Portable Graphics (BPG)

1 Introduction

When developing mission concepts and systems for satellites, mission designers have to take design considerations into account, balancing data storage, data processing, and communication bandwidth for their solutions. The spacecraft platform has specific size, weight, power and cost (SWaP-C) requirements that determine general constraints for these considerations. Then, it is necessary to determine the best combination and balance of these qualities for the mission. For example, the use of more efficient codecs for compressing images can significantly reduce the size of the transferred or stored data, but such codecs are often complex. However, without the use of encoders, on-board sensors quickly saturate the downlink bandwidth of the transceivers, and also preclude on-board data storage. While codecs can significantly reduce the amount of data transmitted or stored, many of them are computationally intensive and require fast processing to maintain the required data rate.

M. Nechyporuk et al. (Eds.): ICTM 2021, LNNS 367, pp. 911–923, 2022.
https://doi.org/10.1007/978-3-030-94259-5_71

Traditional large spacecraft often rely on slow, radiation-hardened processing technology due to the harsh space conditions [1, 2]. However, with the advent of more capable onboard processing systems, including hybrid radically hard/commercial architectures or commercial off-the-shelf (COTS) [3], developers are focusing on more onboard processing and compression to solve mission problems. These systems can provide significant increases in processing power to enable more sophisticated data processing including intelligent image compression techniques for scientific and other applications.

Historically, space-based data compression systems have used a variety of lossless and lossy compression schemes [4, 5]. For many satellite systems, lossy compression is used if either (1) the system throughput was too low to support lossless data transmission, or (2) scientific value was not compromised by the distortion that could have been introduced by lossy compression, or (3) additional sensors were included that did not contribute to the primary data product.

It is well known [6, 7] that (reversible) lossless compression provides a low compression ratio (CR), which often does not meet the requirements of storage and/or transmission of visual data; and the use of lossy compression may be generally acceptable in practice, but only under certain conditions.

The main condition is that the distortion introduced by lossy compression should not degrade the value of the image (i.e., the distortion should be "invisible" and/or there should be no compression artifacts). Other conditions relate to the efficiency and safety of the computation. Compression and decompression should be fast; sometimes it is necessary to use standard encoders [7], while using non-standard encoders can provide additional security for the compressed data.

Another problem is that in all real images, regardless of their type and acquisition conditions, there is noise, which can be of different types and intensities. There are applications or environments where noise is almost invisible, such as optical images captured in good lighting conditions. For other applications, such as radar imagery, noise is always visible [8] and its presence limits the possibilities and efficiency of solving problems of secondary processing (detection and localization of objects [9], classification [10]). In this regard, the noise present in the image is often necessary to take into account when developing methods for compressing such images and choosing the compression parameters [11].

The authors of [11] were the first to draw attention to these points. They showed that lossy compression of images with noise leads, in general, to the positive effect of partial noise suppression – a kind of filtering. In addition, they demonstrated the possible presence of an optimal operating point (OOP) – such lossy compression parameters that the compressed image appears to be closer to the true image (image without noise) than the original (primary) image, subjected to compression, in accordance with some metric.

On the one hand, the analysis of metrics calculated between compressed and true images and their dependences on compression parameters is possible only in the presence of a true image, i.e. when carrying out numerical simulations, when the user has a "pure" (true) image, he/she adds noise to it and performs compression. Therefore, one might get the impression that compression in OOP in practice is impossible due to the absence of a true image. However, studies carried out in recent years have shown [12–14] that lossy compression is possible in the neighborhood of the OOP in the absence of a true image by either using iterative processing [8] or by choosing the parameters of the encoder [14], taking into account the characteristics of the noise assumed to be known a priori or estimated with acceptable accuracy [15].

Since OOP is not always observed (it may be absent), it is advisable to make a decision with what parameters to compress in such a way as to ensure acceptable quality and compression ratio (CR), or at least find an acceptable compromise between them. Then, it is desirable to be able to predict the compression parameters (some points of the curve describing the value of the metric between the true and compressed image depending on the parameter that controls the compression). In [16], we have shown that it is possible to predict the values of the peak signal-to-noise ratio (PSSR) (or its differences from the PSSR for an uncompressed image) for the vicinity of the OOP. In addition, in [17] it was shown that it is possible to predict the values of the CR.

Fabrice Bellard, founder of both the QEMU and FFmpeg projects, recently unveiled his creation of a new lossy image file format. Called BPG (for "Better Portable Graphics"), the format is based on the compression algorithm from the High Efficiency Video Coding (HEVC) video codec. That makes it possible to achieve performance considerably better than the now 20-year-old JPEG format.

Advantages of BPG Compression Format:

- open source accessed
- higher compression ratio with similar quality compared to JPEG.
- wide support for different chroma formats such as grayscale, RGB, YcbCr, and so on.
- provides JavaScript decoder which is light for web browsers and other Portable applications, without requiring sophisticated hardware.
- supports lossless and lossy compression, users can choose one of options freely according to their needs, or use a combination of the two.

The BPG page [18] includes links to several pages of sample images. Viewing them requires a moderately current browser with JavaScript enabled; the page fetches each .bpg image file, decompresses it in JavaScript, and renders it as a lossless PNG images on the page inside of a <canvas> element.

As advertised, the images do exhibit fewer visual artifacts than comparably sized JPEG reference images, and BPG file sizes are noticeably smaller for the same (roughly) equivalent visual quality. Bellard provides a source code bundle for download [19] containing a command-line encoder, decoder, and a library.

The main goal of this study is to check whether OOP might take place if BPG compression is applied to noisy images with different intensities of additive Gaussian noise. We would also like to analyze positions of OOP if it occurs.

2 Test Images for Simulation

2.1 Image Properties

The main performance characteristics of any image compression method are highly dependent on the properties of an image subject to compression, whether this image is noisy [12] or can be considered noise-free [20]. Therefore, when choosing test images, it is necessary to choose them of different complexity.

Images of a simple structure are better compressed without losses, they have a higher quality for a given CR for a selected compression method, and they have better noise suppression [21]. Images of a complex structure have opposite properties.

Six images were selected as test images in PNG format, of which Frisco can be classified as a simple type of image, Diego – as a complex type of images, and Fr01, Fr02, Fr03, Fr04 are images of medium complexity. Examples of images are shown in Fig. 1

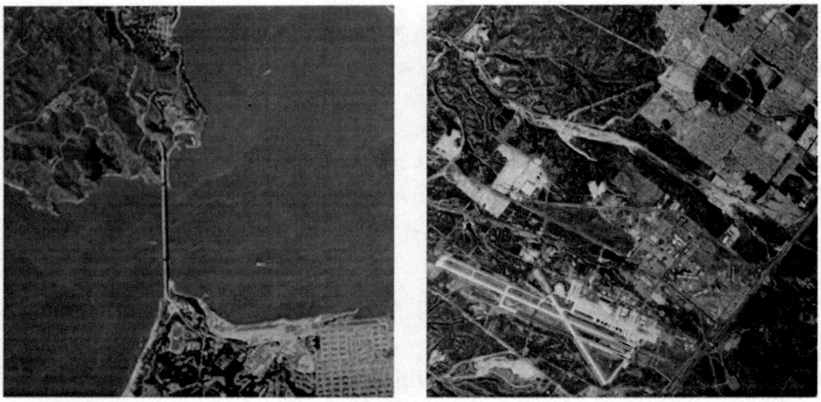

a b

Fig. 1. Noise-free test images used in simulations: a – Frisco, b – Diego

2.2 Noise Models

As it has been mentioned above, we rely on the simplest model of additive white Gaussian noise (AWGN)

$$I_{ij}^{n} = I_{ij}^{true} + n_{ij}, \ i = 1, ..., I, \ j = 1, ..., J \quad (1)$$

where I_{ij}^{n} is the ij-th sample of noisy (original) image, I_{ij}^{true} is the ij-th sample of the true (noise-free) image, n_{ij} is the noise in the same pixel, I, J define the image size. It is assumed that the noise is zero mean and its variance is equal to σ^2. Recall that, in the case of AWGN, the noise can be hardly noticed in an original image if its variance σ^2 is not greater than 15 (for images represented as 8-bit data arrays). Moreover, it is difficult to notice the noise with variance about 25 if an image is highly textural. This property can be treated in another way. Noise is practically invisible if PSNR is larger than 36 dB for simple structure images and larger than 33 dB for highly textural images where.

$$PSNR_{inp} = 10\log_{10}(255^2/\sigma^2).$$

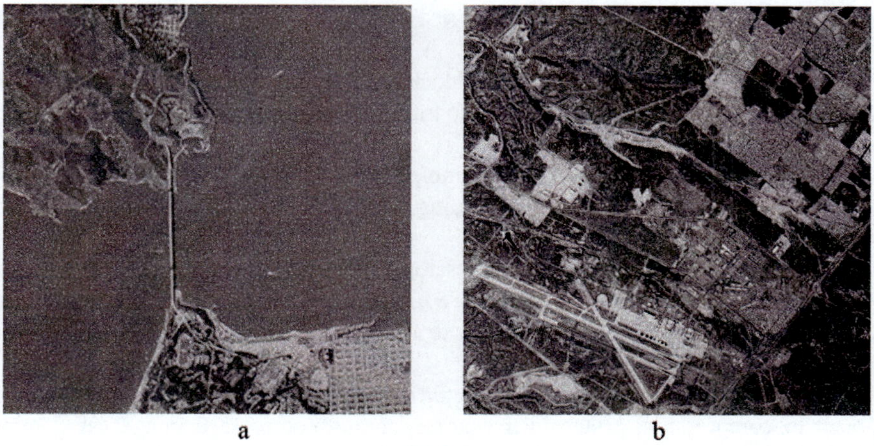

Fig. 2. Noisy test images Frisco (a) and Diego (b), $\sigma^2 = 64$

Figure 2 presents examples of such cases. Both images are corrupted by the noise with $\sigma^2 = 64$. Noise is invisible in textural image Diego whilst it can be noticed in homogeneous regions of the noisy test image Frisco.

Dependences of the traditional PSNR calculated between I^{true} (i.e., noise-free) and compressed image I^c on the parameter Q for these test images are given in Fig. 3 (note that Q is only integer and it can be varied from 1 to 51. As it can be seen from the given curves, the best (largest) PSNR is observed for the Frisco test image with large homogeneous areas for Q > 13. The test images of medium and high complexity in the main area of interest (Q > 20) have PSNR values that are less by 3–7 dB compared to PSNR for Frisco. For Q > 40, the compressed images are mostly characterized by low quality (PSNR is smaller than 27 dB) and the introduced distortions are annoying. Thus, we have to focus on image compression with Q < 40.

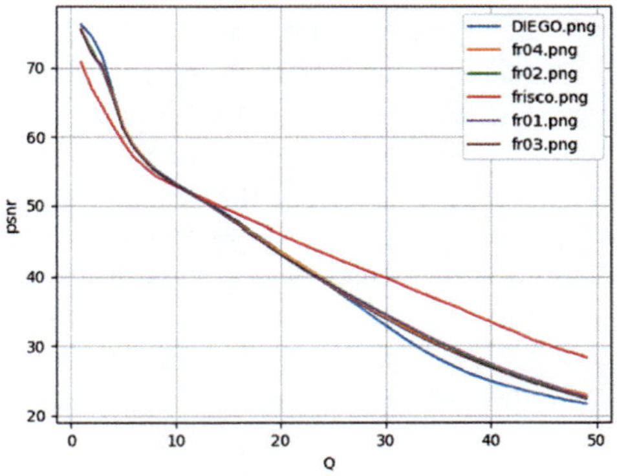

Fig. 3. Rate-distortion curves for six test images without noise

3 BPG Compression Method Analysis

The coder BPG performs DCT in 64 × 64 blocks and minimum recursive block partitions of 4 × 4 pixel blocks and uses DCT as the transformation mechanism to the frequency domain.

After the process of transform, the sample is quantized and transformed to entropy coding. The main objective of entropy coding is to eliminate redundancy which has not been removed by the prediction stage.

The main requirement to lossy compression of images is to obtain acceptable quality with the highest CR possible. Thus, we are faced with the traditional task of solving a compromise, since an increase in CR leads to a decrease in the quality of the compressed image.

A traditional way to define the rate-distortion curve in to "compare" original (noisy, subject to compression) image to the corresponding compressed image using some criterion (metric) for different values of PCC. In our case, this can be Q or CR (let us start our analysis from Q). Having a compressed image $I^c_{ij},\ i = 1, ..., I,\ j = 1, ..., J,$ it is easy to calculate

$$PSNR_{nc} = 10\log_{10}(255^2/\mathrm{MSE}_{nc})$$
$$\mathrm{MSE}_{nc} = (1/IJ)\sum_{i=1}^{I}\sum_{j=1}^{J}(\mathrm{I}^n_{ij} - \mathrm{I}^c_{ij})^2.$$

In this work, the dependences are calculated for six test images noisy with additive Gaussian noise of different intensities. They are shown in Fig. 4 for noise variance values of 25, 64 and 100, respectively. In general sense, these dependences are typical rate-distortion curves where quality (characterized by $PSNR_{nc}$) makes worse ($PSNR_{nc}$ reduces) if Q increases. However, there are two points to pay attention to.

First, for certain Q (larger than a certain threshold T_1 and less than a certain thresh-old T_2), the curves for all test images "go together" with a sufficiently high absolute value of the first derivative. For the curves in Fig. 4, this takes place for $6 \leq Q \leq 20$, while for the second case (Fig. 4, b) there is a larger interval of such behavior ($5 \leq Q \leq 30$) and in the third case (Fig. 4, c) there is a larger interval of such behavior ($5 \leq Q \leq 32$).

It is also possible to see that the curves $PSNR_{nc}$ on Q have the same behavior until $PSNR_{nc}$ is larger than $PSNR_{inp}$ which is about 34 dB for data in Fig. 4, a, about 30 dB for the curves in Fig. 4, b and about 28 dB for the curves in Fig. 4, c. This means that if noise variance is known in advance or pre-estimated with appropriate accuracy, one can predict $PSNR_{nc}$ for a certain range of Q values for BPG encoder using the curves in Fig. 4. This can be also treated in another way. It is possible and easy to set Q to provide a desired $PSNR_{nc}$. For example, if one would like to provide $PSNR_{nc} = 40$ dB, it is recommended to use $Q = 23$. Note that this recommendation is valid for most test images that are practically noise free except the test image Frisco (Fig. 3).

Second, for $Q > T_2$, the curves start to diverge (Fig. 4) and the difference between the values of $PSNR_{nc}$ for the considered test images becomes larger for larger Q. These differences can reach 5…7 dB (see data for Q about 40 in Fig. 4, a. $PSNR_{nc}$ for simpler structure images is larger, the smallest $PSNR_{nc}$ are observed for the most complex structure images.

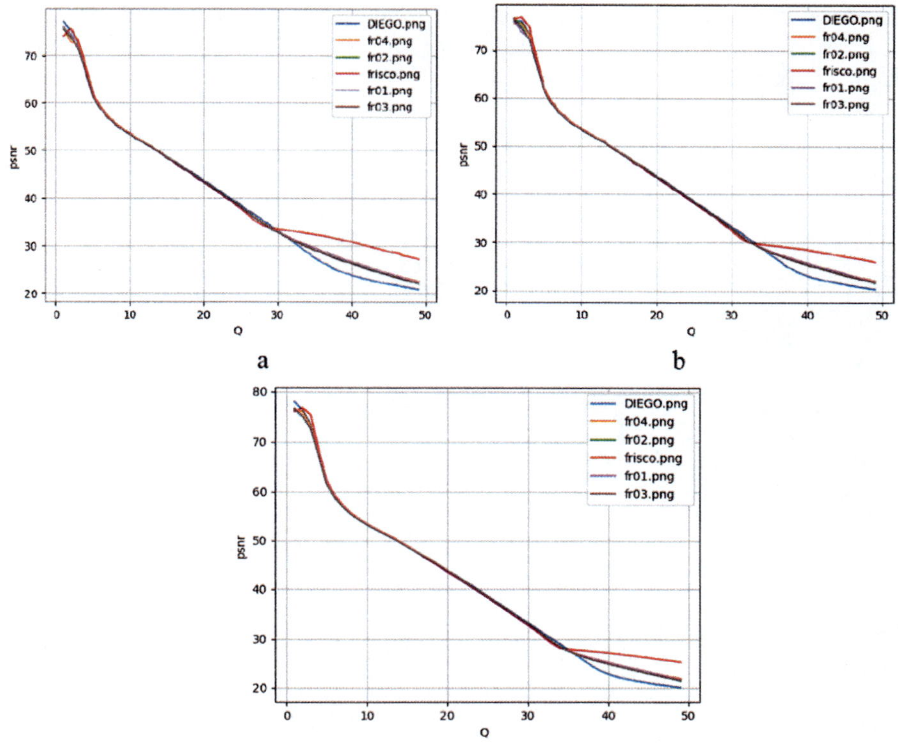

a b

c

Fig. 4. Rate-distortion curves for three test images for $\sigma^2 = 25$ (a), 64 (b) and 100 (c)

In [22, 23], the threshold for the invisibility of distortions for PSNR was approximately determined. Thus, if the PSNR between the distorted and the corresponding reference image is greater than 36 dB, then it is very likely that the distortion is not visible. In all three cases, Fig. 4, this occurs at $Q \leq 25$ (and this is in good agreement with the data in [20]). Otherwise, the original (noisy) and compressed images look different. Let us explain what the difference is.

For this purpose, let us analyze

$$PSNR_{tc} = 10\log_{10}(255^2/\mathrm{MSE}_{tc})$$
$$\mathrm{MSE}_{tc} = (1/IJ)\sum_{i=1}^{I}\sum_{j=1}^{J}(\mathrm{I}_{ij}^{true} - \mathrm{I}_{ij}^{c})^2.$$

Dependences $PSNR_{tc}$(Q) for the same test images and the noise variances equal to 25, 64 and 100 are represented in Fig. 5. Their analysis shows the following.

First, the dependences behave in a very different manner depending on the test image properties and noise variance. The only common thing is that they start from $PSNR_{tc}(Q \to 0) = PSNR_{inp}$, "go together" with small decreasing of $PSNR_{tc}$ till a certain Q_{com}, and then diverge. For $Q > Q_{com}$, $PSNR_{tc}$ for simpler structure images is larger than for complex structure ones and this difference can reach 5…8 dB for large values of noise variance. Note that Q_{com} is about 16 for $\sigma^2 = 25$, about 18 for $\sigma^2 = 64$ and 20 for $\sigma^2 = 100$.

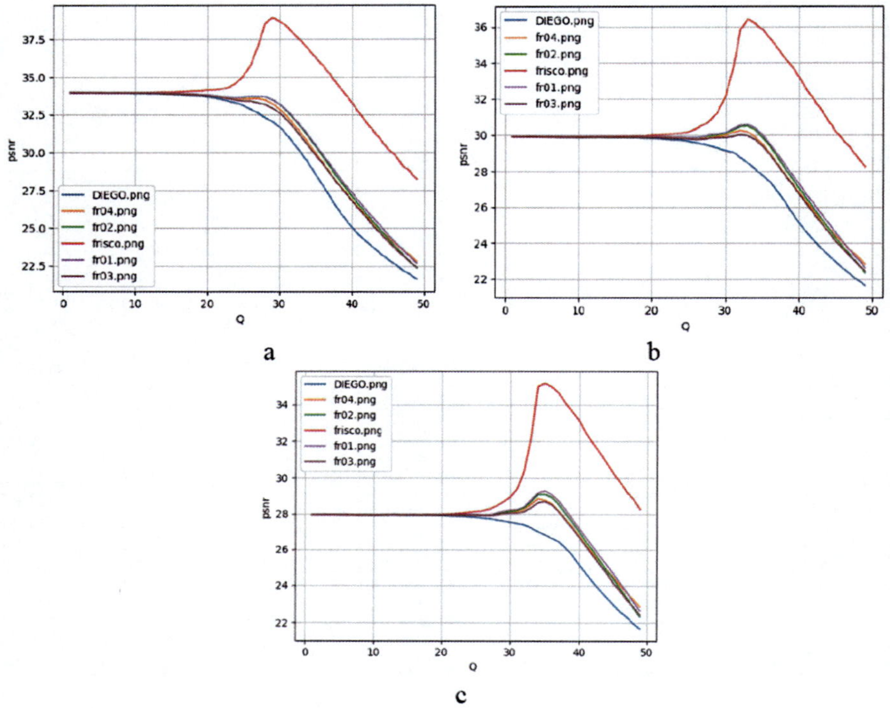

Fig. 5. Rate-distortion curves for three test images for $\sigma^2 = 25$ (a), 64 (b) and 100 (c)

Second, it can be assumed that for the lossy compression standard BPG, as for other lossy compression standards, three practical situations are possible: favorable, neutral and unfavorable. For a favorable situation, there is an obvious OOP, i.e. there is a maximum of $PSNR_{tc}(Q)$ and $PSNR_{tcmax}$ is greater than $PSNR_{inp}$ (dependencies for the Frisco test image in Fig. 5, a–c are examples of such situations). Then, there is an obvious reason to perform lossy compression in OOP (or in its neighborhood, if OOP for some reason cannot be precisely established). Two advantages are achieved simultaneously: the quality of the compressed image is better than the original noisy one, and a rather high CR is achieved.

A neutral situation is if local maximum of $PSNR_{tc}$(Q) exists but $PSNR_{tcmax}$ is about or smaller than $PSNR_{inp}$ (the examples are the dependences for many test image including Diego and noise variance equal to 100, Fig. 5, c). Then, a question is what is the difference $PSNR_{inp} - PSNR_{tcmax}$ and what is the priority of requirements to lossy compression.

Suppose that $PSNR_{inp} - PSNR_{tcmax} < \Delta$ and the main requirement to lossy compression is to provide as large CR as possible with an acceptable degradation of image quality where a preset Δ just controls this possible degradation (for example, Δ can be equal to 1 or 2 dB). In the opposite case, i.e. if $PSNR_{inp} - PSNR_{tcmax} \geq \Delta$ and it is not so necessary to provide a large CR, it is necessary to decide what to do. One way possible is to use Q equal to 25 to provide invisibility of introduced distortions. This guarantees that the compressed image looks close to original noisy one.

Finally, the situation called "unfavorable" consists in the following. The dependence $PSNR_{tc}$(Q) steadily decreases (the examples are the dependences for most test image including Diego for $\sigma^2 = 25$, Fig. 5, a). This means that the compressed image becomes more and more different with respect to both original and noise-free if Q increases. In this case, we recommend to apply Q equal to 25 or slightly smaller to avoid visible degradations.

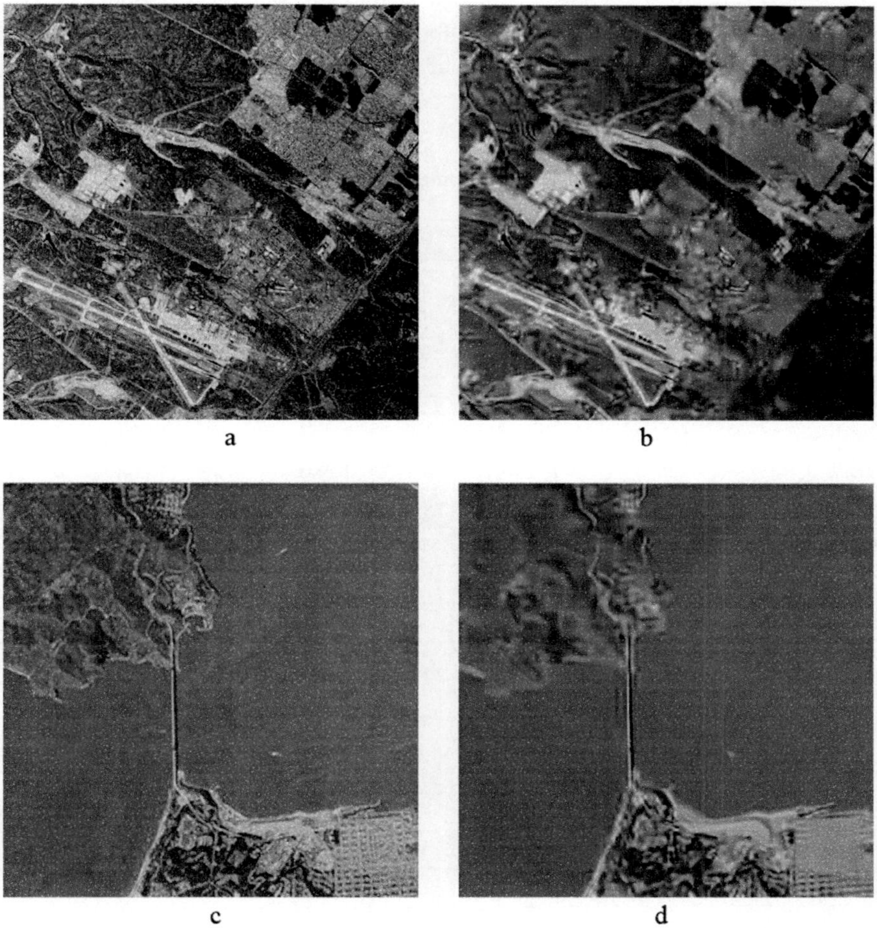

Fig. 6. Compressed test images with noise $\sigma^2 = 64$: Diego.png Q = 33 (a), Q = 46 (b), Fisco.png Q = 33 (c), Q = 46 (d).

Currently, we cannot give an idea how to distinguish aforementioned situations before starting the compression by BPG. This is one task for the future work.

Comparing the curves in Fig. 5 for favorable situations, i.e. when OOP exists, the following observation has been also noticed. If noise variance increases, OOP moves to the area of larger Q. For $\sigma^2 = 25$, OOP is observed for Q about 29 (Fig. 5, a), for $\sigma^2 = 64$ OOP takes place for Q about 33 (Fig. 5, b), and for $\sigma^2 = 100$ OOP is observed for Q about 35. This shows that, most probably, parameters of OOP can be determined under condition that noise variance is a priori known or pre-estimated in advance with appropriate accuracy. This is one more direction of future research.

As an example, we present the results of compression two test images: Diego.png and Frisco.png with noise ($\sigma^2 = 64$). Two values of Q are used: Q = 33 that corresponds to OOP and Q = 46 that provides a larger CR but introduces more distortions. As it can be seen from Fig. 6 compression at the optimal operating point preserves the details of the noisy test image and removes noise while compression with Q = 46 significantly distorts the images.

Table 1. CR for different test images and noise variance values

Test image	Q	CR		
		$\sigma^2 = 25$	$\sigma^2 = 64$	$\sigma^2 = 100$
Frisco	1	1.283	1.135	1.070
Fr01		1.578	1.476	1.420
Fr04		1.511	1.428	1.374
Diego		1.501	1.457	1.424
Frisco	5	1.447	1.280	1.208
Fr01		1.781	1.662	1.599
Fr04		1.704	1.600	1.545
Diego		1.681	1.624	1.589
Frisco	10	1.787	1.553	1.456
Fr01		2.155	1.997	1.914
Fr04		2.058	1.920	1.848
Diego		2.003	1.924	1.879
Frisco	15	2.339	1.944	1.772
Fr01		2.648	2.415	2.304
Fr04		2.521	2.318	2.222
Diego		2.396	2.296	2.236
Frisco	20	3.376	2.593	2.334
Fr01		3.446	3.070	2.898
Fr04		3.279	2.939	2.783
Diego		2.947	2.800	2.739

(*continued*)

Table 1. (*continued*)

Test image	Q	CR		
		$\sigma^2 = 25$	$\sigma^2 = 64$	$\sigma^2 = 100$
Frisco	25	6.051	3.890	3.322
Fr01		4.943	4.187	3.859
Fr04		4.681	4.017	3.722
Diego		3.871	3.619	3.488
Frisco	30	27.596	8.976	5.887
Fr01		9.541	6.833	5.886
Fr04		8.582	6.502	5.638
Diego		5.639	5.125	4.833
Frisco	40	99.902	96.686	95.750
Fr01		39.318	38.057	37.065
Fr04		42.312	40.746	39.377
Diego		38.652	33.982	30.268

One more important aspect is dependence of CR on Q. Some data are presented in Table 1 for three test images and noise variance values equal to 25, 64 and 100. Analysis of the obtained results allows concluding the following:

1. As it can be expected, CR increases if Q increases, but very individually depending on image complexity and noise variance; the values of CR for the same Q can differ by several times (by almost one order, see data for Q = 30);
2. A larger noise variance leads to a smaller CR for the same test image and parameter Q of the BPG encoder.
3. For a more complex images (e.g., Diego), CR is smaller and vice versa;
4. For the case of invisibility of introduced distortions, CR varies from 3 to 6. i.e. it is considerably larger than for lossless compression.

4 Conclusions

This article discusses the lossy BPG compression of single-channel images corrupted by AWGN. The features of such compression are described. Performance criteria are discussed. It is shown that favorable, neutral and unfavorable situations are possible.

Basic analysis is performed using 8-bit png images, but basic recommendations are also valid for images presented in different formats.

Analysis of the data obtained indicates that some parameters characterizing the efficiency of lossy compression can be predicted. This allows giving some recommendations for choosing the compression options for each specific image. In further work, it is planned to obtain rules of thumb for choosing the optimal parameters for compressed images.

References

1. Lovelly, T.M., George, A.D.: Comparative analysis of present and future space-grade processors with device metrics. AIAA J. Aerospace Inf. Syst. **14**(3), 184–197 (2017)
2. 2015 NASA Technology Roadmaps, Washington, D.C., USA: NASA Office of the Chief Technologist (2015) [Online]. Available: https://www.nasa.gov/offices/oct/home/roadmaps/index.html
3. George, A.D., Wilson, C.M.: Onboard processing with hybrid and reconfigurable computing on small satellites. Proc. IEEE **106**(3), 458–470 (2018). https://doi.org/10.1109/JPROC.2018.2802438
4. Beser, N.D.: Space data compression standards. John Hopkins APL Technical Digest. **15**(3), 206–223 (1994)
5. Yu, G., Vladimirova, T., Sweeting, M.N.: Image compression systems on board satellites. Acta Astronaut. **64**(9–10), 988–1005 (2009). https://doi.org/10.1016/j.actaastro.2008.12.006
6. Fidler, A., Likar, B.: What is wrong with compression ratio in lossy image compression? Radiology **245**(1), 299–300 (2007). https://doi.org/10.1148/radiol.2451062005
7. Braunschweig, R., Kaden, I., Schwarzer, J., Sprengel, C., Klose, K.: Image data compression in diagnostic imaging: international literature review and workflow recommendation. RöFo – Fortschritte auf dem Gebiet der Röntgenstrahlen und der bildgebenden Verfahren **181**(7), 629–636 (2009). https://doi.org/10.1055/s-0028-1109341
8. Lee, J.-S., Pottier, E.: Polarimetric Radar Imaging: From Basics to Applications, p. 422. CRC Press (2009)
9. Marques, R., Medeiros, F., Ushizima, D.: Target detection in SAR images based on a level set approach. In: IEEE Trans. on Systems, Man and Cyber Kussul N., Lemoine G., Gallego F. J., Skakun S.V.
10. Kussul, N., et al.: Parcel-based crop classification in Ukraine using Landsat-8 data and Sentinel-1A data. IEEE J. Sel. Top. Appl. Earth Observ. Remote Sens. **9**(6), 2500–2508 (2016). https://doi.org/10.1109/JSTARS.2016.2560141
11. Al-Chaykh, O.K., Mersereau, R.M.: Lossy compression of noisy images. IEEE Trans. Image Process. **7**(12), 1641–1652 (1998)
12. Zemliachenko, A.N., et al.: Lossy compression of hyperspectral images based on noise parameters estimation and variance stabilizing transform. J. Appl. Remote Sens. **8**(1), 083571 (2014). https://doi.org/10.1117/1.JRS.8.083571
13. Ponomarenko, N., Lukin, V., Zriakhov, M., Egiazarian, K.: Lossy compression of images with additive noise. In: Proceedings of International Conference on Advanced Concepts for Intelligent Vision Systems, Antwerpen, Belgium, pp. 381–386 (2005)
14. Ponomarenko, N., Krivenko, S., Lukin, V., Egiazarian, K., Astola, J.T.: Lossy compression of noisy images based on visual quality: a comprehensive study. EURASIP J. Adv. Signal Process. **1**, 976436 (2010). https://doi.org/10.1155/2010/976436
15. Uss, M.L., Vozel, B., Lukin, V.V., Chehdi, K.: Image informative maps for component-wise estimating parameters of signal-dependent noise. J. Electron. Imaging **22**(1), 013019 (2013). https://doi.org/10.1117/1.JEI.22.1.013019
16. Zemliachenko, A.N., Abramov, S.K., Lukin, V.V., Vozel, B., Chehdi, K.: Lossy compression of noisy remote sensing images with prediction of optimal operation point existence and parameters. J. Appl. Remote Sens. **9**(1), 095066 (2015). https://doi.org/10.1117/1.JRS.9.095066
17. Zemliachenko, A., Kozhemiakin, R., Vozel, B., Lukin, V.: Prediction of compression ratio in lossy compression of noisy images. In: Proceedings of TCSET 2016, Lviv-Slavske, Ukraine, pp. 693–697 (2016)
18. http://xooyoozoo.github.io/yolo-octo-bugfixes/#moscow&jpg=s&bpg=s

19. https://bellard.org/bpg/
20. Zemliachenko, A., Lukin, V., Ponomarenko, N., Egiazarian, K., Astola, J.: Still image/video frame lossy compression providing a desired visual quality. Multidimens. Syst. Signal Process. **27**(3), 697–718 (2015). https://doi.org/10.1007/s11045-015-0333-8
21. Chatterjee, P., Milanfar, P.: Is denoising dead? IEEE Trans. Image Process. **19**(4), 895–911 (2010). https://doi.org/10.1109/TIP.2009.2037087
22. Ponomarenko, N., Lukin, V., Astola, J., Egiazarian, K.: Analysis of HVS-metrics' properties using color image database TID2013. In: Battiato, S., Blanc-Talon, J., Gallo, G., Philips, W., Popescu, D., Scheunders, P. (eds.) ACIVS 2015. LNCS, vol. 9386, pp. 613–624. Springer, Cham (2015). https://doi.org/10.1007/978-3-319-25903-1_53
23. Lukin, V., Zriakhov, M., Krivenko, S., Ponomarenko, N., Miao, Z.: Lossy compression of images without visible distortions and its applications. In: CD ROM Proceedings of ICSP, Beijing, 4 p., October (2010)

Powder Mixtures Analysis for Laser Cladding Using OpenCV Library

Dmitriy Kritskiy[1], Olha Pohudina[1(✉)], Mykhailo Kovalevskyi[1], Yevgen Tsegelnyk[2], and Volodymyr Kombarov[2]

[1] National Aerospace University "Kharkiv Aviation Institute", 17 Chkalova Street, Kharkiv 61070, Ukraine
olga.pogudina@khai.edu

[2] O. M. Beketov National University of Urban Economy in Kharkiv, 17 Marshala Bazhanova Street, Kharkiv 61002, Ukraine
y.tsegelnyk@kname.edu.ua

Abstract. Powders and their mixtures are used in innovative additive technologies for aerospace industry. The laser cladding modes required to provide quality parts depend on many parameters, including the fractional composition of powders and their mixtures. The use of computer vision systems for processing images' graphic information and recognizing the contour of powder granules for laser cladding is an actual problem. The paper discusses the use of computer vision methods for processing graphical information of images obtained with an electron microscope. Powder image processing includes the elimination of noise in the original image, obtaining a binary image with a binarization threshold sufficient to identify the contours of granules, and the implementation of an algorithm for determining these contours. The paper proposes to use the OpenCV library to create software for the analysis of the powder image. The comparison of the quality of contour recognition was carried out using the example of three types of powders with different sizes of fraction, the error of the analysis was calculated. For powder of medium and large fractions, where the error was 5% and 6% respectively.

Keywords: Computer vision · OpenCV · Laser cladding · Freeman chain code · Gaussian filter · Threshold transformation

1 Introduction

Powders are used in various technologies, for example, in plasma technologies [1, 2], spraying coatings [3–5], aerospace technologies [6, 7], and most often in recent times, additive manufacturing [8–10]. Powder characteristics are particularly important in the additive manufacturing of parts [11]. In some papers the approaches of modeling the filling of the volume with powder particles in layers, which allows estimating the porosity of the material deposited from powders having different fractional compositions are described [12–14]. The input data for such modeling problems are the results of an experimental analysis of the characteristics of the powder, which can be obtained using computer vision technologies [15].

M. Nechyporuk et al. (Eds.): ICTM 2021, LNNS 367, pp. 924–937, 2022.
https://doi.org/10.1007/978-3-030-94259-5_72

Computer vision-based systems for the analysis of powder mixtures already apply in additive manufacturing [16, 17]. The field of computer vision is quite popular now. Interest in it arose at the dawn of attempts to create artificial intelligence [18]. Currently, the number of new solutions and relevant applications for computer vision and artificial intelligence, in general, continues to grow.

With an increase in the number of proposed methods and algorithms for recognizing contours within an image, it becomes unclear the choice of any specific method for certain needs, since each separately taken method is good for some specific narrow range of tasks.

There are many approaches to edge selection in contour recognition algorithms, but almost everything can be divided into two categories: methods based on finding maximums and methods based on finding zeros. Methods in the first category isolate boundaries by calculating the "edge strength", represented as an expression of the image gradient, and then finding the local maximum of the edge strength using the assumed direction of the boundary, usually perpendicular to the gradient vector. Methods of the second category look for the intersections of the abscissa with the second derivative expression, which is usually the zeros of the Laplacian or the zeros of a nonlinear differential equation. As a preprocessing step, image smoothing is almost always applied to edge selection. As a rule, a Gaussian filter is used for these purposes [19].

Although many of the known boundary extraction techniques are based on the computation of the gradient of an image, they differ in the types of filters used to compute gradients in different directions. The complexity of this problem is significant heterogeneity of the contour boundaries on images of the powder granules.

The aim of the paper is to create tools that improve the quality of laser cladding by determining the parameters of the fractional composition from the images of the used powder mixtures. The paper discusses methods of computer vision for processing graphic information of images obtained with an electron microscope. The solution to the problem of determining the fractional composition can be considered successful if the algorithm allows detecting the maximum number of image objects and identifying at least 80% of the images of granules in the image and evaluating their sizes. To achieve this problem, it is necessary to eliminate the noise of the original image, obtain a binary image with a binarization threshold sufficient to identify the contours of granules, and implement an algorithm for determining these contours. In the considered problem, the following are applied: the method of Gaussian filtering, the method of threshold transformation of a binary image, the method for determining the contours using the approximation of horizontal, vertical, and diagonal segments. This paper proposed to use the OpenCV library to create software that allows analyzing images of powder mixtures. The developed software has been tested for image processing of powders used for laser cladding with a powder of small, medium, and large fractions.

Thus, the problem of recognizing the contour of powder granules for laser cladding is an actual problem. The obtained characteristics of the contour will be used to assess the quality of the powder and select the deposition mode.

2 Features of Contour Analysis

The powder, considered as an object of recognition, in images obtained under a microscope, has distinguished boundaries and, as a rule, a rounded shape. As a consequence, the application of the contour analysis method is possible.

Significant restrictions on the scope of contour analysis are imposed by problems associated with contour identifying in images, namely:

- due to the same brightness with the background, the object may not have a clear boundary, or maybe noisy with interference, which leads to impossibility to recognize the contour;
- overlapping of objects or their grouping leads to the fact that the contour is recognized incorrectly and does not correspond to the boundary of the object.

However, the transition to considering only the contours of objects allows going from the image space to the space of the contours, which significantly reduces the complexity of algorithms and calculations [20, 21].

We can conclude that the contour analysis has a rather weak immunity to interference, and any intersection or only partial visibility of the object leads either to the impossibility of detection or to false positives. However, the simplicity and speed of the contour analysis make it possible to apply this approach quite successfully (with a clearly defined object against a contrasting background and no interference).

The resulting images of granules have the necessary parameters for the application of contour analysis; therefore, we will consider the method of their processing.

3 Step-By-Step Image Processing

Since it is known that a contour is a kind of boundary of an object that separates it from the background (other objects), in order to select it, as a rule, several stages are necessary.

The first step is to blur the image. This operation is performed for the reason of more precise detection of the boundaries since the step between the color gradations will be higher than in the original image. There are quite a lot of filters that provide smoothing (Gaussian filter, median filter, etc.).

But how to get the contours directly? Of course, Canny's edge detector is remembered first, and then any other methods of obtaining a binary image can be cited. For example, threshold transformation and detection of an object by color.

In all these cases, a binary image that explicitly sets the boundaries of the object to us. This collection of pixels that make up the boundary of the object is the object contour.

To operate on the resulting contour, it must be somehow represented (encoded), i.e. to put in correspondence to each contour element a certain number. The sequence of such numbers is called the contour code.

Let's enter the concept of an elementary vector. Elementary vector (EV) $\gamma(n)$ is a vector connecting the centers or nodes of adjacent contour cells of the retina, drawn in the direction of the bypass; n is the number of this EV, counted from the detection point a_0 of the contour, $n = 0, 1, \ldots, k - 1$, k is the number of EV in the contour of this image (Fig. 1). Such EVs will be called standard in contrast to EVs, the length and argument of which are given by random numbers. The resulting code in the form of a sequence of EVs connecting adjacent contour cells is called a chain code. A random contour Γ consisting of EV $\gamma(n)$, $n = 0, 1, \ldots, k - 1$, let's write in the form

$$\Gamma\{\gamma(n)\}_{0,k-1} = \{\gamma(0), \gamma(1), \ldots, \gamma(k-1)\}.$$

There are eight different standard EVs on a square retina that connect adjacent pixels (Fig. 2,*a*). Consider some ways to encode contours.

1. Coding on three features: the length of the current EV, the direction of rotation at the transition to the next EV and the angle between adjacent EV.

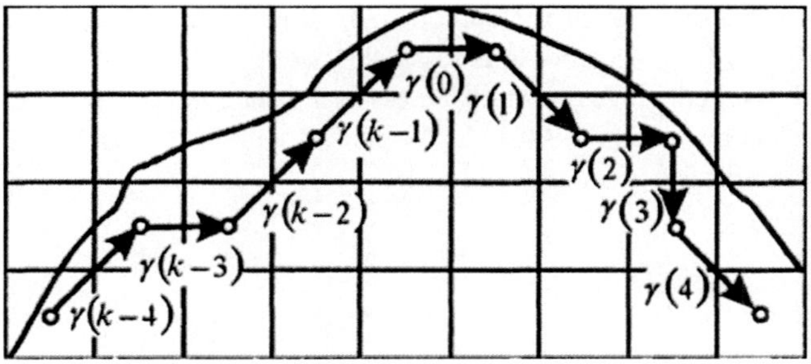

Fig. 1. Definition of contour by elementary vectors

2. Coding of the current EV by three-bit binary code (numbers from 0 to 7) (Fig. 2,*b*). This code was proposed by Freeman and has become widespread in image processing.

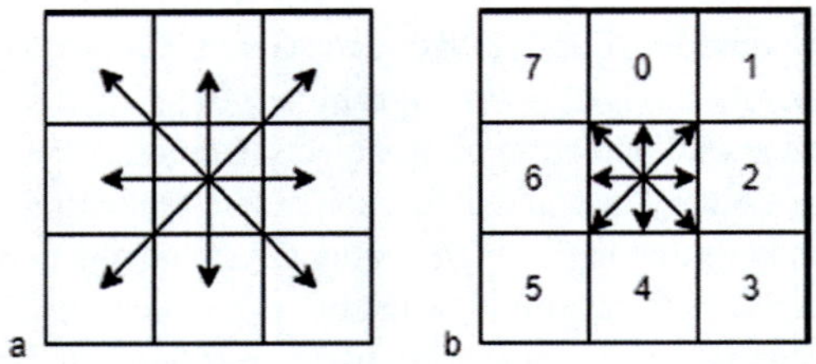

Fig. 2. Varieties of possible EVs on the square retina: without numbering (a) and with numbering of EVs (b)

3. Coding of the current EV by its two projections on the coordinate axes with the starting point, combined with the beginning of EV – two-dimensional code (Fig. 3).

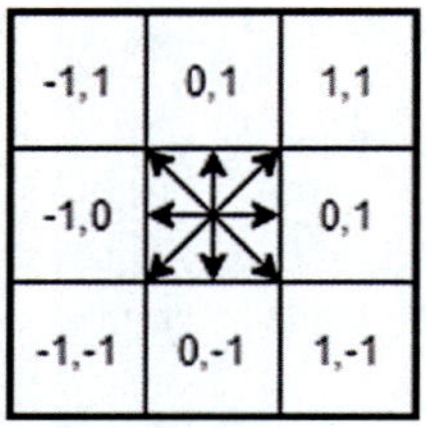

Fig. 3. Coordinates of EV in two-dimensional coding

4. Polygonal representation of the contour, which is obtained by approximating the linear segments (Fig. 4). Coding is to fix the coordinates of the ends of these segments. This method due to the compactness of the obtained descriptions has become widespread. In real cases, it is usually associated with the loss of information about the shape of the images.

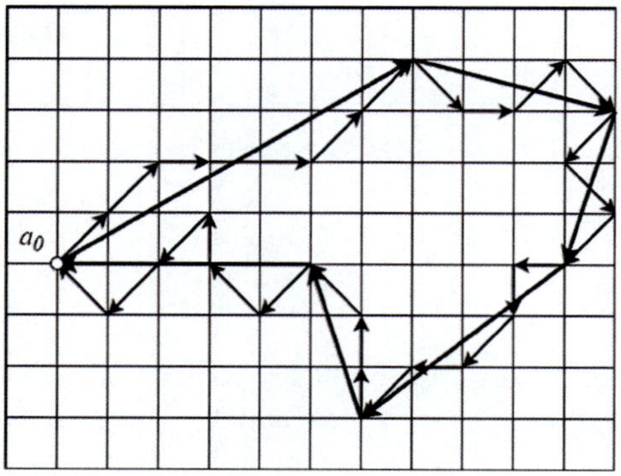

Fig. 4. Polygonal representation of the contour

5. Definition of EV of the contour by eight complex numbers $\{1, 1-i, -i, -1-i, -1, -1+i, i, 1+i\}$. Let's define the complex-valued contour code $\{\gamma(n)\}_{0,k-1}$ as a generalization of the Freeman chain code $\Delta\Phi$ to the plane of a discrete complex variable with a square grid. Instead of the numbers encoding the current contour EVs $\gamma(n)$, each EV corresponds to the coordinate of its end point in the frame of reference, combined with the vector beginning (Fig. 5). These EVs are standard. The modules of their real and imaginary parts are equal to either one or zero.

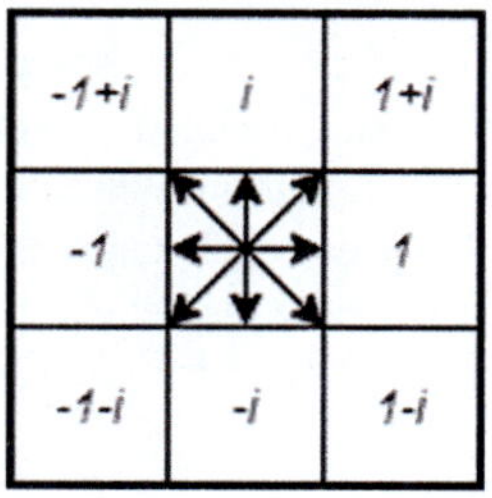

Fig. 5. Coordinates of EV at complex-valued coding

6. *P*-representation of the contour is based on the drawing in the image a number of equidistant sections parallel to the coordinate axis and determining the coordinates of the points of intersection with the contour. The contour is given by a vector, which components are the coordinates of the boundary points of the image in each intersection. The specified representation is useful at an output of the image on the indicator, at definition of the fact of intersection of images, calculation of some geometrical characteristics.
7. Representation of a contour line by a polar code. In the image $\omega(m_1, m_2)$ the pole is selected – the starting point (*P*) of natural (own) coordinate system, i.e. the frame of reference associated with this image. The centers of all boundary points of the image are connected with *P*. The result is a sequence of radius vectors $\beta(n)$, which uniquely define the contour of the image (Fig. 6). Often the starting point *P* is combined with the center of gravity of the image.

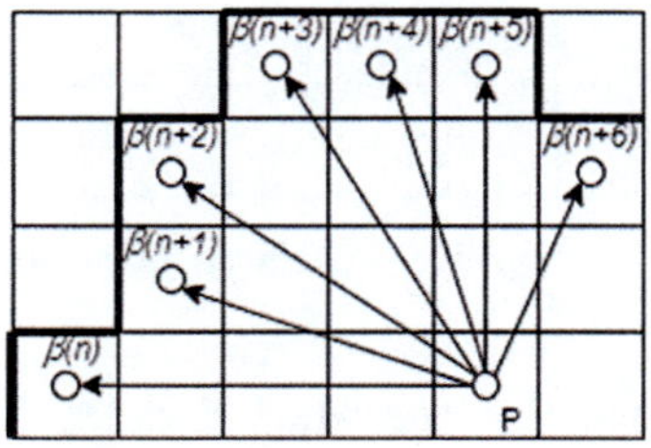

Fig. 6. Example of setting a contour fragment by a polar code: $B = \{\ldots; -4; -3+i; -3+2i; -2+4i; -1+4i; 4i; 1+2i; \ldots\}$

8. Representation of the contour line as a function of a complex variable. The contour as a continuous closed curve in the complex plane is given by its coordinate $z(l)$, which depends on the current length l. An analytical description of the contour of the angular function in the form of the dependence $\theta(l)$ of the tangent angle on the current length l is used, which is invariant to the group of linear transformations. Economical and invariant contour descriptions can be obtained on the basis of Fourier coefficients for the angular function.

These methods of encoding the contours of the images are only a small part of the known and used by various experts in image processing.

To solve problem of coding the contour is proposed and considered Freeman chain code method. In this method chain codes are used to represent a boundary as a sequence of straight line segments of a certain length and direction. This representation is based on a 4- or 8-connected grid. The length of each segment is determined by the resolution of the grid, and the directions are given by the selected code (to represent all directions in a 4-connected grid, 2 bits are enough, and for an 8-connected grid of the chain code (Fig. 7), 3 bits are required).

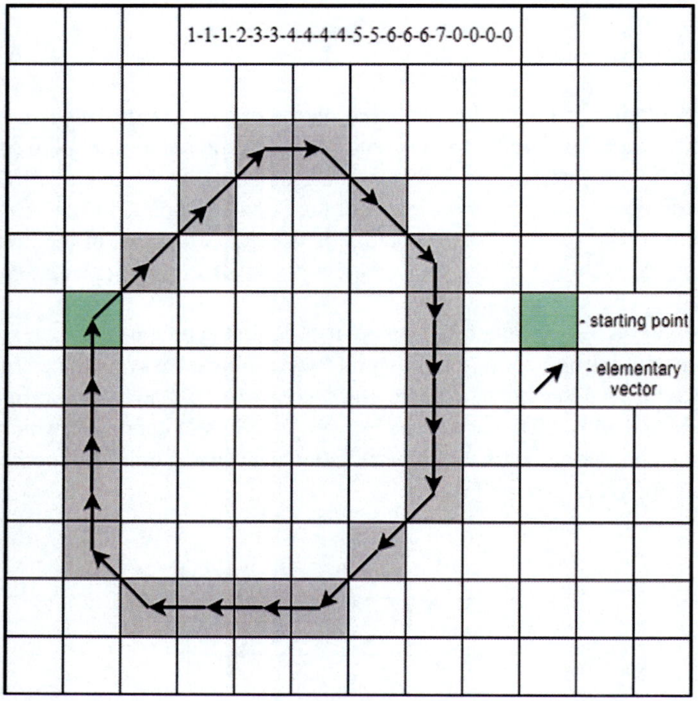

Fig. 7. Freeman chain code in an 8-connected grid

First, a point is fixed on the contour, which is called the starting point. Then, the contour is bypassed, and each offset vector is written by the complex number $a + ib$, where a is the offset of the point on the X axis, and b is the offset on the Y axis. The offset is taken relative to the previous point. In this approach, the contour is defined as a set of elementary vectors represented by their two-dimensional coordinates. Changing the starting point will lead to a cyclic shift of the vector contour, and changing the scale of the image can be considered as multiplying the elementary vector by the scale factor.

The scalar product of the contours $K1$ and $K2$ is the following complex number:

$$\eta = (K1, K2) = \sum_{n=0}^{k-1} (\gamma_n, \nu_n), \tag{1}$$

where k is the dimension of the vector-contour, γ_n is the n-th elementary vector of the contour $K2$, ν_n is the n-th elementary vector of the contour $K1$, (γ_n, ν_n) is the scalar product of complex numbers calculated as

$$(a + ib, c + id) = (a + ib)(c - id) = ac + bd + i(bc - ad). \tag{2}$$

The real part of the scalar product of complex numbers coincides with the scalar product of the corresponding vectors, i.e. includes a vector scalar product. The greater the value of the scalar product, the "closer" the vectors to each other. For perpendicular vectors it will be zero, will be negative for vectors directed in different directions, and for collinear will have the maximum value.

The normalized scalar product (NSP), which is determined by Eq. (3) in the space of complex numbers, is also a complex number. In this case, the one is the maximum possible value of the NSP module and it is achieved only if condition (5) is fulfilled

$$\eta = (K2, K1)/|K2||K1|, \tag{3}$$

where $|K1|$ and $|K2|$ are norms (lengths) of contours.

The norms of the contours are calculated as

$$|K2| = (\sum_{n=0}^{k-1} |\gamma_n|^2)^{1/2}, \tag{4}$$

$$K2 = \mu K1, \tag{5}$$

where μ is the random complex number.

Based on the physical meaning of the multiplication of complex numbers, it turns out that the module of NSP will reach the maximum value – one – only if the contour $K2$ is the same contour as $K1$, even if it is rotated at an angle and scaled by a certain factor.

But equality in (5) is achieved only when the starting points of the contours coincide. Otherwise, even if the contours are the same, the module of NSP of such contours will not be equal to one. To solve this problem, you can use the intercorrelation function (ICF) of the two contours, which is determined by Eq. (6), the values of which show the degree of similarity of the contours, if you shift the starting point N by m positions.

$$\tau(m) = (K2, K1(m)), \tag{6}$$

where $K1(m)$ is the contour obtained from $K1$ by cyclic shifting its elementary vector by m elements. ICF is defined on the whole set of integers, but since the cyclic shift by

k will lead to the original contour, ICF has a period k and it is needed to consider the value of this function only in the range from 0 to $k - 1$.

The maximum module among the values of ICF in (7) shows the degree of similarity of the contours (reaches one for the same contours), and the argument arg(τ_{max}) gives the angle of rotation of one contour relative to another.

$$\tau_{\max} = \max(\frac{\tau(m)}{|K2||K1|}), \quad (7)$$

where $\tau(m)$ is the module of ICF, $|K1|$ and $|K2|$ are norms (lengths) of contours determined by Eqs. (4) and (5).

The chosen method for describing the contours was applied to create the algorithm and software. To operate the program, template contours of powder particles of different diameters were used. These templates were compared with the found contours of the powder particles in the image. A significant limitation in the operation of the algorithm is the need for the same length of the contours. Therefore, the method of approximation of the found contour was additionally applied. In addition, small contours (less than 90 in length) were removed from consideration. Since the contours of the powder have a rounded shape, the angle of rotation was not taken into account, the minimum value of the ICF was set to 0.85.

4 Description of Created Software Product

To solve the problem, namely to find and analyze powder granules on the image, a software product was developed using the OpenCV computer vision library. The object recognition algorithm includes the following steps:

- creation of variables for image storage. The *IplImage* structure is used;
- declarations and initialization of the pointer variable to the name of the image file;
- image loading. The *cvLoadImage* function is used;
- creating *storage* variables (container for storing all contours, type *CvMemStorage*), *first_contour* (pointer to the first contour in the *CvSeq* structure);
- preliminary processing of the image for the correct identification of the necessary objects, namely, the correct translation into a black and white image. Therefore, during finding a contour in OpenCV, it is important to take into account what will be black and white in the image, since black spots in the image appear as the presence of an object, and it is on these objects that the contour is located, and the white area of the image is considered a "hole" (the absence of anything). For such transformations, can used a number of standard functions, as well as splitting the image into channels and transforming each channel using different functions. The type of transformation is determined experimentally by doing experiments on several images with different locations of the same object;
- using the *cvFindContours* contour search function;
- removal of unnecessary contours if the size of the object has some specific value (knowing that the object along the length of the contour cannot be greater or less than any value). To do this, can used a loop that organizes by the list stored in the *cvSeq*

structure and each contour can be accessed using *h_next*, *h_prev* element references. In this case, the length of the contour is determined by the *total* parameter;
- using the contour drawing function *cvDrawContours*() to confirm the correct operation of the contour search function;
- creating program windows in which images will be displayed using the *cvNamedWindow* function;
- loading images using the *cvShowImage*() function, displaying the number of counted powder grains.

The input to the software product is an image of the powder that must be analyzed. Powder, in turn, can be classified into 3 categories: small fraction (Fig. 8,*a*), medium fraction (Fig. 8,*b*), and large fraction (Fig. 8,*c*).

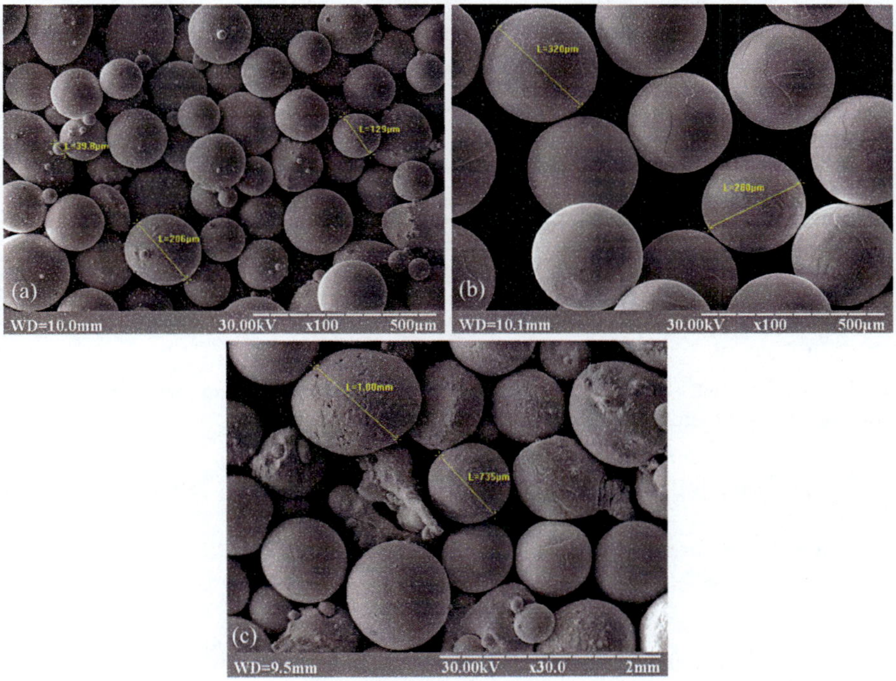

Fig. 8. SEM photo of powder for analysis: small fraction (a), medium fraction (b), and large fraction (c)

After the image has been submitted to the input, the program processes it according to the above described algorithm and analyzes it. As a result, two windows open:

- window with the input image and the contours found on it (Fig. 9,*a*);
- window with analysis results (Fig. 9,*b*).

As can be seen from Fig. 9, the program recognized the powder granules, calculated their number and the average diameter of one granule. Despite the blur and thresholding there are still some noise and darkening on the images, so the recognition algorithm has some error. This error depends not only on the quality of the image received at the input, but also on the quality of the powder, since on some granules the program was able to recognize the internal contours of the granule.

For example, when analyzing powder of small fraction, it was possible to recognize 105 granules, while the actual value is 175. Thus, the analysis error for powder of small fraction is:

$$\varepsilon_s = \frac{|175 - 100|}{175} = 40\%.$$

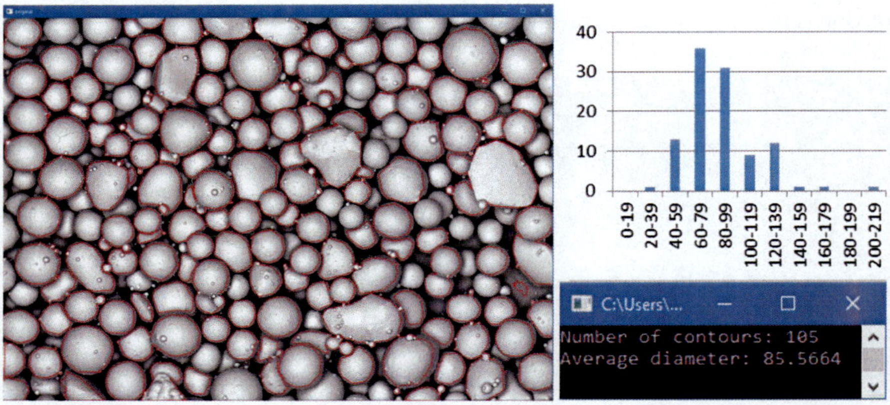

(a) (b)

Fig. 9. Found contours for powder of small fraction (a): red color – external contours, green color – internal contours, diameter distribution and analysis results (b)

The medium powder is analyzed in the same way. The analysis results are shown in Fig. 10.

Analysis error for powder of medium fraction:

$$\varepsilon_m = \frac{|40 - 42|}{40} = 5\%.$$

Now let's also analyze the large powder. The analysis results are shown in Fig. 11.
Analysis error for large powder:

$$\varepsilon_l = \frac{|31 - 29|}{31} \approx 6\%.$$

So the average error of the algorithm:

$$\varepsilon = \frac{\varepsilon_s + \varepsilon_m + \varepsilon_l}{3} = 17\%.$$

To improve the quality of the algorithm in the future, it is necessary to analyze the image binarization parameters, remove contours with a length less than 40, and consider other minimum values for the ISF. We will also look at the HoughCircles function, which uses the Hough transform to detect circles and determine their parameters.

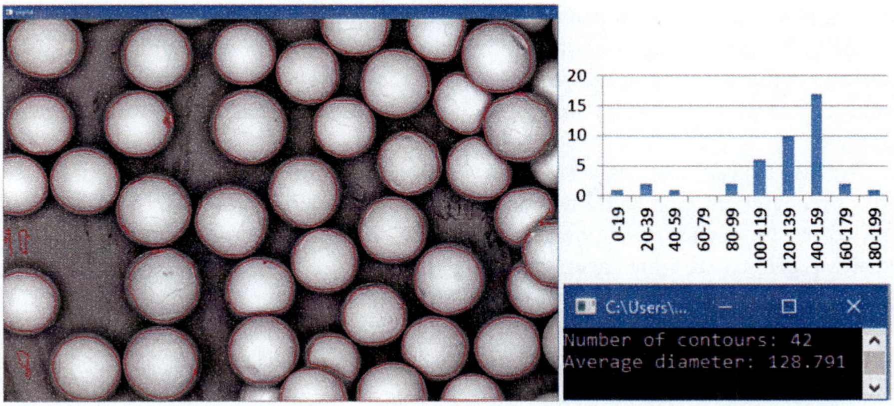

(a) (b)

Fig. 10. Found contours for powder of medium fraction (a): red color – external contours, green color – internal contours, diameter distribution and analysis results (b)

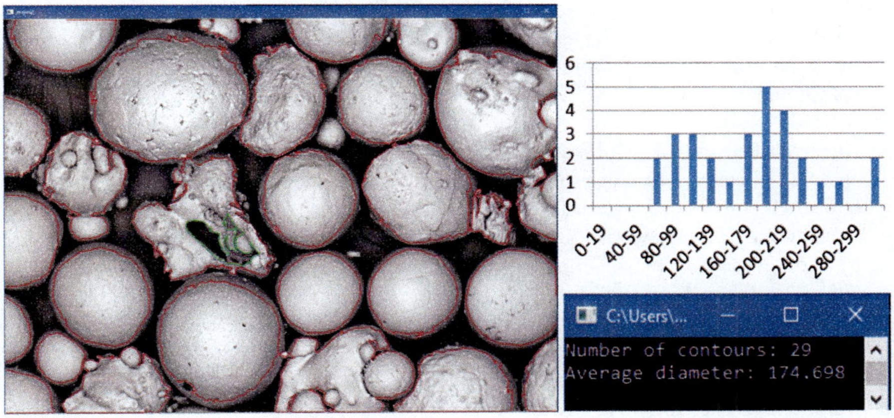

(a) (b)

Fig. 11. Found contours for powder of large fraction (a): red color – external contours, green color – internal contours, diameter distribution and analysis results (b)

5 Conclusions

The article deals with the problem of assessing the quality of powder granules for laser cladding. The possibilities of computer vision methods are considered, namely, the contour processing method using the OpenCV library.

The original images were analyzed to select methods for eliminating noise and preparing contours for recognition. The existing methods of storing information about contours are analyzed. A method has been developed for counting the number of granules and estimating their diameter for three types of powder images.

The software was developed on the basis of the proposed method for processing the contours. The result of processing the initial data will be the image with the outlined contours of the powder granules, the diagram of the distribution of the granule diameters and the general results of the analysis (the number and average diameter of the found granules).

We also analyzed the results of a software product built on the basis of the proposed object recognition approach. The comparison of the quality of contour recognition is carried out for each type of powder: small, medium and large fractions. The analysis error was calculated, which average value is 17%.

Thus, the proposed solution showed quite good results for powder of medium and large fractions, where the error was 5% and 6%, respectively. However, it still needs further improvement when analyzing powder of small fraction.

References

1. Taran, A., Plankovskyy, S., Voronovich, D., Abashin, S.: Emission properties of Re-W dispenser cathodes. In: 2009 IEEE International Vacuum Electronics Conference, pp. 407–408. IEEE, Rome (2009). https://doi.org/10.1109/IVELEC.2009.5193581
2. Taran, A., Plankovskyy, S., Ostrovsky, E., Ordanjan, S.: High-current-density cathodes based on Barium Hafnate with Tungsten. In: 2007 IEEE International Vacuum Electronics Conference, pp. 1–2. IEEE, Kitakyushu (2007). https://doi.org/10.1109/IVELEC.2007.4283309
3. Kuzmych-Ianchuk, I.K., Borisov, Y.S., Bernatskyy, A.V.: Laser-microplasma reactive powder spraying of titanium coatings with nitride phases. Appl. Mech. Mater. **682**, 276–281 (2014). https://doi.org/10.4028/www.scientific.net/AMM.682.276
4. Dolmatov, A.I., Sergeev, S.V., Kurin, M.O., et al.: Kinematics of the solid particle accelerated by a flow of gas in a supersonic nozzle and work hardening of the processed surface. Metallofiz. Noveishie Tekhnol. **37**(7), 871–885 (2016). https://doi.org/10.15407/mfint.37.07.0871
5. Plankovskyy, S., Breus, V., Voronko, V., et al.: Review of methods for obtaining hardening coatings. In: Nechyporuk, M., Pavlikov, V., Kritskiy, D. (eds.) ICTM 2020. LNNS, vol. 188, pp. 332–343. Springer, Cham (2021). https://doi.org/10.1007/978-3-030-66717-7_28
6. Korol', V.N., Bychkov, S.A., Molyar, A.G., et al.: Sintered powder materials for friction units of Antonov aircrafts. Mater. Sci. **39**(1), 132–135 (2003). https://doi.org/10.1023/A:1026194900607
7. Pavlenko, D., Dvirnyk, Y., Przysowa, R.: Advanced materials and technologies for compressor blades of small turbofan engines. Aerospace **8**(1), 1 (2021). https://doi.org/10.3390/aerospace8010001
8. Duriagina, Z.A., Filimonov, O.S., Kulyk, V.V., et al.: Investigation of structural-geometric parameters and elemental composition of spherical VT20 alloy powders. J. Achiev. Mater. Manuf. Eng. **95**(2), 49–56 (2019). https://doi.org/10.5604/01.3001.0013.7914
9. Plankovskyy, S., Shypul, O., Tsegelnyk, Y., et al.: Simulation of surface heating for arbitrary shape's moving bodies/sources by using R-functions. Acta Polytech. **56**(6), 472–477 (2016). https://doi.org/10.14311/AP.2016.56.0472
10. Romanova, T., Stoyan, Y., Pankratov, A., et al.: Optimal layout of ellipses and its application for additive manufacturing. Int. J. Prod. Res. **59**(2), 560–575 (2021). https://doi.org/10.1080/00207543.2019.1697836

11. Duriagina, Z.A., Lemishka, I.A., Trostianchyn, A.M., et al.: The effect of morphology and particle-size distribution of VT20 titanium alloy powders on the mechanical properties of deposited coatings. Powder Metall. Met. Ceram. **57**(11–12), 697–702 (2019). https://doi.org/10.1007/s11106-019-00033-8
12. Stoyan, Y., Yaskov, G., Romanova, T., et al.: Optimized packing multidimensional hyperspheres: a unified approach. Math. Biosci. Eng. **17**(6), 6601–6630 (2020). https://doi.org/10.3934/mbe.2020344
13. Duriagina, Z., Lemishka, I., Litvinchev, I., et al.: Optimized filling of a given cuboid with spherical powders for additive manufacturing. J. Oper. Res. Soc. China **9**(4), 853–868 (2020). https://doi.org/10.1007/s40305-020-00314-9
14. Kartashov, O., Korobchynskyi, K., Yakovleva, I., et al.: Synthesis of spherical object configurations: models and information technologies. CEUR Workshop Proc. **2864**, 287–298 (2021)
15. Pohudina, O., Kritskiy, D., Bykov, A.N., Szalay, T.: Method for identifying and counting objects. In: Nechyporuk, M., Pavlikov, V., Kritskiy, D. (eds.) Integrated Computer Technologies in Mechanical Engineering. AISC, vol. 1113, pp. 161–172. Springer, Cham (2020). https://doi.org/10.1007/978-3-030-37618-5_15
16. Caggiano, A., Zhang, J., Alfieri, V., et al.: Machine learning-based image processing for online defect recognition in additive manufacturing. CIRP Ann. **68**(1), 451–454 (2019). https://doi.org/10.1016/j.cirp.2019.03.021
17. Zhang, J., Habibnejad-korayem, M., Liu, Z., et al.: A computer vision approach to evaluate powder flowability for metal additive manufacturing. Integr. Mater. Manuf. Innovation **10**(3), 429–443 (2021). https://doi.org/10.1007/s40192-021-00226-3
18. Klestov, R.A., Stolbov, V.Y.: Complex recognizing contours method for an image based on computer vision technology. In: GraphiCon 2017 – 27th International Conference on Computer Graphics and Vision, pp. 208–211. Perm (2017)
19. Shih, F.Y.: Image Processing and Pattern Recognition: Fundamentals and Techniques. John Wiley & Sons Inc, Hoboken, NJ, USA (2010). https://doi.org/10.1002/9780470590416
20. Kritskiy, D., Pohudina, O.: Object recognition to refine drone positioning. In: 2020 IEEE 15th International Conference on Computer Sciences and Information Technologies (CSIT), vol. 2, pp. 82–85. IEEE, Zbarazh (2020). https://doi.org/10.1109/CSIT49958.2020.9321945
21. Aksonov, Y., Kombarov, V., Tsegelnyk, Y., et al.: Visualization and analysis of technological systems experimental operating results. In: 2021 IEEE 16th International Conference on Computer Sciences and Information Technologies (CSIT), vol. 2, pp. 141–146. IEEE, Lviv (2021). https://doi.org/10.1109/CSIT52700.2021.9648592

Smart Energy and Grids

Selection Methodology of Ecological Safety Priorities of Sustainable Development Goals of Urban Agglomerations

Natalia Teliura[1(✉)], Nataliia Tsapko[2], Hanna Khabarova[2], Olga Lomakina[1], Olena Pshenichnova[2], and Tetiana Klochko[3]

[1] O. M. Beketov National University of Urban Economy in Kharkiv, 17 Marshala Bazhanova Street, Kharkiv 61002, Ukraine
natalya.telyura@kname.edu.ua

[2] Ukrainian Scientific and Research Institute of Ecological Problems, 6 Bakulina Street, Kharkiv 61165, Ukraine

[3] National Aerospace University "Kharkiv Aviation Institute", 17 Chkalova Street, Kharkiv 61070, Ukraine

Abstract. Paper treats the theoretical and methodological approach to the choice of ecological safe direction of agglomerations development. Groups of indicators and indexes for the selection procedure implementation have been defined. The proposed approach using allows to involve local authorities specialists of different profiles to manage of environmental safety of agglomerations from the position of their sustainable development. The methodological approach essence is to use the Analytic Hierarchy Process (AHP). It has been proposed and implemented groups of indicators and indexes, formulated as sustainable development components – environmental, social and economic. Relevant specialists, as experts, based on information of different types (statistical, forecast, direct measurement data) on the particular agglomeration, give their own judgments about the priority of the advantages of the group characteristics of indicators and indexes. The results of judgments processing of experts according to the formal procedure of AHP are the basis for decision-making in choosing the environmental safe direction of agglomeration development. The multi-criteria hierarchical structure of the choice of environmental safe direction of agglomeration development has been represented by the sequence of actions, which includes three stages: construction of the hierarchical model of features comparison; formation of matrices of elements pairwise comparisons of each level of hierarchy and their local weights determination; global weights determination, consistency index and selection of the best development direction. The advantage of the proposed multi-criteria methodological approach is the ability to combine to the one decision-making algorithm the original data, which differ in content (environmental, social and economic-technological) and the presentation form (statistical, forecast, direct measurement data, expert assessments).

Keywords: Sustainable development · Environmental safety · Agglomeration · Analytic Hierarchy Process

M. Nechyporuk et al. (Eds.): ICTM 2021, LNNS 367, pp. 941–950, 2022.
https://doi.org/10.1007/978-3-030-94259-5_73

1 Introduction

Awareness of the necessity in balance and stabilize in the use and development of resources is crucial modern condition for ensuring the development of the state and the agglomeration, during which the life quality increases while reducing the impact on nature. This is the strategic goal of environmental safety policy that encourages to create the environment conducive to human health. Providing the favorable environment for human life should be based on the criteria of sustainable development [1, 2].

Currently, methodological (program-analytical) approaches creation are important to support management decisions by the highest management section of the settlements.

In the conditions of necessary energy and resource saving, and balanced development, which should be based on the functioning of three system components, namely: economic, social and environmental [3, 4], it becomes important to create the methodological approach to support management decisions in choosing environmental safe direction of agglomeration development. This approach, firstly, based on many factors and criteria will help to solve the issue of evaluating the decision end result. Secondly, in the case of selecting and implementing priority drainage technologies, they will help to analyze the alternative or determine the effectiveness of individual stages of the decision-making process [5].

In such setting in previously published studies, the task has not been set.

2 Problem Statement

Analyzing the treatments of urban agglomerations, and in particular residential complexes it should be noted, that these issues have not been covered in [6]. The paper [6] elucidates, based on study of two residential complexes in Abu Dhabi, the access to sustainability measurements used at both the environmental and social levels within the accredited national rating system by Estidama. Although the study focuses on the application of new technologies that increase the level of sustainability in future housing projects, but the accounting mechanism of economic indicators and indexes have not been identified.

Studies [7] determine that sustainable urban development is the major activity in the world in general, and in India in particular. Researchers in the study [7] have used this concept mainly for sustainable urban development in the concept of the importance of sustainable development.

According to the authors, sustainable urban development has been intended to reduce the harmful effects of climate change, depletion of non-renewable resources and urban degradation, but the optimal ways and directions of development have not been identified based on sustainability, there have been no indicators for thorough further research.

There in the [8] it has been defined that over the last few years the surge of interest in developing and implementing tools, frameworks and indicator sets to assess the smart city. This demonstrates the growing recognition of the evaluation schemes importance to better inform the planning and design of smart and sustainable cities. The outcomes show that different approaches have been applied, but there were some common features. Index is the dominant format in the schemes, and the most common topics are: economy, people, governance, environment, mobility, housing and data, but there has been identified no approaches and methods, hierarchies for further analysis and selection and decision-making on the use of evaluation results.

The global trends of smart city [9] show the focus shift from sectoral interventions to integrated decision-making informed by Big Data. This step to integration manifested in the emergence of Integrated Urban Management Platforms (IUMP). Despite the deepening of the data obtained by IUMP and the concomitant increasing of computing power, limited research has been conducted to study the use of this data to develop quantitative tools for integrated smart planning and urban management. Recognition of the necessity to use model bases in the practice of agglomeration management is the crucial and necessary condition for further development based on sustainability, but there is no definition for the tools using that allow to apply different data, both qualitative and quantitative.

Paper [10] presents the overview of the abandoned military facilities condition on the Zagreb agglomeration area based on mapping of preceded expertise of official digital ortho-photographs. The authors have analyzed previous practices of transformation of deposited territories and optimal use of future military sites depending on their specific position in the urban agglomeration. The analysis has been based on the study included semi-structured interviews with key stakeholders on the revitalization of underdeveloped areas in the Zagreb agglomeration area. Recognition of the need for optimal development of agglomerations and their priority in the implementation and realization requires the approach application, which will allow in the best way to take into account the all available indicators and indexes and determine the reasonable management decision.

One of the levers of regulated and reasonable decision-making on the choice of environmental safe direction of agglomeration development is the methods of mathematical analysis.

The research [11] has been devoted to the substantiation of the decision-making method, based on comparative analysis and evaluation of different methods of choice to determine what one method is better than another, and it has been introduced 16 criteria that could be used to judge and evaluate different methods. The special levers and weights of each of the 16 criteria have been identified, but the expert role, his experience and practical knowledge in the interpretation and analysis of results has been especially defined. It should be noted that the authors have not determined how to take into account and interpret the different views of experts and the approach to coordinate them.

The author of [12] has applied the decision-making method at the level of the settlement for the Odessa city (Ukraine). It has been determined that the implementation of the proposed method should take into account such indicators as treatment efficiency, energy efficiency and other criteria and factors for eutrophied water bodies. However, despite the advantages of the adapted Analytic Hierarchy Process (AHP) T. Saati, for the management system of environmental safety of individual objects, the issue of its application in the agglomeration plane remains actual. Criteria, factors and sources of influence in the cross-border context have been defined, but the mechanisms of interaction of influence levels have not been defined.

This indicates the necessity of expand the list of additional indicators of sustainable development components, adapted in accordance to the object of study, in forming the criteria base of the hierarchy of environmental safe direction of agglomeration development and it is advisable to conduct the treatment to develop the methodological approach. In such setting in previously published studies, the issue has not been set.

3 The Aim and Objectives of the Study

Aim is to develop methodical approach to determine the environmental safe direction of agglomeration development due to the Analytic Hierarchy Process (AHP). The follow tasks have been put forwarded to achieve the aim:

- to develop the main groups of indicators and indexes of the methodological approach
- to form the hierarchical model of indicators and indexes groups that influence on the choice of the priority direction of agglomerations development, and to reveal the order of their analysis by AHP.

4 Development of the Main Groups of Indicators and Indexes of the Methodological Approach of Environmental Safe Direction of Agglomeration Development Definition

Implementation of the methodological approach to determine the priority environmental safe direction of agglomeration development is impossible without reliable and systematic information. To build the hierarchical model, there was the need to create the system of indicators of socio-economic and environmental status of the urban agglomeration, which should be aimed at statistical support and support of urban development processes, be relatively technically simple and, at the same time, solve the necessary issues.

Indicators of socio-ecological and economic development of urban agglomerations have been chosen as such indicators. Indicators are relatively simple specific and structural indicators obtained by calculation from primary indicators, which in the simplified form represent the certain situation that is part of the complicated complex or system.

Indicators of socio-ecological and economic development of urban agglomerations are indicators that characterize the change in the economy state, social sphere and the environment over time. They are the tool for measuring, visualizing and discussing important development issues. They give the quantitative and qualitative description of the issue and allow to assess the situation, to note its changes in the positive or negative direction. They provide the opportunity to timely adjust socio-economic and environmental development.

There has been considered the information base of the study and, in particular, the formation of the information-based initial (priori) list of the indicators and indexes main groups of the methodological approach of urban agglomerations development.

Considering the urban agglomerations development as the complex process that leads to solve urban problems, to improve living conditions of citizens by achieving balanced socio-economic and environmental development, many researchers [9, 12, 13] have identified three basic components (subsystems) that create the system ensuring of socio-ecological and economic urban agglomerations development.

1. Economic development integrates the main indicators of the manufacturing sector, intangibles, financial, investment and foreign economic activity, trade and transport. In conditions of limited funding, it regulates development processes and the level of investment to environmental protection.

2. Social development reflects the demographic situation, education and health, living standards (public associations, labor market, state of infrastructure, etc.).
3. Ecological development accumulates data on the state and level of environment pollution. It also includes indicators of the local environmental management level.

5 Hierarchical Model Formation of Indicators and Indexes Groups that Influence on the Choice of the Priority Direction of Agglomeration Development, and Disclosure of the Order of its Analysis by AHP

According to [11, 12], it has been proposed to implement three stages:

- construction of the hierarchical model of elements (signs) comparison of the problem;
- matrices formation of pairwise elements comparisons of each level of hierarchy and definition of their local weights;
- determination of global weights, consistency index and the best solution selection.

Stage 1. Hierarchical model development for elements comparing of the problem. To develop the hierarchical model the concept of sustainable development has been used adopted by the UN in September 2015 list of Sustainable Development Goals, which includes 17 key positions, one of the goals is to ensure openness, protection, sustainability and environmental sustainability of cities and towns and urban agglomerations (UA). There is the consideration of social, economic and ecological balanced development, aimed to create their economic potential, full living environment for modern and future generations, namely the goal #11, entitled "Sustainable development of cities and communities", adapted for Ukraine. This goal based on the results work of expert groups under the auspices of the UN Development Program in Ukraine together with the Ministry of Economic Development of Ukraine [13–16].

At the 1st level, the goal has been defined (Fig. 1): "Improving the environmental, socio-economic protection due to implementation the goal of sustainable development" and, accordingly, six hierarchical levels. Achieving this goal has been stipulated by the solution of the following general tasks (elements of the 2nd level), which are formulated as components of sustainable development: K1 – environmental, K2 – social and K3 – economic subcriteria of cities, settlements and urban agglomerations safety. Level 3 is the state factors (SF 1,2,3), which detail the safety criteria in terms of the environmental component. At the 4th level there are social indicators (SI 1,2,3,4), which characterize the social component. At the 5th level the economic indicators (EI 1,2,3,4) are presented, directed on improvement of estimation signs of economic subcriteria of cities, settlements and urban agglomerations safety. At the 6th (last) level there is the list of options and objectives of goal #11 "Sustainable development of cities and communities" UN Sustainable Development Goals, adapted for Ukraine:

- G11.1: ensuring the availability of housing;
- G11.2: ensuring the development of settlements and territories exclusively based on integrated planning and management with public participation;

- G11.3: ensuring the preservation of cultural and natural heritage with the involvement of the private sector;
- G11.4: ensuring timely notification of the population about emergencies with the use of innovative technologies;
- G11.5: reducing the negative impact of pollutants, including on the urban environment by the innovative technologies implementation;
- G11.6: reducing the development and implementation of local development strategies aimed at economic growth, job creation, tourism development, recreation, local culture and local production.

The formed hierarchical model of comparison (Fig. 1) at each level of the hierarchy contains no more than nine elements for research that meets the requirements [11].

Stage 2. Matrices formation of pairwise elements comparisons of each level of hierarchy and definition of their local weights. For this purpose, the experts group who are the specialists of the relevant professional field of municipal authorities has been involved. Each expert according to his profile (Fig. 1), forms the square inversely symmetric matrix of dominance (judgments) which is written in the form:

$$A = \left(a_{ij}\right)_{n,n}, a_{ji} = \frac{1}{a_{ij}}, i = \overline{1, n}; j = \overline{1, n}, a_{ij} = 1 \text{ at } i = j, \tag{1}$$

where $a_{ij} = w_i/w_j$; w_i, w_j are weights of relevant elements (criteria and factors).

Pairwise comparison $w_1, w_2, \ldots, w_n$ of elements has been performed using subjective judgments of the expert, numerically evaluated from 1 to 9 on the special scale of relative importance [11, 17].

During the description performing of dominance and assessing the degree of consistency of expert judgments, the eigenvector (i.e. the priority vector w) and, accordingly the maximum eigenvalue $\lambda_{\max}$ has been determined. For each matrix the matrix equation relatively to $\lambda_{\max}$ and vector w_i has been solved.

$$A\begin{pmatrix} w_1 \\ w_n \end{pmatrix} = \lambda_{\max}\begin{pmatrix} w_1 \\ w_n \end{pmatrix}. \tag{2}$$

The solution of this equation has been performed by reducing the matrix A to a sufficiently high degree, followed by summation of rows and normalization (dividing the sum of each row by the sum of all elements of the matrix). The priority vector $w = (w_1, w_2, \ldots, w_n)^T$ has been obtained as result.

Stage 3. Determination of global weights, consistency index and the best solution selection. The completed dominance matrices (1) has been used to determine the weights and global priorities of local criteria and factors.

The global factor priority has been calculated by the formula:

$$w_{ij} = \sum_{j=1}^{m} u_{ij} v_{(i-1),j}, \tag{3}$$

where $v_{(i-1),1}, v_{(i-1),2}, \ldots, v_{(i-1),m}$ are weights or global priorities of local goals or level factors $(i-1)$; $u_{i1}, u_{i2}, \ldots, u_{im}$ are weights or local priorities of the j-th criterion or i-level factor relative to local goals.

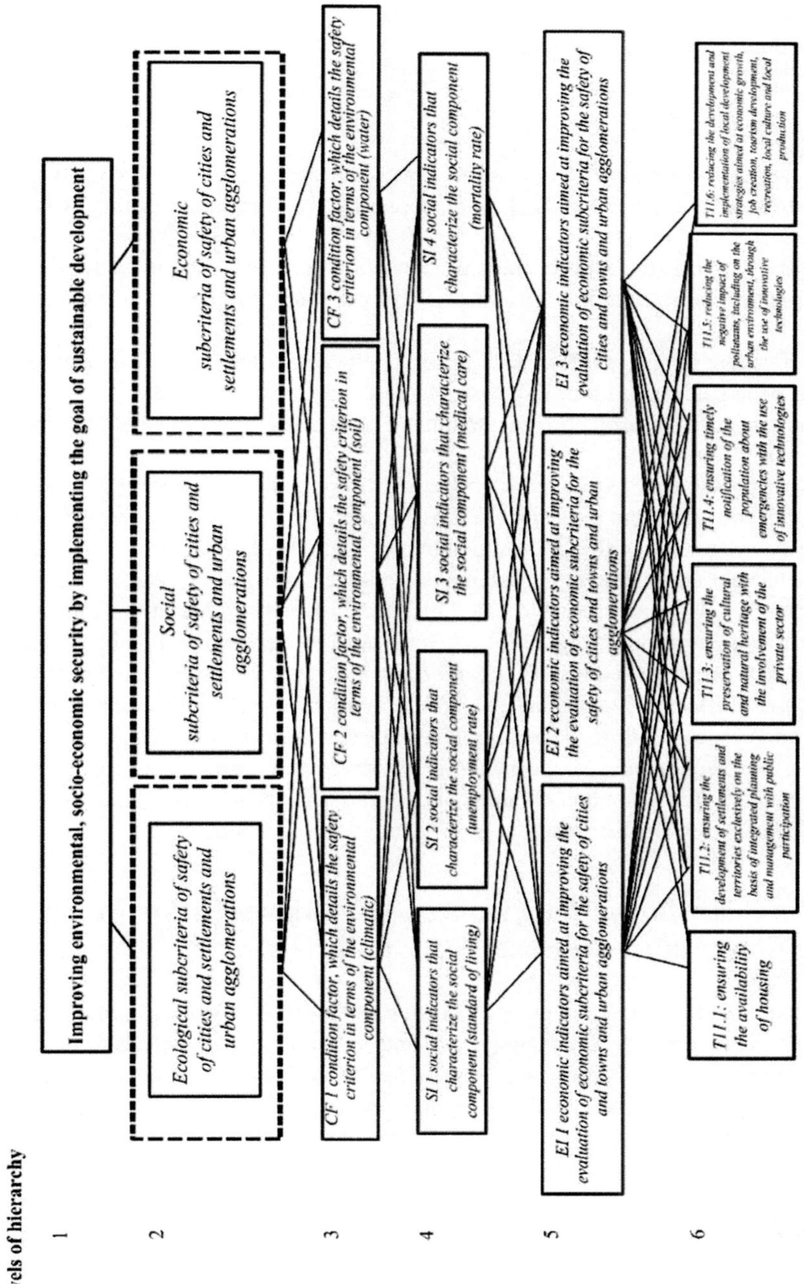

Fig. 1. Decision-making hierarchy.

The consistency of the whole hierarchy has been checked. It could be evaluated by multiplying each consistency index on the priority of the corresponding criterion, summing the obtained numbers and comparing the result with the average consistency index of random matrices of the same order:

$$IY = \frac{\lambda_{\max} - n}{n - 1}. \tag{4}$$

The consistency index has been compared with the average consistency index for random matrices of the same order [17], and if necessary, the quantitative estimates have been refined.

Consistency ratio is the ratio of the consistency index to the average value of the consistency index in the randomly selecting case of the comparison matrix coefficients [17].

For each variant of tasks G11.1…G11.6 (Fig. 1) the quantitative value of the global priority has been calculated according to the largest value of this indicator, which has been recommended for priority implementation in the particular city, settlement or urban agglomeration.

Achieving the Goal 11 "Sustainable Urban and Community Development" implies that sustainable development ensuring of cities, settlements and communities as favorable living environments will require the shift from administratively approved plans to territorial development strategies. This requires partnership, coherence and inclusive participation of local communities, local authorities, civil society organizations and businesses. Development strategies at the local level should be formed in compliance with the principles of openness, inclusiveness and scientific validity with the simultaneous introduction of a system of indicators for further monitoring and public control in the implementation process [18, 19]. Therefore, the chosen option will be the best from the standpoint of real requirements for sustainable development in the particular city, settlement or urban agglomeration and obtained taking into account information of various types (statistical, forecast, direct measurement data, expert assessments).

6 Results Discussion of the Proposed Methodological Approach to the Choice of Development Priority Direction

The developed methodological approach allowed specialists of local authorities with professional orientation (environmental, social and economic-technological) to participate in the preparation of consolidated recommendations for management decisions on the choice of priority direction of development in the particular city, settlement and urban agglomeration.

The advantage of the proposed multi-criteria methodological approach is the ability to combine to the one decision-making algorithm the original data, which differ in content (environmental, social and economic-technological) and the presentation form (statistical, forecast, direct measurement data, expert assessments).

In addition, the next advantage of the developed methodological approach is that it is based on the well-developed and well-established Analytic Hierarchy Process. In particular, there are several software products for this method that allow to effectively calculate the initial indicators. The program MPriority 1.0 has been used for treating the paper materials, which allowed to perform calculations and obtain consistency indexes for the problem with high accuracy.

The limitation of this research is the real limits of application of the proposed approach, as there is the possibility of external factors and conditions, which could lead to stability loss.

The development of this study should be carried out by the following areas:

- development of numerical multi-criteria hierarchies of choice of development priority direction in the conditions of the particular city, settlement, agglomeration based on methods of the systems analysis (SA), to increase the stability and coherence of the methodical approach;
- approbation on the particular settlements and recommendations development on adaptation of the obtained tools at the improvement substantiation or construction of the new drainage system of the settlement located on the eutrophied water object.

7 Conclusion

The AHP methodical approach at definition of priority technologies of drainage from the settlements territories has been developed. Criteria formulated as components of sustainable development – environmental, social and economic-technological for the decision-making method at technological measures choosing for environmental safe drainage in settlements located on eutrophied water bodies, have been formulated and implemented. The hierarchy elements have been determined based on the management strategy in the system of ecological and social safety of the settlements territories.

The proposed method implementation allows to streamline, algorithmize and adjust the procedure of expert evaluation of various factors and improve the quality of the obtained results in the formation of the decision-making process.

Despite the rather large dimension of the elements array of the multi-criteria hierarchical structure of technological measures choice, the correct pairwise comparison with the achievement the given level of consistency ($IY \leq 10\%$) for all three drainage basins of the investigated settlement has been provided. This indicates the correctness of the issue and the reliability of the results.

References

1. Lelechenko, A.P., et al.: Mechanisms of inter-state communications for solving sustainable development problems. Asia Life Sci. Supp. **22**(2), 1–14 (2020)
2. Papagiannis, F., Gazzola, P., Burak, O., Pokutsa, I.: An intelligent environmental plan for sustainable regionalisation policies: the case of Ukraine. Environ. Sci. Policy **108**, 77–84 (2020). https://doi.org/10.1016/j.envsci.2020.03.010
3. Dotsenko, N., Chumachenko, D., Chumachenko, I.: Project-oriented management of adaptive teams' formation resources in multi-project environment. CEUR Workshop Proc. **2353**, 911–923 (2019)

4. Babaiev, V.M., Kadykova, I.M., Husieva, Y., Chumachenko, I.V.: The method of adaptation of a project-oriented organization's strategy to exogenous changes. Naukovyi Visnyk Natsionalnoho Hirnychoho Universytetu **2**, 134–140 (2017)
5. Bobrovska, O.Y., et al.: Management decisions to support and maintain the socio-economic development of the territories. Estudios de Economia Aplicada. **39**(6), (2021). https://doi.org/10.25115/eea.v39i6.5266
6. Ibrahim, I.A.: Sustainable housing development: role and significance of satisfaction aspect. City, Territory and Architecture **7**(1), 1–13 (2020). https://doi.org/10.1186/s40410-020-00130-x
7. Bera, S.: Making a sustainable city through urban development: the Indian scenario. Indian J. Appl. Res. **10**(3), 56–58 (2020). https://doi.org/10.36106/ijar/1417974
8. Sharifi, A.: A typology of smart city assessment tools and indicator sets. Sustain. Cities Soc. **53**, 101936 (2020). https://doi.org/10.1016/j.scs.2019.101936
9. Westraadt, L., Calitz, A.: A modelling framework for integrated smart city planning and management. Sustain. Cities Soc. **63**, 102444 (2020). https://doi.org/10.1016/j.scs.2020.102444
10. Matković, I., Jakovčić, M.: Conversion and sustainable use of abandoned military sites in the Zagreb Urban Agglomeration. Hrvatski Geografski Glasnik. **82**(2), 81–106 (2020). https://doi.org/10.21861/HGG.2020.82.02.03
11. Saaty, T.L., Ergu, D.: When is a decision-making method trustworthy? criteria for evaluating multi-criteria decision-making methods. Int. J. Inf. Technol. Decis. Mak. **14**(06), 1171–1187 (2015). https://doi.org/10.1142/S021962201550025X
12. Teliura, N.O.: Development of the methodological approach to the selection of technologies for environmentallysafe water drainage in populated areas. Eastern-European Journal of Enterprise Technologies. **6**(10–96), 55–63 (2018). https://doi.org/10.15587/1729-4061.2018.148689
13. Baluyeva, O.V.: Method of calculating the index of socio-ecological-economic development of a city. Economic Bulletin of the National Mining University **3**, 137–145 (2013). [in Ukrainian]
14. Boström, M., et al.: Conditions for transformative learning for sustainable development: a theoretical review and approach. Sustainability **10**(12), 4479 (2018). https://doi.org/10.3390/su10124479
15. Kozyk, V., Vorobets, S., Musiiovska, O.: Research of the place of Ukraine in implementation of the goals of the sustainable development model. Technology Audit and Production Reserves. **3**(4), 19–27 (2019). https://doi.org/10.15587/2312-8372.2019.169381
16. Rasoolimanesh, S.M., Badarulzaman, N., Jaafar, M.: City development strategies (CDS) and sustainable urbanization in developing world. Procedia Soc. Behav. Sci. **36**, 623–631 (2012). https://doi.org/10.1016/j.sbspro.2012.03.068
17. Lubentsov, A.V., Bobrov, V.N., Desytov, D.B., Noev, A.N.: The advantage of the method of hierarchy analysis, the statistical methods of decision support. J. Phys: Conf. Ser. **1203**(1), 012079 (2019). https://doi.org/10.1088/1742-6596/1203/1/012079
18. Medvid, M., Komisarov, O., Merdova, O.: Review of the tasks of formation and progress of human resources in the sustainable development strategy of Ukraine in the light of central place theory. Baltic Journal of Economic Studies. **4**(2), 134–140 (2018). https://doi.org/10.30525/2256-0742/2018-4-2-134-140
19. Dmitrieva, O.O., Teliura N.O., Vasilenko V.P.: Implementation environmentally safe wastewater as an element of sustainable human settlements development Ukraine. Municipal Economy of Cities. (7), 174–179 (2018). https://doi.org/10.33042/2522-1809-2018-7-146-174-179 [in Ukrainian]

Information-Analytical System of Landfill Environmental Safety Assessment

Ihor Honcharenko[1](✉), Leonid Pisnia[2], Oleksandr Targonskyi[2], and Hanna Khabarova[3]

[1] Ecology and Environmental Management Department of the V. N. Karazin Kharkiv National University, 4 Svobody Sq., Kharkiv 61022, Ukraine
ihorhoncharenko@gmail.com

[2] Laboratory for Environmental Impact Assessment and Environmental Expertise, Kharkiv, Ukraine

[3] Laboratory for Radioecological Safety and Radiation Monitoring, Ukrainian Scientific Research Institute of Ecological Problems, Kharkiv 61166, Ukraine

Abstract. According to Eurostat [1], the generation of solid waste increases by 2% annually and approaches 300.0 million tons. At the same time, the landfill rate in the EU is about 30% (from 3% for developed countries to 90% – for developing countries). Investigation of the waste management system in Ukraine [2] found that the most dangerous source of environmental impact is the final consumers of waste – landfills for solid waste. Therefore, research on such objects requires special attention, in particular the possibility of further safe operation, identification factors and conditions of local hazards for components of the environment or population, including decommissioning, closure and reclamation. The proposed approach to the development of recommendations is universal and takes into account the scientific environmental assessment of waste disposal sites, local conditions and environmental factors, environmental impact assessment, justification of measures to reduce the negative impact on the environment and ensure operational reliability. Consequently to the calculations carried out in the research, the distribution of hazards for the components of the environment is for C1 – Air – 50,12%; C2 – Water – 26,06%; C3 – Soil – 12,96%; C4 – Biota – 10,86%. Among the measures assessed following the requirements of EU Directives 2008/98/EU, 96/61/EU, 1999/31/EU, the largest contribution will be made by SM4 – Implementation the best available technologies (without reconstruction) – 41,24%.

Keywords: Landfill · Solid waste · System · Analytic hierarchy process · Best available technologies · Expert-analytical assessment · Environmental safety · Management process

1 Introduction

In Ukraine, as a developing country, the problem of waste management, including solid waste (MSW), remains unresolved. The activities of executive bodies at all levels in this area are characterized as critical, which led to a deepening environmental crisis and

M. Nechyporuk et al. (Eds.): ICTM 2021, LNNS 367, pp. 951–975, 2022.
https://doi.org/10.1007/978-3-030-94259-5_74

exacerbated the socio-economic situation in society. The complexity of the situation is due to the lack of a systemic vision and a comprehensive infrastructure for MSW management. As a consequence, the disposal of MSW occurs without comprehensive consideration of possible dangerous consequences, their impact on the state of the environment and human health, low using level of waste as a secondary raw material. The dominant method of MSW management in Ukraine is removal and disposal at landfills and dumps.

In 2018 [3] only 5,8% of generated MSW was recycled, including 2,71% (1,3 million m^3) – incinerated, 3,09% (1,53 million m^3) – sent to other waste processing complexes and about 0,003% (2000 m^3) – composted. The remaining 94% was disposed to landfills and dumps, of which in 2016 there were 5,470 units in Ukraine, of which 305 (5.6%) were overloaded, and 1,646 (30%) doesn't meet domestic environmental safety standards, 99% doesn't meet European requirements. In addition, due to the insufficient level of control or lack of a proper system of MSW management, more than 27 thousand natural landfills are formed every year [4].

Landfills, especially dumps, are potential sources of environmental pollution – atmosphere, hydrosphere, soil. In its body is the formation of a significant amount of toxic filtrates and gases, including explosive, development of pathogenic microorganisms, insects and pests [5]. It is difficult to estimate their chemical composition due to the diversity of the mixture of waste entering landfills and dumps.

Adverse environmental situation, associated with the accumulation of waste and its negative impact on the environment, inherent in all regions of Ukraine [6]. This especially applies to medium and large settlements, which include the city of Kharkiv, where is generated annually about 1,5 million m^3 of MSW [7].

Based on the above is an urgent scientific task development of a systematic approach to hazard assessment and choosing the appropriateness of measures to improve the environmental safety of landfills and dumps (special sites for wastes disposal (SSWD). This approach is the basis of the information-analytical system of support management decisions and an effective mechanism for achieving the goals of sustainable development territorial communities of Ukraine in the conditions of decentralization.

2 The Procedure of Expert-Analytical Assessment of Hazards of the MSW Management Process

The task of supporting making an effective decision takes place when it is necessary to choose the best option means of action among a given set of alternatives to achieve this goal. Solve decision-making tasks with many criteria is possible using different methods: methods of reduction of criteria to one (methods of the main component, complex criterion, fair compromise, Pareto analysis), methods of psychological person features decision-maker (multicriteria utility theory, analytic hierarchy process (AHP), methods of ranking multicriteria alternatives).

The problem of choosing effective methods of management decisions in local authorities devoted to research [8]. The results of in-depth analysis of scientific papers allowed to identify problems (limitations, shortcomings and contradictions) in terms of developing a methodology for making agreed management decisions [9]. According to the conclusions [10, 11], traditional methods of examination practically don't work in the research of organizational and socio-economic systems.

Among the specified block of methods of multicriteria ecological assessment expedient to single out a group of expert analytical methods: GIS, cartographic, checklists, matrix, statistical, adaptive, modelling, etc. [12, 13]. Each of those methods has many advantages (visualization of environmental information, easy to perceive information, reflect the relationship between anthropogenic activities and the impact of second-order impacts, risk accounting, adapted for forecasting and analysis of impacts, etc.) and a lot of shortcomings (do not distinguish between direct and indirect impacts, do not take into account the duration and probability of impacts, do not establish a link between anthropogenic activity and impact, require large amounts of information, poorly formalized sets of indicators and factors difficult to apply to practice, etc.).

Among the considered research methods the expert methods deserve special attention which intends to predict qualitative and quantitative characteristics, the development of which is not fully or partially subject to mathematical formalization due to the lack of sufficient and reliable statistics. Even in the absence of strictly mathematical justifications the level of uncertainty can be reduced through the use of expert assessments [14, 15]. Methodological bases of ecological assessment of objects using AHP in Ukraine are set out in particular in [16, 17]. AHP is highly structured and clear, which allows you to use it in decision-making at different levels of government [18]. The uniqueness of this method is that it is both qualitative and quantitative. AHP allows in step-by-step mode to determine the transitivity and consistency of expert assessments [19].

Issues of support for management decisions are especially relevant in the conditions of decentralization in Ukraine, including the achievement of the Sustainable Development Goals.

The use of independent experts for evaluation will help democratize governance, increase the transparency of the decision-making system with minimization of the corruption component.

Experience in the use of AHP in matters of environmental safety in Ukraine has become widely used [20], including complex-formalized processes, confirming adequacy and acceptability of its use as a mechanism to support decision-making [21, 22].

Schematically the procedure of expert-analytical evaluation hazards of the MSW management process, which was used in the research, presented in a generalized form in Fig. 1.

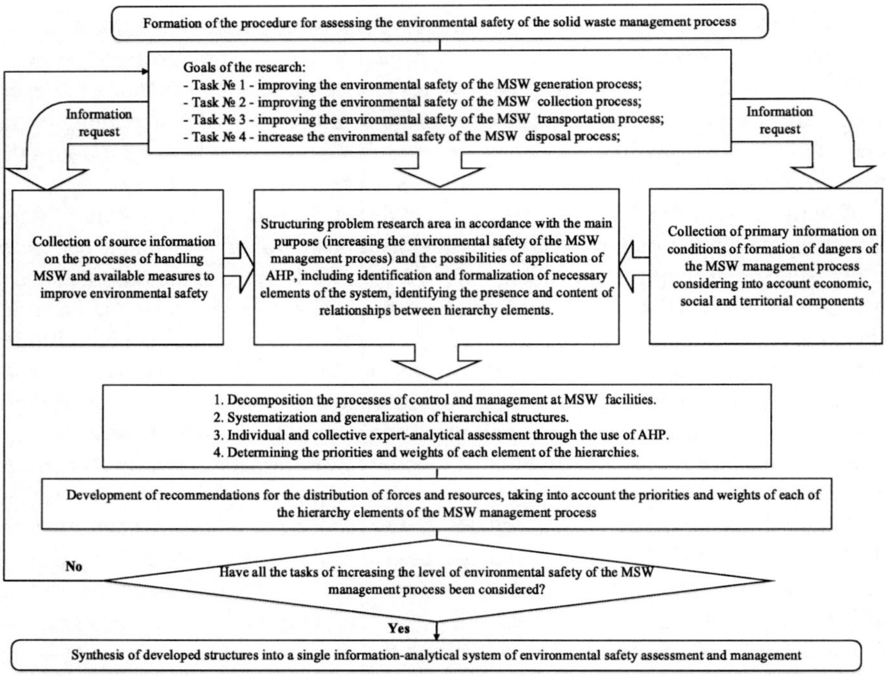

Fig. 1. The procedure of expert-analytical assessment of hazards of the MSW management process

In order to achieve the goal of the research, a step-by-step method of ecological assessment of the situation and operating conditions of the current SSWD was used, which is given in the form of the block diagram in Fig. 2.

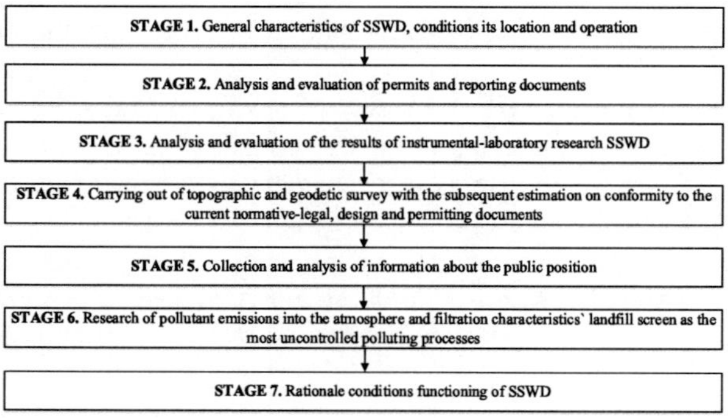

Fig. 2. Block diagram of the assessment of environmental safety SSWD

Practically, the research was conducted at a typical Ukraine high-load landfill which is located near Kharkiv and is called Rogan landfill. The territory of the landfill is near the ring road at the junction of several administrative districts of the Kharkiv region, which makes it promising for the implementation of its functions. At the same time, the Rogan landfill is a potentially hazardous object on which can arise emergencies of technogenic character [23].

As part of the research analysis was performed and evaluation of permits regulatory and technical documents for compliance with the requirements of current urban planning and environmental legislation, as well as a full-scale survey of the landfill. The main positions are given regarding the compliance of the operational characteristics of the landfill with the current regulatory documents urban planning and environmental legislation of Ukraine. According to the research scientifically substantiated conclusions are made and suggestions regarding the environmentally-safety operating conditions of Rogan landfill. The algorithm and methodology of this research can be used for any MSW disposal site, namely is universal [24].

2.1 STAGE 1. General Characteristics of SSWD, Conditions Its Location and Operation

Landfill situated in the village Rohan Kharkiv region in the ravine Pisarenkiv Yar (an abandoned clay quarry) in the north of the village on the lands of the former educational farm «Communist» (Fig. 3).

Fig. 3. Rogan landfill

Rogan landfill is a highly loaded SSWD, the landfill area is 21,7 hectares. Sanitary protection zone is – 500 m. The work area is designed to remove solid household and other waste equated to them III-IV hazard classes. Waste buried by cards, they are compacted by special equipment up to a height of 2 m with subsequent layer-by-layer reclamation with insulating soil and hazard class IV waste, which can be used as an insulating material. Some industrial III-IV hazard classes wastes are accepted with restriction and stored together with solid household waste in compliance with sanitary requirements. The landfill belongs to the engineering and ecological complex, which is designed for centralized collection of MSW and their disposal on compliance with environmental safety requirements, protection of atmospheric air, water bodies, soils from contamination by waste components. Annual volume of MSW, which is accepted at the landfill, is about 100,0 thousand tons.

The inspection revealed that are located on the territory of the landfill waste storage place, auxiliary technological objects: roads and platforms with a hard covering; warehouses; fire tanks; practical zone; existing upland ditches, biogas collection system; filtrate collection system; observation wells. The listed auxiliary objects for the period of researches are in a proper condition for operation. It should be noted, during the summer period, the survey did not reveal areas with the manifestation of exogenous processes (talus, landslides, floods, etc.). Slopes of the MSW storage body was overgrowny with vegetation, have clear slopes, indicating the proper storage of MSW in accordance with the design technology. Ditches which are intended for drainage of surface waters was dried, due to the prolonged absence of rain, upland ditches and system for extinguishing the energy of the water flow of surface waters, are in satisfactory condition. The bottom of the ditch is filled with dried silt. Slopes of ditches are formed and was overgrown with weeds and shrub vegetation. There is a spare volume of the ditch to receive surface runoff, which does not preclude the need for gradual removal of accumulated sludge and debris. Observation wells are in proper conditions: areas around the branch concreted to prevent surface runoff during rains and unforeseen floods. To prevent access by unauthorized persons wells are fenced with closed metal containers. As a result of field research conducted directly on the landfill and analysis of the conditions of its operation found the absence of clusters of birds and insects.

Based on the above, a conclusion is made on the sufficient effectiveness of the existing preventive measures and the satisfactory condition of the environment.

2.2 STAGE 2. Analysis and Evaluation of Permits and Reporting Documents

Landfill project and operation documentation near the village Rogan is available and consists of twenty-eight volumes. At the Rogan landfill, it was planned to use an existing career with a depth of 9–12 m, taking into account two technological stages of operation. After filling the career, the second stage of filling the landfill begins, which provides high-altitude storage of MSW. Landfill in accordance with the working projects includes stage I (the area is 2 hectares) and stage II (the area is 19,2 hectares) for waste disposal. According to project decisions:

1. Closing of each site of MSW disposal is carried out after full filling of cards with waste.

2. The displacement of MSW unloaded by garbage trucks on the working map is carried out by bulldozers, creating a layer up to 0,5 m high. Due to 15–20 compacted layers, a shaft with a gentle slope of 2 m above the level of the garbage truck unloading site is created. The shaft of the next working card is pushed to the previous one. The compacted layer of MSW 2 m high is isolated by a soil layer of 0,25 m.
3. Sealing of MSW laid on the working card in layers up to 0,5 m is carried out by heavy rollers weighing over 14 tones. Sealing in layers of more than 0,5 m is not allowed.
4. After the filling of the career, the second stage of filling the landfill begins, which provides for high-altitude storage. In this area, on the waste covered with a layer of clay, it is planned to lay a slag-clay mixture in a layer of 30 cm, moistened with water to create a base with a slope to the drainage tray for collecting filtrate.
5. On the built basis the anti filtration screen from a soil-bituminous mix of grades of acid tars with the maintenance no more than 18% category III = 0,01 m and the condensed soil of 0,3 m is created.
6. Waste disposal is carried out on the prepared site by creating a compacted working layer of waste – 2.0 m and filling it with soil 0,25 m. The total height of the backfill is 17,0 m, from the screen mark (137,0–138,0 m). In the future, it is planned to fill up to 20,0 m in height under the conditions of compliance with the basic design conditions.
7. Works on recultivation of the landfill will be performed after the landfill is completely filled with waste.

The provided permits and reporting documents, passports, waste register cards, air emissions, wastewater discharges, etc., sufficiently cover the activities of the Rogan landfill for utilization and processing of MSW for the period from 2002 to 2017. Documents meet the requirements of the current legislation of Ukraine at the time of their preparation.

2.3 STAGE 3. Analysis and Evaluation of the Results of Instrumental-Laboratory Research SSWD

Laboratory researches of the Rogan landfill were conducted by territorial structural subdivisions of the State Consumer Service regularly during 2015–2017, in accordance with the requirements of permits and regulations provided for in the working drafts. Reporting documents of those research certify that results tests of air, water, soil samples, noise level measurements are not exceeded of those indicators which are given in the materials of these projects (Table 1).

Table 1. Chemical composition of waste water, groundwater and surface water.

pH	Biogenic elements, mg/dm^3				Hardness mg eq/dm^3	Macrocomponents, $\frac{mg\ eq/dm^3}{mg/dm^3}$						Mineralization/Dry residue, mg/dm^3
	NNH_4	NNO_2	NO_3	PO_4		HCO_3	Cl	SO_4	Ca	Mg	Na	
Sample 1. Groundwater from the observation well №2, located 50 m from the drainage ditch, on the side of the southern slope of the landfill												
7,05	0,65	<0,03	0,89	2,39	12,0	5,4 / 329,4	5,1 / 181,0	5,6 / 268,8	2,6 / 52,1	9,4 / 114,7	4,1 / 102,5	948,5 / 784,0
Sample 2. Landfill infiltrate												
10,75	568,75	2,63	–	36,34	120,0	220 / 13420	380 / 13490	40 / 1920	80 / 1603,2	40 / 488,0	520 / 13000	43921,2 / 37211
Sample 3. River water (Studenok river, 500 m downstream from the intersection of the ravine Pisarenkiv Yar)												
8,48	2,63	<0,03	2.32	3,41	3,8	1.9 / 67,4	1,9 / 67,4	8.2 / 393,6	1,6 / 32,1	2,2 / 26,8	10,1 252.5	1004,2 / 888,0
Standard according to the State Sanitary Rules 2.2.4–171–10 «Tap water»												
6,5–8,5	0,5 (2,6)	≤ 0,5 (0,1)	≤ 50	3,5	≤ 7,0 (10)	No standard-ization	≤ 250 (350)	≤ 250 (500)	No standardization			1000

2.4 STAGE 4. Carrying Out of Topographic and Geodetic Survey with the Subsequent Estimation on Conformity to the Current Normative-Legal, Design and Permitting Documents

In order to comply with the requirements for the height of waste storage, as provided by the permit and design documentation engineering and geodetic surveys were carried out, including field research, and prepared a technical report from topographic and geodetic survey of Rogan landfill. The survey was performed by GC GEOPROM in a polar manner using a certified and certified electronic total station «TOPCON» GTS-239 №ON4653. The survey was performed to obtain a topographic basis in order to develop a land plan. The starting points were determined in real time by RTK using a Trimble R4 GPS receiver. Coordinate system – SK-63. The system of heights is Baltic. According to the terms of reference created a digital vector plan of the territory at a scale of 1:2000 in DXF format (Fig. 4).

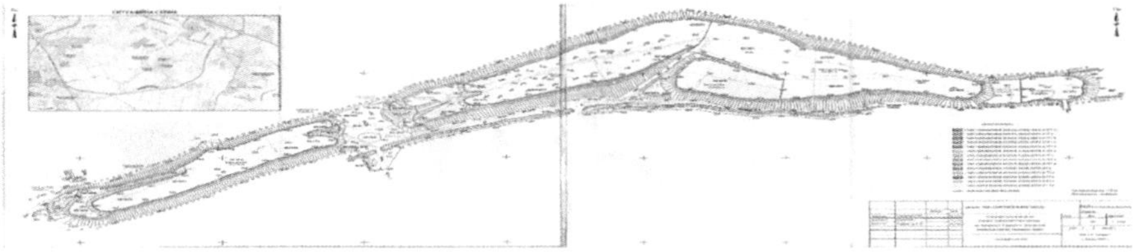

Fig. 4. Topographic-geodetic map of Rogan landfill, made by GEOPROM, 2017

According to the results of topographic and geodetic works the landfill is conditionally divided into 14 zones indicating the heights of the dump. The performed works sufficiently characterize the current state of development of the project area of the researched object as satisfying design conditions for further operation of the facility, as the design height of MSW storage (17 m) not actually achieved in all areas of the landfill, namely: in zone 1 the maximum height of the garbage dump is up to 16,5 m, in zone 2 – up to 16 m, in zone 3 – up to 15,5 m, in zone 4 – up to 15,1 m, in zone 5 – up to 15,5 m, in zone 6 – up to 14,6 m, in zone 7 – up to 11 m, in zone 8 – up to 10,9 m, in zone 9 – up to 10,1 m, in zone 10 – up to 9 m, in zone 11 – up to 7,8 m, in zone 12 – up to 5,9 m, in zone 13 – up to 4,5 m, in zone 14 – up to 1,5 m.

2.5 STAGE 5. Collection and Analysis of Information About the Public Position

The analysis was performed available in free access, including the Internet, information on the activities of the Rogan landfill. Additionally the availability of information in the permit documents is taken into account, on public coverage of the activities of the SSWD.

Analysis of information on available public appeals regarding environmental safety and threats to public health in the area of operation of the Rogan landfill during the period of operation indicates the neutral position of citizens regarding the activities of the SSWD. The biggest concern of the public caused fires of MSW on the territory of Rogan landfill, which arose almost annually.

2.6 STAGE 6. Research of Pollutant Emissions into the Atmosphere and Filtration Characteristics' Landfill Screen as the Most Uncontrolled Polluting Processes

Pollution research in the area of impact of landfills includes calculations of emissions of pollutants into the atmosphere (emission) taking into account biogas emissions, as well as an assessment of the migration of pollutants, generated in the investigated place of waste disposal. To account for and calculate the number of pollutants is important both the implementation of field research (measurements) and theoretical research (mathematical modeling).

Calculation and evaluation of filtration coefficient' for landfill screen [25]. Based on the results of laboratory tests of loams and clays, were set the values of the number of plasticities: for loams 0,16; for clays – 0,20. Loams with a plasticity number of 16 are characterized by a filtration coefficient of less than $8{,}5 \times 10^{-5}$ m/day or less than $8{,}5 \times 10^{-8}$ cm/s, which corresponds to $8{,}5 \times 10^{-10}$ m/s. Clays with a plasticity number of 20 are characterized by a filtration coefficient equal to $7{,}5 \times 10^{-6}$ m/day or $7{,}5 \times 10^{-9}$ cm/s, which corresponds to $7{,}5 \times 10^{-11}$ m/s. In accordance with paragraph 3.12 of CN 2811400–83 filtration coefficient of the base of the landfill should not exceed $1{,}0 \times 10^{-9}$ m/s ($1{,}0 \times 10^{-7}$ cm/s) or should not exceed $8{,}6 \times 10^{-5}$ m/day. Substantiation of the possibility of extending the service life of the landfill, depending on the filtration characteristics of the water permeability of the clay screen at the base of the landfill, was carried out by assessing the values of the filtration coefficient of the clay screen to the regulatory requirements CN 2811400–83.

Comparing the values of the filtration coefficients of loams and clays depending on the number of ductilities obtained during the determination of physical and mechanical properties by laboratory methods, according to the requirements of p.3.12 CN 2811400–83 to the filtration coefficient of the anti-filtration screen, it can be stated that the filtration characteristics, meet the regulatory requirements with a margin of safety because $7{,}5 \times 10^{-11}$ m/s (Kf clay) $<1{,}0 \times 10^{-9}$ m/s (regulatory requirements for the screen), $7{,}5 \times 10^{-6}$ m/s day (Kf clay) $<8{,}6 \times 10^{-5}$ m/day (regulatory requirements for the screen).

It follows that the filtration coefficient of clays at the base of the landfill is two orders of magnitude less than the filtration coefficient of the anti-filtration screen in accordance with regulatory requirements, ie clays at the base of the landfill have more stringent waterproof properties than modern requirements. Thus, the evaluation of the filtration properties of clay rocks indicates the reliability of the protective effect of the anti-filtration screen at the base of the landfill.

According to the results of engineering and geological surveys performed on the landfill in 2004 by the company «Maxitrade» (Kyiv), the composition and physical and mechanical properties were identified a layer of loam, designated as a layer of IGE-2. Using the engineering-geological characteristics of the rocks, the filtration coefficients of the loams lying above the clays were determined by the calculation method, which together with the clays form the upper part against the filtration screen of the Rogan landfill base. To calculate the filtration coefficient of loam was used Zunker's formula:

$$K = C(n/1 - n) * 1/U^2, \quad (1)$$

where U – the «specific surface area», which means a number indicating how many times the total surface area of the particles is greater than the total surface area of the particles with a diameter of 1,0 mm (m/day); C – grain shape factor (400–800 depending on the breed); n – the porosity.

It should be noted that in hydrogeological practice, Zuncker's formula is used mainly to determine the filtration coefficient of clay fine-grained sands and its application to calculate the filtration coefficient of loams is approximate.

As a result of research, it was found that the available filtration properties of clay rocks (clays and loams) of the anti-filtration screen of the Rogan landfill base by calculations of filtration coefficients using the empirical Zunker formula based on the physical and mechanical properties of these rocks indicate reliable water from the possible flow of pollutants that may be contained in the filtrate, which is formed in the body of the landfill.

Determination of the qualitative composition of pollutants and calculation of landfill biogas distribution [26]. The quantitative characterization of emissions of pollutants and greenhouse gases emitted from the SSWD is influenced by several factors, among which the following can be distinguished: climatic characteristics; working area of the landfill; terms of landfill operation; the amount of landfilled waste; waste layer capacity; morphological composition of waste; waste moisture; the content of the organic component in the waste; waste disposal technology. To estimate the volume of emissions at the Rogan landfill field was research conducted and obtained practical data on the spatial distribution of biogas. The results of emissions of biogas components are presented in Fig. 5.

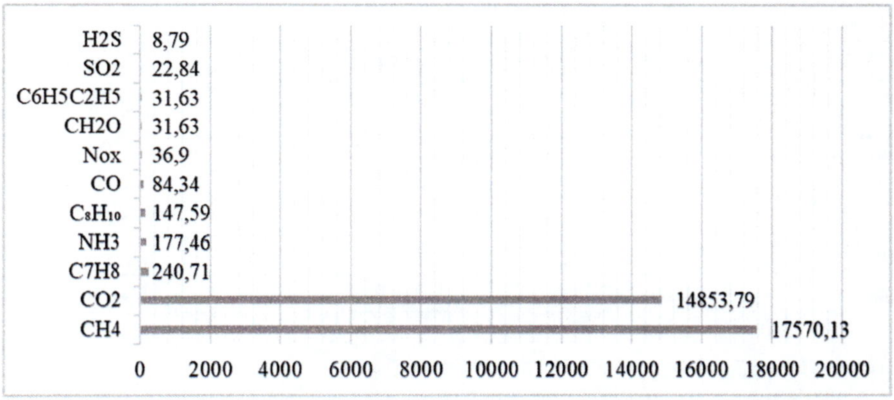

Fig. 5. Emissions of greenhouse gases and pollutants into the atmosphere, tons/year

A model of the convective rise in the atmospheric air of a heated cloud of biogas formed over a landfill has been developed. The results of the calculations are presented in the Table 2.

Table 2. Dependence of the main parameters describing the convective rise of biogas.

Parameter	ϑ_0				
	10^{-3}	$3*10^{-3}$	10^{-2}	$3*10^{-2}$	10^{-1}
$R_0 = 10$ м					
υ_{ch}, м/с	0,50	0,86	1,57	2,71	5,00
υ_{max}, м/с	0,36	0,63	1,14	1,98	3,61
t_0, с	111	64,6	35,4	20,5	11,1
t_υ, с	55,6	31,75	17,54	10,10	5,54
t_ϑ, с	94,3	52,9	29,2	16,8	9,2
t_R, с	383	158,75	87,7	50,50	27,7
t_{max}, с	1,1*104	$6{,}5*10^3$	$3{,}5.10^3$	$2{,}1.10^3$	$1{,}1.10^3$
$R_0 = 100$ м					
υ_{ch}, м/с	1,50	2,71	5,00	8,57	15,00
υ_{max}, м/с	1,10	1,98	3,61	6,26	11,00
t_0, с	370,3	205	111,1	64,8	37,0
t_υ, с	181,8	101,0	55,4	31,9	18,2
t_ϑ, с	302,2	168,3	92	53,2	30,2
t_R, с	906,5	505	277	159,5	90,7
t_{max}, с	3,7.104	$2{,}05.10^4$	$1{,}1.10^4$	$6{,}5.10^3$	$3{,}7.10^3$
$R_0 = 1000$ м					
υ_{ch}, м/с	5,00	8,57	15,00	27,11	50,00
υ_{max}, м/с	3,61	6,26	11,00	19,80	36,1
t_0, с	1111	648,2	370,3	204,9	111,1
t_υ, с	554	319,5	181,8	101	55,4
t_ϑ, с	923	532,5	303	168,3	92,3
t_R, с	2770	1597,5	909	505	277
t_{max}, с	$1{,}1.10^5$	$6{,}5.10^4$	$3{,}7.10^4$	2.10^4	$1{,}1.10^4$

Calculations were made for different buoyancy of biogas ($\vartheta_0 \approx 10^{-3-}$ 10^{-1}) and different sizes ($R_0 = 10$, 100 и 1000 m) of the source of danger (the Rogan landfill).

The obtained results show that with the increase of this radius the maximum height of biogas rise proportionally increases, reaching 1–10 km. The maximum rate of biogas rise varies from values of 0,36–3,6 m/s at $\vartheta_0 = 10^{-3}$ to values of 3,6–36,1 m/s at $\vartheta_0 = 10^{-1}$. The rise time of biogas decreases from 3–24 h at $\vartheta_0 = 10^{-3}$ to 0,3–3 h at $\vartheta_0 = 10^{-1}$.

It is accepted that pop-up gases have the shape of a sphere with a gradually increasing radius, much smaller than the height of a homogeneous atmosphere (the frequency characterizing the oscillations of the atmosphere is neglected). It is also assumed that for heat sources with a radius of 1 km or more, the volume of the heated biogas cloud is determined only by the external scale of turbulence. The change in atmospheric air temperature depending on the height and the removal of heated gas formation horizontally under the action of wind in the mathematical modeling of the researchied process were not taken into account. With the help of the developed model, the estimation of dynamics of the basic parameters of the heated cloud of biogas for several typical cases is executed. Altitude and time dependencies are established. Numerical velocities, characteristic dimensions, excessive relative temperature and cloud buoyancy were estimated numerically.

It is accepted that pop-up gases have the shape of a sphere with a gradually increasing radius, much smaller than the height of a homogeneous atmosphere (the frequency characterizing the oscillations of the atmosphere is neglected). It is also assumed that for heat sources with a radius of 1 km or more, the volume of the heated biogas cloud is determined only by the external scale of turbulence. The change in atmospheric air temperature depending on the height and the removal of heated gas formation horizontally under the action of wind in the mathematical modeling of the researchied process were not taken into account. With the help of the developed model, the estimation of dynamics of the basic parameters of the heated cloud of biogas for several typical cases is executed. Altitude and time dependencies are established. Numerical velocities, characteristic dimensions, excessive relative temperature and cloud buoyancy were estimated numerically.

The estimation of changes of parameters of the heated cloud of biogas executed using mathematical modeling characterizes dynamics of its convection rise in atmospheric air over the landfill of firm household waste. It is shown that in the process of raising this formation its radius increases (due to the addition of cold air). The excess temperature in the heated formation, as well as its buoyancy, are reduced. The rate of ascent first increases and then gradually decreases.

2.7 STAGE 7. Rationale Conditions Functioning of SSWD

Based on the research and with the application of the developed procedure, the decomposition of the tasks of environmental safety management of the Rogan landfill with the development of hierarchical structures of the negative impact. Subsequently, was synthesized the system and was carried out expert-analytical researches of the components of the MSW management process at the Rogan landfill using a systematic multi-criteria integrated approach with AHP evaluation. The hierarchical structure of the environmental safety of the Rogan landfill and measures to reduce the negative impact on environment are shown in Fig. 6.

The connections between the elements of the hierarchy established based on characteristics of elements interaction adjacent levels of hierarchies. In parentheses are the values of generalized weights and priorities in percent, obtained by applying expert-analytical procedures using the AHP program. The overall consistency of expert opinions for the whole scheme is 0,0299, which satisfies the requirements of the AHP.

The formulation of the links content characteristics between the elements of the upper and lower adjacent levels hierarchical structure of environmental safety Rogan

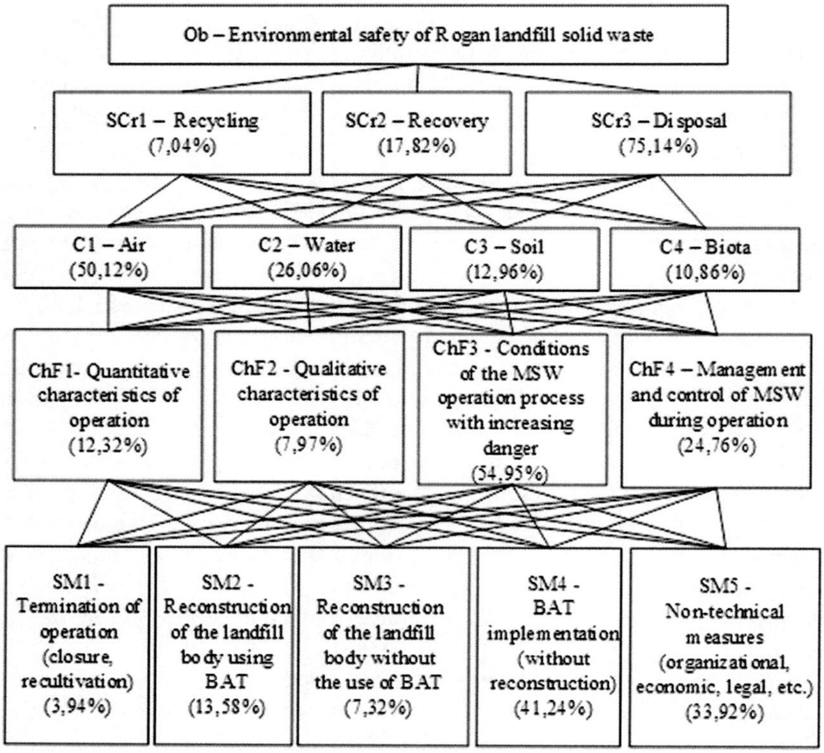

Fig. 6. Hierarchical structure of environmental safety of Rogan landfill and measures to reduce the negative impact on environment

landfill and measures to reduce the negative impact on environment are given in Table. 3.

Summarizing the results of expert-analytical research and in order to bring the activities of the Rogan landfill in line with modern requirements of Ukrainian and EU legislation regarding existing landfills, in particular, in the framework of the implementation of the National Waste Management Strategy in Ukraine until 2030, on the implementation in Ukraine of EU Directives 2008/98/EU and 1999/31/EU (about waste disposal), recommendations (suggestions) were provided to the balance holder of the landfill, the main ones are below:

1. bring in line with the project documentation and put into operation on the technological part of the landfill the existing facilities for sorting and processing of household waste for the purpose of extraction for further utilization of waste for which there are processing technologies in Ukraine (author's note: the proposal meets and solves the problem of improving the level of environmental safety element SM4);
2. carry out waste disposal in compliance with BAT in accordance with the annual developed technological plans for the organization of work on the landfill on work cards.

Table 3. Characteristics of the links between the elements of the MSW management process to assess the environmental safety of the Rogan landfill.

The title of the hierarchy level	Lower level elements	Connection characteristics, that reveals the essence of the interaction or influence between the elements of the assessed level with the higher level and aspects of the general goal
Connections between elements of Level 1 (Objective) and Level 2 (Sub-criteria)		
Level 2. SCr (Sub-criteria) Hazardous MSW management processes	SCr1 – Recycling	The contribution of treatment type to the overall process impact. According to Art. 1 of the Law of Ukraine «On Waste» recycling is defined as minimazing or destruction the hazard of waste by mechanical, physico-chemical or biological treatment. Occurs before disposal at the landfill, for example, sorting of mixed MSW
	SCr2 – Recovery	The contribution of treatment type to the overall process impact. According to Art. 1 of the Law of Ukraine «On Waste» recovery – the use of waste as a secondary material or energy resource. Occurs directly at the landfill, for example, the use of inert waste as a fortification of internal transport routes

(*continued*)

Table 3. (*continued*)

The title of the hierarchy level	Lower level elements	Connection characteristics, that reveals the essence of the interaction or influence between the elements of the assessed level with the higher level and aspects of the general goal
	SCr3 – Disposal	The contribution of treatment type to the overall process impact. According to Art. 1 of the Law of Ukraine «On Waste» disposal is defined as the implementation of operations with waste that does not lead to their recovery, such as final disposal of waste in specially designated places or facilities so that long-term adverse effects of waste on environment and human health did not exceed the established standards. It takes place directly at the landfill, which is a specialized place for that – passport of the SSWD № 05–16
Connections between elements of Level 2 (Sub-criteria) and Level 3 (Criteria)		
Level 3. C (Criteria) – environmental components that are adversely affected	C1 – Air	The possible impact of the MSW management process on the safety of atmospheric air as a component of the environment is assessed
	C2 – Water	The possible impact of the MSW management process on the safety of surface and groundwater as a component of the environment is assessed

(*continued*)

Table 3. (*continued*)

The title of the hierarchy level	Lower level elements	Connection characteristics, that reveals the essence of the interaction or influence between the elements of the assessed level with the higher level and aspects of the general goal
	C3 – Soil	The possible impact of the MSW management process on the safety of soil as a component of the environment is assessed
	C4 – Biota	The possible impact of the MSW management process on the safety of living nature (biota) as a component of the environment is assessed
Connections between elements of Level 3 (Criteria) and Level 4 (Characterization of factors)		
Level 4. ChF (Characterization of factors). Complex factors of influences (characteristics of influence of waste)	ChF1– Quantitative characteristics of operation	The impact (dependence) of the amount of waste on the environmental safety of the environmental component, expressed in quantitative characteristics (weight, volume, etc.). As the number of waste increases, its impact (danger) on the environmental components

(*continued*)

Table 3. (*continued*)

The title of the hierarchy level	Lower level elements	Connection characteristics, that reveals the essence of the interaction or influence between the elements of the assessed level with the higher level and aspects of the general goal
	ChF2 – Qualitative characteristics of operation	The impact (dependence) of the waste quality on the environmental safety of the environmental component, expressed in physical, chemical or biological properties of waste with increasing or emerging hazards. When the quality of waste changes (with an increase or the appearance of danger) increases its impact (danger) on the environmental components
	ChF3 – Conditions of the MSW operation process with increasing danger	Influence (dependence) of waste accumulation conditions with increasing danger on the environmental components (technical characteristics of the landfill, compliance of its operation conditions with environmental safety requirements, territorial features of the landfill, compliance with norms and rules during MSW operations, etc.)

(*continued*)

Table 3. (*continued*)

The title of the hierarchy level	Lower level elements	Connection characteristics, that reveals the essence of the interaction or influence between the elements of the assessed level with the higher level and aspects of the general goal
	ChF4 – Management and control of MSW during operation	Influence (dependence) management and control mechanisms to the process of MSW management, which affects the ecological safety of the environmental components. For example, the absence or weak functioning of management and control systems for the MSW management process leads to an increase in the negative impact on the environment
Connections between elements of Level 4 (Characterization of factors) and Level 5 (SM)		
Level 5. SM (Security measures) Measures to increase the level of environmental safety	SM1–Termination of operation	Termination of operation is defined as a set of works on the termination of operation of the waste disposal site, including the implementation of land recultivation works

(*continued*)

Table 3. (*continued*)

The title of the hierarchy level	Lower level elements	Connection characteristics, that reveals the essence of the interaction or influence between the elements of the assessed level with the higher level and aspects of the general goal
	SM2 – Reconstruction of the landfill body using BAT	The measure which includes a set of works on recycling the existing body of the landfill (according to paragraph 17 of Article 3 of EU Directive 2008/98/EU) with the subsequent arrangement of waste on the same site in accordance with modern environmental safety requirements with the application the best available techniques (BAT) for waste disposal and operation of SSWD (within the meaning of EU Directive 96/61/EU)
	SM3 – Reconstruction of the landfill body without the use of BAT	The measure which includes a set of works on recycling of the existing body of the landfill (according to paragraph 17 of Article 3 of EU Directive 2008/98/EU) with a subsequent arrangement waste on the same site without the application of BAT for waste disposal and operation of SSWD (within the meaning of EU Directives 96/61/EU, 1999/31/EU)

(*continued*)

Table 3. (*continued*)

The title of the hierarchy level	Lower level elements	Connection characteristics, that reveals the essence of the interaction or influence between the elements of the assessed level with the higher level and aspects of the general goal
	SM4 – BAT implementation (without reconstruction)	The measure which includes the introduction of the BAT for waste disposal and operation of SSWD (within the meaning of EU Directives 96/61/EU, 1999/31/EU) without recycling works of the landfill existing body
	SM5 – Non - technical measures (organizational, economic, legal, etc.)	The measure which consisting of a set or separate application of legal, economic, social, organizational, etc., mechanisms to promote environmentally sound waste disposal, such as reducing the cost of disposal for treated (recycled) MSW or construction waste. The list of measures is defined (but not limited to) by EU Directives 2008/98/EU, 96/61/EU, 1999/31/EU, etc.)

The size of work cards can be increased: working card width up to 15 m, length – from 50 to 120 m (sizes may vary depending on storage technology) (author's note: the proposal meets and solves the problem of improving the level of environmental safety element SM4);

3. develop a program of measures to bring the landfill in line with the requirements of the current version of State Building Codes B.2.4–2–2005 and Directive 1999/31/EU about waste disposal, in which to provide, in particular, reconstruction of the fence of the territory, maintenance of stormwater drainage network, development of the

Emergency Response Plan, etc. (author's note: the proposal meets and solves the problem of improving the level of environmental safety element SM5);
4. in the framework of the implementation National Waste Management Strategy apply to local authorities for the inclusion of the existing Rogan landfill in the network of regional landfills, and provide prepared proposals for inclusion in the draft Regional Waste Management Plan in Kharkiv region (author's note: the proposal meets and solves the problem of improving the level of environmental safety element SM5);
5. in order to extend the service life of SSWD, environmentally safe processing of resource-intensive types of waste, accumulated during the existence of the Rogan landfill, to reconstruct the body of the landfill with the subsequent arrangement of the filtrate and biogas collection system in accordance with the requirements of EU Directives 2008/98/EU, 96/61/EU, 1999/31/EU, etc. (author's note: the proposal meets and solves the problem of improving the level of environmental safety element SM2 and SM3).

Requirements for the operation of the landfill can be implemented both by partial revision of existing design documentation and in the form of new design developments.

3 Summary

Research environmentally safe conditions further operation of the Rogan landfill experimentally confirmed the adequacy of the developed methodical apparatus multicriteria expert-analytical assessment with the use of AHP. Based on the results obtained was formed scientifically substantiated recommendations, which were provided to the balance holder of the landfill as part of conducted research work. In particular, as a result of the conducted research established that:

1. Project documentation submitted for expert-ecological analysis and assessment met the requirements of the current legislation of Ukraine to ensure environmental safety at the time of its development. The analyzed permits and reporting documents for the period 2012–2017 meet the requirements of the current legislation of Ukraine at the time of their execution. The results of laboratory tests in 2015–2017 (air, water, soil, measurements of noise levels) show that the indicators given in the project materials are not exceeded. The results of field research of the landfill and analysis of the conditions of its operation indicate the sufficient effectiveness of preventive measures and a satisfactory state of the environment within the sanitary protection zone of the landfill. Analysis of information on public comments regarding environmental safety and threats to public health in the area of Rogan landfill operation indicates the neutral position of citizens.
2. Existing filtration properties of clay rocks (clays and loams) of the anti-filtration screen of the Rogan landfill base by calculating the filtration coefficients of loams using the empirical formula of Zunker and clays, indicate reliable protection of groundwater from possible contaminants, contained in the filtrate, which is formed in the body of the landfill.

3. Materials from the topographic and geodetic survey of the landfill sufficiently characterize the current state of development of the project area of the researched object as one that satisfies the technical conditions for further operation of the object (the design height of MSW storage was not reach). The period of final achievement of the projected maximum height of waste storage depends not so much on the total amount of MSW, as on the sorting and implemented technologies of primary waste processing and need constant updating directly on the results of field measurements of the actual filling of warehousing cards.
4. The proposed approach to the development of recommendations is universal to address the issue regarding the possibility of further operation of the SSWD and takes into account the scientific environmental and expert assessment of the state of landfill, local conditions and environmental factors, involving comprehensive surveys (engineering, environmental, sanitary, etc.), environmental impact assessment, including impacts on human life and health, justification of measures to reduce or eliminate the negative impact on emergencies, development of dangerous geological processes and phenomena and ensuring operational reliability.
5. Among the evaluated measures to improve environmental safety, the highest priority 41,24% is SM – 4 (introduction of the best available technologies for waste disposal and operation of SSWD without changing the existing state of the body of the Rogan landfill).

References

1. Eurostat database. Homepage. https://ec.europa.eu/eurostat/web/waste/data/database
2. Honcharenko, I., Anishchenko, L., Pisnia, L.: Expert-analytical assessment of environmental safety of solid household waste management. East. Eur. J. Enterp. Technol. **1**(103), 63–76 (2020). https://doi.org/10.15587/1729-4061.2020.197007
3. Prokopenko, O.: Statistical collection «Environment of Ukraine 2017». State Statistics Service of Ukraine, Homepage. http://www.ukrstat.gov.ua/druk/publicat/kat_u/2018/zb/11/zb_du2017.pdf (2018)
4. The state of the sphere of household waste management in Ukraine in 2017. Retrieved from: http://www.minregion.gov.ua/napryamki-diyalnosti/zhkh/terretory/
5. Alam, P.: Impact of solid waste on health and the environment. In: Alam, P., Ahmade, K. (eds.) Special Issue of International Journal of Sustainable Development and Green Economics, vol. 2, pp. 165–168 (2013)
6. Yakovliev, E.: Assessment of ecological and man-made safety of cities and towns of Ukraine. Strategic Priorities. **4**, 146–157. Homepage, http://nbuv.gov.ua/UJRN/spa_2013_4_22 (2013)
7. National report on the state of environmental protection in Ukraine in 2018. Homepage. https://mepr.gov.ua/timeline/Nacionalni-dopovidi-pro-stan-navkolishnogo-prirodnogoseredovishcha-v-Ukraini.html
8. Buryanina, O., Abramkina, S.: Management decisions of local authorities: criteria for choosing effective management methods. Society and Power, Homepage. **1**(75), 91–101. https://cyberleninka.ru/article/n/upravlencheskie-resheniya-mestnyh-organov-vlasti-kriterii-vybora-effektivnyh-metodov-upravleniya/ (2019)
9. Podinovsky, V., Podinovskaya, O.: Once again about the incorrectness of the method of analysis of hierarchies. Management Problems, Homepage. **4**, 75–78. https://cyberleninka.ru/article/n/esche-raz-o-nekorrektnosti-metoda-analiza-ierarhiy (2012)

10. Shageev, D.: Management of socio-economic systems. Bulletin of SUSU. Series "Economics and Management". **13**(2), 145–164 (2019)
11. Dorofeyuk, A., Pokrovskaya, I., Chernyavsky, A.: Expert methods of analysis and improvement of control systems. Automation and Telemechanics, Homepage. **10**, 172–188. http://mi.mathnet.ru/rus/at/y2004/i10/p172 (2004)
12. Chumachenko, S.: Comparative analysis of environmental assessment methods and features of their application to assess the impact of military landfills on the environment. Information processing systems **3**, 203–209 (2006)
13. Anischenko, L.: Theoretical substantiation of the combined method of decision-making in problems of multicriteria complex assessment of influence and management of ecological safety of the extended hydraulic engineering constructions. Eastern European Journal of Advanced Technology. **2**(21–28) 2009. Homepage. http://journals.uran.ua/eejet/article/view/3191
14. Yakovliev, E., Chumachenko, S.: Ecological Threats in Donbas, Ukraine. Center for Humanitarian Dialogue, Geneva (2017). Homepage, https://ceobs.org/wp-content/uploads/2020/04/Ecological-Threats-in-Donbas.pdf
15. Bondarenko, I., Anischenko, L., Rudyk, Y.: Substantiation for enhancement of environmental safety of waste management systems through forecasting efficiency of specialized equipment. Visnyk of Lviv State University of Life Safety. **2**(16), 119–129, 20 (2017)
16. Chumachenko, S., Teslenko, O., Mikhailova, A.: Use of expert assessment methods to assess the characteristics of integrated monitoring and alert systems for the threat or occurrence of an emergency. Technogenic and ecological safety and civil protection **12**, 44–51 (2017)
17. Bondarenko, I., Anischenko, L., Rudyk, Y.: Substantiation for enhancement of environmental safety of waste management systems through forecasting efficiency of specialized equipment. Visnyk of Lviv State University of Life Safety. **2**(16), 119–129 (2017)
18. Buryanina, O., Abramkina, S.: Management decisions of local authorities: criteria for choosing effective management methods. Society and Power. **1**(75), 91−101 (2019). Homepage, https://cyberleninka.ru/article/n/upravlencheskie-resheniya-mestnyh-organov-vlasti-kriterii-vybora-effektivnyh-metodov-upravleniya
19. Ogurtsov, A., Staroverova, N.: Algorithm for increasing the consistency of expert assessments in the method of hierarchy analysis. IGEU Bulletin, 5 (2013). Homepage, https://citeseerx.ist.psu.edu/viewdoc/download?doi=10.1.1.1048.1991&rep=rep1&type=pdf
20. Teliura, N.: Development of the methodological approach to the selection of technologies for environmentally safe water drainage in populated areas. Eastern European Journal of Enterprise Technologies. **6**, 10(96), 55–63 (2018)
21. Kachynski, A.: Indicators of national security: definition and application of their limit values: a monograph. NISS, Kyiv (2013)
22. Benatov, D.: System analysis of natural and man-made safety of hydro units of Ukraine: dis. ... cand. tech. Science.: 21.06.01/Nat. Technical University of Ukraine "KPI", Kyiv (2016)
23. Musilli, A.: Landfill elevated internal temperature detection and Landfill Fire Index assessment for fire monitoring. In: Musilli, A. (ed) Rowan University, ProQuest Dissertations Publishing (2016)
24. Honcharenko, I., Anishchenko, L., Pisnia, L.: Development of priority measures (solutions) for the environmentally safe management of municipal waste at the community level. Technology Audit and Production Reserves. **2/3**(52), 17–25 (2020). https://doi.org/10.15587/2706-5448.2020.203687

25. Anishenko, L., Honcharenko, I., Markina, N., Dotsenko, O., Pisnia, L., Sverdlov, B.: Conditions for ecologically safe operation of landfills operating in Ukraine. Collection of scientific works «Problems of environmental protection and environmental safety» UKRNDIEP. **40**, 25–41 (2018)
26. Rashkevich, N., Goncharenko, I., Anishenko, L., Pisnia, L., Petruhin S., Serikova, E.: Mathematical modeling of biogas lifting the municipal solid waste polygon. Scientific Journal «Science Rice. **9**(50), 39—42 (2018). https://doi.org/10.15587/2313-8416.2018.143412

Software Engineering and It-Infrastructure

Digital Finance and Fintech Applications in Ukraine: Current Status and Development Prospects

Olena Zhykhor[1(✉)], Valeriy Ryeznikov[1], Nataliia Pohribna[2], and Vitalina Delas[2]

[1] V.N. Karazin Kharkiv National University, Kharkiv, Ukraine
{olena.zhykhor,valeriy.reznikov}@karazin.ua
[2] Taras Shevchenko National University of Kyiv, Kyiv, Ukraine
{nataliapohribna,delasvitalina}@knu.ua

Abstract. This paper is intended to provide an insight into the importance of success in implementing FinTech as a prospective environment and means for communications in finance. The discussion will call upon the variety of stakeholders in financial markets to review and improve their strategies and detailed measures taking into account the international experience and the current situation. The paper builds up on the approach to show how countries can adapt to the changing conditions of the coronavirus pandemic in the context of Ukraine. It provides an important source to analyze different types of support for solving problems of FinTech during challenging times.

Keywords: FinTech · Digitalization · Digital transformation · Digital finance · Innovation Hub · FinTech Sandbox · RegLab · Cybersecurity

1 Introduction

The quality of information and communication support is a determining factor in the evolutionary development of the state and, to some extent, is its characteristic, because ICT affects all processes related to the functioning of a society. Global experience shows that information support should be treated as a strategically important mechanism that can raise the country to a higher level in all its activities.

Digitalization of the economy is considered to be the 4-th industrial revolution. Technological revolutions are featuring the long-term impact on all sectors of the economy and social life, through diffusion and self-improvement of basic technology. The creation of digital platforms is based on modern breakthrough technologies, namely on cloud computing technologies that have actually created the possibility of the global communication platforms development. Moreover, digital platforms would not be possible without the development of highly loaded big data processing platforms. However, successful development of digital platforms requires big data analysis tools and hybrid

M. Nechyporuk et al. (Eds.): ICTM 2021, LNNS 367, pp. 979–991, 2022.
https://doi.org/10.1007/978-3-030-94259-5_75

processing systems, where the role of artificial intelligence technologies in digital transformation is outstanding, going along with the effective information analysis. At the same time, technology alone is not enough to successfully promote digital platforms, and it requires technology developers, businesses and regulators to be connected into a single ecosystem, contributing to the digital platforms development.

A service-oriented approach, used to integrate different information systems, is the basis for digital platforms at the level of organization, requiring both appropriate flexible technologies and development practices. Digital platforms are designed to communicate with a large number of people and organizations, sometimes tens of millions or more. To be able to handle a large number of transactions, such platforms must be resistant to both high workloads and previously unknown cyber-attacks. This cannot be done in the traditional approach, when security is provided by system administrators, the digital platform itself must identify possible problems and eliminate them, and the human task is to improve the algorithms of such self-protection.

The latest ICT transforms not only the infrastructure and business processes in finance, but also such categories as time and environment, which are crucial for valuation, pricing, etc. Under digitalization, the information between the financial market participants is instantly moved in the global environment, thus significantly reducing the transaction costs of participants associated with the search and analysis of information. The change in communication technologies has qualitatively transformed the role of transnational capital and the impact of foreign affiliates on the national countries. Technology has no borders, it creates the conditions for the instantaneous transfer of financial capital to those jurisdictions that are less regulated. Thus, in the global digital economy, transnational capital has become virtually beyond the control of financial institutions.

Nowadays, the infrastructure of financial markets is forced to continuously transform under the influence of new technological solutions, the study of which is an important task. Given this, the **objective of this paper** is to look into the breakdown of traditional business models with emergence of new, digitally enabled solutions in finance for greater insight and organizational efficiency via new technologies like cloud, robotics, artificial intelligence, and automation.

Analysis of recent research and publications. Digital financial services have enormous potential, both for the ordinary people (saving money, making payments, receiving credit and insurance, controlling irregular cash inflows, recovering from economic crises, etc.), governments (ensuring the availability of social services), and businesses (accepting digital payments, getting new business intelligence, and accessing credit lines). Moreover, digitalization of financial services is the enabler of the United Nations Sustainable Development Goal 1 "End poverty in all its forms everywhere" [1].

The ITU Task Force on Digital Financial Services is the initiative to bring together all participants working to ensure the coverage of financial services, including more than 60 organizations from more than 30 countries. The task force has issued 85 policy recommendations on digital financial services and 28 supporting thematic reports.

The World Bank Group and the Payments and Market Infrastructures Committee under the guidance of ITU launched the Global Financial Services Initiative (FIGI), which is a 3 year collaborative program to advance research into digital finance and accelerated digital financial services in developing countries [2].

The IMF and the World Bank have developed the Bali Fintech Agenda in 2018, bringing together and focusing on the key issues for elaborating the national policies by individual countries according to the broad principles for the safe development of fintech. The experiences of member countries resulted in developing a framework for the consideration of cutting-edge issues within the following 12 elements: embrace the promise of fintech, enable new technologies to enhance financial service provision, reinforce competition and commitment to open, free, and contestable markets, foster fintech to promote financial inclusion and develop financial markets, monitor developments closely to deepen understanding of evolving financial systems, adapt regulatory framework and supervisory practices for orderly development and stability of the financial system, safeguard the integrity of financial systems, modernize legal frameworks to provide an enabling legal landscape, ensure the stability of domestic monetary and financial systems, develop robust financial and data infrastructure to sustain fintech benefits, encourage international cooperation and information-sharing, enhance collective surveillance of the international monetary and financial system [3].

There is a number of recent studies, which closely examined the impact of digitalization of various dimensions in finance. Namely, the research by Loukoianova and Yang (2018) emphasizes the feasibility of adopting fintech applications in the Pacific through a coordinated regional approach, anchored by appropriate regulation and infrastructure [4].

Lashitew van Tulder, and Liasse (2019) aimed to understand the factors that affect the development and diffusion of mobile money services and revealed, how the power and interest dynamics of key actors in the innovation system can shape the emergence of appropriate technologies that aim to address social issues [5].

Patwardhan, Singleton, and Schmitz (2018) studied the emergence of financial technology players focused on enabling greater financial inclusion across four main lines of impact: payments, lending and related ecosystem, savings and financial planning, and insurance. The Fintech companies are increasing the financial capacities of households and organizations worldwide through fostering new approaches to mitigating frictions by designing novel products, or following innovative business strategies, with the common end goal of enhancing financial inclusion [6].

Citi. 2020 explore several financial inclusion initiatives, with a particular focus on digital initiatives, which are helping to increase the number of 'banked' individuals, where the unprecedented growth in financial inclusion is driven by innovative models and technology [7].

Macchiavello and Siri aimed at connecting the Fintech Action Plan and the Sustainable Finance Strategy seeking to find out the common features and highlight the potential of Fintech to respond to existing shortcomings in the current sustainable finance framework [18].

Mitha, Zadek, and Arner raise some considerations of governance in the global digital finance bringing upon the SDG aspects, risks and opportunities, aiming to enhance the governance initiatives to support innovation and developments in fintech [19].

Methodology. Methodologically, the viewpoint on investigation the trends of digital transformation, as well as problems hindering its development, is based on the analysis of secondary data to fulfil the aim of describing and analyzing FinTech. The retrospective data collection was performed via analysis of legal, professional and regulatory texts, providing an important source to analyze different types of support for solving problems of Digital Finance and FinTech. The authors documented the state of FinTech in the world by keeping a research diary, updating the information based upon mass media and information available on the websites of financial market regulators, with analytical reports on Global FinTech Adoption Index around the world and strategies developed by countries. Methods of theoretical generalization were used to study and systematize the current trends of the digital transformation in finance. Systematic approach was applied to examine the variety of stakeholders in financial markets in order to structure their strategies and detailed measures using digital technologies. The data sources included relevant legislature, official reports from international agencies, scientific papers, research materials, etc.

2 Research Results

2.1 The Essence of Digital Finance and FinTech

Digital innovations stimulate the development of the digital economy and society, make it possible to apply developments in many areas and lead to the transformation of the economy. Despite the great interest shown recently in financial technologies, often associated with the development of IT startups, innovative solutions in the field of finance are most often associated with the transformation of the form in which traditional financial products and services are offered to customers.

Modern level of digital technology allows companies to accumulate a huge amount of various information (big data), which can be used to analyze various processes and subsequently monetize. First of all, this is personal data, the analysis of which provides information about human habits, preferences, and models of economic behavior. Analysis of large data arrays based on the use of mathematical methods makes it possible to predict the behavior of demand and further marketing strategy of business entities. Such analytics is at the same time an expensive product in demand on the market.

Using this data for a business marketing strategy or reselling is inherently an effective profit opportunity. Modern digital technologies make it possible to reduce the number of intermediaries and the time for transferring information, and are also characterized by a high level of speed and security. The technology of distributed ledgers and analysis of big data, acting as breakthrough technologies, is already significantly transforming the existing system of the sphere of financial interaction of economic entities.

Distributed ledger technology has clearly demonstrated its practical importance in the issuance of non-fiduciary money, which includes cryptocurrencies. Deloitte developed predictions for the next decade on the possible trends of the digital transformation in finance (Table 1).

Digital platforms are changing what is done and the way it is done in countless ways (phone apps, home automation, cashless commerce, etc.). Today, leading finance organizations are conducting experiment with new technologies (Appendix 1), but the scalable and transformational change is yet to come.

Table 1. Trends of the digital transformation in finance

No	Trend	Explanation
1	The finance factory	Touchless transactions as automation and blockchain reach deeper into finance operations
2	The role of Finance	Finance will double down on business insights and service with largely operations automated. The skills required by finance professionals will change with new combinations of technology and human workforces
3	Finance cycles	Financial operations and decisions will be taken according to the real time reporting
4	Self-service	Self-service will become the prevailing form of service
5	Operating models	New service-delivery models will emerge as robots and algorithms join a more diverse finance workforce. Companies will assess the benefits of automation against onshore and offshore operations
6	Enterprise resource planning	Finance applications and micro services challenge traditional ERP
7	Data	The proliferation of APIs will drive data standardization
8	Workforce and workplace	Employees will be doing new things in new ways

Source: compiled according to [11].

2.2 National Practices in Developing Digital Finance and FinTech

The development of the digital economy has become possible thanks to targeted government policies. In the early 2020s, the priority of the Cabinet of Ministers and Deputies was the structural transformation of the Ukrainian economy, where innovation and digitalization became key drivers. At the same time, this sector is not without drawbacks preventing it from functioning effectively. Figure 1 shows problems hindering the development of digital transformation in Ukraine.

The Strategy for the Development of the Digital Economy was adopted and the basic conditions for its development were created – working legislation, access to a sufficient amount of capital, in particular venture capital, the education system, allowed university graduates to be competitive and meet the challenges of the current technological system.

The main steps made by Ukrainian government over the past years are as follows:

1. ***Creation and implementation of E-government.*** The digitalization of the public sector did not require additional funding from the budget. In fact, funds that had previously been spent on supporting analogue systems were used. During the first year of reforms, the E-government system was formed, which caused a transition to a full electronic document flow for all state bodies and citizens' interaction with them.

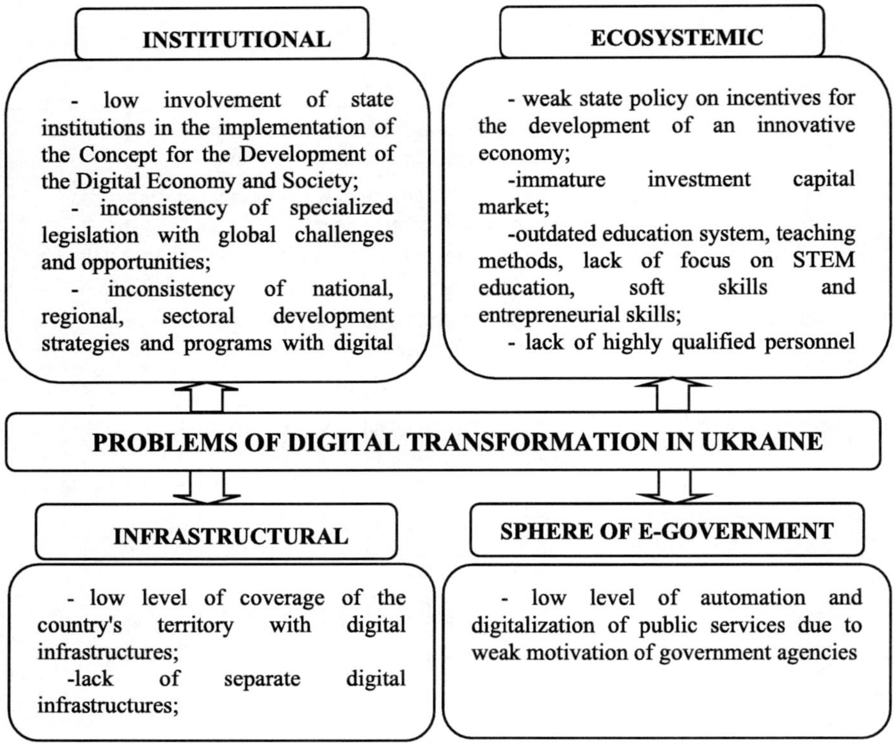

Fig. 1. Problems of digital transformation in Ukraine. Source: [12]

The next step was to open access for all digitally signed citizens to all public registers through personal electronic accounts and all government bodies. This made it possible not only to submit documents in electronic form with a digital signature, but also to track the stages of their passage. The terms of consideration of documents were reduced, the human factor was levered.

All interactions of citizens with the state takes place through an electronic cabinet. There you can see the statement of bank accounts, make a request to government agencies with an electronic signature and receive a prompt response, generate a tax return, sign contracts. Voting and participation in referendums also take place through an electronic cabinet [12].

2. ***Improvement of financial literacy and consumer protection in financial services.*** It should be noted that by the beginning of 2020 the National Bank received new powers to improve financial literacy and protect the rights of consumers of financial services. Back in 2018, the National Bank determined that financial inclusion is one of the strategic goals of the central bank.

An important prerequisite for engagement is improving Ukrainian financial literacy. Therefore, the National Bank, as the leader of the financial sector reform, developed and presented a vision of the Strategy with financial literacy in June 2019. The most important goal of the strategy is the development of a financial culture and a financially educated Ukrainian society.

The Ministry of Digital Transformation, together with the National Bank and PrivatBank, have launched the Digital Money educational course. Recently, the Ministry of Finance and the National Bank agreed on the development of digital financial skills of citizens, which became of great importance during the quarantine. The educational course "Digital Money" is the result of joint work to improve Ukrainian financial literacy.

Constant communication with citizens, explanations in simple language how financial services work, how to compare the services of several financial institutions, what are the foundations of "financial self-protection" – this is one of the most important tasks of the central bank today. Last year, the National Bank approved the "Strategy for Financial Literacy" and the action plan until 2022. Among the main areas of work is the creation of a website on financial literacy, where simple answers to complex financial questions will be provided, a distance learning platform for teachers and educators is already being developed, and all-Ukrainian information campaigns will be conducted in the regions [13].

Improving financial literacy is the key to greater involvement of citizens in the financial system, to the formation of correct financial habits, in particular the habits of saving, planning a budget, making balanced financial decisions.

3. ***Creation of the Ukrainian Association of FinTech and Innovation Companies.*** The Ukrainian Association of FinTech and Innovation Companies is an NGO that aims to foster the development of the domestic financial technology market, create a strong FinTech ecosystem and enhance financial inclusion in Ukraine. The Association was founded in June 2018 and its mission is as follows: research (analyze financial innovations to facilitate market development), advocate (create the best conditions for the growth and to advance financial inclusion), collaborate (encourage cooperation and knowledge sharing in the community), educate (bring forth the next generation of talents). It should be noted that two Memorandums were signed during 2018–2020:

- *Memorandum of Cooperation between the Ukrainian Association of FinTech and Innovation Companies and Fintech Galaxy.* Fintech Galaxy was founded in the United Arab Emirates in 2018 as a regional platform that fuels innovation in financial services, drives ecosystem collaboration and facilitates integration between financial institutions and FinTech companies. From scouting to deployment, it aims to build the future of financial services across the 22 Arab Countries.

The value of this collaboration is the Ukrainian FinTech Community will be reliable partners in Arab countries. This will facilitate the development of effective strategies, expert support and help Ukrainian startups to enter the eastern market. The signed Memorandum will strengthen the representation of Ukrainian FinTech companies in the Arab countries, and this partnership will help to attract the best international practices for the development of the national ecosystem.

- *Memorandum of Cooperation between UAFIC and Ministry of Digital Transformation.* Ukrainian Association of FinTech and Innovation Companies and the Ministry of Digital Transformation of Ukraine signed a Memorandum of Cooperation. This is another important stage in the development of the Ukrainian FinTech Ecosystem. The combining efforts and experience of the Association and the Ministry will provide a right course of development of digital transformation strategy, which, given the spread of COVID-19 in Ukraine and the world, is a particularly pressing issue [14].

During 2020–2025, the government of Ukraine is planning to implement the following steps:

- to identify the state as the main player in the sector;
- to legislatively secure the basic digital rights of a citizen;
- to implement infrastructure projects (connection to fixed broadband Internet for all households, physical infrastructure smartization, etc.);
- to initiate and implement numerous digital transformation projects (e-government, "smart" cities, electronic identification of citizens, electronic customs);
- to create conditions for the appearance of new jobs in the sectors of the digital economy;
- to introduce publicly available universal digital services for citizens: education, medicine, transport, security.

2.3 International Experience in Implementing Digital Finance and FinTech

The experience of individual countries in the field of FinTech shows the importance of balancing regulatory oversight and flexibility in innovation, as well as the need to manage cyber risks. Drivers for growth include supportive regulations and policies, dedicated incubators and accelerators, strong engagement with industry organizations, clear guidelines, availability of start-up and growth capital, quality of Internet and mobile infrastructure, availability of local talent, market structure and maturity financial sector.

It should be noted that international practices for regulating and implementing support measures for FinTech include the following:

1. *Creation of Fintech Bridges* allows to effectively exchange information on innovations and new trends in the financial services market both domestically and with other countries. The most favorable conditions for the development and implementation of financial technologies are offered by UK regulators. In particular, thanks to the creation of such initiatives as the FinTech Delivery Panel and the FinTech Steering Group, which bring together banks, Fintech companies and government agencies to discuss and develop programs for the development of the financial technology market [15].
2. *Creation of FinTech Sandboxes for financial startups.* India has introduced a Fintech sandbox mechanism for financial startups that have enjoyed regulatory requirements and government support for financial experiments.
3. *Creation of innovation hubs.* The Australian Securities and Investment Commission has created an innovation hub with the main mission of helping FinTech start-ups adapt to current regulations in financial markets.

4. *Creation of non-profit organizations and associations* in order to attract financing for FinTech. In Germany, a special non-profit organization has been created to improve the ecosystem for FinTech startups. In the same year, another association began its activity, searching for sources of financing for FinTech projects. Currently, Germany is taking the leading position in Europe in terms of the number of FinTech startups and the amount of attracted funding [16].
5. *Creation of regulatory laboratories (RegLab).* In the United Arab Emirates, this organization selects projects, which will be able to use mentoring, legal advice and other accompanying support necessary for the development and testing of a FinTech product. The most promising areas are blockchain and smart contracts.
6. *Creation of Special Groups to support the development of FinTech.* In Hong Kong, a special group was created to stimulate the development of the industry – Fintech Stiring Group. This association includes representatives of government authorities, the private sector, as well as research institutes, which corresponds to one of the mechanisms of innovative development – the "triple helix" model. It is expected that the group's vigorous activity will stimulate the development of financial technologies in Hong Kong, until it becomes one of the leading centers on the world stage [17].
7. *Strengthening the cybersecurity.* The experience of developed countries confirms that this is an important component, which requires special attention and financial support. Thus, China introduced stricter cybersecurity laws following a global ransomware attack by Wannacry that affected some banking operations. The UK has also invested significant resources in cybersecurity, but a recent cyber-attack by Wannacry has exposed vulnerabilities in a number of financial and non-financial companies, highlighting the need to strengthen defences.

3 Concluding Remarks

The rapid development of information technology leads to the emergence of new financial products that include a number of important advantages. Among them, one should note the high speed of settlements, low cost of operations and high potential for further development. That is why the government of Ukraine needs to pay attention to supporting the FinTech and creating tools for its development and protection.

We believe that due to the coronavirus pandemic, the transition to advanced technologies can be expected to proceed much faster. Indeed, residents of Europe have become more likely to use applications under the pandemic conditions. While the coronavirus has plunged the global economy into recession, the digital industry, by contrast, is booming. The rise in popularity of FinTech applications is a consequence of how society is trying to adapt to the changing conditions of the coronavirus pandemic. Moreover, even after its dying out, society will continue to rely on various digital services.

Now companies work at the intersection of industries and segments, becoming the platforms where they offer their customers not just a product, but a solution to any problems. Over the past 10 years, digitalization has become a key driver of the growth of the Ukrainian economy and the well-being of citizens. In future, digitalization will allow the following: to optimize business processes and create new products and services thanks to IT technology, virtual reality, cloud services, artificial intelligence; to reduce costs; to offer completely new business solutions (new insurance models, P2P systems, alternative banking services, mobile applications for training, an individual approach to online shoppers, personalized advertising).

As a result, we analysed the different types of support for solving problems of FinTech during the adaptation to the changing conditions of the coronavirus pandemic in the case of Ukraine, such as establishment and development of e-government, targeting the customer financial literacy and consumer protection in financial services, institutional elaborations to foster the development of the domestic financial technology market (UAFIC organization).

Appendix 1. Selected examples of digital disruptors in transforming finance

Essence	Problem	Solution	Result
Natural language generation for a global client	The country performance reports for a global client were generated monthly by a bot, right on schedule – but delivered little value. These reports had the potential for deep analysis and insights into market performance, but the bot could not deliver that level of commentary. Instead, each user drew data from the tabulated report and derived their own root-cause analysis for planning to leaders, a task that took nearly half the month to perform. Leadership saw the deep inefficiencies in this process, and wondered if natural language generation (NLG) technology could be applied to the process, helping standardize review, expedite analysis, and deepen insights	A data model that could enable report review was built, and a detailed catalogue of business rules that captured the "what-if" scenarios successfully was created. Touchless, dynamically generated country performance reports were delivered that were nuanced for global audiences and the necessary currency conversions. A modular approach was used, which simplified moving the data, and made implementation particularly nimble and scalable. With automation, these reports give Finance a better understanding of the business, as well as real-time availability of analysis, allowing for quicker decision making and action	The client has realized the return on investment in this engagement, and has found more use cases for this technology across the organization. It is also looking for ways to expand this technology globally. The modular structure of their new technology means that they possess a sort of asset library that can be deployed swiftly and at scale. Natural language generation has helped them make better sense of the data they already had – faster, more accurately, and with significant cost savings

(continued)

(*continued*)

Essence	Problem	Solution	Result
Natural Language Generation speeds financial reports, but human understanding helps them make a difference	The internal controls group at a large European financial institution was tasked with writing monthly progress reports for more than 100 branch offices. That meant a lot of manual labor, but also the potential for repetition and error. The underlying data was growing in size and complexity. The data going into the cloud was sensitive	A cloud-based Natural Language Generation (NLG) system was implemented that could securely and automatically translate the high-volume branch performance data into plain-language expressions of key metrics, then deliver it in useful visual formats. Agile prototyping helped manage validation of the software and speed the growth of its learning capacity. In addition to implementing one tool in one department, an NLG Center of Excellence was set up that is poised to extend the benefits to other parts of the institution	With cloud-based tools at their disposal and more free time due to automation, the control group members began asking for more detail and more analysis, which enriched the reports they created. Overall, the 15-member team now spends 94% less time compiling the monthly reports, while errors have been reduced and employee experience improved

(*continued*)

(*continued*)

Essence	Problem	Solution	Result
Digital budgeting helps Finance project better and faster	One of the largest banks in the world was going through an ambitious digital transformation across the enterprise, and had been working with Finance to disrupt its operations and bring them into the future. Not only did the bank want to update its operating model, it was transforming its entire technology landscape. As part of this sweeping change, a digital budgeting solution was needed that would improve the accuracy and quality of their non-interest, fee-based income projections. The bank was facing a few key roadblocks: a wide variance of forecasting, a dependence on human judgment, and a lack of tools that could provide accurate projections	The digital budgeting tool, which was ultimately delivered, uses 5 billion data points drawing from client risk profiles, industry indicators, governmental policies, and much more. The tool was able to improve the accuracy of non-interest income from 12% to 2%, and it now takes about 5 min to create a projection that previously took the bank up to a day to build. Even for the most complex model that depends on billions of data points, the tool takes less than 2 h. With this effort, the bank was able to lay the groundwork for a phase 2 project: monthly and quarterly rolling forecast models	The second phase of this initiative is in progress examining other potential benefits. The bank is also looking at ways to expand its use across the organization, and the tool is also a way for the bank to truly understand and see their customers. For example, the tool was able to profile high-value credit card customers, a key target audience for the bank, and enable highly targeted marketing campaigns

Source: compiled using [8–10].

References

1. Sustainable development goals UN. https://www.un.org/sustainabledevelopment/poverty/
2. Digital financial inclusion. https://www.itu.int/en/mediacentre/backgrounders/Pages/digital-financial-inclusion.aspx (2019)
3. The Bali FinTech Agenda. IMF policy paper. https://www.imf.org/en/Publications/Policy-Papers/Issues/2018/10/11/pp101118-bali-fintech-agenda (2018)
4. Elena, L., Davidovic, S., Sullivan, C., Tourpe, H.: Strategy for Fintech Applications in the Pacific Island Countries. IMF APD Departmental Paper No. 19/14. International Monetary Fund, Washington, DC (2019)
5. Lashitew, A.A., van Tulder, R., Liasse, Y.: Mobile phones for financial inclusion: what explains the diffusion of mobile money innovations? Res. Policy **48**(5), 1201–1215 (2019). https://doi.org/10.1016/j.respol.2018.12.010
6. Patwardhan, A.: Financial inclusion in the digital age. In: Handbook of Blockchain, Digital Finance, and Inclusion, pp. 57–89. Elsevier (2018). https://doi.org/10.1016/B978-0-12-810441-5.00004-X

7. Citi: "Banking the Next Billion: Digital Financial Inclusion in Action". https://ir.citi.com/pN%2BOlBahjzmphVliXxcaHdn%2BlkONN7NB7l3YKzlQvpsbYl%2Bril0LJyscLIG8hM%2FoOLl0CKhz8l0%3D (2020)
8. Natural language generation for a global client. Deloitte 2020. https://www2.deloitte.com/content/dam/Deloitte/us/Documents/finance-transformation/us-natural-language-generation-for-global-client.pdf
9. Translating data, navigating change. Deloitte 2020. https://www2.deloitte.com/content/dam/Deloitte/us/Documents/finance-transformation/us-translating-data-navigating-change.pdf
10. A big transformation, on a billion data points. Deloitte 2019. https://www2.deloitte.com/content/dam/Deloitte/us/Documents/finance-transformation/us-role-of-finance.pdf
11. Finance 2025: Digital transformation in finance. Deloitte 2018. https://www2.deloitte.com/us/en/pages/finance-transformation/articles/finance-digital-transformation-for-cfos.html
12. 2030E is a country with a developed digital. https://strategy.uifuture.org/kraina-z-rozvinutoyu-cifrovoyu-ekonomikoyu.html
13. Educational series "Digital Money" was launched in Ukraine. https://bank.gov.ua/ua/news/all/v-ukrayini-zapustili-osvitniy-serial-tsifrovi-groshi
14. Ukrainian association of FinTech and Innovation companies. https://fintechua.org/
15. Semenoh, A.: Comparative analysis of financial technology market regulation systems: foreign and domestic experience. Visnyk Sumskoho derzhavnoho universytetu – Bull. Sumy State Univ. **4**, 7–13 (2019)
16. Global FinTech Investment Market 2016 – 2020. Technavio (2016)
17. Report of the Steering Group on Financial Technologies. https://www.fstb.gov.hk/fsb/ppr/report/doc/Fintech_Report_for%20publication_e.pdf
18. Macchiavello, E., Siri, M. Sustainable finance and fintech: can technology contribute to achieving environmental goals? A preliminary assessment of 'Green FinTech'. European Banking Institute Working Paper Series 2020 – No. 71 (2020)https://doi.org/10.2139/ssrn.3672989
19. Mitha, A., Zadek, S., Arner, D.W.: Governing Global Digital Finance. University of Hong Kong Faculty of Law Research Paper No. 2020/045 (2020). https://doi.org/10.2139/ssrn.3678518

Values in Visual Practices of Instagram Network Users

Ekaterina Bataeva(✉) and Anastasiia Chumakova-Sierova

Kharkiv University of Humanities "People's Ukrainian Academy", Lermontovska str. 27, Kharkiv 61000, Ukraine
bataevaekaterina72@yahoo.com

Abstract. In this paper, we consider visual practices of Instagram users to identify the most popular models of social online communication associated with dominant value patterns. This paper builds upon the post-semiotic approach to show that visual images can be studied simultaneously as an intralingual and extra-language phenomenon. We present a methodology for studying visualized values in social practices of Instagram users employing the theory of postmodernization of values by R. Inglehart, which describes three value patterns – traditional, modern, and postmodern. A content analysis of 2,657 photos of Instagram youth audience was carried out. It was revealed that modern and postmodern values are most often visualized in the social practices of Instagram network users in Ukraine. It is worth noting that visualized value practices on Instagram can reflect the habitualized preferences of social actors, which are regulators of their behavior, but at the same time may not be realized. It is concluded that the visualized values in the social practices of Instagram users can be the predictors of subsequent value transformations in real social life.

Keywords: Visualized values · Instagram · Instagram network users · Content analysis

1 Introduction

Visual practices are the subject of various disciplines such as cultural studies, visual studies, visual sociology, social visualistics, visual anthropology, etc., which have an interest in various visual forms (art, cinema, photography, visual everyday life, visual media, Instagram, etc.) [1–12]. In a modern mediatized society, visual practices of social actors can be observed in various social network sites (SNSs), among which Instagram stands out as an image-based platform specializing in photo and video sharing. The study of the visual practices of social actors on Instagram can help to identify the popular models of social behavior online associated with dominant value patterns.

The authors express their deep gratitude for participation in the empirical part of the study to Anastasiia Drannik, Bachelor of Sociology, Department of Sociology and Humanities, Kharkiv University of Humanities "People's Ukrainian Academy".

M. Nechyporuk et al. (Eds.): ICTM 2021, LNNS 367, pp. 992–1002, 2022.
https://doi.org/10.1007/978-3-030-94259-5_76

The goals of this paper are 1) development of the methodology for studying the values visualized in online practices of social actors (which will be tested on Instagram[1]); 2) identification of the dominant values visualized on Instagram based on the R. Inglehart's theory of postmodernization of values.

2 Methodology

The basic thesis of our article is the following: the values and attitudes of social actors can be studied not only with the help of verbal methods but also through the analysis of visual practices implemented via photos and videos. Since Instagram content is visualized and also taking into account the popularity of this social network site among young people, we conducted the study of its visual content to identify the most popular value patterns among young users of the Ukrainian pool of Instagram.

Social visuality can be interpreted based on two principles: 1) media visuality can be an "effect" of the social unconscious (visual media content can reveal latent unconscious mass preferences and values of social actors); 2) media visuality 'constructs' social practices and attitudes by transmitting the most popular value patterns in a particular society; images are constructed as a product of the consciousness or imagination of their authors. As a result, the study of the visual content of modern media (Internet, television) can help to identify massively demanded value and social behavior models.

In the study, we distinguish between the concepts of values and attitudes of social actors. As Z. Toshchenko emphasizes, "if values are material or ideal objects that are significant for all social actors, and these values can satisfy the needs of a society, then attitudes are value orientations of individuals, groups, communities to their chosen values" [17, p. 104]. The attitudes of a social actor can be interpreted as accepted values that have become regulators of his/her behavior (in our case, regulators of the visual practices of Instagram network users). The connection between the values and behavior of social actors is emphasized by many sociologists. Many sociologists note that there is a connection between the values and behavior of social actors. For example, K. Thomas and F. Znaniecki believe that values are life guidelines for a person, for the sake of which he/she will perform particular actions. Values are "objective, social elements which impose themselves upon the individual as given and provoke his reaction" [18, p. 1131]. F. Adler emphasizes that values can be studied by examining the actions of social actors in which values manifest themselves, although there may be no absolute dependence between values and actions [19, p. 276].

Let us make some theoretical assumptions that justify the possibility of studying the value content using the analysis of Instagram visuality.

1. In verbal communication, under the influence of basic conformism or inertia, a social actor can indicate those values that are not truly important for him/her. For example,

[1] The most popular social networks in Ukraine are Facebook (58%), YouTube (41%), Instagram (28%), Telegram (14%) and Vkontakte (7%). As for Facebook, YouTube, Instagram, and Vkontakte, the leading positions in them are occupied by people from 18 to 49 years old [15]. "Women make up 60% of the Ukrainian audience (6.8 million) of Instagram" [15]. "The number of Ukrainian users of the social network Instagram in 2019 increased by 4.5% – to 11.5 million (about 11 million in 2018)" [16].

L. Sokuryanskaya distinguishes between declared, internalized, and realized values [20, p. 130], which may not coincide at the behavioral level. L. Bevzenko emphasizes that there are deep value preferences of an individual, which may not be realized and not verbalized during sociological polls, while articulated values may no longer reflect his/her value priorities. L. Bevzenko determines the deep values as "habitualized values that do not appear in the verbal representation, but work as a deep structure that structures the social practices" [21, p. 146]. In her opinion, unconscious values can be studied by observing everyday social practices and the lifestyle of a social actor, onto which they can be projected. Accepting these premises, we assume that photos and videos can represent 1) conscious values of a social actor, which he/she, due to their importance and priority, would like to visualize, 2) unconscious values that can be involuntarily projected onto the preferred visual online behavior of a person.

2. Since a person is prone to offer others an idealized image of himself/herself [13], it can be assumed that Instagram network users present in photos and videos both actual and idealized self-images, on which value priorities can be displayed precisely because of their idealization.
3. Given the fact that Instagram network users in their self-performances can take into account, copy and reproduce the preferences and values of the intended audience [14], even if we assume that this reproduction is carried out on a superficial level, without affecting the inner beliefs of a person, its studying could provide the information on popular Instagram value patterns and visual behavior.

In the study of visualized values on Instagram, we used content analysis, which is effective when working with various texts (including visual media texts) since it allows identifying the peripheral and dominant tendencies in them and revealing the statistical patterns. "Content analysis is a technique for examining the content or information and symbols contained in written documents or other communication media (e.g., photographs, movies, song lyrics, advertisements)" [22, p. 49]. Traditional, modern, and postmodern values are code categories of our content analysis. The study of visualized values in the social practices of Instagram network users has been carried out using the theory of postmodernization of values by R. Inglehart, which describes three value patterns – traditional, modern, and postmodern. The traditional values include family, health, religiosity, respect for elders, respect for authority, discipline, traditional morality, and public order. The modern values include material well-being, social career, physical security, and physical survival. The postmodern values include self-expression, self-realization, quality of life, aesthetic development, environmental protection [23].

The units of analysis are visual contexts in which a particular value is presented. The units of analysis and units of count have coincided: visualized value in the photo (a unit of analysis) has been recorded as a unit of count. The research procedure included the counting of photographic practices in which some values were manifested. Since the object of our research was the social photographic practices of Instagram network users, we did not analyze those photos that did not contain images of people. Let us consider the units of analysis that reveal the content of the research categories.

The units of analysis that reveal traditional values are the following visual contexts: 1) family context (for example, images of marriage partners; mother or father with children; a whole family with children on vacation, during a walk, at home); 2) religious context (images of a person or a group near the church; images of participation in church services/sacraments and holidays, such as Epiphany, Easter); 3) context of the preparation and consumption of food (images of a person or a group at the table, with the focus on food; an image of a person who demonstrates the enjoyment of food intake). The latter context is included in the methodology since food consumption is a vital need of the human body associated with health, which is very important in traditional communities.

The units of analysis correlated with modern values are the following visual contexts: 1) context of a career and achievements (images of people demonstrating their achievements (diplomas, medals, sports cups), prestigious professional status (for example, an image of a person in a rich room in a boss chair); 2) context of comfort and material well-being (images of people in a luxurious environment, in front of an expensive car or a luxurious house, with expensive gadgets, watches); 3) context of prestigious resort recreation and entertainment (images of people in rich hotels, against the background of swimming pools, in restaurants, in prestigious resort countries; at parties, discos, in clubs, bars); 4) context of glamor (images of people where the focus is on their appearance – clothes, jewelry, makeup, manicure, hairstyle, naked body, presented in a selfie, displayed in a mirror). The latter context is included in the block of "modern values", since the cult of glamorous appearance and imitation of celebrities' life is one of the manifestations of consumer culture, to a greater extent associated with materialistic value preferences.

The units of analysis that reveal postmodern values are the following visual contexts: 1) context of friendly relations (several friends are shown together in the photo; they can hug each other, communicate, spend leisure time together); 2) context of professional self-development (images of people in the process of work, reading books, studying, attending pieces of training or seminars); 3) context of environmentalism and aesthetic perception of life (images in which a person admires a beautiful landscape, flowers at home, sunset/dawn, sunrays, a rainbow; images in which the focus is not on a person, but the sea, rivers, waterfalls, mountains); 4) context of cultural tourism (images of a person while traveling to study other cultures; photos of a person against the background of historical buildings, museum exhibitions, cultural centers).

The aims of our study were 1) to find out which visualized values dominate in the Ukrainian pool of Instagram; 2) conduct a gender analysis of visualized values in social practices of Ukrainian Instagram network users; conduct a comparative analysis of male/female preferences of values in visual Instagram content. To reach these aims, the following codes were presented in the data matrix: 1) the code of a particular value related to some value pattern; 2) the gender code of a visual image (male, female).

3 Research Results

Since the majority of Instagram network users are young [15], data collection was carried out in accounts of users at the age of 14 to 35 (this age interval was determined visually, by the appearance of users since there is no option "user age" on Instagram). Although the majority of Ukrainian Instagram network users are women, we used gender sampling criterion to make a comparative analysis of the values visualized in the social practices of men and women (1,418 (52%) photos presented in women accounts), and 1,303 (48%) photos submitted to men accounts). We also applied the criterion of geographic localization of Instagram network users in our study. By geographic tags, the accounts of Instagram network users living in various regions of Ukraine were selected: West (Volyn, Transcarpathian, Ivano-Frankivsk, Lviv, Rivne, Ternopil, Chernivtsi regions) – 770 (28%) photos; Center (Vinnytsia, Zhytomyr, Kyiv, Kropyvnytsky, Poltava, Sumy, Khmelnytsky, Cherkasy, Chernihiv regions) – 1,013 (37%) photos; South (Zaporizhzhia, Mykolaiv, Kherson, Odesa regions) – 333 (12%) photos; East (Dnepropetrovsk, Kharkiv regions) – 609 (20%) photos. Since these regions were disproportionally represented in the sample, and the differences in the data by regions turned out to be insignificant (the range of discrepancies is 1–4%), we did not conduct a comparative analysis of the value preferences of Instagram network users by geographical regions of Ukraine.

The relevant data were collected within a period from May 2020 to December 2020; we selected those photos on Instagram network user accounts that dated from 2018 to March 15, 2020 (before the introduction of quarantine in Ukraine in connection with the Covid-19 pandemic since during quarantine the values of people could change). The photo selection algorithm was as follows: the locations of settlements in each of the four regions of Ukraine were used; in these locations, approximately 24 accounts were randomly selected, in which every fifth photo was taken (a total of 5–6 photos in one account). The total number of Instagram pictures analyzed was 2,657. Since one photo could represent several visualized values (and, accordingly, several units of count), we identified 2,721 visual value contexts.

Microsoft Office Excel was used to analyze all the collected data. Descriptive statistics were used to identify some trends in the visual content of the Ukrainian pool of Instagram. Tables 1 and 2 present the results of the content analysis of Instagram pictures, in which three indicators are: 1) the number of visual contexts in which a particular value is manifested, 2) a percentage of these visual contexts to all analyzed value contexts, and 3) the rank of a particular visualized value in the overall ranking.

Modern and postmodern values are most often visualized in the social practices of Instagram network users in Ukraine (38.8% and 41.8%, respectively) (see Table 2). These data are inconsistent with the results of the European Values Study conducted in 2008–2010 in 43 European countries using the method of R. Inglehart when only 1% of Ukrainians stated that the values of initiative autonomy and self-expression are important for them (that is, postmodern values), and 33% preferred the modern values of material well-being and social career [24, p. 18]. However, if we take into account that the majority of Instagram network users are young, and in any case, in our study, we analyzed photos only of the youth audience, we can see a partial consistency of our results with the data of other Ukrainian sociological studies that were conducted using the survey method (for example, L. Sokuryanskaya revealed modern and postmodern preferences of Ukrainian student youth in her survey [25]). It can be assumed that the visual content represents the value preferences of social actors that have not been yet censored by the conscious and have not been aligned with public expectations. Instagram network users can unconsciously disclose in their visual practices those value preferences that have not yet been verbalized and rationalized in everyday life, and therefore may differ from the value standards articulated in a particular society. The visualized values can be "below the level of consciousness and language" (which is one of the characteristics of the "habitus" by P. Bourdieu), to be unconscious and unarticulated, but nevertheless, be regulators of human behavior. It can also be assumed that the presence of discrepancies between the data of visual studies and the results of verbal surveys may indicate that the maturation of habitualized values is ahead of the process of comprehension and articulation of value preferences at the individual and mass levels, and may have an inverse effect on the dynamics of social change.

Table 1. Visualized values in the pictures of users of the Ukrainian Instagram

№	Value pattern	Visualized value	N	Percentage	Rank
1	Traditional values	Family context	455	16.7	2
2		Religious context	16	0.6	11
3		Context of the preparation and consumption of food	57	2.1	10
4	Modern values	Context of a career and achievements	136	5.0	8
5		Context of comfort and material well-being	230	8.5	5
6		Context of prestigious resort recreation and entertainment	141	5.2	7
7		Context of glamor	548	20.0	1
8	Postmodern values	Context of friendly relations	444	16.3	3
9		Context of professional self-development	171	6.3	6
10		Context of environmentalism and aesthetic perception of life	108	4.0	9
11		Context of cultural tourism	415	15.3	4

Table 2. Value patterns in visual practices of Instagram network users

№	Value pattern	N	Percentage	Rank
1	Traditional values	528	19.4	3
2	Modern values	1055	38.8	2
3	Postmodern values	1138	41.8	1

If we compare not value patterns, but individual values visualized in the social practices of Instagram network users, we can highlight the following trends. The most popular on Ukrainian Instagram are glamorous images (548 (20%) of visual contexts) (see Table

1), in which women are most often represented (425 female photos versus 123 male photos) (see Table 3). The main message of these photos is a demonstration of attractive appearance, clothes, jewelry, makeup, manicure, hair, body. According to J. Baudrillard, this message is the main in consumer society. The emphasized attention to the material marking of social status, to the visual harmony of the appearance, creates a new hero of our time, Narcissus, a narcissistic esthete that looks for his/her reflection everywhere (in the shop windows, mirrors, in personal photos and videos). The narcissism of a modern person is manifested in an almost hypnotic passion for portraying oneself in photos and videos, in an almost manic desire to capture themselves in various interiors and exteriors. In such visual practices, the desire of a modern person to perpetuate himself/herself (in the ephemeral eternity of photos), to overcome the decay of life is manifested [26].

Table 3. Gender analysis of visualized values on Instagram

№	Visualized value	Female, N	Female, %	Male, N	Male, %
1	Family context	256	18.1	199	15.3
2	Religious context	7	0.5	9	0.7
3	Context of the preparation and consumption of food	34	2.4	23	1.7
4	Context of a career and achievements	48	3.4	88	6.7
5	Context of comfort and material well-being	119	8.4	111	8.5
6	Context of prestigious resort recreation and entertainment	66	4.7	75	5.7
7	Context of glamor	425	30.0	123	9.4
8	Context of friendly relations	146	10.3	298	23.0
9	Context of professional self-development	46	3.2	125	9.6
10	Context of environmentalism and aesthetic perception of life	71	5.0	37	2.8
11	Context of cultural tourism	200	14.0	215	16.6

The second place in the ranking of visualized values is the family (455 (16.7%) visual contexts) (see Table 1). At this point, one can notice a discrepancy between the data obtained from the results of the World Values Survey (WVS) Wave 7 (2017–2020), which recorded the significant importance of the family for Ukrainians (86.0%) [27]. It can be assumed that the family continues to be the main declared value of Ukrainians, but in the daily life of young people, it is no longer dominant, as evidenced by its infrequent representation in the visual content on Instagram.

The postmodern value of friendly relations is slightly inferior to the family (444 (16.3%) of visual contexts) (see Table 1). These data are almost two times different from the results of World Values Survey (WVS) Wave 7 (2017–2020), where 34.0% of Ukrainians consider friends to be a very important value [27]. Interestingly, the context of friendship is twice as common in male photos on Instagram (298 visualized values of friendly relations in male photos versus 146 female ones) (see Table 3).

The postmodern context of cultural tourism (415 (15.3%) of visual contexts) is approximately equally represented in male and female photos on Instagram (200 female photos and 215 male photos) (see Table 3). The popularity of the tourism context is consistent with the conclusion of Z. Bauman, for whom the tourist is one of the central figures in postmodern society. "The tourist is a conscious and systematic seeker of experience, of a new and different experience, of the experience of difference and novelty – as the joys of the familiar wear off and cease to allure. The tourists want to immerse themselves in a strange and bizarre element (a pleasant feeling, a tickling and rejuvenating feeling, like letting oneself be buffeted by sea waves) on condition, though, that it will not stick to the skin and thus can be shaken off they wish" [28, p. 29]. The tourist is driven by the desire to see something new, to get rid of the boredom of daily life.

The context of comfort and material well-being ranked only fifth in the overall rating of visualized values (230 (8.5%) of visual contexts) (see Table 1); women and men in equal proportions share photos on this topic on Instagram (119 female photos and 111 male) (see Table 3). The next popular visualized values are professional self-development (171 (6.3%)), prestigious resort recreation and entertainment (141 (5.2%)), career and achievements (136 (5.0%)), environmentalism and aesthetic perception of life (108 (4.0%)) (see Table 1). The postmodern context of professional self-development is most often used by men (125 male photos versus 46 female photos); the postmodern context of environmentalism and aesthetic perception of life is more often used by women (71 female photos versus 37 male) (see Table 3). The modern context of career and achievements is most often presented in male photos (88 male photos versus 48 female); the modern context of prestigious resort recreation and entertainment is presented in male and female photos in approximately the same way (66 female and 75 male photos) (see Table 3). Religious context (16 (0.6%)) and the context of food preparation and consumption (57 (2.1%)) are rarely found on Ukrainian Instagram (see Table 1). As for the latter context, one can often see images of food without the presence of a person in pictures on Instagram. However, in our study, we did not analyze pictures of objects without human involvement. The data for the "religion" category is significantly different from the results of the World Values Survey (WVS) Wave 7 (2017–2020), where 22.0% of Ukrainians stated the importance of religion for them [27]. This difference can be explained by the fact that religious faith is a private affair of each person, accordingly, believers may not want to demonstrate their inner convictions on display.

If we summarize male and female value preferences in Ukrainian Instagram, we can see the following tendencies. In male photos, one can often see postmodern visualized values (52.0%), followed by modern (30.3%) and traditional (17.7%) values. In female photos, modern visualized values dominate (46.5%), followed by postmodern (32.5%) and traditional (21.0%) values (see Table 4).

Traditional visualized values are the least popular in two groups of Ukrainian Instagram network users, while modern and postmodern values are reversed in male and female accounts. In our study, we did not aim to explain the different dynamics of value changes in visual practices of male and female Instagram network users in Ukraine.

Table 4. Value patterns in visual practices of male/female Instagram network users

№	Value pattern	Female, N	Female, %	Male, N	Male, %
1	Traditional values	297	21.0	231	17.7
2	Modern values	658	46.5	397	30.3
3	Postmodern values	463	32.5	675	52.0

4 Conclusions

Despite numerous Instagram studies have been conducted, little attention has been paid to the analysis of value content visualized in social practices of Instagram network users. However, the study of visualized value content on Instagram is important for forecasting future value changes in some societies since it influences the popularization and formation of particular value patterns at the mass level of social interaction. Consequently, it is necessary to provide monitoring of value transformations in social network sites that are popular (as it was mentioned above, the number of Ukrainian Instagram network users in 2019 was 11.5 million).

In this paper, we presented a methodology for studying visualized values in social practices of Instagram network users employing the theory of postmodernization of values by R. Inglehart, which describes three value patterns – traditional, modern, and postmodern. A content analysis of 2,657 photos of Instagram youth audience was carried out. The key findings are as follows.

1. Modern and postmodern values are most often visualized in the social practices of Instagram network users in Ukraine (38.8% and 41.8%, respectively).
2. The most popular on Ukrainian Instagram are glamorous images (20%); family (16.7%); friendly relations (16.3%); cultural tourism (15.3%); comfort and material well-being (8.5%); professional self-development (6.3%); prestigious resort recreation and entertainment (5.2%); career and achievements (5.0%); environmentalism and aesthetic perception of life (4.0%).
3. Compared to male value preferences, the most popular in female accounts are the visualized values of glamor, environmentalism, and aesthetic perception of life.
4. Compared to female value preferences, the most popular in male accounts are the visualized values of friendship, professional self-development, career, and achievement.
5. In male photos, one can often see postmodern visualized values (52.0%), followed by modern (30.3%) and traditional (17.7%) values. In female photos, modern visualized values dominate (46.5%), followed by postmodern (32.5%) and traditional (21.0%) values.

Following the attitudes of the post-semiotic theoretical approach, we can assume that the visual content of Instagram presents the value preferences of social actors that have not yet been censored by the conscious, have not been verbalized, that is, they can be at the pre-sign, translinguistic level. As a result, they can outstrip value changes in real life and influence the formation of the new value patterns in a particular society. Visualized value practices on Instagram can reflect the habitualized preferences of social actors, which are regulators of their behavior, but at the same time may not be realized. Thus, the visualized values in the social practices of Instagram network users can be the predictors of subsequent value transformations in real social life.

References

1. Mirzoeff, N.: An Introduction to Visual Culture. Routledge, New York (1999)
2. Bataeva, E.V.: Flanering and video mania: modern and postmodern visual practices. Vopr. Filos. **11**, 61–68 (2012). [in Russian]
3. Bataeva, E.V.: An ethno-methodological analysis of on-line communications: a crisis experiment in chats. Sotsiologicheskie Issledovaniya **12**, 88–97 (2011). [in Russian]
4. Bataeva, E.V., Polyakova, Y.A.: Motive-analysis of TV-advertising visual content. Sotsiologicheskiy Zhurnal **24**(2), 66–89 (2018). [in Russian]
5. Lukin, V.V., Zriakhov, M.S., Ponomarenko, N.N., Krivenko, S.S., Zhenjiang, M.: Lossy compression of images without visible distortions and its application. In: International Conference on Signal Processing Proceedings, ICSP, Beijing, China, pp. 698–701 (2010)
6. Elkins, J.: Visual Studies: A Skeptical Introduction. Routledge, New York (2003)
7. Dikovitskaya, M.: Visual Culture. The Study of the Visual After the Cultural Turn. MIT Press, Massachusetts (2005)
8. Jones, A.: The Feminism and Visual Culture Reader. London, New York (2002)
9. Pink, S.: The Future of Visual Anthropology: Engaging the Senses. Routledge, London, New York (2006)
10. Pink, S.: Doing Visual Ethnography: Images, Media, and Representation in Research. Sage, London (2007)
11. Nakamura, L.: Digitizing Race. Visual Cultures of the Internet. University of Minnesota Press, Minneapolis, London (2007)
12. Smith, M.: Visual Culture Studies: Interviews with Key Thinkers. Sage Publication, Los Angeles, London, New Delhi, Singapore (2008)
13. Goffman, E.: The Presentation of Self in Everyday Life. Anchor Books, New York (1959)
14. Boyd, D.: It's Complicated: The Social Lives of Networked Teens. Yale University Press, New Haven, CT (2014)
15. What social networks are the most popular in Ukraine and do we trust them: a survey. https://dengi.informator.ua/2020/02/27/kakie-sotsseti-samoe-populyarnye-v-ukraine-i-doveryayut-li-im-opros/#:~:text. Accessed 21 Feb 2021 [in Russian]
16. The number of Facebook users in Ukraine in 2019 increased by 7.7%, Instagram – by 4.5%, https://interfax.com.ua/news/telecom/647342.html. Accessed 30 Apr 2021 [in Russian]
17. Toshchenko, Z.: Values and Value Orientations. Sociology Thesaurus. UNITY-DANA, Moscow, pp. 103–107 (2013) [in Russian]
18. Thomas, W.I., Znaniecki, F.: The Polish Peasant in Europe and America. Alfred A. Knopf, New York (1927)
19. Adler, F.: The value concept in sociology. Am. J. Sociol. **62**(3), 272–279 (1956)
20. Sokuryanskaya, L.: Students on the way to another society: value discourse of the transition. KhNU im. V. N. Karazin, Kharkiv (2006) [in Russian]
21. Bevzenko, L.: Life success, values, life styles. Sociology **4**, 132–151 (2007). [in Russian]
22. Neuman, W.: Social Research Methods: Qualitative and Quantitative Approaches. Pearson Education Limited, Edinburgh (2014)

23. Inglehart, R., Welzel, C.: Modernization, Cultural Change, and Democracy. The Human Development Sequence. Cambridge University Press (2005)
24. Magun, V., Rudnev, M.: Heterogeneity of the values of the European countries' population: typology on indicators of R. Inglehart. Vestnik obshhestvennogo mnenija. Dannye. Analiz. Diskussii, pp. 12–24 (2012) [in Russian]
25. Sokuryanska, L.: "Modernization" and "postmodernization" of students' value orientations as a factor in the modernization of society. In: Methodology Theory Practice of Sociological Analysis of Modern Society, pp. 36–45 (2012) [in Russian]
26. Baudrillard, J.: The Consumer Society: Myths and Structures. SAGE Publications Ltd. (1998)
27. Haerpfer, C., et al. (eds.): World Values Survey: Round Seven – Country-Pooled Datafile. JD Systems Institute and WVSA Secretariat, Madrid, Spain and Vienna, Austria (2020) https://www.worldvaluessurvey.org/WVSDocumentationWV7.jsp. Accessed 30 Apr 2021
28. Bauman, Z.: From pilgrim to tourist – or a short history of identity. In: Hall, S., Du Gay, P. (eds.) Questions of Cultural Identity, pp. 18–36. Sage Publications Inc. (1996)

Lossy Compression of Remote Sensing and Dental Images Corrupted by Spatially Correlated Noise

Vladimir Lukin[1], Sergey Krivenko[1(✉)], Ihor Kaluzhinov[1], Olha Krylova[2], and Liudmyla Kryvenko[2]

[1] National Aerospace University, Chkalova Street 17, Kharkiv 61070, Ukraine
lukin@ai.kharkov.com, krivenkos@ieee.org, kaluzhynov@i.ua
[2] Kharkiv National Medical University, 4 Nauky Avenue, Kharkiv 61022, Ukraine
krylovaol@ukr.net, ls.kryvenko@knmu.edu.ua

Abstract. Remote sensing and medical imaging are very important in human's life. The former deals with many applications from weather monitoring to military purposes and is performed by sensors mounted onboard of satellite and aerospace carriers. Medical imaging assists in disease detection and diagnostics. Images acquired by synthetic aperture radar or ultrasound medical imagers, alongside with well-known advantages have several drawbacks, particularly, spatially correlated speckle which corrupts images and, simultaneously, restricts possibilities of lossless image compression of remote needed due to a large size typical for modern applications. Therefore, a very important task is lossy compression of such images with acceptable quality that can be described in different ways including visual quality metrics. Three lossy coders are considered in our study: AGU, BPG, and JPEG2000. Three quality metrics are used: PSNR, PSNR-HVS-M, and MSSIM where two latter ones are visual quality metrics. Performance comparisons are performed. Recommendations are given for remote sensing as well as medical imaging applications.

Keywords: Remote sensing · Synthetic aperture radar · Ultrasound · Spatially correlated noise · Lossy compression · Visual quality

1 Introduction

Systems for acquisition and processing of images are an important part of various spheres of human life. These tools allow people to obtain information about the characteristics and condition of various objects in remote sensing, environmental monitoring, biomedical applications, production control, etc. [1]. It is also worth mentioning the digital cameras built into wearable gadgets. All of these systems generate billions of images every day [2]. Thus, image compression is desired. At the same time, image processing should often be carried out automatically, since manual and semi-automatic processing is not able to "manage" a huge amount of data.

M. Nechyporuk et al. (Eds.): ICTM 2021, LNNS 367, pp. 1003–1014, 2022.
https://doi.org/10.1007/978-3-030-94259-5_77

Image processing can be done in several stages. These stages include determination of the type of interference and their characteristics [3], filtering [4], lossy [5] or lossless compression [6], segmentation [7], classification [8], etc. In practice, some of these steps can be skipped. However, the parameters and characteristics of images and interference obtained at some stages of processing can serve as input parameters for subsequent stages. In addition, the expediency of carrying out a certain stage of processing, based on a final goal, has to be taken into account.

Various sensors are used for Earth remote sensing (RS), in particular, synthetic aperture radars (SAR) are widely used. Many modern SAR systems generate high resolution images. High resolution (and, accordingly, a large amount of data) is obtained due to the large size of the synthesized antennas [9], wide swaths, frequent observations. Similar tendencies take place in medical, in particular, dental imaging where hundreds and thousands of images can be acquired each day or week even in one diagnostic center.

The images generated by radar and medical systems have two specific features. In addition to the large size mentioned above, this is distortion of images by a special type of noise which is signal-dependent, for example, speckle, which occurs due to coherent processing of the received signals. At the same time, a decrease in the number of looks or exposition time leads to an increase in the intensity of noise in the generated image, which ultimately worsens the visual quality of images and complicates their classification and analysis.

A large size of acquired images makes necessary or desirable their compression. There are two commonly used approaches to compressing SAR and ultrasound images. The first concerns the compression of raw signals. This is possible only on board the remote sensing system or inside a medical imager, which limits the flexibility of this approach. The second approach is based on the application of various compression methods to an already formed image [10]. In principle, two types of compression are possible – lossless and lossy. However, lossless image compression is often not satisfactory because it reduces the noisy data volume by less than 2 times only [11]. Lossy image compression is a way out in this situation, because it often allows significant increasing the compression ratio.

Lossy compression is common in digital image processing. For example, one of the most common messengers, Viber, which had over 800 million users in 2016 [12] and has more nowadays, already uses unconditional (i.e. not controlled by a user) lossy compression of images when transferring them between users. However, the main problem is to control the quality of the resulting images. The higher the image compression rate, the worse the output quality.

Another problem is the fact that images are usually viewed as undistorted by noise, which is often not true in the considered cases. It is obvious that almost all image processing stages introduce distortions into the original signal to varying degree, including distortions due to complex nonlinear operations [13, 14]. In the considered case, noise is the dominant factor determining their quality. The reasons for this are unsatisfactory shooting conditions (e.g., due to poor lighting) and the formation of images by coherent systems in radars and medicine (X-Ray, CT, MRI) [15, 16] etc. These factors are critical to the efficiency of lossy image compression [17] where the noise observed in different types of images has a different nature and characteristics [18]. In addition, noises can be visually noticeable or unnoticeable [19].

This determines further scenarios for image processing. In the case of visually invisible interference, it makes no sense to determine the parameters of the noise and filter it out. In such situations, the noise has little effect on further processing steps. The quality of segmentation and classification, as a rule, does not change in the absence of filtering of visually imperceptible noise. The features of lossy image compression using the noise factor (compression in the vicinity of the optimal operating point (OOP) [20]) also do not manifest themselves in the case of visually imperceptible distortions.

The situation changes radically with visually noticeable noise. Note that the texture of the image also affects the visual perception, because it masks the noise [21].

An important aspect of image quality analysis is a criterion according to which quality is assessed. These criteria (metrics) can take into account visual quality (PSNR-HVS-M, MSSIM, etc.). Other standard metrics (e.g., MSE, PSNR) do not take peculiarities of human vision system (HVS) into account. However, a common problem with the aforementioned situation is a priori assumption that the noise distorting an image is additive, white and Gaussian. At the same time, SAR and coherent medical systems form images with noise, which have a more complex structure and nature. The main place in such systems is occupied by signal-dependent and spatially correlated noise. Figure 1 shows the fr01 remote sensing image distorted by spatially correlated noise simulating speckle with a different number of looks and a noise parameter. One way to model spatially-correlated noise is to modify the spectrum of AWGN realization by low-pass filtering and transformation back into the spatial domain. Noise parameter is a 'tuning' parameter describing the degree of spatial correlation. The value of 0.5 corresponds to almost AWGN case (very small spatial correlation) whilst the parameter value equal to 1.2 relates to the speckle noise present in Sentinel-1 SAR images [22]. It is easy to see that in the case of a small number of looks and a large value of the noise parameter, the distortions are more noticeable and less pleasant for human perception. This is especially noticeable in homogeneous image areas and regions with minimal texture. Our goal is to analyze opportunities of visually lossless compression of images corrupted by spatially correlated noise.

2 Image and Noise Models. Quality Metrics. Optimal Operation Point

In most situations, an additive model of the effect of noise on an image is considered. Such a model can be represented as:

$$I_{ij}^{noisy} = I_{ij}^{true} + n_{ij}, \quad (1)$$

where I_{ij}^{noisy} is the noisy pixel value, I_{ij}^{true} denotes the true pixel value, n_{ij} is the value of additive Gaussian noise with zero mean and variance σ^2. This formula is also valid for the case of additive spatially correlated noise (SCN). In this case, for n_{ij}, a spatially correlated nature of the noise is assumed. Multiplicative model of noise can be represented as $I_{ij}^{noisy} = \mu I_{ij}^{true}$, where μ is multiplicative noise.After applying lossy image compression, each point of the resulting image can be represented as:

$$I_{ij}^{c} = f(I_{ij}), \tag{2}$$

where I_{ij}^{c} is the pixel value of the compressed image, $f(.)$ is compression algorithm.

The classic metric used to estimate distortions is mean square error (MSE) and peak signal-to-noise ratio (PSNR) associated to MSE:

$$\begin{aligned} MSE^{c} &= \sum_{i=1}^{I} \sum_{j=1}^{J} \left(I_{ij}^{c} - I_{ij}^{true}\right)^{2} / (IJ), \\ PSNR^{c}(QS) &= 10log_{10}\left(255^{2} / MSE^{c}(QS)\right), \end{aligned} \tag{3}$$

where *I, J* – image dimensions, *QS* – quantization step (*QS* is a special case of a parameter controlling compression (PCC)).

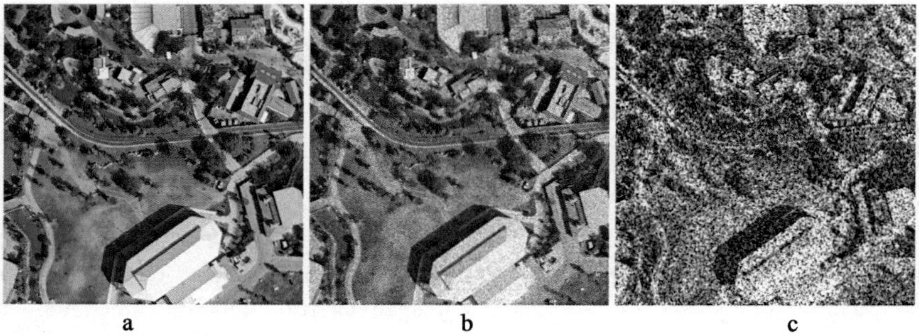

Fig. 1. Image distorted by spatially correlated noise: original (a), 5 looks, noise parameter 0.5 (b), 1 look, noise parameter 1.5 (c).

In addition to the listed classical metrics, the HVS quality metrics as PSNR-HVS, PSNR-HVS-M, and MSSIM, which provide good results for filtering and compression problems, are of particular interest. The PSNR-HVS-M metric, based on the PSNR-HVS metric, is calculated using the quantization Table (its size is 8 × 8), adopted in the JPEG standard for quantizing the Y component of color images [23]. This Table is characterized by a general tendency to increase the quantization step of the spectral DCT components with increasing spatial "frequency". Average shift and contrast correction operations can be performed in a sliding window, as described in [24] when calculating the UQI metric. The size of the used sliding window is 64 × 64 pixels. In addition to taking into account the different sensitivities of a person's vision to different spatial frequencies, masking effects are also taken into account. Without going into details, note that the masking effect is as follows. If the DCT spectrum in a block contains one or two frequencies, the amplitudes of which significantly (by several times) exceed the amplitudes of other frequencies, then human vision practically does not react to distortions corresponding to frequencies with relatively small amplitudes and a small power concentrated in them for each block. This is well illustrated in Fig. 2. The homogeneous area in the upper right corner contains visually noticeable noise (Fig. 2b), which is well observed in comparison with the noisy image (Fig. 2a). In this case, the texture area in the center of the image does not visually differ in both images. The masking power is higher for low frequencies and less for high frequencies. PSNR-HVS-M values are measured in decibels. The PSNR-HVS-M metric is:

$$PSNR - HVS - M = 10log_{10}\left(255^2/MSE_{hvs-m}\right), \tag{4}$$

where MSE_{hvs-m} calculated taking into account the peculiarities of human visual perception in accordance with the approach described in [25].

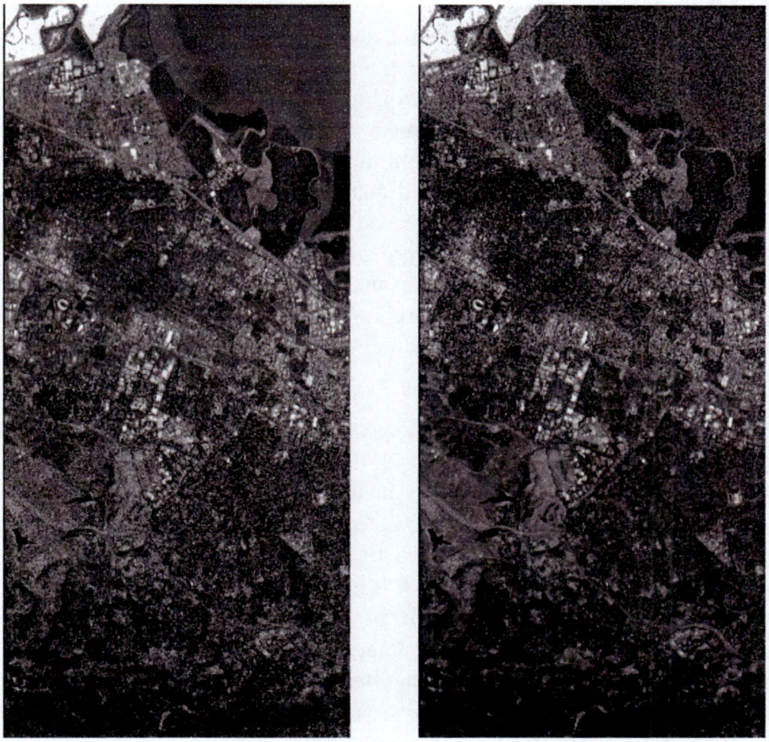

a b

Fig. 2. The effect of masking the RS image noise: a – the noise-free image, b – the image distorted by additive white Gaussian noise with variance equal to 100

The use of the PSNR-HVS-M metric made it possible, in particular, to "identify" different degrees of visual distortions of the same image by spatially correlated and uncorrelated additive noise with the same variance. An image distorted by spatially correlated noise is visually perceived as having a lower quality. Similarly, since distortions in the low frequency region are perceived by the human eye as more noticeable, with the same PSNR and MSE, an image distorted by blur is visually perceived worse than a distorted one by spatially uncorrelated noise (additive or multiplicative). PSNR-HVS-M is expressed in dB, higher values correspond to better visual quality.

A key feature of the MSSIM metric [26] is the determination of the similarity measure of two images for several quantities, such as contrast, brightness, and structure. In this case, the metric is robust to changing the image resolution. The adequacy of the metric is also due to the fact that when measuring the parameters, the image is divided into local blocks and the estimation of the similarity parameters is performed in a sliding window inside the block. The metric can range from 0 (the worst quality) to 1 (the best quality).

For noise-free images, it is true that the higher the compression ratio (CR), the worse the quality. However, this is not always the case for images distorted by noise. Due to the fact that lossy compression often has a filtering effect, it is possible that for a compressed image the MSE calculated relative to the true image is less than the variance of the distorting noise σ^2. The meaning of MSE^c depends on many factors. Among the key factors is the parameter that controls the compression – QS (it can also be the scaling factor (SF), the compression parameter Q [27], etc.). The QS values at which the MSE (or PSNR) vs QS curve has a local minimum (maximum) are called the optimum operating point (OOP). It should be noted that OOP may not be observed for a given image. The existence of the OOP is influenced by several factors, including the type of coder, the complexity of an image, the type and characteristics of noise [20], etc. The existence of OOP is also determined by the analyzed metric [28, 29]. However, here we are interested in lossy but visually lossless compression that allows carrying out analysis of original images without visible distortions.

3 Determining Thresholds for Lossy Compression Metrics

There are methods of data compression of data with losses, which, in general, are capable of providing CR of the order of tens or even more than 100 [30]. However, with such a CR, the introduced distortions can lead to irreversible loss of useful information. Therefore, when compressing with losses, it is advisable to use such a CR, at which, on the one hand, the CR values are significantly higher than with lossless compression, and, on the other hand, the quality of the images provided to users is acceptably high. In this case, the decision about the CR that has to be used should be made preferably automatically. For most practical situations, when compressing an image, quantitative criteria that characterize the quality of the processed image are used as feedback for the coder. However, despite many existing mathematical models, human visual perception remains a subjective characteristic. To determine the threshold values of several HVS metrics described above, we conducted an experiment. The purpose of the experiment was to determine the value of a specific metric, above which visual distortions are invisible.

Three common test images of varying texture complexity were used: Baboon, Goldhill, Peppers. Each of these images was compressed using lossy compression techniques such as AGU. At the same time, the output quality was controlled in accordance with the PSNR metric. A set of a large number of decoded images was obtained with PSNR values in the range of 28.5...36.5 dB for each of the test images. For each set, PSNR-HVS-M and MSSIM metrics were also calculated. Each of the three sets of images was randomly arranged. A user was then asked to compare the original image and each of the compressed versions of a particular test image. The user had to decide if he/she found the difference in the images visually noticeable. In this case, if a user believed that the differences were visually noticeable, the result was taken equal to 1, otherwise – 0. The appearance of the experimental software is shown in Fig. 3. The experiment involved 22 people – 12 men and 10 women. The experimental results for all three images for HVS-metrics (Fig. 4) were presented in the form of dependencies averaged over this metric value, which can be considered an analogy of the probability P of seeing distortions in the image.

Analyzing the data, we can say the following. The dependences of P on PSNR are monotonically decreasing, but they differ for different images. A half of the observers did not notice distortion for the Goldhill and Peppers images, but only a quarter noticed this distortion for the Baboon image. At PSNR $\approx$ 35 dB, almost everyone sees distortion in the Goldhill and Peppers images, but only a half sees them in the Baboon image.

As the PSNR-HVS-M metric increases, the probability P also decreases. At the same time, it can be argued that at PSNR-HVS-M = 37 dB, a half of the observers no longer see the differences in the Baboon image. At a PSNR-HVS-M of about 38 dB, about a third of observers see distortion in the Baboon image, with less than 20% of respondents finding differences in the Goldhill and Peppers images. With PSNR-HVS-M > 40 dB, it can be argued that the distortions in the decompressed image are visually invisible with a high degree of probability. With a value for the MSSIM metric of 0.98, 50% of observers detect differences in the Baboon image. With the same MSSIM for Goldhill and Peppers images, the probability of detecting artifacts is less than 30%. With MSSIM > 0.995, almost all observers do not notice the differences in the presented images.

Thus, with a high degree of probability, it can be argued that the threshold values corresponding to visually imperceptible distortions are the PSNR-HVS-M metric values exceeding 40 dB and MSSIM values exceeding 0.995.

4 Methods for Compressing Images Without Visually Noticeable Distortion

Initially, the compression algorithms are aimed at increasing the amount of data, the number of images that can be stored in a compressed form in the memory of the carrier and transmitted to ground stations. In contrast to lossless compression, lossy compression restores the original signal with varying degrees of information loss. There are three types of redundancy in the image [28]:

- spatial;
- statistical;
- visually dependent.

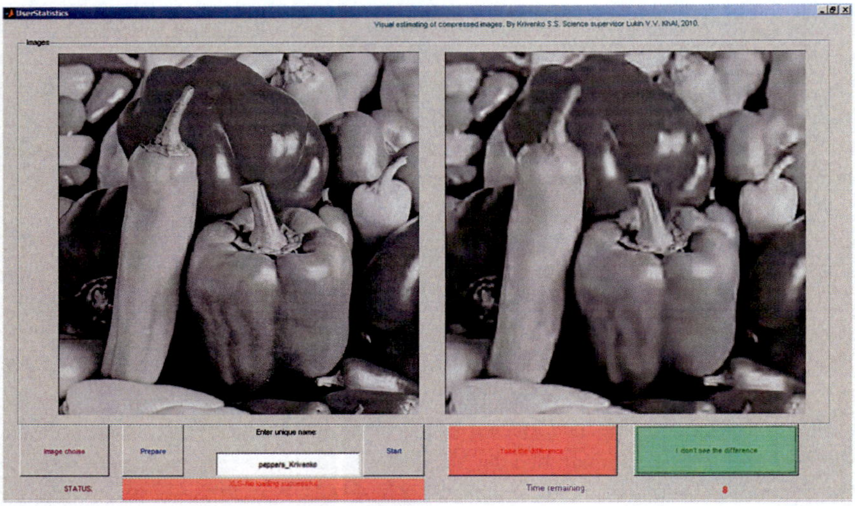

Fig. 3. Appearance of the experimental software workspace

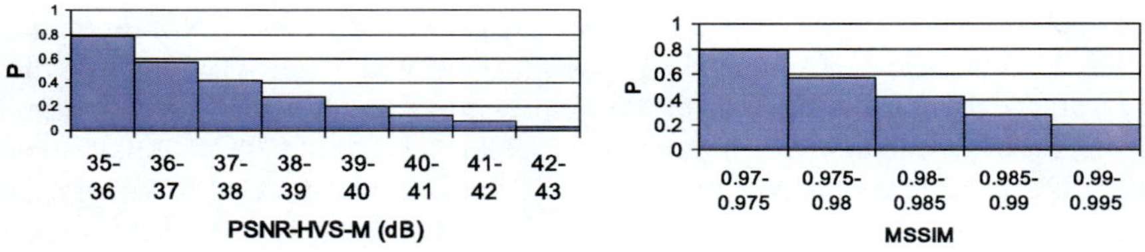

a b

Fig. 4. Probability of recognizing visually noticeable distortions according to the PSNR-HVS-M (a) and MSSIM (b) metrics

While spatial and statistical redundancies are fairly well studied, visual-dependent redundancy is less studied, but it is very important. Visual redundancy is primarily associated with a feature of human vision, which is less sensitive to high spatial frequencies compared to low sensitivity. Elimination of this type of redundancy is carried out in lossy compression at the quantization stage, where high frequencies are either quantized with greater discreteness, or are removed completely. In our work, we will consider three coders – AGU, JPEG2000, BPG.

The DCT coder AGU [31] operates with a fixed 32 × 32 pixel window. This coder has two distinctive features. The first is bit-plane efficient coding of the quantized DCT coefficients. The second is post-filtering to remove blocking artifacts. There is a number of works that show the better efficiency of this coder in comparison with the well-known standard JPEG2000 in terms of rate/distortion curve (RDC). This is especially true for highly textured images. The research version of this coder works with grayscale (8-bit pixel representation) 512 × 512 images.

The properties of the AGU coder contradict the well-known opinion that DWT has better decorrelation properties than DCT. In fact, JPEG2000 has the distinct advantage that the compression ratio is controlled by the number of bits per pixel (bpp), which is important for some telecommunications systems. There are two reasons why the RD curves for JPEG are worse than for JPEG2000. First, JPEG uses 8 × 8 blocks, which is not optimal for data compression. Second, the JPEG standard was developed over 30 years ago and did not use modern lossless coding methods.

The compression ratio for the AGU coder is controlled by a variable QS. This coder does not provide for a simple setting of the required quality (characterized, for example, by PSNR), or the number of bits per pixel. However, with the primary requirement to ensure high retention of visual quality or increase the compression ratio while maintaining the established quality, this is not a significant drawback. Note that the original AGU coder is not adapted for visual quality.

The DWT coder JPEG-2000 uses the wavelet transform technology instead of the discrete cosine transform used in the JPEG format [32]. As a result of such compression, the image is sharper, and the file size compared to JPEG with the same quality turns out to be smaller. JPEG-2000 is completely free from the main drawback of its predecessor: thanks to the use of wavelets, images saved in this format, at high compression rates, do not contain artifacts in the form of a "lattice" of blocks of 8 × 8 pixels. The JPEG-2000 format supports so-called "progressive compression", which allows you to see first a blurry, but then sharper image as it loads. As of 2020, JPEG-2000 is still not widely adopted. This is due to many reasons that simultaneously influenced the adoption of the standard. These include implementation complexity and memory requirements. The parameter that controls the compression is the number of bits per pixel (bpp).

Better Portable Graphics (BPG) is a free lossy image compression format based on HEVC. The format was developed by Fabrice Bellard in 2014 as a replacement for JPEG with a better size-quality ratio [27]. Images in this format show better quality compared to JPEG, JPEG XR and WebP with the same final file size. To display BPG images in browsers that do not support it, the author of the format wrote a JavaScript script. The main parameter that controls compression is the Q parameter, the maximum value of which in the version available to users is 51.

A common difficulty is the fact that the parameters that control the compression in the coders listed above characterize the specific parameters of the algorithms used. This does not allow using these parameters in a single comparative coordinate system. Instead, for the purpose of comparing coders with each other, one has to use the compression ratio as the argument of the used RDCs.

Having the distortion visibility thresholds determined for PSNR, MSSIM, and PSNR-HVS-M in the previous Section, we have compared CR values attained by the considered three coders for several test images like aforementioned fr01 (Fig. 1). Some data are presented in Table 1 for the test image fr01 corrupted by two types of spatially correlated noise. As one can see, CR for AGU and, especially, BPG are essentially (by about 20%) larger than for JPEG2000. The results for visual metrics PSNR-HVS-M and MSSIM metrics practically coincide. The CR values for the more spatially correlated noise (σ = 1.5) are better (larger) than for the spatially correlated noise with less correlation (σ = 0.5). Meanwhile, CR values are considerably larger than for any lossless coder (CR is about 1.05).

Table 1. CR for different coders and metric thresholds

Noise parameters	PSNR = 36 dB			PSNR-HVS-M = 40 dB			MSSIM = 0.995		
	AGU	JPEG 2000	BPG	AGU	JPEG 2000	BPG	AGU	JPEG 2000	BPG
1 look, σ = 0.5	2.0	1.8	2.0	2.5	2.1	2.6	2.6	2.2	2.6
1 look, σ = 1.5	2.7	2.4	2.7	3.7	3.0	4.1	4.3	3.6	4.5

5 Conclusions

Compressing images without visually noticeable distortions or providing high visual quality is not an easy task. The presence of spatially correlated noise complicates this task (especially with signal-dependent interference). The focus of our study was on speckled images. We examined the work of three coders: AGU, BPG, JPEG2000. The results of the work were evaluated using three metrics for assessing image quality: PSNR, PSNR-HVS-M, MSSIM. The focus of our research was on the visual quality of images. Our analysis has showed that for images with one look (high-intensity noise) from the point of view of the PSNR-HVS-M metric, the best is to use the BPG and AGU coders. The MSSIM metric indicates almost equal efficiency in the use of AGU and BPG coders. The classic PSNR metric, which does not take into account visual quality, indicates less benefits of AGU and BPG coders compared to JPEG2000. In the future, we plan to analyze multilook images (with medium intensity noise). Visual analysis of remote sensing and medical images confirms the efficiency of image compression without visually noticeable distortions. At the same time, it is obvious that additional studies are needed, for example, to predict the presence of OOP and select parameters for compression in its vicinity.

References

1. Sharma, G., Vrhel, M.J., Trussell, H.J.: Color imaging for multimedia. Proc. IEEE **86**, 1088–1108 (1998). https://doi.org/10.1109/5.687831
2. Khorram, S., van der Wiele, C.F., Koch, F.H., Nelson, S.A.C., Potts, M.D.: Future trends in remote sensing. In: Principles of Applied Remote Sensing, pp. 277–285. Springer, Cham (2016). https://doi.org/10.1007/978-3-319-22560-9_9
3. Colom, M., Buades, A., Morel, J.-M.: Nonparametric noise estimation method for raw images. J. Opt. Soc. Am. A **31**, 863 (2014). https://doi.org/10.1364/JOSAA.31.000863

4. Chatterjee, P., Milanfar, P.: Is denoising dead? IEEE Trans. Image Process. **19**, 895–911 (2010). https://doi.org/10.1109/TIP.2009.2037087
5. Krivenko, S.S., Krylova, O., Bataeva, E., Lukin, V.V.: Smart lossy compression of images based on distortion prediction. Telecom. Rad. Eng. **77**, 1535–1554 (2018). https://doi.org/10.1615/TelecomRadEng.v77.i17.40
6. Bovik, A.C.: Towards video processing. In: The Essential Guide to Image Processing, pp. 833–834. Elsevier (2009). https://doi.org/10.1016/B978-0-12-374457-9.00028-7
7. Kaarna, A.: Compression of spectral images. In: Goro Obinata, Ashish Dutt (eds.) Vision Systems: Segmentation and Pattern Recognition. I-Tech Education and Publishing (2007). https://doi.org/10.5772/4964
8. Blanes, I., Zabala, A., Moré, G., Pons, X., Serra-Sagristà, J.: Classification of hyperspectral images compressed through 3D-JPEG2000. In: Lovrek, I., Howlett, R.J., Jain, L.C. (eds.) KES 2008. LNCS (LNAI), vol. 5179, pp. 416–423. Springer, Heidelberg (2008). https://doi.org/10.1007/978-3-540-85567-5_52
9. Ager, T.: An introduction to synthetic aperture radar imaging. Oceanography **26**(2), 20–33 (2013). https://doi.org/10.5670/oceanog.2013.28
10. Lukin, V.V., et al.: Lossy compression of images corrupted by mixed Poisson and additive Gaussian noise. In: 2009 International Workshop on Local and Non-Local Approximation in Image Processing, pp. 33–40. IEEE, Tuusula, Finland (2009). https://doi.org/10.1109/LNLA.2009.5278407
11. Jang, H., Kim, C., Lee, J.W.: Practical speculative parallelization of variable-length decompression algorithms. SIGPLAN Not. **48**, 55–64 (2013). https://doi.org/10.1145/2499369.2465557
12. Viber. https://en.wikipedia.org/w/index.php?title=Viber&oldid=1018671172 (2021)
13. Krivenko, S.S., Krivenko, S.A.: Many-to-many linear-feedback shift register. In: 2014 IEEE 34th International Scientific Conference on Electronics and Nanotechnology (ELNANO), pp. 176–178. IEEE, Kyiv, Ukraine (2014). https://doi.org/10.1109/ELNANO.2014.6873939
14. Krivenko, S.A., Pulavskyi, A.A., Krivenko, S.S., Kryvenko, L.S.: Many-to-many linear-feedback shift model for training of artificial neural network in dentistry. In: 2019 IEEE 39th International Conference on Electronics and Nanotechnology (ELNANO), pp. 429–434. IEEE, Kyiv, Ukraine (2019). https://doi.org/10.1109/ELNANO.2019.8783543
15. Nechyporenko, A.S., et al.: Uncertainty of Measurement results for anatomical structures of paranasal sinuses. In: 2019 8th Mediterranean Conference on Embedded Computing (MECO), pp. 1–4. IEEE, Budva, Montenegro (2019). https://doi.org/10.1109/MECO.2019.8760032
16. Krivenko, S., Pulavskyi, A.: Accuracy improvement of noninvasive determination of glucose concentration in human blood. In: 2013 12th International Conference on the Experience of Designing and Application of CAD Systems in Microelectronics (CADSM), pp. 181–183 (2013)
17. Krivenko, S., Lukin, V., Krylova, O., Kryvenko, L., Egiazarian, K.: A fast method of visually lossless compression of dental images. Appl. Sci. **11**, 135 (2021). https://doi.org/10.3390/app11010135
18. Krivenko, S., Lukin, V., Vozel, B., Chehdi, K.: Prediction of DCT-based denoising efficiency for images corrupted by signal-dependent noise. In: 2014 IEEE 34th International Scientific Conference on Electronics and Nanotechnology (ELNANO), pp. 254–258. IEEE, Kyiv, Ukraine (2014). https://doi.org/10.1109/ELNANO.2014.6873920
19. Cedric Valade, J.-M.N.: Homomorphic wavelet transform and new subband statistics models for SAR image compression. In: IEEE International IEEE International IEEE International Geoscience and Remote Sensing Symposium, 2004. IGARSS 2004 Proceedings, pp. 285–288. IEEE, Anchorage, AK, USA (2004). https://doi.org/10.1109/IGARSS.2004.1369017
20. Al-Shaykh, O.K., Mersereau, R.M.: Lossy compression of noisy images. IEEE Trans. Image Process. **7**, 1641–1652 (1998). https://doi.org/10.1109/83.730376

21. Froehlich, J., Su, G.-M., Daly, S., Schilling, A., Eberhardt, B.: Content aware quantization: requantization of high dynamic range baseband signals based on visual masking by noise and texture. In: 2016 IEEE International Conference on Image Processing (ICIP), pp. 884–888. IEEE, Phoenix, AZ, USA (2016). https://doi.org/10.1109/ICIP.2016.7532484
22. Rubel, O., Lukin, V., Egiazarian, K.: Additive spatially correlated noise suppression by robust block matching and adaptive 3D filtering. J. Imaging Sci. Technol. **62**(6), 60401-1–60401-11 (2018). https://doi.org/10.2352/J.ImagingSci.Technol.2018.62.6.060401
23. Hudson, G., Léger, A., Niss, B., Sebestyén, I., Vaaben, J.: JPEG-1 standard 25 years: past, present, and future reasons for a success. J. Electron. Imag. **27**, 1 (2018). https://doi.org/10.1117/1.JEI.27.4.040901
24. Hossny, M., Nahavandi, S.: Image fusion algorithms and metrics duality index. In: 2009 16th IEEE International Conference on Image Processing (ICIP), pp. 2193–2196. IEEE, Cairo, Egypt (2009). https://doi.org/10.1109/ICIP.2009.5413850
25. Toprak, S., Yalman, Y.: A new full-reference image quality metric based on just noticeable difference. Comput. Stand. Interfaces **50**, 18–25 (2017). https://doi.org/10.1016/j.csi.2016.08.003
26. Wang, Z., Simoncelli, E.P., Bovik, A.C.: Multiscale structural similarity for image quality assessment. In: The Thrity-Seventh Asilomar Conference on Signals, Systems and Computers, 2003, pp. 1398–1402. IEEE, Pacific Grove, CA, USA (2003). https://doi.org/10.1109/ACSSC.2003.1292216
27. Albalawi, U., Mohanty, S.P., Kougianos, E.: A hardware architecture for better portable graphics (bpg) compression encoder. In: 2015 IEEE International Symposium on Nanoelectronic and Information Systems, pp. 291–296. IEEE, Indore, India (2015). https://doi.org/10.1109/iNIS.2015.12
28. Sklar, B.: Digital Communications: Fundamentals and Applications. Prentice Hall PTR, Upper Saddle River, NJ (2016)
29. Yea, S., Pearlman, W.A.: Critical encoding rate in combined denoising and compression. In: IEEE International Conference on Image Processing 2005, p. III–341. IEEE, Geneva, Italy (2005). https://doi.org/10.1109/ICIP.2005.1530398
30. Tang, X., Pearlman, W.A.: Three-dimensional wavelet-based compression of hyperspectral images. In: Motta, G., Rizzo, F., Storer, J.A. (eds.) Hyperspectral Data Compression, pp. 273–308. Kluwer Academic Publishers, Boston (2006). https://doi.org/10.1007/0-387-28600-4_10
31. Ponomarenko, N.N., Egiazarian, K.O., Lukin, V.V., Astola, J.T.: High-quality DCT-based image compression using partition schemes. IEEE Signal Process. Lett. **14**, 105–108 (2007). https://doi.org/10.1109/LSP.2006.879861
32. Taubman, D.S., Marcellin, M.W.: JPEG2000: image compression fundamentals, standards, and practice (2013)

Development of Programming Interface for Integrating Assembly Language with Python

Dmytro Kolmogorov and Ievgen Meniailov(✉)

V.N. Karazin Kharkiv National University, Kharkiv, Ukraine
kolmogorov.dmitro@gmai.com

Abstract. The paper analyzes the programming interface for using the Assembler language inserts in the C language, considers the use of C functions in the Python language and describes the use of the programming interface for integrating the Assembler language with Python. The library for integration of Assembly language with Python has been developed. Examples of using functions implemented using PyXAsm are given.

Keywords: Python · C · Assembly language · Assembly-insertions

1 Introduction

Digitalization is a general term for the digital transformation of society and economy [1]. It describes the transition from the industrial era and analog technology to an era of knowledge and creativity, characterized by digital technology and digital business innovation.

In the past two years, due to the global COVID-19 pandemic, the speed of digitalization in all areas of life has accelerated significantly [2, 3]. The introduction of information technology has changed areas such as finance and digital commerce [4, 5], governmental processes [6, 7], project and personnel management [8, 9], green technologies [10], insurance [11], education [12], decision making [13], healthcare [14, 15], epidemiology [16, 17], medical diagnostics [18], etc.

When developing large medical systems, there are problems of integrating a software product at different levels [19]. For example, the collection of statistics and its processing in real time usually occurs using web technologies [20], the collection of personal medical data is implemented using the Internet of Things and low-level programming languages [21, 22], and the processing of big data using machine learning using the Python language [23]. To integrate all components and create a unified information system, it is proposed to use the Assembly language with its integration with the Python programming language.

Assembly language is a low-level machine-oriented programming language [24]. It is a notation system used to represent programs written in machine code in an easy-to-read form. Its commands correspond directly to the individual commands of the machine or their sequences. It is essentially platform-dependent: assembly languages for different hardware platforms are incompatible, although they may be generally similar.

M. Nechyporuk et al. (Eds.): ICTM 2021, LNNS 367, pp. 1015–1026, 2022.
https://doi.org/10.1007/978-3-030-94259-5_78

The assembly language allows the programmer to use alphabetic mnemonic opcodes, assign symbolic names to the computer and memory registers at his own discretion, and also set convenient addressing schemes (for example, index or indirect). In addition, it allows you to use different number systems (for example, decimal or hexadecimal) to represent numeric constants and makes it possible to mark program lines with labels with symbolic names so that they can be accessed (by name, not by address) from others. parts of the program (for example, to transfer control) [25].

Moreover, different compilers have different syntax for using such inserts. For example, the Visual Studio compiler for C++ has the following syntax:

```
int add(int a, int b)
{
    int res;
    asm(
        mov eax, a;
        mov ebx, b;
        add eax, ebx;
        mov res, eax;
    );
    return res;
}
```

The g++ compiler from the GNN project [26], when implementing the same function, has the following syntax:

```
int add(int a, int b)
{
    int res;
    asm(
        "mov eax, %1;"
        "mov ebx, %2;"
        "add eax, ebx;"
        "mov %2, ebx;"
        : "=r" (res)
        : "r" (a), "r" (b)
    );
    return res;
}
```

In addition, we need to compile the source file using the command:

g++ -o main.exe main.cpp -masm=intel

The *-o* flag indicates that the next command line argument is the name of the copied executable file. The *-masm=intel* instruction indicates that the Intel syntax is used for assembly language inserts (the default is *AT&T* syntax).

The **aim** is to develop a PyXAsm library for reason to implement a common programming interface that will not depend on the compiler, which uses an external compiler to create shared libraries (files with the extension .so).

2 Materials and Methods

All the source code of the PyXAsm library can be found at the link: https://github.com/dmitriyKolmogorov/PyXAsm/.

The library was created for the Python language, because:

- Python is widespread among developers and continues to become more and more popular;
- Python has an intuitive syntax that reduces development time;
- It is possible to call functions implemented in C and C++ using the standard library module *ctypes* and shared libraries.

ctypes is a Python standard library module that allows you to interact with a programming interface implemented in C or C++. PyXAsm uses this module as follows: It implements a C function using Intel syntax for Assembly language inserts. The source code is compiled into a shared library that is accessed in a Python session.

For example, let's implement the function of calculating the Pierce arrow (OR-NOT) in C:

```
int pier_arrow(int a, int b)
{
   int result = 0;
   asm(
      "mov eax, %[a];"
      "mov ebx, %[b];"
      "or eax, ebx;"
      "not eax;"
      "mov %[res], eax;"
      :[res]"=r"(result)
      :[a]"r"(a), [b]"r"(b)
   );
   return result;
}
```

Now let's compile the source file (called *main.c*) into a shared library (called *lib.so*) using the command:

gcc -fPIC -shared -o lib.so -m32 -masm=intel main.c

It is important to note here that *ctypes* only works with 32 bits. Therefore, we instruct the compiler to use 32 bits by using the -m32 flag. It is also worth checking that the user is using the 32-bit Python Interpreter.

We are able to use the resulting library as follows:

```
from ctypes import cdll, c_int
import os
lib = cdll.LoadLibrary(os.path.abspath("lib.so"))
lib.pier_arrow.argtypes = (c_int, c_int)
lib.pier_arrow.restype = c_int
lib.pier_arrow(-1, -1)
0
```

Using the *cdll* submodule of the *ctypes* module, the library is loaded and ready to use. If it was saved in a variable named *lib*, and the function is called *pier_arrow*, then the function will be available as a *lib* attribute called *pier_arrow*, that is, *lib.pier_arrow*. To avoid conflicts, we specify the types of the function arguments and the return type (*c_int* is an alternative to int in C).

In the same way, PyXAsm uses *ctypes*: it generates the function code in C, compiles to a *.so* file, loads the function and stores it in a separate variable, deletes the function code file and the *.so* file.

3 Results

In order to install the PyXAsm library, you need to download the source from the link https://github.com/dmitriyKolmogorov/PyXAsm/. Then, in the directory where the code was saved, run *setup.py* using the command:

```
python3 -r setup.py
```

In order to use the register, we need to import them by name from *asm.registers*.

```
from asm.registers import eax, ebx
```

It is possible to import a case under a different name using the as keyword:

```
from asm.registers import eax as e
```

For this purpose, the operation is also possible:

```
from asm.registers import eax
e = eax
```

However, in order to avoid unreadable code and logical errors, registers should be used by their original name.

Also the user has to create a register using the Register class:

```
from asm.registers import Register
eax: Register = Register ('eax')
```

If you specify an unknown register name when initializing an instance of the *Register* class, an *asm.ArgumentValueError* exception will be thrown.

>>> from asm.registers import Register
>>> reg: Register = Register ('aaa')
asm.Argument Value Error: Uknown name for register. Use Register.available_names () to get list of available names.

The *Register* class has a static method *.available_names ()* that returns a list of all possible names for a register.

>>> from asm.registers import Register
>>> Register.available_names ()
['eax', 'ebx', 'edx', 'ecx', 'ebp', 'esp', 'esi', 'edi', 'ax', 'bx', 'dx', 'cx', ' ah ',' bh ',' dh ',' ch ',' al ',' al ',' dl ',' cl ',' esi ',' si ',' edi ',' di ',' esp ', 'sp', 'ebp', 'bpcs',' ss', 'ds',' fs', 'gs',' es', 'mmx0', 'mmx1', 'mmx2', 'mmx3', ' mmx4 ',' mmx5 ',' mmx6 ',' mmx7 ']

Internally, all registers are instances of the *Register* class.

To use the instructions, we need to import them by name from *asm.instructions*. For example, to import *mov, jmp* and *cpuid* instructions, you need to execute the following piece of code:

from asm.instructions import mov, jmp, cpuid

Also, it is possible to import a statement under a different name using the *as* keyword.

from asm.instructions import mov as mv

For this purpose, the following operation is also possible:

from asm.instructions import mov
mv = mov

However, in order to avoid unreadable code and logical errors, it is worth using instructions by their original name.

Internally, instructions without parameters (for example, *cpuid*) are instances of the *InstructionWithoutParameters* class, instructions with one argument (for example, *jmp*) are instances of the *InstructionWithOneParameter* class, and instructions with two parameters (for example, *mov*) are instances of the *InstructionWithTwoParameters* class. All three classes are inherited from the abstract *BaseInstruction* class.

To specify the arguments of the instruction, it is enough to use the call operator () (the magic method *__call__* is responsible for the call).

from asm.instructions import mov
from asm.registers import eax
2_to_eax = mov(eax, 2)

If you pass the wrong number of parameters to the instruction, an *ArgumentsNumberError* exception will be thrown:

>>> mov(2)
asm.ArgumentsNumberError: mov: invalid number of arguments (got 1, expected 2).

If you pass arguments unsupported for the statement, an *ArgumentTypeError* exception will be raised:

>>> mov(2, eax)
asm.ArgumentTypeError: mov: unsupported type argument with index 0.).

The results of invoking an instruction are instances of the *InstructionInstance* class.

Types in PyXAsm are built on top of the types from *ctypes* (for example, c_int) and are used in the definition of variables (see below for the *Variable* class). To use them, you need to import types from *asm.types*:

from asm.types import int_c, short_c

It is also possible to import a type under a different name using the as keyword:

from asm.types import short_c as short

The following operation is also available for this purpose:

```
from asm.types import short_c
short = short_c
```

Internally, all types from *asm.types* are instances of the *Type* class.

It is also possible to create arrays using the *Array* class. To work with this class, it must also be imported from *asm.types*.

from asm.types import Array

During initialization, two parameters are set:

- *ctype* (required) – An instance of the *Type* or *Array* class from *asm.types*.
- *size* (optional) – the size of the array. The value must be an instance of the *int* class and must also be greater than 0.

Further work with the *Array* class will be described in several examples.

To create an array of *int* numbers with a size equal to 5, you need to execute the following code snippet:

```
from asm.types import int_c, Array
int_arr = Array(int_c, 5)
```

To create a two-dimensional array of *short* numbers of 10 × 5 size, you need to execute the following code snippet:

```
from asm.types import short_c, Array
short_arr:Array = Array(short_c, 5)
short_2d_arr:Array = Array(short_arr, 10)
```

If the first argument during initialization is not an instance of type *Type* or *Array*, an *ArgumentTypeError* exception will be thrown:

```
>>> from asm.types import int_c, Array
>>> int_arr:Array = Array(2)
asm.ArgumentTypeError: Unsupposed type of first argument (got 'int', expected Type from asm.types or Array).
```

If the value of the *size* parameter is not an instance of the *int* class, an *ArgumentTypeError* exception will be thrown:

```
>>> from asm.types import int_c, Array
>>> int_arr:Array = Array(int_c, 5.3)
asm.ArgumentTypeError: Unsupposed type of size argument (got 'float', expected 'int').
```

If the *size* parameter is less than or equal to zero, an *ArgumentValueError* exception will be thrown:

```
>>> from asm.types import int_c, Array
>>> int_arr:Array = Array(int_c, -1)
asm.ArgumentValueError: Invalid value of size argument (got -1, expected size > 0).
```

The *Type* and *Array* classes raise 17 different exceptions, which are detailed in the *PyXAsm* library repository.

The *Variable* class is used to define variables used in *PyXAsm* functions (see the *Function* class below). In order to start working with it, you need to import it from *asm*:

```
from asm import Variable
```

When initializing an instance of a class, two parameters are used:

- *ctype* (required) – An instance of the *Type* or *Array* class from *asm.types.*
- *value* (optional, *default = None*) – the value for the variable.

Further work with this class will be described in several examples.

In order to create a variable of the *char* type with the value *'a'*, you need to execute the following code snippet:

```
from asm import Variable
from asm.types import char_c
v:Variable = Variable(char_c, value='a')
```

In order to create a variable of type *int* without a value, you need to execute the following piece of code:

```
from asm import Variable
from asm.types import int_c
v:Variable = Variable(int_c)
```

or

```
v:Variable = Variable(int_c, value=None)
```

In order to create a variable, which is a two-dimensional array of *double* with a size of 10x10, you need to execute the following code snippet:

```
from asm import Variable
from asm.types import double_c, Array
double_2d_ arr:Array = Array(Array(double_c, 10), 10)
v:Variable = Variable(double_2d_arr)
```

Create a variable that is an array of *int* numbers with values from 0 to 9:

```
from asm import Variable
from asm.types import int_c, Array
int_arr_type:Array = Array(int_c, 10)
v:Variable = Variable(int_arr_type, range(10))
```

All of the above classes are used in the main class of the *asm* module – *Function.* Its constructor accepts an iterable with instances of *InstructionInstance* or *Instruction.* For clarity, an example is given:

```
from asm.instructions import mov
from asm.registers import eax, ebx
from asm import Function
func:Function = Function([mov(eax, 2),
                          mov(ebx, 3)])
```

In case the passed object is not iterable, an exception will be thrown *asm.ArgumentTypeError: Can not iter object with instructions.*

If one of the elements in the passed iterable is not an instance of *InstructionInstance* or *Instruction*, an *asm.ArgumentTypeError* exception will be thrown:

```
>>> func = Function([1])
asm.ArgumentTypeError: Object with index 0 in list of instructions is not of type Instruction or InstructionInstance.
```

To use the function further, you need to compile it. To do this, call the *Function().compile()* method, which has 4 named parameters – *input_vars, output_vars, local_vars* and *delete_source.* The first three parameters are responsible for specifying input, local and output variables (an important feature is that when using a function, it is still possible to return a tuple with values). The *delete_source* parameter, which is a boolean value, is responsible for deleting the file with the*.c* extension that stores the function with assembly language insertion. This opportunity was implemented for students who want to understand the structure of the code in the C programming language.

The *Function().compile()* method can throw many exceptions. A complete list can be found in the *_function.py* file or in the description of the method using the *help()* function.

To implement the *chain API*, the *Function().compile()* method returns the current state of the class.

The function is called in the standard way – by passing parameters using parentheses. Internally, the *"magic" method__call__()* is responsible for this. The *Function()._call__()* method takes a number of parameters equal to the number of instances of the *Variable* class passed as *input_vars* to the *Function().compile()* method. If it works correctly, the call returns *None* if the number of output variables is zero; value of the specified type, if the number of output variables is equal to one; a tuple with all output values otherwise.

An example of using *asm.Function* for the task given at the beginning of the post:

```
from asm.instructions import mov, add
from asm.registers import Register
from asm import Variable, Function
from asm.types import int_c
a:Variable = Variable(int_c)
b:Variable = Variable(int_c)
result:Variable = Variable(int_c)
eax:Register = Register('eax')
ebx:Register = Register('ebx')
func:Function = Function([mov(eax, a),
                          mov(ebx, b),
                          add(eax, ebx),
                          mov(result, eax)])
func.compile(input_vars=[a, b],
             ouptut_vars=[result])
>>>func(1, 2)
3
```

4 Conclusions

In this article, the use of assembly language inserts when programming in C was discussed, the use of functions implemented in C in Python using the ctypes module. We also reviewed the PyXAsm library for integrating assembly language with Python.

In practice, in some cases, only fragments of programs are encoded in assembly language, while the rest of the program is developed in one or another high-level language. In this case, the fragments written in assembly language must be linked to the rest of the software system. This is accomplished in two main ways.

At the compilation stage – insertion of inline assembler fragments into the source code of a program in a high-level language using special directives of a high-level language. The method is convenient for simple data transformations, but full-fledged assembly code with data and subroutines, including subroutines with many inputs and outputs, not supported by the high-level language, cannot be done with it.

At the link stage when compiled separately. For the interaction of composable modules, it is sufficient that the imported functions (defined in some modules and used in others) support a certain calling conventions. Individual modules can be written in any languages, including assembly language.

The use of assembly language provides the programmer with a number of features that are usually not available when programming in high-level languages. Most of them are related to the proximity of the language to the hardware platform.

The ability to make the most of all the features of the hardware platform allows, in theory, write the fastest and most compact code of all possible for a given processor.

Due to more rational use of processor resources. For example, the most efficient placement of all or frequently used source data in the internal registers of the processor, you can exclude unnecessary calls to external RAM.

Due to intelligent optimization of calculations, including more efficient use of intermediate results, the amount of code can be reduced and the speed of the program can be increased.

The ability to directly access hardware, and, in particular, I/O ports, specific memory addresses, processor registers (in many operating systems, this feature is significantly limited by the fact that direct access from application programs for writing to peripheral equipment registers is blocked to increase the reliability system).

At the same time, the assembly language also has disadvantages, one way or another limiting its use or making it less profitable.

Acknowledgement. The study was funded by the National Research Foundation of Ukraine in the framework of the research project 2020.02/0404 on the topic "Development of intelligent technologies for assessing the epidemic situation to support decision-making within the population biosafety management" [27].

References

1. Cabitza, F., Magni, M., Batini, C.: Organizing for the digital world: an overview of current IT solutions to support individuals, communities and societies. In: Cabitza, F., Batini, C., Magni, M. (eds.) Organizing for the Digital World. LNISO, vol. 28, pp. 1–6. Springer, Cham (2019). https://doi.org/10.1007/978-3-319-90503-7_1
2. Abd-Alrazaq, A., et al.: Artificial intelligence in the fight against COVID-19: scoping review. J. Med. Internet Res. **22**(12), e20756 (2020). https://doi.org/10.2196/20756

3. Li, J.O., et al.: Digital technology, tele-medicine and artificial intelligence in ophthalmology: a global perspective. Prog. Retinal Eye Res. **82**, 100900 (2021). https://doi.org/10.1016/j.preteyeres.2020.100900
4. Piletskiy, P., Chumachenko, D., Meniailov, I.: Development and analysis of intelligent recommendation system using machine learning approach. In: Nechyporuk, M., Pavlikov, V., Kritskiy, D. (eds.) Integrated Computer Technologies in Mechanical Engineering. AISC, vol. 1113, pp. 186–197. Springer, Cham (2020). https://doi.org/10.1007/978-3-030-37618-5_17
5. Mehrotra, A., Menon, S.: Second Round of FinTech – Trends and Challenges. In: 2nd International Conference on Computation, Automation and Knowledge Management (ICCAKM), pp. 243–248 (2021). https://doi.org/10.1109/ICCAKM50778.2021.9357759
6. Cabitza, F., Batini, C., Magni, M. (eds.): Organizing for the Digital World. LNISO, vol. 28. Springer, Cham (2019). https://doi.org/10.1007/978-3-319-90503-7
7. Fraillon, J., Ainley, J., Schulz, W., Friedman, T., Duckworth, D.: Preparing for Life in a Digital World: IEA International Computer and Information Literacy Study 2018 International Report. Springer International Publishing, Cham (2020). https://doi.org/10.1007/978-3-030-38781-5
8. Dotsenko, N., et al.: Modeling of the processes of stakeholder involvement in command management in a multi-project environment. In: IEEE 13th International Scientific and Technical Conference on Computer Sciences and Information Technologies, CSIT 2018 – Proceedings, vol. 1, pp. 29–32 (2018). https://doi.org/10.1109/STC-CSIT.2018.8526613
9. Dotsenko, N., et al.: Project-oriented management of adaptive teams' formation resources in multi-project environment. In: CEUR Workshop Proceedings, vol. 2353, pp. 911–923 (2019)
10. Kharchenko, V., Gorbenko, A., Sklyar, V., Phillips, C.: Green computing and communications in critical application domains: Challenges and solutions. In: International Conference on Digital Technologies; DT 2013, p. 191 (2013)
11. Bazilevych, K., et al.: Stochastic modelling of cash flow for personal insurance fund using the cloud data storage. Int. J. Comp. **17**(3), 153–162 (2018)
12. Chumachenko, D., et al.: Intelligent expert system of knowledge examination of medical staff regarding infections associated with the provision of medical care. In: CEUR Workshop Proceedings, vol. 2386, pp. 321–330 (2019)
13. Chumachenko, D., et al.: On Intelligent Decision Making in Multiagent Systems in Conditions of Uncertainty. In: 11th International Scientific and Practical Conference on Electronics and Information Technologies, pp. 150–153 (2019). https://doi.org/10.1109/ELIT.2019.8892307
14. Mazorchuck, M., et al.: Web-application development for tasks of prediction in medical domain. In: IEEE 13th International Scientific and Technical Conference on Computer Sciences and Information Technologies, vol. 1, pp. 5–8 (2018). https://doi.org/10.1109/STC-CSIT.2018.8526684
15. Chumachenko, D., Chumachenko, T.: Intelligent agent-based simulation of HIV epidemic process. In: Lytvynenko, V., Babichev, S., Wójcik, W., Vynokurova, O., Vyshemyrskaya, S., Radetskaya, S. (eds.) ISDMCI 2019. AISC, vol. 1020, pp. 175–188. Springer, Cham (2020). https://doi.org/10.1007/978-3-030-26474-1_13
16. Polyvianna, Y., et al.: Computer aided system of time series analysis methods for forecasting the epidemics outbreaks. In: 15th International Conference on the Experience of Designing and Application of CAD Systems, CADSM 2019 – Proceedings, 8779344 (2019). https://doi.org/10.1109/CADSM.2019.8779344
17. Chumachenko, D., et al.: On agent-based approach to influenza and acute respiratory virus infection simulation. In: 14th International Conference on Advanced Trends in Radioelectronics, Telecommunications and Computer Engineering, pp. 192–195 (2018). https://doi.org/10.1109/TCSET.2018.8336184
18. Chumachenko, D., Sokolov, O., Yakovlev, S.: Fuzzy recurrent mappings in multiagent simulation of population dynamics systems. Int. J. Comput. **19**(2), 290–297 (2020)

19. Chen, F., Chen, Y.: Research on information integration architecture of virtual enterprise oriented to e-commerce platform. In: 12th International Conference on Intelligent Computation Technology and Automation (ICICTA), pp. 539–542 (2019). https://doi.org/10.1109/ICICTA49267.2019.00121
20. Chumachenko, D., et al.: On-line data processing, simulation and forecasting of the coronavirus disease (COVID-19) propagation in Ukraine based on machine learning approach. Commun. Comput. Inf. Sci. **1158**, 372–382 (2020)
21. Lu, S., et al.: A study on service-oriented smart medical systems combined with key algorithms in the IoT environment. China Commun. **16**(9), 235–249 (2019). https://doi.org/10.23919/JCC.2019.09.018
22. Cui, Y., et al.: Research on data communication between intelligent terminals of medical internet of things. In: International Conference on Computer Science and Applications (CSA), pp. 357–359 (2015). https://doi.org/10.1109/CSA.2015.39
23. Meniailov, I., Krivtsov, S., Ugryumov, M., Bazilevich, K., Trofymova, I.: Application of parallel computing in robust optimization design. In: Nechyporuk, M., Pavlikov, V., Kritskiy, D. (eds.) Integrated Computer Technologies in Mechanical Engineering. AISC, vol. 1113, pp. 514–522. Springer, Cham (2020). https://doi.org/10.1007/978-3-030-37618-5_44
24. Freudenthal, E.A., et al.: Work in progress – combined introduction of C and assembly with a focus on reduction of high-level language constructs. In: 38th Annual Frontiers in Education Conference, pp. S2H-3–S2H-4 (2008). https://doi.org/10.1109/FIE.2008.4720618
25. Lingling, M., et al.: An assessment tool for assembly language programming. Int. Conf. Comput. Sci. Software, 882–884 (2008)
26. Chen, X.: Application of GNN in Urban computing. Int. Conf. Commun. Inf. Syst. Comput. Eng. 14–17 (2020)
27. Yakovlev, S., et al.: The concept of developing a decision support system for the epidemic morbidity control. In: CEUR Workshop Proceedings, vol. 2753, pp. 265–274 (2020)

Classification of Patients with Diabetes Mellitus Using Neural Networks

Mykola Butkevych[1], Kseniia Bazilevych[1(✉)], Serhii Krivtsov[1], and Anton Tkachenko[2]

[1] National Aerospace University "Kharkiv Aviation Institute", Kharkiv, Ukraine
[2] Kharkiv National Medical University, Kharkiv, Ukraine

Abstract. The work is devoted to the actual problem of determining the diagnosis of a patient according to his biological data. The paper examines the relationship between biological data and the incidence of diabetes. An automated decision support system for analyzing and processing data based on a neural network has been developed. A neural network has been built for the classification of patients with diabetes mellitus. The simulation results show that the use of a software product based on the proposed neural network architecture will improve the accuracy of the diagnosis of diabetes mellitus by doctors.

Keywords: Diabetes mellitus · Diagnostics · Machine learning · Neural network

1 Introduction

Due to the global COVID-19 pandemic, over the past two years, the entire world has accelerated the digitalization of all industries [1]. The introduction of intelligent information technologies has affected all branches of human activity: government [2], finance [3], insurance [4], decision making [5], commerce [6], personnel management [7, 8], urban projecting [9], education [10], etc. Digitalization has not spared the medical sphere [11]. In recent years, many solutions have been implemented and implemented in the medical Internet of Things [12], epidemic modeling [13–15], healthcare applications [16, 17], medical data processing [18, 19] medical diagnostics [20, 21].

Diabetes mellitus is a chronic disease that develops due to insufficient production of insulin by the body or its ineffective use. This pathology is characterized by elevated blood glucose levels, which over time leads to serious damage to many body systems, especially nerve endings and blood vessels [22].

Diabetes mellitus is characterized by the development of micro- and macroangiopathy, neuropathy and diabetic foot syndrome. Diabetic retinopathy is also a dangerous complication, the development of which can lead to complete blindness. According to WHO, 2% of patients with diabetes mellitus, which lasts about 15 years, are diagnosed with complete blindness, and another 10% develop severe visual impairment [23]. Diabetes mellitus is one of the main causes of renal failure, from which 10–20% of patients with diabetes die.

M. Nechyporuk et al. (Eds.): ICTM 2021, LNNS 367, pp. 1027–1037, 2022.
https://doi.org/10.1007/978-3-030-94259-5_79

An increase in the prevalence of diabetes mellitus has also been recorded in Ukraine. According to data published on the website of the Ministry of Health of Ukraine, at the beginning of November 2011, 1.2 million patients with diabetes mellitus were registered in Ukraine. Of these, 184 thousand are insulin-dependent, including 8 thousand children under the age of 18 [24]. That is why overcoming the problem of increasing the incidence of diabetes mellitus in Ukraine is impossible without combining the efforts of government agencies, public organizations, pharmaceutical companies, practicing endocrinologists and, of course, the patients with diabetes themselves [25].

One of the effective solutions to reduce the incidence of diabetes mellitus is the introduction of intelligent information technologies for the classification of patients.

The **aim** of the paper is to develop an information decision support system for the classification of patients with diabetes mellitus based on neural networks.

This research is part of a comprehensive epidemic forecasting and epidemic containment decision-making system development study, which concept is presented in [26].

2 Data Processing

The population analyzed for this study was the Indian population of Pima near Phoenix, Arizona. The population has been in continuous study since 1965 by the National Institute of Diabetes and Digestive and Kidney Diseases for the high incidence of diabetes.

Given the medical data we can collect about people, we should be able to better predict how likely a person is to have diabetes, and therefore act in a specific way to help. We can begin to analyze the data and experiment with algorithms that will help us study the occurrence of diabetes in the Pima Indians.

The dataset is rather small, it consists of 768 rows of 9 columns. Each line contains data collected about a specific patient. Accordingly, each column corresponds to a certain characteristic:

- Pregnancies is Number of pregnancies
- Glucose is plasma glucose concentration a 2:00 in oral glucose tolerance test
- BloodPressure is diastolic blood pressure in mmHg.
- SkinThickness is thickness of the triceps skin component in mm
- Insulin is 2 h serum insulin in mu/ml
- BMI is Body Mass Index, measured as weight in kg/(height in m) ^ 2
- DiabetesPedigreeFunction is Type of diabetes function
- Age is Age in years
- Outcome is Diagnosis, where 1 means the patient is sick, 0 means healthy

Biological parameters used in the dataset:

Body mass index (BMI) is a value that allows you to assess the degree of correspondence between a person's mass and his height and thereby indirectly assess whether the mass is insufficient, normal, overweight (obesity).

Blood pressure is the force with which blood presses against the walls of the arteries. Systolic blood pressure corresponds to the pressure during the contraction of the heart (systole), diastolic blood pressure – at the time of relaxation of the heart (diastole).

The thickness of the triceps subcutaneous fat fold is measured with a caliper.

Insulin is a hormone secreted by the beta cells of the pancreas. Regulates the metabolism of carbohydrates, proteins, fats in the body.

A particularly interesting attribute used in the study was the bloodline function of diabetes. He provided some data on the history of diabetes mellitus in relatives and the genetic relationship of these relatives with the patient. This indicator of genetic influence gave us an idea of the hereditary risk that can arise with the onset of diabetes mellitus. Based on observations in the section continues, it is unclear how well this function predicts the onset of diabetes.

Figure 1 shows the result of a sample of the first 5 patients.

	Pregnancies	Glucose	BloodPressure	SkinThickness	Insulin	BMI	DiabetesPedigreeFunction	Age	Outcome
0	6	148	72	35	0	33.6	0.627	50	1
1	1	85	66	29	0	26.6	0.351	31	0
2	8	183	64	0	0	23.3	0.672	32	1
3	1	89	66	23	94	28.1	0.167	21	0
4	0	137	40	35	168	43.1	2.288	33	1

Fig. 1. Sample of the first 5 patients.

Figure 2 shows the result of the analysis of the distribution of the corresponding analyzes.

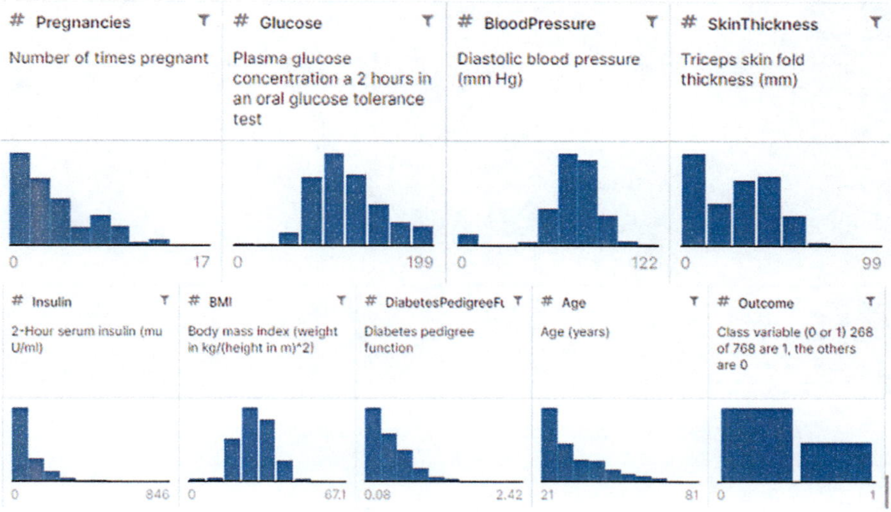

Fig. 2. Distribution of input data.

Having examined the distributions, we can draw the following conclusions:

- It should be noted that the data is based on survey data of at least 21 years and a maximum of 81 years. But most of the sample is between 21 and 50 years old.

- Zero values were also found in columns where such values cannot be obtained as a result of inspection. This means that the data in these columns has not been defined.
- As a result of the analysis of diseases, it was revealed that there are 500 negative cases (65.1%) and 258 positive cases (34.9%).
- Some attributes appear normally distributed (Glucose, BloodPressure, SkinThickness, and BMI).
- Some of the attributes look like they might have an exponential distribution (Pregnancies, Insulin, DiabetesPedigreeFunction, Age).
- Age is likely to be of normal distribution, data collection restrictions may have skewed the distribution.

The correlation matrix is built according to Spearman criterion is shown in Fig. 3.

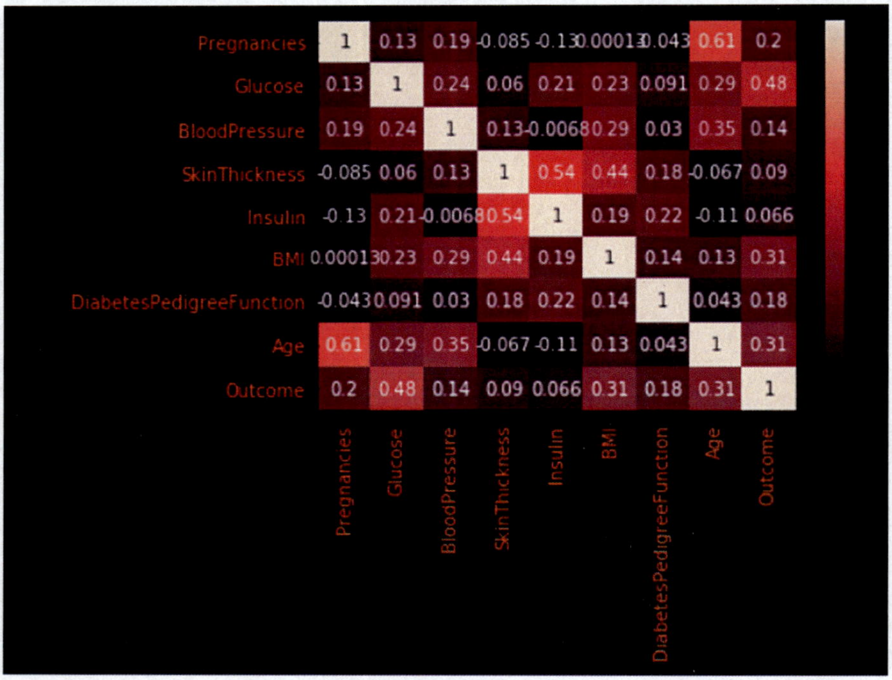

Fig. 3. Correlation matrix according to Spearman criterion.

The correlation matrix is built according to the Pearson criterion is shown in Fig. 4. Figure 5 is the average value of all criteria, only sick and healthy, the difference between them and the correlation with the results of medical research.

Based on the data obtained, conclusions can be drawn:

- criteria BloodPressure, SkinThickness have almost no effect on the result
- Glucose and BMI criteria affect more output

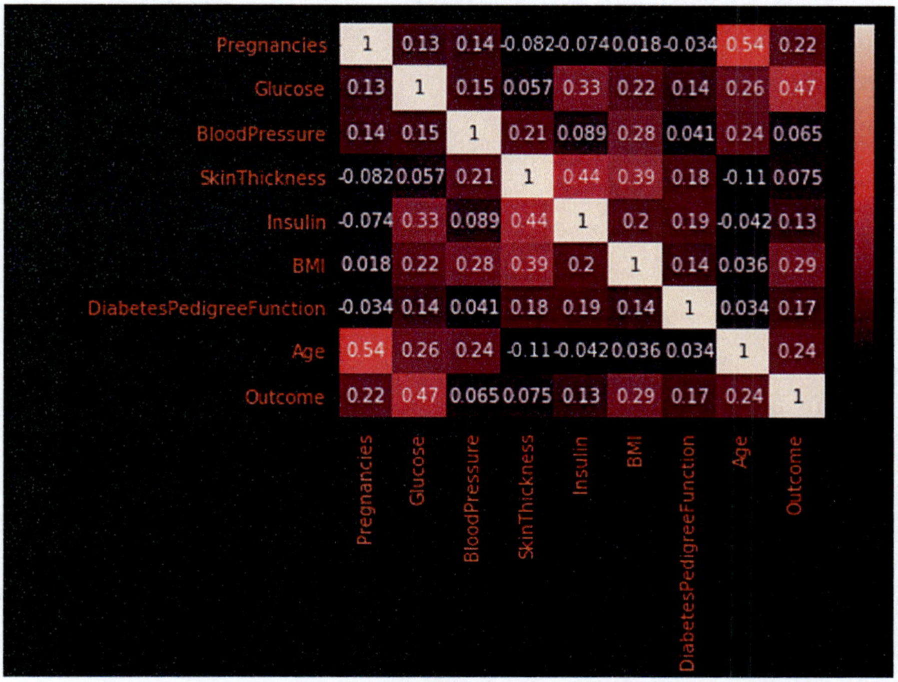

Fig. 4. Correlation matrix according to Pearson criterion.

0	Pregnancies	3.845052	4.865672	3.298000	0.407711	0.221898
1	Glucose	120.894531	141.257463	109.980000	0.258717	0.466581
2	BloodPressure	69.105469	70.824627	68.184000	0.038212	0.065068
3	SkinThickness	20.536458	22.164179	19.664000	0.121743	0.074752
4	Insulin	79.799479	100.335821	68.792000	0.395289	0.130548
5	BMI	31.992578	35.142537	30.304200	0.151233	0.292695
6	DiabetesPedigreeFunction	0.471876	0.550500	0.429734	0.255927	0.173844
7	Age	33.240885	37.067164	31.190000	0.176805	0.238356

Fig. 5. Table of averages analysis.

As a result of the analysis, it was revealed that the input data have the so-called missing values, that is, they are the result of the fact that the measurement was not carried out. We can say that the criterion for pregnancy can have zero values. When solving this problem, you can choose between several strategies:

- "mean", then replace missing data using the mean along each column.
- "median", replace missing values with the median along each column.
- "the most often value", replacing missing data, using most often the value along each column.
- "constant", then replace the missing values with some values.

3 Results

Supervised learning model usually refers to the mathematical structure by which the prediction $y_i y_i$ and is made on the input $x_i x_i$. A common example is the linear model, where predictions are given as linear combination of weighted input features.

$$(y)^i = \sum_{j=0}^{n} \theta_i x_{ij} (y)^i = \sum_{j=0}^{n} \theta_i x_{ij} \quad (1)$$

The prediction value can have different interpretations, depending on the task, that is, regression or classification. For example, it can be turned into logistics to get a positive class in logistic regression, and it can also be used as a rating score if we want to rank the results.

The parameters are the undefined part that we need to know about in the data. In linear regression problems, the parameters are the coefficients θ. Usually we will use θ to denote parameters (there are many parameters in the model, our definition is sloppy here.

By making reasonable choices for $y_i y_i$, we can accomplish various tasks such as regression, classification, and ranking. The problem of training the model is reduced to finding the best parameters θ that correspond to the training data $x_i x_i$ and labels $y_i y_i$. In order to teach a model, we need to define an objective function to measure how well the model matches the training data.

A characteristic feature of target functions is that they consist of two parts: training loss and regularization time:

$$\mathrm{obj}(\theta) = L(\theta) + \Omega(\theta)\mathrm{obj}(\theta) = L(\theta) + \Omega(\theta) \quad (2)$$

where L is the training loss function and Ω is the regularization time. Training loss measures how predictive our model is based on training data. A common choice of L is the root mean square error, which is given as

$$L(\theta) = \sum_{i=0}^{n} \left(y_i - y^i\right) 2 L(\theta) = \sum_{i=0}^{n} \left(y_i - y^i\right) 2 \quad (3)$$

Another commonly used loss function is logistic loss, which is used for logistic regression:

$$L(\theta) = \sum_{i=0}^{n} \left[y_i \ln\left(1 + e - y^i\right) + (1 - y_i)\ln\left(1 + e y^i\right)\right] L(\theta) = \sum_{i=0}^{n} \left[y_i \ln\left(1 + e - y^i\right) + (1 - y_i)\ln\left(1 + e y^i\right)\right] \quad (4)$$

The regularization term controls the complexity of the model and helps us avoid retooling.

The elements introduced above form the core building blocks of supervised learning, and they are the natural building blocks of machine learning toolkits. For example, you should be able to describe the differences and similarities between trees with a higher gradient and random forests. Understanding the process in a formalized way also helps us understand the purpose we are studying and the reason for heuristics such as cropping and anti-aliasing.

Now that we've introduced the elements of supervised learning, let's start with real trees. To get started, let's first learn about choosing an XGBoost model: decision tree ensembles. A tree ensemble model consists of a set of classification and regression trees (CART). Here is a simple example of a CART that classifies whether someone will like a hypothetical PC game X.

CART is somewhat different from decision trees, in which the sheet contains only the decision value. In CART, a real score is associated with each of the leaves, giving us rich interpretations that go beyond classification. It also allows for a principled, unified approach to optimization, as we'll see in the next part of this tutorial.

Usually, one tree is not strong enough to be used in practice. In fact, an ensemble model is used, which encloses the predictions of several trees together.

Here is an example of a woody ensemble of two trees. The estimates of the predictions of each individual tree are summed up to arrive at the final result. Looking at the example, the important fact is that the two trees are trying to complement each other. Mathematically, we can write our model in the form

$$y^i = \sum \mathrm{k} = \mathrm{K} f_k(x_i), f_k \in \mathrm{F} y^i = \sum \mathrm{k} = \mathrm{K} f_k(x_i), f_k \in \mathrm{F} \tag{5}$$

where K is the number of trees, $\mathrm{f_k}$ is a function in the functional space F, and F is the set of all possible CARTs. Objective function to be optimized, given by the formula.

$$\mathrm{obj}(\theta) = \sum_{i=0}^{n} \ln(y_\mathrm{i}, y^i) + \sum \mathrm{k} = \mathrm{K}\Omega(f_\mathrm{k})\mathrm{obj}(\theta) = \sum_{i=0}^{n} \ln\left(y_\mathrm{i}, y^i\right) + \sum \mathrm{k} = \mathrm{K}\Omega(f_k) \tag{6}$$

In this case, wood ensembles are used, so random scaffolding and raised trees are really the same model; the difference arises in how we train them. This means that if you are writing a prediction service for wood ensembles, you only need to write one, and it should work for both random forests and trees with a boosted gradient. One example of why there are elements of supervised learning.

Now that we have introduced the model, we turn to training as always for all learning models that are monitored: the objective function is determined and optimized in a certain way.

Let the next one be the objective function (remember that it should always contain losses and training regularization):

$$\mathrm{obj} = \ln(\mathrm{y_i}, \mathrm{y^t i}) + \sum \mathrm{i} = \mathrm{t}\Omega(\mathrm{f_i})\mathrm{obj} = \ln(\mathrm{y_i}, \mathrm{y^t(t)}) + \sum \mathrm{i} = \mathrm{t}\Omega(\mathrm{f_i}) \tag{7}$$

The first question we want to ask is: tree parameters? You may find that we need to examine these $\mathrm{f_i}$ functions, each containing a tree structure and leaf scores. Studying a tree structure is much more difficult than a traditional optimization problem where you can simply take a gradient. It is unfortunate to study all the trees at once. Instead, we use an additive strategy: we fix what was learned and add one new tree at a time. We add the tree at each step to the one obtained in the previous step, which optimizes our goal.

$$\text{obj(t)} = \ln(y_\text{i}, y^t\text{i}) + \sum \text{i} = \text{t}\Omega(f_i) = \ln(y_\text{i}, y^{t-1}\text{i} + f_t(y_\text{i})) + \Omega(f_t) + \text{const} \tag{8}$$

$$\text{obj(t)} = \ln(y_\text{i}, y^\text{i}(\text{t})) + \sum \text{i} = \text{t}\Omega(f_i) = \sum \text{i} = \ln(y_\text{i}, y^{t-1}\text{i} + f_t(y_\text{i})) + \Omega(f_t) + \text{const} \tag{9}$$

If we consider using the mean squared error (MSE) as our loss function, the goal becomes

$$\text{obj(t)} = \ln(y_\text{i} - (y^{t-1}\text{i} + f_t)((x_\text{i})))2 + \sum \text{i} = \text{t}\Omega(f_t) = \ln[2(y^{t-1}\text{i} - y_\text{i})f_t(x_\text{i}) + f_t(x_\text{i})2] + \Omega(f_t) + \text{const} \tag{10}$$

$$\text{obj(t)} = \ln(y_\text{i} - (y^i(\text{t}-1) + f_t(x_\text{i})))2 + \sum \text{i} = \ln[2(y^i(\text{t}-1) - y_\text{i})f_t(x_\text{i}) + f_t(x_\text{i})2] + \Omega(f_t) + \text{const} \tag{11}$$

The MSE form is friendly, with a first-order term (commonly called a remainder) and a quadratic term. For other interest losses (like logistic losses) it is not easy to get such a good shape. So, in the general case, the Taylor expansion of the loss function is accepted up to the second order.

This becomes the optimization target for the new tree. One of the important advantages of this definition is that the value of the objective function depends only on g_i and h_i. This is how XGBoost supports native loss functions. It is possible to optimize each loss function, including logistic regression and pairwise ranking, using exactly the same solver that accepts g_i and h_i as input.

A learning step has been introduced, you need to add a regulation term. We need to determine the complexity of the tree Ω(f). To do this, we first refine the definition of the tree f(x) as

$$(\text{x}) = \text{wq(x)}, \text{w} \in \text{RT}, \text{q} : \text{Rd} \rightarrow \{1, 2, \cdots, \text{T}\}. \tag{12}$$

$$\text{f}_\text{t}(\text{x}) = \text{wq(x)}, \text{w} \in \text{RT}, \text{q} : \text{Rd} \rightarrow \{1, 2, \cdots, \text{T}\}. \tag{13}$$

Here w is the vector of points on leaves, q is the function that assigns each data point to the corresponding leaf, and T is the number of leaves. XGBoost defines complexity as

$$\Omega(\text{f}) = \gamma\text{T} + 12\lambda \sum \text{j} = 1\text{Tw}2\text{j}\Omega(\text{f}) = \gamma\text{T} + 12\lambda \sum \text{j} = 1\text{Twj}2 \tag{14}$$

Of course, there is more than one way to define complexity, but this one works well in practice. With one part of regularization, most tree packages are less careful or simply ignored. This was because the traditional treatment of studying trees only emphasized the improvement of impurities, while the control of complexity remained with the heuristic. By defining this formally, we can better understand what we are learning and get models that work well in the wild.

4 Conclusions

In this paper, an automated decision support system, sampling, analysis and processing of data using a neural network was considered.

Based on the work done, the following conclusions can be drawn:

For a more accurate analysis, identifying patterns and building a behavioral model, it is necessary:

- Reliable source of information
- More data

Use optimal algorithms as data processing and analysis modules to achieve optimal execution time and best results.

Development of an automated decision support system is real and implemented.

Based on the functional description of the object, its functions were determined, where the main ones are to ensure data input by the user, sampling statistical data according to specified parameters, analyzing and processing this data, as well as providing advice regarding the patient's diagnosis.

Based on this study, it can be noted that this system can be used in the field of medicine.

Acknowledgement. The study was funded by the National Research Foundation of Ukraine in the framework of the research project 2020.02/0404 on the topic "Development of intelligent technologies for assessing the epidemic situation to support decision-making within the population biosafety management".

References

1. Watts, G.: COVID-19 and the digital divide in the UK. Lancet Digital Health **2**(8), e395–e396 (2020). https://doi.org/10.1016/S2589-7500(20)30169-2
2. Ranchordas, S.: The digitalization of Government and digital exclusion: setting the scene. SSRN Electron. J. (2020). https://doi.org/10.2139/ssrn.3663051
3. Mei, S., Peiguang, L., Xiushan, N.: Research on Fintech Industry Development and Talent Training Status. In:13th International Conference on Computer Science & Education (ICCSE), pp. 1–4 (2018). https://doi.org/10.1109/ICCSE.2018.8468796
4. Bazilevych, K., et al.: Stochastic modelling of cash flow for personal insurance fund using the cloud data storage. Int. J. Comput. **17**(3), 153–162 (2018)
5. Chumachenko, D., et al.: On Intelligent Decision Making in Multiagent Systems in Conditions of Uncertainty. In: 11th International Scientific and Practical Conference on Electronics and Information Technologies, pp. 150–153 (2019). https://doi.org/10.1109/ELIT.2019.8892307
6. Piletskiy, P., Chumachenko, D., Meniailov, I.: Development and analysis of intelligent recommendation system using machine learning approach. In: Nechyporuk, M., Pavlikov, V., Kritskiy, D. (eds.) Integrated Computer Technologies in Mechanical Engineering. AISC, vol. 1113, pp. 186–197. Springer, Cham (2020). https://doi.org/10.1007/978-3-030-37618-5_17
7. Dotsenko, N., et al.: Modeling of the processes of stakeholder involvement in command management in a multi-project environment. In: IEEE 13th International Scientific and Technical Conference on Computer Sciences and Information Technologies, CSIT 2018 – Proceedings, vol. 1, pp. 29–32 (2018). https://doi.org/10.1109/STC-CSIT.2018.8526613

8. Dotsenko, N., et al.: Project-oriented management of adaptive teams' formation resources in multi-project environment. In: CEUR Workshop Proceedings, vol. 2353, pp. 911–923 (2019)
9. Davidich, N. et al.: Advanced Traveller Information Systems to Optimizing Freight Driver Route Selection. In: Proceedings – International Conference on Developments in eSystems Engineering, DeSE, pp. 111–115, 9450763 (2020)
10. Chumachenko, D., et al.: Intelligent expert system of knowledge examination of medical staff regarding infections associated with the provision of medical care. In: CEUR Workshop Proceedings, vol. 2386, pp. 321–330 (2019)
11. Hudes, M.K.: Digital health: Hype to reality. 2016 Pan Pacific Microelectronics Symposium (Pan Pacific), pp. 1–5 (2016). https://doi.org/10.1109/PanPacific.2016.7428386
12. Kharchenko, V., Gorbenko, A., Sklyar, V., Phillips, C.: Green computing and communications in critical application domains: challenges and solutions. Int. Conf. Digital Technol. DT 2013, p. 191 (2013)
13. Polyvianna, Y., et al.: Computer aided system of time series analysis methods for forecasting the epidemics outbreaks. In: 15th International Conference on the Experience of Designing and Application of CAD Systems, CADSM 2019 – Proceedings, pp. 8779344 (2019). https://doi.org/10.1109/CADSM.2019.8779344
14. Chumachenko, D., Chumachenko, T.: Intelligent agent-based simulation of HIV epidemic process. In: Lytvynenko, V., Babichev, S., Wójcik, W., Vynokurova, O., Vyshemyrskaya, S., Radetskaya, S. (eds.) ISDMCI 2019. AISC, vol. 1020, pp. 175–188. Springer, Cham (2020). https://doi.org/10.1007/978-3-030-26474-1_13
15. Chumachenko, D., et al.: On agent-based approach to influenza and acute respiratory virus infection simulation. In: 14th International Conference on Advanced Trends in Radioelectronics, Telecommunications and Computer Engineering, pp. 192–195 (2018). https://doi.org/10.1109/TCSET.2018.8336184
16. Mazorchuck, M., et al.: Web-application development for tasks of prediction in medical domain. In: IEEE 13th International Scientific and Technical Conference on Computer Sciences and Information Technologies, vol. 1, pp. 5–8 (2018). https://doi.org/10.1109/STC-CSIT.2018.8526684
17. Chumachenko, D., Sokolov, O., Yakovlev, S.: Fuzzy recurrent mappings in multiagent simulation of population dynamics systems. Int. J. Comput. **19**(2), 290–297 (2020)
18. Harabagiu, S.: Big mechanisms for processing big data in medical informatics. In: IEEE Signal Processing in Medicine and Biology Symposium (SPMB), pp. 1–2 (2015). https://doi.org/10.1109/SPMB.2015.7405440
19. Chumachenko, D., et. al.: On-line data processing, simulation and forecasting of the coronavirus disease (COVID-19) propagation in Ukraine based on machine learning approach. In: Communications in Computer and Information Science, vol. 1158, pp. 372–382 (2020).
20. Skitsan, O., Meniailov, I., Bazilevych, K., Padalko, H.: Evaluation of the informative features of cardiac studies diagnostic data using the Kullback method. CEUR Workshop Proceedingsthis link is disabled, 2917, pp. 186–195 (2021)
21. Dudkina. T.: Classification and prediction of diabetes disease using decision tree method. CEUR Workshop Proceedingsthis link is disabled, 2824, pp. 163–172 (2021)
22. Kerner, W., Brückel, J.: Definition, classification and diagnosis of diabetes mellitus. Exp. Clin. Endocrinol. Diabetes **122**(7), 384–386 (2014). https://doi.org/10.1055/s-0034-1366278
23. Schmidt, A.M.: Highlighting diabetes mellitus: the Epidemic continues. Arterioscler. Thromb. Vasc. Biol. **38**(1), e1–e8 (2018). https://doi.org/10.1161/ATVBAHA.117.310221
24. Mankovsky, B., Halushko, O.: COVID-19 in diabetes patients in Ukraine: lessons for doctors and patients. Georgian Med News, (301), 105–112. PMID: 32535573 (2020)

25. Stuart, R.M., et al.: Diabetes care cascade in Ukraine: an analysis of breakpoints and opportunities for improved diabetes outcomes. BMC Health Serv. Res. (2020). https://doi.org/10.1186/s12913-020-05261-y
26. Yakovlev, S., et al.: The concept of developing a decision support system for the epidemic morbidity control. In: CEUR Workshop Proceedings, vol. 2753, pp. 265–274 (2020)

Transport Systems and Logistics

Theoretical Prerequisites for the Development of Multimodal Transport-Logistic Technologies in Cooperation With Ten-T Network

Oleh Strelko(✉), Oksana Yurchenko, Halyna Vasilova, Ievgen Gusar, and Maryna Rudiuk

State University of Infrastructure and Technologies, 9, Kyrylivska Street, Kyiv 04071, Ukraine

Abstract. The world transport system is undergoing a process of fundamental changes that are associated with the development of world markets for goods and services. The tendency for the further development of a single trading market on different continents significantly simplifies the procedures for crossing the borders, which is reflected in the reduction of tariffs for the handling of goods in international traffic.

Ukraine's inclusion in the Trans-European Core Transport Network makes it possible for Ukraine to qualify for financial assistance from the European Union for the development of its infrastructure.

The article describes the general structure of TEN-T, in particular, two multimodal corridors that have access to the border of Ukraine. The paper also analyzes the problems and prospects of Ukraine's integration into the European transport system. The priorities of the state policy in terms of deepening Ukraine's cooperation with the European Union in the field of transport have been identified. Suggestions for the development of international cooperation in the field of transport services are given.

Keywords: Trans-European Core Transport Networks (TEN-T) · Ukraine's transport system · Investment projects · Pan-European Corridors · Multimodal Transport Networks

1 Introduction

1.1 Problem Statement

Strengthening the European integration processes in which Ukraine participates leads to an increase in the role of transport as an important component of international cooperation. Favorable geographical location allows our country to integrate into the world transport network and effectively use the transport potential. However, during the years of independence, Ukraine has not been able to become an important transit country. Only pipeline transport performed the transit function relatively efficiently, but it is rapidly losing it due to the construction of the South and North Stream pipelines and, accordingly, a significant reduction in transit under contractual terms. [1].

M. Nechyporuk et al. (Eds.): ICTM 2021, LNNS 367, pp. 1041–1050, 2022.
https://doi.org/10.1007/978-3-030-94259-5_80

The current situation has created significant risks for Ukraine's economy, but at the same time will facilitate the search for and implementation of infrastructure projects in the transport sector, in particular the integration of Ukraine's transport system into trans-European transport networks (TEN-T).

World prices for transit cargo transportation are high enough, which greatly affects the value of goods.

Improving transportation processes, reducing costs at border crossings is an impulse that will reduce the overall cost of transportation of freight and, consequently, the cost of the final product.

Over the past 20–30 years, considerable attention has been paid to the project of creating a unified Eurasian system based on the integration of rational transport communications of different states [2]. Why is Europe-Asia the most promising direction?

It is in this direction that international statistics records a steady increase in freight flows, which characterizes the level and state of international trade [3].

The concept of ensuring a single market and freedom of movement of goods, capitals, people, and services in the European states, which arose in 1986, has been continually improved.

To ensure a stable market, national and regional transport and communications networks, multi-branched and existing infrastructure must be well developed. Taking into account that the transport network of Ukraine becomes a single whole with the infrastructure of the European Union, it is necessary to radically modernize it. Both at its own expense and at the expense of the European Union [4].

1.2 Analysis of Recent Research

The papers of such authors [5–9] are devoted to the issues concerning the Euro-Asian multimodal transport-logistic freight transportation technology and the problems of the practical implementation of this system.

Ukraine's transport integration problems and logistics infrastructure into the European transport network are investigated in the works [10–16].

Most authors considered the opportunities for the development of rail transport between Europe and Asia using transport infrastructure of Ukraine, international transport corridors as a special priority for Ukraine on the way of integration into the world economic system, problems of technical modernization of international transport corridors. The agreement concluded between the countries of the European Union has accelerated the consideration of issues and laid the legal basis for trans-European networks as one of the key elements for creating a single market, strengthening economic and social cohesion.

1.3 The Purpose and Tasks of the Research

The purpose and task are to study the organization of trans-European transport operations, to determine problematic issues in the organization of such transportation, and the role of the transport network of Ukraine in the further development of multimodal transport-logistic technologies in cooperation with the TEN-T network.

2 Presenting Main Material

Rail transport of Ukraine is the leading industry in the country's road transport system, which provides 82% of freight and almost 50% of passenger traffic. The operational network of railways of Ukraine is almost 22 thousand km, of which 45% is electrified.

In terms of freight traffic, Ukraine's railways rank fourth in the Euro-Asian continent, being second only to the railways of China, Russia, and India [17].

The developed transport network of Ukraine is a prerequisite for the provision of home and international transport and economic ties in Europe-Asia [18].

According to the sub-index "Infrastructure" of the Global Competitiveness Index 2017–2018 of the World Economic Forum, Ukraine ranked 78th out of 137 countries with a score of 3.9 points, showing a decrease of 3 points from the previous year and 10 points compared to 2014, being at the level of Vietnam (3.9 points), Armenia (3.9 points), Argentina (3.9 points) and Tunisia (3.8 points) [19] (Fig. 1).

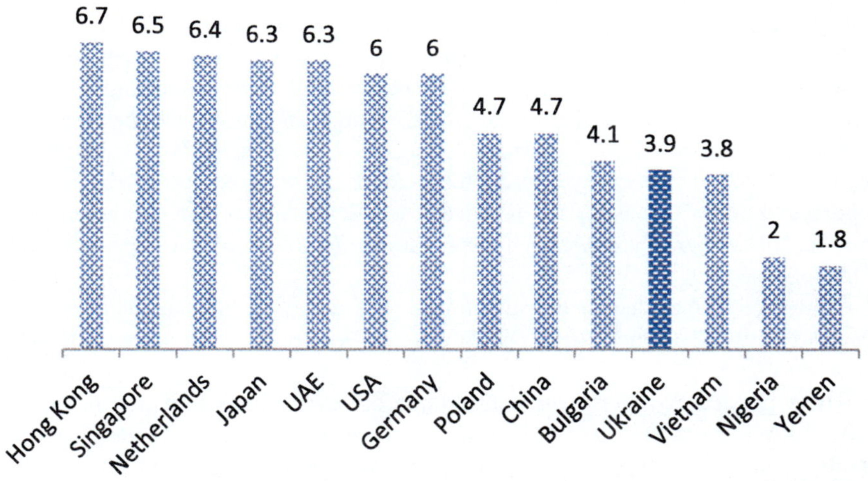

Fig. 1. Infrastructure indicator in GSC 2018, points. Source of information: Global Competitiveness Index 2017–2018.

Undoubtedly, the worst Ukraine's infrastructural indicator is the quality of roads – 130th place out of 137 possible – the level of Moldova, Nigeria and Paraguay. The quality of port and aviation infrastructure remains unsatisfactory (93rd and 92nd place respectively).

The only positive component of Ukraine's transport infrastructure, according to the WEF study, is the railway infrastructure – 37th place along with Italy, Australia and Norway [20].

The inclusion of Ukraine in the Trans-European Transport Network, which took place at the Fifth Eastern Partnership Summit of the European Union countries in October 2017, opens up new opportunities for close cooperation between the EU and partner countries, as well as economic integration and trade [18]. Also at the Summit in the area of transport interconnections, the EU and partner countries agreed on indicative maps of the extended Trans-European Network for Europe (TEN-T) for the Eastern Partner countries as the basis for enhancing transport interconnections and determining common infrastructure priorities. The Eastern Partnership is a partnership of people and its purpose is to improve the standard of living, to protect the values, principles, and aspirations of the people of the European Union and its eastern neighbours.

The transformation of the international transport network development process was first determined at the II Pan-European Transport Conference in Crete (March 1994) and supplemented at the Third Conference in Helsinki (1997).

Since 1997, changes to the corridor map have not been introduced [2]. Of the 10 corridors, 4 corridors pass through the territory of Ukraine (# 3 Brussels – Kyiv, # 5 Venice – Kyiv, # 7 Northwest – Southeast, # 9 Helsinki – Alexandroupolis).

The Trans-European Transport Network (TEN-T) is a project of the European Union and covers the main transport routes within the EU in order to equalize the imbalance between the development level of the Member States' transport networks [18]. Creation of the Trans-European Transport Network is a continuation of the project of the Crete Transport Corridors implemented since the beginning of the '90s.

The goal of TEN-T is to connect the continent from both West to East and South to North. After all, improved transport links both in the Eastern Partnership region and between the Eastern Partnership region and the European Union have the potential to support economic growth and job creation. Also, the aim of the project is to emphasize the importance of road safety, as well as digital solutions in new projects and to bring design standards in line with existing EU practice. The TEN-T Action Plan will help identify priorities for strategic investment in transport infrastructure and increase transport safety in the region.

The projects included in the investment plan were identified together with the countries of the Eastern Partnership with the participation of international financial institutions.

These projects will allow the construction of new and rehabilitation of existing highways, railways, ports, airports, as well as logistics centers and border crossings. Ukraine's infrastructure priority projects are presented in the Figs. 2 and 3.

Special attention is paid to the construction of high-speed railways in the new project. The core network of the existing railway infrastructure is the basis of a sustainable multimodal transport network and it is planned to be completed by 2030, and the final date for the completion of a comprehensive network is by the end of 2050. The basic model covers 9 corridors united into three general directions: "East–West", "North–South" and diagonal corridors [18]. Namely [21]:

1) The Scandinavian-Mediterranean corridor is the most important axis for the European economy from North to South. The route of this corridor crosses the Baltic Sea from Finland to Sweden and then passes through Germany, the Alps and Italy;
2) The North Sea-Baltic corridor connects the ports of the East Coast of the Baltic Sea with the ports of the North Sea. This corridor connects the ports of Finland and Estonia by ferry, and then by modern road and rail connects the three Baltic states – Lithuania, Latvia and Estonia, on the one hand, and Poland, Germany, the

Fig. 2. Ukraine's infrastructure priority projects (Aviation, Port and IWW). Source of information: Center for Transport Strategies [20].

Fig. 3. Ukraine's infrastructure priority projects (Road and Rail). Source of information: Center for Transport Strategies [20].

Netherlands and Belgium – on the other. This multimodal corridor includes the Odra River, which is an inland waterway and connects German, Dutch and Flemish ports;

3) The North Sea-Mediterranean corridor stretches from Ireland and the north of Great Britain through the Netherlands, Belgium and Luxembourg to the Mediterranean Sea in the south of France. This multimodal corridor includes the inland waterways of the Benelux countries and France and aims not only to offer the best multimodal services between the ports of the North Sea, the basins of the rivers Meuse, Rhine, Scheldt, Seine, Saone and Rhone, as well as the ports of Foz-sur-Mer and Marseille, but also improved transport accessibility between the British Isles and mainland Europe;
4) The Baltic-Adriatic Corridor is one of the most important trans-European routes. It connects the Baltic and Adriatic Seas through the industrial zones of southern Poland (Upper Silesia), Vienna and Bratislava, the eastern Alpine region and northern Italy;
5) The Orient-East Med corridor connects the seaports of the North and Baltic Seas with the seaports of the Black and Mediterranean Seas, thus optimizing the use of the respective ports and the highways connecting them. The corridor also covers inland waterways, including the Elbe, which allows for improved multimodal connections between northern Germany, the Czech Republic, the Pannonian region (or the Carpathian Basin) and South-Eastern Europe. It stretches by sea from Greece to Cyprus;
6) The Rhine-Alpine corridor is one of the busiest multimodal corridors. It starts in the ports of Rotterdam and Antwerp in the direction of the Mediterranean basin in Genoa, its route passes through Switzerland and some major economic centers of the Rhine-Ruhr basin and the Rhine-Main-Neckar region, then passes through the Milan agglomeration in northern Italy. This multimodal corridor includes the Rhine as an inland waterway;
7) The Atlantic Corridor connects the western part of the Iberian Peninsula with Paris, where its route forks: one branch goes to the seaport of Le Havre, past Rouen, and the other to Mannheim and Strasbourg. Within this corridor, movement also takes place by inland waterways along the Seine;
8) The Rhine-Danube corridor is the main corridor, where the Danube transport waterway occupies a central place. The route of this multimodal corridor connects the central areas around Strasbourg and Frankfurt and then runs through southern Germany to Vienna, Bratislava, Budapest and finally to the Black Sea. In the Munich area, this corridor has branches in the direction of Prague, Žilina, Košice and ends at the border with Ukraine;
9) The Mediterranean corridor connects the Iberian Peninsula with the Hungarian-Ukrainian border. Its route runs along the Mediterranean coasts of Spain and France, then crosses the Alps to the east through northern Italy, passing through the territory of Slovenia and Croatia in the direction of Hungary. Most of the corridor consists of roads and railways, except for the river Po and some other canals in northern Italy.

At the same time, two of the nine multimodal corridors end at the border with Ukraine (Fig. 4): the Rhine-Danube corridor, which leads through the waterways of the Main and Danube with an important branch from Munich to Prague, Zilina and Kosice to the Ukrainian border; Mediterranean corridor leading from the Iberian Peninsula to the Hungarian-Ukrainian border.

Fig. 4. Multimodal corridors' border with Ukraine. Source of information: TENtec Interactive Map Viewer [22].

The Trans-European Network (TEN) covers three key areas of activity:

- Transport TEN; railways, motorways, waterways, including sea and river ports. Also, this category includes information systems and location of goods;
- Telecommunication TEN: electronic services on the basis of telecommunication networks;
- Energy TEN: gas and electricity supply. Creation of a single energy market and security of its activities.

In the countries of the European Union, the development of transport infrastructure for 2014–2020 costs 26 billion Euros. Development projects are divided into short-term until 2020 and long-term by 2030.

Taking into account that Ukraine is included in the trans-European core network and is an important participant in the Europe–Asia transport corridor, it is possible to rely on EU financial assistance for the development of its part of the Trans-European Transport Network.

Of the 16 projects for the development of Trans-European Transport Network, 9 projects should be implemented on the Ukrainian railways' network. These are: an increase in the carrying capacity of the Hrebinka–Poltava divisions, the electrification of the Berdychiv–Korosten division, the track reconstruction of the Kovel–Yagodyn–Derzhkordon division and its electrification, etc. [23].

In 2018, scientists and specialists of the transport industry together with experts of the EU technical assistance project "Support and Implementation of the Association Agreement and the National Transport Strategy of Ukraine" developed a draft plan of measures for the implementation of the National Transport Strategy of Ukraine for the period up to 2030.

In order to bring Ukraine's legislation closer to European Union law in the field of transport policy and infrastructure, the draft Law of Ukraine "On Multimodal Transport" has been worked out, which establishes common rules for certain types of transportation of goods between the states, members of the Trans-European Network [24].

The Association Agreement between Ukraine and the EU defines the obligation to implement 6 directives and 4 EU regulations in the field of rail transport.

The implementation of the aforementioned legislative acts envisages, in particular, the provision of transparency and non-discriminatory access to the railway infrastructure of all railway undertakings, improvement of rail passenger transportation services and promotion of the integration of the Ukrainian railway system into the Community railway system [25].

Today the work is being carried out in two directions:

1. Development of legislative and regulatory framework taking into account legislative acts of the EU
2. Structural reform of JSC "Ukrainian Railways".

Unfortunately, a newly drafted version of the Law of Ukraine "On Railway Transport", which takes into account the recommendations of the EU legislative acts, is under consideration in the Verkhovna Rada for a long time, and not the adoption of this law does not allow for the development of other legislative acts.

Regarding the implementation of structural reform, it is carried out in accordance with the developed Strategy of the development of the JSC "Ukrainian Railways" until 2023.

3 Conclusions and Suggestions

Ukraine's inclusion in the TEN-T plan is a positive step, as it provides the country with guidelines for medium- and long-term infrastructure development.

Also, Ukraine fits under the European Neighborhood Policy [26], in the framework of cooperation in the field of air transport, which is aimed at signing the Agreement on the Common Airspace of Ukraine-EU. At the same time, this Support Package for Ukraine [27] does not provide for the development of transport infrastructure from the relevant funds.

Investment projects for the development of the TEN-T network included in the strategic plan of the National Transport Strategy of Ukraine give Ukraine an opportunity to claim financial assistance from the EU in building up its part of the Trans-European Transport Network. The adoption of the draft law on "Railway Transport of Ukraine" in the Verkhovna Rada should be accelerated; its adoption will make it possible to accelerate the work on normative and legal documents to implement real reforms in the field of implementation of the Association Agreement. It is important that for Ukraine, as a neighbor of the EU, since the signing of the Association Agreement with the EU opens opportunities for financing the development of transport infrastructure in the framework of TEN-T, which will allow Ukraine to accelerate modernization and integration of its transport infrastructure into EU transport infrastructure.

References

1. Surilova, O.: Common transport policy of the european union and development of the transport system of Ukraine. J. Sci. Pap. National Univ. Odessa Law Academy **26**, 122–133 (2020)
2. Kukharska, N.: Ukraine in Modern Integration Processes. SOIU, Odessa (2015)
3. Association of American Railroads: Rail Traffic Data. Geography and dynamics of the multi-modal transportation market, https://www.aar.org/data-center/rail-traffic-data/. Accessed 15 Aug 2021
4. International cooperation of regions in the context of rapprochement of Ukraine and the EU: foreign economic, investment and management aspects. Analytical note, http://old.niss.gov.ua/. Accessed 15 Aug 2021
5. Harris, I., Wang, Y., Wang, H.: ICT in multimodal transport and technological trends: Unleashing potential for the future. Int. J. Prod. Econ. **159**, 88–103 (2015)
6. Butko, T., Muzykin, M., Prokhorchenko, A., Nesterenko, H., Prokhorchenko, H.: Determining the rational motion intensity of train traffic flows on the railway corridors with account for balance of expenses on traction resources and cargo owners. Transp. Telecommun. **20**(3), 215–228 (2019)
7. Jiang, Y., Qiao, G., Lu, J.: Impacts of the new international land-sea trade corridor on the freight transport structure in China, Central Asia, the ASEAN countries and the EU. Res. Trans. Bus. Manage. **35**, 100419 (2020)
8. Inkinen, T., Hämäläinen, E.: Reviewing truck logistics: solutions for achieving low emission road freight transport. Sustainability (Switzerland) **12**(17), 6714 (2020)
9. Ambrosino, D., Sciomachen, A.: Impact of externalities on the design and management of multimodal logistic networks. Sustainability (Switzerland) **13**(9), 5080 (2021)
10. Strelko, O., Kyrychenko, H., Berdnychenko, Y., Isaienko, S., Tverdomed, V.: Analysis of the intra-day irregularity of intaken empty wagon-flows during the day periods. In: Ostaševičius, V. (ed.) Proceedings of the 23-rd International Conference TRANSPORT MEANS 2019, part I, pp. 16–21. Kaunas University of Technology, Kaunas, (2019). https://transportmeans.ktu.edu/wp-content/uploads/sites/307/2018/02/Transport-means-2019-Part-1.pdf. Accessed 15 Aug 2021
11. Kozachenko, D., Skalozub, V., Gera, B., Hermaniuk, Y., Korobiova, R., Gorbova, A.: A model of transit freight distribution on a railway network. Transp. Prob. **14**(3), 17–26 (2019)
12. Fesovets, O., Strelko, O., Berdnychenko, Y., Isaienko, S., Pylypchuk, O. Ya.: Container transportation by rail transport within the context of Ukraine's european integration. In: Ostaševičius, V. (ed.) Proceedings of the 23-rd International Conference TRANSPORT MEANS 2019, part I, pp. 381–386. Kaunas University of Technology, Kaunas, (2019). https://transportmeans.ktu.edu/wp-content/uploads/sites/307/2018/02/Transport-means-2019-Part-1.pdf. Accessed 15 Aug 2021
13. Strelko, O., Kyrychenko, H., Berdnychenko, Y., Sorochynska, O., Pylypchuk, O.: Application of information technologies for automation of railway and cargo owner interaction. IOP Conf. Ser.: Mater. Sci. Eng. **582**(1), 012029 (2019)
14. Perevozova, I., Minakova, S., Obelnytska, K., Mykhailyshyn, L., Morozova, O.: Mathematical modeling of multimodal transportation in ukraine using methods of the graph theory. E3S Web Conf. **255**, 01030 (2021)
15. Severino, A., Martseniuk, L., Curto, S., Neduzha, L.: Routes planning models for railway transport systems in relation to passengers' demand. Sustainability (Switzerland) **13**(16), 8686 (2021)
16. Kyrychenko, H., Strelko, O., Berdnychenko, Y.A.: Influence of existing operational conditions on compliance with car handling standards. IOP Conf. Ser.: Earth Environ. Sci. **666**(4), 042054 (2021)

17. Myronenko, V.: Transport interoperability and multimodal solutions. Meeting of the national round table on updating the National Transport Strategy of Ukraine until 2030. https://mtu.gov.ua/files/4_Transport_Interoperability_and_multimodal_solutions_Myronenko.pdf. Accessed 15 Aug 2021
18. Proceedings of the Eastern Partnership Summit 2017: Stronger together, https://ec.europa.eu/commission/presscorner/detail/en/IP_17_4845. Accessed 15 Aug 2021
19. Zhmerenetskyi, O.: Transport infrastructure: light at the end of the tunnel. Kyiv International Economic Forum 2018, https://forumkyiv.org/uk/analytics/transportna-infrastruktura:-svitlo-u-kinci-tunelyu. Accessed 15 Aug 2021
20. Center for Transport Strategies: Rating of Ukraine's infrastructure in the world, https://cfts.org.ua/infographics. Accessed 15 Aug 2021
21. Fediai, N.: Features of the integration of Ukrainian transport infrastructure into the trans-European network transport. Efektyvna ekonomika (2018). https://doi.org/10.32702/2307-2105-2018.12.93
22. Trans-European Transport Network (TEN-T) Homepage https://ec.europa.eu/transport/themes/infrastructure/ten-t_en. Accessed 15 Aug 2021
23. Sadlovska, I.: Development of the national transport network of Ukraine and its integration into the priority transport networks of the EU. Collect. Sci. Works Econ. Technol. Univ. Transp.: Econ. Manage. **31**, 16–110 (2015)
24. Railway transport reforms: Ministry of Infrastructure of Ukraine, https://mtu.gov.ua. Accessed 15 Aug 2021
25. 6-th Europe-Ukraine Forum. Budapest, 12–14 may, 2013. Gorshenin institute. Ukraine 2012. Report on transformation. Institute for Eastern Studies, pp. 35–38 (2013)
26. The official site of Enterprise Europe Network: «Ukraine». https://ec.europa.eu (2018). Accessed 15 Aug 2021
27. The official site of Enterprise Europe Network: Support package for Ukraine, https://ec.europa.eu/archives/commission_2010-2014/president/news/archives/2014/03/pdf/20140306-ukraine-package_en.pdf (2014). Accessed 15 Aug 2021

Contract Structures in the Field of Air Carriage

Nataliia Fedosenko[1(✉)], Andriy Starodubtsev[1], Anton Stiebieliev[1], Valentyn Fesiunin[2], and Svitlana Andrenko[3]

[1] Department of Law at National Aerospace University "Kharkiv Aviation Institute", Kharkiv, Ukraine
n.fedosenko@khai.edu

[2] Laboratory of Forensic Economics, Forensic Merchandising, Forensic Psychology and Intellectual Property Analysis of National Scientific Center «Hon. Prof. M. S. Bokarius Forensic Science Institute», Kharkiv, Ukraine

[3] Department of Professional Development at National Scientific Center «Hon. Prof. M. S. Bokarius Forensic Science Institute», Kharkiv, Ukraine
sweta-a@ukr.net

Abstract. The research subject is to determine the concept, characteristics of contracts as well as relations between them in the field of air carriage. The article is dedicated to the analysis of legal nature and basic components of aviation contracts, as well as specificity of their theoretical and practical application, classification, and as a result is aimed at creating a holistic concept of the system of contract-based regulation of air transportation. This analysis is rooted in the study of the context of civil and scientific literature and forensic practice on the specified issue. Apart from general scientific research methods, interpretation legal methods, formal legal, comparative law methods are implemented. Reference to the provisions of civil law doctrine enabled to introduce an individual view on our research subject and suggest grounds for better understanding and improvement of air legislation. The main conclusion of the performed study is that civil law regulation of contracts in the field of air transportation is characterized by the presence of special contracts with its special characteristics determining the content of air legislation. Concerning each of the contract structures, certain questions arise not only in relation to legal regulation and law enforcement (overbooking in the passenger air carriage contract), but also to scientific approaches to determine the concept and nature of a specific contract (status of an air carriage contract as bilateral and trilateral, legal nature of aircraft charter agreement).

The lack of a single understanding and interpretation of concepts, contradictions in legislation, existence of legal gaps do not contribute to air legislation uniformity and provoke difficulty in its study and application. Obtained results can be used in theoretical and legal researches as well as in legislative and practical activities.

Keywords: Air charter · Aircraft charter agreement · Cargo air carriage · Luggage air carriage · Passenger air carriage · Air carriage contract air legislation · Air carriage · Transportation contracts

M. Nechyporuk et al. (Eds.): ICTM 2021, LNNS 367, pp. 1051–1066, 2022.
https://doi.org/10.1007/978-3-030-94259-5_81

1 Introduction

Air (aerial) carriage is a specific field of civil law regulation which role is to provide certain services aimed at moving goods or a person in space. Formation and development of the transport system of Ukraine ensures timely, full and better response to the needs of population and social production in air carriage, creation of comfortable conditions for the development of such activities at transport enterprises of all ownership forms.

Carriage contract is the main category that makes the system of civil law relations while carrying out the transportation process. This air law section keeps evolving and advancing. In turn, as scientists-practitioners believe, our legislative framework lacks practical development of many problems associated with modernization of air transportation, namely peculiarities of regulating certain types of air carriage, ensuring guaranteed execution of air carriage contracts, and their legislative support, including ensuring the protection of rights of air carriage contract chief participant: a passenger or customer of transportation.

Today, air transport takes a leading place in the world in relation to international passenger transportation, which legal regulation keeps constantly developing and advancing. According to the statistics, the popularity of air transport in Ukraine is growing. First of all, it is associated with geographical expansion of air transportation and tourists' aspiration to reduce travel time period to places of recreation. Consequently, the research on the procedure and peculiarities of social relations legal regulation in the field of inland and international air (aerial) carriage of passengers and luggage is relevant.

2 Literature Review

Such domestic and foreign scientists and practitioners as M. Bohuslavskyi, V. Bordunov, V. Butylin, E. V. Horian, K. V. Horian, B. Yeliseiev, V. Kubynskyi, M. Puchkova, P. Rabynovych, O. Stremoukhov, H. Shmelov etc. are among civilian scientists who dedicated their researches and research papers to studying issues of civil law transportation regulation within the regulatory framework of Ukraine and the EU, which governs the mechanism for ensuring passengers' rights according to the air carriage contract.

Innovations concerning the air carriage contract were studied in research papers of such scientists as O. Aksamentov, O. Batalov, O. M. Vereshchahina, N. Vierietielnikova, O. Vyhodianskyў, A. P. Hetman, V. Horobiets, V. Hriaznov, T. Kozlova, Yu. Malieiev, V. Molchanov, V. Mosashvyly, M. Ostroumov, P. Riemishevskyў, A. Smirnov, L. Taltseva, I. Khliestova, K. Kholopov, Ye. Yeriashov, N. Mamchenko, O. Nasadiuk, O. Onishchenko, O. Stoliarskyi, O. Chernenko, A. Filippov, H. V. Tsirat, O. Sheremietieva and others.

Despite the increased attention of scientists to this contract, many theoretical and practical issues are still debatable. What is more, the study of an air carriage contract as a civil law document and in foreign sources is also underdeveloped.

3 Presentation of the Main Content

An integral part of a unified transport system, in conformity with the Law of Ukraine: *On transport*, is an aviation transport which, according to Article 32, includes air transport enterprises performing carriage of passengers, cargo, luggage, mail [1].

In national transport legislation, there is a special legal regulation of civil law relations associated with the contract of carriage. In most general sense, transportation is a kind of economic activity aimed at moving cargoes, passengers and their luggage on the basis of the drawn contract between carriers and their customers.

The important fact is that air transportation is carried out, as a rule, privately, based on the principles of parties' free will, freedom of compiling a contract and formulating its conditions, however, given certain security requirements.

In view of the fact that the air carriage contract is a civil law contract, standards of the Civil Code of Ukraine are applicable to it [2].

Legislative regulation of carriage contracts has a certain feature not inherent in other civil law institutions. In particular, apart from the Civil Code of Ukraine, the existence of individual codified acts considering the specifics of transportation by certain types of transport, including the Air Code of Ukraine [3], is stipulated.

The Civil Code of Ukraine is central in the regulation of transportation relations, so all the regulations contained in other legislative and regulatory acts, in particular, regulations and the Civil Code of Ukraine, should comply with it, except for the cases when the Civil Code of Ukraine stipulates the possibility of transportation relations settlement by means of law in another way.

It should be stressed that the Civil Code of Ukraine has a complex nature and contains private law and public-law regulations governing relations regarding control over air transportation by the state.

A civil-law contract for air carriage, as the main means of transportation legal regulation, has a specific nature conditioned by a particular type of transport. To determine the status of a contract, its legal nature should be studied based on the analysis of current legislation regulations.

Thus, according to Article 1, paragraph 78, of the Air Code of Ukraine, air carriage refers to transportation of passengers, luggage, cargo or mail by an aircraft under a corresponding air carriage contract.

According to Article 98 of the Air Code of Ukraine, air transportation is carried out on the basis of a contract concluded between the air carrier and passenger or consignor. Each air carriage contract and its conditions are certified by a contract of carriage issued by an air carrier or authorized organizations (agents) [3].

An air carriage contract is a civil law contract, the provisions of the Civil Code of Ukraine fully apply to it. According to the passenger carriage contract, one party (carrier) undertakes to transport the other party (passenger) to a destination, and in case of luggage hand-in: also to deliver luggage to a destination and return it to a person who has the right to receive luggage. A passenger is obliged to cover the stipulated travel fee, and in case of luggage hand-in: also for its transportation. (Article 910, part 1, of the Civil Code).

Taking into consideration the announced by Ukraine course for integration into the European Union, priority measures to adapt Ukrainian air transport legislation to EU standards [4] should be implemented, in which a single European aviation market is formed in accordance with the Single European Act [5]. The European Common Aviation Area (ECAA) is a bilateral agreement between the EU and third countries on the establishment of general security standards and initiation of market relations in aviation field. The European Common Aviation Area can use airspace more efficiently and safely. The international organization: The European Organization for the Safety of Air Navigation (EUROCONTROL) is responsible for air transportation and coordination and planning of air traffic control throughout Europe [6].

The corresponding action plan contains many priority directions and provisions of the EU. These provisions need to be translated in Ukrainian, introducing necessary technical terms that are not currently available in Ukrainian legislative framework. At the same time, the plan on preparing for ECAA implementation presupposes the amendment of the current wording of the Air Code of Ukraine: regarding the adoption of aviation rules, certification of aerodromes: on the basis of the Commission regulations (EU) by No. 139/2014 [7]; for operation of aircrafts: Regulations by No. 965/2012 [8]; for civil aviation crews: Regulations by No. 1178/2011) [9].

Preparation process of signing of the European Common Aviation Area agreement continues. For both Ukraine and the European side, the most rapid conclusion of this Agreement is one of the principal common tasks of the present. By signing the ECAA agreement, Ukraine is obliged to align Ukrainian legislation with European standards in the field of aviation safety, air traffic organization and environmental protection. In the future, both the Ukrainian and the European side will have unlimited commercial rights to perform flights within Ukraine and between Ukraine and the EU. This will allow not only to significantly increase competition in the market, but also will contribute to decline in prices for tickets.

The Single European Sky is a European Commission initiative based on which airspace distribution and its management will be integrated for most European countries (European Common Aviation Area) [10].

Furthermore, the procedure for implementation of international passenger transportation in Ukraine is governed by the regulations of Ukrainian national legislation, the provisions of international conventions and regulations [11], to which Ukraine acceded.

Among the basic ones: the Convention for the Unification of Certain Rules for International Carriage adopted on May 28, 1999 (hereinafter referred to as the Montreal Convention) [12, 15]; The Convention for unification of certain rules relating to international air transports dated on October 12, 1929 [13], ratified by the Ukrainian SSR dated on August 14, 1959 by No. 995–181 (hereinafter referred to as the Warsaw Convention dated on 1929); Protocol for amendment to the Convention to unify certain rules relating to international air transportation (The Hague Protocol 1955) [14] and Regulation (EU) 261/20042: *On the establishment of common rules in the area of compensation and assistance to passengers in case of denied boarding, cancellation or prolonged flight delay* [18].

Aviation transportation by Ukrainian air carriers are also governed by air transportation rules and passenger and luggage services (approved by the Order of the State Aviation Administration of Ukraine dated on November 26, 2018, by No. 1239, hereinafter referred to as Air transportation rules) [19], provisions of the Air Code of Ukraine, the Law of Ukraine: *On Consumer Rights Protection* [20], Civil Code of Ukraine and air carrier rules (rules, instructions and technologies established by the carrier used during passengers and/or luggage transportation, including passenger and luggage air transportation rules, rules for applying tariffs and standards for servicing passengers and luggage, the procedure for considering claims).

At the international level, leading organizations in the field of civil aviation: International Civil Aviation Organization (ICAO) [21] and International Air Transportation Association (IATA) [22] are studying the prospects for governing the air contract law.

In conditions of reforming relations in the system of international aviation, the ICAO draws particular attention to legal regulation. Thus, to implement its conceptual vision for sustainable development of civil aviation, the ICAO defines an individual strategic goal, one of which is legal regulation, namely to strengthen legal provisions for regulating international civil aviation activity [24].

In the scientific literature, transportation contracts include the air carriage contract itself as well as contracts that are closely relate to and mediate the process of transportation, including transportation itself. In particular, it concerns obligations on provision of vehicles, company's contracts on transportations organization, chartering, company's contracts while transportation in direct mixed traffic (various types of transport), transport forwarding agreement, etc. The group of contracts associated with air transport activity includes contracts on provision of aviation services and an insurance contract when using air transport.

Accordingly, passenger air carriage contract relate to civil-law contracts in the field of aviation (air) transportation according to the general rule; luggage air carriage contract, cargo (or mail) carriage contract; aircraft charter agreement (aircraft charter).

Each contract has its own specific features and individually defined both in Tax Code of Ukraine and the Aviation Rules of Ukraine: *Rules of air carriage and servicing of passengers and luggage* approved by the Order of The State Aviation Administration of Ukraine No. 1239 dated on November 26, 2018 [25].

The legal nature of aviation contracts also relate to the combination of public and private law provisions, which manifests itself in the presence of imperative rules on the order of conclusion, change and termination of a contract, responsibility of a passenger for aviation rules violation, responsibility of the air carrier as the owner of the source of increased danger. At the same time, the air carriage contract is bilateral, bill paying, consensual (carriage of passengers, aircraft charter) or real (carriage of luggage, cargo), public (transportation of passengers).

In accordance with the general rule, an air carriage contract is bill paying, bilateral, mutual, fixed-term, and the subject of an air carriage is services of the air carrier on transporting passengers, luggage or cargo to a destination, as well as counter-payment of services and insurance of contract while air transport use.

Given the above, traditional aviation (air) contracts and other contracts related to transportation and other civil aviation contracts should be emphasized, as well as contracts on the other use of civil aviation.

At first glance, the mentioned characteristics of contract provoke certain difficulties while defining their precise classification. Let's consider in more detail the legal nature and understanding of particular aviation (air) contracts.

Air carriage is implemented under the which general provisions are contained in the provisions of Chapter 64 of the Civil Code of Ukraine. By analyzing provisions contained in Chapter 64 of the Civil Code of Ukraine, it could be concluded that the concept of *carriage contract* is disclosed through a series of contracts aimed at governing different relations associated with transportation of goods, passengers and luggage, mail.

The mentioned contracts differ from each other by different features: parties involved, subject matter, etc. A majority of scientists do not consider it to be a legal presumption: air carriage takes place only when it is based on the contract until the opposite is proved [26, C. 54].

This principle is inherent in the old Warsaw Convention of 1929, and the new Montreal Convention of 1999 [13] and proceeds from the fact that parties conclude and implement a carriage contract.

Thus, according to the passenger carriage contract, one party (carrier) undertakes to transport the other party (passenger) to a destination, and in case of luggage hand-in: also to deliver luggage to destination and return it to a person who has the right to receive luggage. A passenger is obliged to cover the stipulated travel fee, and in case of luggage hand-in: also for its carriage. (Article 910, part 1, of the Civil Code of Ukraine).

The air carriage contract stipulates contract obligations belonging to the group of service contracts. That is, provisions of Chapter 63 of the Civil Code of Ukraine: *Services. General provisions*: provisions of this chapter, according to Article 901, part 1, of the Civil Code, may apply to all service contracts, provided it does not contradict the essence of obligation. In addition, Chapter 64 of the Civil Code of Ukraine: *Transportation* is dedicated to special regulation of transportation services.

Reference in legislation to such passengers who are transported on preferential terms or for free does not presuppose the contract free nature. Thus, preferential conditions for carriage of cargo, passengers, luggage, mail could be established by an organization, transport enterprise at its expense or financed from a corresponding budget in cases stipulated by law and other legal regulations.

However, for carriage contracts specified by Chapter 64 of the Civil Code of Ukraine, payment is a constitutive feature that follows directly from legal definitions of certain types of a contract. Thus, according to a carriage contract, "the sender is obliged to pay a set fee for carriage of cargo" (Article 909, part 1, of the Civil Code of Ukraine), according to passenger and luggage carriage contract, " a passenger undertakes to pay an established fee" (Article 909, part 1, of the Civil Code of Ukraine), according to the contract of charter (chartering), "one party (charterer) is obliged to provide the other party (charterer) with transportation services for the established fee (Article 912, part 1, the Civil Code of Ukraine)".

In addition, a separate article in Chapter 64 of the Civil Code of Ukraine is dedicated to carriage charge. Thus, as stated in Article 916, part 1, transportation fee is charged for carriage of cargo, passengers, luggage, mail determined by the contract between parties, unless otherwise is established by law or other legal regulations.

According to Article 915 of the Civil Code of Ukraine, carriage carried out by a legal entity is considered to be transportation by public transport. The contract of carriage by public transport is a public contract. Accordingly, a passenger air carriage contract is a public contract, since an aviation company since an airline as a commercial organization has no right to refuse a passenger in contract conclusion. The peculiarity of public contracts is that the customer (passenger) does not possess such scope of rights as the air carrier, since it is impossible to formulate contract terms. In fact, he is deprived of the opportunity to amend terms that would provide him with additional guarantees for implementation of rights belonging to him, their protection and defense. As a matter of fact, the passenger buying a ticket accepts the carrier's offer and agrees to terms and rules of passenger carriage. At the same time, the passenger air carriage contract is a contract of adhesion, since the passenger concludes a contract on conditions determined by the carrier. Terms for concluding a passenger air carriage contract are not only ticket payment by a passenger, but also an obligatory fulfillment of aviation security terms: registration of a passenger, preliminary inspection, and while international air transportation: implementation of the terms of customs and migration control.

Given the above, the air carrier possesses the right as well as obligation to carry out carriage according at the same fare for all consumers (the difference is only in service class) which forbids the unjustified refusal in air carriage contract conclusion and preference of one person over the other. The air carrier is obliged to conclude an air carriage contract with any person who will appeal to him on contract conclusion, but the refusal to sell an aircraft ticket for a passenger is allowed only in the absence of free places and capacities or due to restriction in a particular passenger carriage.

When concluding the air carriage contract, oftentimes problems associated with unilateral refusal of the carrier in passenger's boarding for a flight arise. According to Article 100, part 1, of the Civil Code the air carrier can refuse in transportation of a passenger or consignor in cases stipulated by Ukrainian aviation regulations. At the same time, in practice, other grounds not provided by law also exist. For example, these are the lack of seats on the plane, which is often the result of overselling (overselling of airline tickets: sales and booking by the carrier of a greater number of tickets for a flight than seats provided in a plane), or notify the carrier that a passenger has limited capacity. Overselling in its essence is a one-way refusal of the carrier to fulfill terms of a contract, which is fully complied with provisions of the Civil Code of Ukraine on unilateral refusal to fulfill a contract.

In professional civilized literature, unfortunately, not sufficient attention is paid to the protection of civil rights under the passenger air carriage contract. As a rule, it is only about the responsibility of the carrier for non-fulfillment or improper fulfillment of the terms of passenger air carriage contract Moreover, the issue regarding the use of civil-law liability in case of voluntary compensations payment is also controversial. Thus, according to I.V. Venedyktova, voluntary payment of certain compensations for infringement of rights relates to civil-law liability. At the same time, as R.B. Shyshka notes, erroneous is understanding of responsibility as any sanctions for an offense with their negative consequences without the procedural order of their use by court. According to his view, contractual liability arises precisely as a result of transformation or overgrowth of the violated contractual obligation into an additional (accessory) obligation. Sanction when implementing it by court makes up the concept of civil-law liability [27, P. 154].

Besides, responsibility of the carrier for non-compliance or inappropriate compliance with the contract terms on a passenger carriage by air transport is stipulated in Air transport regulations. Regulations have been developed taking into account provisions of the Montreal Convention, Warsaw Convention, Hague Protocol. While developing Regulations, provisions of Regulation (EC) 261/20042 applicable within the territory of the European Union were also taken into account, and are followed by all the airlines registered in the EU or which operate flights from European airports. Also, other air carriers during flights to / or from Europe are also required to comply with its requirements.

The concept of *flight cancellation* does not include Air carriage rules. Given the fact that in accordance with Article 11, part 4, of the Civil Code of Ukraine all definitions shall be interpreted taking into account the legislation of the European Union in the field of civil aviation, it is possible to return to the definition of the concept of *flight cancellation* contained in Regulations (EU) 261, where Article 2, part 1, of the mentioned Regulation specifies cancellation as non-fulfillment of previously planned flight in which at least one place was booked [28].

Regulation 261/2004 stipulates three types of delay: depending on the number of hours of delay: from the scheduled time of departure of flight and flight range. In the event of its origin, the passenger under Regulation 261/2004 has the right to standard assistance measures defined by the Regulation as "the right to care", which, in particular, is understood as provision of a passenger with food, cool drinks, ability to use communication means, etc. during this delay. Accordingly, regulations on air transportation define only two types of delay, depending on the direction of a passenger transportation: a) in case of a passenger carriage to/from airports of the European Union (Passengers of the EU), the concept of delay is the same as in Regulation 261/2004, that is the carrier is guided by objective criteria; b) in case of delay of flights within Ukraine and of international flights (apart from flights operated from/to airports within the territory of the European Union), where delay is assessed: "in reasonable time regarding the time defined in timetable or a ticket". The rate on calculation from the time established in a schedule is uncoordinated, since the rules of carriage in any world airline, including Ukrainian, presuppose that the time of departure (arrival) indicated in a ticket, traffic schedule or other published timetables of airline flights, is not guaranteed and is not an obligatory prerequisite for a carriage contract as the timetable is not binding. Thus, it cannot be used as a reference point to understand the concepts of *in time /not in time* [29, p. 462].

Article 7 of the EU Regulation No. 261/2004 also stipulates reimbursement in relevant cases. At the same time, this Directive defines cancellation as "non-fulfillment of a flight that was scheduled for which at least one place was reserved". In the Regulation, also the time period after which a long delay is considered to be canceled is mentioned. In the European Commission notification addressed to the European Parliament and the EU Council dated on April 4, 2011, it is stated: "It is difficult to determine, detain a flight or cancel it, as a different approach to the classification of these terms is used by airlines, which obviously affect money compensation payment". Also, data relating to cancellation of flights are provided by airlines, and not by independent sources of information.

Among most important issues in the aviation industry a special place is taken by the control over luggage aviation transportation. Thus, the Ukrainian government pursue many efforts to ensure conformity of national legislation in this field with world standards. As already mentioned above, Ukraine has signed main international contracts regulating a contract on carriage of goods and passengers by aviation transport. The Tax Code of Ukraine defines luggage as subject, property and other personal property of a passenger transported by the aircraft. There are several kinds of luggage: unregistered luggage, unaccompanied and transfer ones. Unregistered luggage is a passenger's luggage which is placed in the aircraft salon by the agreement of the carrier and under control of a passenger and is certified by a special tag. Luggage is unaccompanied: it is a luggage that is accepted by a carrier for air carriage separately from a passenger and is documented with a cargo waybill. Transferred luggage: registered luggage of a transfer passenger which was accepted for transportation from the point of export to the final destination with overload at transfer point [3].

Transportation of luggage by air is a contractual relationship governed, in particular, by Articles 908 and 910 of the Civil Code of Ukraine, Article 98 of the Air Code. Regarding international legal regulation of air carriage, the Warsaw Convention dated on 1929 has become the basis of it. One of the features of the Warsaw Convention is that it regulates both the contract of the international carriage of goods and the contract of the international carriage of passengers and luggage. In this sense, the Convention is unique in the system of legal regulation of international transportation. International legal regulations which also govern international air carriage are, in particular, the Convention for the Unification of Certain Rules for International Carriage by Air [13] and The Hague Protocol dated on 1955.

The subject of air carriage contract of luggage is transportation service, maintenance and dispensing of luggage to a passenger and payment for this service. In terms of its characteristics, this contract is real, paid and bilateral.

Let's note that in the scientific community there is a discussion about the subject of the ratio between the passenger carriage contract and the luggage carriage contract: some authors single them out as separate contracts, others consider them a single contract, and others believe that the contract of luggage carriage is additional to the passenger carriage contract.

In addition, we believe that different types of luggage should be governed by different types of contracts: if the luggage is registered, it is governed by the luggage carriage contract; and in case the luggage is unregistered (hand luggage, passenger's personal belongings), it is the subject of the passenger carriage contract regulation.

There exists a generally accepted and grounded approach as to accessory nature of luggage delivery pledge, and the luggage carriage contract itself is also of an auxiliary, supplementary, optional nature. However, transportation of luggage, despite its supporting nature, has special legal regulation (the presence of a separate luggage ticket, which is issued and attached to a ticket for a passenger carriage, the same route of a passenger and luggage carriage, the existence of a particular civil-law liability of the carrier for damage, loss and delay of luggage).

Thus, the following view is considered to be right: the luggage air-carriage contract is of an accessory nature since this contract is closely interlinked with the passenger air-carriage contract (which is indirectly confirmed by the legislator due to the lack of independent definition and instructions for this Contract in Article 98 of the Tax Code of Ukraine on passenger air-carriage contract, and Article 910 of the Civil Code of Ukraine on a passenger and luggage carriage contract). Although, nevertheless the luggage air-carriage contract has certain individual features which are mentioned in the legislation, certain measures of civil-law liability, independent transport document (luggage ticket) of a real nature of the contract.

Paragraph 78 of the Tax Code of Ukraine provides definition for air carriage as well as for carriage of passengers and luggage, including carriage of cargo or mail carried out by an aircraft according to a corresponding air carriage contract. In the Commercial Code of Ukraine, the term: *economic luggage carriage contract* is used, however, there is no definition for cargo air-carriage contract. However, the definition for the concept of *cargo air-carriage contract* is not provided. The Civil Code of Ukraine contains only a general definition of cargo carriage contract. Thus, according to Article 909 of the Civil Code, one party (carrier) is obliged to deliver cargo to the second party (sender) to a destination and hand it over to a person who possesses the right to receive cargo (recipient), and the sender undertakes to pay the established fee for cargo carriage.

The lack of a legislative framework for the cargo air-carriage contract repeatedly attracted attention of scientists. Thus, based on the findings of I.O. Bezliudko, cargo air-carriage contract is a contract on the basis of which one party (air carrier) is obliged to deliver by air cargo entrusted to him by other party (consignor or customer) to the destination airport within agreed terms in a contract, and hand it to a person who is eligible for cargo (recipient) receipt, and the customer undertakes to pay the cargo set fee for cargo carriage [30].

Such interpretation of a cargo air-carriage contract helps to identify significant terms of the contract, to establish subjective matter and its content.

Thus, it can be assumed that essential terms of a cargo air-carriage contract may be the following: terms on the subject, carriage time period and price. The subject is services of delivery of material assets (cargo, luggage, mail) (Article 909, part 1, of the Civil Code of Ukraine) entrusted to the carrier. At the same time, the positive effect of such a service is in movement (carriage) of material assets and in provision of services associated with carriage, which include the following: movement of cargo and luggage without losses, loading, unloading, temporary storage of cargo, delivering cargo to the recipient and its exporting, etc.

In accordance with the Tax Code, entities of cargo air-carriage are: aviation carrier (air carrier) and airport (aerodrome) user.

A broader list of entities of these relations is provided in the Law of Ukraine: *On Approval of the Rules for Cargo Air Carriage* [31], namely agent; consignor; consignee; cargo owner (or trustee).

Cargo air-carriage contracts can be both consensual and real, though the legislator does not directly tackle this issue. However, if the provision of Article 909, part 1, of the Civil Code of Ukraine specifying that the carrier is obliged to deliver cargo "entrusted" to him by the second party (sender) to the destination is understood as an obligation and only takes place after cargo delivery, then the conclusion may be drawn on the establishment of a legislative presumption as to reality of cargo carriage contract.

Cargo air-carriage contracts can be both paid and free. Payment for a contract depends on the following type of a contract: service contract. Paragraph 6.2, Sect. 6, of the Rules of Cargo Air Carriage, which indicates that "the carrier may issue an air waybill by a consignor's declaration and charge a corresponding fee for this from the consignor". It is not a public contract, taking into account that cargo carriage is not a public transport, and the carrier has the right to independently choose with whom to conclude a contract.

There is a discussion about a three-sided nature of a cargo air-carriage contract in science. Proponents of the position on the bilateral nature of cargo air carriage support it with the fact that the consignee is not a full contracting party of cargo carriage, considering that he does not agree the terms of the contract and does not conclude a carriage contract, does not express his willingness and consent to conclude a contract. Also, this contract is not a contract in favor of the third party, as on the grounds of general principles of civil law it is unacceptable to impose the duty on the third person if he does not participate in an obligation (Article 636 of the Civil Code). Foreign researchers [32] adhere to a similar legal position, in particular, in the law of Switzerland [33] and the law of France (Article 1165 of the Civil Code of France) [34, 35].

The carrier has the right to independently choose with whom to conclude a contract. The subject of cargo air-carriage contract (mail) is a service for transportation, maintenance and delivery of cargo (mail) to the consignee, as well as paying fee for this service.

Proponents of the view on the bilateral nature of cargo air-carriage refer to the presence of third independent parties: consignors, carrier and consignee who become contract parties from the moment of submitting a claim by the third-party beneficiary, and therefore a cargo air-carriage contract is fully considered a contract in favor of the third party. An example of a bilateral nature of a carriage contract is recognized when the consignor and the consignee act as one person.

In the context of air carriage, the above-mentioned views should be taken into account and complement each other. We believe that the criterion for distinguishing and, accordingly appropriateness of inclusion of the third party (consignee) to the contract is the nature of activities performed by him, such as economic or foreign economic activities. It is believed that the cargo air-carriage contract becomes a contract in favor of the third person only when the consignor indicates another person as a consignee in a cargo waybill.

If the sender is indicated as a recipient in the registered cargo waybill, then there is no third party in the legal relationship and a contract is bilateral. The concept of a tripartite cargo air-carriage contract is more acceptable since it considers the existence of a consignee and his role in the right to require the delivery of cargo.

Regarding a mail air-carriage contract, we consider it to be a type of contract on cargo carriage with a special entity: communication organization of both a consignor and a consignee, and a special object: mail cargo and mail dispatches with specific labeling.

Aircraft charter agreement (aircraft charter) governing conditions of irregular air traffic is particularly noteworthy. Aircraft charter agreement establishes special conditions for organizing carriage with provision by a charterer of one or more aircrafts, including all or some parts of the aircraft capacity for one or more flights for transportation of passengers, luggage, cargo or mail.

Article 912, part 1, of the Civil Code provides a general definition for an aircraft charter agreement (aircraft charter). Thus, according to the aircraft charter agreement, one party (charter) at extra fee is obliged to provide the other party (charter) with the capacity or some of its part in one or more vehicles for one or more flights to transport goods, passengers, luggage, mail or for any other purpose, in case it does not contradict the law and other legal regulations [2].

According to Article 1 of the Air Code of Ukraine, aircraft charter carriage is an irregular air transportation, carried out according to an aircraft charter agreement (chartering), by which the air carrier at extra fee provides a charter or charters with a certain number of seats or all aircraft capacity for one or more flights to transport passengers, luggage, cargo or mail indicated by the charterer.

According to Article 61 of the Air Code of Ukraine and in compliance with aircraft charter agreement (aircraft charter) one party (charterer) at extra fee undertakes to provide the other party (charterer) with the entire capacity of one or more aircrafts for one or more flights to transport passengers, luggage, cargo and mail or for other purposes if it does not contradict the current Ukrainian legislation [3].

Based on its characteristics, the aircraft charter agreement (aircraft charter) is consensual, paid, bilateral. It should be stressed that since charter flights are non-scheduled, they are not public transport, and accordingly the air charter agreement is not public.

Despite the fact that both cargo and passenger air carriage contracts can implemented on the grounds of an air charter agreement, it should be emphasized that a certain number of passenger charter transportation by air is much higher compared to the number of cargo carriage.

Cargo charters are used in cases when the demand for transportation cannot be met by regular flights (transportation of vegetables, fruits, flowers, chemicals, etc.). The development of passenger charters is conditioned by the development of tourism and the emergence of air transport users, for whom transportation cost is significant.

In conformity with an air charter agreement an aircraft may be chartered both for the charter's own needs (individual charter agreement) and for carriage of other persons (group charter agreement).

For charter carriage, the charterer can charter the aircraft capacity fully or partially. In the latter case, a contract is called a "block charter", because only a block of seats is chartered according to this contract.

The air carriers act as charters, that is any legal or natural persons who perform air transportation, possess the rights of an aircraft operator (Article 59, part 1, the Air Code of Ukraine). In compliance with this contract, legal entities who don't have their own aircraft park act as charterers, their main activity is not directly aimed at providing transport services, but is associated with the necessity to use vehicles. We refer to travel agencies, sports, professional clubs, etc.

Services for provision of an aircraft capacity and transportation are the subject matter of the aircraft charter agreement. The content of an air charter agreement comprises of the rights and obligations of its parties. The rules for charter flights performance are enshrined by the Order of the Ministry of Transport of Ukraine No. 297 dated on 18.05.2001 [36].

In the English language, the *charter* term is interpreted as rental (loan) [37] of an aircraft or a special-purpose vehicle. In the American Black's Law Dictionary, the *charter* term denotes: "rental or loan of the aircraft to perform a flight", and the air charter agreement ("charter-party") is interpreted as: "a contract by which an aircraft or some of its main part is transferred to a business entity to transport goods by a certain flight to one or several places, "a special and direct contract, on the basis of which the owner transfers an aircraft or some of its part to another person for a certain time or for use", "the charter agreement may be a an aircraft rental contract or a special service provided by the aircraft owner. If the aircraft owner is obliged to carry cargo that should be provided by a charterer according to the established flight, the contract is simply the aircraft charter agreement [38].

Synonymous in the German Language: *fracht* is understood as cargo or cargo carriage [39]. In the German Law Dictionary edited by Kobler H., the term: *fracht* denotes fee for goods transportation, and the air charter agreement is referred to as an agreement aimed at transporting goods for a fee and is considered to be a subtype of the service contract [40].

In the legal doctrine, one of the issues provoking arguments is the problem of legal nature of the aircraft charter which has a complex nature. It's no accident that in the authors' view, the aircraft charter agreement is equivalent to a rental agreement [41], O. D. Keilin expresses the opposite view on the fact that economic content of the aircraft charter agreement for a particular time period, and accordingly its legal authority remain the same, as well as the agreement concluded in case of cargo carriage is not a rental agreement [43].

According to other authors, the aircraft charter agreement is deemed to be one type of carriage contract [44, p. 93]. In the legal doctrine of the number of European countries, the air charter, in this particular case, should be viewed as a type of rental agreement, and in the other case: a type of carriage contract [45, p. 262].

We should agree with the above-mentioned views that it is expedient to attribute individual charter agreements (chartering) by their legal nature to rental agreements (loan), however the purpose of the agreement, namely transportation of cargo from one place to another, plays an important role, since the sender purpose is not to obtain the entire or part of the aircraft capacity, but cargo carriage to a specified place.

Summarizing the above, it is expedient to accept the fact that optimal approach to the legal nature of the charter (chartering) is that the charter agreement (chartering) is a mixed civil-law contract as it includes elements of different types of contracts, depending on the purpose and leased aircraft volume.

4 Conclusions

In conclusion, it should be emphasized that the air carriage contract (by air transport) is a bilateral agreement on services provision, in which one of the parties is the air carrier, and another: the passenger or sender. A studied agreement may be both real and consensual, formal, paid and public agreement.

Civil law regulation of air carriage is characterized by the existence of independent contracts with their own features and characteristics determining the content of air legislation.

The provided analysis of contracts is only a part of the civil law aspect of air legislation and does not reflect all problematic aspects in the field of air transportation, since there is still a huge number of understudied and unresolved issues in this area.

Given the current difficult economic and epidemiological situation in Ukraine and in the world, further adaptation and development of the aviation field requires analysis and reconsideration of existing contractual structures and regulations of air law to reactivate air transportation activity, as well as involvement of unmanned aerial vehicles in its work. The last of the mentioned aspects has not been studied extensively in the legal literature and may be subject of further researches.

References

1. About transport. Law of Ukraine of November 10, 1994. Information of the Verkhovna Rada of Ukraine of December 28, 2015. http://zakon5.rada.gov.ua/laws/show/232/94-BP [in Ukrainian]
2. Civil Code of Ukraine: Law of Ukraine of January 16, 2003 № 435-IV. Voice of Ukraine dated 12.01.2003 № 45. https://zakon.rada.gov.ua/laws/show/435-15#Text [in Ukrainian]
3. Air Code of Ukraine: Law of Ukraine of 19.05.2011 № 3393-VI. Information of the Verkhovna Rada of Ukraine. 2011. № 48–49. St. 536. [in Ukrainian]
4. Single European Act: International document dated 17.02.1986. https://zakon.rada.gov.ua/laws/show/994_028#Text [in Ukrainian]
5. Single European Act (SEA). https://www.europarl.europa.eu/about-parliament/files/in-the-past/ep-and-treaties/single-european-act/en-resolution-on-the-single-european-act-19861211.pdf [in English]
6. Rezo, Z., Steiner, S., Mihetec, T.: European airspace (de) fragmentation assessment model. Traffic-Transp. Transp. **33**(2), 309–318 (2021). http://traffic.fpz.hr/index.php/PROMTT/article/view/3645/561561896 [in English]
7. Commission Regulation (EU) № 139/2014 of 12 February 2014 laying down technical requirements and administrative procedures related to civil aviation aircrew pursuant to Regulation (EC) No 216/2008 of the European Parliament and of the Council. https://eur-lex.europa.eu/LexUriServ/LexUriServ.do?uri=OJ:L:2014:044:0001:0034:EN:PDF [in English]
8. Commission Regulation (EU) No 965/2012 of 5 October 2012 laying down technical requirements and administrative procedures related to civil aviation aircrew pursuant to Regulation (EC) № 216/2008 of the European Parliament and of the Council. https://eur-lex.europa.eu/LexUriServ/LexUriServ.do?uri=OJ:L:2012:296:0001:0148:EN:PDF [in English]
9. Commission Regulation (EU) № 1178/2011 of 3 November 2011 laying down technical requirements and administrative procedures related to civil aviation aircrew pursuant to Regulation (EC) (216/2008 of the European Parliament and of the Council. https://eur-lex.europa.eu/legal-content/EN/TXT/HTML/?uri=CELEX:32011R1178&from=EN [in English]
10. Single European Sky. European Parliament Think Tank. http://www.europarl.europa.eu/thinktank/en/document.html?reference=EPRS_BRI(2015)554173 [in English]
11. See the U.S. Supreme Court's decision in El Al Israel Airlines Ltd. v. Tsui Yuan Tseng (1999) 525 US 155 and Abnett v. British Airways Plc. (Scotland), Sidhu v. British Airways Plc. [1997] A.C. 430; [1997] 2 Lloyd's Rep. 76, a decision by the U.K. House of Lords [in English]
12. Carriage of goods by air: a guide to the international legal framework. https://unctad.org/system/files/official-document/sdtetlb20061_en.pdf [in English]
13. ICAO Convention for the Unification of Certain Rules for International Carriage by Air (ukr / ros): Convention, International Document dated 28.05.1999. https://zakon.rada.gov.ua/laws/show/995_594#Text [in Ukrainian]

14. Protocol of Amendments to the Convention for the Unification of Certain Rules Relating to International Carriage by Air, signed at Warsaw on 12 October 1929 by the United Nations; Protocol, International document of September 28, 1955. https://zakon.rada.gov.ua/laws/show/995_562#Text [in Ukrainian]
15. Convention for the Unification of Certain Rules for International Carriage by Air. https://www.icao.int/secretariat/legal/list%20of%20parties/mtl99_en.pdf [in English]
16. Convention for the Unification of Certain Rules Relating to International Carriage by Air. League of Nations; Convention, International document, Protocol of 12.10.1929. https://zakon.rada.gov.ua/laws/show/995_181#Text [in Ukrainian]
17. Protocol to Amend the Convention for the Unification of Certain Rules Relating to International Carriage by Air, Signed at Warsaw on 12 October 1929, Done at The Hague On 28 September 1955 The Hague Protocol to the Warsaw Convention 1955. https://www.jus.uio.no/lm/air.carriage.warsaw.convention.hague.protocol.1955/doc.html [in English]
18. Regulation (EC) No 261/2004 of the European Parliament and of the Council of 11 February 2004 establishing common rules on compensation and assistance to passengers in the event of denied boarding and of cancellation or long delay of flights, and repealing Regulation (EEC) No 295/91 (Text with EEA relevance) Commission Statement. https://eur-lex.europa.eu/legal-content/en/ALL/?uri=celex:32004R0261 [in English]
19. On approval of the Aviation Rules of Ukraine "Rules of air transportation and servicing of passengers and luggage" by the State Aviation Service of Ukraine. Order, Rules of 26.11.2018 № 1239. https://zakon.rada.gov.ua/laws/show/z0141-19#Text [in Ukrainian]
20. On consumer protection. Law of Ukraine of 12.05.1991 № 1023-XII. https://zakon.rada.gov.ua/laws/show/1023-12#Text [in Ukrainian]
21. Joan H. Stacy: International Civil Aviation Organization (ICAO). J. Air L. Com. **17**, 454 (1950) [in English]
22. International Civil Aviation Organization (ICAO). http://dobrica.savic.ca/pubs/Fundamentals_of_law_ICAO.pdf [in English]
23. IATA Effective 1 Jun 2008, paper tickets are eliminated from IATA BSPs and BSP travel agencies will only have access to e-ticket. IATA. 2008. http://www.iata.org/Pages/default.aspx [in English]
24. Access to Air Transport by Persons with Disabilities [Text] / ICAO. - Montreal: ICAO, 1999. (iii), 8 p. (ICAO Circular; 274). http://www.icao.int/Meetings/FALP/Documents/Falp7-2012/Report/FALP-7_Report_Ru.pdf [in English]
25. Aviation rules: Order of 06.02.2017 № 66/73. https://zakon.rada.gov.ua/laws/show/z0141-19#Text. Accessed 12 Apr 2021 [in Ukrainian]
26. Shulzhenko FP, Gaidulin OO, Kundrik RS Transport law. textbook manual. K .: KNEU, 2005. 244 s. http://chitalka.net.ua/zmist/178.html. Accessed: 10 Apr 2021 [in Ukrainian]
27. Khokhlenko, V.M.: Legal regulation of passenger transportation by air. DICTUM FACTUM **1**(6), 151–157 (2020). [in Ukrainian]
28. Regulation [EC] 261/2004 establishing common rules on compensation and assistance to passengers in the event of denied boarding and of cancellation or long delay of flights, and repealing Regulation (EEC) No 295/91. https://ec.europa.eu/transport/sites/transport/files/2004_261_national_enforcement_bodies.pdf [in English]
29. Tsirat, G.V.: Regulation of flight delay under the 1999 Montreal Convention and the rules of carriage of passengers and luggage. State Law Vip. **56**, 462–467 (2012). [in Ukrainian]
30. Bezlyudko, I.O.: Contract of carriage of goods by air under the civil law of Ukraine: author's ref. dis. for science. degree of Cand. jurid. Science: special. 12.00.03 "Civil law and civil procedure; family law; international private law" . I.O. Inhuman. Kyiv. p. 20 (2005) [in Ukrainian]

31. On approval of the Rules of air cargo transportation State Aviation Service: Order, Rules of 14.03.2006 № 186. https://zakon.rada.gov.ua/laws/show/z0705-06#Text. Accessed 12 Apr 2021 [in Ukrainian]
32. Giemulla, E., Schmid, R., Muller-Rostin, W., Dettling-Ott, R., Margo, R.: Montreal Convention. Art pp. 1–18 (1996). https://www.collectionscanada.gc.ca/obj/s4/f2/dsk2/ftp04/mq29831.pdf [in English]
33. Gronfors, K.: Air Charter and Warsaw Convention. The Hague (1956) p.117. https://link.springer.com/content/pdf/bbm%3A978-94-017-6762-0%2F1.pdf [in English]
34. Katarzyna Sójka-Zielińska. Napoleon's Code. History and community. Wydanie 2. Warszawa: Wydawnictwo Prawnicze LexisNexis Polska, 2008. 585 s. [in Poland]
35. Civil Code of France (Napoleon's Code) VN Zakhvataev, lane. with French, applications VN Zakhvataev. Kyiv: Alerta, p. 1156 (2016) [in Ukrainian]
36. On approval of the Rules of charter flights. Order of the Ministry of Transport of Ukraine, Rules of 18.05.2001 № 297. https://zakon.rada.gov.ua/laws/show/z0461-01#Text. Accessed 16 Apr 2021 [in Ukrainian]
37. The Oxford English Dictionary. https://en.oxforddictionaries.com/definition/charter; Cambridge Business English Dictionary. https://dictionary.cambridge.org/dictionary/english/charter; The Lexico Oxford Dictionary. https://www.lexico.com/definition/charter. Accessed 12 Apr 2021 [in English]
38. Black, H.C.: Black's Law Dictionary: Definitions of the Terms and Phrases of American and English Jurisprudence, Ancient and Modern. Revised Forth Edition. St. Paul, Minn: West Publishing Co. pp. 298–299 (1968). https://ruor.uottawa.ca/bitstream/10393/36717/1/Obel_Andrew_Kinarah_2017_Thesis.pdf. Accessed 12 Apr 2021 [in English]
39. The online dictionary Duden.de. https://www.duden.de/rechtschreibung/Fracht. Accessed 12 Apr 2021 [in German]
40. Köbler G. Juristisches Wörterbuch. Lawyer for German. 12., neubearb. Aufl. Munich: Vahlen, 2003. (Vahlen-Studienreihe Jura). S.188–189. http://www.koeblergerhard.de/beckjurwb14.pdf. Accessed 14 Apr 2021 [in German]
41. Streltsova, E.D.: Contract of sea transportation of goods (comparative aspect): dis. Cand. jur. Sciences: 12.00.03 / Streltsova ED Odessa, p. 51 (2002). [in Ukrainian]
42. Vakulovich, E.V.: Vehicle lease agreement: dis. Cand. jur. Sciences: 12.00.03 / Vakulovich EV Kharkiv, 2002. C. 94. [in Ukrainian]
43. Kailin A.D.: Transport law of the USSR and capitalist countries. In: International Book, Issue. 1: Contract of carriage, p. 120. Moscow (1938)
44. Popov, V.A., Ruban, O.O.: The place of the charter agreement (chartering) in the civil law of Ukraine. Legal Sci. Electron. J. **2**, 92–94 (2021) [in Ukrainian]
45. Abraham, H.J.: The right of air travel. Commentary and Sources Ammlung, pp. 262–266 (1960). [in German]
46. Sundberg: Aircharter: a Study of Legal Development. Stocholm, 1961, p. 272 ff; Rodiere R. Droit des transports. Paris, 1977, pp. 845–847; Shawcross and Beaumont. Air law, 4 Edition, London, 1977, p. 566. [in English]

Author Index

M. Nechyporuk et al. (Eds.): ICTM 2021, LNNS 367, pp. 1067–1070, 2022.
https://doi.org/10.1007/978-3-030-94259-5